PYRIDINE AND ITS DERIVATIVES

Part Five

D1650835

Edited by

George R. Newkome

Louisiana State University
Baton Rouge, Louisiana

AN INTERSCIENCE® PUBLICATION

John Wiley and Sons

NEW YORK • CHICHESTER • BRISBANE • TORONTO • SINGAPORE

An Interscience® Publication

Copyright © 1984 by John Wiley & Sons, Inc.

Library of Congress Cataloging in Publication Data:
(Revised for volume 5)

Klingsberg, Erwin.
    Pyridine and its derivatives.

    (The Chemistry of heterocyclic compounds: a series
of monographs, v. 14)
    Vol. 5-   edited by George R. Newkome.
    Vol. 5-   has imprint: New York: Wiley.
    "An Interscience publication"—Vol. 5, t.p.
    Includes bibliographies.
    1. Pyridine.   I. Newkome, George R. (George Richard).
II. Title.   III. Series: Chemistry of heterocyclic
compounds; v. 14.

    QD401.K712   547′.593   59-13038
    ISBN 0-471-05072-5 (v. 5)

Printed in the United States of America

10 9 8 7 6 5 4 3 2 1

Contributors

T. D. BAILEY
Reilly Tar and Chemical Corporation
Indianapolis, Indiana

G. L. GOE
Reilly Tar and Chemical Corporation
Indianapolis, Indiana

V. K. GUPTA
Department of Chemistry
Louisiana State University
Baton Rouge, Louisiana

G. R. NEWKOME
Department of Chemistry
Louisiana State University
Baton Rouge, Louisiana

J. D. SAUER
Ethyl Corporation
Baton Rouge, Louisiana

E. F. V. SCRIVEN
Reilly Tar and Chemical Corporation
Indianapolis, Indiana

R. P. THUMMEL
Department of Chemistry
University of Houston
Houston, Texas

PYRIDINE AND ITS DERIVATIVES

Part Five

This is the fourteenth volume in the series

THE CHEMISTRY OF HETEROCYCLIC COMPOUNDS

THE CHEMISTRY OF HETEROCYCLIC COMPOUNDS

A SERIES OF MONOGRAPHS

ARNOLD WEISSBERGER AND EDWARD C. TAYLOR

Editors

The Chemistry of Heterocyclic Compounds

The chemistry of heterocyclic compounds is one of the most complex branches of organic chemistry. It is equally interesting for its theoretical implications, for the diversity of its synthetic procedures, and for the physiological and industrial significance of heterocyclic compounds.

A field of such importance and intrinsic difficulty should be made as readily accessible as possible, and the lack of a modern detailed and comprehensive presentation of heterocyclic chemistry is therefore keenly felt. It is the intention of the present series to fill this gap by expert presentations of the various branches of heterocyclic chemistry. The subdivisions have been designed to cover the field in its entirety by monographs which reflect the importance and the interrelations of the various compounds, and accommodate the specific interests of the authors.

In order to continue to make heterocyclic chemistry as readily accessible as possible, new editions are planned for those areas where the respective volumes in the first edition have become obsolete by overwhelming progress. If, however, the changes are not too great so that the first editions can be brought up-to-date by supplementary volumes, supplements to the respective volumes will be published in the first edition.

Research Laboratories   ARNOLD WEISSBERGER
Eastman Kodak Company
Rochester, New York

Princeton University   EDWARD C. TAYLOR
Princeton, New Jersey

Preface

The original four volumes of this pyridine series were published between 1960 and 1964 under the guidance of Dr. Erwin Klingsberg. In 1974–1975, Professor Rudy Abramovitch edited a four-volume supplemental series, which followed the general format of the initial work. These herculean tasks covered most of the important research in pyridine chemistry up to 1970–1972.

As with most areas of organic chemistry, proliferation has occurred at an incredible rate, especially in heterocyclic chemistry. The need for a topical update in key research areas is essential; thus, the supplemental series has changed format in order to keep the interest in pyridine chemistry as current as possible.

In 1977, Professor Abramovitch and I discussed the creation of this expansion of *Pyridine and Its Derivatives* and decided to abandon the difficult-to-organize chapter order of the previous volumes in this series. Also, new topics and directions caused duplication and a need for new chapters to meet the ever-expanding field of pyridine chemistry. As the task started, Professor Abramovitch's writing and editing obligations in other areas of interest prevented his devotion to this series; his efforts are sorely missed.

This and all future supplementary volumes in the *Pyridine and Its Derivatives* series will be devoted to specific areas of interest and will attempt to remain as the comprehensive repository of pyridine chemistry.

I express my thanks to the authors for their contributions and patience as well as to Rudy Abramovitch for his initial guidance and support in this project.

Baton Rouge, Louisiana GEORGE R. NEWKOME
October 1983

Contents

I. SYNTHETIC AND NATURAL SOURCES OF THE PYRIDINE RING   1

T. D. Bailey, G. L. Goe, and E. F. V. Scriven

II. CARBOCYCLIC ANNELATED PYRIDINES   253

R. P. Thummel

III. MACROCYCLIC PYRIDINES   447

G. R. Newkome, V. K. Gupta, and J. D. Sauer

IV. THE REVIEWS OF PYRIDINE CHEMISTRY—1968–1982   635

G. R. Newkome

AUTHOR INDEX   659

SUBJECT INDEX   703

PYRIDINE AND ITS DERIVATIVES

Part Five

This is the fourteenth volume in the series

THE CHEMISTRY OF HETEROCYCLIC COMPOUNDS

CHAPTER I

Synthetic and Natural Sources
of the Pyridine Ring

T. D. BAILEY, G. L. GOE, and E. F. V. SCRIVEN

Reilly Tar and Chemical Corporation, Indianapolis, Indiana

I. Pyridines from Natural Sources	3
1. Pyridines in Nature	3
A. Enzymes, Vitamins, Amino Acids, and Their Biogenesis	3
B. The Tobacco Alkaloids	7
C. Other Pyridine Alkaloids and Related Compounds	11
a. Simple Pyridine Alkaloids	11
b. Monoterpenoid Alkaloids	12
c. Sesquiterpenoid Alkaloids	16
i. Derivatives of Nicotinic Acid	16
ii. Pyridone and Pyridinol Alkaloids	17
iii. Other Sesquiterpenoid Alkaloids	22
d. β-Carboline and Related Alkaloids That Contain a Pyridine Ring	23
2. Degradation of Natural Products	24
A. Coal	24
B. Petroleum	28
C. Shale	29
D. Degradation and Transformation of Alkaloids	31
E. Flavors, Odors, and Volatile Constituents of Food and Beverages	33
F. Miscellaneous Sources	36
II. Pyridines by Synthetic Methods	36
1. From Other Ring Systems	36
A. Carbocyclic Compounds	37
B. Three-Membered Ring Heterocycles	43
C. Four-Membered Ring Heterocycles	46
D. Five-Membered Ring Heterocycles	47
a. Five-Membered Rings Containing One Heteroatom	48
i. Furans, Dihydrofurans, and Tetrahydrofurans	48
ii. Pyrroles	55
b. Five-Membered Rings Containing Two Heteroatoms	60
i. Oxazoles	60
ii. Miscellaneous Five-Membered Ring Heterocycles	68
E. Six-Membered Ring Heterocycles	75
a. One Heteroatom	75
i. Pyrones	75

ii. Pyrans	83
iii. Pyrylium Salts	83
b. Two Heteroatoms	93
i. Pyrimidines	93
ii. Pyridazines	97
iii. Pyrazines	98
iv. Oxazines	99
v. Miscellaneous Six-Membered Heterocycles Containing Two Heteroatoms	101
c. Six-Membered Ring Heterocycles with Three Heteroatoms	102
F. Seven-Membered Ring Heterocycles	104
a. Azepines	104
b. Diazepines	105
G. Pyridines from Reduced Pyridines	106
a. Dihydropyridines	106
b. Tetrahydropyridines	111
c. Piperidines	113
H. Condensed Rings	113
a. Oxidation	113
b. Reductions	117
c. Ring-Opening Reaction	117
2. From Acyclic Compounds	118
A. Cyclization of a 5-Carbon Chain	119
a. 1,5-Dioxo Compounds and Derivatives	119
b. Oxocarboxylic Acids and Derivatives	125
c. 1,5-Dicarboxylic Acids and Derivatives	132
d. Compounds Having Terminal Unsaturation	136
e. Miscellaneous 1,5-Bifunctional Compounds	139
B. 4-1 Condensations	141
a. Dienes with Nitriles	141
b. Other Reactions of Nitriles	142
c. Reactions of Isocyanates	144
d. Reaction of Other Acid Derivatives	144
e. Miscellaneous	146
C. 3-2 Condensations	147
a. 1,3-Dicarbonyl Compounds and Their Derivatives with Methylenic Compounds	147
b. α,β-Unsaturated Carbonyl Compounds and Their Derivatives with Methylenic Compounds	166
c. Condensation of α,β-Unsaturated Carbonyl Compounds with Ammonia	172
d. Miscellaneous 3-2 Condensations	173
D. 1-3-1 Condensations	175
E. 2-2-1 Condensations	176
a. Acetylenes and Nitriles	176
b. Aldehydes with Ammonia	177
i. Acetaldehyde with Ammonia–Vapor Phase	177
ii. Acetaldehyde with Ammonia–Liquid Phase	177
iii. Other Aldehydes with Ammonia–Liquid Phase	178
c. Aldehydes, Ketones, and Mixtures with Ammonia–Gas Phase	179
i. Acetaldehyde and Formaldehyde with Ammonia	179
ii. Other Mixtures of Aldehydes and Ketones with Ammonia	179
d. Other Oxygenated Compounds with Ammonia–Vapor Phase	181
e. Miscellaneous 2-2-1 Condensations	181
F. 2-1-2 Condensations	185

a. Mixtures of Aldehydes and Ketones with Ammonia 185
b. Carbonyl Compounds with Active Methylene Compounds 185
   i. Aldehydes with Active Methylene Compounds 185
   ii. Carboxylic Acid Derivatives with Active Methylene Compounds . . 186
c. Miscellaneous 2-1-2 Condensations 187
G. Cyclization Not Involving the Ring Nitrogen 188
a. Cyclization of Isocyanates 188
b. Cyclization of Imines 190
c. Reactions Related to the Gould–Jacobs Reaction 191
d. Miscellaneous Ring Closures 193
Acknowledgment 194
References 194

I. PYRIDINES FROM NATURAL SOURCES

1. Pyridines in Nature

A. Enzymes, Vitamins, Amino Acids, and Their Biogenesis

The wealth of information derived from isotopic labeling studies over the last few decades has established pathways for the biogenesis of many pyridines. Only the ones that lead to some important pyridines will be outlined.

The pyridine ring of nicotinic acid and the pyridine nucleotides is synthesized by two routes. In the tissues of higher animals and *Neurospora* it is derived from tryptophan, but another pathway starting from aspartic acid and a C_3-fragment is preferred in bacteria (e.g., *E. coli* and *B. subtilis*), green algae, and higher plants (e.g., corn or tobacco). The yeast *S. cerevisiae* has the ability to synthesize pyridine nucleotides by both routes. Under aerobic conditions the tryptophan pathway is favored, but in the absence of oxygen the aspartic acid route (*de Novo* pathway) predominates. The two pathways, which converge at a common intermediate, quinolinic acid, are illustrated in Scheme I-1. Quinolinic acid is decarboxylated and converted to nicotinic acid mononucleotide by phosphoribosyltransferase to provide entry into the Pyridine Nucleotide Cycle (Scheme I-2). Conversion of tryptophan to nicotinic acid has been well studied in both animals, principally the rat (1, 2) and, fungi (3).

The Pyridine Nucleotide Cycle is responsible for the production of nicotinic acid adenine dinucleotide, NAD, nicotinamide and nicotinic acid, and the alkaloids (**I-1**) and (**I-2**) in plants. Several reviews are available on biosynthesis (4–9) and other aspects (10–12) of pyridine nucleotides. The importance of this cycle in the generation of pyridine enzymes and vitamins has excited interest in its control mechanism (13–18). Before giving examples of isolation of members of the cycle from specific sources, the biogeneses of the B_6 vitamins, pyridoxine (**I-3**), pyridoxal (**I-4**), and pyridoxamine (**I-5**) are mentioned.

Evidence from comprehensive ^{14}C labeling work indicated that all the carbon

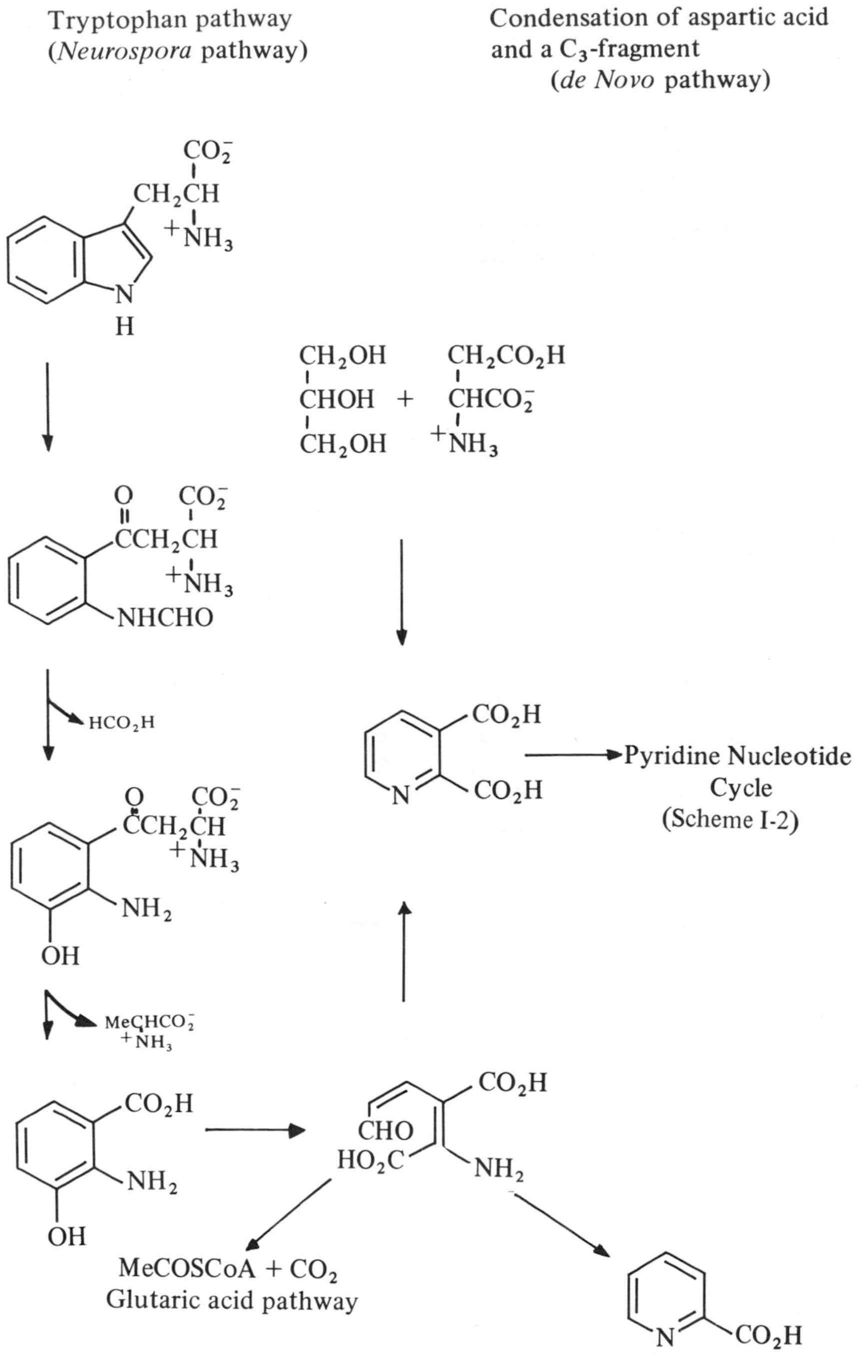

Tryptophan pathway
(*Neurospora* pathway)

Condensation of aspartic acid
and a C_3-fragment
(*de Novo* pathway)

Scheme I-1.   Two pathways for the biogenesis of quinolinic acid.

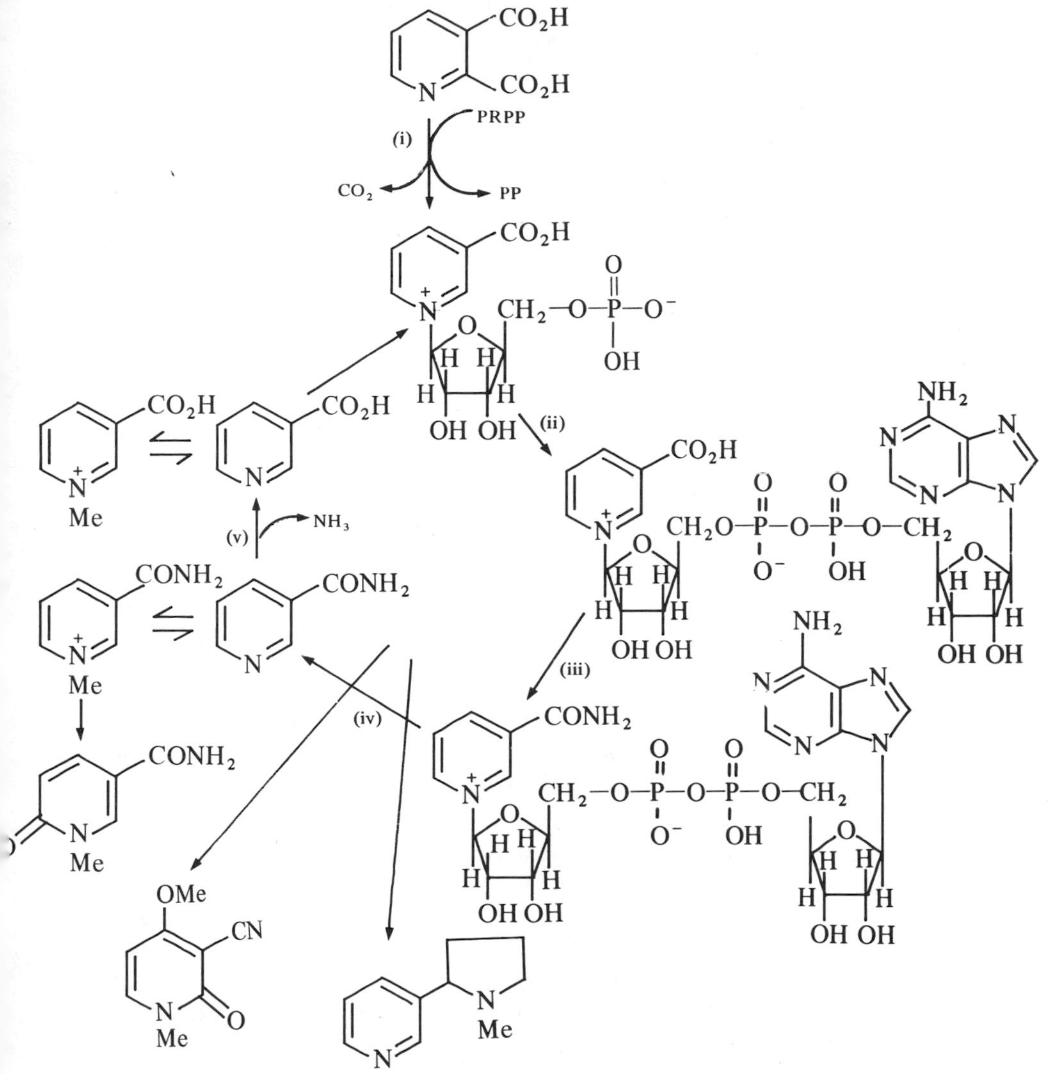

Scheme I-2. Pyridine Nucleotide Cycle. (i) Quinolinic acid phosphoribosyltransferase. (ii) Nicotinic acid mononucleotide adenyltransferase. (iii) NAD + synthetase. (iv) NAD + glycohydrolase. (v) Nicotinamide deamidase. PRPP = phosphoribosyl pyrophosphate.

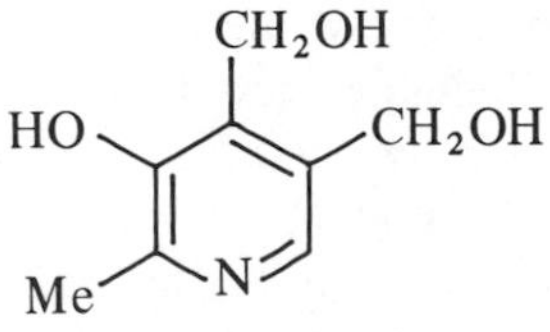

I-3

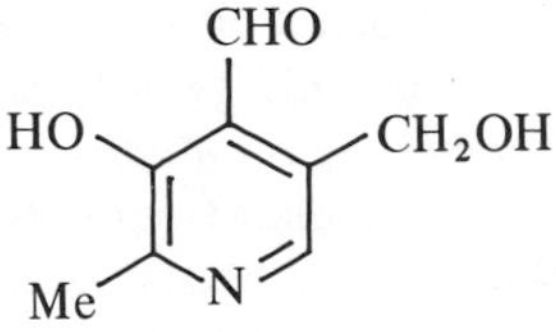

I-4

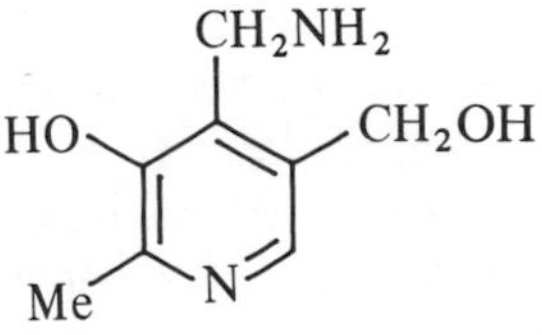

I-5

TABLE I-1.   SYNTHESIS OF NICOTINIC ACID AND B₆
                      VITAMINS BY MICROORGANISMS

Organism	Reference
Hydrogenomonas eutropha	47, 48
Hydrogenomonas pantotropha	47
Hydrogenomonas thermophilus	47
Azotobacter chroococcum	49
Bacillus megaterium	49
Urobacteria	50
Arthrobacter simplex	51
Arthrobacter variabilis	51
Arthrobacter tumescens	51
Rhizobial bacteria	52
Spirulina platensis	53
Fusarium moniliforms	54
Fusarium gibbosum	54
Actinomycetes	55, 56
Promyxobacterium johnsonii	57
Micrococcus freudenreichii	58
Kloeckera apiculata	59
Saccharomyces ellipsoideus etc.	60
Zygosaccharomyces bailii etc.	61
Candida tropicalis	62

atoms in pyridoxal biosynthesized by *E. coli* (B mutant WG2) originate from glycerol (19–23). Nicotinic acid is believed to be the precursor of pyridoxine from a feeding study of a pyridoxine-less mutant of *Aspergillus nidulans* (24).

The yeast *Rhodotorula glutinis* is able to convert *n*-alkanes to nicotinic acid (25). The tryptophan pathway is followed in this organism (26–28) and *Chlamydomonas eugametos* (29). Nicotinic acid has also been produced by soil microorganisms (30), baker's yeast (31), the fungus *Penicillium digitatum* (32), and cotton seeds *Gossypium barbadense* (33). A review has appeared on the biogenesis of nicotinic acid in plants and microbes (34). Other reviews concerning its chemical properties, pharmacology and nutritional aspects as well as occurrence and synthesis have more general interest (35–37).

Pyridoxal phosphate is the coenzyme responsible for transamination and is the chief form of the B₆ vitamins in animal tissues (38). Pyridoxine, the form in which pyridoxal is stored, has been found among the products of the fermentation of methanol, with *Methanomonas methylovora*, which has been claimed to have industrial potential (39). Pyridoxal phosphate has also been synthesized by symbiotic bacteria of nonleguminous plants (40), aerobic cellulose bacteria (41), and all six strains of *Rhizobium leguminosarum* (42). The synthesis of both pyridoxine and nicotinic acid by bacteria from soils has been studied extensively (43–45). Some of these bacteria have been found to produce pyridoxine and nicotinic acid when ethanol is the sole carbon source (46). Other microorganisms producing nicotinic acid and pyridoxine are listed in Table I-1.

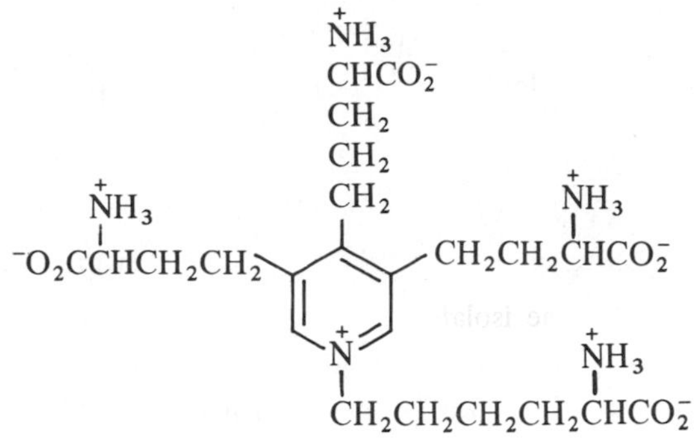

I-6

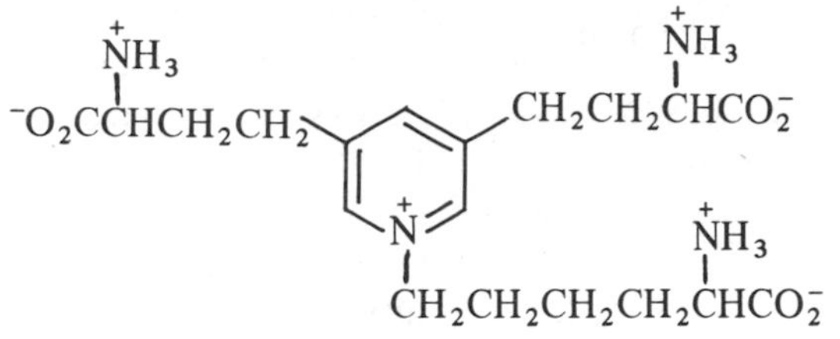

I-7

The unusual amino acids desmosine (**I-6**) and isodesmosine (**I-7**) have been isolated from the protein elastin, in which they act as crosslinking centers of peptides (63).

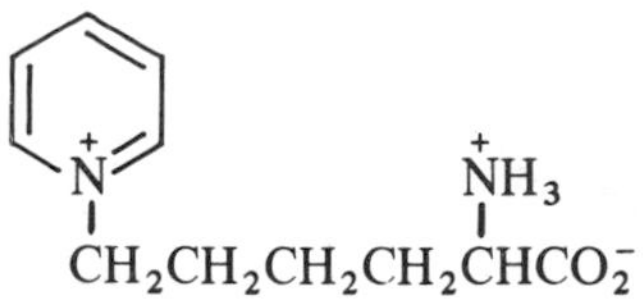

I-8

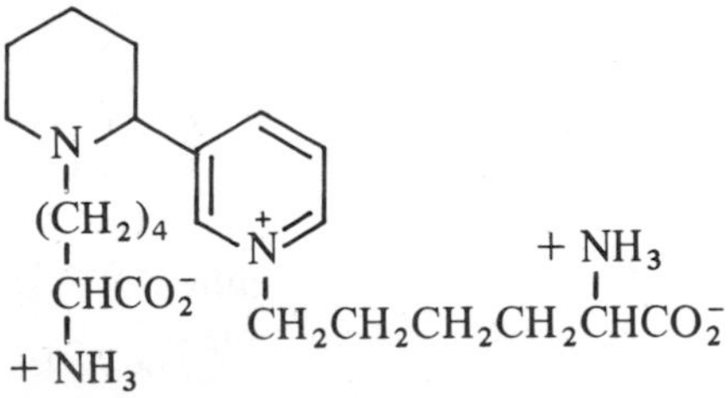

I-9

Recently, two more pyridine-containing amino acids, N-(5-amino-5-carboxy-pentyl)pyridinium chloride (**I-8**) and anabilysine (**I-9**), have been found important in the crosslinking of proteins by glutaraldehyde (64).

B. The Tobacco Alkaloids

Several new methods for the extraction and separation of these alkaloids have appeared (65, 66). Separations on alumina sintered glass plates have been described (67) and new TLC solvent systems have been found which allow the separation of 10 to 13 component mixtures on a semiquantitative basis (68).

I-10

I-11

Two new terpenoid alkaloids have been isolated from Burley tobacco (*Nicotiana tabacum*), 1,3,6,6-tetramethyl-5,6,7,8-tetrahydroisoquinolin-8-one (**I-10**) and 3,6,6-trimethyl-5,6-dihydro-7*H*-pyrindan-7-one (**I-11**) (69). Remarkably, **I-10** may be obtained from the scent gland of the Canadian beaver (*Castor fiber*) (70), or by a synthetic method (71). **I-10** has also been used to improve the aroma of tobacco!

New variants of nicotine isolated from *N. tabacum*, apart from 2,3'-bipyridyl, have been shown to be *N*-acylated derivatives of the pyrrolidine ring (**I-12**) (72). The roots and stems of *N. tabacum*, *N. affinis*, and *N. sylvestris* have all been found to contain *cis* or *trans* nicotine *N*-oxide, which may be reduced to the parent alkaloid with titanous chloride (73). The leaves of *Anthocercis tasmanica* provided a new nontobacco source of nicotine (74). Nicotine has been observed spectroscopically in extracts of the plant *Sedum acre* (75).

$R = CHO, Ac, COC_5H_{11},$
COC_7H_{18}

I-12

Biosynthesis and metabolism of the pyridine alkaloids have been reviewed (76). Many studies of the biosynthesis of nicotine have utilized ^{15}N and ^{14}C labeled precursors (77–79), but more recently ^{13}C labeling has found increasing favor because of the ease of site determination of the label in the product by ^{13}C NMR (80–81). Abnormal synthetic reactions that occur in biological systems, referred to as "aberrant biosynthesis" (81), are currently of particular interest (76). Aberrant biosynthesis may be divided into two types. *Type I is depicted by the formation of a natural compound from an unnatural precursor*, such as production of nicotine when δ-*N*-methylornithine (not normally a component of tobacco) is administered to *N. tabacum* plants (82). *Type II involves the conversion of an unnatural precursor to an unnatural product.* Examples of this type are provided by the conversion of 5-fluoronicotinic acid to either 5-fluoronicotine in *N. tabacum* (83) or 5-fluoroanabasine in *N. glauca* (81).

TABLE I-2.   SIMPLE PYRIDINES FOUND IN TOBACCO LEAF AND SMOKE

Pyridine Bases		Reference	
		Leaf	Smoke
Pyridine		108	108
2-Picoline		109	108
3-Picoline		–	108
4-Picoline		109	108
Lutidines	(only 2,6-)	109	108
2,4,6-Collidine		–	110
2,3,6-Collidine		109	–
2-Ethylpyridine		109	–
3-Ethylpyridine		–	108
Methylethylpyridine(s)		–	111
2-Methyl-5-isopropylpyridine		112	111
2,4-Dimethyl-5-isopropylpyridine		–	111
2-Phenylpyridine		–	108
3-Phenylpyridine		113	108
2-Vinylpyridine		–	111
3-Vinylpyridine		–	108
3-Propenylpyridine		113	–
3-Formylpyridine		113	108
2-Acetylpyridine		113	–
3-Acetylpyridine		108	111
3-Propionylpyridine		113	108
3-Butyrylpyridine		108	–
6-Methyl-3-hydroxypyridine		–	41
Nicotinic acid		108	114
Methyl nicotinate		115	–
Nicotinamide		108	114
N-Methylnicotinamide		116	–
3-Cyanopyridine		113	108
Methylcyanopyridine(s)		–	111
Dimethylcyanopyridine(s)		–	111
3-Methylaminopyridine		–	108

Regulation of the nicotine content of tobacco has been of great importance to the tobacco industry and has been reviewed (84). Such interest has led to the determination of alkaloid content during the course of ontogeny of *N. tabacum*, *N. glutinosa*, and *N. sylvestris* (85–88). Studies have been made of the alkaloid spectrum during germination of seeds (89, 90) and in the roots of seedlings (91, 92). Genetic effects on alkaloid content have also received attention (83–96). Flue-curing and aging of Virginia tobacco have been found to lower nicotine content but increase the amount of simple pyridine components (97). It has been claimed that nicotine may be removed from tobacco by rapid drying of an aqueous alkaline tobacco dispersion (98).

Various aspects of smoking concerned with the occurrence and role of nicotine have been reviewed (99–104). Various metabolites of nicotine have been reported,

TABLE I-3.   ALKALOIDS FOUND IN TOBACCO LEAF AND SMOKE

Alkaloid	Reference	
	Leaf	Smoke
N'-Acetylnornicotine	117	–
Anabasine	108	108
Anatabine	108	108
Anatalline **(I-15)**	118	–
2,2'-Bipyridyl	–	119
N'-Carbomethoxyanabasine	–	120
N'-Carbomethoxynornicotine	–	120
Cotinine	108	108
2',3'-Dehydronicotine	–	108
Dihydrometanicotine	–	108
Dihydronicotyrine (*N*-methylmyosmine)	–	116
N'-Formylnornicotine	117	–
N'-Hexanoylnornicotine	72	–
Isonicoteine (2,3'-bipyridyl)	108	108
Metanicotine	108	108
N'-Methylanabasine	122	119
N'-Methylanatabine	116	–
5-Methyl-2,3'-bipyridyl	117	–
N-Methylnicotone	123	124
Myosmine **(I-16)**	123	123
Nicotelline	125	–
Nicotine	108	108
Nicotyrine	108	108
N'-Nitrosonicotine	126	126
Nornicotine	108	108
Nornicotyrine	108	108
N'-Octanoylnornicotine	72	–
Oxynicotine (nicotine *N*-oxide)	108	127
1,3,6,6-Tetramethyl-5,6,7,8-tetrahydroisoquinolin-9-one **(I-10)**	69	–
3,6,6-Trimethyl-5,6-dihydro-7*H*-pyrindan-7-one **(I-11)**	69	–

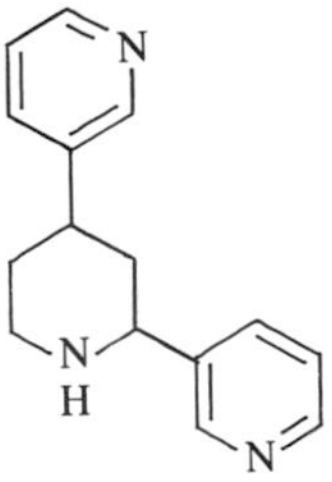

I-15

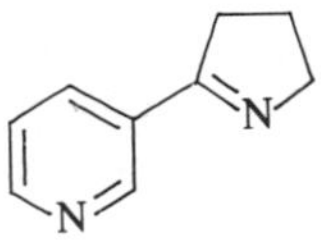

I-16

such as: hydroxycotinine (**I-13**) from urine of smokers (105) and diastereomeric *N*-oxides (**I-14**) from hepatic supernatants of mice, rats, hampsters, rabbits, and guinea pigs (106).

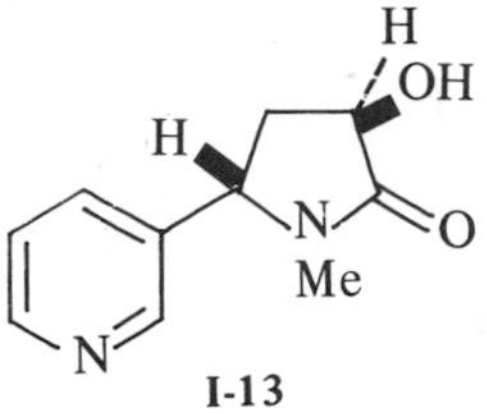

Work on the contents of tobacco leaf and smoke has been reviewed (107), and thus will be dealt with here in a cursory way. Pyridine bases that have been found in tobacco and tobacco smoke are listed in Tables I-2 and I-3. Although bases isolated from tobacco smoke are not strictly speaking alkaloids, they are considered alongside those found in leaf for comparative purposes.

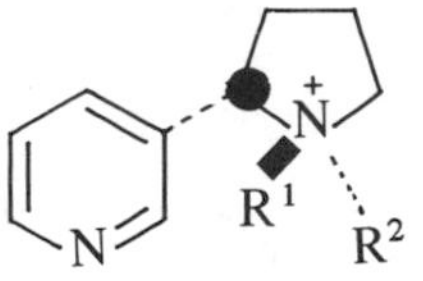

A few points of general interest regarding tobacco smoke are mentioned below. Cigar smoke contained a higher amount of pyridines relative to total alkaloids than did cigarette smoke (128). Cigar butt "head-space-vapors" contained some of the tabulated pyridines (129). Puff frequency has been found to have a greater effect than puff volume on the alkaloid content of smoke (130). Smoke from Cytrel smoking products has been compared with that from flue-cured tobacco (131, 132); nicotine could not be detected in the smoke from 100% Cytrel samples (132).

C. Other Pyridine Alkaloids and Related Compounds

A great expansion in the knowledge of pyridine alkaloids has taken place in the last decade since the appearance of two reviews on the subject (133, 135). Current work is reported in *Alkaloids* (London) in the Chemical Society Specialist Periodical Report Series (442), and another review has appeared (134).

a. SIMPLE PYRIDINE ALKALOIDS

Ricinine (**I-17**) is a well-known 2-pyridone derivative that is found in the castor bean *Ricinus communis*. Recent interest has been centered on the relationship between the pyridine nucleotide cycle and ricinine biogenesis (136, 137). The isomeric pyridones ricinidine (**I-18**) and nudifluorine (**I-19**) have been isolated from the leaves of *Trewia nudiflora* (138, 139).

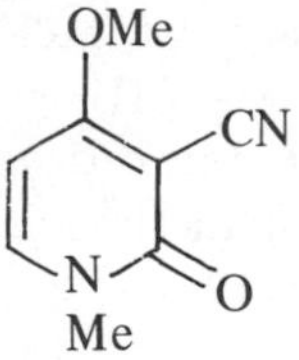

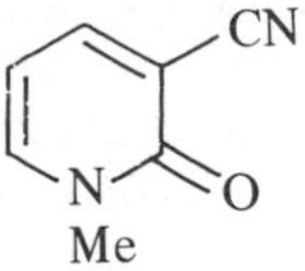

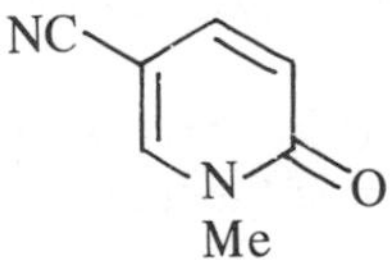

I-20   I-21

I-22

Fusaric acid (**I-20**), a systemic wilt toxin found particularly in cotton plants (140, 141), was produced by different species of *Fusaria* (142–148) and other fungi (149). Dehydrofusaric acid (**I-21**) and (+)-*S*-fusarinolic acid (**I-22**), metabolites of fusaric acid, have been obtained from the mycelium of various *Fusaria*, *S. cerevisiae*, and *Gibberella fujikuroi* (149–151).

Dipicolinic acid was produced by aerobic spore-forming bacteria during sporulation and its calcium salt is a major constituent of endospores. Its biosynthesis and occurrence have been a popular field of study; sources of dipicolinic acid include: *Bacillus megaterium* (152–153), *B. subtilis* (153–157), *B. sphaericus* (158), *B. cereus* (159–161), *B. stearothermophilus* (162, 163), *Penicillium* NRRL 3114 by patented processes (164, 165), and *Chlostridium roseum* (166, 167). An iron complex of pyridine-2,6-di-(monothiocarboxylic acid), which has reported antibiotic activity, has been isolated from a culture medium of a *Pseudomonas* strain (168). Some other naturally occurring simple pyridine alkaloids and their source of origin are listed in Table I-4.

b.   MONOTERPENOID ALKALOIDS

A great deal of progress has been made on the isolation and structure determination of pyridine monoterpenoid alkaloids. These alkaloids have been subdivided into those related to actinidine, mainly pyrindanes Table I-5; and those resembling gentianine which usually have a lactone ring annelated to pyridine Table I-6.

Indicaine and boschniakine were thought originally to differ in stereochemistry at C-7 but have now been shown to be identical (189). The confusion arose because of the formation of a diethylacetal during the preparation of a picrate derivative in ethanol. Indicaine has been found as its *N*-ethyl quaternary salt, indicainine, in *Pedicularis olgae* (187). Actinidine (R = H) and tecostidine (R = OH) occur as their *N*-[β-(4-hydroxyphenyl)-ethyl] quaternary salts (**I-44**) in *Valeriana officinalis* (216). Cantleyine has been shown to be an artifact formed during the treatment of the extraction of the trunk bark of *C. corniculata* with ammonia (198).

Gentianine is now known to be an artifact and has been attributed to the reaction of ammonia, used in extraction, with swertiamarin (**I-56**) or gentiapicrin (**I-57**) found in the plant sources (215).

TABLE I-4.   SIMPLE NATURALLY OCCURRING PYRIDINE ALKALOIDS

Pyridine	Source	Reference
Phenopicolinic acid (I-23)	*Paecilomyces* AF2562	169
Melochinine (I-24)	*Melochia pyramidata*	170
Anibine (I-25)	*Aniba duckei*	171
Duckein (I-26)	*Aniba duckei*	171
Proferrorosamine (I-27)	*Pseudomonas roseus fluorescens*	172
Uvitonic acid (I-28)	*Pseudomonas roseus fluorescens*	173
Caerulomycin (I-29)	*Streptomyces caerulus*	174
1-Methylpyridinium iodide	*Vandopsis longicaulis*	175
1-Methylpyridinium[a]	The oyster, *Crassostrea gigas*	176
1-Methyl-2-picolinium[a]	The oyster, *Crassostrea gigas*	176
3-Butylpyridine	*Fusarium* species	145
2-Heptylpyridine	Bontebok, *Damaliscus dorcas dorcas*	177

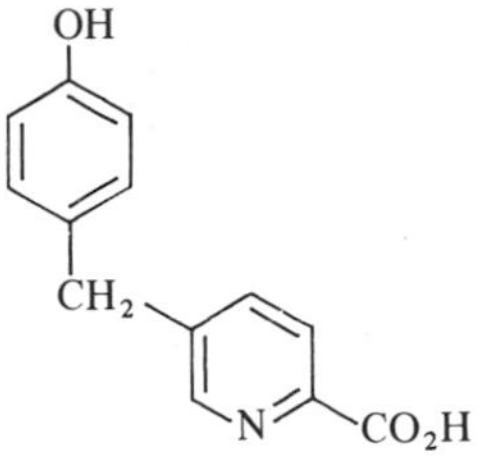

I-23

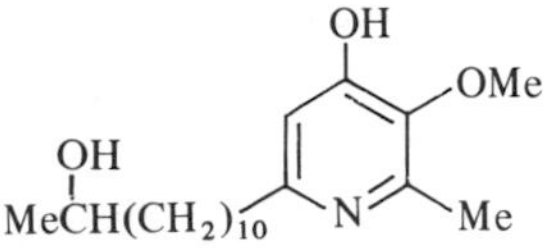

I-24

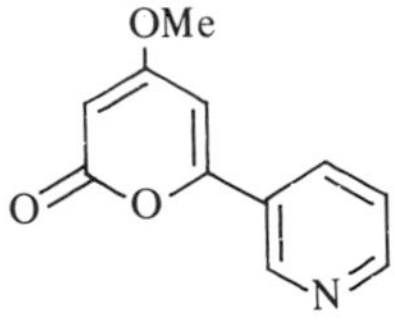

I-25

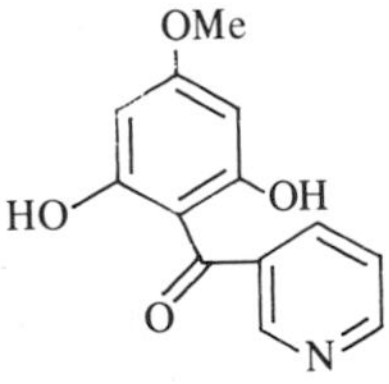

I-26

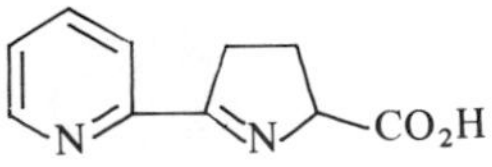

I-27

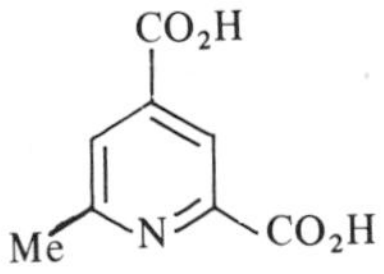

I-28

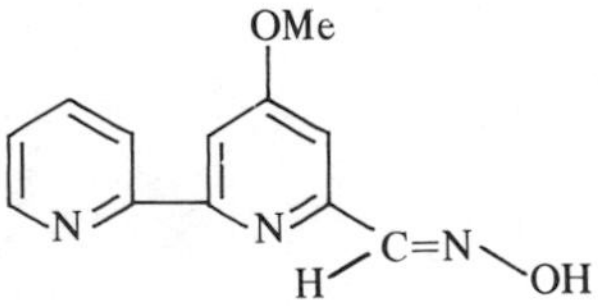

I-29

[a]Counterion not quoted

TABLE I-5.   ACTINIDINE AND RELATED ALKALOIDS

Alkaloid	Species	$[\alpha]_D$	Reference
Actinidine (I-30)	*Actinidia polygama*	−7.2°	178–180
	A. arguta		181
Noractinidine (I-31)	*Tecoma stans*	+3.0	182
	Pedicularis macrochila		183
Valerianine (I-32)	*Valeriana officinalis*	−10.5	184
Tecostidine (I-33)	*Tecoma stans*	−4.0	185, 186
Indicaine (I-34)	*Pedicularis olgae*	+21.02	187
(boschniakine)	*Tecoma stans*		182
	Boschniaka rossica		189
Plantagonine (I-35)	*Plantago indica*	+30.8	190
	P. psyllium		191
	Pedicularis olgae		192
	Verbascum songaricum		193
	Pedicularis macrochila		183
Venoterpine (I-36)	*Alstonia venenata*	+27.0°	194, 195
(RW47)	*Rauwolfia verticillata*		196
Unnamed (I-37)	Jasminum sp. NGF29929	−34°	197
Cantleyine (I-38)	*Cantleya corniculata*	−40 ± 2	198
	Lasianthera austrocaledonica		199
Pedicularidine (I-39)	*Pedicularis olgae*		200
Pedicularine (I-40)	*P. olgae*	−15.3	201
Pediculine (I-41)[a]	*P. olgae*	+61.5	202
Pediculidine (I-42)	*P. olgae*	−	203
Pediculinine (I-43)	*P. olgae*	−	204

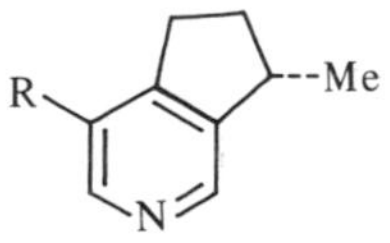

I-30 R = Me
I-31 R = H
I-32 R = CH$_2$OMe
I-33 R = CH$_2$OH
I-34 R = CHO
I-35 R = CO$_2$H

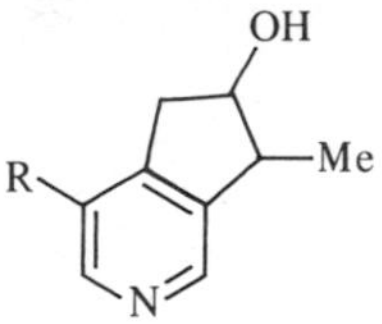

I-36 R = H
I-37 R = CO$_2$Me

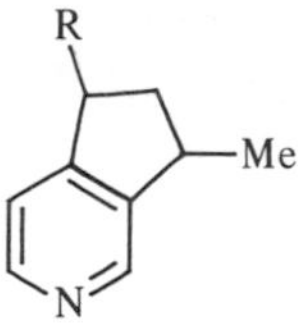

I-39 R = CHO
I-40 R = CO$_2$H

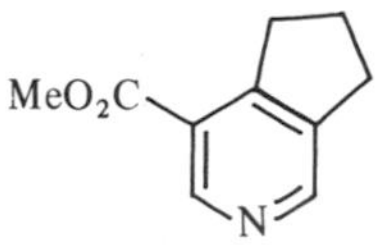

I-38

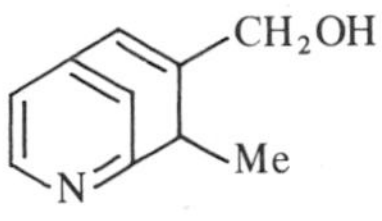

I-41

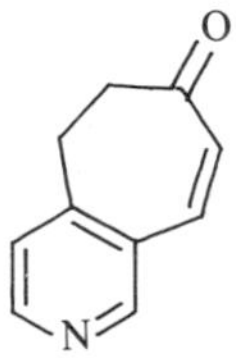

I-42

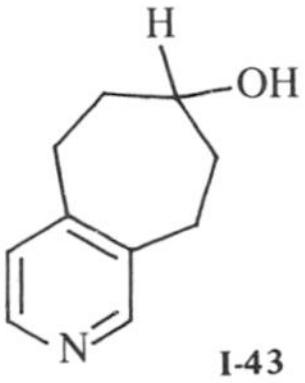

I-43

[a]The proposed structure I-41 appears unlikely.

TABLE I-6. GENTIANINE AND RELATED ALKALOIDS

Alkaloid	Species	m.p.	Reference
Gentianine (I-45)	*Gentiana asclepiadea*	81–82°	121
	G. olgae		205
	G. olivieri		206
	G. tianshanica		205
	G. vvedenskyi		205
	Swertia connata		205
	Dipsacus azurcus		207
	Erythraea centaurium		208
Gentianidine (I-46)	*G. olgae*	131–132°	205
	G. asclepiodae		121
Gentianadine (I-47)	*G. olgae*	77–78°	205
Gentianamine (I-48)	*G. olivieri*	149–150°	206
Gentioflavine (I-49)	*G. tianchanica*	218–220°	205
	G. olivieri		206
	G. olgae		205
	Swertia connata		205
Fontaphilline (I-50)	*Fontanesia phillyreoides*	80–81° and 121–122°	209
Gentiatibetine (I-51)	*G. tibetica*	161–165°	214
Oliveridine (I-52)	*G. olivieri*	260° (dec.)	206
Oliverine (I-53)	*G. olivieri*	206–207°	210
Oliveramine (I-54)	*G. oliveri*		211
Jasminine (I-55)	*Olea paniculata*	174.5–176°	212
	Jasminum sp. NGF29929		197
	Ligustrum novoguineense		213
	Cantleya corniculata		198

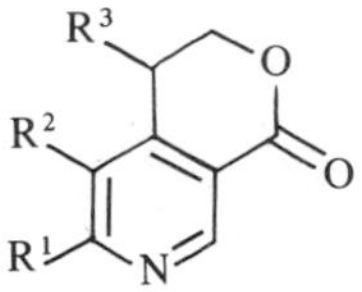

I-45 (R^1 = R^3 = H; R^2 = CHCH$_2$)
I-46 (R^1 = Me; R^2 = R^3 = H)
I-47 (R^1 = R^2 = R^3 = H)
I-48 (R^1 = H; R^2 = CHCH$_2$; R^3 = CH$_2$OH)

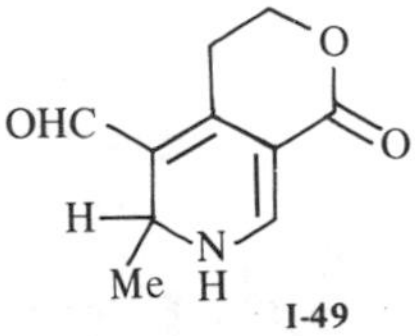

I-49

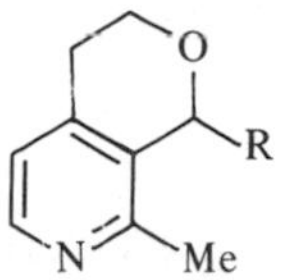

I-51 (R = OH)
I-52 (R = OMe)
I-53 (R = OMe + additional OMe)

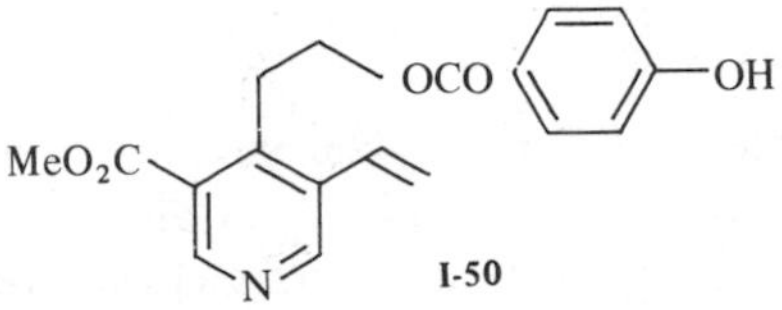

I-50

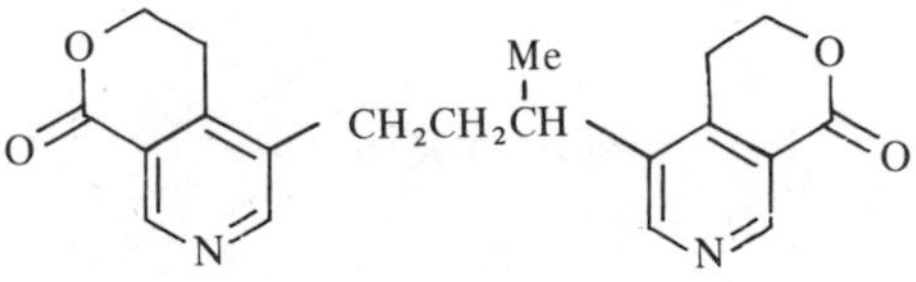

I-54

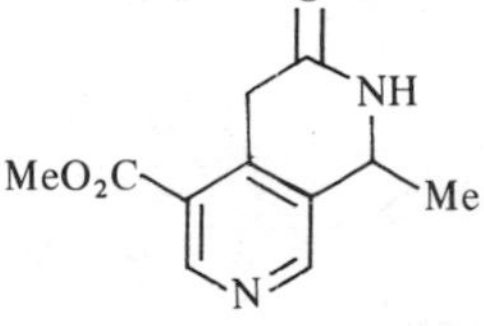

I-55

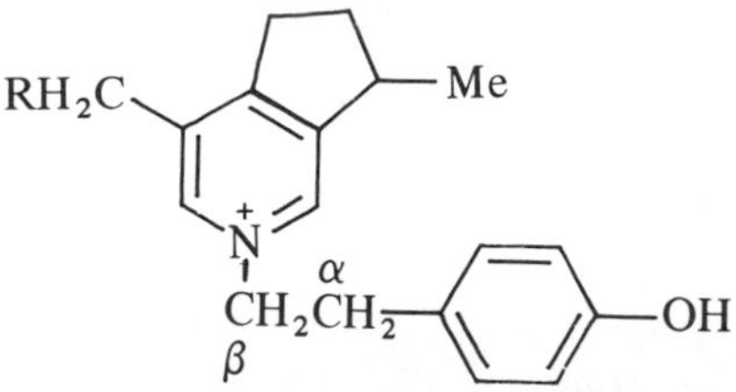

I-44

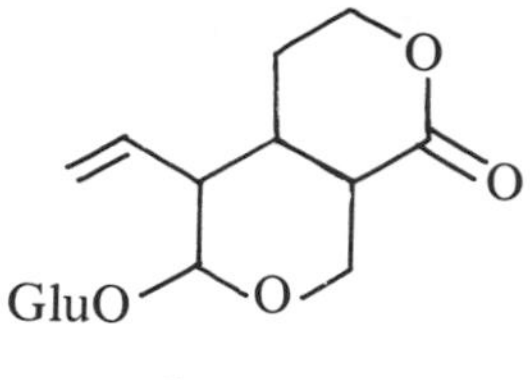

I-56 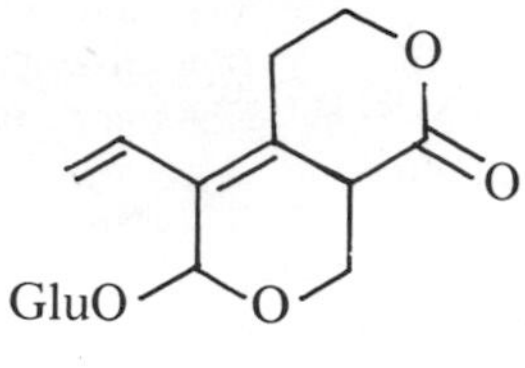   I-57

Fontaphilline, which underwent cyclization to gentianine on treatment with acid, has been shown also to be an artifact (212).

c.   SESQUITERPENOID ALKALOIDS

i.   DERIVATIVES OF NICOTINIC ACID.   Several steroidal nicotinates are known for example, rostratine (I-58) from *Marsdenia rostrata* (217) and two alkaloids kondurangamin *A* and *B* from *Cortex condurango* (218).

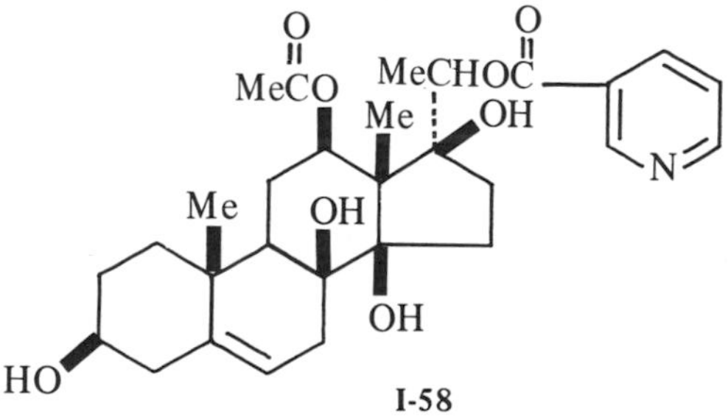

I-58

Halfordinols (I-59), which contain nicotinate masked as an oxazole, have been isolated from rutaceous plants, for example, *Halfordia scleroxyla* (219), *Aeglopsis chevalieri* (220), and *Amyris plumieri* (221). *A. plumieri* has also yielded two nicotinamides (I-60, I-61), that are related to halfordinols (221).

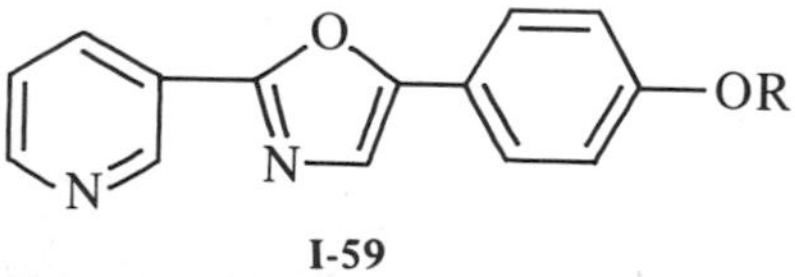

I-59

R = CH(OH)CMe$_2$OH
  = H
  = CHCMe$_2$
  = CH$_2$CMeCH$_2$

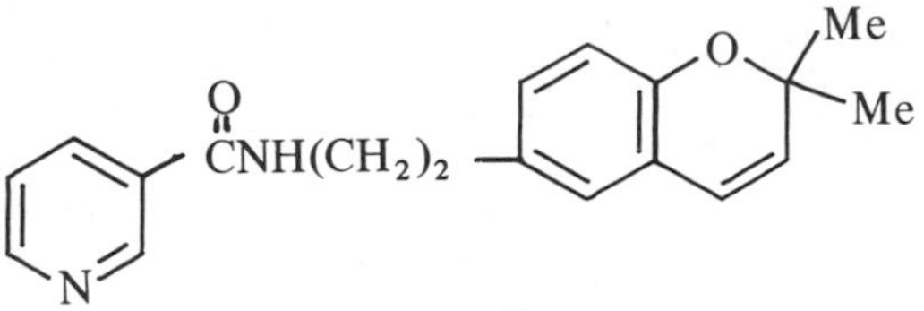

I-60

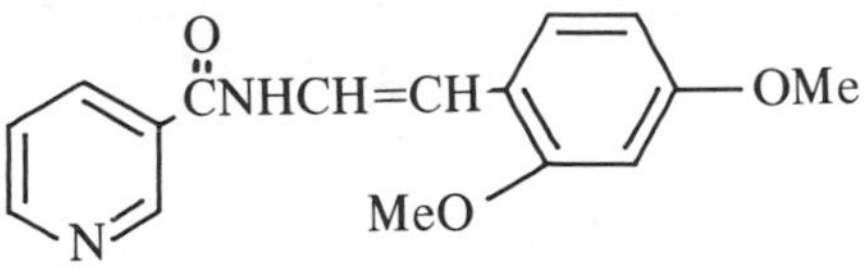

I-61

Sesquiterpenoid pyridine alkaloids, which have been receiving greatest attention, are the euonyminol (I-62) esters of evoninic (I-63), wilfordic (I-64), and hydroxy-wilfordic (I-65) acids, which occur in plants of the Celastraceae family (Table I-7).

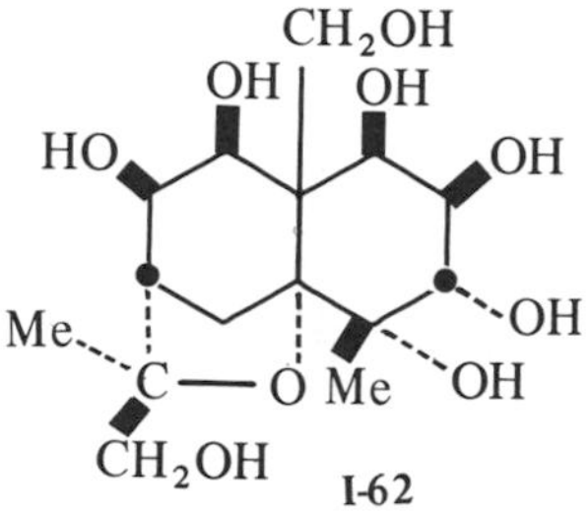

I-62

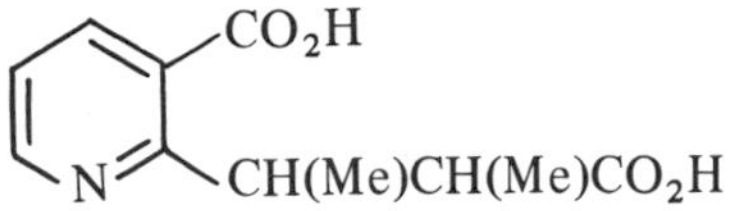

I-63

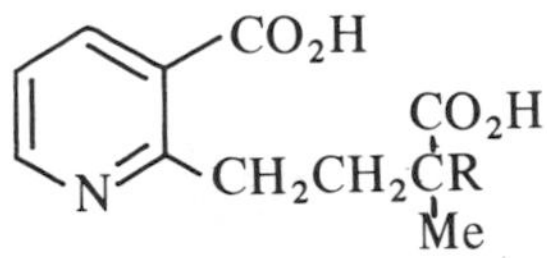

I-64 (R = H)
I-65 (R = OH)

Of topical interest are 11 new celastraceous alkaloids that have been isolated from *Catha edulis* "(khat)", a tree found in parts of East Africa and the Yemen. "Khat" is a drug, permissible to Islam, that has an action similar to that of amphetamine. Bizarre behavior patterns as a result of its use have stimulated a search for the identity of its active ingredients. Extracts so far characterized fall into three categories: (a) cathedulins E1 and E2, which are esters of pentahydroxy-dihydroagarofuran (222); (b) cathedulins K1 (Y1), K2, (I-66) K6, and K15, esters of euonyminol containing one lactone bridge (223); (c) cathedulins E3 (K11) (I-67) E4, E5, E6, and K12, which are more complex esters of euonyminol containing two dilactone bridges (224). The cyphers K (Kenya), E (Ethiopia), and Y (Yemen) refer to the nation of origin of the khat.

Clitidine (I-68), a toxic component of the dokusasako toadstool (*Clitocybe acromelalga*), has been extracted with hot water and it exhibited vasodilator action in dogs (225).

ii. PYRIDONE AND PYRIDINOL ALKALOIDS. The observation of a broad spectrum of biological activity among sesquiterpenoid pyridones has provided a *raison d'etre* for extensive study of compounds of the type shown in Table I-8.

TABLE I-7.   THE CELASTRACEAE ALKALOIDS

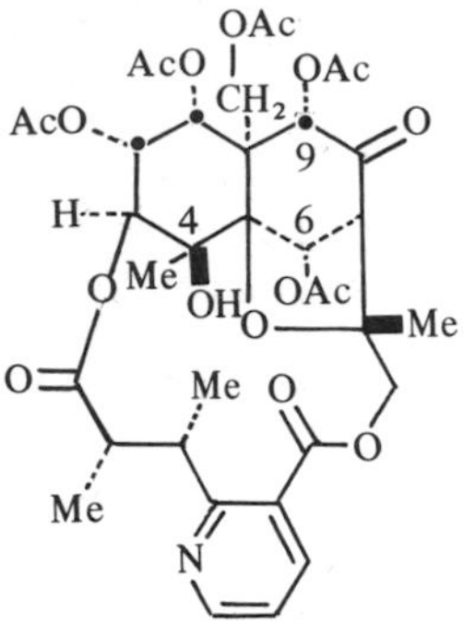

Alkaloid	Species	Reference

Alkaloid	Species	Reference
Evonine	*Euonymus sieboldiana*	226–227
	E. europaeus	228–230
	E. alatus f. striatus	231
Neoevonine (evorine) [O^6-deacetyl]-	*E. sieboldiana*	226–227
	E. europaeus	228, 230
	E. alatus f. striatus	231
Euonymine [8-(acetyloxy)-8-deoxo]-	*E. alatus*	231, 232
	E. sieboldiana	227, 233
Neoeuonymine [O^6-deacetyl-8-(acetyloxy)-8-deoxo-]	*E. sieboldiana*	227, 233
Evozine [O^6, O^9-didacetyl]-	*E. europaeus*	230, 234
Evonoline [4-deoxy]	*E. europaeus*	228

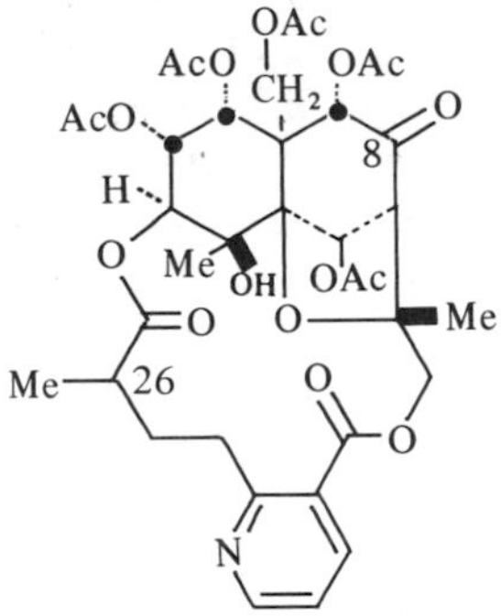

Alkaloid	Species	Reference
Evonimine	*E. sieboldiana*	235
Euonine [8-(acetyloxy)-8-deoxo]-	*E. sieboldiana*	235
Wilfordine [8-(acetyloxy)-O^2-benzoyl-O^2-deacetyl-8-deoxo-26-hydroxy-]	*E. alatus*	232
Alatamine [O^2-benzoyl-O^2-deacetyl-26-hydroxy-]	*E. alatus*	232

TABLE I-7. *(CONTINUED)*

Alkaloid	Species	Reference

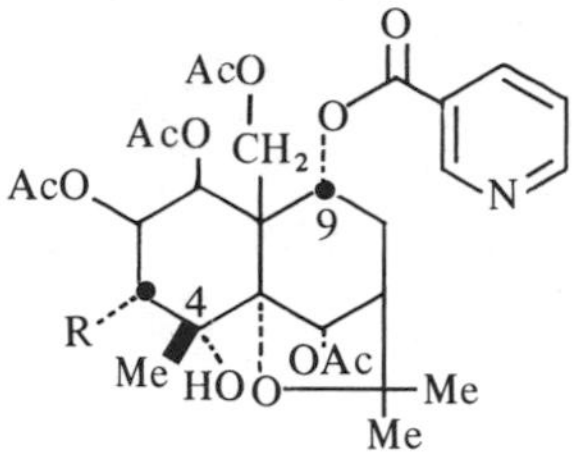		
Maytine (R = OH)	*Maytenus ovatus*	236
Maytoline (R = H)	*M. ovatus*	236

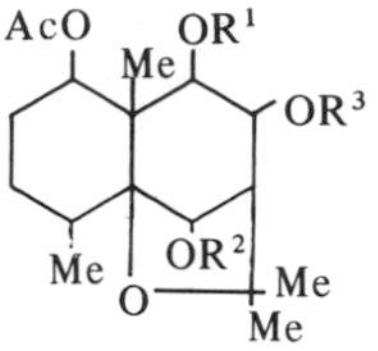		
Celapanin [R¹ = nicotinyl, R² = β-furoyl, R³ = Ac]	*Celastrus paniculatus*	237
Celapanigin [R¹ = nicotinyl, R² = benzoyl, R³ = Ac]	*Celastrus paniculatus*	237
Celapagin [R¹ = nicotinyl, R² = benzoyl, R³ = H]	*Celastrus paniculatus*	237
Cathadulins E2, E8	*Catha edulis*	222
Cathadulins K1, K3, K6, K15	*Catha edulis*	223
Cathadulins E3, E5, E6, E12	*Catha edulis*	224

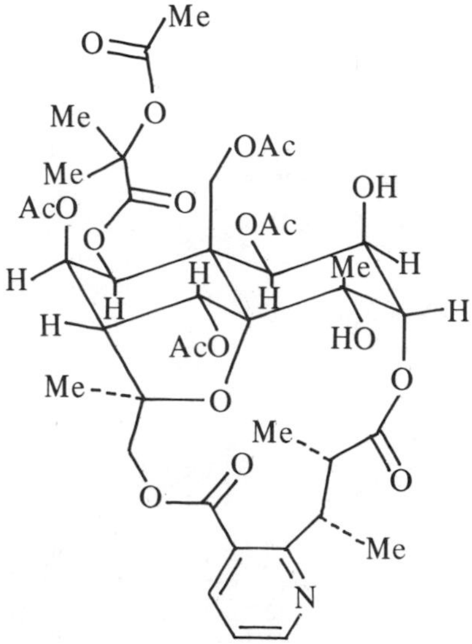

I-66

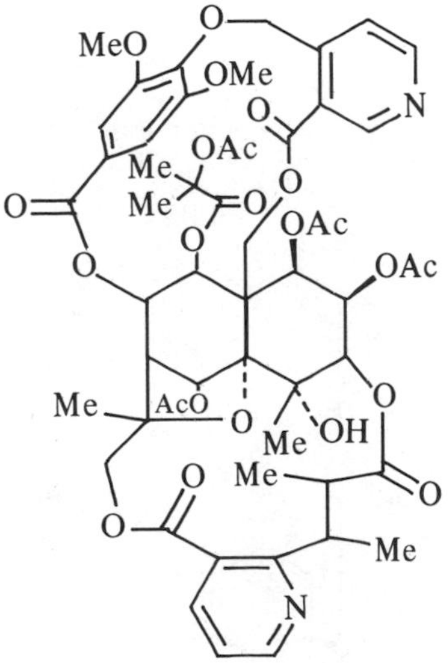

I-67

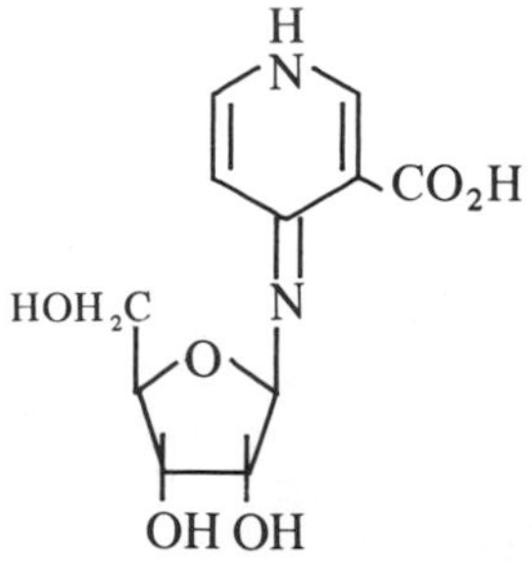

I-68

TABLE I-8.   PYRIDONE AND PYRIDINOL SESQUITERPENOID ALKALOIDS

Alkaloid	Species	Activity	Reference
Tenellin (I-69)	*Beauveria tenella* and *B. bassiana*	–	238–240
Bassianin (I-70)	*Beauveria tenella* and *B. bassiana*	–	238–240
Ilicicolin H (I-71)	*Cylindrocladium ilicicola*	Antifungal, antibiotic	241, 242
Funiculosin (I-72)		Antifungal, antibiotic	243, 244
Mocimycin (I-73)	*Streptomyces ramocissimus*	Antibiotic	245–248
Innovanamine (I-74)	*Evodiopanax innovans*	–	249, 250
Piericidin A (I-75)	*Streptomyces* species	Antibiotic, herbicidal	251–254
Piericidin B (I-76)	*Streptomyces* species	Insecticidal	251–254
Pyridomycin (I-77)	*Streptomyces albidofuscus*	Antibiotic	255
Amylocyanin (I-78)	*Streptomyces coelicolor*	Antibiotic	256
Flavipucine (I-79) (glutamicine)	*Aspergillus flavipes*	Antibiotic	280

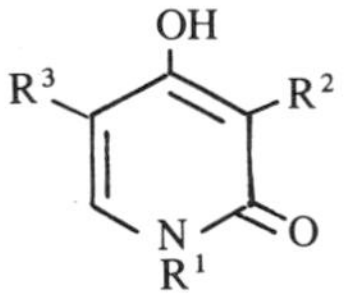

I-69 ($R^1 = -OH$; $R^2 = -COCH=CHCMe=CHCHMeEt$; $R^3 = -C_6H_4OH$-p)
I-70 ($R^1 = -OH$; $R^2 = -CO(CH=CH)_2CMe=CHCHMeEt$; $R^3 = -C_6H_4OH$-p)

I-71 ($R^1 = H$;

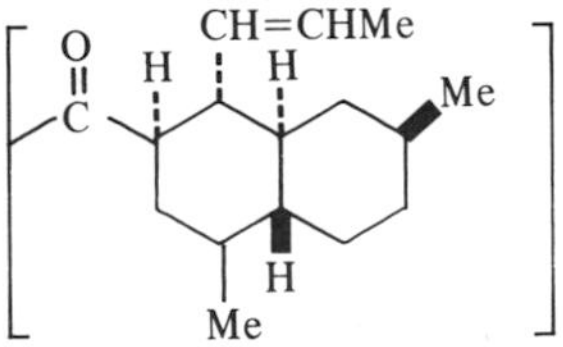

; $R^3 = -C_6H_4OH$-p)

I-72 ($R^1 = Me$

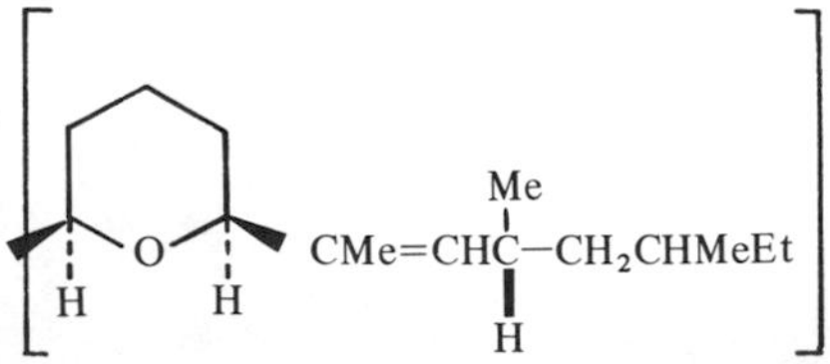

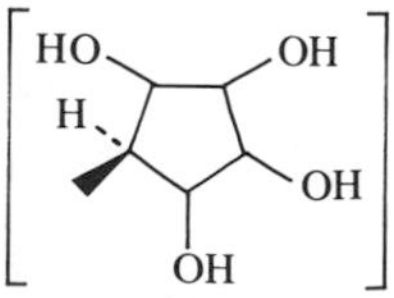

TABLE I-8. *(CONTINUED)*

I-73 $(R^1 = R^3 = H$

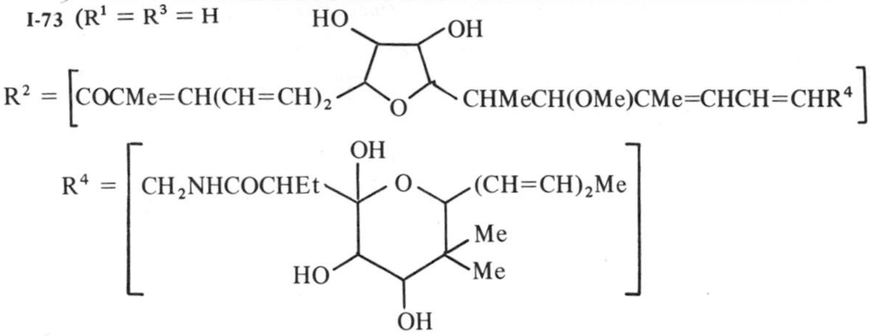

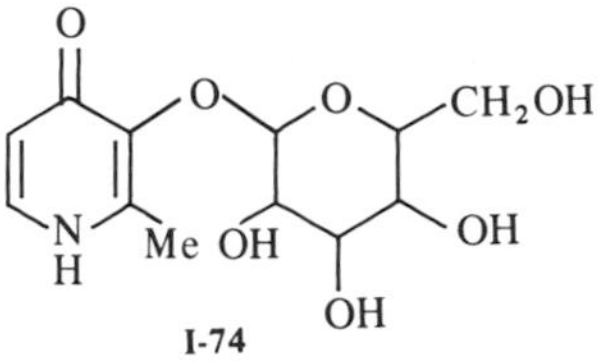

$$R^2 = \left[COCMe=CH(CH=CH)_2 \underset{}{\overset{HO\quad OH}{\bigcirc}} CHMeCH(OMe)CMe=CHCH=CHR^4 \right]$$

$$R^4 = \left[CH_2NHCOCHEt \underset{HO\quad\quad OH}{\overset{OH}{\bigcirc}} (CH=CH)_2Me \quad Me \quad Me \right]$$

I-74

MeO — OH — Me

MeO — N — $CH_2CH=CMeCH_2CH=CH-CMe=CHCHMeCH(OR)CHMeCH_2Me$

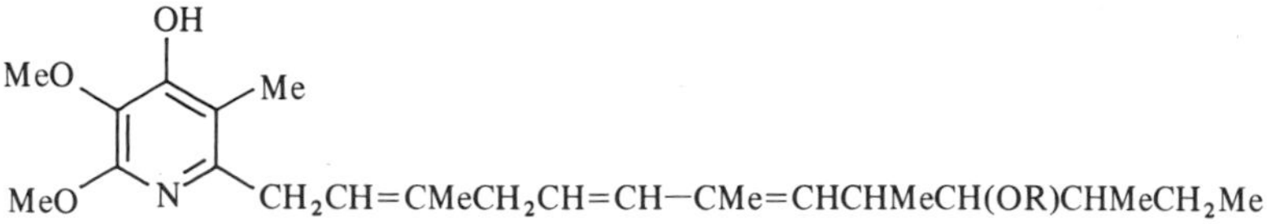

I-75 (R = H)
I-76 (R = Me)

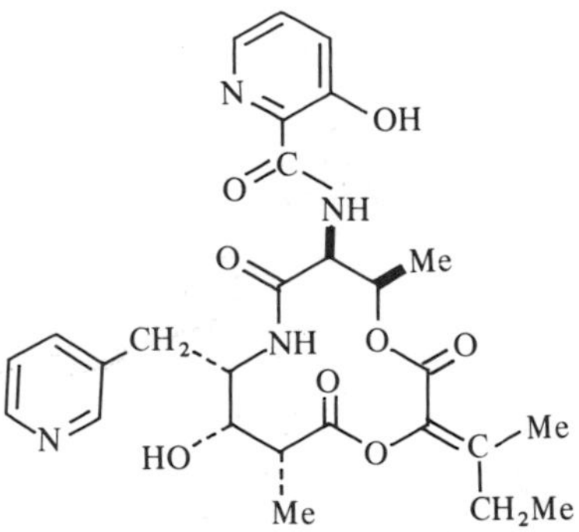

I-77

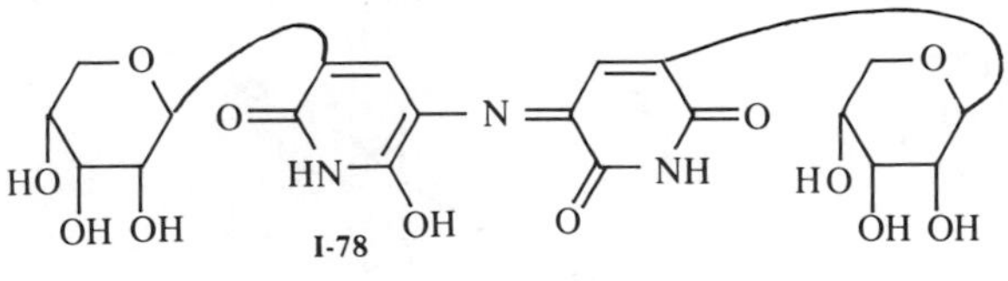

I-78

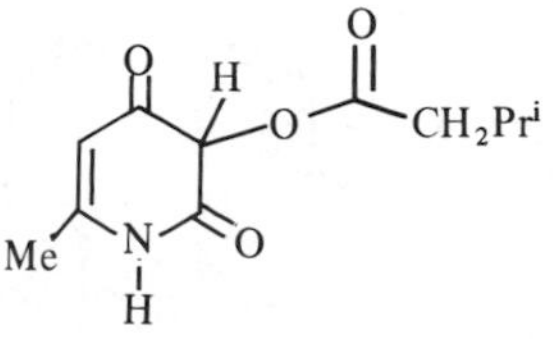

I-79

iii.   OTHER SESQUITERPENOID ALKALOIDS.   Navenone-A **(I-80)** is one of the three pheromones produced by the sea slug *Navanax inermis* (257). Guaipyridine **(I-81)** has now been shown to have the stereochemistry illustrated (258). Sesbanine **(I-82)**, has been obtained from the seeds of *Sesbania drummondii*, which have cytotoxic and antileukemic activity (259).

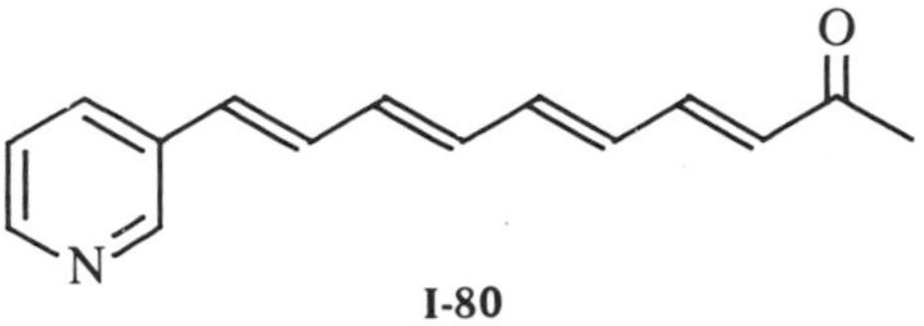

I-80

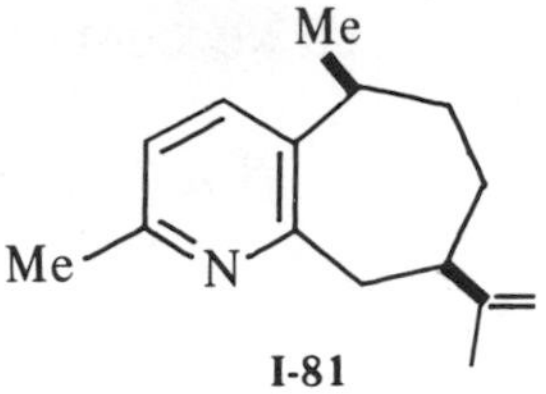

I-81

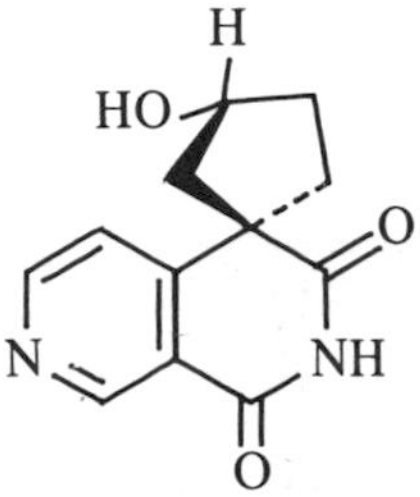

I-82

Pyridine annelated to a cyclohexane ring has been shown to occur in several well-known alkaloid series; namely the sceletium alkaloids, A4 **(I-83)** (188, 260) and tortuosamine **(I-84)** (261); lycopodium alkaloids lycodine **(I-85)** and *N*-methylcodine **(I-86)** (262, 263); fabianine **(I-87)** (264), and the extractives of *Lobelia syphilitica*, syphilobines-A **(I-88)** and F **(I-89)** (265).

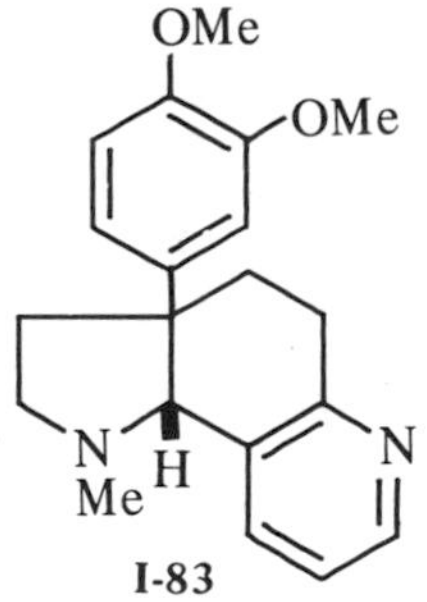

I-83

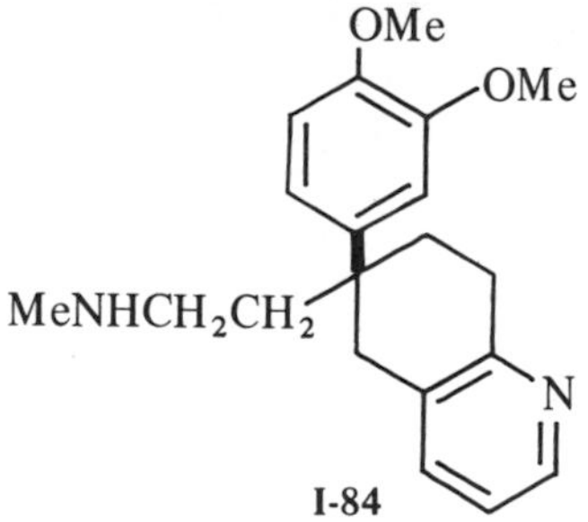

I-84

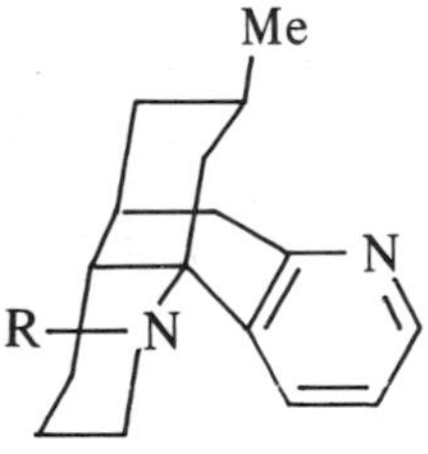

I-85   R = H
I-86   R = Me

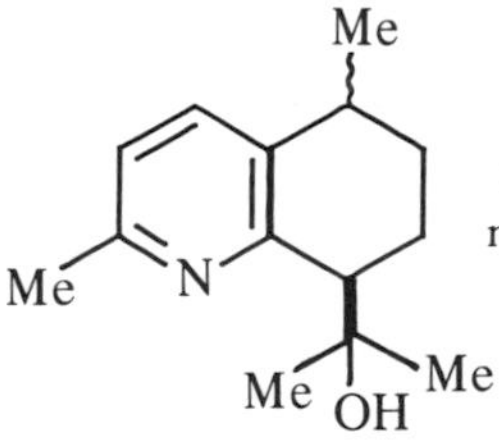

50:50 mixture

I-87

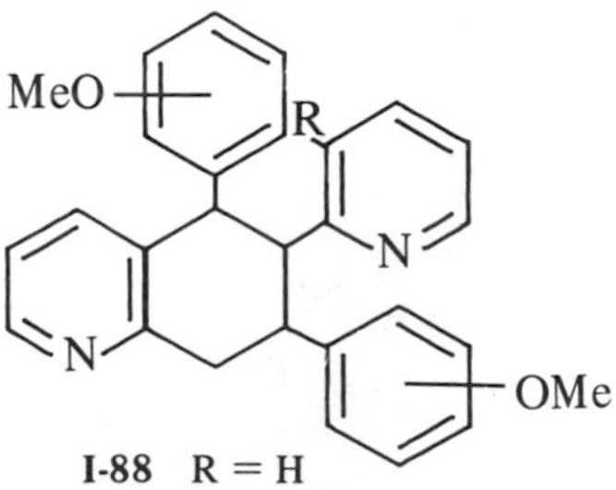

I-88   R = H
I-89   R = OH

Pyridines from Natural Sources 23

d. β-CARBOLINE AND RELATED ALKALOIDS THAT CONTAIN A PYRIDINE RING

Potent antitumor and antileukemic properties of extracts from *Vinca* and *Camptotheca* species have excited great interest in alkaloids from these plants. Some of those that contain pyridine either attached or annelated to another heterocyclic system appear in Table I-9. The structures of vincarpine (**I-90**), naucledine (**I-91**), naufoline (**I-92**), camptothecin (**I-93**), and streptonigrin (**I-94**) are of particular note.

TABLE I-9. β-CARBOLINE AND RELATED ALKALOIDS CONTAINING A PYRIDINE RING

Alkaloid	Species	Reference
Vincarpine[a] (**I-90**)	*Vinca major*	266
18,19-Dihydrovincarpine[a]	*Vinca major*	266
6,7-Dihydroflavopereirine[a]	*Strychnos usambarensis*	267
Cadamine[a]	*Anthrocephalus cadamba*	268
Isocadamine[a]	*Anthrocephalus cadamba*	268
Pauridianthine[a]	*Pauridiantha callicarpoides*	269
Pauridianthinine[a]	*Pauridiantha callicarpoides*	269
Naucledine[a] (**I-91**)	*Nauclea diderichii*	270, 271
Nauclexine[a]	*Nauclea diderichii*	270, 271
Nauclederine	*Nauclea diderichii*	270, 271
Nauclechine	*Nauclea diderichii*	270, 271
Naufoline[a] (**I-92**)	*Nauclea latifolia*	272
Decarbomethoxynauclechine[a]	*Nauclea latifolia*	272
Camptothecin (**I-93**)	*Camptotheca acuminata*	273, 274
Mappicine	*Mappia foetida*	275
Rubrolone	*Streptomyces echinoruber*	276
Onychine	*Onychopetaleum amazonicum*	277
Streptonigrin (**I-94**)	*Streptomyces flocculus*	278, 279

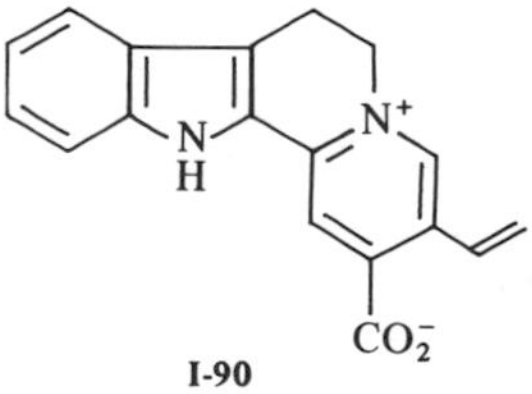

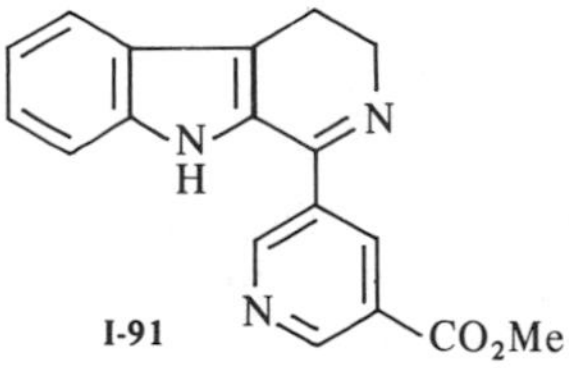

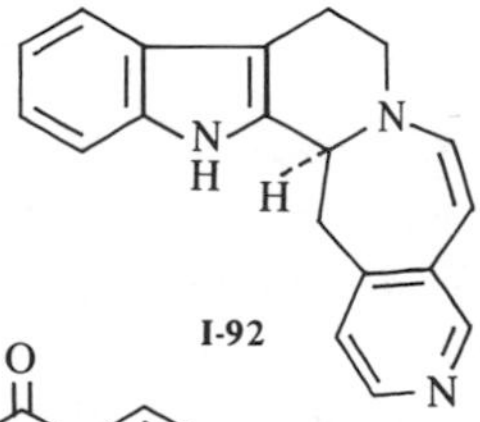

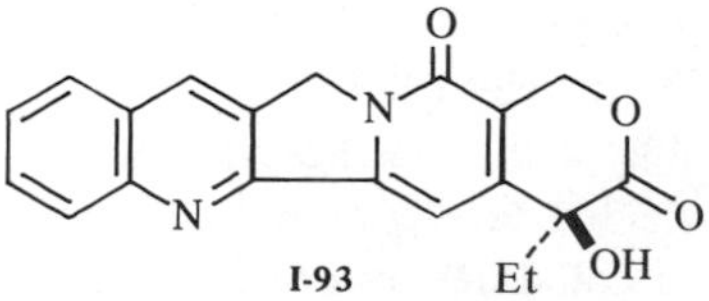

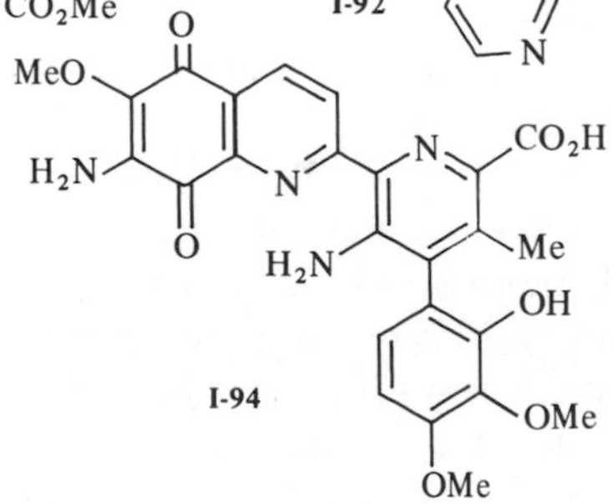

[a]Contains the β-carboline system.

TABLE I-10.   PHYSICAL DATA FOR PYRIDINES EXTRACTED FROM *NAUCLEA*
DIDERICHII

Compound	m.p. ° (solvent)	Derivative m.p. °	[α], (MeOH)

I-95, R = OMe	43–46°C (hexane)		+ 50[a]
I-96, = OH	53–56°C (acetone)	p-nitrobenzoate, 139–142°	+ 23[a,c]
I-97, = NH₂	Syrup	p-nitrobenzoyl, 154–157°	+ 27[b]

I-98	50–75°C (hexane)		
	(sticky needles)		

[a] At 26°C.
[b] At 25°C.
[c] [α] varies with concentration.

Extractives of *N. diderichii*, apart from the alkaloids in Table I-9, are particularly pertinent to this chapter, as they also include four simple pyridines (**I-95–98**). The physical properties of those pyridines are given (Table I-10); surprisingly for such simple compounds, two (**I-96, 98**) have only recently been reported (271), and physical data for the other two (**I-95, 96**) have only appeared in a patent (282).

2.   Degradation of Natural Products

A.   Coal

Coal is formed from plant life that died millions of years ago which subsequently was covered up and subjected to high temperature and pressure in the absence of oxygen. A scheme has been proposed for the formation of pyridines in coal by condensations of amino acids and carbohydrates (283). An examination of 96 Kuznetsk gas coals revealed that they contained 3.25% nitrogen. Under conditions that precluded secondary pyrolysis of gaseous products, 7% of the total nitrogen was found to be present as six-membered heterocycles (284).

Pyridine derivatives are usually obtained from coal by coking (a thermal cracking process). High-temperature coking (900–1200°C) yielded tars containing methylpyridine derivatives, that is, picolines, lutidines, and collidines; use of lower temperature gave tars that contain more aliphatics. Conditions necessary for optimum yields of desired pyridines have received considerable attention (285–289). Oxidation of coal in an alkaline medium yielded naphthalene and pyridinecarboxylic

TABLE I-11.　SEPARATION OF BASES INVOLVING PHYSICAL METHODS

Bases	Method	Separation of:	Reference
3- and 4-Picoline; 2,6-lutidine	Addition of $(NH_4)_2SO_4$ or NH_4OH before distillation	All three	303
3- and 4-Picoline; 2,6-lutidine	Extraction distillation with ethylene glycol	4-Picoline (98.6% pure)	307
3- and 4-Picoline; 2,6-lutidine	Azeotropic distillation with water and H_2SO_4	4-Picoline isolated	305
3- and 4-Picoline; 2,6-lutidine	Distillation on a 55 theoretical plate column	2,6-Lutidine	306
3- and 4-Picoline; 2,6-lutidine; 2-ethylpyridine	(1) Extraction with benzene (2) Azeotropic extractive distillation	2,6-Lutidine (93% yield; 95%, pure)	307
3-Picoline fraction	–	2,6-Lutidine (95% pure)	308
3-Picoline fraction	Azeotropic distillation	3- and 4-Picoline in 87% purity	309
Coking ammonia water	(1) Extraction with benzene (2) 2-stage countercurrent extraction with H_2SO_4 and NH_4OH	Pyridine bases	310
Coke oven gas	(1) H_2SO_4 extraction (2) Neutralization (3) Distillation	Light pyridine bases	311
High-temperature coal carbonization, lutidine fraction	Distillation on an 80-theoretical plate column	2,3-Lutidine (95% pure) 2,4-Lutidine (95% pure) 2,3,6-Collidine (95% pure)	312

acid (290). Alkylpyridines were dealkylated over catalysts containing oxides of vanadium, silver, chromium, molybdenum, or tungsten (291). Pyridine, quinoline, and related heterocycles were reduced further to ammonia without hydrogenation of aromatic hydrocarbons by passage over Co—Mo, WS_2—NiS, and MoS_2—NiS catalysts on alumina at 360–400°C (292).

A list of all the organic compounds detected in high-temperature coal tar was published in 1967 and it includes 242 references (293). In a more recent study, additional high boiling pyridines (viz. 4-methyl-2-propyl-; 5-methyl-2-propyl-; 2,5-diethyl-; 3-methyl-2,6-diethyl-; 4-methyl-2,6-diethyl-; 2,3-dimethyl-6-propyl-pyridine), have been found in the aniline fraction of coal tar (294, 295). GLC analyses of the liquid and gaseous products from the slow burning of coal in an oven at 200–250°C indicated that over 800 compounds were formed; 17 alkyl-pyridines (296) were among the 397 identified.

Tar from the pressure gasification of brown coal was separated into four fractions by distillation, and the first two fractions were found to contain picolines,

TABLE I-12.   SEPARATION OF BASES BY FORMATION OF SALTS OR COMPLEXES

Bases	Method	Separation of:	Reference
Isomeric pyridines	(1) Treated with o-cresol, $-10°$C (2) Centrifuged (3) Caustic added, steam distilled	sym-Collidine	313
Mixture or pyridines	(1) $CaCl_2$, ppted. (2) Filtered, steam distilled	sym-Collidine	314
Coal tar collidine fraction	Complex formed with $CuCl_2$ in MeOH	sym-Collidine; 3,4 and 3,5-lutidine; 3-ethylpyridine	315
Brown coal tar collidine fraction	(1) Treated with HCl (2) Azeotropic distillation (3) NaOH	2,3,6-Collidine (pure); sym-collidine (technical grade)	316 317
Mixture of coal tar bases	$CuCl_2$ in MeOH	2-Picoline	318
Pitch coal tar pyridine fraction	$CuCl_2$ in MeOH	3,4-Lutidine ppted.	319
Coal tar collidine fraction	(1) $CaCl_2$ (2) Azeotropic distillation (3) 50% NaOH	3,5-Lutidine (96% pure)	320
Coal tar 3-picoline fraction	(1) 80% aq. urea soln. (2) 30% NaOH	2,6-Lutidine	321
Coal tar lutidine fraction	(1) HCl/BuOH, (2) base (3) CuCl, (4) base	2,6-Lutidine (pure)	322
Coal tar lutidine fraction	(1) Oxalic acid (2) Azeotropic distillation (3) 50% NaOH	2,4-Lutidine (96% pure)	323 324
Mixture of pyridines	(1) $NiCl_2 \cdot 6H_2O$/KSCN/ 4-picoline/H_2O, 98°C (2) Cool to 20°	2,5-Lutidine separated as a clathrate	325
Mixture of picolines and xylenes	$NiSO_4$/NH_4SCN	3-Picoline (89% pure); 4-Picoline (76% pure)	326 327
Crude pyridine fraction	(1) Diluted with EtOH (2) 85% H_3PO_4 at 65°C (3) NaOH (4) Rectified	2,3-Lutidine	328
2,6-Lutidine, 3- and 4-picolines	(1) Countercurrent extraction with aq. KCNS and with hexane CHCl$_3$ (2) Countercurrent extraction with PhH and aq. NaH_2PO_4	2,6-Lutidine, 4-picoline (aq. phase) 3-picoline (organic phase)	329

lutidines, and collidines (297). Several pyridines have been identified from tar products produced by underground coal gasification, and the effect of gasification conditions on their yield has been discussed (298). The alkaline fraction of soft coal generated tar has been found to contain mainly 2-picoline, 2,6-lutidine, and sym-collidine (299). The content of recycle solvent used in solvent-refined coal processing has been determined, and alkylpyridines were among the constituents

TABLE I-13.   PURIFICATION OF BASES

Base	Method	Reference
Pyridine	(1) pH adjusted to 5 with H_2SO_4 (2) Active carbon (3) Filtered and neutral KOH	332
Pyridine or 2-picoline	(1) Cl_2 gas (2) Distillation	333
2,6-Lutidine	Preparative GC using glycerol and squalene as stationary phases gives product 99.99% pure	334
Pyridine sulfates	Tar removed by extraction with benzene and simultaneous neutralization with ammonia	335
2,6-Lutidine	Picoline impurities removed by complex formation with $CuCl_2$ or $ZnCl_2$ and GLC	336

identified (300). GC–MS analysis has shown picolines and lutidines to be constituents of soot (301).

The techniques of separation and methods of recovery of pyridines from coal tar have been of undiminished interest; the recent ones are listed in Tables I-11 and I-12. This topic has also been reviewed (302).

Two new extraction procedures for the separation of 3-picoline from two phase mixtures have been reported (330, 331). Some new purification methods appear in Table I-13.

Gas chromatography is usually the method of choice for both qualitative and quantitative analyses of tar base fractions containing pyridines. Relative retention times for C_1–C_{15} alkyl- and alkenylpyridines have been determined on different columns at various temperatures (337). Conditions have been established for the estimation of pyridines in coal tar light oil using capillary glass columns coated with either polyethylene glycol 400, Apiezon K-Slovamin 20, Amine 220, or Reoplex 400, as stationary phase (338). Another method involved the conversion of bases into their hydrochloride salts, hydrogenation with Adams catalyst, and chromatography as their pentafluoropropionic amides (339). Combined GC–MS has been used for analysis of the naphthalene and absorption fractions of coal tar (340). This technique has also been found to provide a fast, accurate method for the determination of pyridines (first precipitated as their hydrochloride salts) in the light oil produced by the catalytic hydrodesulfurization of coal (341). Mass spectroscopy has been used to identify pyridines among the photochemical oxidation products of coal (342).

Pyridines and ammonia in coke gas have been determined by a potentiometric method (343). An improved spectrophotometric procedure for the estimation of pyridines in the working area of a coking plant has been developed (344). Infrared spectrophotometry has allowed the estimation of pyridine bases in cane peat (345) and established their distribution in high-temperature coking products (346). A comprehensive study of the pyridine constituents of the aniline fraction of coal tar has utilized GC, IR, and NMR (347, 348). Partition coefficients (between aqueous buffer solutions of pH 2.87–6.10) and hexane, chloroform, butanol, and ethyl acetate have been measured for pyridine, 3- and 4-picoline, 2,4-lutidine, and 2,4,6-collidine obtained from coal liquefaction (349).

Reviews have appeared on problems of carbolic acid processing, which include the recovery of pyridines (350); and trends and prospects for the use of pyridine and its derivatives (351).

B. Petroleum

Methods for the extraction of bases from Khandag and Dehar–Kurgan petroleums have been described (352, 353). The darkening of some light petroleum distillates and formation of resinous tars in medium distillates on storage prompted an investigation of the nitrogen compounds in these fractions (354, 355). Pyridine, alkylpyridines, anilines, quinoline, and pyrrole were characterized. The nitrogen base fraction of gasoline from hydrogenation of the deasphaltate of Arlan petroleum

TABLE I-14.   PYRIDINES IDENTIFIED IN PETROLEUM

Base	Method	Origin	Reference
2,4-Lutidine, sym-collidine	GC–MS	Gasoline fraction of Arlan petroleum	361
2,4-Lutidine, 2,3,6-collidine, 2-picoline, other methyl- and ethylpyridines	GC	Kerosene and gasoline fractions of Emba and Mangyshlak crudes	362
Alkylpyridines	GC, MS, IR, UV, NMR	Wilmington petroleum	363

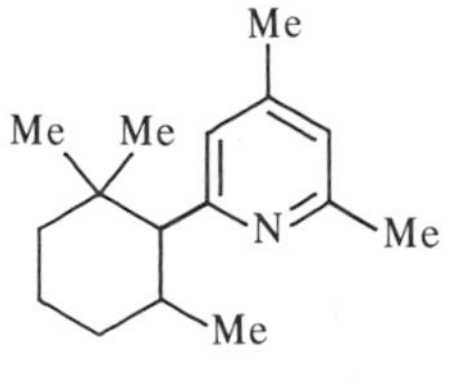

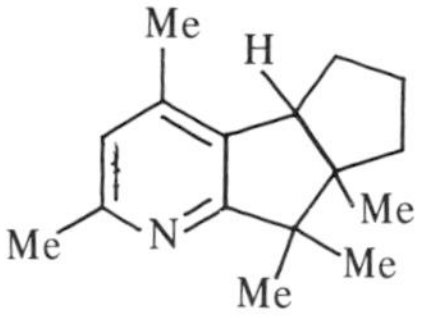

Base	Method	Origin	Reference
Pyridines, pyridones, and carboxylic acids	GC, MS, IR, UV	400–700°C fraction of California crude	364
Alkylpyridines and ethylpyridine	Cation exchange, GC	Naphtha	365
Alkylpyridines	GC, UV, IR, MS	Gasoline fraction of Arlansk petroleum	366
Alkylpyridines	GC–IR	Gasoline and gas oil fractions	367
Pyridines, cycloalkylpyridines, phenylpyridines	GC–MS, IR, UV	Hydrocracking of diesel fuel and coker gas oil	368

contains 22.2% of alkylpyridines (354). It has been suggested that the pyridine content of subsurface waters may provide a guide when prospecting for oil. Surface waters in productive regions of the USSR have a higher pyridine content than the nonproductive regions (356).

Analytical techniques used to determine pyridine contents of oils are the same as those used in the coal tar industry. A combination of GC, IR, MS, and fluorescence methods has been employed to study the distribution in a series of crudes with different geological characteristics (357–359). A GC method has been developed which can detect the principle nitrogen bases in light catalytic cycle oil and vacuum gas oil (360). Some pyridines that have been identified in petroleum are listed in Table I-14.

C. Shale

The potential use of shale oil, produced by combustion retorting of oil shale, as an energy source has created an impetus for the study of its composition and properties. The nitrogen content of crude shale oils (usually more than twice that in high nitrogen petroleum crudes) caused problems during refining. Therefore, crude shale oil was upgraded by catalytic hydrogenation to produce a suitable refinery feedstock, "syn-crude", the characteristics of which have been reported (369).

Tar base concentrate from hydrocracked shale oil naphtha was separated by GC, and the fractions obtained were analysed by NMR, IR, and mass spectrometry (370). The basic portion of the tar consisted of alkylpyridines (64%), alkylanilines (33%), quinoline, and small amounts of cycloalkylpyridines. The individual pyridines occurring in each fraction are given in Table I-15. Part of this study involved an extensive investigation of the NMR spectra of many of the pyridines listed (371–373).

A light distillate (190–310°C), which constituted 15% of a crude shale oil, contained 1.37% total nitrogen. The weak base fraction of this oil was composed mainly of pyridine bases (374).

A high-sulfur shale oil from the Tyrol yielded many simple alkylpyridines (*viz.* 2-, 3-, 4-picolines, 2,3-, 2,4-; 2,5-; 2,6-; 3,4-lutidine, 2-ethyl-4-methyl- and 2-ethyl-6-methylpyridines, and 2,3,6- and 2,4,6-collidines); in addition three sulfur-containing heterocycles (**I-99–101**) were identified (two tentatively **I-100 and 101**) (375).

I-99, (R^1 = Me, R^2 = H)
I-100, (R^1 = R^2 = Me)

I-101

TABLE I-15.   PYRIDINES IDENTIFIED IN TAR BASES FROM HYDROCRACKED SHALE OIL NAPHTHA

	Cut	Wt.% of Tar Bases	-Pyridine
Fraction 1	1	0.19	2-Me-6-Et
(10% of tar bases)	2	0.04	2,6-Me$_2$
	3	0.26	2-Et^a,
			2-Me-6-Pr^i
	4	0.31	2-Me,
			2,4-Me$_2$-6-Et^a
	5	4.05	2-Et-5-Me,
			2,3,6-Me$_3$,
			2,4,6-Me$_3$
	6	1.57	2-Et-4-Me
	7	0.37	2,5-Me$_2$,
			3-Me,
			2,4-Me$_2$,
			2-Me-5-Et
	8	1.76	2-Me-4-Et
	9	0.84	3,5-Me$_2$
	10	0.44	3,4-Me$_2$
Fraction 2	1	1.89	2,4-Me$_2$-6-Pr^i
(26.5% of tar bases)	2	7.09	2-Me-6-Pr
			2,4-Me$_2$-6-Et
	3	0.25	2-Bu
	4	4.44	2,6-Me$_2$-4-Et
	5	1.38	2,4-Et$_2$
	6	0.71	2-Me-4-Pr^i
	7	1.40	4-Pr^a
	8	1.09	3-Me-5-Et
	9	1.80	2,4,5-Me$_3$
	10	0.79	2,3,4-Me$_3$
Fraction 3	1	16.32	2,6-Et$_2$-4-Me,
(42.6% of tar bases)			2,4-Me$_2$-6-Pr$^{i\,a}$,
			2,6-Me$_2$-4-Pr^a,
			Mixed 2,4,6-Me$_3$-Et
	2	11.89	Mixed C$_4$ and C$_5$ pyridines,
			(2-Me-6-ethylaniline)
			2,3,4,6-Me$_4$
	7	0.82	(3,4-dimethylaniline)
			3,4,5-Me$_3$
Fraction 4	1	2.15	2,4-Et$_2$-6-Pr^i
(20.0% of tar bases)			2,3-Et$_2$-6-Pr^i
			2-Et-4-Me-6-Pr^a
	2	6.16	2-Me-4-Et-6-Pr^a
			2,4-Me$_2$-6-pentyl
			2,4,5-Me$_3$-6-Et
			2,5,6-Me$_3$-4-Et^a
			Mixed C$_6$ and C$_7$ pyridines

^{a} Proposed compound.

Other shales from various parts of the world [*viz.* Boltyshsk (376), Baisun (377), other Russian deposits (378, 379), Colorado Green River (380, 381), Japan (382)] have also been found to contain alkylpyridines. The degassing of oil shales using tunnel or shaft furnaces has been reviewed (383). Recently a method linking GC with vapor phase IR spectrophotometry has proven promising for the examination of phenolic and pyridine components of shale oil (384).

D. *Degradation and Transformation of Alkaloids*

Although the characterization of alkaloids today does not rely as heavily on information from chemical degradation as formerly, still routes leading to pyridines are numerous. Only a few examples will be given herein, since this topic is covered in texts on alkaloids.

Many of the alkaloids that contain a pyridine ring yielded pyridine derivatives on degradation. Oxidation of halfordine (**I-102**) with aqueous potassium permanganate gave nicotinamide and 4-carboxyphenoxyacetic acid (**I-103**) (385). Other examples can be found in references quoted in Section I.1.C.

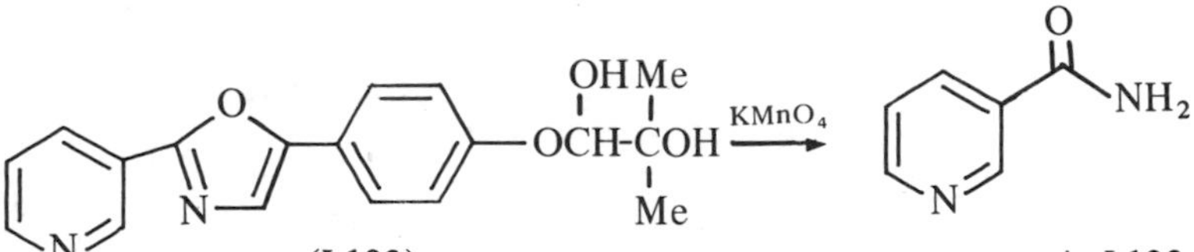

(I-102)   + I-103

Dehydrogenation of the alkaloids **I-104–109** furnished pyridines of varying complexity (386–391). Palladium on charcoal was usually the reagent of choice. The structure of tecomanine (**I-110**) has been assigned on the basis of spectral data and its degradation to actinidine (**I-104–110**) by successive reduction of the double bond and the ketone group, followed by dehydrogenation (392).

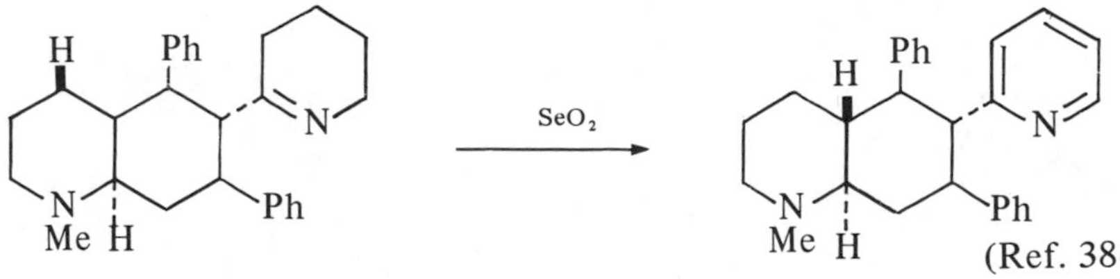

I-104   (Ref. 386)

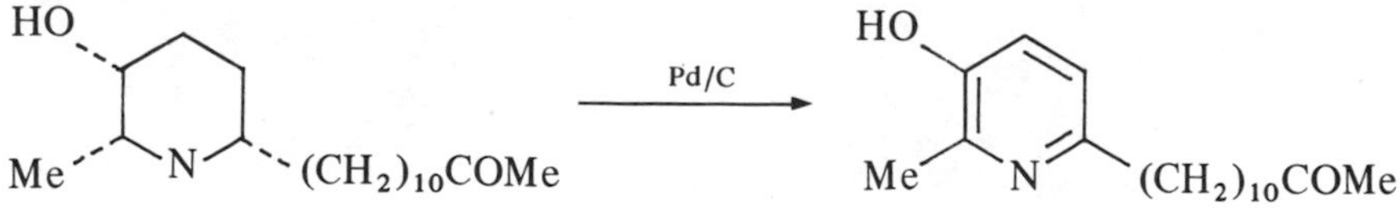

I-105   (Ref. 387)

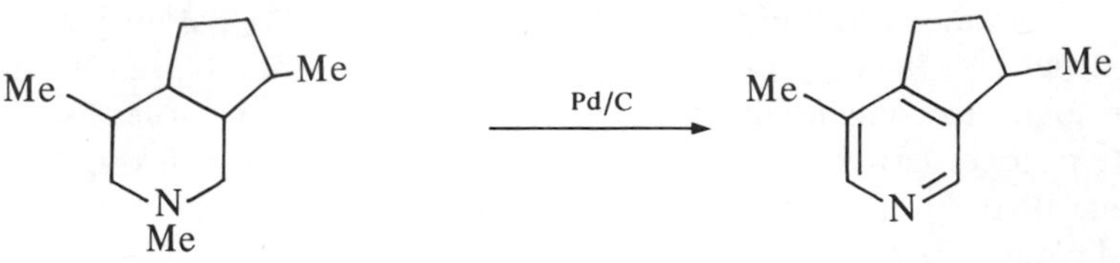

(Ref. 388)

(Ref. 389)

(Ref. 390)

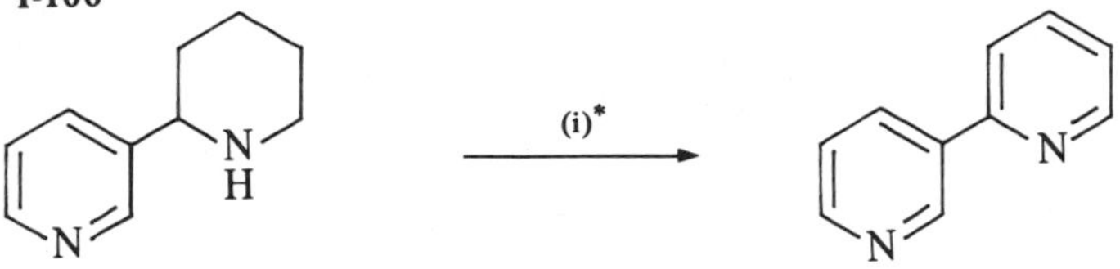

(Ref. 391)

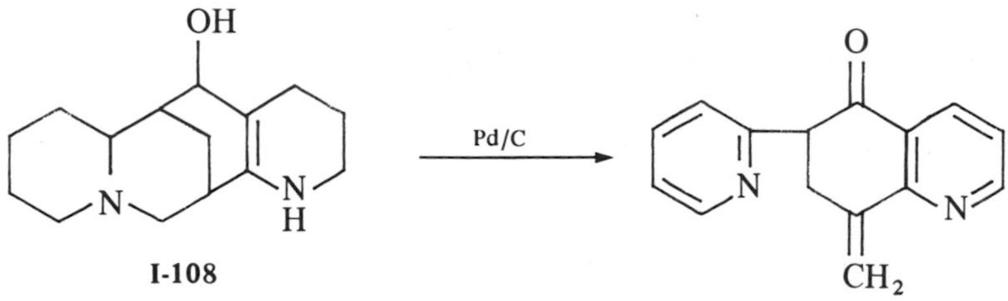

I-106

I-30

I-107

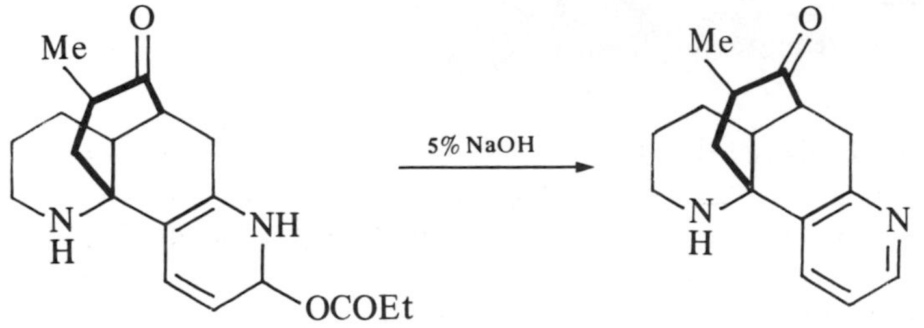

I-108

I-109

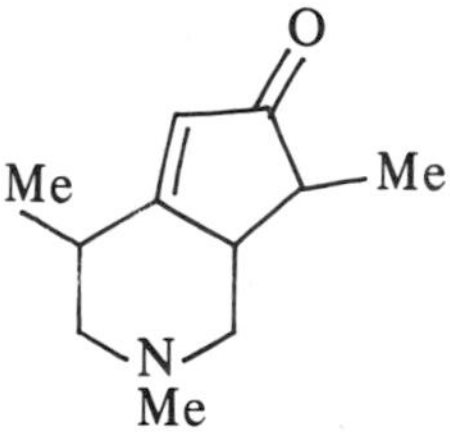

I-110

*(i) — Heating with pyridine 1-oxide at 250° in α-bromonaphthalene.

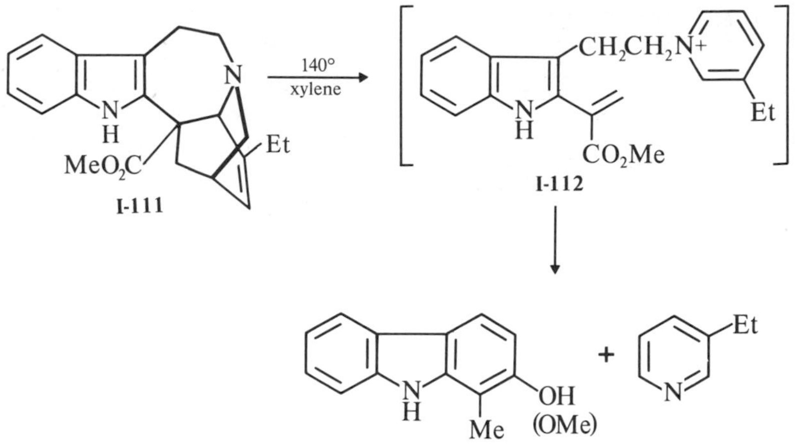

The related indole alkaloids tabersonine, catharanthine (**I-111**), pseudo-catharanthine all yield 3-ethylpyridine on thermolysis in xylene through a common intermediate (**I-112**) (393).

E. Flavors, Odors, and Volatile Constituents of Foods and Beverages

Detection of a large number of pyridines (Table I-16), often in minute amounts, in food and beverages has relied almost entirely upon the recent general availability of GC–MS methods. The new pyridine-containing amino acid, E-[3-hydroxy-6-methylpyridone(1,4)yl]-L-norleucine, has been found in the acid hydrolysates of different dehydrated foods such as carrots, asparagus, leek, mushrooms, and tomatoes (432). A comprehensive list of pyridines found in foodstuffs and their toxicological evaluation has been published by the U.S. Government (441).

TABLE I-16. PYRIDINES IDENTIFIED IN FOODS AND BEVERAGES

Pyridines	Source	Reference
2-Ethylpyridine	Roasted turkey volatiles	394
2- and 3-Pentylpyridine 5-Methyl-2-pentylpyridine 5-Ethyl-2-pentylpyridine 2-Hexylpentylpyridine	Roasted lamb fat	395
2-Ethylpyridine 2-Pentylpyridine 3,4-Dimethylpyridine	Roasted beef aroma	396
Pyridine	Smoked fish	398
Pyridine 2-Methylpyridine	Voltaile components of caviar	399 400

TABLE I-16.   (*CONTINUED*)

Pyridines	Source	Reference
Pyridine	Roasted green coffee beans	401 402 403
Pyridine 2-Methylpyridine 3-Vinylpyridine 2-Methyl-5-ethylpyridine 2-Acetylpyridine 3-Phenylpyridine	Roasted cocoa	404
Pyridine 2-Methylpyridine 2-Pentylpyridine 2-Acetylpyridine Methyl nicotinate	Roasted peanuts	405 406
Pyridines	Roasted filberts	407
Pyridine 2- and 3-Methylpyridine	Cooked rice volatiles	408
Pyridine 2-Methylpyridine 4-Ethylpyridine 2,3-Dimethylpyridine 2,3,6-Trimethylpyridine 2,4,6-Trimethylpyridine Dimethylethylpyridine (3 isomers)[a] 4-Formylpyridine 4-Methyl-2-pyridone 1-Ethyl-2-pyridone	Volatile constituents of wild rice grain *Zizania aquatica*	409
Nicotinic acid	Cauliflowers	410
Pyridine	Volatile constituent of leek (*Allium porrum*)	411
Nicotinic acid	Banana powders	412
Pyridine	Canned and/or frozen corn	413
Pyridine 2,6-Dimethylpyridine 3- and 4-Methylpyridine 2,4,6-Trimethylpyridine	Basic fraction from pyrolysis of sea weed *Fucus serratus*	414
Pyridine 2-Methylpyridine 2- and 3-Acetylpyridine 2-Formylpyridine	White bread crust volatiles	415
2-Methylpyridine 2-Methyl-5-ethylpyridine	Odor of white bread	415
Pyridine Picolines 4-Ethylpyridine 2,3-Dimethylpyridine	Volatile components of white bread	417 400

TABLE I-16.   (*CONTINUED*)

Pyridines	Source	Reference
2,4-Dimethylpyridine 2,6-Dimethylpyridine		
Pyridine Picolines[a] 2-Acetylpyridine	Roasted barley	418 419 420
Pyridine 2- and 3-Methylpyridine	Rum and whisky	421
2- and 3-Acetylpyridine	Beer flavor	422
Pyridine 3-Methylpyridine 2,6-Dimethylpyridine	Raw *shoyu* (soy sauce)	423
Pyridine 2-Ethylpyridine 2,6-Dimethylpyridine	Heated *shoyu*	423
Pyridine 2-Acetylpyridine	Potato chips	424
Pyridine 4-Methylpyridine	Cooked beets	425
Pyridine Picolines[a] Dimethylpyridines[a] Trimethylpyridines[a] 3-Methoxy-2-propylpyridine 2-Methyl-5-ethylpyridine	Potato-like off odor of Gruyère de Comté	426
Pyridine Picolines 2- and 3-Ethylpyridines 2,5-Dimethylpyridine 2,6-Dimethylpyridine 2-Methyl-5-ethylpyridine 2-Methyl-6-ethylpyridine 3-Methoxypyridine 4-Vinylpyridine 2-Acetylpyridine 2-*n*-Butylpyridine 2- and 3-Phenylpyridine	Black tea aroma	427
Pyridine Picolines	Rossiikii cheese	400 428 429
Pyridine Trimethylpyridines[a]	Beaufort cheese	430

[a] The position of the substituent(s) was not quoted.

TABLE I-17.   PYRIDINES DERIVED FROM PYROLYSIS OF AMINO ACIDS

Pyridines	Source	Reference
2-Methylpyridine 2-Methyl-5-ethylpyridine	Pentane extract of the cysteine/ cystine-ribose browning system	436
3-(4-Phenoxy)pyridine	Pyrolysis of tyrosine	437
3-Phenylpyridine 2-Amino-5-phenylpyridine	Pyrolysis of phenylalanine	438
2-Methylpyridine 2-Ethylpyridine 2,4-Dimethylpyridine 3,4-Dimethylpyridine	Pyrolysis of cysteine and xylose in tributyrin	439
3-Dodecylpyridine 3-Tridecylpyridine 3-Tetradecylpyridine 2-Nonyl-5,6,7,8- tetrahydroquinoline	Egg tar	440
Pyridine 2-Methylpyridine	Pyrolysis of egg white and ovalbumin solution	431

F.   Miscellaneous Sources

A search for interstellar pyridine has proved negative (433), but pyridine has been synthesized from HCN and $(CN)_2$ using an electric discharge in a simulated Jovian atmosphere (434). Nicotinonitrile, 2-, and 4-cyanopyridine can be synthesized under primitive earth conditions by the action of electric discharges on ethylene and ammonia (435).

Pyridines have been obtained from pyrolysis of amino acids and egg tar. See Table I.17.

II.   PYRIDINES BY SYNTHETIC METHODS

1.   From Other Ring Systems

Syntheses of pyridine compounds from cyclic materials are considered in this section. These will include ring transformation of carbo- and heterocyclic systems, degradation of condensed pyridine compounds, and dehydrogenation of reduced pyridines. For purposes of continuity the reactions of carbohydrates will be included with the furans as in the past editions of this series.

A. Carbocyclic Compounds

Recently small strained ring carbocycles have been converted into pyridines. Diphenylcyclopropenone (X = O) and the corresponding thione on treatment with glycine derivative I-113 gave substituted pyridines *via* the bicyclic derivative I-114. The mesoionic derivative I-115 was a proposed intermediate (443).

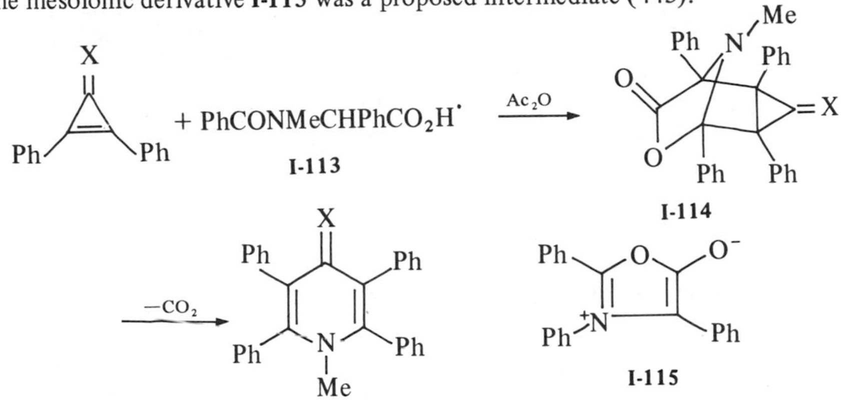

Cyclopropenylazides I-116 with 1-(diethylamino)propyne gave I-117 (444). 1,3-Dipolar addition of the nitrile ylide I-118 with methylene-cyclopropenes gave I-119 *via* the isolated intermediate I-120 (R'=Ph, R=CN). It was shown that under the reaction conditions the bicyclic intermediate I-120 was converted to I-119.

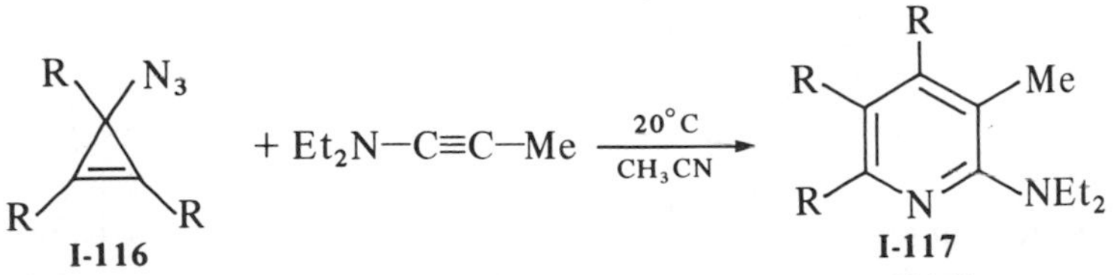

a R = Ph (44%)
b R = p-CH$_3$C$_6$H$_4$— (41%)

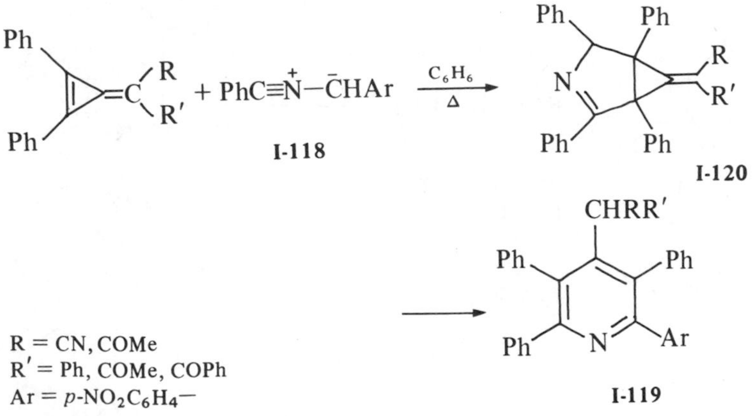

R = CN, COMe
R' = Ph, COMe, COPh
Ar = p-NO$_2$C$_6$H$_4$—

When allowed to react with imines (445), in a similar manner, **I-121** gave pyridones **I-122**. Condensation of diphenylcyclopropenone (**I-123**) with amidines yielded pyrrolinones **I-124**, which when treated with dimethylacetylenedicarboxylate gave 2,5,6-triphenyl-3,4-dicarbomethoxypyridine (**I-125**) (446).

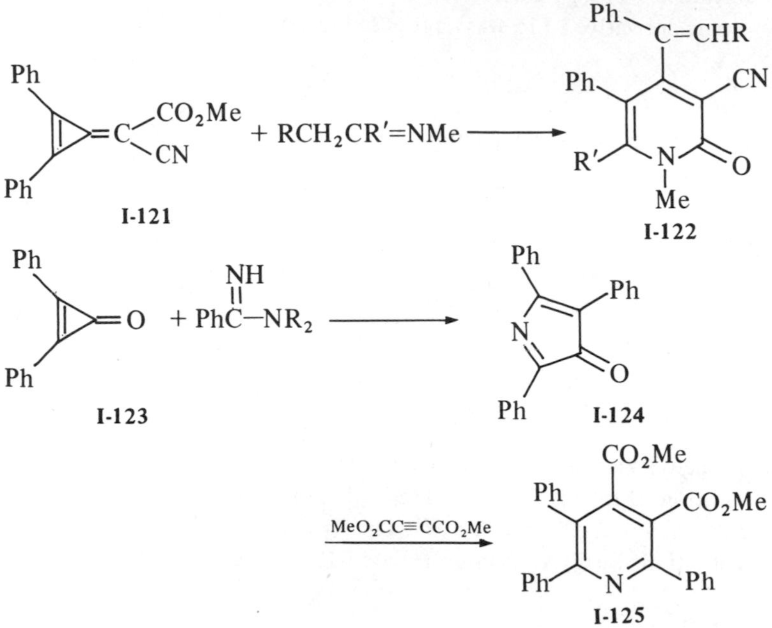

Ph
\
C=CHR

Ph
\
Ph—⌷—CN
Ph
‖
R'—N—C=O
|
Me
I-122

Ph
\
⊳=C<CO₂Me / CN + RCH₂CR'=NMe ⟶
/
Ph
I-121

Ph
\
⊳=O + NH‖ / PhC—NR₂ ⟶
/
Ph
I-123

Ph
\
Ph
N
\
O
Ph
I-124

(MeO₂CC≡CCO₂Me) ⟶

CO₂Me
Ph
CO₂Me
Ph—N—Ph
I-125

In a series of papers, Reid and Baetz (447–449) reported the reactions of phenyl-cyclobutenedione (**I-126**) with a number of enamines to give azabicycloheptenes (**I-127**), which subsequently rearranged in base to give pyridones **I-128**. Catalytic hydrogenation of **I-127** gave the debenzylated pyridine, while treatment with thionyl chloride caused the cyclopropyl ring to open in a different manner to give **I-129**.

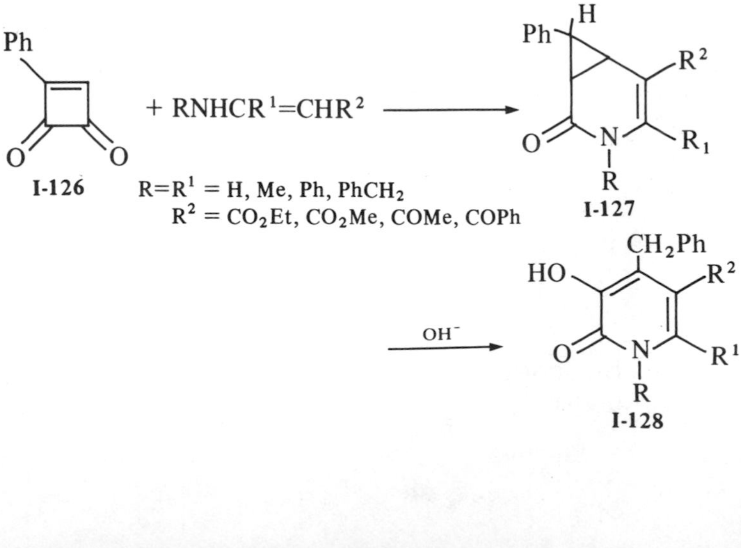

Ph

O   O
I-126   + RNHCR¹=CHR² ⟶

R=R¹ = H, Me, Ph, PhCH₂
R² = CO₂Et, CO₂Me, COMe, COPh

Ph—H
\
R²
O=C—N—R₁
|
R
I-127

(OH⁻) ⟶

CH₂Ph
HO
R²
O=C—N—R¹
|
R
I-128

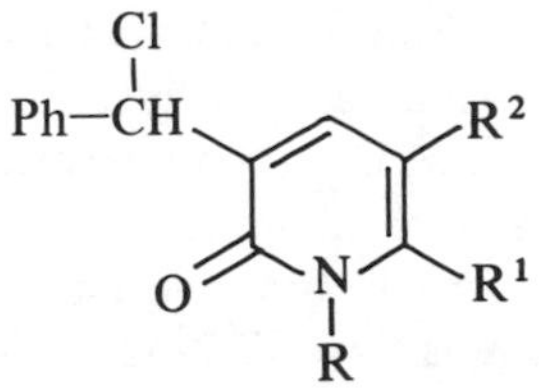

I-129

δ-Complexes (**I-130**) of cyclobutadienes reacted with 2 moles of ethyl cyanoformate under very mild conditions to give highly substituted pyridines **I-131** (450). The Dewar pyridine intermediate **I-132** was proposed (450).

The condensation of phenylisothiocyanate with diphenyl*bis*(diethylamino)-cyclobutadiene (**I-133**) gave a stable Zwitterion intermediate, which on thermolysis gave the 2-pyridylthione **I-134** (451).

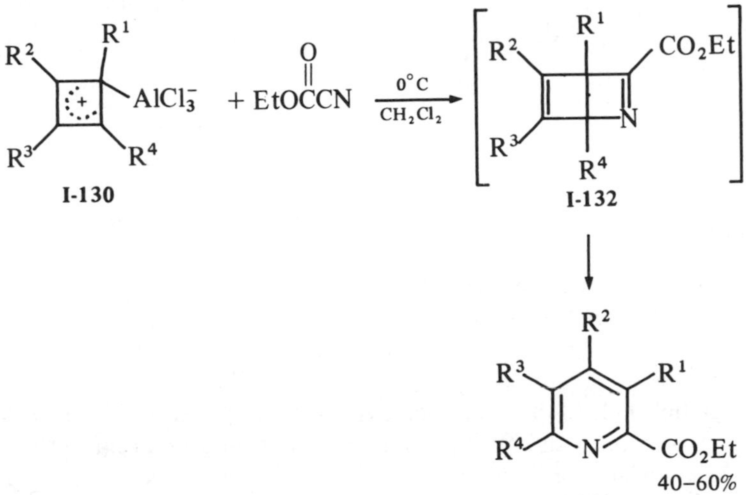

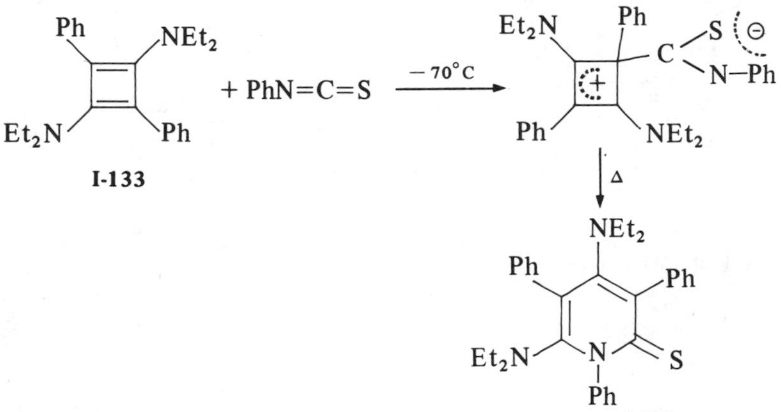

Thermolysis of chloramine and cyclopentadiene in toluene gave (20–25%) pyridine presumably *via* a N-chloroazirine intermediate (452). Similar chemistry has been observed in the indene series. The vapor phase reaction of cyclopentanone with ammonia at 650°C on a silica–alumina catalyst gave (16.5%) pyridine (453). Similarly, the catalytic amination of 2-cyclohexen-1-ol with ammonia using nickel, copper, or platinum on alumina gave (55–62%) 2-picoline (454). Vapor phase chlorination of cyclohexanone oxime was reported to give polychlorinated pyridines. The degree of chlorination in the pyridine products was found to be temperature dependent (455).

Perfluorophenylnitrene was the proposed intermediate in the thermal rearrangement of I-135 to give pyridines I-136–138, along with a number of ring-contraction and open-chain products (456). These pyridines were presumed to arise from the nitrene undergoing a ring-expansion to an azepine, followed by ring-contraction.

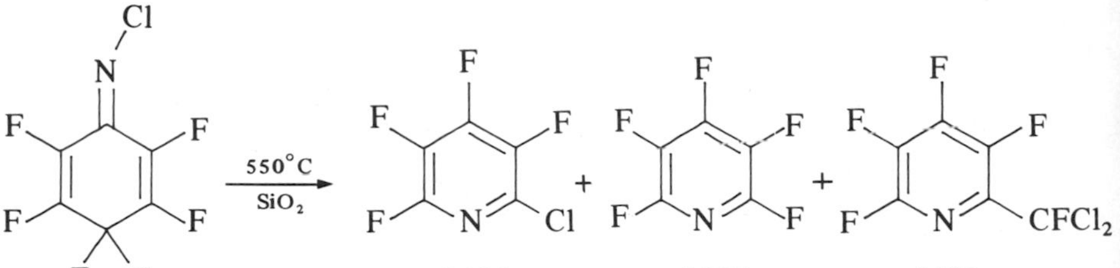

In a different manner, nitrenes have been utilized to give pyridines in a number of cases. Reaction of methanesulfonyl azide or o-bis-benzenesulfonyl azide in the presence of tetracyclone gave the tetraphenylpyridone I-139 (457) *via* a 1,4-addition of the nitrene to tetracyclone. Similarly, 3,4-dinitrophenylazide and 2,5-diethyl-3,4-diphenylcyclopentanone gave the N-dinitrophenylpyridone I-140. Treatment of tetracyclone with phthalimidonitrene gave the isolated 1,2-addition product I-141, which, under acidic conditions such as acetic acid, was converted to pyridone I-142 (458).

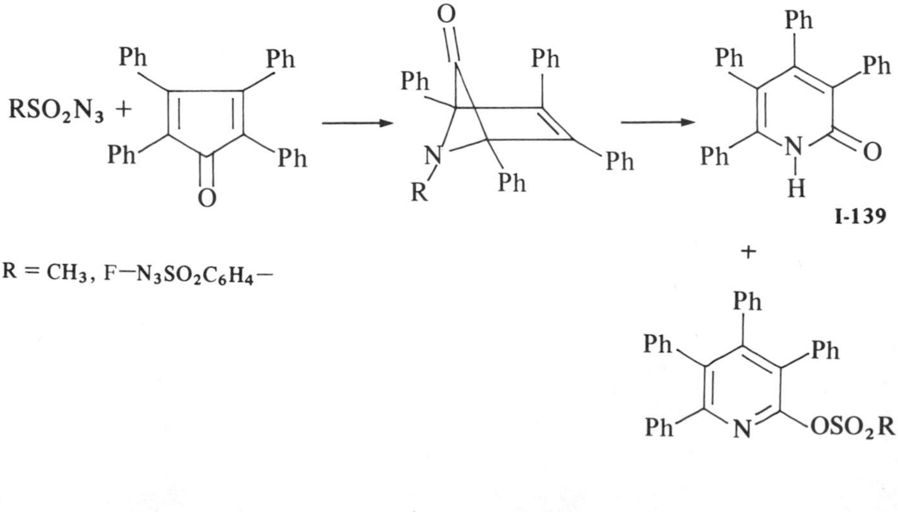

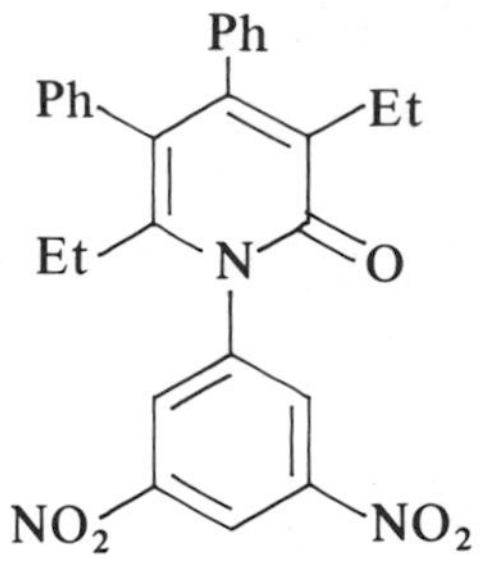

I-140

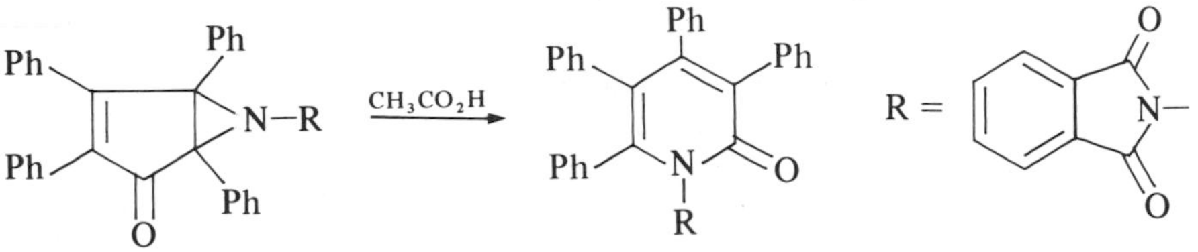

I-141   **I-142**   R =

1,2-Oxazines, such as **I-143**, prepared from cyclopentadienes and α-halo oxime derivatives, are smoothly converted to 2-arylpyridines at 200°C (459).

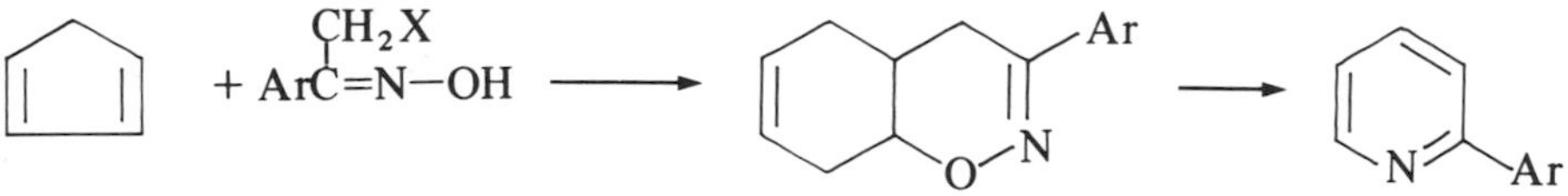

I-143

Reinvestigation of the structure of the product from the reaction of cyanoallene with 1-(N-morpholino)cyclohexene (**I-144**) found that the correct structure was **I-145**, which on treatment with sodium bicarbonate yielded the unexpected tetrahydroquinolone **I-146** (460).

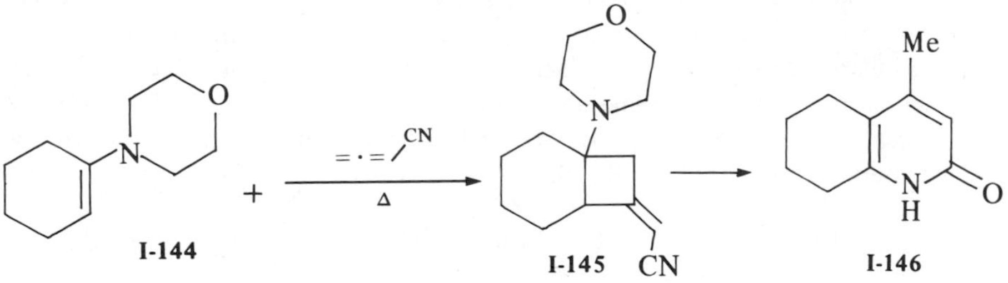

I-144   **I-145** CN   **I-146**

Diels–Alder reaction of carbonyl and sulfonyl nitriles with cyclopentadienes gave the pyridines **I-147** and **I-148**, respectively (461, 462). Similarly, condensation of phenyl cyanoformate (**I-149**) with tetracyclone and tetrachloropentadieneone dimethylketal gave pyridines **I-150** and **I-151**, respectively (463).

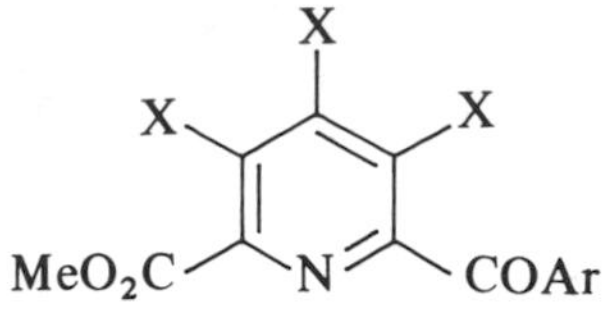

I-147 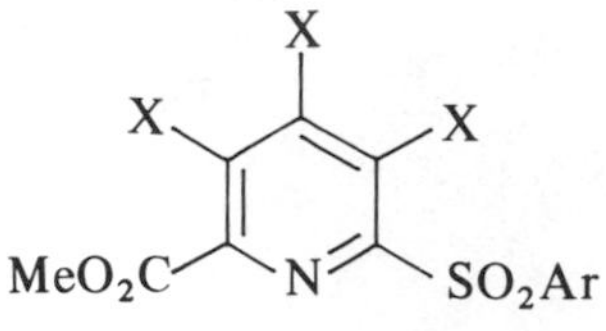   **I-148**

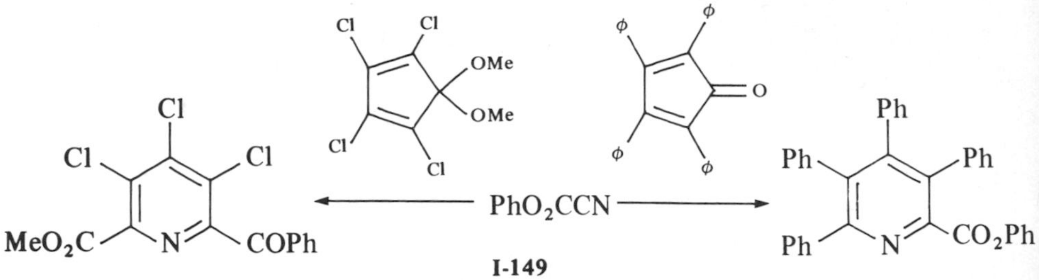

I-151   PhO$_2$CCN   **I-150**

I-149

Thermolysis of cyclopentadiene in the presence of ammonia gave pyridine presumably through glutacondialdehyde (464). The thermal arrangement of the tricyclic decanone oxime **I-152** gave the 9-pyridyloctahydroacridines **I-153** (465).

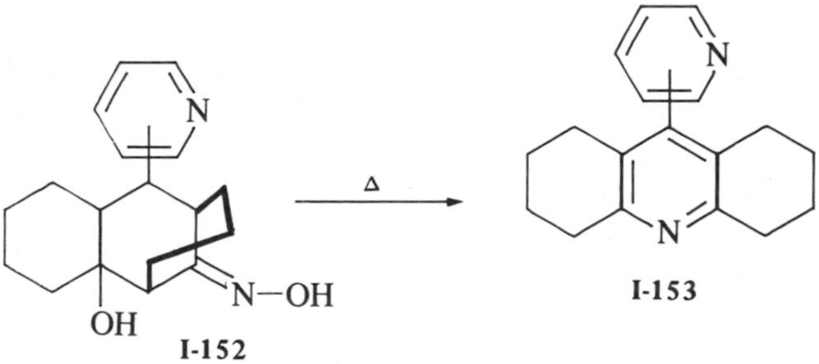

I-152   **I-153**

There have been numerous reports in the last few years on the intramolecular rearrangement of nitrenes to pyridines and these have been reviewed (466). Thermolysis of phenylazide gives pyridine as well as aniline and cyanocyclopentadiene (467). Similarly, 2,6-dimethylphenylazide gave 6-methyl-2-vinylpyridine (468). The vapor phase thermolysis of some alkylsulfonyl azides bearing an aryl substituent has, however, given some novel pyridines **I-154** (469). The pyridine arose from a nitrene insertion, followed by sulfur dioxide extrusion.

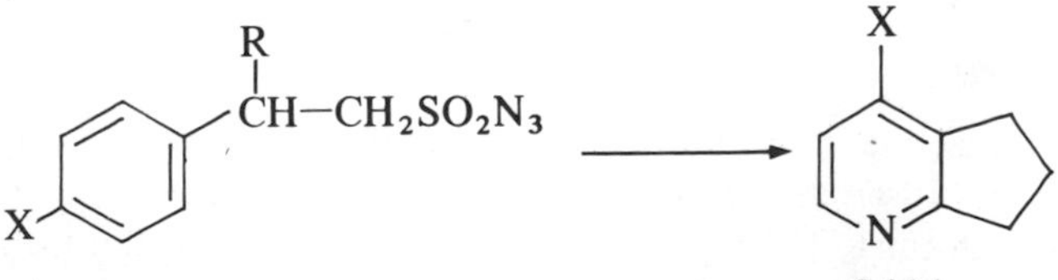

I-154

B.   Three-Membered Ring Heterocycles

Since the first supplement appeared, there have been a large number of reports on the synthesis of pyridines from three-membered ring heterocycles.

One area, which is of interest from both a synthetic as well as a theoretical viewpoint, is the conversion of allyl-substituted 2H-azirines to pyridines. Thermolysis of **I-155** gave the azabicyclo[3.1.0]-hex-2-ene (**I-156**), which was smoothly converted to the pyridine **I-157** (470). Photolysis of **I-155** on the other hand gave pyridines **I-158** *via* the intermediate **I-159** (471). Table I-18 summarizes a number of examples of this procedure.

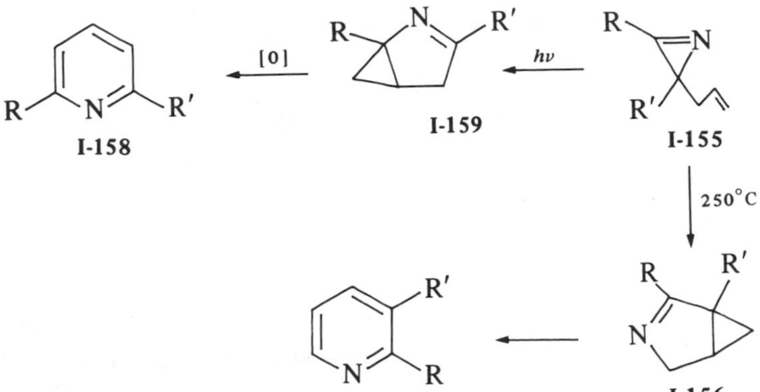

TABLE I-18.   PYRIDINES FROM REARRANGEMENTS OF 2-ALLYL-SUBSTITUTED 2H-AZIRINES

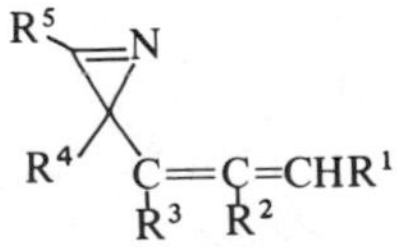

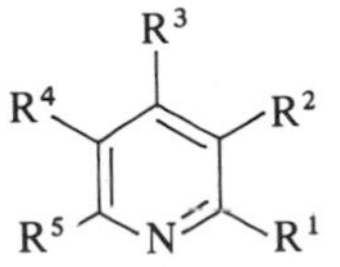

		Substitutent					Product			
R^1	R^2	R^3	R^4	R^5	R^1	R^2	R^3	R^4	R^5	Reference
H	H	H	CH_3	Ph	H	H	H	CH_3	Ph	470
H	H	CH_3	CH_3	Ph	H	H	CH_3	CH_3	Ph	470
CH_3	H	H	CH_3	Ph	CH_3	H	H	CH_3	Ph	470
CO_2Me	H	H	CH_3	Ph	CO_2Me	H	H	CH_3	Ph	470
Ph	H	H	CH_3	Ph	CH_3	H	H	Ph	Ph	471
H	H	H	CH_3	Ph	CH_3	H	H	H	Ph	471
CH_3	H	H	CH_3	Ph	Ph	CH_3	H	H	CH	471
H	H	CH_3	Ph	CH_3	CH_3	Ph	CH_3	H	H	470
CO_2Me	H	H	Ph	CH_3	CH_3	Ph	H	CO_2Me	H	470
CN	H	H	Ph	CH_3	CH_3	Ph	H	CN	H	470

I-160 **I-161** **I-162**

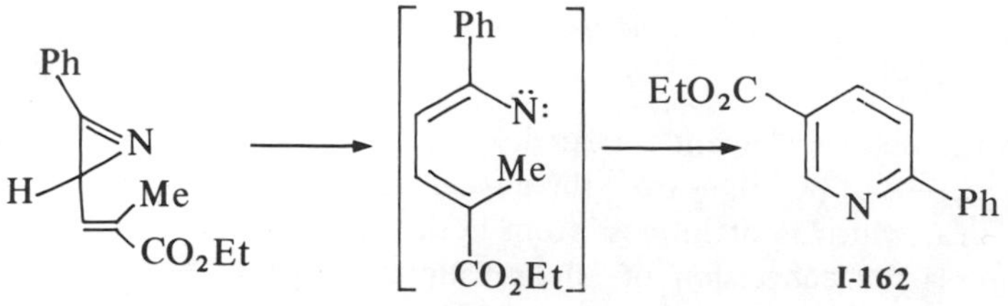

In the case of azirine **I-160**, the intermediate nitrene **I-161** inserted into the neighboring allylic methyl group to give **I-162** directly (472). Thermolysis of allylazirine **I-163** in the presence of a metal carbonyl gave 2,3,5,6-tetraphenylpyridine **(I-164)** *via* a nitrenoid formulated as **I-165** (473, 474).

I-163 **I-165** **I-164**

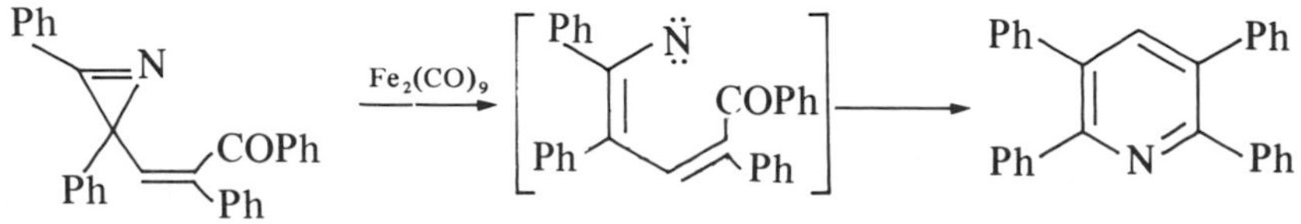

Several reports have appeared on the synthesis of pyridine derivatives from the reaction of azirines **(I-166)** with cyclopropenones **(I-167)** or cyclopropenium salts. Pyridones were also the products of azirines with diphenylcyclopropenone (475).

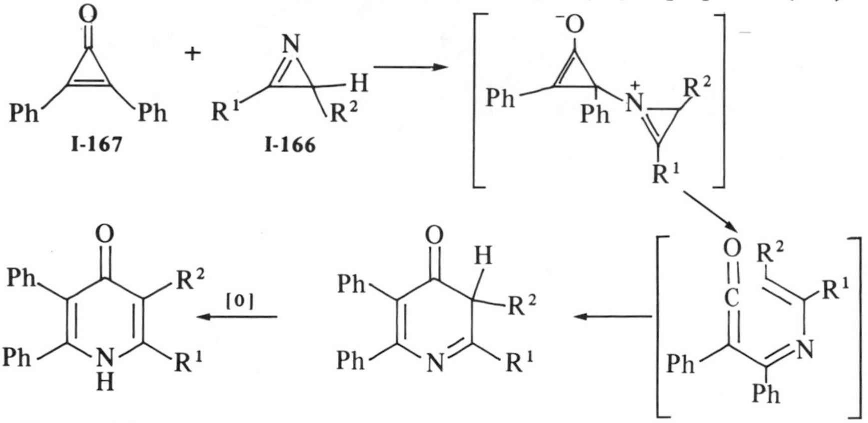

When triphenylcyclopropenyl bromide **(I-168)** and azirines were allowed to react, substituted pyridines **I-169** were isolated (476).

I-168
R = H, Ph, CO_2Et

I-169 (16–84%)

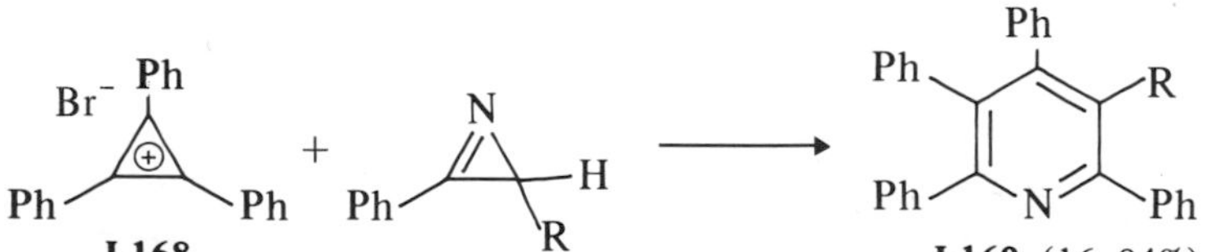

TABLE I-19.   REACTION OF 2-DIMETHYLAMINO-1-AZIRINES WITH OLEFINS AND ACETYLENES (477)

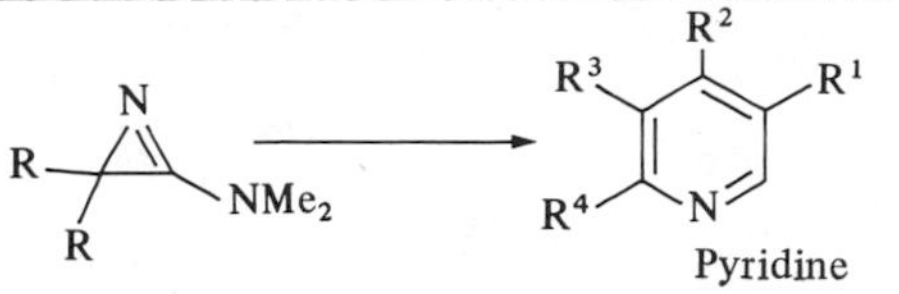

R	R	Unsaturated Reagent	R¹	R²	R³	R⁴	Yield
Me	Me	$MeO_2CC{\equiv}CCO_2Me$	CO_2Me	CO_2Me	H	CH_3	58%
Me	Me	$HC{\equiv}C{-}CO_2Me$	CO_2Me	H	H	CH_3	72%
Me	Me	napthaquinone	$-C(O)-C_6H_4-C(O)-$		H	CH_3	35%
$-(CH_2)_5-$		$H-C{\equiv}C-CO_2Me$	CO_2Me	H	$-(CH_2)_4-$		30%

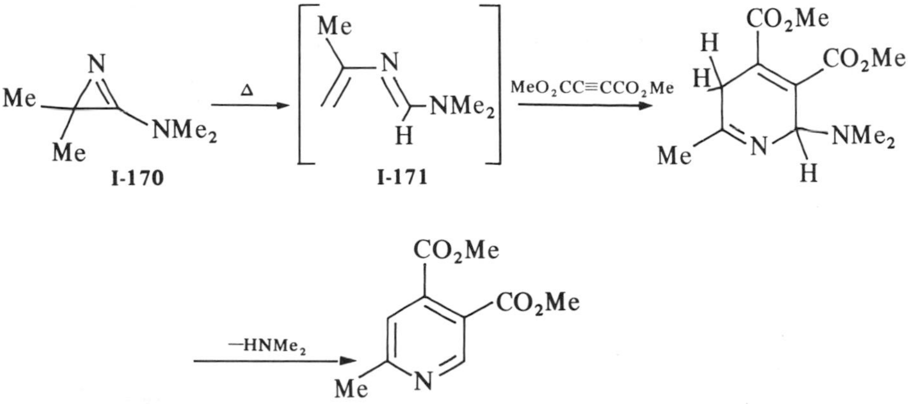

Aminoazirines **I-170** were thermally isomerized to give azabutadiene **I-171**, which can be trapped by olefins and acetylenes to give pyridines (477). The spiro system **I-172** in the presence of an acetylene similarly gave the tetrahydroquinoline **I-173**, see Table I-19.

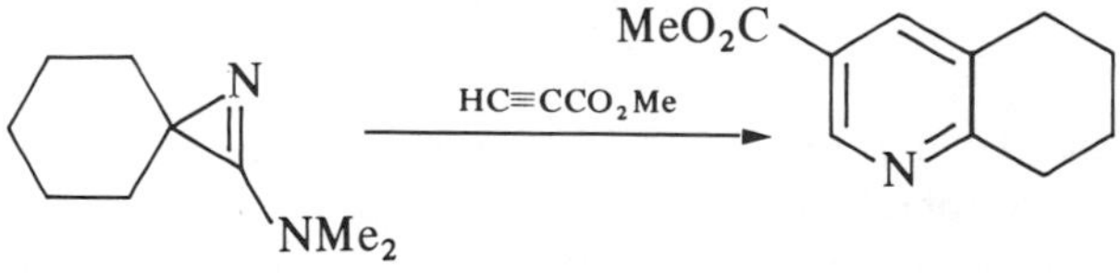

The diaziridine ring system has been used on occasion to synthesize a pyridine. The dialkyldiaziridines **I-174** gave pyridones **I-175** on condensation with acetylene **I-176** (478).

I-174 I-176 I-175

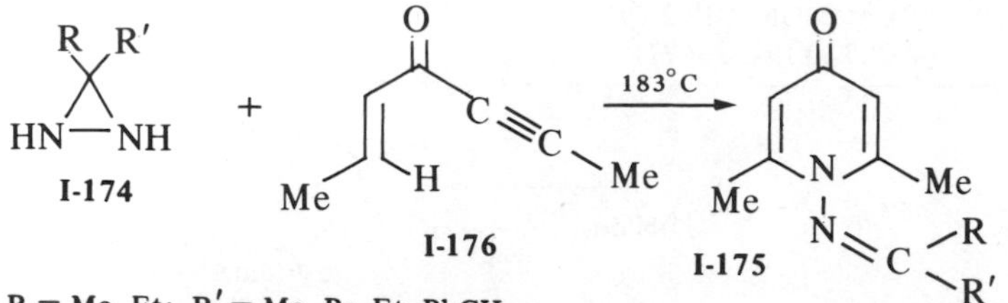

R = Me, Et; R′ = Me, Pr, Et, $PhCH_2$
R, R′ = $-(CH_2)_5-$

Diaziridine imines **I-177** were condensed with dimethyl acetylenedicarboxylate to give the spiroimidazole dihydropyridine **I-178**, which gave pyridine **I-179** upon treatment with acid (479).

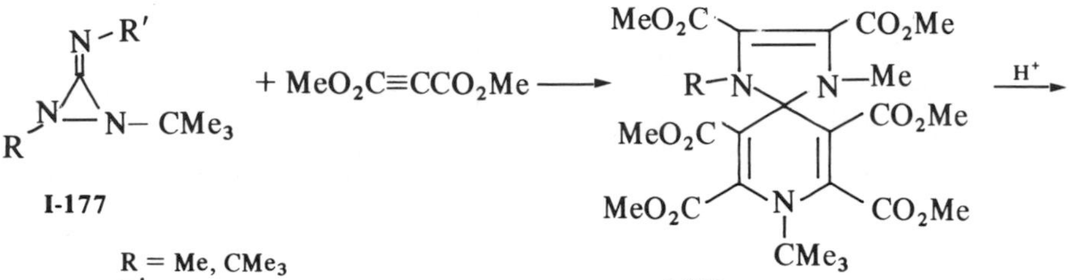

I-177

R = Me, CMe_3
R′ = Me

I-178

I-179

C. *Four-Membered Ring Heterocycles*

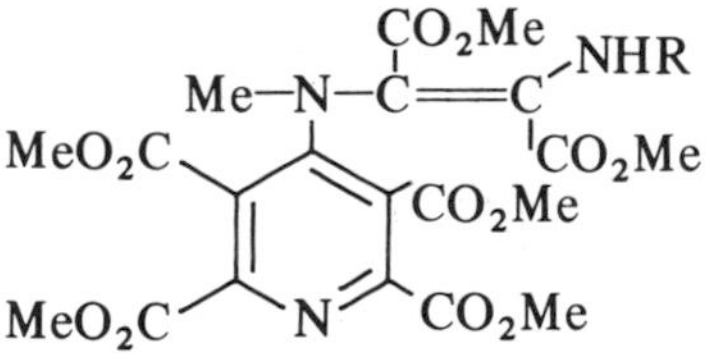

Transformation of four-membered ring heterocycles into pyridines has been a rare occurrence. As one can surmise, such transformations must include a condensation reaction to furnish two of the five carbons needed to complete the pyridine ring. Some workers have accomplished this by rearrangement of bicyclic systems.

Thermolysis of azetine **I-180** in the vapor phase gave the azabutadiene **I-181**, which condensed in a Diels–Alder reaction with dimethyl acetylenedicarboxylate to give **I-182** (480).

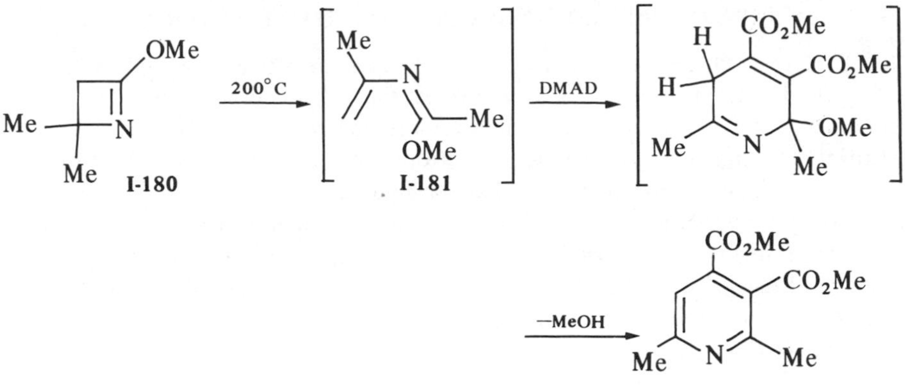

I-180 I-181 I-182

2-Aminopyridines have been prepared by the thermal rearrangement of azetinones **I-183** (481).

I-183

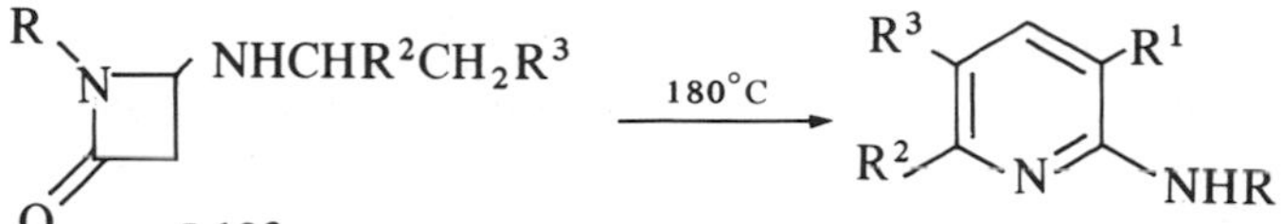

A bicyclic example has been reported in the case of diazetidine **I-184**, which upon treatment with acid or base yielded pyridones **I-185**; the bridged intermediate **I-186** has been proposed to explain these transformations (482).

I-184 I-186 I-185

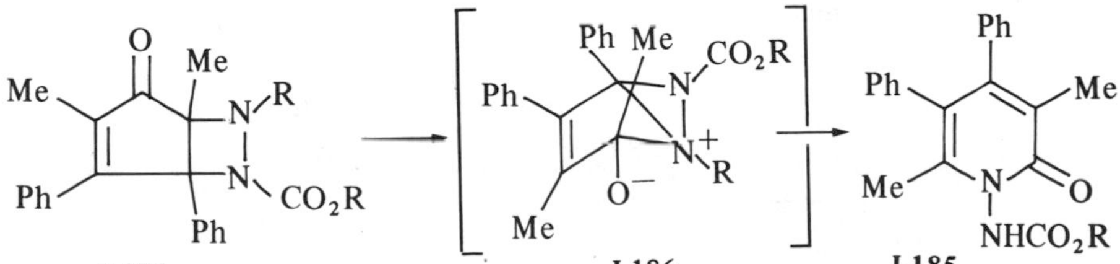

D. *Five-Membered Ring Heterocycles*

Since the publication of the original volume of this work a large number of syntheses of pyridines from five-membered ring heterocycles have appeared, many involving heterocycles containing more than one heteroatom such as the oxazoles and isoxazoles.

a. FIVE-MEMBERED RINGS CONTAINING ONE HETEROATOM

i. FURANS, DIHYDROFURANS, AND TETRAHYDROFURANS.   Syntheses of pyridines from furans generally involve the ring-opening of a 2-substituted furan so that the fifth carbon of the pyridine ring comes from an exocyclic substituent. The source of nitrogen is either from the substituent at the 2-position or from an external reagent such as ammonia or its salts.

The vapor phase thermolysis of tetrahydrofurfuryl alcohol (I-187) with ammonia to give pyridine is well known; however, little definative mechanistic work had been done until recently. A careful study has been conducted which demonstrated that the conversion of tetrahydrofurfuryl alcohol went *via* the intermediate 2,3-dihydro-4*H*-pyran (I-188) (483, 484).

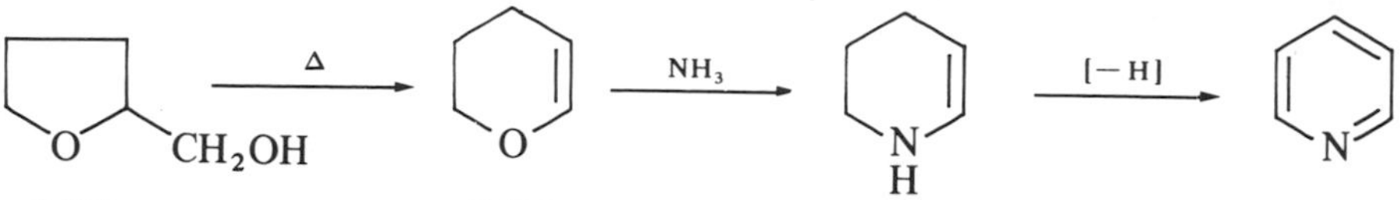

I-187   I-188

An unusual example of furans being converted to pyridines was demonstrated with dihydrofuranones I-189. The pyridine formed contains 3-carbons which were from the original exocyclic substituents. Table I-20 summarizes the known transformations in this series.

Oxidative ring-opening of furfural (I-190) followed by treatment with sulfamic acid gave the 2-iminopyridinesulfonic acid I-191, which on hydrolysis gave 2-amino-3-hydroxypyridine (I-192) (486, 487). The oxidizing agent was either chlorine or hypochlorite. The reaction has been reported to be pH dependent; at pH < 0.5, 5-chloro-2,3-dihydroxypyridine and 2,3-dihydroxypyridine (488) were isolated. If the nitrogen source is not sulfamic acid, *N*-substituted pyridones (I-193) were formed (489). Table I-21 summarizes this conversion.

Table I-20.   4-PYRIDINECARBOXYLIC ACID DERIVATIVES FROM FURAN-ONES I-189 (485)

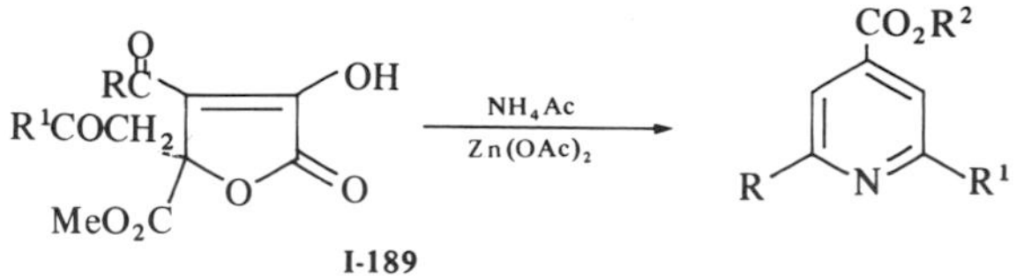

I-189

R	R^1	R^2
Me	Me	H
Et	Et	H
Me	CH$_2$OMe	H
Ph	Ph	H
Me	Ph	Me
CMe$_3$	CMe$_3$	Me

TABLE I-21. OXIDATIVE RING EXPANSION OF FURFURAL TO PYRIDONES I-193 (489)

Nitrogen Source	Conditions	R	Yield
$H_2N(CH_2)_3NMe_2$	$Cl_2/pH = 2$	$(CH_2)_3NMe_2$	24
$H_2NCH_2CO_2H$	$Cl_2/pH = 2$	CH_2CO_2H	8
H_2NOH	$NaClO_4$	OH	14
H_2NSO_3H	$Cl_2/pH = 2$	SO_3Na	43
H_2NNH_2	$Cl_2/pH = 2$	NH_2	1
H_2NNMe_2	$Cl_2/pH = 2$	NMe_2	33
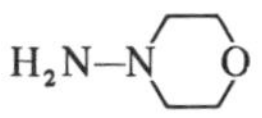 $H_2N-N\!\!\bigcirc\!\!O$	$Cl_2/pH = 2$	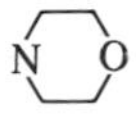 $N\!\!\bigcirc\!\!O$	40

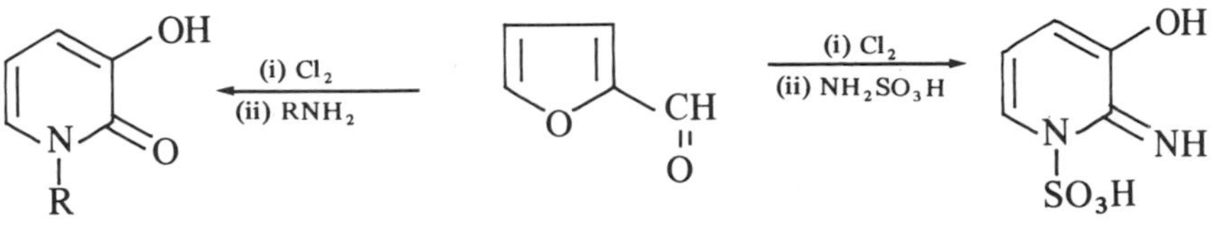

I-193   I-190   I-191

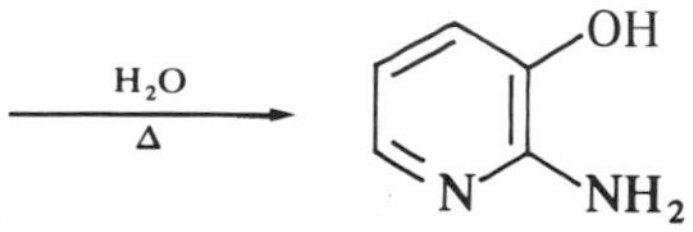

I-192

If excess chlorine was used, the open-chained derivative I-194 was prepared, which on treatment with various amines gave pyridines I-195 (489); see Table I-22.

Unlike furfural, the 2-acylfurans I-196 underwent ring-expansion on simply heating with ammonia or an ammonia salt. Table I-23 summarizes the results of a number of such reactions. It is of interest to note that several acylfurans failed to

TABLE I-22. CHLOROPYRIDONES I-195 FROM FURFURAL (489)

Amine Reagent	Position of Cl	R	% Yield
NH_4OH	4 + 5	H	26 and 57
$MeNH_2$	4	Me	21
$PhNH_2$	5	Ph	21
$o\text{-}CH_3C_6H_4NH_2$	5	$o\text{-}CH_3C_6H_4$	6
$m\text{-}CH_3C_6H_4NH_2$	5	$m\text{-}CH_3C_6H_4$	16
$p\text{-}MeOC_6H_4NH_2$	5	$p\text{-}MeOC_6H_4$	14
$o\text{-}ClC_6H_4NH_2$	5	$o\text{ClC}_6H_4$	6
$m\text{-}ClC_6H_4NH_2$	5	$m\text{-}ClC_6H_4$	23
NH_2OH	4 and 5	OH	15 and 17
NH_2NMe_2	4	NMe_2	12

TABLE I-23.   FORMATION OF 3-PYRIDINOLS I-196 FROM 2-ACYLFURANS

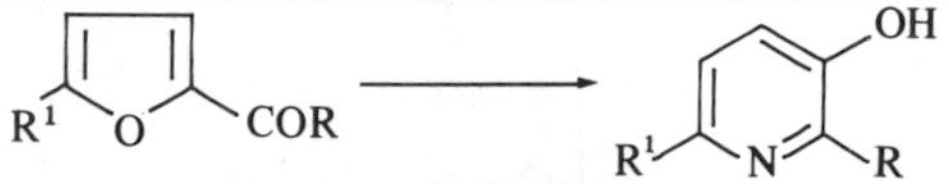

R	R¹	Conditions	3-Hydroxypyridine Product (Yield)	Reference
CH$_2$Ph	H	NH$_4$OAc, 230°C	2-benzyl- (70)	928
CH$_2$C$_6$H$_4$pCl	H	MeOH, NH$_4$OH, 150°C	2-p-chlorobenzyl- (48)	490
CHPhC$_6$H$_4$p-Cl	H	MeOH, NH$_4$OH, 150°C	2-(p-chlorophenyl)benzyl- (15)	490
CHPh$_2$	H	MeOH, NH$_4$OH, 150°C	2-diphenylmethyl- (34)	490
CHPh$_2$	Me	MeOH, NH$_4$OH, 150°C	2-diphenylmethyl-6-methyl-(3.5)	490
CHPhC$_6$H$_4$p-OMe	H	MeOH, NH$_4$OH, 150°C	2-α-p-methoxyphenyl)benzyl- (8)	490
p-tBuC$_6$H$_4$	H	NH$_4$OAc, 230–235°C	2-(p-t-butyl)phenyl-	929
p-EtC$_6$H$_4$	H	NH$_4$OAc, 230–235°C	2-(p-ethyl)phenyl-	929
p-iso-PrC$_6$H$_4$	H	NH$_4$OAc, 230–235°C	2-(p-isopropyl)phenyl-	929

I-190 I-194 I-195

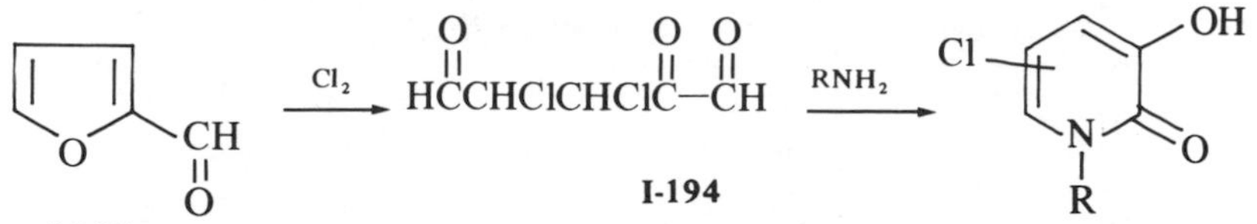

undergo the reaction. Furan **I-197** quantitatively gave fluorene **I-198** on treatment with ammonium hydroxide in methanol at 150–160° C, the 9-xanthenyl derivative gave xanthene (490).

I-197 I-198

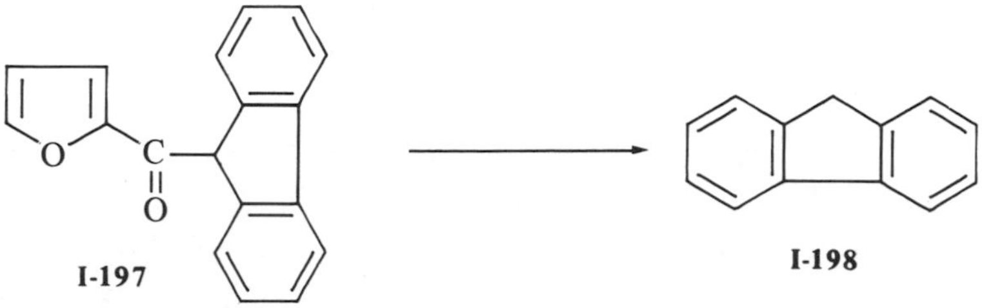

Many derivatives of 2-furoic acid (**I-199**; R^1 = CO_2H) give 2-amino-3-hydroxypyridines (**I-200**) on treatment with ammonia at 200–240°C (491). Table I-24 gives a summary of this reaction.

Furfurylamines **I-201** were rearranged to 3-hydroxypyridines by oxidative ring-opening. If the amino side chain was substituted, then pyridinium salts were obtained, which on pyrolysis gave 3-hydroxypyridines. The summary in Table I-25 outlines a number of such transformations.

TABLE I-24. PYRIDINES FROM 2-FUROIC ACID DERIVATIVES (**I-199**) AND AMMONIA (491)

I-199 I-200

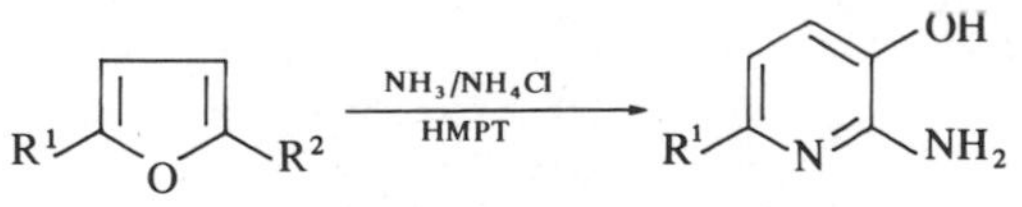

R^1	R^2	Conditions	Yield
H	$CONH_2$	240° C/5 hr	55
H	CO_2Et	240° C/8 hr	50
H	CO_2H	240°C/11 hr	45
H	CN	220°C/5 hr	35
CH_3	CN	240° C/8 hr	30
H	$CONHCH_2Ph$	240° C/10 hr	20
CH_3	CO_2Me	240° C/10 hr	30
H	HN^+=CH—OEt Cl^-	200° C/5 hr	35
H	CH=NOH	220° C/10 hr	1

TABLE I-25. PYRIDINES FROM AMINOFURANS **I-201**

I-201

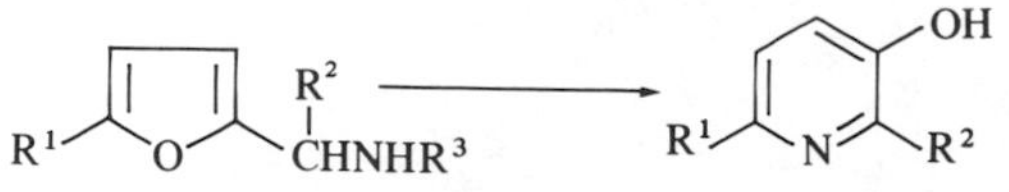

R^1	R^2	R^3	Conditions	R^1	R^2	Reference
Me	Me	H	Pt/250°	Me	Me	492
Et	Me	H	Pt/250°	Et	Me	492
H	CN	Me	(i) Cl_2/HCl (ii) 250°C	H	$CONH_2$	483, 494
H	CN	$PhCH_2$	(i) Cl_2/HCl (ii) 250°C	H	$CONH_2$	493, 494
H	CN	Et	(i) Cl_2/HCl (ii) Δ	H	$CONH_2$	493, 494

An interesting extension of this reaction was the synthesis of optically active pyridinium salts **I-202** from the furans **I-203**. The furans were obtained by condensation of the furfural with amino acids followed by catalytic hydrogenation (495).

I-203

I-202

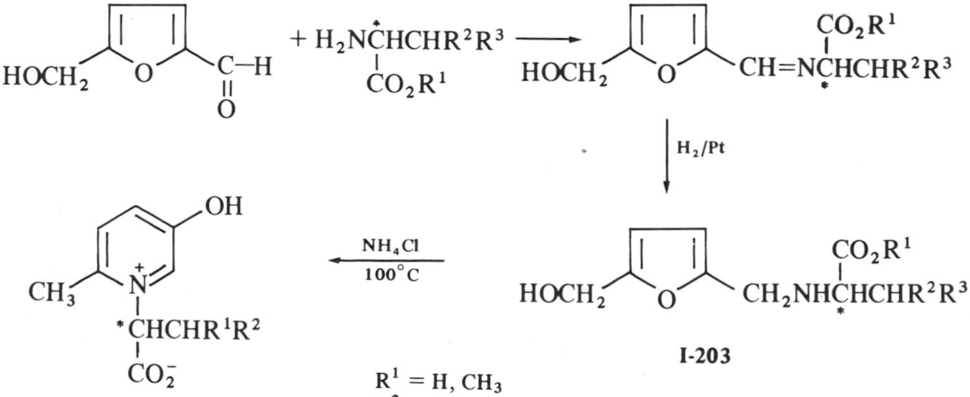

R^1 = H, CH_3
R^2 = H, CH_3, Ph, SCH_2Ph, SH

Norpyridoxine (**I-204**) was prepared from the furfuralamine **I-205** by an electrochemical oxidation (496).

I-205 **I-204**

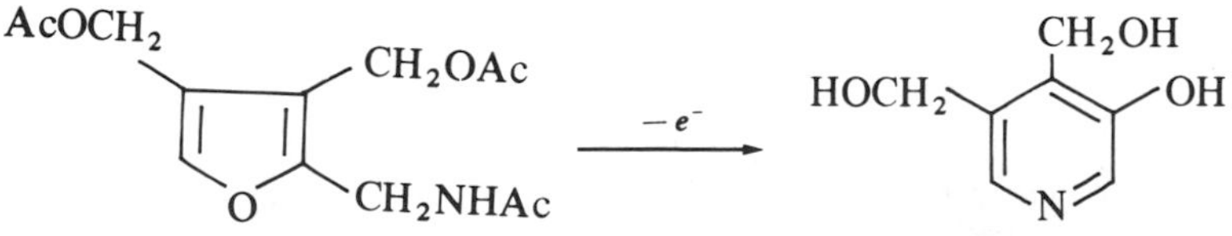

TABLE I-26.   PYRIDINES I-212 FROM MALTOL GLUCOSIDES (501)

R	Solvent	Time (hr)	Yield
H	H_2O	1.5	85
CH_3	30% MeOH—H_2O	1.5	90
Et	95% MeOH—H_2O	1.5	80
n-Pr	40% MeOH—H_2O	1.5	80
n-Bu	40% MeOH—H_2O	1.2	85
iso-Bu	40% MeOH—H_2O	1.0	85
n-pentyl	40% MeOH—H_2O	1.0	90
n-hexyl	40% MeOH—H_2O	1.4	80
n-C_7H_{15}	40% MeOH—H_2O	1.5	90
C_8H_{17}	40% MeOH—H_2O	0.5	–
n-$C_{12}H_{25}$	EtOH	20.0	30
$-(CH_2)_2OH$	H_2O	1.0	80
cyclohexyl	40% MeOH—H_2O	4.0	50
Ph	40% MeOH—H_2O	16	16
p-MeOPh	40% MeOH—H_2O	10	–
$-CH_2CO_2H$	40% MeOH—H_2O	20	70

The 2-furylhydantoin **I-206** was oxidized by chlorine to give 3-hydroxypicolinic acid amide (497). These hydantoins are easily prepared from furfural (498).

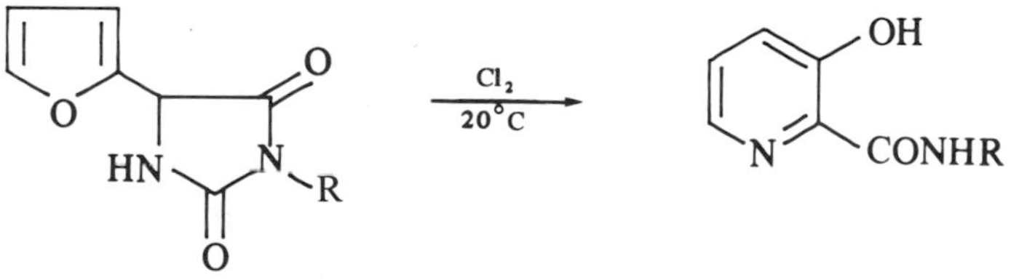

I-206

When furfurylamine was treated with formaldehyde in hydrochloric acid, 6-methyl-3-hydroxypyridine (**I-208**) was generated presumably through the hydroxy-methylfurfurylamine (**I-209**) (499). The isotopically labeled 6-methylpyridine was obtained when this reaction was conducted using ^{14}C-labeled formaldehyde (500).

The reaction of the naturally occurring maltol glucoside (**I-210**) with a wide variety of primary amines gave pyridones **I-211** which upon hydrolysis gave the corresponding 3-hydroxypyridones **I-212** (501). The data from this reaction are summarized in Table I-26.

A maltol derivative **I-213** gave 4-pyridones **I-214** upon treatment with ammonia or methylamine (502).

Acid-catalysed rearrangement of xylose **I-215** gave the aryl quaternary salt **I-216** (503).

TABLE I-27.   MISCELLANEOUS TRANSFORMATIONS OF CARBOHYDRATES TO PYRIDINES

Carbohydrate	Conditions	Products	Reference
Glucose	NH_3/H^+ hydrolysis	3-Hydroxypyridines	508, 509
Glucose + glycine	Acid solution	3-Hydroxypyridines	509
Rhamnose	NH_4OH	2-Amino-5-methyl-pyridine	510
Glucose	NH_3	3-Hydroxypyridine and 2-Methyl-5-hydroxypyridine	511
Glucose	NH_3	2-Methyl-5-hydroxy- and 2-Hydroxymethyl-5-hydroxy-pyridine	512
Glucose and cysteamine	H_2, 100°C	Pyridine	513

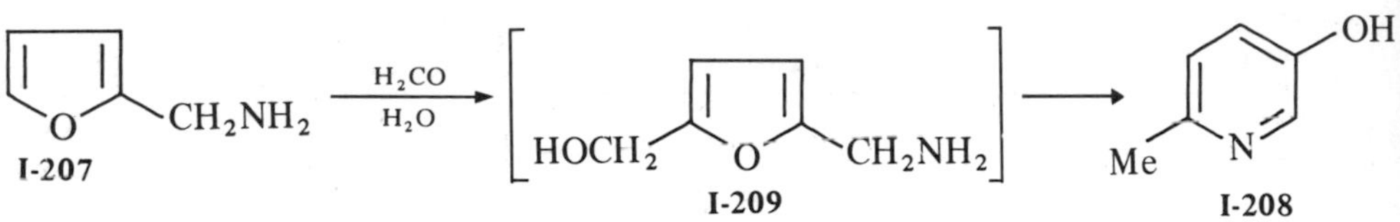

I-207 → [I-209] → I-208 (64% yield)

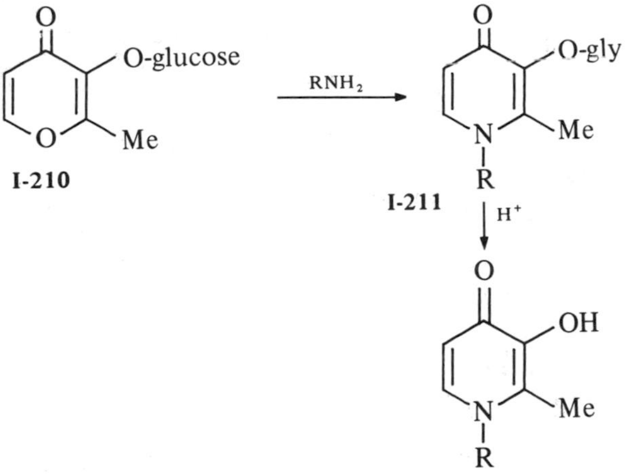

I-210 → I-211 → I-212

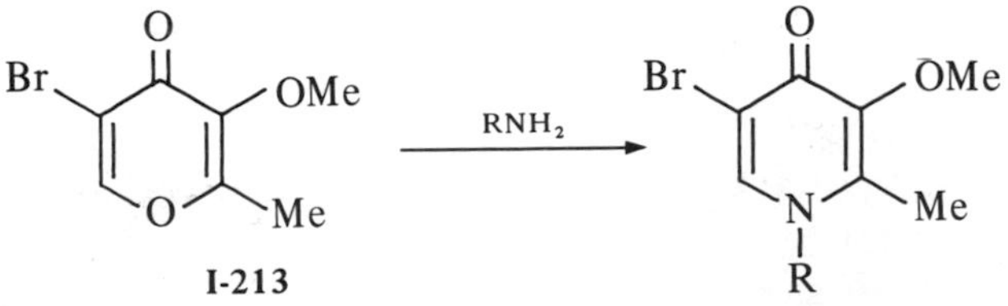

I-213 → I-214

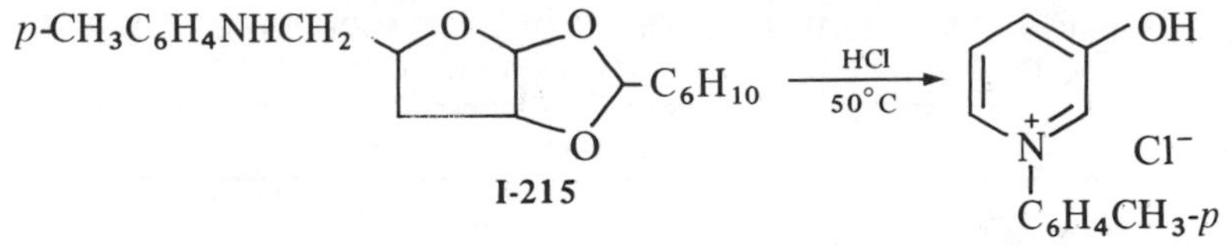

I-215

I-216 (100%)

Several examples of the conversion of D-glucaro-δ-lactams **I-217** to the corresponding pyridones **I-218** have been reported (504–506). Pyridones have been formed when maltose and lactose, maltol, and isomaltol, and isomaltol and glycine were treated with methylammonium acetate (507). In a number of cases, carbohydrates have afforded pyridines, as by-products in low yields. These reactions are not of general synthetic utility but are added here for completeness (Table I-27).

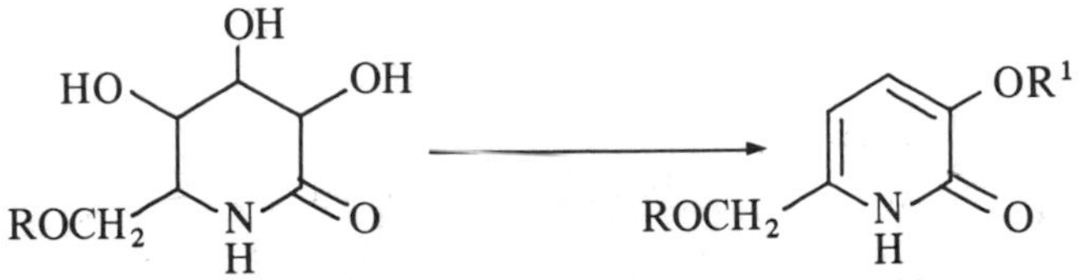

I-217   **I-218**

ii. PYRROLES. Ring-expansion of pyrroles **I-219** into **I-220** is an old and well-studied reaction. The classical reaction involved interaction of an alkyl halide with a strong base under conditions favorable for carbene generation, that is, CX_2 from a haloform and CHX from methylene dihalides. The reaction has been of little synthetic utility due to low yields; however, recent advances have reversed this trend. A recent report details the use of a phase transfer catalyst to improve the yields of the pyridines **I-220** and related indoles as shown in Table I-28 (514). Other sources of carbenes have been utilized and have led to high yields of pyridines (e.g., **I-223**) (Table I-28).

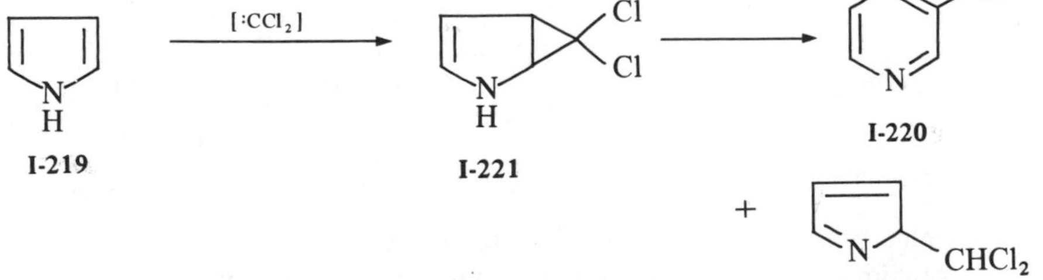

I-219   **I-221**   **I-220**

I-222

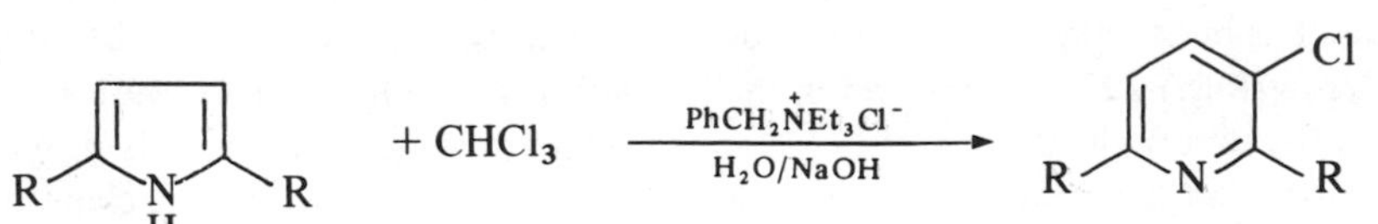

I-223

TABLE I-28.   RING-EXPANSION REACTIONS OF FIVE-MEMBERED HETEROCYCLIC COMPOUNDS USING PHASE TRANSFER CATALYSIS (514)

Substrate	Product	Yield
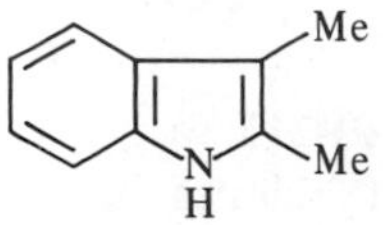	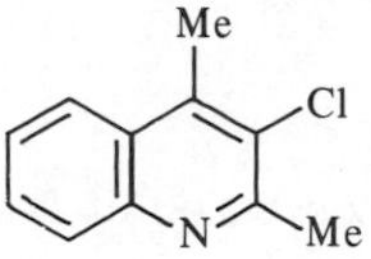	63%
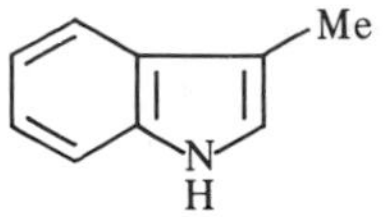	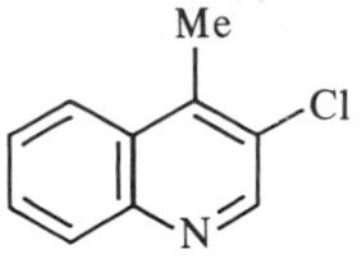	68%
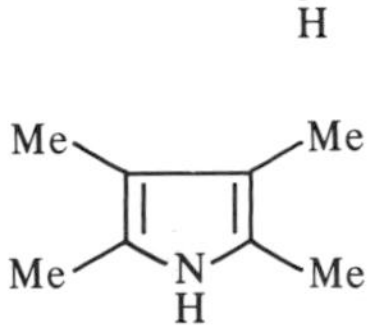	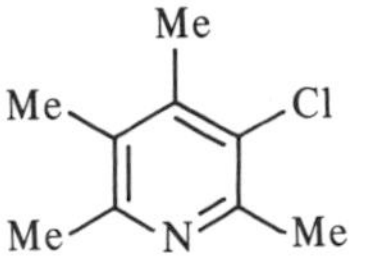	37%
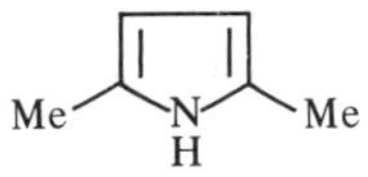	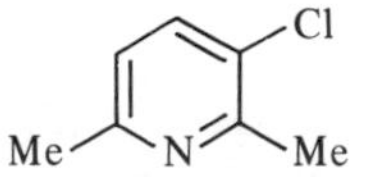	49%

Thermolysis of sodium trichloroacetate in dimethoxyethane gave pyridines (**I-224**) (515).

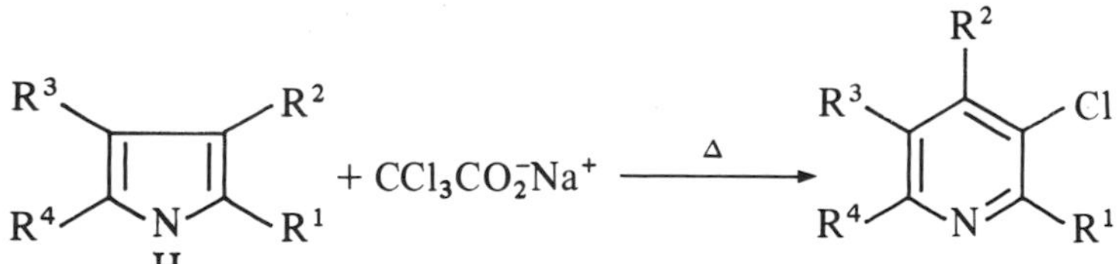

I-224 (50–90%)

The vapor phase reaction of pyrrole with chloroform has been reported to give 3-chloropyridine; however, a recent improvement has been reported in which use of a substrate preheater has increased the yields to 86% (516)!

From a mechanistic viewpoint a great deal of knowledge about this ring-expansion process has been gained over the last decade. In early reports, it was assumed that the 2H-pyrrole **I-222** was involved in the formation of the pyridine derivative **I-220**. Numerous reports have appeared (517–519) which have shown that this proposed transformation is not correct. The most recent paper has shown the mechanism to be as shown in Scheme I-3 (520).

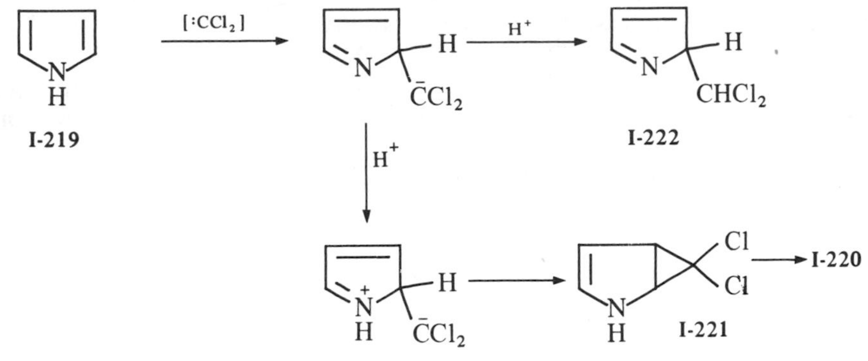

Scheme I-3

Vapor phase reaction of *N*-methylpyrrole in the presence of iodine gave pyridine in 58% yield (521). This is an improvement over the reaction in the absence of iodine where yields were only 2–5%.

The carbene generated from bromoallene I-225 gave I-226 in the presence of 2,5-dimethylpyrrole (552). Yields were low in the pyrrole series but were higher when indoles were used as substrates.

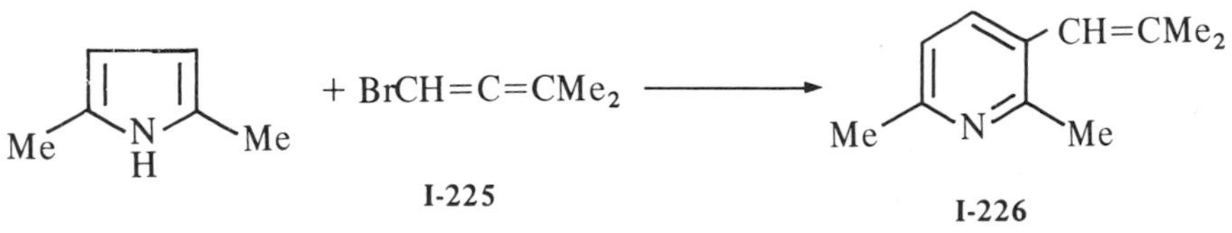

The homopyrroles I-227, prepared from pyrroles and ethyl diazoacetate, gave pyridines I-228 *via* the intermediary 1,2-dihydropyridines I-229 (523).

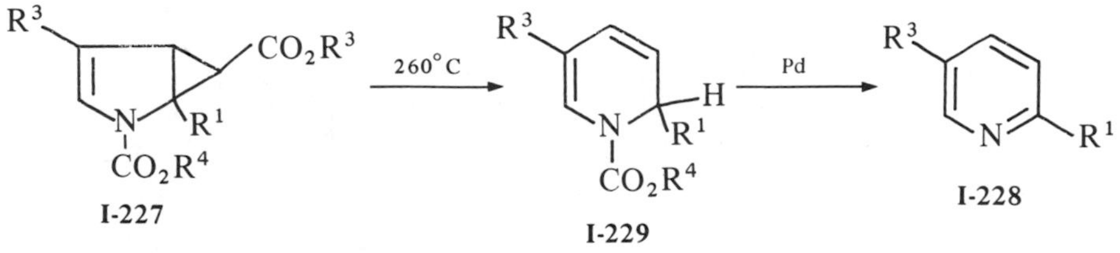

Pyrrolinediones I-230 underwent ring-expansion with diazoalkanes to give pyridones I-231–233 (524) (Table I-29). The bicyclic analogs I-234 gave mainly the enol ethers I-235, the pyridines I-236 and I-237 being minor products (525).

2,4,5-Triphenyl-3*H*-pyrrol-3-one 1-oxide (I-238) underwent cycloadditions with acetylenic dipolarophiles to give pyridines I-239 and I-240 (526).

TABLE I-29. PYRIDONES FROM PYRROLIDONES (524)

I-230 I-231 I-232 I-233

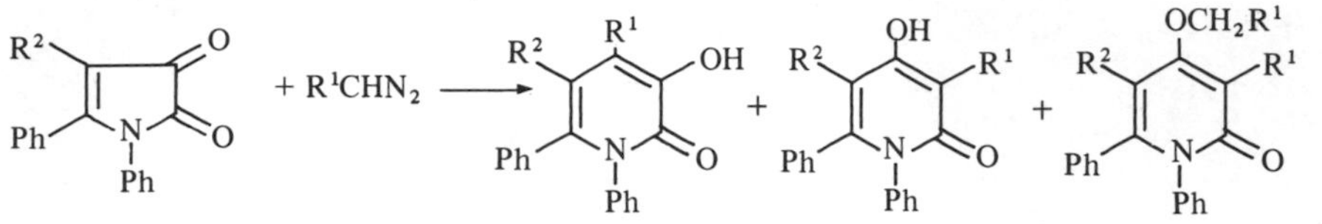

I-230	R^1	R^2	I-231 (%)	I-232 (%)	I-233 (%)
a	H	H	40	26	18
b	H	CH_3	32	25	21
c	H	Ph	14	14	33
d	CH_3	H	52	40	6
e	CH_3	CH_3	52	44	3
f	CH_3	Ph	54	43	2
g	$CO_2C_2H_5$	H	35	60	–
h	$CO_2C_2H_5$	CH_3	58	41	–
i	$CO_2C_2H_5$	Ph	54	45	–
j	Ph	H	59	41	–
k	Ph	CH_3	48	52	–
l	Ph	Ph	49	50	–

I-234 I-235

I-236

I-237

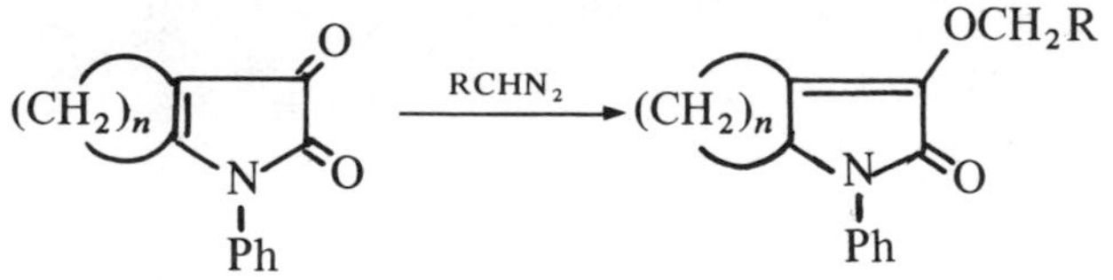

A number of reports have appeared on the electrochemical oxidation of pyrroles (**I-241**) to yield pyridines **I-242** (527–529). The carbon at C-5 appears to be furnished by the nitromethane, since nitromethane-d_3 gave the pyridinium salt (**I-242**) with a deuterium on C-5 (528). When R=H the free pyridine base is obtained.

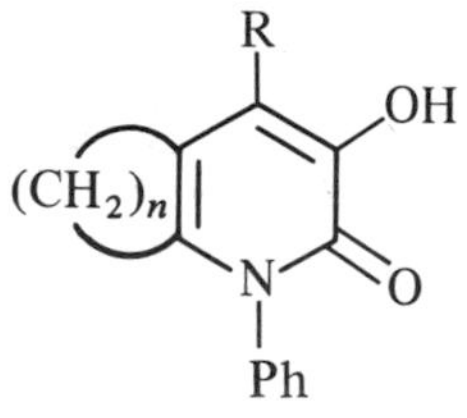

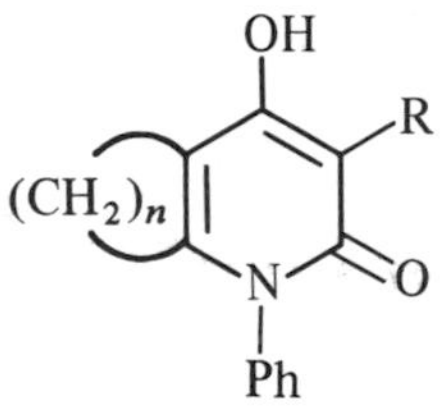

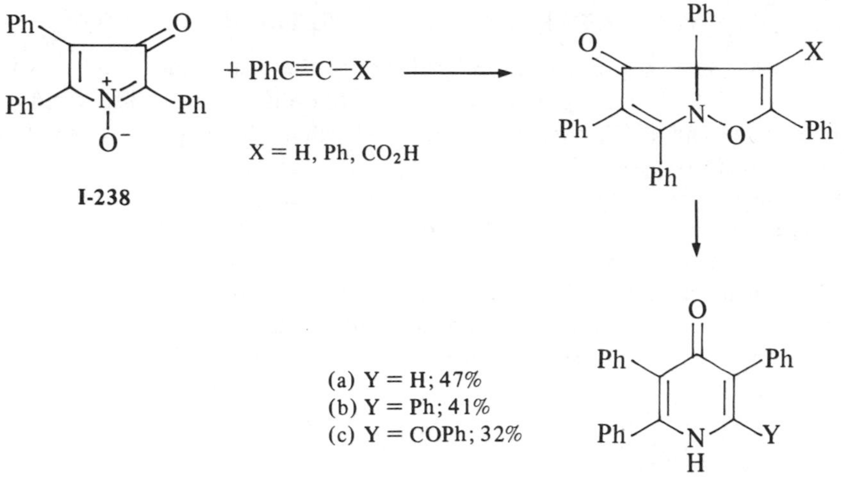

I-238

+ PhC≡C—X $\longrightarrow$

X = H, Ph, CO$_2$H

(a) Y = H; 47%
(b) Y = Ph; 41%
(c) Y = COPh; 32%

I-239

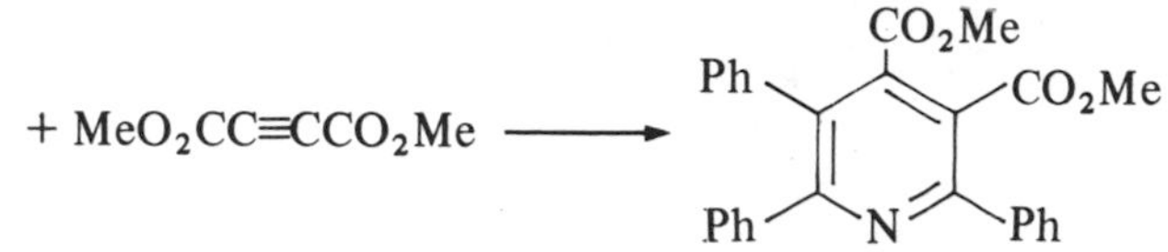

I-238 + MeO$_2$CC≡CCO$_2$Me $\longrightarrow$

I-240 (58%)

+ CH$_3$NO$_2$ $\xrightarrow[\text{NaClO}_4]{-e^-}$

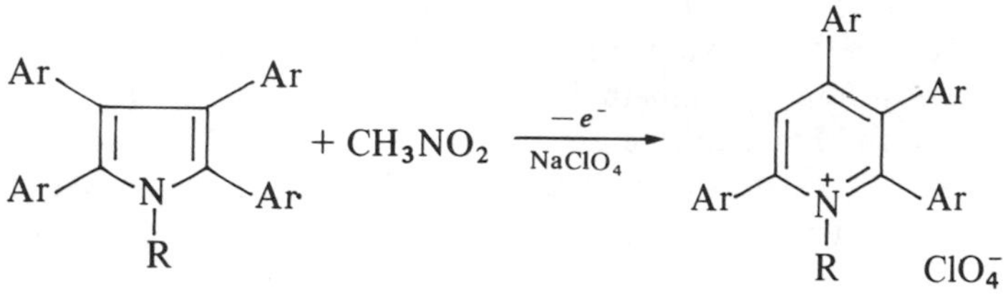

I-241

I-242 (100%)

R = H, Me, Et
Ar = Ph, p-ClC$_6$H$_4$,
p-MeC$_6$H$_4$, p-MeOC$_6$H$_4$

Reaction of the dihydropyrrole **I-243** with diketene (**I-244**) gave the bicyclic pyridone **I-245** in addition to **I-246** (530).

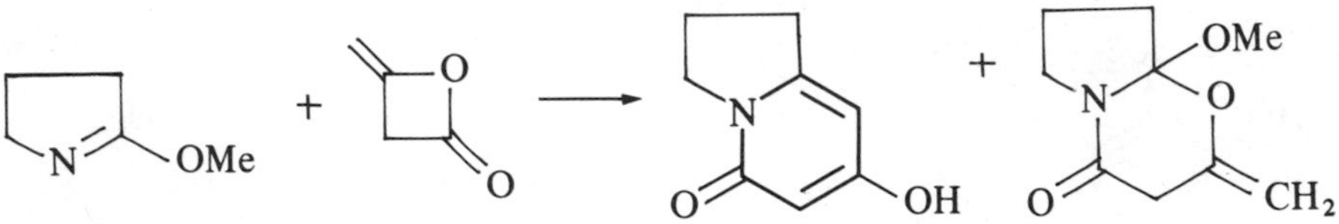

I-243 + I-244 $\longrightarrow$ I-245 + I-246

b. FIVE-MEMBERED RINGS CONTAINING TWO HETEROATOMS

i. OXAZOLES.   Oxazoles react as azadienes in the Diels–Alder reaction with olefinic dienophiles to yield pyridine derivatives. Since Kondrat'eva discovered the reaction in the late 1950s over 100 papers and patents have appeared as well as several reviews (531–534).

When asymmetric dienophiles (I-247) (x ≠ y) are utilized the reaction shows a high degree of regioselectivity with the orientation of the adduct I-248 being such that the more electron withdrawing substituent of the dienophile ends up at the 4-position of the adduct. HMO calculations have been utilized to predict the orientation in several examples (535) although some failure to preduct the correct orientation in the oxazole–dienophile systems has been observed (536). Localization energy calculations have been found to predict the orientation in the oxazole–dienophile systems (536).

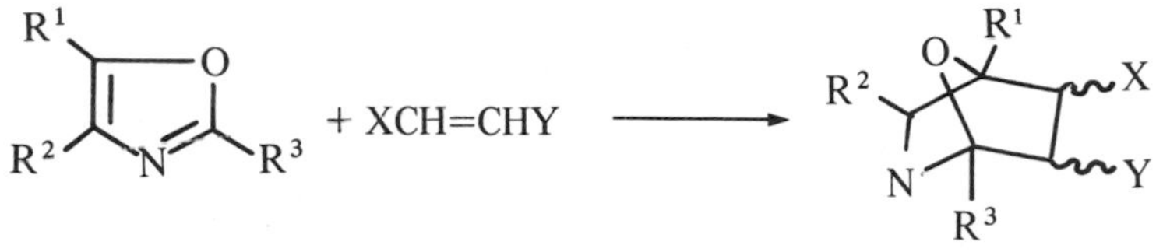

I-247   I-248

The course of the reaction was dependent on subtle changes in the oxazole and dienophile. Use of simple alkyloxazoles illustrates this point. Type A product was obtained from 4,5-dimethyloxazole and fumaronitrile (537) while 2,4-dimethyloxazole and fumaronitrile gave a Type B product (538). When the dienophile was diethyl fumarate the product with 2,4-dimethyloxazole was of Type C (538) (Schemes I-4, I-5).

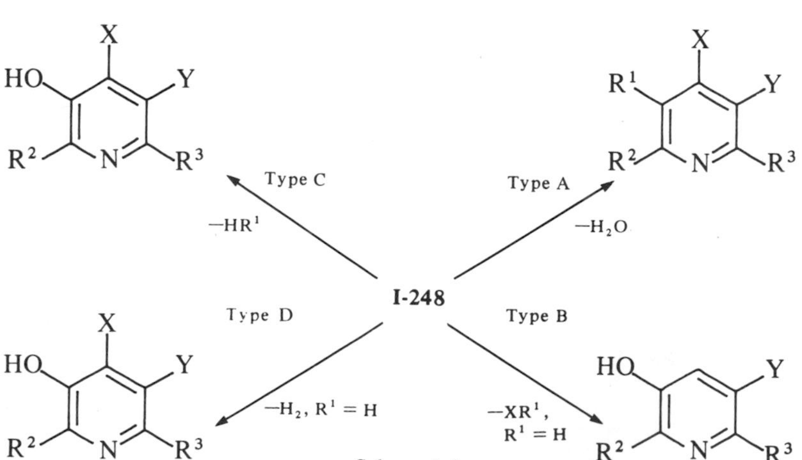

Scheme I-4

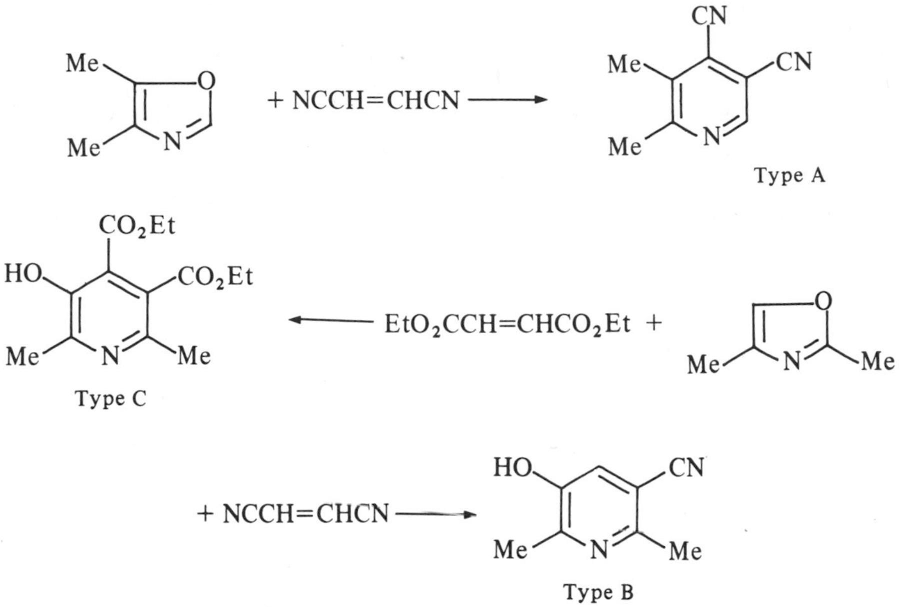

Scheme I-5

A Type A reaction (Scheme I-4) occurred when R^1 was a poor leaving group as in the examples of alkyloxazoles (see Table I-30). If on the other hand when $R^1 = H$ and X was a good leaving group, such as cyanide, then the Type B reaction occurred (Table I-31). When R^1 was a good leaving group such as alkoxy then the Type C pathway predominated. The Type D reaction required the loss of hydride when $R^1 = H$ and X was not a good leaving group. This later process was rarely observed; however, the addition of good oxidizing agents, such as hydrogen peroxide or aromatic nitro compounds, aided the reaction and yields of pyridines from this reaction have been as high as 50–60% (539).

A special case of Type B was the reaction of α,α-disubstituted olefinic dienophiles I-249 [XZC=CHY], where Z is a good leaving group (see Table I-31).

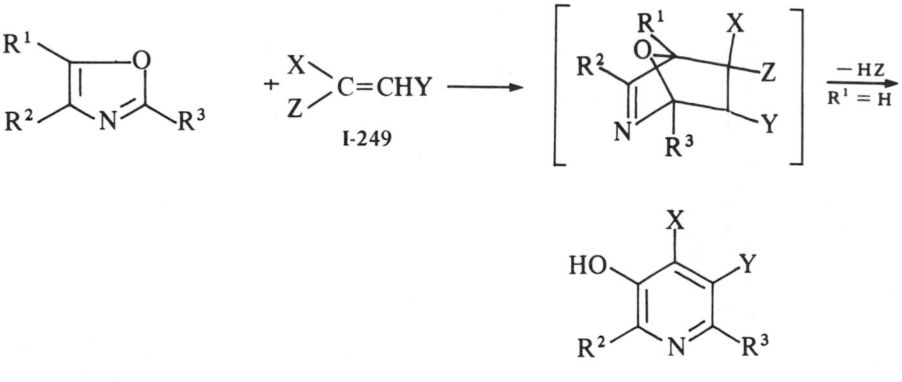

TABLE I-30.   TYPE A REACTIONS OF OXAZOLES TO YIELD PYRIDINES

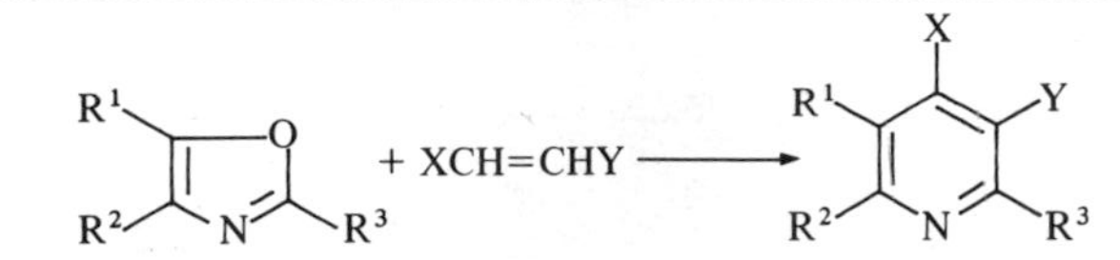

	Dienophile		Pyridine Product						
Substituent in Oxazole	X	Y	R^1	R^2	R^3	X	Y	Yield (%)	Reference
$R^1 = $ 1-(indol-3-yl)ethyl	CN	H	1-(indol-3-yl)ethyl			CN	H	16	541
$R^2 = $ H, $R^3 = $ H									
$R^1 = $ H, $R^2 = $ Ph, $R^3 = $ H	—CONPhCO—		H	Ph	H	—CONPhCO—			542
	—CONEtCO—		H	Ph	H	—CONEtCO—			542
	—CONHCO—		H	Ph	H	—CONHCO—			542
$R^1 = $ Me, $R^2 = $ Me, $R^3 = $ H	CN	CN	Me	Me		CN	CN	58	537
$R^1 = $ Me, $R^2 = $ Me, $R^3 = $ Me	CN	CN	Me	Me	Me	CN	CN	50	537

TABLE I-31.   TYPE B REACTIONS OF OXAZOLES TO YIELD PYRIDINES

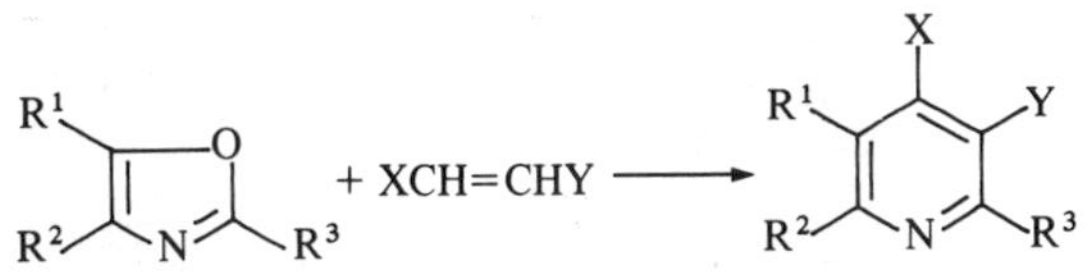

Substituents on Oxazole	Dienophile			Pyridine Product				Yield (%)	Reference
	X	Y	Z	R^2	R^3	X	Y		
R^1 = H, R^2 = Me, R^3 = CH$_2$Ph	—CH$_2$OCH$_2$—		SO$_2$Me	Me	CH$_2$Ph	—CH$_2$OCH$_2$—		19.3	543, 544
R^1 = H, R^2 = Me, R^3 = H	—CH$_2$OCH$_2$—		SO$_2$Me	Me	H	—CH$_2$OCH$_2$—		38	545
	CH$_2$OMe	CH$_2$OMe	SO$_2$Me	Me	H	CH$_2$OMe	CH$_2$OMe		545
	—CH$_2$OCH$_2$OCH$_2$—		SO$_2$Me	Me	H	—CH$_2$OCH$_2$OCH$_2$—			545
	—CH$_2$OCH$_2$—		SO$_2$Ph	Me	H	—CH$_2$OCH$_2$—			545
	CH$_2$OAc	CH$_2$OAc	SO$_2$Ph	Me	H	CH$_2$OAc	CH$_2$OAc		545

TABLE I-32.   TYPE C REACTIONS OF OXAZOLES TO YIELD 3-HYDROXYPYRIDINES

$$R^2\text{-oxazole} + XCH=CHY \longrightarrow \text{3-hydroxypyridine (Pyridine Product)}$$

Dienophile (left) → **Pyridine Product** (right)

$R^1 = OEt,\ R^2 = Me,\ R^3 = H$

Substituents in Oxazole		Pyridine Product					
X	Y	X	R^2	R^3	Y	Yield (%)	Reference
CN	CH_2OH	CN	Me	H	CH_2OH	48.8	546
H	CH_2OH	H	Me	H	CH_2OH	59.8	547, 548
H	CH_2OAc	H	Me	H	CH_2OAc	13.8	549
H	$PhCO_2CH_2$	H	Me	H	$PhCO_2CH_2$		549
H	$PhOCH_2$	H	Me	H	$PhOCH_2$		549
H	CH_2Ph	H	Me	H	CH_2Ph		549
CO_2H	PhCO	CO_2H	Me	H	PhCO	58	550
CO_2H	$o\text{-}MeOC_6H_4CO$	CO_2H	Me	H	$o\text{-}MeOC_6H_4CO$		550
CO_2H	$2,5\text{-}(MeO)_2C_6H_3CO$	CO_2H	Me	H	$2,5\text{-}(MeO)_2C_6H_3CO$		550
CO_2Me	CO_2Me	CO_2Me	Me	H	CO_2Me	57	551, 552
$^{13}CO_2Et$	$^{13}CO_2Et$	$^{13}CO_2Et$	Me	H	$^{13}CO_2Et$		553
$^{14}CO_2Et$	$^{14}CO_2Et$	$^{14}CO_2Et$	Me	H	$^{14}CO_2Et$		554
COMe	CO_2Et	COMe	Me	H	CO_2Et	57.6	555
COMe	Me	COMe	Me	H	Me		556
$-CO-O-CHOEt-$		H	Me	H	CHO	11	557
CH_2OAc	CH_2OAc	CH_2OAc	Me	H	CH_2OAc	70	558
CH_2OMe	CH_2OMe	CH_2OMe	Me	H	CH_2OMe	69.3	559, 561
CH_2OCH_2OMe	CH_2OCH_2OMe	CH_2OCH_2Me	Me	H	CH_2OCH_2OMe		562
CH_2OH	CH_2OH	CH_2OH	Me	H	CH_2OH	78	563, 564
$-CH_2OCH(CH_2CHMeSMe)OCH_2-$		$-CH_2OCH(CH_2SHMeSMe)OCH_2-$	Me	H			565
$-CH_2OCH(CHMe_2)OCH_2-$		$-CH_2OCH(CHMe_2)OCH_2-$	Me	H			566
$-CH(OMe)OCH(OMe)-$		$-CH(OMe)OCH(OMe)-$	Me	H			567, 569
$-CH_2OCH(OEt)CH_2-$		$-CH_2OCH(OEt)CH_2-$	Me	H			570
$-CH_2OCH_2OCH_2-$		$-CH_2OCH_2OCH_2-$	Me	H			571

64

Continuation table (substituted compounds, with yields and literature references). The R¹, R², R³ definitions at the left head successive groups of rows; the two central columns (H and Me) are constant. Bridging substituents (written with dashes) span the two substituent positions.

R¹, R², R³	Substituent	Substituent			Yield (%)	References
R¹ = OPr, R² = CH₃, R³ = H						
	NO₂	Me	H	Me	3	572
	CN	CN	H	Me	37	573
	—CONHCO—		H	Me	55	573
	H	C≡CCMe₂OH	H	Me	40	574
R¹ = OEt, R² = CH₂CO₂Et, R³ = H						
	H	CN	H	Me	88	575, 576
	H	OEt	H	Me		575
	H	CO₂Et	H	Me		577
	Me	CN	H	Me	52	575, 576
	CH₂OAc	CN	H	Me		575
	CH₂OMe	CH₂OMe	H	Me		577
	—CH₂OCH₂OCH₂—		H	Me		577
R¹ = OEt, R² = CH₂CO₂H, R³ = H						
	H	CH₂OH	H	Me	85	548
	H	CN	H	Me	92	578, 579
	H	CO₂H	H	Me	64	578, 579
	H	CO₂Me	H	Me		579
	H	OMe	H	Me		578
	CN	CH₂OH	H	Me	70	580, 581
	CN	CN	H	Me	90	579, 581, 582, 583
	CO₂Me	CO₂Me	H	Me	85	581, 579
	CO₂Et	CO₂Et	H	Me	70	581, 579
	CO₂Bu	CO₂Bu	H	Me	72	581
	CO₂H	CH₃	H	Me	50	579
	CN	CH₃	H	Me	62	579
	CO₂Me	Me	H	Me		579
R¹ = CN, R² = Me, R³ = H						
	—CO—O—CO—		H	Me	24	580, 579
	—CH₂OCH₂OCH₂—		H	Me	20	580, 579
	—CH₂OCH(CHMe₂)OCH₂—		H	Me	77	580, 584, 579
	—CH₂OCH(CHMe₂)OCH₂—		H	Me	82.2	585
	—CH₂OCH(n-Pr)OCH₂—		H	Me		586
	—CH₂OCO₂CH₂—		H	Me		563
R¹ = OMe, R² = Me, R³ = H						
	—CH(OMe)OCH(OMe)—		H	Me		587

TABLE I-32. (CONTINUED)

Substituents in Oxazole	Dienophile		Pyridine Product				Yield (%)	Reference
	X	Y	R^2	R^3	X	Y		
R^1 = OEt, R^2 = CH_2Ph, R^3 = H	CO_2Me	CO_2Me	CH_2Ph	H	CO_2Me	CO_2Me		588
R^1 = SEt, R^2 = CH_3, R^3 = H	—$CH_2OCH(CHMe_2)OCH_2$—		CH_3	H	—$CH_2OCH(CHMe_2)OCH_2$—			589
R^1 = OCH_2CH_2OMe, R^2 = Me, R^3 = H	—$CH_2OCH(CHMe_2)OCH_2$—		CH_3	H	—$CH_2OCH(CHMe_2)OCH_2$—			590
R^1 = OCO_2Et, R^2 = CH_3, R^3 = H	CO_2Me	CO_2Me	CH_3	H	CO_2Me	CO_2Me		591
	COSMe	COSMe	CH	H	COSMe	COSMe		591
	—CO_2CO—		CH_3	H	H	CO_2Et		591
	—CONHCO—		CH	H		—CONHCO—		591
R^1 = $OSiMe_3$, R^2 = CH_2Ph R^3 = Me	—CONPhCO—		CH_2Ph	H		—CONPhCO—		592
R^1 = CO_2H, R^2 = Me, R^3 = H	CO_2Et	CO_2Et	CH_3	H	CO_2Et	CO_2Et		593
	H	CN	CH_3	H	H	CN		593
	CH_2OAc	CH_2OAc	CH_3	H	CH_2OAc	CH_2OAc		593
R^1 = CO_2Et, R^2 = CH_3, R^3 = H	CO_2Me	CO_2Me	CH_3	H	CO_2Me	CO_2Me	56.2	594
	COSMe	COSMe	CH_3	H	COSMe	COSMe		594
	CN	CN	CH_3	H	CN	CN		594
R^1 = OEt, R^2 = Me, R^3 = CO_2H	CO_2Et	CO_2Et	Me	H	CO_2Et	CO_2Et	86.2	595
	CN	CN	Me	H	CN	CN	56.6	595
	—$CH_2OCH_2OCH_2$—		Me	H		—$CH_2OCH_2OCH_2$—	48.7	595, 596
R^1 = OEt, R^2 = CH_2CO_2H R^3 = CO_2H	CO_2Et	CO_2Et	Me	H	CO_2Et	CO_2Et	69	595
R^1 = H, R^2 = Me, R^3 = Me	2-pyridyl	H	Me	H	2-pyridyl	H	60	597, 598
R^1 = H, R^2 = Ph, R^3 = Me	2-pyridyl	H	Ph	Me	2-pyridyl	H	66	597
R^1 = H, R^2 = Ph, R^3 = H	2-pyridyl	H	Ph	H	2-pyridyl	H	50	597

$R^1 = H$, $R^2 = Ph$, $R^3 = Me$	3-(6-methyl)-pyridyl	H	Ph	Me	3-(6-methyl)-pyridyl	H	21	597
	4-pyridyl	H	Ph	Me	4-pyridyl	H	25	597
$R^1 = H$, $R^2 = Me$, $R^3 = H$	$-CH_2OCH_2OCH_2-$		Me	H	$-CH_2OCH_2OCH_2-$			599
	$-CH_2OCHPhOCH_2-$		Me	H	$-CH_2OCHPhOCH_2-$			599
	$-CH_2OCH(CH_2CH_2CH_3)OCH_2$		Me	H	$-CH_2OCH(CH_2CH_2CH_3)OCH_2-$			599
	$-CH_2OCH_2-$		Me	H	$-CH_2O-CH_2-$			599
	$-CH_2OCHMeOCH_2-$		Me	H	$-CH_2OCHMeOCH_2-$			599

TABLE I-33.   PRODUCTION AND PRICE OF PYRIDOXINE HYDROCHLORIDE (B_6) (540)

Year	$/kg	U.S. Consumption (metric tons)
1940	7000	1.9
1947	670	5.9
1955	460	13.6
1963	65	
1967	30	
1976	36^a	$240-260^b$
1978	42^a	264^c
1979	43^a	204^c

a Price rise due to inflation, historic low of $21/kg in the mid-1970s.
b Estimated, Reilly Tar & Chemical Marketing Department.
c Import tonnage only.

Among the four classes, the Type C reaction has been studied most extensively due to its importance in the synthesis of pyridoxine (**I-250**; vitamin B_6). Numerous dienophiles have been utilized (Table I-32); however, the oxazoles (e.g., **I-251**) utilized were generally 5-alkoxyoxazoles with a methyl group at position 5 or a substituent readily transformed into a methyl group.

$$RO \diagdown \quad + XCH=CHX \longrightarrow HO \diagdown \quad (X = CH_2OH)$$

I-251   I-250

Numerous dienophiles have been used which contain substitutions that can be readily converted to CH_2OH. These included esters, nitriles, ethers, and cyclic dienophiles, such as dihydrofurans, dioxapines, and cyclic anhydrides. Although several commercial routes to vitamin B_6 were historically important prior to the discovery of the oxazole route, it was not until its introduction in the early 1960s that the price and availability of vitamin B_6 made it of commercial importance (Table I-33).

ii.   MISCELLANEOUS FIVE-MEMBERED RING HETEROCYCLES.   Various five-membered ring heterocycles containing two heteroatoms have been transformed into pyridines. Usually the transformation involved cycloaddition reactions to give bicyclic intermediates which ring-open to yield a pyridine product. Among the most widely studied systems have been the mesoionic oxazolones (**I-252**) and thiazolones (**I-253**), and their respective isomers **I-254** and **I-255**. Table I-34 and I-35 summarize the reactions of these mesoionic heterocycles which yield pyridines.

TABLE I-34.   PYRIDINES FROM MESOIONIC HETEROCYCLES AND CYCLOPROPENE DERIVATIVES

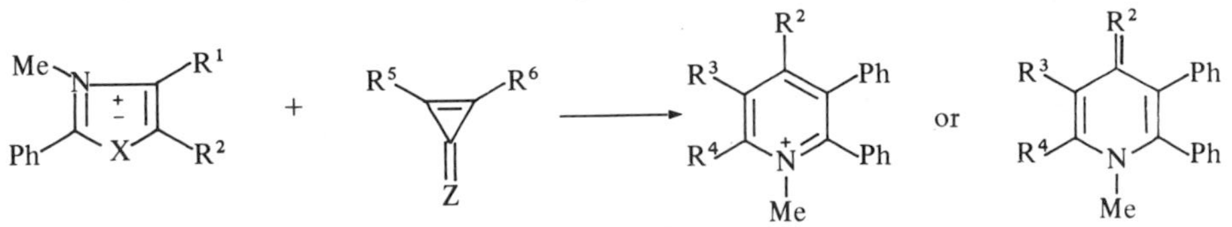

Mesoionic Compounds			Cyclopropene			Pyridinium Product				
X	R^1	R^2	Z	R^5	R^6	R^2	R^3	R^4	Yield	Reference
O	Ph	O	=O	Ph	Ph	=O	Ph	Ph	61%	601–603
			=S	Ph	Ph	=S	Ph	Ph	61%	602–604
			=NTs	Ph	Ph	=NTs	Ph	Ph	64%	604, 602
			=C(CN)$_2$	Ph	Ph	=C(CN)$_2$	Ph	Ph	41%	604, 603
			=C(CN)(CO$_2$Et)	Ph	Ph	=C(CN)CO$_2$Et	Ph	Ph	65%	604
S	O	H	=O	Ph	Ph	Ph	H	—OH	21%	604
S	H	O		Ph	Ph	S	Ph	H	20%	601
S	Ph	O	=O	Ph	Ph	O	Ph	Ph	55%	601, 604
			=S	Ph	Ph	S	Ph	Ph	48%	601, 604
			=NTs	Ph	Ph	=NTs	Ph	Ph	39%	604
			=C(CN)$_2$	Ph	Ph	=C(CN)$_2$	Ph	Ph	74%	604
			=C(CN)CO$_2$Et	Ph	Ph	=C(CN)CO$_2$Et	Ph	Ph	73%	604
O	Ph	O	=O	Ph	CH$_3$	O	Me	Ph	61%	603
			=O	CH$_3$	CH$_3$	O	Me	Me	35%	603
			=C(COMe)$_2$	Ph	Ph	=C(COMe)$_2$	Ph	Ph	37%	603
			=C(COPh)$_2$	Ph	Ph	=C(COPh)$_2$	Ph	Ph	53%	603
			=C(CN)(COPh)	Ph	Ph	=C(CN)(COPh)	Ph	Ph	56%	603
			=C(CN)(CO$_2$Me)	Ph	Ph	=C(CN)(CO$_2$Me)	Ph	Ph	60%	603

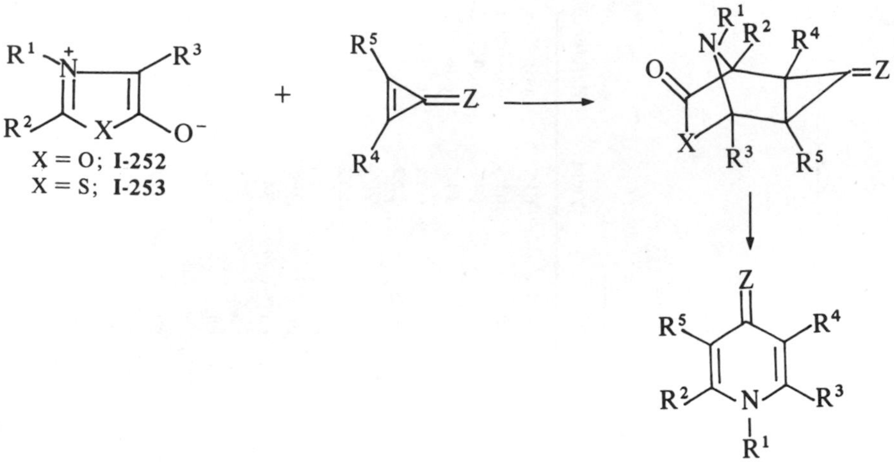

$X = O$; **I-252**
$X = S$; **I-253**

N-Anilinopyridinium tetrafluoroborates (**I-256**) were formed on treatment of pyrazole–boron trifluoride (**I-257**) complexes with dimethyl acetylenedicarboxylate (**600**). The reaction failed when in **I-257** R was acetyl, benzoyl, or carboxamide.

TABLE I-35.   PYRIDINES FROM MESOIONIC HETEROCYCLES AND DIENOPHILES

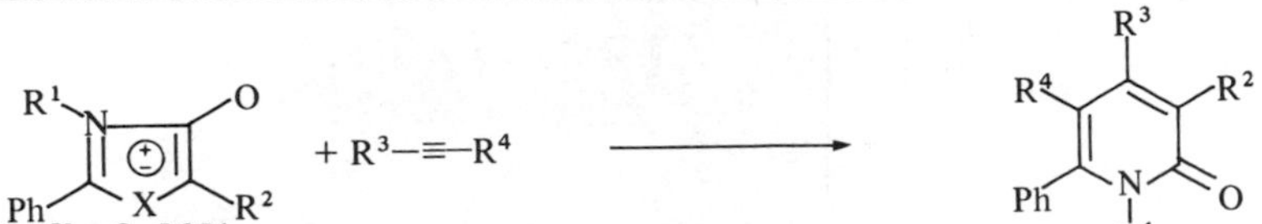

$X = O$; **I-254**
$X = S$; **I-255**

Mesoionic Compound			Dienophile		Product	
X	R^1	R^2	R^3	R^4	Yield	Reference
S	Ph	H	CO_2Me	CO_2Me	70%	605, 606
S	p-ClC_6H_4	H	CO_2Me	CO_2Me	35%	606
			PhCO	PhCO	37%	606
S	Ph	Ph	CN	CN	44%	606

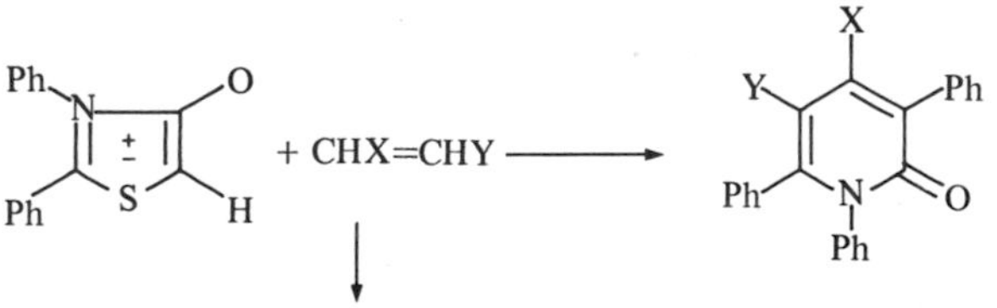

Dienophile ($X = Y$)	Yield	Reference
CN	58%	607
COPh	87%	607
CONPhCO	54%	607
CO_2Me	17%	607

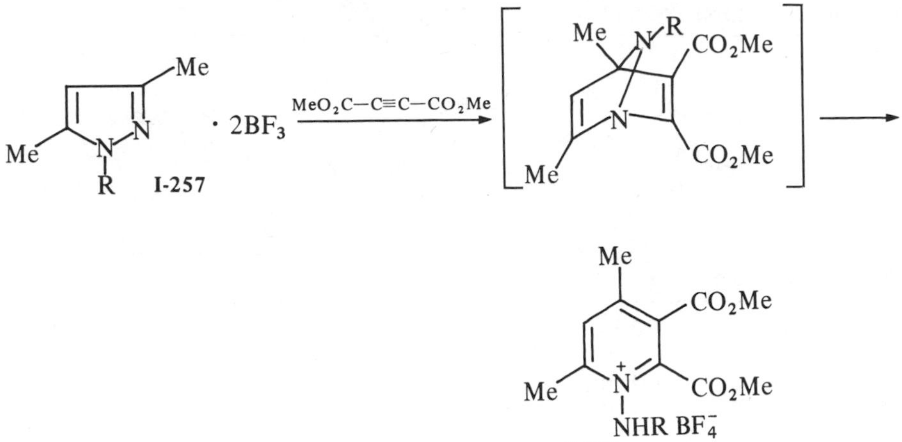

I-256

Ring transformations of isoxazoles into pyridines have been reported frequently and the literature has been reviewed (608, 609). It is well established that the iso-xazole ring can be used as a masked 1,3-dicarbonyl or enamino ketone and as such has become useful in numerous syntheses (610). Catalytic hydrogenation of I-258 gave ring-opened intermediates, such as I-259. When R was an alkyl ketone, then dihydropyridines such as I-260 were isolated (611–614). Isoxazoles I-258 have been shown to react as nucleophiles with the electrophilic diphenylcyclopropenone (I-261) to give pyridones I-262 (615).

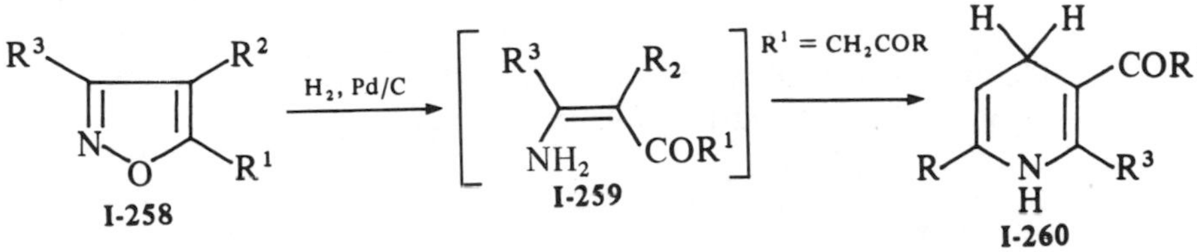

I-258   I-259   I-260

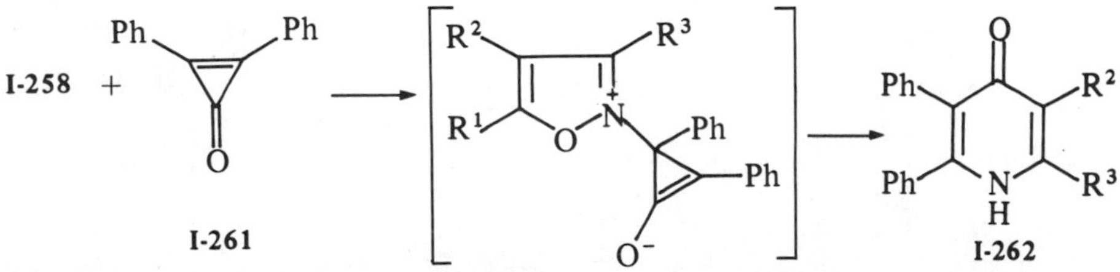

I-258   I-261   I-262

Isoxazoles form quaternary salts (I-263) with some difficulty; however, I-263 is useful in many instances for the preparation of pyridines. Treatment of isoxazolium salts, substituted at C-3 and C-4, with an active methylenic compound in the presence of base leads to I-265 *via* the ketene I-264 (616) (Table I-36). In a similar

TABLE I-36.　PYRIDINES FROM ISOXAZOLIUM SALTS (616)

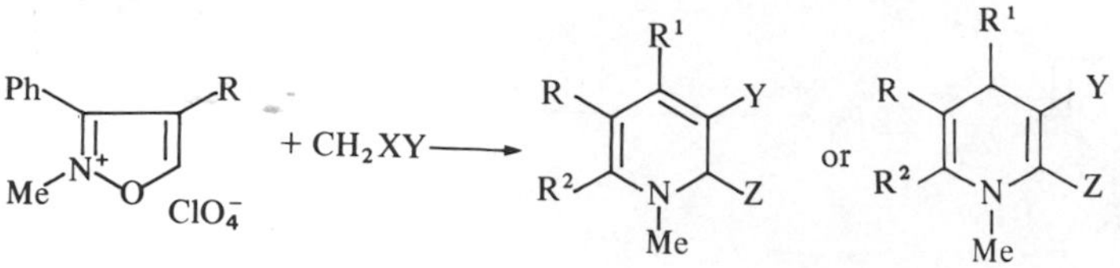

R	X	Y	Base	Product R^1	R^2	Z	Yield
Me	CO$_2$Et	CO$_2$Et	NaOEt	H	Ph	=O	22.7
				OH	Ph	=O	16.0
Me	CN	CONH$_2$	NaOH	H	Ph	=NH	15.8
Me	COCH$_3$	CO$_2$Et	NaOEt	=O	Ph	CH$_3$	5.3

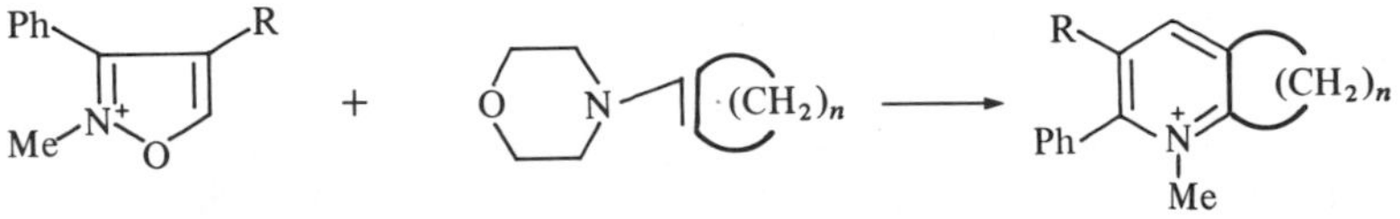

Conditions	R	n	Yield
CH$_3$CN/50°C	Me	3	38.4
CH$_3$CN/50°C	Ph	3	10.7
CH$_3$CN/50°C	H	3	2.3
CH$_3$CN/50°C	Me	4	27.4
CH$_3$CN/50°C	Ph	4	8.9

fashion, these isoxazolium salts gave pyridinium salts on treatment with cyclic enamines.

5-Arylthiazoles condensed with dimethyl acetylenedicarboxylate to afford pyridones I-266 *via* the intermediacy of ylide I-267 (617).

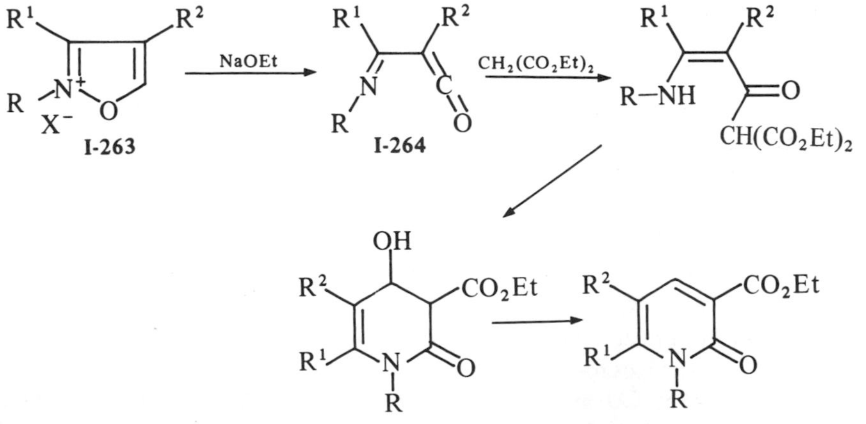

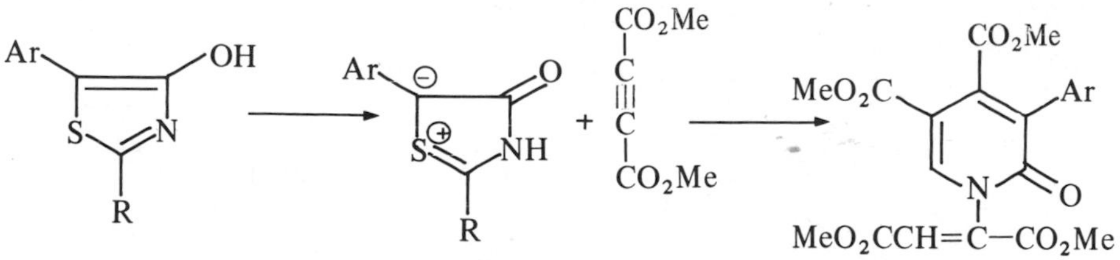

I-267   **I-266**

1,3-Dipolar addition of the oxazolone **I-268** with triphenylcyclopropene gave the 1,4-dihydropyridine **I-269** (618).

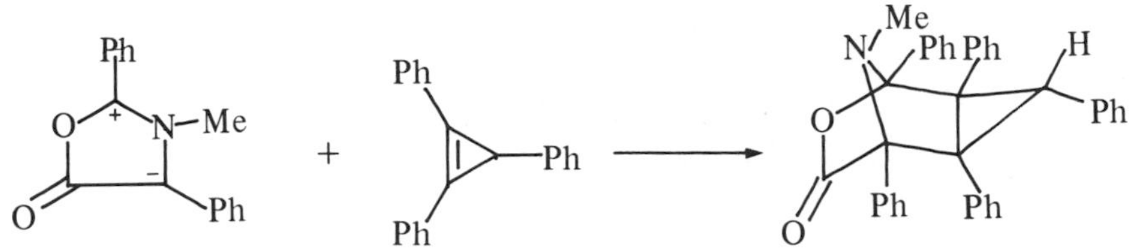

I-268

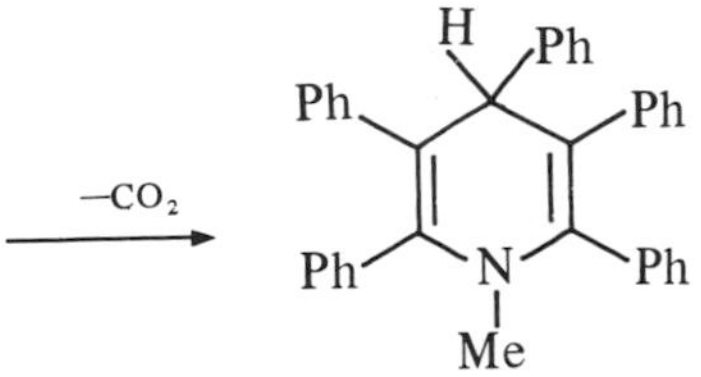

I-269

Hetero–Cope rearrangement of 4-allyl-2-oxazolin-5-ones (**I-270**) gave 2,6-disubstituted pyridines upon the elimination of carbon dioxide (619). When $R^1 =$ Ph, $R^2 = CH_2CO_2CH_3$, 2-phenylpyridine is obtained in 70% yield.

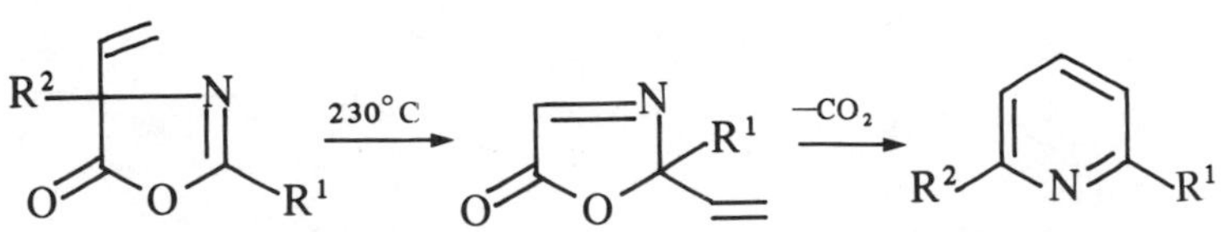

I-270   41–70% yield

The acyldiazabicyclic ketones **I-271** (R = CH_3, Ph) give pyridinium salts **I-272** on treatment with acid, while with methanol gave pyridines **I-273**. Treatment of **I-271** with base gave **I-274**, which eliminated ammonia on standing to give the pyridine. The pyridine products apparently were generated *via* the azetinone **I-275** (620).

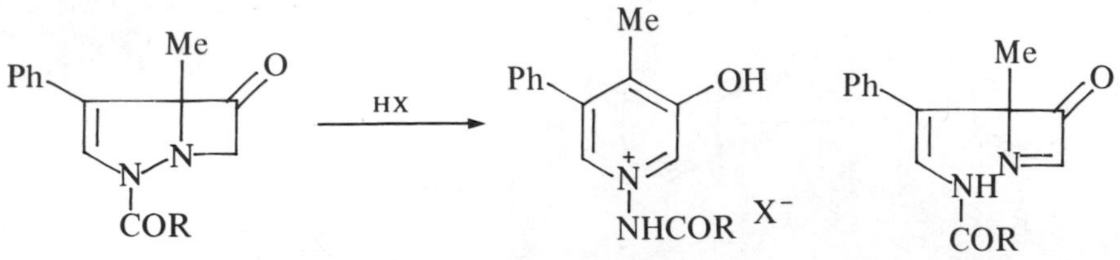

I-271

I-272

I-275

I-271 —MeOH→

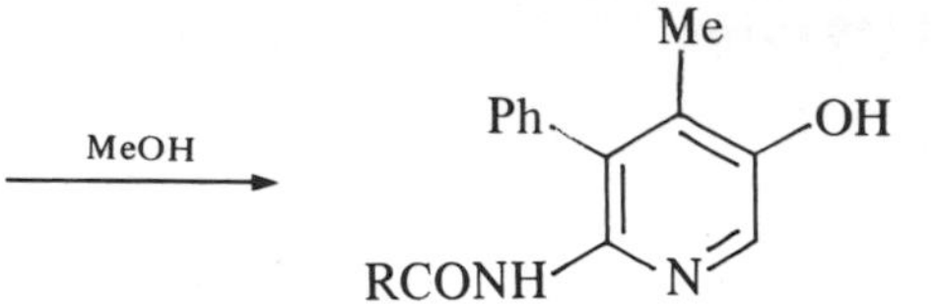

I-273

I-271 —OH⁻→

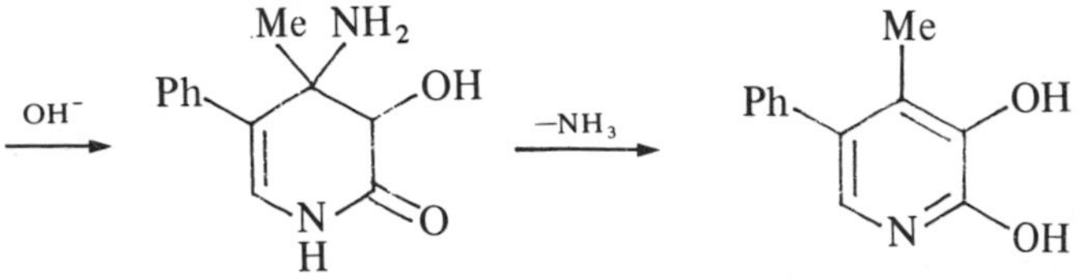

I-274 —(-NH₃)→

N-Carbomethoxy-2-azabicyclo[3.1.0]hex-3-ene (I-276) in the gas phase (285°C) afforded the dihydropyridine I-277 (621). I-276 was prepared easily from the N-carbomethoxypyrrole and ethyl diazoacetate.

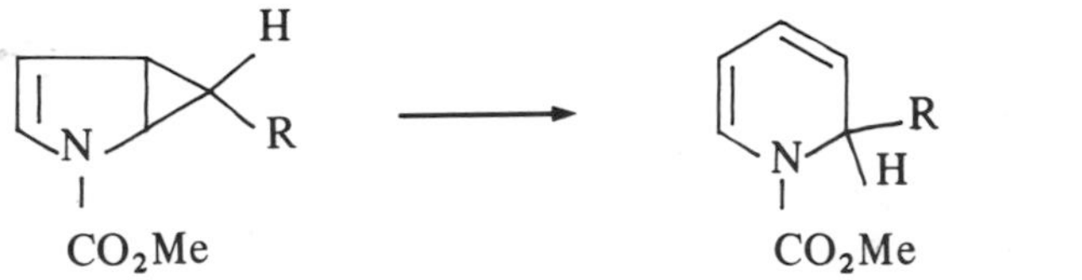

I-276 (R = H; CO₂Et)

I-277 100% yield

Diphenylcyclopropene thione (I-278) with an enamine gave the bicyclic adduct I-279 which can be readily S-alkylated to give the quaternary salt I-280. Treatment of I-280 with potassium hydroxide in methanol or thermolysis at 140°C gave 2,4,5-triphenylpyridine (I-281) (622).

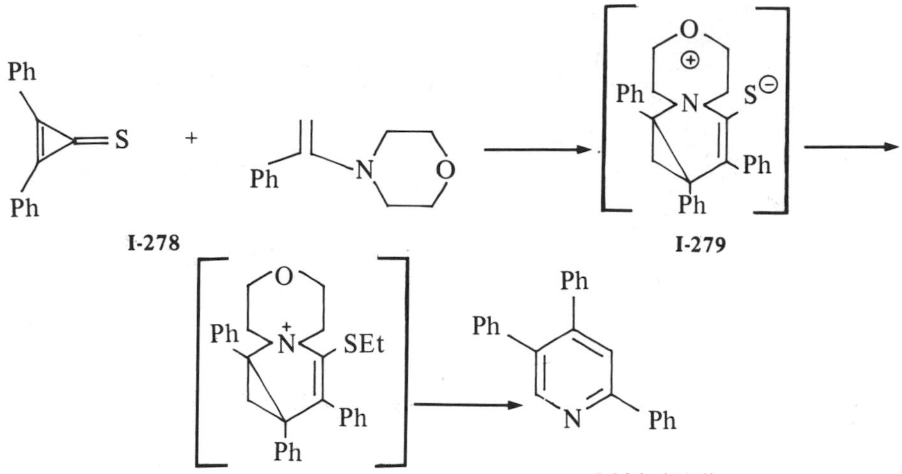

I-278   **I-279**

I-280   **I-281** (70%)

E. Six-Membered Ring Heterocycles

a. ONE HETEROATOM

i. PYRONES. Pyrones continue to be of interest in the synthesis of pyridones and a recent review has appeared (623).

Ammolysis of 2-(**I-282**) and 4-pyrones (**I-283**) gave the corresponding pyridones in high yield. The reaction has been presumed to proceed *via* the open-chained intermediate **I-284**, which cyclized to the pyridone; however recently it has been suggested that 2 moles of amine are involved (623). The open-chained intermediates, for example, **I-285**, were isolated at room temperature; **I-285** (R = Me) was thermally cyclized to pyridone **I-286** (624). Table I-37 summarizes the ammonolysis reaction of 2-pyrones while Table I-38 contains the 4-pyrone examples.

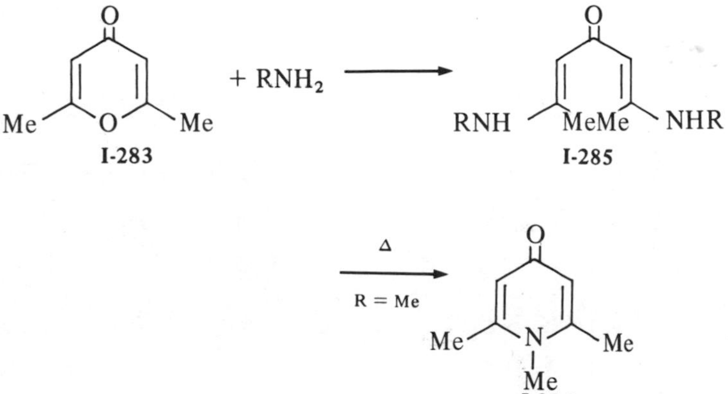

I-283   **I-285**

I-286

TABLE I-37.   2-PYRIDONES FROM 2-PYRONES

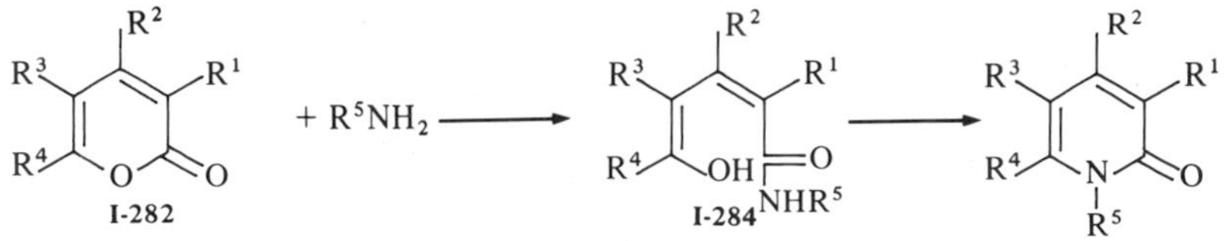

Substituents in Pyrone | Pyridone

R^1	R^2	R^3	R^4	R^1	R^2	R^3	R^4	R^5	Yield	Reference
H	Ph	PhO	p-BrC$_6$H$_4$	H	Ph	PhO	p-BrC$_6$H$_4$	H	—	625
H	Ph	p-MeC$_6$H$_4$O	p-BrC$_6$H$_4$	H	Ph	p-MeC$_6$H$_4$O	p-BrC$_6$H$_4$	H	—	625
H	OH	H	Me	H	N Ph	H	Me	Ph	—	626
H	Ph	H	Ph	H	Ph	H	Ph	Ph	80%	627
				H	Ph	H	Ph	2-pyridyl	60%	627
				H	Ph	H	Ph	5-Me-2-pyridyl	70%	627
				H	Ph	H	Ph	4-Me-2-pyridyl	60%	627
H	Ph	Ph	p-BrC$_6$H$_4$	H	Ph	Ph	p-BrC$_6$H$_4$	H	—	628
				H	Ph	Ph	p-BrC$_6$H$_4$	Me	—	628
H	Ph	Ph	pMeC$_6$H$_4$	H	Ph	Ph	p-MeC$_6$H$_4$	H	—	628
				H	Ph	Ph	p-MeC$_6$H$_4$	Me	—	628
H	NHCOCH$_2$CN	COPh	Ph	H	OH	COPh	Ph	H	92%	629
H	NHCOCH$_2$CN	COPh	Me	H	OH	COPh	Me	H	71%	629
Cl	Cl	H	Me	Cl	Cl	H	Me	H	—	630
Cl	Cl	H	CH$_2$Cl	Cl	Cl	H	CH$_2$Cl	H	—	630
Cl	Cl	H	Ph	Cl	Cl	H	Ph	H	67.1	630, 631
Ph	Ph	H	Ph	Ph	Ph	H	Ph	Me	80.95	632, 633
Ph	Ph	H	2-thienyl	Ph	Ph	H	2-thienyl	Me	—	632
Ph	Ph	H	1-methyl-pyrrole-2-yl	Ph	Ph	H	1-methyl-pyrrole-2-yl	Me	—	632
H	H	H	Ph	H	H	H	Ph	H	66	634
H	H	H	3,4-dimethoxy-phenyl	H	H	H	3,4-dimethoxy-phenyl	H	49	634, 635
H	Ph	Ph	p-morpholino-phenyl	H	Ph	Ph	p-morpholino-phenyl	NH$_2$	79	636

									OH	77	636
H	Me	H	Me	H	Me	H	Me	CH$_2$CO$_2$Et	13	637
H	oClC$_6$H$_4$	Ph	p-morpholino-phenyl	H	oClC$_6$H$_4$	Ph	p-morpholino-phenyl	H	81%	636
Cl	Me	H	CCl=CCl$_2$	Cl	Me	H	CCl=CCl$_2$	H	88.8%	639
H	H	CO$_2$Me	H	H	H	CO$_2$H	H	2-(3-ethylindoyl)	82%	640
Cl	Ph	Ph	Ph	Cl	Ph	Ph	Ph	NH$_2$	—	640
B$_2$	Ph	Ph	Ph	B$_2$	Ph	Ph	Ph	NH$_2$	—	640
I	Ph	Ph	Ph	I	Ph	Ph	Ph	NH$_2$	—	
NO$_2$	Ph	Ph	Ph	NO$_2$	Ph	Ph	Ph	NH$_2$	—	
C$_{10}$H$_7$	Ph	H	Ph	C$_{10}$H$_7$	Ph	H	Ph	NH$_2$	77%	641
C$_{10}$H$_7$	Ph	H	p-CH$_3$C$_6$H$_4$	C$_{10}$H$_7$	Ph	H	pCH$_3$C$_6$H$_4$	NH$_2$	80%	641
C$_{10}$H$_7$	Ph	H	m-ClC$_6$H$_4$	C$_{10}$H$_7$	Ph	H	m-ClC$_6$H$_4$	NH$_2$	74%	641
C$_{10}$H$_7$	Ph	H	pCH$_3$OC$_6$H$_4$	C$_{10}$H$_7$	Ph	H	pMeOC$_6$H$_4$	NH$_2$	77%	641
H	CH$_2$CO$_2$Me	—COCH$_2$CH$_2$CH$_2$—		H	CH$_2$CO$_2$Me	—COCH$_2$CH$_2$CH$_2$—		CH$_2$Ph	80%	642
H	CH$_2$CO$_2$Me	—COCH$_2$CMe$_2$CH$_2$—		H	CH$_2$CO$_2$Me	—COCH$_2$CMe$_2$CH$_2$—		CH$_2$Ph	45%	642
H	CH$_2$CO$_2$Me	—COCH$_2$CH$_2$CH$_2$—		H	CH$_2$CO$_2$Me	—COCH$_2$CH$_2$CH$_2$—		H	92%	642
H	CH$_2$CO$_2$Me	—COCH$_2$CMe$_2$CH$_2$—		H	CH$_2$CO$_2$H	—COCH$_2$CMe$_2$CH$_2$—		H	74%	642
Ph	Ph	H	Ph	Ph	Ph	H	Ph	NH$_2$	67%	643
Ph	Ph	H	p-CH$_3$C$_6$H$_4$	Ph	Ph	H	p-CH$_3$C$_6$H$_4$	NH$_2$	82%	643
Ph	Ph	H	p-ClC$_6$H$_4$	Ph	Ph	H	p-ClC$_6$H$_4$	NH$_2$	77%	643
Ph	Ph	H	p-CH$_3$OC$_6$H$_4$	Ph	Ph	H	p-CH$_3$OC$_6$H$_4$	NH$_2$	82%	643
Ph	Ph	H	3,4-(OCH$_2$O)C$_6$H$_3$	PH	Ph	H	3,4-(OCH$_2$O)C$_6$H$_3$	NH$_2$	67%	643
H	Me	H	PhSCH$_2$	H	Me	H	PhSCH$_2$	OH	43	644
H	Me	H	2-furyl	H	Me	H	2-furyl	OH	26	644
H	Me	H	2CH=CHPh	H	Me	H	2CH=CHPh	OH	34	644
H	Ph	H	Ph	H	Ph	H	Ph	OH	75	645
H	Ph	H	Ph	H	Ph	H	Ph	Me	66.5	645
H	p-MeOC$_6$H$_4$	H	Ph	H	p-MeOC$_6$H$_4$	H	Ph	H	—	645
H	p-MeOC$_6$H$_4$	H	Ph	H	p-MeOC$_6$H$_4$	H	Ph	Me	—	645
H	Ph	H	Ph	H	Ph	H	Ph	NH$_2$	37	645
H	OH	H	Me	H	OH	H	Me	Me	—	646
H	OH	H	Me	H	OH	H	Me	Et	—	646
H	OH	H	Me	H	OH	H	Me	Allyl	61	646

TABLE I-38. 4-PYRIDONES FROM 4-PYRONES

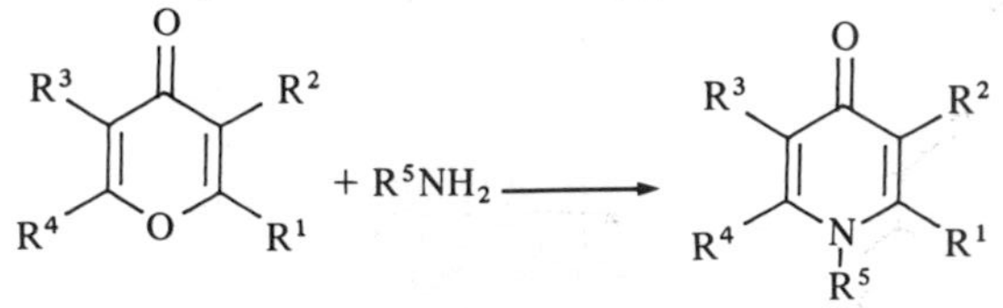

Substituents on Pyrone and Pyridone

Amine	R¹	R²	R³	R⁴	R⁵	Yield	Reference
NH$_3$	CH$_3$	H	H	CH$_3$	H	84	648, 649
NH$_3$	CH$_3$	H	H	Et	H	–	650
NH$_3$	CH$_3$	H	H	Ph	H	–	650
NH$_3$	CH$_3$	H	H	CO$_2$H	H		651
NH$_2$Me	CH$_3$	CH$_3$	H	CH$_3$	CH$_3$	58	652
H$_2$NCH$_2$CH$_2$NH$_2$	CO$_2$H	H	H	H	CH$_2$CH$_2$NH$_2$	–	653
H$_2$NC$_6$H$_4$p-CO$_2$Et	CO$_2$H	H	H	CO$_2$H	p-CO$_2$EtC$_6$H$_4$	–	654
NH$_3$	CO$_2$H	H	H	CO$_2$H	H	–	655
NH$_3$	i-amyl	H	Me	Me	H	75	656, 657
NH$_3$	CH$_3$	CONHPh	H	Me	H	60	658
pClC$_6$H$_4$NH$_2$	CO$_2$H	H	H	CO$_2$H	p-ClC$_6$H$_4$	–	659
BuNH$_2$	CH$_3$	H	H	CH$_3$	Bu	55	660
NH$_3$	H	OMe	H	—C(CN)NHAcCO$_2$Et	H	–	661
	NEt$_2$	Me	H	Me	H	–	662
	NEt$_2$	Ph	H	Me	H	–	662
	NEt$_2$	H	H	Me	H	–	663
N$_2$H$_4$	CH$_2$OH	H	OMe	H	NH$_2$	20	664
NH$_3$	Me	COMe	H	Me	H	–	665
NH$_3$	Me	COMe	H	Ph	H	–	665
NH$_3$	Me	COMe	H	p-BrC$_6$H$_4$	H	–	665
PhNH$_2$	Me	OH	H	H	Ph	–	666
p-MeC$_6$H$_4$NH$_2$	Me	OH	H	H	p-MeC$_6$H$_4$	–	666
p-NO$_2$C$_6$H$_4$NH$_2$	Me	OH	H	H	p-NO$_2$C$_6$H$_4$	–	666

1-amino-1,3,5-triazole	Me		H	H	Me	1,3,5-triazole	40%	667
1-amino-2-pyridone	Me		H	H	Me	N-2-pyridone	15%	667
NH_3	Me		COPh	H	Ph	H	73%	668
NH_3	Me		COMe	H	Ph	H	10%	668
H NNH$_2$	CH_2OH		H	OMe	H	NH_2	—	669
PhNH$_2$	Me		OH	H	H	Ph	—	670
NH_2OH	H		Ph	Ph	H	OH	65	671
MeNH$_2$	H		Ph	Ph	H	Me	60	671
PhNH$_2$	H		Ph	Ph	H	Ph	80	671
NH_3	OEt		H	H	Me	H	72	672
NH_3	OEt		Me	H	Me	H	13	672
NH_3	OEt		Ph	H	Me	H	84	672
NH_3	OEt		Cl	H	Me	H	75	672
NH	—CHOHCH$_2$NH iso-propyl		H	OCH$_2$Ph	H	H	93	673
MeNH$_2$	—CHOHCH$_2$NH iso-propyl		H	OCH$_2$Ph	H	Me	51	673
o-toluidine	Me		CO$_2$Et	CO$_2$Et	Me	o-tolyl	70	674
m-ClC$_6$H$_4$NH$_2$	Me		CO$_2$Et	CO$_2$Et	Me	m-ClC$_6$H$_4$	100	674
p-ClC$_6$H$_4$NH$_2$	Me		CO$_2$Me	H	Me	p-ClC$_6$H$_4$	47	675, 676
p-ClC$_6$H$_4$NH$_2$	Me		CO$_2$H	H	Me	p-ClC$_6$H$_4$	—	675, 676

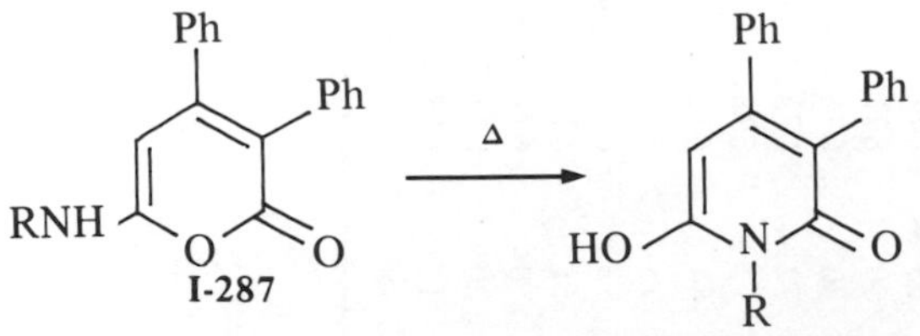

I-287

I-288 (R = Me, Et)

6-Amino-2-pyrone **I-287** rearranged thermally to give pyridone **I-288** (677).

The pyrano[3,4-*b*]pyran **I-289** gave the pyranopyridine **I-290** *via* the pyridone **I-291** (678).

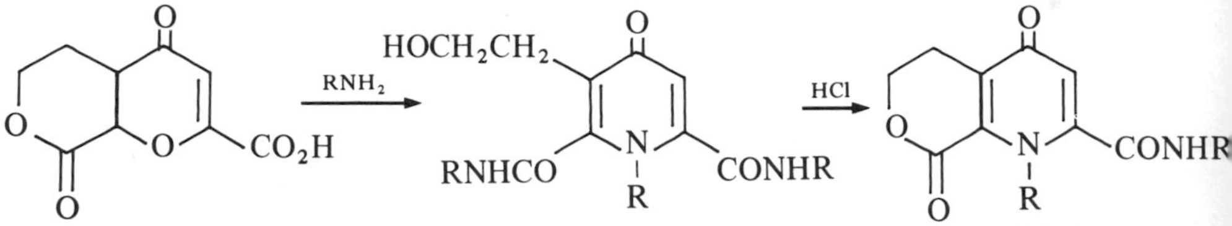

I-289

I-291

I-290

The aminopyranodioxins **I-292** underwent base-catalyzed rearrangement to give the pyridones **I-293** (Table I-39). Similarly the pyranoxazines yielded the carboxy-anilides (680).

TABLE I-39. REARRANGEMENTS OF PYRANODIOXINS
TO PYRIDINES (679, 680)

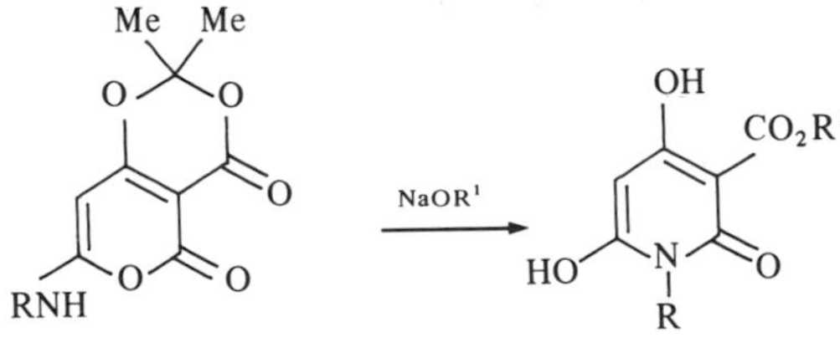

I-292

I-293

R	Base R^1	Yield (%)
m-tolyl	Me	63
Me	Me	51
m-tolyl	Et	60
o-MeOC$_6$H$_4$	Me	47
o-MeOC$_6$H$_4$	Et	50
CH$_2$Ph	Me	47
Ph	Me	62
α-naphthyl	Me	72
α-naphthyl	Et	78

5-Methyl-2*H*-pyran-2,4(3*H*)-dione (**I-294**) gave 4-hydroxy-6-methyl-2-pyridone (**I-295**) upon treatment with ammonia (681).

I-294 **I-295**

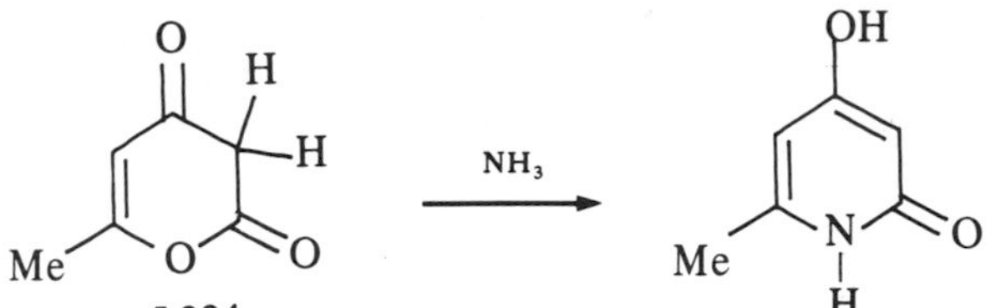

A novel thermal reaction took place between pyrone **I-296** and benzonitrile to give the 6-phenylpyridine **I-297** (682).

I-296 **I-297**

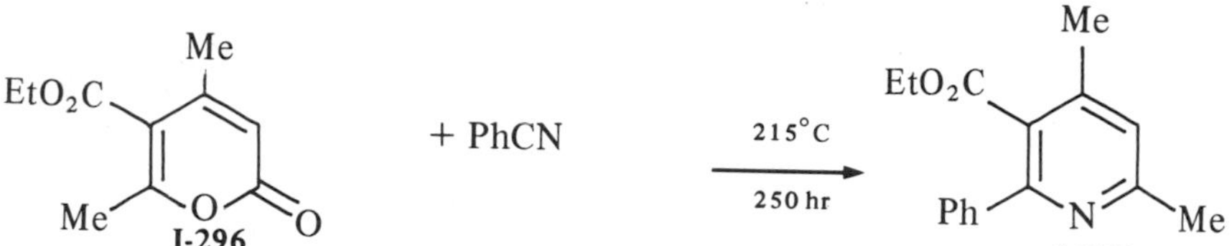

There has been a growing interest in *N,N'*-linked *bis*-heterocycles in recent years due to their interesting chemistry. Numerous *bis*-pyridones have been prepared by the reaction of 1-aminopyridines (**I-298**) with pyrones (683); see Table I-40. (See also the section on pyridines from pyrylium salts.)

I-298

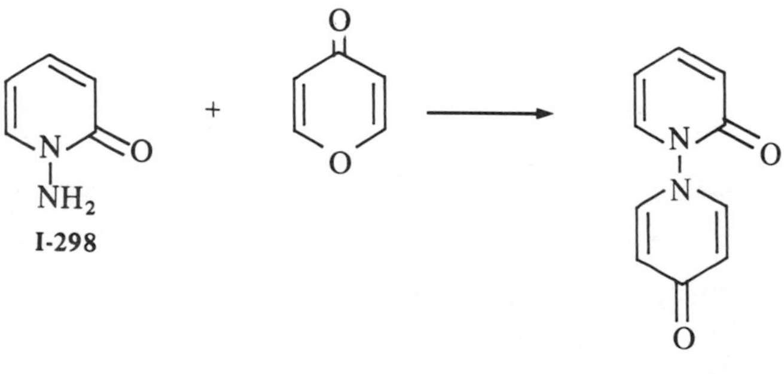

Dehydroacetic acid (**I-299**) gave products upon ammonolysis which appeared to be derived from the rearranged pyrone. Thus heating **I-299** with ammonium hydroxide at 110–130°C gave 2,6-lutidone (**I-300**) and 3-carboxy-2,6-lutidone (**I-301**) (684–687).

TABLE I-40.   *N,N'-BIS*-HETEROCYCLES FROM PYRONES

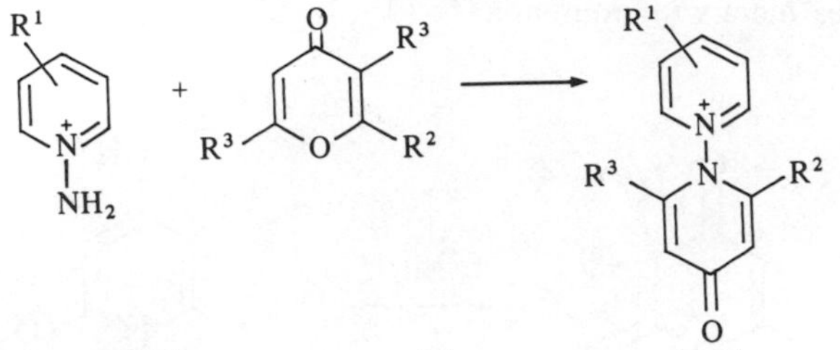

R^1	R^2	R^3	Yield	Reference
2-OH	H	H	50%	683
4-OH	H	H	30%	683
H	CO_2H	CO_2H	78%	690
2-Me	CO_2H	CO_2H	45%	690
3-Me	CO_2H	CO_2H	67%	690
4-Me	CO_2H	CO_2H	70%	690
2-Me-5-Et	CO_2H	CO_2H	13%	690
2-OH	Me	Me	15%	689
2-OH-4,6-diPh	Me	Me	50%	689
4-OH	Me	Me	18%	689

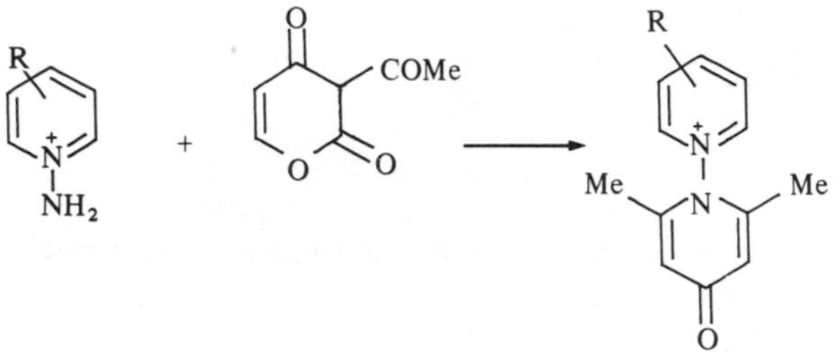

R	Yield	Reference
H	79	683
2-Me	39	683
3-Me	75	683
4-Me	87	683

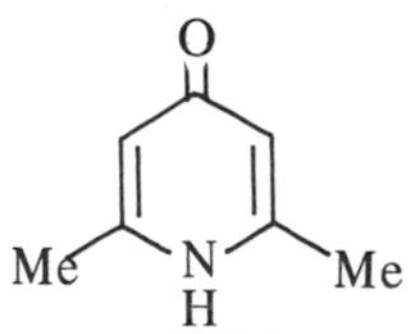

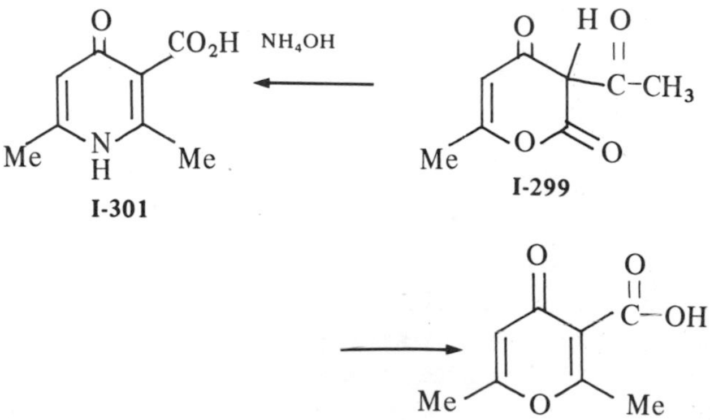

Heating the dehydroacetic acid **I-299** with butylamine gave the open-chained **I-302**, which upon treatment with sodium ethoxide gave pyridone **I-303** (688).

I-299 + *n*-BuNH$_2$ ⟶

I-302 **I-303** 55%

Recently the conversion of pyrones to pyridones has been extended to the thiapyrones yielding pyridylthiones (Table I-41). The thiapyrones are readily obtained from the pyrones and phosphorus pentasulfide.

TABLE I-41. PYRIDONES FROM THIAPYRONES

Amine	R^1	R^2	R^3	R^4	R^5	Yield	Reference
MeNH$_2$	H	Ph	H	Ph	Me	71%	691
MeNH$_2$	H	Ph	OPh	p-BrC$_6$H$_4$	Me	94%	692
CH$_2$=CHCH$_2$NH$_2$	H	Ph	OPh	p-BrC$_6$H$_4$	CH$_2$=CHCH$_2$—	69%	692
H$_2$NCH$_2$CH$_2$NH$_2$	H	Ph	OPh	p-BrC$_6$H$_4$	—CH$_2$CH$_2$NH$_2$	–	692
MeNH$_2$	H	Ph	Ph	Ph	Me	–	693
MeNH$_2$	H	Ph	Ph	p-CH$_3$C$_6$H$_4$	Me	–	693
MeNH$_2$	H	Ph	PhO	Ph	Me	–	693

ii. PYRANS. Pyridines have been prepared from tetrahydro- and dihydro-pyranes by oxidation in the presence of ammonia or by hydroxylamine. Table I-42 presents examples of 3,4-dihydropyran derivatives being converted to pyridines, while Table I-43 lists some reactions of miscellaneous dihydropyrans.

4-Methylene-4*H*-pyran derivatives have been converted to pyridine derivatives by the action of ammonia or ammonium salts. Use of primary amines led to numerous interesting compounds; thus when the pyrans **I-304** were treated with primary amines, the major product in most cases was the corresponding **I-305**, which arose *via* intermediate **I-306** (Table I-44). In several cases, a major by-product, and in certain cases, the only products are the rearranged 2-pyridones, for example, **I-307** (694). These rearrangement products arose from the cyclization of **I-306** where the X or Y substitutent was condensable with the amine (694–696).

iii. PYRYLIUM SALTS. The conversion of pyrylium salts to yield pyridines, pyridine *N*-oxides, and pyridine quaternary salts has been of continuing interest for many years. The accessibility of pyrylium salts from pyrones, 1,5-diketones, and

TABLE I-42.   PYRIDINES FROM 3,4-DIHYDRO-2H-PYRANS

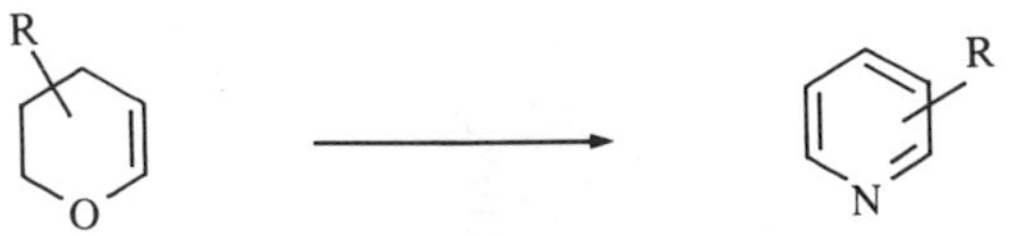

Pyran	Conditions	Pyridine	Yield	Reference
R = H	$NH_4OAc/Cu(OAc)_2$	R = H		852
R = 4-OH	$NH_3/H_2O/O_2/Al_2O_3$	R = H	12.5%	853
R = 2-morpholino-3-Me	$NH_4OAc/Cu(OAc)_2/O_2$	R = 3-CH_3	35%	854
R = 2-OMe	$NH_3/Pt-Al_2O_3/300°C$	R = H	36%	855, 856
R = 4-Me	$NH_3/500°C/PbF_2-Al_2O_3-SiO_2$	R = 4-Me	88%	857
R = 2-OAr	NH_4OH	R = H	—	858
R = 3-OEt	$NH_3/Pt-Al_2O_3/220°$	R = 3-OEt	67%	859
R = 3-OPr	$NH_3/Pt-Al_2O_3/220°C$	R = 3-OPr	58%	859
R = 3-OEt, 4-Me	$NH_3/Pt-Al_2O_3/220°C$	R = 3-OEt, 4-Me	71%	859
R = 2-OBu, 4-Ph, 6-Me	NH_2OH	R = 2-acetyl-4-phenyl	39%	860
R = 2,3-dialkoxy-4-allyl	$NH_3/Pd/200°C$	R = 3-alkoxy-4-alkyl	—	861
R = 2-piperidino-3-benzyl-6-methyl	NH_2OH	R = 3-benzyl-6-methyl	—	862
R = 2-OBu-5-isobutyl	NH_2OH	R = 3-isobutyl	—	863
R = 3-OEt	$NH_3/O_2/H_2O/Bi_8Mo_{12}O_{53}P$	R = H	81%	864, 865
R = [2,3-b]furano	$NH_4OAc/Cu(OAc)_2$	R = CH_2CH_2OH		866
R = 2-OEt	$NH_3/O_2/H_2O/Fe-Al_2O_3$	R = H	51%	867

TABLE I-43.   PYRIDINES FROM VARIOUS PYRAN DERIVATIVES

Pyran	Conditions	Pyridine (% yield)	Reference
4,4'-bi-*4H*-pyran	NH_3, air, NiO—Al_2O_3, 350°C	4,4'-bipyridine	868, 867
2,2'-bi-*4H*-pyran	NH_3, air, NiO—Al_2O_3, 350°C	2,2'-bipyridine	867, 868
5,6-dihydro-*2H*-pyran-3-carboxaldehyde	NH_3, air, V_2O_5—Al_2O_3, 300—400°C	3-cyanopyridine (50%)	869
5,6-dihydro-*2H*-pyran	NH_3, air, V_2O_5—Al_2O_3	pyridine (78%)	870
2-methoxy-5,6-dihydro-*2H*-pyran	NH_3, steam, Pt—Al_2O_3, 300°C	pyridine (36%)	871
4-(4'-pyridyl)-tetrahydropyran	NH_3, air, catalyst, 300°C	4,4'-bipyridine	867, 872–875
4-(2'-pyridyl)-tetrahydropyran	NH_3, air, catalyst, 360°C	2,4-bipyridine	875
2,6-dimethyl-*4H*-pyrano-[3,4-*d*]oxazol-4-one	NH_3	2,6-dimethyl-*5H*-oxazole[4,5-c]pyridine-4-one (84%)	876
6-ethoxy-5,6-tetramethylene-5,6-dihydro-*4H*-pyran	$NH_2OH \cdot HCl$	5,6,7,8-tetrahydroquinoline (76%)	877
3-benzyl-6-methyl-2-(1'-piperidino)-3,4-dihydro-*2H*-pyran	$NH_2OH \circ HCl$	2-methyl-5-benzylpyridine	862
3-*sec*-butyl-6-methyl-2-(1'-piperidino)-3,4-dihydro-*2H*-pyran	$NH_2OH \circ HCl$	2-methyl-5-*sec*-butylpyridine	878
2,6-diethoxy-3-(2'-ethoxyethyl)-5,6-tetramethylene-tetrahydropyran	NH_3, Al_2O_3	3-ethyl-5,6,7,8-tetrahydroquinoline	879

TABLE I-44.   PYRIDINES FROM 4-METHYLENE[4H]PYRAN DERIVATIVES

Amine	R^1	R^2	R^3	R^4	R^5	X	Y	Yield	Reference
n-BuNH$_2$	Me	H	H	Me	n-Bu	NO$_2$	COMe	34	694
n-BuNH$_2$	Me	H	H	Me	n-Bu	NO$_2$	H	98	694
n-BuNH$_2$	Me	H	H	Me	n-Bu	NO$_2$	CO$_2$Me	32	694
n-BuNH$_2$	Me	H	H	Me	n-Bu	NO$_2$	CN	98	694
n-BuNH$_2$	Me	H	H	Me	n-Bu	CN	H	94	694
n-BuNH$_2$	Me	H	H	Me	n-Bu	CN	CO$_2$Me	79	694
n-BuNH$_2$	Me	H	H	Me	n-Bu	CN	CONH$_2$	62	694
MeNH$_2$	Ph	H	H	Ph	Me	CN	CN	86	695
n-BuNH$_2$	Ph	H	H	Ph	n-Bu	CN	CN	40	695
PhCH$_2$NH$_2$	Ph	H	H	Ph	CH$_2$Ph	CN	CN	23	695
(CH$_3$)$_2$CHNH$_2$	Me	H	H	Me	(CH$_3$)$_2$CH	CN	CN	45	696
cyclohexylamine	Me	H	H	Me	C$_6$H$_{11}$	CN	CN	49	696
MeNH$_2$	Me	H	H	Me	Me	2,3-dicyanocyclo-pentadienylidene		70	880
MeNH$_2$	Me	H	H	Me	Me	2,3-dicyanocyclo-pentadienylidene		87	880
MeNH$_2$	Ph	H	H	Ph	Me	p-NO$_2$C$_6$H$_4$		80	881

I-307   X = NO$_2$

various condensation reactions has generated a large body of literature concerning their conversions to pyridines. Several reviews have appeared (697–699).

The ring transformation was usually carried out by treating the pyrylium salt **I-308** with alcoholic ammonia, by passing gaseous ammonia into a solution of the salt or by treating the pyrylium salt with an appropriate amine in alcoholic solutions. Tables I-45, I-46, and I-47 list pyridines and pyridinium salts prepared by these methods.

The mechanism of the ring transformation has been postulated as nucleophilic attack by ammonia to give the vinylogous amide **I-309**, which cyclized to give the pyridine upon loss of water. The reaction with both methylamine and n-butylamine have been studied extensively and the vinylogous amide **I-310** has been isolated in the case of methylamine (700–701).

TABLE I-45. PYRIDINES FROM PYRYLIUM SALTS

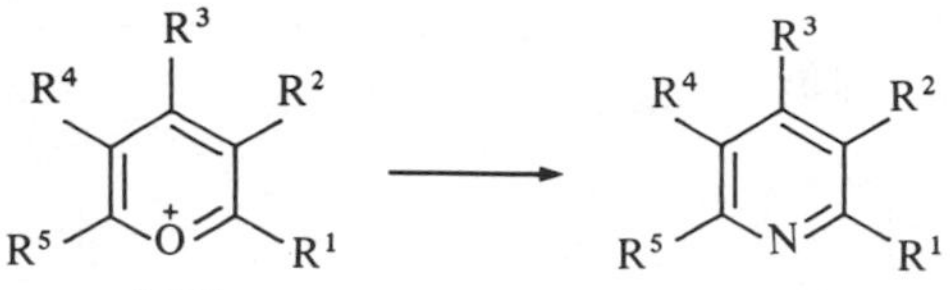

I-308

Substituents	Conditions	Yield	Reference
R^1 = Me, R^2 = R^4 = H, R^3 = R^5 = 3,4(MeO)$_2$C$_6$H$_3$	NH$_4$OAc, AcOH	80	882
R^1 = Me, R^2 = R^4 = H, R^3 = R^5 = 4-PhC$_6$H$_4$	NH$_4$OAc, AcOH	71	882
R^1 = R^5 = Ph, R^2 = R^4 = H, R^3 = PhCOCH$_2$	(NH$_4$)$_2$CO$_3$, EtOH	65	883
R^1 = R^3 = R^5 = Ph, R^2 = OMe, R^4 = H	NH$_4$OAc, AcOH	–	884
R^1 = R^3 = Ph, R^2 = R^4 = R^5 = H	NH$_4$OH, H$_2$O	83	885
R^1 = R^5 = Ph, R^2 = R^3 = R^4 = H	NH$_3$	81	886, 887
R^1 = R^5 = p-MeOC$_6$H$_4$, R^2 = R^3 = R^4 = H	NH$_3$	–	886
R^1 = R^5 = t-Bu, R^2 = R^4 = H, R^3 = Me	NH$_3$	80	888
R^1 = R^2 = R^3 = Me, R^4 = H, R^5 = Et	NH$_3$	–	889
R^1 = R^5 = Me, R^2 = R^4 = H, R^5 = n-Pr	NH$_3$	–	889
R^1 = R^4 = R^5 = Me, R^2 = Et, R^3 = H	NH$_3$	–	889
R^1 = R^5 = Ph, R^3 = C$_6$F$_5$, R^2 = R^4 = H	(NH$_4$)$_2$CO$_3$, AcOH	87	890, 891
R^1 = R^5 = C$_6$F$_5$, R^2 = R^4 = H, R^3 = Ph	(NH$_4$)$_2$CO$_3$, AcOH	94	890
R^1 = Et, R^2 = Me, R^3 = R^5 = Ph, R^4 = H	NH$_3$	–	892
R^1 = R^3 = R^5 = Ph, R^2 = COPh, R^4 = H	NH$_3$	–	892
R^1 = R^5 = Ph, R^2 = H, R^3 = 4-Me$_2$NC$_6$H$_4$, R^4 = H	NH$_3$	⊢	893
R^1 = R^5 = p-MeOC$_6$H$_4$, R^2 = R^3 = R^4 = H	NH$_4$OAc, AcOH	81	887
R^1 = R^5 = p-PhC$_6$H$_4$, R^2 = R^3 = R^4 = H	NH$_4$OAc, AcOH	95	887
R^1 = R^5 = Ph, R^2 = R^4 = H, R^3 = p-MeOC$_6$H$_4$	NH$_3$	96	894
R^1 = R^5 = Ph, R^2 = R^4 = H, R^3 = CH$_2$CO$_2$Me	NH$_3$	–	895
R^1 = R^5 = Me, R^2 = R^4 = H, R^3 = carbazolyl	NH$_3$	44	896
R^1 = R^5 = Me, R^2 = R^4 = H, R^3 = indolinyl	NH$_3$	100	897
R^1 = R^2 = Me, R^3 and R^4 = —(CH$_2$)$_4$—	NH$_3$	64	898
R^1 = R^5 = Me, R^2 = H, R^3 and R^4 = —(CH$_2$)$_4$—	NH$_3$	44	898
R^1 = Ph, R^2 = H, R^3 and R^4 = —(CH$_2$)$_4$—, R^5 = Me	NH$_3$	100	899
R^1 = R^5 = Me, R^2 = R^4 = H, R^3 = 4-pyranylidene	NH$_3$	–	900

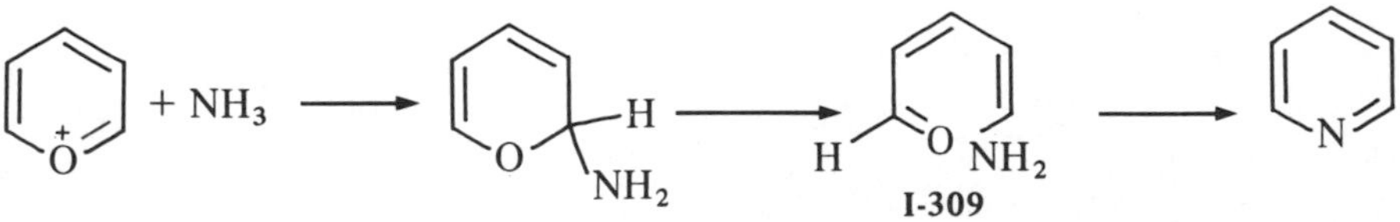

I-309

In the butylamine case, ^{13}C NMR was used to detect the intermediates involved, and the following scheme (Scheme I-6) was proposed (701).

UV studies of the reaction supported the ^{13}C NMR results, with amine to pyrylium salt ratios of at least 2:1. The reaction rate was dependent on the amine concentration and first order in pyrylium salt. This has been interpreted that the rate-determining step was the ring-closure. Solvent dependency, relative rate studies with various amines, and studies on the nature of the amine have been reported (701).

TABLE I-46.　2,4,6-TRIPHENYLPYRYLIUM　SALTS　TO　THE
CORRESPONDING TRIPHENYLPYRIDINIUM SALTS

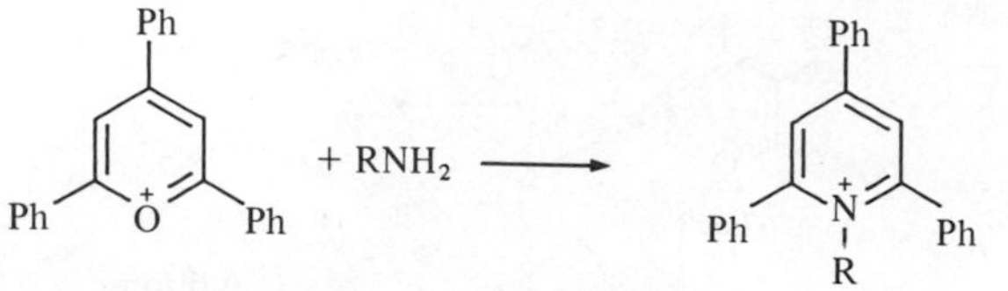

R	Yield of Pyridinium Salt	Reference
CH_3	75%	901
CH_2Ph	87%	902
CH_2CH_2Cl	70%	903
CH_2CH_2OH	90%	903
CH_2CMe_2OH	60%	903
$CH_2CH_2NH_2$	90%	903
$CH_2CH_2CH_2OH$	80%	903
$CH_2CH_2CH_2NH_2$	92%	903
$CH_2CH=CH_2$	85%	902
$n\text{-}C_4H_9$	87%	904
CH_2-2-pyridyl	95%	902, 905
CH_2-4-pyridyl	90%	902, 905
CH_2-2-furyl	45%	902
2-pyridyl	77%	904–906
3-Me-2-pyridyl	80%	904
4-Me-2-pyridyl	78%	904
$2\text{-}HO_2CC_6H_4$	75%	905
$4\text{-}HO_2CC_6H_4$	92%	905
$2\text{-}BrC_6H_4$	88%	905
$4\text{-}BrC_6H_4$	94%	905
$4\text{-}MeC_6H_4CH_2$	68%	904

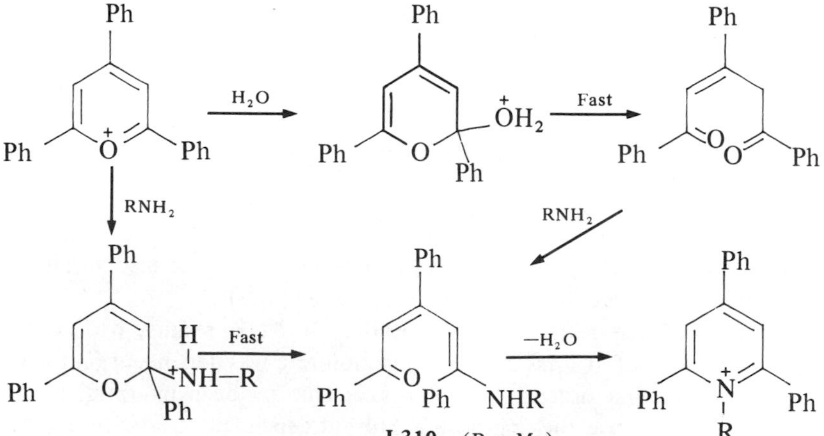

I-310　(R = Me)

Scheme I-6

TABLE I-47. PYRIDINIUM SALTS FROM MISCELLANEOUS PYRYLIUM SALTS

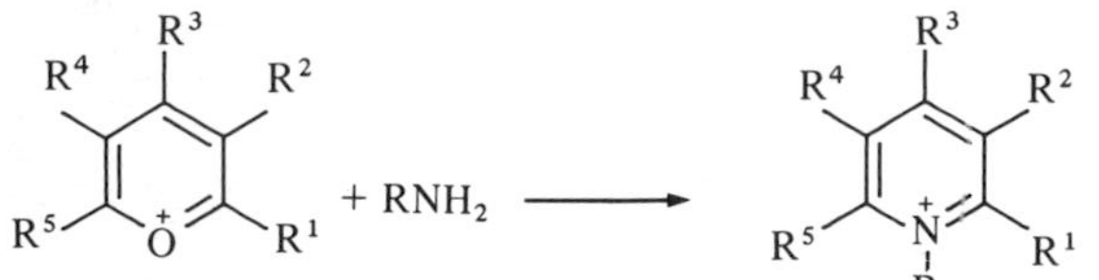

I-308

Substituents	Base	Yield	Reference
$R^1 = R^5 = Me$, $R^3 = H$, R^3 and $R^4 = -(CH_2)_9-$	R = Me or Ph		914
$R^1 = CO_2H$, $R^2 = R^4 = H$, $R^3 = R^5 = Ph$	R = o-phenylenediamine	95	915
$R^1 = R^5 = Me$, $R^2 = R^4 = H$, $R^3 = $ 5-methyl-2-furanyl	R = Ph		916
$R^1 = CHMe_2$, $R^2 = R^4 = H$, $R^3 = R^5 = Me$	R = Ph		917
$R^1 = R^5 = Me$, $R^2 = R^4 = H$, $R^3 = $ 2-benzothiazoyl	R = n-C$_4$H$_9$	53	918
	R = Ph	80	918
	R = H$_2$N	80	918
	R = NHPh	68	918
$R^1 = R^3 = R^5 = Me$, $R^2 = Ph$	R = H	−	919
$R^1 = R^3 = R^5 = Me$, $R^2 = R^4 = CO_2H$	R = N-phthalimido	−	920
$R^1 = -CH = CHOEt$, $R^2 = R^4 = H$, $R^3 = R^5 = Ph$	R = Me	−	921
R^1 and $R^2 = R^4$ and $R^5 = -(CH_2)_4-$, $R^3 = H$	R = Ph, o-, m-, p-tolyl	44–77%	922, 923
	R = Me, Et, Pr, Bu, allyl		924
	R = o-, m-, p-HOC$_6$H$_4$		925
	R = o-, m-, p-MeC$_6$H$_4$		925
	R = HOCH$_2$CH$_2$		925
$R^1 = R^5 = Ph$, $R^2 = R^4 = H$, $R^3 = n$-MeC$_6$H$_4$	R = 4-OH-3,5-diphenyl C$_6$H$_2$		926
$R^1 = R^2 = R^3 = R^4 = R^5 = Ph$	R = Et	65	927
	R = Bu	82	927
	R = Allyl	87	927
	R = PhCH$_2$	60	927
	R = PhCH$_2$CH$_2$	64	927
	R = 2-pyridyl	38	927

Several ring transformations have been carried out where a reagent other than an amine has been utilized, such as triphenylpyrylium tetrafluoroborate **I-311** with sulfinamide **(I-312)** to give the desired pyridine **I-313** and an aryl disulfide (702).

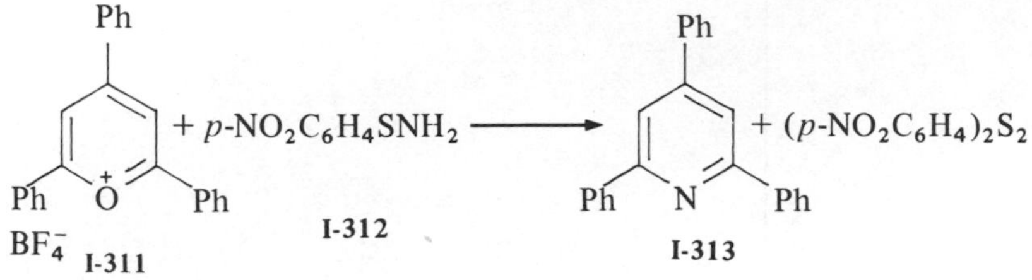

Sulfinylamines **(I-314)** underwent reaction with pyrylium salts to yield pyridinium salts **I-315** with the evolution of sulfur dioxide (703). The reaction is postulated to proceed through the open-chained intermediate **I-316**, which cyclized to **I-317**, followed by elimination of SO_2.

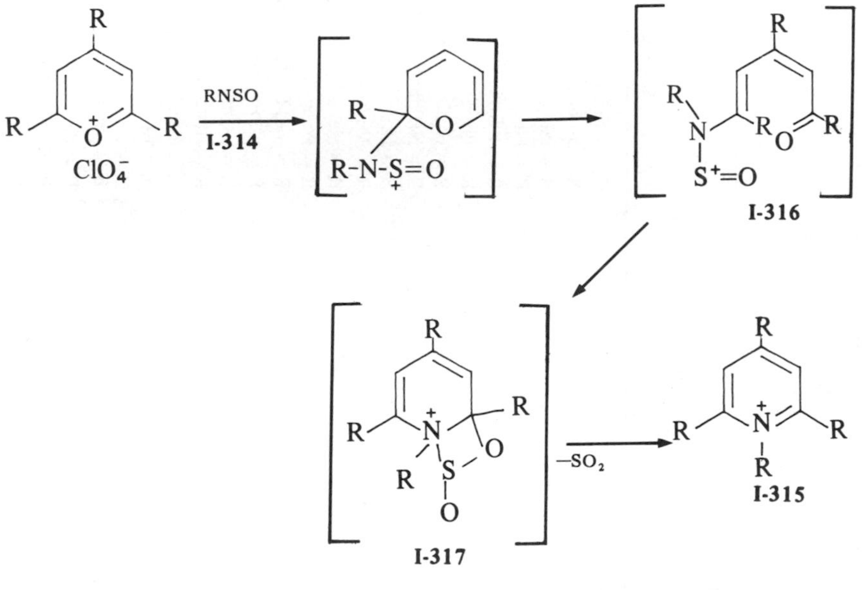

In a similar fashion, imines **I-318** when allowed to react with pyrylium salts gave the corresponding pyridinium salts (704, 705). When an aldoxime was the reagent, the product was the free pyridine base **I-313** believed to arise *via* the N-oxide, which is deoxygenated under the reaction conditions. Indeed the N-oxide can be isolated in low yields if acetic acid is used as the solvent (704). When methyl-hydrazine is used as the nucleophile the expected diazepine **I-319** was not the major product but rather the diazole **I-320** or the pyridinium salt **I-321** was obtained, depending upon the nature of R-group (706).

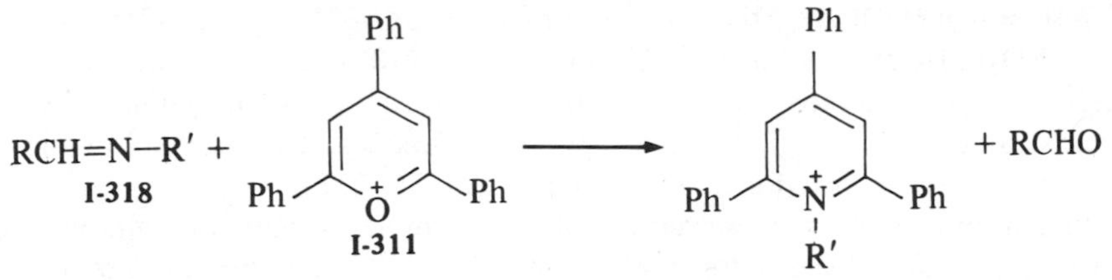

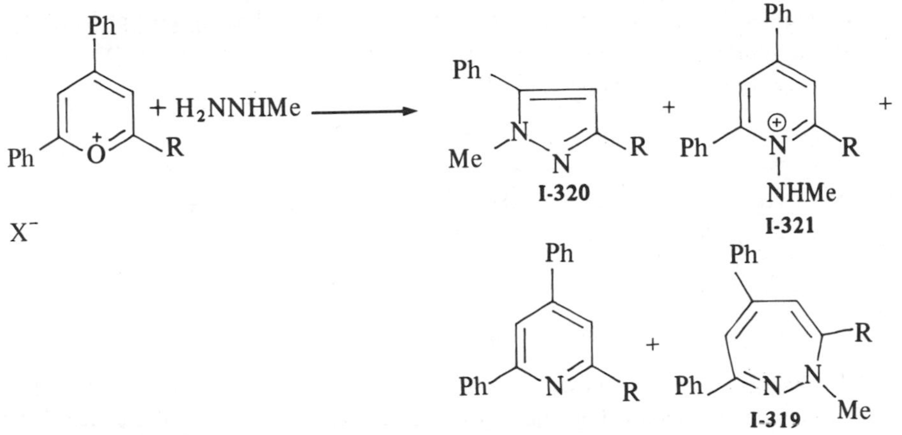

There has been a report of an intramolecular conversion of a pyrylium salt to a pyridine; thus, when the 2-aminopyrylium salt **I-322** was treated with sodium ethoxide in ethanol pyridone **I-323** was obtained. Treatment of **I-322** with pyridine in methanol gave the pyrone derivative **I-324** (707). The reaction with a thiopyrylium salt failed to give the related pyridylthione. When the pyrylium salt contains a good leaving group in either the 2- or 4-position, complications arose in the ring transformation. For example, when the 4-methoxypyrylium salt **I-325** was treated with aniline the iminopyran **I-326** was formed along with **I-327**. The product ratio was found to be dependent on the basicity of the amine.

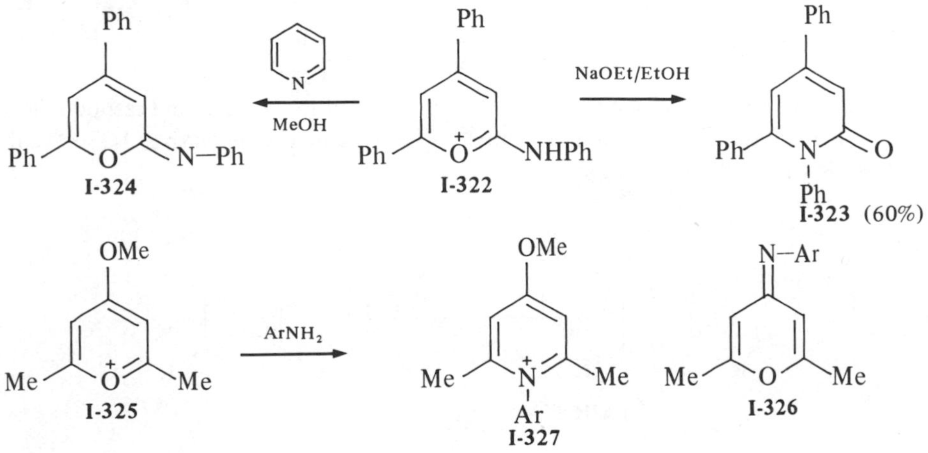

With $R = p\text{-MeOC}_6\text{H}_4$, then the ratio of **I-326** to **I-327** was $68:32$ while when $R = p\text{-NO}_2\text{C}_6\text{H}_4$ then **I-326** was the sole product (708). Similar results have been observed with 2-methoxypyrylium salts (709). Reactions of pyrylium salts with hydroxylamines gave high yields of pyridine *N*-oxides. The mechanism and the general scope of the reaction have been reported (710, 711).

Recent interest in the conversion of pyrylium to pyridinium salts has been spawned by Katritzky and his co-workers in which the trisubstituted pyridinium salts, obtained from pyrylium salts, have been utilized as leaving groups in a wide variety of synthetic reactions. This work has very recently been reviewed and will only be briefly summarized here. (712)

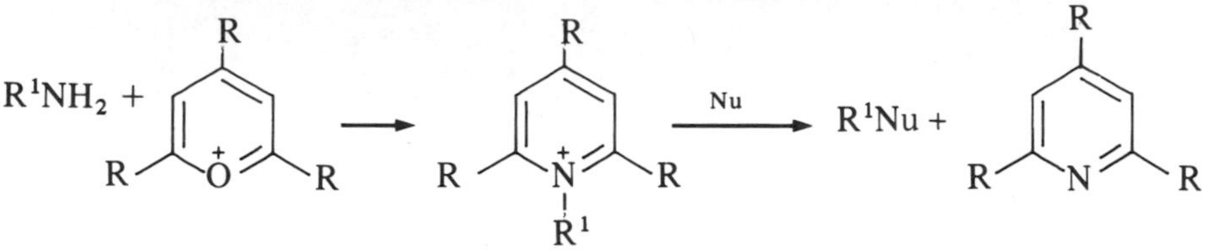

Scheme I-7

The reaction sequence shown in Scheme I-7 has the effect of replacing an amine with various nucleophiles. The nucleophiles that have been successfully utilized are: I^-, Br^-, Cl^-, F^-, RCO_2^-, $^-\text{NO}_3$, RO^-, RNH_2, ^-SCN, $^-\text{S}-\text{C}(=\text{S})\text{OEt}$, and RSO_2^-. Yields in most cases are quite high and a wide range of ring substituents have been utilized including alkyl, allyl, benzyl, aryl, and heteroaryl. This method has also been applied to the deamination of amines *via* the dihydropyridines **I-328**.

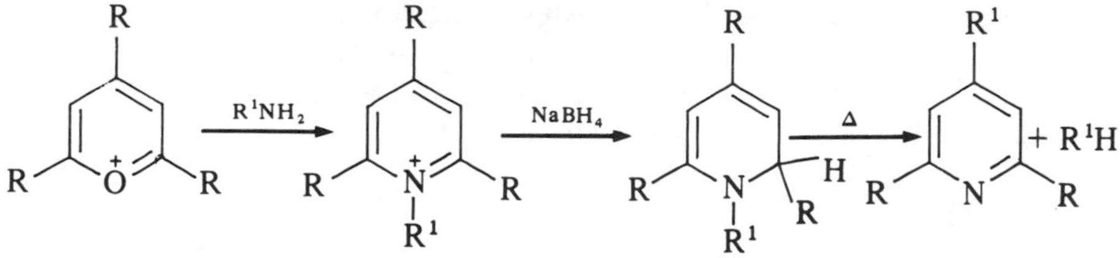

I-328

This concept has been extended to the synthesis of isocyanates and carbodiimides by thermolysis of pyridinium acylimides (**I-329**) and the derivatives **I-330**. Table I-48 lists a number of examples.

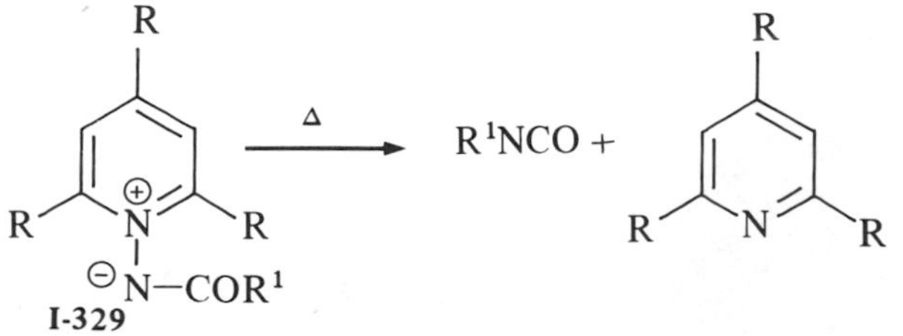

I-329

TABLE I-48. SYNTHESIS OF PYRIDINIUM SALTS FROM PYRYLIUM SALTS AND REAGENTS OTHER THAN AMINES

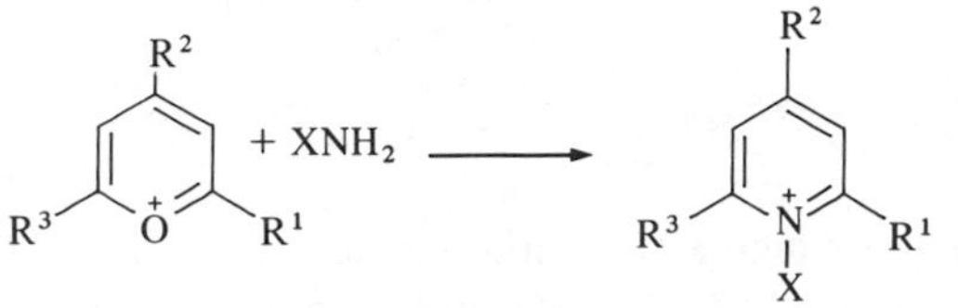

R^1	R^2	R^3	X—NH$_2$	Yield	Reference
Ph	Ph	Ph	H$_2$NNCPhNHPh	92	713, 714
			H$_2$NNHCOMe	76	715
			H$_2$NNHCOPh	82	715
			H$_2$NNHCOCH$_2$PH	85	715
			H$_2$NNHCOCH=CHPh	66	715
			H$_2$NSO$_2$NH$_2$		716
			H$_2$NNHCSPh	93	715
			H$_2$NNHCSC$_6$H$_4$-p-OMe	95	715
			1-NH$_2$-4,6-diphenyl-2-pyridone		717
			H$_2$NNHCONH$_2$	–	718
			H$_2$NNHCONH$_2$	–	718
			H$_2$N-1,2,4-triazol-4-yl		719
			H$_2$N-pyrrol-1-yl		719
Me	Me	Me	H$_2$NNHCONHPh		720
			H$_2$NNHCSNH$_2$		720
			H$_2$NNHCSNHPh		720
			H$_2$NNH$_2$	24	721, 722
			H$_2$NNHCOPh	99	721
			H$_2$NNHSO$_2$Ph		723

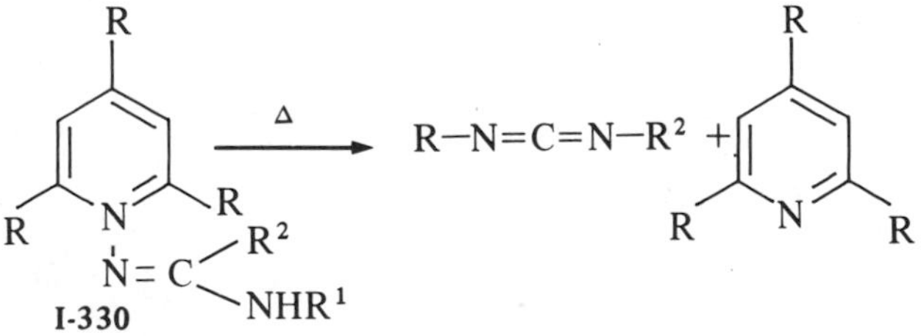

b. TWO HETEROATOMS

i. PYRIMIDINES. The synthesis of pyridines from pyrimidines can be divided into four reaction types. Type I involves nucleophilic attack on pyrimidine, followed by a ring-opening–ring-closure reaction. Type 2 involves thermolysis of tetrahydropyrimidines to yield pyridines. Type 3 is the base-catalyzed hydrolysis of uracils and Type 4 is thermolysis of Diels–Alder adducts from pyrimidines and dienophiles.

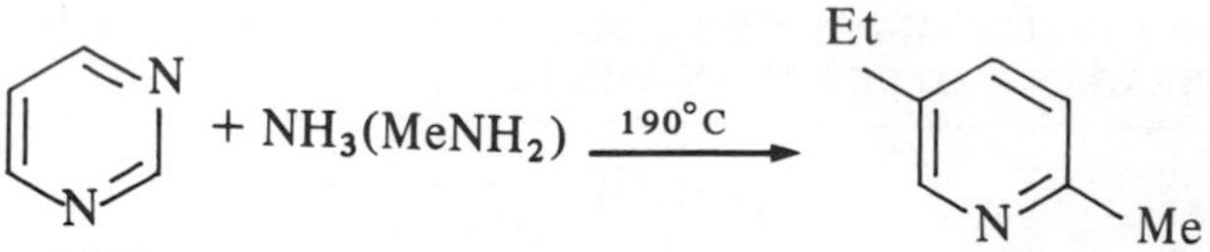

I-331 I-332

Pyrimidines with both amine and carbon nucleophiles to give pyridines have been reported. Thus, pyrimidine **I-331** with methylamine or ammonia gave 5-ethyl-2-methylpyridine **I-332** (724). The reaction mechanism has been postulated to involve ring-opening to give a 2-carbon unit (most likely acetaldehyde), which undergoes the known aldol-type reaction to give **I-332** (725). This would also explain the observation that pyrimidine **I-333** with ammonia gave the diarylpyridine **I-334** (726); the intermediate would be the arylacetaldehyde **I-335**.

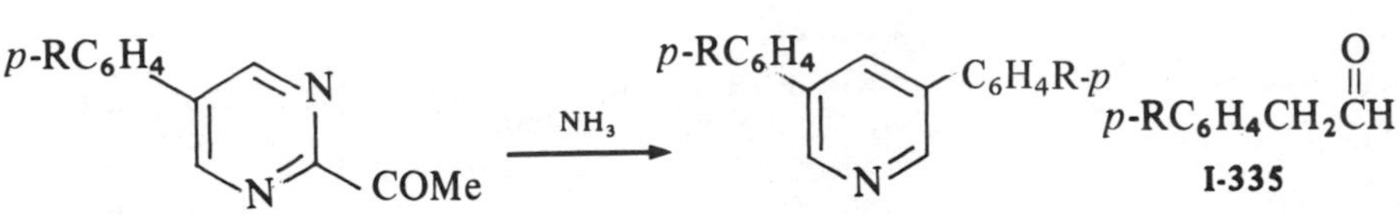

I-333 I-334 I-335

When a pyrimidine was activated by the introduction of an electron-withdrawing group in the five position (e.g., **I-336**) or by quaterization, (e.g., **I-337**) ring transformation occurred on treatment with active methylene compounds or ketones (Scheme I-8) (727–729).

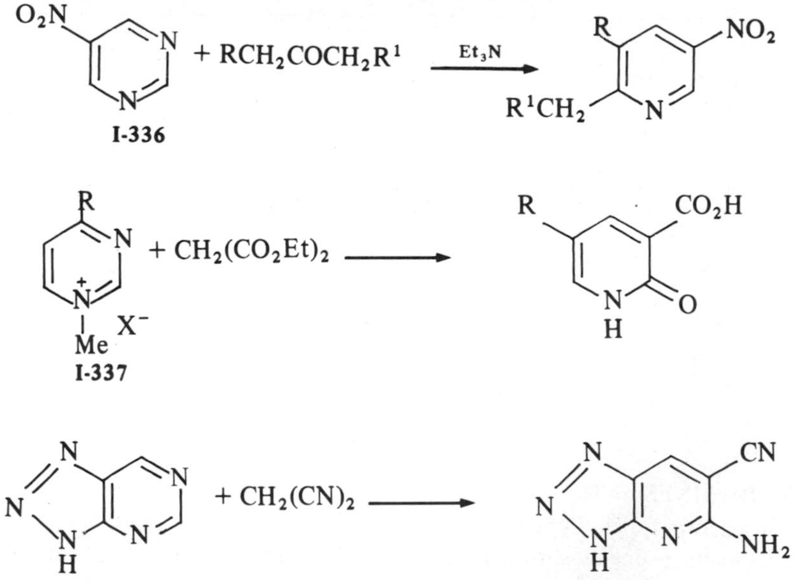

Scheme I-8

Surprisingly, 5-nitropyrimidine (**I-336**) on heating with aqueous acetic acid gave 3,5-dinitropyridine. The postulated mechanism necessitated formation of a 2-carbon unit (**I-338**) and a 3-carbon unit (**I-339**), which condensed to give the pyridine (730).

I-336 **I-339** **I-338**

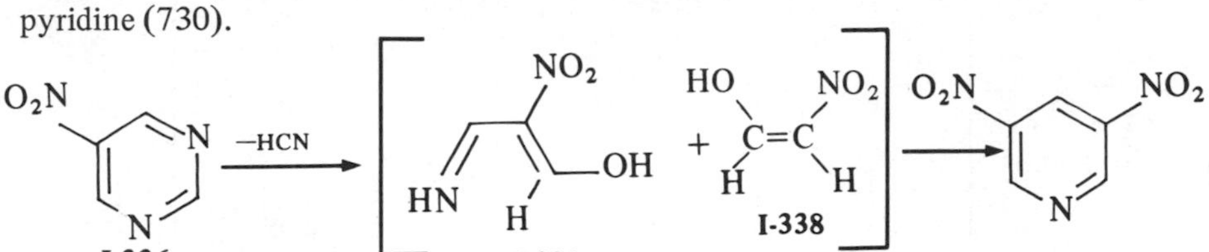

Thermolyses of tetrahydropyrimidines to give pyridines have been reported (731, 732). In general, the tetrahydropyrimidines were easily obtained from an aldehyde and ammonia. For example, butyraldehyde and ammonia in the presence of ammonium chloride at 40°C gave 2,4-dipropyl-5-ethyl-2,3,4,5-tetrahydropyrimidine (**I-340**). Heating **I-340** to 235°C for 2.5 hr gave 2-propyl-3,5-diethylpyridine (**I-341**) (731). The mechanism of this reaction has not been studied but it is assumed that **I-340** cleaved to butyraldehyde imine, which underwent a normal 3–2 condensation to give the desired pyridine (see Section II.2.C). A similar reaction has been reported for the transformation of 4-oxo-1,2,3,4-tetrahydropyrimidines to give pyridones (733).

I-340 (97%) **I-341** (96%)

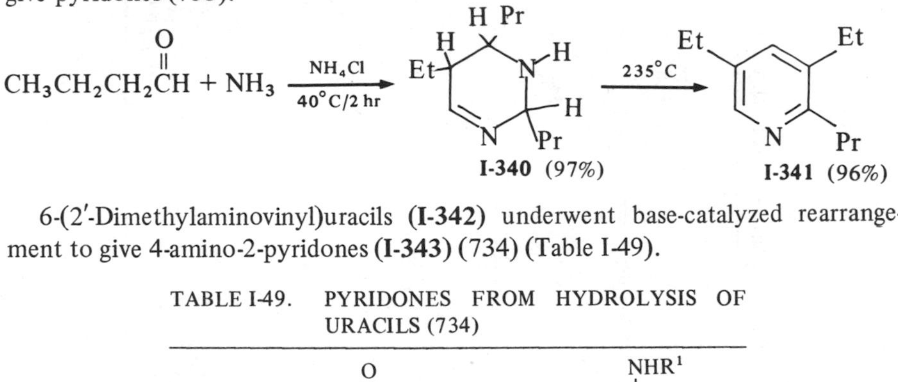

6-(2′-Dimethylaminovinyl)uracils (**I-342**) underwent base-catalyzed rearrangement to give 4-amino-2-pyridones (**I-343**) (734) (Table I-49).

TABLE I-49. PYRIDONES FROM HYDROLYSIS OF URACILS (734)

I-342 **I-343**

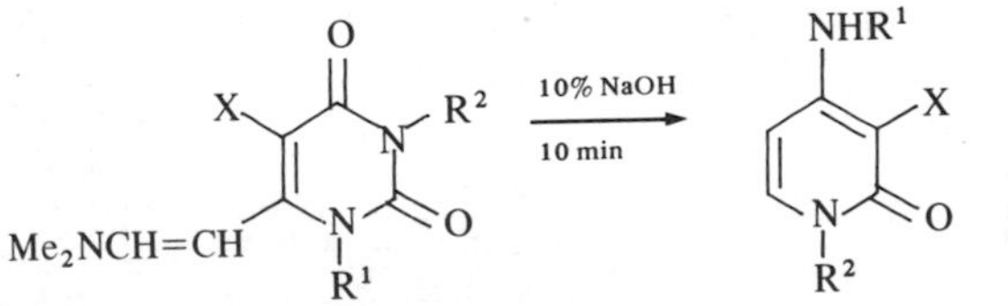

Substituents	Yield (%)
$R^1 = R^2 = Me, X = NO_2$	62
$R^1 = Me, R^2 = C_6H_{11}, X = NO_2$	26
$R^1 = Me, R^2 = p\text{-}NO_2C_6H_4, X = NO_2$	21
$R^1 = R^2 = Me, X = CN$	70
$R^1 = C_6H_5, R^2 = Me, X = CN$	85
$R^1 = CH_2C_6H_5, R^2 = Me, X = CN$	35
$R^1 = C_6H_5, R^2 = Me, X = H$	56

TABLE I-50. PYRIDINES OBTAINED FROM THE DIELS–ALDER REACTIONS OF PYRIMIDINES

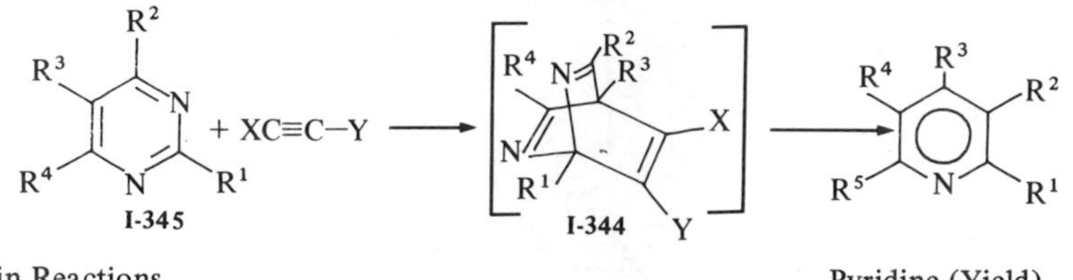

Substituents in Reactions	Pyridine (Yield)	Reference
$R^1 = R^2 = R^4 = N(Me)_2$, $R^3 = H$, $X = Y = CO_2Me$	$R^1 = R^5 = N(Me)_2$, $R^2 = R^3 = CO_2Me$ (53)	736
$R^1 = R^2 = N(Me)_2$, $R^3 = R^4 = H$, $X = Y = CO_2Me$	$R^1 = R^5 = N(Me)_2$, $R^2 = R^3 = CO_2Me$ (19)	736
$R^1 = R^2 = N(Me)_2$, $R^3 = H$, $R^4 = OMe$, $X = Y = CO_2Me$	$R^1 = R^5 = N(Me)_2$, $R^2 = R^3 = CO_2Me$ (17)	736
$R^1 = OMe$, $R^2 = R^4 = N(Me)_2$, $R^3 = H$, $X = Y = CO_2Me$	$R^1 = OMe$, $R^2 = R^3 = CO_2Me$, $R^5 = N(Me)_2$ (30)	736
$R^1 = R^4 = OMe$, $R^2 = N(Me)_2$, $R^3 = H$, $X = Y = CO_2Me$	$R^1 = OMe$, $R^2 = R^3 = CO_2Me$, $R^5 = N(Me)_2$ (55)	736
$R^1 = R^2 = R^4 = OMe$, $R^3 = H$, $X = Y = CO_2Me$	$R^1 = R^5 = OMe$, $R^2 = R^1 = CO_2Me$ (12)	736
$R^1 = N(Me)_2$, $R^2 = R^4 = OMe$, $R^3 = H$, $X = Y = CO_2Me$	$R^1 = R^5 = OMe$, $R^2 = R^3 = CO_2Me$ (40)	736
$R^1 = R^3 = R^4 = H$, $R^2 = CO_2Me$, $X = N(Et)_2$, $Y = Me$	$R^1 = R^3 = H$, $R^2 = Me$, $R^4 = N(Et)_2$, $R^5 = CO_2Me$ (10)	737
$R^1 = R^2 = R^4 = H$, $R^3 = CO_2Me$, $X = N(Et)_2$, $Y = Me$	$R^1 = R^5 = H$, $R^2 = Me$, $R^3 = N(Et)_2$, $R^4 = CO_2Me$ (90)	737
$R^1 = R^4 = H$, $R^2 = R^3 = CO_2Me$, $X = N(Et)_2$, $Y = Me$	$R^1 = H$, $R^2 = Me$, $R^3 = N(Et)_2$, $R^4 = R^5 = CO_2Me$ (80)	737
$R^1 = R^2 = CO_2Me$, $R^3 = R^4 = H$, $X O N(Et)_2$, $Y = Me$	$R^1 = R^5 = CO_2Me$, $R^2 = N(Et)_2$, $R^3 = Me$, $R^4 = H$ (81)	737
$R^1 = R^3 = H$, $R^2 = R^4 = CO_2Me$, $X = N(Et)_2$, $Y = Me$	$R^1 = R^4 = H$, $R^2 = N(Et)_2$, $R^3 = Me$, $R^5 = CO_2Me$ (64)	737
$R^1 = Me$, $R^2 = R^4 = OH$, $R^3 = H$, $X = Y = CO_2Me$	$R^1 = OH$, $R^2 = H$, $R^3 = R^4 = CO_2Me$, $R^5 = Me$ (62)	738

In a similar manner 1,3-disubstituted uracils gave pyridones in moderate yields (735).

Adducts (I-344) obtained from the Diels–Alder reaction of pyrimidines (I-345) with acetylenes can be thermally rearranged to yield pyridines in moderate to high yields (Table I-50). The reaction proceeded well when electron-withdrawing substituents were on the pyrimidine ring and the dienophile was electron-rich [i.e., X = Et_2N] or the converse.

A novel intramolecular Diels–Alder reaction in the pyrimidine series has been reported (739). When the pyrimidines I-346 (n = 3, R = Me or Ph) were subjected to thermolysis at 180–200°C, the annulated pyridines I-348 were formed. The intermediate I-347 was not isolated nor observed spectroscopically. When the acetylene analog was used, the pyridine (I-348) was obtained in 60% yield.

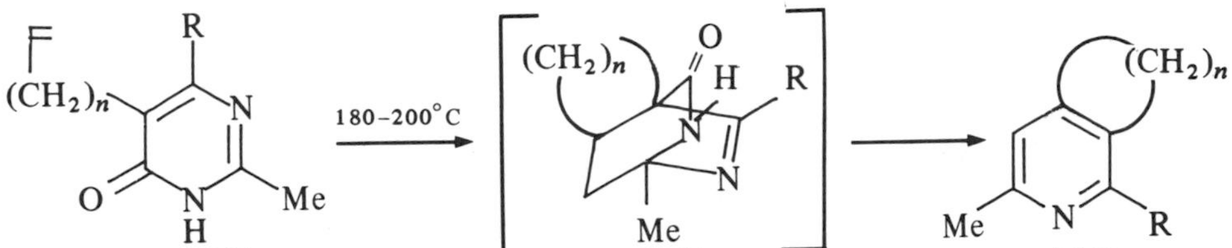

R = Me, 65%
R = Ph, 51%

ii. PYRIDAZINES. Pyridazines underwent a Diels–Alder addition with 1-diethylaminopropyne to give two possible adducts (I-349 and I-350) depending on the nature of the substituents on the pyridazine. I-349 on further heating lost HCN to give the pyridine I-351, while I-350 lost N_2 to afford I-352 (740, 741). Examples of pyridazines leading to pyridines are given in Table I-51.

TABLE I-51. PYRIDINES FROM PYRIDAZINES *VIA* A DIELS–ALDER REACTION (740)

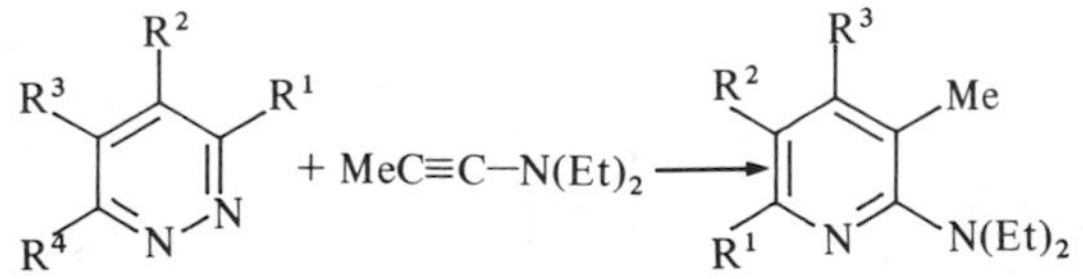

Substituents	Pyridine (Yield)
$R^1 = R^3 = H, R^2 = CO_2Me$	$R^1 = R^3 = H, R^2 = CO_2Me$ (30)
$R^1 = R^4 = Ph, R^2 = CO_2Et, R^3 = H$	$R^1 = Ph, R^2 = CO_2Et, R^3 = H$ (95)
$R^1 = R^4 = CO_2Me, R^2 = R^3 = H$	$R^1 = CO_2Me, R^2 = R^3 = H$ (75)
$R^1 = R^4 = CO_2Me, R^2 = Ph, R^3 = H$	$R^1 = CO_2Me, R^2 = Ph, R^3 = H$ (89)
$R^1 = R^2 = R^4 = CO_2Me, R^3 = H$	$R^1 = R^2 = CO_2Me, R^3 = H$ (85)

I-349 I-351

I-350

I-352

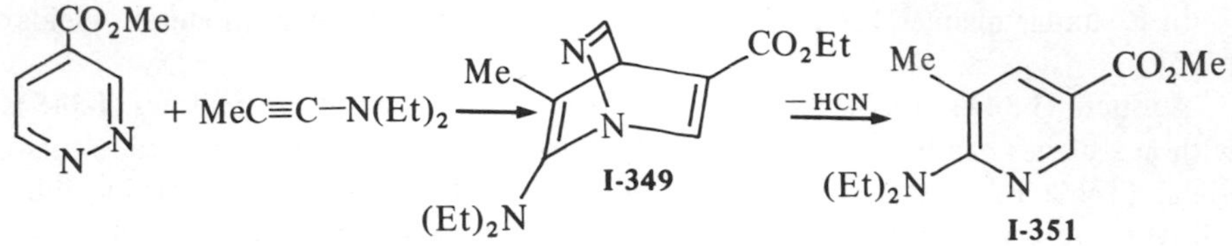

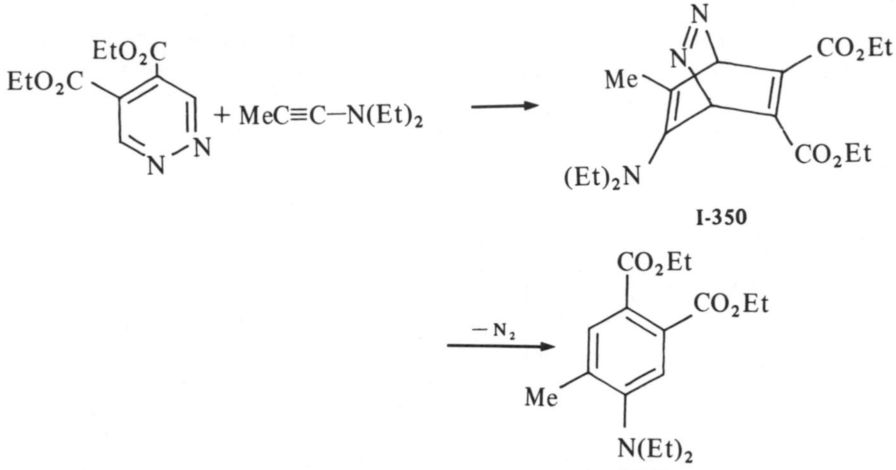

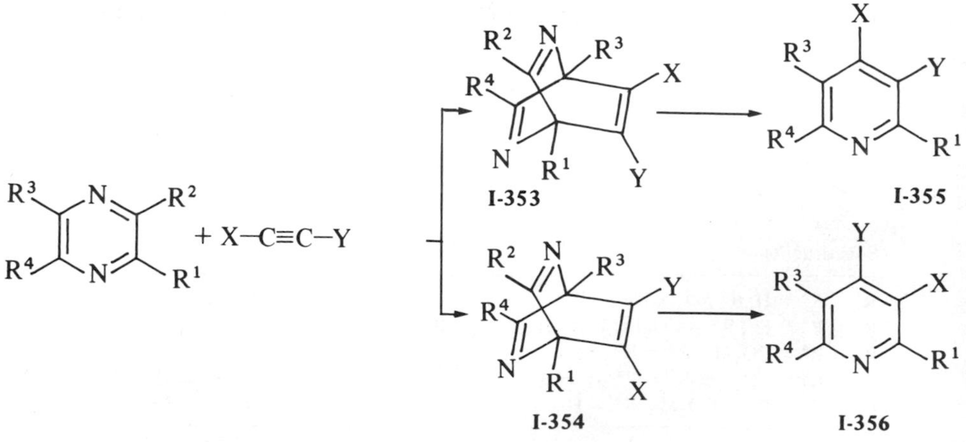

iii. PYRAZINES. Pyrazines like the pyridazines undergo cycloadditions with dienophiles to yield pyridines. With an unsymmetrical dienophile two possible bicyclic intermediates **I-353** and **I-354** were possible each leading to different pyridines (**I-355** and **I-356**, respectively). Thus, when pyrazine and 1-diethylamino-propyne [$X = NEt_2$, $Y = Me$] were allowed to react in acetonitrile containing boron trifluoride-etherate, 3-diethylamino-4-methylpyridine (**I-357**) along with **I-358** and **I-359** (742) were isolated. It is yet unclear how **I-358** and **I-359** arose,

I-353 I-355

I-354 I-356

I-357 I-358

I-359

however, **I-358** is composed of two molecules of the acetylene and HCN, and did not arise from the Diels–Alder adduct since **I-358** and **I-359** were also formed from 2-methyl- and 2-carbomethylpyrazine along with the expected pyridines.

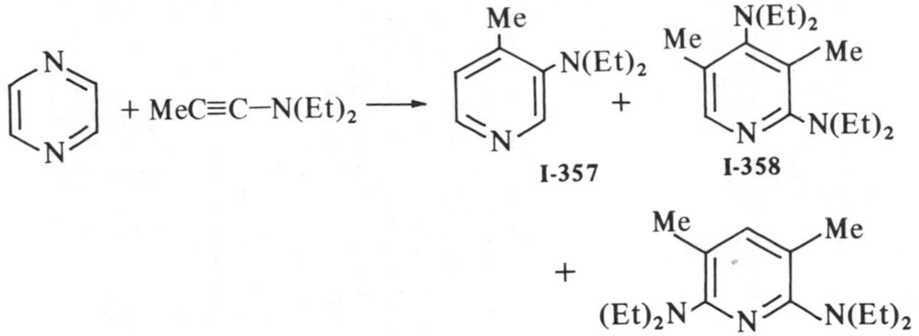

2,5-Dihydroxypyrazines underwent a similar reaction with dimethyl acetylenedicarboxylate to give the isolated adduct **I-360**, which on further heating gave pyridones **I-361** and **I-362**.

I-360

I-361 I-362

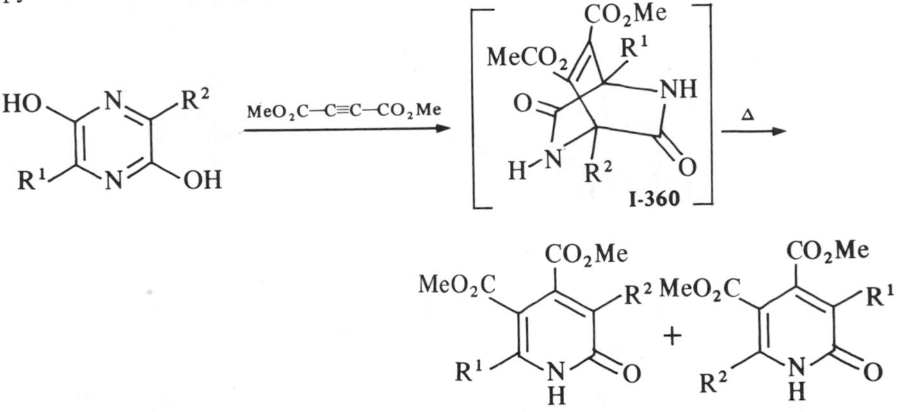

iv. OXAZINES. Oxazines like the pyridazines and pyrazines underwent cycloadditions with dienophiles to afford pyridines. For example, the 1,3-oxazin-6-ones **I-363** when treated with 1-diethylaminopropyne gave pyridines **I-364** (744, 745). (Table I-52)

I-363 I-364

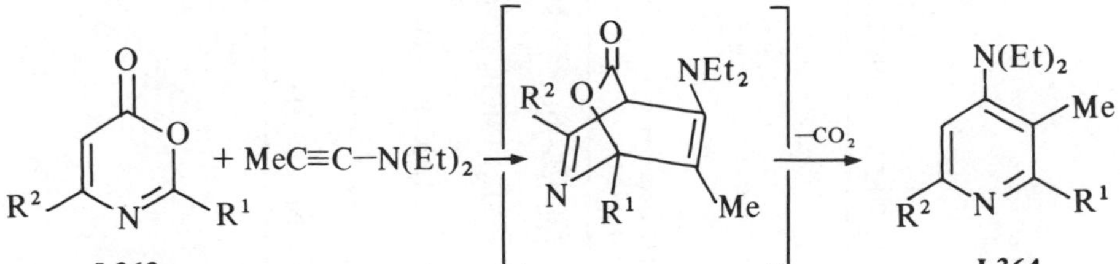

TABLE I-52.   PYRIDINES FROM 1,3-OXAZIN-6-ONES (744, 745)

I-296 + dienophile ⟶ (pyridine structure with substituents R^3, R^2, R^1, R^4)

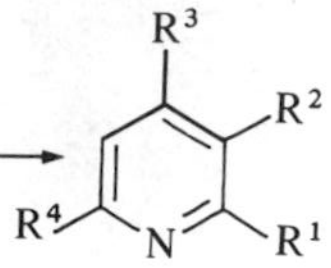

1,3-Oxazin-6-one Substituents	Dienophile	Pyridine (Yield)
$R^1 = (CH_3)_2CH_2$, $R^2 = Me$	$MeC\equiv CN(Et)_2$	$R^1 = (CH_3)_2CH_2$, $R^2 = Me$, $R^3 = N(Et)_2$, $R^4 = Me$ (79)
	$H_2C=COEt[N(Me)_2]$	$R^1 = (CH_3)_2CH$, $R = H$, $R^3 = N(Me)_2$, $R^4 = Me$ (80)
$R^1 = (CH_3)_2CHCH_2$, $R^2 = Me$	$MeC\equiv CN(Et)_2$	$R^1 = (CH_3)_2CHCH_2$, $R^2 = Me$, $R^3 = N(Et)_2$, $R^4 = Me$ (89)
	$H_2C=COEtN(Me)_2$	$R^1 = (CH_3)_2CHCH_2$, $R^2 = H$, $R^3 = N(Me)_2$, $R^4 = Me$ (90)
$R^1 = (CH_3)_3C$, $R^2 = Me$	$MeC\equiv CN(Et)_2$	$R^1 = (CH_3)_3C$, $R^2 = Me$, $R^3 = N(Et)_2$, $R^4 = Me$ (65)
	$H_2C=COEtN(Me)_2$	$R^1 = (CH_3)_3C$, $R^2 = H$, $R^3 = N(Me)_2$, $R^4 = Me$ (94)
$R^1 = CF_3$, $R^2 = Me$	$MeC\equiv CN(Et)_2$	$R^1 = CF_3$, $R^2 = Me$, $R^3 = N(Et)_2$, $R^4 = Me$ (67)
	$H_2C=COEtN(Me)_2$	$R^1 = CF_3$, $R^2 = H$, $R^3 = N(Me)_2$, $R^4 = Me$ (80)
$R^1 = Ph$, $R^2 = Me$	$MeO_2C-C\equiv C-CO_2Me$	$R^1 = Ph$, $R^2 = R^3 = CO_2Me$, $R^4 = Me$ (30)

However, unlike pyridazine and pyrazines, oxazines (e.g., **I-365**) with active methylene compounds gave pyridines **(I-366)** under base catalysis. Active methylenes used include malonates, acetylacetone, ethyl cyanoacetate, and similar compounds (Table I-53). Similarly, the dihydrooxazines gave similar results (746). Spiro-1,3-oxazines **(I-367)** upon treatment with phosphorus oxychloride gave pyridones **I-368**. The mechanism proposed involves a ring-opening to give the intermediate **I-369**, which subsequently cyclized to give **I-368** (747).

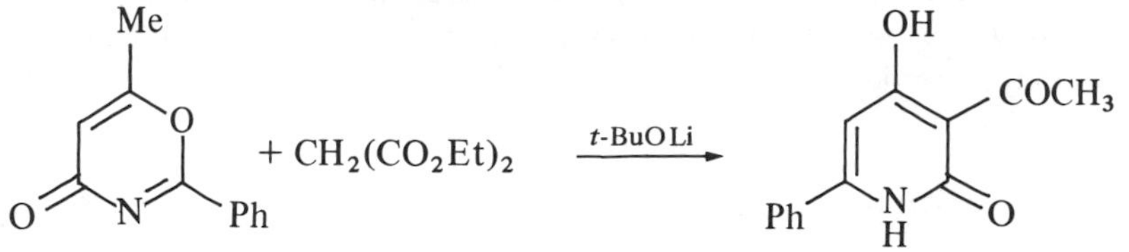

I-365   + $CH_2(CO_2Et)_2$ $\xrightarrow{t\text{-BuOLi}}$ I-366   60%

TABLE I-53.   PYRIDONES FROM OXAZINES (748)

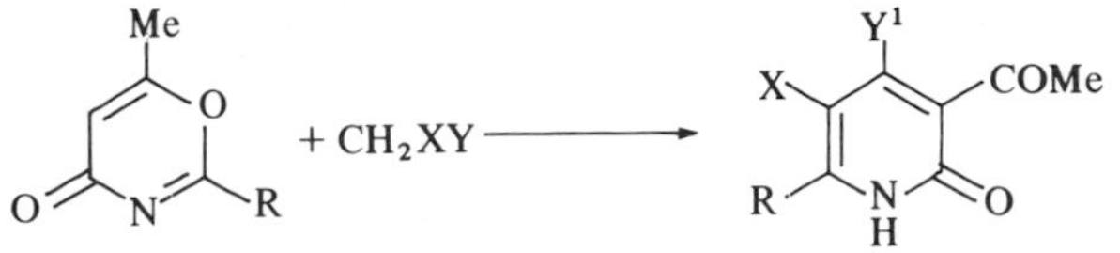

$+ CH_2XY \longrightarrow$

Active Methylane Compound	R	X	Y^1	Yield of Pyridone
$CH_2(CO_2Et)_2$	Ph	CO_2Et	OH	60
$CH_3COCH_2COCH_3$	Ph	$COCH_3$	Me	80
1,3-cyclohexanedione	Ph	$-(CH_2)_3CO-$		90
$CH_2(CN)_2$	Ph	CN	NH_2	70
$CH_3COCH_2CO_2Et$	Ph	CO_2Et	CH_3	83
$NCCH_2CO_2Et$	Ph	CN	OH	80
$NCCH_2COPh$	Ph	CN	Ph	90

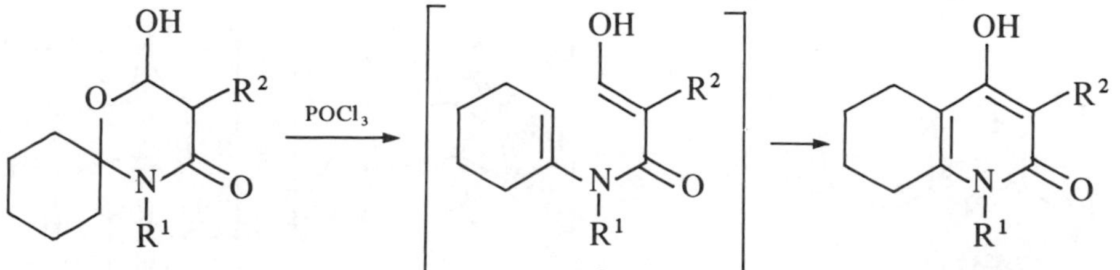

I-367 $\xrightarrow{POCl_3}$ I-369   I-368

v.   MISCELLANEOUS SIX-MEMBERED HETEROCYCLES CONTAINING TWO HETEROATOMS.   4,4-Dimethyl-1,3-dioxane **(I-370)** and acetaldehyde in the presence of ammonia over $SiO_2-Al_2O_3-CdSO_4$ at $400°C$ gave a mixture of pyridine and 3-methylpyridine (749).

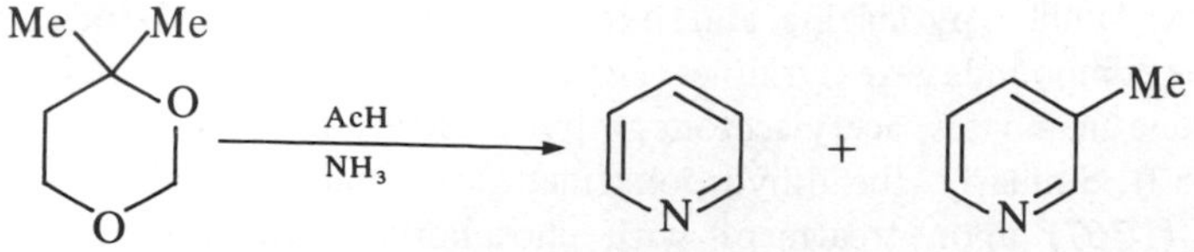

Me Me

I-370 $\xrightarrow[\text{NH}_3]{\text{AcH}}$

The cycloadditions of azides with 2,3-diazabicycloheptenes (e.g., **I-371**) gave aziridino adducts **I-372**, which on hydrolysis and oxidation produced the azo compounds **I-373**. Subsequent thermolysis of **I-373** gave the dihydropyridines **I-374**. These dihydropyridines, where $R^1 = R^2SO_2$ upon treatment with base, gave pyridine (750).

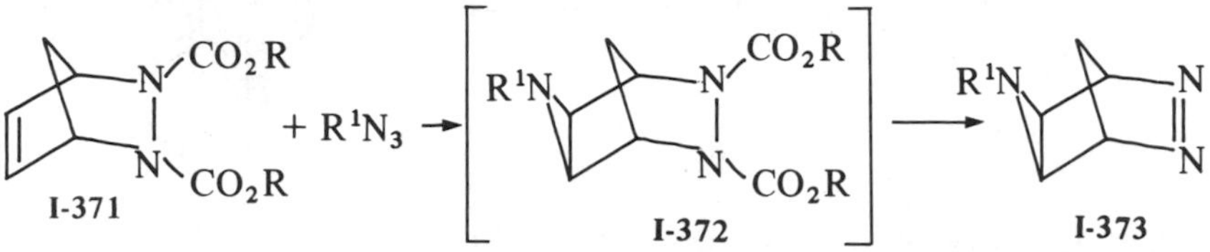

$$\text{I-371} + R^1N_3 \rightarrow [\text{I-372}] \rightarrow \text{I-373}$$

$$\text{I-373} \xrightarrow[-N_2]{\Delta} \xrightarrow{RO^-}$$

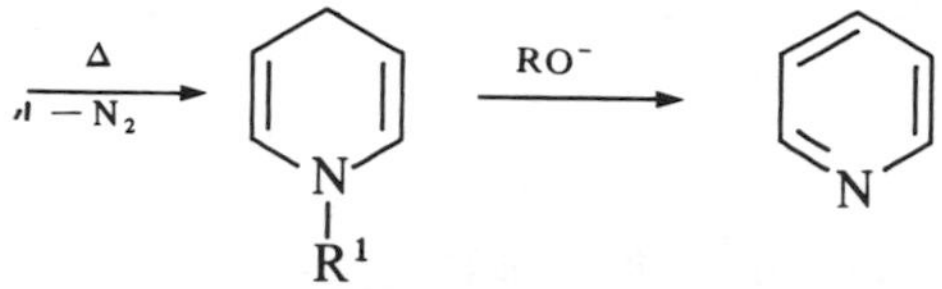

I-374 $(R^1 = R^2SO_2^-)$

c. SIX-MEMBERED RING HETEROCYCLES WITH THREE HETEROATOMS

Cycloadditions of electron-rich dienophiles with triazines (**I-375**) gave pyridines in good to moderate yields. The ratios of the two possible pyridines **I-376** and **I-377** were dependent on the nature of substituents and on experimental conditions, such as solvent (751–754). Table I-54 details some examples of this reaction. In a similar manner the triazine (R = Me) gave 2-methylpyridine (756).

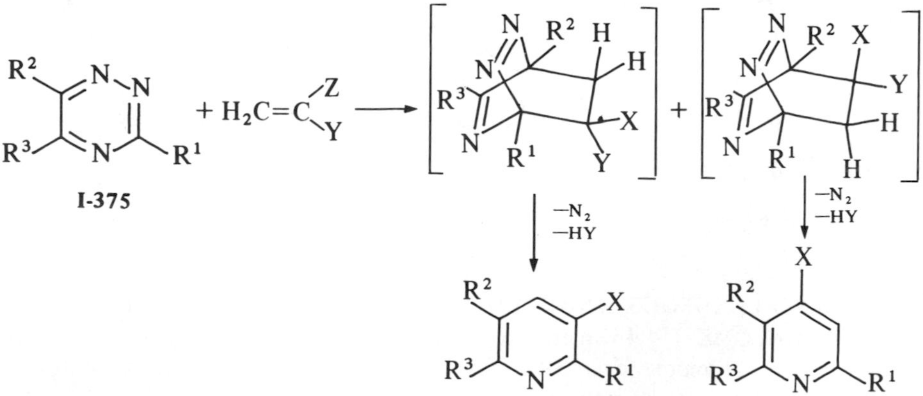

$$\text{I-375} + H_2C=C\begin{smallmatrix}Z\\Y\end{smallmatrix} \longrightarrow [\cdots] \xrightarrow[-HY]{-N_2} + [\cdots] \xrightarrow[-HY]{-N_2}$$

I-376   **I-377**

TABLE I-54.   PYRIDINES FROM TRIAZINES WITH DIENOPHILES

Triazine (I-375) Substituents	Dienophile	Pyridine (Yield)	Reference
$R^1 = CO_2Me$, $R^2 = R^3 = H$	$H_2C=C(OEt)_2$	I-376; $R^1 = CO_2Me$, $X = OEt$, $Y = R^2 = R^3 = H$ (57)	751
	$H_2C=C[N(Me)_2](OEt)$	I-376; $R^1 = CO_2Me$, $X = N(Me)_2$, $Y = R^2 = R^3 = H$ (74)	751
	$H_2C=C[N(Me)_2](SMe)$	I-376; $R^1 = CO_2Me$, $X = N(Me)_2$, $Y = R^2 = R^3 = H$ (82)	751
	$H_2C=C[N(Me)_2]_2$	I-376; $R^1 = CO_2Me$, $X = N(Me)_2$, $Y = R^2 = R^3 = H$ (17.5)	751
		I-377; $R^1 = CO_2Me$, $Y = H$, $X = N(Me)_2$, $R^2 = R^3 = H$ (74.5)	
$R^1 = CO_2Me$, $R^2 = H$, $R^3 = Ph$	$H_2C=C(OEt)_2$	I-376; $R^1 = CO_2Me$, $X = N(Me)_2$, $Y = R^2 = H$, $R^3 = Ph$ (85)	751
	$H_2C=C[N(Me)_2](OEt)$	I-376; $R^1 = CO_2Me$, $X = N(Me)_2$, $Y = R^2 = H$, $R^3 = Ph$ (58)	751
		I-377; $R^1 = CO_2Me$, $X = N(Me)_2$, $Y = R^2 = H$, $R^3 = Ph$ (5)	
$R^1 = CO_2Me$, $R^2 = H$, $R^3 = Ph$	$H_2C=C(OEt)_2$	I-376; $R^1 = CO_2Me$, $X = N(Me)_2$, $Y = H$, $R^2 = R^3 = Ph$ (5)	751
		I-377; $R^1 = CO_2Me$, $X = N(Me)_2$, $Y = H$, $R^2 = R^3 = Ph$ (95)	
$R^1 = CO_2Me$, $R^2 = H$, $R^3 = Ph$	$CH_3-C\equiv C-N(Et)_2$	I-376; $R^1 = CO_2Me$, $X = N(Et)_2$, $Y = Me$, $R^2 = H$, $R^3 = Ph$ (85)	753
	$CH_3-C\equiv C-N(Me)_2$	I-376; $R^1 = CO_2Me$, $X = N(Me)_2$, $Y = Me$, $R^2 = H$, $R^3 = Ph$ (89)	753
$R^1 = CO_2Me$, $R^2 = R^3 = Ph$	$CH_3-C\equiv C-N(Et)_2$	I-376; $R^1 = CO_2Me$, $X = N(Et)_2$, $Y = Me$, $R^2 = R^3 = Ph$ (16.9)	753
		I-376; $R^1 = CO_2Me$, $X = N(Et)_2$, $Y = Me$, $R^2 = R^3 = Ph$ (30.2)	
$R^1 = R^2 = R^3 = CO_2Me$	$CH_3-C\equiv C-N(Et)_2$	I-376; $R^1 = R^2 = R^3 = CO_2Me$, $X = N(Et)_2$, $Y = Me$ (85)	753
$R^1 = R^2 = R^3 = CO_2Me$	1-pyrrolidinocyclopentane	I-376; $R^1 = R^2 = R^3 = CO_2Me$, X & Y = $-(CH_2)_3-$ (91)	754
$R^1 = CO_2Me$, $R^2 = R^3 = Ph$	1-pyrrolidinocyclopentane	I-376; $R^1 = CO_2Me$, $R^2 = R^3 = Ph$, X & Y = $-(CH_2)_3-$ (86)	756

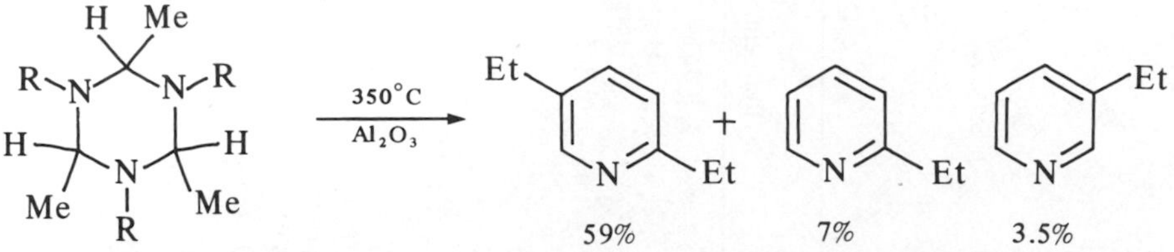

I-378

R = Allyl

59%   7%   3.5%

F.   Seven-Membered Ring Heterocycles

a.   AZEPINES

Ring-contractions of azepines leading to pyridines have been reported. These have involved 1H-azepines, dihydroazepines, azepinediones, and 3H-azepines.

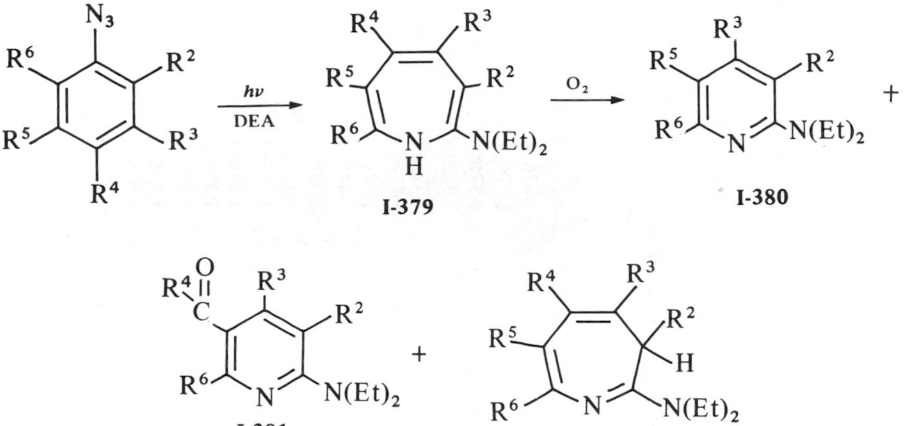

I-379   I-380

I-381   I-382

The primary products of the photolysis of substituted arylazides in the presence of diethylamine were the 2-diethylamino-1H-azepines I-379. Air oxidation of I-379 led to secondary products including pyridines I-380 and I-381 (757). The yields of these pyridines were generally low; however in some cases (R^2 = Me, R^3 = R^4 = R^5 = R^6 = H), I-380 was found in 20% yield. It was shown that 3H-azepine I-382 did *not* rearrange to either I-380 or I-381. This is of interest in light of the report that photolysis of 3H-azepines I-383 led to pyridines I-384 and I-385 (758).

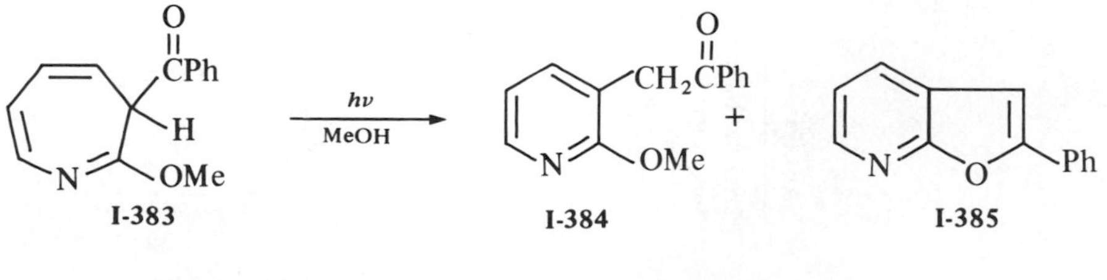

I-383   I-384   I-385

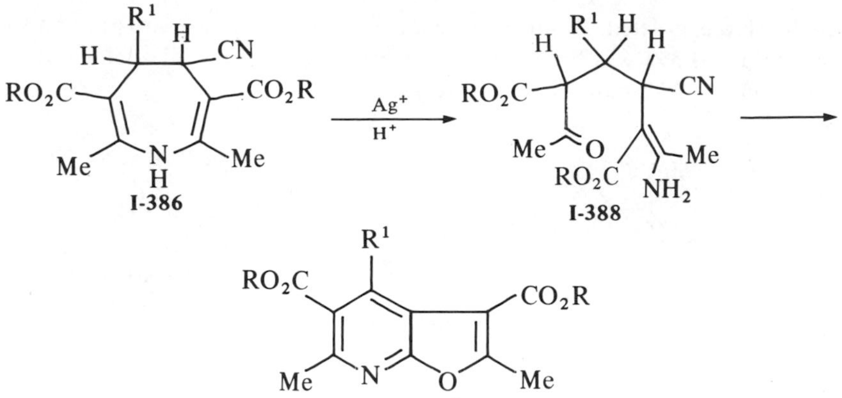

I-386 I-388

$R = Me, R^1 = H; 66\%$

I-387

Silver-catalyzed rearrangement of 4,5-dihydroazepines **I-386** in aqueous acid gave the furo[2,3-*b*]pyridines **I-387** in high yields (759). The open-chained intermediate **I-388** was shown to be involved in the ring-contraction.

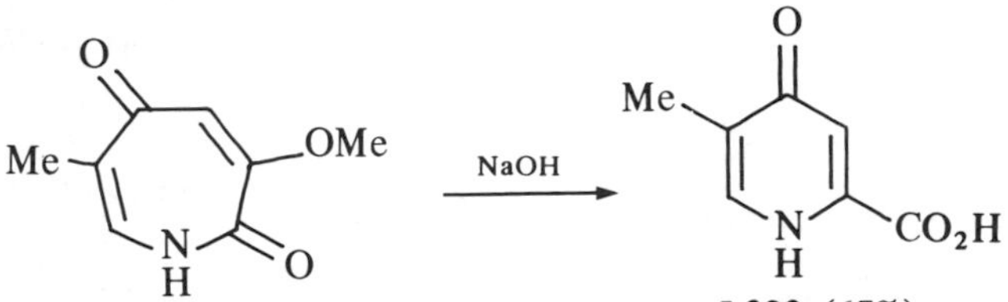

I-389 **I-390** (67%)

Azepinediones, such as **I-389**, in the presence of base underwent a ring-contraction to give pyridone **I-390** (760, 761).

b. DIAZEPINES

Photolysis of the pyridinium ylide **I-391** gave the 1,2-diazepine **I-392**. It was postulated that intermediate **I-393** was involved in this ring-expansion reaction. The reaction was reversible, since when **I-392** was treated with acetic acid under reflux,

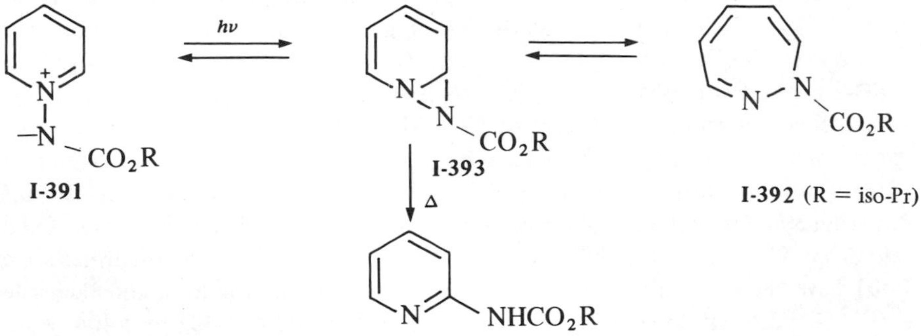

I-391 **I-393** **I-392** (R = iso-Pr)

I-395

ylide **I-391** was isolated. When the diazepine was heated in the absence of acid the aminopyridine **I-395** was found in high yield (762). Similarly, diazepine **I-396** gave aminopyridine **I-397** upon treatment with base (763).

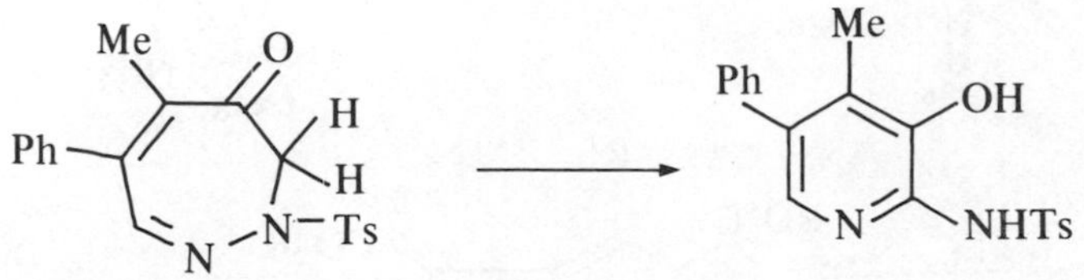

Acid-catalyzed ring-contractions of diazepines of type **I-398** gave modest yields of pyridines **I-399** with pyrazoles being the major products. The *N*-methyldiazepines **I-400** gave **I-399** in yields of synthetic utility (764, 765). A novel ring-contraction was reported to occur when the anion of the diazepine **I-398** was treated with sodium–potassium amalgam at low temperatures (766).

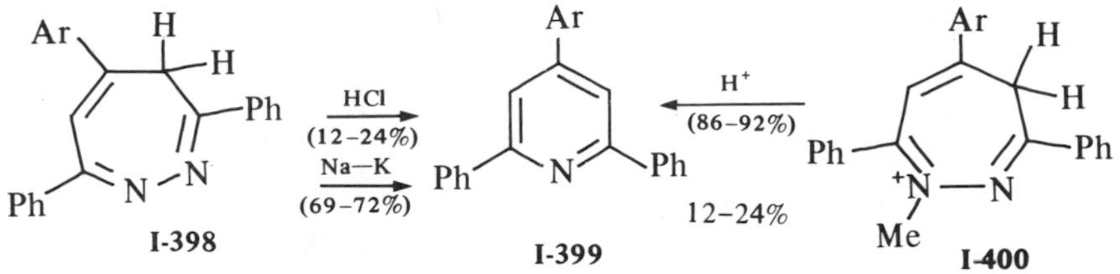

G. Pyridines from Reduced Pyridines

a. DIHYDROPYRIDINES

The aromatization of dihydropyridines to pyridines is an important synthetic procedure since many methods of preparing dihydropyridines are known (see Section II.2.E), the Hantzsch synthesis being one of the oldest and most well-known examples.

The methods of oxidation can be classified as chemical oxidizing agents, catalytic dehydrogenations, and electrochemical methods. Table I-55 outlines the various inorganic reagents which have been utilized to carry out such oxidations while Table I-56 lists the organic compounds that have been used to oxidize 2,6-dimethyl-3,5-dicarbethoxy-1,4-dihydropyridine, the so-called Hantzsch ester. Other dihydropyridines have also been oxidized by organic reagents. Octahydroacridines **I-401** have been aromatized by azobenzenes, nitrobenzenes, imines, and aldehydes (767). Dihydropyridine **I-402** was aromatized to the pyridine **I-403** by **I-404**, which also is a precursor to **I-402** (768).

TABLE I-55. OXIDATIONS OF DIHYDROPYRIDINES TO PYRIDINES USING INORGANIC REAGENTS

(Dihydropyridine structure: R⁴, R³, H, R², R⁵, R¹, N–R⁶) → (Pyridine structure: R³, R⁴, R², R⁵, N–R¹, (R⁶))

Substituents	Conditions	Pyridine	Reference
$R^1 = R^5 = R^6 = H$, $R^2 = R^4 = CO_2Me$ $R^3 = m\text{-}NO_2C_6H_4$	HNO_3	$R^1 = R^5 = H$, $R^2 = R^4 = CO_2Me$ $R^3 = m\text{-}NO_2C_6H_4$	774, 775
$R^1 = R^5 = Ar$, $R^2 = R^4 = CN$, $R^3 = H$, $R^6 = H$	$NaNO_3/AcOH$	$R^1 = R^5 = Ar$, $R^2 = R^4 = CN$	776
$R^1 = R^5 = Me$, $R^2 = R^4 = CO_2Et$ $R^6 = H$, $R^3 = 2\text{-}, 3\text{-}, \text{ or } 4\text{-pyridyl}$	$3\ HNO_3$	$R^1 = R^5 = Me$, $R^2 = R^4 = CO_2Et$ $R^3 = 2\text{-}, 3\text{-}, \text{ or } 4\text{-pyridyl}$	777
$R^1 = R^5 = R^3 = H$, $R^2 = R^4 = CO_2CH_3$ $R^6 = H$	i. dil NHO_3 ii. NH_3	$R^1 = R^5 = R^3 = H$, $R^2 = R^4 = CONH_2$	778, 779
$R^1 = Me$, $R^2 = CO_2Et$, $R^3 = R^5 = R^6 = H$, $R^4 = Et$	HNO_3	$R^1 = Me$, $R^2 = CO_2H$, $R^3 = R^5 = H$ $R^4 = Et$	780
$R^1 = R^5 = R^6 = H$, $R^2 = R^4 = COCH_3$ $R^3 = Me$	$NaNO_2\text{–}H_2SO_4$	$R^1 = R^5 = H$, $R^2 = R^4 = COCH_3$, $R^3 = Me$	781
$R^1 = R^5 = R^3 = Me$, $R^2 = R^4 = CN$ $R^6 = H$	$NaNO_2\text{–}HOAc$	$R^1 = R^5 = R^3 = Me$, $R^2 = R^4 = CN$	782
$R^1 = Me$, $R^2 = COCH_3$, $R^3 = Ph$, $R^4\ \&\ R^5 = -CO-CH_2CHMe_2CH_2-$ $R^6 = H$	CrO_3	$R^1 = Me$, $R^2 = COCH_3$, $R^3 = Ph$ R^4 and $R^5 = -COCH_2CHMe_2CH_2-$	783
$R^1 = R^2 = R^4 = R^5 = R^6 = H$, $R^3 = 4\text{-pyridyl}$	SO_2	$R^1 = R^2 = R^4 = R^5 = H$ $R^3 = 4\text{-pyridyl}$	784
$R^1 = R^3 = R^4 = R^5 = H$	$H_2O_2\text{–}Cu^{+2}$	$R^1 = R^3 = R^4 = R^5 = H$	785
$R^2 = CONH_2$, $R^6 = CH_2Ph$	$H_2O_2\text{–}MoO_4-$ Anhy. HBr	$R^3 = CONH_2$, $R^6 = CH_2Ph$	786
$R^1 = R^2 = R^4 = R^5 = H$, $R^3 = CH_2COAr$, $R^6 = Ar$		$R^1 = R^2 = R^4 = R^5 = H$, $R^3 = CH_2COAr$, $R^6 = Ar$	787
$R^1 = R^5 = Aryl$, $R^6 = H$, $R^3 = Aryl$ $R^2 = R^4 = N\text{-pyridinium}$	$N_2O_3/AcOH$	$R^1 = R^5 = Aryl$, $R^5 = Aryl$ $R^2 = R^4 = NH_2$	788

107

TABLE I-56.   DEHYDROGENATIONS OF THE HANTZSCH
             ESTER UTILIZING ORGANIC OXIDIZING AGENTS

Oxidizing Agent	Yield	Reference
Quinoline N-oxide	91%	789
Pyridine N-oxide	80%	789–792
PhNOMe$_2$	–	791
Azoxybenzene	–	791
Nicotinamide N-oxide	67%	793
Picolinic acid N-oxide	96%	794
4-Nitropyridine N-oxide	–	794
2,2′,-4,4′-, 3,3′-bipyridine N,N'-dioxides	–	795
2-Picoline N-oxide	–	796
Diacetylperoxide	–	797–798
Triphenylverdazyls	–	799–802
Fluorenone	–	803
Benzophenone	–	803
Iminium salts	–	804, 809
Indolenine salts		806
Nitrosobenzene	–	807
N-Tosylimides	70–92%	808
Indan-1,3-diones	–	809
Chloranil	–	809
Pyridinioacylamidates	78–99%	810
Phenacylonium salts	–	811

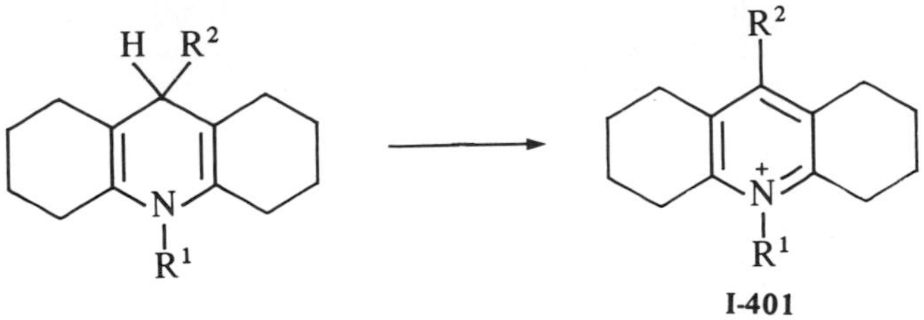

I-401

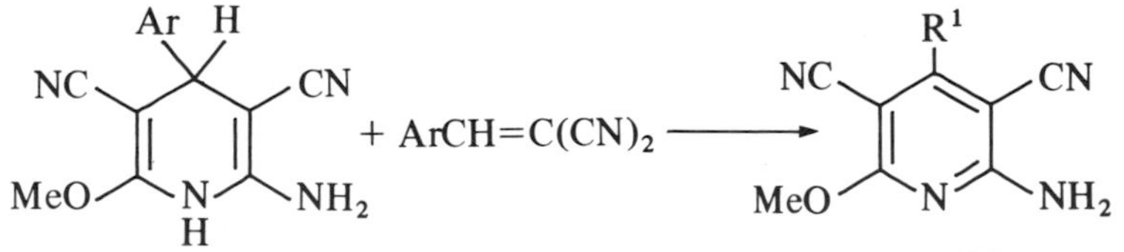

I-402   I-404   I-403

Catalytic dehydrogenation has most often been utilized in the aromatization of tetrahydropyridines and piperidines, however only occasionally on dihydropyridines. Table I-57 lists some recent examples.

There have been reports of the thermal aromatization of dihydropyridines. Thermolysis of I-405 gave lactone I-406 as well as the pyrrole I-407 (812, 813). When the ketone at the 3 and 5 positions is replaced by ester functions, pyridine I-408 was obtained along with the pyrrole (814).

TABLE I-57.   CATALYTIC DEHYDROGENATION OF DIHYDROPYRIDINES USING PALLADIUM ON CARBON

Dihydropyridine	Conditions	Pyridine	Reference
3,4-dihydro-2-pyridone	Xylene, reflux	2-pyridone	
2-carbomethoxy-3-methyl-4-aryl-5-acetamido-2,5-dihydro-6-pyridone	Pd-carbon	2-carbomethoxy-3-methyl-4-aryl-5-acetamido-6-pyridones	820
2-phenyl-4-benzoyl-5-oxo-7,7'-dimethyl-1,4,5,6,7,8-hexahydroquinoline	Pd-carbon xylene, reflux	2-phenyl-4-benzoyl-5-oxo-7,7'-dimethyl-5,6,7,8-tetrahydroquinoline	821
4-phenyl-1,4-dihydro-3,5-pyridine-dicarboxamide	Pd-carbon	4-phenyl-3,5-dicarboxamidopyridine	822
2-methyl-3-(β-acetoxyethyl)-4,5-dihydro-pyridine	5% Pd-carbon	2-methyl-3-(β-acetoxyethyl)pyridine	819, 823

I-405 I-406 I-407

I-408

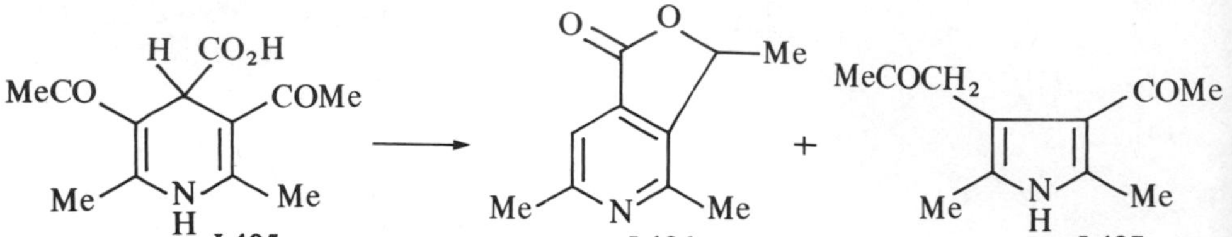

Electrochemical oxidation has become of interest in recent years and has been utilized to probe the NADH, NADPH, and NMNH biological systems (815). Preparatively, the Hantzsch ester was oxidized at a graphite electrode to give the pyridine in $> 65\%$ yield with a current efficiency of 80% (816). It has been shown that, when dihydropyridines **I-409** were oxidized at a rotating platinum electrode, the pyridine product was obtained by a two-electron process, except $R^2 = R^3 = Me$, then the reaction stopped after the loss of one electron (817, 818).

I-409

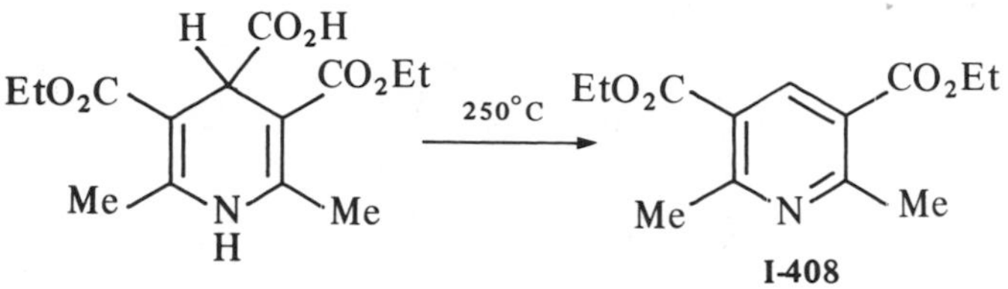

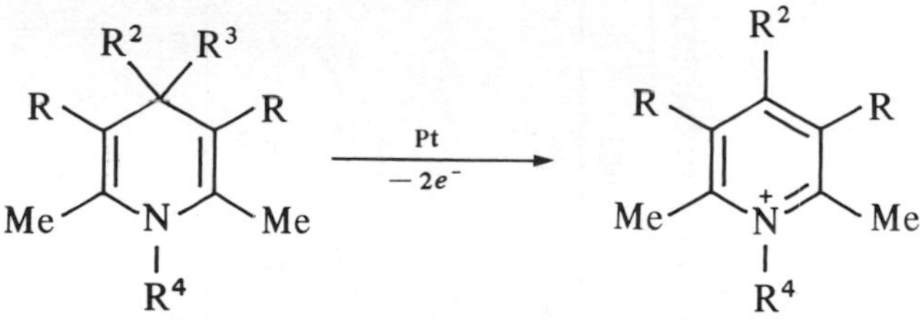

As in the past edition of this work the aromatizations of piperidinediones (glutarimides) have been included in the section on dihydropyridines.

Chlorinations of glutarimides, for example, **I-410** with phosphorus oxychloride gave pyridine **I-411** in high yields (825–831, 837). Similarly **I-412** gave **I-413** upon treatment with phosphorus pentachloride (832).

I-410 I-411

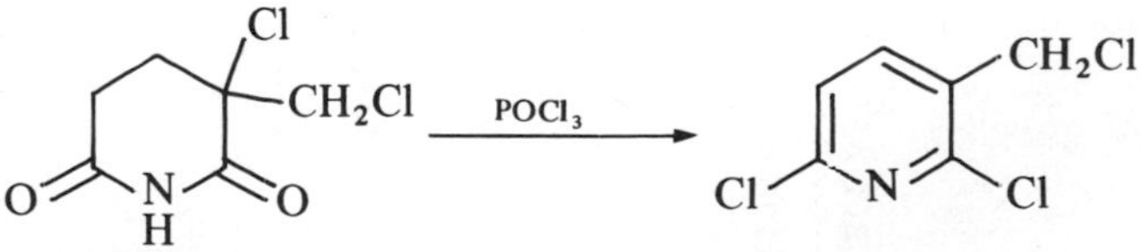

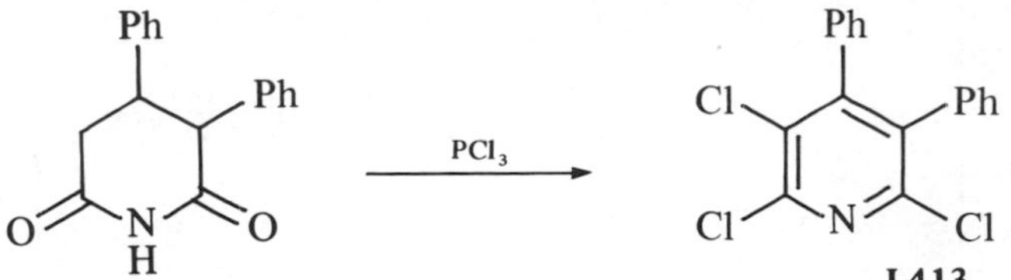

I-412   →(PCl₃)→   **I-413**

Dehydrogenation of tetrahydropyridones **I-414** over palladium on carbon gave the pyridones **I-415** (833, 834). In a similar manner N-phenyltetrahydro-4-pyridones gave the N-phenyl-4-pyridones (835).

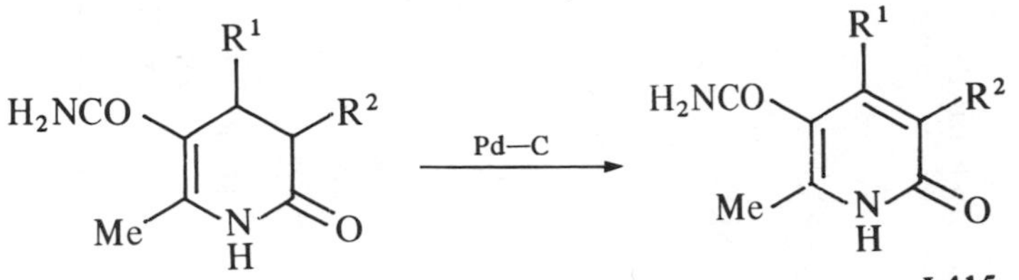

I-414   →(Pd—C)→   **I-415**

Bromination followed by dehydrohalogenation of the tetrahydropyridone **I-416** afforded the pyridone **I-417** (836, 837).

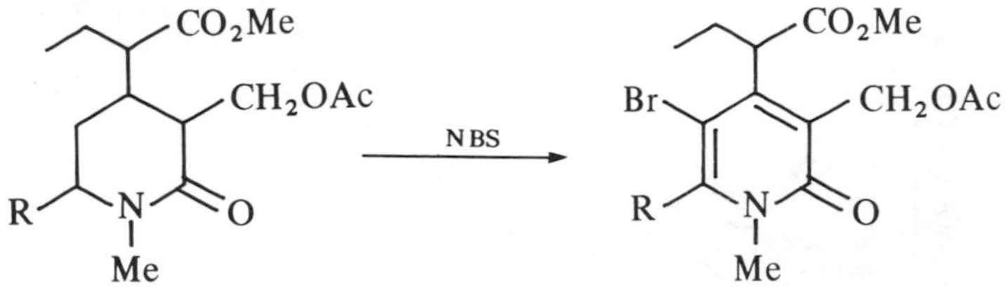

I-416   →(NBS)→   **I-417**

b.   TETRAHYDROPYRIDINES

Tetrahydropyridines are aromatized by chemical oxidation, catalytic dehydrogenation over palladium, vapor phase thermal dehydrogenation, and by elimination reactions on substituted tetrahydropyridines. Table I-58 delineates the vapor phase and catalytic methods. These methods have been extended to the aromatization of 2-piperidone to give pyridine (838–840). Elimination of suitable substituents from tetrahydropyridines gave pyridines: **I-418** to **I-419** is a representative example (841–843).

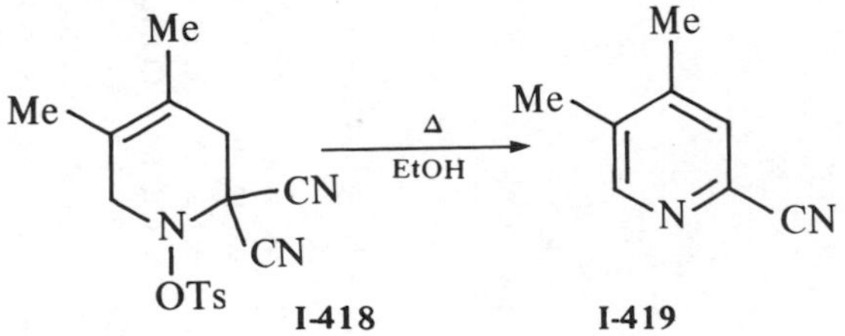

I-418   →(Δ, EtOH)→   **I-419**

TABLE I-58. CATALYTIC AND VAPOR PHASE DEHYDROGENATIONS OF TETRAHYDROPYRIDINES

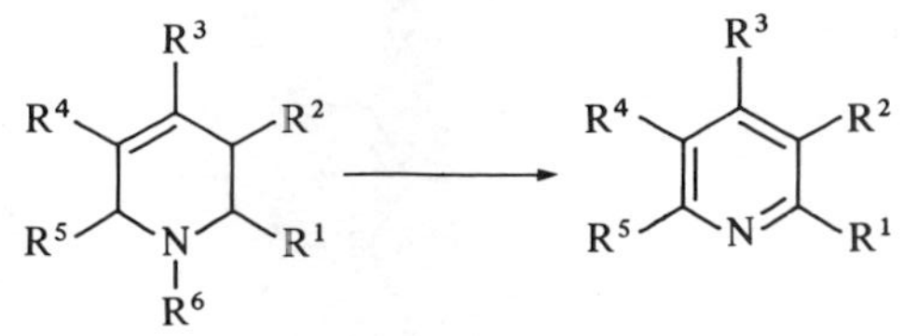

Substituents	Conditions	Pyridine (Yield)	Reference
$R^1 = R^2 = R^3 = R^4 = R^5 = R^6 = H$	SiO_2–Al_2O_3, O_2, 300°C	Pyridine	844
	$Cu(OAc)_2$–O_2		851
$R^1 = R^5 = H$, $R^2 = R^4 = R^6 = Me$, $R^3 = Ph$	K–2 catalyst, 400°C	$R^2 = R^4 = Me$, $R^3 = Ph$ (73%)	845
$R^1 = R^2 = R^4 = R^5 = H$, $R^3 = Alkyl$, $R^6 = CH_2Ph$	K–2 catalyst, 400°C	$R^3 = Alkyl$	846
$R^1 = R^3 = R^4 = R^5 = H$, $R^2 = CN$, $R^6 = COMe$	10% Pd–C, PhCH=CHPh	$R^3 = CN$	847
$R^1 = R^4 = R^5 = H$, $R^2 = COMe$, $R^3 = Me$, $R^6 = CH_2Ph$	(i) Pd–C (ii) hydrogenation	$R^2 = Me$, $R^3 = Et$	848
$R^2 = R^4 = R^5 = R^6 = H$, $R^1 = n\text{-Pr}$, $R^3 = CN$	4% Pd–MgO	$R^1 = n\text{-Pr}$, $R^3 = CN$	849, 850

c. PIPERIDINES

Catalytic dehydrogenation has historically been the main technique for the conversion of piperidines to pyridines. Table I-59 lists some recent examples. Chemical oxidations have been reported less frequently; however, this method has been utilized more recently. Oxidations of piperidines by heterocyclic *N*-oxides (930–932), pyridine *N*-imines (933, 934), and hydrogen peroxide (935) have been reported.

Recently, an interesting synthetic route to pyridines substituted in the 4-position has been based on the dehydrogenation of piperidines. The following example of the synthesis of **I-420** illustrates the method (936). This method has been used in several synthetic schemes (955–958).

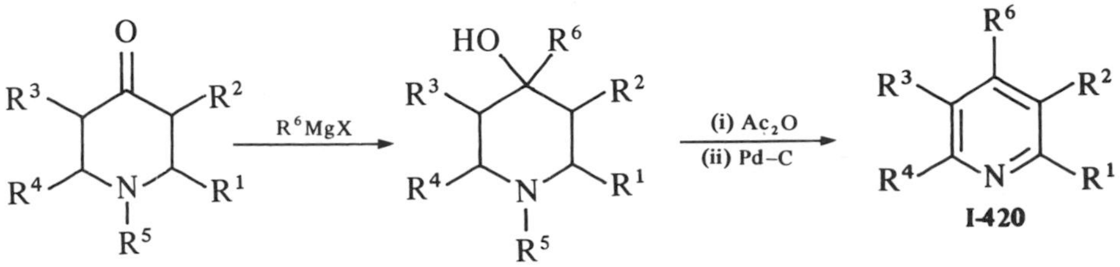

Defluorination of perfluoropiperidine (**I-421**) in the presence of **I-422** gave pyridine **I-423**. The intermediate **I-424** has been proposed (959).

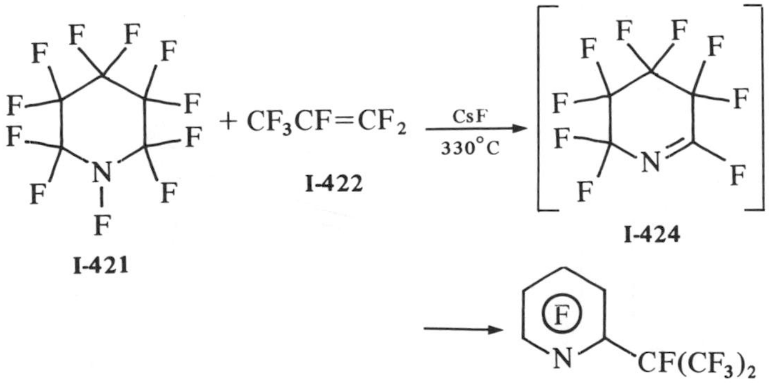

H. Condensed Rings

a. OXIDATION

Oxidation of quinoline and related condensed pyridines gave pyridinecarboxylic acids and similar derivatives. Classically, oxidations by nitric acid, potassium permanganate, and other common oxidizing agents have been utilized. More recently ozone, ruthenium oxides, and vapor phase methods have been used (Table I-60). In

TABLE I-59.   DEHYDROGENATIONS OF PIPERIDINES TO PYRIDINES *VIA* CATALYTIC DEHYDROGENATIONS

Piperidine	Conditions	Pyridine	Reference
Piperidine	$Pd-SiO_2$, 325°C	Pyridine	937–939
	$Pd-Al_2O_3$, 400		940
	$MoO_3-Al_2O_3$		941
	$Pt-Al_2O_3$, 280°C		942
N-methylpiperidine	$Pt-Al_2O_3$, 280°C	Pyridine	942
3-methylpiperidine	$V_2O_5-SiO_2-Al_2O_3$	3-methylpyridine	943, 944
	$Pd-Al_2O_3$, 300°C		945
	$V_2O_5-SiO_2-Al_2O_3$, NH_3, O_2	3-cyanopyridine	946
3,3-dimethylpiperidine	5% $Pd-C$, H_2, 485°C	3-methylpyridine	947
4-(4'-pyridyl)piperidine	Al_2O_3, O_2, NH_3, 340°C	4,4'-bipyridine	948
1,4-dialkoxy-3-cyanopiperidine	$Pd-C$, 300°C	3-cyanopyridine	949
1,2,5-trimethyl-4-alkenyl-4'-alkoxypiperidine	K–16, 420°C	2,5-dimethyl-4-alkenylpyridine	950, 951
1-acetyl-4-acetoxy-3-(*N*-butyl)carboxamidopiperidine	30% $Pd-C$, Decalin	3-cyanopyridine	952, 953
1,3-dimethyl-2,4,6-triphenyl-4-hydroxypiperidine	K–16, 550°C	3-methyl-2,4,6-triphenylpyridine	954

TABLE I-60.   OXIDATION OF CONDENSED RING HETEROCYCLES TO PYRIDINES

Substrate	Conditions	Pyridine (Yield)	Reference
Quinoline	O_3	2,3-Diformylpyridine	960
	$O_3-H_2SO_4$	Quinolinic acid	961
	O_3-AcOH	Nicotinic acid	962–964
	$V_2O_5-SnO_2$, air, 420°C	Nicotinic acid, nicotinaldehyde	965
	$V_2O_5-Al_2O_3$, air	Nicotinic acid	966
	$V_2O_5-Al_2O_3$, air, NH_3	Nicotinonitrile	966
	$Co(OAc)_3-HOAc$, air	Quinolinic acid (90%)	967, 968
	$V_2O_5-MoO_3$, air, NH_3	Nicotinonitrile	969
	RuO_4	Quinolinic acid	970
	Electrochemical oxidation–PdO_2 electrode	Quinolinic acid	971, 972
	KOH, air (250°C)	Nicotinic acid (54%)	973
	HNO_3, 210°C	Nicotinic acid (76%)	974
Isoquinoline	O_3	3,4-Diformylpyridine	925
Perfluoroquinoline	HNO_3	2,3,4-Trifluoro-5,6-dicarboxypyridine	976
8-Hydroxyquinoline	O_3	Nicotinic acid	977
5-Bromo-6-methylmerimine	$KMnO_4$	5-Bromo-6-methyl-4-cinchomeronamic acid	978
Perfluoroisoquinoline	$KMnO_4$	2,5,6-Trifluoropyridine-3,4-dicarboxylic acid (52%)	979

TABLE I-61.   REDUCTION OF QUINOLINES AND ISOQUINOLINES TO YIELD PYRIDINES

Substrate	Conditions	Product (Yield)	Reference
Quinoline	Li–MeOH, liq. NH$_3$	5,6,7,8-tetrahydro	980, 981
	4 N MeOH–HCl, PtO$_2$, H$_2$	5,6,7,8-tetrahydro (52%)	982
		1,2,3,4-tetrahydro (39%)	
	Liq. HF, PtO$_2$, H$_2$	5,6,7,8-tetrahydro (75%)	983
		1,2,3,4-tetrahydro (8%)	
	12 N HCl, PtO$_2$, H$_2$	5,6,7,8-tetrahydro (70%)	984
2-methylquinoline	12 N HCl, PtO$_2$, H$_2$	2-methyl-5,6,7,8-tetrahydro (95%)	984
3-methylquinoline	12 N HCl, PtO$_2$, H$_2$	3-methyl-5,6,7,8-tetrahydro (98%)	984
6-methylquinoline	12 N HCl, PtO$_2$, H$_2$	6-methyl-5,6,7,8-tetrahydro (53%)	984
		6-methyl-1,2,3,4-tetrahydro (24%)	
8-methylquinoline	12 N HCl, PtO$_2$, H$_2$	8-methyl-5,6,7,8-tetrahydro (55%)	984
		8-methyl-1,2,3,4-tetrahydro (22.5%)	
Isoquinoline	5 N MeOH–HCl, PtO$_2$, H$_2$	5,6,7,8-tetrahydro (87%)	982
		1,2,3,4-tetrahydro (13%)	
	PrNH$_2$, Li, liq. NH$_3$	5,6,7,8-tetrahydro	985
	Liq. HF, PtO$_2$, H$_2$	5,6,7,8-tetrahydro (66%)	983
	12 N HCl, PtO$_2$, H$_2$	5,6,7,8-tetrahydro (95%)	984
7,8,9,10-tetrahydro-phenanthridine	PtO$_2$, H$_2$, AcOH	1,2,3,4,7,8,10-octahydro	986
Acridine	(i) Li-EtOH	1,2,3,4,5,6,7,8-octahydro	987
	(ii) (PH$_3$P)$_3$RH$_1$RH$_2$		
2-methyl-4-hydroxyquinoline	Raney Ni, H$_2$	2-methyl 5,6,7,8-tetrahydro-4-hydroxy	988–990
2-methyl-4-aminoquinoline	PtO$_2$, AcOH, H$_2$	2-methyl-4-amino-5,6,7,8-tetrahydro (95.2%)	991
2-methyl-4-methoxyquinoline	Pt, AcOH, H$_2$	2-methyl-4-methoxy-5,6,7,8-tetrahydro	992
Dimethyl-4-quinolylmethanol	Zn, HCO$_2$H	4-isopropyl-5,6,7,8-tetrahydro	993
		4-isopropyl-1,2,3,4-tetrahydro	

general, the vapor phase oxidations of quinolines afforded nicotinic acid, since quinolinic acid was unstable to the reaction conditions. Oxidation with ozone gave the diozanide, which was converted to 2,3-diformylpyridine (960).

b.   REDUCTIONS

Reductions of fused-ring pyridines, such as quinoline and isoquinoline, gave two possible tetrahydro compounds **I-425** and **I-426**. When the pyridine portion was highly substituted, i.e. 2,3- or 2,4-dimethylquinoline, the predominate product was **I-426**. Recently, the reductions have been conducted in the presence of acid to yield the 5,6,7,8- tetrahydro derivatives (Table I-61).

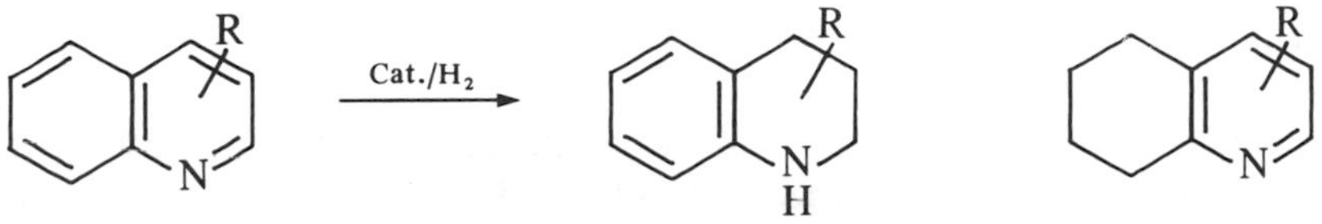

I-425   **I-426**

Similar to the catalytic hydrogenations of quinolines, the Birch reductions normally gave the 1,2,3,4-tetrahydro derivatives. It has recently been found that Li—NH$_3$ reductions of quinoline and isoquinoline in the presence of methanol or *n*-propylamine gave the 5,6,7,8-tetrahydro derivative, as the major product (Table I-61). An interesting example was the reduction of 6-methoxyquinoline to yield the mixture of **I-427** and **I-428**, which on hydrolysis gave ketone **I-429**, which has been difficult to obtain by classical methods (994).

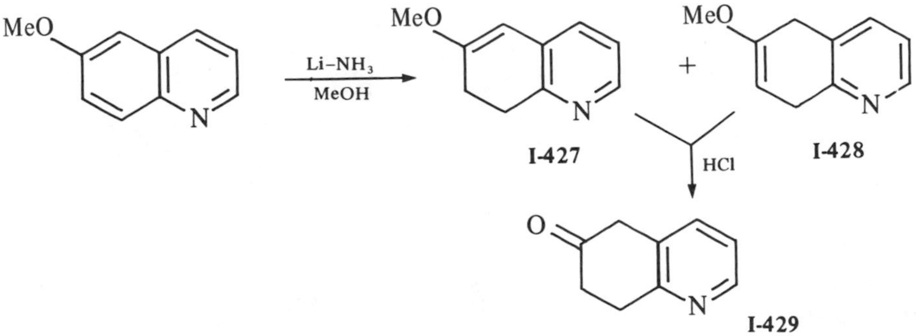

I-427   **I-428**

I-429

c.   RING-OPENING REACTION

Bicyclic heterocyclic rings, in which one ring is a pyridine nucleus, undergo ring-opening reactions to form substituted pyridines.

2-(3-Pyridyl)pyrido[2,3-*d*]pyrimidine (**I-430**), prepared from 3-cyanopyridine and ammonium sulfamate, underwent a ring-cleavage catalyzed by acid to give pyridine **I-431** plus 3-cyanopyridine (995).

The pyridopyrimidinones **I-432** were reductively opened to pyridines **I-433**. When LiAlH$_4$ was the reducing agent, the carbonyl function at the two position was reduced to **I-434** (996).

I-430 **I-431**

I-432 **I-433** **I-434**

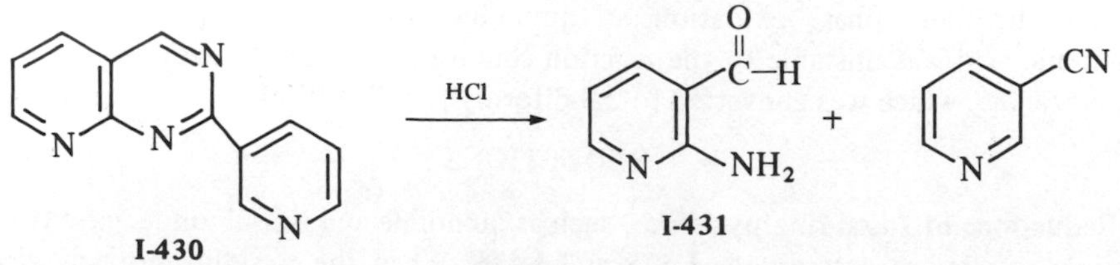

An interesting method for the synthesis of 3,4-disubstituted pyridines utilized such a ring-opening. 2-Methylthiazolium salts **I-435** underwent condensation with 1,2-diketones to give thiazole[3,2-*a*]pyridinium salts, which were cleaved by Raney nickel in ethanol to give the disubstituted pyridine **I-436** (997).

I-435

I-436

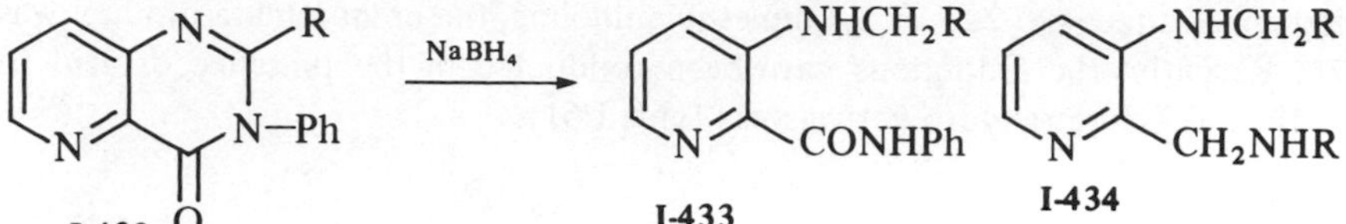

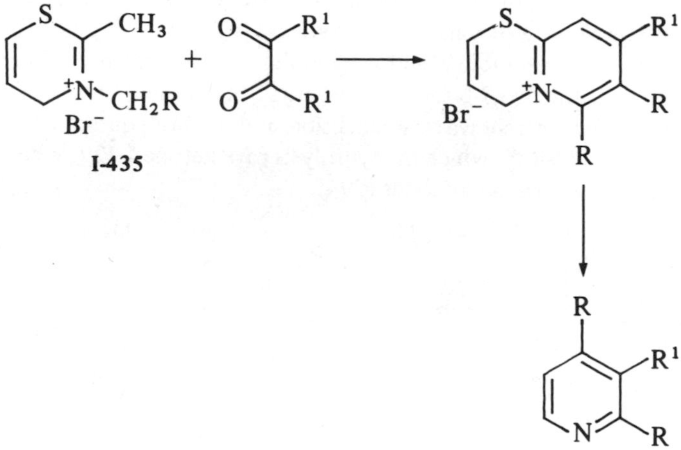

(R = H, $CONH_2$)
(R^1 = Alkyl, aryl)

2. From Acyclic Compounds

The organization of Brody and Ruby (998), as shown in Scheme I-9, has been followed. Care should be taken in using this scheme since condensations in which an intermediate is isolated, whether or not it is fully characterized, are classified by the intermediate rather than by the starting materials. Thus nearly identical reactions can be widely separated in this Section, even by the whims of the experimentalist.

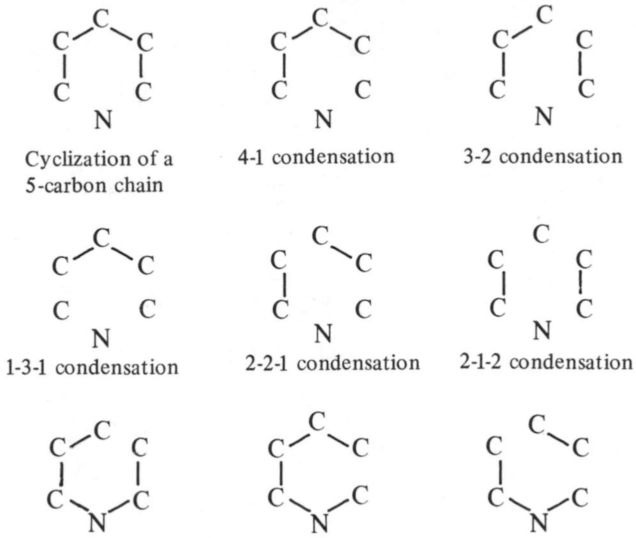

Cyclization of a 5-carbon chain   4-1 condensation   3-2 condensation

1-3-1 condensation   2-2-1 condensation   2-1-2 condensation

Cyclizations not involving the ring nitrogen

Scheme I-9

A. Cyclization of a 5-Carbon Chain

a. 1,5-DIOXO COMPOUNDS AND DERIVATIVES

Use of hydroxylamine for the transformation of a saturated 1,5-diketone to a pyridine is usually the preferred route, since neither disproportionation nor oxidation is required. Ketals, enol ethers, and so on, may be used, as shown for the optically active **I-437a** and **b** (999).

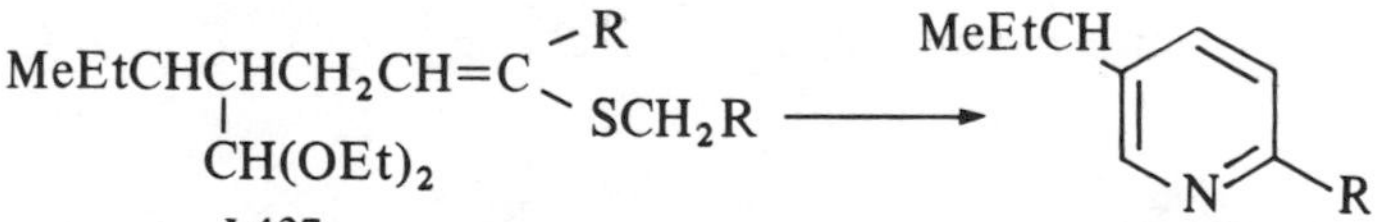

$$\text{MeEtCHCHCH}_2\text{CH=C} \overset{\text{R}}{\underset{\text{SCH}_2\text{R}}{\diagdown}}$$
$$\underset{\text{CH(OEt)}_2}{|}$$

I-437

I-438

I-438a R = H (41%)
     b R = Me (62%)

The hydroxylamine method has frequently been used for the preparation of 5,6,7,8-tetrahydroquinoline and related compounds (Table I-62). Since the starting 1,5-dioxo compounds are often prepared by Michael addition of a substituted 3-carbon unit to a cyclic ketone, the 3-2 condensation using the same or similar starting materials can be regarded as a synthetic equivalent combining the carbon skeleton construction and ring formation steps. In addition, many 5,6,7,8-tetrahydroquinolines are accessible by catalytic hydrogenation of the corresponding quinolines.

TABLE I-62.   CYCLIZATION OF 1,5-DIOXO COMPOUNDS YIELDING TETRAHYDROQUINOLINES AND RELATED COMPOUNDS

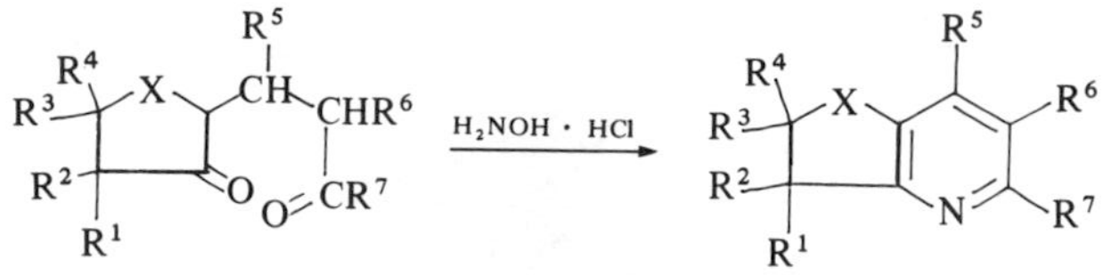

R^1	R^2	R^3	R^4	R^5	R^6	R^7	X	Yield	Reference
i-Pr	H	H	H	H	H	Me	$-CH_2-\overset{\underset{\displaystyle CH_3}{\displaystyle CH_3}}{CH}-$	—	1000
$(CH_3)_2C=$		H	H	H	H	CH_3	$-CH_2\overset{\underset{\displaystyle CH_3}{\displaystyle \mid}}{CH}-$	—	264
H	H	H	H	C_6H_5	CH_2NMe_2	C_6H_5	$-(CH_2)_2-$	58% $(R^6 = H)$	1001
				C_6H_5	CH_2NMe_2	C_6H_5	$-(CH_2)_2-$	11% $(R^6 = H)$ 50% $(R^6 = -CH_2NMe_2)$	1001
$-CH_2NMe_2$	H	H	H	C_6H_5	H	C_6H_5	$-(CH_2)_2-$	77%[a]	1001
H	H	CH_3	CH_3	C_6H_5	H	C_6H_5	$-OCH_2-$	94%	1002
H	H	H	H	H	H	$-CH=CHC_6H_5$	$-CH_2-$	28	1003
						$-CH=CHC_6H_5$	$-(CH_2)_2-$	34	1003
						$-CH=CHC_6H_5$	$-(CH_2)_3-$	28	1003
						C_6H_5	$-(CH_2)_2-$	64	1004
						$-CMe_3$	$-(CH_2)_2-$	64	1004
						CH_3	$-(CH_2)_2-$	60	1004
H	H	H	H	H	$-(CH_2)_4-$		$-CH_2-$	65	1005
H	H	H	H	Ph	H	Ar	$-(CH_2)_2-$	—	1006
H	H	Me	Me	Ph	R	Ph	$-SCH_2-$	—	1007
R	H	H	H	H	H	Ph	$-(CH_2)_2-$	—	1007
R	H	H	H	H	H	Ph	$-CH_2-$	—	1007
$-(C_4H_4)-$				Ph	H	Ph	$-(CH_2)_2-$	—	1007

[a] Mono-oxime (79%) cyclized with dry HCl (98%).

TABLE I-63. PREPARATION OF REDUCED ACRIDINES

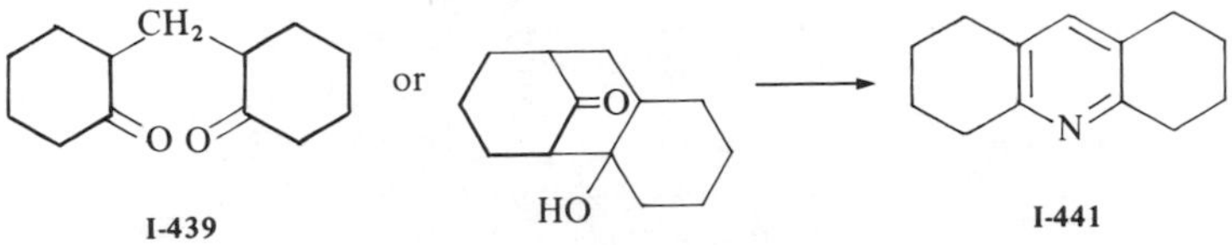

Reactant	Nitrogen Source	Products (Yields)	Reference
I-439	$NH_3(H_2, Ni)$	I-441 (60), perhydroacridine (9)	1008a
I-439	$H_2NOH \cdot HCl$	I-441	1008b
I-439	NH_3	I-441 (53.8)	1009
I-439	$N_2H_4 \cdot H_2O$	I-441	1010
I-439	$PhNHNH_2$, $RCONHNH_2$	N-Amino-tetrahydroacridinium and perhydroacridines	1010
I-439 or I-440	NH_4OAc	I-441 (7.6), others	1011
I-439	RNH_2	N-R-Tetrahydro-acridinium (40–85), N-R-Perhydroacridine (9–46), Decahydroacridines	1012–1018
I-439	H_2NCHO	I-441, others	1019
Substituted I-439	–	Substituted I-441, others	1020–1024

Extensive work on the cyclization of 2,2'-dimethylenecyclohexanone (I-439) or its internal ketol (I-440) to octahydroacridine (I-441) and related compounds has been reported (Table I-63).

Additional cyclizations of 1,5-dioxo compounds using ammonia, amines, and other nitrogen sources are shown in Table I-64.

Reaction of glutaraldehyde with labeled iminium salt I-442 and aqueous ammonia at pH 10.3 gave labeled nicotine I-443 in 21.2% radiochemical yield; the biomimetic mechanism shown was suggested (1034).

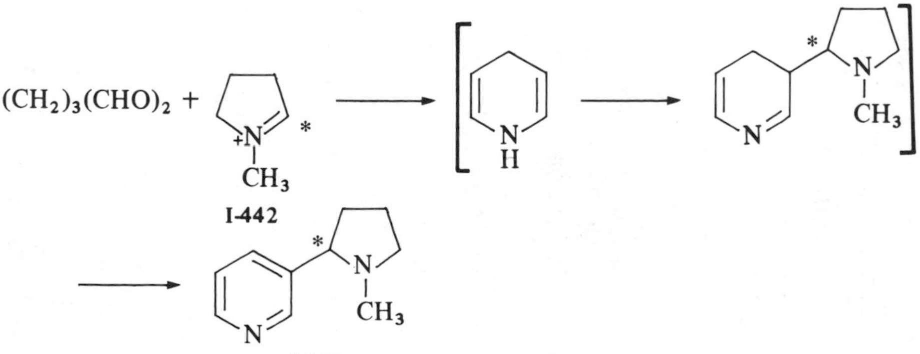

TABLE I-64.   PREPARATION OF PYRIDINES FROM 1,5-DIOXO COMPOUNDS

Reactant	Nitrogen Source	Product (Yield)	Reference
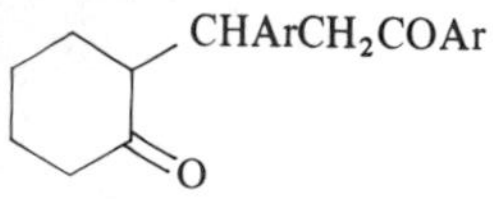 CHArCH$_2$COAr (cyclohexanone structure with =O)	HCONH$_2$	Tetrahydro- and decahydroquinoline derivatives	1025
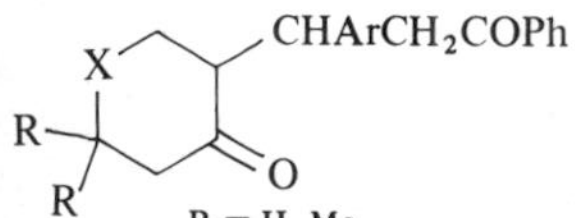 (ring structure with X, CHArCH$_2$COPh, R, R', =O) R = H, Me; X = CH$_2$, O	PhNHNH$_2$	Pyridines (8–19), indoles	1026
PhCOCH=CPhCH$_2$COPh	PhNH$_2$	1,2,4,6-tetraphenyl-pyridinium (95)	1027
4-C$_5$H$_4$NCH[CH$_2$CH(OEt)$_2$]$_2$	NH$_4$OAc	4,4'-bipyridyl (25)	1028
PhCO(CH$_2$)$_3$COPh	NH$_3$ (H$_2$, Ru/C)	2,6-diphenylpyridine (24), 2,6-diphenylpiperidine (45)	1029
Various 1,5-diketones	NH$_3$, RNH$_2$, H$_2$, Ru or Ni	Pyridines (11–60), piperidines (0–12)	1030, 1031
PhCOCH$_2$CHRCH(CH$_2$NR$_2$)COPh	H$_2$NOH·HCl	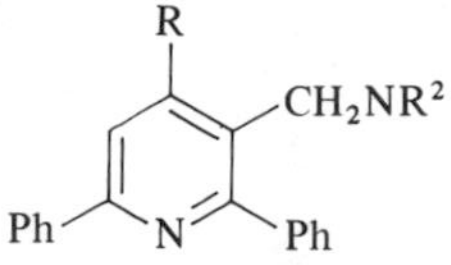 (pyridine ring: R, CH$_2$NR^2, Ph, N, Ph)	1032
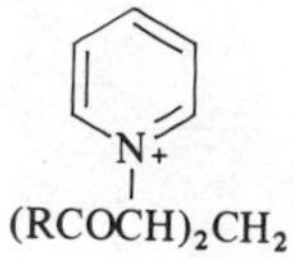 (N$^+$ pyridinium) (RCOCH)$_2$CH$_2$	NH$_4$OAc	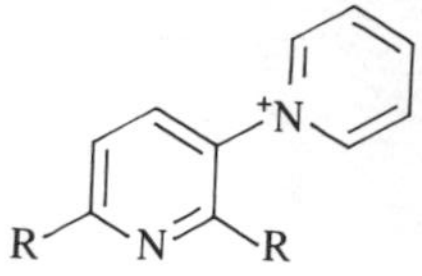 (bipyridinium product: R, N$^+$ pyridinium, R, N, R)	1033
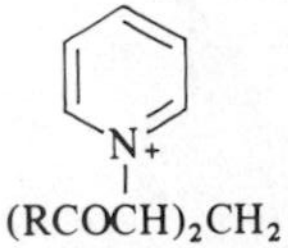 (N$^+$ pyridinium) (RCOCH)$_2$CH$_2$	NH$_4$OAc, O$_2$ or H$_2$NOH	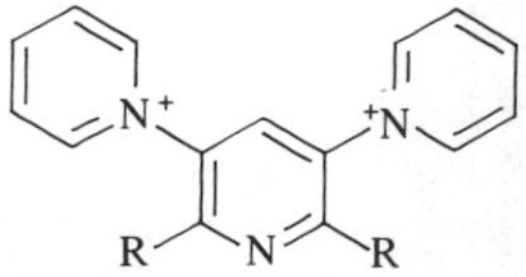 (di-pyridinium product: two N$^+$ pyridinium groups, R, N, R)	1033

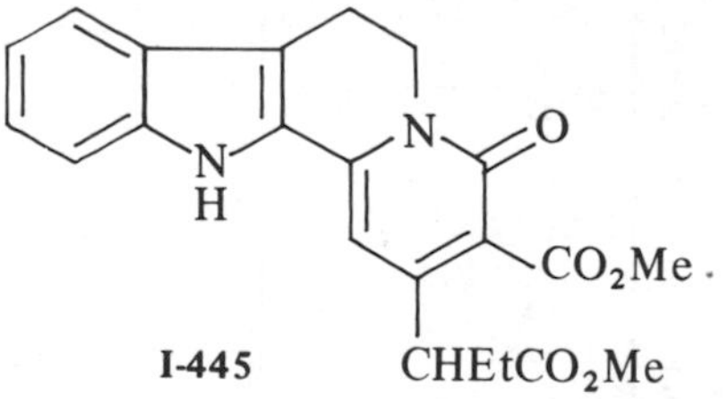

I-445 CHEtCO$_2$Me

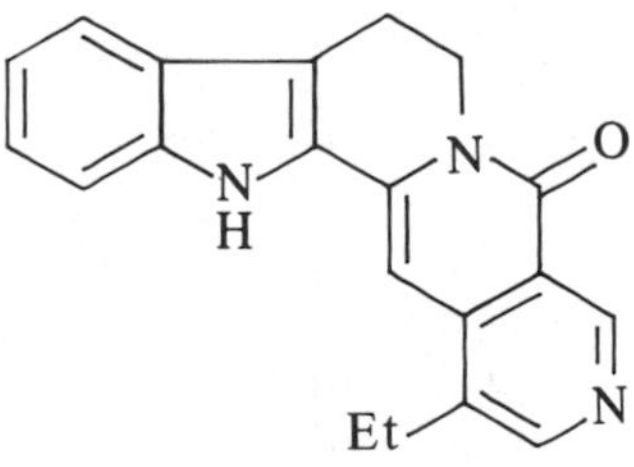

I-444

Dihydroangustine **I-444** was synthesized by reduction of ester **I-445** with diisobutylaluminum hydride and cyclization with aqueous ammonia (1035).

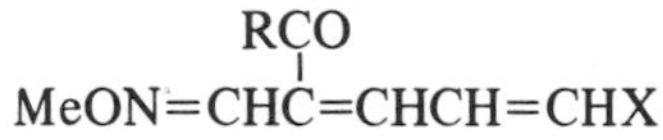

$$MeON=CHC=CHCH=CHX$$

I-446

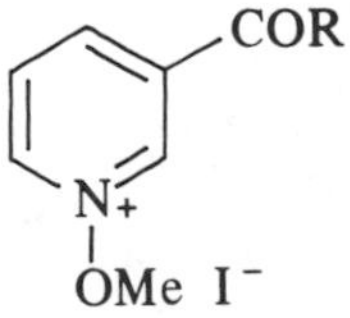

I-447

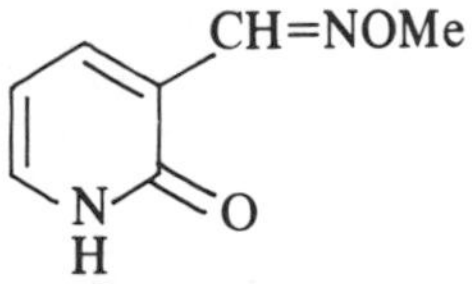

I-448

Oxime ethers **I-446** cyclize at the oxime nitrogen to give N-methoxypyridinium derivatives **I-447** or pyridones **I-448** depending on reaction conditions (1036).

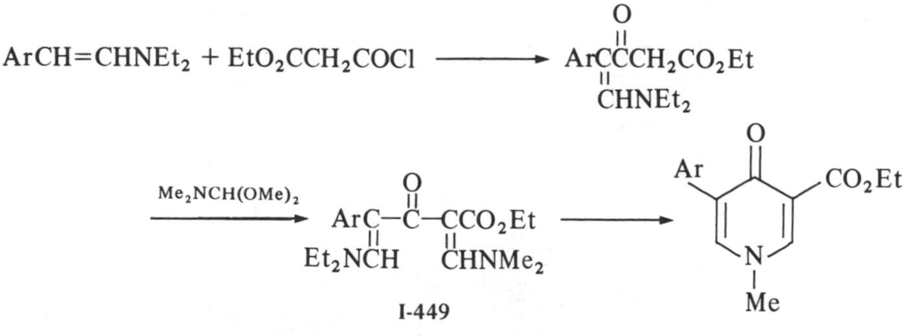

$$ArCH=CHNEt_2 + EtO_2CCH_2COCl \longrightarrow ArCCH_2CO_2Et$$

I-449

Cyclizations of 1,3,5-trioxo compounds and their derivatives to give 4-pyridones or related compounds are shown in Table I-65. Noteworthy is the construction of 5-carbon units **(I-449)** by reactions using amide acetals (1044). Similar 4-pyridones can be prepared by 4-1 condensations which combine the last two steps.

Heating cyanine **I-450** resulted in displacement of the alkoxy group in addition to cyclization; a 4-alkoxypyridinium salt may be an intermediate (1045).

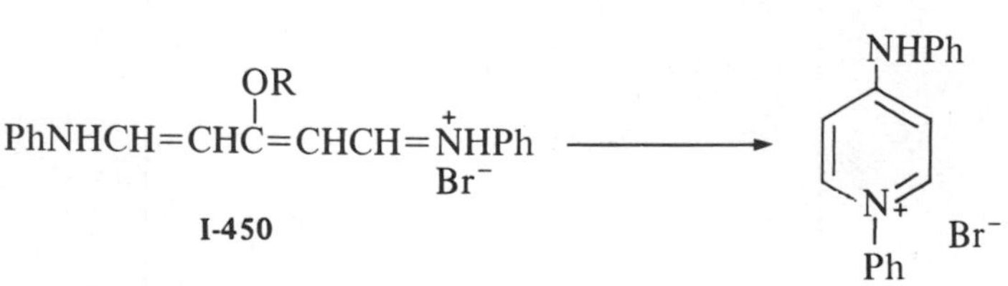

$$PhNHCH=CHC=CHCH=\overset{+}{N}HPh \quad Br^-$$

I-450

TABLE I-65.   PREPARATION OF 4-PYRIDONES FROM 1,3,5-TRIOXO COMPOUNDS AND DERIVATIVES

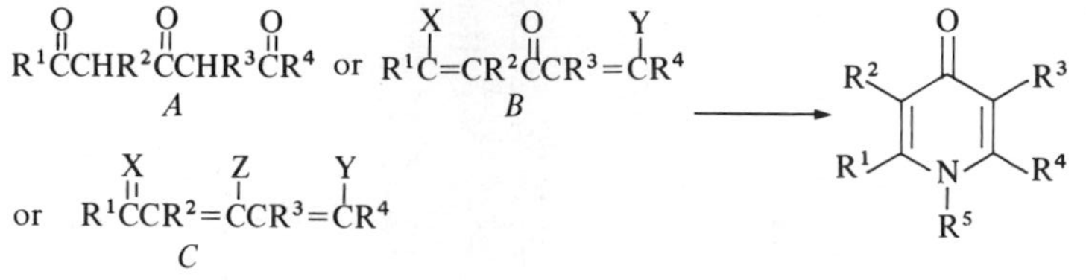

Reactant	R¹	R²	R³	R⁴	R⁵	X	Y	Z	Yield	Reference
A	H	Ar	Ar′	H	Alkyl	—	—	—	—	1037
B	H	Ph	Ph	H	Ar	ArNH	ArNH			1038
A	CF₃	H	H	CF₃	H	—	—	—		1039
A	Ph	Me	Me	Me	H	—	—		86	1040
C	H	H	CO₂Et	H	H	NMe₂⁺	NH₂	Cl	54	1041
B	H	Ar	Ar	H	R	ONa	ONa	—	—	1042
B	Me	H	H	Ph	Ar	ArNH	OH	—	89–95	1043

I-451

a R = H, X = OH, Y = Ph
b R = Me, XY = O

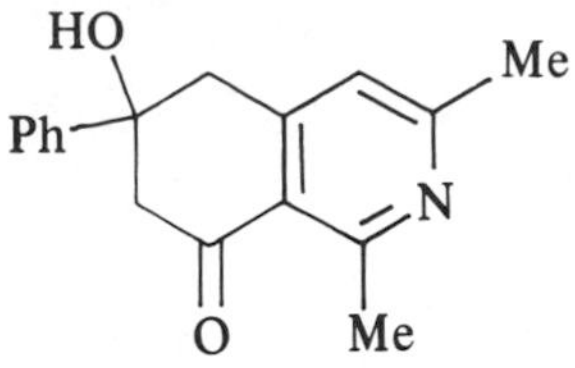

I-452

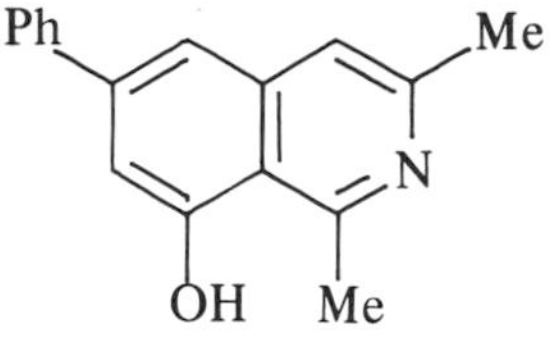

I-453

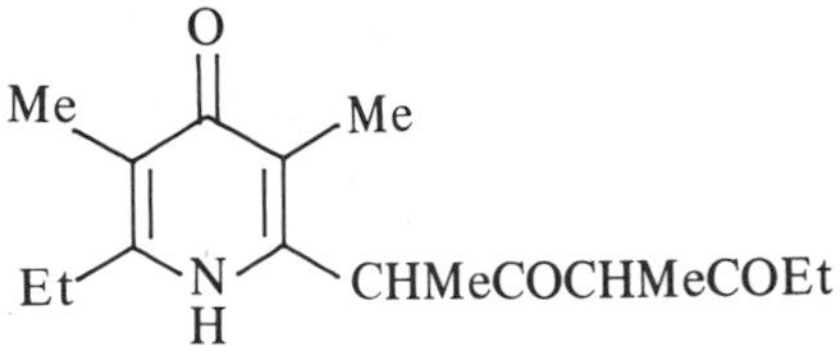

I-454

The polyfunctional **I-451a** underwent cyclization in dilute acid to isoquinoline derivatives **I-452** (46.6%) and **I-453** (18.1%), while its more hindered derivative **I-451b** gave pyridone **I-454** (1046).

b.   OXOCARBOXYLIC ACIDS AND DERIVATIVES

Ring-closure of unsaturated oxo-acids **I-455** or related derivatives gives 2-pyridones **I-456** or related compounds such as 2-amino- or 2-halo-pyridines (Table I-66). The 5-carbon unit is frequently constructed by reactions similar to the 3-2 or 4-1 condensations; therefore many related syntheses are classified under those headings.

I-455   I-456

Thus malonaldehyde acetal **I-457** condensed with methyl cyanoacetate (zinc chloride, acetic anhydride) to give **I-458** (90%, as a single stereoisomer), which was cyclized by hydrogen bromide in acetic acid to give methyl 2-bromonicotinate (**I-459**, 95%); construction of the 5-carbon unit is related to the 3-2 condensation. Malononitrile reacted similarly to give 2-bromonicotinonitrile (1066).

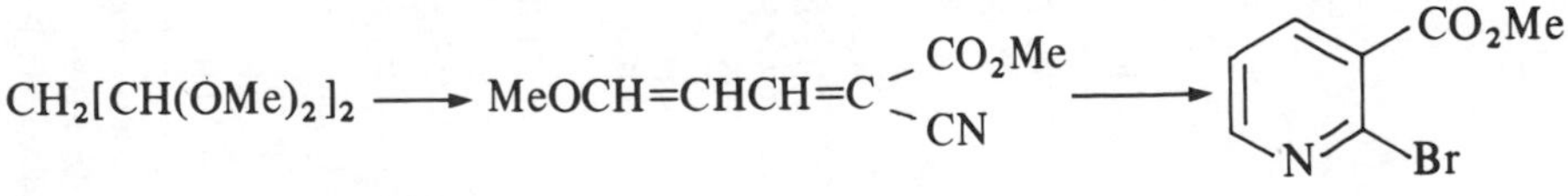

$CH_2[CH(OMe)_2]_2 \longrightarrow MeOCH=CHCH=C \begin{smallmatrix} CO_2Me \\ CN \end{smallmatrix} \longrightarrow$

I-457   I-458   I-459

TABLE I-66.   CYCLIZATION OF OXO-ACID DERIVATIVES

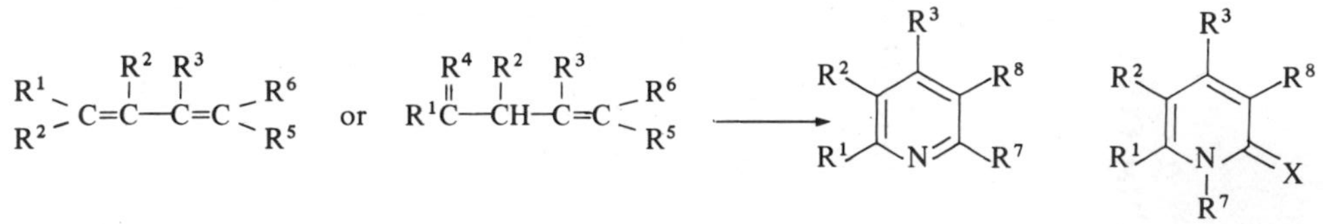

Reactant	R¹	R²	R³	R⁴	R⁵	R⁶	R⁷	R⁸	X	Product	Yield	Reference
A	Me, Ph	H	Me	NMeNMe₂	H	CN	H	H	O	D	52–68	1047
A	H, Me, Ph	H	Me	NR₂	H	CN	NR₂	H	–	C	–	1048
B	Me, Bu, Ph	H	Me	O	H	CN	H	H	O	D	10–55	1049
A	—(CH₂)₄—		H	N(CH₂)₄	CN	CN	Cl, Br	CN	–	C	–	1050
A	Alkyl	Alkyl	H	N(CH₂)₄	CN	CN	NH₂	CN	–	C	–	1051
A	H, Me	H, Me	H	RO	CO₂Et	CO₂Et	Ar	CO₂Et	O	D	67–100	1052
A	ArCH=CH—	H	H	NR₂	CONH₂	CN	H	CN	O	D	82–100	1053
B	R	R	R	O	R	CO₂R	OH	R	O	D	34–50	1054, 1055
A	H	H	H	RNH	CO₂Et	CO₂Et	R	CO₂H	O	D	–	1056
A	R	CO₂Et	H	RNH	H	CO₂Et	R	H	O	D	66–88	1057
B	—CH₂CMe₂CH₂CO—		SMe	O	CN	CO₂Me	Me	H	O	D	–	1058
B	—CH₂CMe₂CH₂CO—		SMe	O	CN	CO₂Me				Dᵃ	–	1058
B	H	H	OEt	(OEt)₂	H	CO₂Et				Dᵇ	93	1059
A	Ar	H	H	NHMe	CN	CN, CONH₂, CO₂R				ᶜ	87–100	1060, 1061
A	R	R	SMe	NR₂	CN	CN	H	CN	O	D		
							Cl, Br	CN	–	D		
							NH₂	CN	–	C	–	1062
A	H, Ar	H	CO₂Et	NMe₂	CO₂R	CO₂R	PhCH₂	CO₂R	O	D	26–90	1063
A	H	H, R	H	OR	CN	CN	NH₂	CN	–	C	–	1064, 1065

ᵃ Product is 1-methyl-3-cyano-4-methylamino-7,7-dimethyl-7,8-dihydro-1H,6H-quinoline-2,5-dione.

ᵇ Product is 4-amino-2-pyridone.

ᶜ Products are 1-methyl-2-aminopyridinium salts.

Alternatively, the same compound can be constructed by a variation of the 4-1 condensation. Ethyl ethylidenecyanoacetate **(I-460)** was condensed with the dimethylformamide acetal **I-461** to give enamine **I-462** (81%), which was cyclized to **I-463** in 95% yield (1067). The varied choice of starting materials in these two sequences allows preparation of nicotinic acid derivatives isotopically labeled in specific positions.

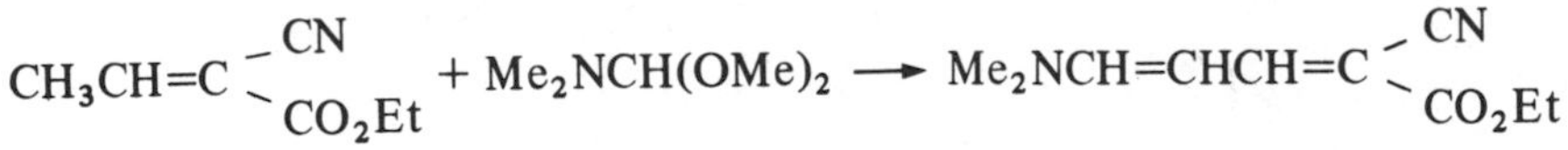

$$CH_3CH=C \overset{CN}{\underset{CO_2Et}{}} + Me_2NCH(OMe)_2 \longrightarrow Me_2NCH=CHCH=C \overset{CN}{\underset{CO_2Et}{}}$$

I-460	I-461	I-462

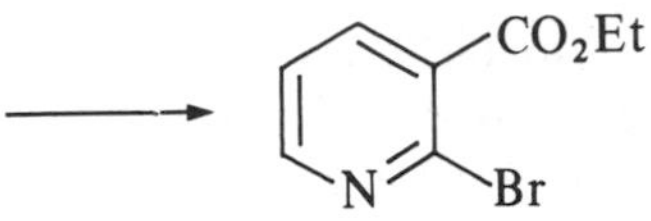

I-463

The pyrrolidine enamine of cyclohexanone **(I-464)** reacted with methoxymethylenemalononitrile to give **I-465** (80%), which was cyclized by ammonium hydroxide in tetrahydrofuran to afford **I-466** in 94% yield (1068). Since the 3-carbon unit is prepared from malononitrile and methyl orthoformate, the synthesis is related to the 2-1-2 condensation.

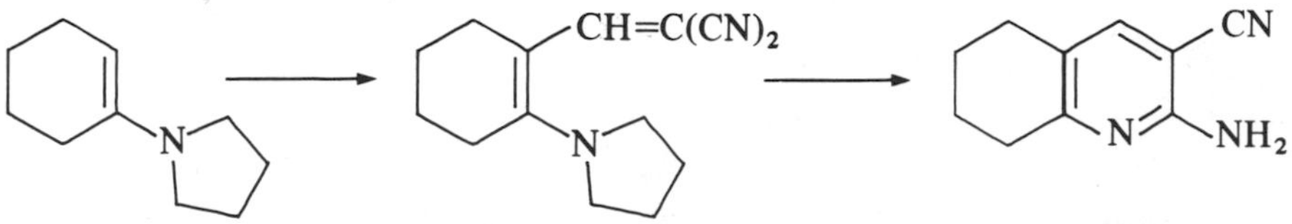

I-464	I-465	I-466

A pyridine ring can be constructed fused to another heterocyclic ring which is later destroyed, as shown in the acid-catalyzed cyclization of **I-467** to **I-468**, followed by hydrogenolysis of the isoxazole ring to give **I-469** (1069, 1070).

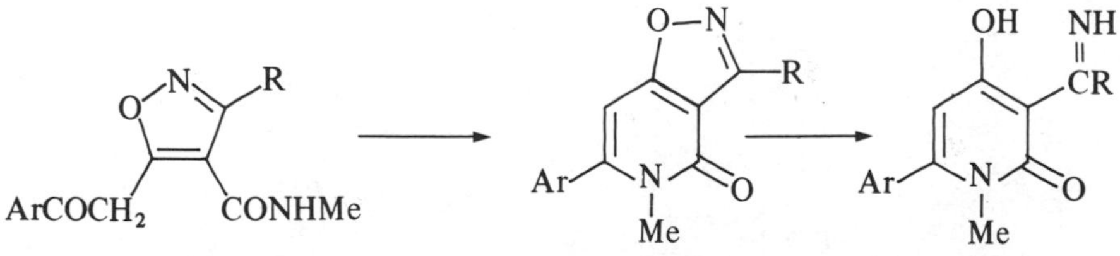

I-467	I-468	I-469

Cyclization of **I-470** in 50% sulfuric acid gave pyridone **I-471** in 94% yield (1071).

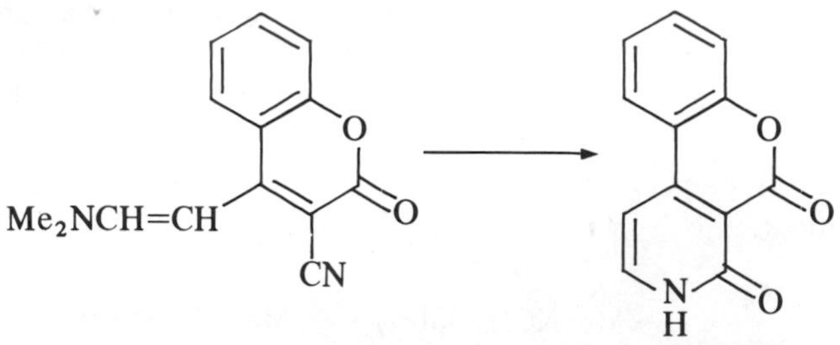

I-470   I-471

Addition of diethylamine to **I-472** gave enamine **I-473**; the latter cyclized in aqueous acetic acid to give a mixture of pyrone **I-474** (50%) and pyridone **I-475** (15%), but to **I-475** alone (80–90%) in aqueous dimethylamine (1072, 1073) (cf. Section II2.A.d).

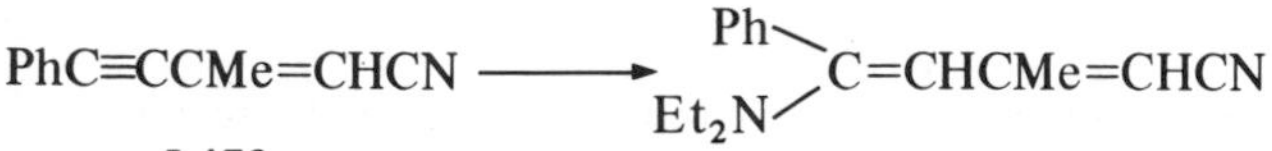

$$PhC{\equiv}CCMe{=}CHCN \longrightarrow \underset{Et_2N}{\overset{Ph}{}}C{=}CHCMe{=}CHCN$$

I-472   I-473

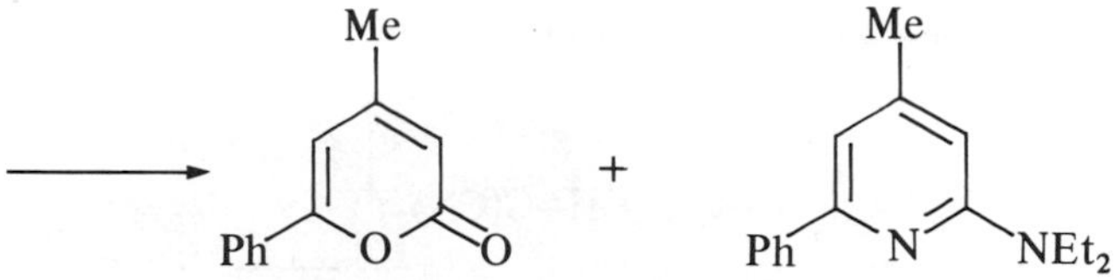

I-474   I-475

A carbethoxy group was lost during the cyclization of **I-476** to **I-477** (90%) using 80% sulfuric acid (1074). Similarly, **I-478** was decarboxylated during cyclization to **I-479** in 95% sulfuric acid (1075).

$$\underset{\underset{CO_2Et}{|}}{PhCOCH{=}CHCCN} \longrightarrow$$

I-476

I-477

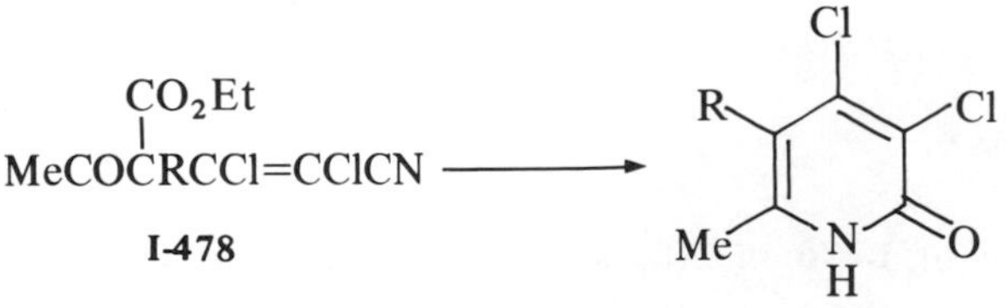

$$\underset{I-478}{\overset{\overset{CO_2Et}{|}}{MeCOCRCCl{=}CClCN}} \longrightarrow$$

I-479

Treatment of 4-phenylpyrimidine *N*-oxide (**I-480**) with acetic anhydride and ethyl cyanoacetate gave a mixture of ring-opened compounds (56%), which underwent solvolysis in ethanolic sulfuric acid to generate **I-481** (82%). The latter compound was cyclized by heating in ethanolic sulfuric acid to the aminopyridine **I-482**, but in ethanolic sodium hydroxide to the pyridone **I-483** (1076). The ring-opening and cyclization reactions are relevant to the one-step synthesis of pyridines from pyrimidines.

I-480 **I-481** **I-482** **I-483**

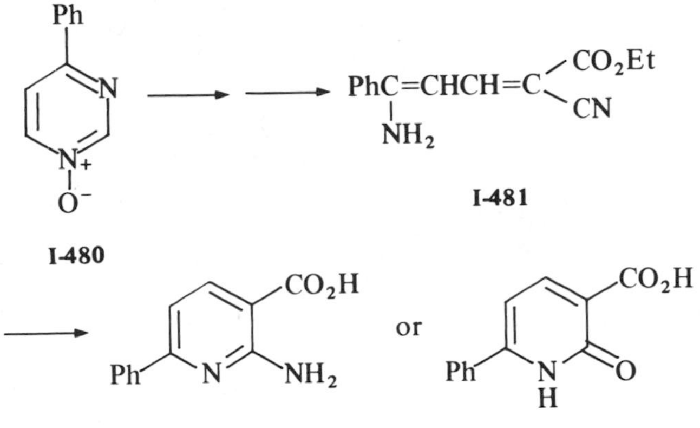

Treatment of **I-484** with hydrazine gave the pyridine **I-485** in 36% yield, accompanied by 6.2% of the pyrazole **I-486** (1077).

I-484 **I-485** **I-486**

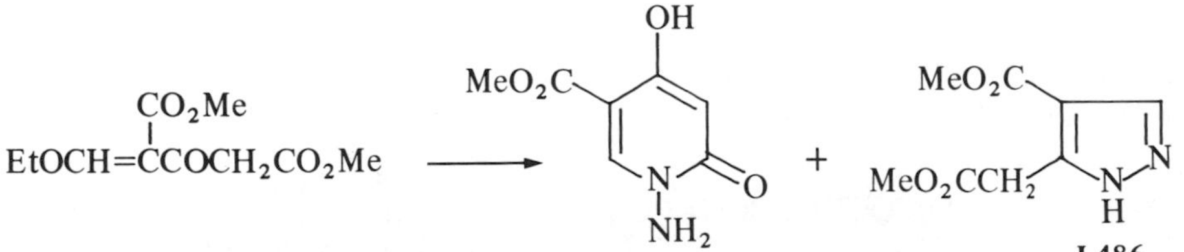

While additional unsaturation in the 5-carbon unit as shown in **I-455** is usually necessary for formation of the fully aromatized pyridine, **I-487** was formed from **I-488** (57%) on heating in hydrochloric and acetic acids mixture, with migration of the exocyclic unsaturation into the pyridine ring and decarboxylation (1078).

I-488 **I-487**

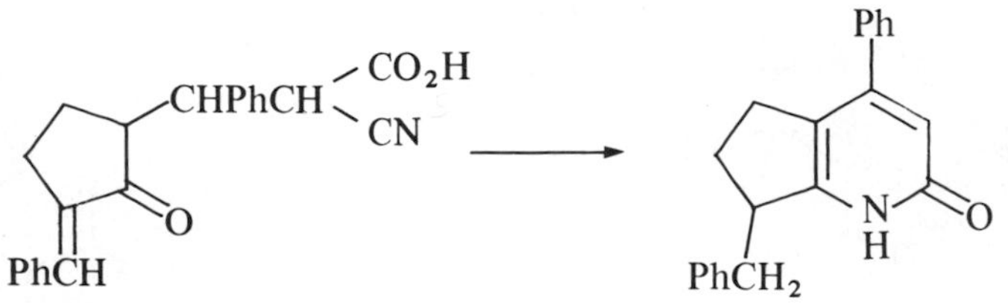

On standing in methanol, **I-489** cyclized to the expected pyridone **I-490** and aldol product **I-491**; the former predominated in aqueous methanol (48–57%), and the latter in anhydrous methanol (59–65%). Crossover experiments established that the aldol condensation occurred between the methyl group of **I-490** and the ketone carbonyl of **I-489** (1079, 1080).

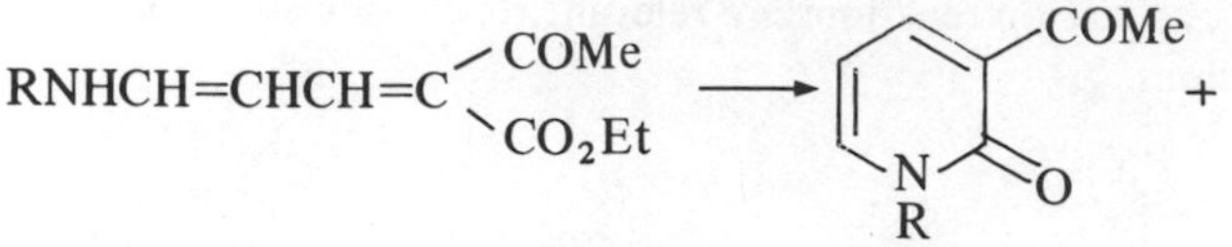

$$RNHCH=CHCH=C \overset{COMe}{\underset{CO_2Et}{\big\langle}} \longrightarrow$$

I-489   **I-490**   +

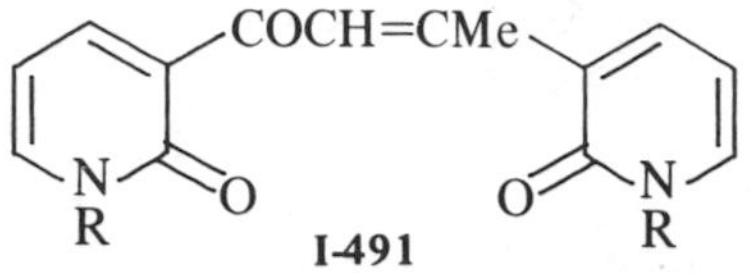

I-491

Irradiation of **I-492**, prepared *in situ* from **I-493** in aqueous ammonia, gave 6-aminonicotinonitrile (**I-494**); these prebiotic conditions are similar to those used for the synthesis of purines from diaminomaleonitrile (hydrogen cyanide tetramer). **I-493** was prepared by hydrolysis of cyanoacetylene and aldol condensation (1081).

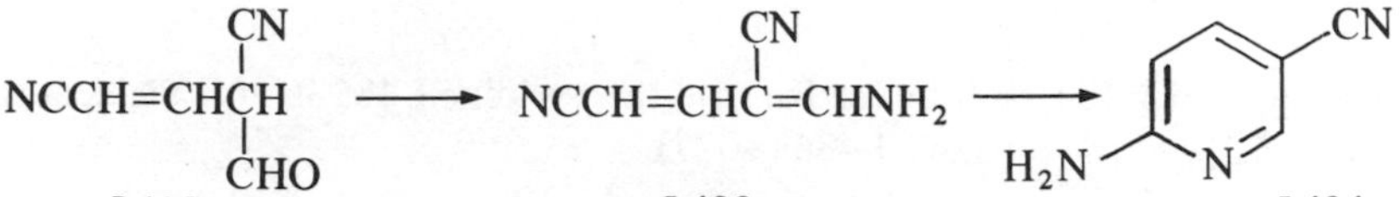

$$NCCH=CH\overset{CN}{\underset{CHO}{C}}H \longrightarrow NCCH=CH\overset{CN}{C}=CHNH_2 \longrightarrow$$

I-493   **I-492**   **I-494**

The *gem*-dichloride **I-495** cyclized to **I-496** (57%) in the presence of formamide and sodium bicarbonate (1082).

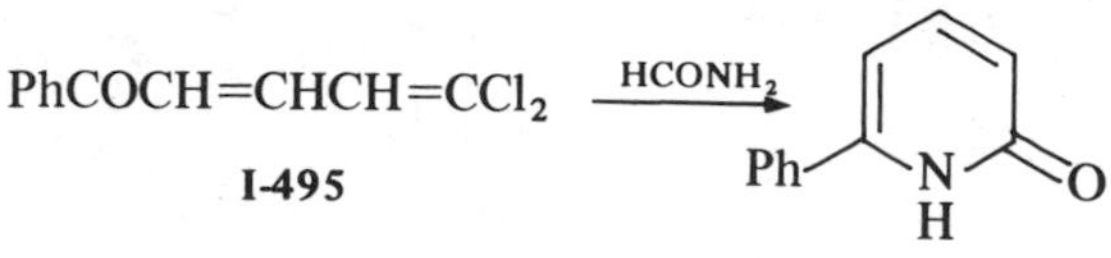

$$PhCOCH=CHCH=CCl_2 \xrightarrow{HCONH_2}$$

I-495

I-496

In the condensed ring series, **I-497** formed the 1,8-naphthyridine **I-498** on treatment with hydrochloric acid (1083, 1984). Similarly, the enamine **I-499** formed the 2,7-naphthyridine **I-500** (1085); other related compounds were prepared in a similar manner.

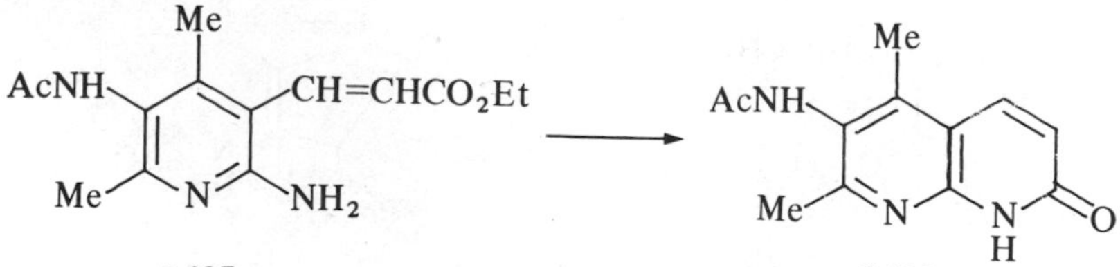

I-497   **I-498**

I-499   I-500

A reaction of commercial importance is the preparation of 2-picoline by cyclization of the ketonitrile **I-501**. The intermediate is prepared by cyanoethylation of acetone, using a large excess of acetone to reduce dicyanoethylation and an amine salt as catalyst; the reported yields of **I-501** are 83–91% based on 56–83% conversion of acrylonitrile, and 71–85% based on acetone converted (1086). Earlier work (1087) showed that imines could be more cleanly monocyanoethylated, albeit at higher temperatures, than carbonyl compounds. Basic by-products are removed from **I-501** (1088), and cyclization is performed in the vapor phase in the presence of hydrogen at 230° over a palladium-containing catalyst to give 2-picoline in 73–86% yield, accompanied by 5–21% of 2-methylpiperidine (1089). When nickel catalysts are used, 2-methylpiperidine and 2-methyltetrahydropyridine are major products. Earlier work described the preparation of pyridines and quinolines by similar methods in lower yields (1090).

$$Me_2CO + CH_2{=}CHCN \longrightarrow MeCO(CH_2)_3CN \longrightarrow$$

I-501

By similar reactions, cyanoethylation of methyl ethyl ketone (1091–1093) followed by cyclization gave 2,3-lutidine (1089); cyanoethylation of cyclohexanone followed by cyclization gave a mixture of partially reduced quinolines (1090) which were further dehydrogenated by a second pass over the same catalyst at a higher temperature to give quinoline in high yield (1094).

$$CH_3CH{=}NR \longrightarrow NC(CH_2)_3CH{=}NR \longrightarrow$$

I-502   I-503

In a related series of reactions, an imine of acetaldehyde (**I-502**) was cyanoethylated (1095, 1096) and the resulting imine (**I-503**), or the aldehyde obtained from it by hydrolysis, was hydrogenated in the presence of ammonia to give piperidine (1097–1100).

$$CH_3CO(CH_2)_3CONH_2 \longrightarrow$$

I-504   I-505

Vapor phase reaction of **I-504** over a catalyst containing palladium gave the pyridone **I-505** in 80% yield at 50% conversion (1101).

TABLE I-67.   CYCLIZATION OF CYANO ESTERS

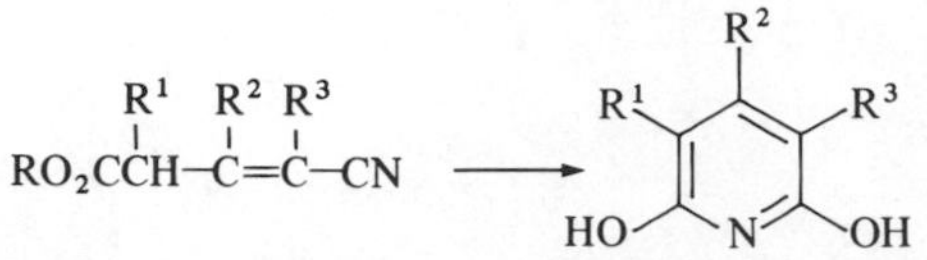

R	R^1	R^2	R^3	Conditions	Yield	Reference
Et	H	Me	CN	$HClO_4$	56	1102
Et	H	Me	CN	H_2O	10	1103
Et	H	Ph	CN	$HClO_4$	19.5	1102
Et	$-(CH_2)_3-$		CN	H_2O	46	1103

c.   1,5-DICARBOXYLIC ACIDS AND DERIVATIVES

Glutaconic acid derivatives typically cyclize with ease to the imides, here shown as 2,6-pyridinediols. Details of the cyclization of cyano esters (Table I-67) have come from work leading to 5,6,7,8-tetrahydroisoquinoline-1,3-diols.

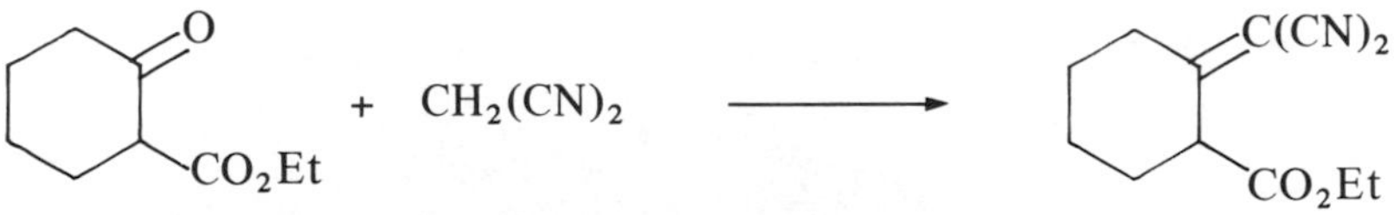

Knoevenagel condensation of carbethoxycyclohexanone I-506 with malononitrile I-507 (ammonium acetate, acetic acid, benzene, with continuous separation of water) gave I-508, the expected product (1104). When ethanol comprised part of the solvent, and the reaction was stopped after 0.5 hr, I-508 was isolated in 80% yield. However, carrying out the condensation for 24 hr under the latter conditions, or treating I-508 similarly, or warming I-508 in ethanol, gave the pyridine I-509 (1103–1105). Warming I-508 in water gave I-510 (1103). A long-wavelength band in the ultraviolet spectrum of I-508 in hydroxylic solvents was attributed to the tautomer I-511, which was postulated as an intermediate in the cyclization (1103).

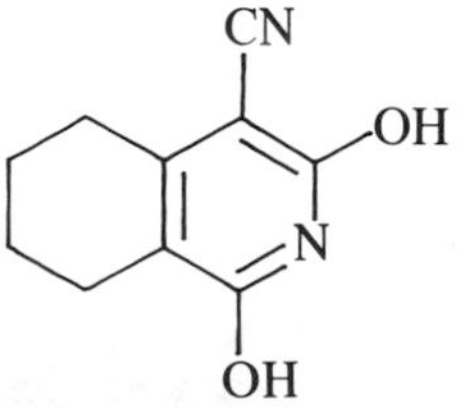

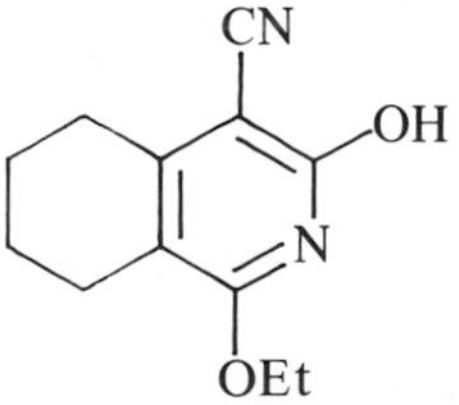

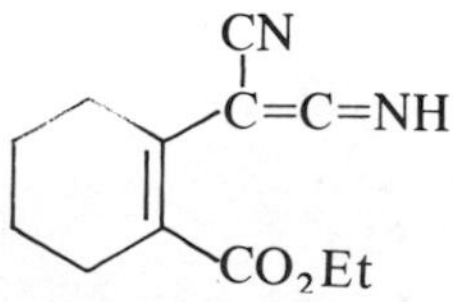

Treatment of the related intermediate **I-512** with hydrogen chloride in ethanol, or aqueous sodium hydroxide followed by acidification, also gave **I-510**. However, reaction in concentrated sulfuric acid gave an isomer for which the tautomeric structure **I-513** was proposed (1106).

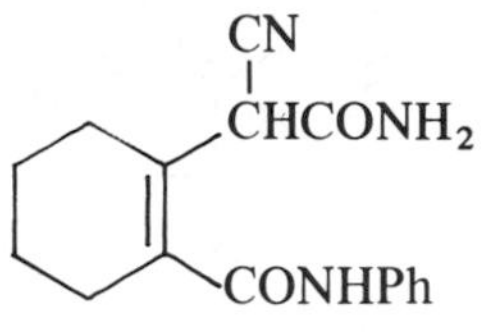

I-512 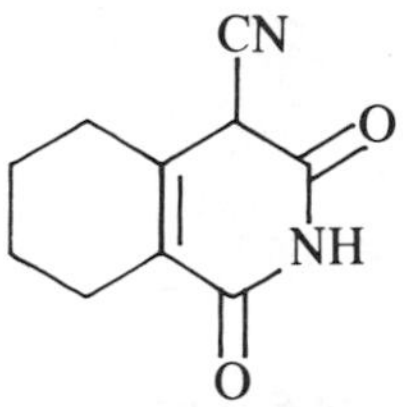   I-513

For preparative purposes, the 3-2 condensation of **I-506** with **I-507** using methanolic potassium hydroxide followed by acidification (1107) gave **I-510** in 88.4% isolated yield (1108).

Treatment of **I-514** (R = Et) or the crude condensation product of **I-506** with cyanoacetic acid (**I-514**, R = H, and its decarboxylation product) with concentrated sulfuric acid gave **I-515** in 83–87% and 90% yield, respectively (1109–1111).

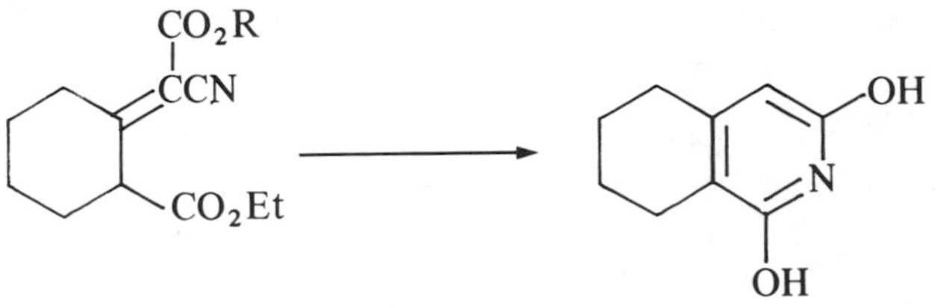

I-514   I-515

Heating **I-516** in 50% sulfuric acid gave **I-517** in 26% yield, accompanied by 63% of **I-518** (1112).

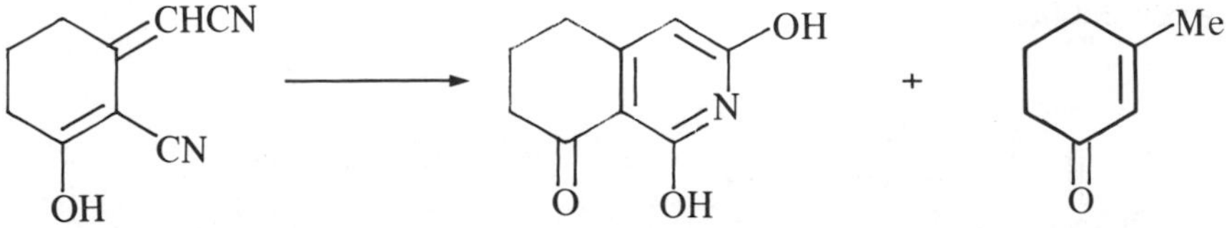

I-516   I-517   I-518

Knoevenagel condensation of diethyl acetonedicarboxylate with malononitrile gave an intermediate presumed to be **I-519**, which cyclized to **I-520** in 77% yield overall on treatment with perchloric acid (1102).

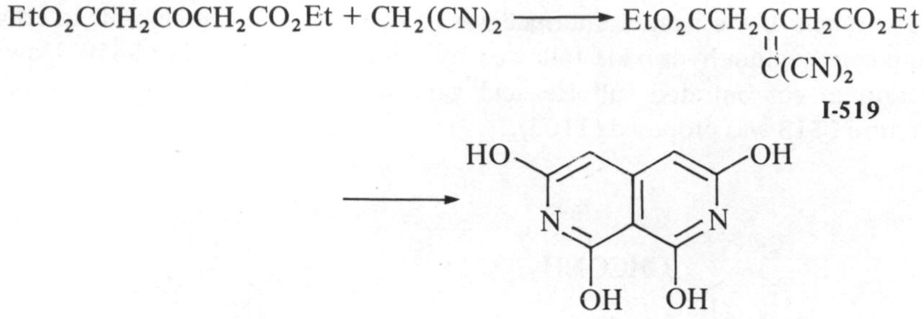

$$EtO_2CCH_2COCH_2CO_2Et + CH_2(CN)_2 \longrightarrow EtO_2CCH_2CCH_2CO_2Et$$

$$\overset{\|}{C}(CN)_2$$

I-519

I-520

Treatment of **I-521** with ammonia at 110° gave only a 9% yield of **I-522** (1077).

Closure of **I-523** to **I-524** (81–87%, R = H, Me, Et) occurred on treatment with concentrated hydrochloric acid; the intermediates were prepared from malononitrile and ortho esters (45–68%). Alternatively, the preparation of **I-524** (R = H) could be carried out without isolation of **I-523** in 75% yield (1113) (cf. Sections II.2.C.b and II.2.E.B.ii).

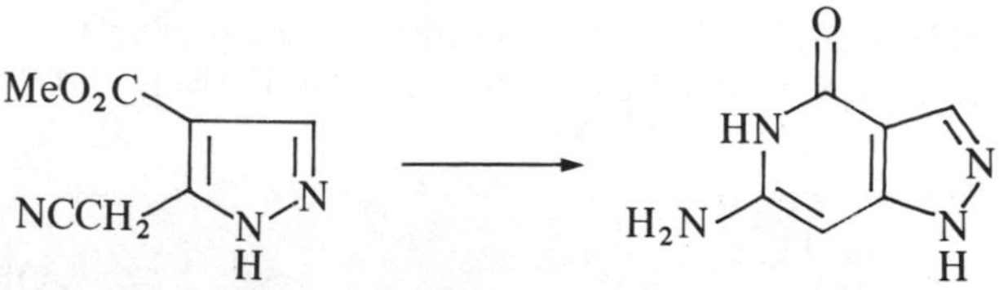

I-521　　　　　　　　　　**I-522**

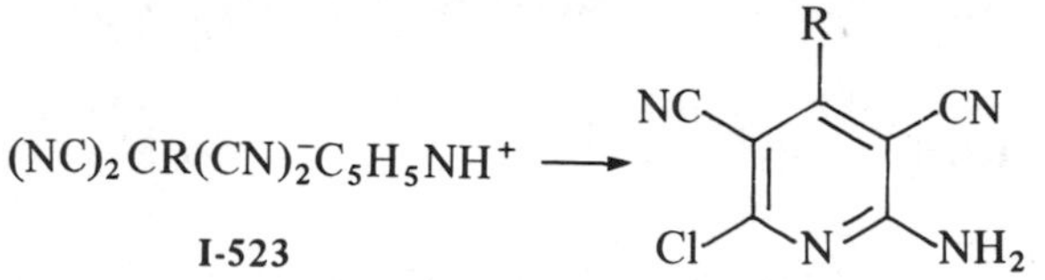

$$(NC)_2CR(CN)_2^-C_5H_5NH^+ \longrightarrow$$

I-523

I-524

Similarly, cyclization of **I-525** to **I-526** using anhydrous hydrogen bromide was carried out in 79.5% yield (1114).

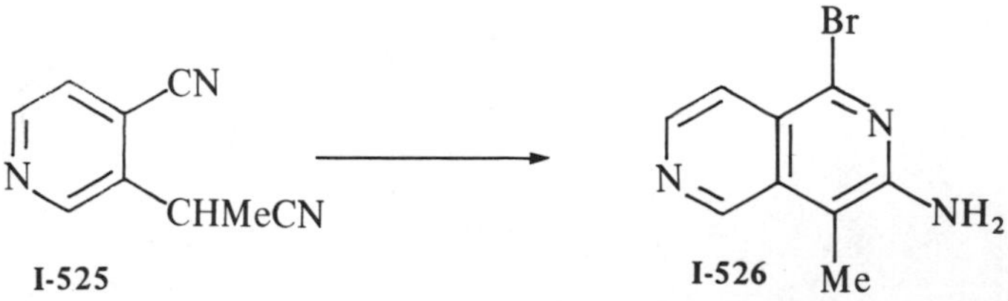

I-525　　　　　　　　　　**I-526**　Me

Treatment of **I-523** (as the potassium salt) with a trichloromethylamine gave **I-527**, which cyclized to **I-524** (R = H, 62%) with hydrochloric acid (1115).

$$\underset{\text{I-527}}{(NC)_2C=CHC=CClN=CClNR_2}$$
with CN above the central carbon.

Dimethylcrotonamide **I-528** was alkylated with the phosgeniminium salt **I-529** to form the cyanine **I-530** (95%), which gave the pyridineimine **I-531** (80%) on treatment with aniline (1116).

$$CH_3CH=CHCONMe_2 + Me_2\overset{+}{N}=CCl_2Cl^- \longrightarrow Me_2\overset{+}{N}=CClCH=CHCH=CClNMe_2$$

I-528   I-529   Cl⁻   I-530

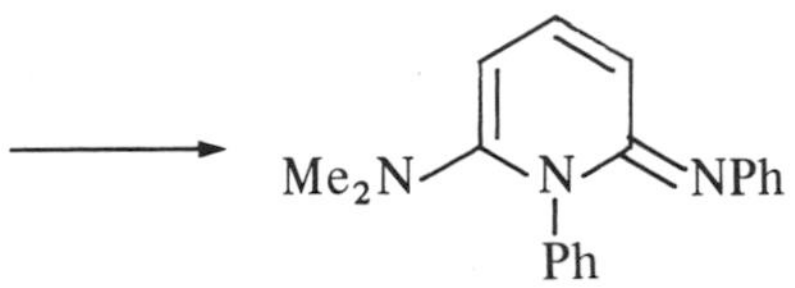

I-531

The acid chlorides **I-532** [R_1, R_2 = Me, H; Ph, H; $(CH_2)_3$; $(CH_2)_4$] cyclized to form **I-533** in 35–63% yield on heating with hydrogen chloride in dibutyl ether (1117).

$$NCCH=CR^1CHR^2COCl \longrightarrow$$
I-532

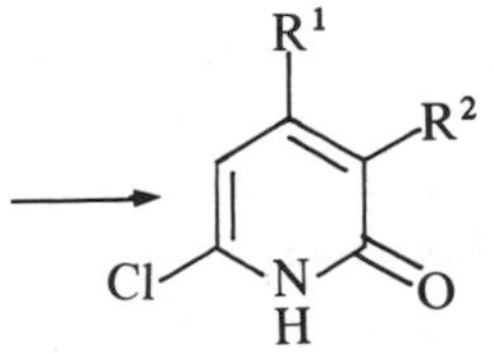

I-533

Vapor-phase chlorination of glutaronitrile gave pentachloropyridine in 40% yield (1118).

Acylation of **I-534** with methyl isocyanate gave amide **I-535** (48%), which cyclized at 170° (89%) to **I-536**. Isothiocyanates behaved similarly (1119).

$$(Me_2N)_2C=C-C=CH_2 \longrightarrow (Me_2N)_2C=C-C=CHCONHMe$$
with CN and OMe above the carbons.

I-534   I-535

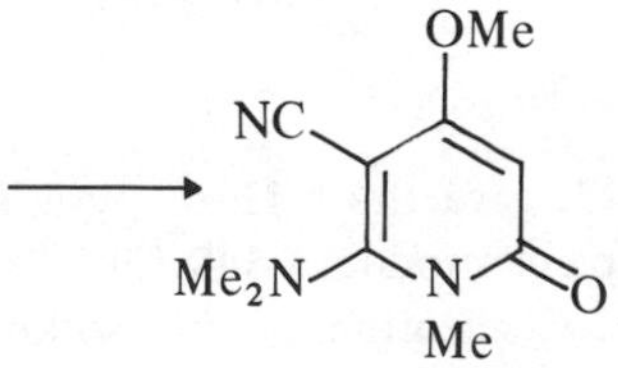

I-536

d. COMPOUNDS HAVING TERMINAL UNSATURATION

Amides **I-537** (R = H, Me, Ph) were cyclized to the pyridones **I-538** by a stoichiometric quantity of lithium tetrachloropalladate in 60–65% yield (1120, 1121). Similarly, the *o*-allyl- (**I-539**, R = H, Me) and *o*-vinylbenzamides (**I-540**, R′ = Ar′, PhCO, CN) were cyclized to the isocarbostyrils **I-541** in 37–91% yield (1122, 1123).

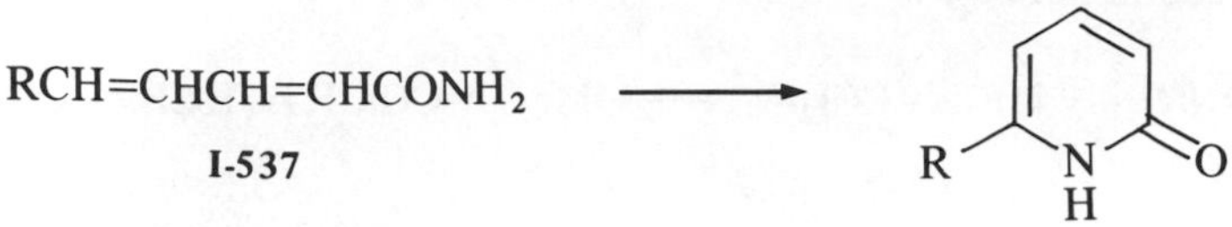
RCH=CHCH=CHCONH₂
R
N
H
O

I-537 **I-538**

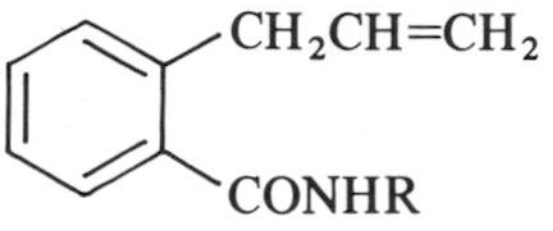
CH₂CH=CH₂
CONHR

I-539

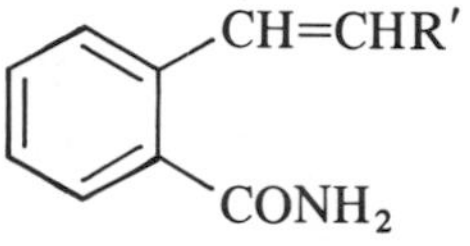
CH=CHR′
CONH₂

I-540

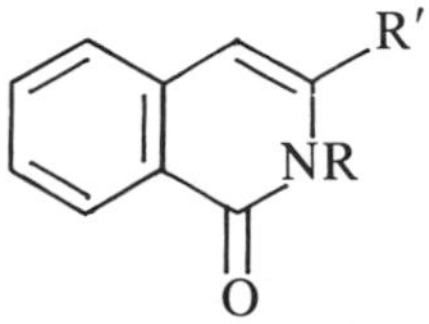
R′
NR
O

I-541

Oximes **I-542**, **I-543**, or **I-544** were cyclized by a stoichiometric amount of $PdCl_2(PPh_3)_2$ in the presence of sodium phenoxide to give the pyridines **I-545** in fair yields. However, with sodium carbonate as the base, isoxazoles **I-546** were formed from **I-542** or **I-543** (1124, 1125).

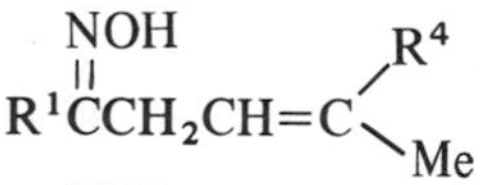
NOH
R¹CCH₂CH=C
R⁴
Me

I-542

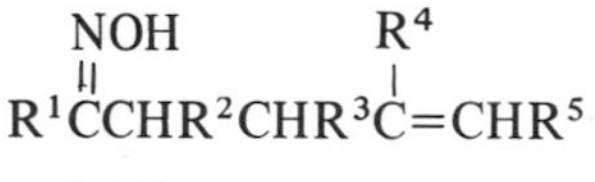
NOH
R⁴
R¹CCHR²CHR³C=CHR⁵

I-543

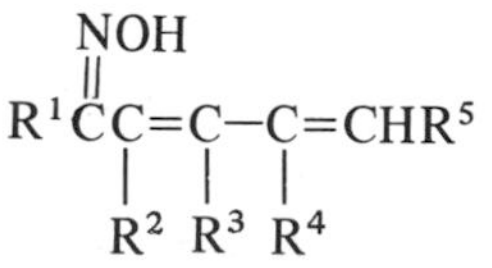
NOH
R¹CC=C−C=CHR⁵
R² R³ R⁴

I-544

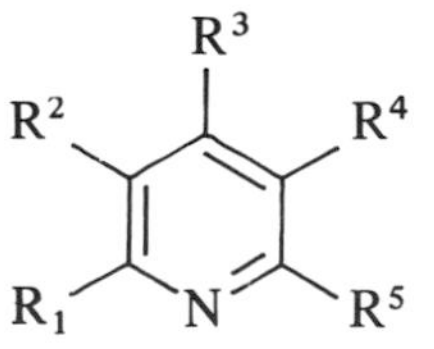
R³
R²
R⁴
R₁
N
R⁵

I-545

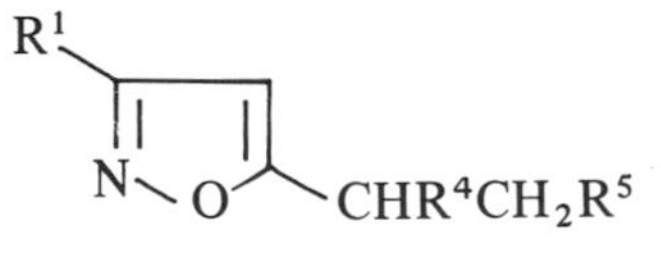
R¹
N
O
CHR⁴CH₂R⁵

I-546

A structurally related oxime, **I-547**, gave the bridged pyridine **I-548** (25%) in addition to the Beckmann rearrangement product **I-549** (20%) on treatment with phosphorus oxychloride in pyridine. Cyclization of the Beckmann intermediate **I-550** was postulated (1126).

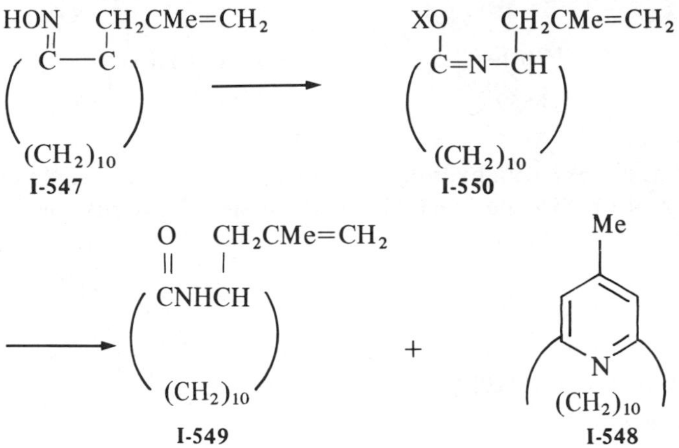

HON CH$_2$CMe=CH$_2$

XO CH$_2$CMe=CH$_2$

I-547

I-550

O CH$_2$CMe=CH$_2$

I-549

+

I-548

Reaction of **I-551** (R = Me, Ph, *t*-Bu) with ammonia or secondary amines gave aminopyridines **I-552**, or the pyridoneimines **I-553** (R = Me, Et, Bu) with saturated primary amines (1127–1130) (cf. Section II.2.A.b). Allylamine gave **I-553** (R^1 = —CH=CHCH$_3$, 54%), while propargylamine gave the pyridoimidazole **I-554** in 47% yield (1131). With 1,1-dimethylhydrazine **I-551** gave **I-555**, which cyclized to **I-556** on hydrolysis (1132).

RC≡CMeC=CHCN
I-551

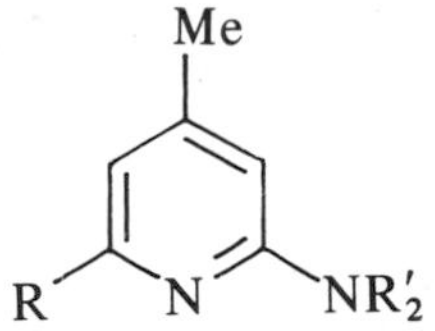

I-552

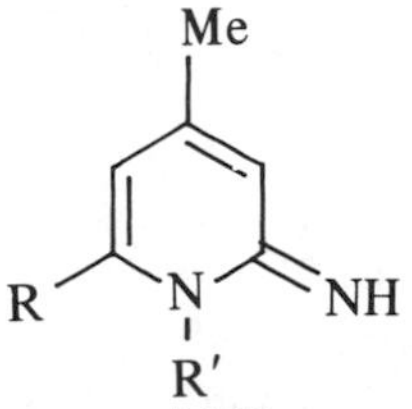

I-553

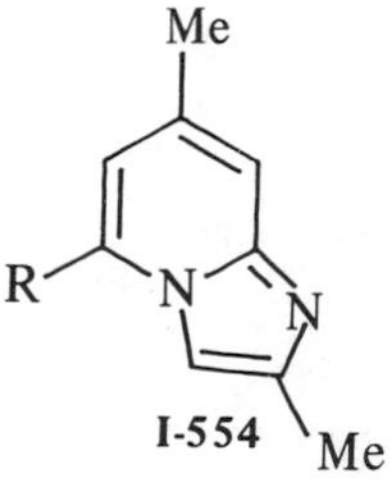

I-554

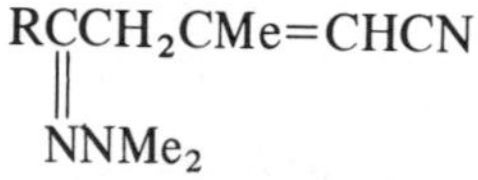

RCCH$_2$CMe=CHCN

NNMe$_2$

I-555

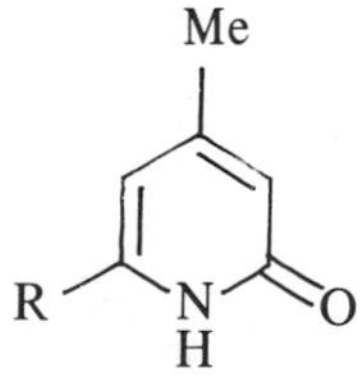

I-556

Treatment of **I-557** with butylamine at 10° gave **I-558** (30%), **I-559** (18%), and **I-560**; at − 10°, **I-560** (58%) was the sole isolated product (1133).

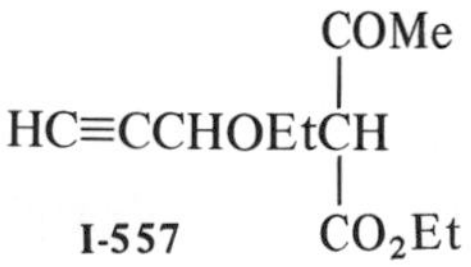

HC≡CCHOEtCH

I-557   CO$_2$Et

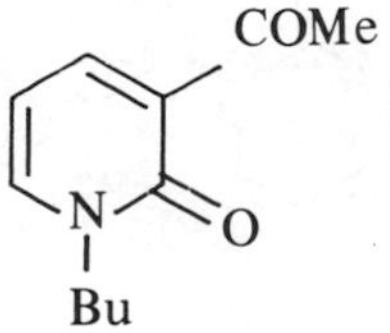

I-558

MeC(NHBu)=CHCO$_2$Et   BuNHCH=CHCH=C(COMe)(CO$_2$Et)

I-559   **I-560**

Treatment of **I-561** with hydroxylamine gave **I-562** in 58% yield (1134).
Reaction of the dienone **I-563** with ammonia in the gas phase gave 2,6-lutidine (1135).

PhC≡CCOCH$_2$COC$_6$H$_4$Ph-p ⟶

I-561

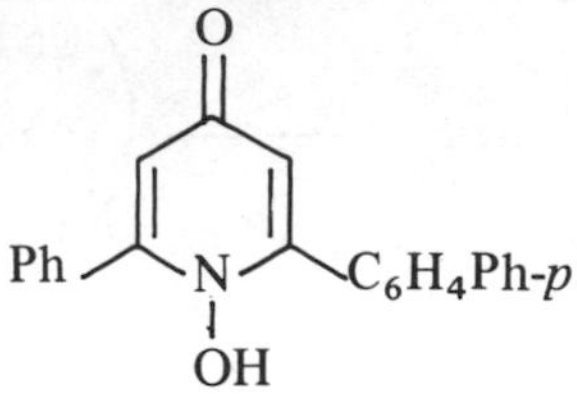

I-562

MeCH=CHCH=CHCOMe ⟶

I-563

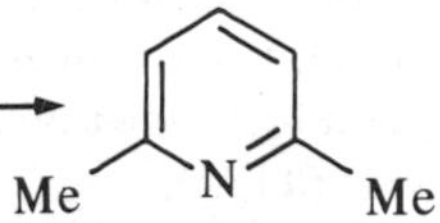

Ammonia or primary amines add to diethynyl ketones **I-564** to give pyridones **I-565**; the intermediates **I-566** are sometimes isolable (1136–1138). A by-product from **I-566** (R^1 = R^2 = R^3 = Ph), originally identified as **I-567** (1139), was reassigned the structure **I-568** (1140). Heating **I-566** (R^1 = R^2 = R^3 = Ph) in xylene gave 45% each of **I-565** and **I-568**, while in Me$_2$SO 80% of **I-568** and only 15% of **I-565** were formed.

R^{1}C≡CCOC≡CR^2 R^{1}C=CHCOC≡CR^2 ($\overset{|}{N}$HR^3)

I-564   **I-566**

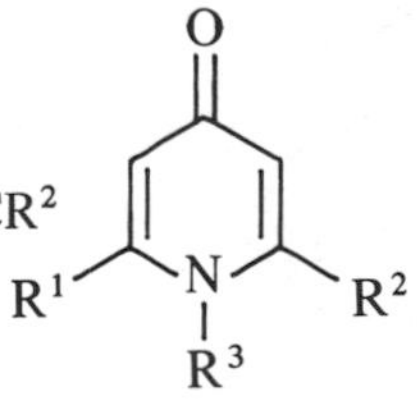

I-565

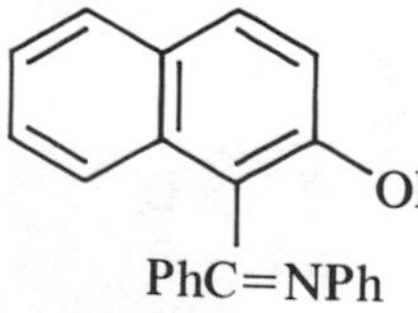

PhC=NPh
I-567

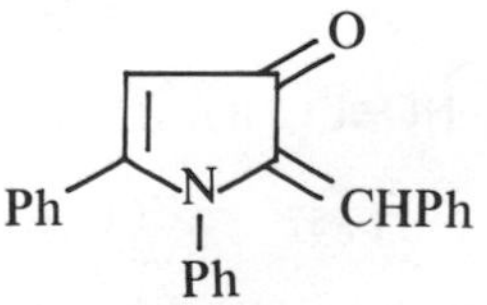

I-568

Cl$_2$C=CClCCl=CClX=CXCHO   PhCCl=CClCCl=CClCHO

I-569   **I-570**

Phenylhydrazones, dinitrophenylhydrazones, and semicarbazones of **I-569** (X = Cl or Br) and **I-570** cyclized to the pyridones **I-571** or **I-572** in high yields (1141, 1142). Hydroxylamine with **I-570** gave the pyridine *N*-oxide **I-573** in 65–70% yield (1142).

I-571   **I-572**   **I-573**

e. MISCELLANEOUS 1,5-BIFUNCTIONAL COMPOUNDS

Oxidation of *n*-hexylamine with iodine in the vapor phase at 500° gave 2-picoline in 33% yield, along with 18% of hexane and lesser amounts of other by-products (1143). Amines and imines with five carbons (*n*-amylamine, *N*-ethyl-*n*-propylamine, *N*-butylidinemethylamine, etc.) gave pyridine (10–30%) under similar conditions, the yield being improved by packing the reactor with calcium oxide to remove the hydrogen iodide, feeding oxygen, or both (1144) (cf. Section II.2.G.b).

Cadaverine (1,5-pentanediamine) gave 41–84% pyridine and 4–56% piperidine in the vapor phase over 0.5% palladium on alumina at 200–260°, with more of the dehydrogenated product at higher temperatures; using 30% nickel on silica at 168–180°, only piperidine was formed (1145). Using copper chromite catalyst at 450°, 1,5-pentanediol with ammonia gave pyridine in 64% yield (1146).

Ammoxidation of 1,3-pentadiene or cyclopentene gave pyridine, acetonitrile, and pentadienenitrile in low yields (1147).

Chlorination of 2-methyleneglutaronitrile **I-574** (produced by dimerization of acrylonitrile) gave dichloride **I-575**, which on cyclization with stannic chloride afforded 3-cyanopyridine (**I-576**) in 77% yield. Alternatively, addition of hydrogen chloride to **I-574** gave a monochloride which could be similarly cyclized to a dihydro derivative of **I-576**. The latter was catalytically dehydrogenated to **I-576** (1148, 1149).

$$\overset{\overset{\displaystyle CH_2}{\|}}{NCCH_2CH_2CCN}$$

I-574

$$\overset{\overset{\displaystyle CH_2Cl}{|}}{NCCH_2CH_2CClCN}$$

I-575

I-576

Acidification of the salts of trinitro compounds **I-577** (R^1, R^2, R^3 = CN, SO_2Me, CO_2Me) gave pyridine *N*-oxides **I-578** in poor to good yields in a reaction shown to be autocatalytic in the nitrous acid liberated. Isoxazoles **I-579** and pyridine **I-580** were by-products (1150, 1151) (cf. Section II.2.E.e).

I-577 **I-578** **I-579** **I-580**

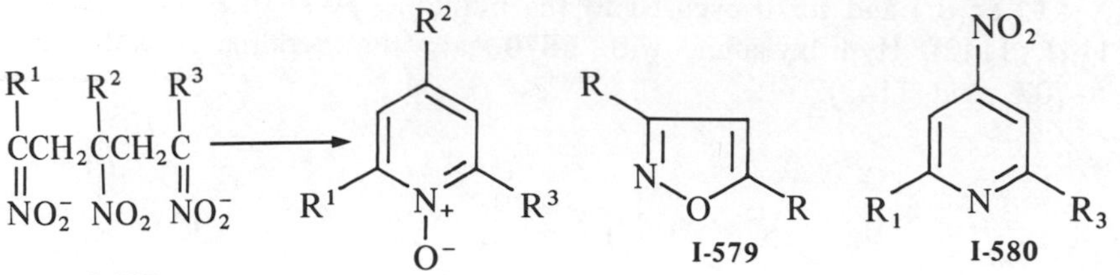

Compound **I-581**, obtained from malononitrile dimer and a dialkylaminoacrolein, was cyclized to **I-582** by treatment with alcoholic potassium hydroxide followed by acidification (1152).

I-581 **I-582**

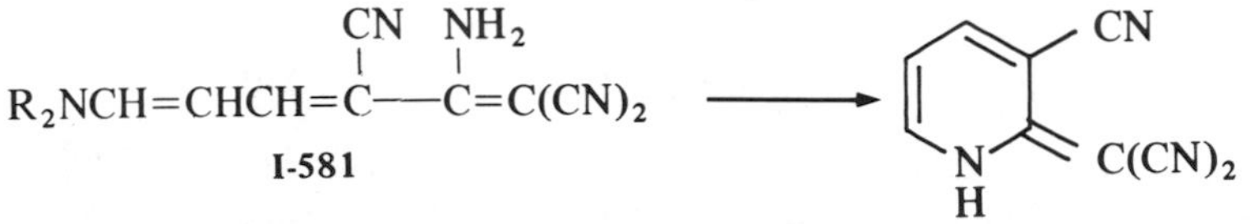

Methylthienopyridines **I-583** and **I-584** were prepared in 71% and 83% yield by cyclization of **I-585** and **I-586**, respectively, in 2 *N* hydrochloric acid (1153).

I-585 **I-583** **I-586** **I-584**

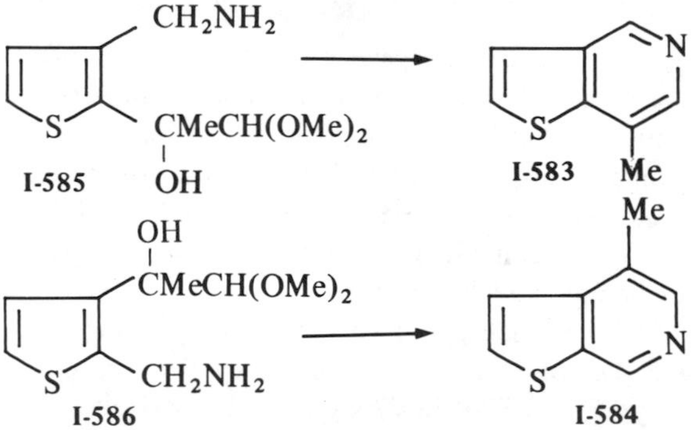

Cyclization of indolepropionamides **I-587** with phosphorus pentachloride gave isocarbolines **I-588** in 26–94% yield (1154).

I-587 **I-588**

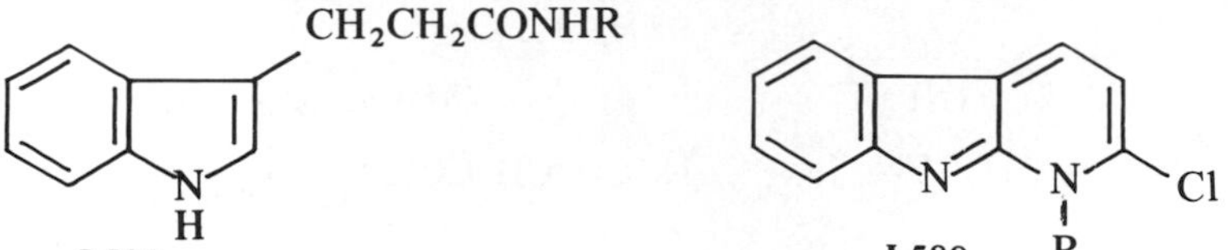

B. 4-1 Condensations

a. DIENES WITH NITRILES

The cobalt metallocycles **I-589** reacted with nitriles to give pyridines **I-590** in 20–72% yield (1155–1157), and with isocyanates to give pyridones **I-591** in 37–87% yield (1158). The cobaltocycles are probably intermediates in the synthesis of pyridines from acetylenes and nitriles (cf. Section II.2.E.a).

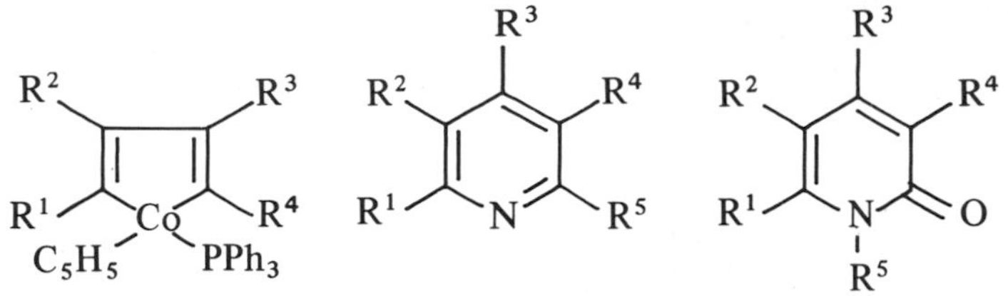

The reaction of butadiene with cyanogen (1159) has been optimized to give 69.4% yield of 2-cyanopyridine at 90.1% conversion of butadiene (1160–1163).

Sulfonyl cyanides **I-592**, particularly tosyl cyanide, were found to react with dienes under relatively mild conditions to give pyridyl sulfones **I-563** (6–38%), accompanied by the dihydropyridones **I-594** (0–55%) formed by hydrolysis of the intermediate dihydropyridines (1164).

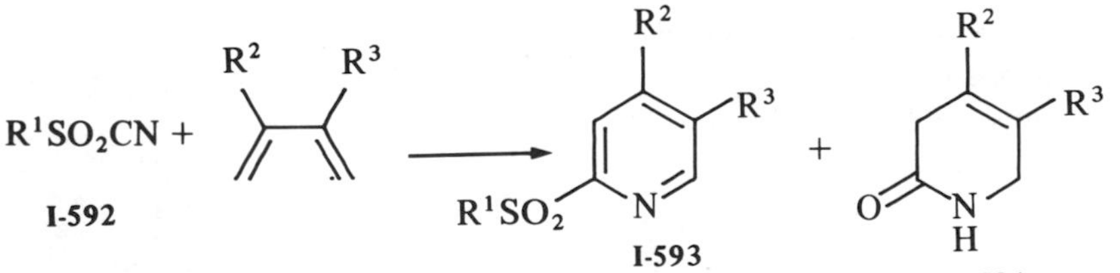

Related reactions of cyclopentadienones and other cyclic dienes are discussed in Section II.1.

Vapor-phase catalytic reactions of butadiene with formaldehyde and ammonia (1165) or methylamine (1166) gave pyridine in 9–44% yield at 15–23% conversion of the butadiene.

Diels–Alder reaction of **I-595** with **I-596** gave as the major regioisomer the tetra-hydropyridine **I-597**, which was hydrolyzed, esterified, and dehydrogenated to give **I-598** (1167).

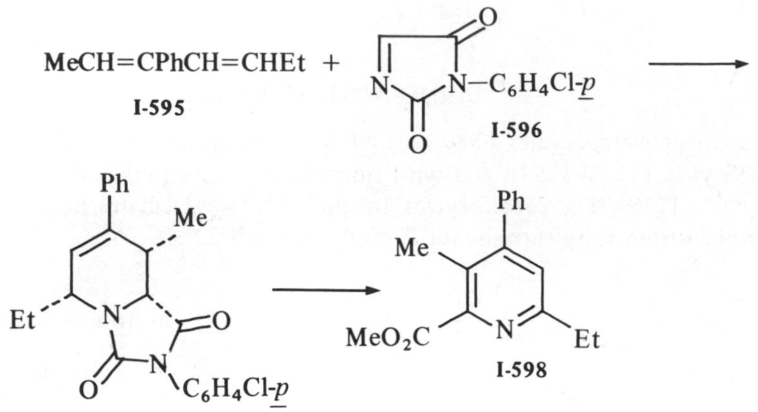

MeCH=CPhCH=CHEt +

I-595

I-596

I-597

I-598

Reaction of tosylisonitrosomalononitrile **I-599** with various dienes gave dihydro-pyridines **I-600**, which on heating gave pyridines **I-601** (1168).

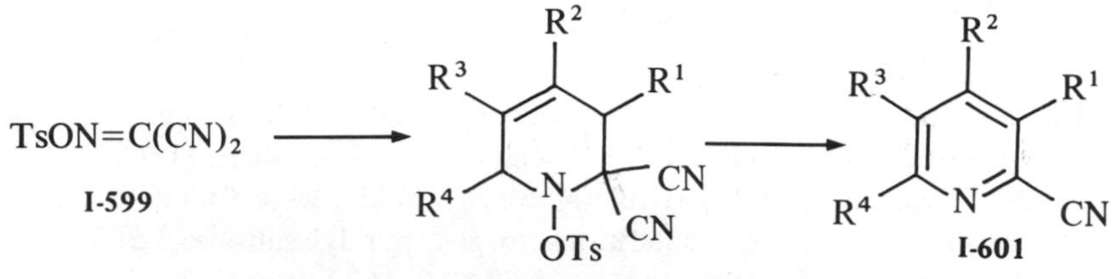

TsON=C(CN)$_2$ ⟶

I-599

I-600

I-601

b.   OTHER REACTIONS OF NITRILES

Reaction of *N-tert*-butylimines of α,β-unsaturated aldehydes **I-602** with lithium diisopropylamide followed by reaction with alkyl or aryl nitriles gave pyridine derivatives **I-603** in 22–87% yield (1169, 1170).

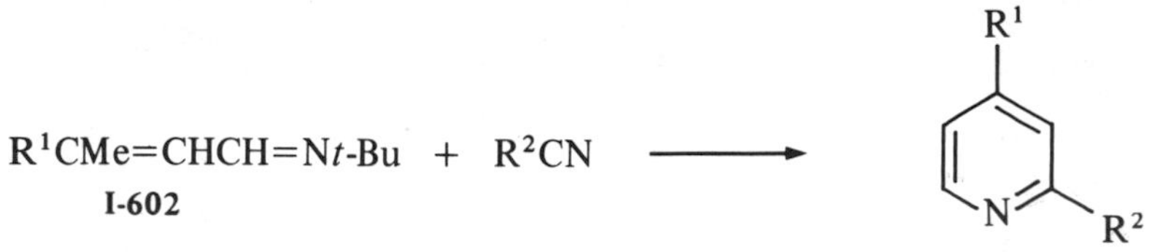

R^{1}CMe=CHCH=N*t*-Bu + R^{2}CN ⟶

I-602

I-603

Dicarbonyl compounds **I-604** (X = Me, Ph, PhCH$_2$CH$_2$, OEt) reacted with benzonitrile and excess sodium or potassium amide to give pyridones **I-605** (Y = Me, Ph, PhCH$_2$CH$_2$, OH) in 20–51% yield; in the case of X = OEt, **I-606** was the major product (1171, 1172). Dianion **I-607**, prepared by reaction of isoxazole **I-608** with lithium diisopropylamide, was found to react with nitriles **I-609** to yield pyridones **I-610** (1173).

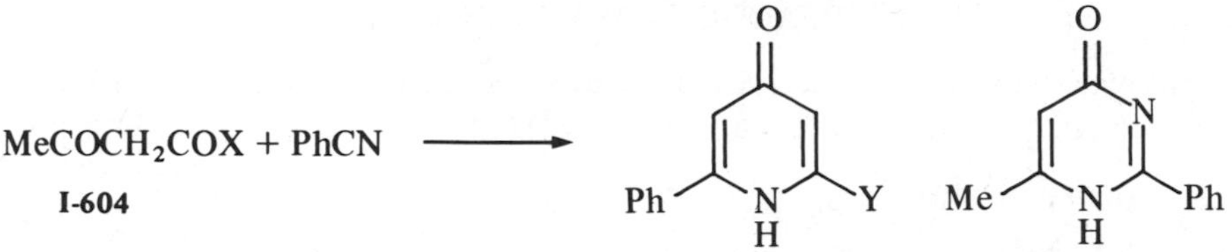

MeCOCH$_2$COX + PhCN $\longrightarrow$

I-604

I-605   **I-606**

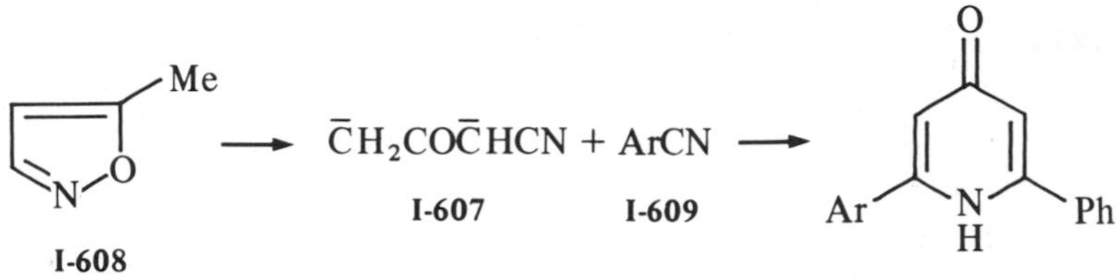

I-608 $\longrightarrow$ $\bar{C}H_2CO\bar{C}HCN + ArCN \longrightarrow$

I-607   **I-609**

I-610

Glutarimide **I-611** reacted with benzonitrile and sodium to give pyridone **I-612** in 35% yield (1174).

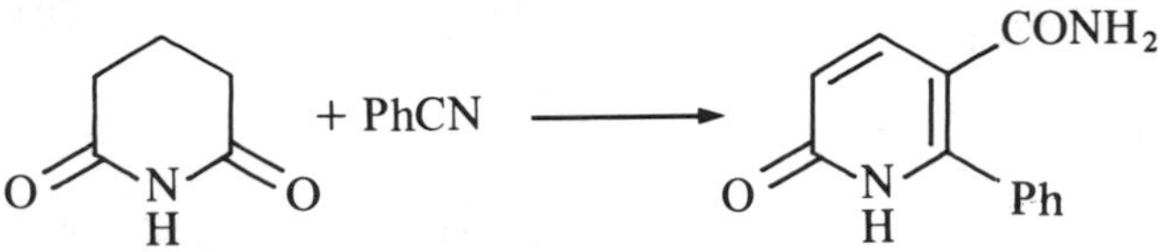

+ PhCN $\longrightarrow$

I-611   **I-612**

Cyanopyridines **I-613** condensed with identical or different (**I-614**) cyano compounds in the presence of strong base to give naphthyridines **I-615** (1175, 1176).

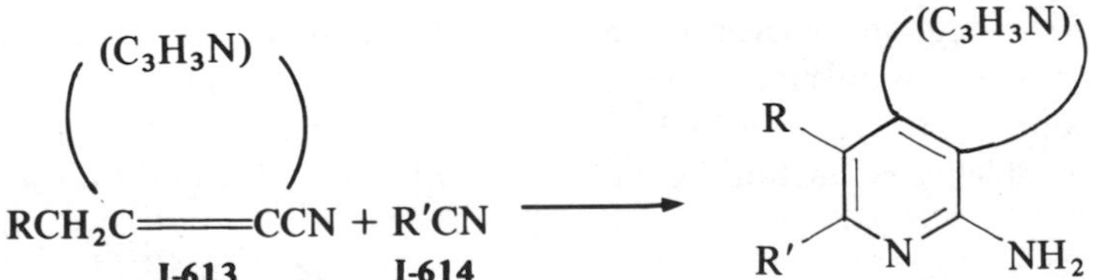

RCH_2C=$=$$CCN + R'CN \longrightarrow$

I-613   **I-614**

I-615

The indoleacetonitrile **I-616** reacted with aluminium chloride in acetonitrile to give **I-617**, a potent mutagent from tryptophan pyrolyzate, in 19% yield (1177).

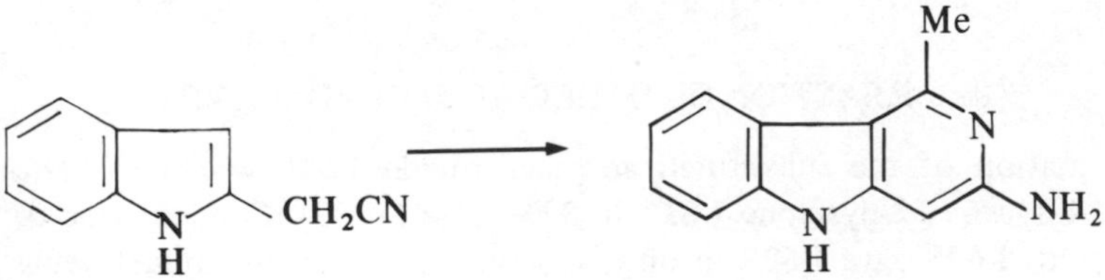

CH$_2$CN

I-616   **I-617**

c.   REACTIONS OF ISOCYANATES

Enamine derivatives of acetylacetone **I-618** reacted with phenyl isocyanate to give pyridones **I-619** (27–43%) and **I-620** (6–8%) (1178).

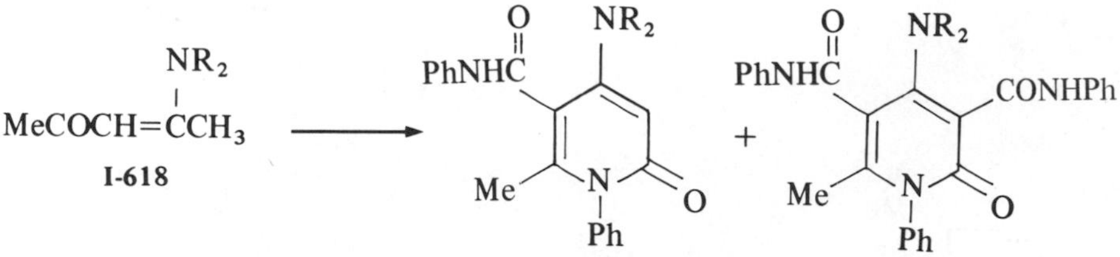

I-618   I-619   I-620

Similarly, the sodium salt of diethyl acetonedicarboxylate **I-621** reacted with aryl isocyanates to give **I-622** in 50–85% yield. With phenyl isothiocyanate, the carbethoxy group was not lost and **I-623** was formed in 87% yield (1179). Alkyl isocyanates reacted with **I-621** to give intermediates that cyclized to **I-622** in 62–70% yield on warming with phosphorus oxychloride (1180).

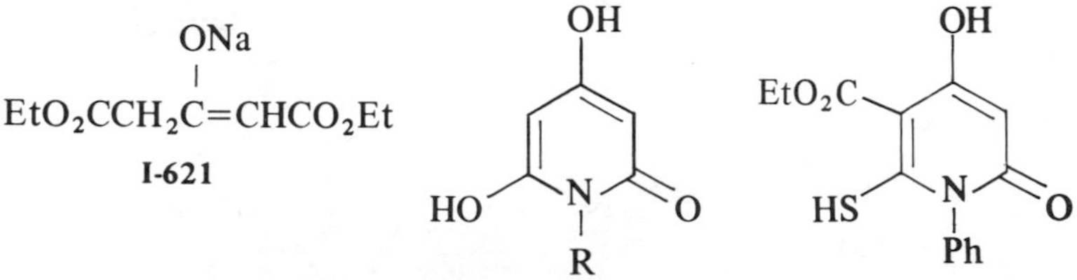

I-621   I-622   I-623

Analogously, the sodium salt of glutacondialdehyde **I-624** reacted with aryl or alkyl thiocyanates to give the pyridinethiones **I-625** (X = S) in 10–98% yield (1181–1183). Acyl isothiocyanates except for dimethylcarbamyl did not give pyridine rings, but isoselenocyanates and isocyanates gave **I-625** (X = Se, O), the latter in poor yield. Pyridones **I-625** (X = O) were prepared in high yield by protecting the aldehyde function of **I-625** (X = S) and oxidation with a peracid (1184).

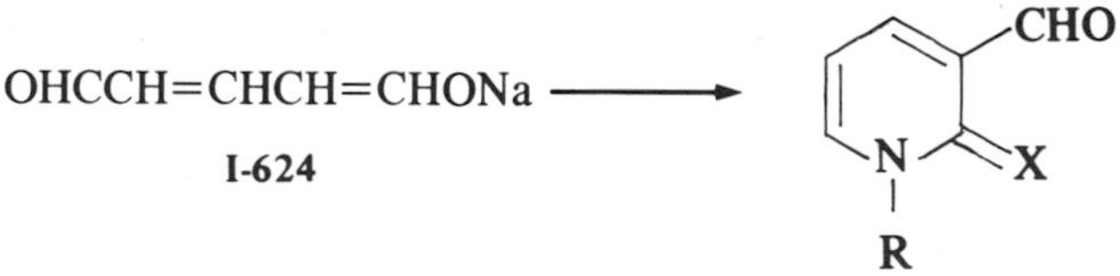

I-624   I-625

d.   REACTION OF OTHER ACID DERIVATIVES

Condensation of the substituted acetoacetanilide **I-626** with methyl formate gave the 4-hydroxy-2-pyridone **I-627** in 52% yield (1185). Condensation of ethyl formate with **I-628** gave **I-629** in only 10% yield, the major product being **I-630** (1186).

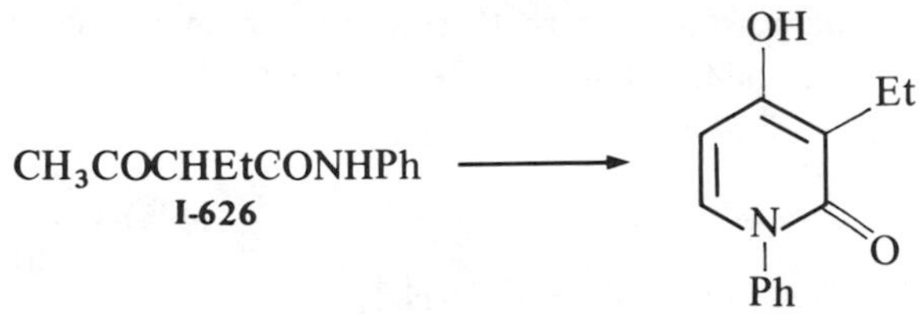

$CH_3COCHEtCONHPh$
I-626

I-627

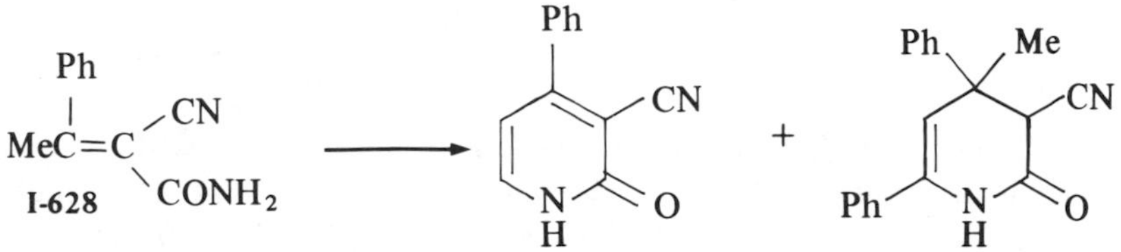

I-628

I-629

I-630

The related **I-631** condensed with the amide acetal **I-632** to give **I-633** (1187). Similarly, **I-634** condensed with the corresponding ethyl acetal **I-635** to give 85% of **I-636**, which was elaborated to prepare an aza-steroid (1188, 1189).

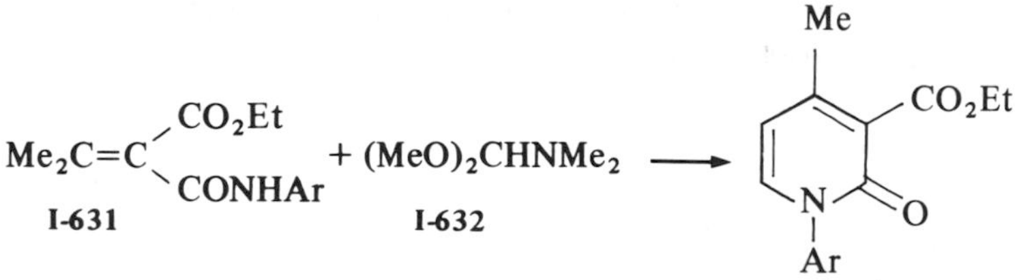

$Me_2C=C$ with CO_2Et and $CONHAr$ **+ (MeO)_2CHNMe_2**

I-631

I-632

I-633

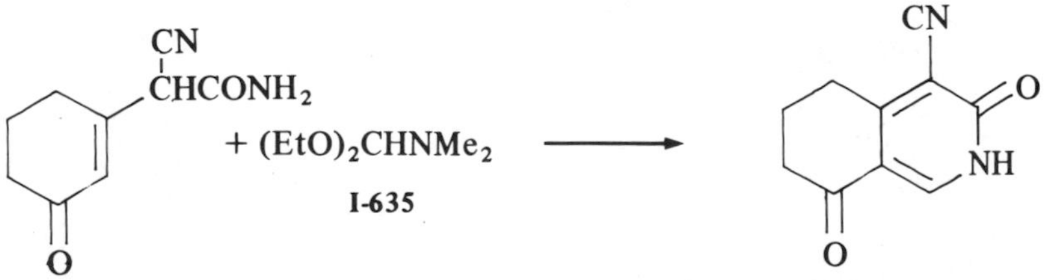

+ (EtO)_2CHNMe_2

I-635

I-634

I-636

Diketene (**I-637**) was found to react with thioamides **I-638** to yield pyridones **I-639** (1190).

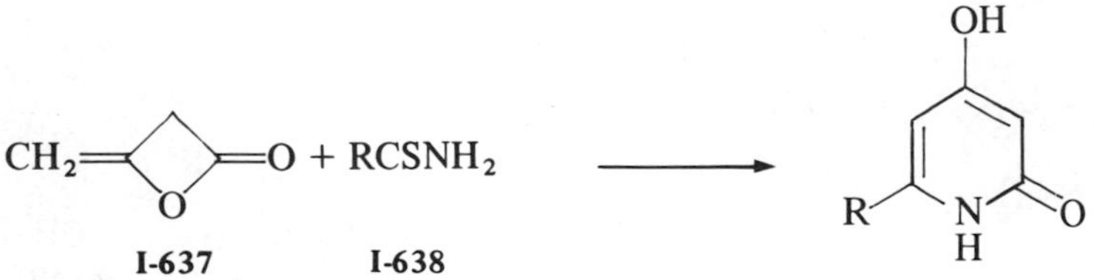

$CH_2=$ ⬜ $=O + RCSNH_2$

I-637

I-638

I-639

Dimethyl acetonedicarboxylate **I-640** condensed with a series of imidates **I-641** to give 4-hydroxy-2-pyridones **I-642** in 17–66% yield (1191–1193).

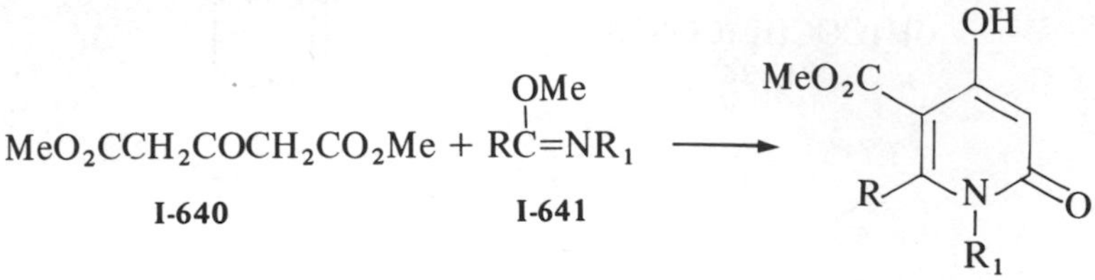

$MeO_2CCH_2COCH_2CO_2Me + RC(OMe){=}NR_1$
OH
MeO_2C
R
N
O
R_1

I-640 **I-641** **I-642**

Cyclohexenylcyclohexanone **I-643** has been found to react with a variety of compounds to give derivatives of octahydrophenanthridines **I-644**. Thus aldehydes, ammonia, and **I-643** in the vapor phase gave **I-644** (R = alkyl, aryl) in 12–55% yield, accompanied by **I-644** (R = n-C_5H_{11}), the expected product from condensation of cyclohexanone with ammonia, in 9–38% yield (1194, 1195). Amides also reacted with **I-643** to give **I-644** in 13–70% yield (1194, 1196). Urea reacted with the conjugated isomer of **I-643** to give the pyridone **I-644** (R = OH) in 59% yield (1197, 1198); other unsaturated ketones reacted similarly. The octahydrophenanthridone could also be prepared from cyclohexanone and urea in 64% yield (1197), or from the cyclohexanone–urea intermediate condensation product **I-645** (1199) (cf. Section II.2.G.d).

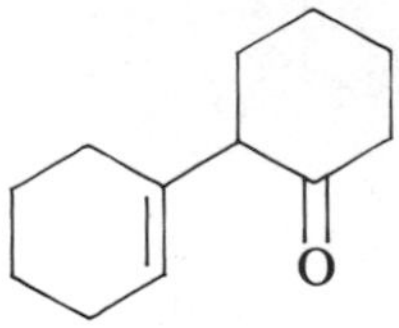

O

I-643

N
R

I-644

HN
NH
O

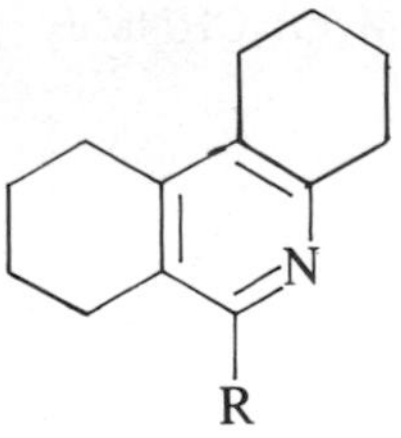

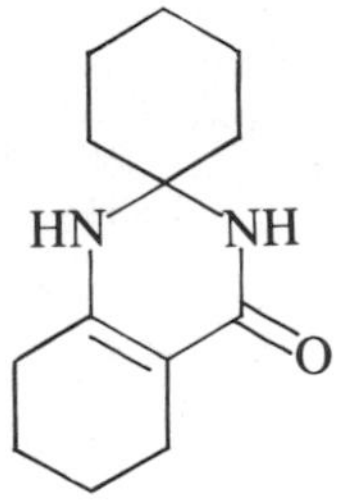

I-645

e. MISCELLANEOUS

Diacetylenes **I-646** heated with primary amines **I-647** gave pyridines **I-648** sometimes accompanied by the corresponding *N*-oxide; with large amounts of cuprous chloride catalyst, pyrroles **I-649** were formed instead (1200, 1201).

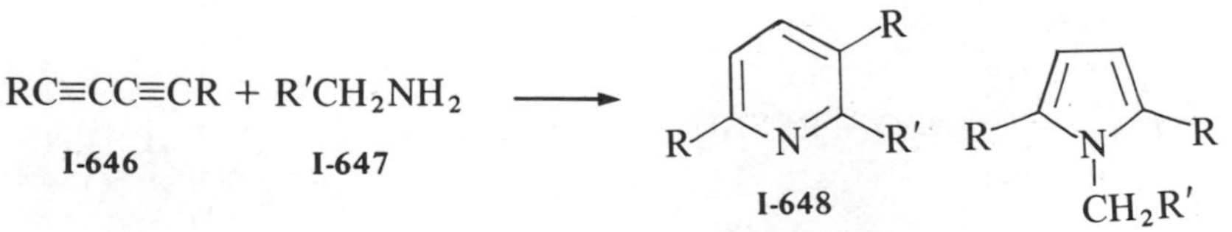

$RC{\equiv}CC{\equiv}CR + R'CH_2NH_2$
R
R
N
R'
R
N
R
CH_2R'

I-646 **I-647** **I-648** **I-649**

Further examples of the vapor-phase reaction of crotonaldehyde, formaldehyde, and ammonia to give pyridine and lesser amounts of 3-picoline were reported (1202–1206). A similar reaction using 2-methyl-2-butenal gave mostly 3-picoline (1207).

C.   3-2 Condensations

a.   1,3-DICARBONYL COMPOUNDS AND THEIR DERIVATIVES WITH METHYLENIC COMPOUNDS

This reaction continues to be exploited as the most common laboratory pyridine-ring synthesis. We use the term Guareschi condensation to refer to all variations, not only Guareschi's original reaction of β-keto esters with cyanoacetamide or cyanoacetic ester and ammonia or amines. Variations include as the 3-carbon component β-diketones, nitro compounds, enol ethers and esters, enamines, chlorovinyl compounds, diketene, and so on. The 2-carbon component includes various derivatives of malonic acid (Table I-68), β-keto acid derivatives (Table I-69), and more complex or exotic groups.

Guareschi's reaction of ethyl acetoacetate with cyanoacetamide (Scheme I-10) has been improved by conducting the reaction in the presence of piperidine or potassium hydroxide, and acidifying the isolated salt (I-650) thus formed (1107). The products of this kind have found patentable utility because diazonium salts couple at the 5-position to give azo dyes (1208a, 1294).

The generalizations made by Brody and Ruby (998) regarding the "normal" course of Guareschi-type reactions hold true:

1. The most reactive carbonyl group (or carbonyl group equivalent) in the 3-carbon unit becomes attached to the methylene group of the 2-carbon unit, by an apparent Knoevenagel condensation (which is not distinguishable from Michael addition of the methylene group to the enol of the carbonyl group of the 3-carbon unit).

2. If there are more than two carbonyl groups or equivalents in the 3-carbon unit, the second most reactive group becomes attached to the nitrogen to close the ring.

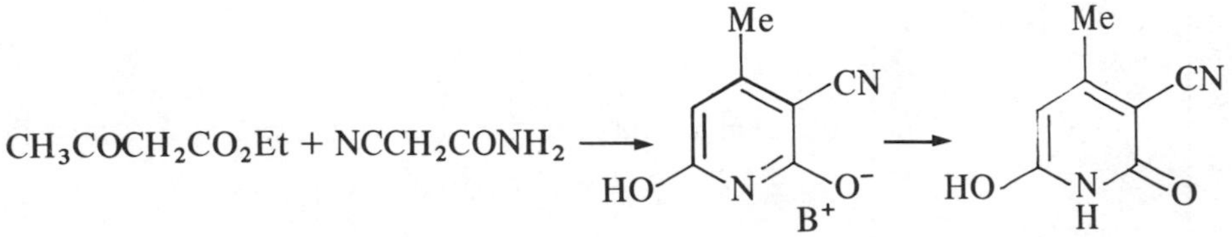

$$CH_3COCH_2CO_2Et + NCCH_2CONH_2 \longrightarrow$$

I-650

Scheme I-10

TABLE I-68.   GUARESCHI CONDENSATIONS – MALONIC ACID DERIVATIVES

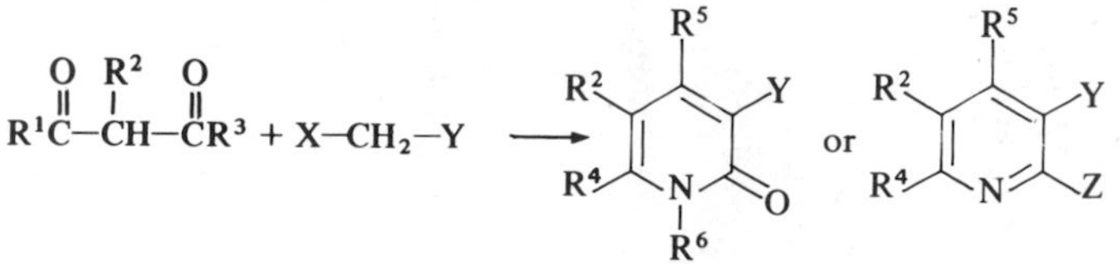

R^1	R^2	R^3	X	Y	R^4	R^5	R^6	Z	Reference
Me	H	OEt	CO_2Et	CN	OH	Me	$-(CH_2)_3OMe$	–	1208
Me	H	NHEt	CO_2Et	CO_2Et	OH	Me	Et	–	1209
Alkyl	H	OEt	$CONR_2$	$CONR_2$	OH	Alkyl	R	–	1210
Me	H	H^a	CONHAr	CN	Me	H	Ar	–	1052
Me	H	Me	CONHR	CN	Me	Me	Ar	–	1056
$-(CH_2)_4-$		OEt	CN	CN	OEt	R_2	H	–	1211
Me	R	OEt	CO_2Et	CN	OH	Me	R'	–	1212, 1213
Me	H	Me	CONHNHCOR	CN	Me	Me	NHCOR	–	1214
n-Pr	H	CO_2Et	$CONH_2$	CN	n-Pr	CO_2Et	H	–	1215
H	Ph	H^b	$CONH_2$	CN	H	H	H	–	1216
Ph	H	H^b	CO_2Et	CN	H	Ph	H	–	1186
Me	AcNH	Me	$C(=NH)OEt$	CO_2Et	Me	Me	–	NH_2	1217–1219
Me	H	Me	CN	CN	Me	Me	H	–	1220, 1221
Ph	H	Ph	CN	CN	Ph	Ph	H	–	1221
Ar	H	H	CONHR	CN	Ar	H	R	–	1222
$-(CH_2)_3CO-$		H^b	CN	CN	R_1	H	Ar	–	1223
CF_3	H	CF_3	$C(=NH)OEt$	CO_2Et	CF_3	CF_3	–	NH_2	1224
OEt	CO_2Et	H^a	$CONH_2$	CN	OH	H	H	–	1225
Me	CH_2OMe	H	CO_2Et	CN	Me	CH_2OMe	H	–	1226
H	NO_2	H	CONHR	CN	H	H	R	–	1227
H	NO_2	H	CSNHR	CN	H	H	R	$-^c$	1227
Ar	H	H^b	$CONH_2$	CN	Ar	H	H	–	1228
Me	H	CO_2Et	$CONH_2$	NO_2	Me	CO_2Et	H	–	1229
H^b	Ar	H	$CONH_2$	CN	H	H	H	–	1230–1232
Me	H	CO_2Et	$CSNH_2$	CN	Me	CO_2Et	H^c	–	1233, 1234

148

Me	H	OEt	CONHR	COMe	HO	Me	R	—	1235
Me[b]	H	CH$_2$Ph	CONH$_2$	CO$_2$Et	Me	CH$_2$Ph	H	—	1236
R	H	CO$_2$Et	CONH$_2$	CN	R	CO$_2$Et	H	—	1237
Me	H	Me	CONHR	CN	Me	Me	R	—	1238
OEt	H	H	CONH$_2$	CN	OH	H	H	—	1239
Me	\-(CH$_2$)$_4$\-		CONH$_2$	CN	Me	R$_3$	H	—	1240
\-(CH$_2$)$_4$\-		CO$_2$Et	CO$_2$Et	CO$_2$Et	R$_1$	OH	H	—	1241
Me	H	Me	CONHR	CN	Me	Me	R	—	1242, 1243
Me	H	Me	CN	C$_9$H$_6$NS[d]	Me	Me	H	—	1244
Me	H	H[b]	CN	CO$_2$Et	Me	H	H	—	1245
Me[b]	H	ArCH$_2$	CO$_2$Et	CO$_2$Et	Me	ArCH$_2$	H	—	1246–1248
3- or 4-C$_5$H$_4$N	H	OEt	CONH$_2$	CN	OH	3- or 4-C$_5$H$_4$N	H	—	1249, 1250
H	Ar	H[e]	MeCO	CO$_2$Et	H	H	—	Me	1251
MeOCH$_2$	H	CH$_3$	CN	CO$_2$Et	MeOCH$_2$	Me	H	—	1252
					Me	MeOCH$_2$			
EtO	H	Me	CONHR	CONHR	OH	Me	R	—	1253
Me	H	Me	CO$_2$Et[f]	CN	Me	Me	OH	—	1254
Me	H	OEt	CONHR	CN	OH	Me	R	—	1255
H[g]	ArCO	H	CONH$_2$	CONH$_2$	H	H	H	—	1256
R	\-(CH$_2$)$_4$\-		CONH$_2$	CN	R	R$_3$	H	—	1257
OEt	H	OEt	CO$_2$Et	CONHMe[h]	OH	OH	Me	—	1258
H[b]	NO$_2$	H	CO$_2$Et	CO$_2$Et	H	H	R	—	1259
H[b]	NO$_2$	H	CO$_2$Et	NO$_2$	H	H	R	—	1259
Ph	H	OEt	CONH$_2$	CN	OH	Ph	H	—	1260
Ar	H	CO$_2$Et	CONH$_2$	CN	Ar	CO$_2$Et	H	—	1261
Me	CO$_2$Et	H[i]	CONH$_2$	CN	Me	H	H	—	1262
Ar	Me	Me[b]	CO$_2$Et	CN	Me	Ar	H	—	1263
R	H	H[j]	CONHR'	CN	R	H	R'	—	1264–1267
Me	R	H[b]	CO$_2$Et	CN	Me	H	H	—	1245
Me[b]	H	ArCH$_2$	CO$_2$Et	CO$_2$Et	Me	ArCH$_2$	H	—	1268
H	C$_5$H$_4$N	H[b]	CN	CONH$_2$	H	H	H	—	1269
Me	C$_5$H$_4$N	H[b]	CN	CONH$_2$	Me	H	H	—	1270

^{a} As acetal. ^{c} Product is the thiopyridone. ^{e} As vinyl chloride. ^{g} As chromone. ^{i} As enol ether.

^{b} As enamine. ^{d} 4-Phenylthiazol-2-yl. ^{f} Plus hydroxylamine. ^{h} —CO$_2$Et and MeNH$_2$. ^{j} As Na enolate.

149

TABLE I-69. CONDENSATIONS IN WHICH THE ACTIVE METHYLENE COMPOUND IS A β-KETO ACID OR β-DIKETONE DERIVATIVE

$$\text{A}\!-\!\underset{\underset{\text{R}^1}{|}}{\text{CH}}\!-\!\text{B} + \text{X}\!-\!\text{CH}_2\!-\!\text{Y} \longrightarrow$$

(pyridine product with substituents R^1, R^2, R^3, Y, Z)

A	R^1	B	X	Y	R^2	R^3	Z	Reference
—COCOCHRCO—		COCO$_2$Et	MeCO^a	CO$_2$Et	—COCHRCO—	CO$_2$Et	Me	1271
CHO^a		—CHR(CH$_2$)$_n$CHRCO—	MeCO	MeCO	—CHR(CH$_2$)$_n$CHR—	H	Me	1272, 1273
PhCH=CHCO	H	CO$_2$Et	MeCO^a	CO$_2$Et	PhCH=CH	CO$_2$Et	Me	1274
RCO	H	RCO	CONH$_2$	MeCO	R	R	OH	1275
CO$_2$Et	H	MeCO	PhCO^a	CN	OH	Me	Ph	1276
MeCO	H	CO$_2$Et	MeCO^a	CO$_2$Et	Me	OH	Me	1277
3-C$_4$H$_3$OCO	H	CHO	MeCO^a	CO$_2$Et	3-C$_4$H$_3$O^c	H	Me	1278
R	R	R	—CO(CH$_2$)$_2$CO—^{a}		R	R	—(CH$_2$)$_2$CO—	1279
MeCO	H	CHO^b	—CO(CH$_2$)$_3$CO—		Me	H	—(CH$_2$)$_3$CO—	1280
CN	H	ArCO	ArCO	CN	R$_2$N	Ar	Ar	1281
Ar	R	CHO	MeCO^a	MeCO	Ar	H	Me	1282
H^d	CO$_2$Et	CO$_2$Et	MeCO^a	CN	H	OH	Me	1283
—COC$_6$H$_4$(CH$_2$)$_2$—		CHO	MeCO	CO$_2$Et	—C$_6$H$_4$(CH$_2$)$_2$—	H	Me	1284
n-PrCO	H	CHO^a	n-PrCO	CHO^a	n-Pr	H	H	1285
CO$_2$Ar	R	CO$_2$Ar	MeCO^a	CN	OH	OH	Me	1286, 1287
CHO^d	MeCO	COCO$_2$Et	MeCO^a	CONH$_2$	H	CO$_2$Et	Me	1288
MeCO	H	MeCO^a	CONH$_2$	NO$_2$	Me	H	H	1289
CHO^a	R	CHO^e	MeCO^a	CO$_2$Et	H	H	Me	1290
MeCO^a	H	CN	MeCO	CN	NH$_2$	Me	Me	1291
RCO	H	COCH$_2$Cl^f	MeCO^a	CO$_2$Et	R	CH$_2$Cl	Me	1292
PhCH=CHCO	H	CHO	CONH$_2$	CN	PhCH=CH	H	OH	1293

^{a} As enamine. ^{c} 3-furyl-. ^{e} As iminium salt.
^{b} As acetal. ^{d} As enol ether. ^{f} As vinyl chloride.

3. Products with an "abnormal" orientation are often formed if an imine or
    enamine having a free N—H is part of the 3-carbon unit; in that case, the
    apparent sequence of steps is frequently reversed, the imine or enamine
    nitrogen becoming the nitrogen of the pyridine ring.

4. If one of the carbonyl groups of the 3-carbon unit is present as an enol
    ether, enol ester, or chlorovinyl group, Michael addition frequently appears
    to dominate as the first step; "abnormal" products may be intentionally
    prepared in this manner.

Although no irrefutable evidence for the mechanism implied by these generalizations is available, acyclic intermediates have been isolated on occasion and
cyclized to pyridines (cf. Section II.2.A) in a manner entirely consistent with the
mechanism. For preparative purposes, the one-step 3-2 cyclization is preferred.

Thus the methyl ketone I-651 condensed with malononitrile to yield the expected
"normal" product I-652, but the enamine I-653 yielded the expected "abnormal"
isoquinoline derivative I-654 (1295, 1296).

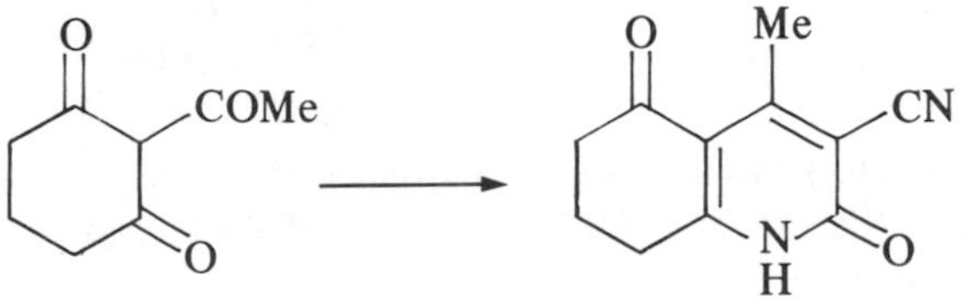

I-651   I-652

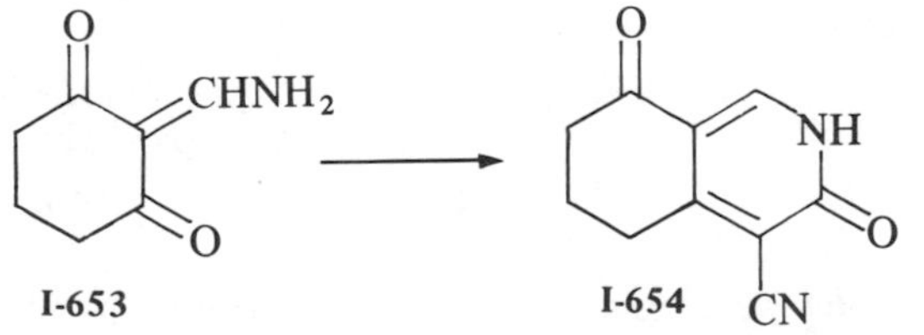

I-653   I-654   CN

The enaminoketone I-655 yielded the "normal" product I-656 on reaction with
cyanoacetamide, but an "abnormal" product (I-657) on reaction with the enamine
I-658 (1228); Michael addition of the amino group of I-658 to I-655 could be
suspected as a first step.

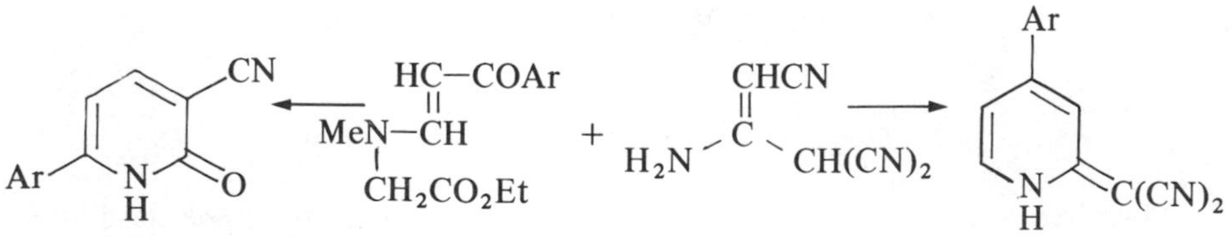

I-656   I-655   I-658   I-657

However, the enaminonitrile **I-659** yielded the "normal" product **I-660** on condensation with acetoacetic ester; the "abnormal" product **I-661** was formed on reaction with diketene **(I-637)** (1276, 1297). The latter reaction may have occurred *via* acylation at carbon of **I-659** by diketene as the first step.

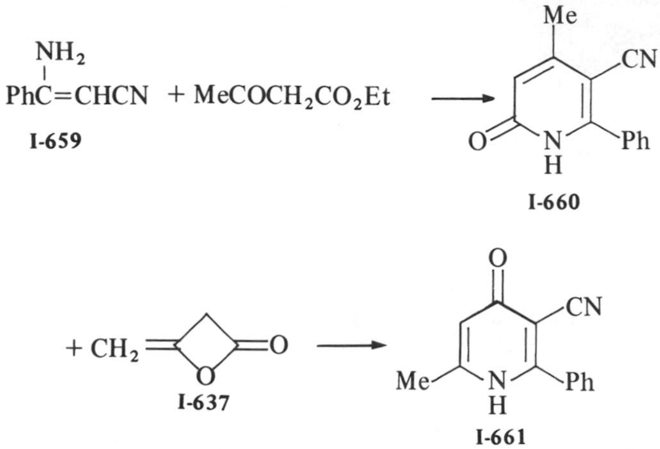

A 4-pyridone **(I-662)** was also formed by the condensation of the aminocrotonic ester **I-663** with acetoacetic ester (1277); reaction of the amino group of **I-663** with the ketone carbonyl is a reasonable first step.

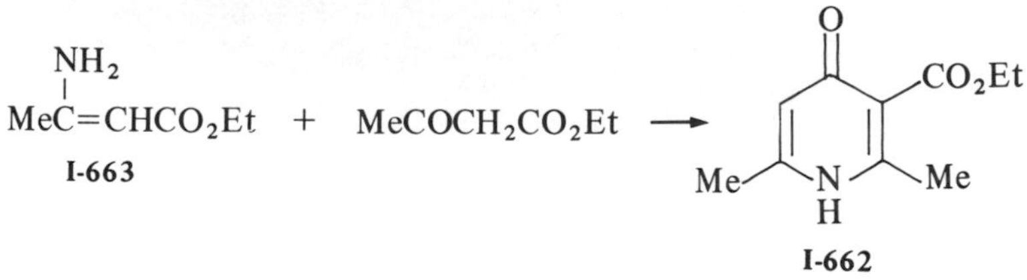

The enamines **I-664** were found to yield 2,5-substituted pyridinium salts **I-665** on cyclization in strong acid; however, when R = tertiary butyl, the 3,4-substituted product **I-666** was also observed (1298).

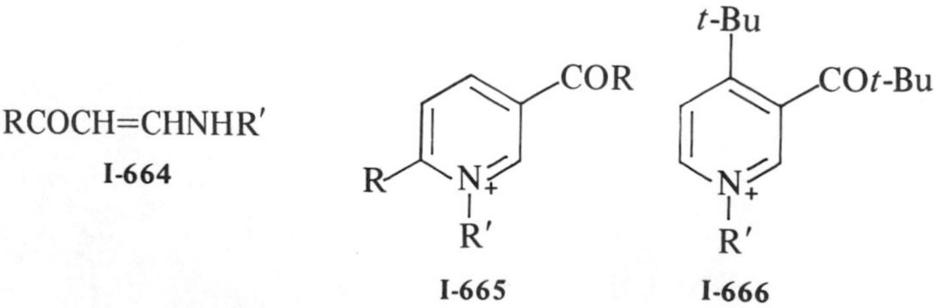

Pyrolysis of acetoacetamide **(I-667)** yielded pyridine derivative **I-668** in addition to the major product, **I-669** (1299).

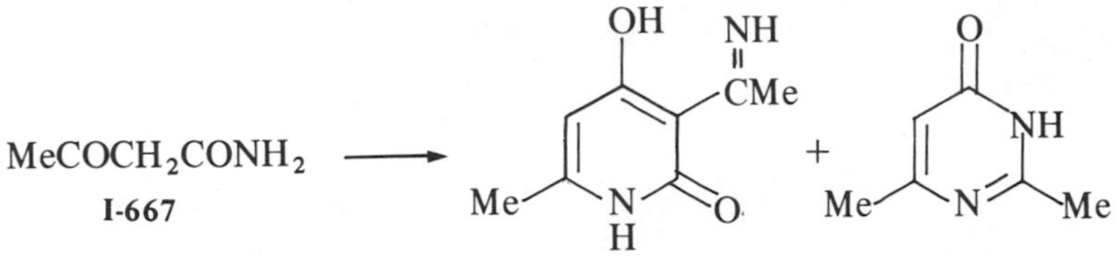

MeCOCH$_2$CONH$_2$ $\longrightarrow$

I-667

I-668 + **I-669**

It has been claimed that condensation of N-substituted acetoacetamides **I-670** gave the tautomer **I-671** of the expected product **I-672** (1300).

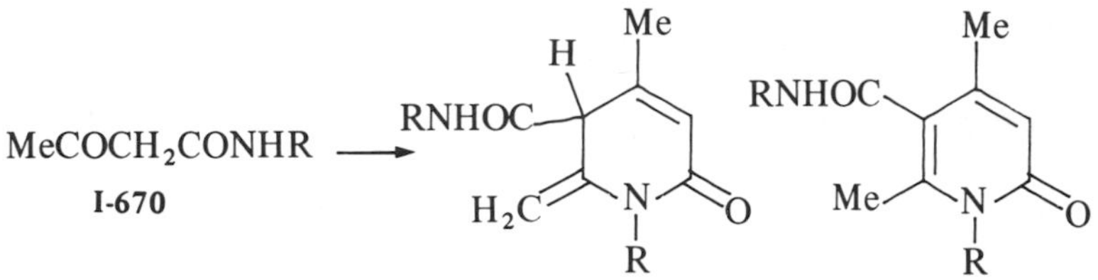

MeCOCH$_2$CONHR $\longrightarrow$

I-670

I-671   **I-672**

Self-condensation of cyanoacetamide using sodium ethoxide yielded **I-673** (1301).

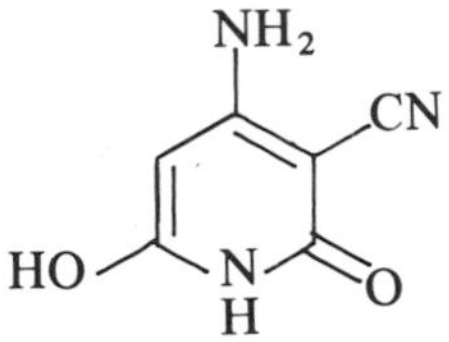

I-673

The reactions discussed in the remainder of this section are viewed as variations of the Guareschi condensation.

Ketene S,S-acetals have been condensed with active methylene compounds to yield pyridines (Table I-70).

Diketene has been found generally to yield derivatives of 4-hydroxy-2-pyridone or of 4-pyridone on reaction with amines or with β-dicarbonyl compounds, respectively; examples are shown in Table I-71. An exception is hydroxylamine, which yielded a 6-hydroxy-2-pyridone derivative on reaction with diketene (1307).

Derivatives of 4-hydroxy-2-pyridone were formed by reactions of malonyl chloride with compounds containing various nitrogen-containing functional groups (Table I-72).

Similarly, the related chloroiminium salts **I-674** yielded 2,4-diaminopyridine derivatives on reaction with imines (1325). Such intermediates may be used to explain the self-condensation of cyanoacetamides **I-675** in phosphoryl chloride to give chloropyridines **I-676** (1326–1328) or **I-677** (1329–1332). Malononitrile has been converted to 2-bromo-3-cyano-4,6-diaminopyridine on reaction with hydrogen bromide (1333).

TABLE I-70.   CONDENSATION OF KETENE S,S-ACETALS

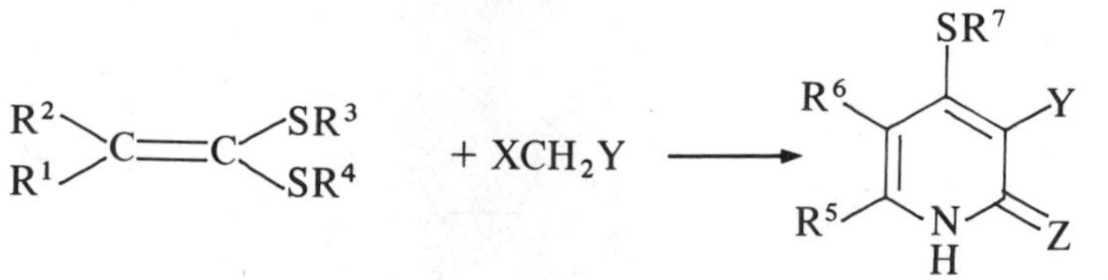

R^1	R^2	R^3	R^4	X	Y	R^5	R^6	R^7	Z	Reference
$CONH_2$	CN	Me	Me	CHO[a]	Et	H	CN	Me		1302
$CONH_2$	CN	Me	H	CN	CN	MeNH[b]	H	Me	NH	1303
RCO	R'	Me	Me	$CONH_2$	CN	R	R'	Me	O	1304–1306

[a] As enamine.
[b] Reaction in $MeNH_2$.

TABLE I-71.   REACTIONS OF DIKETENE

Reactant	Product	Reference
MeC(NH$_2$)=CHCONH$_2$	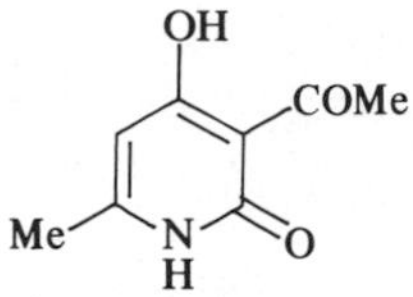	1308
H$_2$NCHRCO$_2$H	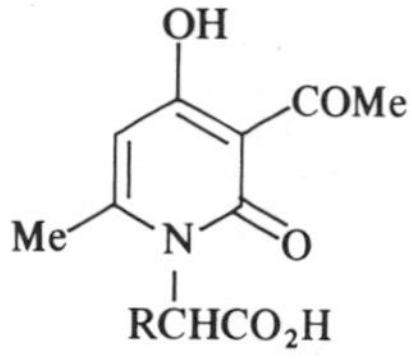	1309
RNH$_2$	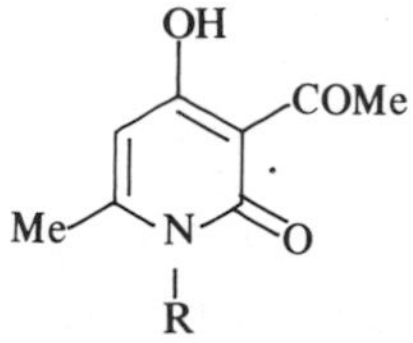	1310–1312
R'COCH=CMeNHR	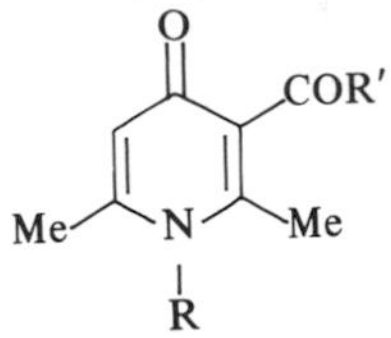	1313
RC(=NPh)CH$_2$R'	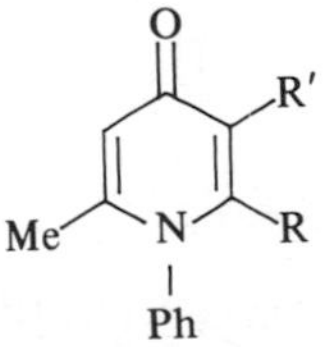	1314, 1315
H$_2$NOH	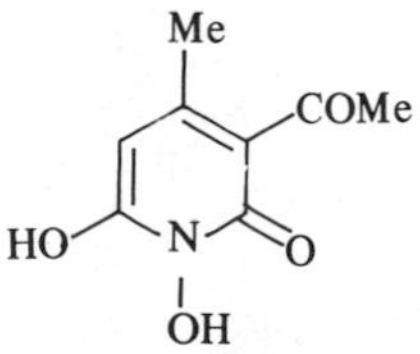	1307
MeCOCH$_2$CONHR	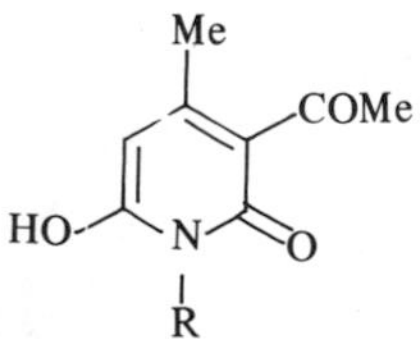 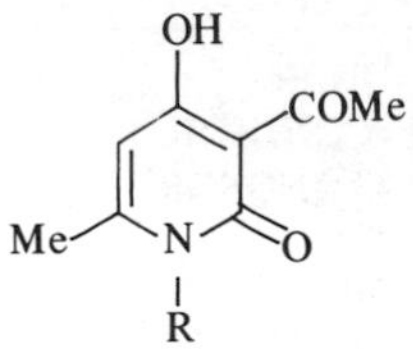	1316

TABLE I-71. *(CONTINUED)*

Reactant	Product	Reference
$RC(NH_2)=CHCO_2Et$	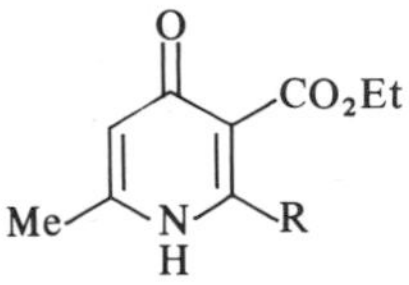	1317
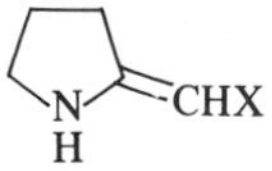	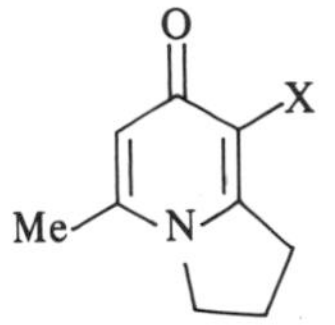 $X = COMe, CN, CO_2Et$	1318

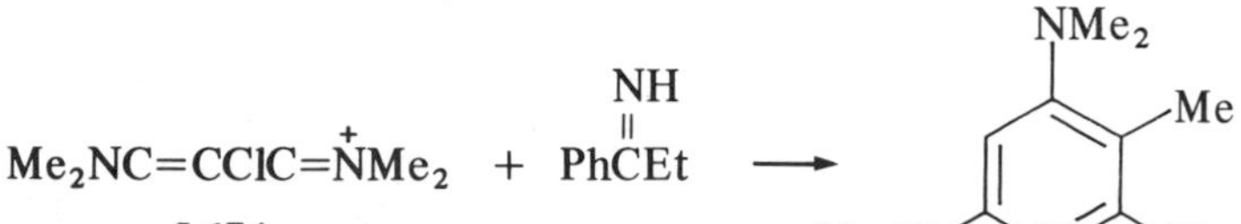

Me$_2$NC=CClC=$\overset{+}{N}$Me$_2$ + PhCEt $\longrightarrow$

I-674

NCCH$_2$CONHR $\longrightarrow$

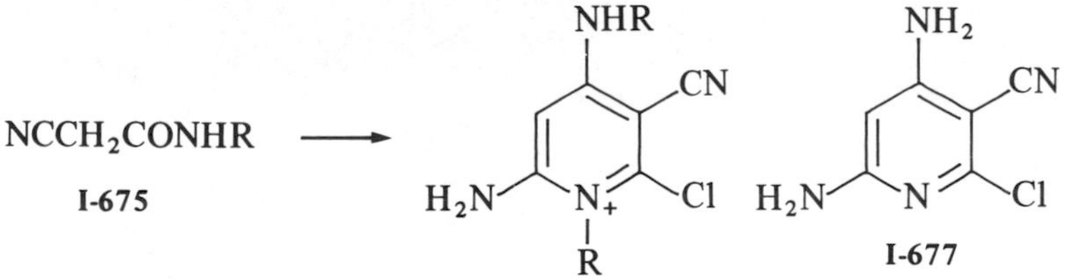

I-675

I-676   **I-677**

TABLE I-72. REACTIONS OF MALONYL CHLORIDES

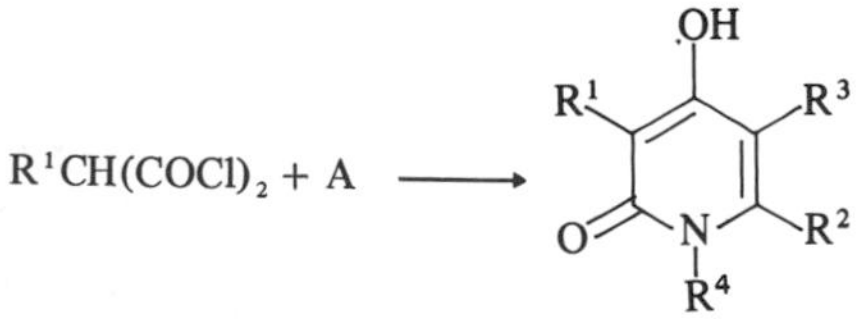

$R^1CH(COCl)_2 + A \longrightarrow$

R^1	A	R^2	R^3	R^4	Reference
Et, PhCH$_2$	PhC(=NH)Et	Ph	Me	H	1319
–	$R^3CH=C(NHR^4)R^2$	–	–	–	1320
–	$R^3CH_2C(=NOR')R^2$	–	–	OR'	1321
H	R^3CH_2CN	Cl	–	H	1322, 1323
H	ArNHCSCH$_2$C(=NAr)Ar	SH	C(=NAr)Ar	Ar	1324

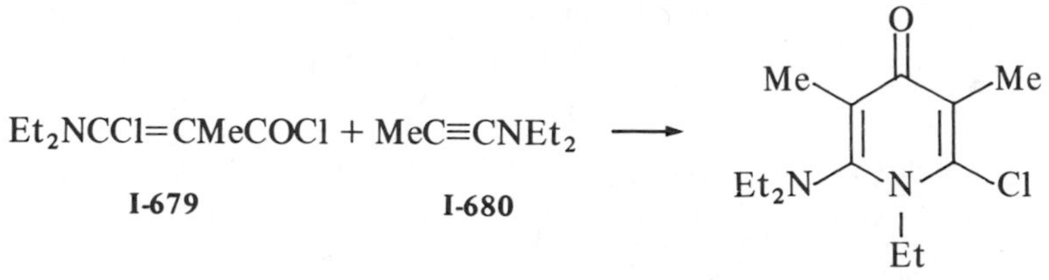

Et$_2$NCCl=CMeCOCl + MeC≡CNEt$_2$ ⟶

I-679   **I-680**

I-678

Chloropyridone **I-678** was formed from reaction of **I-679** with **I-680** (1334).

Enamine derivatives of cyclic β-diketones **I-681** yielded tetrahydroquinoline derivatives **I-682** or related compounds on reaction with malonaldehyde derivatives **I-683** (1335–1340). Cyclic ketones without a second carbonyl group were found to condense in fair yield with enamino ketones to yield fused pyridines such as **I-684** (1335, 1336, 1341, 1342).

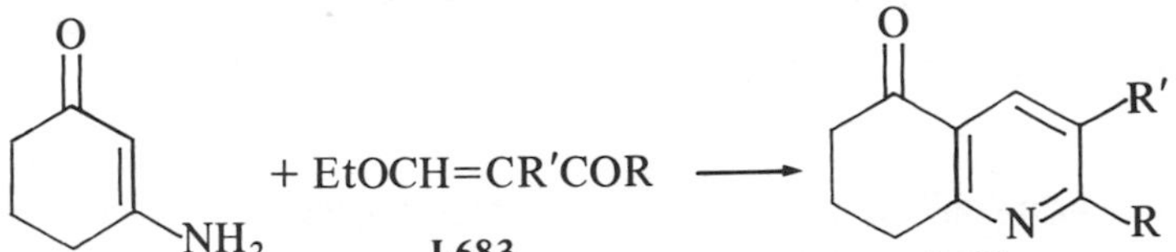

+ EtOCH=CR′COR ⟶

I-681   **I-683**   **I-682**

+ H$_2$NCH=CHCHO ⟶

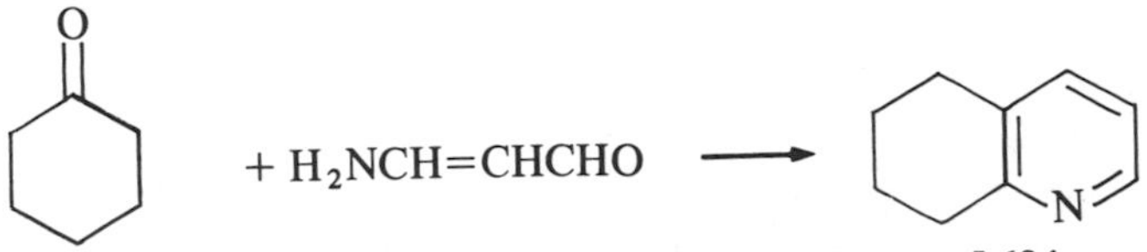

I-684

Enaminonitrile **I-685** yielded fused pyridines **I-686** on reaction with ketones **I-687** in polyphosphoric acid (1343, 1344).

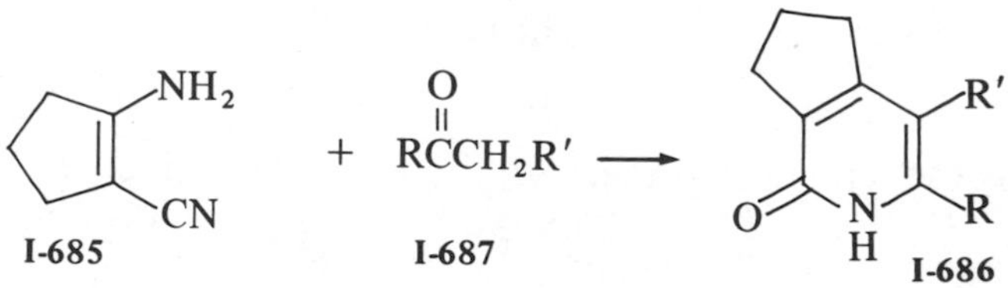

+ RCCH$_2$R′ ⟶

I-685   **I-687**   **I-686**

Some ketones **I-687** were found to condense with acetoacetic ester and ammonium acetate to give pyridones **I-688** in low yield (1345).

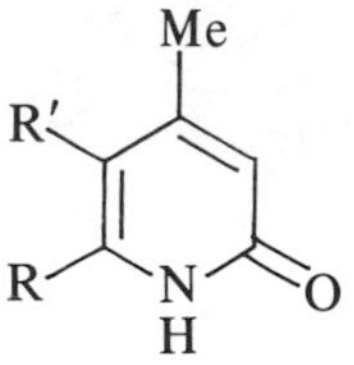

I-688

The ketene–ester **I-689** was found to react with cyclohexanone anil to yield **I-690** (30%), or with acetylacetone imine to yield **I-691** (51%) (1346).

I-689

I-690

I-691

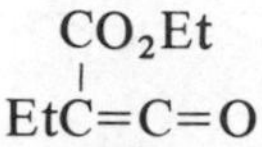

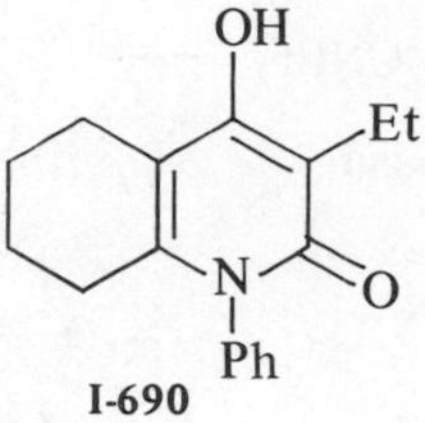

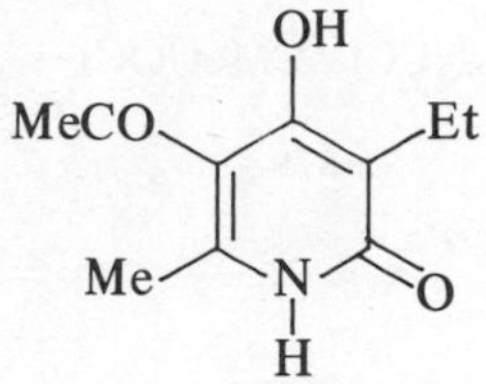

Reaction of lithium imide **I-692** with ketenimine **I-693** yielded pyridine **I-694** (1347–1348).

I-692 **I-693** **I-694**

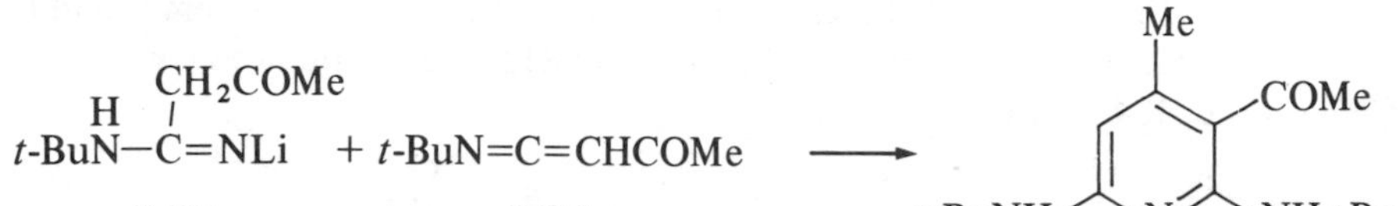

Acylaminoacetamidines **I-695** were found to react with acetylacetone to yield pyridine derivatives **I-696** (1349), demonstrating the typical preparation of 2-aminopyridine derivatives from amidines or imino ethers (see Table I-68).

I-695 **I-696**

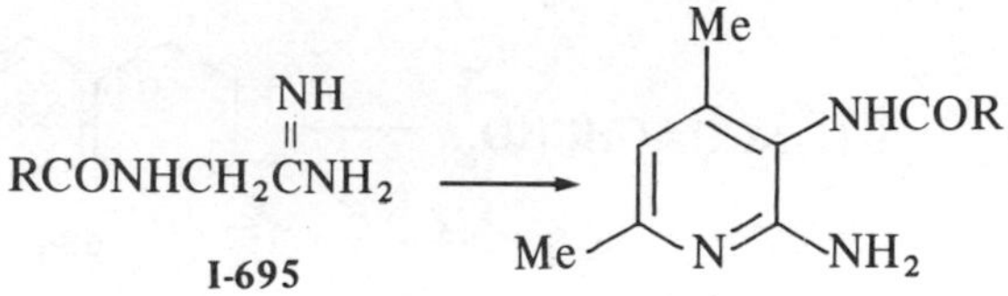

Similarly, ethoxymethylenemalononitrile **I-697** on reaction with amidine **I-698** yielded 2,6-diaminopyridine derivative **I-699** (1350).

I-697 **I-698** **I-699**

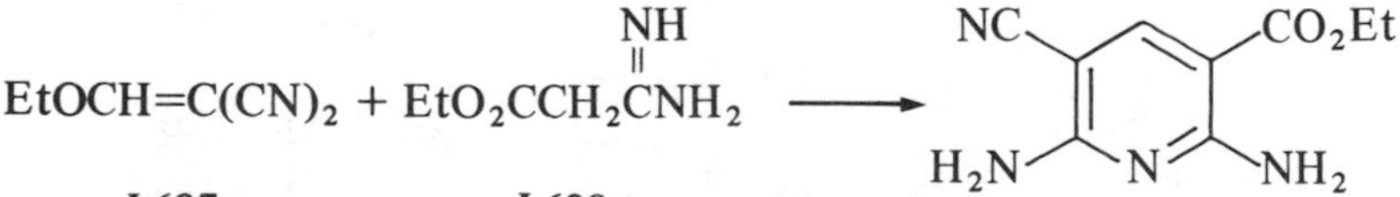

Reaction of **I-697** with picolyl ketone **I-700** gave pyridone **I-701** (1351).

I-700

I-701

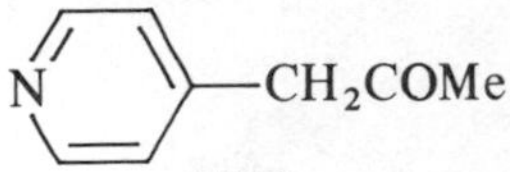

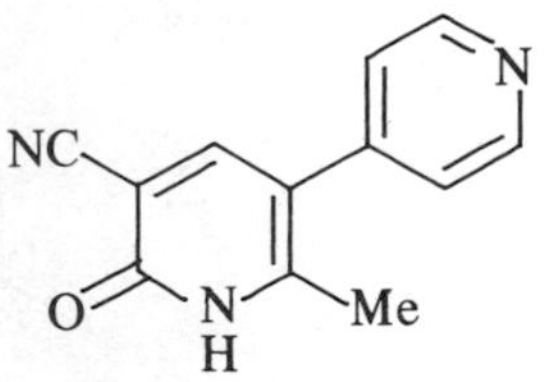

The malondiamidines **I-702** and **I-703** yielded pyridine derivatives **I-704** and **I-705**, respectively, on reaction with β-diketones (1352).

$H_2NC(=NH)CH_2C(=NH)NH_2$

I-702

I-703

I-704

I-705

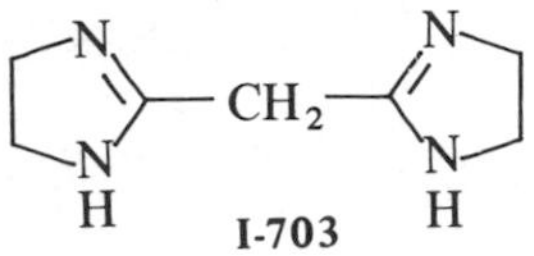

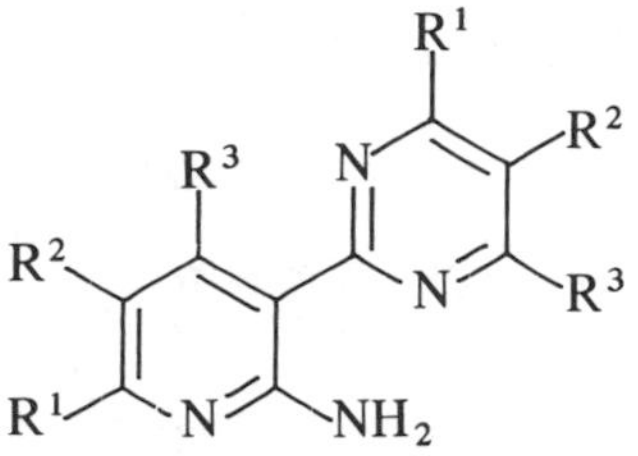

Condensation of amidine **I-706** with a β-ketoester yielded pyrimidine **I-707**, as the major product; pyridine **I-708** was also formed (1353).

$$H_2NCOCH_2C(=NH)NH_2 \cdot HCl + RCOCH_2CO_2Et \longrightarrow$$

I-706

I-707

I-708

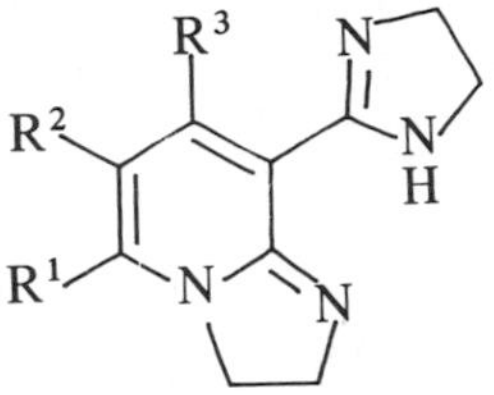

Ethoxyacrylic ester **I-709** was found to form **I-710** on reaction with one-half equivalent of ammonia, but dinicotinic ester **I-711** on reaction with excess ammonia (1354).

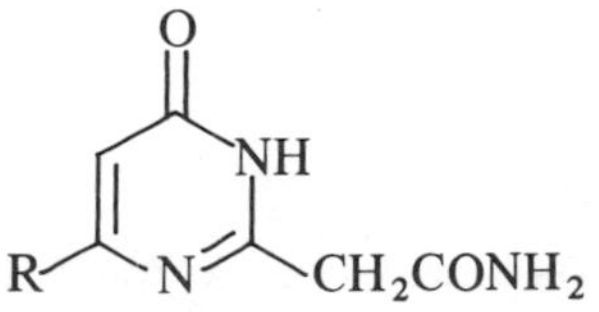

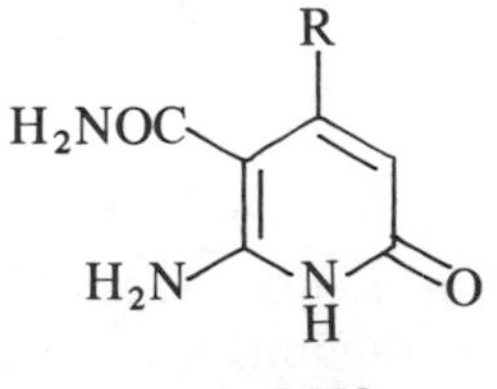

$$EtOCH{=}CHCO_2Et \longrightarrow$$

I-709

I-710

or

I-711

The β-ketonitriles **I-712** formed pyridones **I-713** on reaction with acetophenone in polyphosphoric acid (1355).

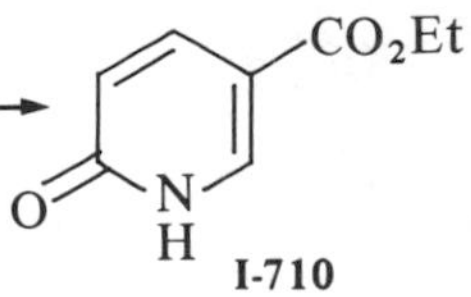

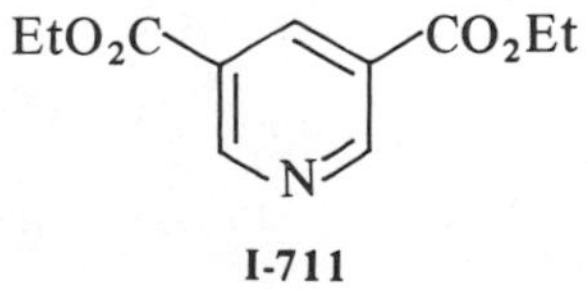

I-712 I-713

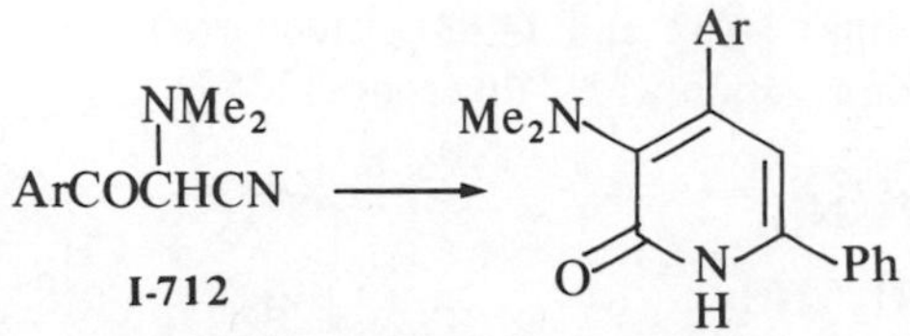

Ketoanilides **I-714** were condensed in the presence of succinic acid dichloride to yield pyridones **I-715** and **I-716** (1356).

$RCOCH_2CONHPh$

I-714

I-715

I-716

Carbon suboxide C_3O_2, prepared by thermolysis of *bis* (trimethylsilyl) malonate, formed pyridone **I-717** on reaction with *bis*(trimethylsilyl)amine (1357).

I-717

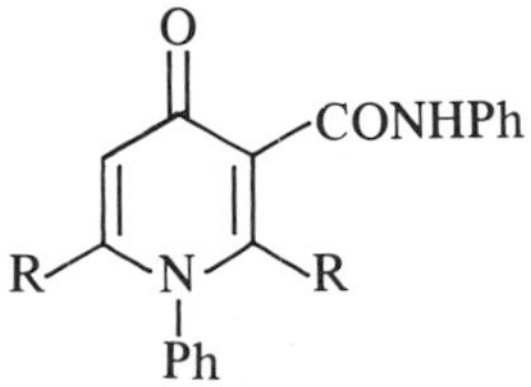

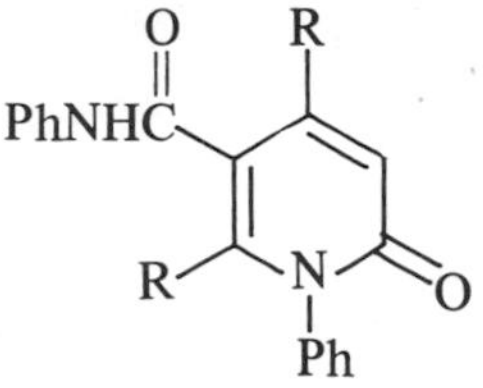

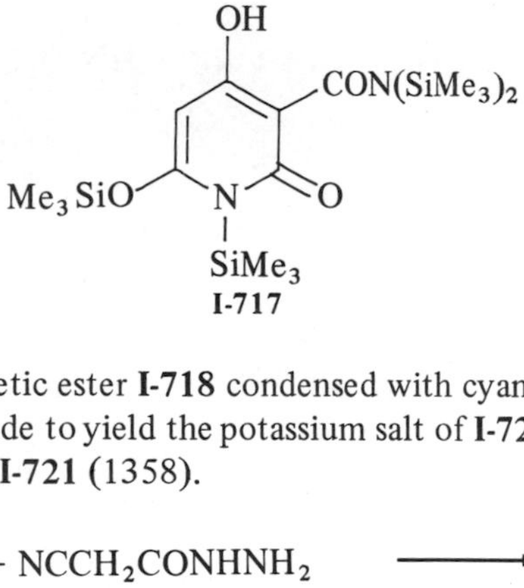

The trifluoroacetoacetic ester **I-718** condensed with cyanoacetic hydrazide **I-719** using potassium hydroxide to yield the potassium salt of **I-720**; acidification brought about rearrangement to **I-721** (1358).

$CF_3COCH_2CO_2Et + NCCH_2CONHNH_2 \longrightarrow$

I-718 **I-719**

I-720 **I-721**

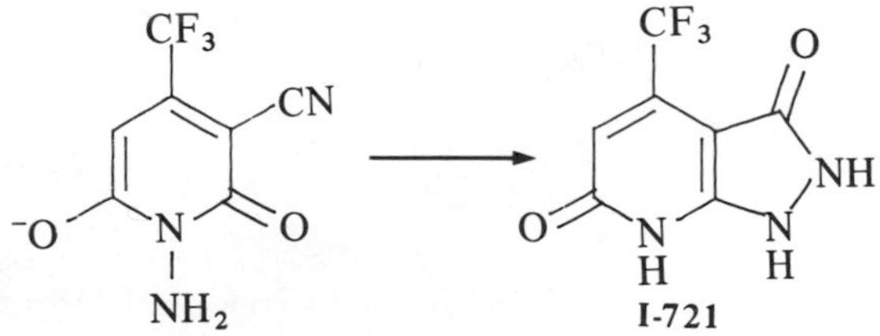

Reaction of 3-pyridylacetonitrile with phenylpropiolic ester **I-722** yielded the intermediate **I-723** and as a major product, the 2 : 1 adduct **I-724** in addition to the pyridone **I-725** (1359).

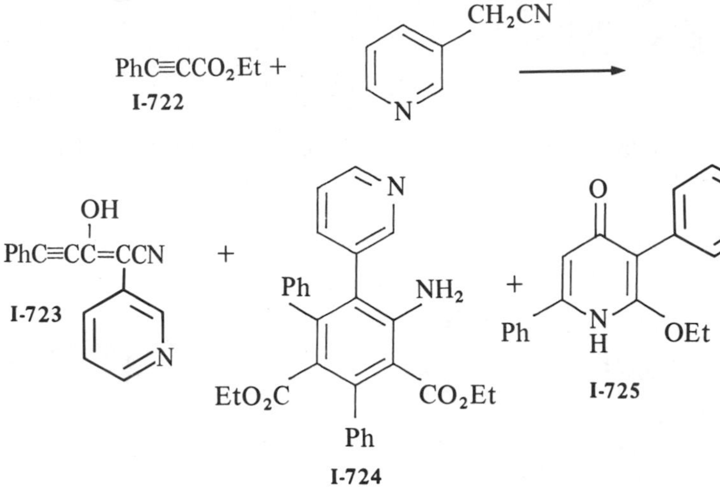

In general, acetylenic carbonyl compounds may be substituted for β-dicarbonyl compounds as the 3-carbon unit in a Guareschi-type condensation. Thus methyl propiolate **I-726** reacted with enamino ketone **I-681** to yield pyridone **I-727** (1360). Use of acetylenic aldehydes or ketones instead of propiolic ester gave pyridines **I-728** (1361–1365).

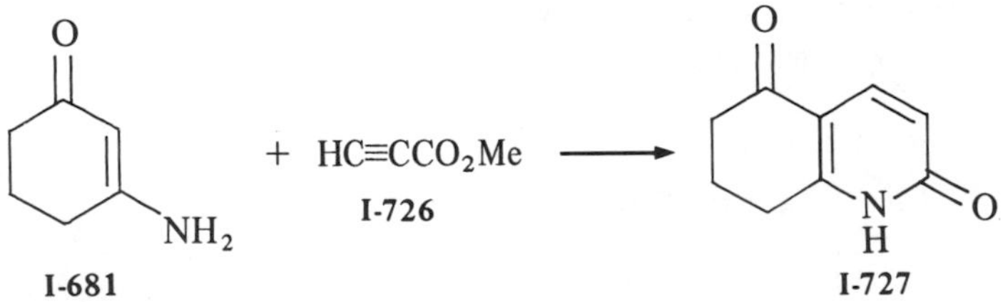

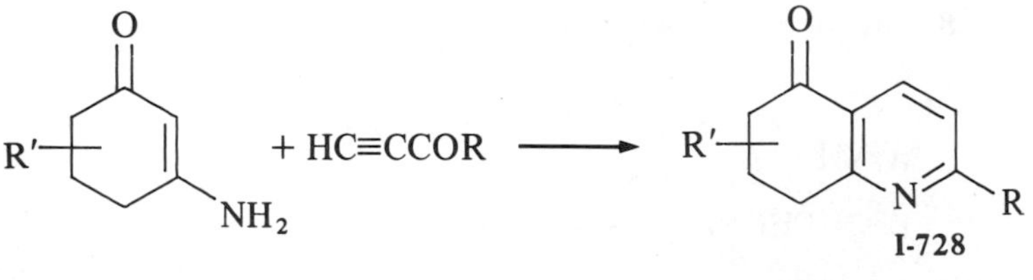

Reaction of the hydrazine **I-729** with dimethyl acetylenedicarboxylate yielded pyridone **I-730** (1366).

I-729 I-730

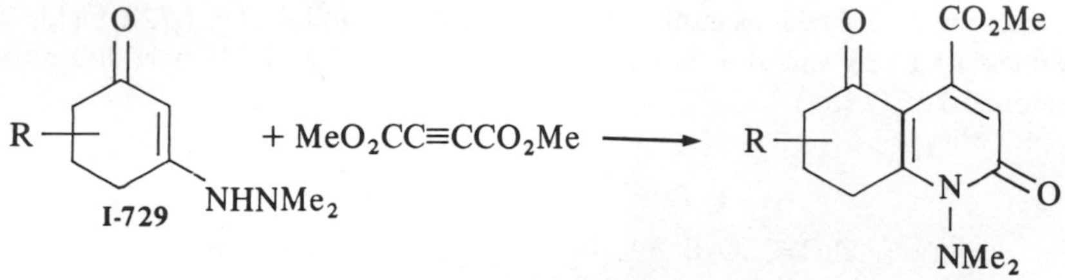

Similarly, arylpropiolic esters **I-731** condensed with substituted amides **I-732** to yield pyridinediones **I-733**, R = Ar (1367–1369), 1-indolyl (1370).

I-731 I-732 I-733

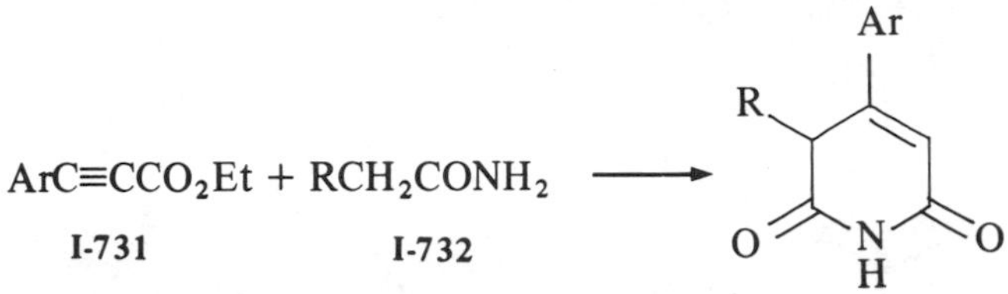

Propiolic esters or amides reacted similarly with electronegatively substituted amides **I-734** (X = CN, NO_2, RCO, etc.) to yield pyridinediones **I-735** (1371, 1372).

I-734 I-735

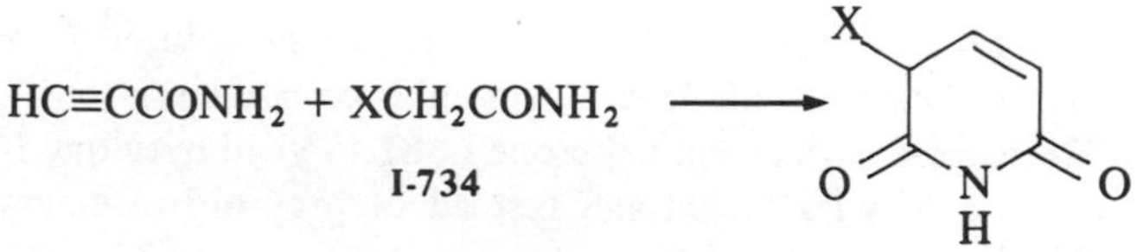

The indole salt **I-736** on reaction with diethyl acetylenedicarboxylate yielded the 4-pyridone **I-737**, with decarboxylation (1373).

I-736 I-737

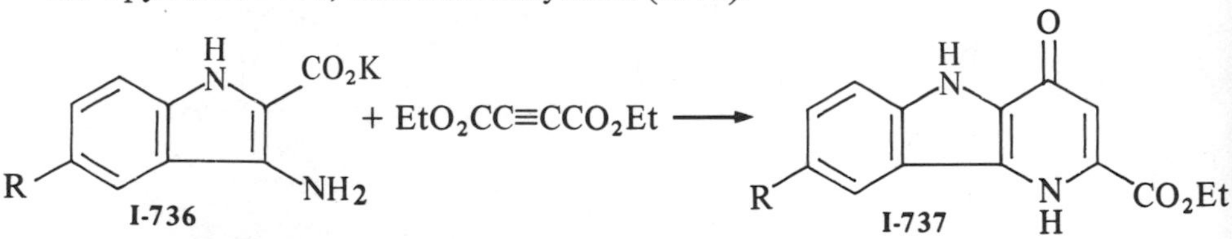

Acid-catalyzed self-condensation of anilides **I-738** (Z = O) or thioanilides **I-738** (Z = S) yielded pyridones or thiopyridones **I-739** respectively (1374).

I-738 I-739

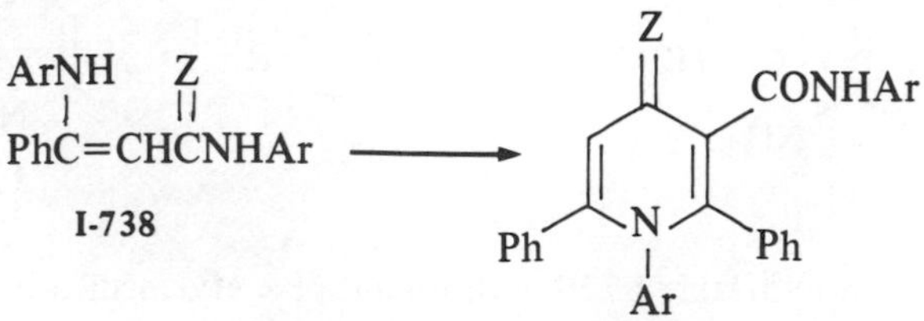

Condensation of excess alkoxymethylenecyanoacetic ester **I-740** with amino-crotonitrile **I-741** yielded **I-742**; the intermediate **I-743** could also be isolated (1375, 1376).

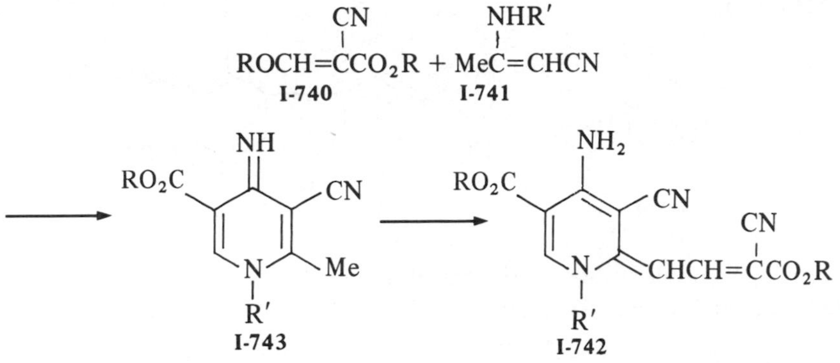

$$ROCH=CCO_2R + MeC=CHCN$$
I-740   **I-741**

I-743   **I-742**

Heating **I-744** with an arylamine tosylate salt (160°) gave low yields of pyridone imines **I-745** in addition to the major products **I-746** (1377).

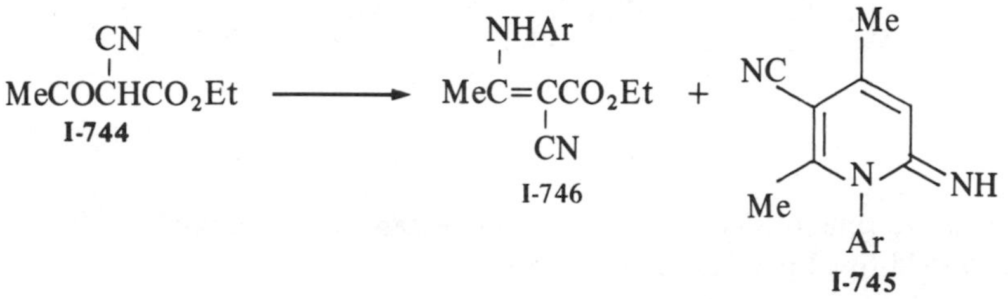

$$MeCOCHCO_2Et \longrightarrow MeC=CCO_2Et +$$
I-744   **I-746**   **I-745**

Nitro compound **I-747** reacted with ethoxymethylene derivatives **I-748** (X = CN, CO$_2$Et) to yield pyridines **I-749** (1378).

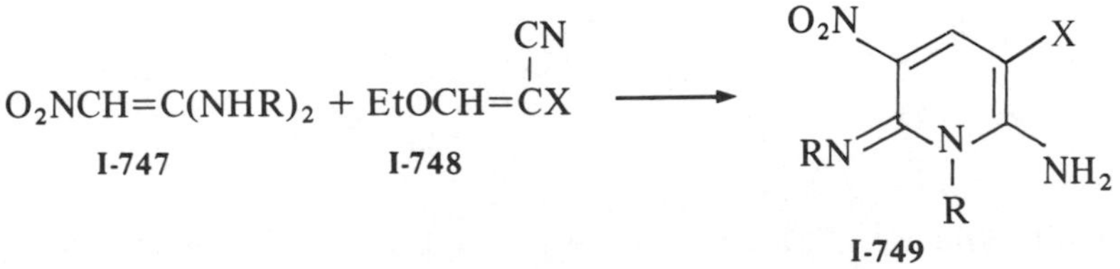

$$O_2NCH=C(NHR)_2 + EtOCH=CX \longrightarrow$$
I-747   **I-748**   **I-749**

Reaction of **I-750** (X = NHMe, NMe$_2$, ONa) with malononitrile dimer **I-751** yielded bipyridyls **I-752** among other products (1379, 1380); although the loss of malononitrile is unexceptional, the reductive coupling is unusual.

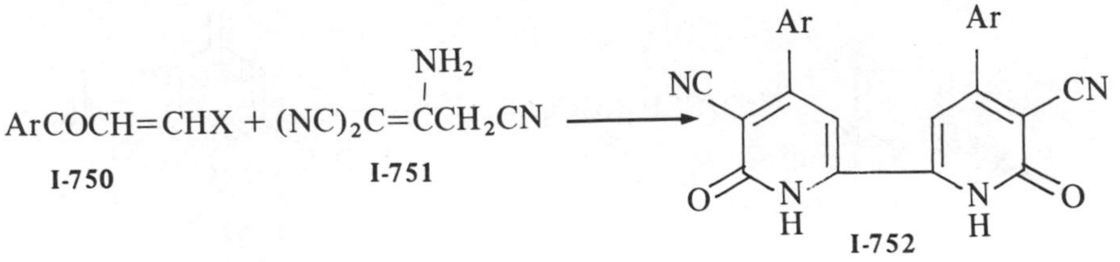

$$ArCOCH=CHX + (NC)_2C=CCH_2CN \longrightarrow$$
I-750   **I-751**   **I-752**

Further variations of the Guareschi-type reaction have been used to prepare condensed pyridine derivatives. Thus, α-carbolines **I-753** were obtained on reaction of *N*-substituted aminoindoles **I-754** (X = —SO$_2$Ar, —CO$_2$CH$_2$Ph) with β-diketones (1381–1383).

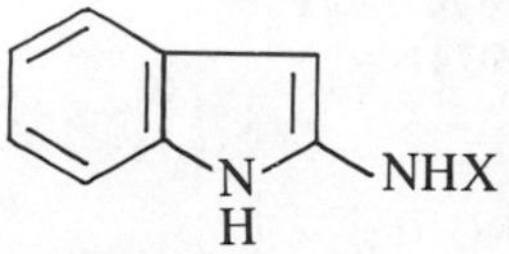

I-754 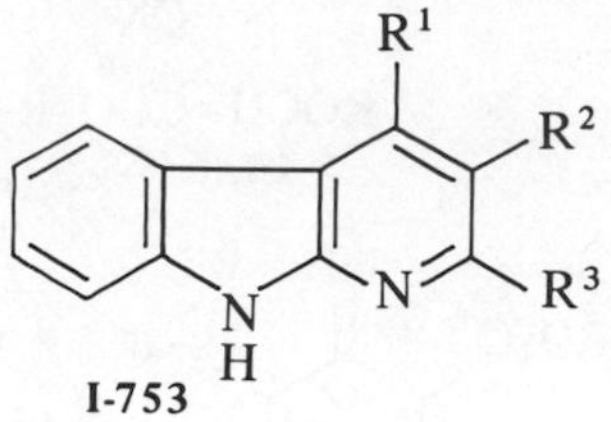   I-753

Aminopyrrole derivative **I-755** formed azaindole **I-756** on reaction with acetal **I-757**, or the isomeric **I-758** on reaction with methyl vinyl ketone (1384).

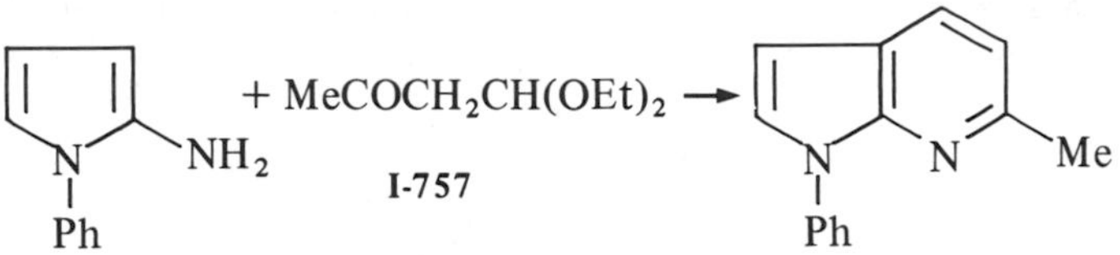

I-755   + MeCOCH$_2$CH(OEt)$_2$ →   I-756 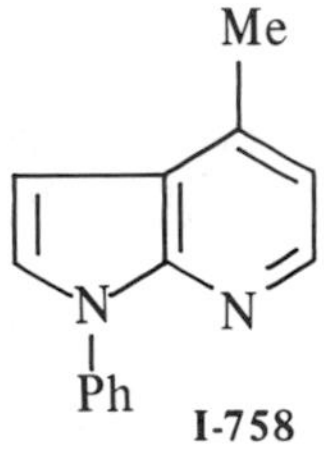   I-758

I-757

Similarly, amino derivative **I-759** condensed with ethoxymethylenemalonic ester to yield fused pyridone **I-760** (1385).

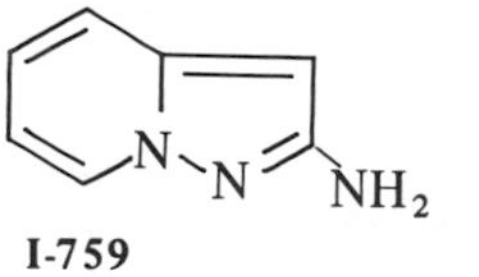

I-759 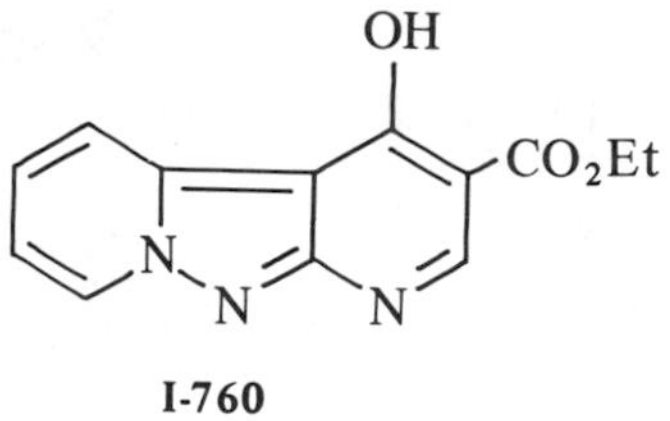   I-760

Pyrazolopyridines **I-761** were formed by the reaction of aminopyrazoles **I-762** with keto ester **I-763** (1386, 1387). Similarly, aminopyrazolinones **I-764** reacted with enamino ketones **I-765** to form **I-766** (1388).

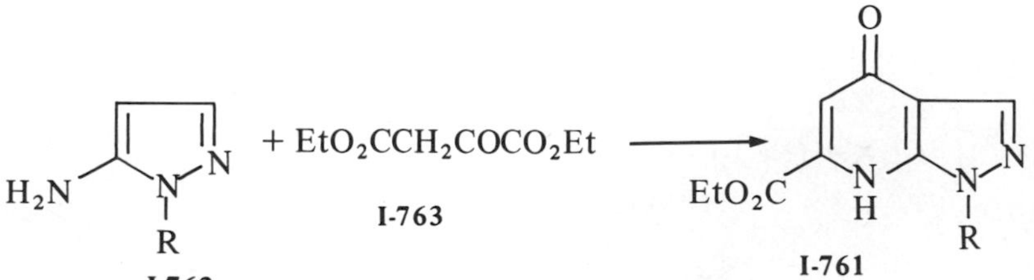

I-762   + EtO$_2$CCH$_2$COCO$_2$Et →   I-761

I-763

I-764 **I-765** **I-766**

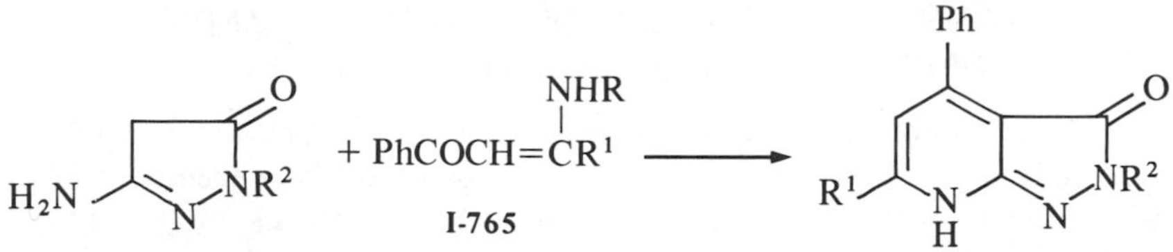

Isoxazole derivative **I-767** was found to react with malonic ester to yield isoxazolopyridine **I-768** (1389).

I-767

I-768

I-769

I-770

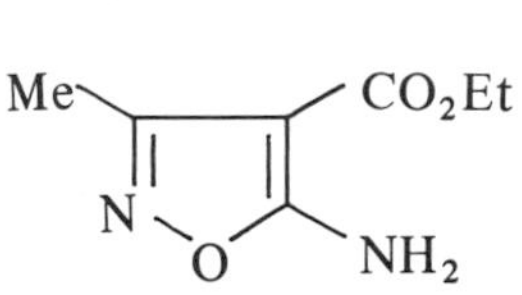

Aminoisoxazolone **I-769** formed **I-770** on reaction with acetylacetone (1390).

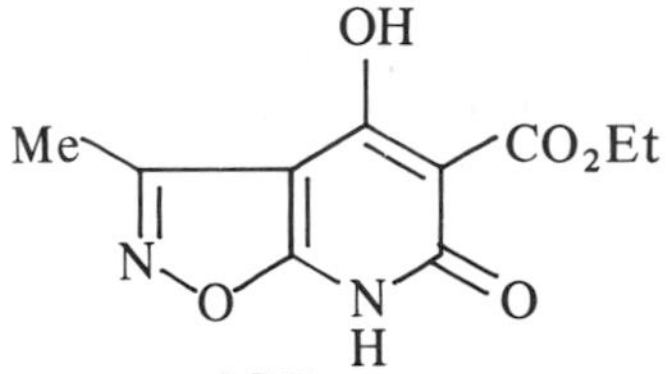

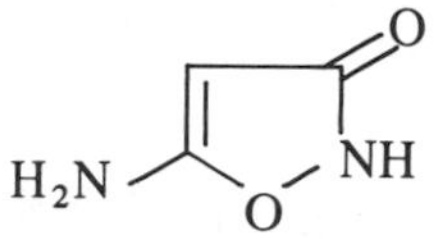

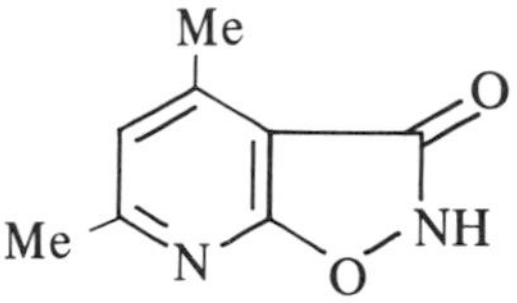

Aminoselenophenes **I-771** and **I-772** (X = Se) formed selenolopyridines **I-773** and **I-774** (X = Se), respectively, on reaction with 1,1,3,3-tetramethoxypropane and zinc chloride; **I-774** (X = S) was formed similarly from **I-772** (X = S) (1391).

I-771

I-772

I-773

I-774

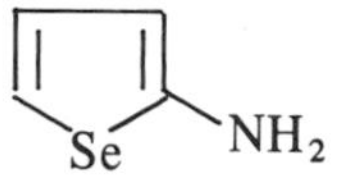

The Friedlander condensation of *ortho*-amino aldehydes with methylenic ketones may be thought of as another variation on the Guareschi synthesis (1392). Interesting uses of pyridine derivatives include reaction of 2-aminonicotinaldehyde with cyclic ketones to form **I-775** (1393), and of 2,6-diaminopyridine-3,5-dicarboxaldehyde with acenaphthenone to form **I-776** (1394).

I-775

I-776

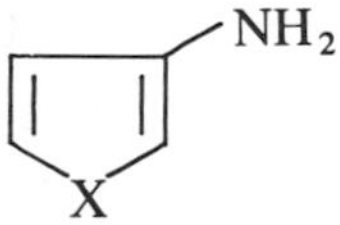

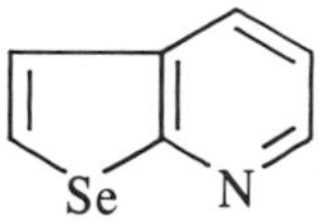

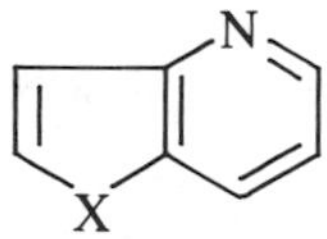

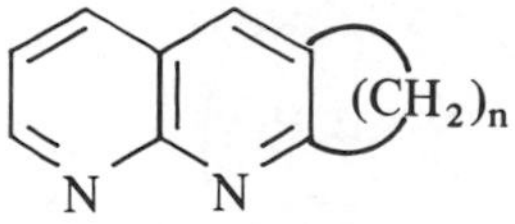

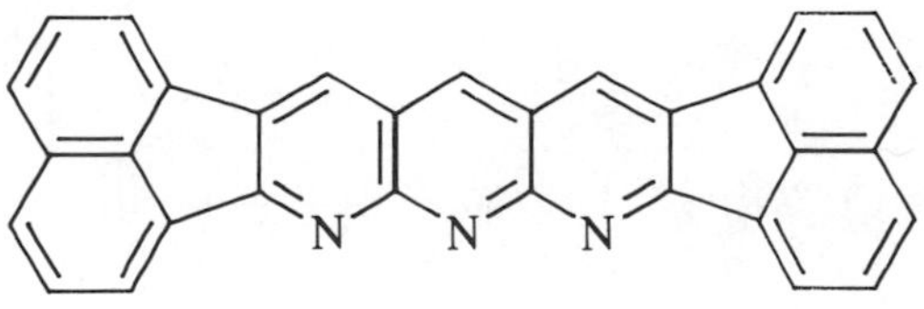

b.   α,β-UNSATURATED CARBONYL COMPOUNDS AND THEIR DERIVATIVES WITH METHYLENIC COMPOUNDS

Since an α,β-unsaturated carbonyl compound **I-777** is of a lower oxidation state than a similar β-dicarbonyl 3-carbon unit, the product of its condensation with an active methylene **I-778** or similar 2-carbon unit should be a dihydropyridine **I-779** (or a tautomer) rather than the pyridines produced by the Guareschi-type condensations. Often, the fully aromatic pyridine **I-780** is the only isolated product; it is usually presumed that the aromatization has taken place by disproportionation or air oxidation. In some cases, variation in reaction conditions can allow isolation of either the dihydropyridine or the pyridine product. Examples of this kind of reaction are given in Table I-73.

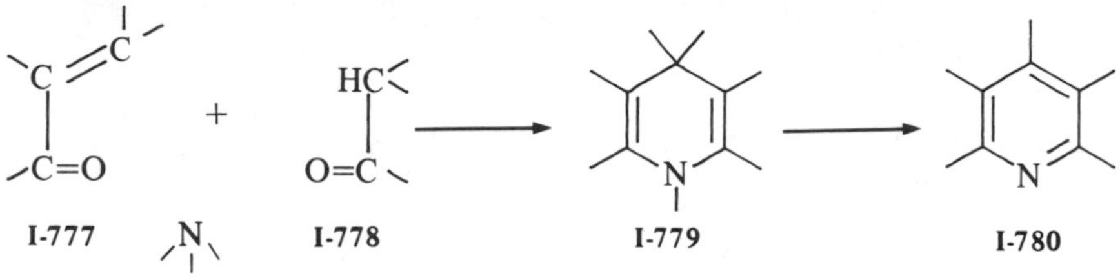

The mechanism seems generally to be Michael addition of **I-778** to **I-777**; the orientation of the 1- and 3-substituents of **I-777** has most often been demonstrated (or frequently only presumed) to appear in the product in a manner consistent with this mechanism. The reversed orientation, following from 1,2- rather than 1,4-addition to **I-777**, has been noted in by-products (998), or in a single product (1422).

Extensive variation from the typical functional groups is common; thus the nitriles **I-781** were found to react with malononitrile in methanol to yield **I-782** or the fully aromatic derivative (1423).

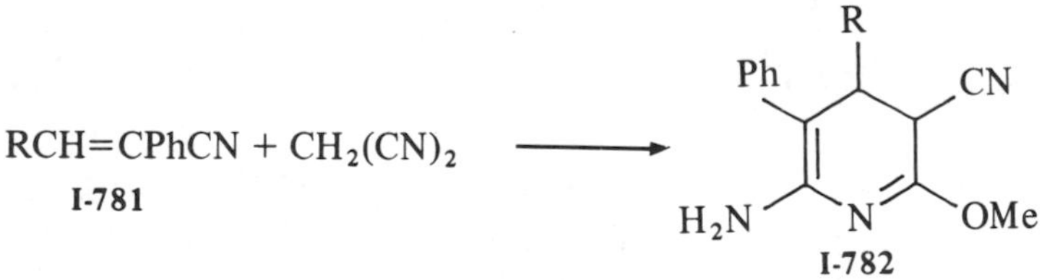

$$RCH=CPhCN + CH_2(CN)_2 \longrightarrow$$
I-781

Compound **I-783** was formed from reaction of **I-784** and malononitrile using sodium methoxide; **I-785** was also formed (1424).

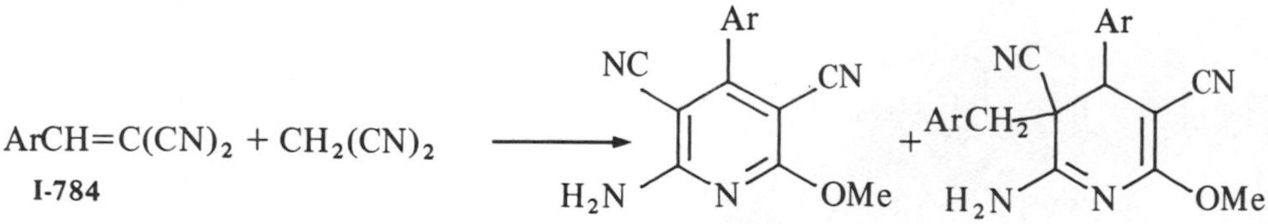

$$ArCH=C(CN)_2 + CH_2(CN)_2 \longrightarrow$$
I-784

TABLE I-73.   CONDENSATION OF ACTIVE METHYLENIC COMPOUNDS WITH α,β-UNSATURATED CARBONYL COMPOUNDS

$$R^1COCR^2=CHR^3 + XCH_2Y \longrightarrow$$

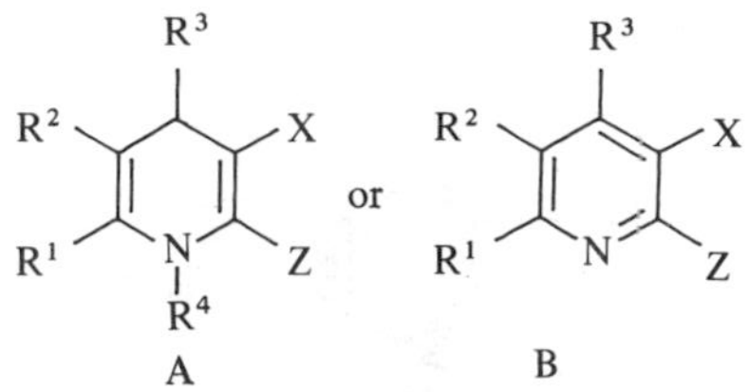

A   B

R^1	R^2	R^3	X	Y	Product A	Product B	Reference
NHR	CN	Ar	CN	CO_2Et		X	1395, 1396
NHR	CN	Ar	CN	CN		X	1397
Ar	H	Ar	CN	CO_2Et	X or	X	1398–1402
Ar	H	Ar	CN	CN		X	1398
Ar	H	Ar	CO_2Et	CN		X	1399
Ar	H	Ar	$CONH_2$	$CONH_2$		X	1403
Me	H	Pr	CN	CO_2Et		X	1401
Ph	H	Ph	$CONH_2$	$MeCO^a$		X	1404
RCO_2	H, Me	H, Me	$CONH_2$	$MeCO^a$	X		1405
Me	$-CO_2C_6H_4-$		CO_2Et	Me		X	1406
Ar	H	$2-C_4H_3S^c$	CN	CO_2Et		X	1407
Ar	Alkyl	H^b	RCO	R^a	X and	X	1408–1410
OEt	CN	Ar	CN	CN		X	1411, 1412
OEt	CN	Ar	Alkyl	COMe		X	1413
$2-C_4H_3O^d$	H	Ar	H	$2-COC_4H_3O^d$		X	1414
$2-C_7H_5N_2{}^e$	H	Ar	CN	CO_2Et	X or	X	1415
$2-C_{10}H_{11}{}^f$	H	Ph	CN	CO_2Et	X or	X	1416
Ar	H	Ar	CONHR	CONHR	X or	X	1417
$-(CH_2)_6-$		H^b	$-(CH_2)_6CO-$			X	1418
Ar	H	$-CH=CHPh$	CN	CO_2Et, CN	X and	X	1400, 1419, 1420
$-CH_2CR_2CH_2CO-^a$			PhCO	PhCO	X or	X	1421

^{a} As enamine. ^{c} Thienyl. ^{e} Benzimidazolyl.
^{b} As Mannich base precursor. ^{d} Furyl. ^{f} Tetrahydronaphthyl.

The coumarin **I-786** was found to yield fused pyridines **I-787** or dihydropyridines on reaction with ketones **I-788**; the same products could be prepared by reaction of salicyclaldehydes, malononitrile, and ketones (1425).

I-786 I-787

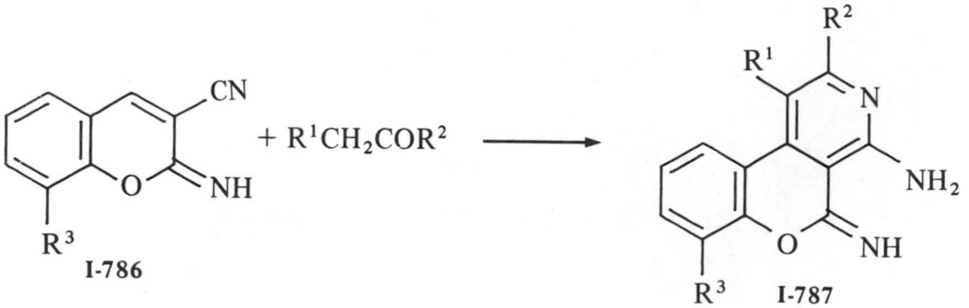

Lead tetraacetate oxidation of enamine **I-789** (R = Me, Et) yielded pyridone **I-790** among other products (1426).

I-789 I-790

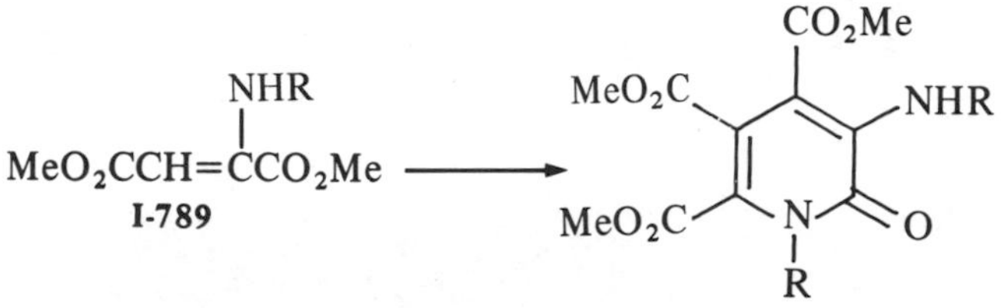

Some α,β-unsaturated carbonyl compounds undergo a retro-aldol or -Knoevenagel reaction under the condensation conditions; the resulting active methylene can sometimes enter the condensation reaction instead of, or in competition with, an added active methylene unit.

Thus chalcones **I-791** were found to react with ammonium formate to yield triarylpyridines **I-350** (1427, 1428) by condensation of the chalcone with its acetophenone fission product **I-793**.

I-791 I-792 I-793

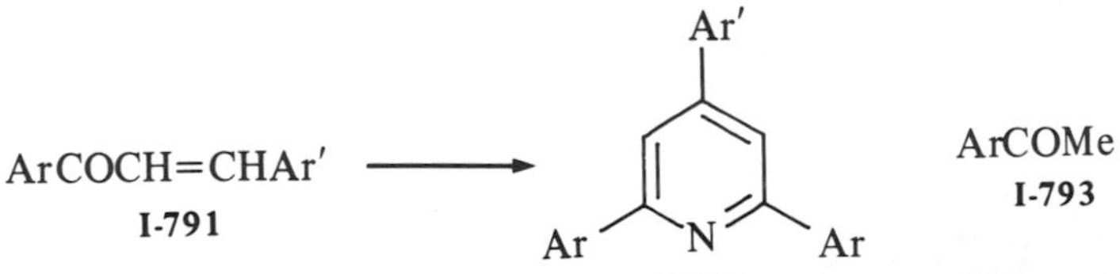

Condensation of the Knoevenagel product **I-794** with aldehyde **I-795** yielded pyridone **I-796**; metathesis of the reactants to **I-797** and **I-798** no doubt preceded the condensation (1413).

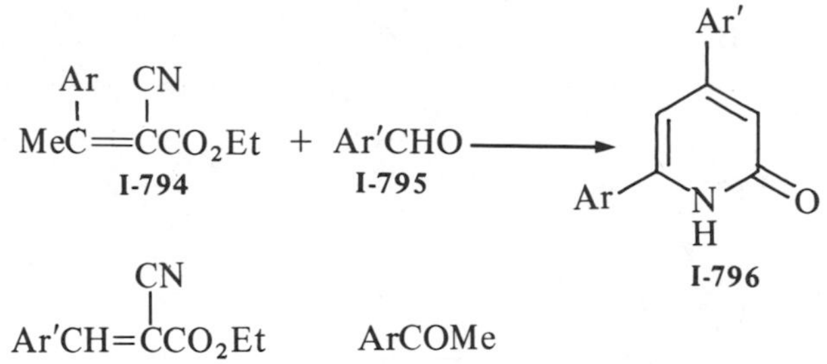

I-794 I-795 I-796 I-797 I-798

Loss of a substituent during reaction is illustrated by formation of **I-799** from reaction of **I-800** with **I-801** (1429), and of **I-802** from reaction of **I-803** with acrylonitrile (1430).

I-800 I-801 I-799

I-803 I-802

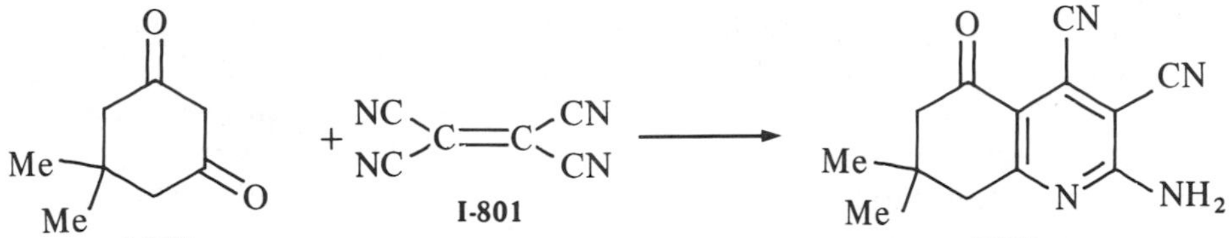

Application in fused-ring systems is possible; for example, condensation of indole derivatives **I-804** with **I-805** gave low yields of **I-806** (1431).

I-804 I-805 I-806

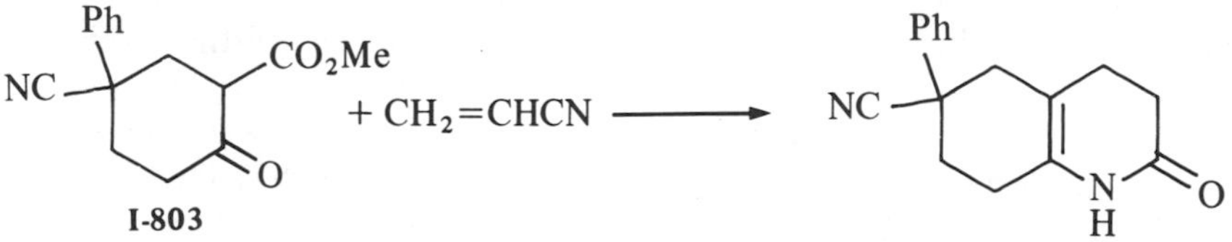

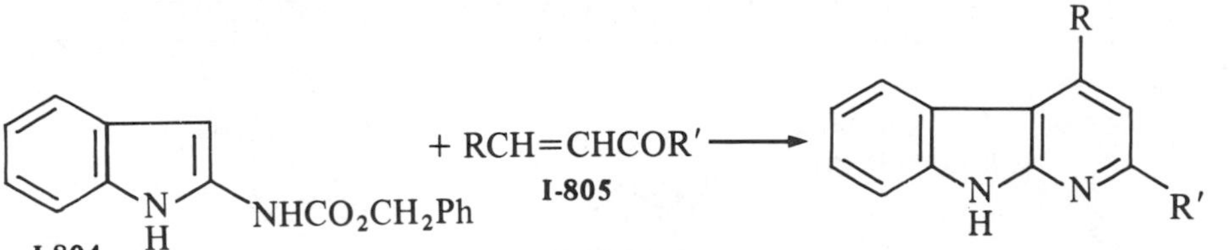

Avoidance of the problem of dihydropyridine by-products, and improved yield overall, may be simply accomplished by increasing the oxidation state of the 2-carbon unit, thereby incorporating a leaving group as in **I-807**. The most useful leaving group seems to be pyridine; the pyridinium salt **I-809** is easily prepared from an α-halocarbonyl compound. The additional advantage of this route is the stability of the betaine (**I-808**) formed from the pyridinium salt **I-809**, facilitating the Michael reaction step of the ring synthesis.

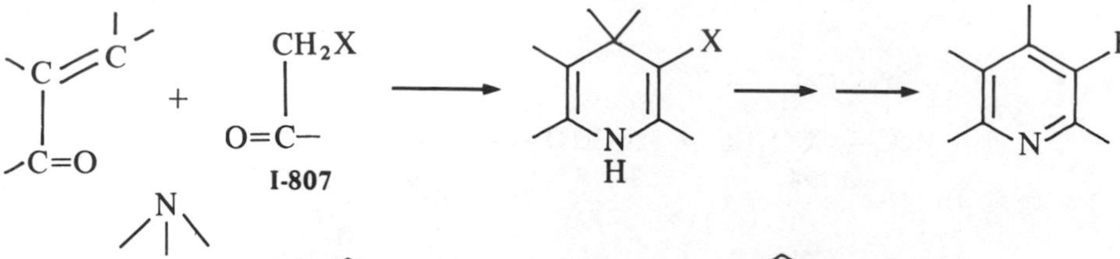

The first example of the utility of this route was the pyridone synthesis of Thesing, using N-carbamolymethylpyridinium chloride **I-810**. This reaction was reviewed in earlier volumes of this series; recent examples appear in Table I-74.

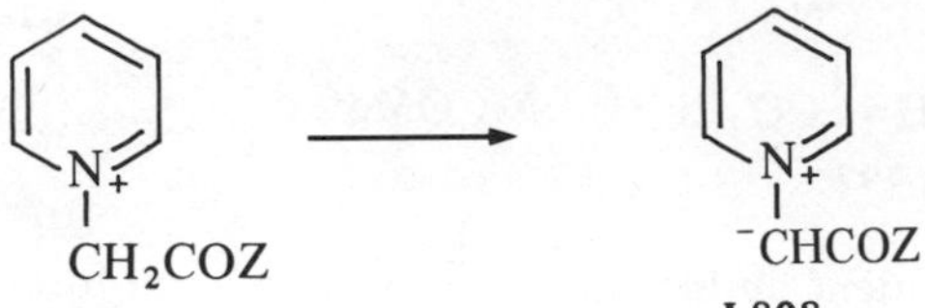

Kröhnke demonstrated the use of ketomethylpyridinium salts **I-811** for preparation of pyridines **I-812**, and noted the utility of the Ortoleva–King reaction for preparing the salt **I-811** from a methyl ketone directly without isolation of an α-halo ketone. The latter reaction, and the high yields in the ring synthesis, have made the Kröhnke route preferred for the synthesis of 2,4,6-triarylpyridines and related compounds. The reaction has been reviewed (1434), and further examples are given in Table I-75.

The reaction has also been carried out in poor to fair yield by generation of the chalcone *in situ*; thus reaction of **I-813**, **I-814**, and **I-815** gave **I-816** (1446).

Other leaving groups (X in **I-807**) that also stabilize the intermediate anion or betaine are, of course, possible, but have been rarely explored. The sulfonium group as in **I-817** can be used (1447).

TABLE I-74.   REACTIONS OF N-CARBOMOLYMETHYLPYRIDINIUM CHLORIDE

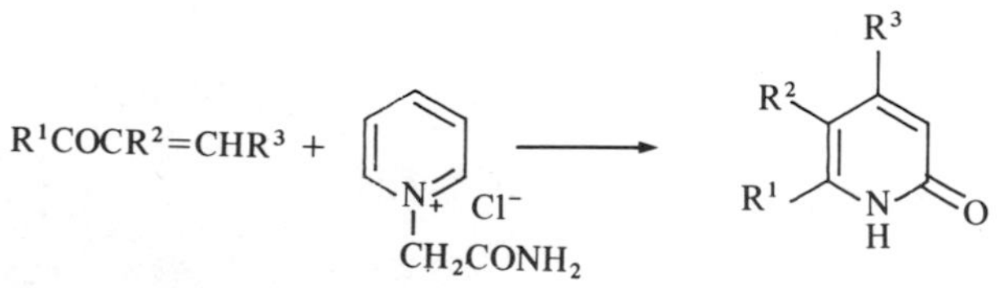

R^1	R^2	R^3	Reference
H	H	$o\text{-}NO_2C_6H_4$	1432
Ar	R	$3\text{-}C_8H_6N\text{--}$[a]	1433
Ar	H	$2\text{-}C_9H_6N\text{--}$[b]	1433

[a] Indolyl.
[b] Quinolyl.

TABLE I-75. REACTIONS OF KETOMETHYLPYRIDINIUM SALTS

I-811 I-812

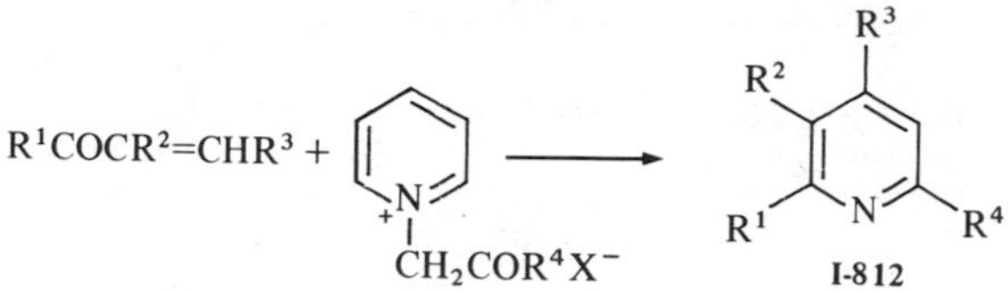

R^1	R^2	R^3	R^4	Reference
Ar	H	2-C_9H_6N—[a]	Me	1433
Ar	H	CO_2H	Ar	1435–1437
H	H	o-$NO_2C_6H_4$	Me	1438–1441
Ar	H or Br	Ar	Ar	1442
Ar	H	Ar	Ar	1443, 1444
Ar	H, OMe	Ar	Ar	1445

[a] Quinolyl.

$PhCOCH_2OMe$ (I-813) $PhCHO$ (I-814)

I-815

I-816

I-817

The gas-phase reaction of acrolein with carbonyl compounds remains an interesting method for the synthesis of relatively simple compounds. The reaction of acrolein with acetaldehyde and ammonia to produce pyridine has been extensively patented (1448–1456).

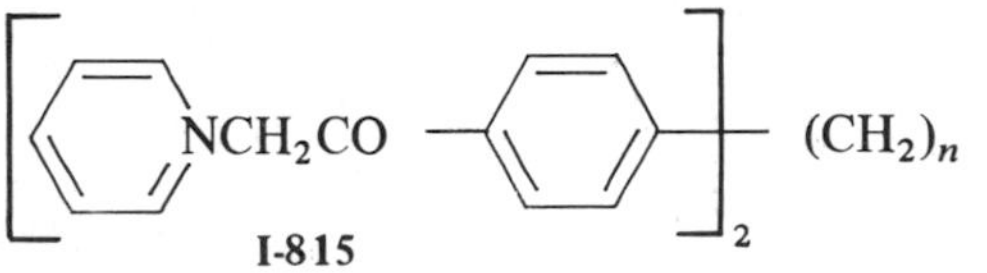

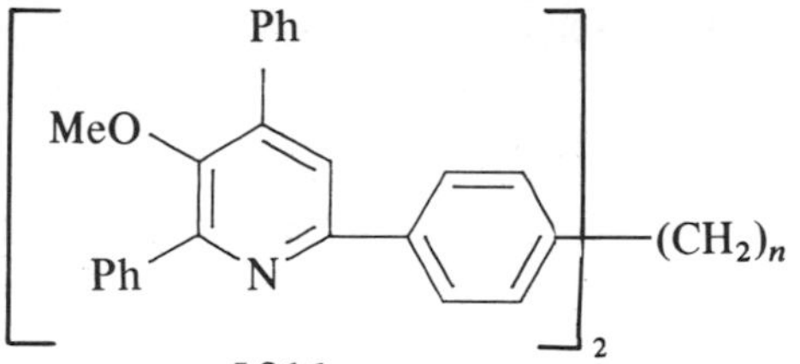

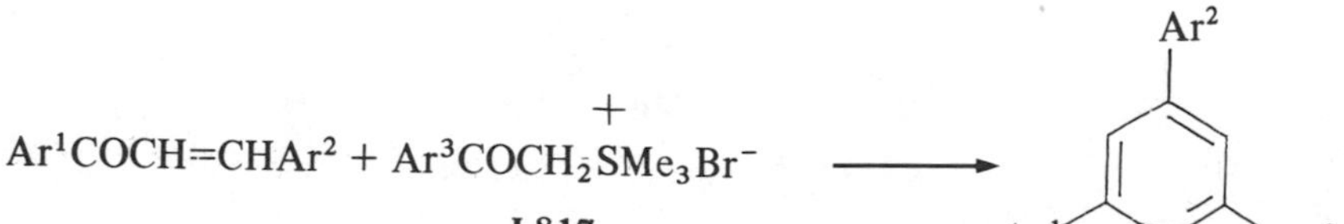

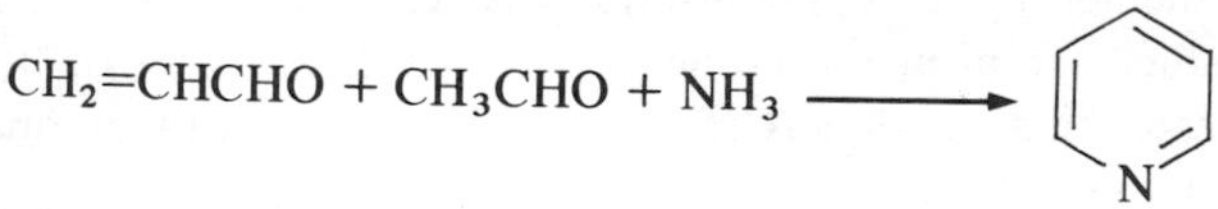

TABLE I-76.   GAS-PHASE CONDENSATIONS OF CARBONYL
COMPOUNDS WITH ACROLEIN AND AMMONIA

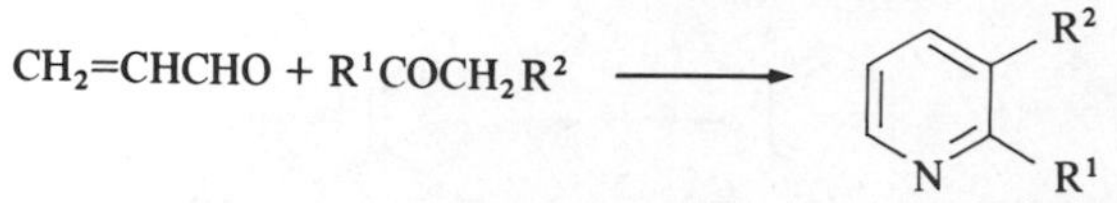

$$CH_2=CHCHO + R^1COCH_2R^2 \longrightarrow$$

R^1	R^2	Reference
$PhCH_2$	Ph	1457, 1458
CH_3	H	1459
H	CH_3	1460
Ph	H	1461
$-(CH_2)_n-$		1462

The gas-phase reaction of acrolein with more highly substituted compounds has yielded substituted pyridines, as shown in Table I-76.

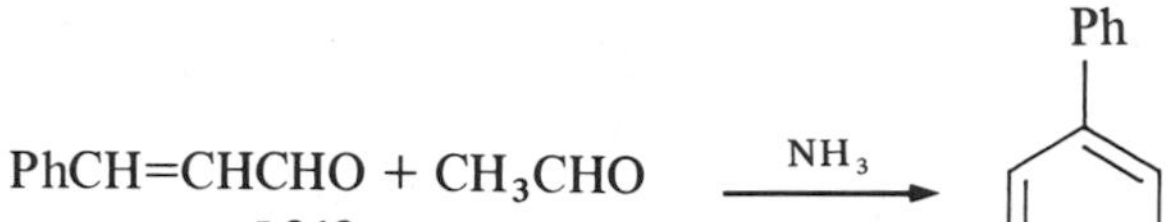

$$PhCH=CHCHO + CH_3CHO \xrightarrow{NH_3}$$
I-818

Naturally, substitution in the 3-carbon component is also allowed; gas-phase reaction of cinnamaldehyde **I-818** with acetaldehyde and ammonia produced 4-phenylpyridine (1463); crotonaldehyde with phenylacetaldehyde gave poor yields of **I-819** and **I-820** (1464).

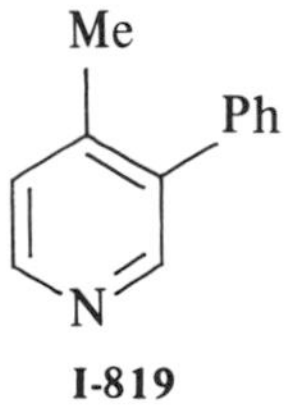

I-819 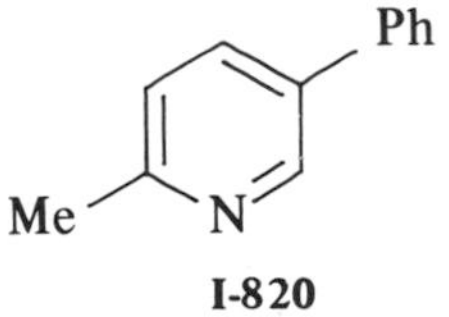   **I-820**

c.   CONDENSATION OF α,β-UNSATURATED CARBONYL COMPOUNDS
WITH AMMONIA

Gas-phase reaction of acrolein with ammonia is a well-known reaction yielding 3-methylpyridine and pyridine; the reaction has been reviewed (1465). Comparison with the products from the reaction of the acrolein Diels–Alder dimers with ammonia indicates that the dimers are not intermediates in the normal reaction of acrolein with ammonia (1466).

Several patents have been concerned with the catalyst for the acrolein–ammonia reaction (1467–1476). Addition of air or oxygen, with catalyst variation, has been claimed to increase the yield of pyridine (1477–1485). Reaction of propylene and air over oxidation catalysts, followed by reaction of the crude gas stream with ammonia over a condensation catalyst, has given pyridine and 3-methylpyridine (1486–1489); acrolein is certainly implicated as an intermediate. Direct combination of propylene, air, and ammonia has been claimed to give low yields of pyridine and methylpyridines (1490–1497). Reaction of isobutylene, acetaldehyde, air, and ammonia has yielded pyridine and methylpyridines (1498). Propylene oxide may be included with the acrolein (1499, 1500). Substitution of methylamine for ammonia in reaction with acrolein has given decreased yields of pyridine but increased amounts of 3-methylpyridine and 3,5-dimethylpyridine (1501). Liquid-phase reaction of acrolein and ammonium acetate has yielded pyridine (1502).

The protected acrolein derivative, 1,1,3-trimethoxypropane, has given 3-methylpyridine and pyridine on reaction with ammonia in either the liquid (1503) or gas (1504) phase.

Gas-phase reaction of diethyl ketone with ammonia, using a platinum on alumina catalyst, gave pyridine **I-821**; methyl ethyl ketone similarly yielded **I-822** and **I-823** (1505). Dehydrogenation of the starting ketones to **I-824** and normal 3-2 reaction can be suspected.

Liquid-phase reaction of the trimethylol **I-825** or its cyclic derivative **I-826** with ammonia or amines yielded 3,5-dialkylpyridines **I-827**. Loss of formaldehyde and decomposition to the intermediate **I-828** was suggested (1506–1513).

d. MISCELLANEOUS 3-2 CONDENSATIONS

The allene **I-829** was found to react with several imines or enamines **I-830** to give pyridones **I-831** (1514, 1515); one of these products was used as an intermediate in a total synthesis of camptothecine (1514b).

The cyclobutenedione **I-832** yielded 3-hydroxypyridones **I-833** in reaction with enamines **I-830** (R = H, R^2 = —CO) (1516).

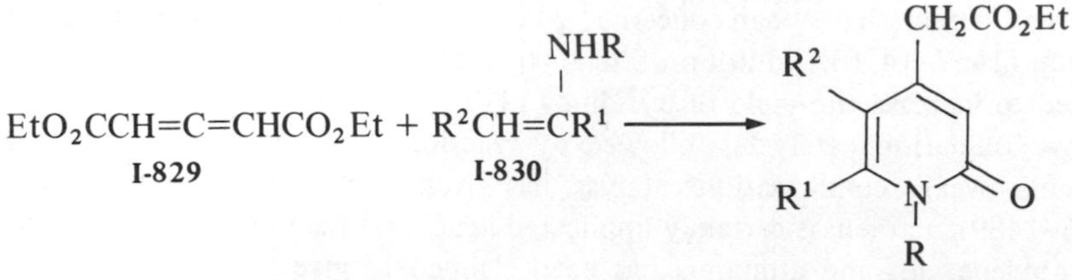

$$EtO_2CCH=C=CHCO_2Et + R^2CH=CR^1 \longrightarrow$$

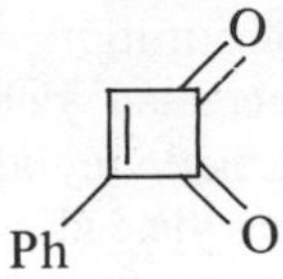

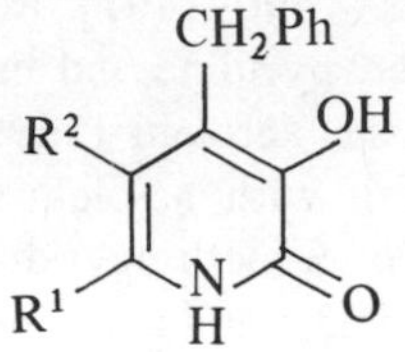

I-829   **I-830**

(with NHR on the first reactant)

I-831   (product pyridone bearing CH_2CO_2Et, R^2, R^1, R, $=O$)

I-832   (cyclobutenedione with Ph)

I-833   (pyridone bearing CH_2Ph, OH, R^2, R^1, N-H, $=O$)

Reaction of acetylenedicarboxylic esters with *bis*-imine **I-834** yielded pyridines **I-835** (1517).

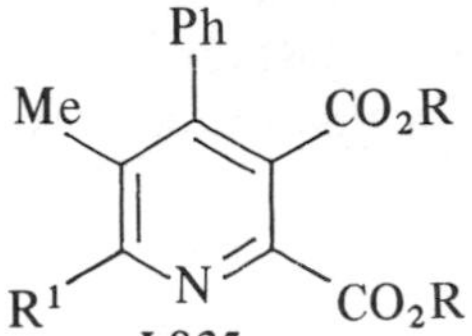

$$\underset{\textbf{I-834}}{R^1\overset{\overset{NH}{\|}}{C}CHMe\overset{\overset{NR}{\|}}{C}Ph}$$

I-835   (pyridine with Ph, Me, CO_2R, R^1, CO_2R)

Dichloroketene **I-836**, generated from dichloroacetyl chloride, reacted with cinnamaldehyde imine **I-837** to yield **I-838** and subsequently **I-839** (1518).

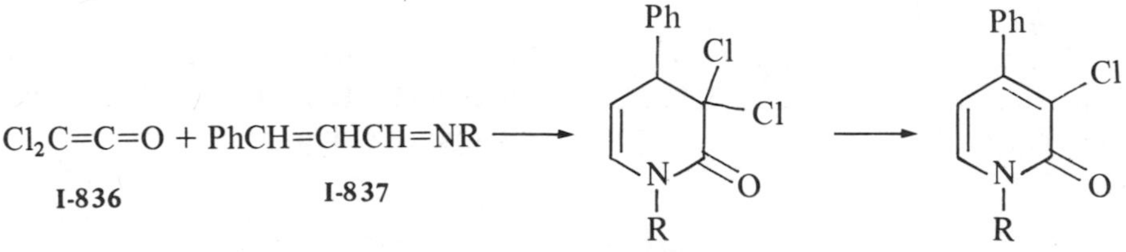

$$Cl_2C=C=O + PhCH=CHCH=NR \longrightarrow$$

I-836   **I-837**

I-838   (dihydropyridone with Ph, Cl, Cl, N-R, $=O$) $\longrightarrow$ **I-839** (pyridone with Ph, Cl, N-R, $=O$)

Dienamines **I-840** were found to react with enamines **I-841** ($R' = Me$, OEt) to yield dihydropyridines **I-842** (or tautomers), or pyridines **I-843** (1519). The related enynamines **I-844** similarly yielded mixtures of **I-843** and the isomeric **I-843a**, the former predominating (1519, 1520).

$$CH_2=CHCH=CHNR_2$$

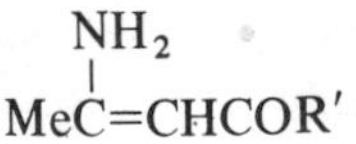

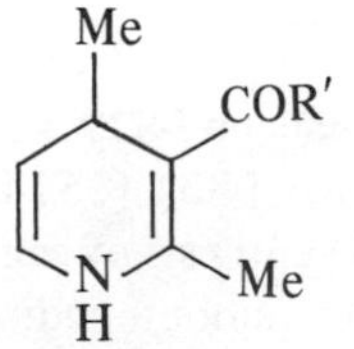

I-840

$$\underset{\textbf{I-841}}{Me\overset{\overset{NH_2}{|}}{C}=CHCOR'}$$

I-842   (dihydropyridine with Me, COR', Me, N-H)

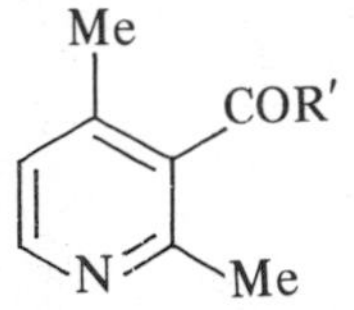

I-843 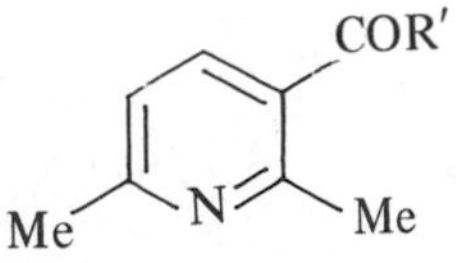   I-843a   HC≡CCH=CHNR₂   I-844

Selenoles **I-845** and **I-846** (R = H, Me) were used to prepare selenolopyridines **I-847** and **I-848**, respectively, by condensation with α-aminoacetaldehyde dimethyl acetal and acid-catalyzed ring closure (1521).

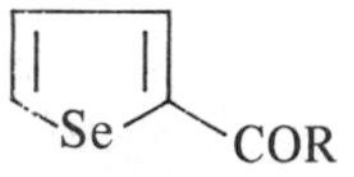

I-845 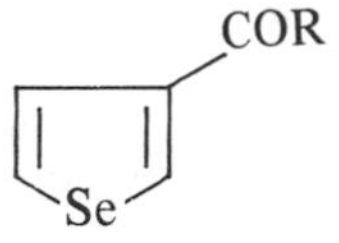   I-846 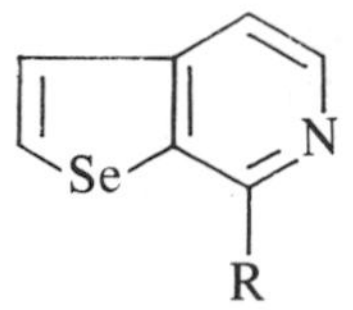   I-847 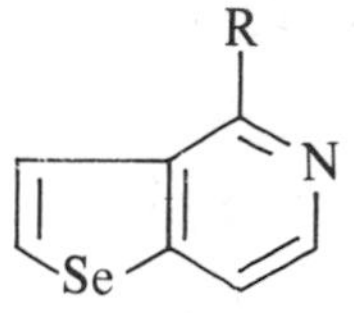   I-848

D. *1-3-1 Condensations*

This class of reaction is one of the most rare kinds of pyridine ring-forming reactions. However, it should be noted that in the cyclization of 1,3,5-triketones to 4-pyridone derivatives (Section II.2.A.a), the construction of the intermediate corresponds essentially to a 1-3-1 condensation.

Reaction of vinylogous amidinium salt **I-849** with imine anion **I-850** was found to yield the pyridine **I-851** (1522).

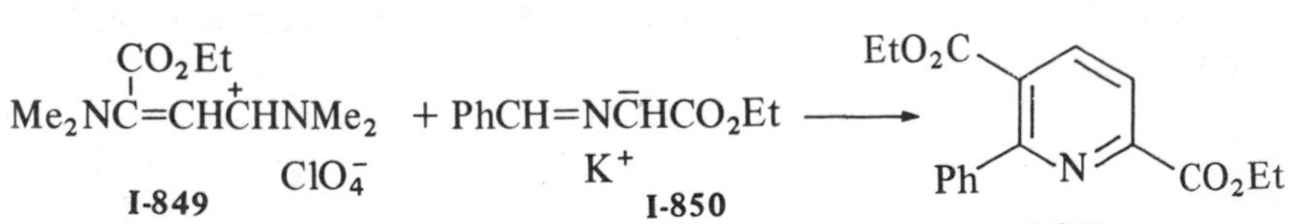

Condensation of dibenzyl ketone **I-852** with excess formamide gave the pyridine **I-853**; the same product could be formed from phenylacetic anhydride and s-triazine (1523).

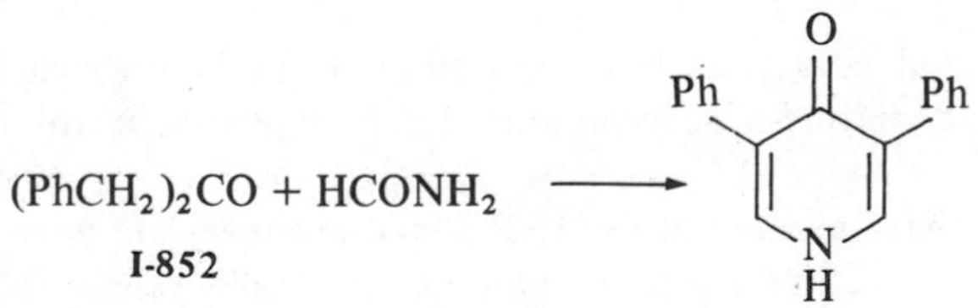

$$(PhCH_2)_2CO + HCONH_2 \longrightarrow$$

I-852

I-853

TABLE I-77.   PYRIDINES AND ACETYLENES AND NITRILES

$$R^1C{\equiv}CR^2 + R^3CN \longrightarrow (R^1)_2 \underset{\text{N}}{\overset{(R^2)_2}{\diagdown}} R^3$$

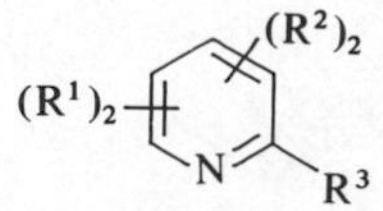

R^1	R^2	R^3	Reference
H, Me	H, Me	Me, Et, C_3H_5	1525
H, Ph, CH_2OMe	H, Ph, CH_2OMe	Me, Ar, $PhCH_2$, MeO_2CCH_2	1526
$(CH_2)_n$	H	R	1427
H, alkyl, Ph	H	Alkyl, C_2H_3, Ar	1528, 1529
H, alkyl, Ph	H, alkyl, Ph	Me, Ar	1530, 1531
H, Me, Ph	H, Me, Ph	$(CH_2)_n$, $-C_6H_4-$	1532
H, alkyl, Ph	H, alkyl	Alkyl, $PhCH_2Ar$, $(CH_2)_n$, $-C_6H_4-$	1533
H, alkyl	H	Alkyl	1534
H	H	Alkyl, C_2H_3, C_5H_4N	1535
$(CH_2)_n$	H, alkyl	R	1536, 1537
H	H	$-CH_2CH_2NR_2$, $-CH_2CH_2OR$	1538–1540
H	H	R_2N	1541

E.   2-2-1 Condensations

a.   ACETYLENES AND NITRILES

Various acetylenes **I-854** have been shown to react with nitriles **I-855** in the presence of homogeneous cobalt catalysts to yield pyridines **I-856**; benzene derivatives were formed as by-products by trimerization of the acetylene. Reactions of acetylene itself have been conducted at 80–200° and about 12 atm; special safety precautions are required.

$$2RC{\equiv}CR + R'CN \longrightarrow$$

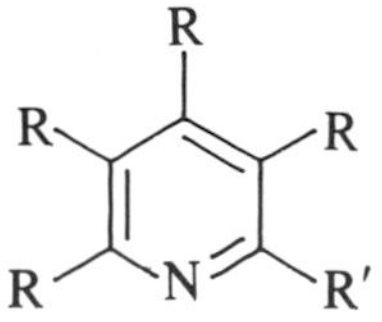

I-854   **I-855**

I-856

Monosubstituted acetylenes can form isomeric pyridines including **I-857** and **I-858**; typically a mixture of such products has been observed, with **I-857** being the major product.

The reaction has been reviewed (1524), and examples are given in Table I-77. References to the reaction of nitriles with cobalt metallocycles, the probable intermediates in this cyclization, are given in Section II.2.B.a.

I-857   I-858   I-859   I-860

In a similar manner, isocyanates have been found to react with acetylenes to yield **I-859**, and carbodiimides to yield pyridone imines **I-860** (1542–1544).

b. ALDEHYDES WITH AMMONIA

i. ACETALDEHYDE WITH AMMONIA — VAPOR PHASE.   The commercial Chichibabin synthesis of 2- and 4-methylpyridine has received considerable attention, mostly directed toward changing the balance of products to favor the more valuable 2-methylpyridine by variations in catalyst or reaction conditions. A well-established reaction mechanism, focused on the partitioning between the 2-2-1 and 2-1-2 paths to 2- and 4-methylpyridine, respectively, would be an aid to these attempts, but no details of the mechanism have yet been rigorously established. Much of the published work has been reviewed (1545, 1546), and several of the reaction by-products, which should be of aid in discerning the reaction path, have been identified (1547, 1548).

Catalyst variations have included silica–alumina (1549, 1550) promoted by fluoride ion (1551–1553), metal chlorides and fluorides (1554–1557), metal oxides (1558–1562) which have also been applied by ion-exchange to silica–alumina (1563, 1654) or zeolites (1565), and metal phosphates (1566–1571) or phosphoric acid (1572). Very high ratios of 2- to 4-methylpyridine have been obtained using molten salts such as zinc chloride as catalysts (1573–1575). Effects of catalyst pores (1576, 1577), temperature (1578), and other reaction variables (1579) have been determined.

Acetylene also reacts with ammonia (1580–1584) or formamide (1585) to produce the same mixtures of 2- and 4-methylpyridine as are derived from acetaldehyde. The observation that this reaction occurs in the absence of water supports the proposition that the aldehyde condensation can occur by way of enamine and imine intermediates, but does not clearly require that this path be favored over that occurring by way of aldehyde and enol intermediates. The self-condensation of crotonaldehyde with ammonia yields 2- and 4-methylpyridine in addition to 3-ethyl-4-methylpyridine (1586), confirming the reversibility of the first condensation step in the acetaldehyde reaction.

Reaction of propargyl alcohol with ammonia yielded 2- and 4-methylpyridine and 5-ethyl-2-methylpyridine (1587), from which products dissociation of the starting alcohol to acetylene can be inferred.

ii. ACETALDEHYDE WITH AMMONIA — LIQUID PHASE.   The commercial synthesis of 5-ethyl-2-methylpyridine (MEP) from paraldehyde has been reviewed

(1588) and numerous by-products have been identified (1589, 1590). The major variation of the process described has been addition of copper salts to the reaction, which has had the effect of increasing the yield of 2-methylpyridine (1591–1598); inclusion of formaldehyde in some of the reaction mixtures described yielded additional pyridine derivatives (1599). Addition of oxygen to the reaction mixture with copper salts also increased the yield of 2-methylpyridine (1600, 1601). Silica-alumina has been used to catalyze the reaction (1602), and the continuous operation of the process was described (1603). Hydroxylamine has been used as the source of nitrogen (1604).

Reaction of ethylene and ammonia in the presence of the typical Wacker reaction catalyst (palladium chloride–copper chloride) has yielded MEP and 2-methylpyridine; in accord with the previously mentioned effect of copper salts on the MEP reaction, increased amounts of 2-methylpyridine were observed (1605–1619). Aspects of the process have been reviewed (1612, 1614).

iii.   OTHER ALDEHYDES WITH AMMONIA – LIQUID PHASE.   Addition of oxygen and copper salts to reaction mixtures has increased yields of pyridine products, for instance, in the preparation of 3,5-diethyl-2-propylpyridine from butyraldehyde (1600, 1601). Preparation of aldehydes *in situ* by the oxo reaction has yielded pyridines; thus reaction of ethylene, carbon monoxide, and ammonia in the presence of rhodium salts yielded 2-ethyl-3,5-dimethylpyridine (1620–1622).

Reaction of aliphatic nitro compounds, for example, 1-nitrobutane **I-861**, with carbon monoxide in the presence of palladium and ferric chloride has yielded pyridines, for example, **I-862**, whose structures correspond to those of the apparent reaction of aldehydes with ammonia. Nitroethane produced only ethyl carbamate under these reaction conditions (1623). The aldehyde–ammonia from butyraldehyde (**I-863**) also formed **I-862** on heating with acetic acid (1624).

$n\text{-}C_4H_9NO_2 \longrightarrow$

I-861   I-862   I-863

The condensation product of butyraldehyde with aniline, originally assigned the structure **I-864** (1625), was correctly identified as the 1,2-dihydropyridine **I-865** (1626).

I-864   I-865

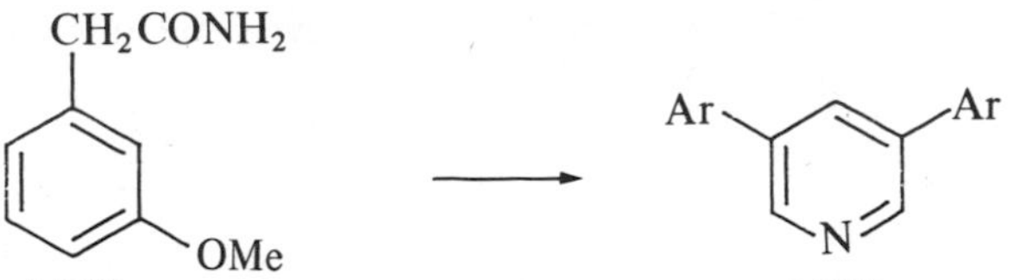

CH$_2$CONH$_2$

I-866   **I-867**

Reduction of the amide **I-866** with lithium aluminium hydride yielded the pyridine **I-867** (10%) in addition to the expected primary amine (1627).

c.   ALDEHYDES, KETONES, AND MIXTURES WITH AMMONIA – GAS PHASE

i.   ACETALDEHYDE AND FORMALDEHYDE WITH AMMONIA.   The commercial synthesis of pyridine and 3-methylpyridine has naturally received considerable study. Comments on the mechanism have been made (1628–1631), and criticized in a review (1545). The process has been compared to the acrolein process (1632). Optimization of the process has been carried out in a kinetic model (1633). While the course of the reaction is typically unclear, it should be noted that acrolein, acetone, propionaldehyde, and methyl ethyl ketone have been detected as products from the gas-phase condensation of acetaldehyde with formaldehyde, in the absence of ammonia, catalyzed by fluoridated silica–alumina (1634).

Catalysts for the condensation have included silica–alumina (1635–1645) or silica–alumina impregnated with fluoride (1646) or other halides (1647, 1648), or metal oxides (1649–1655) or metal halides (1656–1662).

ii.   OTHER MIXTURES OF ALDEHYDES AND KETONES WITH AMMONIA. Although many of these reactions are clearly 2-1-2 condensations, they are all collected in this section because of their clear relationship to other gas-phase condensations.

Reaction of aldehydes **I-868** (R = Ph, pyridyl) with an excess of acetaldehyde and ammonia yielded 4-substituted pyridines **I-869** in addition to 2- and 4-methylpyridine (1663–1666).

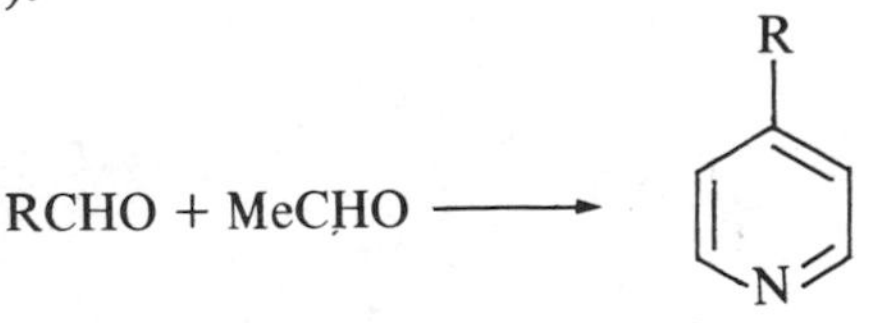

RCHO + MeCHO $\longrightarrow$

I-868   **I-869**

In contrast, condensation of propionaldehyde or butyraldehyde with aromatic aldehydes yielded the 2-2-1 products **I-870** (R = Me, Et) (1667).

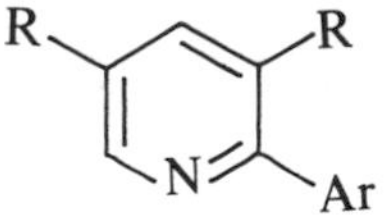

I-870

With aliphatic methyl ketones **I-871**, benzaldehyde and furfural were found to condense to yield pyridines **I-872**; by-products **I-873** were also formed (1668, 1669).

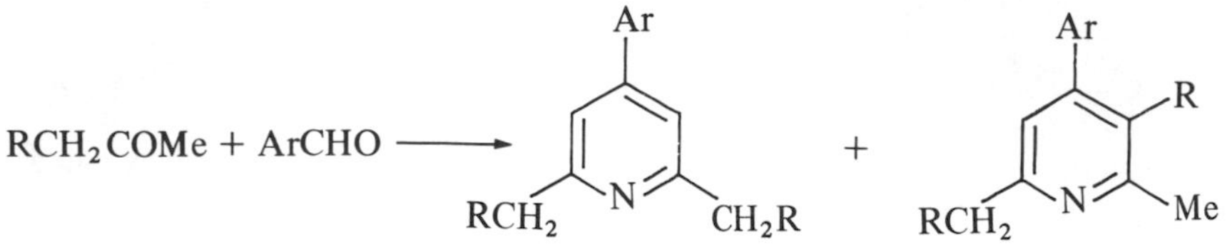

$$RCH_2COMe + ArCHO \longrightarrow$$

I-871   **I-872**   **I-873**

Acetophenone and related compounds reacted with formaldehyde and ammonia to yield 2,6-diarylpyridines (1670). In the liquid phase, acetophenone condensed with p-aminobenzaldehydes and ammonium acetate yielding 4-aryl-2,6-diphenyl-pyridines (1671).

Condensation of 1-hexanol with acetaldehyde and ammonia using dehydrogenating catalysts yielded pyridines **I-874** and **I-875** in addition to 2- and 4-methyl-pyridine (1672).

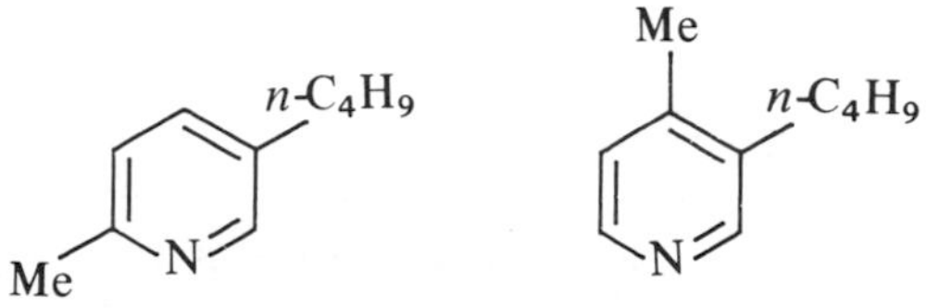

I-874   **I-875**

1-Butanol and formaldehyde reacted with ammonia using dehydrogenating catalysts to yield 3,5-diethylpyridine (1673).

Cyclic ketones **I-876** were found to react with ammonia to yield the 2-2-1 fused product **I-877** (1674).

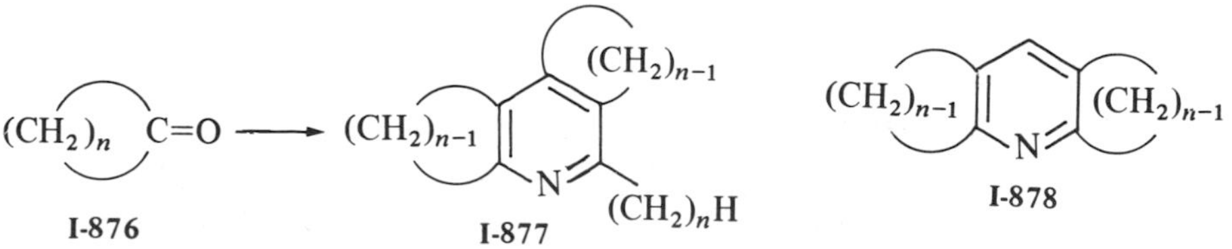

I-876   **I-877**   **I-878**

By contrast, cycloalkanones **I-876** reacted with formaldehyde and ammonia to yield **I-878** (1675).

Products from the reaction of acetone with ammonia have been identified as 2,4- and 2,6-dimethylpyridine and 2,4,6- and 2,3,6-trimethylpyridine (1676).

Variations of the well-known synthesis of 2,6-dimethylpyridine from acetone, formaldehyde, and ammonia have been documented (1677–1681).

d. OTHER OXYGENATED COMPOUNDS WITH AMMONIA – VAPOR PHASE

Reaction of ethylene oxide with ammonia (1682–1684) or other amines (1685) has yielded pyridine and 2- and 4-methylpyridine; numerous by-products have been identified (1686). Propylene oxide has similarly yielded mixtures of alkylpyridines (1687–1689). Methylpyridines have been identified among the products of the reactions of ethylene glycol with ethylenediamine (1690) or ammonia (1691), and of the self-condensation of ethanolamine (1692).

e. MISCELLANEOUS 2-2-1 CONDENSATIONS

Liquid-phase condensations of cyclic ketones I-876 with formaldehyde and ammonia yielded 2-2-1 products I-879 (1693, 1341). The contrasting 2-2-1 and 2-1-2 condensations in the liquid and gas phases, respectively, have been noted previously. However, reaction of cyclohexanone with aldehydes and amines in strong acid has yielded the 2-1-2 product I-880 (1694).

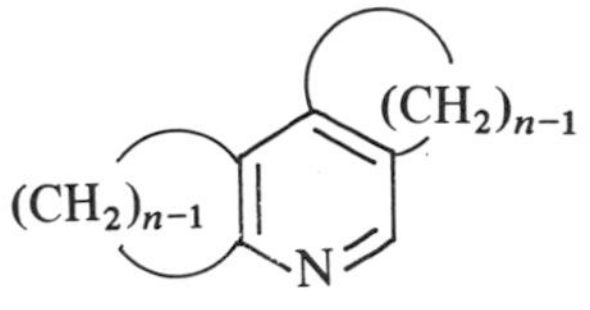

I-879

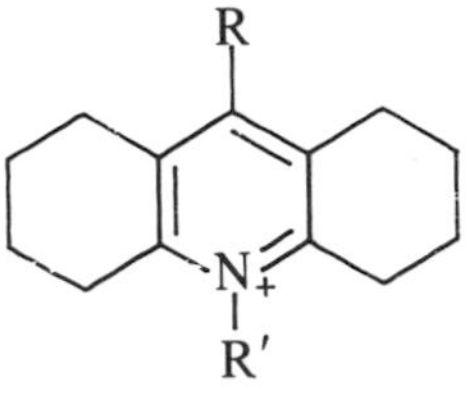

I-880

Reaction of aldehydes I-881 with the ethanolamine complex I-882 gave pyridines I-883 in 20–40% yield (1695).

$$RCH_2CHO \qquad [(H_2NCH_2CH_2O)_2Cr(H_2NCH_2CH_2OH)(H_2O)]\,Cl \longrightarrow$$

I-881 I-882

Phosphatidylethanolamine on warming with aldehydes yielded the N-hydroxyethyl quaternary salts of I-883 (1696).

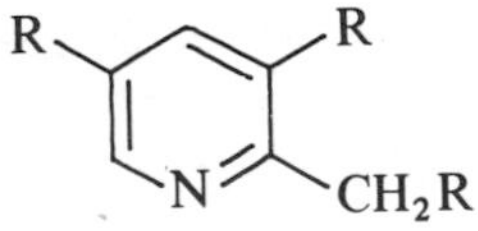

I-883

The formation *in vivo* of desmosine (I-6) and isodesmosine (I-7), believed to be responsible for the crosslinking of the polypeptide chains in elastin, appears to have come about by 2-1-2 and 2-2-1 condensation of lysine with the aldehyde derived from lysine (1697).

In vitro, several aliphatic aldehydes were shown to react with amino acids to give low yields of the quaternary salts I-884 (1698).

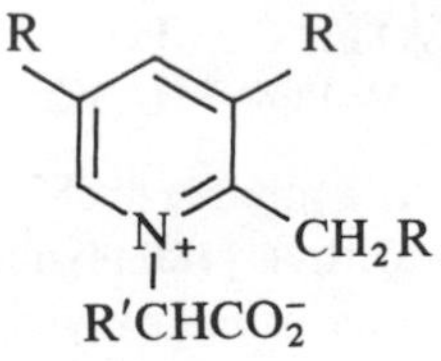

I-884

Reaction of the 1,5-diketone **I-885** with malononitrile yielded the dihydro-quinoline **I-886** (1699, 1700).

PhCO(CH$_2$)$_3$COPh

I-885

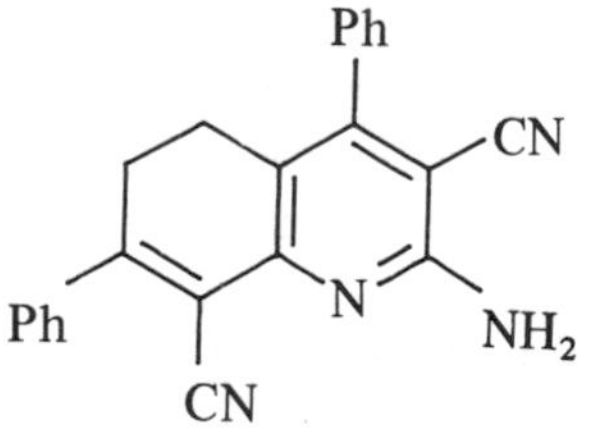

I-886

Trimers formed by treatment of malononitrile with base have been identified as **I-887**, **I-888**, and **I-889**; dimer **I-751** is believed to be a precursor (1701–1704). Dimer **I-751** and trimers **I-887** and **I-888** were also identified from high-pressure reactions (1705).

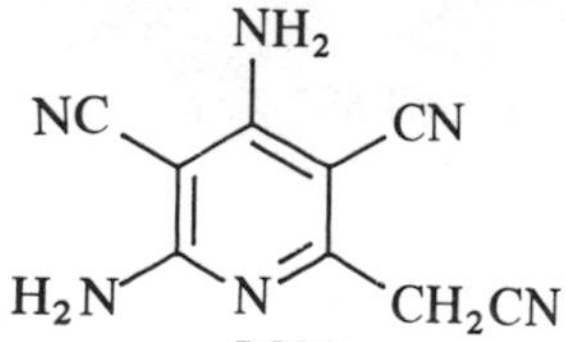

I-887

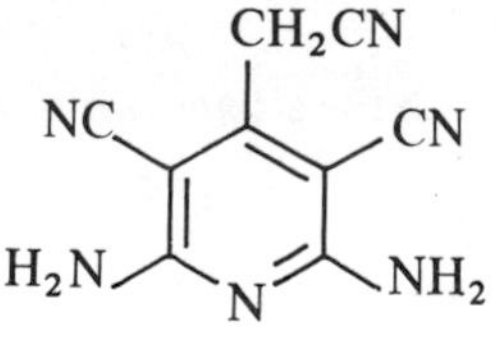

I-888

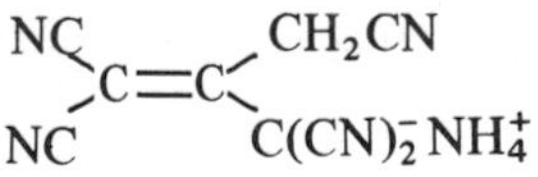

I-889

Reaction of malononitrile with aliphatic methyl ketones RCOMe gave pyridines **I-890** in 4–34% yield (1706).

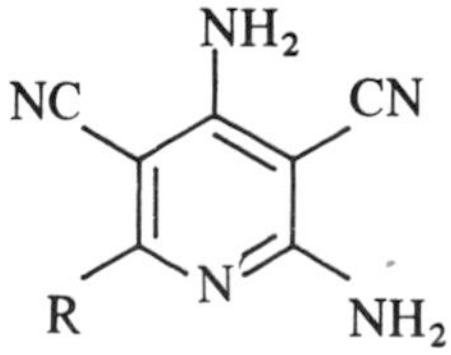

I-890

The malonic acid derivative **I-891** was found to react with methyl formate to yield the pyridine derivative **I-892** (1707).

Reaction of acetone oxime with acetylene (generated *in situ* from calcium carbide) gave 2,4,6-trimethylpyridine in 10% yield (1708–1710).

EtNHCOCH$_2$CSNH$_2$
I-891

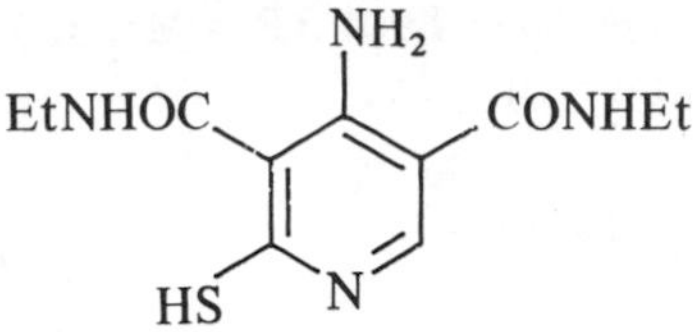

I-892

The potassium salt **I-893**, generated by the reaction of the potassium salt of nitroacetonitrile with formaldehyde, gave the pyridine N-oxide **I-894** on acidification (1711, 1712). The related **I-895** yielded **I-896** on acidification (1711, 1713) (cf, Section II.2.A.e).

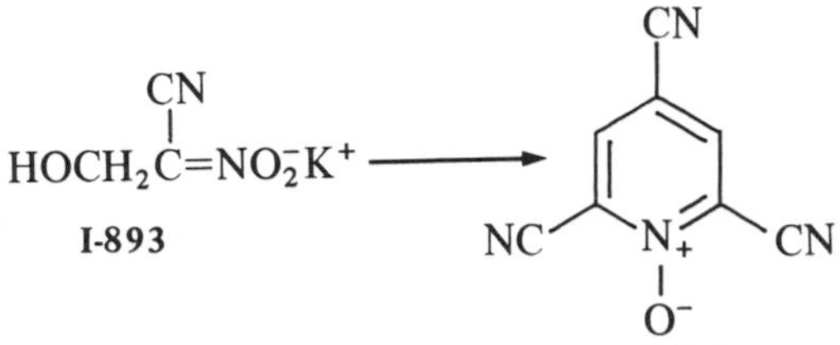

The methylene ketones **I-687** reacted with urea to give pyridones **I-897** (1714).

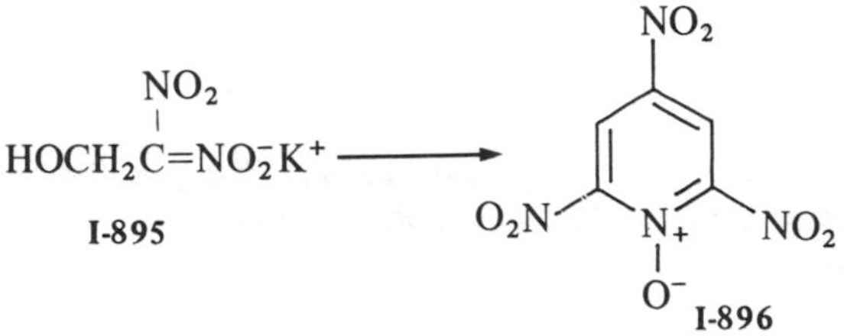

Reaction of acetal **I-898**, ethyl orthoformate, and **I-899** yielded **I-900**. Ethyl aminocrotonate reacted in a similar fashion to give ethyl 2-methylnicotinate (1715).

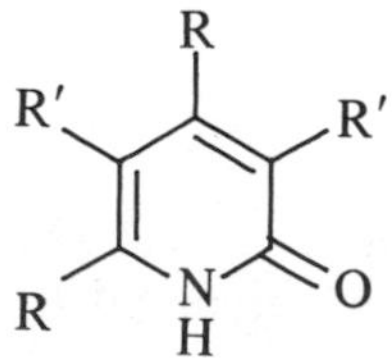

I-897

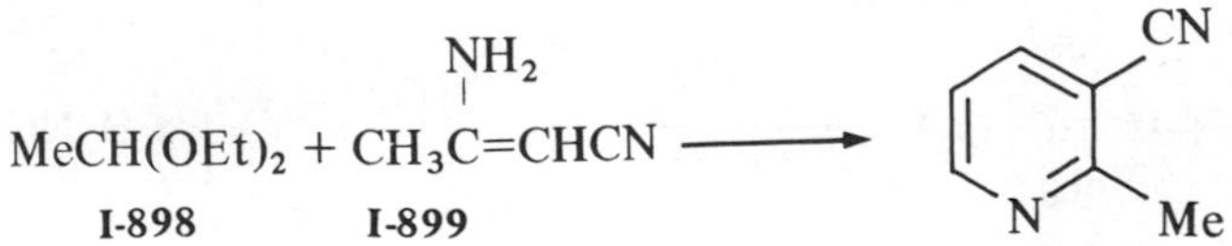

MeCH(OEt)$_2$ + CH$_3$C=CHCN ⟶
I-898   **I-899**

I-900

Alkyl formimidates **I-901** (R = Me, Et) were found to react with dimethyl acetylenedicarboxylate to yield the dihydropyridines **I-902** (1716).

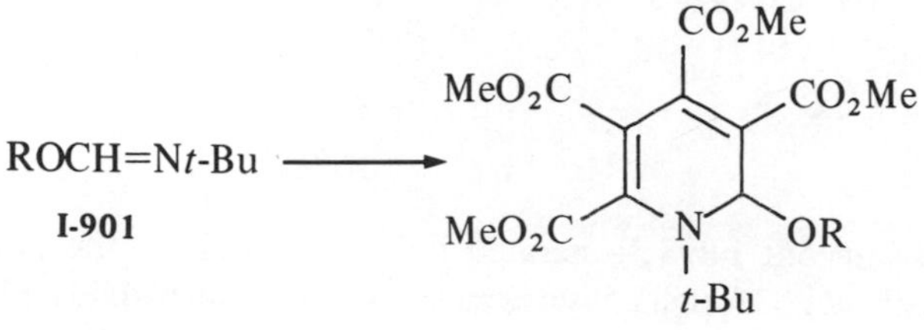

ROCH=N*t*-Bu   ⟶

I-901

CO$_2$Me

MeO$_2$C   CO$_2$Me

MeO$_2$C   N   OR

|

t-Bu

I-902

The reaction of *cis-β-*(methylthio)styrene **I-903** with chlorosulfonyl isocyanate

PhCH=CHSMe   ⟶

I-903

Ph   Ph

N   O

H

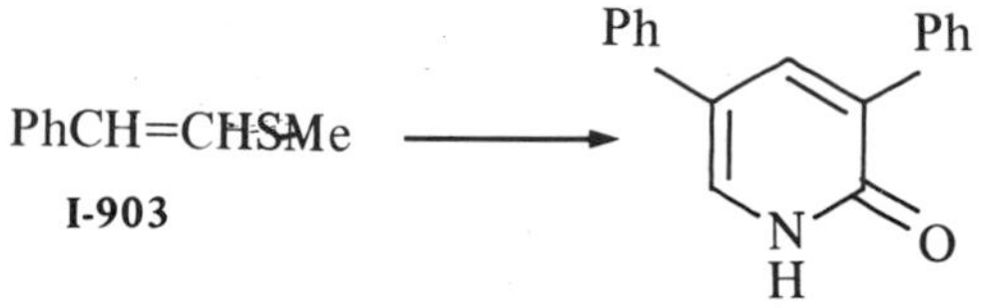

I-904

gave the pyridone **I-904** in 12% yield (1717). With the ynamine **I-905**, chlorosulfonyl isocyanate reacted to produce **I-906** (1718).

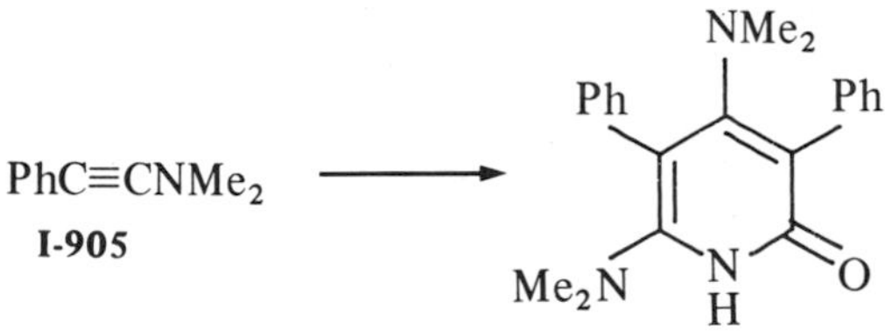

PhC≡CNMe$_2$ ⟶

I-905

NMe$_2$

Ph   Ph

Me$_2$N   N   O

H

I-906

Formaldoxime **I-907** was found to react with ethyl propiolate **I-908** to yield the dinicotinic ester **I-909**; the zwitterion **I-910** was suggested as an intermediate (1719).

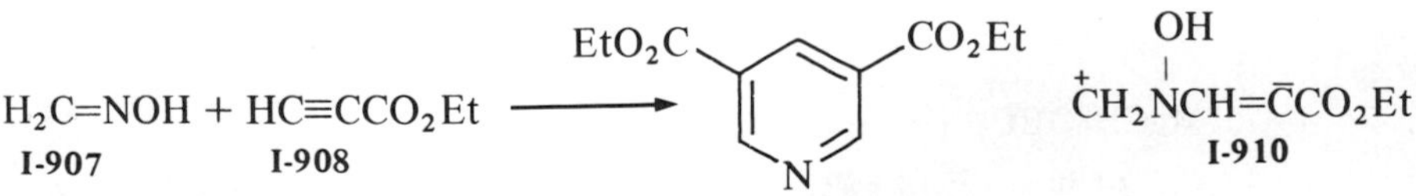

H$_2$C=NOH + HC≡CCO$_2$Et   ⟶

I-907   **I-908**

EtO$_2$C   CO$_2$Et

N

I-909

OH

|

$\overset{+}{C}H_2NCH=\overset{-}{C}CO_2Et$

I-910

F.   2-1-2 Condensations

a.   MIXTURES OF ALDEHYDES AND KETONES WITH AMMONIA

These reactions are treated along with other similar gas-phase reactions in Section 2.E.c.ii.

b.   CARBONYL COMPOUNDS WITH ACTIVE METHYLENE COMPOUNDS

i.   ALDEHYDES WITH ACTIVE METHYLENE COMPOUNDS.   The Hantzsch synthesis, because it produces only 1,4-dihydropyridines, will not be treated here; the reaction has been well reviewed (1720, 1721).

The reaction of aldehydes with ethyl cyanoacetate or malononitrile generally yields aromatic pyridines I-911 or I-912, respectively; these reactions are included in Table I-78. It should be noted that these products may also be prepared in two steps with the ring-closure carried out as a 3-2 condensation of an active methylene compound with a Knoevenagel product, or as cyclization of the 5-carbon unit so constructed.

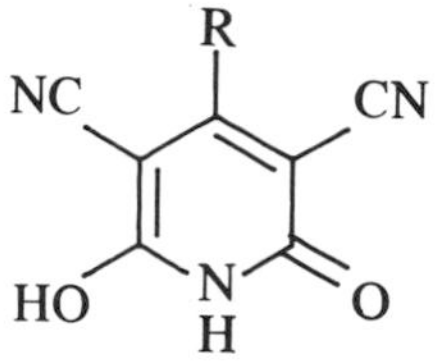

I-911 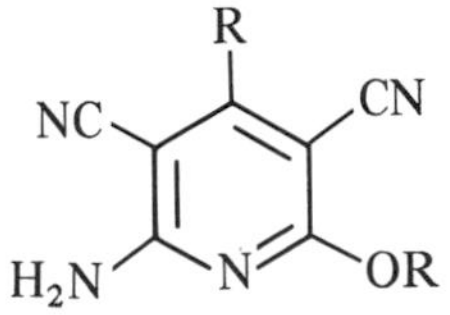   I-912

TABLE I-78.   REACTIONS OF ALDEHYDES WITH MALONONITRILE OR CYANOACETIC ESTER

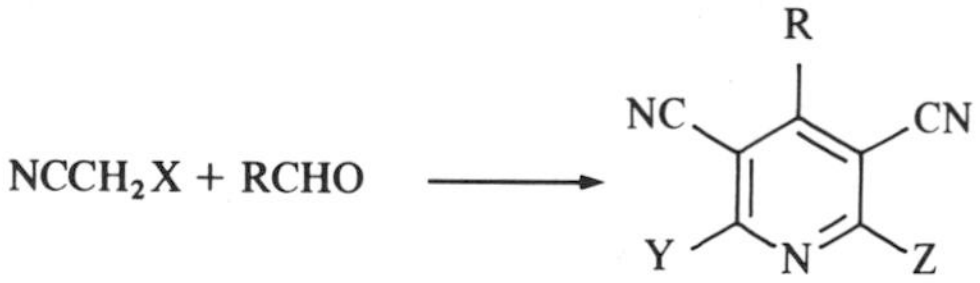

$$NCCH_2X + RCHO \longrightarrow$$

X	R	Y	Z	Reference
CN	Me[a]	H_2N	OEt	1722
CN	Alkyl, aryl	H_2N	OR'	1723, 1724
CN	C_5H_4N-	H_2N	OR'	1725
CO_2Et	Alkyl	OH	OH	1726
CO_2Et	Aryl	OH	OH	1727
CO_2Et	Aryl[b]	OH	OH	1728, 1729
CO_2Et	C_5H_4N	OH	OH	1730

[a] As cyanohydrin.

[b] As imine; N-propylpyridone product.

Aromatic pyridines I-913 were produced by condensation of aldehydes I-914 (R = H, Me) with cyanoacetophenone I-915, but o-methoxybenzaldehyde gave only the 1,4-dihydropyridine (1731).

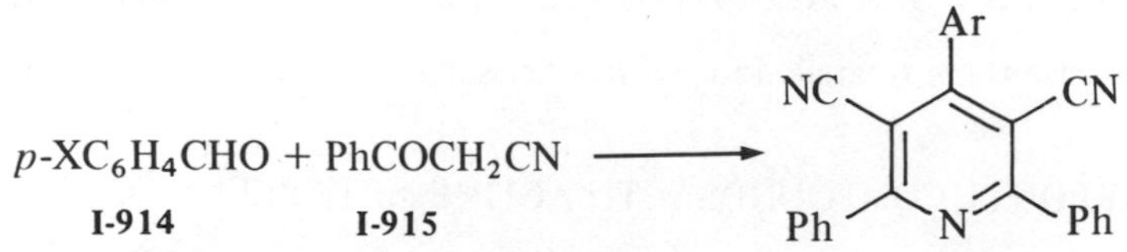

$$p\text{-}XC_6H_4CHO + PhCOCH_2CN \longrightarrow$$

I-914   I-915

I-913

Condensation of desoxybenzoin I-916 with aldehyde I-917 (Ar = p-FC₆H₄) gave the substituted pentaphenylpyridine I-918 (1732).

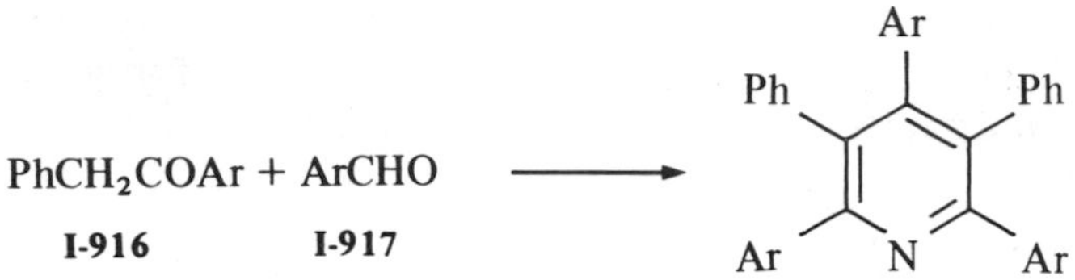

$$PhCH_2COAr + ArCHO \longrightarrow$$

I-916   I-917

I-918

ii. CARBOXYLIC ACID DERIVATIVES WITH ACTIVE METHYLENE COMPOUNDS. By substituting a compound with the oxidation state of a carboxylic acid for the aldehyde normally used in the Hantzsch synthesis, aromatic pyridines may be produced directly in some cases. Thus, reaction of the aminouracils I-919 (R = H, Me) with dimethylformamide (1733), ethyl orthoformate, formamide, or *tris*(formamido)methane (1734) gave the pyridine derivatives I-920. The same reaction occurred with dimethylsulfoxide as the source of the methine carbon (1735).

Vilsmeier reaction of phenylacetonitriles I-921 yielded pyridines I-922 (1736).

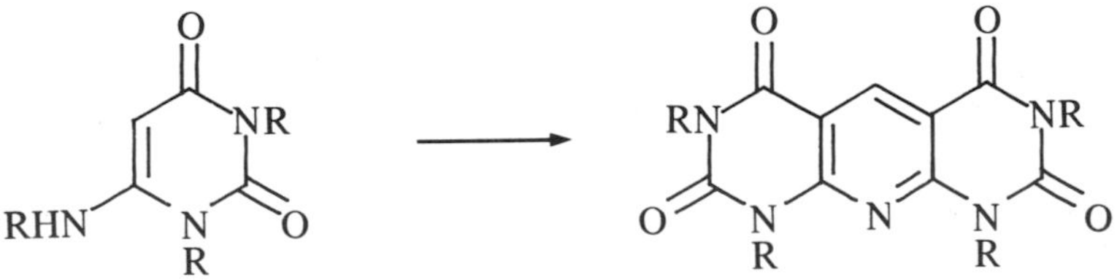

I-919   I-920

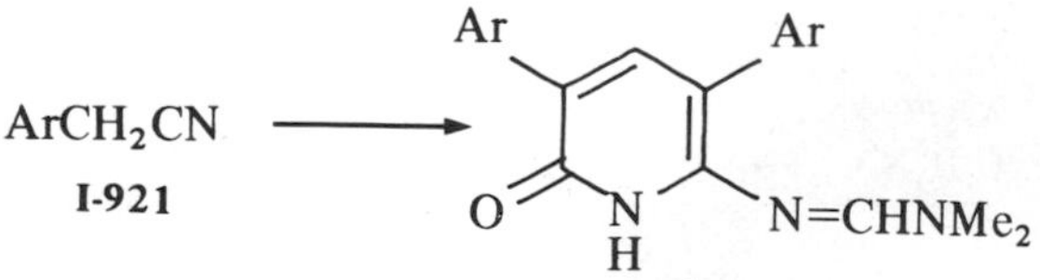

$$ArCH_2CN \longrightarrow$$

I-921

I-922

Condensation of ethyl oxalate, cyanoacetamide, and ketones **I-923** gave pyridones **I-924** (1737).

I-923 **I-924**

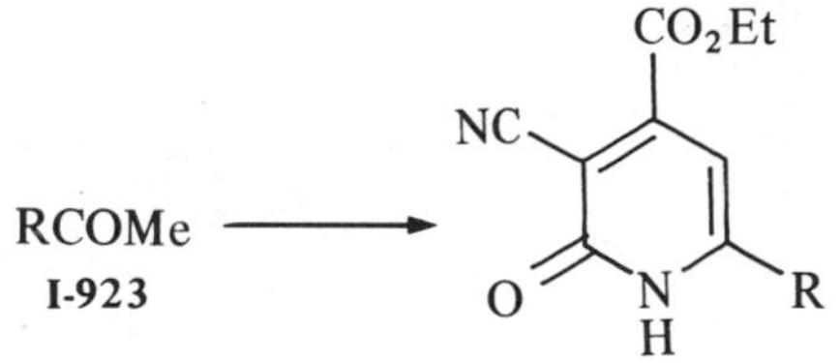

Reaction of ethyl orthoformate with enamines **I-925** yielded the pyridines **I-926** (1738).

I-925 **I-926**

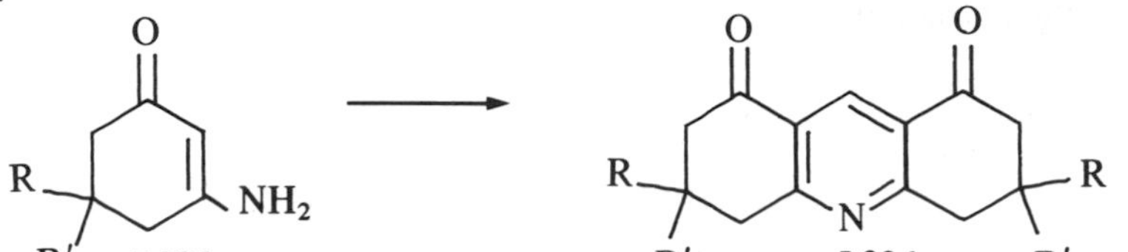

c. MISCELLANEOUS 2-1-2 CONDENSATIONS

Pyrolysis of the hydrazonium salts **I-927** yielded pyridines **I-928** (1739–1741).
Pyrolysis of the ketazine **I-929** gave triphenylpyridine **I-930** (1742). The hydrazone **I-931** gave a low yield of **I-932** in addition to other pyrolysis products (1743).

I-927 **I-928**

I-929 **I-930**

I-931 **I-932**

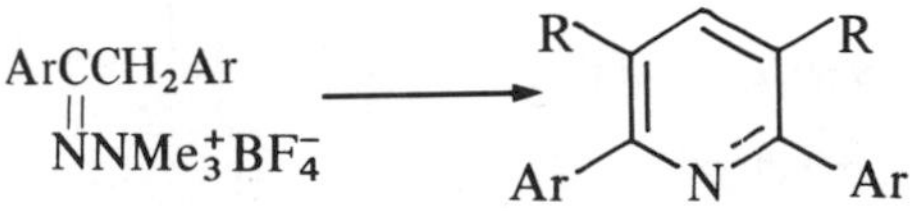

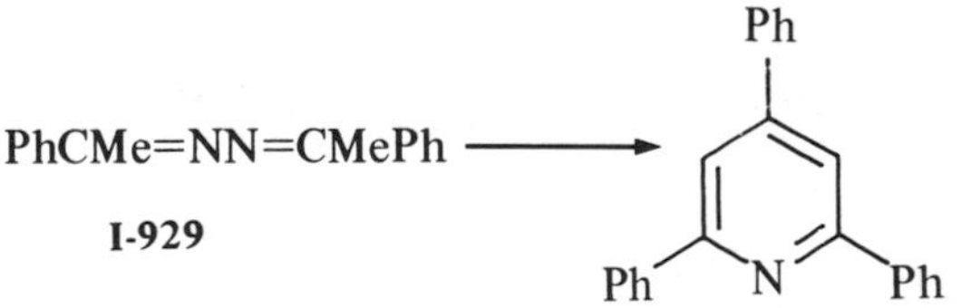

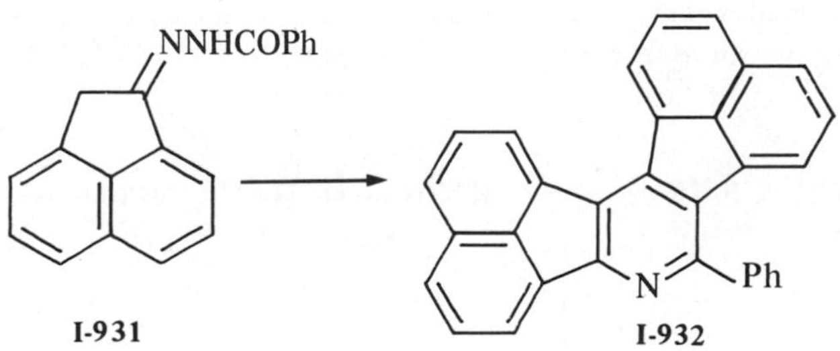

Reaction of triazenes **I-933** with dimethyl acetylenedicarboxylate yielded pyridine derivatives **I-934**; hydrolysis gave pyridones (1744).

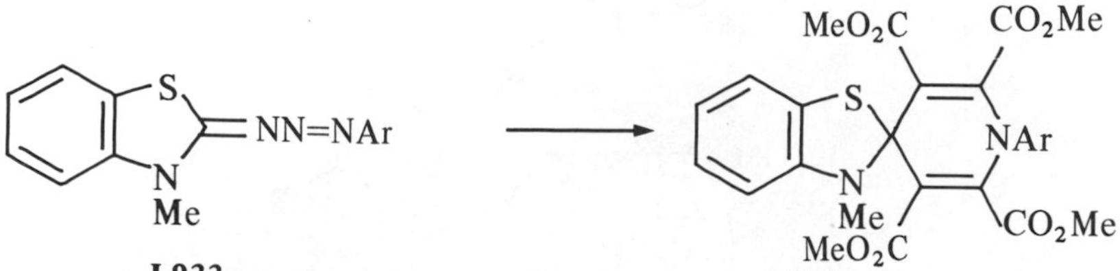

I-933 **I-934**

Ketones **I-935** (X = H, OH) gave modest yields of pyridine **I-936** on heating in hexamethylphosphoric triamide (1745–1748).

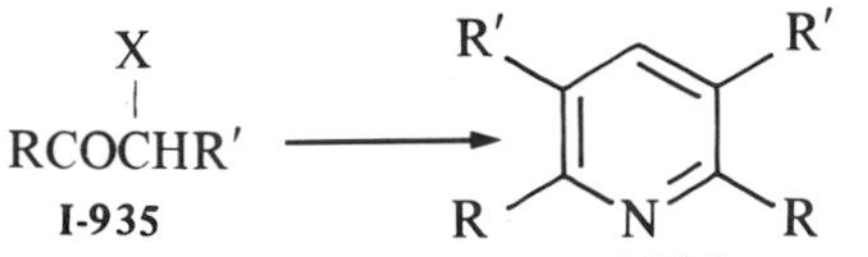

I-935 **I-936**

The ketene dithioacetal **I-937** was found to react with *N*-methylcyanoacetamide to give the naphthyridine derivative **I-938** (1306).

$PhCOCH{=}C(SMe)_2$

I-937

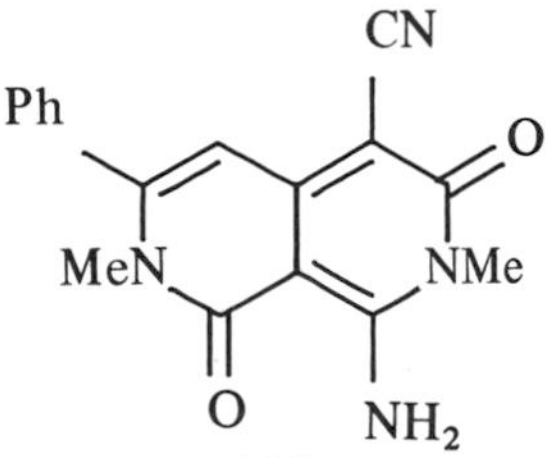

I-938

G. *Cyclization Not Involving the Ring Nitrogen*

a. CYCLIZATION OF ISOCYANATES

Thermolysis of acyl azides **I-939** has given pyridones **I-940** (1749–1751); isocyanates **I-941** are a presumed intermediate, as in the Curtius rearrangement. Similar isocyanates have been postulated as intermediates in the synthesis of pyridones by the route from propargylic pseudoureas **I-942** (1752, 1753).

$R^1CH{=}CHCH{=}CR^2CON_3$ ⟶ $R^1CH{=}CHCH{=}CR^2NCO$ ⟶

I-939 **I-941**

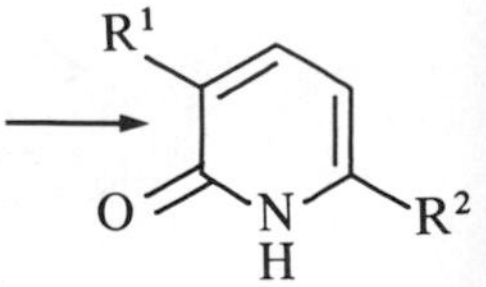

I-940

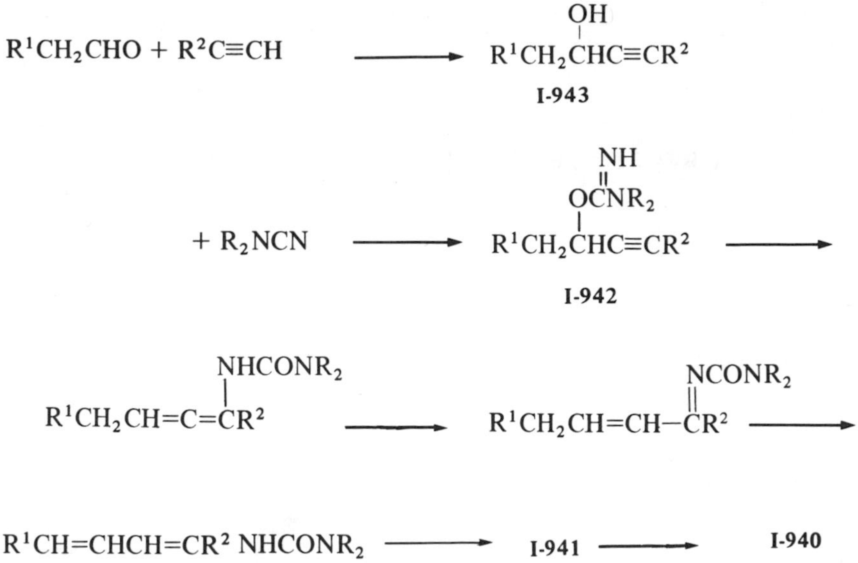
OH
R¹CH₂CHO + R²C≡CH
R¹CH₂CHC≡CR²
I-943
NH
OCNR₂
+ R₂NCN
R¹CH₂CHC≡CR²
I-942
NHCONR₂
R¹CH₂CH=C=CR²
NCONR₂
R¹CH₂CH=CH–CR²
R¹CH=CHCH=CR² NHCONR₂
I-941
I-940

Related reactions include the thermolysis of azide **I-944** to yield the fused pyridone **I-945** (1754).

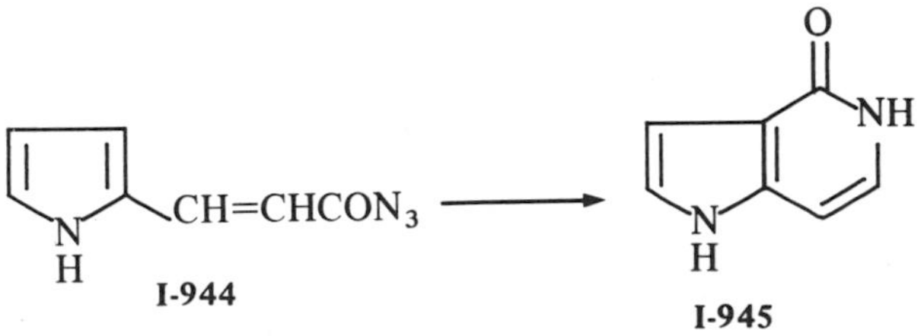
CH=CHCON₃
N
H
I-944
O
NH
N
H
I-945

Styryl isocyanates **I-946** were found to react with ynamines **I-947** to produce *N*-substituted pyridones **I-948**; the latter compounds formed the parent adducts **I-949** on heating (1755).

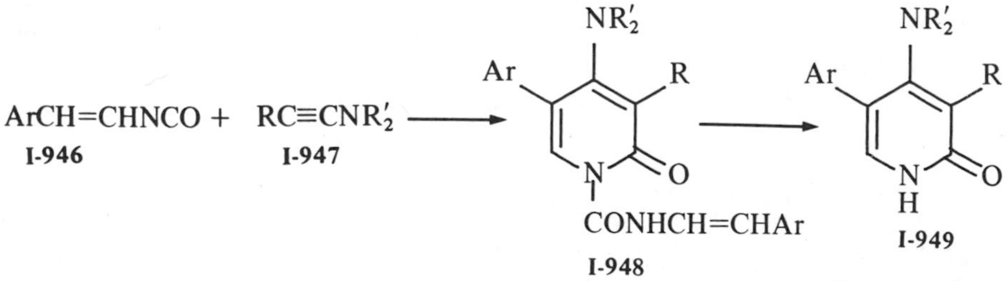
ArCH=CHNCO +
I-946
RC≡CNR′₂
I-947
NR′₂
Ar
R
N
O
CONHCH=CHAr
I-948
NR′₂
Ar
R
N
H
O
I-949

Benzoyl or thiobenzoyl isocyanates **I-950** reacted with enaminoketones **I-951** to yield pyridones or thiopyridones **I-952** (1756, 1757).

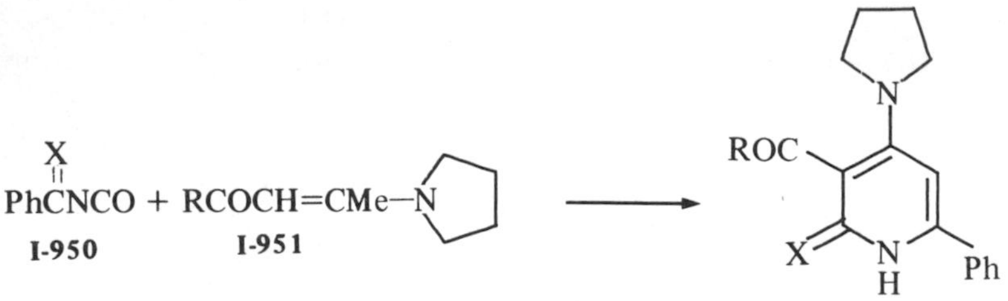

$$\underset{\textbf{I-950}}{\overset{X}{\underset{\|}{PhCNCO}}} + \underset{\textbf{I-951}}{RCOCH{=}CMe{-}N} \longrightarrow \textbf{I-952}$$

b.   CYCLIZATION OF IMINES

A large number of unsaturated nitrogen compounds undergo homogeneous or catalyzed reactions at moderately high temperatures to give pyridine derivatives in varying yields. For example, gas-phase reaction of the *N*-allylimine **I-953** in the presence of potassium oxide on alumina at 350° gave the cyclopentenopyridine **I-954** in 38% yield; a variety of other imines, with or without additional unsaturation, gave pyridines similarly (1758–1764). Substituted ethanolamines and ethylenediamines behaved in a similar fashion. In the presence of acrylonitrile, **I-955** was formed from **I-953**, and a Diels–Alder reaction of intermediate isomer **I-956**, followed by loss of hydrogen cyanide, was postulated (1756–1767). Similar reactions occurred with simple olefins as well (1768).

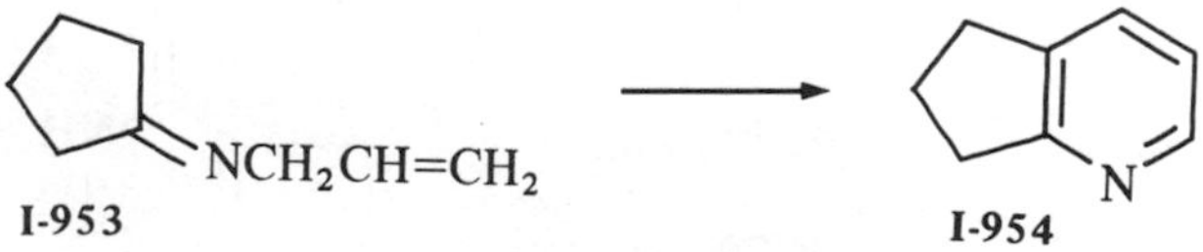

$$\underset{\textbf{I-953}}{\diagdown N CH_2 CH{=}CH_2} \longrightarrow \underset{\textbf{I-954}}{\diagdown}$$

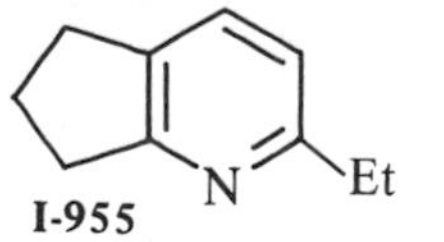

$$\underset{\textbf{I-955}}{\diagdown N\text{—}Et} \qquad \underset{\textbf{I-956}}{\diagdown N{=}CHCH_2CH_3}$$

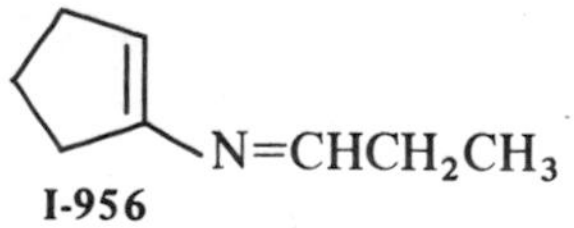

Imines with pendant benzene rings, such as **I-957**, were cyclized in the vapor phase with iodine to phenylpyridines such as **I-958** (1769); the vapor-phase reaction of saturated amines with iodine (Section II.2.A.e) should be compared.

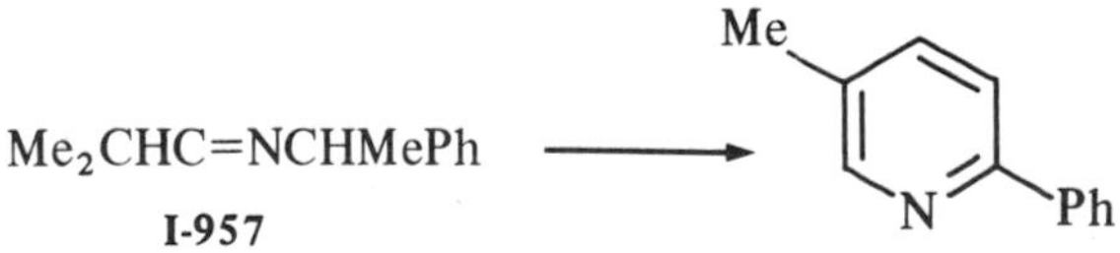

$$\underset{\textbf{I-957}}{Me_2CHC{=}NCHMePh} \longrightarrow \underset{\textbf{I-958}}{\diagdown}$$

In the only study of the reaction mechanism, the pyrolysis of triallylamine (**I-959**) at 260–350°C was investigated and found to be homogeneous and first order. Elimination of propylene by a retro-ene reaction to give **I-960**, followed by rapid intramolecular rearrangements and dehydrogenation to yield 3-methylpyridine, was postulated (1770).

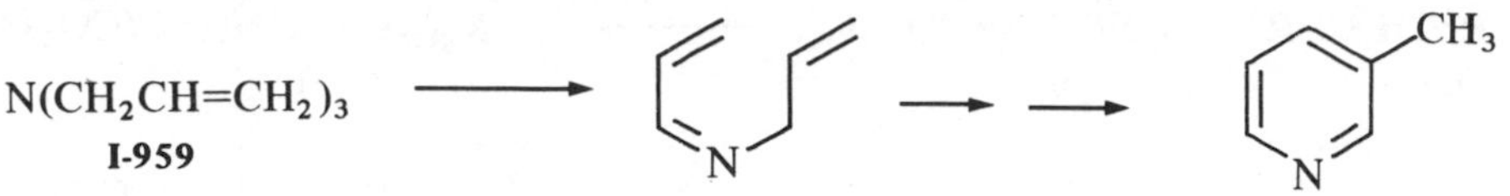

$N(CH_2CH=CH_2)_3$
I-959

I-960

Imines **I-961** were converted to pyridines **I-962** and **I-963** ($R^4 = R^1 = CH_2$) in the presence of sodium and alloocimene (1771, 1772).

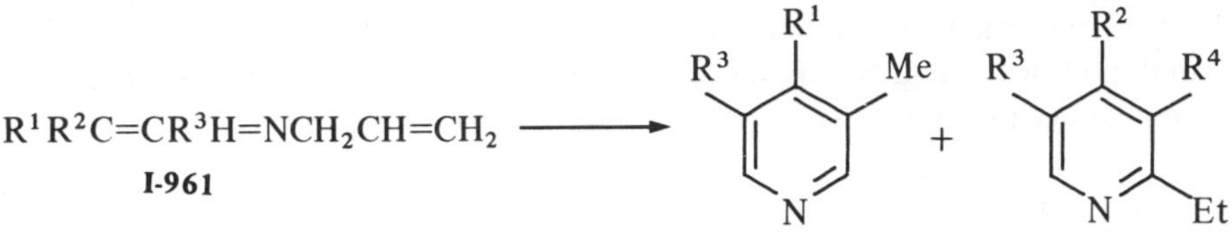

$R^1R^2C=CR^3H=NCH_2CH=CH_2$
I-961

I-962   **I-963**

Reaction of tripropenylamine **I-964** with phosphoric acid yielded pyridine **I-965**; propionaldehyde reacted with ammonia to yield the same product, as expected (1773).

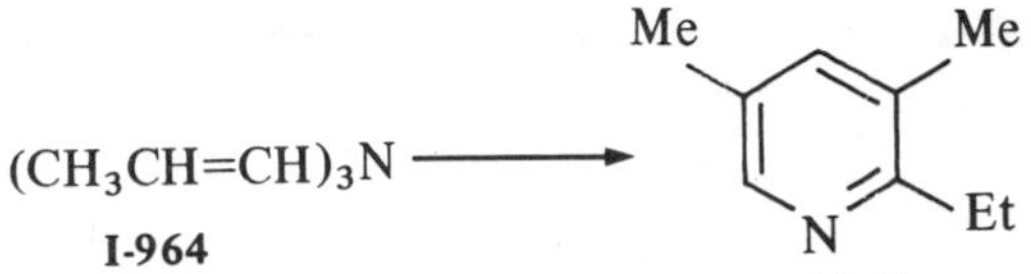

$(CH_3CH=CH)_3N$
I-964

I-965

On heating in nitrobenzene, the propargylaminoketones **I-966** formed fused pyridines **I-967** in 40–68% yield; Claisen rearrangement to **I-968** was suggested as the first step (1774).

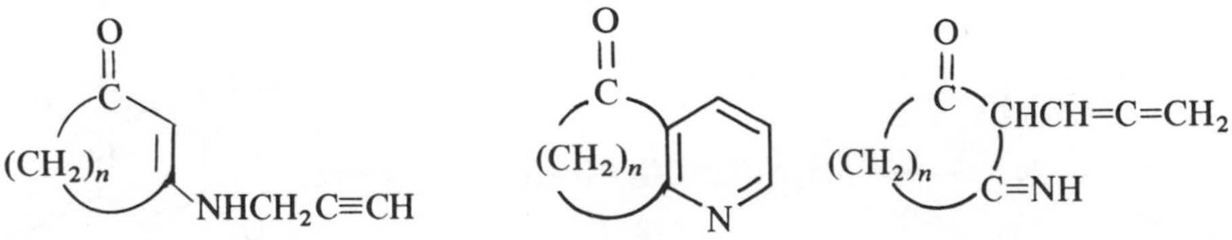

I-966   **I-967**   **I-968**

c.   REACTIONS RELATED TO THE GOULD-JACOBS REACTION

Condensation of aromatic or heteroaromatic amines with ethoxymethylenemalonic ester and thermal cyclization of the resulting anilinomethylenemalonic esters has been a method of choice for the synthesis of 4-quinolones and related fused 4-pyridones. Extension of the reaction to aliphatic derivatives yields 4-pyridones.

Thus, ketones **I-969** (R^1 = alkyl, COR, CO_2R, SO_2R) were found to react with aminomethylenemalonic ester **I-970**, and the resulting enamines **I-971** were cyclized by heating to yield the pyridones **I-972** (1775–1777).

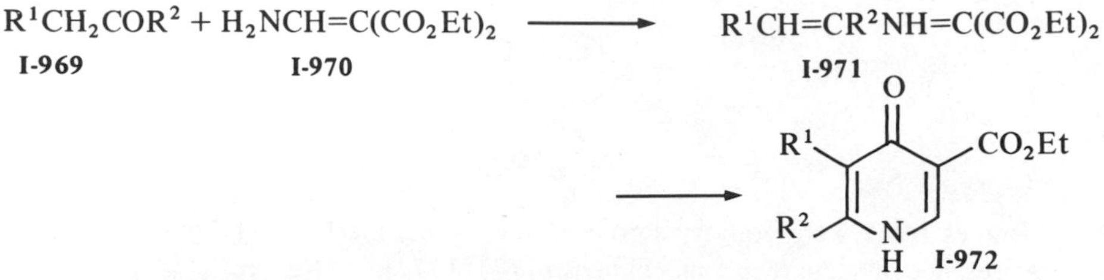

$$R^1CH_2COR^2 + H_2NCH=C(CO_2Et)_2 \longrightarrow R^1CH=CR^2NH=C(CO_2Et)_2$$

I-969   **I-970**   **I-971**

I-972

In a similar kind of reaction, the amide acetals **I-973** [R, R′ = Me, H, or $(CH_2)_n$] condensed with aminocrotonate **I-974** yielding intermediate **I-975**, which was found to cyclize on heating to the pyridone **I-976** (1778).

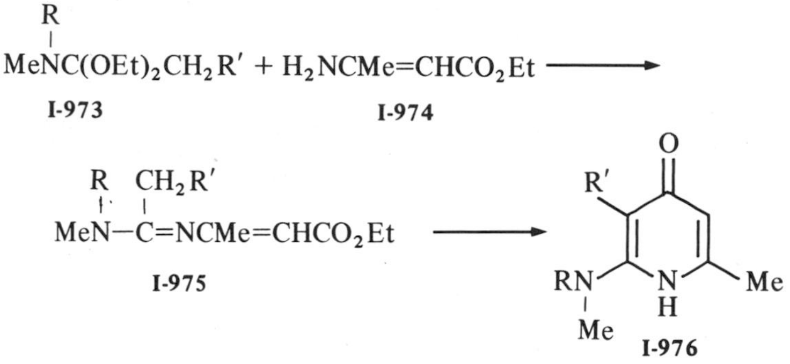

$$\underset{\textbf{I-973}}{MeNC(OEt)_2CH_2R'} + \underset{\textbf{I-974}}{H_2NCMe=CHCO_2Et} \longrightarrow$$

I-975   I-976

The amide acetals **I-973** reacted with the related **I-977** to yield intermediate **I-978**, which again cyclized on heating to **I-979** (1779). Intermediate **I-980**, formed from **I-981** and dimethylformamide diethyl acetal, formed **I-982** on heating (1780).

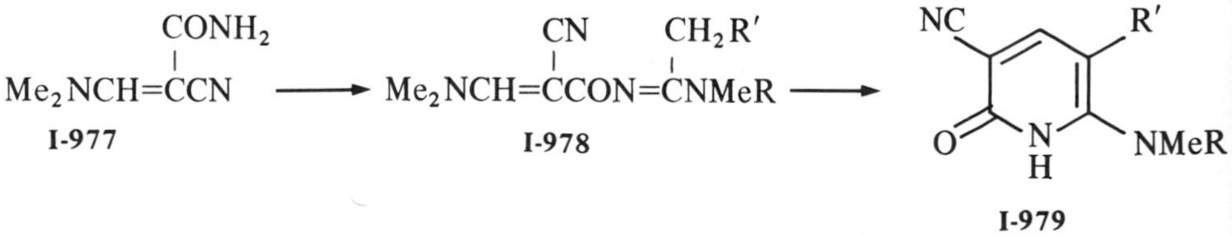

I-977   I-978   I-979

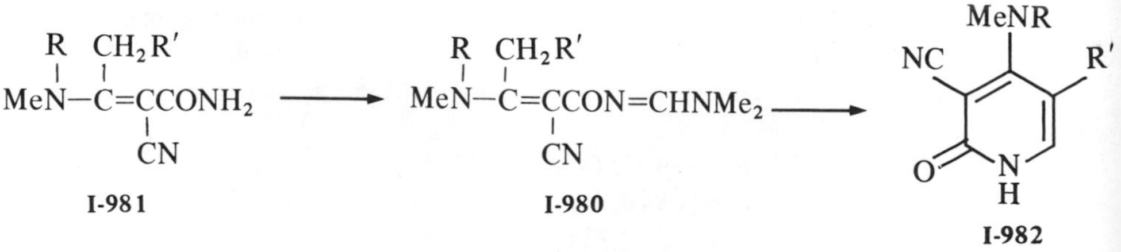

I-981   I-980   I-982

d. MISCELLANEOUS RING CLOSURES

Reaction of the dienamides **I-983** with phosphoryl chloride gave the pyridines **I-984** (1781) (cf. Section II.2.B.d).

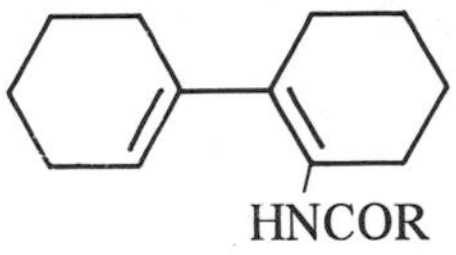

I-983

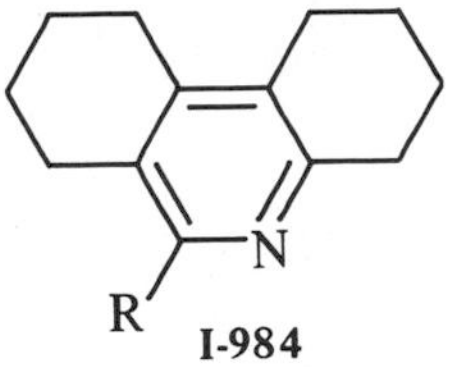

I-984

Irradiation of a solution of diphenylacetylene **I-985** and the aminocrotonic ester **I-986** gave the pyridone **I-987** in 19% yield; intermediates **I-988** and **I-989** were suggested (1782).

$PhC{\equiv}CPh$ + $MeC(NHMe){=}CHCO_2Et$ ⟶

I-985 **I-986**

I-987

I-988

$MeC(NHMe){=}CPhCPh{=}CHCO_2Et$

I-989

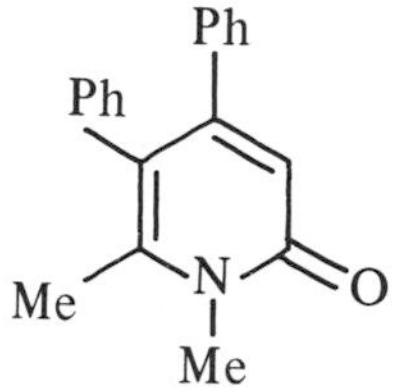

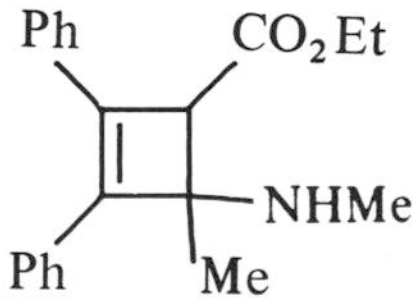

Reaction of diethyl maleate with the isocyanide **I-990** at 170–180° afforded a 17% yield of the pyridine **I-991** (1783).

$MeCH(NC)CO_2Et$ ⟶

I-990

I-991

The enamides **I-992** (R = H, Me) could be cyclized to the pyridones **I-993** in 60–70% yield by reaction with excess sodamide (1784).

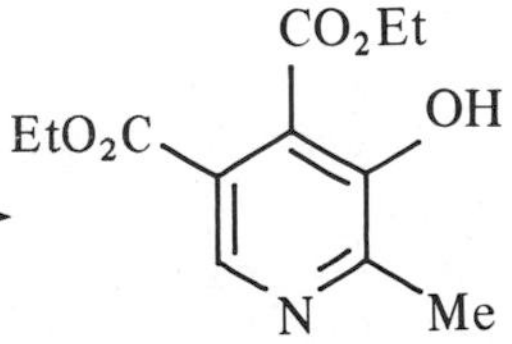

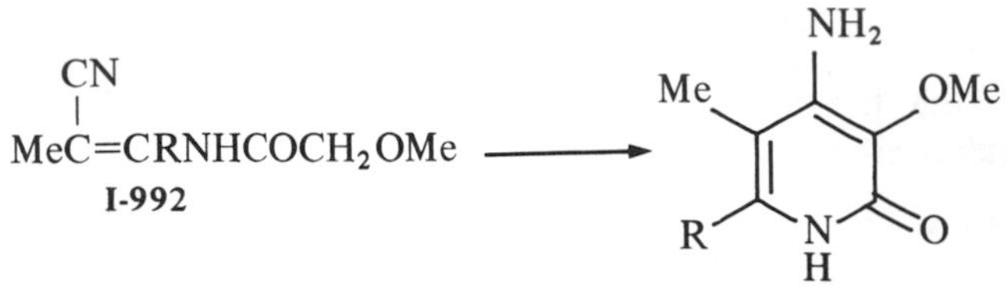

I-992 → I-993

Cyclization of the acetonedicarboxylic acid derivative **I-994** to the pyridone **I-995** took place in 87% yield on heating in acetonitrile containing piperidine (1785).

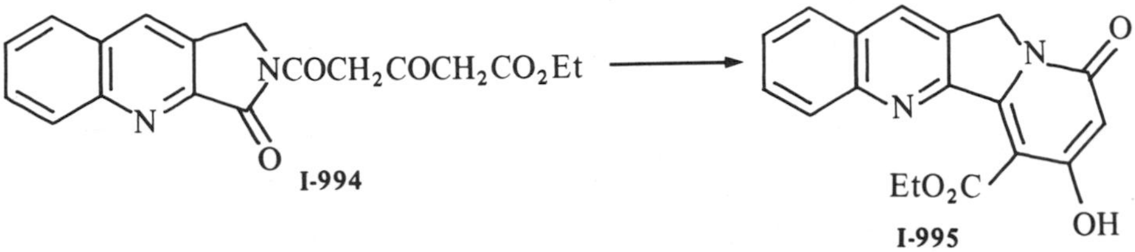

I-994 → I-995

The action of a high-frequently electric discharge on mixtures of methane, ammonia, and water led to the production of detectable amounts of methylpyridines, nicotinic acid, and nicotinamide, among other compounds (1786).

ACKNOWLEDGMENT

The authors thank Mrs. Karen Young for assisting with the literature search, Mrs. Joyce Rawley for typing and preparing the manuscript, and Mrs. LaDonna Ball and Miss Sandi Moore for typing the reference section.

REFERENCES

1. Y. Hagino, S. J. Lan, C. Y. Ng, and L. M. Henderson, *J. Biol. Chem.,* **243,** 4980 (1968).
2. H. H. Loh and C. P. Berg, *J. Nutr.,* **101,** 1601 (1971).
3. H. Buerstell and W. Hilgenberg, *Biol. Zentralbl.,* **94,** 401 (1975).
4. D. Gross, *Fortschr. Chem. Org. Naturst.,* **28,** 109 (1970).
5. E. R. Jaffe, in 'Hematology," W. J. Williams, E. Beutler, A. J. Erslev, eds., 2nd ed., McGraw-Hill, New York, 1977, p. 146.
6. F. M. Chiancone, *Acta Vitaminol. Enzymol.,* **26,** 81 (1972).
7. J. Benes and B. Zicha, *Chem. Listy,* **61,** 203 (1967) [*Chem. Abstr.,* **66,** 62039 (1967)].
8. T. Deguchi, *Tampakushitsu Kakusan Koso,* **14,** 312 (1969) [*Chem. Abstr.,* **70,** 111532 (1969)].
9. V. L. Nemchinskaya, V. M. Bozhkov, and V. P. Kushner, *Tsitologiya,* **13,** 799 (1971) [*Chem. Abstr.,* **75,** 94724 (1971)].

10. K. S. You, L. J. Arnold, Jr., W. S. Allison, and N. O. Kaplan, *Trends Biochem. Sci.*, **3**, 265 (1978).

11. C. Ricci, *Acta Vitaminol. Enzymol.*, **25**, 65 (1971).

12. N. O. Kaplan, *Trends Biochem. Sci.*, **3**, N52 (1978).

13. C. Bernofsky, *FEBS Lett.*, **4**, 167 (1969).

14. J. L. R. Chandler and R. K. Gholson, *Biochim. Biophys. Acta*, **264**, 311 (1972).

15. G. J. Tritz, *Can. J. Microbiol.*, **20**, 205 (1974).

16. H. H. Schott, H. Staudinger, and V. Ullrich, *Hoppe-Seyler's Z. Physiol. Chem.*, **352**, 1654 (1971).

17. J. R. Chandler and R. K. Gholson, *J. Bacteriol.*, **111**, 98 (1972).

18. A. Ichiyama, S. Nakamura, and Y. Nishizuka, *Arzeneim.-Forsch.*, **17**, 1346, 1525 (1967).

19. R. E. Hill, I. Miura, and I. D. Spenser, *J. Am. Chem. Soc.*, **99**, 4179 (1977).

20. R. E. Hill, P. Horsewood, I. D. Spenser, and Y. Tani, *J. C. S. Perkin Trans. I*, **1975**, 1622.

21. W. B. Dempsey, *Biochim. Biophys. Acta*, **264**, 344 (1972).

22. R. Suzue and Y. Haruna, *J. Vitaminol.*, **16**, 161 (1970).

23. Y. Tani and W. B. Dempsey, *J. Bacteriol.*, **116**, 341 (1973).

24. T. M. Vatsala, K. R. Shanmugasundaram, and E. R. B. Shanmugasundaram, *Biochem. Biophys. Res. Commun.*, **72**, 1570 (1976).

25. I. A. Samsonova, D. Birnbaum, and F. Boettcher, *Biol. Rundsch.*, **10**, 353 (1972).

26. D. Birnbaum, F. Boettcher, and I. A. Samsonova, in *Yeasts, Models Sci. Tech., Proc. Spec. Int. Symp., 1st*, A. Kockova-Kratochvilova and E. Minarik, eds., Slovak Acad. Sci., Bratislava, **1972**, 133 [*Chem. Abstr.*, **84**, 102062 (1976)].

27. D. Birnbaum, F. Boettcher, and I. A. Samsonova, *Z. Allg. Mikrobiol.*, **12**, 365 (1972).

28. F. Boettcher, D. Birnbaum, and I. A. Samsonova, *Eur. J. Biochem.*, **29**, 88 (1972).

29. D. J. Uhlik and C. S. Gowans, *Int. J. Biochem.*, **5**, 79 (1974).

30. M. M. Kole and A. K. Banerjee, *Sci. Cult.*, **40**, 76 (1974).

31. K. Wihervaara, *J. Inst. Brew.*, **73**, 167 (1967) [*Chem. Abstr.*, **67**, 10334 (1967)].

32. I. V. Stakheev, I. S. Vazhinskaya, and S. G. Latysheva, *Vestsi Akad. Navuk B. SSR, Ser. Biyal. Navuk*, **1976**, 54 [*Chem. Abstr.*, **85**, 141310 (1976)].

33. N. I. Mansurov, S. G. Manzyuk, and A. Mirzorakhimov, *Izv. Akad. Nauk Tadzh. SSR. Otd. Biol. Nauk*, **1975**, 85 [*Chem. Abstr.*, **84**, 86879 (1976)].

34. R. F. Dawson, D. R. Christman, and R. U. Byerrum, *Methods Enzymology*, **18** (pt. B), 90 (1971).

35. C. Del Rio-Estrada and H. W. Daugherty, "Kirk-Othmer Encyclopedia of Chemical Technology," 2nd ed., Interscience, New York, 1970, Vol. 21, p. 509.

36. W. J. Darby, K. W. McNutt, and E. N. Todhunter, *Nutr. Rev.*, **33**, 289 (1975).

37. M. Weiner, *Drug Metab. Rev.*, **9**, 97 (1979).

38. P. Gyorgy and F. W. Zilliken, *Fermente, Horm., Vitam. Beziehungen Wirkstoffe Zueinander, Dritte Erweiterte Aufl.*, **3**, 662 (1974) [*Chem. Abstr.*, **83**, 143090 (1975)].

39. T. Oki and A. Kitai, *Process Biochem.*, **9**, 31 (1974).

40. I. S. Rodynyuk, in "Obraz. Fiziol. Akt. Veshchestv Mikoorg.," I. L. Klevenskaya, ed., Nauka, Sib. Otd., Novosibirsk, USSR, 1975, 50 [*Chem. Abstr.*, **85**, 188924 (1976)].

41. L. G. Safranova, in "Obraz. Fiziol. Akt. Veshchesty Mikroorg.," I. L. Klevenskaya, ed., 1975, 66 [*Chem. Abstr.*, **85**, 188925 (1976)].

42. T. K. Begishvili, M. Z. Machavariana, and N. M. Ramishvili, *Soobshch Akad. Nauk Gruz. SSR*, **87**, 465 (1977) [*Chem. Abstr.*, **88**, 18836 (1978)].

43. E. G. Vukhrer and S. M. Mirdzhalilova, *Tr. Kirg. Nauch-Issled. Inst. Pochvoved*, **2**, 118 (1969) [*Chem. Abstr.*, **75**, 97640 (1971)].

44. Yu. M. Voznyakovskaya and S. M. Gasanova, in "Vopr. Ekol. Fiziol. Mikroorg. Ispol'z. Sel'sk. Khoz.," O. A. Berestetskii, ed., Nauchno-Issled. Inst. S-kh. Mikrobiol., Leningrad, USSR, 1976, 147 [*Chem. Abstr.*, **88**, 117717 (1978)].

45. I. L. Klevenskaya, in "Obraz. Fiziol. Akt. Veshchestv Mikroorg.," I. L. Klevenskaya, ed., Nauka, Sib. Otd. Novosibirsk, USSR, 1975, 21 [*Chem. Abstr.*, **85**, 188921 (1976)].

46. E. I. Kvasnikov, M. N. Gavrilenko, V. G. Sumnevich, V. V. Stepanyuk, G. S. Eliseeva, and L. P. Stognii, *Mikrobiologiya*, **45**, 944 (1977) [*Chem. Abstr.*, **88**, 18886 (1978)].

47. E. A. Katruk, *Izv. Akad. Nauk Mold. SSR, Ser. Biol. Khim, Nauk,* 1975, 46 [*Chem. Abstr.*, **85**, 18906 (1976)].

48. E. A. Katruk, V. V. Kotelev, E. I. Andreyuk, D. A. Volkova, and I. A. Cheban, *Izv. Akad. Nauk Mold SSR, Ser. Biol. Khim, Nauk,* **1973**, 41 [*Chem. Abstr.*, **88**, 74618 (1974)].

49. A. K. Panosyan, G. S. Babayan, V. G. Nikogosyan, N. M. Sayadyan, D. S. Mkrtchyan, and S. B. Shakhmuradyan, *Vop. Mikrobiol.*, **6**, 145 (1973) [*Chem. Abstr.*, **81**, 48477 (1974)].

50. B. Tolysbsaev and O. Bisenbaev, *Izv. Akad. Nauk Kaz. SSR, Ser. Biol.*, **14**, 22 (1976) [*Chem. Abstr.*, **85**, 44849 (1976)].

51. E. I. Kvasnikov, E. N. Pisarchuk, and T. M. Klyushnikova, *Mikrobiol. Zh. (Kiev)*, **37**, 573 (1975) [*Chem. Abstr.*, **84**, 14412 (1976)].

52. S. A. Samtsevich and L. I. Stefanovich, in "Ispol'z Mikroorganizmov Ikh Metab. Nar. Khoz.," S. A. Samtsevich, ed., Nauka i Tekhnika, Minsk, USSR, 1973, 79 [*Chem. Abstr.*, **81**, 150221 (1974)].

53. N. D. Tupik and S. I. Los, *Ukr. Bot. Zh.*, **32**, 39 (1975) [*Chem. Abstr.*, **83**, 94846 (1975)].

54. V. I. Bilai, S. M. Shcherbina, and I. A. Ellanskaya, *Mikrobiol. Zh. (Kiev)*, **33**, 310 (1971) [*Chem. Abstr.*, **76**, 23797 (1972)].

55. U. A. Sartbaeva, L. F. Frolova, A. K. Balitskaya, B. D. Duhamalieva, and N. K. Zhardetskaya, *Tr. Inst. Mikrobiol. Virusol., Akad, Nauk Kas. SSR,* **17**, 10 (1971) [*Chem. Abstr.*, **76**, 123965 (1972)].

56. N. N. Naplekova and M. V. Smagina, in Ref. 40, p. 29 [*Chem. Abstr.*, **85**, 188922 (1976)].

57. Z. P. Rybakova, in "Mikrobn. Metab. Ikh Ispol'z. Sel'sk. Khoz.," Yu. M. Voznyakovskaya, ed., Inst., S-kh. Mikrobiol; Leningrad, USSR, 1973, 70 [*Chem. Abstr.*, **85**, 18909 (1976)].

58. G. S. Eliseeva and D. M. Isakova, *Mikrobiol. Zh. (Kiev)*, **38**, 156 (1976) [*Chem. Abstr.*, **85**, 31567 (1976)].

59. T. A. Scott and C. Picton, *Biochem. J.*, **154**, 35 (1976).

60. R. M. Akhinyan, F. G. Sarukhanyan, A. G. Sevoyan, and M. L. Stepanyan, *Biol. Zh. Arm.*, **24**, 14 (1971) [*Chem. Abstr.*, **76**, 97903 (1972)].

61. G. I. Mosiashvili, E. V. Mirianashvili, and M. S. Pataraya, *Tr. Inst. Sadovod., Vinograd. Vinodel., Tiflis*, **24**, 356 (1976) [*Chem. Abstr.*, **88**, 188031 (1978)].

62. E. I. Kvasnikov, D. M. Isakova, V. I. Sudenko, and G. S. Eliseeva, *Mikrobiol. Zh. (Kiev)*, **34**, 160 (1972) [*Chem. Abstr.*, **77**, 32622 (1972)].

63. R. A. Anwar, G. E. Gerber, and K. M. Baig, *Adv. Exp. Med. Biol.*, **68B**, 709 (1977) [*Chem. Abstr.*, **88**, 33236 (1978)].

64. P. M. Hardy, G. J. Hughes, and H. N. Rydon, *J. Chem. Soc. Perkin Trans. I*, **1979**, 2282.

65. E. Nurimov and M. Ya. Lovkova, *Prikl. Biokhim. Mikrobiol.*, **9**, 789 (1973) [*Chem. Abstr.*, **80**, 30635 (1974)].

66. L. Jusiak, *Acta Pol. Pharm.*, **30**, 49 (1973).

67. T. Okumura and T. Kadono, *J. Chromatography*, **86**, 57 (1973).

68. R. L. Farkas, *Acta Agron. Acad. Sci. Hung.*, **23**, 11 (1974).

69. E. Demole and C. Demole, *Helv. Chim. Acta*, **58**, 523 (1975).

70. B. Maurer and G. Ohloff, *Helv. Chim. Acta*, **59**, 1169 (1976).

71. E. P. Demole, *Swiss Pat.*, 585523 (1977) [*Chem. Abstr.*, **86**, 168148 (1977)].

72. A. J. N. Bolt, *Phytochemistry*, **11**, 2341 (1972) [*Chem. Abstr.*, **77**, 62199 (1972)].

73. J. D. Phillipson and S. S. Handa, *Phytochemistry*, **14**, 2683 (1975).

74. I. R. C. Bick, J. B. Bremner, J. W. Gillard, and K. N. Winzerberg, *Aust. J. Chem.*, **27**, 2515 (1974).

75. A. Z. Gulubov and I. Z. Bozhkova, *Nauch. Tr. Plovdivski Univ. Mat. Fiz. Chem. Biol.*, **10**, 105 (1972) [*Chem. Abstr.*, **78**, 13763 (1973)].

76. E. Leete, in *Recent Adv. Chem. Comp. Tob., Tob. Smoke, Symp.*, R. J. Reynolds Tob. Co., Winston Salem. N.C., 1977, 365 [*Chem. Abstr.*, **88**, 186075 (1978)].

77. M. Y. Lovkova and G. S. Il'in, *Biokhimiya*, **32**, 812 (1967) [*Chem. Abstr.*, **67**, 88365 (1967)].

78. H. R. Schuette, W. Maier, and U. Stephan, *Z. Naturforsch.*, **23**, 1426 (1968).

79. M. Y. Lovkova and N. I. Kliment'eva, in "Biochem. Physiol. Alkaloide, Int. Symp., 4th," K. Mothes, ed., Akad.-Verlag, Berlin, 1972, 209 [*Chem. Abstr.*, **77**, 98096 (1972)].

80. E. Leete, *Biorg. Chem.*, **6**, 273 (1977).

81. E. Leete, *J. Org. Chem.*, **44**, 165 (1979).

82. T. J. Gilbertson and E. Leete, *J. Am. Chem. Soc.*, **89**, 7085 (1967).

83. E. Leete, G. B. Bodem, and M. F. Manuel, *Phytochemistry*, **10**, 2687 (1971).

84. L. P. Bush and J. L. Saunders, *Recent Adv. Chem. Comp. Tob., Tob. Smoke, Symp.*, Ref. 76, p. 389.

85. M. Y. Lovkova, B. Gyorfi, G. S. Il'in, and N. S. Minozhedinova, *Izv. Akad. Nauk SSSR Ser. Biol.*, 1973, 15 [*Chem. Abstr.*, **78**, 121566 (1973)].

86. G. S. Il'in, B. Gyorfi, M. Y. Lovkova, and N. S. Minozhedinova, *Izv. Akad. Nauk SSSR, Ser. Biol.*, 1972, 756 [*Chem. Abstr.*, **77**, 162093 (1972)].

87. M. Y. Lovkova, G. S. Il'in, B. Gyorfi, and N. S. Minozhedinova, *Izv. Akad. Nauk SSSR, Ser. Biol.*, 1973, 576 [*Chem. Abstr.*, **79**, 123897 (1973)].

88. M. Y. Lovkova, N. S. Minozhedinova, and G. S. Il'in, *Izv. Akad. Nauk SSSR, Ser. Biol.*, 1975, 603 [*Chem. Abstr.*, **83**, 175542 (1975)].

89. W. W. Weeks and L. P. Bush, *Plant Physiol.*, **53**, 73 (1974) [*Chem. Abstr.*, **80**, 575842 (1974)].

90. M. Y. Lovkova, G. S. Il'in, and N. S. Minozhedinova, *Prikl. Biokhim. Mikrobiol.*, **6**, 481 (1970) [*Chem. Abstr.*, **73**, 127917 (1970)].

91. G. S. Il'in, *Izv. Akad. Nauk SSSR, Ser. Biol.*, 1967, 835 [*Chem. Abstr.*, **68**, 47097 (1968)].

92. M. Y. Lovkova and G. S. Il'in, *Prikl. Biokhim. Mikrobiol.*, **3**, 391 (1967) [*Chem. Abstr.*, **67**, 106035 (1967)].

93. T. Ogino, N. Hiraoka, and M. Tabata, *Phytochemistry*, **17**, 1907 (1978).

94. G. V. Izman, N. A. Sherstyannykh, and L. G. Astakhova, *Genetika*, **11**, 50 (1975) [*Chem. Abstr.*, **83**, 40163 (1975)].

95. A. P. Grebenkin, N. A. Sherstyannykh, and Y. F. Sarychev, *Genetika*, **10**, 37 (1974) [*Chem. Abstr.*, **81**, 133025 (1974)].

96. L. P. Bush and J. L. Sims, in "Tob. Health Workshop Conf., Proc. 4th," Univ. of Kentucky, Lexington, Ky., 1973, 742 [*Chem. Abstr.*, **79**, 89672 (1973)].

198   T. D. Bailey, G. L. Goe, and E. F. V. Scriven

97.   I. Wahlberg, K. Karlsson, D. J. Austin, N. Junker, J. Roeraade, C. R. Enzell, and W. H. Johnson, *Phytochemistry,* **16,** 1233 (1977).

98.   W. J. Casey, *U.S. Pat.,* 4,068,671 (1978) [*Chem. Abstr.,* **88,** 101791 (1978)].

99.   G. B. Gori and C. J. Lynch, *J. Am. Med. Assoc.,* **240,** 1255 (1978).

100. L. Buratti, *Riv. Med. Aeronaut. Spaz.,* **40,** 194 (1977) [*Chem. Abstr.,* **88,** 164913 (1978)].

101. M. E. Jarvik, *NIDA Res. Monogr.,* **17,** 122 (1977).

102. F. M. Chiancone and L. Bonollo, *Acta Vitaminol. Enzymol,* **23,** 103 (1969).

103. R. W. Ryall, in "Neuropoisons: Their Pathophysiol. Actions", L. L. Simpson and D. R. Curtis, eds., Vol. 2, Plenum, New York, 1974, 61 [*Chem. Abstr.,* **83,** 72739 (1975)].

104. J. F. Emele, *Proc. World Conf., 3rd, Smoking Health,* **1975,** Vol. 2, N.I.H., Washington D.C., 1977, p. 561 [*Chem. Abstr.,* **88,** 183933 (1978)].

105. E. Dagne and N. Castagnoli, *J. Med. Chem.,* **15,** 356 (1972).

106. P. Jenner and J. W. Gorrod, *Res. Commun. Chem. Pathol. Pharmacol.,* **6,** 829 (1973).

107. I. Schmeltz and D. Hoffman, *Chem. Rev.,* **77,** 295 (1977).

108. G. B. Neurath, *Beitr. Tabakforsch.,* **5,** 115 (1969).

109. C. R. Enzell, I. Wahlberg, and A. J. Aasen, *Fortschr. Chem. Org. Naturst.,* **34,** 1 (1977).

110. I. Schmeltz, R. L. Stedman, W. J. Chamberlan, and D. Burdick, *J. Sci. Food Agr.,* **15,** 744 (1964).

111. G. B. Neurath, in "The Chemistry of Tobacco and Tobacco Smoke," I. Schmeltz, ed., Plenum, New York, 1972, p. 77.

112. S. S. Hecht, R. L. Thorne, R. R. Maronpot, and D. Hoffmann, *J. Natl. Cancer Inst.,* **55,** 1329 (1975).

113. E. Demole and D. Berthet, *Helv. Chim. Acta,* **55,** 1866 (1972).

114. D. A. Buyske, J. M. Flowers, P. Wilder, and M. E. Hobbs, *Science,* **124,** 1080 (1956).

115. T. Fujimori, R. J. Kasuga, H. Matsushita, H. Kaneko, and M. Naguchi, *Agr. Biol. Chem.,* **40,** 303 (1976).

116. R. L. Stedman, *Chem. Rev.,* **68,** 153 (1968).

117. A. H. Warfield, W. D. Galloway, and A. G. Kallianos, *Phytochemistry,* **11,** 3371 (1972).

118. T. Kisaki, S. Mitzusaki, and E. Tamaki, *Phytochemistry,* **7,** 323 (1968).

119. E. V. Brown and I. Ahmad, *Phytochemistry,* **11,** 3485 (1972).

120. S. S. Hecht, R. M. Ornaf, and D. Hoffman, *J. Natl. Cancer Inst.,* **54,** 1237 (1975).

121. F. Rulko and K. Nadler, *Diss. Pharm. Pharmacol.,* **22,** 329 (1970) [*Chem. Abstr.,* **74,** 83989 (1971)].

122. E. Spath and F. Keszther, *Chem. Ber.,* **70,** 404 (1937).

123. T. C. Tso, *Physiology and Biochemistry of Tobacco Plants,* Dowden, Hutchinson, & Ross, Stroudsburg, Pa., 1972.

124. M. Pailer, in "Tobacco Alkaloids and Related Compounds," U. S. von Euler, ed., Macmillan, New York, 1965, p. 15.

125. J. Thesing and A. Muller, *Angew. Chem.,* **68,** 577 (1956).

126. H. Klus and H, Kuhn, *Fachliche Mitt. Oesterr. Tabakregie,* **16,** 307 (1975) [*Chem. Abstr.,* **85,** 30771 (1976)].

127. L. De Salles de Hys, M. L. Renard, and J. Goy, *Ann. Tab, Sect. 1,* **14,** 119 (1976) [*Chem. Abstr.,* **88,** 34694 (1978)].

128. H. Kuhn, *Fachliche Mitt. Oesterr. Tabakregie,* **6,** 89 (1966) [*Chem. Abstr.,* **67,** 29938 (1967)].

129. R. L. Peck, S. F. Osman, and J. L. Barson, *Tob. Sci.,* **13,** 38 (1969).

130. L. P. Bush, C. Grunwald, and D. L. Davis, *J. Agr. Food Chem.*, **20**, 676 (1972).

131. D. G. Vickroy, *Beitr. Tabakforsch.*, **8**, 415 (1976) [*Chem. Abstr.*, **86**, 40370 (1977)].

132. R. K. Mauldin, *Beitr. Tabakforsch.*, **8**, 422 (1976) [*Chem. Abstr.*, **86**, 40371 (1977)].

133. D. Gross, in "Biosyn. Alkaloide," K. Mothes, ed., VEB Deut. Verlag Wiss., Berlin, **1969**, 215 [*Chem. Abstr.*, **71**, 45722 (1969)].

134. R. K. Hill, in "Alkaloids," S. W. Pelletier, ed., Van Nostrand Reinhold, New York, 1970, p. 385.

135. E. H. Eugster, *Heterocycles*, **4**, 51 (1976).

136. R. D. Johnson and G. R. Waller, *Phytochemistry*, **13**, 1493 (1974).

137. D. F. Mann and R. U. Byerrum, *Plant Physiol.*, **53**, 603 (1974).

138. S. N. Ganguly, *Phytochemistry*, **9**, 1667 (1970).

139. M. Mukherjee and A. Chatterjee, *Tetrahedron*, **22**, 1461 (1966).

140. A. Kost, P. B. Terent'ev, and L. V. Modyanova, in "Some Recent Dev. Chem. Nat. Prod.," S. Rangaswami and N. V. Subba Rao, eds., Prentice-Hall, India, New Delhi, India, 1972, p. 153.

141. R. Kalyanasundaram, *Bull. Nat. Inst. Sci. India*, **1966**, 115 [*Chem. Abstr.*, **68**, 75996 (1968)].

142. R. Bhaskaran and N. N. Prasad, *Indian J. Exp. Biol.*, **9**, 516 (1971) [*Chem. Abstr.*, **76**, 83309 (1972)].

143. S. E. Becker, I. D. Pushkareva, V. F. Poletaeva, S. G. Shilina, and E. I. Yasakova, *Bodenkultur*, **23**, 256 (1972) [*Chem. Abstr.*, **77**, 162068 (1972)].

144. V. F. Poletaeva and K. Muraimukhamedov, *Ekol.-Fiziol. Metody Bor'be Fuzarioznym Viltom, Khlop.*, **2**, 75 (1973) [*Chem. Abstr.*, **84**, 42416 (1976)].

145. S. G. Shilina, Z. E. Bekker, and M. G. Goshaev, *Ekol.-Fiziol. Metody Bor'be Fuzarioznym Viltom Khlop.*, **2**, 219 (1973) [*Chem. Abstr.*, **84**, 39502 (1976)].

146. V. I. Bilai, S. N. Shcherbina, L. A. Bogomolova, and N. S. Proskuryakova, *Mikrobiol. Zh. (Kiev)*, **36**, 293 (1974) [*Chem. Abstr.*, **81**, 148113 (1974)].

147. A. M. Bezborodov, T. N. Medvedeva, E. N. Disler, L. F. Goncharova, and V. I. Bilai, *Prikl. Biokhim. Mikrobiol.*, **15**, 40 (1979) [*Chem. Abstr.*, **90**, 150278 (1979)].

148. N. S. Erofeev, A. M. Bezborodov, V. I. Bilai, L. A. Bogomolova, E. A. Zolotareva, T. N. Medvedeva, V. M. Surovtseva, and D. N. Chermenskii, *Russ. Pat.* 516,739 (1976) [*Chem. Abstr.*, **85**, 92209 (1976)].

149. D. W. Pitel and L. C. Vining, *Can. J. Biochem.*, **48**, 623 (1970).

150. D. Steiner, U. Graf, and E. Hardegger, *Helv. Chim. Acta*, **54**, 845 (1972).

151. J. C. Frederiks, *Verh. Schweis. Naturforsch. Ges.*, **1971**, 84 [*Chem. Abstr.*, **77**, 98563 (1972)].

152. G. Leanz and C. Gilvarg, *J. Bacteriol.*, **114**, 455 (1973).

153. S. I. Prigoda, A. Z. Prilepo, V. G. Chalenko, and V. A. Kutuzov, *Probl. Osobo Opasnykh Infek.*, **1970**, 192 [*Chem. Abstr.*, **77**, 150517 (1972)].

154. L. A. Chasin and J. Szulmajster, *Biochem. Biophys. Res. Commun.*, **29**, 648 (1967).

155. K. Kimura, *J. Biochem. (Tokyo)*, **75**, 96 (1974).

156. K. Kimura and T. Sasakawa, *J. Biochem. (Tokyo)*, **78**, 381 (1975).

157. G. Peschke, *German Pat.* 2,300,056 (1974) [*Chem. Abstr.*, **81**, 118549 (1974)].

158. Y. Imae and J. L. Strominger, *J. Biol. Chem.*, **251**, 1493 (1976).

159. A. V. Kulikovskii, *Mikrobiologiya*, **45**, 128 (1976) [*Chem. Abstr.*, **84**, 130892 (1976)].

160. G. R. Germaine and W. G. Murrell, *J. Bacteriol.*, **118**, 202 (1974).

161. M. Forman and A. Aronson, *Biochem. J.*, **126**, 503 (1972).

162. M. R. W. Brown and J. Melling, *J. Pharm. Pharmacol.*, **25**, 478 (1973).

200 T. D. Bailey, G. L. Goe, and E. F. V. Scriven

163. S. Kivilaakso, V. Kauppinen, and J. Halmekoski, *Acta Pharm. Fenn.*, **86**, 187 (1977) [*Chem. Abstr.*, **88**, 166438 (1978)].

164. P. H. Hodson and W. A. Darlington, *U.S. Pat.* 3,334,021 (1967) [*Chem. Abstr.*, **67**, 89809 (1967)].

165. R. B. Forrester and P. W. Vipond, *Brit. Pat.* 1,505,654 (1978) [*Chem. Abstr.*, **90**, 4443 (1979)].

166. M. W. Tabor, J. MacGee, and J. W. Holland, *Appl. Environ. Microbiol.*, **31**, 25 (1976).

167. K. M. Prakasan, D. Sharma, K. G. Gollakota, and B. D. Lakchaura, *Use of Radiation and Radioisotopes in Study of Plant Prod., Proc. Symp.*, Bombay, India, **1974**, 794.

168. W. Ockels, A. Roemer, H. Budzikiewicz, H. Korth, and G. Pulverer, *Tetrahedron Lett.*, **1978**, 3341.

169. T. Nakamura, H. Yasuda, A. Obayashi, O. Tanabe, S. Matsumura, F. Ueda, and K. Ohata, *J. Antibiot.*, **28**, 477 (1975).

170. E. Medina and G. Spiteller, *Chem. Ber.*, **112**, 376 (1979).

171. D. de B. Correa and O. R. Gottlieb, *Phytochemistry*, **14**, 271 (1975).

172. A. M. Helbling and M. Viscontini, *Helv. Chim. Acta*, **59**, 2284 (1976).

173. M. Pouteau-Thouvenot, J. Padikkala, and M. Barbier, *Biochimie*, **54**, 115 (1972).

174. P. V. Divekar, G. Read, and C. C. Vining, *Can. J. Chem.*, **45**, 1215 (1972).

175. S. Brandange and B. Luning, *Acta Chem. Scand.*, **24**, 353 (1970).

176. T. Yasumoto, N. Kawagishi, and H. Asano, *Nippon Suisan Gakkaishi*, **44**, 529 (1978) [*Chem. Abstr.*, **89**, 104127 (1978)].

177. B. V. Burger, M. Le Roux, C. F. Garbers, H. S. C. Spies, R. C. Bigalke, K. G. R. Pachler, P. L. Wessels, V. Christ, and K. H. Maurer, *Z. Naturforsch., C. Biosci.*, **32**, 49 (1977).

178. T. Sakan, A. Fujino, F. Murai, Y. Butsugan, and S. Suzui, *Bull. Chem. Soc. Japan*, **32**, 315, 1155 (1959).

179. T. Sakan, A. Fujino, and F. Murai, *J. Chem. Soc. (Japan)*, **81**, 1320 (1960).

180. T. Sakan, A. Fujino, F. Murai, A. Suzui, Y. Butsugan, and Y. Terashima, *Bull. Chem. Soc. Japan*, **33**, 712 (1960).

181. D. Gross, W. Berg, and H. R. Schuette, *Phytochemistry*, **11**, 3082 (1972).

182. E. M. Dickinson and G. Jones, *Tetrahedron*, **25**, 1523 (1969).

183. A. Abdusamatov, A. Samatov, and S. Y. Yunusov, *Khim. Prir. Soedin.*, **1976**, 122 [*Chem. Abstr.*, **85**, 74887 (1976)].

184. B. Frank, U. Peterson, and F. Hueper, *Angew. Chem. Internat. Ed. Engl.*, **9**, 891 (1970).

185. Y. Hammouda and J. Le Men, *Bull Soc. Chim. Fr.*, **1963**, 2901.

186. G. W. K. Cavill and A. Zeitlin, *Aust. J. Chem.*, **20**, 253 (1967).

187. S. Khakimdzhanov, A. Abdusamatov, and S. Y. Yunusov, *Khim. Prir. Soedin.*, **7**, 126 (1971) [*Chem. Abstr.*, **74**, 142118 (1971)].

188. C. P. Forbes, J. D. Michau, T. van Ree, A. Wiechers, and M. Woudenberg, *Tetrahedron Lett.*, **1976**, 935.

189. T. Sakan, F. Murai, Y. Hayashi, Y. Honda, T. Shono, M. Nakajima, and M. Kato, *Tetrahedron*, **23**, 4635 (1967).

190. K. Torssell, *Acta Chem. Scand.*, **22**, 2715 (1968).

191. M. S. Karawya, S. I. Balbaa, and M. S. Afifi, *U. A. R. J. Pharm. Sci.*, **12**, 53 (1971) [*Chem. Abstr.*, **77**, 13729 (1972)].

192. K. L. Lutfullin, P. K. Yuldashev, and S. Y. Yunusov, *Khim. Prir., Soedin.*, **1965**, 365 [*Chem. Abstr.*, **64**, 3620 (1965)].

193. R. Ziyaev, A. Abdusamatov, and Y. Yunosov, *Khim. Prir. Soedin.*, **7**, 853 (1971) [*Chem. Abstr.*, **76**, 124154 (1972)].

194. L. A. Mitscher, A. B. Ray, and A. Chatterjee, *Experientia*, **27**, 16 (1971).

195. A. B. Ray and A. Chatterjee, *Tetrahedron Lett.*, **1968**, 2763.

196. H. R. Arthur, S. R. Johns, J. A. Lamberton, and S. N. Loo, *Aust. J. Chem.*, **20**, 2505 (1967).

197. N. K. Hart, S. R. Johns, and J. A. Lamberton, *Aust. J. Chem.*, **22**, 1283 (1969).

198. T. Sevenet, B. C. Das, J. Parello, and P. Potier, *Bull. Soc. Chim. Fr.*, **1970**, 3120.

199. T. Sevenet, A. Husson, and H. P. Husson, *Phytochemistry*, **15**, 576 (1976).

200. S. Khakimdzhanov, A. Abdusamatov, and S. Y. Yunusov, *Khim. Prir. Soedin.*, **9**, 132 (1973) [*Chem. Abstr.*, **78**, 159, 944 (1973)].

201. A. Abdusamatov, S. Khakimdzhanov, and S. Y. Yunusov, *Khim. Prir. Soedin.*, **5**, 457 (1969) [*Chem. Abstr.*, **72**, 67156 (1970)].

202. A. Abdusamatov, K. Ubaev, and S. Y. Yunusov, *Khim. Prir. Soedin.*, **4**, 136 (1968) [*Chem. Abstr.*, **69**, 67572 (1968)].

203. A. Abdusamatov and S. Y. Yunusov, *Khim. Prir. Soedin.*, **7**, 306 (1971) [*Chem. Abstr.*, **75**, 110477 (1971)].

204. A. Abdusamatov, M. U. Rashidov, and S. Y. Yunusov, *Khim. Prir. Soedin.*, **7**, 304 (1971) [*Chem. Abstr.*, **75**, 110476 (1971)].

205. T. U. Rakhmatullaev, *Khim. Prir. Soedin.*, **7**, 128 (1971) [*Chem. Abstr.*, **74**, 136454 (1971)].

206. T. U. Rakhmatullaev, S. T. Akramov, and S. Y. Yunusov, *Khim. Prir. Soedin.*, **5**, 608 (1969) [*Chem. Abstr.*, **73**, 84624 (1970)].

207. P. K. Alimbaeva, Z. S. Nuralieva, and M. M. Mukhamedziev, *Fiziol. Aktiv. Soedin. Rast. Kirg.*, **1970**, 88 [*Chem. Abstr.*, **75**, 148459 (1971)].

208. F. Rulko and K. Witkiewicz, *Diss. Pharm. Pharmacol.*, **24**, 73 (1972) [*Chem. Abstr.*, **77**, 2771 (1972)].

209. H. Budzikiewicz, C. Horstmann, K. Pufahl, and K. Schreiber, *Chem. Ber.*, **100**, 2798 (1967).

210. T. U. Rakhmatullaev and S. Y. Yunusov, *Khim. Prir. Soedin.*, **8**, 350 (1972) [*Chem. Abstr.*, **77**, 162,012 (1972)].

211. T. U. Rakhmatullaev and S. Y. Yunusov, *Khim. Prir. Soedin.*, **9**, 64 (1973) [*Chem. Abstr.*, **78**, 159,956 (1973)].

212. N. K. Hart, S. R. Johns, and J. A. Lamberton, *Aust. J. Chem.*, **24**, 1739 (1971).

213. N. K. Hart, S. R. Johns, and J. A. Lamberton, *Aust. J. Chem.*, **21**, 1321 (1968).

214. F. Rulko and K. Witkiewicz, *Rocz. Chem.*, **47**, 1871 (1973).

215. H. G. Floss, U. Mothes, and A. Rettig, *Z. Naturforsch.*, **19B**, 1106 (1964).

216. K. Torsell and K. Wahlberg, *Acta Chem. Scand.*, **21**, 53 (1967).

217. E. Gellert and R. E. Summons, *Aust. J. Chem.*, **27**, 919 (1974).

218. M. Pailer and D. Ganzinger, *Monatsh. Chem.*, **106**, 37 (1975).

219. W. D. Crow and J. H. Hodgkin, *Aust. J. Chem.*, **17**, 119 (1964).

220. D. L. Dreyer, *J. Org. Chem.*, **33**, 3658 (1968).

221. B. A. Burke and H. Parkins, *Tetrahedron Lett.*, **1978**, 2723.

222. R. L. Baxter, L. Crombie, D. J. Simmonds, and D. A. Whiting, *J. Chem. Soc., Perkin Trans. I*, **1979**, 2972.

223. L. Crombie, W. M. L. Crombie, D. A. Whiting, and K. Szendrei, *J. Chem. Soc., Perkin Trans. I*, **1979**, 2976.

224. R. L. Baxter, W. M. L. Crombie, L. Crombie, D. J. Simmonds, D. A. Whiting, and K. Szendrei, *J. Chem. Soc., Perkin Trans. I*, **1979**, 2982.

225. I. Ushizawa, N. Katagiri, T. Kato, and N. Taira, *Igaku To Seibutsugaku*, **94**, 251 (1977) [*Chem. Abstr.*, **89**, 141568 (1978)].

226.   Y. Shizuri, H. Wada, K. Sugiura, K. Yamada, and Y. Hirata, *Tetrahedron,* **29,** 1773 (1973).

227.   K. Sugiura, Y. Shizuri, H. Wada, K. Yamada, and Y. Hirata, *Tetrahedron Lett.,* **1971,** 2733.

228.   H. Budzikiewicz, A. Roemer, and K. Taraz, *Z. Naturforsch.,* **27,** 800 (1972).

229.   L. Dubravkova, J. Tomko, and L. Dolejs, *Phytochemistry,* **12,** 944 (1973).

230.   A. Klasek, F. Santavy, A. M. Duffield, and T. Reichstein, *Helv. Chim. Acta,* **54,** 2144 (1971).

231.   K. Yamada, Y. Shizuri, and Y. Hirata, *Tetrahedron,* **34,** 1915 (1978).

232.   Y. Shizuri, K. Yamada, and Y. Hirata, *Tetrahedron Lett.,* **1973,** 741.

233.   K. Yamada, K. Sugiura, Y. Shizur, H. Wada, and Y. Hirata, *Tetrahedron,* **33,** 1725 (1977).

234.   A. Klasek, Z. Samek, and F. Santavy, *Tetrahedron Lett.,* **1972,** 941.

235.   K. Sugiura, K. Yamada, and Y. Hirata, *Tetrahedron Lett.,* **1973,** 113.

236.   S. M. Kupchan, R. M. Smith, and R. F. Bryan, *J. Am. Chem. Soc.,* **92,** 6667 (1970).

237.   H. Wagner, E. Heckel, and J. Sonnenbichler, *Tetrahedron,* **31,** 1949 (1975).

238.   A. G. McInnes, D. G. Smith, C. Wat, L. C. Vining, and J. L. C. Wright, *J. Chem. Soc., Chem. Commun.,* **1974,** 281.

239.   A. G. McInnes, D. G. Smith, J. A. Walter, L. C. Vining, and J. L. C. Wright, *J. Chem. Soc., Chem. Commun.,* **1974,** 282.

240.   C. K. Wat, A. G. McInnes, D. G. Smith, J. L. C. Wright, and L. C. Vining, *Can. J. Chem.,* **55,** 4090 (1977).

241.   M. Matsumoto and H. Minato, *Tetrahedron Lett.,* **1976,** 3827.

242.   M. Matsumoto, *J. Sci. Hiroshima Univ., Ser. A: Phys. Chem.,* **43,** 47 (1979).

243.   K. Ando, I. Matsuura, Y. Nawata, H. Endo, H. Sasaki, T. Okytomi, T. Saehi, and G. Tamura, *J. Antibiot.,* **31,** 533 (1978).

244.   I. Matsuura, H. Endo, T. Okutomi, and K. Hosoda, *Japan Kokai,* **78,** 121,767 (1978) [*Chem. Abstr.,* **90,** 121426 (1979)].

245.   C. Vos and P. E. J. Verwiel, *Tetrahedron Lett.,* **1973,** 5173.

246.   H. Maehr, M. Leach, L. Yarmchuk, and A. Stempel, *J. Am. Chem. Soc.,* **95,** 8449 (1973).

247.   H. Maehr, T. H. Williams, M. Leach, and A. Stempel, *Helv. Chim. Acta,* **57,** 212 (1974).

248.   C. Vos and P. E. J. Verwiel, *Tetrahedron Lett.,* **1973,** 2823.

249.   M. Yasue, N. Kawamura, and T. Kato, *Yakugaku Zasshi,* **87,** 732 (1967) [*Chem. Abstr.,* **67,** 100382 (1967)].

250.   M. Yasue and N. Kawamura, *Chem. Pharm. Bull.,* **14,** 1443 (1966).

251.   N. Takahashi, K. Fukunaga, H. Yonehara, M. Asakawa, and A. Sieno, *Japan Kokai,* 75,132,183 (1975) [*Chem. Abstr.,* **84,** 149203 (1976)].

252.   Y. Kimura, N. Takahashi, and S. Tamura, *Agr. Biol. Chem.,* **33,** 1507 (1969) [*Chem. Abstr.,* **71,** 120674 (1969)].

253.   N. Takahashi, Y. Kimura, and S. Tamura, *Tetrahedron Lett.,* **1968,** 4659.

254.   S. Yoshida, K. Yoneyama, S. Shiraishi, A. Watanabe, and N. Takahashi, *Agric. Biol. Chem.,* **41,** 849 (1977) [*Chem. Abstr.,* **88,** 120936 (1978)].

255.   Y. Okami, K. Maeda, and H. Umezawa, *J. Antibiotics (Tokyo),* **7A,** 55 (1954).

256.   G. Habermehl, B. G. Christ, and H. J. Kutzner, *Z. Naturforsch B. Anorg. Chem. Org. Chem.,* **32B,** 1195 (1977).

257.   H. L. Sleeper and W. Fenical, *J. Am. Chem. Soc.,* **99,** 2367 (1977).

258.   A. Van der Gen, L. M. Van Der Linde, and J. G. Witteveen, *Rec. Trav. Chim.,* **91,** 1433 (1972).

259. R. G. Powell, C. R. Smith, D. Weisleder, D. A. Muthard, and J. Clardy, *J. Am. Chem. Soc.*, **101**, 2784 (1979).

260. R. V. Stevens, P. M. Lesko, and R. Lapoline, *J. Org. Chem.*, **40**, 3495 (1975).

261. F. O. Snyckers, F. Strelow, and A. Wiechers, *Chem. Commun.*, **1971**, 1467.

262. W. D. Marshall, T. T. Nguyen, D. B. Maclean, and I. D. Spenser, *Can. J. Chem.*, **53**, 41 (1975).

263. T. T. Nakashima, P. P. Singer, L. M. Browne, and W. A. Ayer, *Can. J. Chem.*, **53**, 1936 (1975).

264. P. Teisseire, B. Shimizu, M. Plattier, B. Corbier, and P. Rouillier, *Recherches*, **19**, 241 (1974).

265. R. Tschesche, D. Kloeden, and H. W. Fehlhaber, *Tetrahedron*, **20**, 2885 (1964).

266. E. Ali, V. S. Giri, and S. C. Pakrashi, *Tetrahedron Lett.*, **1976**, 4887.

267. L. Angenot, M. Dubois, C. Ginion, W. van Dorsser, and A. Dresse, *Arch. Intern. Pharmacodyn.*, **215**, 246 (1975).

268. R. T. Brown and C. L. Chapple, *Tetrahedron Lett.*, **1976**, 1629.

269. J. L. Pousset, A. Bouquet, A. Cave, and R. R. Paris, *Compt. Rend.*, **272C**, 665 (1971).

270. D. G. Murray, A. Szakolcai, and S. McLean, *Can. J. Chem.*, **50**, 1486 (1972).

271. S. McLean and D. G. Murray, *Can. J. Chem.*, **50**, 1478 (1972).

272. F. Hotellier, P. Delaveau, R. Besselièvre, and J. L. Pousset, *Compt. Rend.*, **282C**, 595 (1976).

273. A. G. Schultz, *Chem. Rev.*, **73**, 385 (1973).

274. T. Sugasawa, T. Toyoda, and K. Sasakura, *Tetrahedron Lett.*, **1972**, 5109.

275. T. R. Govindachari, K. R. Ravindranath, and N. Viswananthan, *J. Chem. Soc., Perkin Trans.*, **1974**, 1215.

276. W. Schuep, J. F. Blount, T. H. Williams, and A. Stempel, *J. Antibiot.*, **31**, 1226 (1978).

277. M. E. L. De Almeida, R. Braz, M. V. von Buelow, O. R. Gottlieb, and J. G. S. Maia, *Phytochemistry*, **15**, 1186 (1976).

278. Y. -Y. H. Chiu and W. N. Lipscomb, *J. Am. Chem. Soc.*, **97**, 2525 (1975).

279. F. Z. Basha, S. Hibino, D. Kim, W. E. Pye, T. -T. Wu, and S. M. Weinreb, *J. Am. Chem. Soc.*, **102**, 3962 (1980).

280. J. A. Findlay and L. Radics, *J. Chem. Soc., Perkin Trans. I*, **1972**, 2071; J. A. Findlay and D. Kwan, *ibid.*, 2962.

281. S. McLean and D. G. Murray, *Can. J. Chem.*, **50**, 1478 (1972).

282. F. E. Cislak, *U.S. Pat.* 2,938,037 (1960) [*Chem. Abstr.*, **55**, 1664 (1961)].

283. V. E. Rakovskii, *Khim. Tverd. Topl.*, **1969**, 148 [*Chem. Abstr.*, **71**, 115005 (1969)].

284. A. P. Bronshtein, G. N. Makarov, P. V. Akulov, and N. P. Girgor'ev, *Khim. Tverd. Topl.*, **1975**, 48 [*Chem. Abstr.*, **82**, 108219 (1975)].

285. G. A. Markus, *Koks Khim.*, **1974**, 29 [*Chem. Abstr.*, **82**, 75213 (1975)].

286. K. Wiszniowski, *Chemik*, **27**, 204 (1974) [*Chem. Abstr.*, **81**, 135828 (1974)].

287. V. Masek and K. Vojtovic, *Chem. Prum.*, **17**, 535 (1967) [*Chem. Abstr.*, **67**, 118982 (1967)].

288. A. P. Bronshtein, G. N. Makarov, N. P. Grigor'ev, and N. A. Bunkina, *Tr. Mosk. Khim.-Tekhnol. Inst.*, **91**, 104 (1976) [*Chem. Abstr.*, **88**, 173280 (1978)].

289. B. D. Litvinov, S. A. Litvinova, P. F. Gurtovnik, V. V. Tonkous, and A. A. Kulakov, *Koks Khim.*, **8**, 27 (1978) [*Chem. Abstr.*, **89**, 179819 (1978)].

290. T. A. Kukharenko, *Khim. Tverd. Topl.*, **1973**, 167 [*Chem. Abstr.*, **78**, 126632 (1978)].

291. T. Yokoyama, H. Kojima, and M. Uragami, and K. Miwa, *German Pat.* 2,736,269 (1976) [*Chem. Abstr.*, **88**, 152439 (1978)].

292. O. Svajgl, *Czech Pat.* 150,077 (1973) [*Chem. Abstr.*, **80**, 50194 (1974)].

204   T. D. Bailey, G. L. Goe, and E. F. V. Scriven

293.   K. F. Lang and I. Eigen, *Fortschr. Chem. Forsch.*, **8**, 91 (1967).

294.   J. Vymetal, *Chem. Prum.*, **24**, 447 (1974) [*Chem. Abstr.*, **82**, 75363 (1975)].

295.   J. Vymetal, *Chem. Listy*, **64**, 65 (1970).

296.   H. Pichler and P. Hennenberger, *Brennst.-Chem.*, **50**, 341 (1969) [*Chem. Abstr.*, **72**, 34138 (1970)].

297.   I. Kavan and K. Tesarik, *Ropa Uhlie*, **19**, 29 (1977) [*Chem. Abstr.*, **87**, 55587 (1977)].

298.   S. B. King, C. F. Brandenburg, and W. J. Lanum, *Energy Sources*, **3**, 263 (1978).

299.   P. Buryan, J. Macak, and J. Lelek, *Chem. Prum.*, **28**, 70 (1978) [*Chem. Abstr.*, **89**, 149210 (1978)].

300.   F. K. Schweighardt, C. M. White, S. Friedman, and J. L. Shultz, *ACS Symp. Ser.*, **71**, 240 (1978).

301.   G. R. Clemo, *Tetrahedron*, **29**, 3987 (1973).

302.   L. Achremowicz, *Wiad. Chem.*, **25**, 491 (1971) [*Chem. Abstr.*, **75**, 142590 (1971)].

303.   A. Maczynski, *Pol. Pat.* 53,526 (1967) [*Chem. Abstr.*, **68**, 14688 (1968)].

304.   C. Zinsstag and B. Righetti, *Swiss Pat.* 564363 (1975) [*Chem. Abstr.*, **84**, 180046 (1976)].

305.   V. A. Kostyuk, I. V. Volgina, S. M. Grigor'ev, and G. A. Markus, *U.S.S.R. Pat.* 393,275 [*Chem. Abstr.*, **79**, 146403 (1973)].

306.   G. A. Markus, V. K. Terent'ev, and V. S. Kirsanova, *Koks Khim.*, **1977**, 29 [*Chem. Abstr.*, **88**, 152376 (1978)].

307.   I. V. Volgina, V. A. Kostyuk, and V. V. Rode, *Khim. Tverd. Topl. (Moscow)*, **1978**, 58 [*Chem. Abstr.*, **90**, 41134 (1979)].

308.   E. M. Gepshtein, *Proc. Koksovaniya Uglei Vostoka, SSSR, Poluch., Obrab., Ispol's., Metody Anal., Vost. Nauch.-Issled. Uglekhim. Inst., Sb. States*, **1967**, 97 [*Chem. Abstr.*, **69**, 2824 (1968)].

309.   S. Neuhaeuser and F. Wolf, *Brennst.-Chem.*, **49**, 355 (1968) [*Chem. Abstr.*, **71**, 30328 (1969)].

310.   R. Swierezek, F. Wanecki, R. Ulatowski, and F. Dziembala, *Pol. Pat.* 94,387 [*Chem. Abstr.*, **90**, 124471 (1979)].

311.   V. V. Markow, N. N. Karacharov, M. N. Chernyavskaya, and L. A. Korzhan, *Otkrytiya, Izobret., Prom. Obraztsy, Tovarnye Znaki*, **4**, 72 (1971) [*Chem. Abstr.*, **75**, 112630 (1971)].

312.   J. Vymetal, *Erdoel Kohle, Erdgas, Petrochem. Brennst.-Chem.*, **25**, 537 (1972) [*Chem. Abstr.*, **78**, 138,593 (1973)].

313.   A. -G. Ruetgerswerke and Teerverwertung, *Ger. Pat.* 1,245,963 (1967) [*Chem. Abstr.*, **68**, 87192 (1968)].

314.   J. Baron and J. Szewczyk, *Pol. Pat.* 56,971 (1969) [*Chem. Abstr.*, **71**, 61223 (1969)].

315.   J. Vymetal and A. Kulhankova, *Chem. Prum.*, **23**, 620 (1973) [*Chem. Abstr.*, **80**, 82593 (1974)].

316.   J. Vymetal, *Czech. Pat.* 147,295 (1971) [*Chem. Abstr.*, **79**, 31891 (1973)].

317.   J. Vymetal and A. Kulhankova, *Chem. Prum.*, **23**, 124 (1973) [*Chem. Abstr.*, **79**, 68456 (1973)].

318.   O. F. Sidorov, E. G. Zaidis, V. V. Mochalov, and G. G. Rusanov, *Koks Khim.*, **1977**, 33 [*Chem. Abstr.*, **87**, 120302 (1977)].

319.   J. Vymetal and P. Tvaruzek, *Czech. Pat.* 133,170 (1969) [*Chem. Abstr.*, **73**, 77069 (1970)].

320.   J. Vymetal and A. Kulhankova, *Czech. Pat.* 163,586 (1976) [*Chem. Abstr.*, **86**, 121174 (1977)].

321.   J. Vymetal and L. Kulhanek, *Chem. Prum.*, **20**, 565 (1970) [*Chem. Abstr.*, **74**, 144311 (1971)].

322. L. A. Litsinova and V. G. Gegele, *Soobshch. Akad. Nauk Grus. SSR*, **55**, 313 (1969) [*Chem. Abstr.*, **72**, 5005 (1970)].

323. J. Vymetal and A. Kulhankova, *Czech. Pat.* 147,239 (1970) [*Chem. Abstr.*, **78**, 159447 (1973)].

324. J. Vymetal and A. Kulhankova, *Chem. Prum.*, **22**, 172 (1972) [*Chem. Abstr.*, **77**, 22632 (1972)].

325. J. Skura, J. Czarnecki, A. Maczynski, and D. Sybilska, *Pol. Pat.* 73,862 (1975) [*Chem. Abstr.*, **86**, 106394 (1977)].

326. J. Menyhart, A. Ujhidy, and O. Borlai, *Hung. Pat.* 9,984 (1975) [*Chem. Abstr.*, **84**, 30625 (1976)].

327. J. Menyhart, K. Domsa, J. Suto, and P. Siklos, *Hung. Pat.* 155,473 (1968) [*Chem. Abstr.*, **70**, 106399 (1969)].

328. J. Vymetal, *Czech. Pat.* 149,514 (1973) [*Chem. Abstr.*, **81**, 49571 (1974)].

329. M. M. Anwar, S. T. M. Cook, C. Hanson, and M. W. T. Pratt, in "Proc. Int. Solvent Extr. Conf.," G. V. Jeffreys, ed., Soc. Chem. Ind., London, 1974, p. 895.

330. D. Wyrzykowska-Stankiewicz and A. Szafranski, *Przem. Chem.*, **53**, 358 (1974) [*Chem. Abstr.*, **81**, 171907 (1974)].

331. M. M. Anwar, C. Hanson, A. N. Patel, and M. W. T. Pratt, *Trans. Inst. Chem. Eng.*, **51**, 151 (1973).

332. A. Tamano, *Japan Pat.* 68 21,545 (1968) [*Chem. Abstr.*, **70**, 57659 (1969)].

333. A. Tamano and Y. Kawai, *Japan Pat.* 68 20,187 (1968) [*Chem. Abstr.*, **70**, 57666 (1969)].

334. V. S. Novoselov and G. D. Kharlampovich, *Khim. Tverd. Topl. (Moscow)*, **1976**, 116 [*Chem. Abstr.*, **86**, 106310 (1977)].

335. A. Skrzecz and A. Macznski, *Pol. Pat.* 76,876 (1975) [*Chem. Abstr.*, **85**, 78013 (1976)].

336. P. Tomasik, A. Woszczyk, and F. Kret, *Koks, Smola, Gaz*, **21**, 330 (1976) [*Chem. Abstr.*, **87**, 134941 (1977)].

337. J. Oszczapowicz, J. Golab, and H. Pines, *J. Chromatogr.*, **64**, 1 (1972).

338. K. Tesarik and S. Ghyczy, *J. Chromatogr.*, **91**, 723 (1974).

339. D. E. Durbin and A. Zlatkis, *J. Chromatogr. Sci.*, **8**, 608 (1970).

340. Y. N. Popov, T. V. Portnova, B. E. Kogan, E. A. Kiva, and E. A. Kruglov, *Tezisy Dokl. Simp. Khim. Tekhnol. Geterotsikl. Soedin. Goryuch. Iskop., 2nd,* Donetsk. Gos. Univ. Donetsk, USSR, **1973**, p. 160 [*Chem. Abstr.*, **86**, 174047 (1977)].

341. C. M. White, F. K. Schweighardt, and J. L. Schultz, *Fuel Process. Technol.*, **1**, 209 (1978).

342. R. Hayatsu, R. G. Scot, L. P. Moore, and M. H. Studier, *Nature (London)*, **257**, 378 (1975).

343. V. M. Kagasov and N. A. Zharkova, *Koks Khim.*, **1966**, 29 [*Chem. Abstr.*, **66**, 12856 (1967)].

344. J. Neiser and V. Masek, *Zentralbl. Arbeitsmed. Arbeitsschutz*, **23**, 111 (1973) [*Chem. Abstr.*, **79**, 57236 (1973)].

345. G. A. Evdokimova, V. E. Rokovskii, and E. A. Yurkevich, *Kompleks. Ispol'z. Torfa*, **1968**, 58 [*Chem. Abstr.*, **73**, 100840 (1970)].

346. J. Suto, *Period. Polytech., Chem. Eng. (Budapest)*, **11**, 283 (1967) [*Chem. Abstr.*, **69**, 60595 (1968)].

347. J. Vymetal, *Chem. Prum.*, **22**, 119 (1972) [*Chem. Abstr.*, **77**, 37387 (1972)].

348. J. Vymetal, *Chem. Prum.*, **22**, 612 (1972) [*Chem. Abstr.*, **78**, 138591 (1973)].

349. R. Zalewski and P. Tomasik, *Koks, Smola, Gaz*, **22**, 203 (1977) [*Chem. Abstr.*, **90**, 171239 (1979)].

206 T. D. Bailey, G. L. Goe, and E. F. V. Scriven

350. A. Bylicki and A. Maczynski, *Koks, Smola, Gaz,* **14,** 306 (1969) [*Chem. Abstr.,* **73,** 37218 (1970)].

351. V. M. Kagasov, G. D. Kharlampovich, and B. V. Suvorov, *Vestn. Akad. Nauk Kaz. SSR,* **23,** 40 (1967) [*Chem. Abstr.,* **67,** 110348 (1967)].

352. T. V. Speshilova, I. U. Numanov, G. L. Tolmacheva, B. I. Gizatova, and N. Yusupova, *Dokl. Akad. Nauk Tadzh. SSR,* **12,** 36 (1969) [*Chem. Abstr.,* **72,** 23089 (1970)].

353. I. U. Numanov, T. V. Speshilova, G. L. Tolmacheva, and I. I. Rakitin, *Izv. Akad. Nauk Tadzh. SSR, Otd. Fiz.-Mat. Geol.-Khim. Nauk.,* **1972,** 51 [*Chem. Abstr.,* **78,** 149468 (1973)].

354. A. I. Stekhun, T. S. Ivanova, V. P. Kubrak, and Z. F. Kuz'mina, *Neftepererab. Neftekhim. (Moscow),* **1972,** 7 [*Chem. Abstr.,* **77,** 142052 (1972)].

355. G. F. Bol'shakov, *Izv. Vyssh. Uchebn. Zaved., Neft Gaz,* **18,** 44 (1975) [*Chem. Abstr.,* **84,** 33407 (1976)].

356. G. I. Spiridonova and S. S. Kogan, *Metod. Issled. Obl. Izuch. Org. Veshchestva Poszemn. Vod Neftegazonos. Obl.,* **10,** 60 (1971) [*Chem. Abstr.,* **76,** 88534 (1972)].

357. J. F. McKay, J. H. Weber, and D. R. Latham, *Anal. Chem.,* **48,** 891 (1976).

358. J. F. McKay, J. H. Weber, and D. R. Latham, *Prepr., Div. Pet. Chem., Am. Chem. Soc.,* **20,** 12 (1975).

359. J. F. McKay and D. R. Latham, *ERDA Energy Res. Abstr.,* **2,** Abstr. No. 24017 (1977) [*Chem. Abstr.,* **87,** 186906 (1977)].

360. D. K. Albert, *Anal. Chem.,* **50,** 1822 (1978).

361. V. G. Ben'kovskii, A. Y. Baidova, B. T. Bulatova, N. S. Lyubopytova, and Y. N. Popov, *Neftekhimiya,* **12,** 454 (1972) [*Chem. Abstr.,* **77,** 116548 (1972)].

362. A. V. Kotova, L. N. Tokareva, A. Y. Lanchuk, and I. D. Leonov, *Tr. Inst. Khim. Nefti Prir. Solei Akad. Nauk Kaz. SSR,* **3,** 110 (1971) [*Chem. Abstr.,* **77,** 154651 (1972)].

363. C. F. Brandenburg and D. R. Latham, *J. Chem. Eng. Data,* **13,** 391 (1968).

364. L. R. Snyder, *Anal. Chem.,* **41,** 314 (1969).

365. J. C. Escalier, M. Caude, C. Bollet, R. Rosset, P. Sassiat, and J. P. Massoue, *Analusis,* **5,** 395 (1977) [*Chem. Abstr.,* **88,** 155426 (1978)].

366. V. G. Ben'kovskii, A. Y. Baikova, Y. N. Popov, B. T. Bulatova, and N. A. Alekseeva, *Neftekhimiya,* **13,** 581 (1973) [*Chem. Abstr.,* **80,** 29020 (1974)].

367. A. I. Stekhun, A. I. Sumskaya, and V. P. Zavadskaya, *Tezisy Dokl.-Symp. Khim. Tekhnol Geterotsikl Soedin. Goryuch. Iskop, 2nd,* **1973,** 154 [*Chem. Abstr.,* **86,** 158015 (1977)].

368. V. G. Ban'kovskii, A. Y. Baikova, Y. N. Popova, E. A. Kruglov, A. I. Stekun, N. Lyubopytova, and F. N. Latypova, in *Tezisy Dokl. Nauch. Sess. Khim. i Tekhnol Org. Soedin. Sery Sernisty kh Neftei, 13th,* G. D. Galpern, ed., Zinatne, Riga, USSR, **1974,** 27 [*Chem. Abstr.,* **87,** 25665 (1977)].

369. R. E. Poulson, C. M. Frost, and H. B. Jensen, *Adv. Chem. Ser.,* **1976,** 151.

370. D. Brown, D. G. Ernshaw, F. R. McDonald, and H. B. Jensen, *Anal. Chem.,* **42,** 146 (1970).

371. F. R. McDonald, A. W. Decora, and G. L. Cook, *U.S., Bur. Mines, Bull.,* **1971,** 657 [*Chem. Abstr.,* **76,** 88196 (1972)].

372. F. R. McDonald, A. W. Decora, and G. L. Cook, *Appl. Spectrosc.,* **22,** 325 (1968).

373. F. R. McDonald, A. W. Decora, and G. L. Cook, *Appl. Spectrosc.,* **22,** 329 (1968).

374. R. E. Poulson, H. B. Jensen, and G. L. Cook, *Am. Chem. Soc., Div. Petrol. Chem., Prepr.,* **16,** A49 (1971) [*Chem. Abstr.,* **78,** 6133 (1973)].

375. M. Pailer and W. Jiresch, *Monatsh. Chem.,* **100,** 121 (1969).

376. G. S. Igon'kina and I. M. Ozerov, *Khim. Tverd. Topl.,* **1973,** 101 [*Chem. Abstr.,* **80,** 147507 (1974)].

377. E. V. Glushchenkova, N. D. Dokshina, and S. S. Semenov, *Khim. Tekhnol. Goryuch. Slantsev, Prod. Ikh. Pererab.*, **1968**, 56 [*Chem. Abstr.*, **71**, 83213 (1969)].

378. O. G. Eizen, A. Khallik, and I. R. Klesment, *Esti NSV Teaduste Akad. Toimetised, Fuusik.-Mat.-ja Tehnikateaduste Seer.*, **1966**, 230 [*Chem. Abstr.*, **66**, 4593 (1967)].

379. N. N. Bezinger, G. D. Gal'pern, and V. N. Karicheva, *Metody Anal. Org. Soedin. Nefti. Ikh Smesei Proiszvod.*, **1969**, 121 [*Chem. Abstr.*, **73**, 37153 (1970)].

380. D. E. Anders, F. G. Doolittle, and W. E. Robinson, *Geochim. Cosmochim. Acta*, **39**, 1423 (1975) [*Chem. Abstr.*, **84**, 7372 (1976)].

381. B. R. Simoneit, H. K. Schnoes, P. Haug, and A. Burlingame, *Nature (London)*, **226**, 75 (1970).

382. E. Kitatsuji, E. Yoshii, and T. Iida, *Yakugaku Zasshi*, **92**, 665 (1972) [*Chem. Abstr.*, **77**, 64,362 (1972)].

383. J. Grudzien and H. Zielinski, *Koks, Smola, Gaz*, **19**, 129 (1974) [*Chem. Abstr.*, **82**, 127195 (1975)].

384. P. C. Uden, A. P. Carpenter, Jr., H. M. Hackett, D. E. Henderson, and S. Siggia, *Anal. Chem.*, **51**, 38 (1979).

385. W. D. Crow and J. H. Hodgkin, *Aust. J. Chem.*, **17**, 119 (1964).

386. M. M. Robison, W. G. Pierson, L. Dorfman, B. F. Lambert, and R. A. Lucas, *J. Org. Chem.*, **31**, 3206 (1966).

387. R. J. Highet, *J. Org. Chem.*, **29**, 471 (1964).

388. C. Djerassi, J. P. Kutney, M. Shamma, J. N. Shoolery, and L. F. Johnson, *Chem. and Ind.*, **1961**, 210.

389. Y. V. Kurbatov, A. S. Kurbatova, O. S. Otroshchenko, and A. S. Sadyknov, *Rr. Samarkand. Gos. Univ.*, **167**, 71 (1969) [*Chem. Abstr.*, **75**, 139950 (1971)].

390. N. Y. Novogorodova, S. K. Maekh, and S. Y. Yunusov, *Khim. Prir. Soedin.*, **11**, 435 (1975) [*Chem. Abstr.*, **84**, 59823 (1976)].

391. G. S. Rao, J. E. Sinsheimer, and H. M. McIlhenny, *Chem. and Ind.*, **1972**, 537.

392. G. Jones, H. M. Fales, and W. C. Wildman, *Tetrahedron Lett.*, **1963**, 397.

393. A. I. Scott and P. C. Cherry, *J. Am. Chem. Soc.*, **91**, 5872 (1969).

394. L. Crawford and M. J. Kretsch, *J. Food Sci.*, **41**, 1470 (1976).

395. R. G. Buttery, L. C. Ling, R. Teranishi, and T. R. Mon, *J. Agr. Food Chem.*, **25**, 1227 (1977).

396. K. Watanabe and Y. Sato, *J. Agr. Food Chem.*, **19**, 1017 (1971).

397. G. MacLeod and B. M. Coppock, *J. Agr. Food Chem.*, **25**, 113 (1977).

398. G. P. Ionas, V. A. Alsuf'ev, L. V. Kondakova, M. I. Stepanova, and Y. K. Shaposhnikov, *Rybn. Khoz. (Moscow)*, **1977**, 69 [*Chem. Abstr.*, **87**, 51810 (1977)].

399. R. V. Golovnya, G. A. Mironov, and I. L. Zhuraveleva, *Zh. Anal. Khim.*, **22**, 612 (1967) [*Chem. Abstr.*, **67**, 52845 (1967)].

400. R. V. Golovnya, *IARC Sci. Publ.*, **14**, 237 (1976) [*Chem. Abstr.*, **86**, 169468 (1977)].

401. M. Quijano-Rico, R. E. Bautista, B. F. Chaparro, G. V. Zamudio, P. A. Ortiz, and J. Von Helden, *Adv. Mass. Spectrom. Biochem. Med.*, **2**, 207 (1977).

402. E. N. Suassuna de Oliveira, K. Kizawa, and B. Campos, *Rev. Microbiol.*, **3**, 61 (1972).

403. E. Turchetto, *Minerva Med.*, **63**, 3314 (1972).

404. O. G. Vitzthum, P. Werkhoff, and P. Hubert, *J. Food Sci.*, **40**, 911 (1975).

405. J. P. Walradt, A. O. Pittet, T. E. Kinlin, R. Muralidhara, and A. Sanderson, *J. Agr. Food Chem.*, **19**, 972 (1971).

406. C. K. Shu and G. R. Waller, *J. Food Sci.*, **36**, 579 (1971).

407. T. E. Kinlin, R. Muralidhara, A. O. Pittet, A. Sanderson, and J. P. Walradt, *J. Agr. Food Chem.*, **20**, 1021 (1972).

408. I. Yajima, T. Yanai, M. Nakamura, H. Sakakibara, and T. Habu, *Agr. Biol. Chem.*, **42**, 1229 (1978).

409. D. A. Withycombe, R. C. Lindsay, and D. A. Stuiber, *J. Agr. Food Chem.*, **26**, 816 (1978).

410. E. E. Ter-Manuel'yants and O. V. Studentsov, *Nauchn. Tr. Maikopskoi Opytn. Stn., Vses. Nauchno-Issled. Inst. Rastenievod.*, **7**, 107 (1973) [*Chem. Abstr.*, **83**, 95158 (1975)].

411. L. Schreyen, P. Dirinck, F. Van Wassenhove, and N. Schamp, *J. Agr. Food Chem.*, **24**, 336 (1976).

412. D. L. Patil and N. G. Magar, *Curr. Sci.*, **43**, 249 (1974).

413. A. L. Boyko, M. E. Morgan, and L. M. Libbey, in *"Anal. Foods Beverages [Proc. Symp.]* 1977," G. Charalambous, ed., Academic Press, New York 1978, p. 57.

414. P. J. Morgan and K. Smith, *Analyst (London)*, **103**, 1053 (1978).

415. D. J. Folkes and J. W. Gramshaw, *J. Food Technol.*, **12**, 1 (1977).

416. E. J. Mulders, *Versl. Landbouwk. Onders.*, No. 798 (1973) [*Chem. Abstr.*, **80**, 13787 (1974)].

417. R. V. Golovnya, N. G. Enikeeva, I. L. Zhuravleva, and A. S. Zyuz'ko, *Nahrung*, **18**, 143 (1974) [*Chem. Abstr.*, **81**, 62336 (1974)].

418. R. J. Harding, J. J. Wren, and H. E. Nursten, *J. Inst. Brew.*, **84**, 41 (1978).

419. E. Collins, *J. Agr. Food Chem.*, **19**, 533 (1971).

420. P. Wang, H. Kato, and M. Fujimaki, *Agr. Biol. Chem.*, **33**, 1775 (1969).

421. H. J. Wobben, R. Timmer, R. Ter Heide, and P. J. De Valois, *J. Food Sci.*, **36**, 464 (1971).

422. R. J. Harding, H. E. Nursten, and J. J. Wren, *J. Sci. Food Agr.*, **28**, 225 (1977).

423. N. Nunomura, M. Sasaki, Y. Asao, and T. Yokotsuka, *Agr. Biol. Chem.*, **42**, 2123 (1978).

424. R. G. Buttery, R. M. Seifert, D. G. Guadagni, and L. C. Ling, *J. Agr. Food Chem.*, **19**, 969 (1971).

425. T. H. Parliment, M. G. Kolor, and I. Y. Maing, *J. Food Sci.*, **42**, 1592 (1977).

426. J. P. Dumont, S. Roger, and J. Adda, *Lait*, **55**, 479 (1975) [*Chem. Abstr.*, **84**, 15814 (1976)].

427. O. G. Vitzthum, P. Werkhoff, and P. Hubert, *J. Agr. Food Chem.*, **23**, 999 (1975).

428. R. V. Golovnya, I. L. Zhuravleva, G. A. Mironov, and R. M. Abdullina, *Moloch. Prom.*, **31**, 8 (1970) [*Chem. Abstr.*, **72**, 131194 (1970)].

429. R. V. Golovnya, R. M. Abdullina, I. L. Zhuravleva, and G. A. Mironov, *Izv. Akad. Nauk SSSR, Ser. Khim.*, **1969**, 2570 [*Chem. Abstr.*, **72**, 99289 (1970)].

430. J. P. Dumont and J. Adda, *J. Agr. Food Chem.*, **26**, 364 (1978).

431. Y. Kato, K. Watanabe, and Y. Sato, *Lebensm.-Wiss. Technol.*, **11**, 128 (1978).

432. H. Sulser and W. Buchi, *Lebensm.-Wiss. Technol.*, **2**, 105 (1969).

433. M. N. Simon and M. Simon, *Astrophys. J.*, **184**, 757 (1973).

434. C. Ponnamperuma, in "Chem. Evol. Giant Planets [Invited Lect. Colloq.] 1974," C. Ponnamperuma, ed., Academic Press, New York, 1976, p. 221.

435. N. Friedmann, S. L. Miller, and R. A. Sanchez, *Science*, **171**, 1026 (1971).

436. E. J. Mulders, *Z. Lebensm.-Unters. Forsch.*, **152**, 193 (1973).

437. T. Yazaki, M. Makino, T. Yamamoto, K. Tsuji, H. Zenda, and T. Kosuge, *Yakugaku Zasshi*, **98**, 914 (1978) [*Chem. Abstr.*, **89**, 197912 (1978)].

438. K. Tsuji, T. Yamamoto, H. Zenda, and T. Kosuge, *Yakugaku Zasshi*, **98**, 910 (1978) [*Chem. Abstr.*, **89**, 163925 (1978)].

439. F. Ledl and T. Serverin, *Chem. Mikrobiol., Technol. Lebensm.,* **2,** 155 (1973).

440. K. Tsuji, *Yakugaku Zasshi,* **96,** 479 (1976) [*Chem. Abstr.,* **85,** 99075 (1976)].

441. Scientific Literature Review of Pyridine and Related Substances in Flavor Usage, FDA, Washington, D.C., Jan. **1979,** Vol. 1, p. 92.

442. A. R. Pinder, *Alkaloids (London),* **8,** 37 (1978), and other volumes in the series.

443. K. T. Potts and J. Baum, *J. Chem. Soc.,* **1973,** 883.

444. H. Neunhoeffer and H. Ohl, *Chem. Ber.,* **111,** 299 (1978).

445. T. Eicher, J. L. Weber, and J. Kopf, *Ann. Chem.,* **1978,** 1222.

446. T. Eicher, F. Abdesaken, G. Franke, and J. L. Weber, *Tetrahedron Lett.,* **1975,** 3915.

447. W. Ried and F. Baetz, *Ann. Chem.,* **762,** 1 (1972).

448. W. Ried and F. Baetz, *Angew. Chem. Internat. Ed. Engl.,* **10,** 735 (1971).

449. W. Ried and F. Baetz, *Ann. Chem.,* **725,** 230 (1969).

450. P. B. J. Driessen, H. Hogeveen, H. Jorritsma, and D. S. B. Grace, *Tetrahedron Lett.,* **1976,** 2263.

451. T. Kato, Y. Yamamoto, and M. Kondo, *Heterocycles,* **3,** 927 (1975).

452. E. Schmits, U. Bicker, and K. P. Dietz, *Z. Chem.,* **14,** 230 (1974).

453. A. Murata and H. Suzuki, *Japan. Kokai* 74 49,967 (1974) [*Chem. Abstr.,* **81,** 135972 (1974)].

454. N. S. Kozlov, G. L. Lomako, and L. T. Gurskaya, *Dokl. Akad. Nauk B. SSR,* **1975,** 248 [*Chem. Abstr.,* **82,** 170624 (1975)].

455. R. D. Bowden and T. Seaton, *U.S. Pat.* 3,830,820 (1974) [*Chem. Abstr.,* **81,** 135977 (1974)].

456. B. Al-Saleh, R. E. Banks, M. G. Barlow, and J. C. Hornby, *J. Fluorine Chem.,* **12,** 341 (1978).

457. R. A. Abramovitch and G. N. Knaus, *J. Chem. Soc., Chem. Commun.,* **1974,** 238.

458. L. Hoesch, *Chimia,* **29,** 531 (1975).

459. R. Faragher and T. L. Gilchrist, *J. Chem. Soc., Chem. Commun.,* **1977,** 252.

460. J. E. Baldwin, R. H. Fleming, and D. M. Simons, *J. Org. Chem.,* **37,** 3963 (1972).

461. R. G. Pews, E. B. Nyquist, and F. P. Corson, *J. Org. Chem.,* **35,** 4096 (1970).

462. R. G. Pews and F. P. Corson, *U.S. Pat.* 3,641,004 (1972) [*Chem. Abstr.,* **76,** 140549 (1972)].

463. T. Jaworski and B. Korybut-Daszkiewicz, *Rocz. Chem.,* **41,** 1521 (1967) [*Chem. Abstr.,* **69,** 2822 (1967)].

464. E. K. Fields, *World Petrol. Congr., Proc. 7th 1967,* Vol. 5, Elsevier Publ. Co. Ltd. Barking, Engl., **1968,** p. 23 [*Chem. Abstr.,* **71,** 60374 (1968)].

465. N. Barbulescu, F. Potmischil, and G. Badita, *Chem. Ber.,* **104,** 787 (1971).

466. W. Lwowski, "Nitrenes," in M. Jones and R. A. Moss, eds., *Reactive Intermediates,* Vol. 1, Wiley, New York, 1978, pp. 205–207.

467. W. D. Crow and M. N. Paddon-Row, *Aust. J. Chem.,* **28,** 1755 (1975).

468. C. Wentrup, *J. Chem. Soc., Chem. Commun.,* **1967,** 1387.

469. R. A. Abramovitch and W. D. Holcomb, *J. Am. Chem. Soc.,* **97,** 676 (1975).

470. A. Padwa and P. H. J. Carlsen, *J. Org. Chem.,* **43,** 2029 (1978).

471. A. Padwa and P. H. J. Carlsen, *J. Am. Chem. Soc.,* **99,** 1514 (1977).

472. A. Padwa, J. Smolanoff, and A. Tremper, *J. Org. Chem.,* **41,** 543 (1976).

473. F. D. Bellamy, *J. Chem. Soc., Chem. Commun.,* **1978,** 998.

474. F. D. Bellamy, *Tetrahedron Lett.,* **1978,** 4577.

475. A. Hassner and A. Kascheres, *J. Org. Chem.,* **37,** 2328 (1972).

210 T. D. Bailey, G. L. Goe, and E. F. V. Scriven

476. R. E. Moerck and M. A. Battiste, *Tetrahedron Lett.*, **1973**, 4421.

477. A. Demoulin, H. Gorissen, A. M. Hesbain-Frisque, and L. Ghosez, *J. Am. Chem. Soc.*, **97**, 4409 (1975).

478. V. N. Yandovskii and T. K. Klindukhova, *Zh. Org. Khim.*, **10**, 1510 (1974) [*Chem. Abstr.*, **81**, 105356 (1974)].

479. H. Quast, K. H. Ross, E. Spiegel, K. Peters, and H. G. Von Schnering, *Angew. Chem. Internat. Ed. Engl.*, **16**, 177 (1977).

480. D. H. Aue and D. Thomas, *J. Org. Chem.*, **40**, 1349 (1975).

481. C. Metzger and J. Kurz, *Chem. Ber.*, **108**, 233 (1975).

482. D. Mackay and L. L. Wong, *Can. J. Chem.*, **53**, 1973 (1975).

483. J. D. Butler and R. D. Laundon, *J. Chem. Soc. C,* **1969**, 173.

484. J. D. Butler and R. D. Laundon, *J. Chem. Soc. B,* **1970**, 716.

485. F. Bonadies, F. Savagnone, and M. L. Scarpati, *Gazz. Chim. Ital.*, **108**, 87 (1978).

486. J. R. Geigy, A. -G., *Fr. Pat.* 1,477,998 (1967) [*Chem. Abstr.*, **68**, 29606 (1968)].

487. J. R. Geigy, A. -G., *Brit. Pat.* 1,108,975 (1968) [*Chem. Abstr.*, **69**, 67226 (1969)].

488. J. R. Geigy, A. -G., *Fr. Pat.* 1,538,729 (1968) [*Chem. Abstr.*, **71**, 38810 (1969)].

489. J. B. Petersen, J. Lei, N. Clauson-Kaas, and K. Norris, *Kgl. Dan. Vidensk. Selsk., Mat. Fys. Medd.*, **36**, 23 (1967) [*Chem. Abstr.*, **69**, 59063 (1968)].

490. L. A. Walter, C. K. Springer, J. Kenney, S. K. Galen, and N. Sperber, *J. Med. Chem.*, **11**, 792 (1968).

491. H. Greuter and D. Bellus, *J. Heterocycl. Chem.*, **14**, 203 (1977).

492. I. F. Bel'skii, F. I. Bel'skii, and V. M. Shostakovskii, *Izv. Akad. Nauk SSSR, Ser. Khim.*, **1977**, 1364 [*Chem. Abstr.*, **87**, 152127 (1977)].

493. J. V. B. Petersen, *U.S. Pat.* 3,658,826 (1972) [*Chem. Abstr.*, **77**, 883341 (1972)].

494. J. V. B. Petersen, K. Norris, N. Clauson-Kaas, and K. Svanholt, *Acta Chem. Scand.*, **23**, 1785 (1969).

495. K. Undheim and T. Greibrokk, *Acta Chem. Scand.*, **23**, 2475 (1969).

496. L. Mavougou-Gomes, *Bull. Soc. Chem. Fr.*, **1967**, 1764.

497. J. R. Geigy A. -G., *Swiss Pat.* 415,630 (1967) [*Chem. Abstr.*, **66**, 94915 (1966)].

498. H. R. Henze and R. J. Speer, *J. Am. Chem. Soc.*, **64**, 522 (1942).

499. N. Clauson-Kaas and M. Meister, *Acta Chem. Scand.*, **21**, 1104 (1967).

500. K. Undheim and G. Bye, *Acta Chem. Scand.*, **23**, 695 (1969).

501. M. Yasue, N. Kawamura, and J. Sakakibara, *Yakugaku Zasshi*, **90**, 1222 (1970) [*Chem. Abstr.*, **74**, 23102 (1971)].

502. J. H. Looker, R. L. Prokop, W. E. Serbousek, and M. D. Cliffton, *J. Org. Chem.*, **44**, 3408 (1979).

503. V. J. Veksler, *Zh. Obshch. Khim.*, **38**, 1649 (1968) [*Chem. Abstr.*, **70**, 4460 (1969)].

504. T. Koeda, H. Asaoka, U. Shibata, T. Tsuruoka, S. Inoue, and T. Niida, *Japan. Kokai*, 77 05,774 (1977) [*Chem. Abstr.*, **87**, 53089 (1977)].

505. T. Tsuruoka, K. Kawamura, I. Kenji, S. Inoue, and T. Nuda, *Japan Kokai*, 75 29,568 (1975) [*Chem. Abstr.*, **83**, 97049 (1975)].

506. T. Tsuruoka, T. Niwa, H. Goi, Y. Ogawa, S. Omoto, S. Inoue, K. Saito, and T. Niida, *Japan. Kokai*, 75 36,467 (1975) [*Chem. Abstr.*, **83**, 97050r (1975)].

507. T. Severin and A. Loidl, *Z. Lebensm.-Unters. Forsch.*, **161**, 119 (1976) [*Chem. Abstr.*, **85**, 143377 (1976)].

508. H. Tsuchida, M. Komoto, H. Kato, and M. Fujimaki, *Agr. Biol. Chem.*, **39**, 1143 (1975).

509. K. Olsson, P. A. Pernemalm, and O. Theander, *Acta Chem. Scand. Ser. B*, **32**, 249 (1978).

510. T. Shabamoto and R. A. Bernhard, *J. Agr. Food Chem.*, **26**, 183 (1978).

511. H. Tsuchida, M. Komoto, H. Kato, T. Kurata, and M. Fujimaki, *Agr. Biol. Chem.*, **40**, 2051 (1976).

512. H. Tsuchida, M. Komoto, H. Kato, and M. Fujimaki, *Agr. Biol. Chem.*, **37**, 403 (1973).

513. M. Sakaguchi and T. Shibamoto, *J. Agr. Food Chem.*, **26**, 179 (1978).

514. F. De Angelis, A. Gambacorta, and R. Nicoletti, *Synthesis*, **1976**, 798.

515. R. L. Jones and C. W. Rees, *J. Chem. Soc. C*, **1969**, 2249.

516. F. S. Baker, R. E. Busby, M. Iqbal, J. Parrick, and C. J. G. Shaw, *Chem. Ind. (London)*, **1969**, 1344.

517. A. Gambacorta, R. Nicoletti, and M. L. Forcellese, *Tetrahedron*, **27**, 985 (1971).

518. M. L. Forcellese, A. Gambacorta, and R. Nicoletti, *Atta Acad. Naz. Lincei, Cl. Sct. Fis., Mat. Nat., Rend.*, **53**, 569 (1972) [*Chem. Abstr.*, **81**, 151202 (1974)].

519. R. L. Jones and C. W. Rees, *J. Chem. Soc. C*, **1969**, 2255.

520. A. Gambacorta, R. Nicoletti, S. Cerrini, W. Fedeko, and E. Gavuzzo, *Tetrahedron Lett.*, **1978**, 2439.

521. W. H. Bell and P. M. Quan, *Brit. Pat.* 1,184,244 (1970) [*Chem. Abstr.*, **72**, 132541 (1970)].

522. S. R. Landor, V. Rogers, and H. R. Sood, *J. Chem. Soc., Perkin Trans. I*, **1976**, 2103.

523. J. F. Biellmann and M. P. Goeldner, *Tetrahedron*, **27**, 2957 (1971).

524. B. Eistert, G. W. Mueller, and T. J. Arackal, *Justus Leibigs Ann. Chem.*, **1976**, 1023.

525. B. Eistert, G. W. Mueller, and T. J. Arackal, *Justus Liebigs Ann. Chem.*, **1976**, 1031.

526. R. A. Y. Jones and N. Sadighi, *J. Chem. Soc., Perkin Trans. I*, **1976**, 2259.

527. M. Libert, C. Caullet, and J. Huguet, *Bull. Soc. Chim. Fr.*, **1972**, 3639.

528. M. Libert and C. Caullet, *Bull. Soc. Chem. Fr.*, **1974**, 800.

529. M. Libert and C. Caullet, *C. R. Acad. Sci., Ser. C*, **276**, 1073 (1973).

530. T. Kato, Y. Yamamoto, and M. Kondo, *Heterocycles*, **3**, 927 (1975).

531. M. Y. Karpeiskii and V. L. Florent'ev, *Russ. Chem. Rev.*, **38**, 540 (1969).

532. I. J. Turchi and M. J. S. Dewar, *Chem. Rev.*, **75**, 389 (1975).

533. H. C. van der Plas, "Ring Transformations of Heterocycles," Vol. 1, Academic Press London, 1973, pp. 316–321.

534. R. Lakhan and B. Terna, in "Advances in Oxazole Chemistry," A. R. Katritzky and A. J. Boulton, eds., *Advances in Heterocyclic Chemistry*, Vol. 17, Academic Press, New York, 1974, pp. 99–211.

535. N. D. Doktorova, L. V. Ionava, M. Y. Karpeiskii, N. S. Padyukova, K. F. Turchin, and V. L. Florent'ev, *Tetrahedron*, **25**, 3527 (1969).

536. N. S. Padyukova and V. L. Florent'ev, *Khim. Geterotsikl. Soedin.*, **1973**, 600 [*Chem. Abstr.*, **79**, 52483 (1973)].

537. V. A. Puchnova and E. A. Luk'yanets, *Khim. Geterot. sikl. Soedin., Sb. 2: Kislorod-soderzhashchie Geterotsikly.*, **1970**, 327 [*Chem. Abstr.*, **76**, 140453 (1972)].

538. T. Naito and T. Yoshikawa, *Chem. Pharm. Bull.*, **14**, 918 (1966).

539. T. Yoshikawa, F. Ishikawa, Y. Omura, and T. Naito, *Chem. Pharm. Bull.*, **13**, 873 (1965)].

540. J. W. White, Reilly Tar & Chemical Corp., private communication (1980).

541. A. P. Kozikowski and N. M. Hasan, *J. Org. Chem.*, **42**, 2039 (1977).

542. Y. Usui, Y. Hara, N. Shimamoto, S. Yurugi, and T. Masuda, *Heterocycles*, **3**, 155 (1975).

543. W. Boell and H. Koening, *Ger. Offen.* 2,711,656 (1978) [*Chem. Abstr.*, **90**, 6252 (1979)].

212   T. D. Bailey, G. L. Goe, and E. F. V. Scriven

544. W. Boell, D. Lenke, and G. Von Philipsborn, *Ger. Offen.* 2,711,655 (1978) [*Chem. Abstr.*, **90**, 103837 (1979)].

545. W. Boell and H. Koening, *Ger. Offen.* 2,143,989 (1973) [*Chem. Abstr.*, **78**, 147815 (1973)].

546. T. Maruyama, E. Araki, and N. Tokai, *Japan Pat.* 11, 745 (1967) [*Chem. Abstr.*, **68**, 87177 (1967)].

547. Y. Morisawa, M. Kataoka, and T. Watanabe, *Chem. Pharm. Bull.*, **24**, 1089 (1976).

548. Y. Morisawa, M. Kataoka, and T. Watanabe, *Japan Kobai*, 75 58,075 (1975) [*Chem. Abstr.*, **83**, 178832 (1975)].

549. Y. Morisawa, M. Kataoka, and T. Watanabe, *Japan Kokai*, 75 69,081 (1975) [*Chem. Abstr.*, **84**, 43853 (1976)].

550. M. Kawazu, M. Seto, and M. Watanabe, *Japan. Pat.* 71 02,020 (1971) [*Chem. Abstr.*, **74**, 99897 (1971)].

551. M. V. Balyakin, Z. N. Zhukova, and E. S. Zhdanovich, *Zh. Prikl. Khim. (Leningrad)*, **41**, 2324 (1968) [*Chem. Abstr.*, **70**, 68074 (1969)].

552. G. Y. Kondrat'eva, B. A. Kazanskii, N. G. Proshchina, and N. A. Oshuevea, *U.S.S.R. Pat.* 213,879 (1963) [*Chem. Abstr.*, **69**, 52013 (1968)].

553. J. Wursch, S. F. Schaeren, and H. Frick, *Methods Enzymol.*, **18**, 606 (1970).

554. C. Colombini and E. Celon, *Gazz. Chim. Ital.*, **99**, 526 (1969).

555. N. A. Drobinskaya, L. V. Ionova, M. Y. Karpeiskii, N. S. Padyukova, K. F. Turchin, and V. L. Florent'ev, *Khim. Geterotsikl. Soedin.*, **1970**, 37 [*Chem. Abstr.*, **72**, 121313 (1970)].

556. M. Y. Karpeiskii, N. S. Padyukova, and V. L. Florent'ev, *Tetrahedron Lett.*, **1970**, 4489.

557. V. P. Chekhun, E. N. Zvonkova, M. I. Struchkova, T. P. Belova, and R. P. Evstigneeva, *Khim. Geterotsikl. Soedin.*, **1974**, 100 [*Chem. Abstr.*, **80**, 95791 (1974)].

558. Y. Morita, S. Onishi, and T. Yamagami, *Japan. Pat.* 73 15,949 (1973) [*Chem. Abstr.*, **79**, 53287m (1973)].

559. Y. Morita, S. Onishi, and T. Fujiwara, *Japan. Pat.* 73 30,636 (1973) [*Chem. Abstr.*, **80**, 82700 (1974)].

560. E. E. Harris, J. L. Zabriskie, E. M. Chamberlin, J. P. Crane, E. R. Peterson, and W. Reuter, *J. Org. Chem.*, **34**, 1993 (1969).

561. Merck & Co., Inc., *Neth. Appl.* 6,614,801 (1967) [*Chem. Abstr.*, **68**, 87190 (1968)].

562. E. E. Harris and R. B. Currie, *Ger. Offen.* 1,948,103 (1970) [*Chem. Abstr.*, **72**, 121512 (1970)].

563. Merck & Co., Inc. *Neth. Appl.* 6,614,802 (1967) [*Chem. Abstr.*, **68**, 87191 (1968)].

564. H. Miki and H. Saikawa, *Ger. Offen.* 2,218,739 (1972) [*Chem. Abstr.*, **78**, 43288 (1973)].

565. T. Naito, K. Ueno, and F. Ishikawa, *Japan. Pat.*, 71 11,500 (1971) [*Chem. Abstr.*, **75**, 63622 (1971)].

566. T. Naito, Y. Morita, K. Ueno, S. Shimada, S. Miyazaki, and T. Fujiwara, *Japan. Pat.*, 70 39,259 (1970) [*Chem. Abstr.*, **74**, 125462 (1971)].

567. Daiichi Seiyaku Co., *Fr. Pat.* 1,530,842 (1968) [*Chem. Abstr.*, **71**, 81203 (1969)].

568. T. Naito, K. Ueno, M. Sano, Y. Omura, I. Itoh, and F. Ishikawa, *Tetrahedron Lett.*, **1968**, 5767.

569. K. Ueno, F. Ishikawa, H. Omura, S. Miyazaki, and T. Naito, *Japan. Pat.*, 70 31,180 (1970) [*Chem. Abstr.*, **74**, 100013 (1971)].

570. T. Naito, K. Ueno, T, Miki, and H. Omura, *Japan. Pat.*, 70 36,301 (1970) [*Chem. Abstr.*, **74**, 64211 (1971)].

571. G. Y. Kondrat'eva, *U.S.S.R. Pat.* 196,854 (1968) [*Chem. Abstr.,* **69,** 67240 (1968)].

572. S. V. Stepanova, S. D. L'vova, B. S. El'yanov, and V. I. Gunar, *Khim.-Farm. Zh.,* **11,** 92 (1977) [*Chem. Abstr.,* **83,** 117803 (1977)].

573. Z. L. Itov, S. D. L'vova, and V. I. Gunar, *Khim.-Farm. Zh.,* **10,** 100 (1976) [*Chem. Abstr.,* **85,** 62921 (1976)].

574. S. D. L'vova, S. V. Stepanova, I. S. Cherkosova, and V. I. Gunar, *Zh. Org. Khim.,* **11,** 1537 (1975) [*Chem. Abstr.,* **83,** 164051 (1976)].

575. M. Kawazu, K. Azuma, and M. Wada, *Japan. Pat.,* 70 26,492 (1970) [*Chem. Abstr.,* **74,** 53544q (1971)].

576. T. Matsuo and T. Miki, *Chem. Pharm. Bull.,* **20,** 669 (1972).

577. T. Naito, T. Yoshikawa, and H. Omura, *Japan. Pat.,* 70 11,499 (1970) [*Chem. Abstr.,* **73,** 45360r (1970)].

578. T. Miki and Y. Matsuo, *Japan. Pat.,* 69 21,007 (1969) [*Chem. Abstr.,* **71,** 124263y (1969)].

579. T. Matsuo and T. Miki, *Chem. Pharm. Bull.,* **20,** 806 (1972).

580. T. Miki and T. Matsuo, *Yakugaku Zasshi,* **87,** 323 (1967) [*Chem. Abstr.,* **67,** 32549k (1968)].

581. T. Miki and Y. Matsuo, *Japan. Pat.,* 67 25,664 (1967) [*Chem. Abstr.,* **69,** 43807k (1968)].

582. T. Miki and T. Matsuo, *U.S. Pat.* 3,413,297 (1968) [*Chem. Abstr.,* **70,** 68172 (1969)].

583. Daiichi Seiyaku Co., Ltd., *Neth. Appl.* 6,607,005 (1966) [*Chem. Abstr.,* **67,** 32593v (1967)].

584. S. F. Schaeren, *Ger. Offen.* 2,008,854 (1969) [*Chem. Abstr.,* **73,** 130900m (1970)].

585. D. Szlompek-Nesteruk, A. Rudnicki, and T. Sikorska, *Przem. Chem.,* **54,** 238 (1975) [*Chem. Abstr.,* **83,** 79037n (1975)].

586. D. Szlompek-Nesteruk, A. Rudnicki, K. Woisa, S. Spychala, K. Suwalska, and M. Adamus, *Pol. Pat.* 93,375 (1977) [*Chem. Abstr.,* **89,** 109108 (1978)].

587. E. E. Harris, D. W. Rosenburg, and E. M. Chamberlin, *U.S. Pat.* 3,381,014 (1968) [*Chem. Abstr.,* **69,** 52023 (1968)].

588. T. Miki and T. Matsuo, *Yakagaku Zasshi,* **91,** 1030 (1971) [*Chem. Abstr.,* **76,** 3660v (1972)].

589. T. Naito, K. Ueno, and T. Miki, *Japan. Pat.,* 70 07,747 (1970) [*Chem. Abstr.,* **73,** 25318 (1970)].

590. T. Naito, K. Ueno, Y. Morita, S. Simada, and H. Omura, *Japan. Pat.,* 70 11,906 (1970) [*Chem. Abstr.,* **73,** 25317 (1970)].

591. M. Murakami and M. Iwanami, *Bull. Chem. Soc. Japan.,* **41,** 726 (1968).

592. H. Takagaki, N. Yasuda, M. Assoka, and H. Takei, *Chem. Lett.,* **1979,** 183.

593. T. Yoshikawa and T. Naito, *Japan. Pat.,* 69 32,572 (1969) [*Chem. Abstr.,* **72,** 78886 (1970)].

594. M. Murakamo, M. Iwanami, and I. Ozawa, *Japan. Pat.,* 69 32,574 (1969) [*Chem. Abstr.,* **72,** 66839y (1970)].

595. I. Maeda, M. Takehara, K. Togo, S. Asai, and R. Yoshida, *Bull. Chem. Soc. Japan,* **42,** 1435 (1969).

596. Ajinomoto Co., Inc., *Fr. Pat.* 1,535,817 (1968) [*Chem. Abstr.,* **71,** 101839b (1969)].

597. P. B. Terent'ev, M. Islam, A. A. Zaitsev, and A. N. Kost, *Vestn. Mosk. Univ. Khim.,* **24,** 123 (1969) [*Chem. Abstr.,* **71,** 12970v (1969)].

598. A. N. Kost, P. B. Terent'ev, M. Islam, and A. A. Zaitsev, *U.S.S.R. Pat.* 250,142 (1969) [*Chem. Abstr.,* **72,** 78883w (1970)].

599. J. M. Osbond, *Brit. Pat.* 1,034,483 (1966) [*Chem. Abstr.,* **69,** 106571b (1968)].

214   T. D. Bailey, G. L. Goe, and E. F. V. Scriven

600. F. Abjean, *C. R. Acad. Sci., Ser. C,* **278,** 359 (1974).

601. K. T. Potts, J. Baum, and E. Houghton, *J. Org. Chem.,* **41,** 818 (1976).

602. H. Matsukubo and H. Kato, *J. Chem. Soc., Chem. Commun.,* **1974,** 412.

603. T. Eicher and V. Schaefer, *Tetrahedron,* **30,** 4025 (1974).

604. H. Matsukubo and H. Kato, *J. Chem. Soc., Perkin Trans. I,* **1975,** 632.

605. K. T. Potts, E. Houghton, and U. P. Singh, *Chem. Commun.,* **1969,** 1129.

606. K. T. Potts, E. Houghton, and U. P. Singh, *J. Org. Chem.,* **39,** 3627 (1974).

607. K. T. Potts, J. Baum, and E. Houghton, *J. Org. Chem.,* **39,** 3631 (1974).

608. T. Nishiwaki, *Synthesis,* **1975,** 20.

609. B. J. Wakefield and D. J. Wright, "Isoaxazole Chemistry Since 1963," in *Advances in Heterocyclic Chemistry,* A. R. Katritzky and A. J. Boulton, eds., Vol. 25, Academic Press, New York, 1979, pp. 147–200.

610. A. I. Meyers, "Heterocycles in Organic Synthesis," Wiley, New York, 1974, pp. 180–181.

611. P. Caramella and A. Querci, *Synthesis,* **1972,** 46.

612. G. Stork, M. Ohashi, H. Kamachi, and H. Kakisawa, *J. Org. Chem.,* **36,** 2784 (1971).

613. E. Ajello, *J. Heterocycl. Chem.,* **8,** 1035 (1971).

614. P. Caramella, R. Metelli, and P. Gruenanger, *Tetrahedron,* **27,** 379 (1971).

615. R. Grigg, R. Hayes, J. L. Jackson, and T. J. King, *J. Chem. Soc., Chem. Commun.,* **1973,** 349.

616. I. Adachi, *Chem. Pharm. Bull.,* **17,** 2209 (1969).

617. A. Robert, M. Ferrey, and A. Foucaud, *Tetrahedron Lett.,* **1975,** 1377.

618. H. D. Martin and M. Heckman, *Angew. Chem. Internat. Ed. Engl.,* **11,** 926 (1972).

619. S. Goetze, B. Kuebel, and W. Steglich, *Chem. Ber.,* **109,** 2331 (1976).

620. J. A. Moore, R. L. Winecholt, F. J. Marascia, R. W. Medeiros, and F. J. Creegan, *J. Org. Chem.,* **32,** 1353 (1967).

621. S. R. Tanny, J. Grossman, and F. W. Fowler, *J. Am. Chem. Soc.,* **94,** 6495 (1972).

622. T. Eicher and S. Boehm, *Tetrahedron Lett.,* **1972,** 3965.

623. H. C. van der Plas, "Ring Transformations of Heterocycles," Vol. 22, Academic Press, London, 1973, pp. 34–41.

624. J. A. Van Allan, G. A. Reynolds, J. T. Alessi, S. C. Chang, and R. C. Joines, *J. Heterocycl. Chem.,* **8,** 919 (1971).

625. F. K. Rafia, *J. Chem. Soc. C,* **1971,** 2048.

626. A. K. Kiang, S. F. Tan, and W. S. Wong, *J. Chem. Soc. C,* **1971,** 2721.

627. A. S. Afridi, A.R. Katritzky, and C. A. Ramsden, *J. Chem. Soc., Perkin Trans. I,* **1977,** 1436.

628. I. E. El-Kholy, F. K. Rafla, and M. M. Mishrikey, *J. Chem. Soc. C,* **1969,** 1950.

629. E. Ziegler, F. Raninger, and A. K. Mueller, *Ann. Chem.,* **1976,** 250.

630. A. N. Akopyan and A. M. Saakin, *Dokl. Vses. Konf. Khim. Atsetilena, 4th,* **1,** 127 (1972) [*Chem. Abstr.,* **79,** 104668 (1973)].

631. A. N. Akopyan, A. M. Saakyan, and A. A. Safaryan, *Zh. Org. Khim.,* **9,** 459 (1973) [*Chem. Abstr.,* **79,** 5218t (1974)].

632. N. N. Magdesieve, N. M. Koloskova, and N. Le Nguyen, *Khim. Geterotsikl. Soedin.,* **1977,** 1475 [*Chem. Abstr.,* **88,** 62254m (1978)].

633. T. Eicher, E. Von Angerer, and A. M. Hansen, *Ann. Chem.,* **746,** 102 (1971).

634. Y. V. Maevskii and S. V. Sokolovskaya, *Khim. Geterotsikl. Soedin.,* **7,** 579 (1971) [*Chem. Abstr.,* **76,** 59402e (1972)].

635. Y. V. Maevskii, S. V. Sokolovskaya, and I. P. Komkov, *Khim. Geterotsikl. Soedin.*, **1970**, 1160 [*Chem. Abstr.*, **74**, 87785j (1971)].

636. M. Y. Shandala, A. Y. Al-Khashab, and M. M. Al-Arab, *J. Iraqi Chem. Soc.*, **2**, 53 (1977) [*Chem. Abstr.*, **90**, 103908q (1979)].

637. N. P. Shusherina, O. V. Slavyanova, L. K. Petrova, and R. Y. Levina, *Zh. Org. Khim.*, **1972**, 387 [*Chem. Abstr.*, **76**, 126736g (1972)].

638. A. N. Akopyan, A. M. Saakyan, and A. A. Safarvan, *Arm. Khim. Zh.*, **29**, 323 (1976) [*Chem. Abstr.*, **86**, 5265a (1977)].

639. R. Oehl, G. Lenzer, and P. Rosenmund, *Chem. Ber.*, **109**, 705 (1976).

640. I. E. El-Kholy, M. M. Mishrikey, and H. M. Fuid-Alla, *J. Heterocycl. Chem.*, **12**, 129 (1975).

641. Y. A. Al-Farkh, F. H. Al-Hajjar, N. R. El-Rayyes, and H. S. Hamoud, *J. Heterocycl. Chem.*, **15**, 759 (1978).

642. J. Oehldrich and J. M. Cook, *J. Org. Chem.*, **42**, 889 (1977).

643. F. G. Baddar, F. H. Al-Hajiar, and N. R. El-Rayyes, *J. Heterocycl. Chem.*, **13**, 195 (1976).

644. G. Lohaus and W. Dittmar, *Ger. Offen.* 2,214,608 (1973) [*Chem. Abstr.*, **79**, 146419w (1973)].

645. I. E. El-Kholy, F. K. Rafia, and M. M. Mishrikey, *J. Chem. Soc. C*, **1970**, 1578.

646. P. H. Schroeder, *U.S. Pat.* 3,657,426 (1972) [*Chem. Abstr.*, **77**, 34351v (1972)].

647. T. Gostea and A. Maza, *Rom. Pat.* 56,409 (1974) [*Chem. Abstr.*, **82**, 57575p (1975)].

648. A. Svab, *Czech. Pat.* 144,978 (1972) [*Chem. Abstr.*, **78**, 29628g (1973)].

649. V. Janata, A. Svab, D. Nemcova, L. Bruna, and B. Sevcik, *Czech. Pat.* 149,740 (1973) [*Chem. Abstr.*, **80**, 133276t (1974)].

650. K. S. Banetjee and S. S. Deshapande, *J. Indian Chem. Soc.*, **52**, 41 (1975).

651. L. C. Dorman, *J. Org. Chem.*, **32**, 4105 (1967).

652. S. A. Vartanyan, A. S. Noravyan, and V. N. Zhamagortsyan, *Khim. Geterotsikl. Soedin.*, **1966**, 670 [*Chem. Abstr.*, **66**, 55339k (1966)].

653. G. A. Garkusha, G. A. Khutornenko, and N. A. Kurakina, *Zh. Org. Khim.*, **3**, 1699 (1967) [*Chem. Abstr.*, **68**, 29530n (1968)].

654. M. P. Martes and A. J. Lin, *J. Heterocycl. Chem.*, **6**, 941 (1969).

655. W. Kowitz, *Ger. Offen.* 2,021,872 (1971) [*Chem. Abstr.*, **76**, 72410a (1972)].

656. S. Yoshida, K. Morimoto, and N. Takahashi, *Japan. Kokai*, 77 87,171 (1977) [*Chem. Abstr.*, **88**, 6735g (1978)].

657. S. Yoshida, K. Morimoto, and N. Takahashi, *Japan. Kokai*, 77 151,175 (1977) [*Chem. Abstr.*, **88**, 169977b (1968)].

658. T. Kato and Y. Kubota, *Yakugaku Zasshi*, **87**, 1412 (1967) [*Chem. Abstr.*, **68**, 95644d (1968)].

659. A. W. Douglas, M. H. Fisher, J. J. Fishinger, P. Gund, E. F. Harris, G. Olson, A. A. Patchett, and W. V. Ruyle, *J. Med. Chem.*, **20**, 939 (1977).

660. S. Bimecki, B. Gutkowski, *Acta Pol. Pharm.*, **27**, 1 (1970) [*Chem. Abstr.*, **73**, 87740v (1970)].

661. S. J. Norton and E. Sanders, *J. Med. Chem.*, **10**, 961 (1967).

662. J. Ficini and J. P. Genet, *C.R. Acad. Sci., Ser. C*, **270**, 650 (1970).

663. J. Ficini and J. P. Genet, *Bull. Soc. Chim. Fr.*, **1974**, 2086.

664. I. Ichimoto, K. Fujn, and C. Tatsunn, *Agr. Biol. Chem. (Tokyo)*, **31**, 979 (1967).

665. M. Simalty, H. Strzelecka, and M. Dupre, *C.R. Acad. Sci., Paris, Ser. C*, **265**, 1284 (1967).

666. B. Tambina, F. Zorko, and M. J. Herak, *J. Inorg. Nucl. Chem.*, **39**, 1201 (1977).

667. A. S. Afridi, A. R. Katritzky, and C. A. Ramsden, *J. Chem. Soc., Perkin Trans. I*, 1977, 1428.

668. E. M. Kaiser, S. D. Work, J. F. Wolfe, and C. R. Hauser, *J. Org. Chem.*, **32**, 1483 (1967).

669. E. S. H. El Ashry, *Carbohyd. Res.*, **33**, 178 (1974).

670. B. Tamhina, K. Jakopcic, F. Zorko, and M. J. Herak, *J. Inorg. Nucl. Chem.*, **36**, 1855 (1974).

671. I. El-Kholy, M. M. Mishrikey, and R. F. Atmeh, *J. Heterocycl. Chem.*, **11**, 487 (1974).

672. T. Kato, Y. Yamamoto, and S. Takeda, *Chem. Pharm. Bull.*, **21**, 1047 (1973).

673. H. W. R. Williams, *Can. J. Chem.*, **54**, 3377 (1976).

674. R. R. G. Haber and C. Sominovitch, *Ger. Offen.* 2,217,739 (1972) [*Chem. Abstr.*, **78**, 29634f (1973)].

675. G. R. Carlson, *U.S. Pat.* 4,051,142 (1977) [*Chem. Abstr.*, **88**, 22643f (1978)].

676. R. R. G. Haber and C. Simonovitch, *U.S. Pat.* 3,821,235 (1974) [*Chem. Abstr.*, **82**, 139968h (1975)].

677. T. Eicher, E. Von Angerer, and A. M. Hansen, *Ann. Chem.*, **746**, 102 (1971).

678. K. Yoshioka, G. Goto, and K. Hiraga, *Yakugaku Zasshi*, **93**, 1183 (1973) [*Chem. Abstr.*, **79**, 137003d (1973)].

679. M. A. Butt, I. A. Akhtar, S. A. Qureshi, and M. Akhtar, *Pak. J. Sci. Ind. Res.*, **10**, 240 (1967) [*Chem. Abstr.*, **69**, 35881q (1968)].

680. A. Butt, M. Hasang, and Z. Manawwar, *Tetrahedron*, **23**, 2461 (1967).

681. C. Wang, *J. Heterocycl. Chem.*, **7**, 389 (1970).

682. T. Jaworski and S. Kwiatkowski, *Rocz. Chem.*, **44**, 555 (1970) [*Chem. Abstr.*, **73**, 130845x (1970)].

683. A. R. Katritzky, A. S. Afridi, and C. A. Ramsden, *Pak. J. Sci. Ind. Res.*, **21**, 1 (1978).

684. V. Janata and D. Nemcova, *Czech. Pat.* 144,986 (1972) [*Chem. Abstr.*, **78**, 29626e (1973)].

685. C. S. Wang, J. P. Easterly, and N. E. Skelly, *Tetrahedron*, **27**, 2581 (1971).

686. M. Hruby and M. Ferenc, *Czech. Pat.* 152,394 (1975) [*Chem. Abstr.*, **84**, 4811x (1976)].

687. M. Hruby and M. Ferenc, *Czech. Pat.* 162,232 (1976) [*Chem. Abstr.*, **86**, 72446y (1977)].

688. S. Biniecki and B. Gutkowska, *Acta Pol. Pharm.*, **27**, 1 (1970) [*Chem. Abstr.*, **73**, 87740v (1970)].

689. A. S. Afridi, A. R. Katritzky, and C. A. Ramsden, *J. Chem. Soc., Perkin Trans. I*, 1977, 1428.

690. M. P. Sammes, H. K. Wah, and A. R. Katritzky, *J. Chem. Soc., Perkin Trans. I*, 1977, 327.

691. I. E. El-Kholy, F. K. Rafla, and M. Mishrikey, *J. Chem. Soc. C*, 1970, 1578.

692. F. K. Rafla, *J. Chem. Soc. C*, 1971, 2048.

693. A. E. El-Kholy, F. K. Rafla, and M. M. Mishrikey, *J. Chem. Soc. C*, 1969, 1950.

694. I. Belsky, H. Dodiuk, and Y. Shvo, *J. Org. Chem.*, **39**, 989 (1974).

695. J. A. Van Allan, G. A. Reynolds, C. C. Petropoulos, and D. P. Maier, *J. Heterocycl. Chem.*, **7**, 495 (1970).

696. J. A. Van Allan and G. A. Reynolds, *J. Heterocycl. Chem.*, **8**, 367 (1971).

697. H. C. van der Plas, "Ring Transformations of Heterocycles," Vol. 2, Academic Press, London, 1973, pp. 17–24.

698. K. Dimroth, *Angew. Chem.,* **72,** 331 (1960).

699. A. T. Balaban, W. Schroth, and G. Fischer, in "Advances in Heterocyclic Chemistry," Vol. 10, A. R. Katritzky and A. J. Boulton, eds., Academic Press, New York, 1969, pp. 241–326.

700. A. R. Katritzky, R. T. C. Brownlee, and J. G. Musumarva, *Tetrahedron,* **36,** 1643 (1980).

701. A. R. Katritzky, *Tetrahedron,* **36,** 679 (1980).

702. R. A. Abramovitch and E. P. Kyba, *Org. Prep. Proced. Int.,* **3,** 127 (1971).

703. N. S. Zefirov, G. N. Dorofeenko, and T. M. Pozdnyakova, *Zh. Org. Khim.,* **9,** 387 (1973) [*Chem. Abstr.,* **78,** 124413s (1973)].

704. G. N. Dorofeenko, E. A. Zvezdina, M. P. Zhdanova, V. V. Derbenev, and E. S. Matskovskaya, *Khim. Geterotsikl. Soedin.,* **1974,** 1036 [*Chem. Abstr.,* **81,** 169408x (1974)].

705. G. N. Dorofeenko, E. A. Zvezdina, and V. V. Derbenev, *Zh. Org. Khim.* **9,** 1079 (1973) [*Chem. Abstr.,* **79,** 53145p (1973)].

706. V. Snieckus and G. Kan, *Chem. Commun.,* **1970,** 1208.

707. A. S. Afridi, A. R. Katritzky, and C. A. Ramsden, *J. Chem. Soc., Chem. Commun.,* **1976,** 899.

708. M. P. Sammes and K. L. Yip, *J. Chem. Soc., Perkin Trans. I,* **1978,** 1373.

709. G. P. Safaryan and G. N. Dorofeenko, *Khim. Geterotsikl. Soedin.,* **1976,** 1323 [*Chem. Abstr.,* **86,** 72367 (1977)].

710. C. Pedersen, N. Harrit, and O. Buchardt, *Acta Chem. Scand.,* **24,** 3435 (1970).

711. O. Buchardt, C. L. Pedersen, and N. Harit, *J. Org. Chem.,* **37,** 3592 (1972).

712. A. R. Katritzky, *Tetrahedron,* **36,** 679 (1980).

713. A. R. Katritzky, P. L. Nie, A. Dondoni, and D. Tassi, *J. Chem. Soc., Perkin Trans. I,* **1979,** 1961.

714. A. R. Katritzky, P. L. Nie, A. Dondoni, and D. Tassi, *Synth. Commun.,* **7,** 387 (1977).

715. A. R. Katritzky, J. Lewis, and P. L. Nie, *J. Chem. Soc., Perkin I,* **1979,** 446.

716. R. Neidlein and P. Witerzens, *Monatsh. Chem.,* **106,** 643 (1975).

717. M. Alajarin, A. Arques, P. M. Fresneda, A. Lorenzo, P. Molina, A. Soler, and M. J. Vilaplana, *An. Quim.,* **74,** 625 (1974) [*Chem. Abstr.,* **89,** 197293q (1978)].

718. E. Zvezdina, M. P. Zhdanova, and G. N. Dorofeenko, *Khim. Geterotsikl. Soedin.,* **1979,** 324 [*Chem. Abstr.,* **91,** 5080q (1979)].

719. A. R. Katritzky and J. W. Suwinski, *Tetrahedron Lett.,* **1974,** 4123.

720. R. Neidlein and P. Witerzens, *Arch. Pharm.,* **309,** 649 (1976).

721. G. N. Dorofeenko, A. N. Narkevich, Y. A. Zhdanov, O. E. Shelepin, and T. G. Soroka, *Khim. Geterotsikl. Soedin.,* **1970,** 223. [*Chem. Abstr.,* **76,** 140454v (1971)].

722. G. N. Dorofeenko, A. N. Narkevich, and Y. A. Zhdanov, *Khim. Geterotsikl. Soedin.,* **1967,** 1130 [*Chem. Abstr.,* **69,** 67175d (1968)].

723. G. N. Shibanov, V. A. Palchkov, and M. A. Kuznetsova, *U.S.S.R. Pat.* 239,340 (1969) [*Chem. Abstr.,* **71,** 49784p (1969)].

724. H. C. van der Plas, H. Jongejan, G. Guertsen, and M. C. Vollering, *Recl. Trav. Chim. Pays-Bas,* **90,** 1246 (1971).

725. H. C. van der Plas, "Ring Transformations of Heterocycles," Vol. 2, Academic Press, London, 1973, p. 134.

726. D. J. Brown and B. T. England, *Aust. J. Chem.,* **23,** 625 (1970).

727. P. Barczynski and H. C. van der Plas, *Recl. Trav. Chim. Pays-Bas,* **97,** 256 (1978).

728. E. A. Oostveen and H. C. van der Plas, *Recl. Trav. Chim. Pays-Bas,* **93,** 233 (1973).

729. A. Albert and W. Pendergast, *J. Chem. Soc., Perkin Trans. I*, 1973, 1620.

730. H. C. van der Plas, H. Jongejan, and A. Koudijs, *J. Heterocycl. Chem.*, 15, 485 (1978).

731. B. A. Oude Alink, *U.S. Pat.* 3,931,191 (1976) [*Chem. Abstr.*, 84, 90019 (1976)].

732. B. A. Oude Alink, *U.S. Pat.* 4,113,730 (1978) [*Chem. Abstr.*, 90, 38803 (1979)].

733. A. V. Upadysheva, N. D. Grigor'eva, and A. P. Znamenskaya, *Khim. Geterotsikl. Soedin.*, 1977, 1549 [*Chem. Abstr.*, 88, 105251 (1978)].

734. S. Senda, K. Hirota, T. Asao, and Y. Abe, *Heterocycles*, 9, 739 (1978).

735. E. G. Lovett and D. Lipkin, *J. Org. Chem.*, 42, 2574 (1977).

736. H. Neunhoeffer and B. Lehmann, *Ann. Chem.*, 1975, 1113.

737. H. Neunhoeffer and G. Werner, *Ann. Chem.*, 1974, 1190.

738. P. J. Manchin, A. E. A. Porter, and P. G. Sammes, *J. Chem. Soc., Perkin Trans. I*, 1973, 404.

739. L. B. Davies, P. G. Sammes, and R. A. Watt, *J. Chem. Soc., Chem. Commun.*, 1977, 663.

740. H. Neunhoeffer and G. Werner, *Ann. Chem.*, 1973, 437.

741. H. Neunhoeffer and G. Werner, *Tetrahedron Lett.*, 1972, 1517.

742. H. Neunhoeffer and G. Werner, *Ann. Chem.*, 761, 39 (1972).

743. P. J. Machin, A. E. A. Porter, and P. G. Sammes, *J. Chem. Soc., Perkin Trans. I*, 1973, 404.

744. W. Steglich, E. Buschmann, and O. Hollitzer, *Angew. Chem., Int. Ed. Engl.*, 13, 533 (1974).

745. E. Buschmann and W. Steglich, *Angew. Chem., Int. Ed. Engl.*, 13, 484 (1974).

746. T. Kato, U. Izumi, and N. Katagiri, *J. Heterocycl. Chem.*, 15, 1475 (1978).

747. E. Ziegler, K. Belegratis, and G. Brus, *Monatsh. Chem.*, 98, 555 (1967).

748. T. Kato, Y. Yamamoto, and M. Kondo, *Chem. Pharm. Bull.*, 23, 1873 (1975).

749. A. Tamano and T. Nakamura, *Japan. Pat.*, 73 35,076 (1973) [*Chem. Abstr.*, 80, 82695u (1974)].

750. D. M. Stout, T. Takaya, and A. I. Meyers, *J. Org. Chem.*, 40, 563 (1975).

751. B. Burg, W. Dittmar, H. Reim, A. Steigel, and J. Sauer, *Tetrahedron Lett.*, 1975, 2897.

752. H. Neurhoeffer and H. W. Fruehauf, *Ann. Chem.*, 758, 120 (1972).

753. H. Reim, A. Steighel, and J. Sauer, *Tetrahedron Lett.*, 1975, 2901.

754. W. Dittmar, J. Sauer, and A. Steigel, *Tetrahedron Lett.*, 1969, 5171.

755. Y. S. Dol'skaya, G. Y. Kondrat'eva, N. I. Golovina, A. D. Shashkov, and V. I. Kadentsev, *Izv. Akad. Nauk SSSR. Ser. Khim.*, 1975, 1812 [*Chem. Abstr.*, 83, 206216x (1975)].

756. D. C. Frost, B. Mac Donald, C. A. McDowell, and N. P. C. Westwood, *J. Electron Spectros. Relat. Phenom.*, 14, 379 (1978).

757. R. J. Sundberg, S. R. Suter, and M. Brenner, *J. Am. Chem. Soc.*, 94, 513 (1972).

758. M. Ogata, H. Matsumoto, and H. Kano, *Tetrahedron*, 25, 5217 (1969).

759. B. Gregory, E. Bullock, and T. S. Chen, *Can. J. Chem.*, 55, 4061 (1977).

760. C. G. Hughes, E. G. Lewars, and A. H. Rees, *Can. J. Chem.*, 52, 3327 (1974).

761. T. Sano, Y. Horiguchi, and Y. Tsuda, *Heterocycles*, 9, 731 (1978).

762. J. Streith, J. P. Luttringer, and M. Nastasi, *J. Org. Chem.*, 36, 2962 (1971).

763. J. A. Moore, E. J. Volker, and C. M. Kopay, *J. Org. Chem.*, 36, 2676 (1971).

764. D. J. Harris, M. T. Thomas, V. Snieckus, and E. Klingsberg, *Can. J. Chem.*, 52, 2805 (1974).

765. V. Snieckus, G. Kan, and M. T. Thomas, *Chem. Commun.*, 1971, 1022.

766. R. R. Schmidt and H. Vatter, *Tetrahedron Lett.*, 1972, 4891.

767. A. N. Saverchenko, V. A. Kaminskii, and M. N. Tilichenko, *Khim. Geterotsikl. Soedin.*, 1974, 809 [*Chem. Abstr.*, 81, 120421 (1974)].

768. L. Fuentes and J. L. Soto, *An. Quim.*, 73, 1349 (1977) [*Chem. Abstr.*, 89, 129369u (1978)].

769. Sterling Drug, Inc., *Brit. Pat.* 1,396,681 (1975) [*Chem. Abstr.*, 83, 97055w (1975)].

770. P. M. Carabateas and G. L. Williams, *J. Heterocycl. Chem.*, 13, 927 (1976).

771. P. M. Carabateas and G. L. Williams, *J. Heterocycl. Chem.*, 11, 819 (1974).

772. G. Y. Lesher and P. M. Carabateas, *Ger. Offen.* 2,224,090 (1972) [*Chem. Abstr.*, 78, 84280m (1973)].

773. P. M. Carabateas and G. L. Williams, *U.S. Pat.* 4,008,239 (1977) [*Chem. Abstr.*, 86, 171270t (1977)].

774. P. M. Carabateas and G. L. Williams, *U.S. Pat.* 3,970,662 (1976) [*Chem. Abstr.*, 85, 192571m (1976)].

775. T. Chennat and U. Eisner, *J. Chem. Soc., Perkin Trans. I*, 1975, 926.

776. J. Kuthan and J. Hakr, *Coll. Czech. Chem. Commun.*, 32, 1438 (1967).

777. P. Nantka-Namirski and R. Balicki, *Acta Pol. Pharm.*, 31, 279 (1974).

778. J. Palecek and J. Kuthan, *Czech. Pat.* 125,165 (1967) [*Chem. Abstr.*, 69, 96476g (1968)].

779. J. Palacek and J. Kuthan, *Czech. Pat.* 125,164 (1967) [*Chem. Abstr.*, 69, 96475h (1968)].

780. H. D. Eilhauer and I. Kaempfer, *Z. Chem.*, 9, 188 (1969).

781. J. Palecek and J. Kuthan, *Coll. Czech. Chem. Commun.*, 34, 3336 (1969).

782. J. Palecek, K. Vondra, and J. Kuthan, *Coll. Czech. Chem. Commun.*, 34, 2991 (1969).

783. E. Grinsteins, B. Stankevics, and G. Duburs, *Khim. Geterotsikl. Soedin.*, 1967, 1118 [*Chem. Abstr.*, 69, 77095 (1968)].

784. J. G. Carey and J. E. Colchester, *U.S. Pat.* 3,664,383 (1972) [*Chem. Abstr.*, 76, 140548d (1972)].

785. L. A. Negievich, O. M. Grishin, and A. A. Yasnikov, *Ukr. Khim. Zh.*, 34, 802 (1968) [*Chem. Abstr.*, 70, 28776b (1969)].

786. G. A. Tolstikov, U. M. Dzhemilev, and V. P. Yur'ev, *Izv. Akad. Nauk SSSR, Ser. Khim.*, 3, 670 (1972) [*Chem. Abstr.*, 77, 101354p (1972)].

787. L. A. Khanna, N. S. Pivnenko, F. N. Grigor'eva, and V. F. Lavrushin, *Zh. Org. Khim.*, 12, 1097 (1976) [*Chem. Abstr.*, 85, 46333u (1976)].

788. P. Wild and F. Kroehnke, *Ann. Chem.*, 1975, 849.

789. Y. V. Kurbatov and A. S. Kurbatova, *Rr. Samarkund Gos. Univ.*, 167, 71 (1969) [*Chem. Abstr.*, 75, 139950 (1971)].

790. G. V. Kireev, V. B. Leont'ev, Y. V. Kurbatov, O. S. Otroshchenko, and A. S. Sadykov, *Izv. Akad. Nauk SSSR, Ser. Khim.*, 1978, 1425 (1978) [*Chem. Abstr.*, 89, 107481x (1978)].

791. A. S. Kurbatova, Y. V. Kurbatov, O. S. Ostroshchenko, and A. S. Sadykov, *Tezisy Dokl. Simp. Khim. Tekhnol. Geterotsikl. Soedin. Goryuch. Iskop., 2nd*, 1973, 179 [*Chem. Abstr.*, 86, 43520n (1977)].

792. G. V. Kireev, V. B. Leont'ev, Y. V. Kurbatov, O. I. Kartavtsev, O. S. Otroshchenko, and A. S. Sadykov, *Izv. Akad. Nauk SSSR, Ser. Khim.*, 1978, 807 [*Chem. Abstr.*, 89, 23471f (1978)].

793. Y. V. Kurbatov, A. S. Kurbatova, O. S. Otroshchenko, A. S. Sadykov, and F. Zairov, *Tr. Samarkand. Univ.*, 206, 138 (1972) [*Chem. Abstr.*, 80, 95677q (1974)].

794. A. S. Kurbatova, Y. V. Kurbatov, and D. A. Nivazova, *Khim. Geterotsikl. Soedin.*, 1977, 1651 [*Chem. Abstr.*, 88, 120946b (1978)].

795.   Y. V. Kurbatov, A. S. Kurbatova, O. S. Otroshchenko, and A. S. Sadykov, *Tr. Sarmakand. Gos. Univ.*, **1969**, 167 [*Chem. Abstr.*, **74**, 99820e (1971)].

796.   A. S. Kurbatova, Y. V. Kurbatov, O. S. Otroshchenko, and A. S. Sadykov, *Nauch. Tr. Samarkand. Univ.*, **167**, 26 (1969) [*Chem. Abstr.*, **74**, 141474x (1971)].

797.   E. S. Huyser, J. A. K. Harmony, and F. L. McMillian, *J. Am. Chem. Soc.*, **94**, 3176 (1972).

798.   A. A. Tyshchenko, V. B. Leont'ev, K. A. Aslanov, and A. S. Sadykov, *Dokl. Akad. Nauk Usb. SSR*, **29**, 32 (1972) [*Chem. Abstr.*, **78**, 135355a (1973)].

799.   O. M. Polumbrik, G. F. Dvorko, E. A. Ponomareva, and E. I. Zaika, *Zh. Org. Khim.*, **8**, 2417 (1972) [*Chem. Abstr.*, **78**, 96982f (1973)].

800.   O. M. Polumbrik, E. I. Zaika, and G. F. Dvorko, *Zh. Org. Khim.*, **10**, 1943 (1974) [*Chem. Abstr.*, **82**, 57164d (1975)].

801.   O. M. Polumbrik, E. I. Zaika, and G. F. Dvorku, *Ukr. Khim. Zh.*, **40**, 963 (1974) [*Chem. Abstr.*, **83**, 96005t (1975)].

802.   O. M. Polumbrik and O. I. Zaika, *Reakts. Sposobn. Org. Soedin.*, **11**, 391 (1974) [*Chem. Abstr.*, **83**, 57640b (1975)].

803.   M. L. Khidekel, A. S. Astakhova, N. F. Dmitrieva, S. N. Zelenin, G. I. Kozub, P. A. Kaikaris, and Y. A. Shvetsov, *Zh. Obshch. Khim.*, **37**, 1483 (1967) [*Chem. Abstr.*, **68**, 12169d (1968)].

804.   M. J. De Nie-Sarink and U. K. Pandit, *Tetrahedron Lett.*, **1978**, 1335.

805.   U. K. Pandit, F. R. Mas Cabre, R. A. Gase, and M. J. De Nie-Sarink, *J. Chem. Soc., Chem. Commun.*, **1974**, 627.

806.   R. W. Huffman and T. C. Bruice, *J. Am. Chem. Soc.*, **89**, 6243 (1967).

807.   B. G. Gribov, R. B. Ivanova, D. D. Mozzhuikhin, A. S. Strizhkova, G. A. Tychin, and M. L. Khidekel, *Katal. Reakis, Zhidk, Faze, Tr. Vases. Konf., 2nd; Alma-Ata, Kaz. SSR*, **1966**, 420 [*Chem. Abstr.*, **69**, 18389t (1968)].

808.   Y. V. Kurbatov, S. V. Zalyalieva, O. S. Otoshchenko, and A. S. Sadykov, *Khim. Geterotsikl. Soedin.*, **1975**, 225 [*Chem. Abstr.*, **82**, 170623r (1975)].

809.   G. Duburs and J. Uldrikis, *Khim. Geterotsikl. Soedin*, **1970**, 83 [*Chem. Abstr.*, **72**, 121317d (1970)].

810.   S. V. Zalyalieva, Y. V. Kurbatov, O. S. Otroshchenko, and A. S. Sadykov, *Khim. Geterotsikl. Soedin.*, **1976**, 226 [*Chem. Abstr.*, **85**, 32781b (1976)].

811.   D. M. Hedstrand, W. H. Kruizinga, and R. M. Kellogg, *Tetrahedron Lett.*, **1978**, 1255.

812.   J. F. Biellmann and M. P. Goeldner, *Tetrahedron*, **27**, 1789 (1971).

813.   J. F. Biellmann, R. J. Highet, and M. P. Goeldner, *Chem. Commun.*, **1970**, 295.

814.   J. F. Biellmann and H. J. Callot, *J. Chem. Soc., Chem. Commun.*, **1969**, 140.

815.   R. D. Braun, K. S. V. Santhanam, and P. J. Elving, *J. Am. Chem. Soc.*, **97**, 2591 (1975).

816.   H. Berge, P. Jeroschewski, and K. Tellert, *Ger. (East) Pat.* 106,376 (1974) [*Chem. Abstr.*, **82**, 23757j (1975)].

817.   V. Skala, J. Volke, V. Ohanka, and J. Kuthan, *Coll. Czech. Chem. Commun.*, **42**, 292 (1977).

818.   Y. I. Beilis, G. Duburs, J. Uldrikjis, and A. Sausins, *Nov. Polyarogr., Tezisy Dokl. Vses. Soveshch. Polyarogr., 6th*, **1975**, 124 [*Chem. Abstr.*, **86**, 23365 (1977)].

819.   C. S. Kadyrov, *Khim. Geterotsikl. Soedin.*, **1968**, 840 [*Chem. Abstr.*, **71**, 13080 (1969)].

820.   T. Kametami, S. Tanaka, and A. Kosuka, *Yakugaku Zasshi*, **91**, 1068 (1971) [*Chem. Abstr.*, **76**, 14272w (1972)].

821.   Y. Tamura, T. Sakaguchi, T. Kawasaki, and K. Kita, *Chem. Pharm. Bull.*, **24**, 1160 (1976).

822. T. Kato, H. Yamanaka, S. Konno, and H. Shimomura, *Yakugaku Zasshi,* **90**, 606 (1976) [*Chem. Abstr.,* **73**, 35319 (1970)].

823. C. S. Kadyrov, *U.S.S.R. Pat.* 181,643 (1966) [*Chem. Abstr.,* **66**, 28677n (1966)].

824. R. Harmetz and R. Tull, *U.S. Pat.* 3,644,380 (1972) [*Chem. Abstr.,* **76**, 140558g (1972)].

825. U. Horn, F. Mutterer, and C. D. Weis, *Ger. Offen.* 2,432,793 (1975) [*Chem. Abstr.,* **82**, 156110q (1975)].

826. C. D. Weis, *Ger. Offen.* 2,252,063 (1973) [*Chem. Abstr.,* **79**, 18586m (1973)].

827. C. D. Weis, *Ger. Offen.* 2,341,777 (1974) [*Chem. Abstr.,* **80**, 146030h (1974)].

828. F. Mutterer, *Ger. Offen.* 2,415,748 (1974) [*Chem. Abstr.,* **82**, 72799h (1975)].

829. R. Huff. F. Mutterer, and C. D. Weis, *Helv. Chim. Acta,* **60**, 907 (1977).

830. H. Fritz, C. D. Weis, and T. Winkler, *Helv. Chim. Acta,* **59**, 179 (1976).

831. C. D. Weis, *J. Heterocycl. Chem.,* **15**, 31 (1978).

832. P. I. Mortimer, *Aust. J. Chem.,* **21**, 467 (1968).

833. T. Kato, H. Yamanaka, and J. Kawamata, *Chem. Pharm. Bull.,* **17**, 2411 (1969).

834. Z. Bomika, M. B. Andabu, J. Pelcers, and G. Duburs, *Khim. Geterotsikl. Soedin.,* 1108 (1975) [*Chem. Abstr.,* **83**, 193035k (1975)].

835. N. Sugiyama, M. Yamamoto, and C. Kashima, *Bull. Chem. Soc. Japan,* **42**, 2690 (1969).

836. J. J. Plattner, R. D. Gless, and H. Rapoport, *J. Am. Chem. Soc.,* **94**, 8613 (1972).

837. D. Diller and F. Bergmann, *J. Org. Chem.,* **37**, 2147 (1972).

838. Albright & Wilson (Mfg.) Ltd., *Neth. Appl.* 6,610,329 (1967) [*Chem. Abstr.,* **68**, 29613s (1968)].

839. T. Naito and M. Ogawa, *Japan. Kokai,* 74 116,072 (1974) [*Chem. Abstr.,* **82**, 125282t (1975)].

840. A. A. Liebman, D. H. Malarek, A. M. Dorsky, and H. H. Kaegi, *J. Heterocycl. Chem.,* **11**, 1105 (1974).

841. J. P. Fleury, J. M. Biehler, and M. Desbois, *Tetrahedron Lett.,* **1969**, 4091.

842. J. P. Fleury, M. Desbois, and J. See, *Bull. Soc. Chim. Fr.,* **1978**, 147.

843. A. I. Meyers and S. Singh, *Tetrahedron,* **25**, 4161 (1969).

844. J. A. Corran, *Brit. Pat.* 1,125,735 (1968) [*Chem. Abstr.,* **70**, 3844j (1969)].

845. N. S. Prostakov, O. I. Sotokin, and A. Y. Ismailov, *Khim. Geterotsikl. Soedin.,* **1967**, 674 [*Chem. Abstr.,* **68**, 114376 (1968)].

846. N. S. Prostakov, *Geterogennyi Katal. Reakts. Poluch. Prevrashch. Geterosikl. Soedin.,* **1971**, 195 [*Chem. Abstr.,* **77**, 48181b (1972)].

847. N. L. Wendler, D. Taub, and C. H. Kuo, *U.S. Pat.* 3,551,432 (1970) [*Chem. Abstr.,* **75**, 5717x (1971)].

848. J. A. W. Gutzwiller and M. R. Uskokovic, *Ger. Offen.* 1,902,310 (1969) [*Chem. Abstr.,* **72**, 12583s (1970)].

849. L. N. Yakhontov, V. A. Azimov, T. P. Sycheva, V. M. Aryuzina, T. V. Sakovich, and M. N. Shchukina, *U.S.S.R. Pat.* 510,472 (1976) [*Chem. Abstr.,* **85**, 46410s (1976)].

850. L. N. Yakhontov, V. A. Azimov, T. P. Sycheva, V. M. Aryuzina, T. V. Sakovich, and M. N. Shchukina, *Khim.-Farm. Zh.,* **10**, 96 (1976) [*Chem. Abstr.,* **85**, 32787h (1976)].

851. J. E. Colchester and J. A. Corran, *Brit. Pat.* 1,157,001 (1969) [*Chem. Abstr.,* **71**, 91312h (1969)].

852. J. E. Colchester, *Brit. Pat.* 1,233,643 (1971) [*Chem. Abstr.,* **75**, 63631a (1971)].

853. J. A. Corran and G. Swift, *Brit. Pat.* 1,186,884 (1970) [*Chem. Abstr.,* **73**, 3802f (1970)].

222 T. D. Bailey, G. L. Goe, and E. F. V. Scriven

854. G. Swift, *Brit. Pat.* 1,217,167 (1970) [*Chem. Abstr.*, **74**, 111924b (1971)].

855. D. H. Bartholomew, A. Campbell, D. W. H. Oliver, and B. G. Dutton, *Brit. Pat.* 1,087,279 (1967) [*Chem. Abstr.*, **68**, 87183n (1968)].

856. J. E. Colchester, *Brit. Pat.* 1,110,865 (1968) [*Chem. Abstr.*, **69**, 59100k (1968)].

857. A. Tamano, G. Hosokawa, and T. Nakamura, *Japan. Pat.*, 72 25,352 (1972) [*Chem. Abstr.*, **77**, 75137y (1972)].

858. G. G. Skvortsova, V. G. Kozyrev, and K. V. Zapunnaya, *U.S.S.R. Pat.* 215,214 (1968) [*Chem. Abstr.*, **69**, 59102n (1968)].

859. Y. I. Chumakov and V. P. Sherstyuk, *Tetrahedron Lett.*, **1967**, 771.

860. L. Proevska, A. Orakhovats, and B. Kurtev, *Izv. Old. Khim. Nauki, Bulg. Akad. Nauk.*, **6**, 747 (1973) [*Chem. Abstr.*, **81**, 3737d (1974)].

861. Y. I. Chumakov and V. P. Sherstyuk, *U.S.S.R. Pat.* 201,408 (1967) [*Chem. Abstr.*, **69**, 19027k (1968)].

862. H. Murai, S. Matsumura, T. Okubo, S. Tada, and O. Tanabe, *Japan. Kokai*, 77 48,670 (1977) [*Chem. Abstr.*, **87**, 201320z (1977)].

863. D. Tatone, T. C. Dich, R. Nacco, and C. Botteghi, *J. Org. Chem.*, **40**, 2987 (1975).

864. S. Nagao and E. Omori, *Japan. Pat.*, 72 44,747 (1972) [*Chem. Abstr.*, **78**, 71927d (1973)].

865. S. Nagao and E. Omori, *Japan. Pat.*, 72 44,747 (1972) [*Chem. Abstr.*, **78**, 719227 (1973)].

866. J. G. Carey and J. E. Colchester, *Brit. Pat.* 1,143,994 (1969) [*Chem. Abstr.*, **71**, 2382v (1969)].

867. R. D. Bowden, *Ger. Offen.* 1,913,732 (1970) [*Chem. Abstr.*, **72**, 132535z (1970)]; K. Sakakibara and K. Yasuda, *Japan. Kokai*, 75 37,782 (1975) [*Chem. Abstr.*, **83**, 114222t (1975)].

868. G. J. Moore and M. T. Richardson, *Brit. Pat.* 1,424,679 (1976) [*Chem. Abstr.*, **84**, 180064q (1976)].

869. H. Marketz, *Ger. Offen.* 2,614,204 (1967) [*Chem. Abstr.*, **86**, 55296g (1977)].

870. H. S. Inglis, *Ger. Offen.* 2,038,533 (1971) [*Chem. Abstr.*, **75**, 20202u (1971)].

871. D. H. Bartholomew, A. Campbell, D. W. H. Oliver, and B. G. Dutton, *Brit. Pat.* 1,087,279 (1967) [*Chem. Abstr.*, **68**, 87183 (1968)].

872. R. D. Bowden, *Brit. Pat.* 1,268,195 (1972) [*Chem. Abstr.*, **76**, 126789b (1972)].

873. R. D. Bowden, *Brit. Pat.* 1,268,194 (1972) [*Chem. Abstr.*, **76**, 126790v (1972)].

874. R. D. Bowden, *Brit. Pat.* 1,268,191 (1972) [*Chem. Abstr.*, **76**, 126791w (1972)].

875. R. D. Bowden, *U.S. Pat.* 3,651,071 (1972) [*Chem. Abstr.*, **76**, 153609z (1972)].

876. T. Kato, J. Kawamata, and T. Shibata, *Yakugaku Zasshi*, **88**, 106 (1968) [*Chem. Abstr.*, **69**, 27302m (1968)].

877. Y. I. Chumakov and N. B. Bulgakova, *Ukr. Khim. Zh.*, **36**, 514 (1970) [*Chem. Abstr.*, **73**, 55950 (1970)].

878. C. Botteghi, G. Caccia, S. Gladiali, and D. Tatone, *Syn. Commun.*, **9**, 69 (1979).

879. Y. I. Chumakov and N. B. Bulgakova, *Khim. Geterotsikl. Soedin.*, **7**, 1533 (1971) [*Chem. Abstr.*, **77**, 5306 (1972)].

880. G. Seitz and H. Moenninghoff, *Ann. Chem.*, **732**, 131 (1970).

881. J. A. Van Allen, S. C. Chang, G. A. Reynolds, and D. P. Maier, *J. Chem. Eng. Data*, **20**, 210 (1975).

882. G. N. Dorofeenko, E. I. Demidenko, and S. V. Krivun, *Izv. Vyssh. Ucheb. Zaved., Khim. Khim. Teknol.*, **10**, 304 (1967) [*Chem. Abstr.*, **68**, 29529u (1968)].

883. H. Strzelecka and M. Simalty, *Bull. Soc. Chim. Fr.*, **1968**, 832.

884. H. J. Teuber, H. J. Bader, and G. Schuetz, *Ann. Chem.*, **1977**, 1335.

885. G. N. Dorofeenko and G. P. Safaryan, *Khim. Geterotsikl. Soedin.*, **2**, 278 (1970) [*Chem. Abstr.*, **72**, 111216f (1970)].

886. G. R. Zhungietu and G. V. Lazur'evskii, *Zh. Vses. Khim. Obshchest.*, **13**, 597 (1968) [*Chem. Abstr.*, **70**, 47246v (1969)].

887. V. V. Mezheritskii and G. N. Dorofeenko, *Khim. Geterotsikl. Soedin, Sb. 2: Kislorodsodershashchie Geterotsikl.*, **1970**, 232 [*Chem. Abstr.*, **76**, 140412e (1972)].

888. A. T. Balaban, *Org. Prep. Proced. Int.*, **9**, 125 (1977).

889. A. T. Balaban, A. Bota, F. Chiraleu, E. Sliam, A. Hanes, and C. Draghici, *Rev. Roum. Chim.*, **22**, 1003 (1977) [*Chem. Abstr.*, **88**, 22563e (1978)].

890. V. I. Vlasov, *Izv. S. Otd. Akad. Nauk SSSR, Ser. Khim. Nauk*, **1971**, 96 [*Chem. Abstr.*, **76**, 140411d (1972)].

891. V. M. Vissov, *Zh. Vses. Khim. Obshchest.*, **15**, 476 (1970) [*Chem. Abstr.*, **75**, 35600w (1971)].

892. G. N. Dorofeenko and L. B. Olekhnovich, *Khim. Geterotsikl. Soedin.*, **7**, 883 (1972) [*Chem. Abstr.*, **77**, 164383t (1972)].

893. S. V. Krivun, G. N. Dorofeenko, and A. S. Kovalevskii, *Khim. Geterotsikl. Soedin.*, **1970**, 733 [*Chem. Abstr.*, **73**, 98769n (1970)].

894. G. N. Dorofeenko, E. P. Olekhnovich, and L. L. Laukhina, *Zh. Org. Khim.*, **7**, 1296 (1971) [*Chem. Abstr.*, **75**, 98392a (1971)].

895. S. V. Krivun, A. L. Buryak, and S. N. Baranov, *Khim. Geterotsikl. Soedin.*, **1973**, 1317 [*Chem. Abstr.*, **80**, 47782y (1974)].

896. G. I. Zhungietu, E. A. Revenko, and F. N. Chukhrii, *Khim. Geterotsikl.*, **1973**, 347 [*Chem. Abstr.*, **78**, 147927b (1973)].

897. G. I. Zhungietu and B. P. Sukhanyuk, *Khim. Geterotsikl. Soedin.*, **1972**, 1531 [*Chem. Abstr.*, **78**, 43316k (1973)].

898. G. N. Dorofeenko, Y. A. Zhdanov, and L. N. Etmetchenko, *Khim. Geterotsikl.*, **1969**, 781 [*Chem. Abstr.*, **72**, 111223 (1970)].

899. V. L. Dulenko and G. N. Dorofeenko, *Metody Poluch. Khim. Reaktivov Prep.*, **17**, 91 (1967) [*Chem. Abstr.*, **71**, 12952 (1969)].

900. J. A. Van Allan and G. A. Reynolds, *J. Heterocycl. Chem.*, **13**, 557 (1976).

901. A. B. Susan and A. T. Balaban, *Rev. Roum. Chim.*, **14**, 111 (1969) [*Chem. Abstr.*, **71**, 61150m (1969)].

902. A. J. Boulton, J. Epsztajn, A. R. Katritzky, and P. L. Nie, *Tetrahedron Lett.*, **1976**, 2689.

903. A. R. Katritzky, J. B. Bapat, R. M. Claramunt-Elguero, F. S. Yates, A. Dinculescu, A. T. Balaban, and F. Chiraleu, *J. Chem. Res. (S)*, **1978**, 395.

904. N. F. Eweiss, A. R. Katritzky, P. L. Nie, and C. A. Ramsden, *Synthesis*, **1977**, 634.

905. A. R. Katritzky, A. Krutosikova, C. A. Ramsden, and J. Lewis, *Coll. Czech. Chem. Commun.*, **43**, 2046 (1978).

906. E. A. Zvezdina, M. P. Zhdanova, V. A. Bren, and G. N. Dorofeenko, *Khim. Geterotsikl. Soedin.*, **1974**, 1461 [*Chem. Abstr.*, **82**, 97303e (1975)].

907. M. P. Zhdanova, E. A. Zvezdina, and G. N. Dorofeenko, *Khim. Geterotsikl. Soedin.*, **1975**, 277 [*Chem. Abstr.*, **82**, 156212z (1975)].

908. E. A. Avezdina, M. P. Zhdanova, V. A. Bren, and G. N. Dorofeenko, *Khim. Geterotsikl. Soedin.*, **1978**, 944 [*Chem. Abstr.*, **89**, 179817j (1978)].

909. A. Balaban and F. A. Urseanu, *Rom. Pat.* 57,177 (1974) [*Chem. Abstr.*, **83**, 195216u (1975)].

910. A. R. Katritzky, U. Gruntz, D. H. Kenny, M. C. Rezende, and H. Sheikh, *J. Chem. Soc., Perkin I*, **1979**, 430.

224   T. D. Bailey, G. L. Goe, and E. F. V. Scriven

911. A. R. Katritzky, U. Gruntz, N. Mongelli, and M. C. Rezende, *J. Chem. Soc., Chem. Commun.*, **1978**, 133.

912. A. R. Katritzky, N. F. Eweiss, and P. L. Nie, *J. Chem. Soc., Perkin Trans. I*, **1979**, 433.

913. A. R. Katritzky, J. Lewis, and P. L. Nie, *J. Chem. Soc., Perkin Trans. I*, **1979**, 442.

914. A. T. Balaban, *Rev. Roum. Chim.*, **18**, 1609 (1973) [*Chem. Abstr.*, **80**, 27069t (1974)].

915. Y. P. Andreichikov, N. V. Kholodova, and G. N. Dorofeenko, *Khim. Geterotsikl. Soedin.*, **1975**, 1578 [*Chem. Abstr.*, **84**, 43980w (1976)].

916. A. V. Koblik, T. I. Polyakova, B. A. Tertov, B. V. Mezhov, and G. N. Dorofeenko, *Zh. Org. Khim.*, **11**, 2153 (1975) [*Chem. Abstr.*, **84**, 43782h (1976)].

917. C. Uncuta and A. T. Balaban, *Rev. Roum. Chim.*, **21**, 251 (1976) [*Chem. Abstr.*, **85**, 4707p (1976)].

918. G. N. Dorofeenko, A. V. Koblik, B. A. Tertov, and T. I. Polyakova, *Khim. Geterotsikl. Soedin.*, **1973**, 1016 [*Chem. Abstr.*, **79**, 137021h (1973)].

919. A. Bota, A. T. Balaban, and F. Chiraleu, *Rev. Roum. Chem.*, **21**, 101 (1976) [*Chem. Abstr.*, **85**, 93214r (1976)].

920. A. R. Katritzky and J. W. Suwinski, *Tetrahedron*, **31**, 1549 (1975).

921. G. N. Dorofeenko, V. V. Mezheritskii, and A. L. Vasserman, *Khim. Geterotsikl. Soedin.*, **1974**, 1338 [*Chem. Abstr.*, **82**, 139912k (1975)].

922. N. Barbulescu, G. Nicolae, and V. Niculaita, *An. Univ. Bucuresti, Chim.*, **20**, 37 (1971) [*Chem. Abstr.*, **79**, 66146q (1973)].

923. V. A. Kaminskii and M. N. Tilichenko, *Zh. Org. Khim.*, **5**, 186 (1969) [*Chem. Abstr.*, **70**, 87541 (1969)].

924. N. Barbulescu and G. Nicolae, *Rev. Chim. (Bucharest)*, **22**, 368 (1971) [*Chem. Abstr.*, **75**, 129638 (1971)].

925. N. Barbulescu and G. Nicolae, *Rev. Chim. (Bucharest)*, **23**, 69 (1972) [*Chem. Abstr.*, **77**, 48176 (1972)].

926. C. Reichardt and R. Mueller, *Ann. Chem.*, **1976**, 1937.

927. A. R. Katritzky, U. Gruntz, A. A. Ikigler, D. H. Kenny, and B. P. Leddy, *J. Chem. Soc., Perkin Trans. I*, **1979**, 436.

928. L. D. Smirnov, V. S. Zhuravlev, V. P. Lezina, M. A. Gugunava, and K. M. Dyumaev, *Azv. Akad. Nauk SSSR, Ser. Khim.*, **1972**, 1878 [*Chem. Abstr.*, **77**, 151832p (1972)].

929. L. D. Smirnov, V. I. Kuz'min, L. N. Mikhailova, V. P. Lezina, and K. M. Dyumaev, *Izv. Akad. Nauk SSSR, Ser. Khim.*, **1970**, 1845 [*Chem. Abstr.*, **74**, 76279w (1971)].

930. A. S. Kurbatova, Y. V. Kurbatov, A. Palamar, O. S. Otroshchenko, and A. S. Sadykov, *Nauch. Tr. Samarkand Univ.*, **167**, 17 (1969) [*Chem. Abstr.*, **74**, 141421 (1971)].

931. Y. V. Kurbatov, *Khim. Rast. Veshchestv*, **1972**, 102 [*Chem. Abstr.*, **78**, 136012y (1973)].

932. A. S. Kurbatova, Y. V. Kurbatov, A. Palamar, O. S. Otroshchenko, and A. S. Sadykov, *Tr. Samarkand. Gos. Univ.*, **167**, 17 (1969) [*Chem. Abstr.*, **75**, 35645q (1971)].

933. Y. V. Kurbatov, S. V. Zalyalieva, O. S. Otroshchenko, and A. S. Sadykov, *Nauch. Tr. Samarkand. Univ.*, **167**, 185 (1969) [*Chem. Abstr.*, **74**, 142134 (1971)].

934. Y. V. Kurbatov and S. V. Zalyalieva, *Khim. Geterotsikl. Soedin.*, **1977**, 1535 [*Chem. Abstr.*, **88**, 74275g (1978)].

935. T. M. Nagiev, G. M. Maned'yarov, N. L. Ali-Zade, Y. A. Mardiev, and N. L. Rzabekova, *Zerb. Khim. Zh.*, **1978**, 6 [*Chem. Abstr.*, **90**, 6214c (1979)].

936. H. Inoue, G. Thyagarajan, and E. L. May, *J. Heterocycl. Chem.*, **12**, 709 (1975).

937. R. J. Benzie, *Brit. Pat.* 1,393,088 (1975) [*Chem. Abstr.*, **83**, 97054v (1975)].

938. R. J. Benzie, *Brit. Pat.* 1,393,087 (1975) [*Chem. Abstr.*, **83**, 97053u (1975)].

939. R. J. Benzie, *Brit. Pat.* 1,393,086 (1975) [*Chem. Abstr.*, **83**, 97052t (1975)].

940. M. Inoue, S. Enomoto, and J. Imamaura, *Yakugaku Zasshi*, **95**, 849 (1975) [*Chem. Abstr.*, **83**, 146830 (1975)].

941. J. Sonnemans, J. M. Janus, and P. Mars, *J. Phys. Chem.*, **80**, 2107 (1976).

942. I. C. Nigam, *J. Chromatog.*, **24**, 188 (1966).

943. A. P. Gelbein, P. Janssen, and H. Richtzenhain, *U.S. Pat.* 4,051,140 (1977) [*Chem. Abstr.*, **88**, 6743h (1978)].

944. A. P. Gelbein, P. Janssen, and H. Richtzenhain, *Ger. Offen.* 2,729,072 (1978) [*Chem. Abstr.*, **88**, 89532n (1978)].

945. G. Daum and H. Richtzenhain, *Ger. Offen.* 2,519,529 (1976) [*Chem. Abstr.*, **86**, 72454z (1977)].

946. H. Richtzenhain and P. Janssen, *U.S. Pat.* 4,057,552 (1977) [*Chem. Abstr.*, **88**, 74301n (1978)].

947. P. Bellingham, *Brit. Pat.* 1,308,152 (1973) [*Chem. Abstr.*, **78**, 147810h (1973)].

948. R. D. Bowden, *Brit. Pat.* 1,268,192 (1972) [*Chem. Abstr.*, **76**, 153605v (1972)].

949. N. L. Wendler, D. Taub, and C. H. Kuo, *U.S. Pat.* 3,435,044 (1969) [*Chem. Abstr.*, **71**, 38820w (1969)].

950. N. S. Prostakov, V. G. Pleshakov, V. V. Dorogov, and V. P. Zvolinsku, *Khim. Geterotsikl. Soedin.*, **1970**, 60 [*Chem. Abstr.*, **72**, 111243n (1970)].

951. N. S. Prostakov, S. Y. Govor, N. N. Mikheeva, and R. K. P. Franko, *Khim. Geterotsikl. Soedin.*, **1969**, 1018 [*Chem. Abstr.*, **73**, 35181u (1970)].

952. N. L. Wendler, D. Taub, and C. H. Kuo, *U.S. Pat.* 3,450,706 (1969) [*Chem. Abstr.*, **71**, 81206 (1969)].

953. N. L. Wendler, D. Taub, and H. Kuo, *U.S. Pat.* 3,441,568 (1969) [*Chem. Abstr.*, **71**, 81191d (1969)].

954. N. S. Prostakov, G. A. Vasil'ev, V. P. Zvolinskii, A. V. Varlamov, A. A. Savina, O. I. Sorokin, and N. D. Lopatina, *Khim. Geterotsikl. Soedin.*, **1975**, 1112 [*Chem. Abstr.*, **83**, 193036 (1975)].

955. N. S. Prostakov, V. I. Kuznetsov, V. F. Zakharov, and V. P. Zvolinskii, *Khim. Geterotsikl. Soedin.*, **1976**, 1077 [*Chem. Abstr.*, **87**, 151974t (1977)].

956. N. S. Prostakov, V. I. Kuznetsov, V. F. Zakharov, and V. P. Zvolinskii, *Khim. Geterotsikl. Soedin.*, **1976**, 1077 [*Chem. Abstr.*, **86**, 5293h (1977)].

957. N. S. Prostakov, M. Torres, A. V. Varlamov, and G. A. Vasil'ev, *Khim. Geterotsikl. Soedin.*, **1979**, 648 [*Chem. Abstr.*, **91**, 91480z (1979)].

958. N. S. Prostakov, L. A. Gaivoronskaya, A. O. Tosunyan, and V. N. Madakyan, *V sb., Sintezy Geterotsikl. Soedinenii, Erevan*, **1975**, 27 [*Chem. Abstr.*, **84**, 150467f (1976)].

959. R. E. Banks, K. Mullen, W. J. Nicholson, C. Oppenheim, and A. Prakash, *J. Chem. Soc., Perkin Trans. I*, **1972**, 1098.

960. G. Queguiner and P. Pastour, *Bull. Soc. Chim. Fr.*, **1968**, 4117.

961. B. T. Kalakutskii, M. A. Kosareva, N. D. Rus'yanova, and N. V. Malysheva, *Zh. Prikl. Khim. (Leningrad)*, **50**, 2121 (1977) [*Chem. Abstr.*, **88**, 22560b (1978)].

962. N. V. Malysheva, L. P. Yurkina, and V. K. Kondratov, *U.S.S.R. Pat.* 191,562 (1967) [*Chem. Abstr.*, **68**, 104997a (1968)].

963. N. D. Rus'yanova, L. P. Yurkina, and N. S. Popova, *Koks Khim.*, **1969**, 41 [*Chem. Abstr.*, **71**, 38753b (1969)].

964. L. P. Yurkina, N. D. Rus'yanova, and N. V. Malysheva, *Khim. Prod. Koksovaniya Uglei Vostoka SSSR, Poluch., Obrab., Ispol's Metody Anal., Vost. Nauch.-Issled. Uglekhim. Inst., Sb. States*, **4**, 251 (1967) [*Chem. Abstr.*, **69**, 35876s (1968)].

965. I. B. Chekmareva, E. S. Zhdanovich, and B. V. Suvorov, *Khim. Geterotsikl. Soedin.*, **1968**, 842 [*Chem. Abstr.*, **70**, 96570e (1969)].

966. S. K. Roy, S. C. Roy, and H. S. Rao, *J. Appl. Chem. Biotechnol.*, **23**, 363 (1973).

967. J. D. V. Hanotier and M. Hanotier-Bridoux, *Fr. Demande* 2,193,820 (1974) [*Chem. Abstr.*, **82**, 170704t (1975)].

968. J. D. V. Hanotier and M. Hanotier-Bridoux, *U.S. Pat.* 3,829,432 (1974) [*Chem. Abstr.*, **82**, 43197e (1975)].

969. V. I. Trubnikov, V. V. Petrov, E. S. Zhdanovich, and N. A. Preobrazhenskii, *Khim.-Farm. Zh.*, **3**, 49 (1969) [*Chem. Abstr.*, **72**, 12498 (1970)].

970. D. C. Ayres and A. M. M. Hossain, *J. Chem. Soc., Perkin Trans. I*, **1975**, 707.

971. L. D. Borkhi, V. G. Khomyakov, and I. G. Yakimchuk, *Tr. Vses. Nauch.-Issled. Inst. Khim. Reaktiv. Osobo Chist. Khim. Veshchestv.*, **32**, 122 (1970) [*Chem. Abstr.*, **77**, 108725 (1972)].

972. V. G. Khomyakov, N. A. Dzbanovskii, and L. D. Borkhi, *Tr. Vses. Nauch. Issled. Inst. Khim. Reaktiv. Osobo Chist. Khim. Veshchestv.*, **29**, 304 (1966) [*Chem. Abstr.*, **67**, 116796b (1967)].

973. L. M. Govorova, V. A. Proskuryakov, and A. N. Chistyakov, *Zh. Prikl. Khim. (Leningrad)*, **43**, 2356 (1970) [*Chem. Abstr.*, **74**, 76277 (1971)].

974. J. Bialek, *Przem. Chem.*, **46**, 526 (1967) [*Chem. Abstr.*, **68**, 114393 (1968)].

975. G. Queguiner and P. Pastour, *Bull. Soc. Chim. Fr.*, **1968**, 4117.

976. P. Sartori, K. Ahlers, and H. J. Frohn, *J. Fluorine Chem.*, **7**, 363 (1976).

977. N. D. Rus'yanova, N. V. Malysheva, and L. P. Yurkina, *Khim. Prod. Koksovaniya Uglei Vostaka SSSR, Poluch., Obrab., Ispol's., Metody Anal., Vost. Nauch.-Issled. Uglekhim. Inst., Sb. States*, **4**, 231 (1967) [*Chem. Abstr.*, **69**, 2825w (1968)].

978. M. G. Shreekrishna and J. L. Frederick, *J. Heterocycl. Chem.*, **5**, 125 (1968).

979. R. D. Chambers, B. Iddon, K. R. William, and R. A. Storey, *Brit. Pat.* 1,151,863 (1969) [*Chem. Abstr.*, **71**, 38819c (1969)].

980. W. A. Remers, G. J. Gibs, C. Pidacks, and M. J. Weiss, *J. Am. Chem. Soc.*, **89**, 5513 (1967).

981. W. A. Remers, G. J. Gibs, C. Pidacks, and M. J. Weiss, *J. Org. Chem.*, **36**, 279 (1971).

982. J. Z. Ginos, *J. Org. Chem.*, **40**, 1191 (1975).

983. A. E. Feiring, *J. Org. Chem.*, **42**, 3255 (1977).

984. F. W. Vierhapper and E. L. Eliel, *J. Am. Chem. Soc.*, **96**, 2256 (1974).

985. N. Nagano and K. Okumura, *Japan. Kokai*, 77 97,980 (1977) [*Chem. Abstr.*, **87**, 201346 (1977)].

986. A. C. W. Curran and R. G. Shepherd, *U.S. Pat.* 4,046,895 (1977) [*Chem. Abstr.*, **87**, 184384 (1977)].

987. A. J. Birch and H. H. Mantsch, *Aust. J. Chem.*, **22**, 1103 (1969).

988. E. Ochiai, M. Takahashi, and R. Tanabe, *Chem. Pharm. Bull. (Tokyo)*, **15**, 1385 (1967).

989. G. Salbeck, *Ger. Offen.* 2,426,851 (1976) [*Chem. Abstr.*, **84**, 105422 (1976)].

990. W. Werner, *Tetrahedron*, **25**, 255 (1969).

991. H. Tani, T. Yamaguchi, and T. Teshigawara, *Japan. Pat.*, 69 04,979 (1969) [*Chem. Abstr.*, **70**, 115019a (1969)].

992. H. C. van der Plas, M. Wozniak, and A. van Veldhuizen, *Rec. Trav. Chim., Pays-Bas*, **97**, 130 (1978).

993. M. Erles, O. Kocian, and J. Lavy, *Coll. Czech. Chem. Commun.*, **41**, 758 (1976).

994. W. A. Remers, G. J. Gibs, and M. J. Weiss, *J. Heterocycl. Chem.*, **8**, 1083 (1971).

995. T. G. Majewicz and P. Caluwe, *J. Org. Chem.*, **39**, 720 (1974).

996. W. J. Irwin, *J. Chem. Soc., Perkin Trans. I*, **1972**, 353.

997. O. Westphal, G. Feix, and A. Joos, *Angew. Chem. Internat. Ed. Engl.*, **8**, 74 (1969).

998. F. Brody and P. R. Ruby, in "Pyridine and Its Derivatives," E. Klingsberg, ed., Interscience, New York, 1960, Vol. 1, p. 2.

999. C. Botteghi, G. Caccia, and S. Gladiali, *Syn. Commun.*, **6**, 549 (1976).

1000. M. Plattier, B. Shimizu, and P. J. Teisseire, *Ger. Pat.* 2,437,901 (1975) [*Chem. Abstr.*, **83**, 131481 (1975)].

1001. M. N. Tilichenko, G. V. Pavel, and A. D. Chumak, *Zh. Org. Khim.*, **7**, 704 (1971) [*Chem. Abstr.*, **75**, 35662 (1971)].

1002. G. A. Klimov, V. N. Ovsyannikova, and M. N. Tilichenko, *Khim. Geterotsikl. Soedin.*, **1972**, 1547 [*Chem. Abstr.*, **78**, 58266 (1973)].

1003. J. Epsztajn, W. E. Hahn, and K. Tosik, *Rocz. Chem.*, **43**, 807 (1969).

1004. A. C. W. Curran, R. Crossley, and D. G. Hill, *Ger. Pat.* 2,352,585 (1974) [*Chem. Abstr.*, **81**, 37486 (1974)].

1005. A. C. W. Curran and R. G. Shepherd, *Ger. Pat.* 2,459,631 (1975) [*Chem. Abstr.*, **83**, 131483 (1975)].

1006. A. Sammour, A. Raouf, M. Elkasaby, and M. A. Ibrahim, *Acta Chim. (Budapest)*, **78**, 399 (1973) [*Chem. Abstr.*, **79**, 146100 (1973)].

1007. G. B. Pavel, M. N. Tilichenko, and A. D. Chumak, *Deposited Doc.*, **1976**, VINITI 4253 [*Chem. Abstr.*, **89**, 163461 (1978)].

1008a. O. V. Fedotova and A. P. Kriven'ko, *Issled. Obl. Sint. Katal. Org. Soedin.*, **1975**, 5 [*Chem. Abstr.*, **86**, 106320 (1977)].

1008b. G. A. Klimov and M. N. Tilichenko, *U.S.S.R. Pat.* 572,455 (1977) [*Chem. Abstr.*, **88**, 50829 (1978)].

1009. L. R. Freimiller and J. W. Nemec, *U.S. Pat.* 3,325,498 (1967) [*Chem. Abstr.*, **67**, 73533 (1967)].

1010. T. V. Moskovkina, V. A. Kaminskii, V. I. Vysotskii, and M. N. Tilichenko, *Khim. Geterotsikl. Soedin.*, **1973**, 826 [*Chem. Abstr.*, **79**, 91958 (1973)].

1011. V. I. Vysotskii and M. N. Tilichenko, *Khim. Geterotsikl. Soedin.*, **1969**, 751 [*Chem. Abstr.*, **72**, 31579 (1970)].

1012. N. Barbulescu and F. Potmischil, *Rev. Roum. Chim.*, **15**, 1601 (1970) [*Chem. Abstr.*, **74**, 53483 (1971)].

1013. V. I. Vysotskii and M. N. Tilichenko, *Khim. Geterotsikl. Soedin.*, **7**, 376 (1971) [*Chem. Abstr.*, **76**, 14295 (1972)].

1014. V. A. Kaminskii, A. N. Saverchenko, and M. N. Tilichenko, *Zh. Org. Khim.*, **6**, 404 (1970) [*Chem. Abstr.*, **72**, 111272 (1970)].

1015. V. A. Kaminskii, A. N. Saverchenko, and M. N. Tilichenko, *Khim. Geterotsikl. Soedin.*, **1970**, 1538 [*Chem. Abstr.*, **74**, 53482 (1971)].

1016. N. P. Maslennikova and N. N. Letyushova, *Issled. Obl. Sint. Katal. Org. Soedin.*, **1975**, 15 [*Chem. Abstr.*, **86**, 106259 (1977)].

1017. A. N. Saverchenko, V. A. Kaminskii, and M. N. Tilichenko, *Khim. Geterotsikl. Soedin.*, **1972**, 1232 [*Chem. Abstr.*, **77**, 164426 (1972)].

1018. V. A. Kaminskii and M. N. Tilichenko, *Zh. Org. Khim.*, **5**, 186 (1969) [*Chem. Abstr.*, **70**, 87541 (1969)].

1019. N. Barbulescu and F. Potmischil, *Rev. Roum. Chim.*, **14**, 1427 (1969) [*Chem. Abstr.*, **72**, 132483 (1970)].

1020. V. I. Vysotskii and M. N. Tilichenko, *Khim. Geterotsikl. Soedin.*, **1976**, 383 [*Chem. Abstr.*, **84**, 180006 (1976)].

1021. A. N. Saverchenko, V. A. Kaminskii, and M. N. Tilichenko, *Khim. Geterotsikl. Soedin.*, **1973**, 384 [*Chem. Abstr.*, **78**, 159393 (1973)].

228 T. D. Bailey, G. L. Goe, and E. F. V. Scriven

1022. A. N. Saverchenko, Z. R. Bekkerova, V. A. Kaminskii, and M. N. Tilichenko, *Khim. Geterotsikl. Soedin.*, 1974, 243 [*Chem. Abstr.*, 80, 145982 (1974)].

1023. N. Barbulescu, F. Potmischil, and D. Romer, *Rev. Chim. (Bucharest)*, 21, 677 (1970).

1024. T. L. Akimova, M. N. Tilichenko, V. V. Isakov, and T. A. Budina, *Zh. Org. Khim.*, 13, 2095 (1977) [*Chem. Abstr.*, 88, 62025 (1978)].

1025. T. V. Moskovkina, M. N. Tilichenko, V. M. Kurilenko, and L. P. Fedyaeva-Basova, *Khim. Farm. Zh.*, 7, 3 (1973) [*Chem. Abstr.*, 78, 147761 (1973)].

1026. T. V. Moskovkina and M. N. Tilichenko, *Khim. Geterotsikl. Soedin.*, 1976, 645 [*Chem. Abstr.*, 85, 143008 (1976)].

1027. C. Chadwick, *Anal. Chem.*, 48, 1201 (1976).

1028. R. D. Bowden and T. Seaton, *Brit. Pat.* 1,324,644 (1973) [*Chem. Abstr.*, 79, 126323 (1973)].

1029. V. O. Fedotova and A. P. Kriven'ko, *Issled. Obl. Sint. Katal. Org. Soedin.*, 1975, 5 [*Chem. Abstr.*, 86, 106320 (1977)].

1030. O. V. Fedotova, A. P. Kriven'ko, V. G. Kharchenko, and M. N. Tilichenko, *Deposited Doc.*, 1976, VINITI 2254 [*Chem. Abstr.*, 89, 146728 (1978)].

1031. N. S. Smirnova, A. P. Kriven'ko, A. D. Shebaldova, V. N. Kravtsova, L. I. Markova, O. V. Fedotova, L. G. Chichenkova, G. I. Rybina, and A. A. Terekhin, *Khim. Dikarbonil'nykh Soedin., Tezisy Dokl. Vses. Konf., 4th*, 1975, 155 [*Chem. Abstr.*, 87, 53046 (1977)].

1032. G. V. Pavel and M. N. Tilichenko, *Khim. Geterotsikl. Soedin.*, 1968, 484 [*Chem. Abstr.*, 69, 96405 (1968)].

1033. J. Curtze, P. Wild, and F. Kroehnke, *Justus Liebigs Ann. Chem.*, 1975, 864.

1034. E. Leete, *J. Chem. Soc., Chem. Commun.*, 1972, 1091.

1035. A. Shafiee and E. Winterfeldt, *Synthesis*, 1974, 185.

1036. J. Schnekenburger and D. Heber, *Arch. Pharm. (Weinheim)*, 308, 225 (1975).

1037. F. Abdulla, T. L. Emmick, and H. M. Taylor, *Syn. Commun.*, 7, 305 (1977).

1038. I. El-S. El-Kholy, M. M. Mishrikey, and R. F. Atmeh, *J. Heterocycl. Chem.*, 10, 665 (1973).

1039. H. Rigterink, *U.S. Pat.* 3,748,334 (1973) [*Chem. Abstr.*, 79, 78627 (1973)].

1040. P. J. Wittek, K. B. Hindley, and T. M. Harris, *J. Org. Chem.*, 38, 896 (1973).

1041. K. Tamazawa, K. Takahashi, M. Murakami, and S. Iwanami, *Japan. Pat.* 75 93,976 (1975) [*Chem. Abstr.*, 84, 43862 (1976)].

1042. H. M. Taylor, *Ger. Pat.* 2,537,753 (1976) [*Chem. Abstr.*, 85, 46406 (1976)].

1043. S. Boatman, R. E. Smith, G. F. Morris, W. G. Kofron, and C. R. Hauser, *J. Org. Chem.*, 32, 3817 (1967).

1044a. R. F. Abdulla, K. H. Fuhr, and H. M. Taylor, *Syn. Commun.*, 7, 313 (1977).

1044b. R. F. Abdulla and R. S. Brinkmeyer, *Tetrahedron Report No. 67*, 1979, 1675.

1045. A. F. Vompe, N. V. Monich, N. F. Turitsyna, and L. V. Ivanova, *Zh. Org. Khim.*, 7, 2590 (1971) [*Chem. Abstr.*, 76, 99463 (1972)].

1046. I. Iijima, M. Miyazaki, N. Taga, and T. Tanaka, *Heterocycles*, 8, 357 (1977).

1047. K. G. Golodova and S. I. Yakimovich, *Zh. Org. Khim.*, 8, 2019 (1972) [*Chem. Abstr.*, 78, 71642 (1973)].

1048. N. V. Koshmina and F. Ya. Perveev, *Tezisy Dokl. Vses. Konf. Khim. Atsetilena, 5th*, 1975, 291 [*Chem. Abstr.*, 89, 59819 (1978)].

1049. K. G. Golodova and S. I. Yakimovich, *Zh. Org. Khim.*, 8, 2015 (1972) [*Chem. Abstr.*, 78, 29570 (1973)].

1050. H. Mishima and H. Kurihara, *Japan. Pat.*, 78 68,788 (1978) [*Chem. Abstr.*, 89, 197349 (1978)].

1051. H. Mishima and H. Kurihara, *Japan. Pat.*, 78 68,781 (1978) [*Chem. Abstr.*, 89, 163424 (1978)].

1052. J. McNulty, H. O. Bayer, and M. C. Seidel, *U.S. Pat.* 2,838,155 (1974) [*Chem. Abstr.*, 81, 152008 (1974)].

1053. H. Hartmann and J. Liebscher, *East. Ger. Pat.* 107,033 (1974) [*Chem. Abstr.*, 82, 139971 (1975)].

1054. W. Dittmar, *Ger. Pat.* 1,795,831 (1978) [*Chem. Abstr.*, 89, 197347 (1978)].

1055. C. Lohacs and W. Dittmar, *S. African Pat.*, 69 06,039 (1970) [*Chem. Abstr.*, 73, 120508 (1970)].

1056. R. Weyer, V. Hitzel, and E. Granzer, *Ger. Pat.* 2,637,477 (1978) [*Chem. Abstr.*, 88, 152443 (1978)].

1057. N. Anghelide, C. Draghici, and D. Raileanu, *Tetrahedron,* 30, 623 (1974).

1058. T. Hatada, M. Sone, Y. Tominaga, R. Natsuki, Y. Matsude, and G. Kobayashi, *Yakugaku Zasshi,* 95, 623 (1975) [*Chem. Abstr.*, 83, 206076 (1978)].

1059. P. D. Cook, R. T. Day, and R. K. Robins, *J. Heterocycl. Chem.*, 14, 1295 (1977).

1060. J. Liebscher and H. Hartmann, *Z. Chem.*, 13, 342 (1973).

1061. J. Liebscher and H. Hartmann, *East Ger. Pat.* 106,377 (1974) [*Chem. Abstr.*, 81, 152025 (1974)].

1062. E. Poetsch, *Ger. Pat.* 1,811,973 (1970) [*Chem. Abstr.*, 73, 55978 (1970)].

1063. R. F. Borch, C. V. Grudzinskas, D. A. Peterson, and L. D. Weber, *J. Org. Chem.*, 37, 1141 (1972).

1064. *Ger. Pat.* 2,365,302 (1974) [*Chem. Abstr.*, 81, 49575 (1974)].

1065. M. J. Abu El-Haj and B. W. Dominy, *U.S. Pat.* 3,962,264 (1976) [*Chem. Abstr.*, 85, 123967 (1976)].

1066. T. A. Bryson, J. C. Wisowaty, R. B. Dunlap, R. R. Fisher, and P. D. Ellis, *J. Org. Chem.*, 39, 3436 (1974).

1067. T. A. Bryson, D. M. Donelson, R. B. Dunlap, R. R. Fisher, and P. D. Ellis, *J. Org. Chem.*, 41, 2066 (1976).

1068. T. Nishimura, H. Misawa, H. Kurihara, and H. Yamanaka, *Japan. Pat.*, 78 69,935 (1978) [*Chem. Abstr.*, 90, 22841 (1979)].

1069. J. Nadelson, *U.S. Pat.* 4,122,182 (1978) [*Chem. Abstr.*, 90, 203873 (1979)].

1070. J. Nadelson, *U.S. Pat.* 4,054,653 (1977) [*Chem. Abstr.*, 88, 37809 (1978)].

1071. W. Ried and E. Nyiondi-Bonguen, *Justus Liebigs Ann. Chem.*, 1973, 1.

1072. N. V. Koshmina and F. Ya. Perveev, *Zh. Org. Khim.*, 12, 2074 (1976) [*Chem. Abstr.*, 86, 105889 (1977)].

1073. F. Ya. Perveev and N. V. Koshmina, *U.S.S.R. Pat.* 434,082 (1974) [*Chem. Abstr.*, 81, 77806 (1974)].

1074. D. Muenzner, H. Lattau, and H. Schubert, *Z. Chem.*, 7, 278 (1967).

1075. K. Grohe and A. Roedig, *Chem. Ber.*, 100, 2953 (1967).

1076. V. F. Sedova and V. P. Mamaev, *Zh. Org. Khim.*, 13, 2220 (1977) [*Chem. Abstr.*, 88, 50782 (1978)].

1077. K. W. Ehler, R. K. Robins, and R. B. Meyer, Jr., *J. Med. Chem.*, 20, 317 (1977).

1078. J. P. Schneider and P. Cordier, *Chim. Ther.*, 4, 330 (1969).

1079. Zh. A. Krasnaya, T. S. Stytsenko, E. P. Prokof'ev, and V. F. Kucherov, *Khim. Dikarbonil'nykh Soedin., Tezisy Dokl. Vses. Konf., 4th 1975,* 81 (1976) [*Chem. Abstr.*, 87, 68099 (1977)].

1080. Zh. A. Krasnaya, T. S. Stytsenko, E. P. Prokof'ev, and V. F. Kucherov, *Khim. Geterotsikl. Soedin.,* 1973, 668 [*Chem. Abstr.*, 79, 78541 (1973)].

1081. J. P. Ferris, J. E. Kuder, and A. W. Catalano, *Science,* 166, 765 (1969).

230   T. D. Bailey, G. L. Goe, and E. F. V. Scriven

1082. Yu. V. Maevskii and S. V. Sokolovskaya, *Khim. Geterotsikl. Soedin.*, **7**, 567 (1971) [*Chem. Abstr., 76*, 25048 (1972)].

1083. C. S. Rooney, H. W. R. Williams, and B. K. Wasson, *U.S. Pat.* 3,993,656 (1976) [*Chem. Abstr., 86*, 106555 (1977)].

1084. H. W. R. Williams, C. S. Rooney, and B. H. Wasson, *Fr. Pat.* 2,183,753 (1974) [*Chem. Abstr., 80*, 120895 (1974)].

1085. J. J. Baldwin, K. Mensler, and G. S. Ponticello, *J. Org. Chem.*, **43**, 4878 (1978).

1086. J. J. M. Deumens and S. H. Groen, *U.S. Pat.* 3,780,082 (1973) [*Chem. Abstr., 77*, 4985 (1972)].

1087. H. Krimm, *U.S. Pat.* 2,768,962 (1956) [*Chem. Abstr., 51*, 6684 (1957)].

1088. J. J. M. Deumens, H. J. G. Henskens, J. A. Thoma, and H. Van der Zalm, *Brit. Pat.* 1,378,464 (1974) [*Chem. Abstr., 80*, 120336 (1974)].

1089. *Brit. Pat.* 1,304,155 (1973) [*Chem. Abstr., 78*, 124454 (1973)].

1090. B. D. Simpson, A. M. Schnitzer, and R. L. Cobb, *U.S. Pat.* 3,007,931 (1961) [*Chem. Abstr., 56*, 4738 (1962)].

1091. E. Van Poelvoorde and H. Van der Zalm, *Ger. Pat.* 2,246,284 (1973) [*Chem. Abstr., 78*, 158993 (1973)].

1092. S. H. Groen and J. J. M. Deumens, *U.S. Pat.* 3,686,262 (1972) [*Chem. Abstr., 74*, 124898 (1971)].

1093. J. J. M. Deumens and P. A. M. Stijfs, *U.S. Pat.* 3,780,083 (1972) [*Chem. Abstr., 77*, 151521 (1972)].

1094. J. A. S. Thoma and P. A. M. J. Stijfs, *Ger. Pat.* 2,459,095 (1975) [*Chem. Abstr., 83*, 147401 (1975)].

1095. L. G. E. Beekhuis and J. F. M. Klein, *U.S. Pat.* 3,821,274 [*Chem. Abstr., 78*, 15556 (1973)].

1096. J. A. Thoma, J. F. M. Klein, and L. H. Geurts, *U.S. Pat.* 3,708,515 (1973) [*Chem. Abstr., 73*, 98428 (1970)].

1097. J. J. M. Deumens and J. A. Thoma, *Brit. Pat.* 1,243,327 (1971) [*Chem. Abstr., 73*, 120505 (1970)].

1098. S. H. Groen and A. M. J. Petrus, *U.S. Pat.* 3,728,353 (1973) [*Chem. Abstr., 74*, 64209 (1971)].

1099. S. H. Groen, J. J. M. Deumens, and P. A. M. J. Stijfs, *U.S. Pat.* 3,843,659 (1974) [*Chem. Abstr., 76*, 14353 (1972)].

1100. J. A. Thoma and J. J. M. Deumens, *U.S. Pat.* 3,658,824 (1972) [*Chem. Abstr., 77*, 61828 (1972)].

1101. T. Kiyoura and Y. Kogure, *Japan. Pat.*, 78 18,580 (1978) [*Chem. Abstr., 89*, 129404 (1978)].

1102. E. Gudriniece and B. Rigerte, *Latv. PSR Zinat. Akad. Vestis, Kim. Ser.*, **1974**, 239 [*Chem. Abstr., 81*, 37490 (1974)].

1103. T. R. Kasturi, V. K. Sharma, A. Srinivasan, and G. Subrahmanyam, *Tetrahedron*, **29**, 4103 (1973).

1104. J. L. Van der Baan and F. Bickelhaupt, *J. Chem. Soc. D,* **1970**, 326.

1105. T. R. Kasturi and V. K. Sharma, *Tetrahedron,* **31**, 527 (1975).

1106. K. Bogdanowicz-Szwed, *Rocz. Chem.*, **48**, 641 (1974).

1107. J. M. Bobbitt and D. A. Scola, *J. Org. Chem.*, **25**, 560 (1960).

1108. E. Wenkert, K. G. Dave, and F. Haglid, *J. Am. Chem. Soc.*, **87**, 5461 (1965).

1109. M. I. Murakami, N. Inukai, and N. Nagano, *Chem. Pharm. Bull.*, **20**, 1699 (1972).

1110. M. Murakami, T. Inukai, and N. Nagano, *Japan. Pat.*, 72 14,388 [*Chem. Abstr., 77*, 48285 (1972)].

1111. N. Nagano and M. Murakami, *Japan. Pat.*, 74 66,685 [*Chem. Abstr.*, **82**, 72800 (1975)].

1112. V. Bertini, P. Pelosi, and A. De Munno, *J. Heterocycl. Chem.*, **9**, 741 (1972).

1113. H. W. Schmidt and H. Junek, *Monatsh. Chem.*, **108**, 895 (1977).

1114. A. Taurins and R. T. Li, *Can. J. Chem.*, **52**, 843 (1974).

1115. V. P. Kukhar and N. G. Pavlenko, *Zh. Org. Khim.*, **10**, 36 (1974) [*Chem. Abstr.*, **80**, 120187 (1974)].

1116. M. Huys-Francotte, Z. Janousek, and H. G. Viehe, *J. Chem. Res. (S)*, 1977, 100.

1117. G. Simchen, *Chem. Ber.*, **103**, 389 (1970).

1118. R. D. Bowden and T. Seaton, *Brit. Pat.* 1,334,922 (1973) [*Chem. Abstr.*, **80**, 47857 (1974)].

1119. W. Kantlehner, *Ger. Pat.* 2,718,171 [*Chem. Abstr.*, **90**, 54833 (1979)].

1120. A. Kasahara and T. Saito, *Chem. Ind. (London)*, 1975, 745.

1121. A. Kasahara, *Japan. Pat.*, 76 143,672 (1976) [*Chem. Abstr.*, **87**, 39289 (1977)].

1122. D. E. Korte, L. S. Hegedus, and R. K. Wirth, *J. Org. Chem.*, **42**, 1329 (1977).

1123. A. Kasahara, T. Izumi, and O. Saito, *Chem. Ind. (London)*, 1980, 666.

1124. T. Hosokawa, N. Shimo, K. Maeda, A. Sonoda, and S. Murahashi, *Tetrahedron Lett.*, 1976, 383.

1125. T. Hosokawa, N. Shimo, A. Sonoda, and S. Murahashi, *Hukusokan Kagaku Toronkai Keon Yoshisu, 8th*, 1975, 234 [*Chem. Abstr.*, **85**, 5462 (1976)].

1126. P. Dubs and R. Stuessi, *J. Chem. Soc., Chem. Commun.*, 1976, 1021.

1127. F. Ya. Perveev and N. V. Koshmina, *Zh. Org. Khim.*, **4**, 177 (1968) [*Chem. Abstr.*, **68**, 78086 (1968)].

1128. F. Ya. Perveev and N. V. Koshmina, *Zh. Org. Khim.*, **5**, 1334 (1969) [*Chem. Abstr.*, **71**, 112771 (1969)].

1129. F. Ya. Perveev, M. S. Ivakhnyuk, and N. V. Koshmina, *Zh. Org. Khim.*, **6**, 1116 (1970) [*Chem. Abstr.*, **73**, 35173 (1970)].

1130. F. Ya. Perveev and I. I. Afonina, *Zh. Org. Khim.*, **8**, 2026 (1972) [*Chem. Abstr.*, **78**, 29566 (1973)].

1131. I. I. Afonina and F. Ya. Perveev, *Zh. Org. Khim.*, **9**, 2006 (1973) [*Chem. Abstr.*, **80**, 27171 (1974)].

1132. K. G. Golodova, S. I. Yakimovich, and F. Ya. Perveev, *Dokl. Vses. Konf. Khim. Atsetilena, 4th*, **1**, 310 (1972) [*Chem. Abstr.*, **79**, 78051 (1973)].

1133. Zh. A. Krasnaya, T. S. Stytsenko, E. P. Prokof'ev, and V. F. Kucherov, *Izv. Akad. Nauk SSSR, Ser. Khim.*, 1972, 2213 [*Chem. Abstr.*, **78**, 71347 (1973)].

1134. H. N. Al-Jallo, *J. Chem. Eng. Data*, **17**, 513 (1972).

1135. V. A. Shikhanov, A. P. Ivanovskii, and A. M. Kut'in, *U.S.S.R. Pat.* 791,747 (1980) [*Chem. Abstr.*, **94**, 192158 (1981)].

1136. V. N. Yandovskii and T. K. Klindukhova, *Zh. Org. Khim.*, **10**, 730 (1974) [*Chem. Abstr.*, **81**, 25505 (1974)].

1137. T. Metler, A. Uchida, and S. I. Miller, *Tetrahedron*, **24**, 4285 (1968).

1138. G. L. Isele and K. Scheib, *Chem. Ber.*, **108**, 2312 (1975).

1139. J. Chauvelier, *Ann. Chim.*, **3**, 393 (1948).

1140. J. Chauvelier, M. Chauvin, and J. Aubouet, *Bull. Soc. Chim. Fr.*, 1966, 1721.

1141. A. Roedig, G. Maerkl, and H. Schaller, *Chem. Ber.*, **103**, 1022 (1970).

1142. A. Roedig, H. A. Renk, V. Schall, and D. Scheutzow, *Chem. Ber.*, **107**, 1136 (1974).

1143. R. L. Hodgson, *Tetrahedron*, **24**, 4833 (1968).

1144. W. H. Bell, G. B. Carter, and J. Dewing, *J. Chem. Soc. (C)*, 1969, 352.

1145. J. J. M. Deumens, S. H. Groen, and J. M. J. G. Lipsch, *Neth. Pat.*, 70 05,792 [*Chem. Abstr.*, **76**, 46099 (1972)].

1146. R. D. Bowden, J. H. Marsden, and F. Talor, *Ger. Pat.* 2,013,181 (1970) [*Chem. Abstr.*, **73**, 120504 (1970)].

1147. A. Murata and H. Suzuki, *Japan. Pat.*, 73 64,020 [*Chem. Abstr.*, **80**, 59872 (1974)].

1148. R. Harmetz and R. J. Tull, *U.S. Pat.* 3,644,380 (1972) [*Chem. Abstr.*, **76**, 140558 (1972)].

1149. R. Harmetz and R. J. Tull, *U.S. Pat.* 3,873,597 (1975) [*Chem. Abstr.*, **83**, 43199 (1975)].

1150. A. Garming, D. Redwan, P. Gelbke, D. Kern, and U. Dierkes, *Justus Liebigs Ann. Chem.*, **1975**, 1744.

1151. A. Garming, D. Kern, G. Cohausz, G. Hillert, P. Gelbke, and D. Severin, *Justus Liebigs Ann. Chem.*, **1977**, 1822.

1152. B. T. Gatica and F. J. Bartulin, *Rev. Real Acad. Cienc. Exactas, Fis. Natur. Madrid*, **62**, 259 (1968) [*Chem. Abstr.*, **70**, 57583 (1969)].

1153. E. Sandberg, *Chem. Scr.*, **2**, 241 (1972).

1154. V. V. Kononova, A. L. Vereshchagin, V. M. Polyachenko, and A. A. Semenov, *Khim.-Farm. Zh.*, **12**, 30 (1978) [*Chem. Abstr.*, **90**, 103873 (1979)].

1155. Y. Wakatsuki and H. Yamazaki, *J. Chem. Soc., Dalton Trans.*, **1978**, 1278.

1156. Y. Wakatsuki and H. Yamazaki, *Japan. Pat.*, 74 100,073 (1973) [*Chem. Abstr.*, **82**, 86419 (1975)].

1157. Y. Wakatsuki and H. Yamazaki, *J. Chem. Soc., Chem. Commun.*, **1973**, 280.

1158. P. Hong and H. Yamazaki, *Synthesis*, **1977**, 50.

1159. G. J. Janz, in "1,4-Cycloaddition Reactions," J. Hamer, ed., Academic Press, New York, 1967, Chap. 4.

1160. D. J. Perettie, *Belg. Pat.* 815,514 (1974) [*Chem. Abstr.*, **83**, 97033 (1975)].

1161. D. J. Perettie, *Ger. Pat.* 2,427,162 (1975) [*Chem. Abstr.*, **84**, 164628 (1976)].

1162. *Neth. Pat.*, 74 07,395 (1975) [*Chem. Abstr.*, **85**, 192567 (1976)].

1163. *Japan. Pat.* 75 157,374 (1975) [*Chem. Abstr.*, **85**, 108549 (1976)].

1164. J. C. Jagt and A. M. Van Leusen, *Recl. Trav. Chim. Pays-Bas*, **92**, 1343 (1973).

1165. J. I. Darragh and H. S. Inglis, *Brit. Pat.* 1,495,233 (1977) [*Chem. Abstr.*, **89**, 24155 (1978)].

1166. J. I. Darragh, *Brit. Pat.* 1,493,553 (1977) [*Chem. Abstr.*, **88**, 190596 (1978)].

1167. D. Kim and S. M. Weinreb, *J. Org. Chem.*, **43**, 121 (1978).

1168. J. P. Fleury, M. Desbois, and J. See, *Bull. Soc. Chim. Fr. II*, **1978**, 147.

1169. K. Takabe, H. Fujiwara, T. Katagiri, and J. Tanaka, *Tetrahedron Lett.*, **1975**, 4375.

1170. J. Tanaka, T. Katagiri, and K. Takabe, *Japan. Pat.*, 77 42,882 (1977) [*Chem. Abstr.*, **87**, 84836 (1977)].

1171. C. Kashima, M. Yamamoto, S. Kobayashi, and N. Sugiyama, *Bull. Chem. Soc. Japan*, **42**, 2389 (1969).

1172. N. Sugiyama, M. Yamamoto, S. Kobayashi, and C. Kashima, *Bull. Chem. Soc. Japan*, **45**, 296 (1972).

1173. F. J. Vinick, Y. Pan, and H. W. Gschwend, *Tetrahedron Lett.*, **1978**, 4221.

1174. C. Kashima, M. Yamamoto, S. Kobayashi, and N. Sugiyama, *Bull. Chem. Soc. Japan*, **47**, 1805 (1974).

1175. J. Mayer and M. H. Sherlock, *U.S. Pat.* 4,017,500 (1977) [*Chem. Abstr.*, **87**, 23252 (1977)].

1176. *Belg. Pat.* 835,770 (1976) [*Chem. Abstr.*, **87**, 23251 (1977)].

1177. K. Takeda, T. Ohta, K. Shudo, T. Okamoto, K. Tsuji, and T. Kosuge, *Chem. Pharm. Bull.*, **25**, 2145 (1977).

1178. O. Tsuge and A. Inaba, *Bull. Chem. Soc. Japan*, **46**, 286 (1973).

1179. H. Junek, A. Metallidis, and E. Ziegler, *Monatsh. Chem.*, **100**, 1937 (1969).

1180. A. Metallidis, *Chem. Chron.*, **1**, 151 (1972) [*Chem. Abstr.*, **77**, 126383 (1972)].

1181. J. Becher and E. G. Frandsen, *Tetrahedron Lett.*, **1976**, 3347.

1182. J. Becher and E. G. Frandsen, *Tetrahedron*, **33**, 341 (1977).

1183. J. Becher and E. G. Frandsen, *Acta Chem. Scand., Ser. B*, **B30**, 863 (1976).

1184. J. Becher, E. G. Frandsen, C. Dreier, and L. Henriksen, *Acta Chem. Scand., Ser. B*, **B31**, 843 (1977).

1185. E. Gegner, *Tetrahedron Lett.*, **1969**, 287.

1186. J. C. Powers and I. Ponticello, *J. Am. Chem. Soc.*, **90**, 7102 (1968).

1187. H. O. Bayer and P. J. Nulty, *U.S. Pat.* 3,711,488 (1973) [*Chem. Abstr.*, **78**, 84270 (1973)].

1188. R. J. Chorvat and R. Pappo, *Tetrahedron Lett.*, **1975**, 623.

1189. R. Pappo and R. J. Chorvat, *Ger. Pat.* 2,520,013 (1975) [*Chem. Abstr.*, **84**, 122135 (1976)].

1190. T. Kappe, I. Maninger, and E. Ziegler, *Monatsh. Chem.*, **99**, 85 (1968).

1191. S. Rajappa, B. G. Advani, and R. Sreenivasan, *Indian J. Chem.*, **10**, 323 (1972).

1192. S. Rajappa and R. Sreenivasan, *Indian J. Chem., Sect. B*, **14B**, 400 (1976).

1193. T. Kappe, M. Hariri, and F. S. G. Soliman, *Arch. Pharm. (Weinheim, Ger.)*, **309**, 684 (1976).

1194. E. L. Patmore and H. Chafetz, *Amer. Chem. Soc., Div. Petrol. Chem., Prepr.*, **1972**, B27 [*Chem. Abstr.*, **80**, 3368 (1974)].

1195. H. Chafetz and R. C. Anderson, *U.S. Pat.* 3,336,313 (1967) [*Chem. Abstr.*, **68**, 29621 (1968)].

1196. H. Chafetz and E. L. Patmore, *U.S. Pat.* 3,408,351 (1968) [*Chem. Abstr.*, **70**, 19950 (1969)].

1197. C. Bischoff and H. Herma, *J. Prakt. Chem.*, **318**, 891 (1976).

1198. A. V. Upadysheva, N. D. Grigor'eva, and A. P. Znamenskaya, *U.S.S.R. Pat.* 551,328 (1977) [*Chem. Abstr.*, **87**, 39308 (1977)].

1199. G. Zigeuner and G. Gübitz, *Monatsh.*, **101**, 1547 (1970).

1200. A. J. Chalk, *Tetrahedron Lett.*, **1972**, 3487.

1201. A. J. Chalk, *Tetrahedron*, **30**, 1387 (1974).

1202. A. Tamano and M. Doya, *Japan. Pat.*, 73 22,704 (1973) [*Chem. Abstr.*, **79**, 146406 (1973)].

1203. *Brit. Amended Pat.* 1,053,290 (1969) [*Chem. Abstr.*, **72**, 132531 (1970)].

1204. G. Grigoleit, R. Oberkobusch, and G. Collin, *Ger. Pat.* 1,931,945 (1971) [*Chem. Abstr.*, **74**, 64207 (1971)].

1205. G. Grigoleit, R. Oberkobusch, G. Collin, and K. Matern, *Ger. Pat.* 2,051,316 (1972) [*Chem. Abstr.*, **77**, 19545 (1972)].

1206. J. Veitch, *Brit. Pat.* 1,135,854 (1968) [*Chem. Abstr.*, **70**, 47310 (1969)].

1207. S. Yasuda and S. Nishi, *Japan. Pat.*, 70 08,420 (1970) [*Chem. Abstr.*, **73**, 14699 (1970)].

1208a. A. H. Berrie and N. Hughes, *Brit. Pat.* 1,256,095 (1971) [*Chem. Abstr.*, **76**, 87170 (1972)].

1208b. H. Matsumoto, *Japan. Pat.*, 76 37,919 (1976) [*Chem. Abstr.*, **85**, 34635 (1976)].

1209. A. H. Berrie and N. Hughes, *Brit. Pat.* 1,256,094 (1971) [*Chem. Abstr.*, **76**, 87171 (1972)].

234   T. D. Bailey, G. L. Goe, and E. F. V. Scriven

1210.   E. Heinrich, *Ger. Pat.* 2,150,772 (1973) [*Chem. Abstr., 79*, 32700 (1973)].

1211.   J. L. Van der Baan and F. Bickelhaupt, *J. Chem. Soc. D*, 1970, 326.

1212.   P. W. Austin and A. Crabtree, *Ger. Pat.* 2,202,270 (1972) [*Chem. Abstr., 77*, 151954 (1972)].

1213.   H. J. Quadbeck-Seeger, *Ger. Pat.* 2,531,035 (1977) [*Chem. Abstr., 86*, 139879 (1977)].

1214.   Z. Cojocaru, E. Chindris, D. Ghiran, M. Moga-Iuga, V. Ariesan, and C. Nistor, *Farmacia (Bucharest), 21*, 607 (1973) [*Chem. Abstr., 80*, 133208 (1974)].

1215.   P. Melius and J. L. Green, Jr., *J. Med. Chem., 15*, 206 (1972).

1216.   G. E. Hardtmann, *Ger. Pat.* 2,051,013 (1971) [*Chem. Abstr., 76*, 113242 (1972)].

1217.   C. S. Rooney, H. W. R. Williams, and B. K. Wasson, *U.S. Pat.* 3,993,656 (1976) [*Chem. Abstr., 86*, 106555 (1977)].

1218.   H. W. R. Williams, C. S. Rooney, and B. H. Wasson, *Fr. Pat.* 2,183,753 (1974) [*Chem. Abstr., 80*, 120895 (1974)].

1219.   E. Eichler, C. S. Rooney, and H. W. R. Williams, *J. Heterocycl. Chem., 13*, 841 (1976).

1220.   E. Gudriniece, A. V. Nikitenko, and S. Valtere, *Latv. PSR Zinat. Akad. Vestis, Kim. Ser., 1969*, 627 [*Chem. Abstr., 72*, 55196 (1970)].

1221.   E. Gudriniece and A. V. Nikitenko, *Latv. PSR Zinat. Akad. Vestis, Kim. Ser., 1972*, 569 [*Chem. Abstr., 78*, 29567 (1973)].

1222.   L. Doub, J. S. Kaltenbronn, and D. Schweiss, *U.S. Pat.* 3,954,734 (1976) [*Chem. Abstr., 85*, 108683 (1976)].

1223.   O. S. Wolfbeis and E. Ziegler, *Z. Naturforsch., B: Anorg. Chem., Org. Chem., 31B*, 1519 (1976).

1224.   E. Eichler, C. S. Rooney, and H. W. R. Williams, *J. Heterocycl. Chem., 13*, 43 (1976).

1225.   S. V. Sunthankar and S. D. Vaidya, *Indian J. Chem., 11*, 1315 (1973).

1226.   T. A. Grushina, Z. N. Zhukova, M. V. Balyakina, M. Ts. Yanotovskii, and V. I. Gunar, *Khim.-Farm. Zh., 10*, 141 (1976).

1227.   P. F. H. Freeman, *U.S. Pat.* 3,674,877 (1972) [*Chem. Abstr., 77*, 88314 (1972)].

1228.   S. Hauptmann and J. Weisflog, *J. Prakt. Chem., 314*, 353 (1972).

1229.   J. H. Maguire and R. L. McKee, *J. Heterocycl. Chem., 16*, 133 (1979).

1230.   G. E. Hardtmann, *U.S. Pat.* 3,717,645 (1973) [*Chem. Abstr., 78*, 136080 (1973)].

1231.   G. E. Hardtmann, *U.S. Pat.* 3,660,415 (1972) [*Chem. Abstr., 77*, 48274 (1972)].

1232.   G. E. Hardtmann, *U.S. Pat.* 3,886,161 (1975) [*Chem. Abstr., 83*, 79090 (1975)].

1233.   K. Ito and A. Miyake, *Japan. Pat., 74* 135992 (1974) [*Chem. Abstr., 83*, 43362 (1975)].

1234.   B. Tornetta, F. Guerrera, and G. Ronsisvalle, *Ann. Chim. (Rome), 64*, 833 (1974).

1235.   V. Radtke, *Ger. Pat.* 2,721,888 (1978) [*Chem. Abstr., 90*, 72074 (1979)].

1236.   A. I. Gurevich, M. N. Kolosov, V. N. Omel'chenko, and V. V. Onoprienko, *Bioorg. Khim., 1*, 176 (1975) [*Chem. Abstr., 83*, 114168 (1975)].

1237.   J. Kreidl, L. Szabo, O. Polyak, and J. Felmeri, *Hung. Teljes* 1158 (1970) [*Chem. Abstr., 74*, 111923 (1971)].

1238.   H. Jahine, H. A. Zaher, A. A. Sayed, and M. Seada, *J. Prakt. Chem., 316*, 337 (1974).

1239.   G. Lamm, *Ger. Pat.* 2,307,445 (1974) [*Chem. Abstr., 81*, 169441 (1974)].

1240.   F. Freeman, D. K. Farquhar, and R. L. Walker, *J. Org. Chem., 33*, 3648 (1968).

1241.   D. R. Buckle, B. C. C. Cantello, and H. Smith, *Ger. Pat.* 2,448,389 (1975) [*Chem. Abstr., 83*, 79100 (1975)].

1242.   P. J. McNulty and H. L. Warner, *U.S. Pat.* 4,028,084 (1977) [*Chem. Abstr., 87*, 117785 (1977)].

1243.   W. O. Johnson, M. C. Seidel, and H. L. Warner, *U.S. Pat.* 4,038,065 (1977) [*Chem. Abstr., 87*, 152034 (1977)].

1244. H. Schaefer and K. Gewald, *J. Prakt. Chem.*, **316**, 684 (1974).

1245. Zh. A. Krasnaya, T. S. Stytsenko, E. P. Prokof'ev, and V. F. Kucherov, *Isv. Akad. Nauk SSSR, Ser. Khim.*, **1973**, 2543 [*Chem. Abstr.*, **80**, 59832 (1974)].

1246. G. Y. Lesher and M. D. Gruett, *Brit. Pat.* 1,322,318 (1973) [*Chem. Abstr.*, **79**, 105231 (1973)].

1247. G. Y. Lesher and M. D. Gruett, *Ger. Pat.* 2,125,310 (1972) [*Chem. Abstr.*, **78**, 58383 (1973)].

1248. G. Y. Lesher and M. D. Gruett, *Fr. Pat.* 2,138,216 (1973) [*Chem. Abstr.*, **79**, 32022 (1973)].

1249. P. Nantka-Namirski and R. Balicki, *Acta Pol. Pharm.*, **29**, 131 (1972) [*Chem. Abstr.*, **77**, 114204 (1972)].

1250. P. Nantka-Namirski and R. Balicki, *Pol. Pat.* 72,750 (1974) [*Chem. Abstr.*, **85**, 21120 (1976)].

1251. P. Doyle and G. J. Stacey, *Brit. Pat.* 1,147,068 (1969) [*Chem. Abstr.*, **71**, 38813 (1969)].

1252. S. Saijo, H. Tobiki, and K. Nishio, *Japan. Pat.*, 68 24,178 (1968) [*Chem. Abstr.*, **70**, 57674 (1969)].

1253. E. Schefczik, *Ger. Pat.* 2,701,659 (1978) [*Chem. Abstr.*, **89**, 215406 (1978)].

1254. F. K. Rafla and M. A. Khan, *J. Chem. Soc. C*, **1971**, 2044.

1255. G. Evans and D. L. Farmery, *Ger. Pat.* 2,835,074 (1979) [*Chem. Abstr.*, **90**, 188521 (1979)].

1256. A. Nohara, T. Ishiguro, and Y. Sanno, *Tetrahedron Lett.*, **1974**, 1183.

1257. A. Rosowsky and N. Papathanasopoulos, *J. Med. Chem.*, **17**, 1272 (1974).

1258. K. H. Schuendehuette and K. L. Moritz, *Ger. Pat.* 2,438,497 (1976) [*Chem. Abstr.*, **84**, 166243 (1976)].

1259. Yu. V. Maksimov and S. M. Kvitko, *Nekotor. Vopr. Khimii Redkozemel'n. Elementov*, **1975**, 52 [*Chem. Abstr.*, **87**, 39242 (1977)].

1260. J. Roch, E. Mueller, B. Narr, J. Nickl, and W. Haarmann, *Ger. Pat.* 2,643,753 (1978) [*Chem. Abstr.*, **89**, 6322 (1978)].

1261. T. Kametani, K. Ogasawara, A. Kozuka, and K. Nyu, *Yakugaku Zasshi*, **87**, 1189 (1967) [*Chem. Abstr.*, **68**, 95651 (1968)].

1262. S. V. Sunthankar and S. D. Vaidya, *Indian J. Chem., Sect. B*, **15B**, 187 (1977).

1263. T. K. Liao, P. J. Wittek, and C. C. Cheng, *J. Heterocycl. Chem.*, **13**, 1283 (1976).

1264. L. Doub and J. S. Kaltenbronn, *U.S. Pat.* 3,948,903 (1976) [*Chem. Abstr.*, **85**, 21348 (1976)].

1265. O. P. Goel, *U.S. Pat.* 3,951,982 (1976) [*Chem. Abstr.*, **85**, 143094 (1976)].

1266. L. Doub and J. S. Kaltenbronn, *U.S. Pat.* 3,873,523 (1975) [*Chem. Abstr.*, **83**, 58807 (1975)].

1267. L. Doub, J. S. Kaltenbronn, and D. Schweiss, *Ger. Pat.* 2,502,119 (1975) [*Chem. Abstr.*, **83**, 164214 (1975)].

1268. A. I. Gurevich, M. N. Kolosov, V. N. Omel'chenko, and V. V. Onoprienko, *Izv. Akad. Nauk SSSR, Ser. Khim.*, **1972**, 1452 [*Chem. Abstr.*, **77**, 126392 (1972)].

1269. G. Y. Lesher and C. J. Opalka, *U.S. Pat.* 4,004,012 (1977) [*Chem. Abstr.*, **86**, 189725 (1977)].

1270. G. Y. Lesher, R. E. Philion, D. F. Page, and C. J. Opalka, *Brit. Pat.* 2,065,642 (1981) [*Chem. Abstr.*, **96**, 181150 (1982)].

1271a. A. Ozola, E. I. Stankevich, S. Kalnins, and O. Neilands, *Khim. Geterotsikl. Soedin.*, **1976**, 256 [*Chem. Abstr.*, **85**, 32782 (1976)].

236   T. D. Bailey, G. L. Goe, and E. F. V. Scriven

1271b. A. Ozola, E. Stankevics, and G. Duburs, *Khim. Geterotsikl. Soedin.*, **1973**, 1147 [*Chem. Abstr.*, **79**, 126256 (1973)].

1272. G. Bouchon, K. H. Spohn, and E. Breitmaier, *Chem. Ber.*, **106**, 1736 (1973).

1273. E. Brietmaier, S. Gassenman, and E. Bayer, *Tetrahedron*, **26**, 5907 (1970).

1274. H. M. Mokhtar and L. Rateb, *Pharmazie*, **33**, 782 (1978).

1275. T. Kato and M. Noda, *Chem. Pharm. Bull.*, **24**, 303 (1976).

1276. T. Kato, H. Yamanaka, and T. Hozumi, *Yakugaku Zasshi*, **91**, 772 (1971) [*Chem. Abstr.*, **75**, 129623 (1971)].

1277. V. Janata and D. Nemcova, *Czech. Pat.* 147,252 (1973) [*Chem. Abstr.*, **78**, 159448 (1973)].

1278. J. Szychowski, J. T. Wrobal, and A. Leniewski, *Bull. Acad. Pol. Sci., Ser. Sci. Chim.*, **22**, 335 (1974) [*Chem. Abstr.*, **82**, 31426 (1975)].

1279. C. Ruangsiyanand, H. J. Rimek, and F. Zymalkowski, *Chem. Ber.*, **103**, 2403 (1970).

1280. H. Oehlschlaeger, O. Riester, and H. Meckl, *Ger. Pat.* 2,001,572 (1971) [*Chem. Abstr.*, **75**, 152990 (1971)].

1281. G. Purrello and A. L. Vullo, *J. Heterocycl. Chem.*, **11**, 481 (1974).

1282. H. Pinhas and S. Beranger, *Fr. Pat.* 2,248,027 (1975) [*Chem. Abstr.*, **84**, 59170 (1976)].

1283. T. Denzel and H. Hoehn, *Ger. Pat.* 2,322,073 (1973) [*Chem. Abstr.*, **80**, 37145 (1974)].

1284. S. Beranger and H. Pinhas, *Ger. Pat.* 2,345,924 (1974) [*Chem. Abstr.*, **81**, 25568 (1974)].

1285. T. Koshigoe and H. Hamano, *Japan. Pat.*, 74 07,289 (1974) [*Chem. Abstr.*, **81**, 13398 (1974)].

1286. T. Kappe, M. A. A. Chirazi, H. P. Stelzel, and E. Ziegler, *Monatsh. Chem.*, **103**, 586 (1972).

1287. F. S. G. Soliman and T. Kappe, *Pharmazie*, **32**, 278 (1977).

1288. T. Kurihara and T. Uno, *Heterocycles*, **6**, 547 (1977).

1289. Zh. A. Krasnaya, T. S. Stytsenko, E. P. Prokof'ev, I. P. Yakovlev, and V. F. Kucherov, *Izv. Akad. Nauk SSSR, Ser. Khim.*, **1974**, 845 [*Chem. Abstr.*, **81**, 49608 (1974)].

1290. C. Jutz, H. G. Loebering, and K. H. Trinkl, *Synthesis*, **1977**, 326.

1291. K. Sato, M. Ohashi, T. Amakasu, and K. Takeda, *Bull. Chem. Soc. Japan*, **42**, 2319 (1969).

1292. I. I. Ibragimov, A. N. Kost, A. G. Aliev, S. P. Godzhaev, R. A. Gadzhily, and L. A. Sviridova, *U.S.S.R. Pat.* 515,746 (1976) [*Chem. Abstr.*, **86**, 16543 (1977)].

1293. L. Rateb, G. A. Mina, and G. Soliman, *J. Chem. Soc. C*, **1968**, 2140.

1294. A. H. Berrie and N. Hughes, *Brit. Pat.* 1,256,340 (1971) [*Chem. Abstr.*, **76**, 101205 (1972)].

1295. P. Pastors, *Biol. Akt. Savienojumu Kim. Tehnol. Rigas Politeh. Inst. 1964–1973*, **1**, 75 (1974) [*Chem. Abstr.*, **85**, 32778 (1976)].

1296. A. D. Yukhnevich and E. Gudriniece, *Latv. PSR Zinat. Akad. Vestis, Kim. Ser.*, **1973**, 694 [*Chem. Abstr.*, **80**, 82596 (1974)].

1297. T. Kato, H. Yamanaka, and T. Hozumi, *Yakugaku Zasshi*, **91**, 740 (1971) [*Chem. Abstr.*, **75**, 129619 (1971)].

1298. L. Kozerski and E. Czerwinska, *Tetrahedron*, **33**, 1365 (1977).

1299a. T. Kato, H. Yamanaka, and T. Shibata, *Chem. Pharm. Bull. (Tokyo)*, **15**, 921 (1967).

1299b. T. Kato, H. Yamanaka, J. Kawamata, and T. Shibata, *Chem. Pharm. Bull. (Tokyo)*, **16**, 1835 (1968).

1300. W. Ehm, *Justus Liebigs Ann. Chem.*, **1977**, 1642.

1301. H. Junek and A. Schmidt, *Monatsh. Chem.*, **98**, 1097 (1967).

1302. E. Poetsch, *Ger. Pat.* 1,809,467 (1970) [*Chem. Abstr.*, **73**, 66443 (1970)].

1303. M. Yokoyama, *Bull. Chem. Soc. Japan*, **44**, 3195 (1971).

1304. R. R. Rastogi, H. Ila, and H. Junjappa, *J. Chem. Soc., Chem. Commun.*, **1975**, 645.

1305. R. R. Rastogi, A. Kumar, H. Ila, and H. Junjappa, *J. Chem. Soc., Perkin Trans 1*, **1978**, 549.

1306. R. R. Rastogi, A. Kumar, I. Ila, and H. Junjappa, *J. Chem. Soc., Perkin Trans. 1*, **1978**, 554.

1307. T. Kato, N. Katagiri, and N. Minami, *Chem. Pharm. Bull.*, **20**, 1368 (1972).

1308. T. Kato, H. Yamanaka, J. Kawamata, and H. Shimomura, *Chem. Pharm. Bull.*, **17**, 1889 (1969).

1309. T. Kato and Y. Kubota, *Yakugaku Zasshi*, **87**, 1219 (1967) [*Chem. Abstr.*, **68**, 87536 (1968)].

1310. T. Kato and Y. Kubota, *Yakugaku Zasshi*, **89**, 1477 (1969) [*Chem. Abstr.*, **72**, 31559 (1970)].

1311. S. Nagai, T. Yorie, Y. Hirota, T. Hibi, K. Sato, H. Yamamura, T. Wada, and I. Aoi, *Japan. Pat.*, 77 66,630 (1977) [*Chem. Abstr.*, **88**, 1599 (1978)].

1312. H. Toda and S. Seto, *Chem. Pharm. Bull.*, **19**, 1477 (1971).

1313. E. Ziegler, I. Herbst, and T. Kappe, *Monatsh. Chem.*, **100**, 132 (1969).

1314. T. Kato, H. Yamanaka, Y. Yamamoto, and T. Sakamoto, *Yakugaku Zasshi*, **90**, 613 (1970) [*Chem. Abstr.*, **73**, 45305 (1970)].

1315. T. Kappe, I. Herbst, and E. Ziegler, *Monatsh. Chem.*, **102**, 848 (1971).

1316. V. Radtke, *Ger. Pat.* 2,705,562 (1978) [*Chem. Abstr.*, **89**, 215429 (1978)].

1317. G. Hoerlein, B. Kuebel, A. Studeneer, and G. Salbeck, *Justus Liebigs Ann. Chem.*, **1979**, 371.

1318. T. Kato and T. Hozumi, *Yakugaku Zasshi*, **93**, 1084 (1973) [*Chem. Abstr.*, **79**, 105033 (1973)].

1319. E. Ziegler and K. Belegratis, *Monatsh. Chem.*, **98**, 219 (1967).

1320. E. Ziegler, F. Hradetzky, and K. Belegratis, *Monatsh. Chem.*, **96**, 1347 (1965).

1321. E. Ziegler and K. Belegratis, *Monatsh. Chem.*, **99**, 995 (1968).

1322. J. A. Elvidge and N. A. Zaidi, *J. Chem. Soc. C*, **1968**, 2188.

1323. T. Stensrud, E. Bernatek, and M. Johnsgaard, *Acta Chem. Scand.*, **25**, 523 (1971).

1324. W. Zankowska-Jasinska and J. Eilmes, *Rocz. Chem.*, **50**, 1059 (1976).

1325. H. G. Viehe, G. J. De Voghel, and F. Smets, *Chimia*, **30**, 189 (1976).

1326. A. L. Cossey, R. L. N. Harris, J. L. Huppatz, and J. N. Phillips, *Angew. Chem., Int. Ed. Engl.*, **11**, 1100 (1972).

1327. A. L. Cossey, R. L. N. Harris, J. L. Huppatz, and J. N. Phillips, *Angew. Chem., Int. Ed. Engl.*, **11**, 1098 (1972).

1328. A. L. Cossey, R. L. N. Harris, J. L. Huppatz, and J. N. Phillips, *Aust. J. Chem.*, **29**, 1039 (1976).

1329. J. Dehnert and W. Juenemann, *Ger. Pat.* 2,349,373 (1975) [*Chem. Abstr.*, **83**, 81208 (1975)].

1330. J. N. Phillips, *Austr. Pat.* 491,554 (1978) [*Chem. Abstr.*, **89**, 109114 (1978)].

1331. J. N. Phillips, *Ger. Pat.* 2,345,801 (1974) [*Chem. Abstr.*, **81**, 3775 (1974)].

1332. A. L. Cossey, R. L. N. Harris, J. L. Huppatz, and J. N. Phillips, *Angew. Chem., Int. Ed. Engl.*, **11**, 1099 (1972).

1333. R. Portmann, *Ger. Pat.* 2,809,629 (1978) [*Chem. Abstr.*, **90**, 38792 (1979)].

1334. W. Eberlein, J. Heider, and H. Machleidt, *Justus Liebigs Ann. Chem.*, **738**, 54 (1970).

1335. E. Breitmaier and E. Bayer, *Angew. Chem., Int. Ed. Engl.*, **8**, 765 (1969).

1336.   E. Breitmaier and E. Bayer, *Tetrahedron Lett.*, **1970**, 3291.

1337.   E. Breitmayer, S. Gassenmann, and E. Bayer, *Tetrahedron*, **26**, 5907 (1970).

1338.   A. C. W. Curran, *Brit. Pat.* 1,432,379 (1976) [*Chem. Abstr.*, **85**, 78021 (1976)].

1339.   A. C. W. Curran, *Brit. Pat.* 1,460,457 (1977) [*Chem. Abstr.*, **87**, 23081 (1977)].

1340.   A. C. W. Curran, *J. Chem. Soc., Perkin Trans. 1*, **1976**, 975.

1341.   R. P. Thummel and D. K. Kohli, *J. Org. Chem.*, **42**, 2742 (1977).

1342.   E. Breitmayer and S. Gassenmann, *Chem. Ber.*, **104**, 665 (1971).

1343.   A. V. Upadysheva, N. D. Grigor'eva, G. S. Sergeeva, and A. P. Znamenskaya, *Zh. Org. Khim.*, **12**, 687 (1976) [*Chem. Abstr.*, **85**, 32786 (1976)].

1344.   A. V. Upadysheva, N. D. Grigor'eva, G. S. Sergeeva, and A. P. Znamenskaya, *U.S.S.R. Pat.* 514,828 (1976) [*Chem. Abstr.*, **85**, 159898 (1976)].

1345.   A. Sakurai and H. Midorikawa, *Bull. Chem. Soc. Japan*, **41**, 165 (1968).

1346.   E. Ziegler, H. Wittmann, and V. Illi, *Monatsh. Chem.*, **100**, 1741 (1969).

1347.   G. B. Bennett, W. R. J. Simpson, R. B. Mason, R. J. Strohschein, and R. Mansukhani, *J. Org. Chem.*, **42**, 221 (1977).

1348.   W. R. Simpson and R. J. Strohschein, *U.S. Pat.* 4,055,649 (1977) [*Chem. Abstr.*, **88**, 22655 (1978)].

1349.   H. Uchida, H. Iwasawa, and M. Ohta, *Bull. Chem. Soc. Japan*, **46**, 3277 (1973).

1350.   S. David and H. Hirshfeld, *J. Chem. Soc. C*, **1969**, 133.

1351.   B. Singh, *U.S. Pat.* 4,347,363 (1982) [*Chem. Abstr.*, **97**, 216018 (1982)].

1352.   H. Meyer and J. Kurz, *Justus Liebigs Ann. Chem.*, **1978**, 1491.

1353.   D. J. Brown, J. C. Gamble, and P. Waring, *Nucleic Acid Chem.*, **1**, 67 (1978) [*Chem. Abstr.*, **89**, 215348 (1978)].

1354.   V. D. Adams and R. C. Anderson, *Synthesis*, **1974**, 286.

1355.   R. E. Smith, G. F. Morris, and C. R. Hauser, *J. Org. Chem.*, **33**, 2562 (1968).

1356.   W. Zankowska-Jasinska and M. Reczynska-Dutka, *Zesz. Nauk. Uniw. Jagiellon., Pr. Chem.*, **21**, 141 (1976) [*Chem. Abstr.*, **88**, 22565 (1978)].

1357.   L. Birkofer and P. Sommer, *Chem. Ber.*, **109**, 1701 (1976).

1358.   R. Balicki and P. Nantka-Namiriski, *Acta Pol. Pharm.*, **32**, 129 (1975).

1359.   H. N. Al Jallo and F. Al Azawi, *J. Heterocycl. Chem.*, **14**, 27 (1977).

1360.   M. A. T. Dubas-Sluyter, W. N. Speckamp, and H. O. Huisman, *Recl. Trav. Chim. Pays-Bas*, **91**, 157 (1972).

1361.   F. Zymalkowski and M. Kothari, *Arch. Pharm. (Weinheim)*, **303**, 667 (1970).

1362.   E. Reimann and R. Reitz, *Justus Liebigs Ann. Chem.*, **1975**, 1081.

1363.   E. Schroeder and H. Koch, *Ger. Pat.* 2,449,030 (1976) [*Chem. Abstr.*, **85**, 32877 (1976)].

1364.   H. Klar and F. Zymalkowski, *Arch. Pharm. (Weinheim, Ger.)*, **307**, 577 (1974).

1365.   E. Reimann and R. Reitz, *Justus Liebigs Ann. Chem.*, **1976**, 610.

1366.   U. Wolf, W. Sucrow, and H. J. Vetter, *Z. Naturforsch., B: Anorg. Chem., Org. Chem.*, **34B**, 102 (1979) [*Chem. Abstr.*, **90**, 168414 (1979)].

1367.   I. El-S. El-Kholy, H. N. A. Al-Jallo, M. Y. Shandala, and F. Al-Hajjar, *Bull. Coll. Sci., Univ. Baghdad*, **11**, 90 (1970) [*Chem. Abstr.*, **74**, 12955 (1971)].

1368.   H. N. Al-Jallo, *J. Chem. Eng. Data*, **17**, 513 (1972).

1369.   M. Y. Shandala and N. H. Al-Jobour, *J. Chem. Eng. Data*, **21**, 120 (1976).

1370.   M. Y. Shandala and N. H. Al-Jobour, *Aust. J. Chem.*, **29**, 1583 (1976).

1371.   H. J. Opgenorth and H. Scheuermann, *Ger. Pat.* 2,719,079 [*Chem. Abstr.*, **90**, 54840 (1979)].

1372. W. Juenemann, H. J. Opgenorth, and H. Scheuermann, *Angew. Chem.,* **92**, 390 (1980).

1373. O. N. Boyarintseva, G. N. Kurilo, O. S. Anisimova, and A. N. Grinev, *Khim. Geterotsikl. Soedin.,* **1977**, 82 [*Chem. Abstr.,* **86**, 189761 (1977)].

1374. W. Zankowska-Jasinska, Z. Kamela, and U. Zieba, *Bull. Acad. Pol. Sci., Ser. Sci. Chim.,* **23**, 901 (1975) [*Chem. Abstr.,* **85**, 21037 (1976)].

1375. J. Kuthan, D. Ilavsky, J. Krechl, and P. Trska, *Tetrahedron Lett.,* **1976**, 4763.

1376. J. Kuthan, D. Ilavsky, J. Krechl, V. Kubelka, and P. Trska, *Collect. Czech. Chem. Commun.,* **43**, 2024 (1978).

1377. R. J. Grout, B. M. Hynam, and M. W. Partridge, *J. Chem. Soc. C,* **1969**, 1590.

1378. H. Schaefer and K. Gewald, *Z. Chem.,* **18**, 335 (1978).

1379. H. Junek and A. Schmidt, *Monatsh. Chem.,* **99**, 635 (1968).

1380. H. Junek and G. Stolz, *Monatsh. Chem.,* **101**, 1234 (1970).

1381. A. N. Kost, R. S. Sagitullin, V. V. Men'shikov, and T. Jagodzinski, *Khim. Dikarbonil'nykh Soedin., Tezisy Dokl. Vses. Konf., 4th 1975,* 78, (1976) [*Chem. Abstr.,* 87, 53118 (1977)].

1382. A. N. Kost, V. V. Men'shikov, and R. S. Sagitullin, *Zh. Org. Khim.,* **12**, 2234 (1976) [*Chem. Abstr.,* **86**, 72487 (1977)].

1383. A. N. Kost, R. S. Sagitullin, T. Yagodzin'ski, and V. V. Men'shikov, *Vestn. Mosk. Univ., Khim.,* **17**, 618 (1976) [*Chem. Abstr.,* **86**, 106436 (1977)].

1384. P. I. Abramenko, *Zh. Vses. Khim. Obshchest.,* **18**, 715 (1973) [*Chem. Abstr.,* **80**, 95788 (1974)].

1385. T. Irikura, *Japan. Pat.,* 77 17,497 (1977) [*Chem. Abstr.,* 87, 84992 (1977)].

1386. H. Dorn and A. Zubek, *Pharmazie,* **26**, 732 (1971).

1387. T. Denzel and H. Hoehn, *Ger. Pat.* 2,453,305 (1975) [*Chem. Abstr.,* 83, 114399 (1975)].

1388. H. Junek and I. Wrtilek, *Monatsh. Chem.,* **100**, 1250 (1969).

1389. A. Camparini, F. Ponticelli, and P. Tedeschi, *J. Heterocycl. Chem.,* **14**, 435 (1977).

1390. M. A. Khan and F. K. Rafla, *J. Chem. Soc., Perkin Trans. 1,* **1975**, 693.

1391. F. Outurquin, G. Ah-Kow, and C. Paulmier, *C.R. Acad. Sci., Ser. C,* **277**, 29 (1973).

1392. C. -C. Cheng and S. -J. Yan, *Org. React.,* **28**, 37 (1982).

1393. R. P. Thummel and D. K. Kohli, *J. Heterocycl. Chem.,* **14**, 685 (1977).

1394. P. Caluwe and T. G. Majewicz, *J. Org. Chem.,* **42**, 3410 (1977).

1395. J. L. Soto, A. Lorente, and L. Fuentes, *An. Quim.,* **74**, 648 (1978) [*Chem. Abstr.,* **90**, 6201 (1979)].

1396. C. N. O'Callaghan, *Proc. R. Ir. Acad., Sect. B,* **77B**, 533 (1977) [*Chem. Abstr.,* **89**, 108949 (1978)].

1397. A. Lorente and J. L. Soto, *Afinidad,* **35**, 138 (1978) [*Chem. Abstr.,* 89, 146729 (1978)].

1398. A. Sammour, Y. Akhnookh, and H. Jahine, *U.A.R. J. Chem. 1970,* **13**, 421 (1971) [*Chem. Abstr.,* 77, 101348 (1972)].

1399. A. Sammour, A. F. Fahmy, S. Abd El-Rahman, Y. Akhnookh, and M. S. Abd Elmoez, *U.A.R. J. Chem.,* **14**, 581 (1971) [*Chem. Abstr.,* **79**, 115413 (1973)].

1400. M. Abdalla, A. Essawy, and A. Deeb, *Pak. J. Sci. Ind. Res.,* **20**, 139 (1977) [*Chem. Abstr.,* **90**, 137633 (1979)].

1401. A. Sakurai and H. Midorikawa, *Bull. Chem. Soc. Japan,* **40**, 1680 (1967).

1402. H. Jahine, H. A. Zaher, A. A. Sayed, and O. Sherif, *Indian J. Chem.,* **11**, 1122 (1973).

1403. A. Sammour, M. I. B. Selim, and M. S. Abd Elhalim, *Egypt. J. Chem.,* **15**, 23 (1972) [*Chem. Abstr.,* **79**, 115259 (1973)].

1404. T. Kato and M. Noda, *Chem. Pharm. Bull.,* **24**, 1408 (1976).

1405. T. Kato, H. Yamanaka, and J. Kawamata, *Chem. Pharm. Bull.*, 17, 2411 (1969).

1406. A. Sakurai and H. Midorikawa, *J. Chem. Soc., Perkin Trans. 1*, 1975, 2025.

1407. M. El-Kerdawy, A. M. Samour, and A. G. El-Agamey, *Acta Pharm. Jugosl.*, 26, 135 (1976) [*Chem. Abstr.*, 85, 46571 (1976)].

1408. H. J. Roth and R. Troschuetz, *Arch. Pharm. (Weinheim, Ger.)*, 310, 48 (1977).

1409. R. Troschuetz and H. J. Roth, *Arch. Pharm. (Weinheim, Ger.)*, 311, 400 (1978).

1410. R. Troschuetz and H. J. Roth, *Arch. Pharm. (Weinheim, Ger.)*, 311, 542 (1978).

1411. M. A. Cabrerizo and J. L. Soto, *An. Quim.*, 72, 926 (1976) [*Chem. Abstr.*, 87, 184337 (1977)].

1412. H. H. Otto, O. Rinus, and H. Schmelz, *Synthesis*, 1978, 681.

1413. S. Kambe, K. Saito, A. Sakurai, and T. Hayashi, *Synthesis*, 1977, 841.

1414. P. M. Carbateas and G. L. Williams, *J. Heterocycl. Chem.*, 11, 819 (1974).

1415. H. H. Zoorob and E. S. Ismail, *Z. Naturforsch., B: Anorg. Chem., Org. Chem.*, 31B, 1680 (1976).

1416. H. H. Zoorob and J. Metri, *Indian J. Chem., Sect. B*, 16B, 53 (1978).

1417. Z. Bomika, J. Pelcers, G. Duburs, and M. B. Andaburskaya, *Khim. Dikarbonil'nykh Soedin., Tezisy Dokl. Vses. Konf., 4th 1975*, 30, (1976) [*Chem. Abstr.*, 87, 53054 (1977)].

1418. A. C. W. Curran, *U.S. Pat.* 3,963,722 (1976) [*Chem. Abstr.*, 86, 5334 (1977)].

1419. A. Sammour, A. Raouf, M. Elkasaby, and M. A. Hassan, *Acta Chim. Acad. Sci. Hung.*, 83, 209 (1974) [*Chem. Abstr.*, 82, 31219 (1975)].

1420. A. Sammour, A. Raouf, M. Elkasaby, and M. Hassan, *J. Prakt. Chem.*, 315, 1175 (1973).

1421. Y. Tamura, T. Sakaguchi, T. Kawasaki, and Y. Kita, *Chem. Pharm. Bull.*, 24, 1160 (1976).

1422. D. V. Sokolov, Zh. I. Isin, and N. Yu. Kim, *U.S.S.R. Pat.* 435,237 (1974) [*Chem. Abstr.*, 81, 120505 (1974)].

1423. L. Fuentes, A. Lorente, and J. L. Soto, *An. Quim.*, 73, 1359 (1977) [*Chem. Abstr.*, 89, 146731 (1978)].

1424. L. Fuentes and J. L. Soto, *An. Quim.*, 73, 1349 (1977) [*Chem. Abstr.*, 89, 129369 (1978)].

1425. A. Sakurai, Y. Motomura, and H. Midorikawa, *Bull. Chem. Soc. Japan*, 46, 973 (1973).

1426. R. M. Carr, R. O. C. Norman, and J. M. Vernon, *J. Chem. Soc., Chem. Commun.*, 1977, 854.

1427. M. Del Carmen, G. Barrio, J. Barrio, G. Walker, A. Novelli, and N. J. Leonard, *J. Am. Chem. Soc.*, 95, 4891 (1973).

1428. C. S. Angadiyavar and R. Srinivasan, *U.S. Publ. Pat. Appl. B* 506,916 (1976) [*Chem. Abstr.*, 84, 128682 (1976)].

1429. H. Junek and H. Aigner, *Z. Naturforsch. B*, 25, 1423 (1970).

1430. K. Nomura, J. Adachi, M. Hanai, S. Nakayama, and K. Mitsuhashi, *Chem. Pharm. Bull.*, 22, 1386 (1974).

1431. R. S. Sagitullin, T. Jagodzinski, A. N. Kost, and V. V. Men'shikov, *Khim. Geterotsikl. Soedin.*, 1978, 925 [*Chem. Abstr.*, 89, 215256 (1978)].

1432. M. A. Akhtar, W. G. Brouwer, J. A. D. Jeffreys, C. W. Gemenden, W. I. Taylor, R. N. Seelye, and D. W. Stanton, *J. Chem. Soc. C*, 1967, 859.

1433. G. Kobayashi, Y. Matsuda, R. Natsuki, and H. Yamaguchi, *Yakugaku Zasshi*, 91, 934 (1971) [*Chem. Abstr.*, 75, 151643 (1971)].

1434. F. Kroehnke, *Synthesis*, 1976, 1.

1435a. A. B. Ash, P. Blumbergs, A. Markovac, and M. P. La Montagne, *U.S. Pat.* 3,763,148 (1973) [*Chem. Abstr., 79*, 146417 (1973)].

1435b. A. B. Ash, M. P. La Montagne, and A. Markovac, *U.S. Pat.* 3,940,404 (1976) [*Chem. Abstr., 84*, 180059 (1976)].

1436. A. B. Ash, P. Blumbergs, A. Markovac, and M. P. La Montagne, *U.S. Pat.* 3,753,997 (1973) [*Chem. Abstr., 79*, 126325 (1973)].

1437. A. B. Ash, C. L. Stevens, and A. Markovac, *U.S. Pat.* 3,764,604 (1973) [*Chem. Abstr., 80*, 47858 (1974)].

1438. A. Markovac, A. R. Patel, M. P. La Montagne, and A. B. Ash, *J. Heterocycl. Chem., 14*, 147 (1977).

1439. C. -H. Wang and J. -M. Horng, *Bull. Inst. Chem., Acad. Sin., 25*, 47 (1978) [*Chem. Abstr., 89*, 197295 (1978)].

1440. C. -H. Wang and J. -M. Horng, *Heterocycles, 7*, 759 (1977).

1441. C. -H. Wang, C. -T. Chen, J. -M. Horng, T. -M. Hwang, J. -I. Lee, Y. -I. Chen, J. -M. Luo, and D. -C. Chen, *Bull. Inst. Chem., Acad. Sin., 25*, 93 (1978) [*Chem. Abstr., 89*, 163367 (1978)].

1442. P. S. Kendurkar and R. S. Tewari, *Z. Naturforsch., Teil B, 29*, 552 (1974).

1443. P. S. Kendurkar and R. S. Tewari, *J. Chem. Eng. Data, 19*, 184 (1974).

1444. R. S. Tewari and K. C. Gupta, *Indian J. Chem., Sect. B, 14B*, 829 (1976).

1445. H. J. Teuber, G. Schuetz, and H. J. Bader, *Justus Liebigs Ann. Chem.*, 1977, 1321.

1446. H. J. Teuber and H. J. Bader, *Justus Liebigs Ann. Chem.*, 1978, 1297.

1447. R. S. Tewari and A. K. Awasthi, *Synthesis*, 1981, 314.

1448. S. Suma, A. Nakamachi, and Y. Sakai, *Japan. Pat.*, 70 39,262 (1970) [*Chem. Abstr., 74*, 87857 (1971)].

1449. Y. Watanabe and T. Matsuzawa, *Japan. Pat.*, 72 42,678 (1972) [*Chem. Abstr., 78*, 84264 (1973)].

1450. H. Ito and S. Nakamura, *Japan. Pat.*, 72 44,746 (1972) [*Chem. Abstr., 78*, 71924 (1973)].

1451. S. Takenaka, T. Matsuzawa, M. Kikuchi, and Y. Watanabe, *Japan. Pat.*, 72 49,592 (1972) [*Chem. Abstr., 78*, 124459 (1973)].

1452. H. Ito and Y. Nakamura, *Japan. Pat.*, 73 03,621 (1973) [*Chem. Abstr., 78*, 147818 (1973)].

1453. H. Ito and Y. Nakamura, *Japan. Pat.*, 73 03,624 (1973) [*Chem. Abstr., 78*, 147817 (1973)].

1454. H. Beschke, F. L. Dahm, H. Friedrich, and G. Schreyer, *Ger. Pat.* 2,554,946 (1977) [*Chem. Abstr., 87*, 135094 (1977)].

1455. H. Beschke and H. Friedrich, *Ger. Pat.* 2,703,049 (1978) [*Chem. Abstr., 89*, 146780 (1978)].

1456. *Neth. Pat.*, 78 00,217 (1978) [*Chem. Abstr., 89*, 179866 (1978)].

1457. *Belg. Pat.* 858,389 (1978) [*Chem. Abstr., 89*, 163412 (1978)].

1458. H. Beschke and H. Friedrich, *Ger. Pat.* 2,712,694 (1978) [*Chem. Abstr., 90*, 72061 (1979)].

1459. H. Beschke and H. Friedrich, *Ger. Pat.* 2,703,069 (1978) [*Chem. Abstr., 89*, 146781 (1978)].

1460. H. Beschke and H. Friedrich, *Ger. Pat.* 2,703,070 (1978) [*Chem. Abstr., 89*, 179865 (1978)].

1461. H. Beschke, H. Friedrich, and H. Offermanns, *Ger. Pat.* 2,639,701 (1978) [*Chem. Abstr., 88*, 169982 (1978)].

242 T. D. Bailey, G. L. Goe, and E. F. V. Scriven

1462. H. Beschke, H. Friedrich, and H. Offermanns, *Ger. Pat.* 2,639,702 (1978) [*Chem. Abstr.*, **88**, 190604 (1978)].

1463. A. Avots, I. Lazdins, M. Sile, and E. Lavrinovics, *U.S.S.R. Pat.* 527,425 (1976) [*Chem. Abstr.*, **86**, 29647 (1977)].

1464. N. S. Prostakov, P. K. Radzhan, and A. T. Soldatenkov, *Khim. Geterotsikl. Soedin.*, **1980**, 1516 [*Chem. Abstr.*, **95**, 24768 (1981)].

1465. H. Beschke and H. Friedrich, *Chem.-Ztg.*, **101**, 377 (1977).

1466. M. Muchlstaedt, H. J. Uebel, and K. K. Moll, *J. Prakt. Chem.*, **312**, 849 (1971).

1467. K. K. Moll and H. J. Uebel, *East Ger. Pat.* 58,960 (1967) [*Chem. Abstr.*, **69**, 106559 (1968)].

1468. Y. Watanabe and S. Takenaka, *Japan. Pat.*, 70 39,545 (1970) [*Chem. Abstr.*, **74**, 141558 (1971)].

1469. Y. Watanabe and S. Takenaka, *Japan. Pat.*, 72 44,743 (1972) [*Chem. Abstr.*, **78**, 71925 (1973)].

1470. H. Beschke, H. Schaefer, G. Schreyer, W. Schuler, and W. Weigert, *Ger. Pat.* 2,224,160 (1973) [*Chem. Abstr.*, **80**, 70713 (1974)].

1471. H. Beschke, H. Schaefer, G. Schreyer, W. A. Schuler, and W. Weigert, *Ger. Pat.* 2,151,417 (1973) [*Chem. Abstr.*, **79**, 31897 (1973)].

1472. N. A. Titova, V. V. Antonova, and B. F. Ustavshchikov, *Org. Khim.*, **1973**, 25 [*Chem. Abstr.*, **81**, 120405 (1974)].

1473. H. Beschke, G. Schreyer, and A. Kleeman, *Ger. Pat.* 2,239,801 (1974) [*Chem. Abstr.*, **80**, 120782 (1974)].

1474. H. Beschke, H. Friedrich, and G. Schreyer, *Ger. Pat.* 2,449,340 (1976) [*Chem. Abstr.*, **85**, 32865 (1976)].

1475. T. Kawahito, K. Koga, and M. Kuniyoshi, *Japan. Pat.*, 77 31,078 (1977) [*Chem. Abstr.*, **87**, 135081 (1977)].

1476. *Japan. Pat.*, 81 26,546 (1981) [*Chem. Abstr.*, **95**, 86970 (1981)].

1477. *Fr. Pat.* 1,563,467 (1969) [*Chem. Abstr.*, **72**, 55274 (1970)].

1478. G. Swift, *Brit. Pat.* 1,158,365 (1969) [*Chem. Abstr.*, **71**, 81189 (1969)].

1479. G. Swift, *Ger. Pat.* 1,917,037 (1969) [*Chem. Abstr.*, **72**, 43479 (1970)].

1480. J. A. Corran, G. Swift, and I. S. McColl, *Brit. Pat.* 1,193,341 (1970) [*Chem. Abstr.*, **73**, 45358 (1970)].

1481. R. H. Benson and P. A. E. Whincup, *Brit. Pat.* 1,208,291 (1970) [*Chem. Abstr.*, **74**, 53550 (1971)].

1482. D. G. Parks-Smith, *Ger. Pat.* 2,023,158 (1970) [*Chem. Abstr.*, **74**, 42284 (1971)].

1483. G. Swift, *Brit. Pat.* 1,217,857 (1970) [*Chem. Abstr.*, **74**, 141539 (1971)].

1484. J. A. Corran, *Brit. Amended Pat.* 1,193,341 (1973) [*Chem. Abstr.*, **79**, 126316 (1973)].

1485. Y. Ichikawa, N. Suzuki, and K. Soma, *Japan. Pat.*, 74 75,586 (1974) [*Chem. Abstr.*, **82**, 156082 (1975)].

1486. *Brit. Pat.* 1,134,163 (1968) [*Chem. Abstr.*, **70**, 68173 (1969)].

1487. J. M. Ross and G. Wilbert, *S. African Pat.*, 68 05,247 (1969) [*Chem. Abstr.*, **71**, 101723 (1969)].

1488. J. M. Ross and G. Wilbert, *S. African Pat.*, 68 05,246 (1969) [*Chem. Abstr.*, **72**, 55268 (1970)].

1489. K. Yamagishi, K. Sakakibara, and M. Kawano, *Japan. Pat.*, 70 26,493 (1970) [*Chem. Abstr.*, **74**, 87844 (1971)].

1490. Y. Ichikawa, N. Suzuki, and K. Soma, *Japan. Pat.*, 74 01,570 (1974) [*Chem. Abstr.*, **80**, 146023 (1974)].

1491. Y. Ichikawa, N. Suzuki, and K. Soma, *Japan. Pat.*, 74 01,569 (1974) [*Chem. Abstr.*, 80, 146025 (1974)].

1492. Y. Ichikawa, N. Suzuki, and K. Soma, *Japan. Pat.*, 74 75,585 (1974) [*Chem. Abstr.*, 82, 156083 (1975)].

1493. Y. Ichikawa, Y. Honda, K. Soma, N. Suzuki, and T. Yamaji, *Ger. Pat.* 2,401,103 (1975) [*Chem. Abstr.*, 83, 114224 (1975)].

1494. Y. Ichikawa, N. Suzuki, and K. Soma, *Japan. Pat.*, 73 99,177 (1973) [*Chem. Abstr.*, 80, 95744 (1974)].

1495. L. Forni and M. Stanga, *J. Catal.*, 59, 148 (1979).

1496. L. Forni, M. Tescari, and P. Zambelli, *J. Catal.*, 65, 470 (1980).

1497. L. Forni and G. Gianetti, *J. Catal.*, 75, 375 (1982).

1498. G. Wilbert, *Fr. Pat.* 1,452,433 (1966) [*Chem. Abstr.*, 66, 104910 (1967)].

1499. *Brit. Pat.* 1,192,255 (1970) [*Chem. Abstr.*, 73, 45362 (1970)].

1500. S. Numa, A. Nakamachi, and K. Murakami, *Japan. Pat.*, 71 08,853 (1971) [*Chem. Abstr.*, 75, 35778 (1971)].

1501. H. J. Uebel, M. Muehlstaedt, and K. Moll, *Chem. Tech. (Leipzig)*, 22, 679 (1970).

1502. A. Nicolson, *Brit. Pat.* 1,240,928 (1971) [*Chem. Abstr.*, 75, 98455 (1971)].

1503. E. Catalucci, *Ger. Pat.* 1,965,010 (1970) [*Chem. Abstr.*, 74, 3524 (1971)].

1504. M. Kubo and K. Yasuda, *Japan. Pat.*, 75 111,078 (1975) [*Chem. Abstr.*, 84, 121655 (1976)].

1505. R. Hamilton, M. A. McKervey, and J. J. Rooney, *J. Chem. Soc., Chem. Commun.*, 1976, 1038.

1506. D. Dieterich, H. Reiff, H. Ziemann, and R. Braden, *Justus Liebigs Ann. Chem.*, 1973, 111.

1507. D. Dieterich and R. Braden, *Brit. Pat.* 1,208,707 (1970) [*Chem. Abstr.*, 74, 53552 (1971)].

1508. D. Dieterich and R. Braden, *S. African Pat.*, 70 01,082 (1970) [*Chem. Abstr.*, 75, 5716 (1971)].

1509. H. Reiff, D. Dieterich, and H. Ziemann, *Ger. Pat.* 2,045,884 (1972) [*Chem. Abstr.*, 77, 48267 (1972)].

1510. *Fr. Pat.* 2,001,070 (1969) [*Chem. Abstr.*, 72, 66836 (1970)].

1511. D. Dieterich, H. Reiff, and H. Ziemann, *Ger. Pat.* 2,045,885 (1972) [*Chem. Abstr.*, 77, 48268 (1972)].

1512. H. Reiff, D. Dieterich, and H. Ziemann, *Ger. Pat.* 2,045,880 (1972) [*Chem. Abstr.*, 77, 48269 (1972)].

1513. D. Dieterich and H. Ziemann, *Ger. Pat.* 2,045,886 (1972) [*Chem. Abstr.*, 77, 48276 (1972)].

1514a. S. Danishefsky, S. J. Etheredge, R. Volkmann, J. Eggler, and J. Quick, *J. Am. Chem. Soc.*, 93, 5575 (1971).

1514b. R. Volkmann, S. Danishefsky, J. Eggler, and D. M. Solomon, *J. Am. Chem. Soc.*, 93, 5576 (1971).

1515. J. Quick, *Tetrahedron Lett.*, 1977, 327.

1516. W. Ried and F. Baetz, *Justus Liebigs Ann. Chem.*, 725, 230 (1969).

1517. J. Barluenga, S. Fustero, and V. Gotor, *Synthesis*, 1975, 191.

1518. F. Duran and L. Ghosez, *Tetrahedron Lett.*, 1970, 245.

1519. Yu. A. Zaichenko, I. A. Maretina, and A. A. Petrov, *Zh. Org. Khim.*, 11, 708 (1975) [*Chem. Abstr.*, 83, 28062 (1975)].

1520. M. A. Kirillova, Yu. A. Zaichenko, I. A. Maretina, and A. A. Petrov, *Zh. Org. Khim.*, 8, 1575 (1972) [*Chem. Abstr.*, 77, 139742 (1972)].

1521.   F. Outurquin, C. Paulmier, J. Morel, and P. Pastour, *C.R. Acad. Sci., Ser. C,* **274,** 1696 (1972).

1522.   R. Gompper and R. Sobotta, *Angew. Chem.,* **90,** 808 (1978).

1523.   A. Kreutzberger and D. Abel, *Arch. Pharm. (Weinheim),* **302,** 362 (1969).

1524.   H. Boennemann, *Angew. Chem.,* **90,** 517 (1978).

1525.   R. A. Clement, *U.S. Pat.* 3,829,429 (1974) [*Chem. Abstr.,* **81,** 135962 (1974)].

1526.   Y. Wakatsuki and H. Yamazaki, *Tetrahedron Lett.,* **1973,** 3383.

1527.   K. P. C. Vollhardt and R. G. Bergman, *J. Am. Chem. Soc.,* **96,** 4996 (1974).

1528.   H. Boennemann, R. Brinkmann, and H. Schenkluhn, *Synthesis,* **1974,** 575.

1529.   H. Boennemann and M. Samson, *U.S. Pat.* 4,266,061 (1981) [*Chem. Abstr.,* **93,** 95136 (1980)].

1530.   Y. Wakatsuki and H. Yamazaki, *Japan. Pat.,* 74 126,680 (1974) [*Chem. Abstr.,* **82,** 170714 (1975)].

1531.   Y. Wakatsuki and H. Yamazaki, *Japan. Pat.,* 77 25,780 (1977) [*Chem. Abstr.,* **87,** 68168 (1977)].

1532.   H. Boennemann and R. Brinkmann, *Synthesis,* **1975,** 600.

1533.   H. Boennemann and H. Schenkluhn, *Ger. Pat.* 416,295 (1975) [*Chem. Abstr.,* **84,** 30909 (1976)].

1534.   Y. Wakatsuki and H. Yamazaki, *Synthesis,* **1976,** 26.

1535.   P. Hardt, *Ger. Pat.* 2,615,309 (1977) [*Chem. Abstr.,* **87,** 23068 (1977)].

1536.   A. Naiman and K. P. C. Vollhardt, *Angew. Chem.,* **89,** 758 (1977).

1537.   K. P. C. Vollhardt and A. U. Naiman, *U.S. Pat. Appl.* 886,119 (1978) [*Chem. Abstr.,* **90,** 186806 (1979)].

1538.   P. Hardt, *Ger. Pat.* 2,742,541 (1978) [*Chem. Abstr.,* **89,** 24163 (1978)].

1539.   P. Hardt, *U.S. Pat.* 4,212,978 (1980) [*Chem. Abstr.,* **87,** 23068 (1977)].

1540.   *Belg. Pat.* 859,768 (1978) [*Chem. Abstr.,* **90,** 22819 (1979)].

1541.   P. Hardt, *U.S. Pat.* 4,196,287 (1980) [*Chem. Abstr.,* **93,** 114337 (1980)].

1542.   P. Hong and H. Yamazaki, *Tetrahedron Lett.,* **1977,** 1333.

1543.   P. Hong and H. Yamazaki, *Nippon Kagaku Kaishi,* **1978,** 730 [*Chem. Abstr.,* **89,** 108980 (1978)].

1544.   K. Ko and H. Yamazaki, *Japan. Pat.,* 78 50,179 (1978) [*Chem. Abstr.,* **89,** 109129 (1978)].

1545.   I. Lazdins and A. Avots, *Khim. Geterotsikl. Soedin.,* **1979,** 1011 [*Chem. Abstr.,* **91,** 175087 (1979)].

1546.   I. Lazdins, M. Sile, M. V. Shimanskaya, and A. Avots, *Katalitich. Sintez i Prevrashcheniya Geterotsikl. Soedin. Geterogen. Katiliz.,* **1976,** 93 [*Chem. Abstr.,* **87,** 134958 (1977)].

1547.   A. P. Ivanovskii, V. A. Shikhanov, A. M. Kut'in, and M. A. Korshunov, *Osnovn. Org. Sint. Neftekhim.,* **2,** 64 (1975) [*Chem. Abstr.,* **83,** 178757 (1975)].

1548.   L. G. Sednevets, L. I. Doktorova, A. P. Ivanovskii, V. A. Shikhanov, A. G. Pankov, A. F. Moskvin, and O. P. Yablonskii, *Prom-st' Sintetich. Kauchuka. Ref. Sb.,* **1975,** 1 [*Chem. Abstr.,* **85,** 108499 (1976)].

1549.   A. P. Ivanovskii, V. A. Shikhanov, A. M. Kut'in, and M. A. Korshunov, *Org. Khim.,* **1973,** 42 [*Chem. Abstr.,* **81,** 135911 (1974)].

1550.   A. P. Ivanovskii, V. A. Shikhanov, A. M. Kut'in, and M. A. Korshunov, *Sb. nauch. tr NII monomerov dlya sintetich. kauchuka,* **1973,** 266 [*Chem. Abstr.,* **83,** 28064 (1975)].

1551.   K. K. Moll, H. Baltz, K. Peltzing, H. Barsch, and H. J. Uebel, *East Ger. Pat.* 61,545 (1968) [*Chem. Abstr.,* **70,** 47314 (1969)].

1552. *Fr. Pat.* 1,533,498 (1968) [*Chem. Abstr.*, **71**, 112824 (1969)].

1553. J. Herzenberg, G. Boccato, M. Pieroni, and V. Fattore, *Ger. Pat.* 1,695,434 (1973) [*Chem. Abstr.*, **79**, 5266 (1973)].

1554. *Fr. Pat.* 1,529,418 (1968) [*Chem. Abstr.*, **71**, 61235 (1969)].

1555. Y. Watanabe, S. Takenaka, and K. Koyasu, *Japan. Pat.*, 71 39,873 (1971) [*Chem. Abstr.*, **76**, 34112 (1972)].

1556. M. Pieroni, *Ger. Pat.* 1,695,437 (1972) [*Chem. Abstr.*, **77**, 114252 (1972)].

1557. A. B. Baylis, *U.S. Pat.* 3,970,655 (1976) [*Chem. Abstr.*, **86**, 43568 (1977)].

1558. K. Nakajima and T. Sato, *Japan. Pat.*, 74 36,680 (1974) [*Chem. Abstr.*, **81**, 37482 (1974)].

1559. C. W. Hargis, *U.S. Pat.* 3,829,428 (1974) [*Chem. Abstr.*, **81**, 135963 (1974)].

1560. M. Levi, K. Markov, A. Shcherev, Ch. Ivanov, P. Mileva, M. Dryanska, and O. Zabunova, *Tr. Nauchnoizsled. Khim.-Farm. Inst.*, **9**, 173 (1974) [*Chem. Abstr.*, **83**, 58617 (1975)].

1561. T. Miyake, K. Noguchi, and K. Imamura, *Ger. Pat.* 2,746,177 (1978) [*Chem. Abstr.*, **89**, 43139 (1978)].

1562. A. P. Ivanovskii, V. A. Shikhanov, A. M. Kut'in, and M. A. Korshunov, *Neftepererab. Neftekhim. (Moscow)*, **1978**, 33 [*Chem. Abstr.*, **89**, 108996 (1978)].

1563. Y. Watanabe and S. Takenaka, *Japan. Pat.*, 72 44,745 (1972) [*Chem. Abstr.*, **78**, 71926 (1973)].

1564. T. Miyake, K. Noguchi, and K. Imamura, *Japan. Pat.*, 78 53,659 (1978) [*Chem. Abstr.*, **89**, 129421 (1978)].

1565. C. D. Chang and W. H. Lang, *U.S. Pat.* 4,220,783 (1980) [*Chem. Abstr.*, **93**, 22067 (1980)].

1566. *Brit. Pat.* 1,216,866 (1970) [*Chem. Abstr.*, **74**, 141559 (1971)].

1567. Y. Minato and S. Yasuda, *Japan. Pat.*, 71 09,463 (1971) [*Chem. Abstr.*, **75**, 20211 (1971)].

1568. *Brit. Amended Pat.* 1,216,866 (1970) [*Chem. Abstr.*, **77**, 151962 (1972)].

1569. Y. Minato, S. Yasuda, and T. Yoshida, *Japan Pat.*, 73 72,172 (1973) [*Chem. Abstr.*, **80**, 27108 (1974)].

1570. Y. Minato and S. Yasuda, *U. S. Pat.* 3,932,421 (1976) [*Chem. Abstr.*, **85**, 46402 (1976)].

1571. L. -H. Li, B. -A. Zhang, B. -J. Li, S. -J. Weng, and D. -Q. Wang, *Ta-lien Kung Hsueh Yuan Hsueh Pao*, **1979**, 38 [*Chem. Abstr.*, **92**, 180959 (1980)].

1572. K. Nakajima and T. Sato, *Japan. Pat.*, 74 61,174 (1974) [*Chem. Abstr.*, **81**, 120480 (1974)].

1573. A. Rigo and G. Boccato, *Chim. Ind. (Milan)*, **52**, 653 (1970) [*Chem. Abstr.*, **73**, 81080 (1970)].

1574. G. Boccato and A. Rigo, *Ital. Pat.* 851,021 (1969) [*Chem. Abstr.*, **75**, 110197 (1971)].

1575. G. Boccato and A. Rigo, *Ger. Pat.* 2,017,480 (1971) [*Chem. Abstr.*, **76**, 34117 (1972)].

1576. G. Ferraiolo, F. Donetti, and A. Peloso, *Ann. Chim. (Rome)*, **57**, 250 (1967) [*Chem. Abstr.*, **67**, 94345 (1967)].

1577. A. Reverberi, G. Ferraiolo, and D. Beruto, *Ann. Chim. (Rome)*, **58**, 68 (1968) [*Chem. Abstr.*, **68**, 117457 (1968)].

1578. G. Fabbri and G. Campazzi, *Chim. Ind. (Milan)*, **49**, 458 (1967) [*Chem. Abstr.*, **67**, 90627 (1967)].

1579. E. Just, K. Pelzing, and K. K. Moll, *Chem. Tech. (Leipzig)*, **22**, 210 (1970).

1580. E. V. Lukin, A. P. Musakin, G. N. Nikandrov, and G. N. Semanov, *Zh. Prikl. Khim. (Leningrad)*, **42**, 2109 (1969) [*Chem. Abstr.*, **72**, 12504 (1970)].

246 T. D. Bailey, G. L. Goe, and E. F. V. Scriven

1581. A. Abdurakhmanov, K. M. Akhmerov, and A. B. Kuchkarov, *Uzb. Khim. Zh.*, **18**, 65 (1974) [*Chem. Abstr.*, **80**, 145970 (1974)].

1582. K. M. Akhmerov, A. Abdurakhmanov, and A. B. Kuchkarov, *Uzb. Khim. Zh.*, **18**, 53 (1974) [*Chem. Abstr.*, **81**, 37462 (1974)].

1583. K. M. Akhmerov, D. Yusupov, A. Abdurakhmanov, and A. B. Kuchkarov, *Khim. Geterotsikl. Soedin.*, **1975**, 221 [*Chem. Abstr.*, **82**, 170622 (1975)].

1584. D. Yusupov, K. M. Akhmerov, A. Abdurakhmanov, and A. B. Kuchkarov, *Tr. Tashk. Politekh. Inst.*, **107**, 15 (1973) [*Chem. Abstr.*, **83**, 9723 (1975)].

1585. J. G. Crist and J. O. Hawthorne, *U.S. Pat.* 3,575,986 (1971) [*Chem. Abstr.*, **75**, 35788 (1971)].

1586. A. P. Ivanovskii, V. A. Shikhanov, A. M. Kut'in, and M. A. Korshunov, *Khim. Prom-st., Ser.: Reakt. Osobo Chist. Veshchestva*, **1980**, 33 [*Chem. Abstr.*, **93**, 239168 (1980)].

1587. M. G. Ismatullaeva, K. M. Akhmerov, D. Yusupov, and A. Kuchkarov, *Uzb. Khim. Zh.*, **1979**, 59 [*Chem. Abstr.*, **92**, 58571 (1980)].

1588. A. Nenz and M. Pieroni, *Hydrocarbon Process.*, **47**, 139 (1968).

1589. A. G. Pankov, L. G. Srednevets, L. I. Doktorova, A. F. Moskin, and O. P. Yablonskii, *Prom. Sint. Kauch., Nauchno-Tekh. Sb.*, **1973** (2), 8 [*Chem. Abstr.*, **81**, 135898 (1974)].

1590. J. Vahldieck and G. Buchmann, *Pharmazie*, **24**, 196 (1969).

1591. Y. Kusunoki, H. Okazaki, and E. Sono, *Japan. Pat.*, 71 39,870 (1971) [*Chem. Abstr.*, **76**, 34113 (1972)].

1592. Y. Kusunoki and H. Okazaki, *Japan. Pat.*, 71 39,871 (1971) [*Chem. Abstr.*, **76**, 34109 (1972)].

1593. H. Okazaki, *Japan. Pat.*, 71 41,909 (1971) [*Chem. Abstr.*, **76**, 46091 (1972)].

1594. *Fr. Pat.* 2,161,128 (1973) [*Chem. Abstr.*, **80**, 14850 (1974)].

1595. Y. Kusunoki and H. Okazaki, *Japan. Pat.*, 73 44,262 (1973) [*Chem. Abstr.*, **79**, 66188 (1973)].

1596. Y. Kusunoki and H. Okazaki, *Japan. Pat.*, 73 44,261 (1973) [*Chem. Abstr.*, **79**, 66190 (1973)].

1597. Y. Kusunoki, H. Okazaki, and E. Sano, *Ger. Pat.* 2,155,896 (1973) [*Chem. Abstr.*, **79**, 31895 (1973)].

1598. Y. Kusunoki and H. Okazaki, *Japan. Pat.*, 73 54,074 (1973) [*Chem. Abstr.*, **80**, 27115 (1974)].

1599. R. Dinkel, *Eur. Pat. Appl. EP* 40,698 (1981) [*Chem. Abstr.*, **96**, 104105 (1982)].

1600. H. B. Charman, *Brit. Pat.* 1,208,569 (1970) [*Chem. Abstr.*, **74**, 76335 (1971)].

1601. H. B. Charman and J. M. Rowe, *J. Chem. Soc. D*, **1971**, 476.

1602. H. Okazaki, *Japan. Pat.*, 74 31,671 (1974) [*Chem. Abstr.*, **81**, 91358 (1974)].

1603. C. B. Ross, J. A. Wantuck, and A. Kaufman, *Ger. Pat.* 2,337,087 (1974) [*Chem. Abstr.*, **80**, 120778 (1974)].

1604. Y. Kusunoki and H. Okazaki, *Japan. Pat.*, 72 02,094 (1972) [*Chem. Abstr.*, **76**, 126802 (1972)].

1605. Y. Kusonoki, H. Okazaki, Y. Sato, and E. Sano, *Ger. Pat.* 1,919,810 (1969) [*Chem. Abstr.*, **72**, 55275 (1970)].

1606. Y. Kusunoki, H. Okazaki, Y. Sato, and E. Sano, *Japan. Pat.*, 71 09,586 (1971) [*Chem. Abstr.*, **75**, 20213 (1971)].

1607. Y. Kusunoki, H. Okazaki, and E. Sano, *Japan. Pat.*, 71 39,869 (1971) [*Chem. Abstr.*, **76**, 34110 (1972)].

1608. Y. Kusunoki, H. Okazaki, and E. Sano, *Japan. Pat.*, 71 39,689 (1971) [*Chem. Abstr.*, **76**, 34121 (1972)].

1609. Y. Kusunoki and H. Okazaki, *Japan. Pat.,* 72 22,580 (1972) [*Chem. Abstr.,* 77, 88321 (1972)].

1610. Y. Kusunoki and H. Okazaki, *Japan. Pat.,* 72 44,744 (1972) [*Chem. Abstr.,* 77, 71923 (1972)].

1611. Y. Kusunoki and H. Okazaki, *Japan. Pat.,* 74 76,872 (1974) [*Chem. Abstr.,* 82, 57576 (1975)].

1612. Y. Kusunoki and H. Okazaki, *Hydrocarbon Process.,* 53, 129 (1974).

1613. Y. Kusunoki and H. Okazaki, *Japan. Pat.,* 75 05,384 (1975) [*Chem. Abstr.,* 83, 28115 (1975)].

1614. N. Takahashi, *Chem. Econ. Eng. Rev.,* 7, 34 (1975).

1615. Y. Kusunoki and H. Okazaki, *Japan. Pat.,* 76 75,076 (1976) [*Chem. Abstr.,* 86, 72445 (1977)].

1616. Y. Kusunoki and H. Okazaki, *Nippon Kagaku Kaishi,* 1979, 1520 [*Chem. Abstr.,* 92, 163814 (1980)].

1617. Y. Kusonoki and H. Okazaki, *Japan. Pat.,* 75 101,362 (1975) [*Chem. Abstr.,* 84, 59214 (1976)].

1618. Y. Kusunoki and H. Okazaki, *Japan. Pat.,* 75 101,363 (1975) [*Chem. Abstr.,* 84, 59213 (1976)].

1619. Y. Kusunoki and H. Okazaki, *Nippon Kagaku Kaishi,* 1980, 1734 [*Chem. Abstr.,* 95, 97525 (1981)].

1620. D. M. Fenton, *U.S. Pat.* 3,679,688 (1972) [*Chem. Abstr.,* 77, 101388 (1972)].

1621. D. M. Fenton, *U.S. Pat.* 3,679,689 (1972) [*Chem. Abstr.,* 77, 101389 (1972)].

1622. E. H. Homeier and T. Imai, *U.S. Pat.* 4,204,066 (1980) [*Chem. Abstr.,* 93, 168142 (1980)].

1623. A. G. Mohan, *J. Org. Chem.,* 35, 3982 (1970).

1624. B. A. M. Oude-Alink and N. E. S. Thompson, *U.S. Pat.* 4,022,785 (1977) [*Chem. Abstr.,* 87, 102174 (1977)].

1625. D. Craig, L. Schaefgen, and W. P. Tyler, *J. Am. Chem. Soc.,* 70, 1624 (1948).

1626. G. Krow, E. Michener, and K. C. Ramey, *Tetrahedron Lett.,* 1971, 3653.

1627. D. R. Eckroth, *Chem. Ind. (London),* 1967, 920.

1628. N. A. Titova, V. V. Antonova, B. F. Ustavschikov, and E. V. Olemskaya, *Osnovn. Org. Sint. Neftekhim.,* 2, 60 (1975) [*Chem. Abstr.,* 83, 177752 (1975)].

1629. A. P. Ivanovskii, V. A. Shikhanov, A. M. Kut'in, and M. A. Korshunov, *Sb. Nauch. Tr. VNII Khim. Sredstv Zashchity Rast.,* 1976, 88 [*Chem. Abstr.,* 89, 162671 (1978)].

1630. A. P. Ivanovskii, V. A. Shikhanov, A. M. Kut'in, and M. A. Korshunov, *Khim. Sredstva Zashch. Rast.,* 7, 88 (1976) [*Chem. Abstr.,* 91, 90803 (1979)].

1631. V. V. Vetrova, N. A. Titova, and B. F. Ustavshchikov, *Zh. Prikl. Khim. (Leningrad),* 46, 2735 (1973) [*Chem. Abstr.,* 80, 69971 (1974)].

1632. H. J. Uebel, K. K. Moll, and M. Muehlstaedt, *Chem. Tech. (Leipzig),* 22, 745 (1970).

1633. N. A. Titova, G. N. Abaev, V. V. Vetrova, B. F. Ustavshchikov, and G. K. Denisova, *Zh. Prikl. Khim. (Leningrad),* 46, 1566 (1973) [*Chem. Abstr.,* 79, 105047 (1973)].

1634. St. Malinowski and W. J. Palion, *React. Kinet. Catal. Lett.,* 1, 73 (1974).

1635. V. V. Antonova, T. I. Ovchinnikova, B. F. Ustavshchikov, and V. K. Promonenkov, *Zh. Org. Khim.,* 16, 547 (1980) [*Chem. Abstr.,* 93, 45503 (1980)].

1636. F. E. Cislak and W. R. Wheeler, *Ger. Pat.* 1,255,661 (1967) [*Chem. Abstr.,* 68, 105012 (1968)].

1637. *Brit. Pat.* 1,182,705 (1970) [*Chem. Abstr.,* 72, 132552 (1970)].

1638. H. Terada and M. Nishi, *Japan. Pat.,* 69 32,789 (1969) [*Chem. Abstr.,* 72, 78891 (1970)].

248   T. D. Bailey, G. L. Goe, and E. F. V. Scriven

1639. A. P. Ivanovskii, V. A. Shikhanov, A. M. Kut'in, and M. A. Korshunov, *Khim. Prom. (Moscow)*, **48**, 26 (1972) [*Chem. Abstr.*, **76**, 99465 (1972)].

1640. N. A. Titova, V. V. Antonova, G. K. Denisova, and B. F. Ustavshchikov, *Org. Khim.*, **1973**, 20 [*Chem. Abstr.*, **81**, 120406 (1974)].

1641. V. V. Antonova, T. I. Smirnova, N. A. Titova, K. P. Bespalov, and B. F. Ustavshchikov, *Zh. Prikl. Khim. (Leningrad)*, **50**, 382 (1977) [*Chem. Abstr.*, **86**, 189665 (1977)].

1642. A. P. Ivanovskii, V. A. Shikhanov, A. M. Kut'in, and M. A. Korshunov, *Katal. Sint. Prevrashch. Geterotsikl. Soedin.*, **1976**, 86 [*Chem. Abstr.*, **88**, 22559 (1978)].

1643. K. Becker, H. D. Berrouschot, K. K. Moll, H. J. Uebel, and M. Weber, *East Ger. Pat.* 130,784 (1978) [*Chem. Abstr.*, **90**, 87297 (1979)].

1644. J. I. Darragh, *U.S. Pat.* 4,089,863 (1978) [*Chem. Abstr.*, **91**, 5117 (1979)].

1645. K. P. Bespalov, V. V. Antonova, T. I. Smirnova, N. A. Titova, and B. F. Ustavshchikov, *Deposited Doc. 1977, VINITI 4412-77*, 110 [*Chem. Abstr.*, **91**, 140691 (1979)].

1646. H. Baltz, K. K. Moll, H. J. Uebel, and M. Muehlstadt, *East Ger. Pat.* 71,550 (1970) [*Chem. Abstr.*, **73**, 109701 (1970)].

1647. Y. Minato and S. Nishikawa, *Ger. Pat.* 2,163,524 (1972) [*Chem. Abstr.*, **77**, 139825 (1972)].

1648. Y. Minato and S. Nishikawa, *Japan. Pat.*, 76 44,946 (1976) [*Chem. Abstr.*, **86**, 189727 (1977)].

1649. A. Tamono and M. Doya, *Brit. Pat.* 1,141,526 (1969) [*Chem. Abstr.*, **70**, 106396 (1969)].

1650. A. Tamono and M. Doya, *Brit. Amended Pat.* 1,141,526 (1970) [*Chem. Abstr.*, **75**, 140708 (1971)].

1651. Y. Minato and S. Yasuda, *Japan. Pat.*, 69 32,790 (1969) [*Chem. Abstr.*, **72**, 78892 (1970)].

1652. H. Ito and S. Nakamura, *Japan. Pat.*, 72 20,167 (1972) [*Chem. Abstr.*, **77**, 164521 (1972)].

1653. K. Nakajima and T. Sato, *Japan. Pat.*, 73 86,874 (1973) [*Chem. Abstr.*, **80**, 120785 (1974)].

1654. Y. Wada, S. Yasuda, T. Niwa, Y. Tsuruta, and T. Tagano, *Japan. Pat.*, 76 63,176 (1976) [*Chem. Abstr.*, **85**, 192569 (1976)].

1655. *Japan. Pat.*, 80 151,559 (1980) [*Chem. Abstr.*, **94**, 121346 (1981)].

1656. S. Cane and L. E. Cooper, *Fr. Pat.* 2,000,863 (1969) [*Chem. Abstr.*, **72**, 55267 (1970)].

1657. S. Cane and L. E. Cooper, *Ger. Pat.* 1,903,878 (1969) [*Chem. Abstr.*, **72**, 3386 (1970)].

1658. Y. Minato and T. Niwa, *Ger. Pat.* 2,054,773 (1971) [*Chem. Abstr.*, **75**, 35783 (1971)].

1659. Y. Minato and T. Niwa, *Japan. Pat.*, 71 41,546 (1971) [*Chem. Abstr.*, **76**, 46090 (1972)].

1660. A. Tamano and M. Doya, *Japan. Pat.*, 73 22,705 (1973) [*Chem. Abstr.*, **79**, 146405 (1973)].

1661. G. Grigoleit, R. Oberkobusch, G. Collin, and K. Matern, *Ger. Pat.* 2,203,384 (1973) [*Chem. Abstr.*, **79**, 136985 (1973)].

1662. K. Nakajima and T. Sato, *Japan. Pat.*, 74 01,568 (1974) [*Chem. Abstr.*, **80**, 146024 (1974)].

1663. O. S. Otroshchenko, A. A. Ziyaev, and A. S. Sadykov, *Khim. Geterotsikl. Soedin.*, **1969**, 365 [*Chem. Abstr.*, **71**, 22001 (1969)].

1664. I. Lazdins and A. Avots, *Latv. PSR Zinat. Akad. Vestis, Kim. Ser.*, **1974**, 427 [*Chem. Abstr.*, **81**, 169409 (1974)].

1665. I. Lazdins and A. Avots, *Tezisy Dokl.-Simp. Khim. Tekhnol. Geterosikl. Soedin. Goryuch. Iskop., 2nd*, **1973**, 58 [*Chem. Abstr.*, **85**, 177216 (1976)].

1666. A. Avots, E. Lavrinovics, I. Lazdins, and M. Sile, *U.S.S.R. Pat.* 512,209 (1970) [*Chem. Abstr.*, **85**, 108537 (1976)].

1667. N. S. Prostakov, A. T. Soldatenkov, and V. O. Fedorov, *Zh. Org. Khim.*, **15**, 1109 (1979) [*Chem. Abstr.*, **91**, 56789 (1979)].

1668. A. T. Soldatenkov, V. O. Fedorov, R. Chandra, V. M. Polosin, A. I. Mikaya, and N. S. Prostakov, *Zh. Org. Khim.*, **16**, 188 (1980) [*Chem. Abstr.*, **92**, 215230 (1980)].

1669. N. S. Prostakov, P. K. Radzhan, A. T. Soldatenkov, and A. I. Mikaya, *Khim. Geterotsikl. Soedin.*, **1981**, 383 [*Chem. Abstr.*, **95**, 80663 (1981)].

1670. *Belg. Pat.* 858,390 (1978) [*Chem. Abstr.*, **89**, 163413 (1978)].

1671. H. Baumann and Oberlinner, *Ger. Pat.* 2,227,597 (1974) [*Chem. Abstr.*, **81**, 8432 (1974)].

1672. S. Yasuda and N. Asegawa, *Japan. Pat.*, **74**, 94,679 (1974) [*Chem. Abstr.*, **82**, 72793 (1975)].

1673. V. V. Antonova, N. A. Titova, V. K. Promonenkov, and B. F. Ustavshchikov, *U.S.S.R. Pat.* 652,176 (1979) [*Chem. Abstr.*, **90**, 203890 (1979)].

1674. H. Chafetz and R. C. Anderson, *U.S. Pat.* 3,349,092 (1967) [*Chem. Abstr.*, **68**, 105026 (1968)].

1675. *Belg. Pat.* 858,391 (1978) [*Chem. Abstr.*, **89**, 163427 (1978)].

1676. Y. Tanba, *Japan. Pat.*, 77 00,034 (1977) [*Chem. Abstr.*, **86**, 189736 (1977)].

1677. Y. Minato, Y. Wada, S. Yasuda, T. Niwa, and S. Nishikawa, *Ger. Pat.* 2,064,397 (1971) [*Chem. Abstr.*, **75**, 118236 (1971)].

1678. S. Aoshima, K. Ueda, and S. Miyagawa, *Japan. Pat.*, 72 46,070 (1972) [*Chem. Abstr.*, **78**, 97495 (1973)].

1679. H. Ito and S. Nakamura, *Japan. Pat.*, 72 31,935 (1972) [*Chem. Abstr.*, **78**, 29632 (1973)].

1680. V. A. Shikhanov, A. P. Ivanovskii, A. M. Kut'in, and M. A. Korshunov, *Khim. Prom-st. (Moscow)*, **1975**, 828 [*Chem. Abstr.*, **84**, 59127 (1976)].

1681. Y. Tanba, *Japan. Pat.*, 77 00,034 (1977) [*Chem. Abstr.*, **86**, 189736 (1977)].

1682. M. Sile, A. Avots, M. V. Shimanskaya, I. I. Ioffe, and V. Ulaste, *Latv. PSR Zinat. Akad. Vestis, Kim. Ser.*, **1972**, 218 [*Chem. Abstr.*, **77**, 88175 (1972)].

1683. M. Sile, A. Avots, I. I. Ioffe, and M. V. Shimanskaya, *Geterogennyi Katal. Reakts. Poluch. Prevrashch. Geterosikl. Soedin.*, **1971**, 169 [*Chem. Abstr.*, **76**, 72362 (1972)].

1684. A. Avots, I. I. Ioffe, M. Sile, and M. V. Shimanskaya, *U.S.S.R. Pat.* 366,193 (1973) [*Chem. Abstr.*, **78**, 159449 (1973)].

1685. E. A. Mailey, *U.S. Pat.* 3,428,640 (1969) [*Chem. Abstr.*, **70**, 77798 (1969)].

1686. M. Sile, A. Avots, M. V. Shimanskaya, and G. Strautins, *Latv. PSR Zinat. Akad. Vestis, Kim. Ser.*, **1969**, 567 [*Chem. Abstr.*, **73**, 18907 (1970)].

1687. B. F. Ustavshchikov, V. V. Antonova, and N. A. Titova, *U.S.S.R. Pat.* 478,003 (1975) [*Chem. Abstr.*, **83**, 193096 (1975)].

1688. B. F. Ustavshchikov, V. V. Antonova, N. A. Titova, and L. B. Zyul'kova, *U.S.S.R. Pat.* 490,799 (1975) [*Chem. Abstr.*, **84**, 43874 (1976)].

1689. V. V. Antonova, N. A. Titova, and B. F. Ustavshchikov, *Oznovn. Organ. Sintez i Neftekhimiya*, **1977**, 96 [*Chem. Abstr.*, **90**, 22769 (1979)].

1690. J. Okada and K. Nakano, *Yakugaku Zasshi*, **19**, 416 (1971) [*Chem. Abstr.*, **74**, 141692 (1971)].

1691. Kh. I. Areshidze and G. O. Chivadze, *Khim. Geterotsikl. Soedin.*, **1973**, 937 [*Chem. Abstr.*, **79**, 126213 (1973)].

1692. D. Yusupov, K. M. Akhmerov, A. B. Kuchkarov, Z. Tapilov, and M. G. Ismatullaeva, *U.S.S.R. Pat.* 789,519 (1980) [*Chem. Abstr.*, **94**, 208721 (1981)].

250 T. D. Bailey, G. L. Goe, and E. F. V. Scriven

1693. W. E. Hahn and W. Koziolkiewicz, *Lods. Tow. Nauk. Wydz. III, Acta Chim.*, **15**, 61 (1970) [*Chem. Abstr.*, **75**, 110170 (1971)].

1694. V. A. Kaminskii and M. N. Tilichenko, *Khim. Geterotsikl. Soedin.*, **1974**, 1434 [*Chem. Abstr.*, **82**, 43157 (1975)].

1695. Yu. S. Dol'skaya and G. Ya. Kondrat'eva, *Izv. Akad. Nauk. SSSR, Ser. Khim.*, **1978**, 1207 [*Chem. Abstr.*, **89**, 118426 (1978)].

1696. T. Nakanishi and K. Suyama, *Agr. Biol. Chem.*, **38**, 1141 (1974).

1697. N. R. Davis and R. A. Anwar, *J. Am. Chem. Soc.*, **92**, 3778 (1970).

1698. K. Suyama and S. Adachi, *J. Org. Chem.*, **44**, 1417 (1979).

1699. A. A. Usol'tsev and E. S. Karaulov, *Zh. Org. Khim.*, **13**, 84 (1977) [*Chem. Abstr.*, **86**, 139807 (1977)].

1700. A. A. Usol'tsev, E. S. Karaulov, and M. N. Tilichenko, *Khim. Dikarbonil'nykh Soedin.*, *Tezisy Dokl. Vses. Konf., 4th, 1975*, 169 (1976) [*Chem. Abstr.*, **87**, 67883 (1977)].

1701. M. Taguchi and K. Matsuura, *Yuki Gosei Kagaku Kyokai Shi*, **27**, 1230 (1969) [*Chem. Abstr.*, **74**, 53464 (1971)].

1702. M. Taguchi and A. Matsuura, *Japan. Pat.*, 71 09,701 (1971) [*Chem. Abstr.*, **75**, 35765 (1971)].

1703. T. Takeshima. M. Yokoyama, N. Fukada, and M. Akano, *J. Org. Chem.*, **35**, 2438 (1970).

1704. J. W. Ducker and M. J. Gunter, *Aust. J. Chem.*, **26**, 2567 (1973).

1705. I. Inoue, K. Hara, and J. Osugi, *Rev. Phys. Chem. Japan*, **46**, 64 (1976).

1706. A. Sakurai and H. Midorikawa, *Bull. Chem. Soc. Japan*, **41**, 430 (1968).

1707. K. Sasse, *Justus Liebigs Ann. Chem.*, **1976**, 768.

1708. B. A. Trofimov, A. I. Mikhaleva, A. S. Atavin, and E. G. Chebotareva, *Khim. Geterotsikl. Soedin.*, **1975**, 1427 [*Chem. Abstr.*, **84**, 59121 (1976)].

1709. B. A. Trofimov, A. I. Mikhaleva, and A. S. Atavin, *Zh. Org. Khim.*, **11**, 1141 (1975) [*Chem. Abstr.*, **83**, 42776 (1975)].

1710. B. A. Trofimov, A. I. Mikhaleva, S. E. Korostova, L. N. Balabanova, A. N. Vasil'ev, G. A. Kalabin, and A. S. Atavin, *Tezisy Dokl.-Vses. Konf. Khim. Atsetilena, 5th*, **1975**, 281 [*Chem. Abstr.*, **88**, 169869 (1978)].

1711. K. D. Gundermann and H. U. Alles, *Angew. Chem.*, **78**, 906 (1966).

1712. K. D. Gundermann and H. U. Alles, *Chem. Ber.*, **102**, 3014 (1969).

1713. L. I. Bagal, I. V. Tselinskii, and I. N. Shokhor, *Zh. Org, Khim.*, **5**, 2016 (1969) [*Chem. Abstr.*, **72**, 55180 (1970)].

1714. C. Bischoff and H. Herma, *J. Prakt. Chem.*, **318**, 891 (1976).

1715. H. Rozsa and B. Borivoje, *Ind. Chim. Belge*, **32**, 165 (1967).

1716. D. H. Aue and D. Thomas, *J. Org. Chem.*, **40**, 2360 (1975).

1717. K. Hirai, H. Matsuda, and Y. Kishida, *Chem. Pharm. Bull.*, **21**, 1090 (1973).

1718. K. Hirai, H. Matsuda, and Y. Kishida, *Sankyo Kenkyusho Nempo*, **24**, 108 (1972) [*Chem. Abstr.*, **78**, 159569 (1973)].

1719. M. Ochiai, M. Obayashi, and K. Morita, *Tetrahedron*, **23**, 2641 (1967).

1720. U. Eisner and J. Kuthan, *Chem. Rev.*, **72**, 1 (1972).

1721. F. Bossert, H. Meyer, and E. Wehinger, *Angew. Chem., Int. Ed. Engl.*, **20**, 762 (1981).

1722. A. Alberola, M. A. Gunther, M. Lora-Tamayo, and J. L. Soto, *An. Real Soc. Espan. Fis. Quim., Ser. B*, **63**, 691 (1967) [*Chem. Abstr.*, **67**, 81825 (1967)].

1723. A. S. Alverez-Insua, M. Lora-Tamayo, and J. L. Soto, *J. Heterocycl. Chem.*, **7**, 1305 (1970).

1724. M. A. Cabrerizo and J. L. Soto, *An. Quim.*, **70**, 951 (1974) [*Chem. Abstr.*, **83**, 193024 (1975)].

1725. P. Nantka-Namirski and R. Balicki, *Acta Pol. Pharm.*, **31**, 271 (1974) [*Chem. Abstr.*, **82**, 125251 (1975)].

1726. J. S. A. Brunskill, *J. Chem. Soc. C*, **1968**, 960.

1727. W. Nagai, Y. Hirata, and Y. Miwa, *J. Org. Chem.*, **39**, 3735 (1974).

1728. M. H. Pera, T. Q. Duc, H. Fillion, and L. D. Cuong, *Bull. Soc. Chim. Fr.*, **1975**, 321.

1729. M. H. Pera, H. Fillion, A. Boucherle, and L. D. Cuong, *Ann. Pharm. Fr.*, **33**, 141 (1975).

1730. P. Nantka-Namirski and R. Balicki, *Acta Pol. Pharm.*, **29**, 545 (1972) [*Chem. Abstr.*, **78**, 136021 (1973)].

1731. A. Sakurai and H. Midorikawa, *J. Org. Chem.*, **34**, 3612 (1969).

1732. M. M. Bursey and T. A. Elwood, *J. Org. Chem.*, **35**, 793 (1970).

1733. R. Vonderwahl, *U.S. Pat.* 3,035,061 (1962) [*Chem. Abstr.*, **57**, 11213 (1962)].

1734. H. Bredereck, F. Effenberger, and R. Sauter, *Chem. Ber.*, **95**, 2049 (1962).

1735. R. C. Elderfield and M. Wharmby, *J. Org. Chem.*, **32**, 1638 (1967).

1736. T. Hirota, T. Koyama, T. Nanba, and M. Yamato, *Chem. Pharm. Bull.*, **25**, 2838 (1977).

1737. O. Fuchs, V. Senkariuk, A. Nemes, G. Zolyomi, T. Somogyi, and A. Lazar, *Hung. Pat.* 157,008 (1970) [*Chem. Abstr.*, **72**, 100533 (1970)].

1738. O. S. Wolfbies and H. Junek, *Z. Naturforsch., B: Anorg. Chem., Org. Chem.*, **30B**, 249 (1975).

1739. G. R. Newkome and D. L. Fishel, *J. Heterocycl. Chem.*, **4**, 427 (1967).

1740. G. R. Newkome and D. L. Fishel, *J. Chem. Soc. D*, **1970**, 916.

1741. G. R. Newkome and D. L. Fishel, *J. Org. Chem.*, **37**, 1329 (1972).

1742. O. Tsuge, M. Tashiro, K. Hokama, and K. Yamada, *Kogyo Kagaku Zasshi*, **71**, 1667 (1968) [*Chem. Abstr.*, **70**, 37587 (1969)].

1743. O. Tsuge, K. Hokama, and M. Koga, *Yakugaku Zasshi*, **89**, 789 (1969) [*Chem. Abstr.*, **71**, 81292 (1969)].

1744. E. Fanghoenel, R. Ebisch, and P. Niedermeyer, *Z. Chem.*, **15**, 143 (1975).

1745. R. S. Monson, D. N. Priest, and J. C. Ullrey, *Tetrahedron Lett.*, **1972**, 929.

1746. R. S. Monson and A. Baraze, *J. Org. Chem.*, **40**, 1672 (1975).

1747. R. S. Monson and A. Baraze, *Tetrahedron*, **31**, 1145 (1975).

1748. R. S. Monson and A. Baraze, *Chem. Lett.*, **1976**, 555.

1749. A. M. Derijckere and F. G. F. Eloy, *Ger. Pat.* 2,052,536 (1971) [*Chem. Abstr.*, **75**, 35756 (1971)].

1750. F. Eloy and A. Deryckere, *Chim. Ther.*, **5**, 416 (1970).

1751. F. Eloy and A. Deryckere, *J. Heterocycl. Chem.*, **7**, 1191 (1970).

1752. L. E. Overman and S. Tsuboi, *J. Am. Chem. Soc.*, **99**, 2813 (1977).

1753. L. E. Overman and J. P. Roos, *J. Org. Chem.*, **46**, 811 (1981).

1754. B. A. J. Clark, M. M. S. El-Bakoush, and J. Parrick, *J. Chem. Soc., Perkin Trans. 1*, **1974**, 1531.

1755. R. Fuks, *Tetrahedron*, **26**, 2161 (1970).

1756. O. Tsuge and A. Inaba, *Heterocycles*, **3**, 1081 (1975).

1757. O. Tsuge and A. Inaba, *Bull. Chem. Soc. Japan*, **49**, 2828 (1976).

1758. G. Ya. Kondrat'eva, Yu. S. Dol'skaya, E. A. Aleksandrova, and B. A. Kazanskii, *Khim. Gererotsikl. Soedin.*, **1972**, 970 [*Chem. Abstr.*, **77**, 139753 (1972)].

1759. Yu. S. Dol'skaya, G. Ya. Kondrat'eva, and B. A. Kazanskii, *Izv. Akad. Nauk SSSR, Ser. Khim.*, **1972**, 2263 [*Chem. Abstr.*, **78**, 58216 (1973)].

252   T. D. Bailey, G. L. Goe, and E. F. V. Scriven

1760.   Yu. S. Dol'skaya, G. Ya. Kondrat'eva, and N. I. Golovina, *Izv. Akad. Nauk SSSR, Ser. Khim.*, **1975**, 1809 [*Chem. Abstr.*, **84**, 17087 (1976)].

1761.   Yu. S. Dol'skaya, V. S. Bogdanov, and G. Ya. Kondrat'eva, *Izv. Akad. Nauk SSSR, Ser. Khim.*, **1978**, 1691 [*Chem. Abstr.*, **89**, 146739 (1978)].

1762.   Yu. S. Dol'skaya, G. Ya. Kondrat'eva, and B. Z. Bartkevich, *Izv. Akad. Nauk SSSR, Ser. Khim.*, **1978**, 1446 [*Chem. Abstr.*, **89**, 109018 (1978)].

1763.   B. A. Kazanskii, G. Ya. Kondrat'eva, and Yu. S. Dol'skaya, *U.S.S.R. Pat.* 270,735 (1970) [*Chem. Abstr.*, **73**, 98806 (1970)].

1764.   Yu. S. Dol'skaya, M. A. Kapustin, G. Ya. Kondrat'eva, V. I. Garanin, Kh. M. Minachev, and N. I. Reutova, *Izv. Akad. Nauk SSSR, Ser. Khim.*, **1980**, 655 [*Chem. Abstr.*, **93**, 71483 (1980)].

1765.   B. A. Kazanskii, G. Ya. Kondrat'eva, and Yu. S. Dol'skaya, *Zh. Org. Khim.*, **6**, 2203 (1970) [*Chem. Abstr.*, **74**, 42256 (1971)].

1766.   B. A. Kazanskii, G. Ya. Kondrat'eva, and Yu. S. Dol'skaya, *U.S.S.R. Pat.* 199,890 (1967) [*Chem. Abstr.*, **68**, 95710 (1968)].

1767.   Yu. S. Dol'skaya, G. Ya. Kondrat'eva, and B. A. Kazanskii, *Geterogennyi Katal. Reakts. Poluch. Prevrashch. Geterosikl. Soedin.*, **1971**, 165 [*Chem. Abstr.*, **76**, 85656 (1972)].

1768.   Yu. S. Dol'skaya and G. Ya. Kondrat'eva, *Izv. Akad. Nauk SSSR, Ser. Khim.*, **1970**, 2123 [*Chem. Abstr.*, **74**, 64183 (1971)].

1769.   W. H. Bell and R. A. C. Rennie, *Brit. Pat.* 1,247,361 (1971) [*Chem. Abstr.*, **75**, 151688 (1971)].

1770.   P. Vitins and K. W. Egger, *Helv. Chim. Acta*, **57**, 17 (1974).

1771.   G. Dauphin, B. Jamilloux, A. Kergomard, and D. Planat, *Tetrahedron*, **33**, 1129 (1977).

1772.   G. Dauphin, L. David, B. Jamilloux, A. Kergomard, and H. Veschambre, *Fr. Pat.* 2,057,192 (1971) [*Chem. Abstr.*, **76**, 85704 (1972)].

1773.   B. Adler, C. Burtzlaff, C. Duschek, J. Ohl, H. Schmidt, and W. Zech, *J. Prakt. Chem.*, **320**, 904 (1978).

1774.   K. Berg-Nielsen and L. Skatteboel, *Acta Chem. Scand., Ser. B*, **B32**, 553 (1978).

1775.   H. Agui, H. Tobiki, and T. Nakagome, *J. Heterocycl. Chem.*, **12**, 1245 (1975).

1776.   T. Kametani, K. Kigasawa, M. Hiiragi, K. Wakisaka, O. Kusama, H. Sugi, and K. Kawasaki, *J. Heterocycl. Chem.*, **14**, 477 (1977).

1777.   K. Kigasawa, M. Hiiragi, K. Wakisaka, O. Kusama, H. Sugi, and K. Kawasaki, *Japan. Pat.*, 78 63,382 (1978) [*Chem. Abstr.*, **89**, 129420 (1978)].

1778.   V. G. Granik, N. B. Marchenko, E. O. Sochneva, R. G. Glushkov, T. F. Vlasova, and Yu. N. Sheinker, *Khim. Geterotsikl. Soedin.*, **1976**, 80 [*Chem. Abstr.*, **85**, 108568 (1976)].

1779.   V. G. Granik, O. Ya. Belyaeva, R. G. Glushkov, T. F. Vlasova, and O. S. Anisimova, *Khim. Geterotsikl. Soedin.*, **1977**, 1160 [*Chem. Abstr.*, **88**, 6779 (1978)].

1780.   V. G. Granik, N. B. Marchenko, T. F. Vlasova, and R. G. Glushkov, *Khim. Geterotsikl. Soedin.*, **1976**, 1509 [*Chem. Abstr.*, **86**, 139783 (1977)].

1781.   M. S. Schrider, *U.S. Pat.* 4,006,236 (1977) [*Chem. Abstr.*, **86**, 171283 (1977)].

1782.   M. Kawanisi, K. Matasunaga, and N. Miyamoto, *Bull. Chem. Soc. Japan*, **45**, 1240 (1972).

1783.   K. Togo, A. Maeda, and R. Kyoshida, *Japan. Pat.*, 69 26,104 (1969) [*Chem. Abstr.*, **72**, 12589 (1970)].

1784.   F. P. Schmidtchen and H. Rapoport, *J. Am. Chem. Soc.*, **99**, 7014 (1977).

1785.   T. Sugasawa, T. Toyoda, and K. Saskura, *Tetrahedron Lett.*, **1972**, 5109.

1786.   D. Stefanescu, D. Moraru, I. Predescu, P. Barza, T. Stanciu, and Fl. Coman, *Farmacia (Bucharest)*, **17**, 115 (1969) [*Chem. Abstr.*, **71**, 30331 (1969)].

CHAPTER II

Carbocyclic Annelated Pyridines

R. P. THUMMEL

Department of Chemistry, University of Houston, Houston, Texas

I.	Introduction and Scope	254
II.	Synthesis of Annelated Pyridines	255
	1. Condensation Reactions	255
	A. Condensation with 1,5-Dicarbonyl Compounds	255
	B. Chichibabin Condensation	261
	C. Friedlander Condensation	264
	D. Claisen- and Aldol-Type Condensations	271
	2. Cyclization Reactions	274
	A. Nucleophilic Cyclizations	274
	B. Electrophilic Cyclizations	275
	C. Friedel–Crafts Acylations	278
	D. Nitrile Cyclizations	280
	E. Azatriene Cyclizations	282
	F. Thermal Cyclizations	286
	G. Other Cyclizations	289
	3. Cycloaddition Reactions	290
	A. Diels–Alder Reactions	290
	a. Azadienes	290
	b. Nitriles	293
	c. 2,3-Dehydropyridine	294
	d. 3,4-Dehydropyridine	295
	e. Other [2 + 4] Cycloadditions	297
	B. Cooligomerization	299
	C. Photocycloadditions	300
	4. Rearrangement Reactions	300
	5. Hydrogenations	303
	A. Quinoline and Isoquinoline	303
	B. Acridine	306
	6. Halogenation	306
	7. Pyrylium Salts	307
	8. Malononitrile and Cyanoacetamide Reaction with 1,3-Dicarbonyl Compounds	310
III.	Reactions of Annelated Pyridines	312
	1. Mono-Annelated Pyridines	312
	A. *N*-Oxide Acetylation	312

    B. α-Metallation 313
    C. α-Condensation 314
  2. *Bis*-Annelated Pyridines 316
    A. Reactions of *N*-Oxides 316
    B. α-Condensation 317
    C. Metallation 318
    D. Other Reactions 319
  3. Azafluorenes 319
  4. Other Annelated Pyridines 322
IV. Naturally Occurring Annelated Pyridines 323
V. Biologically Active Annelated Pyridines 325
VI. Physical Properties 327
  1. NMR Spectra 327
  2. Ultraviolet Spectra 331
  3. Basicity 332
  4. Infrared Spectra 333
  5. Pyrindine Tautomerism 334
  6. Tables of Physical Constants 335
Acknowledgments 335
References 436

I. INTRODUCTION AND SCOPE

This chapter represents the first extensive review of carbocyclic annelated pyridines. As a result, it will attempt to cover the pertinent literature available up until the end of 1981. Emphasis will be placed upon more recent work and, although references prior to 1940 are scarce, they will be included wherever possible and relevant.

A cursory review of 2,3-cycloalkenopyridines based heavily on patent literature from one laboratory has recently appeared (1). Related reviews have discussed various other aspects of the topic such as cycloadditions to dehydropyridines (2) and cyclizations of various pyridyl alkenes (3). This series, *The Chemistry of Heterocyclic Compounds*, has published a supplement chapter devoted to Alkyl-pyridines in which some reference is made to carbocyclic annelated derivatives (4). The cyclohexapyridines (5,6,7,8-tetrahydroquinoline and 5,6,7,8-tetrahydro-isoquinoline) will be discussed in separate chapters on the hydrogenated quinolines (5) and isoquinolines (5) and for that reason these compounds will be presented here only as they relate to other homologs or as examples of synthetic approaches. An excellent review on the 1-pyrindines (6) has recently appeared which deals in detail with derivatives of 1*H*-1-, 5*H*-1-, and 6,7-dihydro-5*H*-1-pyrindine. This review will discuss these same compounds as cyclopentapyridines.

Specifically, this article will deal with ortho, monoannelated, or *bis*-annelated neutral pyridines in which the fused ring is composed only of carbon atoms. Not included in the review will be: benzoannelated pyridines (i.e., quinolines or iso-quinolines), annelation on nitrogen, annelated pyridones, or dihydro-, tetrahydro-, or hexahydro-annelated pyridines.

II. SYNTHESIS OF ANNELATED PYRIDINES

The nitrogen atom of pyridine produces an inherent asymmetry which leads to the possibility of two sites for simple monoannelation. Thus one can have 2,3- or 3,4-fused pyridines. With regard to *bis*-annelation, two orientations are also possible; 2,3:4,5-fusion (also referred to as meta or angular fusion) and 2,3:5,6-fusion (also referred to as para or linear fusion).

Most all the possible isomers are known of annelated pyridines where the fused ring or rings contain from four to seven carbons. No cyclopropapyridines have yet been reported although the benzene analog, benzocyclopropene, is a well-known molecule which has been extensively studied (7).

The following section will present synthetic methods that have been employed to prepare annelated pyridines arranged by reaction type. The accompanying tables will summarize detailed information regarding specific methods employed and yields obtained.

1. Condensation Reactions

The most convenient method for constructing the pyridine nucleus is by way of condensation reactions. These have normally involved an amino group combining with two carbonyl groups accompanied by the loss of two or more equivalents of water. Very often a final oxidation step was necessary to obtain the aromatic ring system. The Hantzsch synthesis is probably the best example of such a condensation and many other approaches bear a similarity to this reaction. This section will attempt to make some distinction between various types of condensations leading to annelated pyridines but no strict divisions will be established. In fact, most condensations leading to a pyridine derivative proceed through an intermediate which can be related to a 1,5-dicarbonyl compound. Thus we will begin by considering this most general synthesis.

A. *Condensation with 1,5-Dicarbonyl Compounds*

The simplest example of this reaction would be the treatment of glutaraldehyde (**II-1**) with ammonia, followed by an oxidation step to yield pyridine (**II-3**).

256 R. P. Thummel

The fusion of one or two rings onto the glutaraldehyde backbone should similarly lead to mono- or *bis*-annelated pyridines. This reaction was first observed by Striegler who reacted **II-4** with hydroxylamine in hopes of preparing the corresponding mono- or *bis*-oxime (8). Instead he obtained pyridine **II-5**. Similar condensations were reported by Knoevenagel (9) and later by Stobbe (10), wherein the substitution of NH_2OH for ammonia allowed for a triple dehydration avoiding the necessity for a final oxidation step.

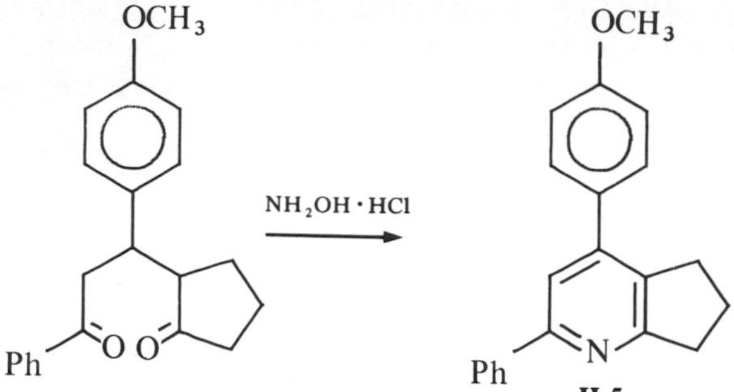

Potts et al. (11) found that the decomposition of ketonic Mannich bases (e.g. **II-6**) by controlled heating with suitable ketones led to β-acylethylation of the ketone and formation, primarily, of a 1,5-diketone (**II-7**), which could then be converted to a variety of annelated pyridines (12). It was likewise found that enamines of cyclic ketones could react with Mannich bases, leading to the formation of diketones, such as **II-7** (13). Utilizing the appropriate Mannich bases, the *bis*-annelated systems **II-9** and **II-10** could also be prepared. Several years later, Colonge et al. found that cyclic ketones (e.g., **II-11**) could be condensed with formaldehyde to provide 1,5-diketones **II-12** (14), which upon treatment with ammonium acetate provided the corresponding *para-bis*-annelated pyridines **II-13**. These workers were able to prepare dioxime and disemicarbazone derivatives of **II-12** and further showed that refluxing these derivatives in aqueous HCl also led to the desired pyridines often in substantially better yield (15).

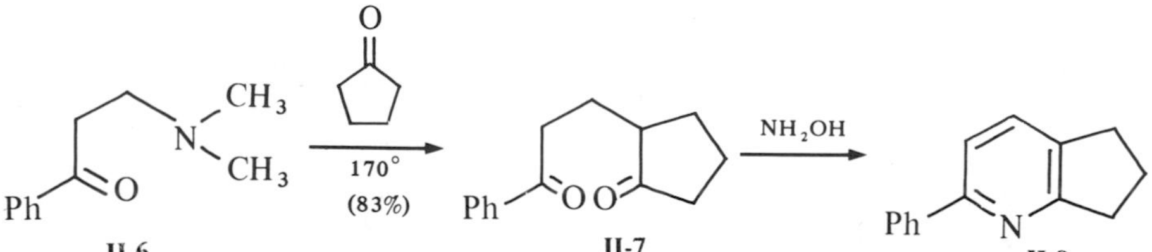

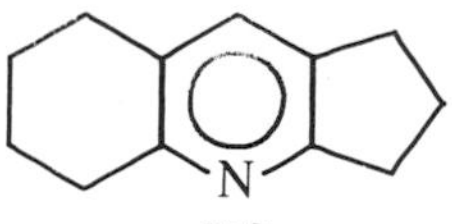

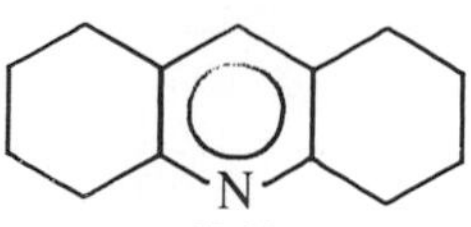

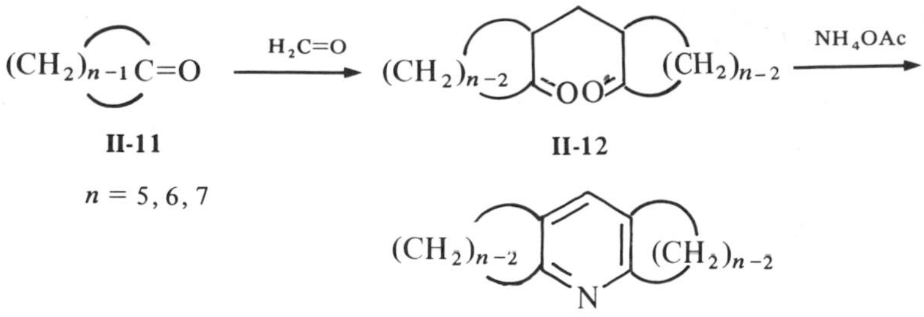

II-11

$n = 5, 6, 7$

II-12

II-13

Ketoaldehydes such as **II-14** and **II-16** could be prepared by treatment of the corresponding cycloalkanone enamine with acrolein (321). Subsequent condensation with ammonium acetate or hydroxylamine provided entry to the simple parent compounds: 2,3-cyclopentapyridine (**II-15**) (16) and 2,3-cycloheptapyridine (**II-17**) (17).

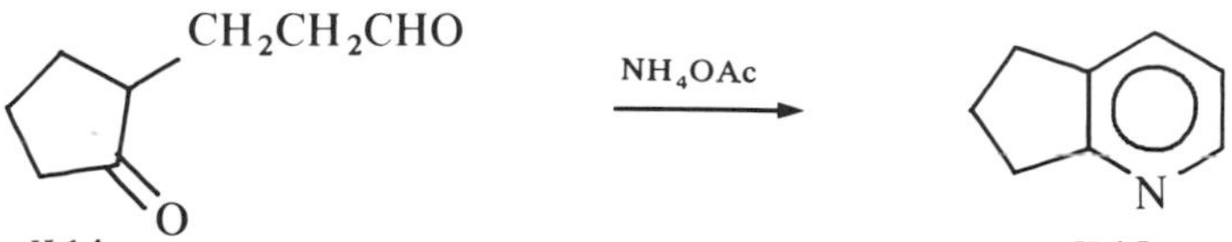

II-14

II-15

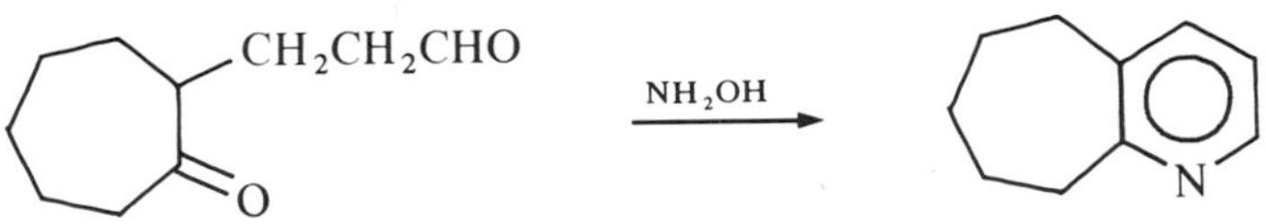

II-16

II-17

Of all the 1,5-diketone condensations, perhaps the most widely studied is that of 2,2'-methylenedicyclohexanone (**II-18**) to give 1,2,3,4,5,6,7,8-octahydroacridine (**II-19**). A large majority of this work has been carried out by Tilichenko's group at Far East State University in Vladivostok. Aside from the usual reagents, **II-19** has also been produced by the action of urea (18) or hydrazine. Hydrazine derivatives led to the formation of N-substituted salts, **II-20** and **II-22** in addition to **II-19** (19). Primary amines in the presence of carbon tetrachloride gave the corresponding octahydroacridinum chlorides (**II-21**) (20).

The result with phenylhyrazine has been interpreted in terms of an initial adduct **II-23**, which loses aniline to give **II-19** *via* an acid-catalyzed oxidation. Aniline can then be reacted with another equivalent of **II-18** to produce **II-20**. The N–N cleavage did not occur with the hydrazides of carboxylic acids, and thus products, such as **II-22**, were obtained.

It was found that a mixture of two products (**II-26** and **II-27**), epimeric at the α-positions, could be prepared by condensation of diketone **II-25** with hydroxylamine hydrochloride (21).

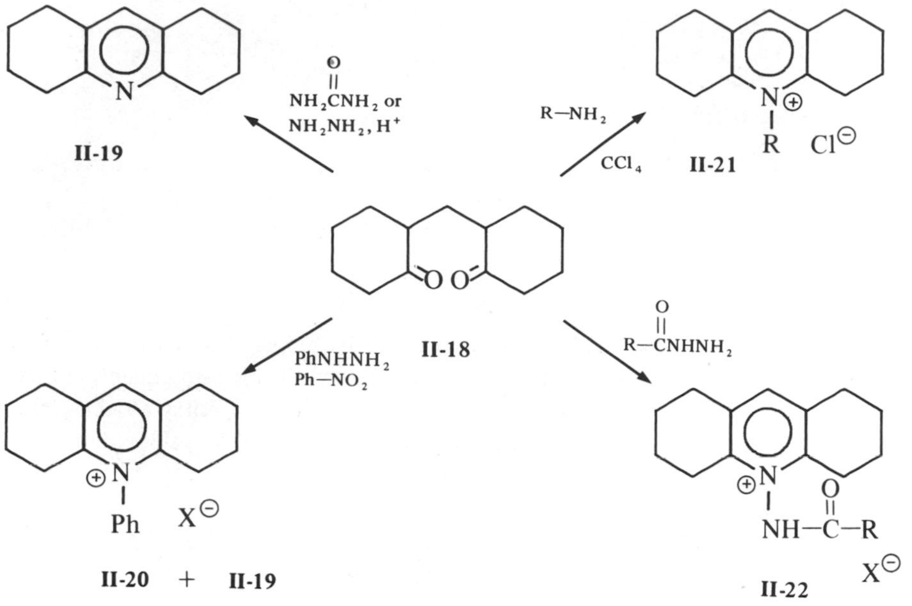

II-19

$$\overset{\overset{\oplus}{\parallel}}{NH_2CNH_2} \text{ or } NH_2NH_2, H^+$$

R—NH$_2$

CCl$_4$

II-21

Cl$^\ominus$

II-18

PhNHNH$_2$
Ph—NO$_2$

$$\overset{O}{\overset{\parallel}{R-CNHNH_2}}$$

II-20   +   II-19

II-22

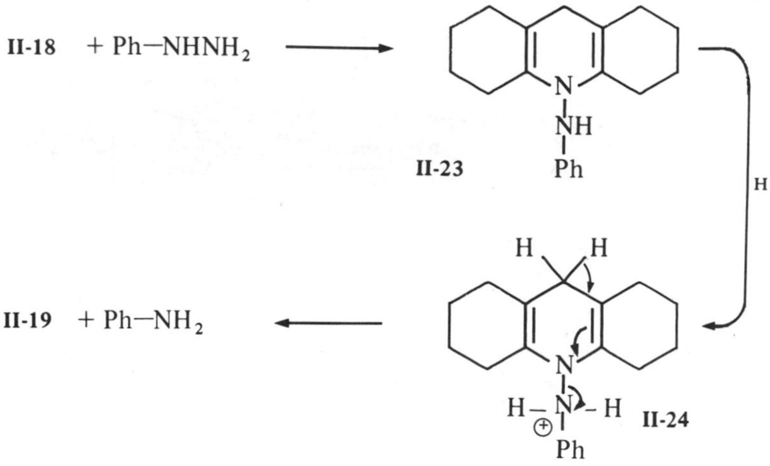

II-18   + Ph—NHNH$_2$ ⟶

II-23

H$^+$

II-19   + Ph—NH$_2$ ⟵

II-24

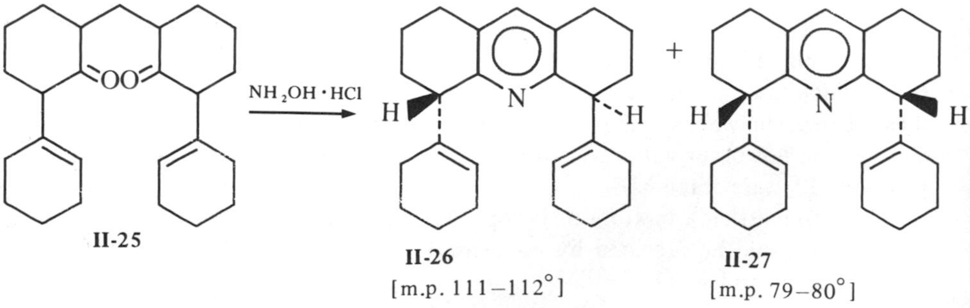

II-25

NH$_2$OH · HCl

II-26
[m.p. 111–112°]

+

II-27
[m.p. 79–80°]

A dibenzo derivative (**II-29**) of **II-19** may be obtained by treating 2,2′-methylene-*bis*-1-tetralone (**II-28**) with ammonium acetate in acetic acid (22).

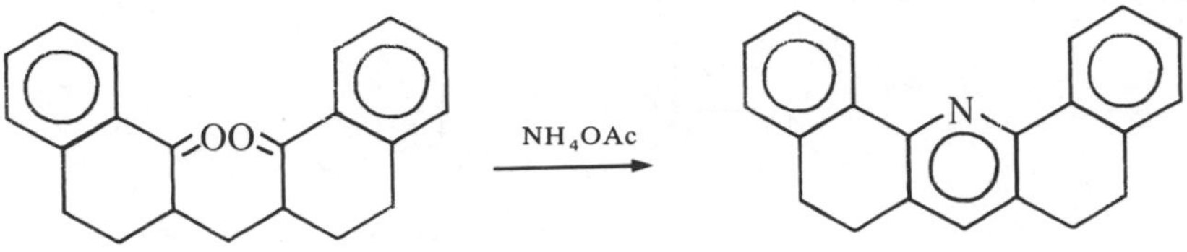

II-28   →(NH₄OAc)→   **II-29**

An interesting reaction leading to 9-methyl-1,2,3,4,5,6,7,8-octahydroacridine (**II-32**) involved a prior fragmentation of keto-alcohol **II-30** to diketone **II-31**, followed by condensation in the normal manner (23).

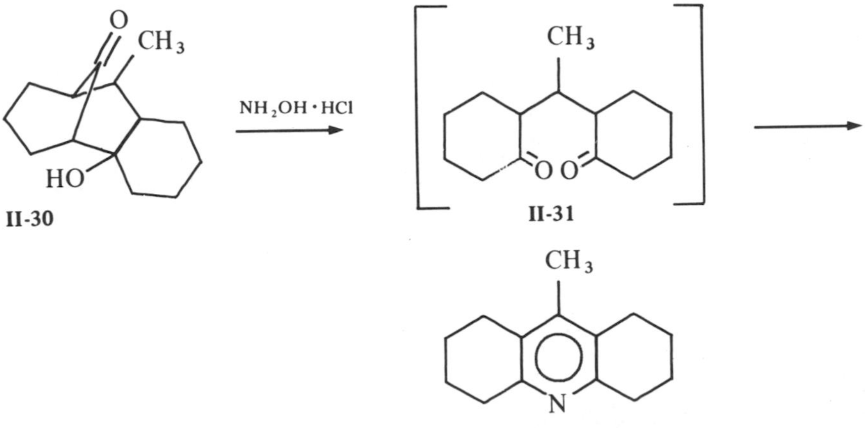

II-30   →(NH₂OH · HCl)→   **II-31**

II-32

Annelated derivatives of glutaraldehyde may be used to prepare monoannelated pyridines; however, the yields were typically poorer than for diketones or keto-aldehydes. These diminished yields were presumably due to the comparably higher reactivity of these dialdehydes. In actual fact, pure glutaraldehyde has been quite difficult to obtain. Iridodial (**II-33**), α-[2-formyl-3-methylcyclopentyl] propional can be prepared from *D*-(+)-citronellal (24). Cavill et al. (25, 26) have reported the conversion of iridodial into the pyridine alkaloid actinidine (**II-34**, 1′,5-dimethyl-3,4-cyclopentapyridine) *via* the *bis*-2,4-dinitrophenylhydrazone intermediate.

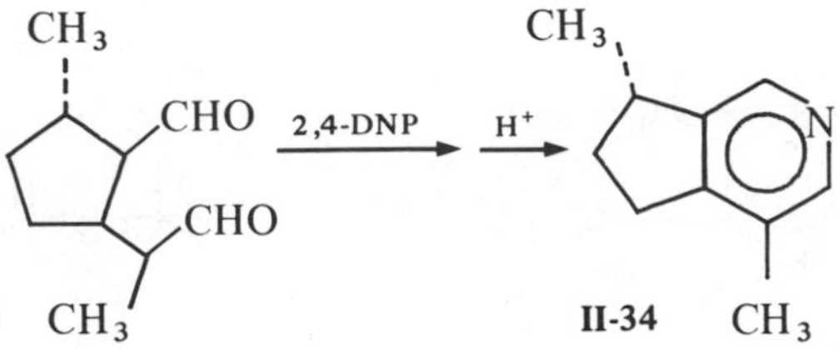

II-33   →(2,4-DNP)→ →(H⁺)→   **II-34**

The preparation of a cyclobutane-fused glutaraldehyde (**II-34a**) by oxidation of bicyclo[3.2.0]hept-2-ene has been attempted utilizing ozone or periodic acid but, thus far, with no success (27).

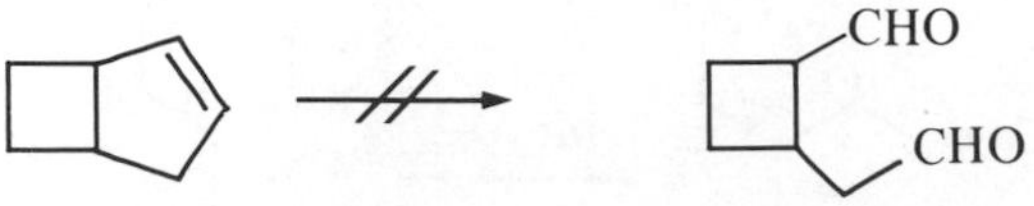

II-34a

Indanedione derivatives have been employed to prepare various 4-aza-9-fluorenones. When a solution of **II-35** and ammonium acetate in acetic acid was heated, the dihydropyridine **II-36** was obtained which could then be readily oxidized to the annelated system **II-37** (28). In fact, a *bis*-indanedione, **II-38**, has been utilized to prepare the pentacyclic symmetrical **II-40**, via oxidation of **II-39**, followed by a photochemical decarboxylation (29).

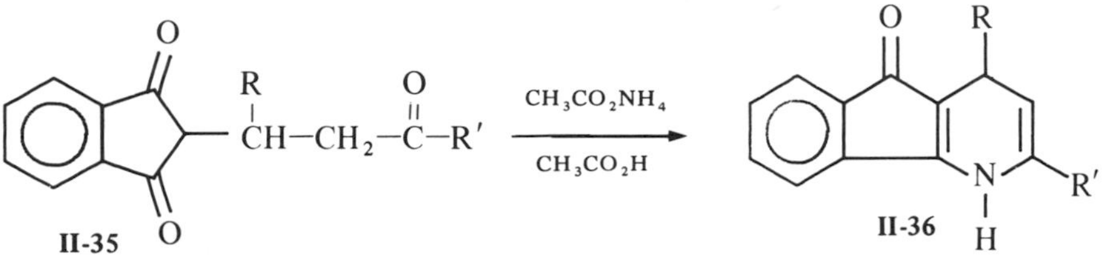

II-35

$R = R' = C_6H_5$
$R = C_6H_4OCH_3 \ (-p), \ R' = C_6H_5$
$R = \alpha$-furyl, $R' = C_6H_5$
$R = C_6H_5, \ R' = C(CH_3)_3$
$R = C_6H_5, \ R' = CH_3$

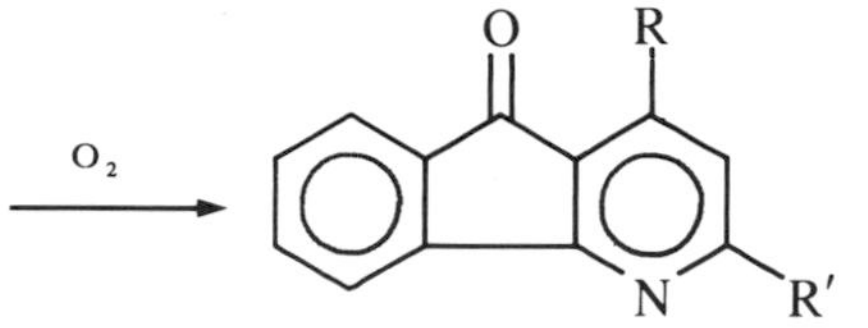

II-36

II-37

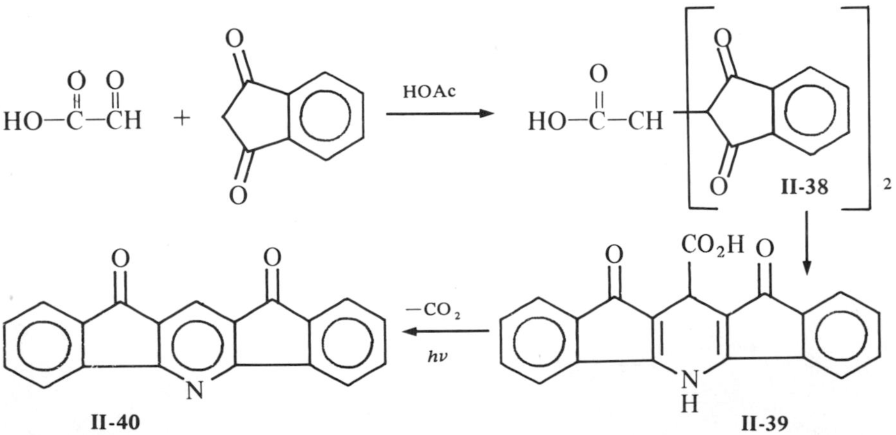

II-40   **II-39**   **II-38**

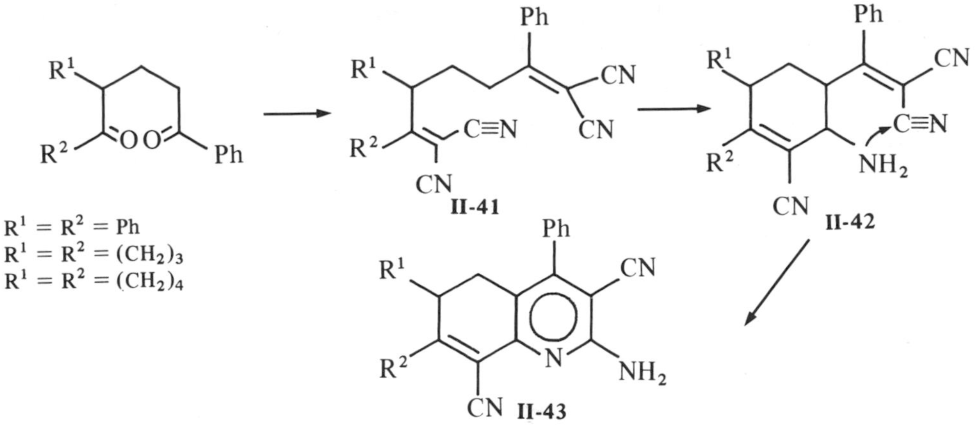

$R^1 = R^2 = Ph$
$R^1 = R^2 = (CH_2)_3$
$R^1 = R^2 = (CH_2)_4$

An interesting deviation from the normal mode of condensation was found upon treatment of 1,5-diketones with malononitrile, whereupon *bis*-dicyanomethylene derivatives **II-41** were initially formed (30). These species then underwent a cyclization to form **II-42** containing what will ultimately become the annelated cyclohexene ring. Finally, attack of the amino group on one of the geminal cyano groups provided the pyridine ring of dihydroquinoline **II-43**. In this sequence the 1,5-dicarbonyl group gave rise to the annelated ring rather than to the pyridine moiety. Therefore, 1,5-diketones can undergo a similar conversion to give cyclopentapyridine **II-44a** (31). By NMR it has been determined that solvent dependent tautomatic equilibrium is established between the pyrindine **II-44a** and the psuedoazulene **II-44b**. Functionalization of the unsubstituted 5-position of **II-44** with aldehydes and nitroso-compounds led to a variety of dye stuffs (32).

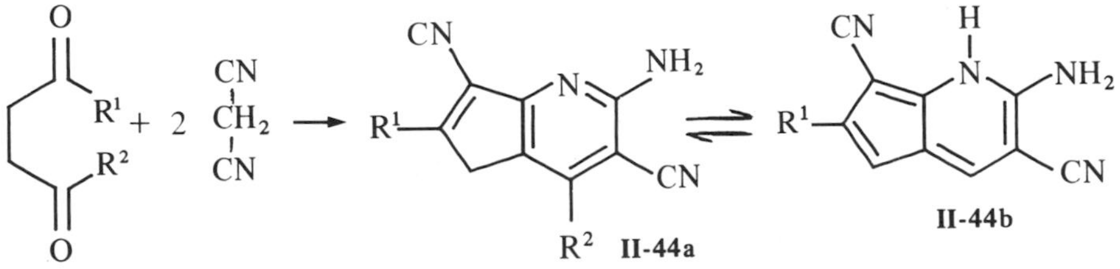

$R^1 = CH_3, R^2 = CH_3$
$R^1 = CH_3, R^2 = CH_2CH_3$
$R^1 = CH_3, R^2 = Ph$
$R^1 = Ph, R^2 = CH_3$
$R^1 = Ph, R^2 = Ph$

B.   Chichibabin Condensation

The Chichibabin condensation involves cyclocondensation of two equivalents of an aldehyde or ketone with one equivalent of each an aldehyde and ammonia. If a cyclic ketone was employed, a *bis*-annelated pyridine would result (33–35). As

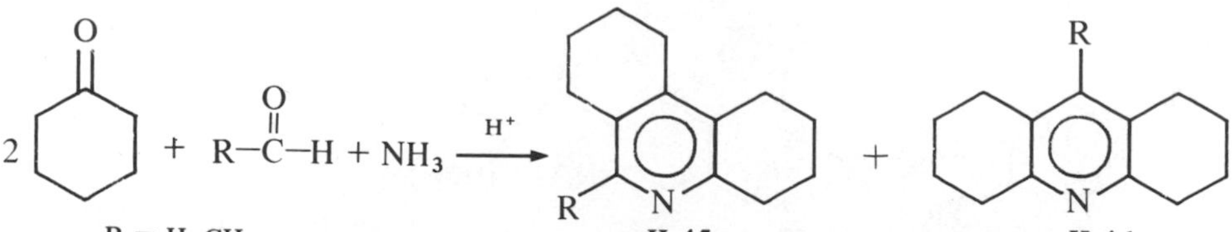

R = H, CH₃ — R = H, CH_3

II-45   II-46

the reaction was originally formulated, it involved cyclohexanone reacting with ammonia and either formaldehyde or acetaldehyde to give a 7:1 mixture of **II-45** and **II-46**. The method has since been adapted as a general procedure to prepare a variety of 6-substituted octahydrophenanthridines by variation of the aldehyde (36). The yields were found to be much lower when carried out with cyclopentanone. Nevertheless, when an equimolar mixture of cyclohexanone and cyclopentanone was treated with formaldehyde and ammonia, a complex mixture was obtained from which the two angularly fused isomers **II-47** and **II-48** were isolated. The structural assignments were made by comparison of ^{13}C NMR spectra with those of homologous compounds (37).

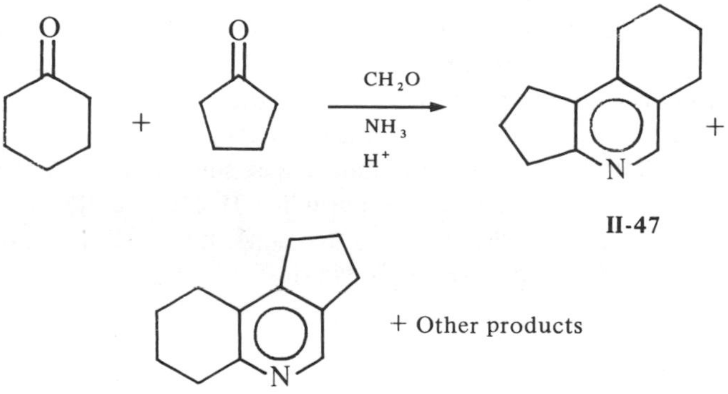

II-47

+ Other products

II-48

Patmore and Chafetz have carried out a fairly extensive investigation of this reaction (38) and discovered that 6-pentyl-1,2,3,4,7,8,9,10-octahydrophenanthridine (**II-49**) was prepared in high yield by the vapor-phase condensation of cyclohexanone and ammonia over metal oxides. Similar results have been observed for cyclopentanone (39).

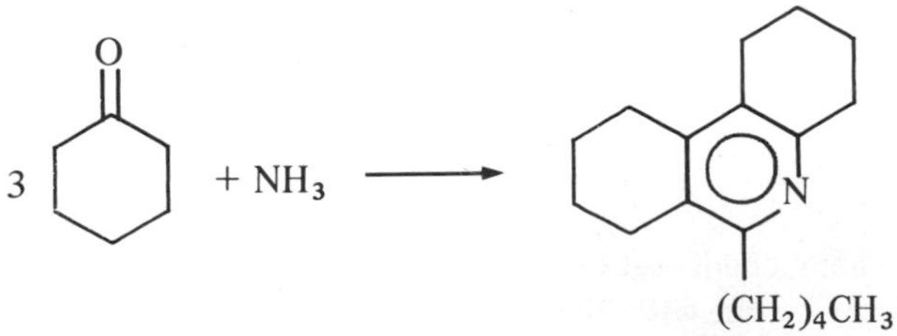

$(CH_2)_4CH_3$

II-49

A mechanism has been proposed (38, 40) to explain the formation of products such as **II-45**. It involved an initial aldol condensation of cyclohexanone to form the β,γ-unsaturated ketone **II-50**, which reacted with ammonia to form a dienamine **II-52**. This dienamine was condensed with an aldehyde to form imine **II-53** which cyclized and was finally dehydrogenated to give the *bis*-annelated pyridine **II-45**. It has been found that treatment of **II-50** under these conditions led to the same products (41, 42), lending credence to its involvement an initial intermediate.

II-50 **II-51** **II-52** **II-53** **II-54** **II-45**

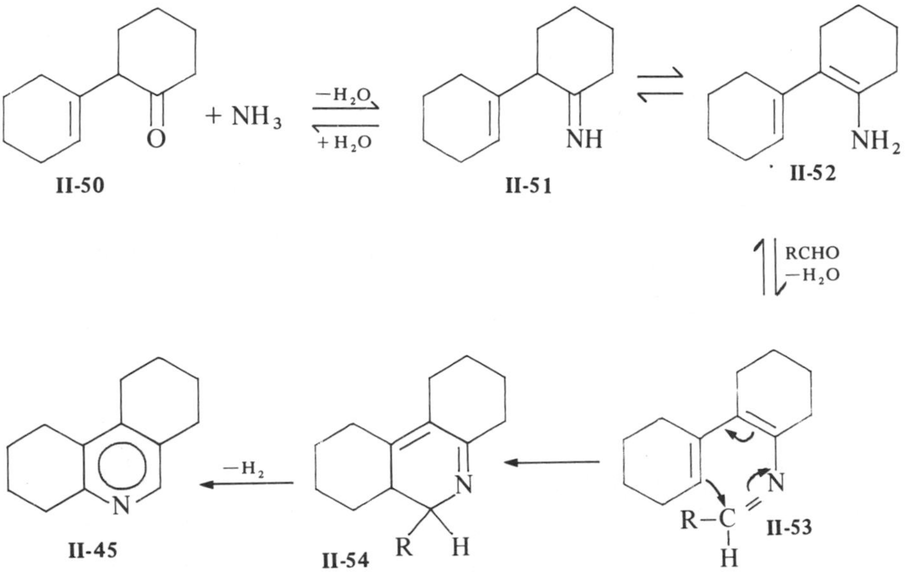

It appears that the final orientation of the two fused rings depends upon the sequence of condensation steps. If the initial condensation occurred between cyclohexanone and the aldehyde RCHO, then the ultimate orientation will be *para* as is observed in the minor product **II-46**. More reactive ketones, such as dimedone (43) and 1,3-cyclohexanedione (44), were observed to prefer this mode of reaction, leading to products such as **II-56**. There was, however, a report of the *para* isomer being formed in a 38% yield when cyclohexanone, formaldehyde, and ammonia were condensed over a catalyst containing Al, Mg, and F in an atomic ratio of 1000:25:50 at 400°C (45).

II-55 **II-56**

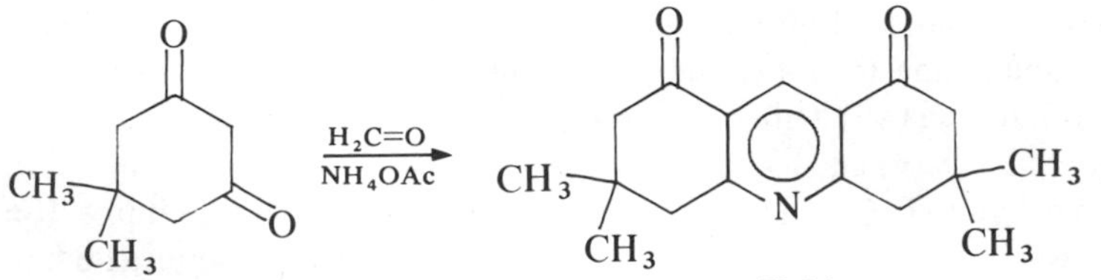

C.   Friedlander Condensation

The Friedlander condensation fundamentally involves condensation of a β-amino-α,β-unsaturated carbonyl compound with a second carbonyl compound such that a pyridine ring is formed. Two molecules of water are normally lost in the process. A simple example was set forth by Breitmaier and Bayer (46) who treated β-aminoacrolein with cyclohexanone to prepare the tetrahydroquinoline **II-61**.

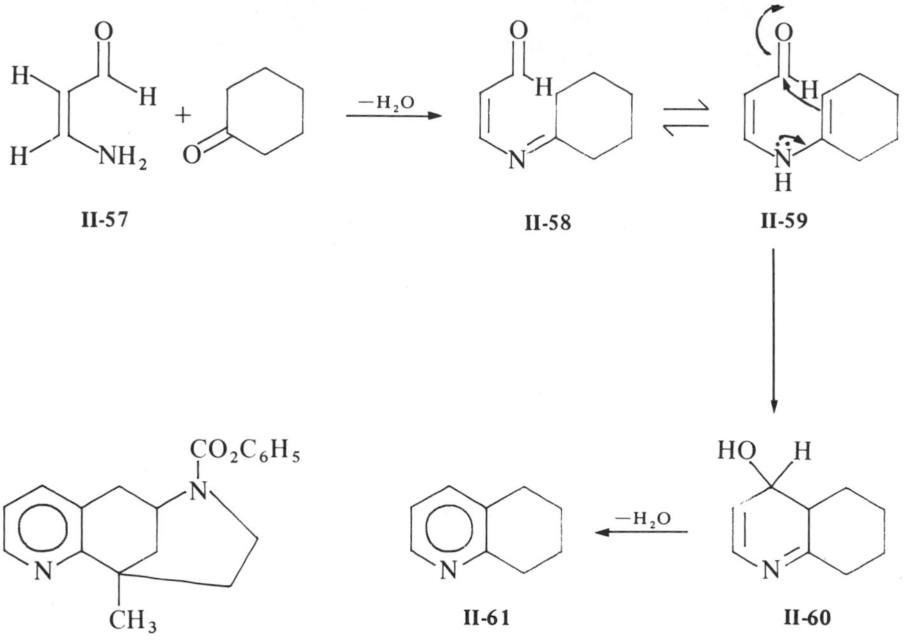

The reaction worked equally well for cyclopentanone (37) but failed with cyclobutanone (27), from which only polymeric material was obtained. The rationale for this may be the high reactivity of the cyclobutanone enamine analogous to **II-59**. Bicyclic ketones may be reacted with β-aminoacrolein and thus pyrido [2,3-f]morphan **II-61a** is prepared in moderate yield (322).

Condensation with cyclobutanone did proceed if the unsaturated portion of **II-57** was incorporated into an aromatic ring. Thus o-aminobenzaldehyde gave 2,3-cyclobutaquinoline (47, 48) and 2-aminonicotinaldehyde afforded 2,3-cyclobuta-1,8-naphthyridine (49). Other aromatic systems were condensed in a similar fashion. Caluwe and coworkers have taken particular advantage of 4-aminopyrimidine-5-carboxaldehyde (**II-62**) in this respect (50, 51). Cyclopentapyridine **II-65**, which was obtained after hydrolysis of the initial condensation product **II-64**, underwent a second Friedlander reaction to generate the bis-annelated naphthyridine **II-66**.

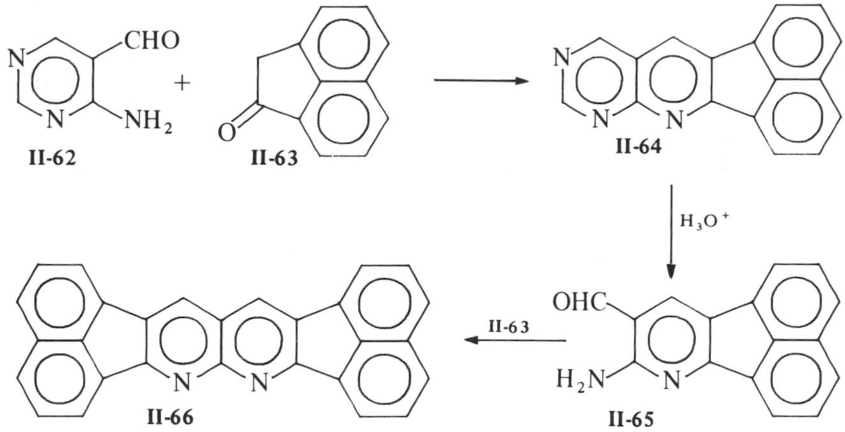

Similarly, 2-amino-1-formylazulene (**II-67**) reacted with acetylacetone to provide azulenopyridine **II-68** (52).

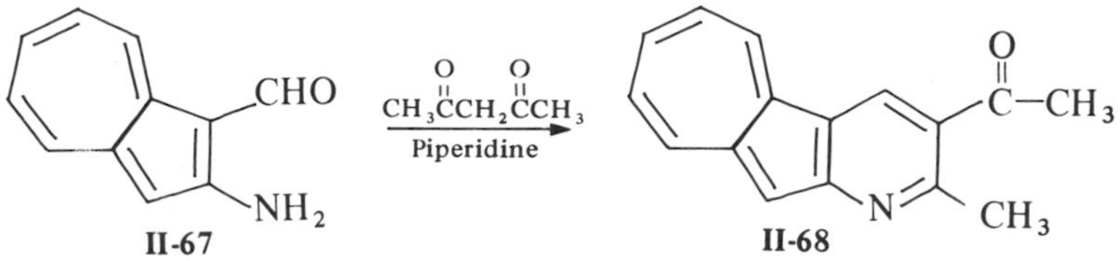

One of the most useful sources of β-aminoacrolein (**II-57**) and various substituted derivatives is *via* reductive cleavage of the N—O bond of isoxazoles (327) by catalytic hydrogenation over Raney–Nickel. Stork et al. (53) have shown that appropriately substituted isoxazoles (e.g., **II-69**), after hydrogenolysis, underwent an abnormal intramolecular condensation in which the carbonyl portion of the β-aminoenone was not involved. Only one dehydration occurred from **II-72**, which then must undergo an additional oxidation step to provide the annelated pyridine **II-73**.

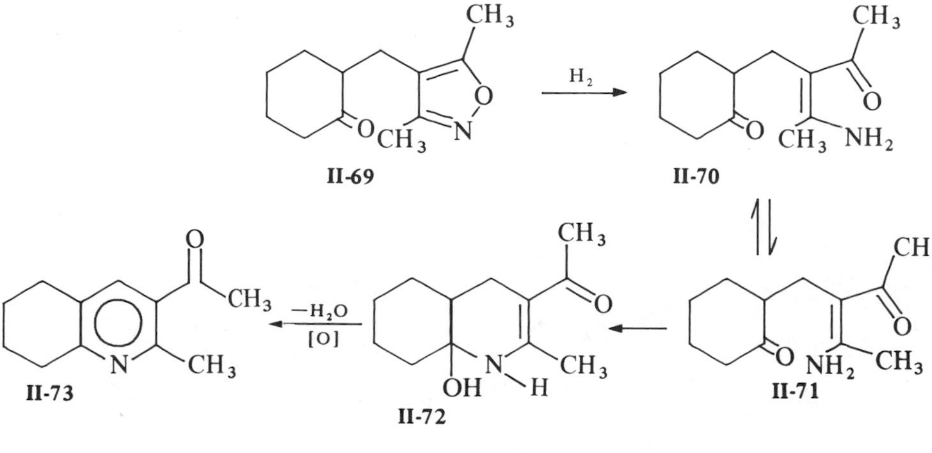

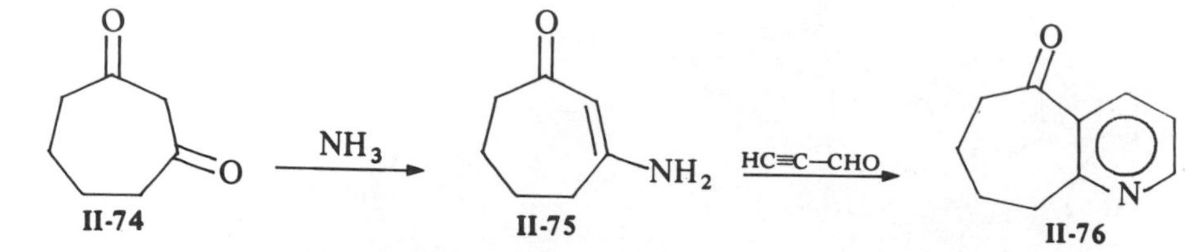

For the direct Friedlander condensation to occur in the normal fashion the amino and carbonyl moieties, as in **II-57**, must be *cis* to one another. In certain cyclic systems this orientation is either impossible or unfavorable and condensation must occur in an alternate fashion. β-Aminoenones derived from cyclic 1,3-diketones by treatment with ammonia fall into this category. Thus, 1,3-cycloheptanedione **(II-74)** gave **II-75**, which was condensed with propargaldehyde to provide annelated pyridine **II-76** (54). Reaction with yneones was preferable since the final oxidation step was not necessary (55, 56). With propargaldehyde, analysis of the product did not allow one to distinguish whether the initial enamine attack occurred in an 1,2- or 1,4-fashion. When 3-butyn-2-one was employed the methyl group was found in the 2-position indicative of a 1,4-addition *via* intermediate **II-80** (57).

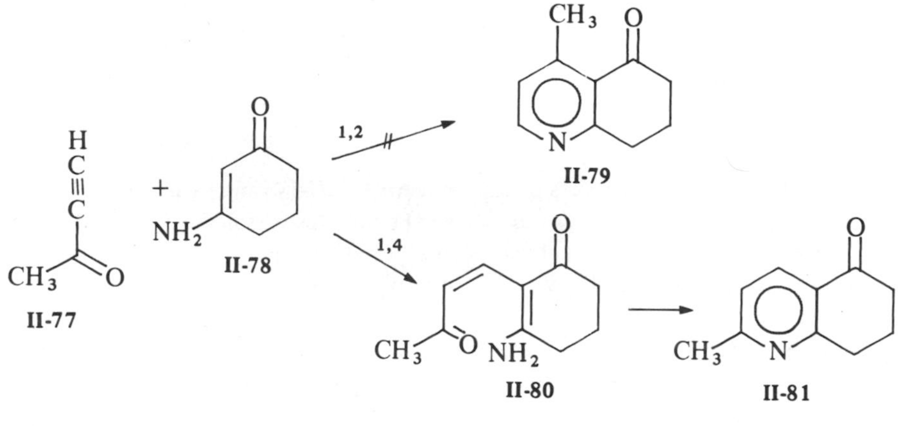

Many 1,3-diketones exist preferentially in their tautomeric β-hydroxy-enone form which can be envisioned to undergo a Friedlander condensation such as illustrated for **II-82**, where annelation can occur at both ends of the molecule (58). Once again, symmetry prevents one from deciding conclusively where initial attack of ammonia occurred.

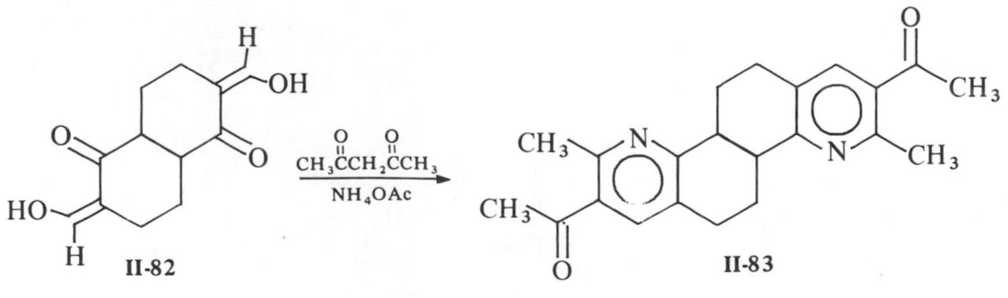

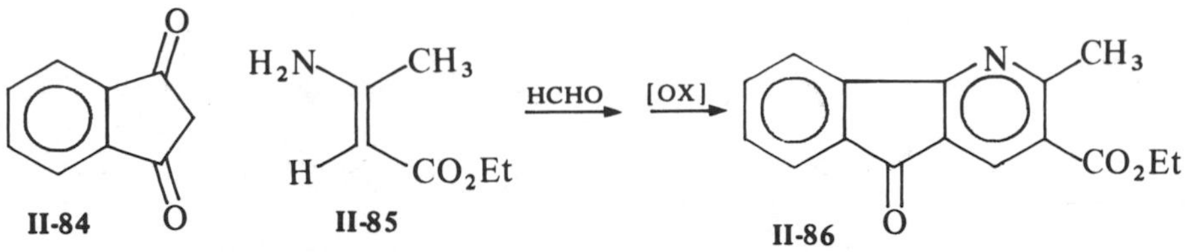

II-84   II-85   II-86

Some condensations appeared to combine features of both the Friedlander and Chichibabin reactions (59, 61). One such example would be condensation of 1,3-indanedione (II-84) with formaldehyde and ethyl β-aminocrotonate (II-85) to afford the azafluorenone II-86, after a final oxidation (59). Breitmayer and Bayer have shown that cyclic analogs of II-85 (e.g., II-87) would condense with activated carbonyl compounds in a more normal fashion such that formaldehyde was not required (62). Functionality may be removed from the pyridine ring of II-88 to provide ultimately the simple 2,3-cycloalkenopyridine II-91. A similar condensation has been observed for β-amino-α, β-unsaturated amides (63).

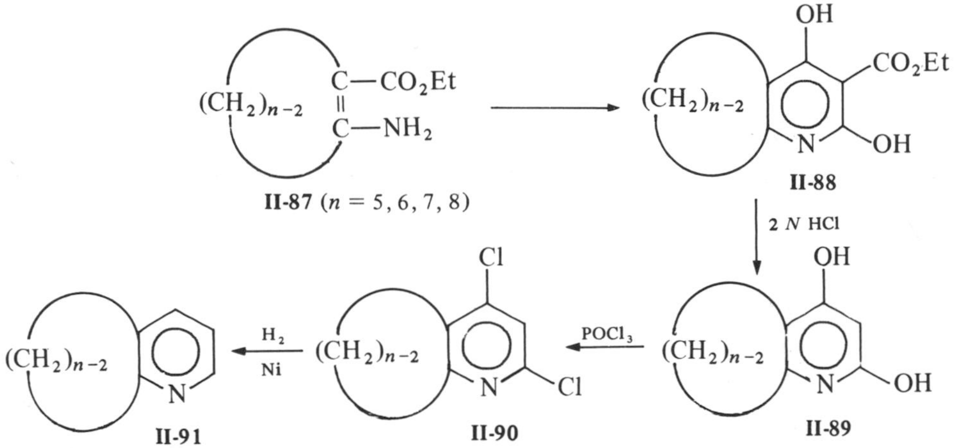

II-87 (n = 5, 6, 7, 8)   II-88   II-89   II-90   II-91

There is an anomaly, which has not yet been resolved, in the mechanism of certain Friedlander condensations. In some cases apparent rearrangements have occurred, while in other cases normal, direct condensation occurred. A good example of normal condensation would be the reaction of the cyclic β-amino-enones II-92 and II-78 with α-formylcycloalkanones II-93 and II-94 to give a series of para-bis-annelated pyridines (64). If the 2-formyl group was replaced by a 2-acetyl group, the condensation took a different course and the angular bis-annelated pyridine II-95 was formed. In a similar fashion pyridines II-96–98 were prepared. It is tempting to claim that the initial attack occurred at the ring carbonyl group rather than at the acetyl group. However, some studies involving other related systems were inexplicable in this fashion (37, 65). The β-aminoenones II-99 and II-100, II-100, prepared by treating II-93 and II-94, respectively, with ammonia, were found to condense with cyclopentanone or cyclohexanone to provide the

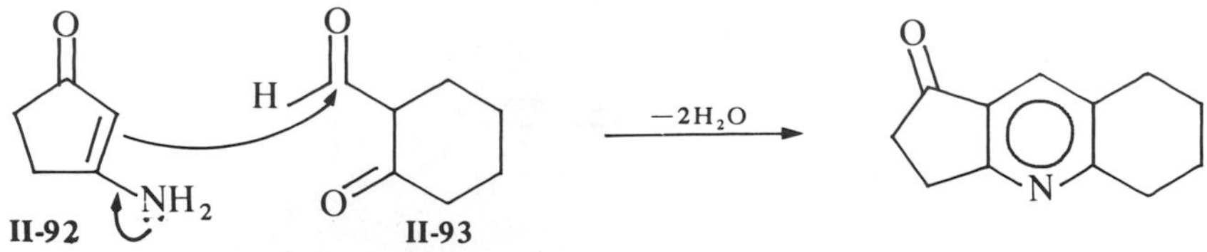

II-92   **II-93**

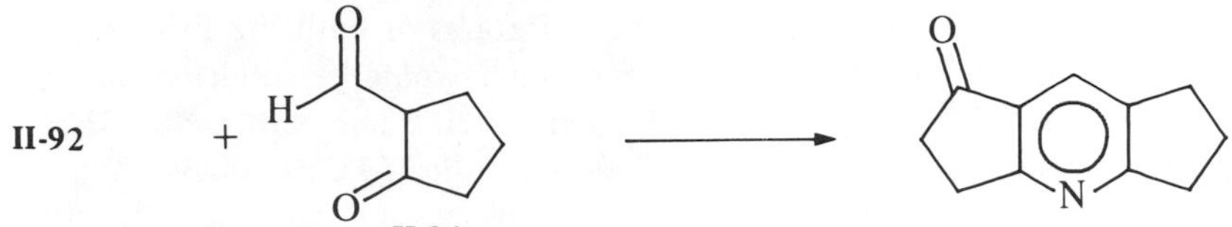

II-92 + **II-94**

II-78  + **II-93**

II-78 + **II-94**

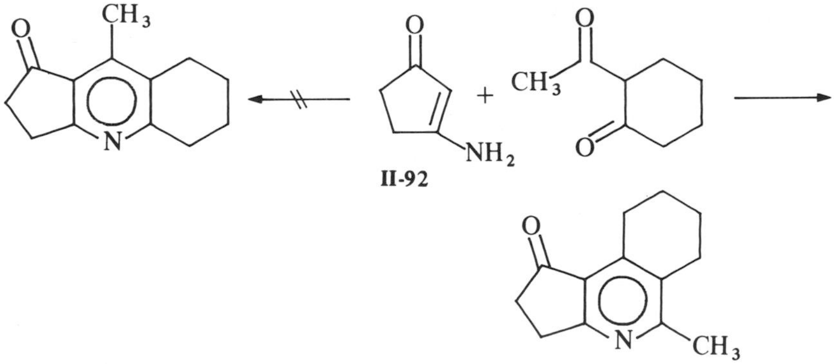

II-92

II-95

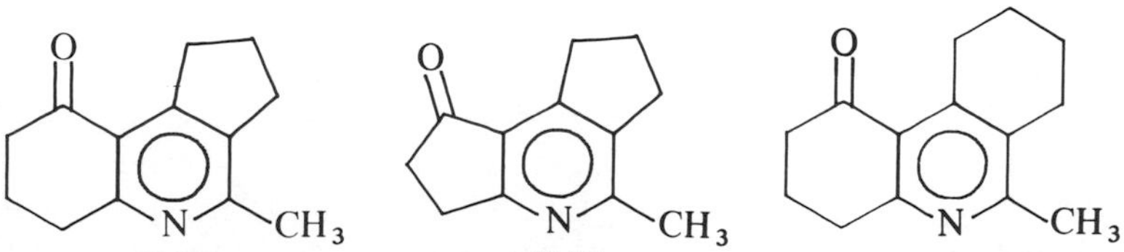

II-96   **II-97**   **II-98**

II-102 II-99 II-101

II-100

II-103

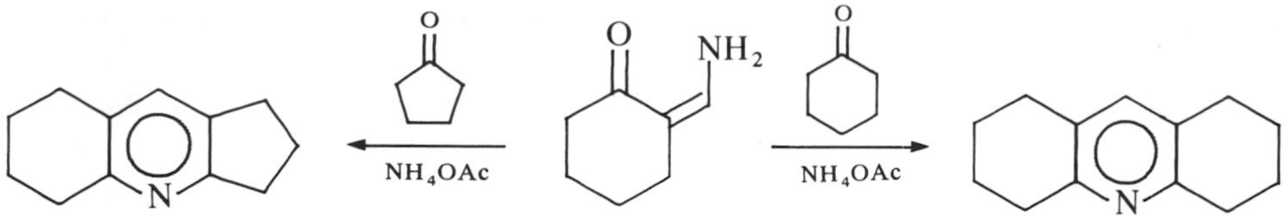

linear isomers **II-101–II-103**, where the angular products were expected. The mode of condensation appeared to depend on the substitution pattern of the β-aminoenone as illustrated for the reaction of cyclopentanone with methyl-substitute enamines **(II-107–II-109)** obtained by hydrogenation of the corresponding isoxazole **(II-104–II-106**, respectively) (66).

II-104 II-107 II-110

II-105 II-108 II-110

II-106 II-109 II-111

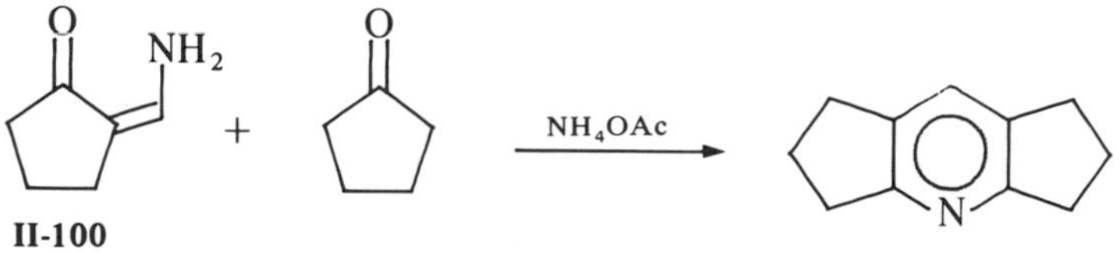

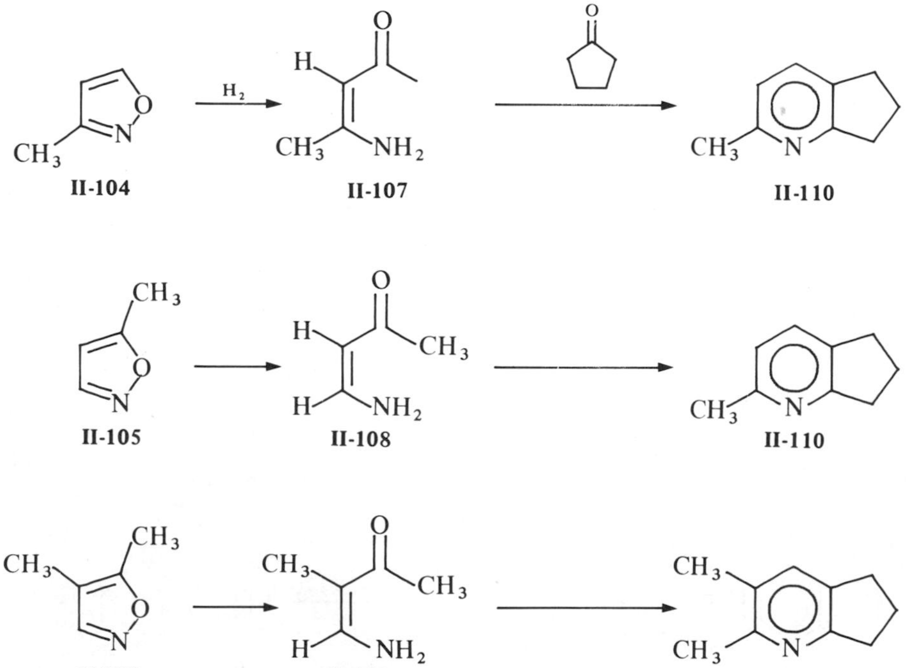

At this stage one could safely conclude that rearrangement occurred for β-amino-α,β-unsaturated ketones but not for analogous aldehydes (compare **II-107** and **II-108**). Curran (67) has made a similar observation for the reaction of **II-108** with 1,3-cyclohexanedione where the rearranged product **II-113** was observed. The mechanism, which he postulated, invoked a prior equilibrium between the two starting materials, giving rise to the enamine (**II-78**) of the 1,3-diketone and the keto-aldehyde precursor of **II-108**. These species then condensed in a normal fashion to form **II-113**.

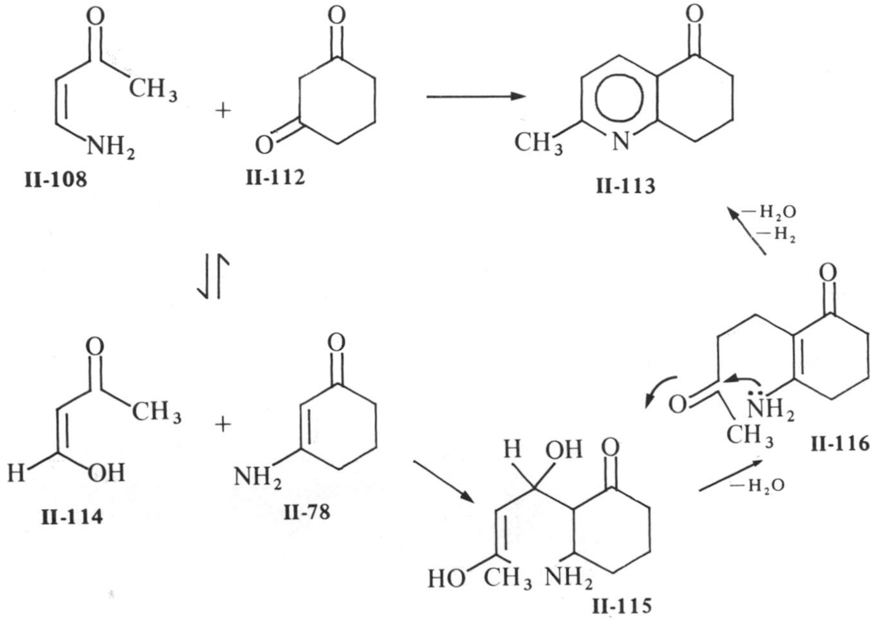

Having seemingly clarified this problem, this section is closed with a flagrant exception. Amine **II-118** should undergo rearrangement during its condensation with ketone **II-117** to give the linear fused isomer, analogous to **II-103**; but instead **II-119** is observed (68) *via* the normal condensation pathway.

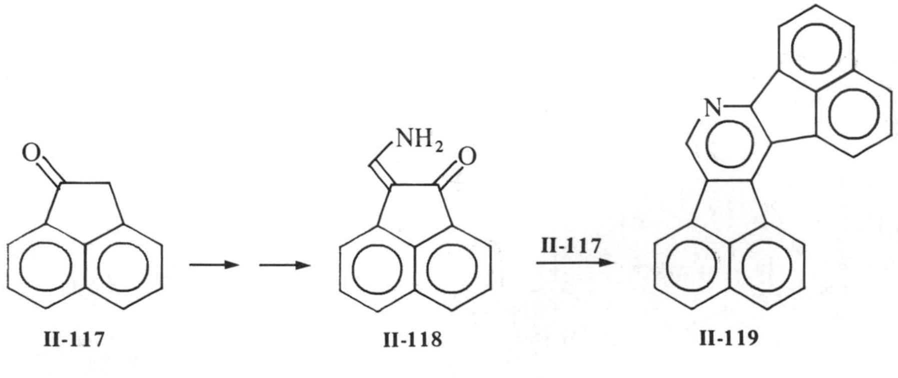

D.   Claisen- and Aldol-Type Condensations

The condensation reactions discussed thus far have involved reactions in which the principle step involved formation of a pyridine ring. A second class of condensation reactions leading to annelated pyridines would be one in which the pyridine ring was preformed and the annelated portion was synthesized in the condensation step.

II-120   **II-121**   **II-122**

Mosher et al. have shown that acylindanediones (**II-121**) may be prepared by reacting dimethyl 2,3-pyridinedicarboxylate (**II-120**) with a wide variety of methyl ketones in the presence of sodium methoxide (69). When R was an aryl group, the reaction was slower and accompanied by side reactions. Further treatment of **II-121** with hydrazine resulted in cyclocondensation at the 5-keto and 6-acyl groups to generate the 1,4-dihydropyrazolo[3′,4′:3,4]cyclopenta(1,2b)pyridine (**II-122**). Excess hydrazine resulted in hydrazone formation at the 7-keto group and Wolff–Kishner-type reduction could subsequently be effected. If **II-121** (R = CH$_3$ or C$_6$H$_5$) was treated instead with 1,2-ethylenediamine or o-phenylenediamine, the corresponding adducts **II-123** and **II-124** were then obtained (70). The methyl ester derivative of **II-121** (R = OCH$_3$) has also been prepared and some of the chemistry of its sodium salt has been investigated (71).

II-123   CH$_3$ **II-124**   C$_6$H$_5$

Jones and Jones have examined the condensation of ethyl pyridine-2,3-dicarboxylate (**II-125**) and -3,4-dicarboxylate (**II-129**) with dialkyl succinates and dialkyl glutarates (72). The reaction with diethyl succinate (**II-126**) led to the formation of quinolines **II-127** and **II-130**, which can be oxidized with lead tetraacetate to the corresponding quinones **II-128** and **II-131**. Diethyl glutarate reacted with **II-125** to give adduct **II-133**, which appeared to exist preferentially as the bis-enol **II-133b**. Acidic hydrolysis of **II-133** led to the formation of cyclohepta-[b]pyridine-5,9-dione (**II-134a**), which can be brominated with NBS to provide

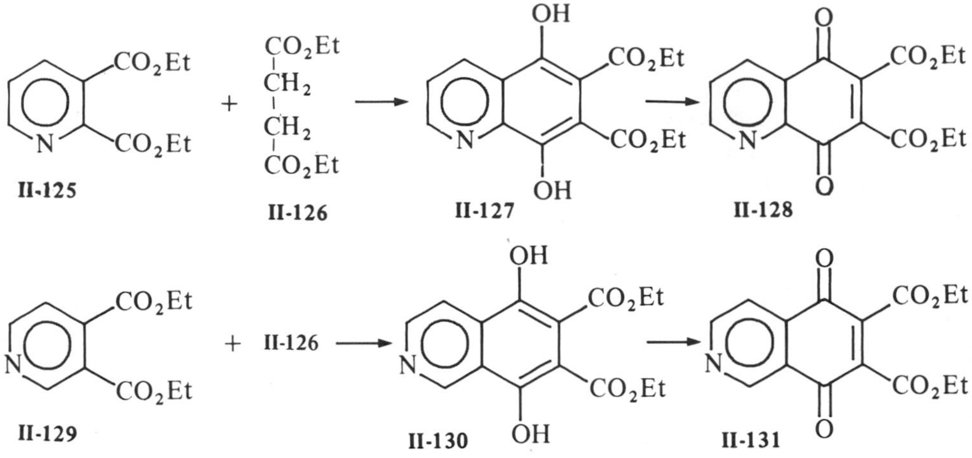

II-125 II-126 II-127 II-128

II-129 II-130 II-131

II-134b and/or II-134c. Finally, triethylamine will effect dehydrobromination with concurrent ring closure to give the cyclopropaquinolinedione **II-135**. Other reactions of the initial adduct **II-133** have also been studied (73). Similar treatment of **II-129** with **II-132** gave rise to the corresponding 3,4-fused cycloadduct **II-136**, which could likewise be converted into diketones **II-137** and **II-138**.

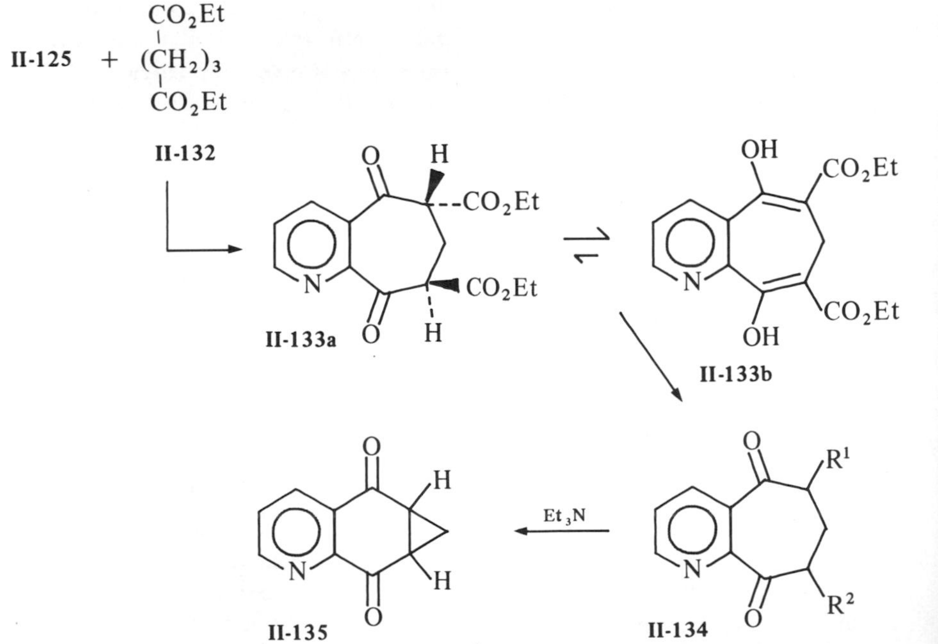

II-125 II-132 II-133a II-133b II-134 II-135

a $R^1 = R^2 = H$
b $R^1 = H$, $R^2 = Br$
c $R^1 = Br$, $R^2 = H$

II-129 + II-132 ——

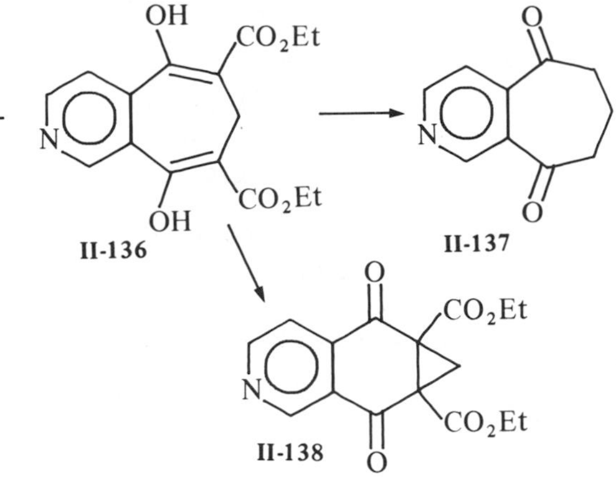

II-136 II-137 II-138

Binder has prepared 5,6-dihydro-2-pyrindin-7-one (**II-141**) by a Dieckmann cyclization of diester **II-139**, followed by hydrolysis and decarboxylation (74). Diester **II-139** was synthesized by the condensation of diethyl malonate with 3,4-pyridinedicarboxylic anhydride followed by reduction, saponification, decarboxylation, and finally esterification.

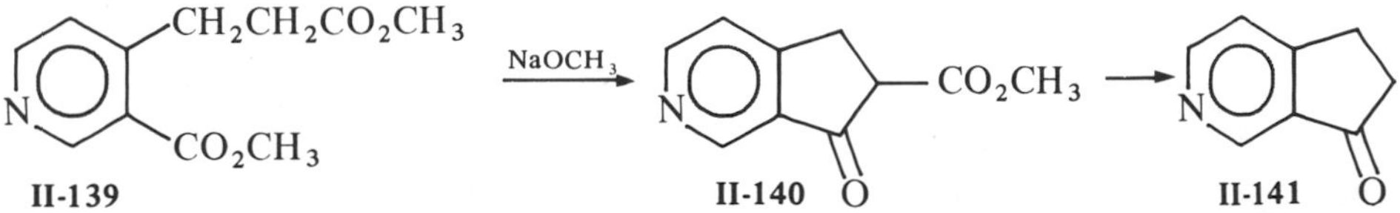

II-139 II-140 II-141

Bis-aldol condensations have been employed to prepare pyridotropolones. Thus, 2,3-pyridinedicarboxaldehyde (**II-142**) reacted with methoxyacetone in the presence of sodium hydroxide to give a mixture of two tropolones: **II-143** (10%) and **II-144**

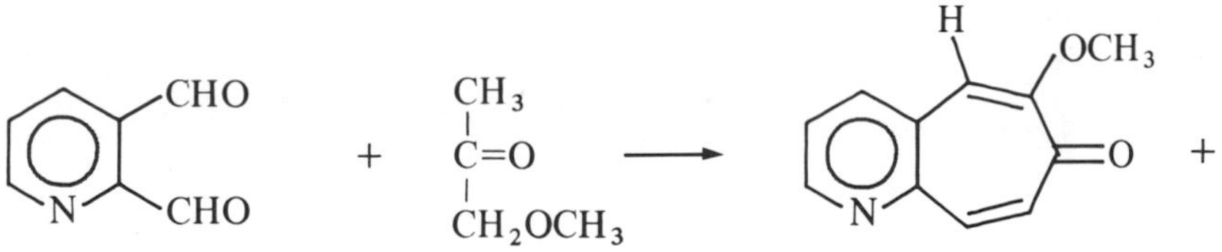

II-142 **II-143** (m.p. 153°)

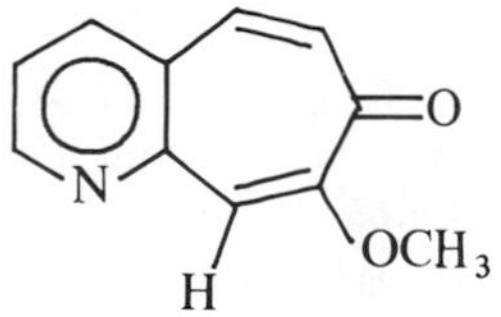

II-144 (m.p. 119°)

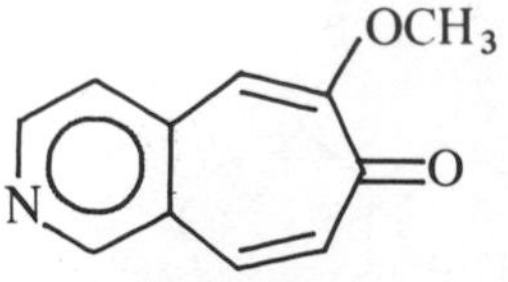

II-145

(90%) (75). These isomers were distinguished and identified by NMR. The analogous reaction with 3,4-pyridinedicarboxaldehyde led only to tropolone **II-145** (76), which was identified by NMR and deuterium-labeling studies and verified by an alternate synthesis of the other possible isomer **II-149** by way of two distinct aldol condensations.

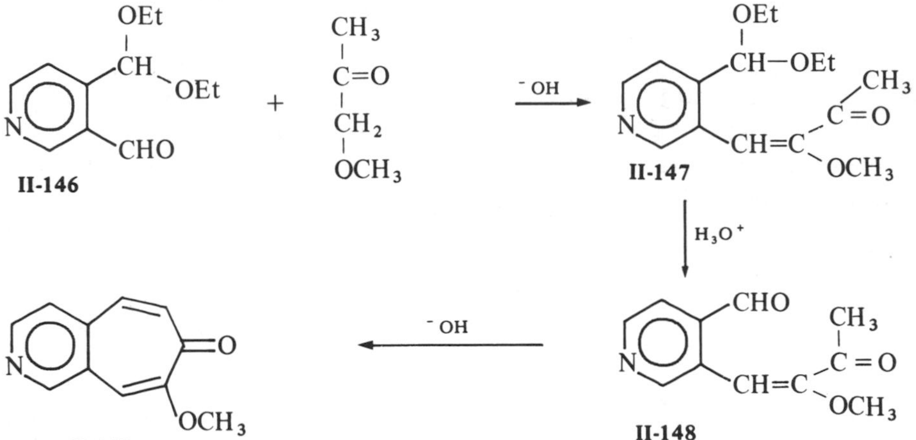

2.   Cyclization Reactions

A.   Nucleophilic Cyclizations

Pyridine is more susceptible to nucleophilic as opposed to electrophilic attack. Thus if a carbanionic center can be generated on a pyridine side chain, it can attack the parent ring preferably at the 2- or 4-position to generate an annelated derivative. Such cyclizations work best when the carbanion is located at a γ-carbon thus giving rise to a fused five-membered ring. Eisch et al. (77, 78) have treated 3-(3′-chloropropyl)-pyridine (**II-150**) with magnesium to prepare **II-151**, which then afforded the cyclized products **II-152**, **II-153**, and **II-154** in a ratio of 20:10:70. If prior to hydrolysis, the excess magnesium was removed, and the solution was diluted and heated; the ratio of the products **II-152**, **II-153**, and **II-154** became 20:50:30, respectively. Eisch et al. take this to be good evidence that **II-153** was produced directly by cyclization of **II-151** (the precursor to **II-154**), while **II-152** was produced only from **II-150** and magnesium metal.

Pines and Kannan (79, 80) found that sodium or potassium catalyzed the reaction of 3-ethylpyridine with ethylene to give a variety of 2,3-pyrindanes (**II-162** and **II-163**), which could be explained by the sequence outlined in Scheme II-1. Carbanion **II-156** can add to ethylene to give **II-157**, which can either cyclize to **II-159** or transmetallate to give **II-158**. Due to its high reactivity, **II-159** was not isolated but subsequently went on to metallate and add ethylene twice more at its 7-position, ultimately providing **II-162**, which could then, more slowly, metallate and add ethylene at the 5-position to provide **II-163**. This same product could arise in a similar stepwise fashion from **II-158**.

Pines et al. have also reported an interesting intramolecular variation on the same process (81). When 6-(3′-pyridyl)-1-hexene (**II-166**) was treated with base a carbanion was generated which then attacked the double bond at the 2-position to form **II-168**, which could cyclize again at either the 2- or 4-position of the pyridine ring in an 88:12 ratio, respectively.

B. Electrophilic Cyclizations

Due to the inductive deactivating influence of the nitrogen atom, pyridine is reluctant to undergo electrophilic aromatic substitution. When it does undergo such reactions, attack occurs preferentially at the 3-position. One of the only examples of this process (82) leading to an annelated pyridine is the reductive cyclization of **II-171** which presumably proceeds *via* reduction of the carbonyl group followed by pyridine capture of the incipient carbonium ion **II-172** to give **II-173**, after loss of a proton.

Electrophilic cyclizations were observed, however, when a phenyl or an alkenyl substituent on a pyridine ring attacked a positive site on an adjacent substituent to close a ring, thus giving an annelated pyridine. When **II-174** was treated with poly-phosphoric acid, cyclization occurred to give the azafluorenone **II-175** with loss of mesitylene (83). A somewhat similar case was the acid-catalyzed formation of **II-179** from the nicotinic acid derivative **II-176** (84). Ramirez and Paul proposed

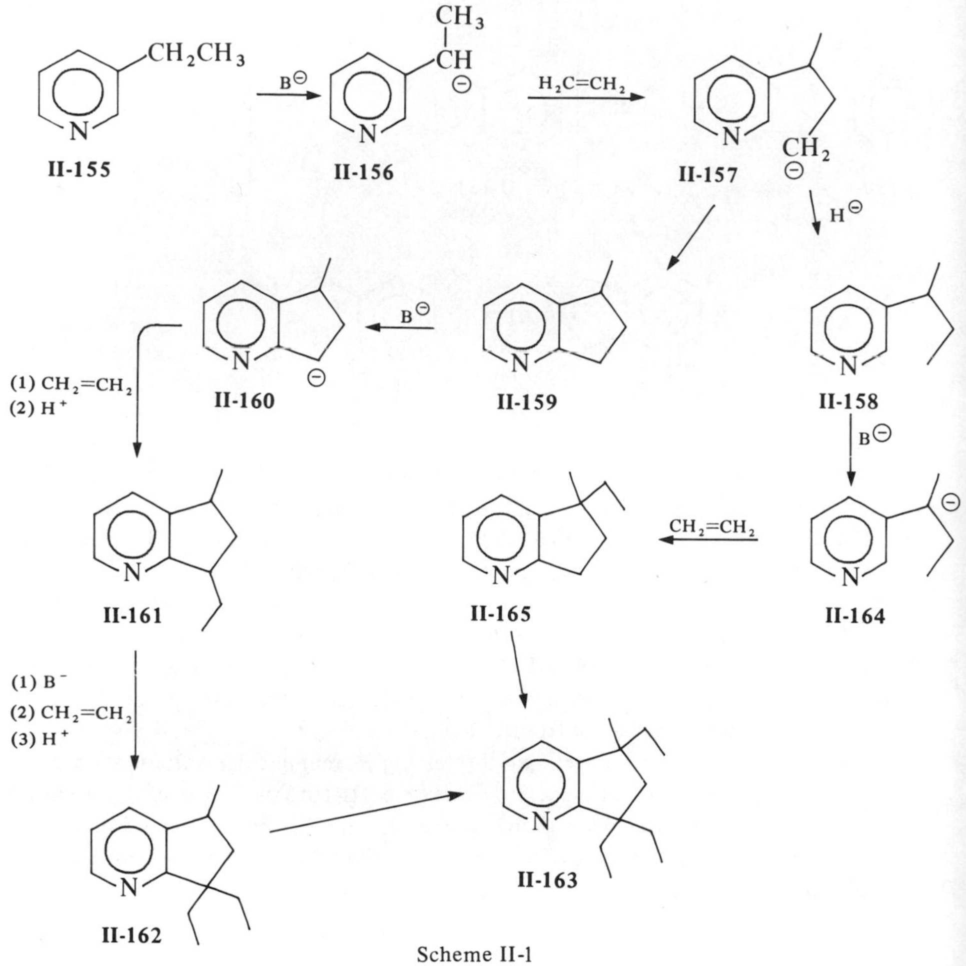

Scheme II-1

II-169 (88%)   +   II-170 (12%)

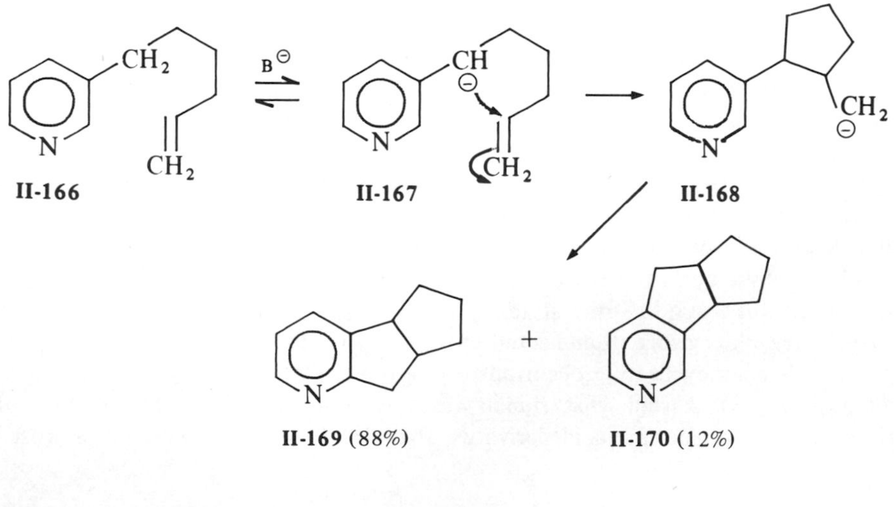

II-171 II-172

II-173

II-174 II-175

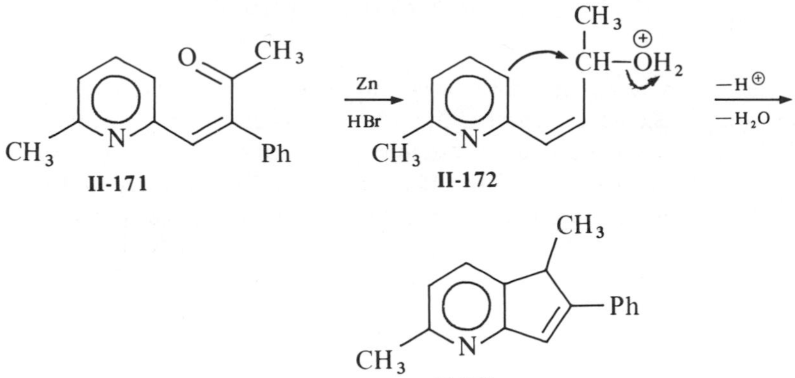

that nucleophilic attack on the carboxyl group occurred through the enol form **II-177**. The intermediate spiro-diketone **II-178** then underwent ring-opening in what was essentially a reversal of the previous step. Under the reaction conditions the 6-chloropyridine was converted to the corresponding 2-pyridone.

II-176 II-177

II-179 II-178

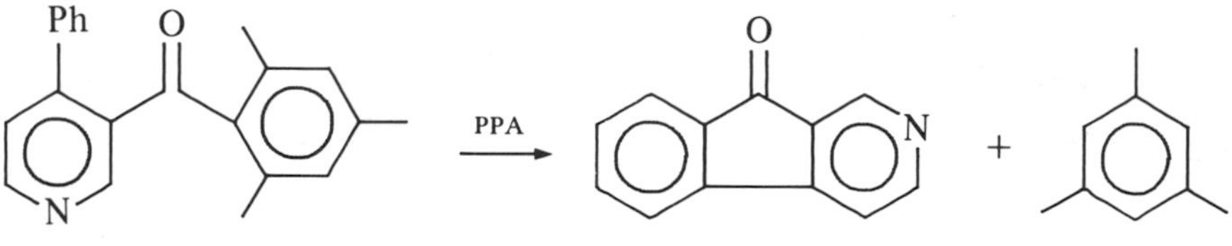

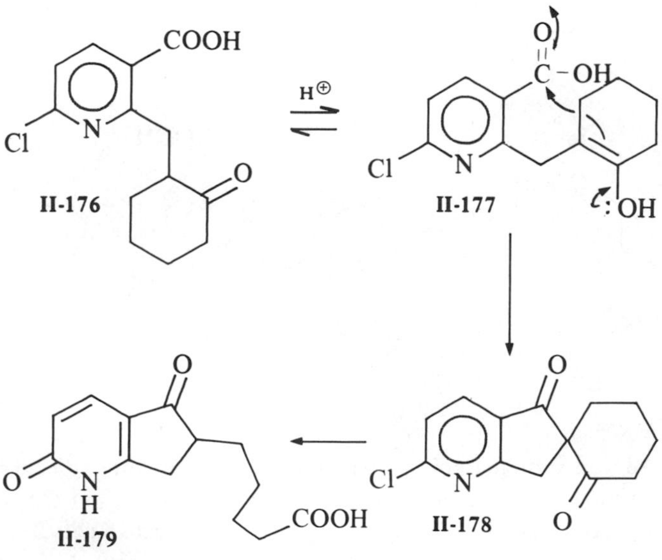

C.   Friedel–Crafts Acylations

Friedel–Crafts acylations have been employed primarily in the synthesis of azadibenzocycloalkanones. The reaction does not occur on the pyridine nucleus, which is deactivated toward electrophilic substitution by the electronegative nitrogen atom. Rather an acyl function on pyridine is typically attacked by a phenyl ring which is bonded to an *ortho* ring carbon by zero, one, or two methylene units. Closure in this fashion forms a central ring containing usually five, six, or seven carbons. The preparation of 2-azafluorenone followed this route (85);

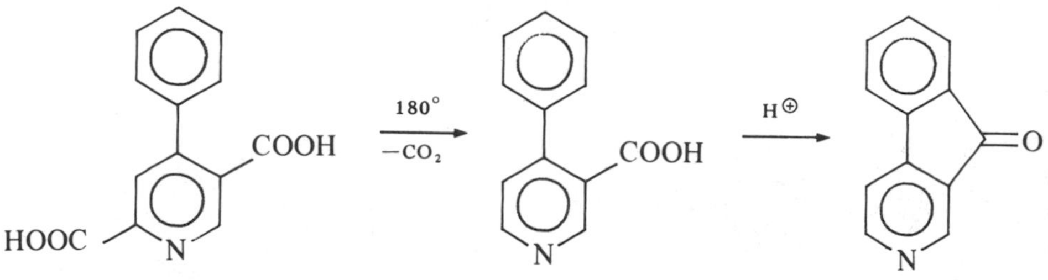

II-180   II-181   II-182

4-phenyl-2,5-pyridinedicarboxylic acid (**II-180**) heated at 180°C decarboxylated to give **II-181**, which was then heated at 100°C with concentrated sulfuric acid to provide **II-182**. A double barreled version of this reaction has been reported for 4-phenyl-3,5-pyridinedicarboxylic acid (**II-183**) (86), which when heated with sulfuric acid gave not only the expected azafluorenone **II-184** but also a low yield of **II-185** resulting from two Friedel–Crafts acylations.

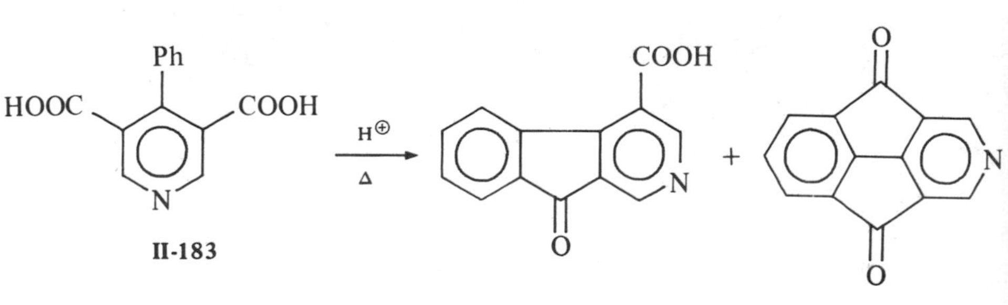

II-183   II-184   II-185

A series of aza-anthraquinones was prepared *via* an intermolecular acylation (87). When reacted with a variety of substituted benzenes, 2,3-pyridinedicarboxylic acid (**II-186**) provided **II-187**, which showed some biological activity.

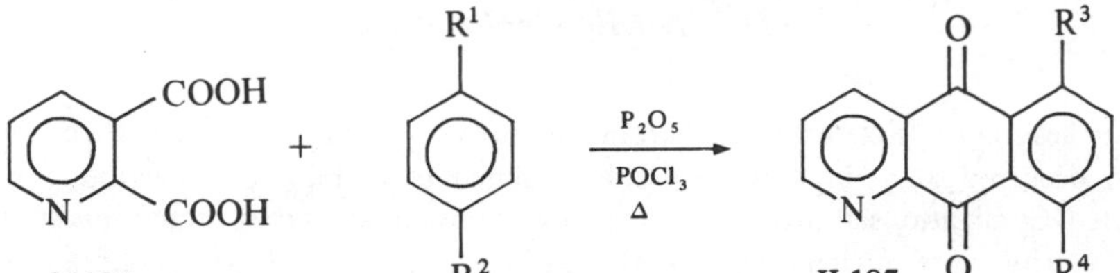

II-186 **II-187**

$R^1 = OCH_3$, $R^2 = OH, OCH_3$
$R^1 = OH$, $R^2 = Cl$
$R^1 = R^2 = Cl$

$R^3 = R^4 = OH$
$R^3 = OH$, $R^4 = Cl$
$R^3 = R^4 = Cl$

Two groups have independently reported preparation of the four isomeric aza-dibenzocycloheptanones **II-188** and aza-dibenzocycloheptenones **II-189** (88–93).

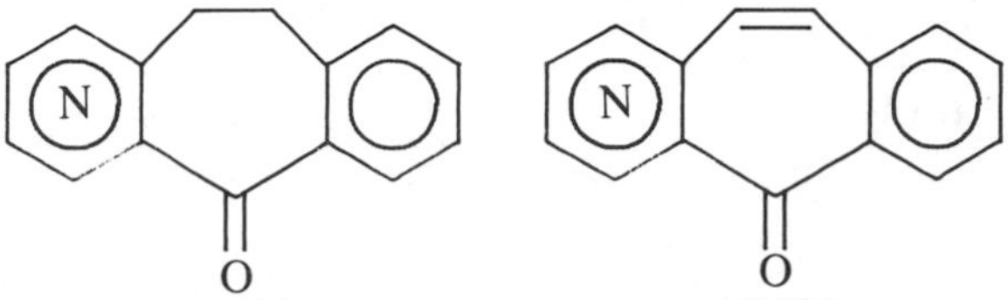

a 1-aza
b 2-aza
c 3-aza
d 4-aza

II-188 **II-189**

The saturated systems **II-188** were all prepared by Friedel–Crafts ring closure on the appropriate β-phenylethyl substituted pyridine carboxylic acids **II-190–93**.

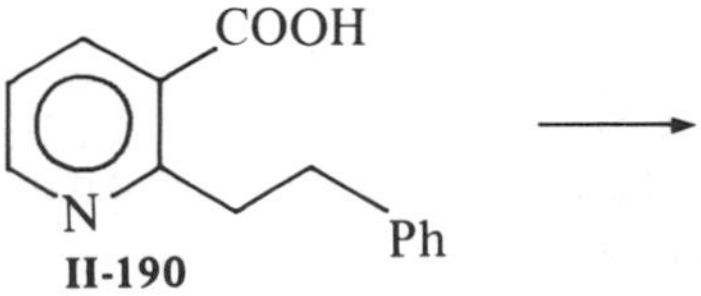

II-190 → **II-188a** (1-aza)

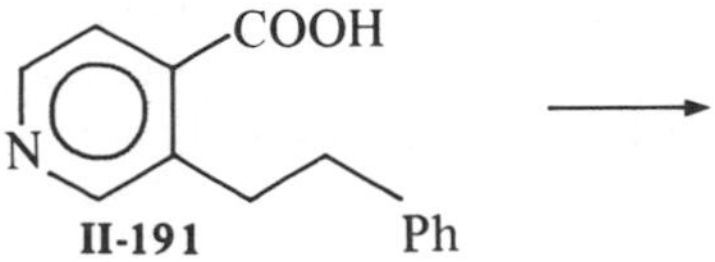

II-191 → **II-188b** (2-aza)

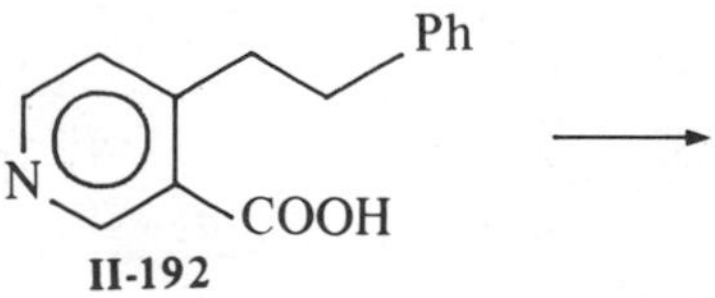

II-192 → **II-188c** (3-aza)

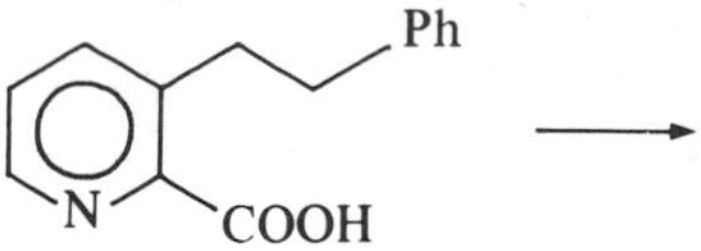

II-193 → **II-188d** (4-aza)

The corresponding unsaturated systems **II-189** were derived from **II-188** by oxidation with selenium dioxide (88) or by treatment first with *N*-bromosuccinimide, followed by triethylamine (88, 90). A variety of similar derivatives have been reported (89, 91).

D. Nitrile Cyclizations

Nitriles may cyclize at either carbon or nitrogen to give pyridines. If the cyclization occurred at a ring site or a site attached to a ring, an annelated pyridine
resulted. Cyclizations at nitrogen were more common and incorporated this nitrogen
into the pyridine nucleus. One of the early syntheses of 2,3-cyclopentapyridine
(94) followed this approach where the ring closure of keto-nitrile II-195 was effected
as a reductive amination.

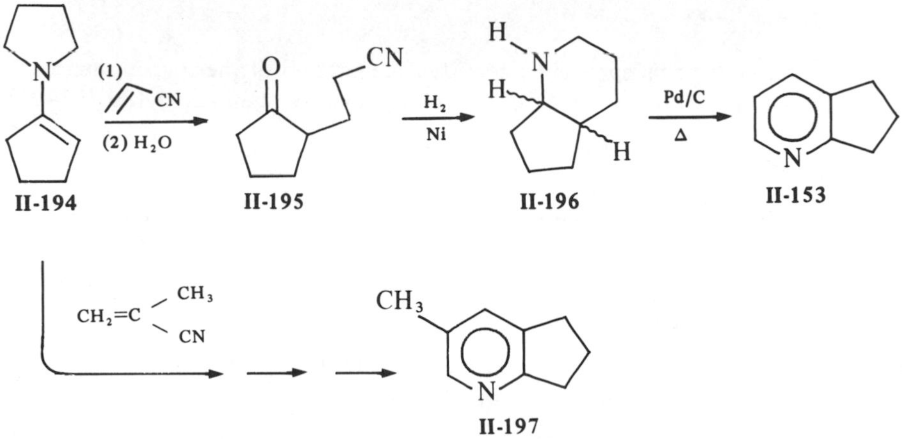

Employing methacrylonitrile in the initial alkylation ultimately led to 2,3-
cyclopenta-5-methylpyridine (II-197) (95). Aside from enamines such as II-194,
β-ketoesters may also be treated with acrylonitrile to synthesize the cyclic
precursors. This route has been used to prepare two methyl-substituted derivatives,
II-201 and II-203 (96). In these cases the final dehydrogenation of the piperidine
ring also resulted in loss of the bridgehead ethoxycarbonyl group.

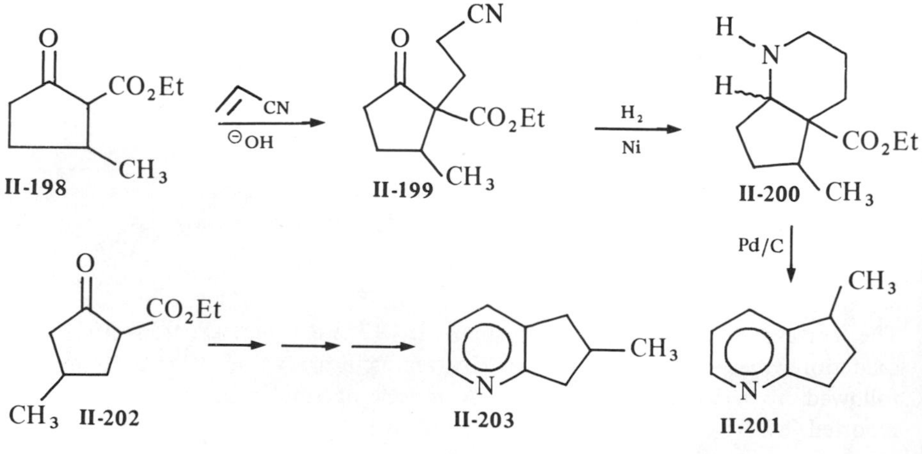

II-204 II-205 II-206

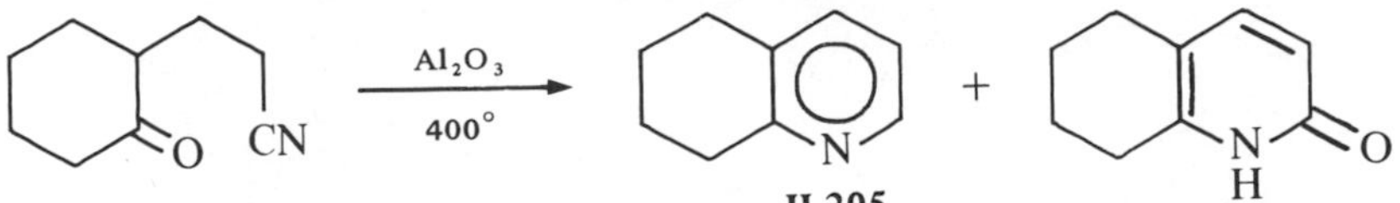

The analogous cyclohexanone systems have been used to prepare 5,6,7,8-tetrahydroquinolines, for example, **II-205** (97). It was found that the cyclization and dehydrogenation could be effected in a single process by heating the keto-nitrile to 400°C in the presence of alumina. Yields were generally low (20–25%), and the product was often contaminated by a nearly equivalent amount of pyridone **II-206** (98).

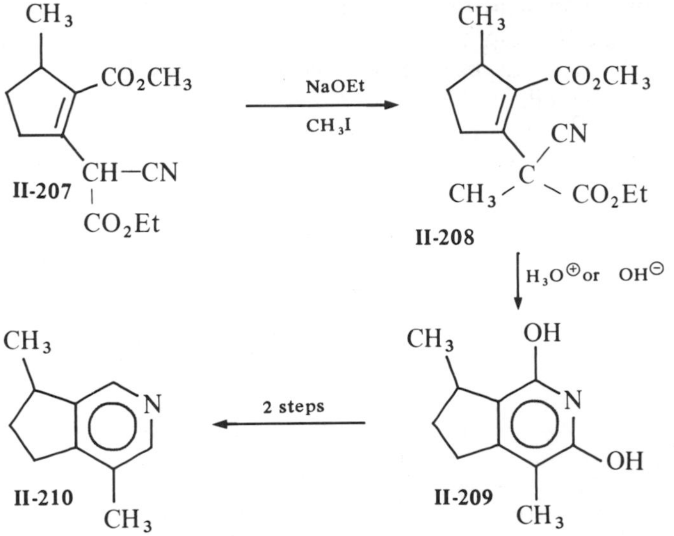

When nitrile cyclization occurred at a carbonyl group attached directly to a carbocyclic ring, a 3,4-annelated pyridine resulted. Most frequently, the carbonyl moiety was an ester group, which gave rise to a hydroxy substituent on the final pyridine (99, 100). An early synthesis of Actinidine (**II-210**) followed this approach (101). A similar cyclization has been observed with acid chlorides, such as **II-211**, which gave pyridine **II-213** *via* pyridone **II-212** (102).

II-212 II-213

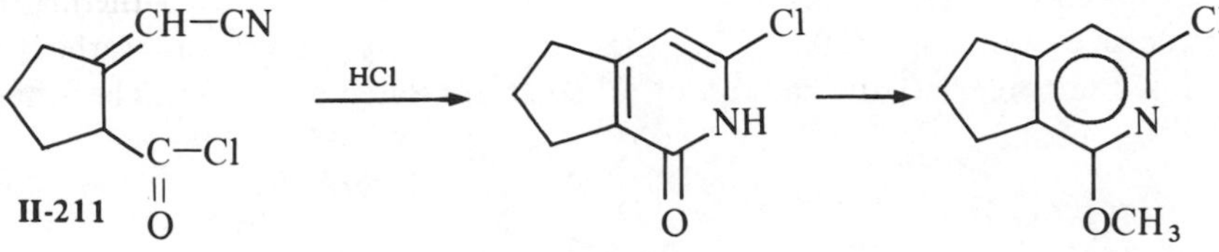

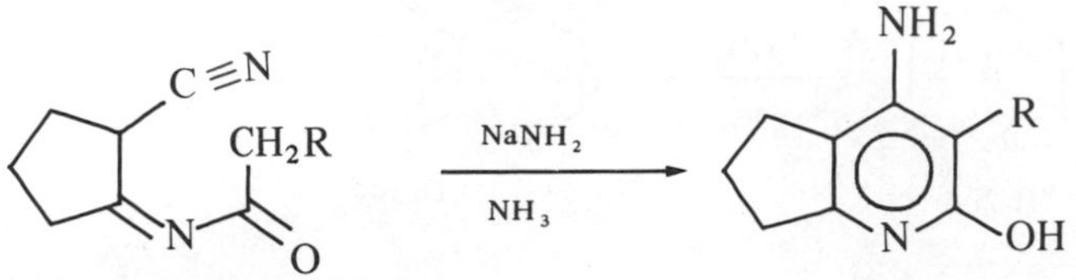

II-214a R = H
    b R = CH₃
    c R = Cl
    d R = COCH₃

II-215

Ring closures leading to annelated pyridines which result from attack at the carbon end of a nitrile are rare and convert the nitrile nitrogen to an amino substituent on the pyridine ring. Thus, the carbanion generated from nitrile **II-214** attacked the cyano group to give pyridine **II-215** after tautomeric shift of two protons (103). When the imine **II-216**, generated from dimedone, reacted with tetracyanoethylene, adduct **II-217** was formed. The imino nitrogen then attacked a cyano group so that after proton transfer and loss of HCN, the annelated pyridine **II-218** was obtained (104).

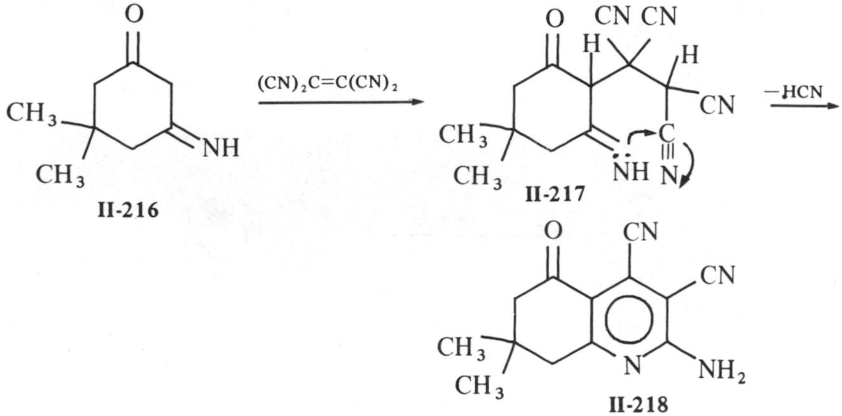

II-216

II-217

II-218

E.   Azatriene Cyclizations

It is known that azatrienes of the types **II-219–221** can cyclize ultimately to form the pyridine nucleus. These reactions have been postulated to proceed either by a carbanionic mechanism under basic conditions or simply by a thermal 6-π electrocyclic process (105). If the azatriene possesses a carbocyclic ring fused to two adjacent carbon atoms, then an annelated pyridine will be formed.

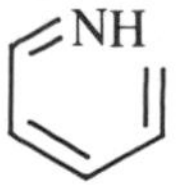

II-219

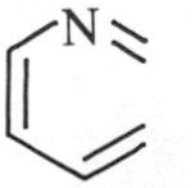

II-220

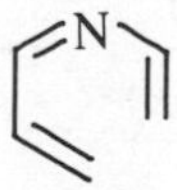

II-221

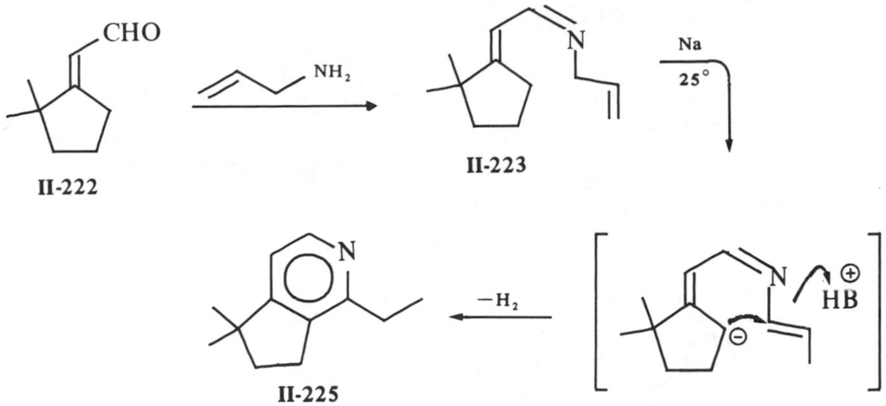

II-222   **II-223**   **II-225**   **II-224**

Thus, aldehyde **II-222** upon treatment with allylamine provided imine **II-223**, which cyclized by a carbanionic intermediate **II-224** to give the annelated pyridine **II-225** after an aromatization step (105).

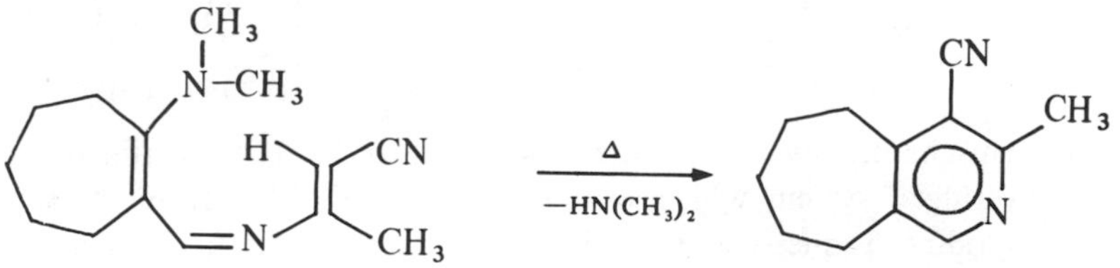

II-226   **II-227**

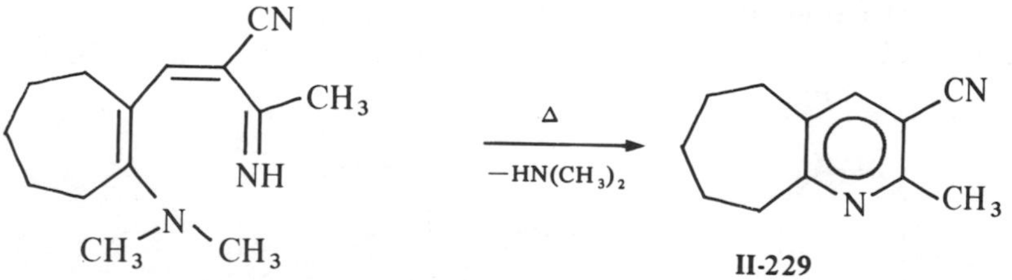

II-228   **II-229**

Other azatrienic systems underwent purely thermal cyclizations. Systems **II-226** and **II-228** loose dimethylamine after ring closure to generate the pyridines **II-227** and **II-229**, respectively (106). Closely related are the cyclizations of some azatriene derivatives of indene and acenaphthalene (**II-230, II-232, II-234**). Upon treatment with ammonium chloride or molten ammonium acetate these molecules underwent electrocyclic ring closure followed by the loss of trimethylamine and finally demethylation to give azafluorenes **II-231** and **II-233**, as well as 7-azafluoranthrene (**II-235**) (107).

II-230 II-231 (70%)

II-232 II-233 (80%)

II-234 II-235 (65%)

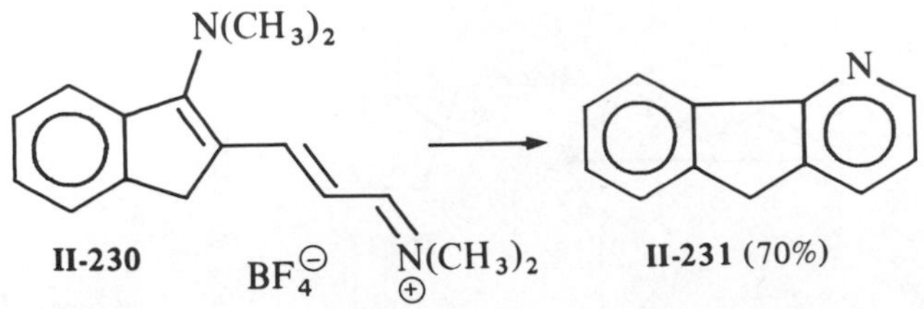

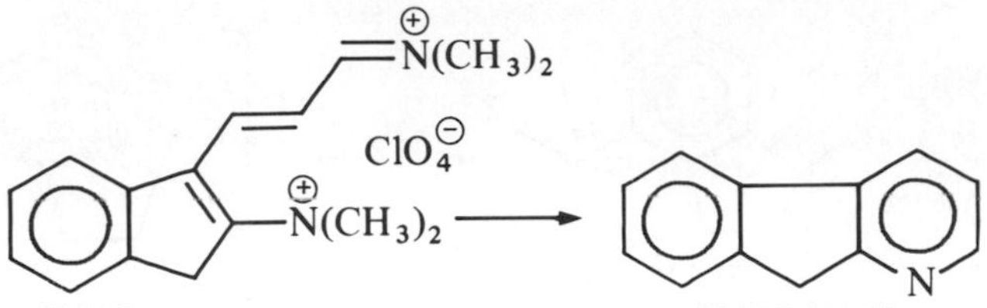

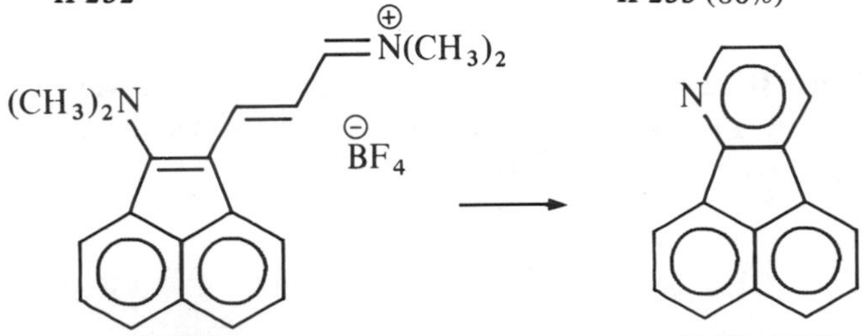

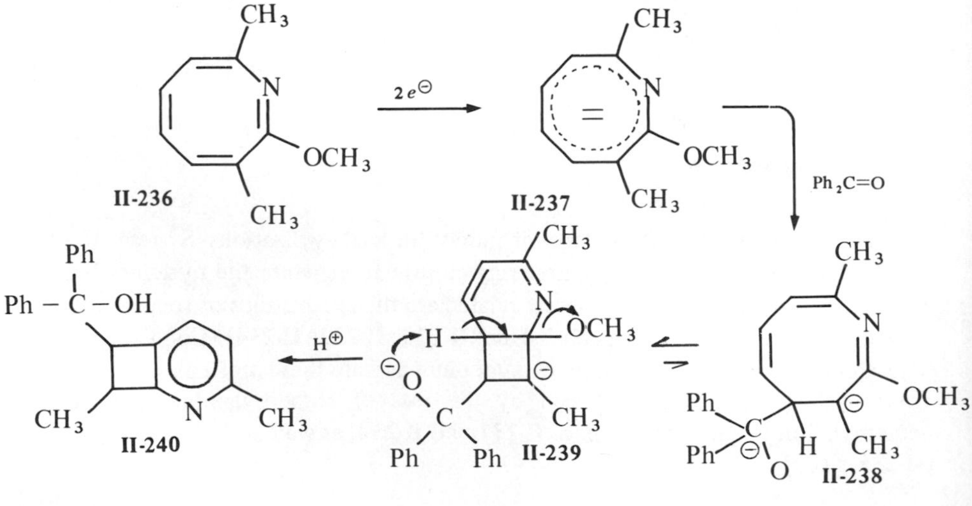

If the azatriene moiety is contained in a ring, then electrocyclic ring closure will generate a bridged system which would simply be the dihydro-precursor to an annelated pyridine. Paquette and Kakihana have reported the first preparation of a cyclobutapyridine by the reaction of an azocinyl dianion with a nonenolizable ketone, such as benzophenone (108). When 2 M equivalents of benzophenone were added to a THF solution of dipotassium 3,8-dimethyl-2-methoxyazocinate (**II-237**), adduct **II-238** was presumably obtained. Valence isomerization of **II-238** to the

II-236 II-237 II-238 II-239 II-240

bicyclic tautomer **II-239** was followed by the intramolecular elimination of methanol to aromatize the pyridine ring, giving **II-240**, after protonation. The proposed mechanism is consistent with the fact that the 3,4-dihydro derivative (**II-241**) of **II-236**, when treated with potassium *t*-butoxide, gave dimethyl-2,3-cyclobutapyridine **II-243** in high yield (107). A similar reaction of the next higher homolog of **II-241** has also been reported (314).

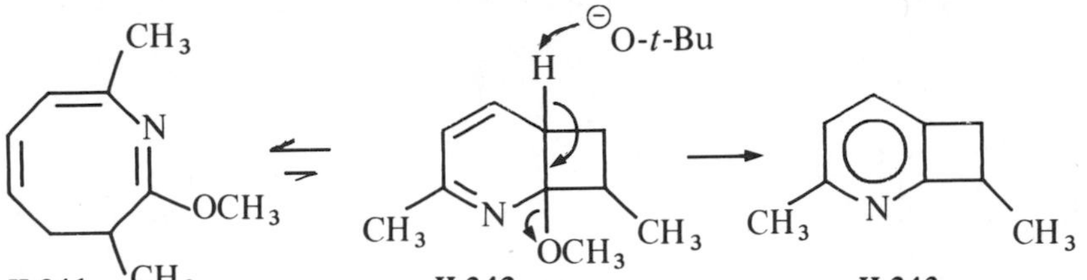

II-241   II-242   II-243

A somewhat related cyclization occurred in the corresponding nine-membered ring triene **II-244** (109, 110). Sensitized irradiation at $-10°C$ produced an intermediate for which the structure **II-245** was suggested. Alumina catalyzed cyclodehydration of **II-245** produced the bicyclic system **II-246**. Two other pathways for the conversion of **II-245** to **II-246** were also reported. Lithium aluminium hydride reaction of **II-246** gave a mixture of 5*H*- and 7*H*-1-pyrindines (**II-247** and **II-248**) (110, 111).

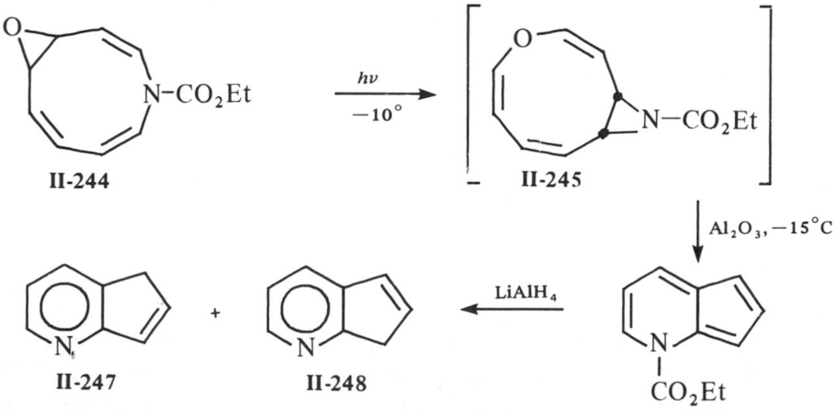

II-244   II-245

II-247   II-248   II-246

When a benzene solution of the *O*-allyl ether of cyclohexanone oxime was heated at 200° in a sealed tube, 5,6,7,8-tetrahydroquinoline (**II-61**) was obtained (303); the mechanism is outlined in Scheme II-2. The *O*-allyl ether rearranged to an *N*-oxide which tautomerized to an *N*-hydroxy species which then dehydrated to an annelated azatriene of type **II-221**. In a similar fashion the analogous cyclopenta-, cyclohepta-, and cyclooctapyridines were prepared in yields of 20%, 44%, and 48%, respectively.

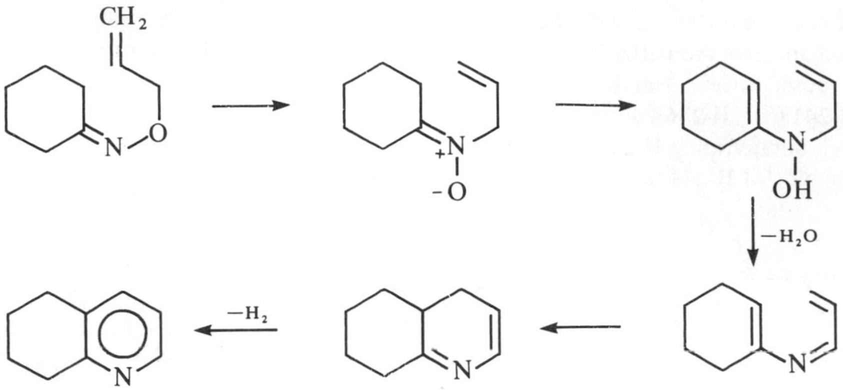

II-61 (43%)   Scheme II-2

F. Thermal Cyclizations

The thermolysis of ketone *N,N,N*-trimethylhydrazonium fluoroborates has been shown to provide pyridine derivatives (112). Hindered cyclic ketones and benzylic cyclic ketones, such as 1-indanone or 1-tetralone, gave hydrazonium salts (**II-249**), which provided *bis*-annelated pyridines after pyrolysis (113). The origin of C-4 of the pyridine ring was not obvious but labeling experiments suggested that this carbon may be derived from the trimethylamine moiety of the quaternary fluoroborate.

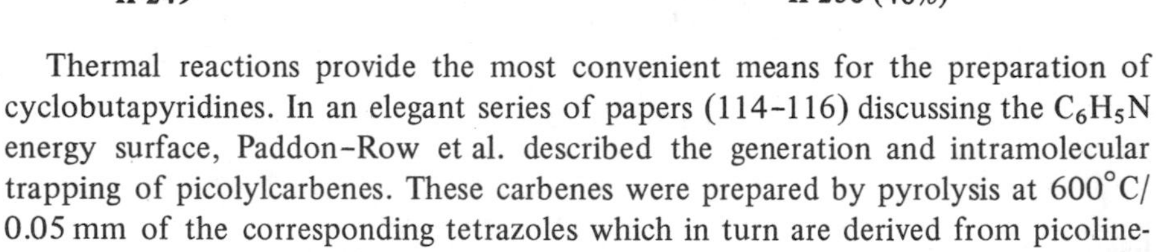

II-249   **II-250** (40%)

Thermal reactions provide the most convenient means for the preparation of cyclobutapyridines. In an elegant series of papers (114–116) discussing the C_6H_5N energy surface, Paddon–Row et al. described the generation and intramolecular trapping of picolylcarbenes. These carbenes were prepared by pyrolysis at 600°C/ 0.05 mm of the corresponding tetrazoles which in turn are derived from picoline-carbonitriles. An interesting feature was that the initially formed carbene could isomerize so that intramolecular trapping by the α-methyl group became possible. Thus, **II-252**, which was generated from **II-251**, could "migrate" the carbene functionality around the pyridine ring to give first **II-253** and finally **II-254**, which then underwent C—H insertion to provide 2,3-cyclobutapyridine. Products were also observed to result from phenylnitrene formation. Azacycloheptatrienylidenes were proposed as possible intermediates in these carbene migrations.

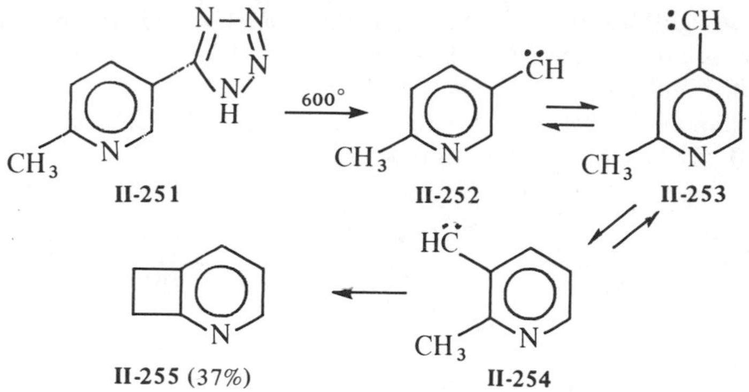

Riemann and Trahanovsky have shown that flash vacuum pyrolysis of phenyl propargyl ether led to the formation of benzocyclobutene along with some 2-indanone (117). The same reaction applied to propargyl 4-pyridyl ether provided both isomers of cyclobutapyridine (II-255 and II-265) (118, 119). A somewhat elaborate mechanistic scheme has been suggested to explain this transformation. An initial Claisen rearrangement gave II-257, which then underwent an intra-molecular [2 + 4] cycloaddition to provide II-258, which decarbonylated to II-259. Fission of two bonds would generate carbene II-263, which could insert in an adjacent C—H bond to give II-264, which ring closed to II-265. If a 1,3-sigmatropic bond shift occurred prior to bond fission, then the same sequence of steps could give rise to the 2,3-fused isomer II-255.

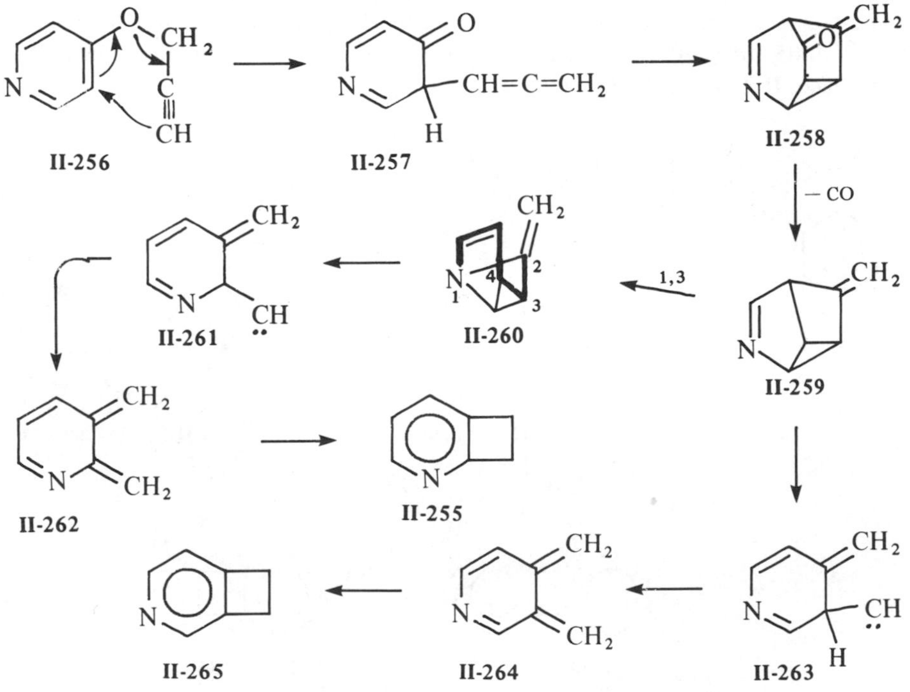

A more straightforward pyrolytic elimination has been employed to prepare [2,3:5,6]dicyclobutapyridine (**II-268**) (120, 304). Lithium aluminum hydride reduction of 3,5-dicarbethoxy-2,6-dimethylpyridine (**II-266**), followed by treatment with thionyl chloride provided the dichloride **II-267**. Sublimation (750°C/0.1 mm) of **II-267** in a quartz pyrolysis tube gave **II-268**.

II-266 **II-267** **II-268** (15%)

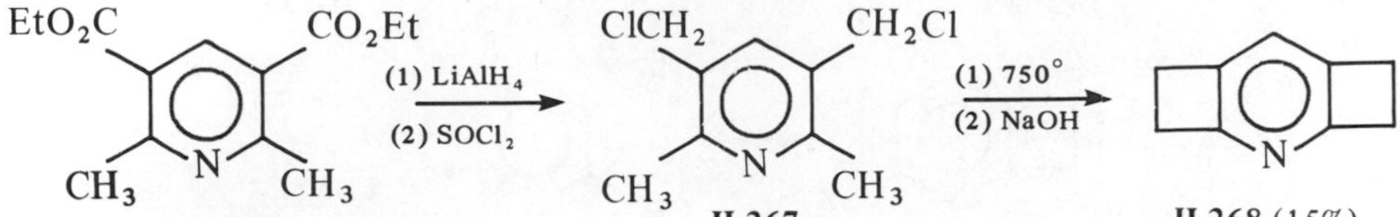

The thermal extrusion of nitrogen from cyclic diazo compounds has been employed to prepare both mono- and diazabiphenylenes. When 2,7,9,10-tetraazaphenanthrene (**II-269**) was passed through a silica tube (900°C/0.04 mm), 2,7-diazabiphenylene (**II-270**) was produced (121).

II-269 **II-270** (60%)

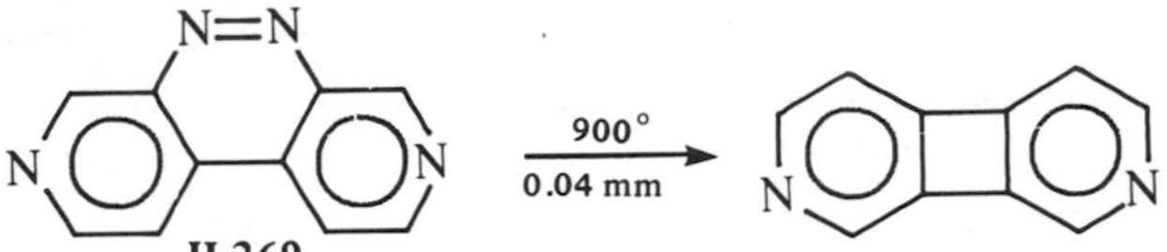

A novel photocyclization reaction has been observed for 3-phenylazopyridine (**II-271**), which after 72 h of irradiation as a 1% solution in concentrated sulfuric acid, provided a mixture of **II-272** (19%) and **II-273** (3%). Vacuum thermolysis (800°C/0.05 mm) of these two materials gave the two isomeric monoazabiphenylenes **II-274** and **II-275**, respectively (122).

II-271 **II-272** (19%) **II-273** (3%) **II-274** (58%) **II-275** (34%)

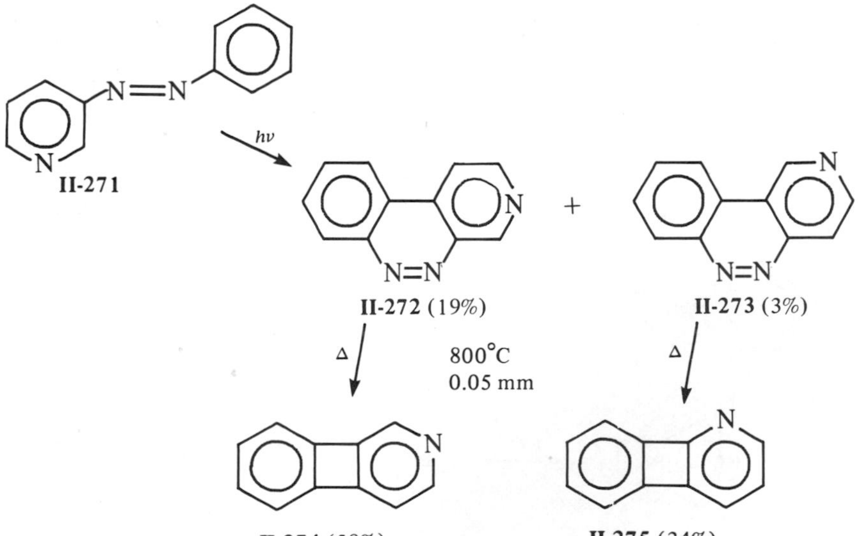

G.   Other Cyclizations

Dehydrocyclization of 3-methyl-4-phenylpyridine derivatives has been shown to lead to 2-azafluorenes (123, 124). Thus, when **II-276** was subjected to a type **K-16** industrial dehydrogenation catalyst at 550°C, the cyclized material **II-277** was obtained along with a good deal of unchanged **II-276**. It has been shown that the same reaction can be accomplished in the presence of iodine or even in the absence of catalyst at elevated temperatures (124).

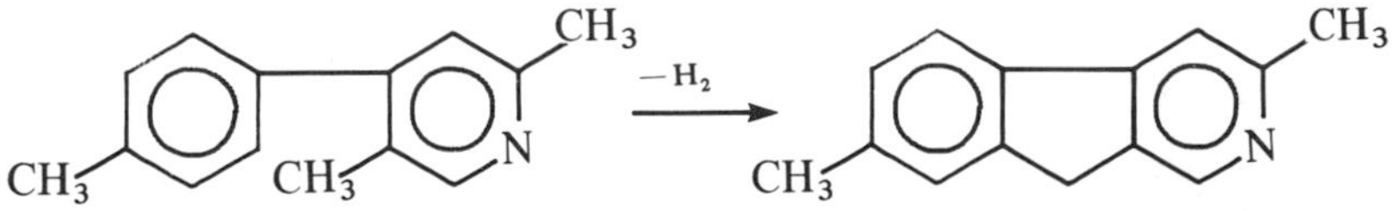

A similar cyclization has been observed for the aza-fluorene derivative **II-278**, which ring closed in both possible directions with loss of HCl to provide **II-279** and **II-280** (125).

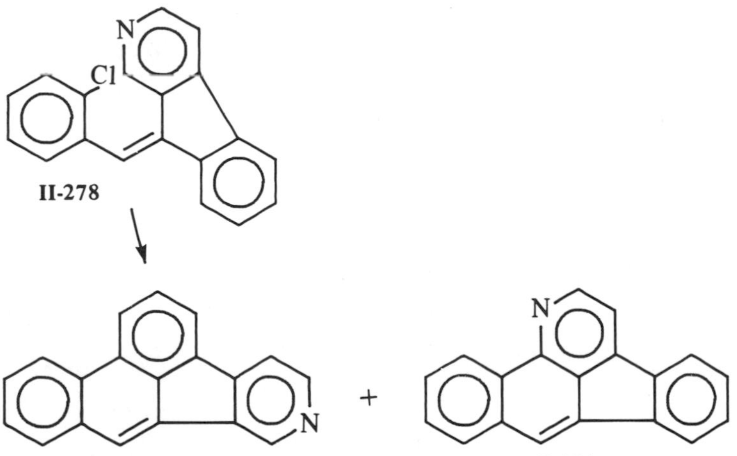

A more unusual reaction involved the chlorination of 2-propylpyridine (**II-281**) in carbon tetrachloride at 550°C (126). The major observed product was heptachloro-5*H*-1-pyrindine (**II-282**) in which the location of the *gem*-dichloro moiety was determined by x-ray diffraction analysis (127).

A simple pyridotropolone synthesis has been reported in which the pyridine portion of **II-287** was formed by intramolecular cyclization of an enamine derivative (128). Reaction of **II-285** with ethyl (α-acetyl-β-ethoxy)acrylate provided the substituted tropolone **II-286**, which cyclized at the ester carbonyl group to provide **II-287**, after loss of ethanol.

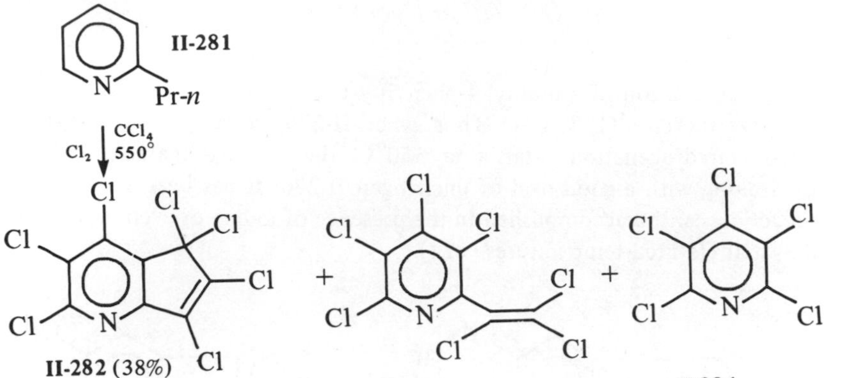

II-281

II-282 (38%)

II-283

II-284

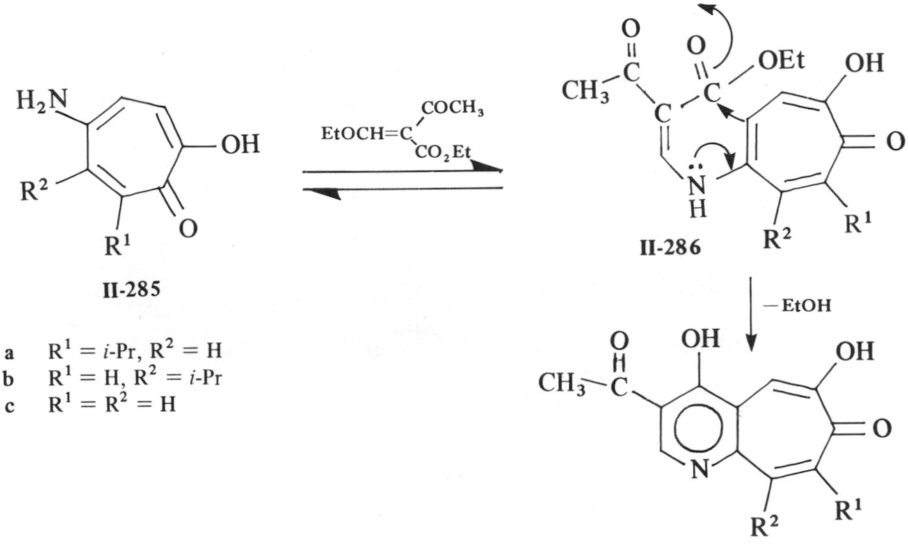

II-285

a $R^1 = i$-Pr, $R^2 = H$
b $R^1 = H$, $R^2 = i$-Pr
c $R^1 = R^2 = H$

II-286

II-287

3. Cycloaddition Reactions

A. Diels–Alder Reactions

a. AZADIENES

In a simple [2 + 4] cycloaddition, which ultimately is to result in formation of a pyridine ring, the nitrogen atom must be incorporated into either of the two reacting partners, the diene or dienophile. Sauer et al. have shown that triazines, such as **II-288**, reacted exothermically with the morpholine or pyrollidine enamine of cyclopentanone to give the 3,4-cyclopentapyridine derivative **II-290** (129). In several cases, intermediate **II-289** was detected spectroscopically by NMR. Reaction of norbornene with **II-288** led to the formation of norbornyl-fused pyridine **II-291**, in which an oxidation step had occurred after cycloaddition and loss of nitrogen. Reaction of the unstable precursor to **II-291** with norbornene resulted in formation of *bis*-adduct **II-292**. Utilizing the unsubstituted 1,2,4-triazine, this approach has been developed into a general synthesis of 3,4-annelated pyridines (323).

II-288 (R = CO_2Et, R′ = CH_3) **II-289** **II-290** **II-291** **II-292**

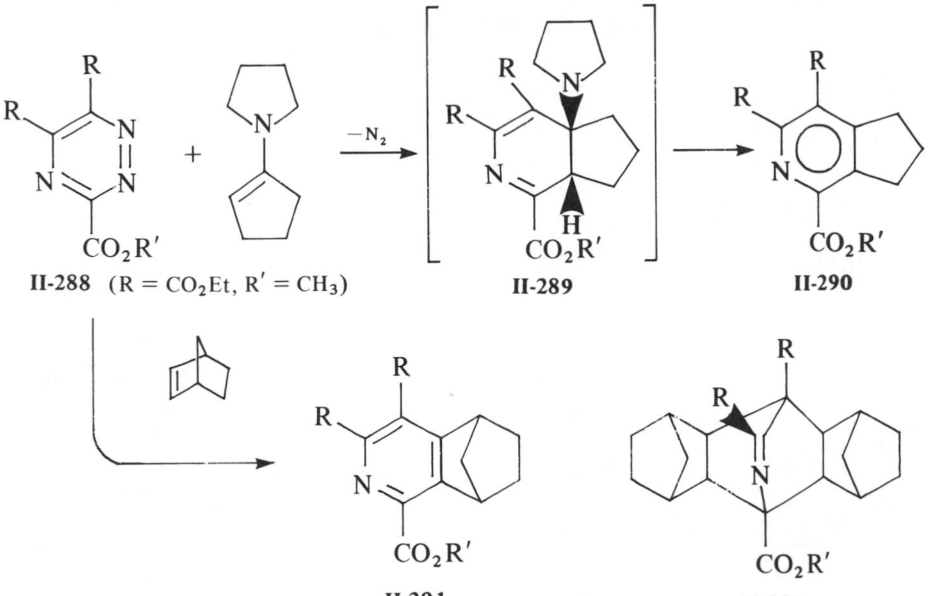

Ghosez et al. have found that thermolysis of 2-amino-1-azirines provides isomeric 2-azabutadiene derivatives, which can undergo Diels–Alder cycloaddition to give pyridines (130). In the case of the spiro system **II-293**, heating at 400°C (0.1 mm) provided azadiene **II-294**, which reacted with acetylenic dienophiles, such as methyl propargylate, to afford the tetrahydroquinoline derivative **II-295**. Once again a final dehydrogenation step must be involved.

II-293 **II-294** (95%) **II-295**

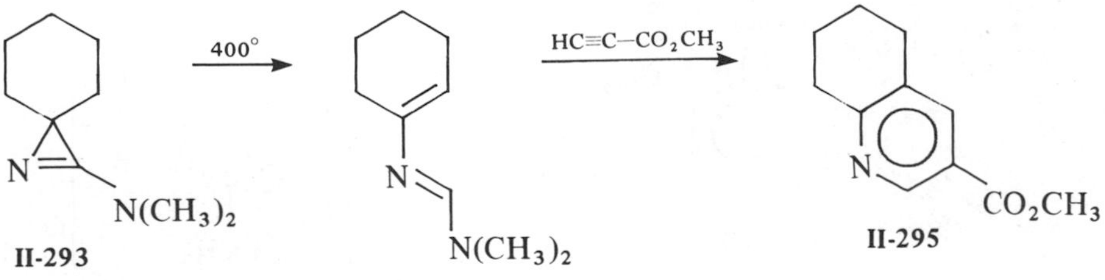

Russian workers have reported an interesting isomerization which makes numerous 2-aza-1,3-butadienes quite accessible (131). When *N*-allylcyclopentylidenimine (**II-296**) was treated with acrylonitrile over alkali-treated aluminum oxide (3–6% K_2O/Al_2O_3) at 300–400°C, 6-ethyl-2,3-cyclopentapyridine (**II-299**) was obtained. It appears that both double bonds in **II-296** migrated to give what is presumably the conjugated diene **II-297**, which was not isolated but rather reacted with acrylonitrile to afford **II-298** which then lost hydrogen and hydrogen cyanide to give **II-299**. In the case of *N*-alkylcyclohexlidenimine further dehydration occurred to provide the fully aromatized 2-ethylquinoline.

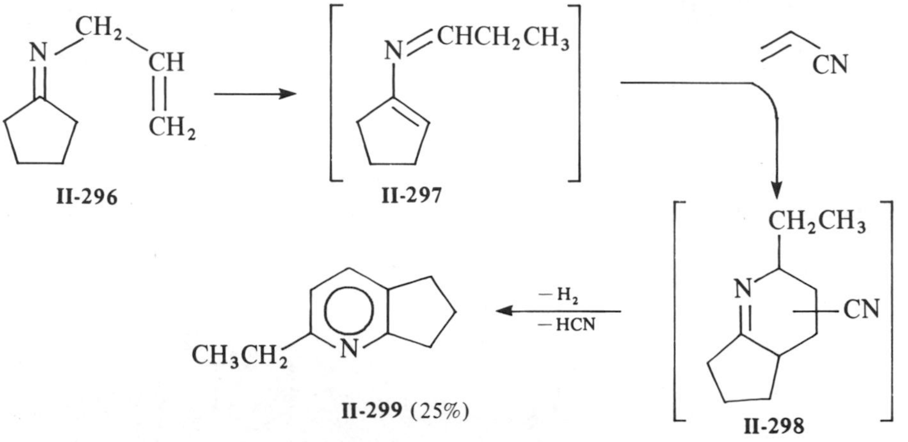

An intramolecular [2 + 4] cycloaddition route to annelated pyridines has been set forth by Sammes et al. (132). Thermolysis of the mono-oxopyrimidine **II-300** in either acetonitrile or dimethylformamide at 180–200°C afforded pyridine **II-302**.

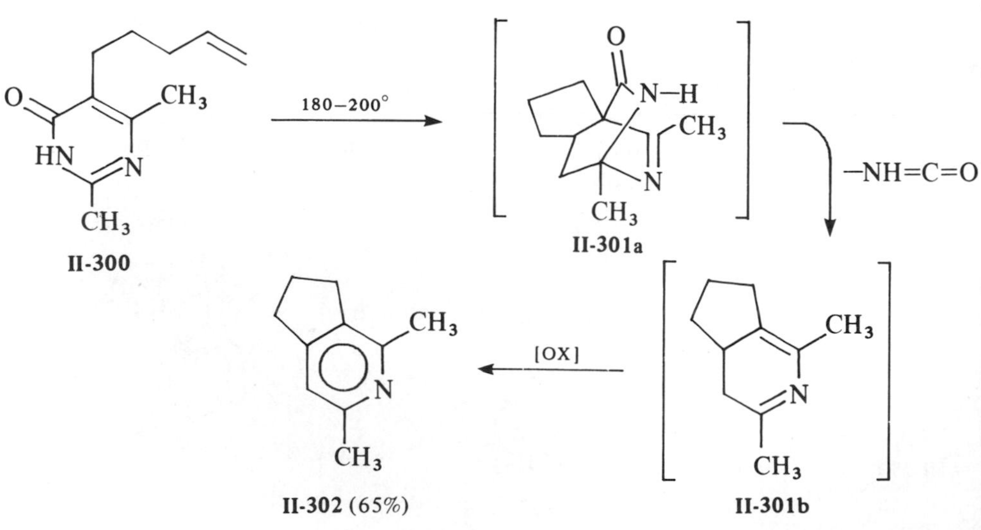

Monitoring the reaction spectroscopically did not reveal any appreciable quantity of the expected intermediate bicyclic adduct **II-301a**, implying that subsequent elimination of 1 mole of cyanic acid must be at least as fast as initial cyclization. Mass spectral analysis of the crude product indicated the presence of traces of the dihydro-precursor **II-301b**, which efficiently underwent dehydrogenation.

b.   NITRILES

Under certain circumstances, a nitrile triple bond can behave as a dienophile and give rise to a cycloaddition which provides a pyridine ring. Intramolecular Diels–Alder reactions are often facilitated by the proximity of the two reacting partners; thus when a mixture of allylic alcohols **II-303** and **II-304** were treated with sulfuric acid, dehydration to diene **II-305** occurred. Pyrolysis of **II-305** at 500°C gave the cyclopentapyridine **II-307** (133).

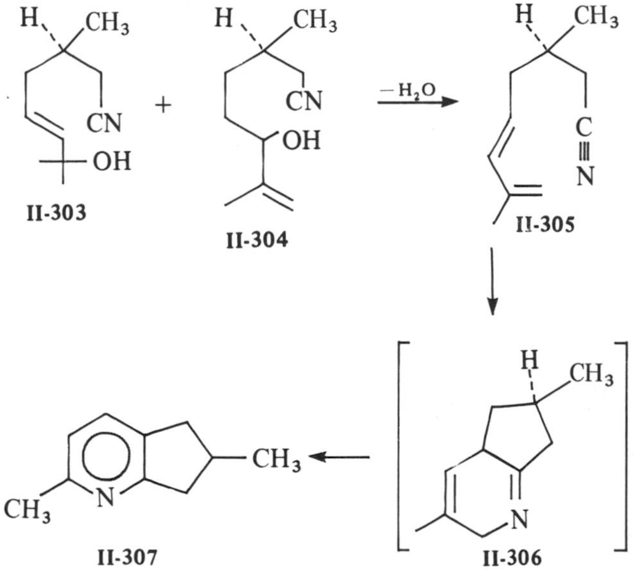

A cyano group has been employed as a dienophile in an intermolecular [2 + 4] cycloaddition where both partners were highly activated. The diene portion was an aluminum trichloride complexed cyclobutadiene and the dienophile was ethyl cyanoformate (134), in which the electron-withdrawing ethoxycarbonyl group was necessary to activate the nitrile moiety. When the reaction was carried out starting with the annelated complex **II-308**, a mixture of two products was obtained consisting of **II-314**, as the major component, and either **II-315** or **II-316**, as the minor component. The formation of these materials may be interpreted as the cycloaddition of ethyl cyanoformate to the cyclobutadienes **II-309** and **II-310** leading to Dewar pyridines **II-311–313**, which valence isomerize to pyridines **II-314–316**. If the same reaction was carried out employing *bis*-annelated complexes **II-317**, then *bis*-annelated pyridines **II-318** were obtained (134).

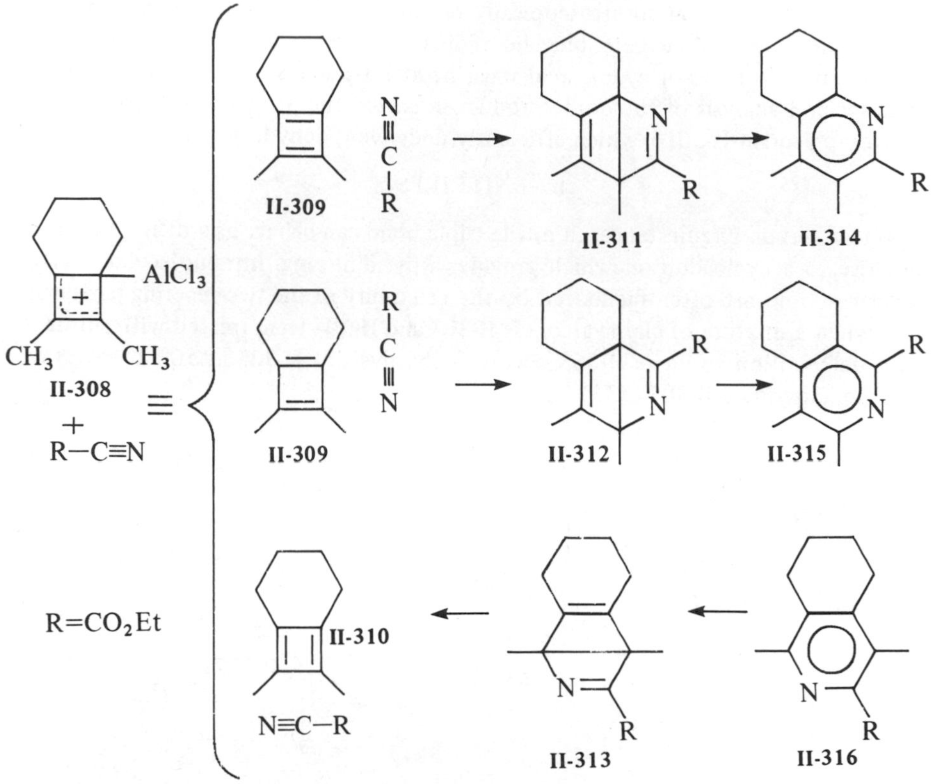

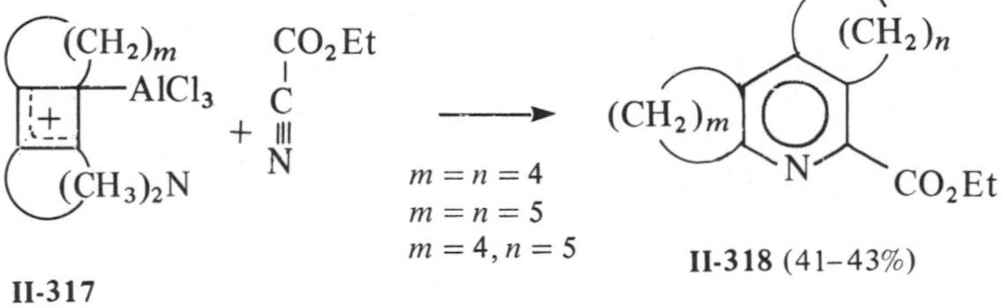

c.   2,3-DEHYDROPYRIDINE

One of the most versatile ways of preparing annelated aromatic systems is by the cycloaddition of a dehydroaromatic species to a diene. There is one report on the generation of 2,3-dehydropyridine (**II-320**) by treatment of 2-chloro-3-bromo-pyridine (**II-319**) with lithium amalgam and its subsequent trapping with furan (135, 136). Unfortunately the Diels–Alder adduct **II-321** was never isolated because under the reaction conditions it was found to readily lose oxygen to afford quinoline in very low yield. A stable adduct **II-324** was obtained when the 3-lithiopyridine derivative **II-323a** was heated in the presence of furan; no adducts were obtained in the case of **II-323b** or **II-323c**. Furthermore, only the adduct derived from a 2,3-pyridyne intermediate was obtained (137).

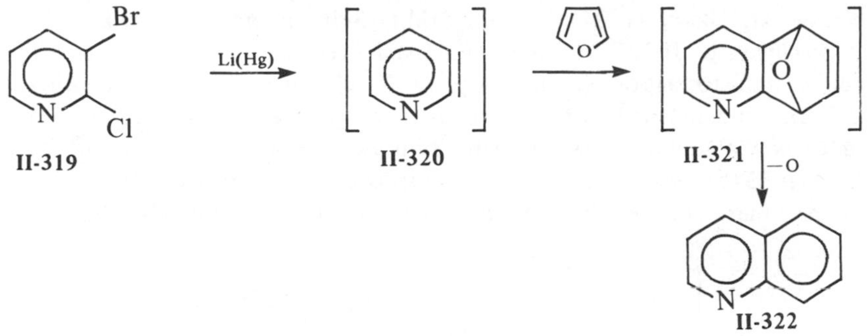

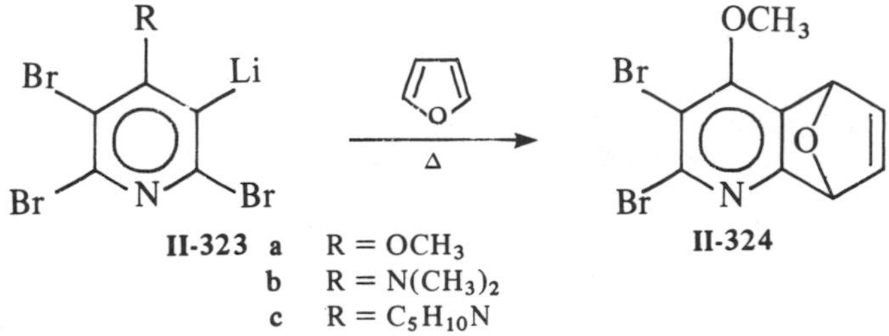

II-323   a   R = OCH₃
         b   R = N(CH₃)₂
         c   R = C₅H₁₀N

d.   3,4-DEHYDROPYRIDINE

The reactions of 3,4-dehydropyridine are more abundant than those of the 2,3-isomer. Kauffmann and Boettcher (138) first generated this species from 3-amino-4-pyridinecarboxylic acid (**II-325**), which was converted to the diazocarboxylate **II-326** and then heated to 60° causing loss of CO_2 and N_2 to provide **II-327**. This unstable species was trapped with furan but the adduct **II-328** was stable and could be isolated. Treatment of **II-327** with phenyl azide provided a mixture of the two possible triazoline adducts. When **II-327** is generated in the presence of cyclopentadiene, however, it appears that [2 + 2] cycloaddition was preferred and a postulated adduct (**II-329** or **II-330**) was obtained (139).

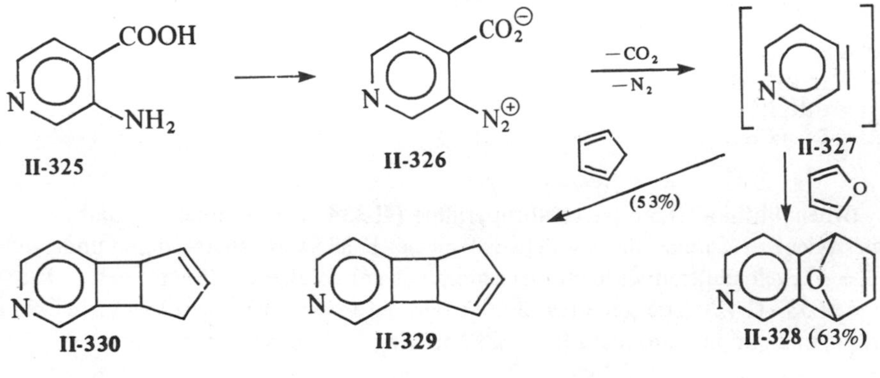

Kramer and Berry (140) also used pyridine-3-diazonium-4-carboxylate (**II-326**) as a precursor to **II-327** but they employed flash photolytic decomposition to effect the transformation. In the gas phase **II-327** dimerized to afford either 2,6- or 2,7-diazabiphenylene depending on whether dimerization occurred in a head-to-head or head-to-tail fashion. The later synthesis by MacBride (121) of the 2,7-isomer (**II-331b**), which had different spectroscopic properties from the 3,4-pyridyne dimer, indicated this latter material to be the 2,6-isomer **II-331a**.

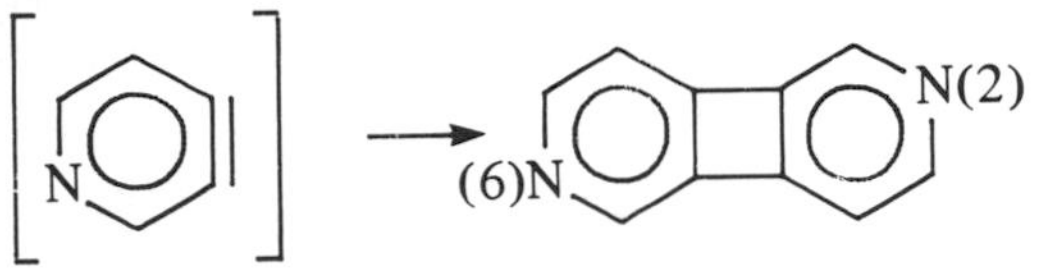

II-327   II-331a

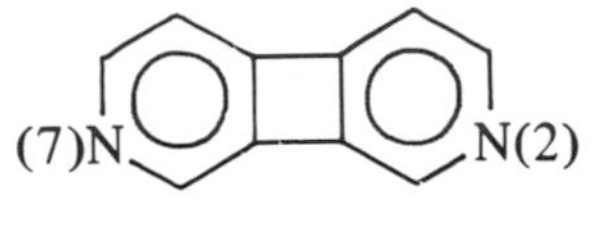

II-331b

Various 3-halopyridines have also been employed as precursors of 3,4-pyridyne. French workers have demonstrated that treatment of 3-bromopyridine (**II-332**) with butyllithium will generate **II-327**, which was trapped with a variety of furans to provide the annelated pyridines **II-333** (141).

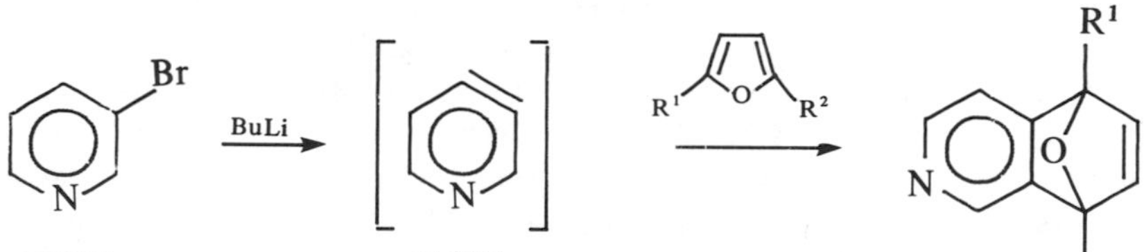

II-332   II-327   II-333

$R^1 = R^2 = H$
$R^1 = H, R^2 = CH_3$
$R^1 = CH_3, R^2 = H$
$R^1 = R^2 = CH_3$

When 4-lithio-2,3,5,6-tetrachloropyridine (**II-334**) was refluxed in solution with mesitylene or durene, the 3,4-dehydro species **II-335** was generated and underwent [2 + 4] cycloaddition with the benzenoid solvent to afford adducts such as **II-336** and **II-337** (142). The same reaction carried out in benzene gave a low yield (about 0.5%) of an impure product reported to be the trichloropyridobarralene.

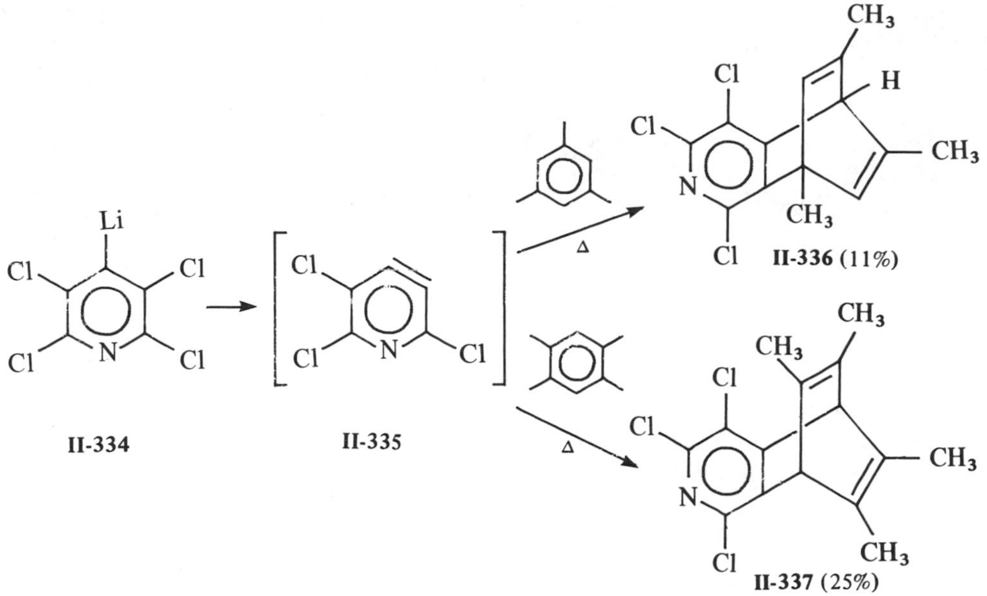

II-334   **II-335**   **II-336** (11%)

II-337 (25%)

Wakefield et al. (137) prepared a similar lithio-derivative **II-338** and found that of the two possible 3,4-pyridyne isomers which could be formed, only furan adducts **II-340**, derived from **II-339**, were detected.

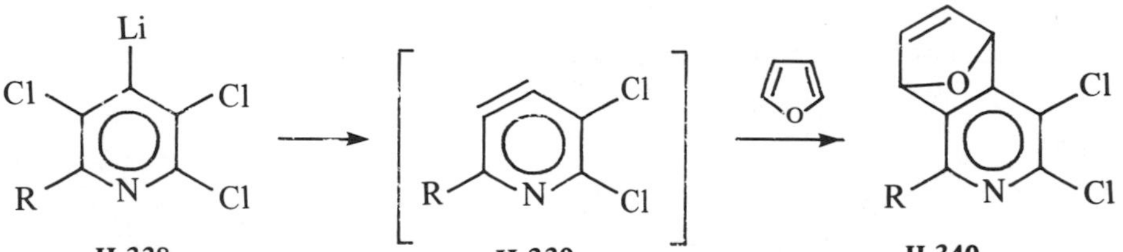

II-338   **II-339**   **II-340**

R = NMe₂, C₄H₄N, C₅H₁₀N

e.   OTHER [2 + 4] CYCLOADDITIONS

A Diels–Alder dimerization has been reported to occur during the poly-phosphoric acid promoted dehydration of 2-(2'-pyridyl)ethanol at 200°C (143). 2-Vinylpyridine (**II-342**), which is initially formed, added to itself in such a way that the vinyl group acts as a dienophile as well as part of the diene moiety. Of the two possible orientations, only the 5-substituted product (**II-343**) is observed. After cycloaddition, a 1,3-hydrogen shift must have occurred to rearomatize the pyridine ring, which contributed to the diene moiety.

II-341   **II-342**   **II-343** (37%)

Annelated dihydropyrans (**II-346**) can be formed by Diels–Alder addition of the cyclic enol ether **II-345** to enone **II-344**. Further reaction of **II-346** in a flow apparatus with ammonia over alumina impregnated with platinum provided annelated pyridine **II-347** (144). The reaction has been demonstrated for several different α,β-unsaturated aldehydes.

II-344   **II-345**   **II-346**   **II-347** (40–79%)

Kametani et al. have heated the pyridoxyl dibromide **II-348** to 150–160° for 4 hr with indole in dimethylformamide and obtained adducts **II-350** and **II-351** (145). Presumably the reaction proceeded with generation of the *ortho*-lutidylene species **II-349**, which then added to indole in either of two fashions. A final oxidation must have occurred to reestablish the indole portion of the product.

II-348   **II-349**

II-350 (4%)   **II-351** (15%)

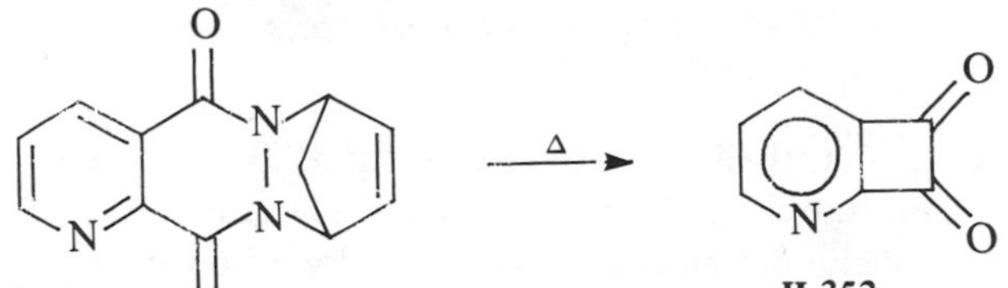

II-352

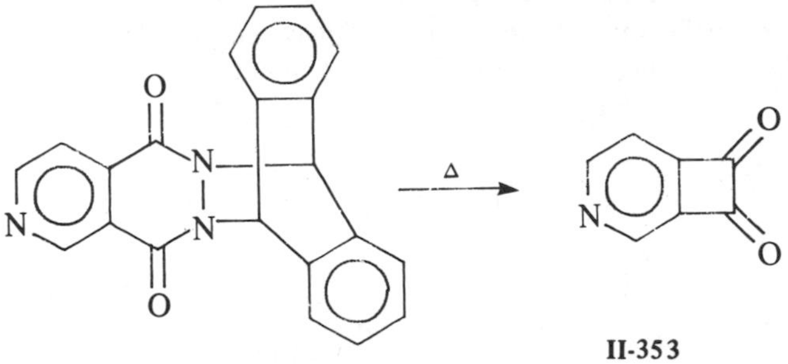

II-353

Infrared spectra have been cited as evidence for the formation of cyclobuta[a]- and cyclobuta[b]-pyridine-1,2-dione (**II-352** and **II-353**). Unlike cyclobutabenzene-1,2-dione, these two substances are too labile to be isolated in a pure state although they are prepared by a pyrolytic retro Diels–Alder reaction at 450–500°C (324).

B.   Cooligomerization

An interesting reaction, which amounts to a cyclotrimerization, has been reported to lead to annelated pyridines under appropriate conditions (145). Thus cyclopentadienyl cobalt dicarbonyl can act as a template for the cooligomerization of two acetylene moieties with a nitrile group to provide a pyridine nucleus. When the two terminal acetylene functions are connected by a polymethylene bridge, annelated pyridines may be prepared in wide ranging yields (3–81%). Unlike other nitrile cycloadditions, the reaction does not appear to depend as heavily on the nature of the R group, although electron-withdrawing substituents do afford the lowest yields. When an ethyl group replaced a terminal hydrogen on the diacetylene, a strong regioselectivity prevailed such that the ethyl and R group (*n*-butyl) did not end up on adjacent carbons in the pyridine ring. Unfortunately the reaction was not successful for $n = 2$, which would lead to a 3,4-cyclobutapyridine.

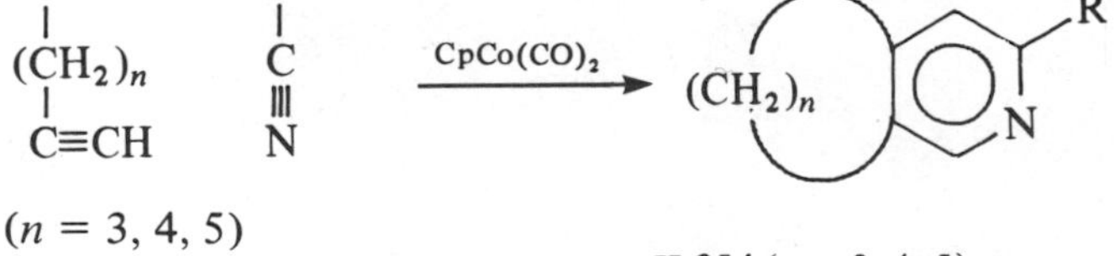

II-354 ($n = 3, 4, 5$)

C.   Photocycloadditions

Irradiation of 2,4-dimethylbenzo[g]quinoline (**II-355**) in methanol with light of greater than 300-nm wavelength gave rise to the photodimers **II-356** and **II-357** (146). The two isomeric products were separated by chromatography on alumina and identified by a combination of NMR and mass spectroscopy. Upon irradiation with 253.7-nm light, these dimers reverted back to starting material (**II-355**). The authors did not make any mention of the other two isomeric dimers in which both the pyridine rings would be oriented in the same direction.

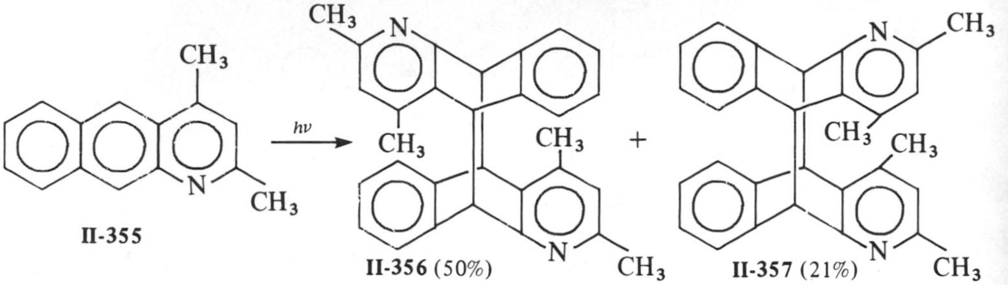

II-355

II-356 (50%)   +   **II-357** (21%)

4.   Rearrangement Reactions

One of the most frequently cited rearrangement reactions which leads to an annelated pyridines is the thermolysis of the oxime derived from the tricyclic ketone **II-358** (148–152). This reaction is best understood when one realizes that **II-358** is simply the intramolecular aldol condensation product of **II-359** which according to earlier discussions (Section II.1.A) is a straightforward precursor to pyridine **II-361**. Ketones **II-358** and **II-359** were obtained by the reaction of cyclohexanone with an aldehyde (RCHO) where numerous different R groups have been employed.

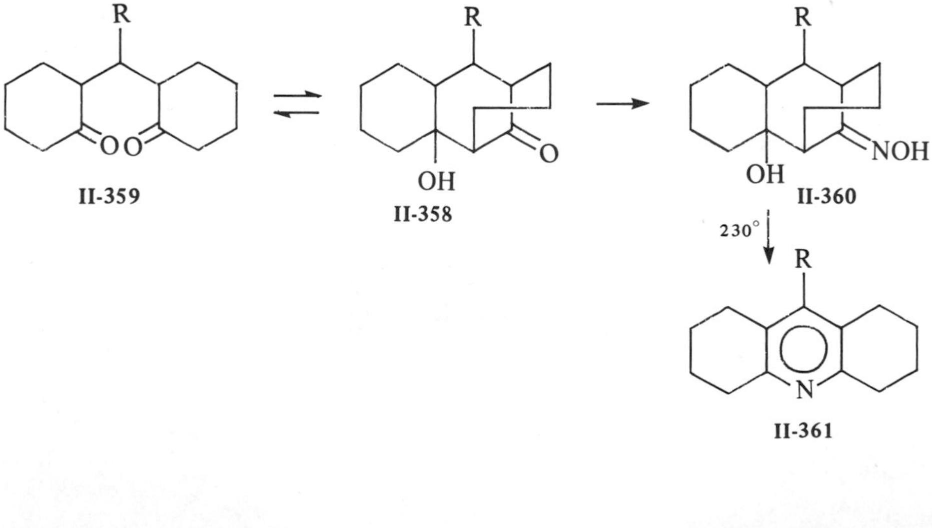

II-359

II-358

II-360

II-361

Treatment of an annelated 4-oxo-1,2,3,4-tetrahydropyrimidine, such as **II-362**, with polyphosphoric acid at 135°C led primarily to 2-pyridones but minor amounts of a 4-aminopyridine (**II-368**) were also observed (153). The mechanism proposed to explain this rearrangement involves protonation of the pyrimidine ring followed by fragmentation of the ring and reclosure after the loss of water.

II-362 **II-363** **II-364** **II-365** **II-366** **II-367** **II-368**

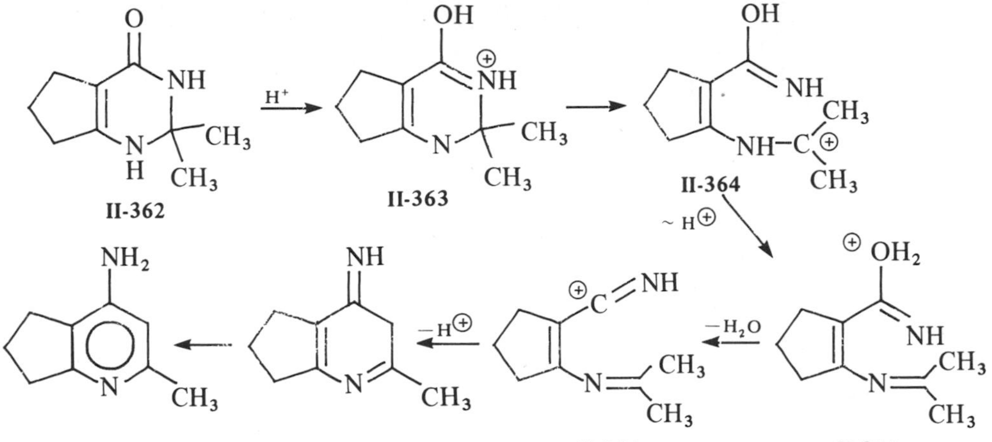

Photochemical reduction of 2-acetylpyridine (**II-369**) in isopropanol provided the pinacol dimer **II-370**. When this diol was treated with concentrated sulfuric acid to effect a pinacol-type rearrangement, the expected ketone **II-372** resulting from 1,2-methyl migration was observed along with **II-373**, which is presumed to arise from a double dehydration proceeding *via* intermediate **II-371** (154).

II-369 **II-370** **II-371** **II-372** **II-373** **II-372**

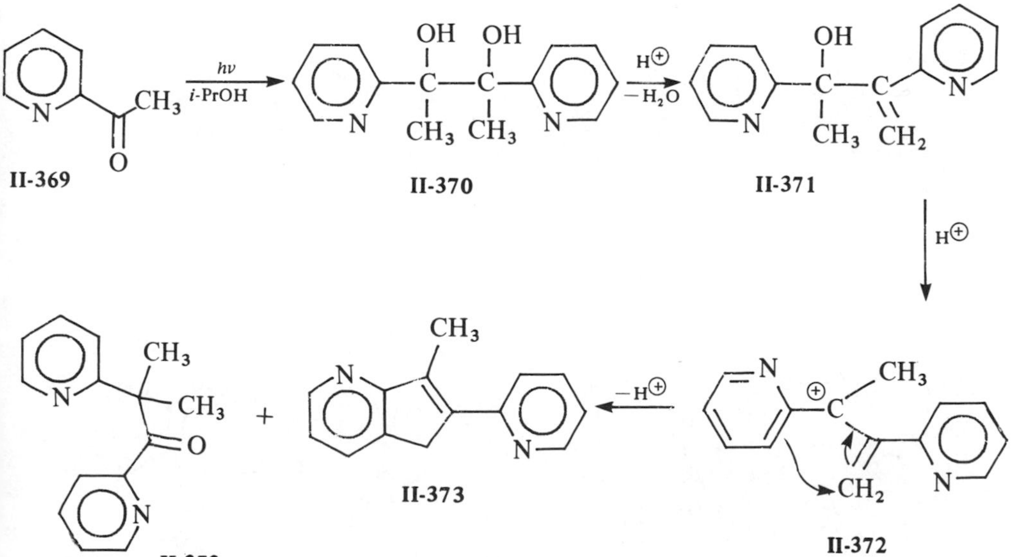

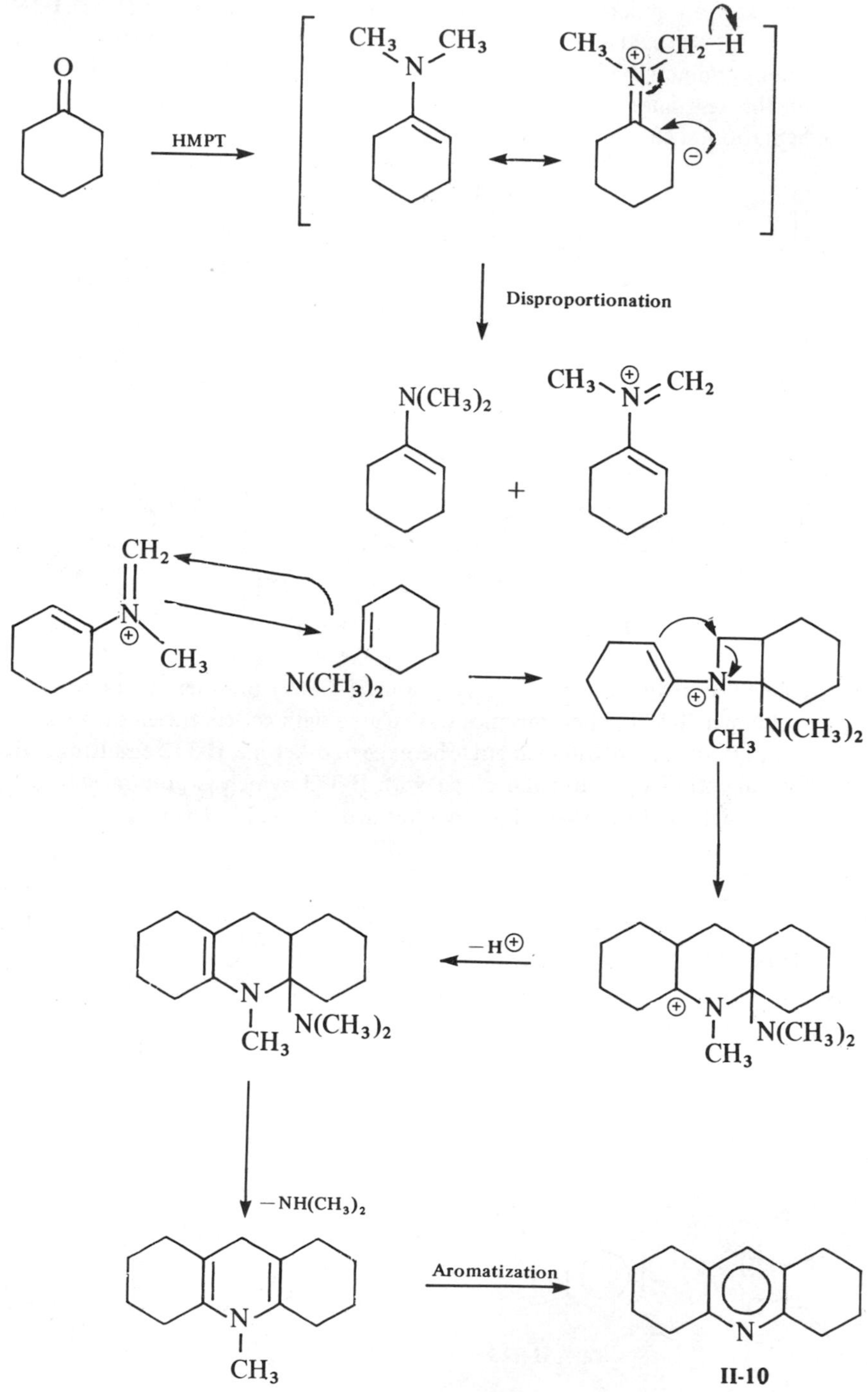

Scheme II-3

TABLE II-1. BISANNELATED PYRIDINES FROM THE REACTION OF CYCLIC KETONES AND HMPT

Ketone	Product	Yield
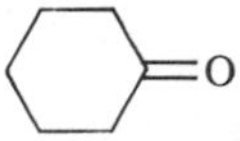 (cyclohexanone)	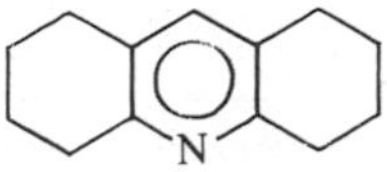 (octahydroacridine)	4.8%
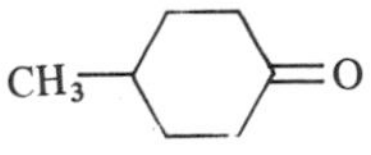 (methylcyclohexanone)	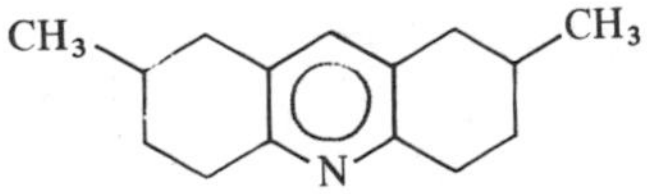 (dimethyl-substituted product)	10.8
 (tert-butylcyclohexanone)	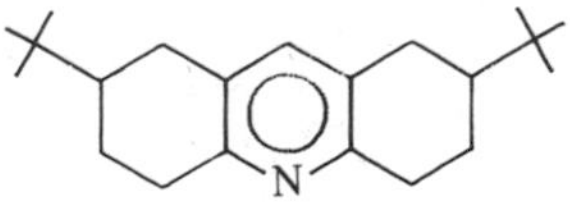 (di-tert-butyl-substituted product)	2.7
 (isopropylcyclohexanone)	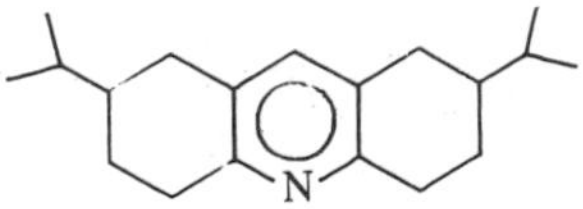 (diisopropyl-substituted product)	4.9
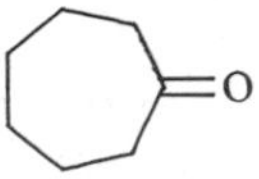 (cycloheptanone)	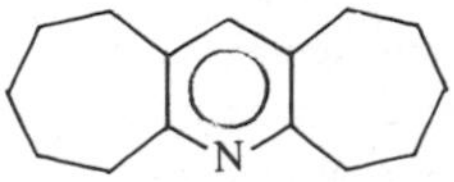 (bis-cycloheptano-fused pyridine)	6.0

Monson et al. have reported an unusual reaction of cyclic ketones with hexa-methylphosphoric triamide (HMPT) by which *para-bis*-annelated pyridines were generated in a single process (155). The yields were unfortunately quite low (see Table II-1). The mechanism for this reaction is shown in Scheme II-3. The key steps involve a [2 + 2] cycloaddition to give an azetidinium salt, followed by a ring-expansion, which establishes the framework of the final product (**II-10**).

5. Hydrogenations

A. *Quinoline and Isoquinoline*

Under the normal conditions for catalytic hydrogenation, it is well known that pyridine will reduce more readily than benzene. Thus, partial hydrogenation of quinoline or isoquinoline leads to the 1,2,3,4-tetrahydro derivatives. In 1923, von Braun and coworkers discovered that the number and position of methyl substituents on a quinoline ring could markedly influence the direction of catalytic hydrogenation (156, 157). Methyl substitution on the benzene nucleus of quinoline inhibits reduction of that portion of the molecule while similar substitution on the pyridine nucleus inhibits its reduction and favors reduction of the benzenoid ring. The results are shown in Table II.2.

TABLE II-2.   DIRECTION OF CATALYTIC HYDROGENATION
OF METHYL-SUBSTITUTED QUINOLINES (156)

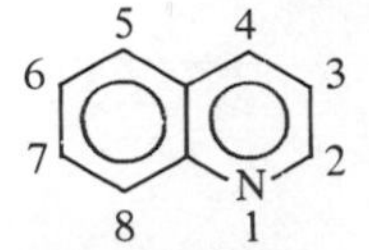

| | | % Reduction | |
		Benzenoid Ring	Pyridine Ring
2-methyl		4	96
3-		33	66
4-		33	66
5-		Not studied	
6-		0	100
7-		0	100
8-		0	100
2,3-dimethyl		44	56
2,4-		80	20
2,3,4-trimethyl		> 80	

Japanese workers, studying the catalytic reduction of 4-amino-, 4-acetamido-, and 4-benzamido-isoquinoline, discovered that the direction of reduction depended on the choice of reagents and reaction conditions (158). Reduction of **II-374b** and **II-374c** over Adams catalyst in acetic acid or over Raney nickel in ethanol afforded the corresponding 1,2,3,4-tetrahydro derivatives **II-375b** and **II-375c**, whereas, their reduction over Adams catalyst in acetic acid-sulfuric acid gave the corresponding 5,6,7,8-tetrahydro derivatives **II-376b** and **II-376c**. Over either Adams catalyst in acetic acid or Raney nickel in ethanol 4-aminoisoquinoline (**II-374a**) was reduced to the 5,6,7,8-tetrahydro derivative **II-367a**.

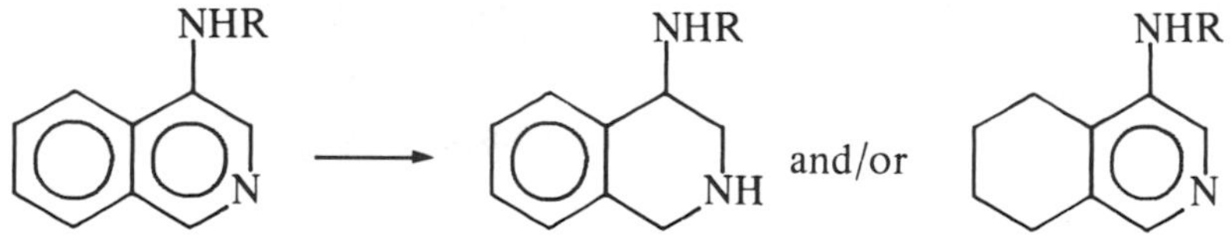

II-374a R = H   II-375b, c   II-376a, b, c
     b R = COCH$_3$
     c R = COC$_6$H$_5$

Ginos has studied the catalytic reduction of quinoline and isoquinoline in an attempt to establish conditions that may be used in the synthesis of precursors of apomorphine and morphinan analogs from the readily synthesized derivatives of 1-benzylisoquinoline (159). He found that the proportion of product resulting from reduction of the benzenoid ring increased as the concentration of hydrochloric acid in methanol was increased (Table II-3). In the case of quinoline, small amounts of decahydro- and octahydroquinoline were also observed.

TABLE II.3. CATALYTIC REDUCTION OF ISOQUINOLINE AND QUINOLINE UNDER ACIDIC CONDITIONS (159)

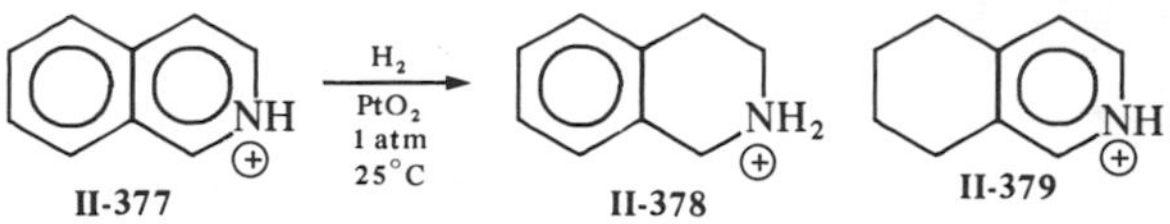

Solvent	Time (min)	II-378	II-379
MeOH	60	87%	13%
1.0 N HCl/MeOH	140	30	70
4.0 N HCl/MeOH	160	13	87

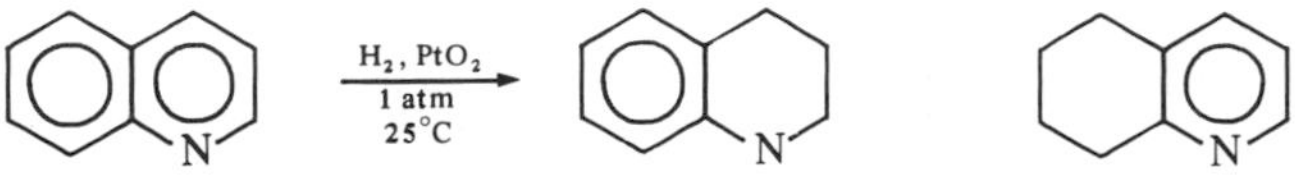

Solvent	Time (min)	II-381	II-382
MeOH	345	54%	34%
1.0 N HCl/MeOH	445	43	37
4.0 N HCl/MeOH	450	39	52

At about the same time Vierhapper and Eliel made a detailed study on the same subject (160–162). They found that hydrogenations utilizing platinum oxide (and, in some instances, palladium or rhodium on charcoal) in strong acids such as $12 N$ HCl, $12 N$ H_2SO_4, or trifluoroacetic acid produced selective hydrogenation of the benzene ring. The fastest procedure employed platinum oxide in trifluoroacetic acid and the results are shown in Table II-4.

The cyclohexapyridines could be further reduced with sodium in ethanol to give the fully saturated ring systems (162); thus, 5,6,7,8-tetrahydroquinoline gave *trans*-decahydroquinoline (**II-383**) as well as the *cis* isomer **II-384**. Reduction of 5,6,7,8-tetrahydroisoquinoline gave $\Delta^{9,10}$-octahydroisoquinoline (**II-386**) and about equal amounts of the two decahydro isomers **II-387** and **II-388**.

TABLE II-4. HYDROGENATION IN CF_3CO_2H WITH PtO_2 AT 50 psi OF HYDROGEN AND ROOM TEMPERATURE (161)

Substrate	% Product Reduced in Benzene Ring
Quinoline	84
Isoquinoline	90.5
2-Methylquinoline	95
2-Isopropylquinoline	95
6-Methylquinoline	74
8-Methylquinoline	56

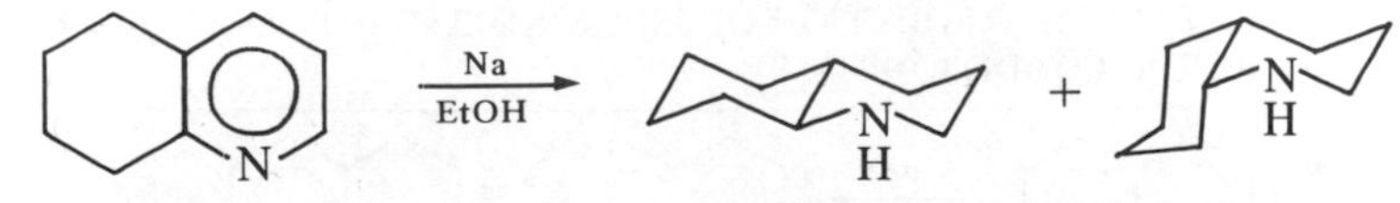

II-382 II-383 (90%) + II-384 (10%)

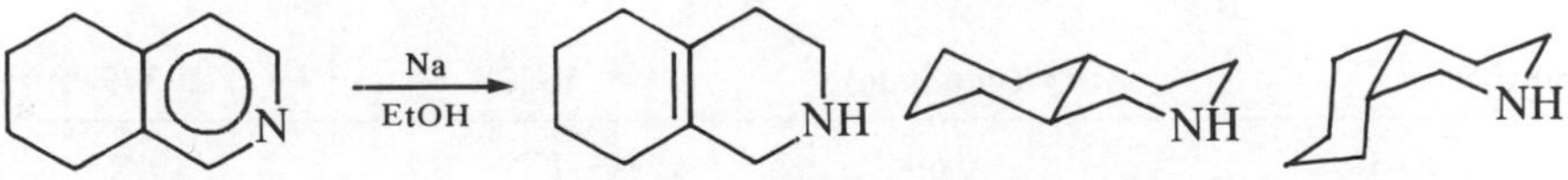

II-385 II-386 (56%) II-387 (22%) II-388 (20%)

B. Acridine

Vierhapper and Eliel have shown that acridine is one of the most cleanly reduced systems affording exclusively octahydroacridine **II-10** (161). Further reduction of **II-10** with sodium in ethanol was similarly very specific giving *trans-syn-trans*-perhydroacridine **(II-390)** (162).

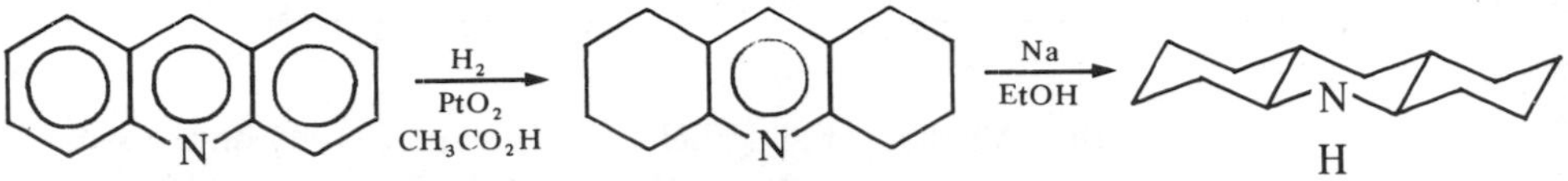

II-389 II-10 II-390 (90%)

Adkins and Coonradt (163) examined the hydrogenation of acridine **(II-389)** over Raney Nickel and copper chromite and found that the nickel reduction at 100°C afforded **II-10** (16%) along with as-octahydroacridine **II-391** (38%) and dodeca-hydroacridine **II-392** (22%); whereas at 190°C, copper chromite gave **II-391** (70%), **II-392** (13%), and no **II-10**. On the other hand, reduction of acridine with lithium and ethanol in ammonia provided 1,4,5,8-tetrahydroacridine **II-393** (60%) (164).

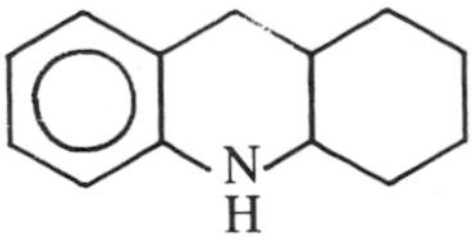

II-391 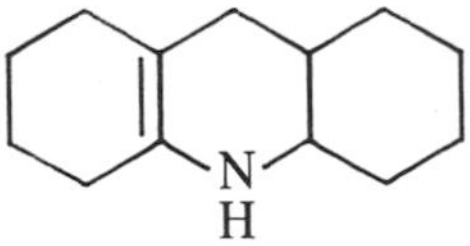 II-392 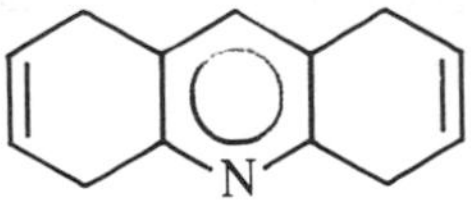 II-393

6. Halogenation

In the previous section we have seen that hydrogenation can be employed to saturate one or more rings of an aza-polynuclear aromatic system to produce an annelated pyridine. It has been shown that addition of halogen, in particular chlorine, can be employed to achieve a similar result (165). When 2,4-disubstituted

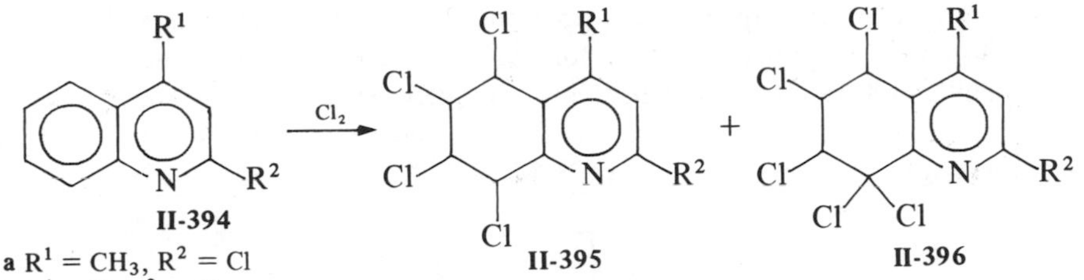

II-394 II-395 II-396

a $R^1 = CH_3$, $R^2 = Cl$
b $R^1 = H$, $R^2 = Cl$
c $R^1 = CH_3$, $R^2 = OC_2H_5$
d $R^1 = Cl$, $R^2 = CH_3$
e $R^1 = CH_3$, $R^2 = H$

quinolines **II-394a**, **b**, and **c**, were treated with chlorine in carbon tetrachloride, **II-395a**, **b**, and **c** along with other products were isolated. In the case of **II-395c**, however, this material spontaneously lost two equivalents of HCl to give 6,8-dichloro-2-ethoxy-4-methylquinoline. For **II-394d** and **e**, the pentachloro-derivatives **II-396d**, and **e** were observed.

7. Pyrylium Salts

When a pyrylium salt, most often a perchlorate, is treated with ammonia, the oxygen is replaced by nitrogen and a pyridine results. Of course if the initial pyrylium salt possesses one or two fused rings, then the corresponding annelated pyridine will result (166–168, 312).

The most general procedure for pyrylium salt formation involves the use of acetic anhydride and perchloric acid. A variety of cyclic precursors, such as the β-hydroxyacetal **II-397**, can be elaborated in this fashion (169). The reaction most likely proceeded through the enol form of **II-398** which was first acetylated and then cyclized to the pyrylium salt **II-401**. After treatment with ammonia, the annelated pyridine **II-402** was obtained in 72% overall yield.

II-397 II-398 II-399

II-402 II=401 II-400

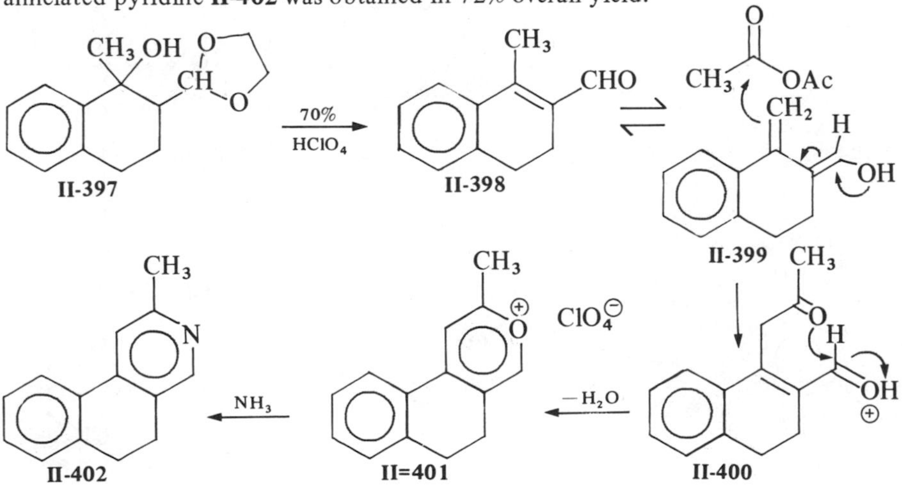

A similar reaction sequence was observed for β,γ-unsaturated ester **II-403** in which an apparently unactivated carbon–carbon double bond can undergo initial electrophilic addition (170). Interestingly, the methoxy group was retained in the pyrylium product. In the ammonolysis step, however, this group was replaced by nitrogen to give the 6-amino derivative **II-407**.

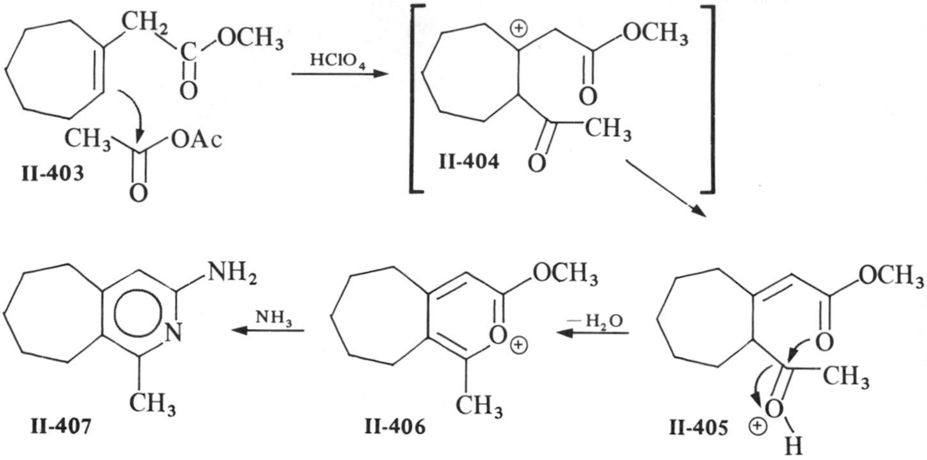

The entire anhydride moiety can be incorporated into a pyrylium salt by means of the *bis*-acylation of cyclic tertiary alcohols (171). The first step would be dehydration of 1-methylcyclopentanol (**II-408**) to a mixture of *endo*- and *exo*-cyclic alkenes followed by *bis*-acylation and then a double dehydration. The process is illustrated for the exocyclic alkene **II-409** which afforded the annelated pyridine **II-413**.

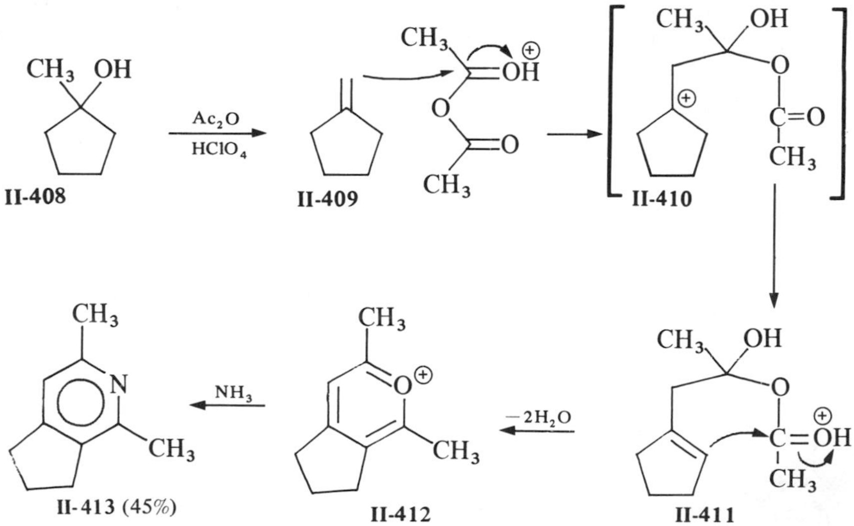

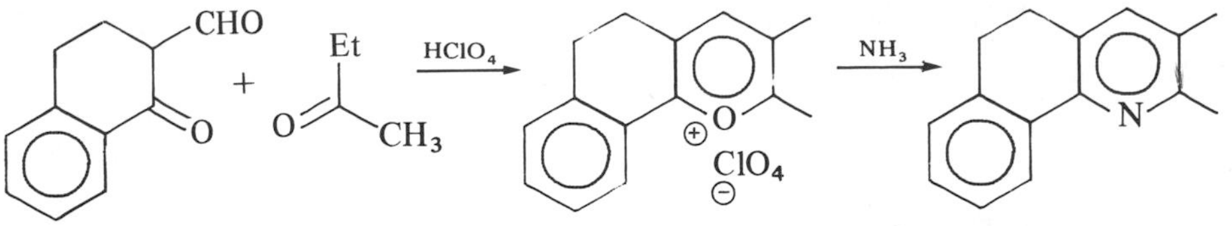

II-414   II-415   II-416

Acetic anhydride is not essential to these cyclizations. Methyl ethyl ketone has been found to react with β-keto aldehyde **II-414** to produce pyrylium salt **II-415**, which in turn afforded pyridine **II-416**, albeit in low yield (172). The mechanism of this reaction is similar to the one involved in the formation of **II-402**. This same type of reaction works equally well for the preparation of *bis*-annelated pyrylium salts and the corresponding *bis*-annelated pyridines. Thus, hydroxymethylenecyclohexanone (**II-417**) reacted with cyclopentanone in the presence of perchloric acid to provide **II-418** (173). In a similar fashion, 2-benzylidenecyclohexanone (**II-419**) reacted with cyclohexanone and sulfuric acid, followed by treatment with ammonia, to produce octahydroacridine **II-421** (174).

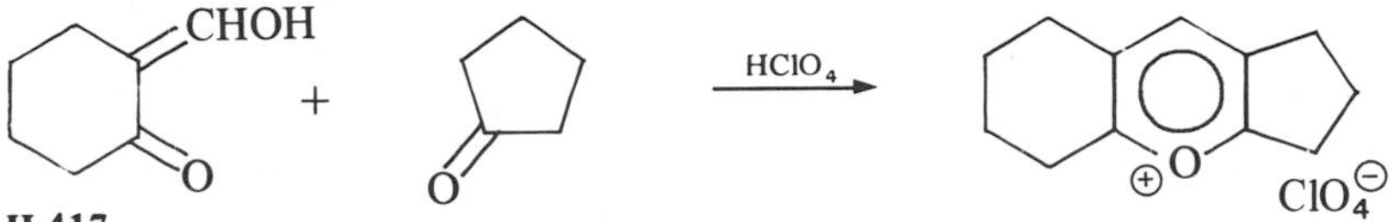

II-417   II-418 (23%)

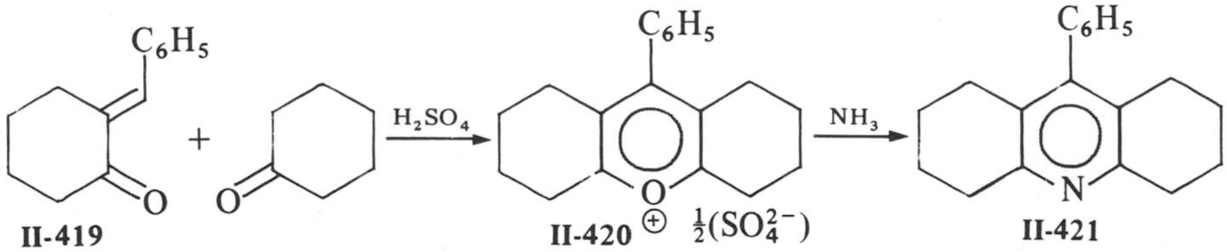

II-419   II-420 $\frac{1}{2}(SO_4^{2-})$ II-421

Treatment of 1,5-diketones (e.g., **II-31**) with triphenylmethyl perchlorate in acetonitrile resulted in a slightly exothermic dehydrogenation to give annelated pyrylium salts, such as **II-422**, in yields of 80–90% (175, 176). Further conversion of these salts into *bis*-annelated pyridines was effected by ammonia in 70–80% yield. If a primary amine was substituted for ammonia, *N*-octahydroacridinium salts may be prepared (177, 178).

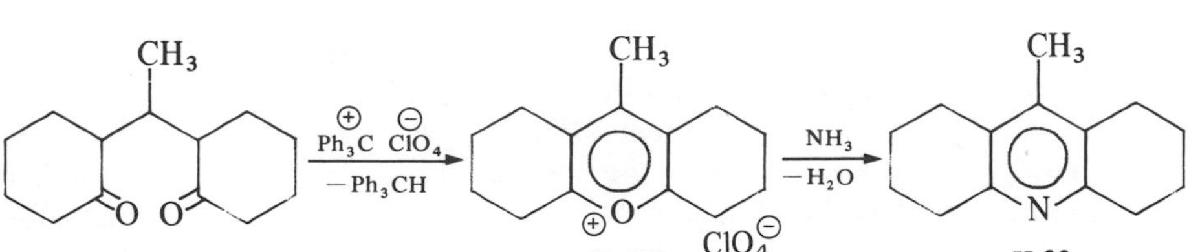

II-31   II-422   II-32

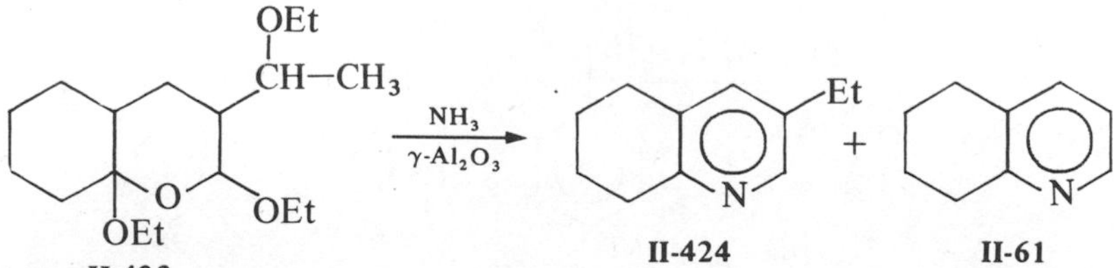

II-423   **II-424**   **II-61**

In a related reaction, when triethoxy-substituted tetrahydropyran **II-423** in ammonia was passed over a catalyst of γ-Al_2O_3 or Pt/Al_2O_3, a mixture of 5,6,7,8-tetrahydroquinolines was formed (179). It is appealing to postulate a mechanism which involves initial loss of three equivalents of ethanol to give an intermediate pyrylium salt, which subsequently would react with ammonia to provide the observed products.

8. Malononitrile and Cyanoacetamide Reaction with 1,3-Dicarbonyl Compounds

In 1969 Freeman and Ito reported that cyanoacetamide reacted with 2-acylclo-alkanones (**II-425**) to afford 3,4-fused pyridines (180). The same reaction also occurred with malononitrile (181). A mechanism was suggested which involved initial attack of the cyanoacetamide anion at the cyclic carbonyl carbon to form the cycloalkylidenecyanoacetamide **II-426**, which cyclized and then eliminated water to provide **II-427**. The reaction was not found to be as straightforward for the cyclopentanone derivative ($n = 1$).

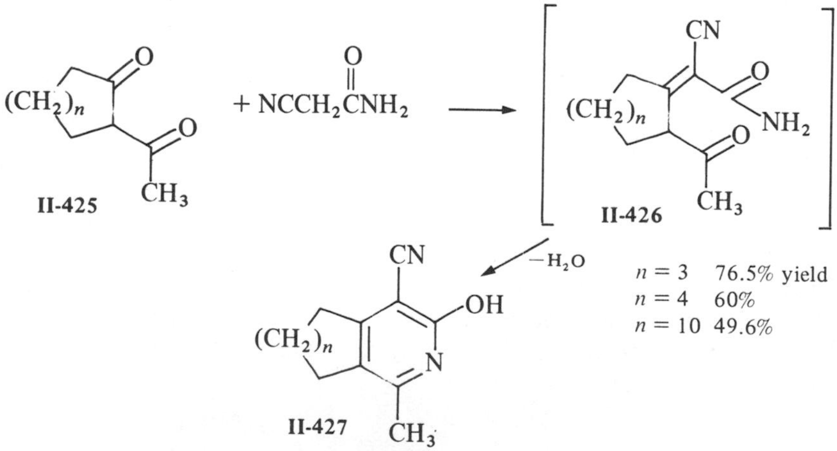

II-425   **II-426**

$n = 3$ 76.5% yield
$n = 4$ 60%
$n = 10$ 49.6%

II-427

At about the same time van der Baan and Bickelhaupt corrected their earlier report (182) regarding a very similar reaction of cyclic β-keto esters. They found that 2-(ethoxycarbonyl)cyclohexanone (**II-428**) condensed with malononitrile to give **II-429** (183, see also 184). In the absence of ethanol, the product was the simple Knoevenagel adduct **II-430**, which could be converted into **II-429** under the original reaction conditions.

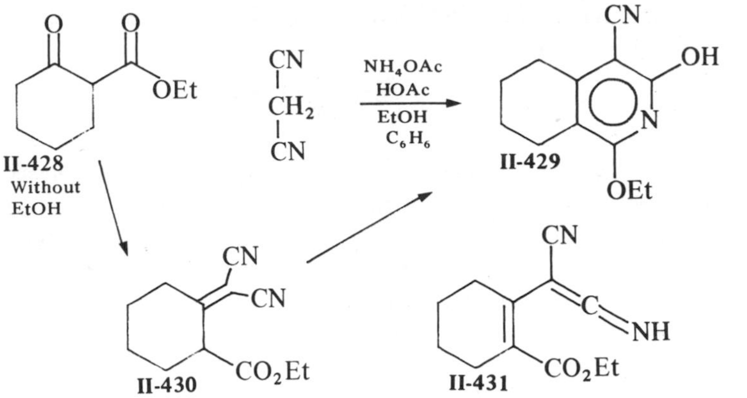

Kasturi et al. (185, 186) were examining the same reaction and proposed a mechanism which invoked the intermediacy of ketenimine **II-431**. They reported that an ethanol solution of **II-430** showed UV absorption maxima at 240 and 356 nm, the intensity of the latter absorption increasing on the addition of sodium hydroxide. The longer wavelength band was attributed to **II-431** and is the sole evidence for their mechanism.

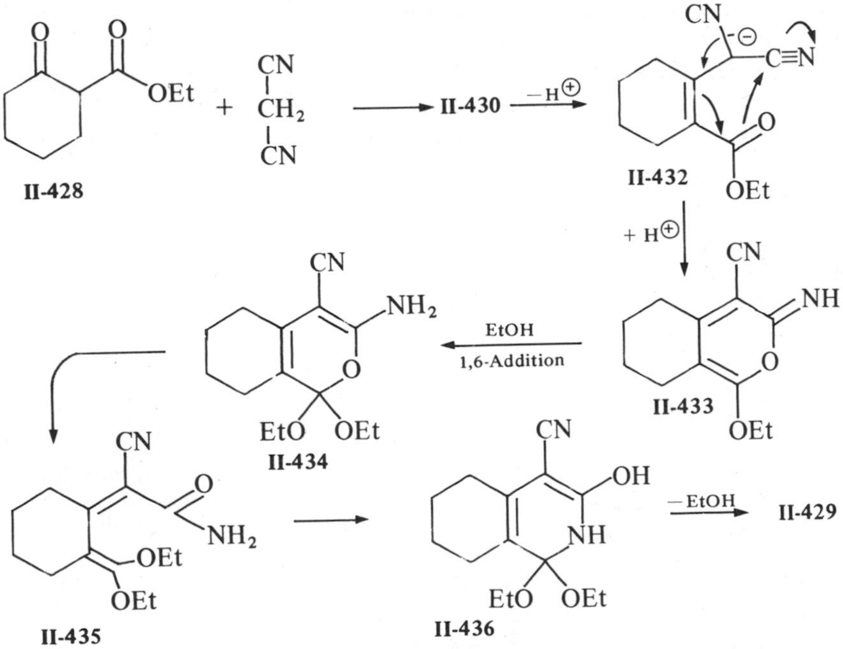

A solution to this mechanistic problem was finally set forth by Ducker and Gunter (187) for the base-promoted process and by van der Baan and Bickelhaupt (188) for the acid-promoted version. Since both processes are quite similar, only the former is illustrated. The involvement of the solvent alcohol as well as intermediates **II-434** and **II-435** was confirmed.

III.   REACTIONS OF ANNELATED PYRIDINES

1.   Mono-Annelated Pyridines

The ability of an aromatic ring to stabilize an adjacent carbon bearing a charge or an unpaired electron explains the high reactivity of such positions in many types of organic reactions. For unsubstituted benzocycloalkenes (i.e., indan, tetralin, etc.) both available α-positions are equivalent. For the corresponding annelated pyridines, these positions are nonequivalent. For 2,3-fused pyridines, the methylene group fused at the 2-position is far more reactive than the one fused at the 3-position.

A.   N-Oxide Acetylation

The first reported preparation of 1,5-pyridine (**II-441**) was accomplished by functionalization of the α-position of 2,3-cyclopentapyridine (**II-437**) followed by an elimination step (189). N-oxide **II-438**, formed from **II-437**, was heated with acetic anhydride to afford **II-439**. The elimination was effected either directly from the acetate or by initial hydrolysis to alcohol **II-440**, followed by dehydration. The two-step process proceeded in higher overall yield. The question of tautomeric equilibrium between **II-441a** and **II-441b** will be treated in a later section. Hahn and Epsztajn have carried out a systematic study of the acetylation of annelated pyridine N-oxides for **II-442**, where the fused-ring was varied from five to seven carbons with either a hydrogen or phenyl group at the 6-position (190).

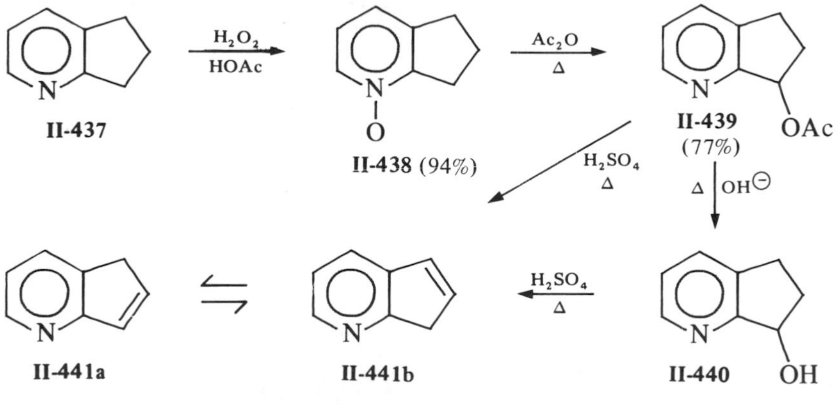

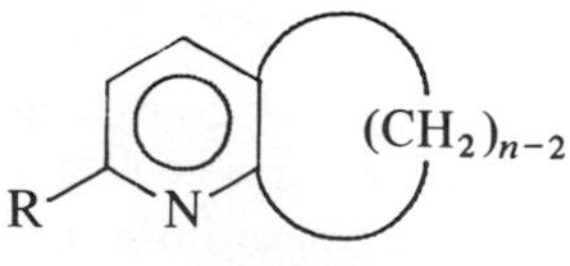

II-437

H_2O_2 / HOAc

II-438 (94%)

Ac_2O / Δ

II-439 (77%) OAc

H_2SO_4 / Δ

Δ | OH⊖

II-441a ⇌ II-441b

H_2SO_4 / Δ

II-440 OH

II-442 $n = 5, 6, 7$
R = H, C_6H_5

An interesting variation occurred with 3,4-fused pyridines. In a thorough study on the preparation of hydroxyactinides, Cavill and Zeitlin have prepared 6-hydroxy-actinidine (**II-447**) by functionalization of **II-443** (191). Attack by acetate on the *N*-oxide **II-444** presumably occurred by intermediacy of **II-445**.

II-443 **II-444** **II-445** **II-446** (56%) **II-447**

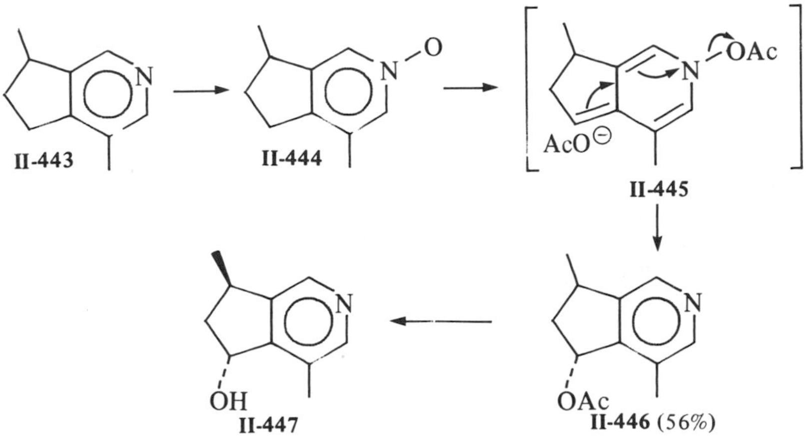

N-oxides may also be formed from annelated pyridines which have already been functionalized, allowing a second substituent, usually an acetate group, to be introduced into the α-position (192, 193). Thus systems, such as **II-448**, may be functionalized to **II-450** in two simple steps. Further interconversions of the two α-substituents of **II-450** make available a wide variety of derivatives.

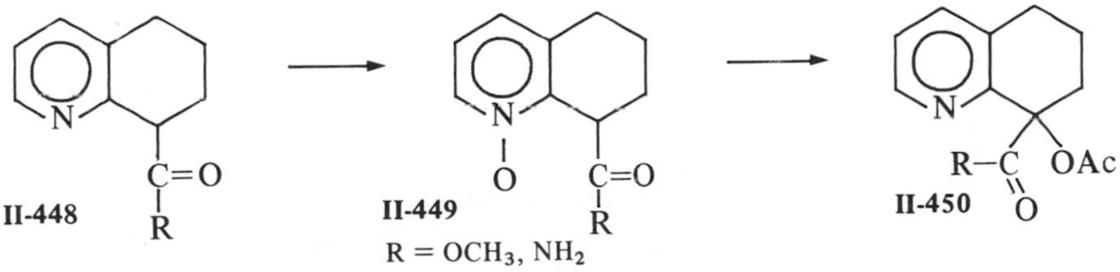

II-448 **II-449** **II-450**

$R = OCH_3, NH_2$

B. α-Metallation

The major effort in α-metallation of annelated pyridines has been made by Curran and his coworkers at Wyeth Laboratories. This group found that the Grignard reagent **II-451b** was best prepared by an exchange reaction involving tetrahydroquinoline **II-451a** and isopropyl magnesium bromide. The lithio species **II-451c** was generated from **II-451a** using either phenyllithium or butyllithium (194). Reaction of these organometallics with carbon dioxide provided the 8-carboxylic acid (**II-452a**) while subsequent esterification provided **II-452b**. These esters were then converted to the corresponding amides, nitriles, and thioamides (**II-452c, d, e**, respectively) (194). The yield of the overall conversion of **II-451a** to **II-452b** varied from 42–68% depending on the method employed.

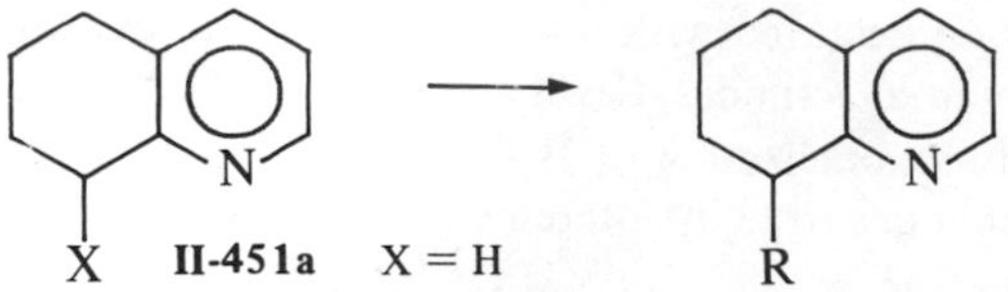

X   II-451a   X = H
    b   X = MgBr   R
    c   X = Li   II-452 a   R = CO$_2$H
                            b   R = CO$_2$CH$_3$
                            c   R = CONH$_2$
                            d   R = CN
                            e   R = CSNH$_2$

Direct conversion of the 8-lithio derivative **II-453** to the thioamide **II-455** was achieved by utilizing trimethylsilyl isothiocyanate, followed by mild hydrolysis (195). The yields varied from 0.5–70%, depending on the solvent (benzene or hexane) and the nature of the substrate.

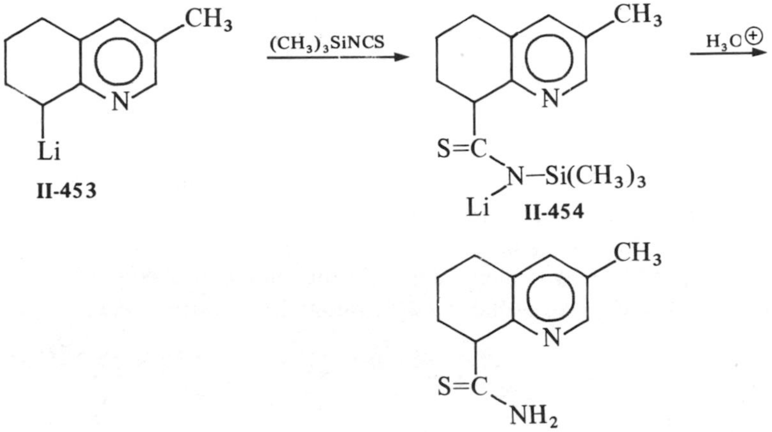

A wide variety of 5,6,7,8-tetrahydroquinolines substituted in the 8-position has been reported by Curran's group and extensively patented. For more detailed descriptions of these specific systems, the patent literature (196–204) should be consulted.

C.   α-Condensation

Hahn, Epsztajn, and coworkers have made a careful study on the condensation of 2,3-annelated pyridines with a variety of aldehydes. They found that when 2,3-cyclopentapyridine or 5,6,7,8-tetrahydroquinoline was heated at 110–120°C in a sealed tube for 25 hr, the mono-condensation product **II-457** was obtained. The same reaction carried out at 150–160°C gave predominantly the *bis*-condensation product **II-458** (205). Treatment of primary alcohols **II-457** with polyphosphoric acid effected dehydration to afford the exomethylene derivative **II-459** (206).

II-456 $n = 2, 3$ II-457 II-458 II-459

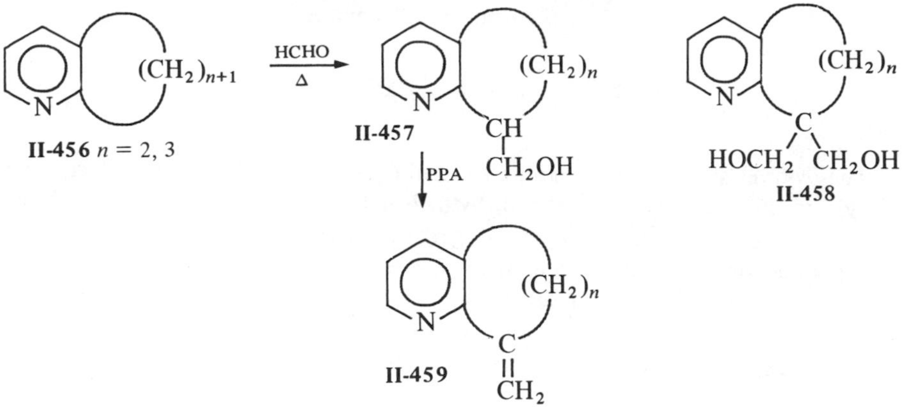

A competition experiment has been carried out to determine the relative reactivity of the α-methylene group of an annelated pyridine versus an α-methyl group. For substrates **II-460a** and **b**, the observed products of condensation with benzaldehyde or *p*-nitrobenzaldehyde were **II-461a** and **b**, where the reaction had occurred exclusively at the ring methylene. For the case of 6-methyl-2,3-cycloheptapyridine **(II-460c)** condensation occurred primarily at the methyl group to afford **II-462** (207). If the condensation product of 5,6,7,8-tetrahydroquinoline and benzaldehyde was ozonized, followed by a reductive workup,

II-460a $n = 3$
b $n = 4$
c $n = 5$

II-461a $n = 3$
b $n = 4$

II-462

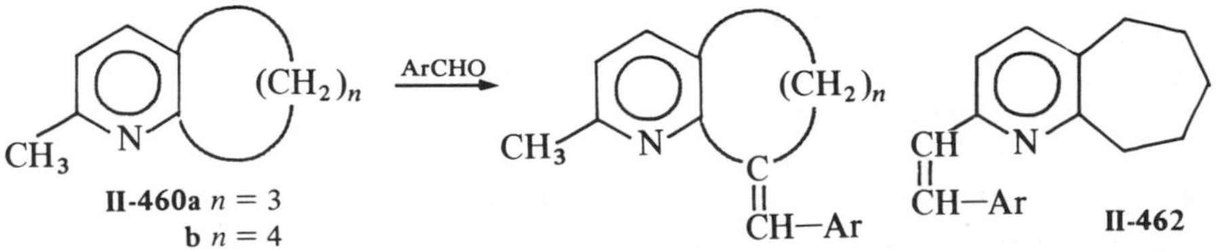

quinolone **II-464** could be obtained (208). Condensation of compounds **II-442** ($n = 5, 6, 7, 8$; $R = C_6H_5$) with benzaldehyde and *p*-nitrobenzaldehyde has been studied (209) with the finding that reactivity of the 2-methylene group diminished as the size of the cycloalkyl ring was increased. Only the systems where $n = 5$ and $n = 6$ showed any appreciable reactivity. The basicity and UV spectra of these compounds were recorded but no direct relationship with reactivity could be observed.

II-463 **II-464** (66%)

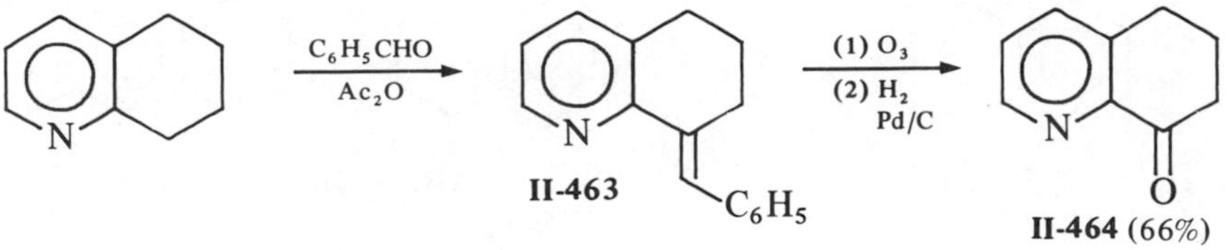

2. *Bis*-Annelated Pyridines

A. *Reactions of N-Oxides*

The *N*-oxides (**II-465a–d**), prepared from *para-bis*-annelated pyridines, were acetylated in acetic anhydride and hydrolyzed to the corresponding alcohols. The symmetrical systems (**II-465a, b**) gave a single product with an OH group incorporated on one of the two equivalent α-carbons. The unsymmetrical systems (**II-465c, d**) were claimed to give mixtures of the two possible alcohol products (210). *Sym*-octahydroacridine *N*-oxide (**II-465b**) can be acetylated, converted to the *N*-oxide, and acetylated again to provide the diacetate **II-466** (211, 212).

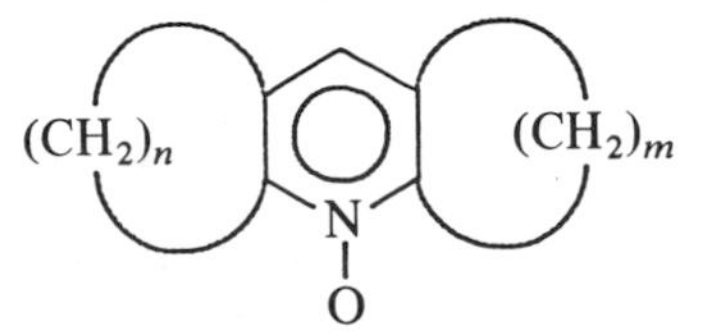

II-465a $m = n = 3$,   b $m = n = 4$
c $m = 3, n = 4$, d $m = 3, n = 5$

Hydrolysis provided diol **II-467** from which can be derived a variety of other functionalized derivatives **II-468–470**. If the acetylation was carried out in the presence of an aromatic aldehyde, the unacetylated α-position can condense with the aldehyde to provide a product such as **II-471** (213). This same material may be obtained by subjecting **II-472** to identical reaction conditions. Hydrolysis of the ester function of **II-471** and dehydration of the resulting alcohol have been reported (213).

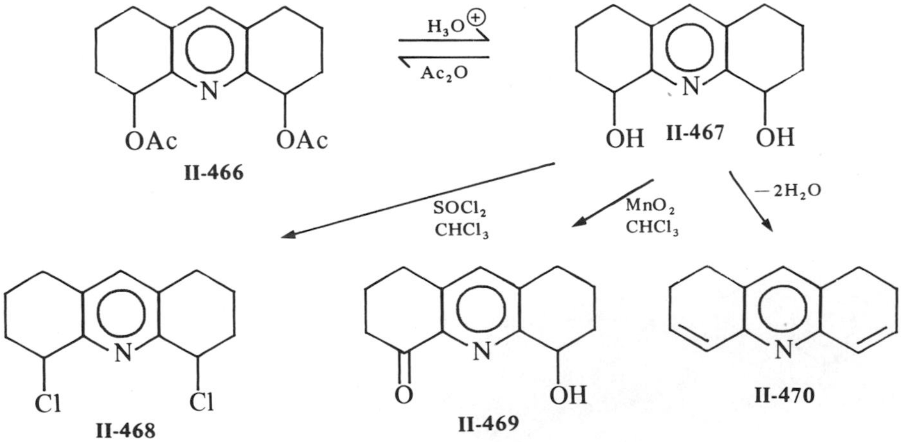

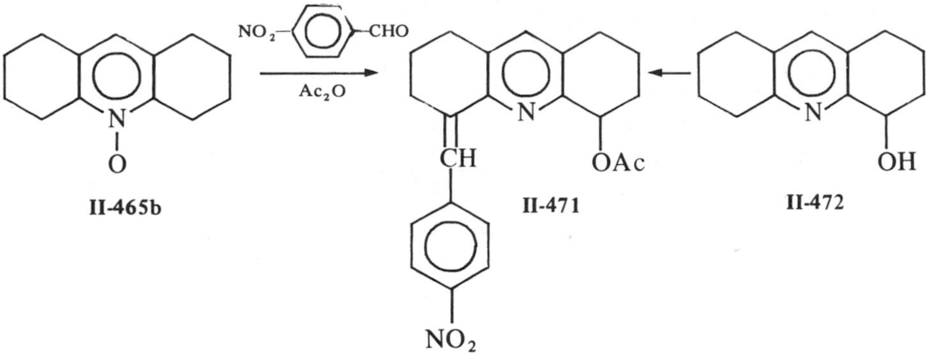

II-465b II-471 II-472

Nitration of **II-465b** with nitric acid and sulfuric acid led to the 9-nitro derivative **II-473**, which can be converted to a variety of other functionalities, such as **II-474a–d** (214).

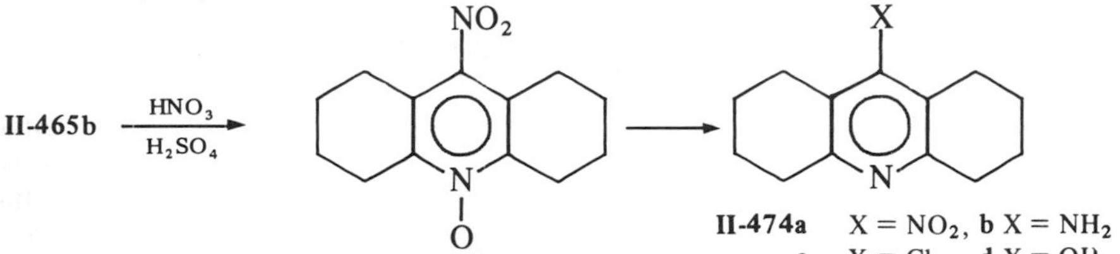

II-474a X = NO$_2$, **b** X = NH$_2$
c X = Cl, **d** X = OH
II-473

B. α-Condensation

Vysotskii and Tilichenko have reported that *sym*-octahydroacridine will condense with two molecules of an aromatic aldehyde to give a diarylideneoctahydroacridine **II-475** (215, 216). If an excess of aromatic aldehyde was treated with cyclopentanone and ammonium acetate in ethanol, pyridines **II-477** may be prepared (217).

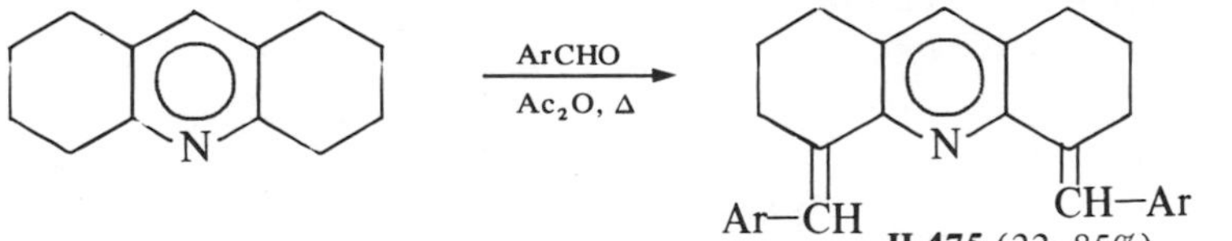

II-475 (22–85%)

Initially a Chichibabin-type condensation occurred to give the *bis*-annelated pyridine **II-476**, which then condensed with two additional molecules of the aldehyde to give **II-477**.

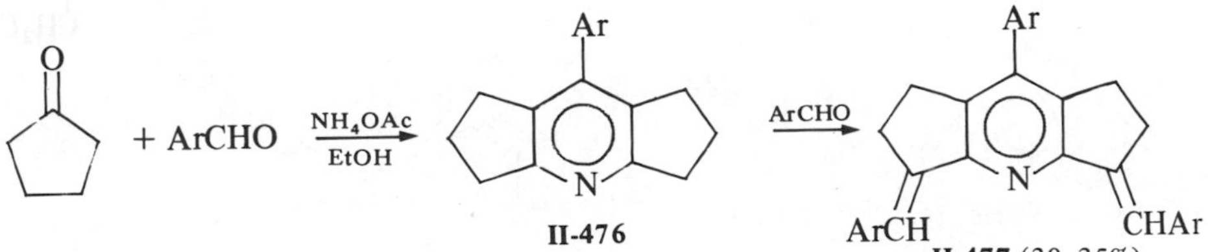

II-476
II-477 (30–35%)

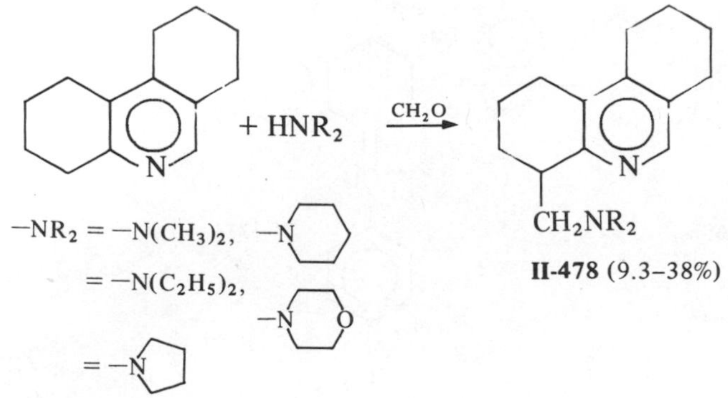

$$-NR_2 = -N(CH_3)_2, \quad -N\langle\text{piperidine}\rangle$$
$$= -N(C_2H_5)_2,$$
$$= -N\langle\text{pyrrolidine}\rangle$$

$-N\langle\text{morpholine}\rangle O$

CH_2NR_2

II-478 (9.3–38%)

Hahn and Koziolkiewicz have found that aminomethylation of *bis*-annelated pyridines occurred only under relatively drastic conditions (218). Their sole success was with 1,2,3,4,7,8,9,10-octahydrophenanthridine, which condensed only at the α-position. With formaldehyde and various secondary amines, the yields of **II-478** ranged from 9.3 to 38%.

C.   Metallation

The principle work in this area has been carried out by the Tilichenko and Curran groups. If *sym*-octahydroacridine was treated with one equivalent of phenyllithium, the metallated species **II-479** was obtained. Addition of an aldehyde or ketone to **II-479** resulted in formation of the expected alcohol by nucleophilic addition to the acyl group (219). If **II-479** was treated with benzyl chloride, a coupling reaction occurred to give **II-480** (220), which reacted with another equivalent of phenyllithium to effect α′-metallation so that reaction with benzyl chloride provided an approximately equal mixture of the two possible stereo-isomers of **II-482** (221). This same material may also be prepared by the catalytic hydrogenation of **II-475**.

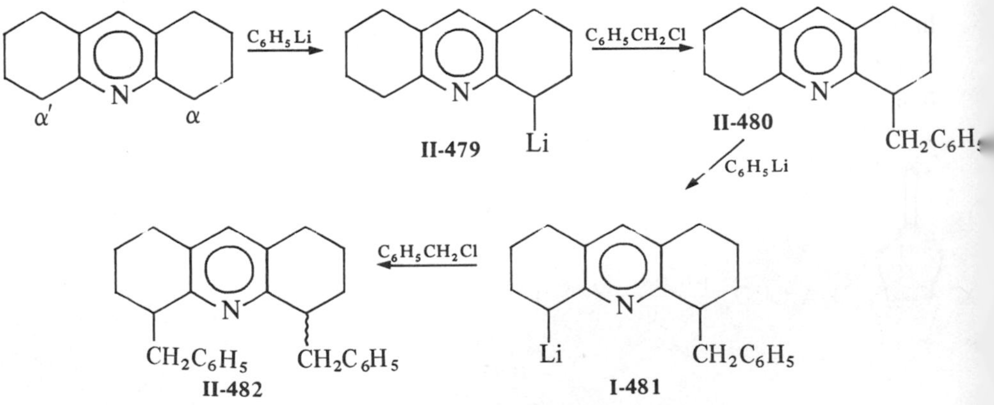

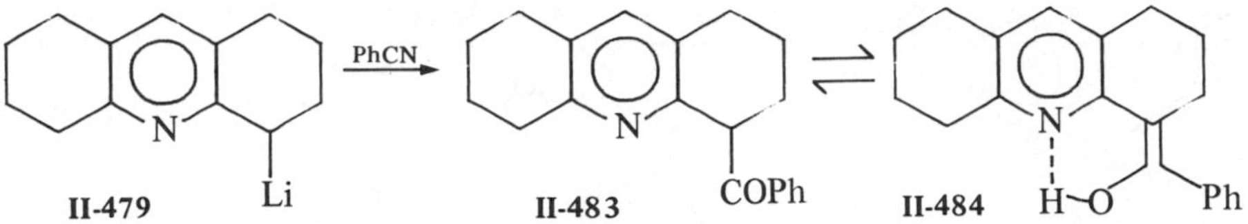

II-479   Li   II-483   COPh   II-484   H–O   Ph

Treatment of **II-479** with benzonitrile provided the benzoyl derivative **II-483**. The infrared spectrum of **II-483** indicated that it exists largely as enol **II-484** (222). A complex may be formed between **II-484** and cupric acetate in which two molecules of the pyridine derivative chelate one atom of copper.

Curran and coworkers have prepared a variety of derivatives from **II-479** and related *bis*-annelated pyridines in which the α-substituent was a cyano, amide, thioamide, or ester group (223, 224).

D. Other Reactions

Sodium borohydride or sodium hydrosulfite reduction of *bis*-annelated pyridine **II-485** has proceeded only in poor yield. Better results were obtained by catalytic hydrogenation of **II-485** in ethanol over platinum oxide to yield dodecahydro-acridinedione **II-486** (225).

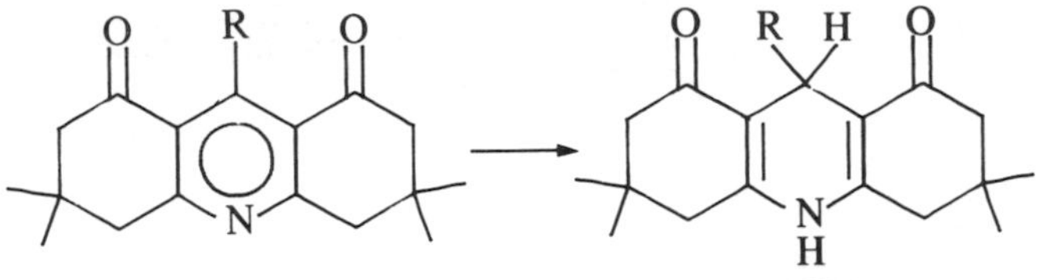

II-485   II-486 (80–90%)

Octahydroacridine and octahydrophenanthridine have been dehydrogenated by the use of chloranil (226) or palladium (227). In both cases the yields were either low or mixtures of products were obtained.

3. Azafluorenes

Fluorene does not react with refluxing acetic anhydride. Under these reaction conditions, however, 3-methyl-2-azafluorene (**II-487**) was acetylated to give **II-488**, a high-melting, red material (228). Depending on the pH of the medium, this acetyl derivative underwent a number of transformations. Most notably, with atmospheric oxygen in ethanol, **II-488** was oxidized to azafluorenone **II-489**.

Condensation of **II-487** with diethyl oxalate provided α-ketoester **II-490** (229), while cyanoethylation gave **II-491a** which can be hydrolyzed to **II-491b** (230). Ethyl acrylate underwent a double Michael addition with **II-487** in the presence of potassium metal to provide diester **II-491c** (231). A similar reaction was observed for ethyl crotonate.

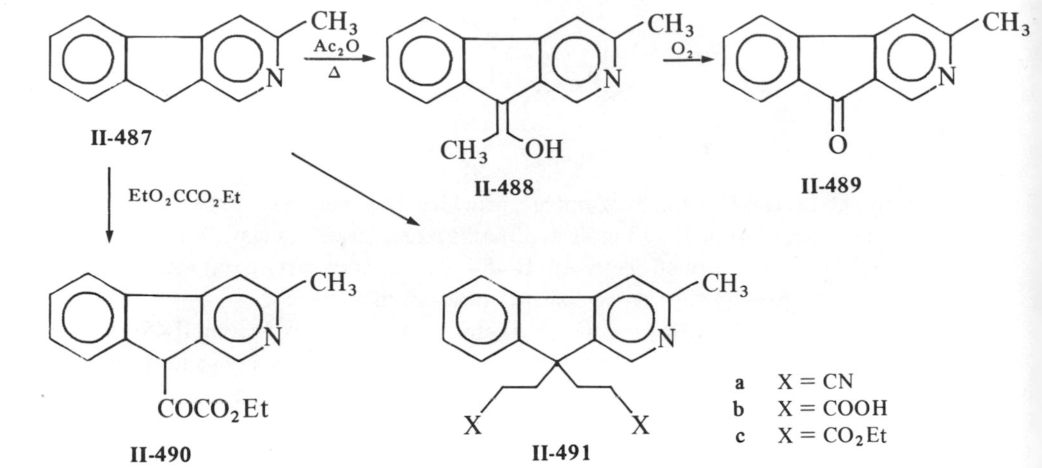
CH3
Ac2O
Δ
N
CH3
O2
CH3
OH
O
II-487
II-488
II-489
EtO2CCO2Et
COCO2Et
X
X
II-490
II-491
a X = CN
b X = COOH
c X = CO2Et

The difunctionalized azafluorene **II-491b** and **c** underwent further cyclizations, as illustrated in Scheme II-4.

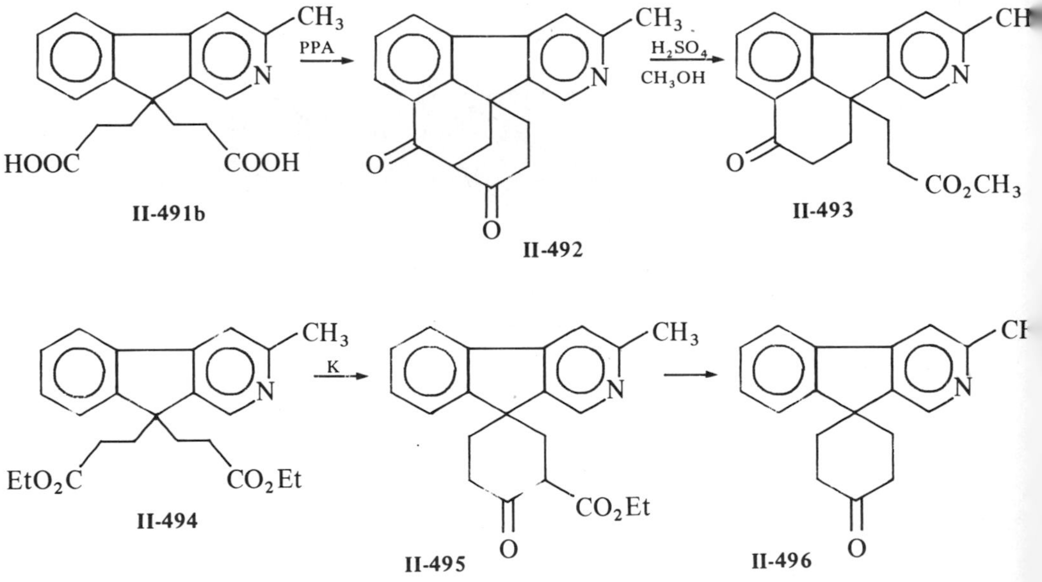
CH3
PPA
H2SO4
CH3OH
N
HOOC
COOH
O
CO2CH3
II-491b
II-492
II-493
K
EtO2C
CO2Et
II-494
II-495
II-496

Scheme II-4

Azafluorenone **II-497** is a *bis*-annelated pyridine similar to **II-485**. When **II-497** was hydrogenated over Adams catalyst in ethanol, the five-membered ring carbonyl group was reduced rather than the pyridine ring. Reaction of **II-497** with zinc dust in glacial acetic acid led to reduction of both carbonyl groups (232).

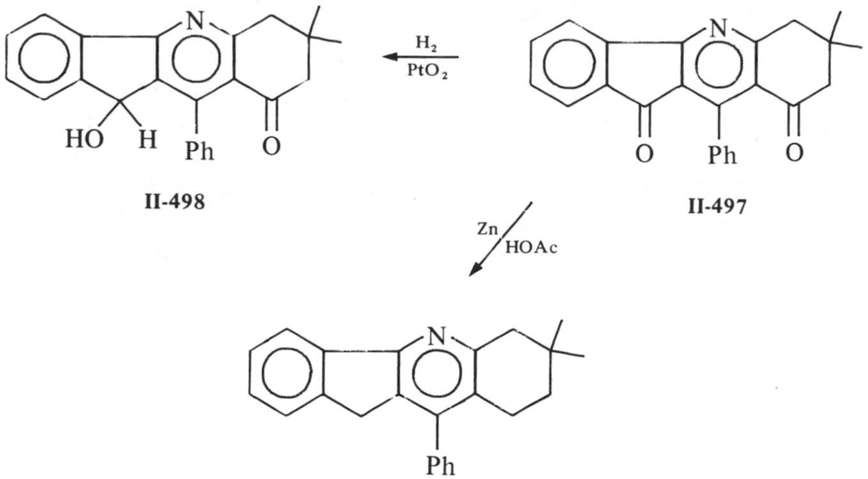

II-498 II-497

II-499

Schönberg and Junghans have prepared a number of derivatives of 1,8-diaza-fluorenone **II-500** (233, 234). Perhaps the most interesting of these compounds was *bis*[1,8-diazafluorenylidene] **II-503**, which was prepared by the photochemical addition of **II-500** to 1,8-diazafluorene followed by dehydration. In some more recent work, Newkome and Roper have continued investigation of compounds of this type (235).

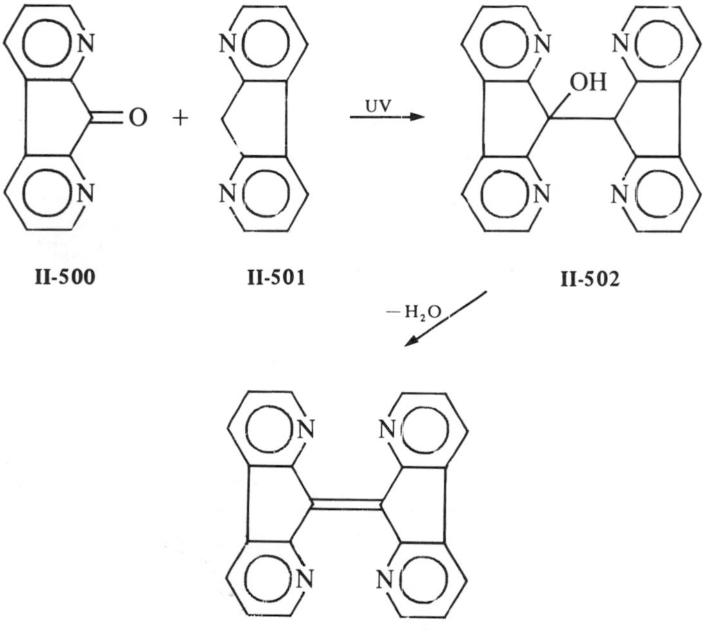

II-500 II-501 II-502

II-503

4. Other Annelated Pyridines

Villiani et al. have prepared a large series of azadibenzocycloheptanes (**II-505a**, **II-506a**) and azadibenzocycloheptenes (**II-505b**, **II-506b**) by treating the appropriate azaketones **II-504** with a Grignard reagent to give the corresponding tertiary carbinols by a reductive alkylation procedure (236). Carbinols **II-505** were subjected to the usual acid dehydrating conditions and gave the unsaturated compounds **II-506**. Derivatives were prepared with the pyridine nitrogen atom in all four possible positions where X was hydrogen, chlorine, or bromine, and R was an alkylamino group.

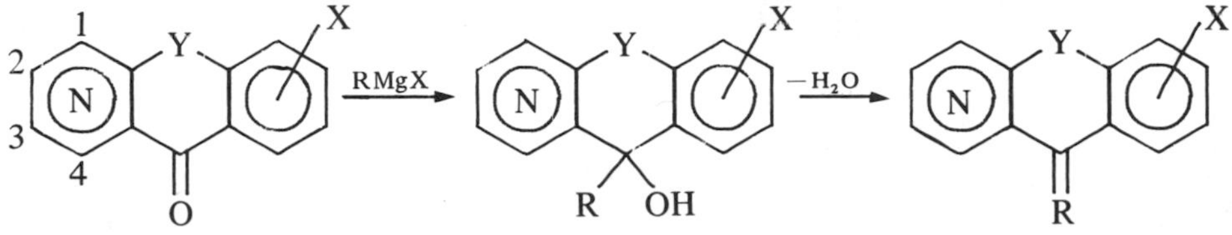

II-504a   Y = CH₂CH₂   **II-505**   **II-506**
    b   Y = CH=CH

Yamane has reported that the nitroso derivative of pyrido[3,2d]-tropolone **II-507** afforded **II-508** upon condensation with o-phenylenediamine (237). A later study revealed that the parent tropolone **II-509** reacted with phenylhydrazine in an unexpected fashion such that one molecule of ammonia was lost during the reaction. It was believed that the fused indole derivative **II-510** was obtained (238).

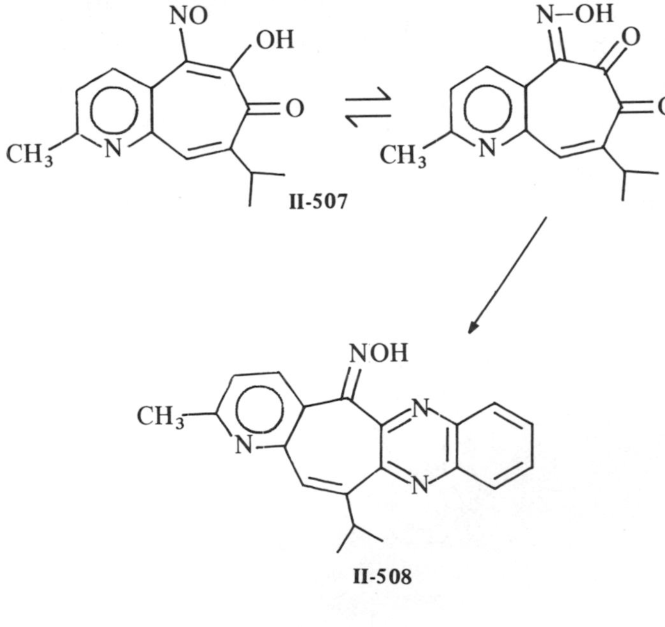

II-507

II-508

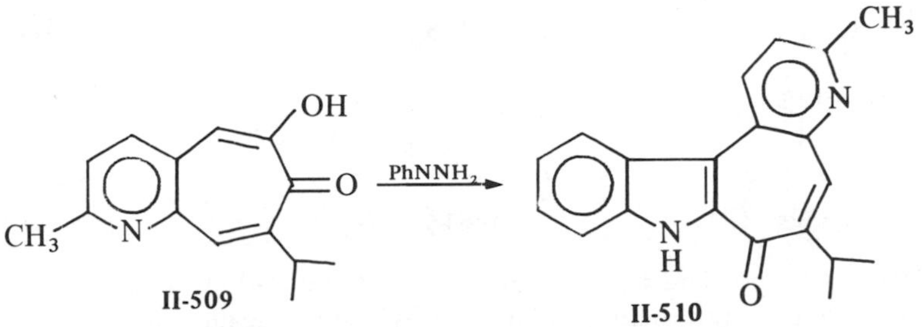

II-509   II-510

Although the chemistry of quinolinediones should be treated in the appropriate chapter of this series devoted to quinolines, one example of an interesting Diels–Alder reaction has to be cited. Cyclohexadiene was added to 5,8-quinolinedione **II-511** (an annelated pyridine) to provide **II-512**, which tautomerized to **II-513** and then further oxidized back to quinone **II-514**. The initial cycloaddition proceeded in higher yields with 2,3-dimethyl-1,3-butadiene, cyclopentadiene, and anthracene (239).

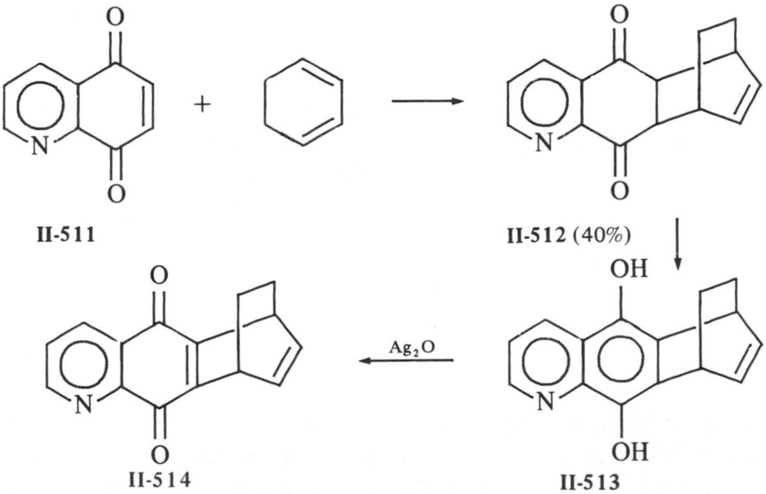

II-511   II-512 (40%)

II-514   II-513

IV. NATURALLY OCCURRING ANNELATED PYRIDINES

In 1928 Eguchi reported the detection of a number of pyridine bases in Fushun shale tar (240). Among these compounds was 2,3-cyclopentapyridine. This finding was verified three years later by Thompson, who accomplished the first authentic synthesis of this material (241). In 1930 Bailey et al. described their investigation of the bases in the kerosene distillate of California petroleum wherein they mentioned the presence of a hydroaromatic base of the formula $C_{16}H_{25}N$ (242). A year later this same group published a paper in which they attempted to define the structure of this material (243), provisionally claiming it to be **II-515**.

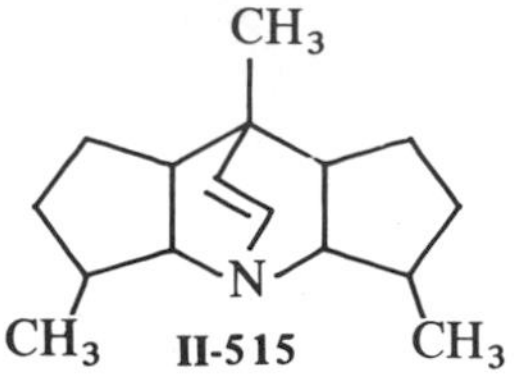

II-515 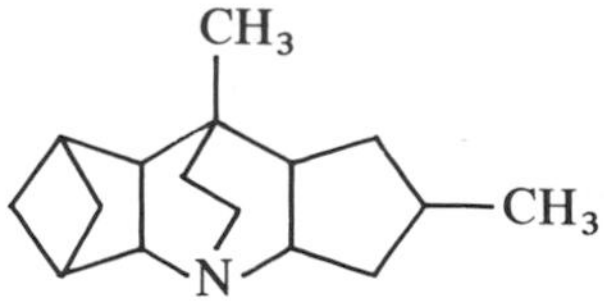   II-516 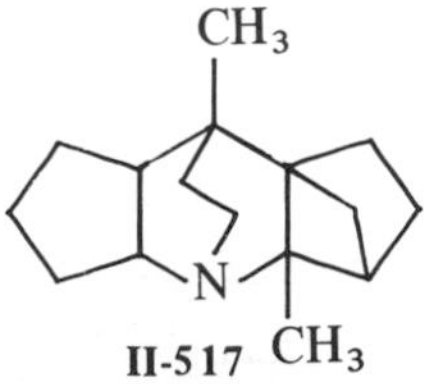   II-517

In 1933, Armendt and Bailey reinvestigated the question of $C_{16}H_{25}N$, settling on the equally unlikely structure **II-516** (244) which was again modified the following year to **II-517** (245). This reviewer remains unconvinced by all three proposals but will not compound Bailey's folly by advancing still another possibility. Twenty years later Arnall continued these investigations of coal tar bases and reported the presence of 3,4-cyclopenta- and 6-methyl-2,3-cyclopentapyridine in addition to the previously described 2,3-cyclopentapyridine (246, 247). These results were further confirmed by Lochte and Pittman (94). At about the same time a group of German chemists identified a variety of bases in lignite coal among which were 2,3-cyclopentapyridine as well as 5,6,7,8-tetrahydroquinoline (248).

There have been scattered reports of annelated pyridines being isolated from natural sources other than petroleum. Hammouda and Le Men identified Tecosfanine (**II-518**) as an alkaloid extracted from the juice of Tecoma Stans (249). The fungus Rhizoctonia leguminicola has afforded the octahydropyrindine **II-519** (250) while five closely related monoterpenoid alkaloids have been found to be produced by the Chilean desert plant *Skytanthus actus M.* (251).

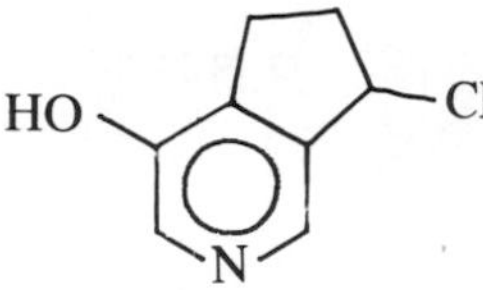

II-518 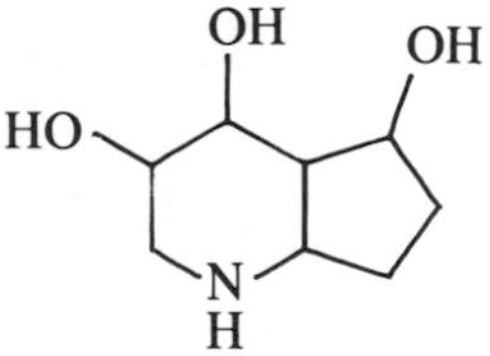   II-519

Examination of the alkaloid constituents of the *Sceletium* species has indicated the presence of two closely related materials containing an annelated pyridine ring, *Sceletium* A_4 (**II-520a**) (252, 253), and tortuosamine (**II-520b**) (254). The structure of **II-520a** has been verified by x-ray crystallography (255). Lycodine (**II-521**) was

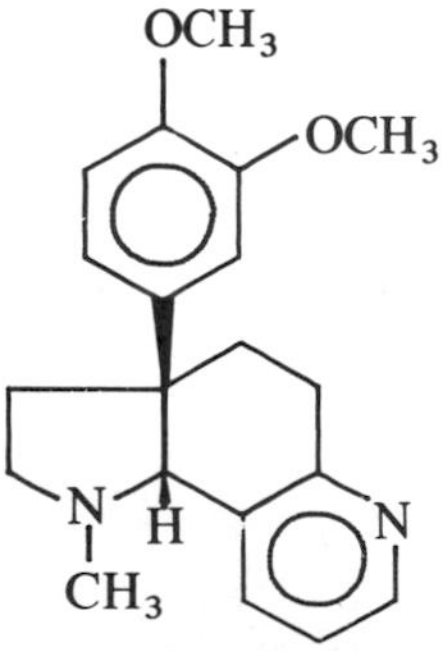

II-520a 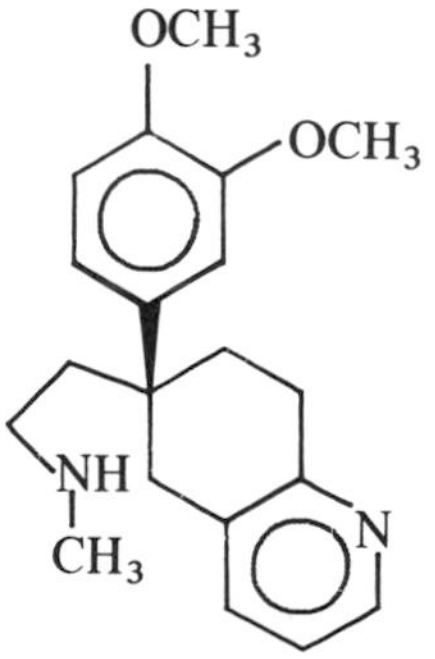   II-520b 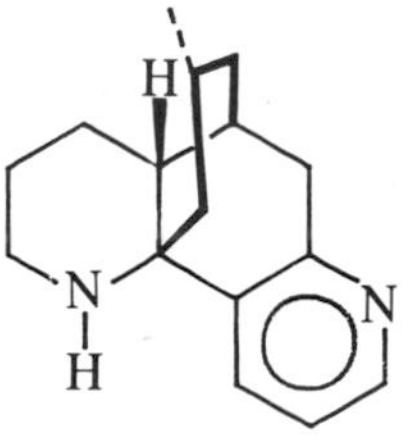   II-521 first isolated from *Lycopodium annotinum* by Anet and Eves (325) and more recently has been synthesized by Kleinman and Heathcock (326).

Several azaquinone-type species have been isolated which, at least formally, incorporate an annelated pyridine ring. Streptonigrin (**II-522**), a metabolite of *Streptomyces flocculus*, in an antibiotic that exhibits striking activity against a variety of animal tumors (256). It is a monobasic acid and is readily susceptible to reversible two electron reduction. *Phoma terrestris Hansen*, the fungus responsible for the "pinkroot disease" of onions, produced an orange pigment when grown on Czapek–Dox medium containing starch. This pigment metabolite has been identified as phomazarin and assigned the structure **II-523** (257). Another pigment, bostrycoidin, was isolated from *Fusarium solani* D$_2$ purple and identified as a β-azaanthraquinone having structure **II-524** (258).

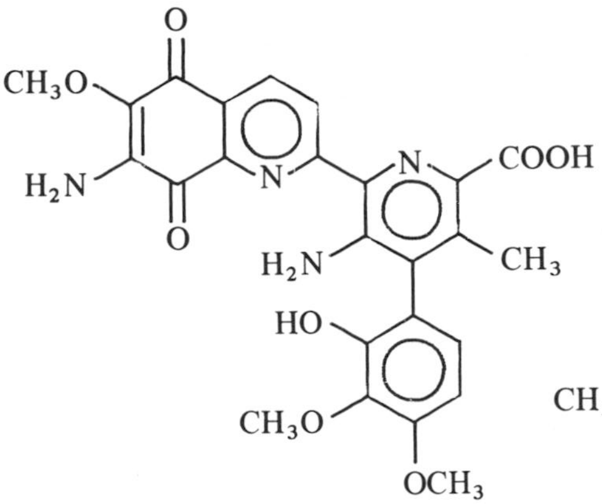

II-522

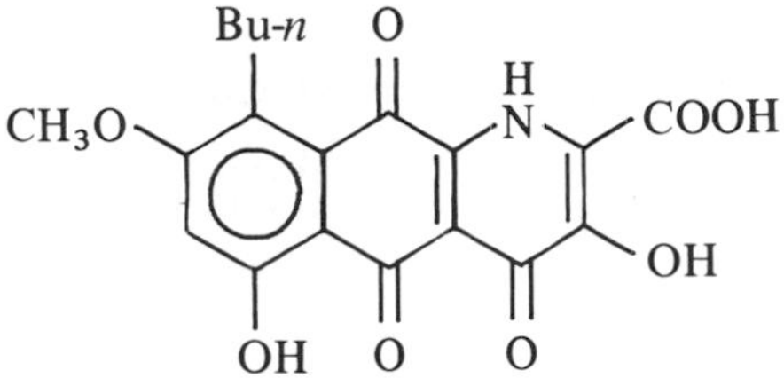

II-523

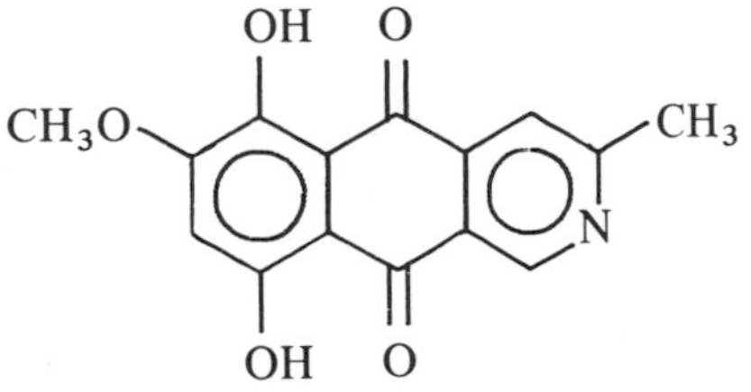

II-524

V. BIOLOGICALLY ACTIVE ANNELATED PYRIDINES

One of the most potent nonanticholinergic antisecretory agents reported in recent years has been pyridyl-2-thioacetamide (**II-525**). On the premise that 5,6,7,8-tetrahydroquinoline-8-thiocarboxamide (**II-526**) resembled **II-525**, Curran et al. undertook a systematic study which revealed that the 8-substituted thioamides as well as the 8-nitriles were potent inhibitors of basal gastric secretion in the pylorus-ligated rat. These compounds also afforded protection against gastric erosions induced in rats by cold-restraint stress (193). A somewhat different structure-activity relationship was exhibited by the analogous 8-thiourea (**II-527**) (259).

The azadibenzocycloheptadienones and related derivatives have been shown to be effective in regulating a variety of biological functions. Thus, **II-528** and several similar compounds substituted differently on the pyridine ring have shown central nervous system activity (260). The tertiary amine **II-529** has proved useful

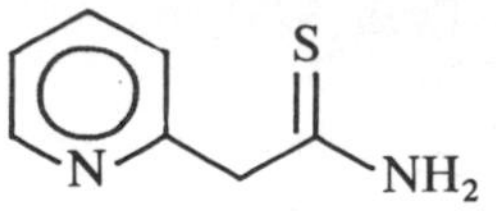

II-525

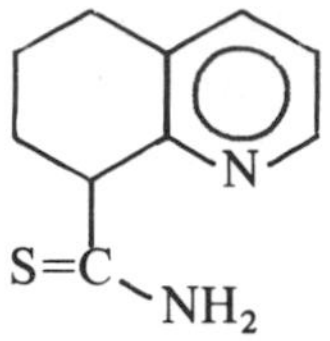

II-526

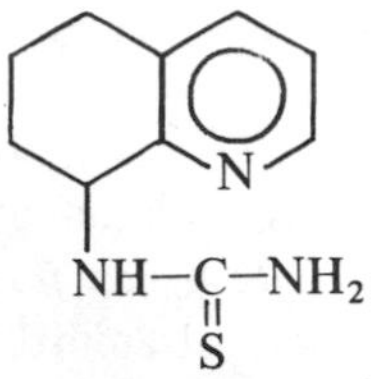

II-527

as a coronary vasodilator and blood pressure heightening agent (261). Villani et al. have examined an extensive series of azadibenzocycloheptadienone derivatives and found that the 4-aza derivatives showed the maximum antianaphylactic and antihistaminic activity where the order of decreasing potency was 4-aza > 2-aza > 1-aza > 3-aza. The most active compound studied was the one possessing a 1-methyl-4-piperidylene group at the 5-position (**II-530**). It showed high activity against fatal anaphylaxis in mice sensitized with horse serum and pertussis vaccine and later challenged with horse serum (236, 262).

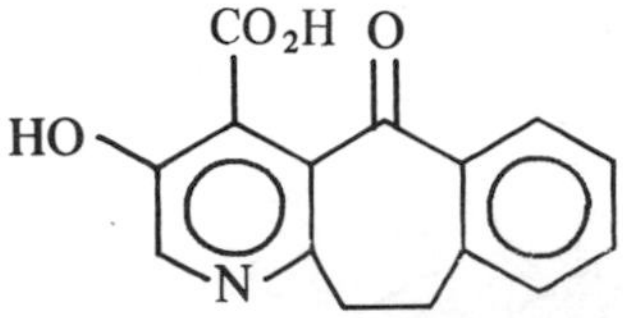

II-528

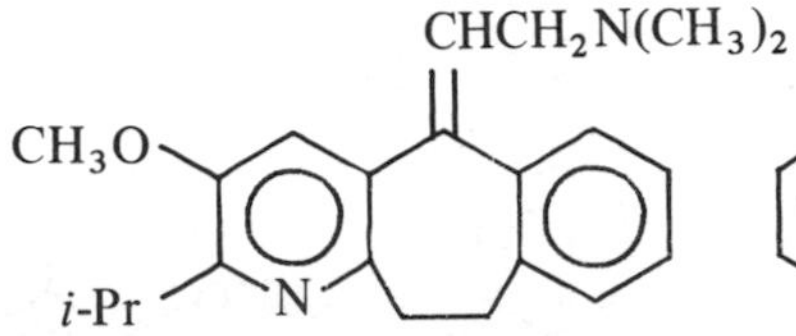

II-529

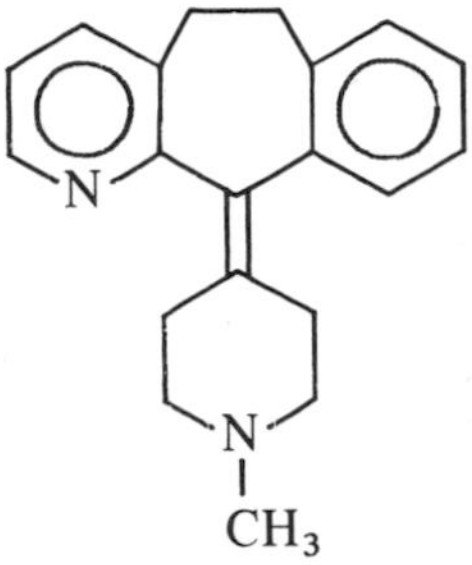

II-530

Tuberculostatic activity has been evidenced by some cyclopentapyridines (263). Numerous isonicotinic thioamides were tested for their antituberculostatic activity and **II-531** was found to be among the least active (264, 265). 2,3-Diaminoazanaphthoquinones **II-532** have been shown to be effective tuberculostatic agents and were also efficacious in inhibiting aerobic glycolysis (266).

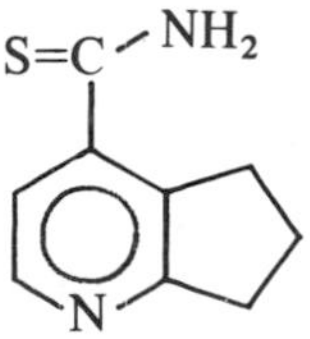

II-531

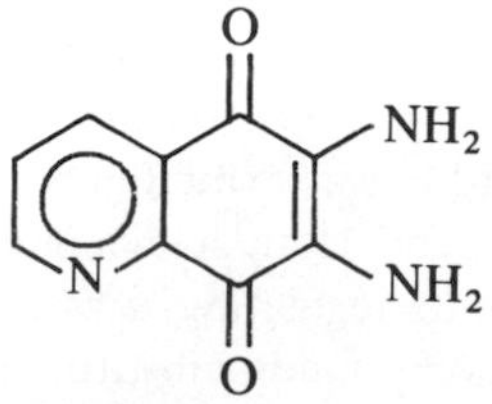

II-532

A total synthesis of (±)-2-azaestradiol 3-methyl ether (**II-533**) has been reported (267). The biological profile of such 2-azaestratrienes has thus far shown this series to possess antiviral properties in assays to determine anti-influenza activity. Several 3,4-cyclopentapyridine derivatives with different substituents on the pyridine ring have been found to be inhibitors of collagen biosynthesis (268, 269). The spiro-hydantoins **II-534** where $n = 0, 1$ inhibit aldose reductase activity, decrease sorbitol accumulation, and inhibit cataract formation in diabetics (270). Several other simple annelated pyridines have been found to be useful as analgesics and anti-inflammatory agents (271–273).

II-533

II-534 $(n = 0, 1)$

There is a report of a fungicide which contains an azanaphthoquinone derivative as its active ingredient (274). Some 5,6,7,8-tetrahydroquinolines have found use as insecticides (275, 276). Various substituted octahydrophenanthridines have found value as pesticides especially in the control of *Acarina*, particularly ixodid ticks such as *Amblyomma* and *Boophilus* and insect pests such as fleas (277).

VI. PHYSICAL PROPERTIES

1. NMR Spectra

Pyridine proton (^{1}H) and carbon-13 (^{13}C) chemical shift assignments are greatly facilitated by the electronic influence which the nitrogen atom exerts on the various ring positions. Thus, the 2- and 6-(α) positions are the most deshielded due to the proximity of the electronegative heteroatom. The 4-(γ) position possesses significant positive character by resonance and is the next most deshielded. Atoms at the 3- and 5-(β) positions are the least deshielded and are found at highest field. An additional feature facilitates assignment of carbon chemical shifts. Aromatic carbons bearing a hydrogen have shorter relaxation times and greater nuclear Overhauser enhancements, thus giving rise to more intense signals.

In Tables II-5 and II-6 the ^{1}H and ^{13}C chemical shifts for *mono-* and *bis*-annelated pyridines as well as appropriate dimethyl and tetramethyl substituted analogs are recorded (37). In the case of 3,4-substituted pyridines, C-2 was assigned to lower field than C-6 by analogy with assignments for 3-methylpyridine. Assignments for an unsymmetrical pyridine such as **II-9** were facilitated by the observation of close

TABLE II.5. NMR AND ULTRAVIOLET SPECTRAL DATA FOR MONO-ANNELATED PYRIDINES – CHEMICAL SHIFTS (ppm) (37)

	H_2	H_3	H_4	H_5	H_6	C_2	C_3	C_4	C_5	C_6	Max (95% EtOH) (e)		
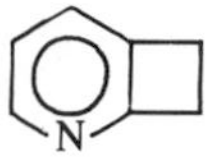 II-255			7.32	7.10	8.39	164.2	140.2	129.6	122.9	148.6	278(3420)	272(4800)	269(4770)
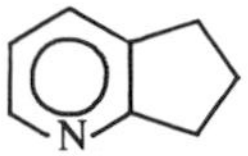 II-15			7.47	7.00	8.31	165.3	136.8	132.0	120.8	147.0	278(2768)	274(3580)	270(3873)	265(3421)
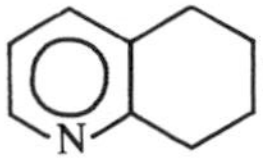 II-61			7.31	6.98	8.31	157.1	132.1	136.7	120.7	146.4	276(2180)	271(2556)	268(2851)	264(2491)
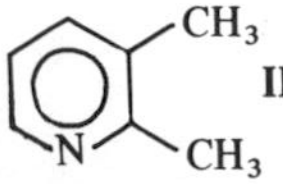 II-535			7.37	7.00	8.29	156.7	131.0	136.6	120.8	146.1	271(2417)	265(3182)	259(2767)
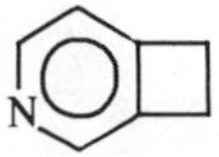 II-265	8.16			6.92	8.39	147.8	142.4	155.3	118.2	142.7	263(1620)	258(1950)	253(1740)
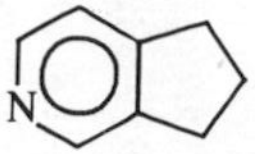 II-152	8.44			7.14	8.33	147.1	139.9	153.3	119.7	145.7	267(1728)	259(2027)	254(1646)
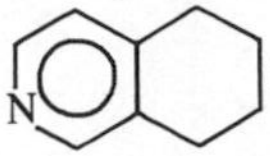 II-385	8.27			6.94	8.24	150.3	132.8	145.9	123.7	146.3	269(1771)	261(2169)	254(1708)
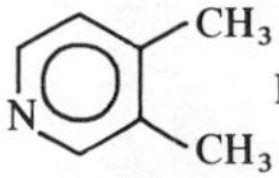 II-536	8.28			6.99	8.26	149.6	131.6	145.0	124.1	146.9	267(1842)	259(2248)	255(1947)

TABLE II-6. NMR AND ULTRAVIOLET SPECTRAL DATA FOR *BIS*-ANNELATED PYRIDINES — CHEMICAL SHIFTS (ppm) (37)

		H_{Ar}	C_2	C_3	C_4	C_5	C_6	$J_{C_{Ar}}$—H	Max (95% EtOH) (e)			
	II-268	7.00	161.7	138.6	124.6	138.6	161.7	163.2 Hz			290(8078)	
	II-103	7.30	163.2	134.1	128.4	134.1	163.2	159.8	297(4875)	292(7200)	287(7925)	282(6800)
	II-9	7.15	162.4	133.9	132.8	129.0	154.6	155.2	295(4805)	289(6603)	285(7000)	281(6217)
	II-10	7.02	153.9	129.1	137.3	129.1	153.9	152.2	291(3980)	286(5073)	281(5520)	277(4625)
	II-537	7.10	153.1	128.1	138.5	128.1	153.1	152.6	281(3388)	275(4518)	272(4556)	267(3743)
	II-538	8.18	162.8	132.5	149.2	137.3	142.9	174.9	278(2804)	274(2942)	270(2927)	
	II-48	8.22	154.1	128.1	152.3	136.8	142.2	174.0	276(2840)	272(2985)	268(3205)	
	II-47	8.06	161.6	135.0	142.0	129.8	148.1	173.7	276(4300)	272(4470)		
	II-539	8.06	153.4	129.6	144.0	129.6	146.9	172.2	279(3064)	276(3321)	271(3423)	267(3013)
	II-540	8.04	153.9	129.1	143.6	129.1	146.3	172.2	275(2552)	271(2958)	267(4000)	263(2682)

TABLE II-7.   CHANGES IN ^{13}C CHEMICAL SHIFTS UPON DECREASING ANNELATED
               RING FROM SIX CARBONS TO FIVE (37)

Compound	Altered Ring Fusion	ΔC_2	ΔC_3	ΔC_4	ΔC_5	ΔC_6
II-61 → II-15	2,3	−8.2	−4.7	+4.7	−0.1	−0.6
II-10 → II-9	2,3	−8.5	−4.8	+4.5	+0.1	−0.7
II-9 → II-103	5,6(2,3)^a	−8.6	−5.1	+4.4	−0.2	−0.8
II-539 → II-47	2,3	−8.2	−5.4	+2.0	−0.2	−1.2
II-48 → II-538	2,3	−8.7	−4.4	+3.1	−0.5	−0.7
II-10 → II-103	2,3 and 5,6	−9.3	−5.0	+8.9	−5.0	−9.3
II-385 → II-152	3,4	+3.2	−7.1	−7.4	+4.0	+0.6
II-539 → II-48	4,5(3,4)^a	+4.7	−7.2	−8.3	+1.5	−0.7
II-47 → II-538	4,5(3,4)^a	+5.2	−7.5	−7.2	+2.5	−1.2
II-539 → II-538	2,3 and 4,5	−9.4	−2.9	−5.2	−7.7	+4.0

^{a} Numbering patterns (C_2 and C_6, C_3 and C_5) in these molecules are reversed to preserve
consistency throughout the table.

agreement with the symmetrical analogs II-10 and II-103. Pyridines II-47 and II-48
were assigned by similar analogy to II-538 and II-539. The fusion of a six-membered
ring onto the pyridine nucleus resulted in chemical shifts very much like those of
the corresponding methyl derivatives as was illustrated by the similarity between
the NMR data for II-61 and II-535, II-385 and II-536, II-10 and II-537, and II-539
and II-540.

As the size of the fused ring was decreased from six to five carbons, a very
consistent variation in pyridine chemical shifts was observed. The ^{13}C chemical
shifts are tabulated in Table II-7. For a ring fused at the 2,3-position, both bridgehead carbons were observed to shift downfield with C-2 shifting 8.2–8.7 ppm and
C-3 shifting 4.4–5.4 ppm. For 3,4-fused systems, the shift was also downfield
with C-3 shifting 7.1–7.5 ppm and C-4 shifting 7.2–8.3 ppm. At the pyridine ring
positions ortho to the bridgehead carbons, the change was in the opposite direction,
shifting upfield 1.5–3.1 ppm when the ortho carbon was bonded to a methylene
and 3.2–5.2 ppm when bonded to a hydrogen. The chemical shift of the hydrogen
bonded to this ortho carbon was found to move downfield as the size of the fused
ring was decreased from six carbons to five. Changes in proton and carbon chemical
shifts at positions meta and para to the bridgehead varied only slightly when the
annelated ring size was decreased, indicating that the influence of the fused ring
was localized around the bridgehead and did not significantly affect the π-sextet
as a whole. It should be noted that similar variations in carbon chemical shifts were
observed between tetralin and indan.

For the cyclobutapyridines, the chemical shift of greatest interest was that of
the aromatic carbon and attached proton located ortho to the point of ring fusion.
In the benzene series, the appearance of this aromatic proton (H-3) at substantially
high field has been pointed out in the case of the benzo[1,2:4,5]dicycloalkenes
(278). The ortho ring protons of II-255, II-265, and II-268 exhibited this same high-
field chemical shift which is nearly identical with the corresponding cyclohexene-

fused analogs **II-61**, **II-385**, and **II-10**. With the exception of C-2 for both cyclo-butapyridines, all of the previously observed ^{13}C chemical shift trends are preserved. The bridgehead carbons move downfield while the pyridine carbon ortho to the bridgehead moves upfield with decreasing size of the fused ring. The failure of C-2 to follow these trends indicated that the electronegative effect of the adjacent nitrogen atom plays an important role even though rehybridization effects can be transmitted through this heteroatom.

The chemical shift effects appear to be additive. In going from **II-10** to **II-103**, C-4 is ortho to two rings both of which decrease by one carbon. The chemical shift of this carbon moves upfield by 8.9 ppm, about twice the value observed when only one ring is altered. Furthermore, the effect of decreasing two rings in going from **II-539** to **II-538** can be reasonably well approximated by summation of the changes observed in going from **II-539** to **II-47** and from **II-539** to **II-48**; therefore one may calculate (37) ΔC-2 = $-$ 8.9, ΔC-3 = $-$ 2.9, ΔC-4 = 6.3, ΔC-5 = $-$ 7.0, and ΔC-6 = + 3.5 ppm.

Most of the evidence indicates that as the size of a fused ring is decreased, there is an increase in electron density at the carbons ortho to the bridgehead, resulting in a significant polarization of this C—H bond. Thus the ortho carbon becomes more shielded and its resonance moves upfield while the proton attached to it becomes more deshielded and moves to lower field. The principal exception would be the proton ortho to a fused cyclobutene ring.

The coupling constant between the ortho carbon and hydrogen increases regularly along the series **II-537**, **II-10**, **II-9**, **II-103**, and **II-268** as well as along the series **II-539**, **II-47**, **II-48**, and **II-538**, indicating increased *s*-character resulting from polarization of this bond. The downfield shift of the bridgehead carbon atoms with increasing strain indicates a decrease in electron density at these positions. These effects are consistent with a rehybridization model earlier proposed. For aryl positions adjacent to a fused strained ring, Streitweiser (179b) claims that for the bridgehead carbons "the atomic orbitals used to construct the strained ring have higher *p*-character. Hence, the remaining orbital has higher *s*-character. The ortho-carbon is thus bound to an orbital of higher electronegativity." As the size of the fused ring is decreased from six carbons to five, the electronegativity of this ortho-carbon should increase. The expected effect of this would be to decrease shielding of the attached proton, shifting the carbon resonance to higher field and the proton resonance to lower field.

2.   Ultraviolet Spectra

As regards the UV data reported in Tables II-5 and II-6, two trends bear comment. The positions of the maxima observed for the *para-bis*-annelated series **II-537**, **II-10**, **II-9**, **II-103**, **II-268** move to longer wavelength with increasing strain while the extinction coefficients increase. The latter effect is analogous to what is observed for *para-bis*-annelated benzenes and is explained by increased planarity leading to better Franck–Condon overlap between the ground and excited states. In the *mono*-annelated systems, the extinction coefficients for **II-265** are nearly identical with those observed for the other 3,4-fused systems **II-152** and **II-385**, a regular trend of increasing extinction coefficients is observed for the series **II-61**, **II-15**, **II-255**.

Thus, the extinction coefficient of pyridine is more sensitive to effects resulting from ring fusion at the position adjacent to the nitrogen atom (37).

3.   Basicity

The basicities for *mono*- and *bis*-annelated pyridines are recorded in Table II-8. The basicities were determined as half-neutralization potentials (HNP) by titration at 25° with 0.10 N perchloric acid in acetic acid with acetic anhydride as the solvent. The HNP's were determined for a series including pyridine and seven methyl-substituted derivatives of known basicity. A plot of HNP versus pK_a for these compounds resulted in a straight line from which the pK_a's of the annelated pyridines could be determined (37).

A very dramatic decrease in basicity was observed for 2,3-cyclobutapyridine **(II-255)** when compared to its higher homologues **II-15** and **II-61** (119). There was a comparable difference of almost 2 pK_a units between the positionally isomeric cyclobutapyridines, with the 3,4-fused system being decidedly more basic. In fact, the size of a ring fused at the 3,4-position has only a minor and apparently inconsistent influence on the basicity of the molecule. Similarly, there was a very regular decrease in basicity with decreasing ring size along the *para-bis*-annelated series. A drop of about one pK_a unit occurred with the loss of each methylene unit. The influence of four-membered ring fusion was also seen to be additive with respect

TABLE II-8.   BASICITIES OF ANNELATED PYRIDINES (37, 119, 120)

	pK_a
2,3-Cyclobutapyridine (II-255)	4.85
2,3-Cyclopentapyridine (II-15)	5.95
5,6,7,8-Tetrahydroquinoline (II-61)	6.65
2,3-Dimethylpyridine (II-535)	6.56
3,4-Cyclobutapyridine (II-265)	6.75
3,4-Cyclopentapyridine (II-152)	6.96
5,6,7,8-Tetrahydroisoquinoline (II-385)	6.83
3,4-Dimethylpyridine (II-536)	6.61
Dicyclobuta[2,3:5,6]pyridine (II-268)	4.40
Dicyclopenta[2,3:5,6]pyridine (II-103)	6.42
2,3-Cyclopenta-5,6,7,8-tetrahydroquinoline (II-9)	7.30
1,2,3,4,5,6,7,8-Octahydroacridine (II-10)	8.09
3,4-Cyclopenta-5,6,7,8-tetrahydroisoquinoline (II-47)	7.35
Dicyclopenta[2,3:4,5]pyridine (II-538)	7.39
1,2,3,4,7,8,9,10-Octahydrophenanthridine (II-539)	7.75
3,4-Cyclopenta-5,6,7,8-tetrahydroquinoline (II-48)	8.05

to pyridine: one four-membered ring caused a drop of 0.45 pK_a units while two caused a drop of 0.90 units (120). The *meta-bis*-annelated pyridines were consistent with these observations. With a 2,3-fused six-membered ring and a 3,4-fused five-membered ring, 3,4-cyclopenta-5,6,7,8-tetrahydroquinoline (II-48) was the most basic (pK_a 8.05) while 3,4-cyclopenta-5,6,7,8-tetrahydroisoquinoline (II-47) with the opposite orientation was the least basic (pK_a 7.35). Pyridines II-538 and II-539 fell inbetween. The 2,7-diazabiphenylene (II-270) (pK_a 4.97 and 2.32) is a weaker base than pyridine but stronger than the most basic bipyridyl (3,4'-bipyridyl) (319).

Streitweiser's arguments (279a) for the rehybridization of bridgehead carbons can be invoked to explain the decrease in basicity for pyridines with a small ring fused in the 2,3-position. Bridgehead C-2 uses orbitals high in p-character in forming bonds to the small ring. This leaves an orbital of higher s-character to bond to the adjacent nitrogen atom causing the lone pair of electrons to be held more tightly and the basicity to be diminished.

4. Infrared Spectra

Godar and Mariella have carried out an extensive study of the infrared spectra of a variety of substituted pyridines in an effort to identify bands which are characteristic of a particular size carbocyclic ring annelated in the 2,3-position of pyridine (280, 281). They found that pyridines with a fused cyclohexane ring showed a band in the ranges 1005–952 cm^{-1} and 1055–1000 cm^{-1}. In a series of 50 compounds, only two exceptions were noted and each of these exhibited one of the two bands. A third band appeared in the narrow range of 832–818 cm^{-1} and

TABLE II-9.   INFRARED BANDS TYPICAL OF CYCLOHEXENO-FUSED PYRIDINES (cm^{-1}) (281)

5,6,7,8-Tetrahydro-2-chloroquinoline	832	988	
5,6,7,8-Tetrahydroquinoline-2-ol	820	990	
5,6,7,8-Tetrahydro-2-chloroquinoline-3-carboxylic acid	822	955	1032
5,6,7,8-Tetrahydro-2-methoxy-3-cyanoquinoline	828	955	1007, 1013
5,6,7,8-Tetrahydro-2-quinoline-3-carboxylic acid	818	965	
5,6,7,8-Tetrahydro-2-chloro-3-cyanoquinoline	828	950	1040
5,6,7,8-Tetrahydro-3-cyano-2-quinolone	820	965	
5,6,7,8-Tetrahydro-2-cyanoquinoline	823		1007
5,6,7,8-Tetrahydro-3-aminomethylquinoline dihydrochloride	828	982	1010
5,6,7,8-Tetrahydro-3-aminomethylquinoline	825		1007
5,6,7,8-Tetrahydro-3-hydroxymethylquinoline hydrochloride	825	997	1034
5,6,7,8-Tetrahydro-3-hydroxymethylquinoline	825	1005	1037
5,6,7,8-Tetrahydroquinoline-3-carboxylic acid	827		1010
5,6,7,8-Tetrahydroquinoline-3-carboxylic acid hydrochloride	822	970	1018
5,6,7,8-Tetrahydro-3-carboethoxyquinoline	828	1005	1027
Di-([5,6,7,8-tetrahydroquinol-β-yl]methyl)amine	825		1007
Di-([5,6,7,8-tetrahydroquinol-β-yl]methyl)amine hydrochloride	828	983	1020
Di-([5,6,7,8-tetrahydroquinol-β-yl]methyl)N-nitrosamine	825	965	1008
5,6,7,8-Tetrahydroquinoline	828	988	1050
5,6,7,8-Tetrahydroquinoline hydrochloride	818	993	1029

334   R. P. Thummel

TABLE II-10.   INFRARED BANDS TYPICAL OF CYCLOHEPTENO-FUSED PYRIDINES (cm^{-1}) (281)

6,7,8,9-Tetrahydro-2-chloro-5H-cyclohepta(b)pyridine	769	828
6,7,8,9-Tetrahydro-5H-cyclohepta(b)pyridine-2-ol	732	825
6,7,8,9-Tetrahydro-5H-cyclohepta(b)-2-pyridone-3-carboxylic acid	718	830
6,7,8,9-Tetrahydro-3-cyano-5H-cyclohepta[b]-2-pyridone	775	830
6,7,8,9-Tetrahydro-2-chloro-3-cyano-5H-cyclohepta(b)pyridine	770	833
6,7,8,9-Tetrahydro-2-methoxy-3-cyano-5H-cyclohepta(b)pyridine	763	832
6,7,8,9-Tetrahydro-2-chloro-5H-cyclohepta(b)pyridine-3-carboxylic acid	768	832
6,7,8,9-Tetrahydro-3-hydroxymethyl-5H-cyclohepta(b)pyridine hydrochloride	760	835
6,7,8,9-Tetrahydro-3-hydroxymethyl-5H-cyclohepta(b)pyridine	774	832
6,7,8,9-Tetrahydro-3-cyano-5H-cyclohepta(b)pyridine	772	840
6,7,8,9-Tetrahydro-3-carboethoxy-5H-cyclohepta(b)pyridine	765	835
6,7,8,9-Tetrahydro-3-aminomethyl-5H-cyclohepta(b)pyridine dihydrochloride	762	835
6,7,8,9-Tetrahydro-3-aminomethyl-5H-cyclohepta(b)pyridine	773	832
6,7,8,9-Tetrahydro-5H-cyclohepta(b)pyridine-3-carboxylic acid	777	838
6,7,8,9-Tetrahydro-5H-cyclohepta(b)pyridine-3-carboxylic acid hydrochloride	757	835
Di-([6,7,8,9-tetrahydro-5H-cyclohepta(b)pyrid-β-yl]methyl)amine hydrochloride	754	835
Di-([6,7,8,9-tetrahydro-5H-cyclohepta(b)pyrid-β-yl]methyl)amine	762	835
Di-([6,7,8,9-tetrahydro-5H-cyclohepta(b)pyrid-β-yl]methyl)N-nitrosamine	745	838
6,7,8,9-Tetrahydro-5H-cyclohepta(b)pyridine	767	837

was exhibited by all the compounds examined without exception (see Table II-9). Pyridines with a cyclohepteno ring fused at the 2,3-position showed two bands in the regions of 840–827 cm^{-1} and 777–718 cm^{-1} (see Table II-10).

5. Pyrindine Tautomerism

An interesting property of 5H-1-pyrindine (**II-441a**) was noted upon its first being prepared by Robison (**II-189**) in 1958 (see Section III.1.A). He observed that the freshly prepared material had an orange color which could be discharged upon dilution with organic solvents. He proposed that the material existed as a tautomeric mixture of **II-441a** and **II-441b** but did not delve further into the problem. The question was taken up by Reese who prepared 1-methyl-1-pyrindine (**II-541**), a dark orange oil (282). He measured the UV absorption spectrum of **II-541** and compared the intensity of its absorption at 456 μm with that of the orange impurity in **II-441**. Assuming equal extinction coefficients for **II-541** and **II-542**, he estimated that pseudoazulene **II-542** was present to the extent of 0.1% in the neat, tautomeric mixture of pyrindines. He confirmed this estimate by carrying out pK_a measurements.

The existence of a tautomeric equilibrium for **II-441** was also consistent with a calculation of the electronic excitation energies and total Π-energies by the standard Hückel method neglecting overlap (283). NMR evidence supporting this same premise has been presented by Anderson and Ammon (111). They observed two signals with dissimilar splittings at δ 3.22 and 3.36 ppm and attributed these peaks to the methylene group at either the 5- or 7-position.

II-441a   **II-441b**   CH₃   H

II-541   **II-542**

Anastassiou et al. have developed a preparation of *N*-carbethoxypyridine (**II-246**) and found that the NMR spectrum of this species, as well as that of **II-541**, evidenced strongly diatropic proton shifts. Thus, the *N*-electrons must contribute heavily to the development of "aromatic" delocalization. The magnitude of this effect was apparently insensitive to variations in the electronegativity of the hetero-atom (110). It appears, therefore, that the relative stability of the two tautomeric pyrindines, **II-441a** versus the aza-azulene **II-542**, will depend primarily on the relative resonance energies of the two types of systems.

6.   Tables of Physical Constants

The following tables (Tables II-11 through II-23) of physical constants are divided into *mono-* and *bis*-annelated pyridines and are arranged according to the size and position of fusion of the carbocyclic ring. An effort has been made to include all derivatives of each ring system which differ with respect to the substituents attached to the parent. In cases where more than one report of a compound has appeared that example has been chosen which gives the most complete and recent physical data.

Melting points and boiling points are given in degrees centigrade at atmospheric pressure, unless otherwise noted. NMR chemical shift data are given in ppm down-field from tetramethylsilane. Infrared bands are reported in reciprocal centimeters and microns. Ultraviolet absorption bands are given in millimicrometers with extinction coefficients included in parentheses.

ACKNOWLEDGMENTS

Financial support from the Robert A. Welch Foundation and the Donors of the Petroleum Research Fund, administered by the American Chemical Society, is gratefully acknowledged. I would also like to especially thank Dr. Dalip Kohli for his research efforts which led us into the field of annelated pyridine chemistry; Mrs. Lynn Sterba for her invaluable help in interlibrary loan; Professor Tom Lemke for reading the entire manuscript; and Dorothy Newman, Wanda Delong, and Virginia Anderson for their excellent typing.

TABLE II-11.   CYCLOBUTAPYRIDINES

	Method of Preparation	Yield	b.p./m.p. (°C)
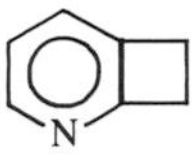	II.2.F	35%	
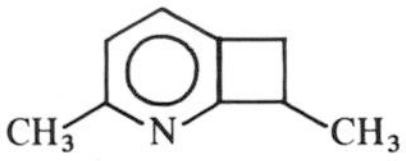	II.2.E	85%	
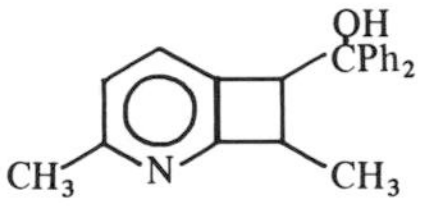	II.2.E	46%	m.p. 185–86°
	II.2.F	17%	
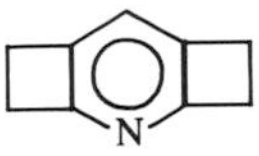	II.2.F	15%	m.p. 117–118°
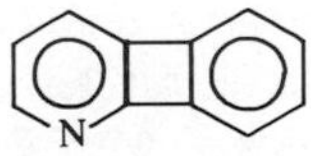	II.2.F	34%	m.p. 70–71°
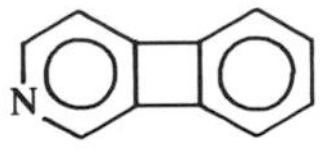	II.2.F	58%	m.p. 88.5–89.5°
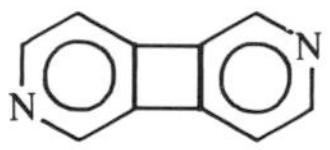	II.3.A.d	70%	m.p. 169–169.5°
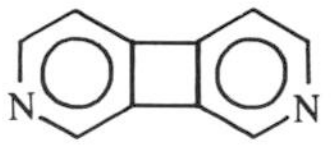	II.2.F	60%	m.p. 192–192.5°
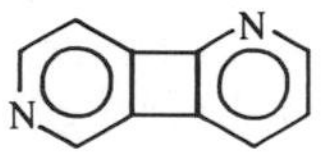	II.2.F	38%	m.p. 106.5–108.

NMR	IR	UV	Derivative (m.p.)	Reference
9(H$_6$), 7.32(H$_4$) 0(H$_5$)		269(4770) 272(4800) 278(3420)		118 119
6(d,1), 6.95(d,1) 7(m,1), 3.26(dd,1) 4(d of m,1), 2(s,3), and 1.43(d,3)	1600, 1580 818	276(6760) 278 sh (6660) 285 sh (5550)		108
7–8.21(m,10) 3(s,2), 4.04(d,1), 9(s,3), 2.31(s,1), 3(d,3)	3330, 1610, 1590 705, and 690 287 sh (5550)	277(8400) 279 sh (8250)		108
9(H$_6$), 8.16(H$_2$) 2(H$_5$)		253(1740) 258(1950) 263(1620)		118 119
)(s,1), 3.32(m,4),)(m,4)	2970, 2930, 1594 1546, 1423, 1365 1250, 1210, 1140 1095	290(8078)		120
7(m,4), 7.66(d,1) 3(d,1), 6.51(d,1)		230(4.27) 236(4.29) 264(3.84) 344(3.66) 358(3.73)		122
5(d), 7.75(s) (d of d), 6.84(m)		241(3.73) 248.5(3.90) 318(3.22) 331(3.43) 345.5(3.49)		122
2), 8.0(2) 2)	1600, 838, 830 7.33, 720	338(3.08) 325(3.15) 312(3.17) 297(3.18) 283(3.16) 241(5.10)		140
(d,2), 7.90(d,2), (dd,2)	834	234.5(4.79) 242(4.93) 284(3.25) 297.5(3.26) 312(3.17) 326.5(2.94)		121
(d,H$_7$), 8.60(d,H$_5$) (dd,H$_2$), (dd,H$_4$), (dd,H$_8$), (dd,H$_3$)		228.5(4.52) 258 (4.17) 289(3.56) 318.5(3.79) 335.5(3.80)		308

TABLE II-12. 2,3-CYCLOPENTAPYRIDINES

	Method of Preparation	Yield	b.p./m.p. (°C)
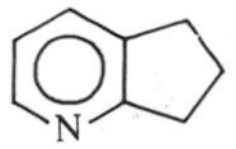	II.1.B	35%	b.p. 87–88° (11 mm)
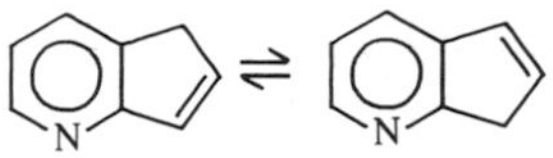	II.1.A	88%	b.p. 79–82° (8 mm)
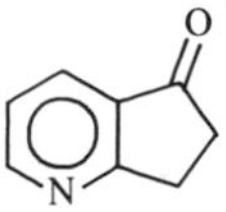	II.1.B	28%	m.p. 57–59°
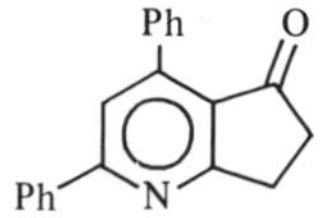	II.1.B	25%	m.p. 166–169°
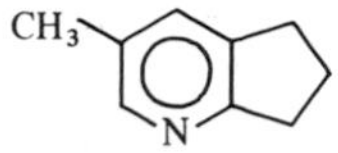	II.1.B	40%	b.p. 94–95° (11 mm) m.p. 39–41°
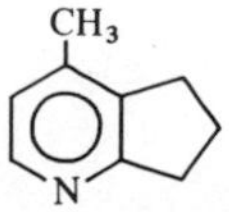	II.1.A	23%	b.p. 105–107° (16 mm)
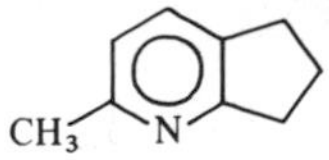	II.1.A	72%	b.p. 105.5–106° (23 mm)
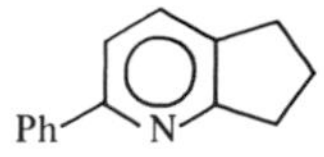	II.1.A	16%	b.p. 200–206° (20 mm) m.p. 79°
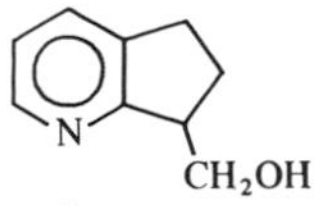	III.1.C	66%	b.p. 76–78° (0.25 mm)
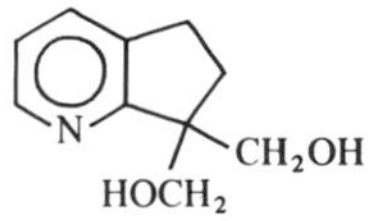	III.1.C	90%	b.p. 120–124° (0.2 mm) m.p. 101–102.5°
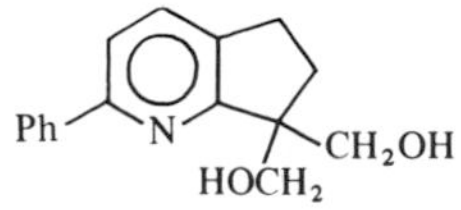	III.1.C	29%	m.p. 125.5–126.5
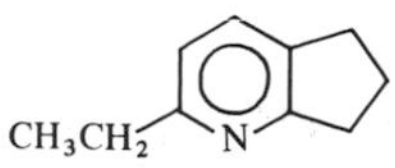	II.3.A.a	36%	b.p. 125–128° (13 mm)

NMR	IR	UV	Derivative (m.p.)	Reference
1(d,1), 7.47(d,1)	3050, 2963, 2853	278(2768)	Picrate	62, 37
0(dd,1), 2.96(t,4)	2853, 1720, 1597	274(3580)	(181–182°)	77, 78
0(m,2)	1586, 1430, 1267	270(3873)		243
	790, 725	265(3421)		315
3(t,1), 7.51(d,1)	1380, 1400, 1570	285(3.7)	Picrate	189
5–7.11(m,3)	1580	247(3.85)	(181.5–184°)	111
5 and 3.32(d,2)				
				64
				64
		278(3.92)	Picrate	62, 94
			(203–204°)	95
		269(3.85)	Picrate (151°)	16
			(169–170°)	94
1,1), 6.7(d,1)			Picrate	12
(m,4), 2.45(s,3)			(149–150.5°)	
n,2)				
		289(4.15)	Picrate (198°)	11
				209
	3250		Picrate	205
			(144.5–145.5°)	
	3280		Picrate	205
	3050		(141–142°)	
	3340		Picrate	205
	3200		(191.5–193°)	
			Picrate	131
			(165–166°)	

TABLE II-12.   (*CONTINUED*)

	Method of Preparation	Yield	b.p./m.p. (°C)
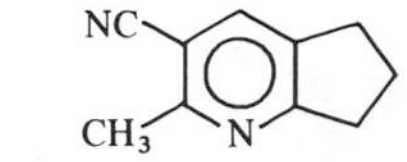	II.2.E	70%	m.p. 95–99°
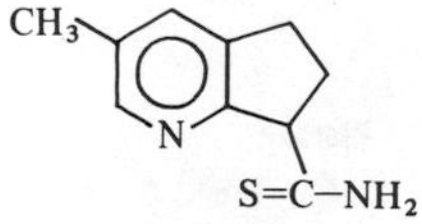		73%	m.p. 64–66°
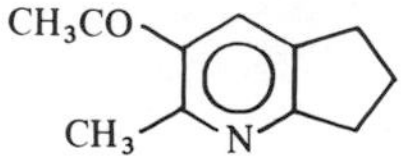	II.1.B	65%	b.p. 77° (0.06 mm) m.p. 28–30°
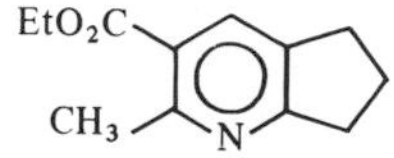	II.1.B	55%	b.p. 78° (0.04 mm)
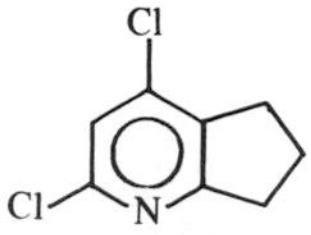		82%	b.p. 134–136° (13 mm)
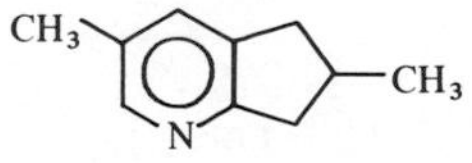	II.3.A.b		b.p. 63–65° (3.5 mm)
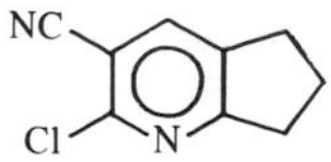			
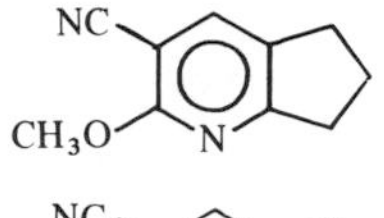		86%	m.p. 114–114.5°
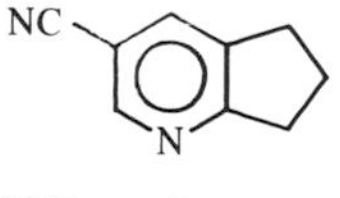		87%	m.p. 88°
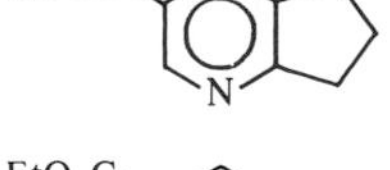		88.5%	m.p. 223–225°
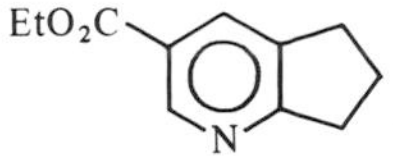		65%	m.p. 42°
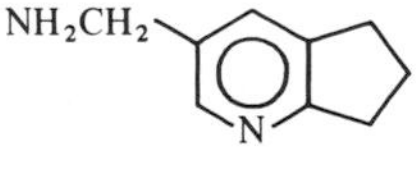		85%	m.p. 208–210°

NMR	IR	UV	Derivative (m.p.)	Reference
	2210		Picrate	107
	2175		(150°)	
			HCl	194
			(198–202°)	203
2(s,1), 2.57(s,3)	1675	243(3.50)	Picrate	65
◀(s,3)		288(3.59)	(146–147°)	53
)(s,1), 2.80(s,3)			Picrate	65
8(q,2), 1.39(t,3)			(123–124°)	
				286
5(s,1), 7.17(s,1)	1568, 871, 715	275(26, 600)	Picrate	133
-3.3(m,5),			(180°)	
8(s,3), 1.20(d,3)				
	1594, 1540, 1157			287
	1020, 938, 699			
	1607, 1570, 1217			280
	1007, 933, 762			
	1600, 1558, 1240			280
	1032, 913, 712			
	1640, 1548, 1180			280
	1045, 910, 760			281
	3090, 1610, 1575			280
	1485, 1185, 1026			281
	950, 778			
	1612, 1580, 1478			280
	1220, 1040, 897			281
	720			

TABLE II-12. (CONTINUED)

	Method of Preparation	Yield	b.p./m.p. (°C)
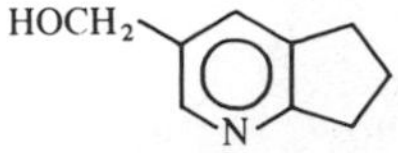		60%	m.p. 130°
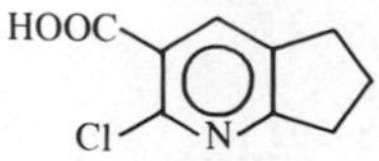		96%	m.p. 206–208°
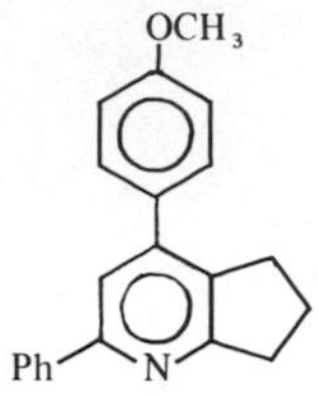	II.1.A		m.p. 144–145°
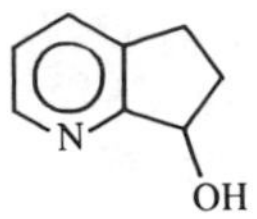	III.1.A	92%	m.p. 83–84°
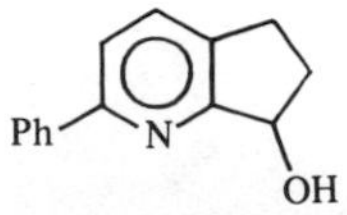	III.1.A	92%	m.p. 121–122°
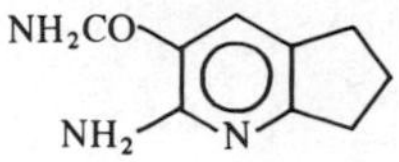	II.1.B	25%	m.p. 254°
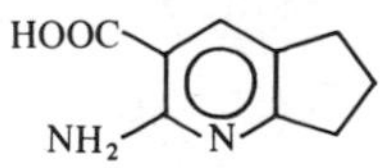		80%	m.p. 311–313°
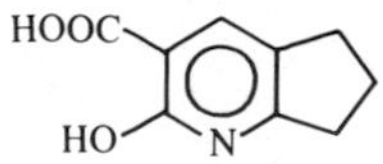		69%	m.p. 270°
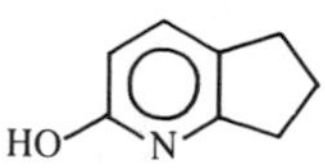			m.p. 100–105°
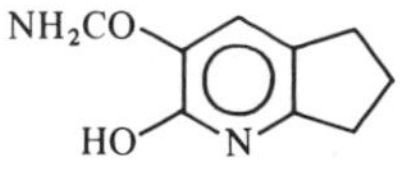		78%	m.p. 266°
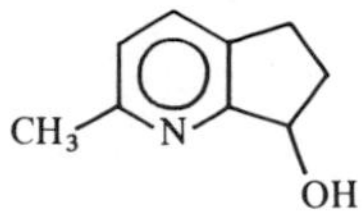	III.1.A		m.p. 79.5–81°
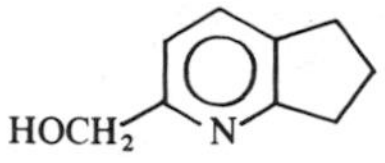	III.1.A		

NMR	IR	UV	Derivative (m.p.)	Reference
	1620, 1590, 1220			280
	1036, 890, 720			281
	3030, 1608, 1571			280
	1468, 1160, 1130			281
	923, 762			
			Picrate	8
			(185–186°)	
			HCl (218°)	
	3150		Picrate	190
			(192–193°)	
	3170		Picrate	190
			(173–175°)	
			HCl (288–290°)	63
				63
				63
			Picrate (160°)	63
			Picrate (159°)	63
				210
				210

TABLE II-12.   (*CONTINUED*)

	Method of Preparation	Yield	b.p./m.p. (°C)
		71%	b.p. 95–96° (0.05 mm) m.p. 45–47°
		81%	b.p. 102–104° (17 mm)
	II.2.A		
	II.2.A		
	II.4		m.p. 66–67° b.p. 110–112° (0.03 mm)
	II.2.E		
	II.2.E	100%	
	II.2.F	3.3%	
	II.2.F	14%	

NMR	IR	UV	Derivative (m.p.)	Reference
				206
	2900, 700		Picrate (172–174°)	206
	1600, 1450 1220, 1120, 880		Picrate (149–151°)	206
?(d,3), 3.10(m,1) ?(m,6), 0.70(t,3) ?(t,3)				79, 80
?(s,3), 1.67(m,8) ?m,9)				79, 80
	790	312–313 (24,690)		154
(d,1), 6.84(d,1) (s,3), 3.4–1.3(m,s) ?(d,3)	2940, 2860 1590, 1450 1380, 1235 1110, 815			314
(d,1), 6.85(d,1) ?(t,2), 2.55(s,3) ?(t,2), 1.28(s,6)	2940, 2870 1585, 1455			314
(d,H$_2$), 8.26(d,H$_4$) ?(dd,H$_3$), 2.4(m,5), 1.25(d,3)			Picrate (137–137.5°)	315
?(d,H$_2$), 7.46(d,H$_4$) ?(dd,H$_3$), –3.00(m,H$_7$), –2.78(m,H$_5$) –2.15(m,1), –1.54(m,1), ?(d,3)			Picrate (128.5–129°)	315

TABLE II-12.   *(CONTINUED)*

	Method of Preparation	Yield	b.p./m.p. (°C)
		64%	m.p. 282–283°
	II.8	57%	m.p. 284–286°
	II.2.D	93%	m.p. 321–322°
	II.2.D	97%	m.p. 287°
	II.2.D	92%	m.p. 324°
	II.2.D	85%	m.p. > 330°
	II.2.B		m.p. 133–134°
	II.1.D	86%	m.p. 263–264°
	II.1.D	72%	m.p. 97–98°
	II 3.C	36%	b.p. 130–135° (20 mm)

NMR	IR	UV	Derivative (m.p.)	Reference
	1650, 1670 3200, 3350, 3470 3100, 2600			316
	1660, 2230 3400, 2600			316
				103
				103
				103
				103
	1560, 1400, 1320 790(s)			82
	5.82, 5.91, 6.13 3.0–4.0(b) μ	275(16,100) 295(11,400) 308(9500) 323(4900)		84
	5.80, 5.85, 608(w), 6.19(w) 630 μ	239(7500) 291(11,600) 297(12,400) 309(8700)		84
.65(dd,1), 7.80(dd 1), .34(dd,1), 3.74(m,1), .98(m,1), .50–2.40(m,6)	1710			309 310

TABLE II-12.   *(CONTINUED)*

	Method of Preparation	Yield	b.p./m.p. (°C)
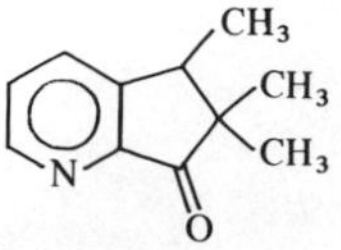	II.3.C	16%	Oil
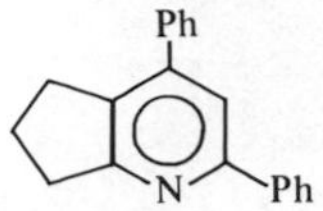	II.1.C	63%	m.p. 144–146°
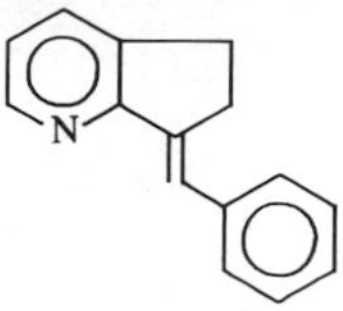	III.2.B	53%	m.p. 74.5–75.5°
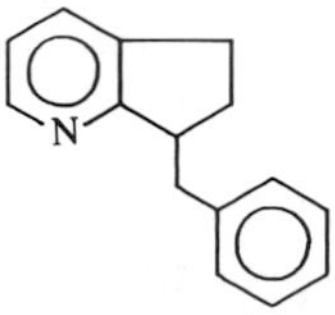		96%	Oil
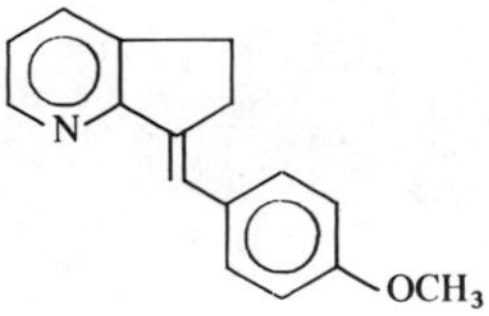		47.6%	m.p. 70–70.5°
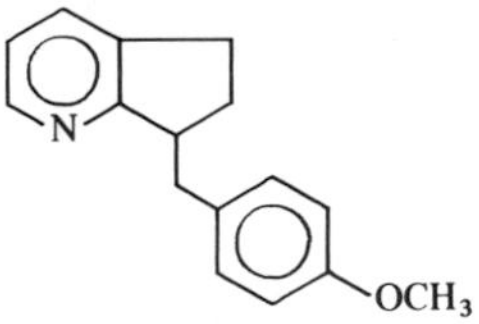		97%	Oil
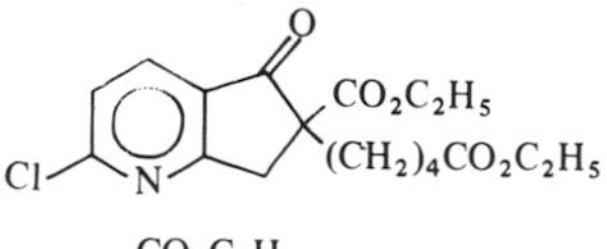	II.1.D		b.p. 190° (0.05 mm)
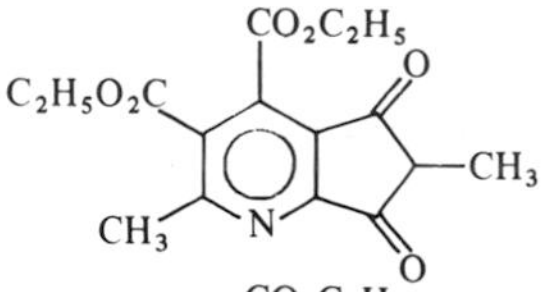	II.1.B	30%	m.p. > 250° (dec)
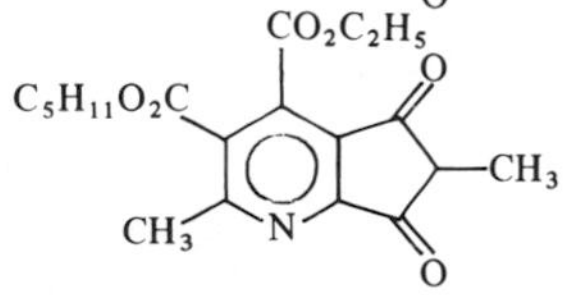	II.1.B	33%	m.p. 230–235° (dec)

NMR	IR	UV	Derivative (m.p.)	Reference
63(*dd*,1), 7.79(*dd*,1) 33(*dd*,1), 3.03(*q*,1) 31(*d*,3), 1.20(*s*,3) 03(*s*,3)	1715			310
				312
52(*dd*,1), 85–7.65(8*H*), 19(*s*,4)				301
40(*dd*,1), 3–7.6(7*H*), –3.8(7*H*),			CH$_3$I (249–251°)	301
43(*dd*,1), 33–7.68(7*H*), 3(*s*,3), 3.18(*s*,4)				301
(*dd*,1), –7.5(6*H*), 3.8(*s*,3) –3.7(7*H*)			CH$_3$I (178–179°)	301
	5.80, 5.85, 6.30 6.35 μ	238(10,300) 289(11,900) 297(10,400)		84
	1547, 1557, 1600 1615, 1677, 1735 3195, 3394			288
	1530, 1589, 1608 1651, 1668, 1727 3270, 3340			288

TABLE II-12. *(CONTINUED)*

	Method of Preparation	Yield	b.p./m.p. (°C)
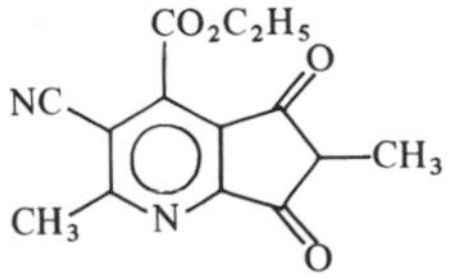	II.1.B	30%	m.p. > 300° (dec)
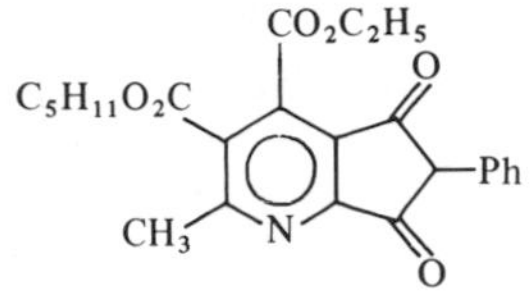	II.1.B	24%	m.p. > 300° (dec)
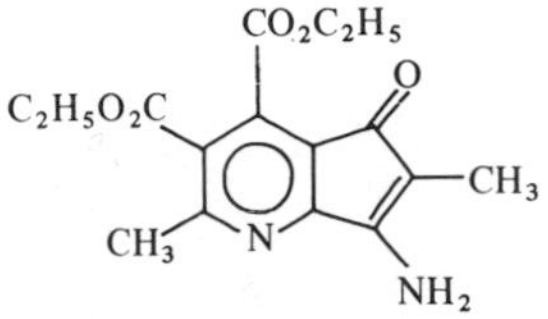	II.1.B	9%	m.p. 140–142°
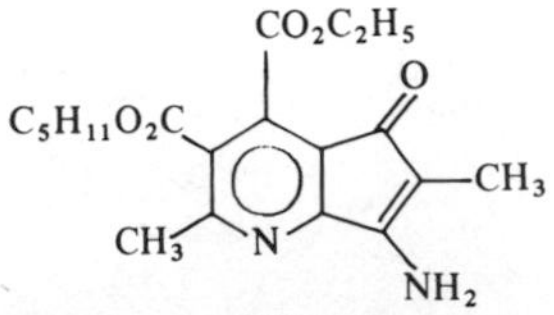	II.1.B	10%	m.p. 101–103°
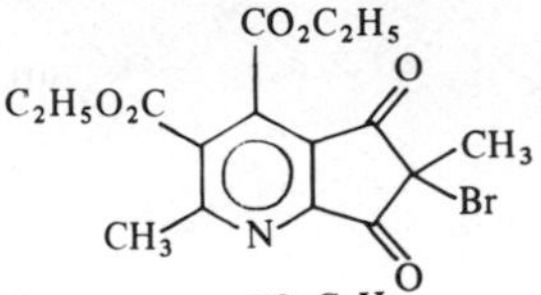		38%	m.p. 83–84°
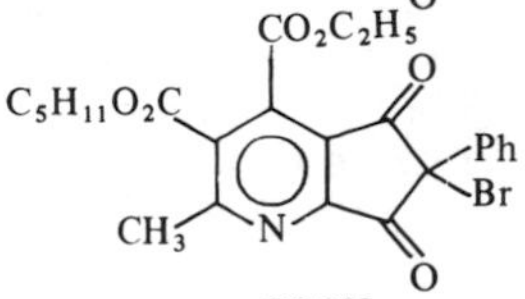		46%	m.p. 67–70°
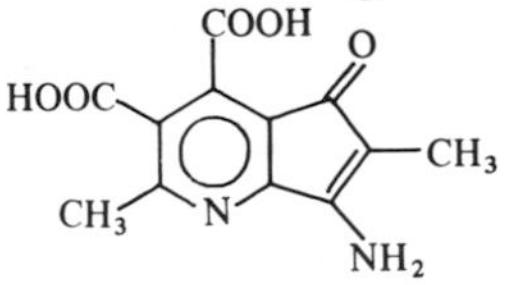		62%	m.p. 236° (dec)
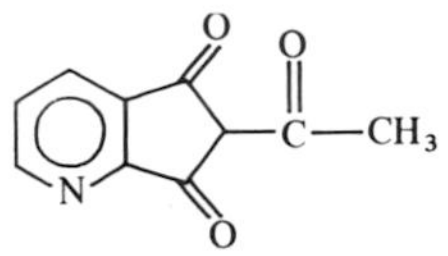	II.1.D	37%	m.p. 148°
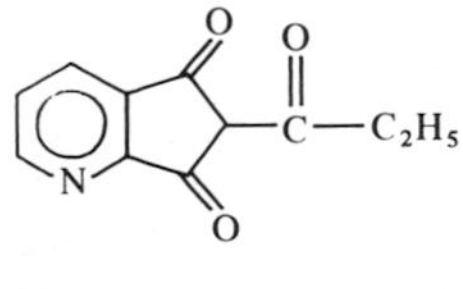	II.1.D	39%	m.p. 149°

NMR	IR	UV	Derivative (m.p.)	Reference
	1553, 1585, 1604 1674, 1740, 2235 3415			288
	1533, 1595, 1650 1718, 3060, 3430			288
	1568, 1598, 1630 1645, 1683, 1720 1742, 3180, 3445			288
	1570, 1600, 1636 1685, 1730, 3196 3415			288
	1568, 1726, 1748 1772			288
	1570, 1734, 1770			288
	1525, 1590, 1630 1700, 1720, 3190 3355			288
	2960, 1680, 1650 1600			69
				69

TABLE II-12. (*CONTINUED*)

	Method of Preparation	Yield	b.p./m.p. (°C)
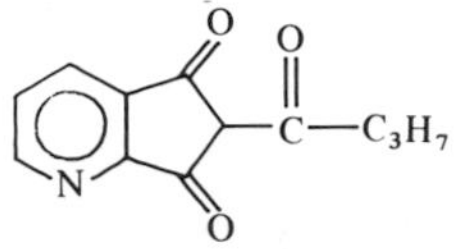 C—C₃H₇	II.1.D	32%	m.p. 89°
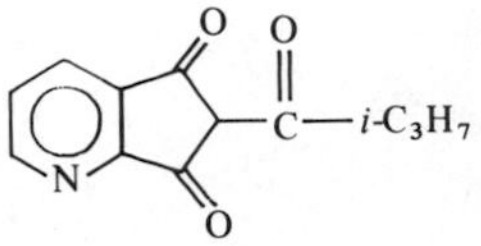 C—i-C₃H₇	II.1.D	36%	m.p. 92°
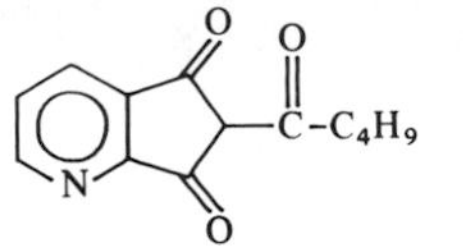 C-C₄H₉	II.1.D	64%	m.p. 88°
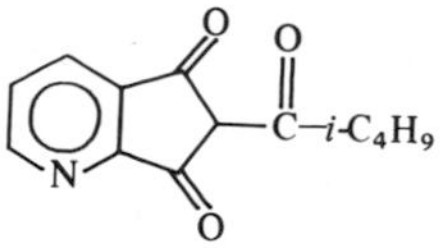 C—i-C₄H₉	II.1.D	46%	m.p. 101°
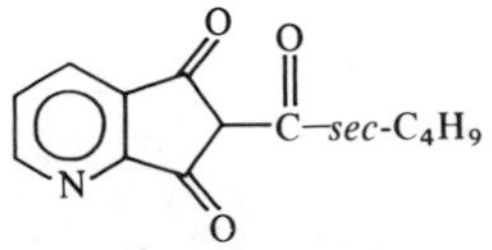 C—sec-C₄H₉	II.1.D	27%	m.p. 77°
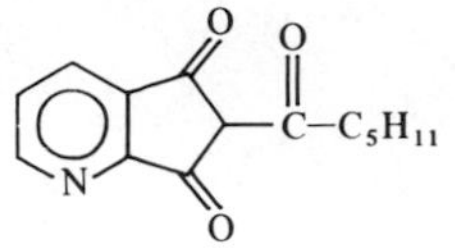 C-C₅H₁₁	II.1.D	50%	m.p. 98°
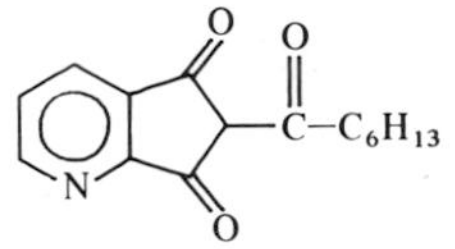 C-C₆H₁₃	II.1.D	20%	m.p. 88°
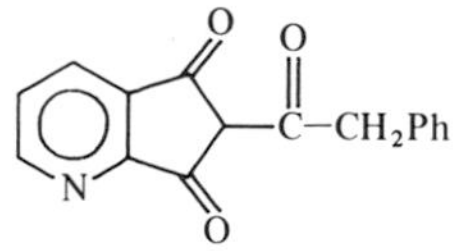 C-CH₂Ph	II.1.D	18%	m.p. 139°
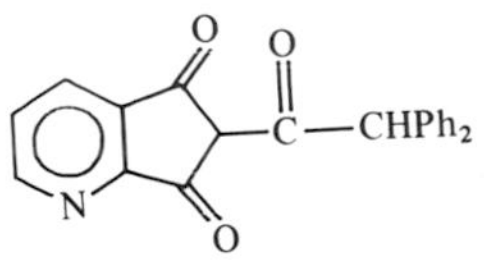 C-CHPh₂	II.1.D	69%	m.p. 170°
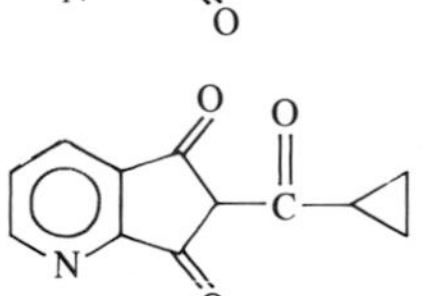 C-	II.1.D	49%	m.p. 177°

NMR	IR	UV	Derivative (m.p.)	Reference
				69
				69
				69
				69
				69
				69
				69
	2950, 1720, 1650 1600			69
	3000, 1710, 1680 1650, 1570			69
				69

TABLE II 12.   *(CONTINUED)*

	Method of Preparation	Yield	b.p./m.p. (°C)
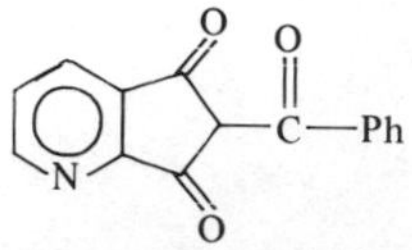	II.1.D	38%	m.p. 188°
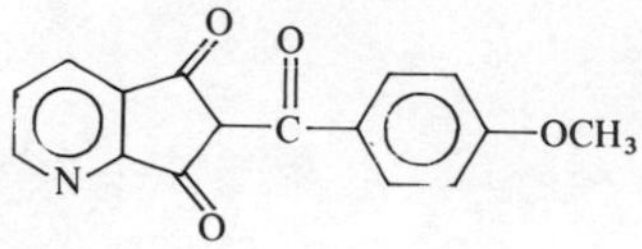	II.1.D	8%	m.p. 161°
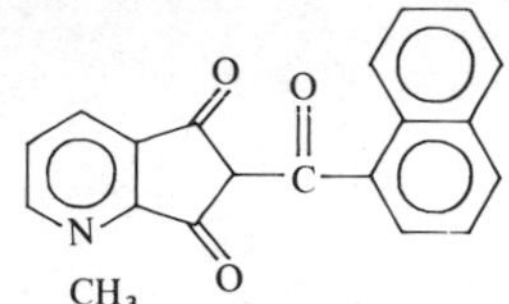	II.1.D	21%	m.p. 110°
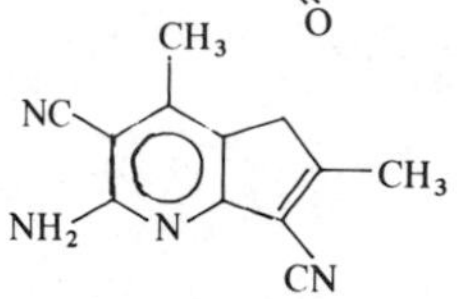	II.2.D	95%	m.p. 280°
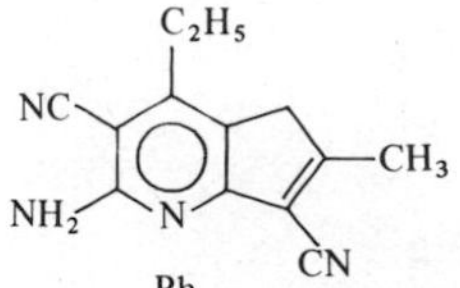	II.2.D	92%	m.p. 216°
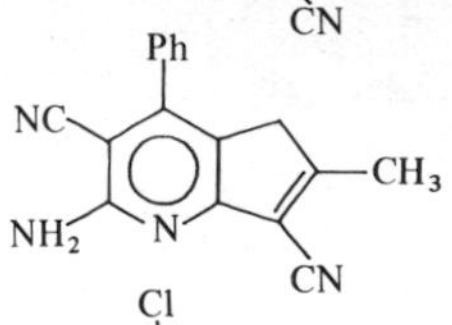	II.2.D	25%	m.p. 268°
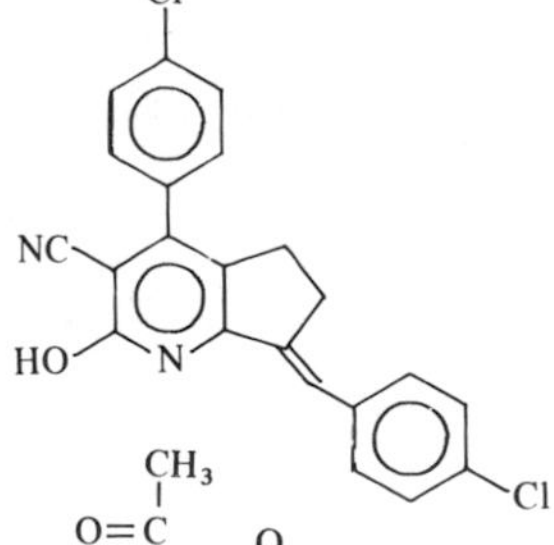	II.1.C	38%	m.p. > 300°
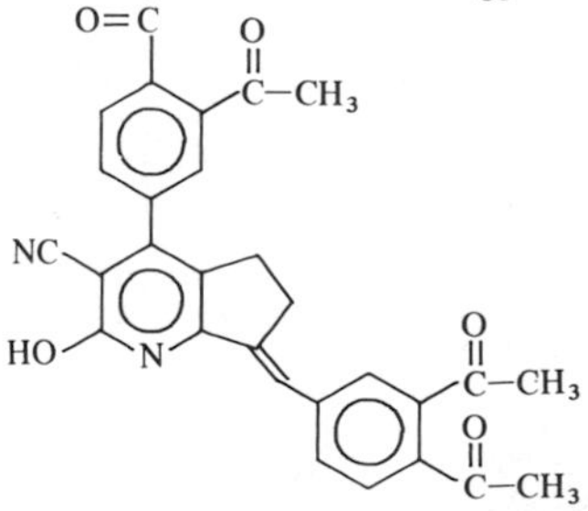	II.1.C	57%	m.p. > 300°

NMR	IR	UV	Derivative (m.p.)	Reference
	3000, 1680, 1650 1580			69
				69
				69
	3472, 3311, 3175 1639, 2227, 2212 1592, 1567, 1560			31
2(*t*), 2.70(*q*) 5(*s*), 3.67(*s*) 1(*s*)	3484, 3322, 3185 1650, 2232, 2217 1597, 1570			31
8(*s*), 3.66(*s*) (*s*), 7.69(*s*)	3497, 3333, 3195 1639, 2242, 2217 1610, 1592, 1567 1560			31
?(*s*,4), 7.53(*s*,1), 4(*s*,4), 3.20(*s*,4)	3600, 2210 1632			318
-7.2(*m*,7) (*s*,12) (*s*,4)	3200, 2200 1680			318

TABLE II-12.　　*(CONTINUED)*

	Method of Preparation	Yield	b.p./m.p. (°C)
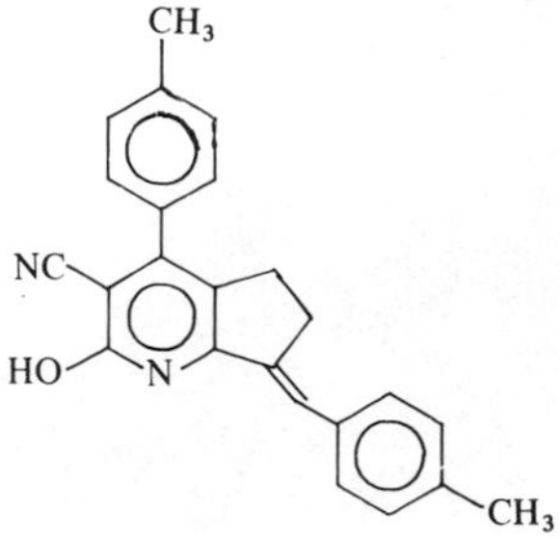	II.1.C	57%	m.p. > 300°
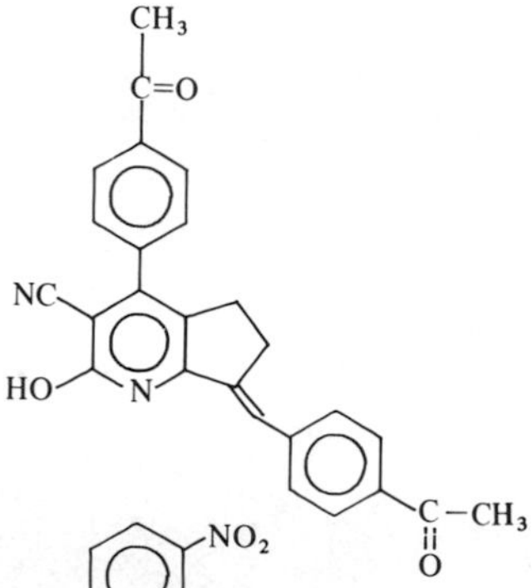	II.1.C	47%	m.p. > 300°
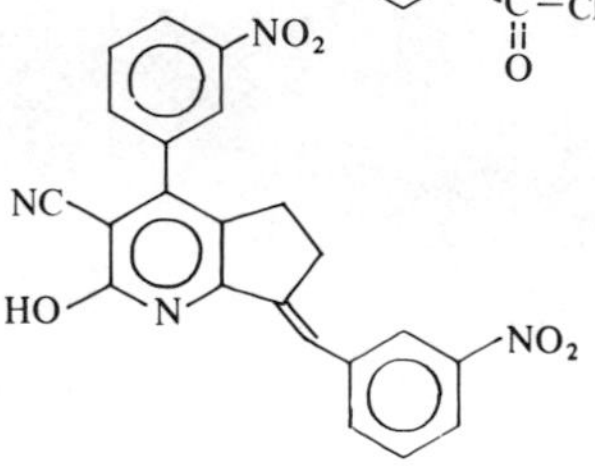	II.1.C	42%	m.p. > 300°
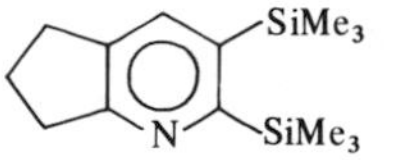	II.3.B	77%	Oil
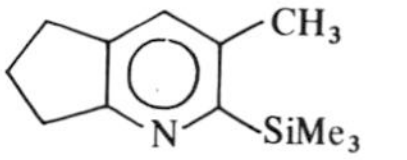	II.3.B	70%	Oil
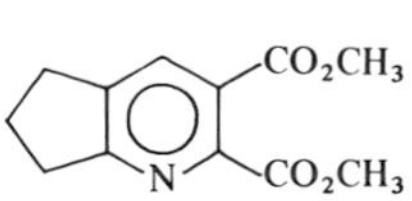	II.3.B	68%	Oil
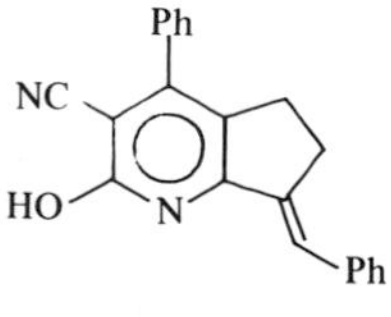	II.1.C	40%	m.p. > 300°

NMR	IR	UV	Derivative (m.p.)	Reference
5(s,8), 7.34(s,1) 2(s,4), 2.49(s,3) 0(s,3)	3650, 2200 1628			318
5(d,4), 7.34(s,1) 4(d,4), 4.09(s,3) 5(s,3), 3.18(s,4)				318
-7.4(m,9) 8(s,4)	3650, 2210 1632			318
				313
				313
(s,11), 3.17(s,4)	3650, 2210 1628			318

TABLE II 12.   (*CONTINUED*)

	Method of Preparation	Yield	b.p./m.p. (°C)
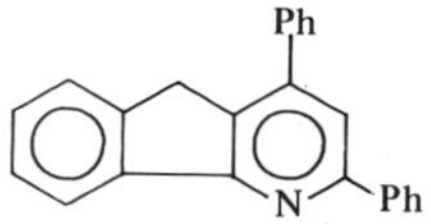	II.1.C	88%	m.p. 158–159°
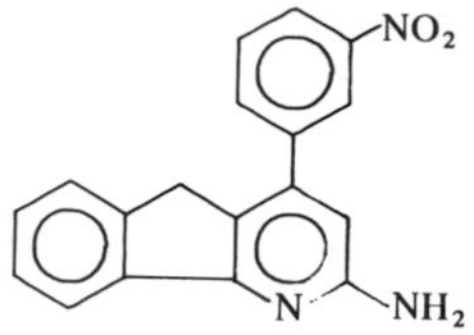			m.p. 199–201°
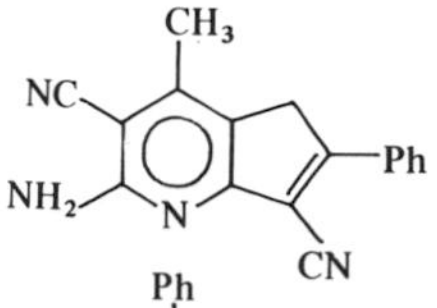	II.2.D		m.p. 308°
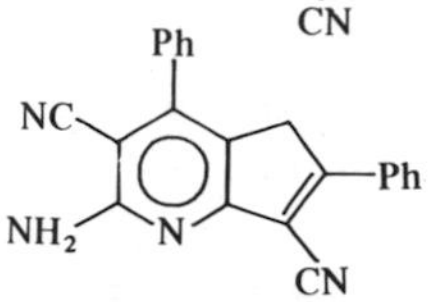	II.2.D	43%	m.p. 305°
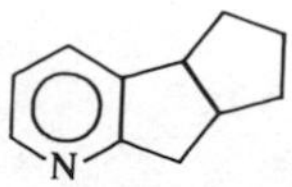	II.2.A	30%	
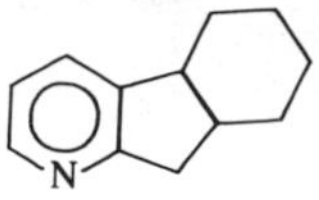	II.2.A	37%	
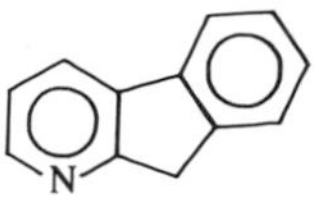	II.2.E	80%	m.p. 79–81°
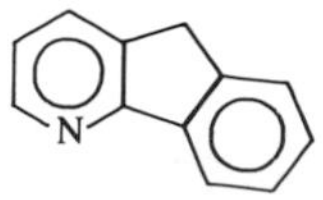	II.2.E	70%	m.p. 96–97°
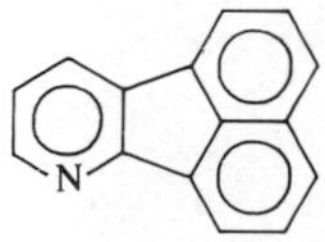	II.2.E	65%	m.p. 96–97°

NMR	IR	UV	Derivative (m.p.)	Reference
				305, 312
				305
3(s), 4.22(s))(s), 7.65(m)	3484, 3333, 3205 1642, 2232, 2208 1608, 1587, 1572			31
6(s)	3436, 3390, 3333 3226, 1647, 2217 2193			31
0–8.26(1) 6–7.39(1) 0–6.96(1)				81
0–8.26(1) 6–7.39(1) 0–6.96(1)				81
3(dd,1), 3(dd,1), –7.67(m,s) 7(s,2)				107
3(dd,1), 8.02(m,1) 9(dd,1), 3–7.33(m,3), 2(dd,1), 3.48(s,2)				107
				107

TABLE II-12.   *(CONTINUED)*

Structure	Method of Preparation	Yield	b.p./m.p. (°C)
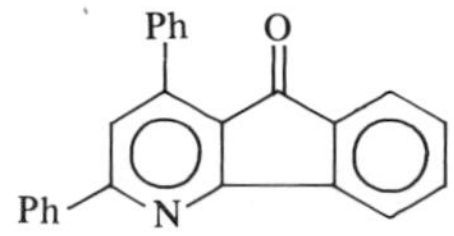	II.1.A		
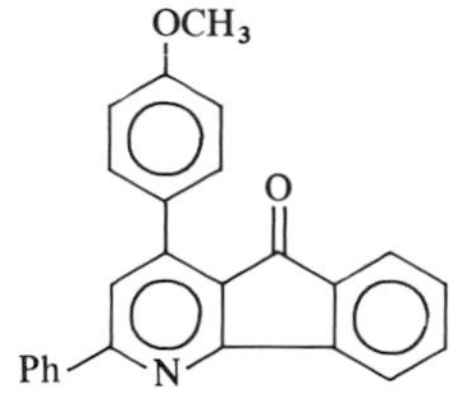	II.1.A	29%	m.p. 206°
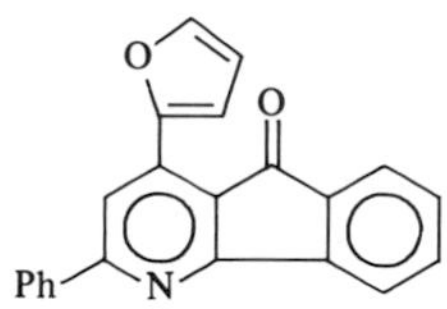	II.1.A	25%	m.p. 184–185°
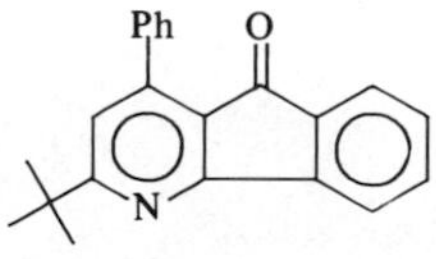	II.1.A	95%	m.p. 110–111°
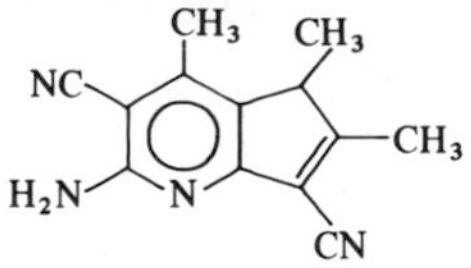	II.1.B		m.p. 208–211°
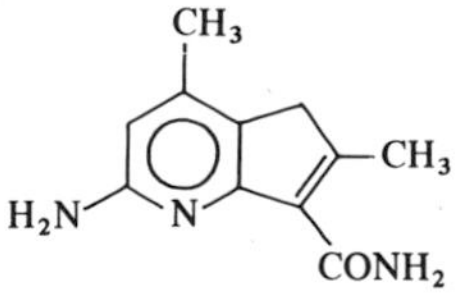	II.1.B		m.p. 196°
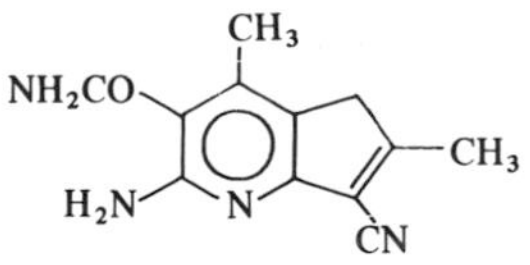	II.1.B		m.p. 295°
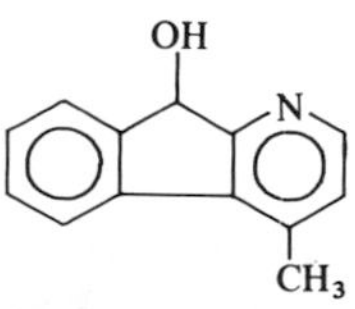			m.p. 154–156°

NMR	IR	UV	Derivative (m.p.)	Reference
	1709, 1590, 1572 1551			28
	1710, 1606, 1586 1575, 1546, 1509			28
	1705, 1591, 1552			28
	1709, 1604, 1582 1555			28
				32
	3430, 3150, 1650 1610, 1590, 1550			32
				32
(d,1), −7.62(m,2), −7.39(m,2) (d,1), 5.64(s,1) (s,3)				320

TABLE II-12. *(CONTINUED)*

	Method of Preparation	Yield	b.p./m.p. (°C)
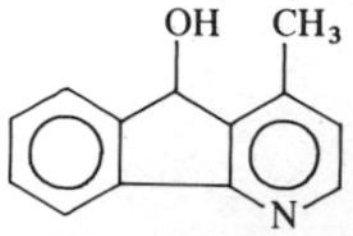			m.p. 156–157°
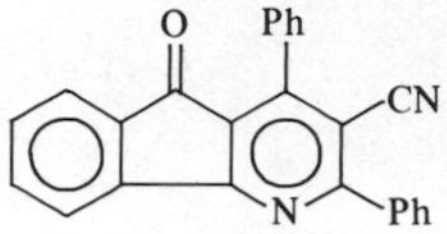			m.p. 274–275°
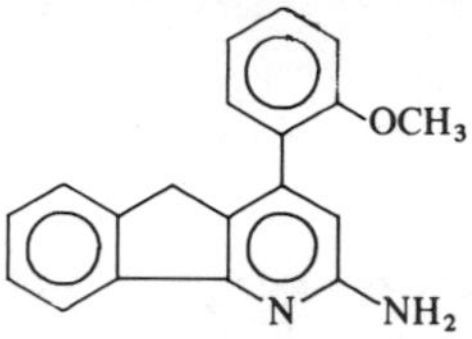	II.1.C		m.p. 138–139°
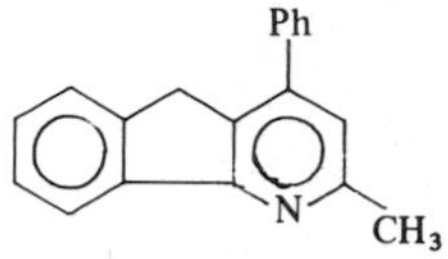			Oil
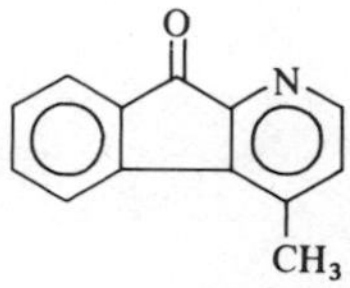			m.p. 158–160°
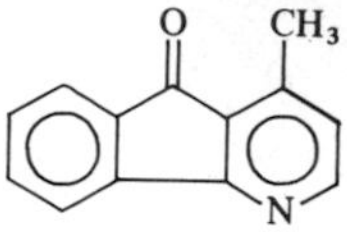	II.2.E		m.p. 133–135°
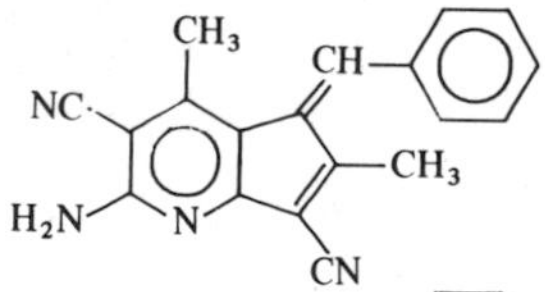			m.p. 255–258°
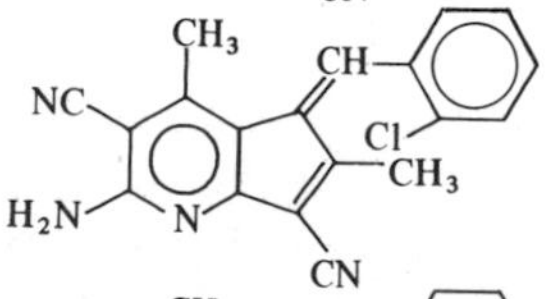			m.p. 280°
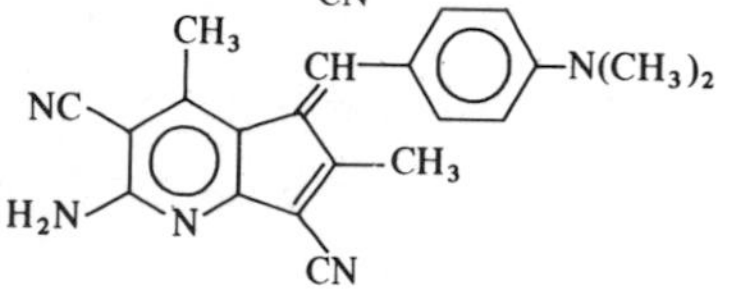			m.p. 244°

NMR	IR	UV	Derivative (m.p.)	Reference
3(d,1), 0–7.56(m,2), 7–7.30(m,2) 4(d,1), 5.53(s,1) 7(s,3)				320
	1712 2208			305
				305
			Picrate 233–234° (dec)	305
(d,1), 7.82, 7.71, , 7.40, 7.18(d,1), 3(s,3)				320
(d,1), 7.86, 7.73 , 7.44, 6.98(d,1) (s,3)				320
		329(4.22) 246(4.41)		32
				32
		495(3.94) 355(4.08) 290(4.05) 260(4.23) 236(4.46)		32

TABLE II-12. (*CONTINUED*)

	Method of Preparation	Yield	b.p./m.p. (°C)
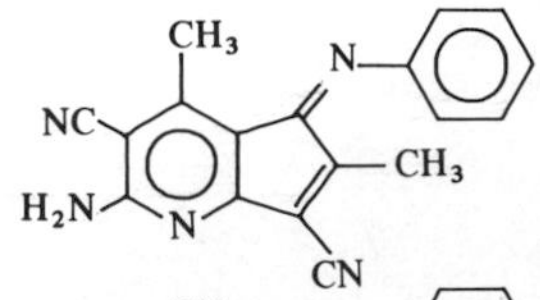			m.p. 255–260°
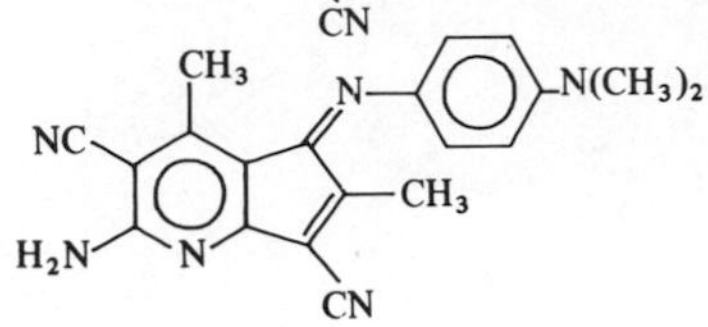			m.p. 255–260°
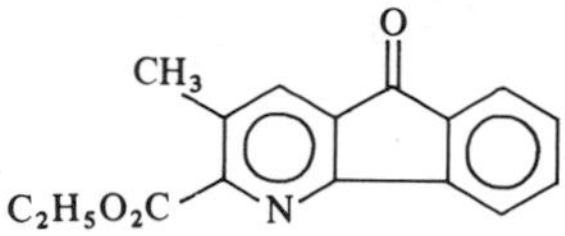	II.1.B	76%	m.p. 125–126°
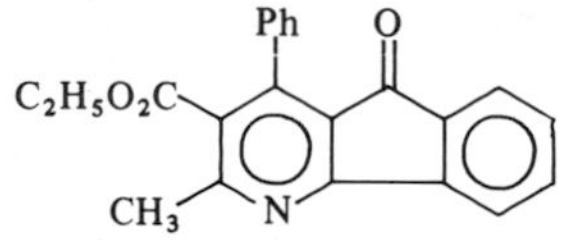	II.1.B	88%	m.p. 174.5–175.5°
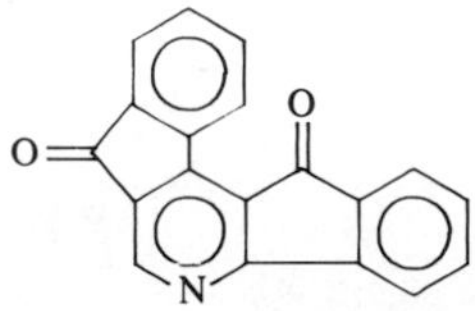			m.p. 263°
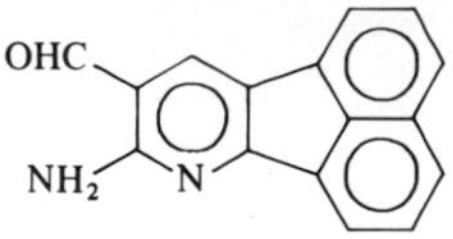	II.1.B	95%	m.p. 215–217°
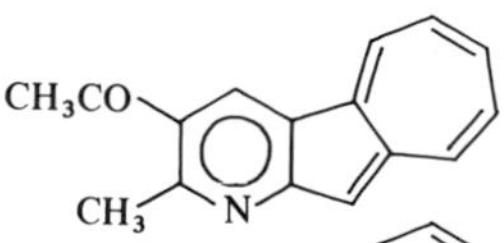	II.1.B	50%	m.p. 145°
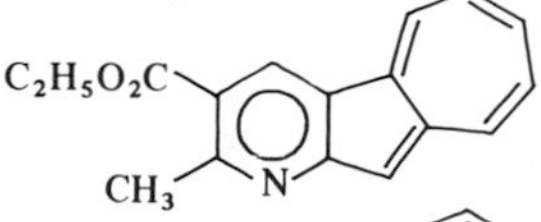	II.1.B		m.p. 138°
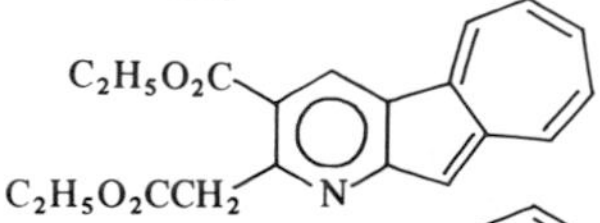	II.1.B		m.p. 217°
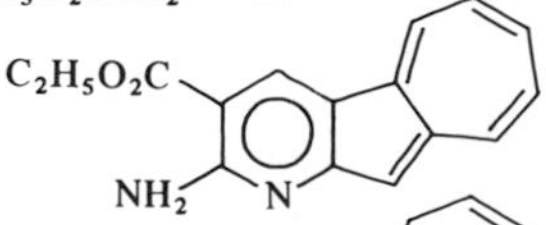	II.1.B		m.p. 190°
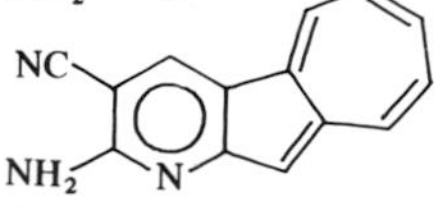	II.1.B		m.p. 268°

NMR	IR	UV	Derivative (m.p.)	Reference
		450(3.49) 336(3.88) 292(4.52) 246(4.46)		32
		610(3.65) 330(3.68) 290(4.03) 250(4.03)		32
				59
			Oxime (281°)	61
				61
	1665	246(4.15)320(4.74) 332(4.76)378(3.77) 400(3.95)424(3.92) 488(2.44)520(2.52) 560(2.51)610(2.33)	HCl (> 280°) Oxime (251°)	50
				52
			HCl (217°, dec)	52
			HCl (200°, dec) Picrate (216°, dec)	52
				52
				52

TABLE II 12. (*CONTINUED*)

	Method of Preparation	Yield	b.p./m.p. (°C)
			m.p. 157°
			m.p. 265°
			m.p. 163°
			m.p. 230°
			m.p. 210°
		74%	m.p. 211–212°
		75%	m.p. 175–177°
		85%	m.p. 201–203°
		95%	m.p. 243°
	II.2.F	38%	m.p. 180.5–181.

NMR	IR	UV	Derivative (m.p.)	Reference
				52
				52
				52
				52
				233, 234
				234
				233
				233
		433(4.31)		233
		385(4.05) 425–435 298–304		233
				126

TABLE II-13.   3,4-CYCLOPENTAPYRIDINES

	Method of Preparation	Yield	b.p./m.p. (°C)
	II.2.A		
n-Bu	II.3.B	67%	
Ph	II.3.B	56%	
		80%	m.p. 84–86°
CO$_2$CH$_3$	II.1.D	82%	m.p. 128–130°
CH$_3$ / CH$_3$	II.1.A	40%	b.p. 92–94° (10 mm)
CH$_3$ / OH / CH$_3$			m.p. 136°
CH$_3$ / CH$_3$	II.2.F	55%	
CH$_3$ / CH$_3$		63%	
CH$_3$ / OH / CH$_3$		15%	m.p. 132°
CH$_2$OH / CH$_3$		56%	

NMR	IR	UV	Derivative (m.p.)	Reference
4(s,1), 8.33(d,1) 4(d,1), 2.91(dt,4) 0(quin, 2)	3040, 2960, 2855 1605, 1577, 1491 1430, 1180 827 721	267(1758) 259(2027) 254(1646)		37 77
				146
				146
8(s,1), 8.70(d,1) 7(d,1), 7–3.10(m,2), 7–2.60(m,2)				74
				74
6(s,1), 8.19(s,1) 5(s,3), –1.4(m,1), 0(d,3)		261.5(2620) 269.5(2420)		191 25 101
1(s,1), 8.08(s,1) 5(s,1), 8–2.05(m,4), 0(s,3), 1.60(s,3)		260(1900) 268(1580)		191
7(s,3), 3.60(m,1) 5(s,2), 2.25(s,3) 4(d,3)	1720			307
			Picrate 146–146.5°	307
2(s,1), 8.07(s,1) 0(s,1), 5.23(t,1) 8(s,3), 1.35(d,3)	3100	265(2410) 272(2140)		191
4(s,1), 8.18(s,1) 4(s,1), 4.67(s,2) 7(s,3)	3200 1600	261(1950) 269(1750)	Picrate (110–111°)	191 249

TABLE II-13.   *(CONTINUED)*

	Method of Preparation	Yield	b.p./m.p. (°C)
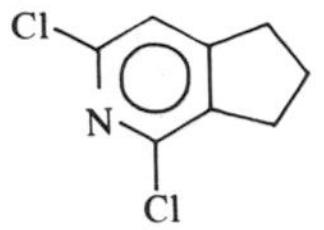	II.2.D	78%	m.p. 36°
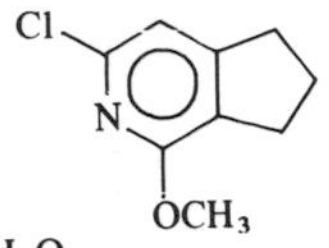	II.2.D		b.p. 145° (14 mm)
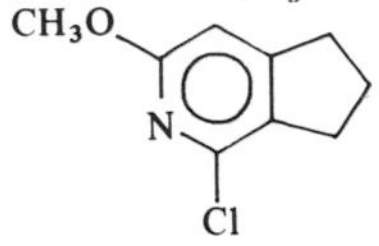	II.2.D	5%	
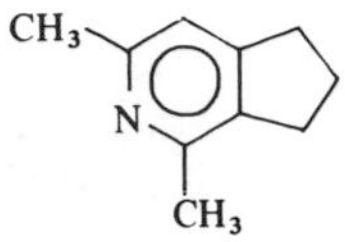	II.3.A.a	65%	
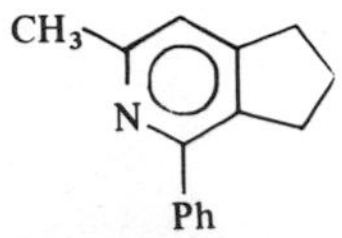	II.3.A.a	81%	
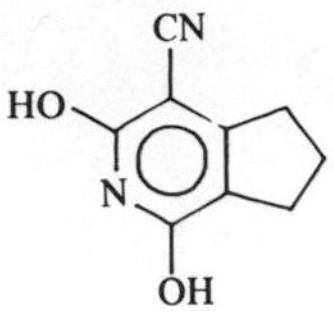		38%	m.p. 276–278°
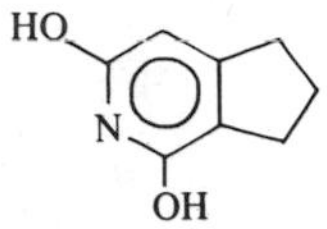		76%	m.p. 264°
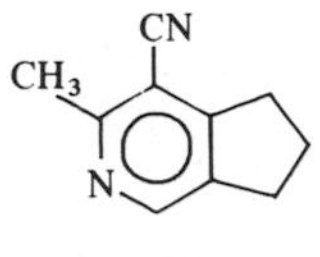	II.2.E	70%	m.p. 95–99°
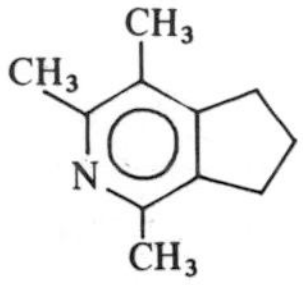	II.7		b.p. 103–104° (8 mm)
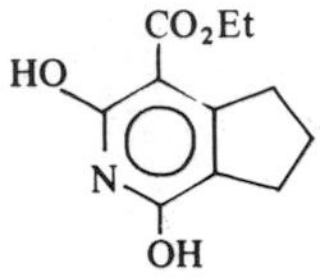	II.2.D		m.p. 241°

NMR	IR	UV	Derivative (m.p.)	Reference
				102, 184
				102
				102
			Picrate (121–122°)	132, 171
				132
				184
				184
				99
	2210 2175		Picrate (150°)	107
	1672, 1656, 1584 1412, 1430		Picrate (146–147°)	171
				99

TABLE II-13. (*CONTINUED*)

	Method of Preparation	Yield	b.p./m.p. (°C)
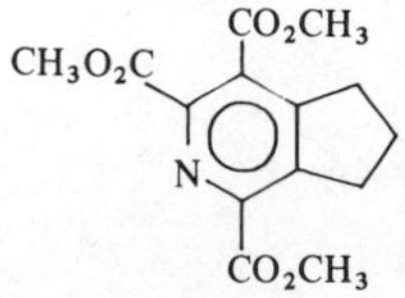	II.3.A.a	95%	m.p. 118–119°
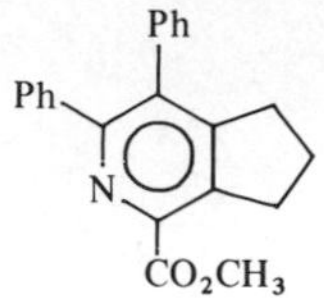	II.3.A.a	86%	m.p. 136–137°
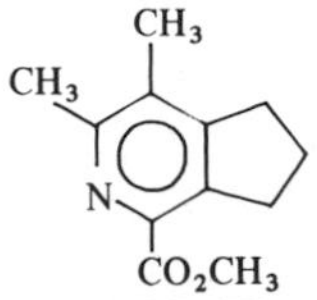	II.3.A.a	49%	m.p. 82–83°
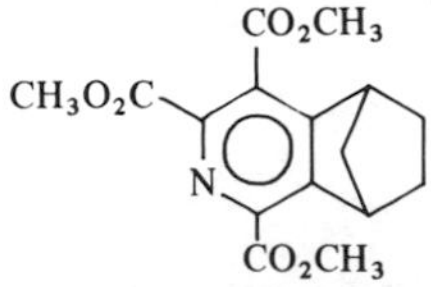	II.3.A.a	76%	m.p. 122–123°
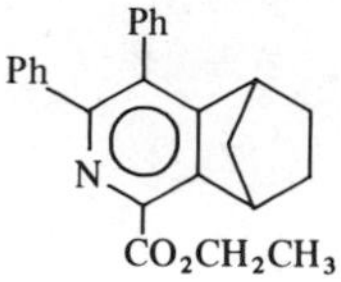	II.3.A.a	74%	m.p. 176–177°
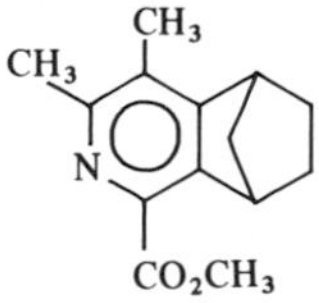	II.3.A.a	17%	m.p. 109–110°
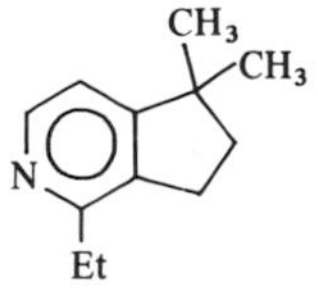	II.2.E		b.p. 50–60° (12 mm)
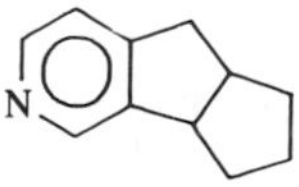	II.2.A	12%	
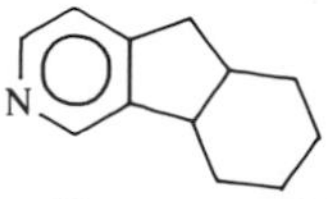	II.2.A	14%	
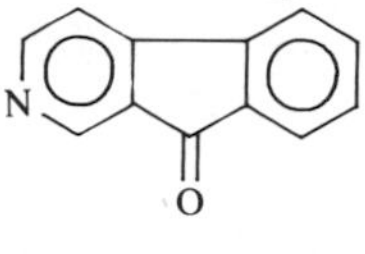		16%	m.p. 78.5–80°

NMR	IR	UV	Derivative (m.p.)	Reference
5(9), 3.27(4))(2)				129
				129
				129
				129
				129
				129
2.5(m,2), 1.5(m,2), 1.25(s,6) (t,3)	1460, 1560, 1580	260(3000)		105
8.20, 7.00–6.93				81
				81
				124 83

TABLE II-13.   (*CONTINUED*)

	Method of Preparation	Yield	b.p./m.p. (°C)
			trans m.p. 165–166° *cis* m.p. 82–83°
		91%	m.p. 140–141°
		75%	*trans* m.p. 129–130.5°
			cis m.p. 115–116.5°
		100%	m.p. 118–119°
			m.p. 147–148.5°
		31%	
		38%	m.p. 204.5–205°
		34%	
		20%	m.p. 224–225°

NMR	IR	UV	Derivative (m.p.)	Reference
				124
			Picrate (217–218°)	124
		228(4.56) 258(4.41) 332(4.24)	Picrate (257–258°)	124
		226(4.64) 258(4.56) 336(4.34)	Picrate (239–240°)	
			Picrate (197–198.5°)	124
			HCl (253–254°)	124
				228
				228
				229
				228

TABLE II-13. *(CONTINUED)*

	Method of Preparation	Yield	b.p./m.p. (°C)
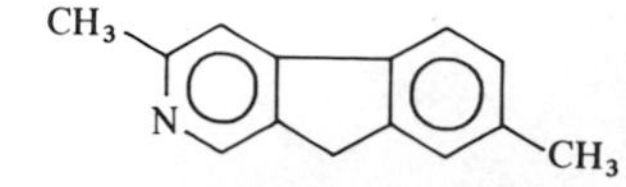			b.p. 148–153° (2 mm) m.p. 101–101.5°
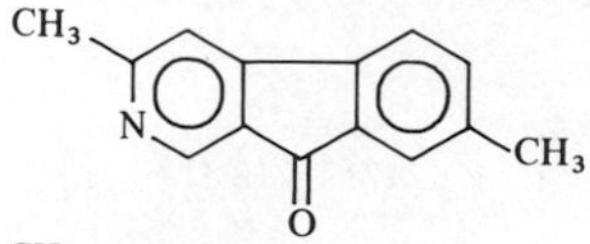			m.p. 173°
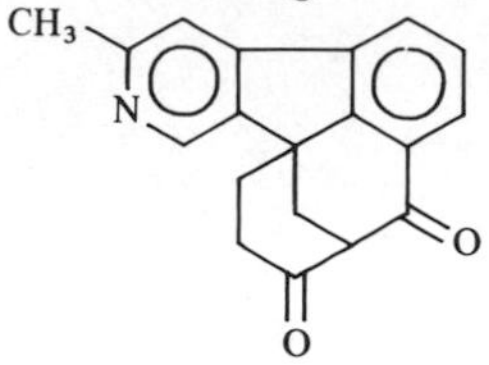	II.2.C	40%	m.p. 202–203°
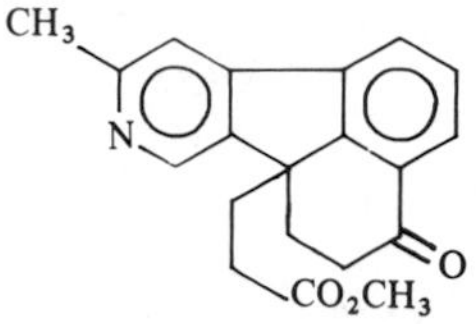		60%	m.p. 86–88.5°
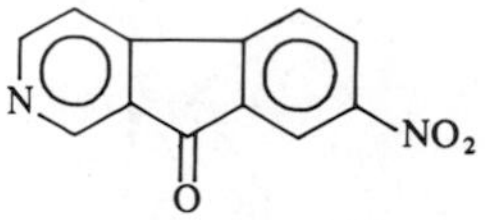		70%	m.p. 210–211°
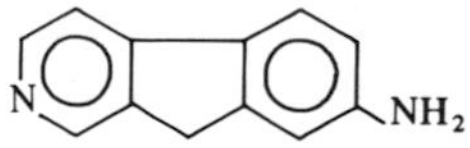		80%	m.p. 219–220°
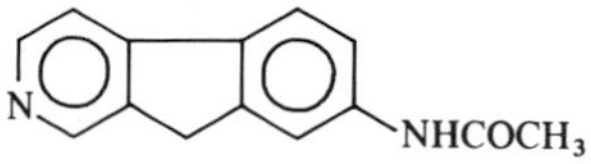		100%	m.p. 367–368°
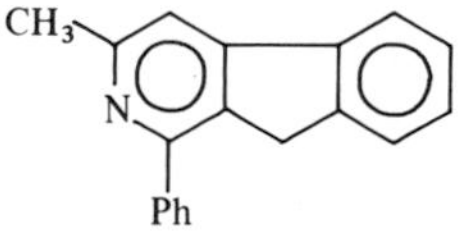		14%	m.p. 123.5–124°
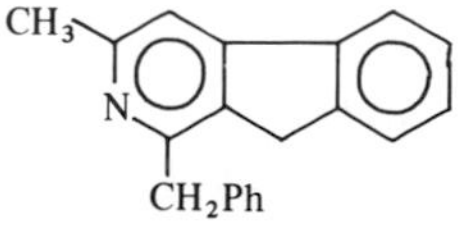		23%	m.p. 94–94°
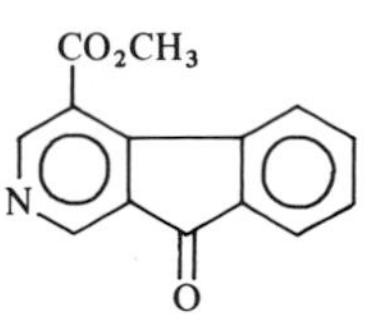			m.p. 195–197°

NMR	IR	UV	Derivative (m.p.)	Reference
			HCl (290°)	123
			HCl (229–230°)	123
			Picrate (220–222°)	230
			HCl (206–208°)	230
				83
			Picrate (231°)	83
				83
s,2), 2.57(s,3)	1560, 695		HCl (265–266°)	284
s,2), 3.5(s,2) s,3)	1570, 700	265(4.37) 286(4.24) 296(4.20)	HCl (263–265°)	284
				86

TABLE II-13. (*CONTINUED*)

	Method of Preparation	Yield	b.p./m.p. (°C)
	II.2.C		m.p. 132–134°
		81%	m.p. 107–108°
		54%	m.p. 210–212°
			m.p. 152°
			m.p. 267–268°

NMR	IR	UV	Derivative (m.p.)	Reference
				86
	3074, 3025			285
1(4), 2.37(4) 5(3)	1730			231
		224(4.58) 278(4.38) 354(4.09)	Picrate (258°)	125
3, 7.55, 8.85, 8.97 5, 8.79, 7.96, 7.53				125

TABLE II-14. 2,3-CYCLOHEPTAPYRIDINES

	Method of Preparation	Yield	b.p./m.p. (°C)
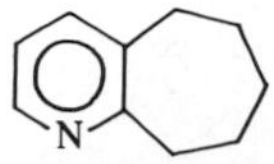	II.1.B	55%	b.p. 97–98° (11 mm)
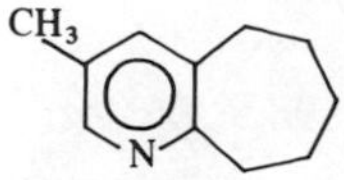	II.1.B	60%	b.p. 46–47° (0.05 mm)
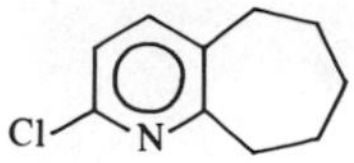		81%	m.p. 194–195°
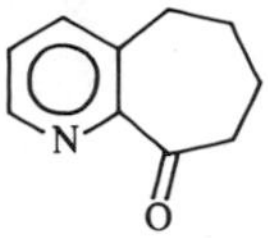		25%	b.p. 125° (0.3 mm)
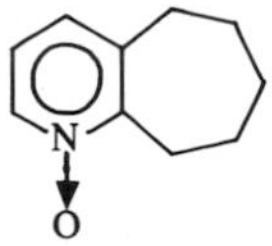		95%	b.p. 124° (0.03 mm) m.p. 107–108°
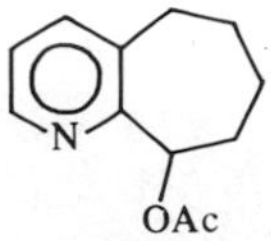		78%	b.p. 106° (0.6 mm)
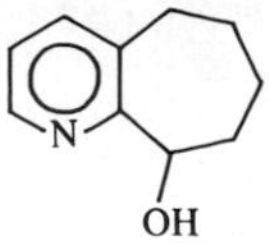		76%	b.p. 91–93° (0.4 mm)
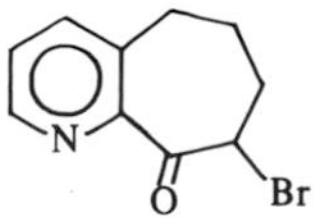		10%	
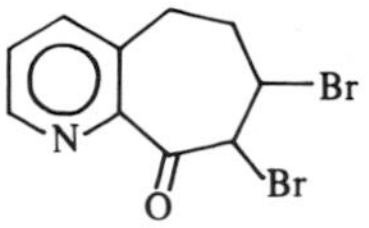		30%	
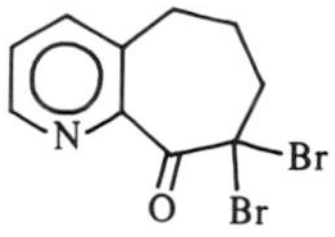		14%	m.p. 83–84°
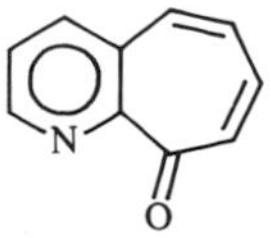		46%	

NMR	IR	UV	Derivative (m.p.)	Reference
			Picrate (138–139°)	62
			Picrate (162–163°)	62
				17
(m,4), 2.84(m,4) 5(dd,1), 7.65(dd,1) 5(dd,1)	1695	209(3.64) 228(3.64) 274(3.51)		17
(m,6), 2.8(m,2) (m,2), 7.05(s,2) 5(t,1)		221(4.36) 264(4.02)		17
(m,9), 2.85(m,2) (m,1), 7.1(dd,1) 5(dd,1), 8.4(dd,1)	1735	207(3.85) 249(3.64) 290(3.46)		17
5(m,2), 1.85(m,4) (m,2), 4.75(d,1) (s,1), 7.0(dd,1) (dd,1), 8.3(dd,1)	3360	210(3.75) 263(3.62) 290(sh)		17
5(m,4), 2.9(m,2) 5(m,1), 7.4(dd,1) (d,1), 8.7(dd,1)	1710			17
5(m,2), 2.95(m,2) 7(m,1), 5.75(d,1) 5(dd,1), 7.5(d,1) dd,1)	1720			17
m,2), 2.8(m,4) m,2), 8.6(m,1)	1718	211(3.85) 264(3.58) 270(sh)		17
q,1), 8.1(q,1), m,1), 7.0–7.2(2) m,1), 7.3(q,1)	1642, 1612, 1588	220(4.57) 256(sh), 321(sh) 338(4.20)		17

TABLE II-14. (*CONTINUED*)

	Method of Preparation	Yield	b.p./m.p. (°C)
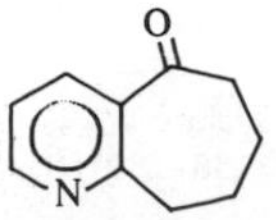	II.1.B	39%	b.p. 135–138° (10 mm)
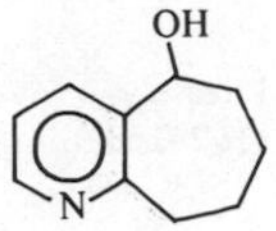		87%	m.p. 73.5–75.5°
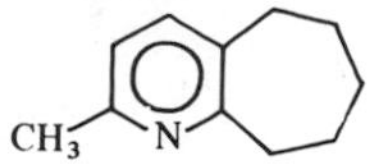	II.1.A	47%	b.p. 116–117° (17 mm)
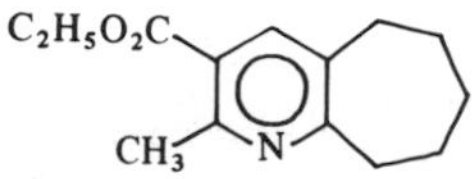	II.1.B	50%	m.p. 49–50° b.p. 106–107° (0.3 mm)
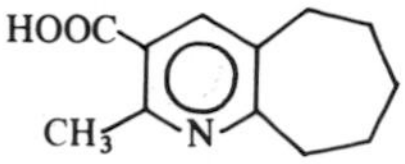		64%	m.p. 225–227°
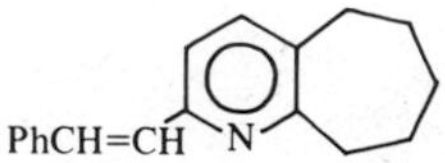	III.1.C	28%	b.p. 168–172° (0.8 mm) m.p. 86.5–87.5°
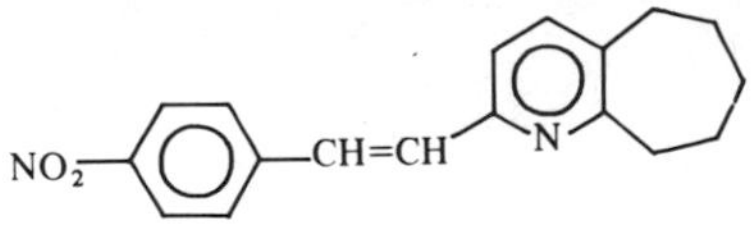	III.1.C	64%	m.p. 147–148.5°
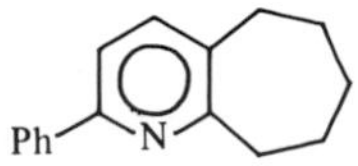		61%	b.p. 136–138° (0.2 mm)
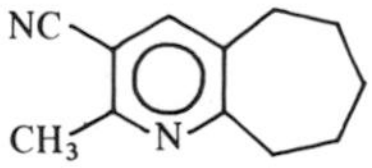	II.2.E	72%	m.p. 112–116°
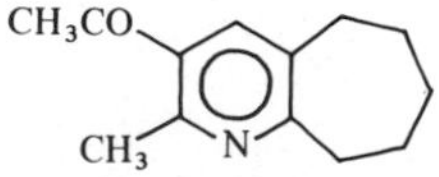	II.1.B	53%	b.p. 88° (0.04 mm)
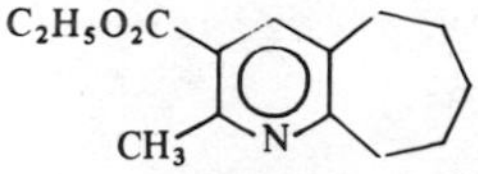	II.1.B	56%	b.p. 94° (0.04 mm)
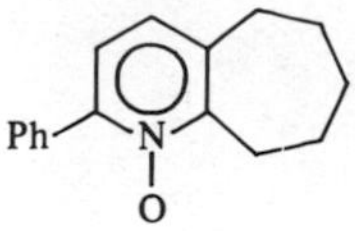		63%	m.p. 102–103°

NMR	IR	UV	Derivative (m.p.)	Reference
3(dd,1), 8.13(dd,1) 1(dd,1), 3.21(t,2) 3(t,2), 1.93(m,4)	1670	235.5(3.62) 274(3.81)	HCl (162°)	54
3(dd,1), 7.90(dd,1) 1(dd,1), 4.98(1) 6(s,1), 3.03(m,2) 6(m,6)	3180			54
(d,1), 6.7(d,1) (m,2), 2.6(m,2) (s,3), 1.7(m,6)			Picrate (159–160°)	12
			Picrate (163–164°)	12
				12
				207
				207
		282(4.14)	Picrate (142–143.5°)	190
	2225		Picrate (191–195°)	107
0(1), 2.55(3) 8(3)			Picrate (140–141°)	65
1(1), 2.78(3) 8(2), 1.39(3)			Picrate (163–164°)	65
				190

TABLE II-14. (*CONTINUED*)

	Method of Preparation	Yield	b.p./m.p. (°C)
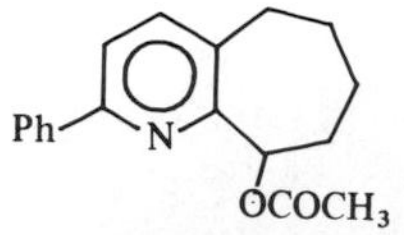	III.1.A	73%	b.p. 151–153° (0.1 mm)
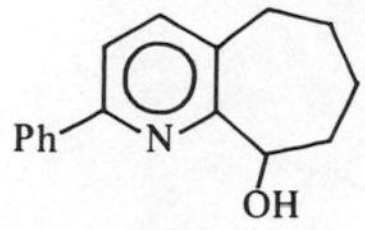		82%	b.p. 135–137° (0.04 mm) m.p. 51–52°
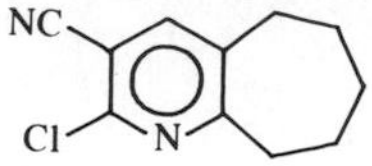			
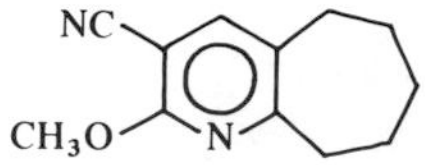		68%	m.p. 82–84°
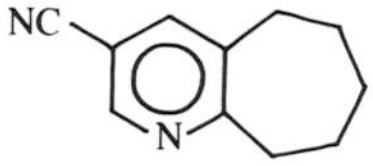		98%	m.p. 89–90°
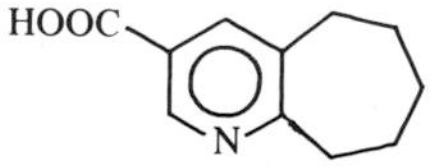		90%	m.p. 218°
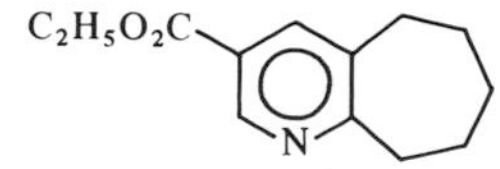		78%	m.p. 52°
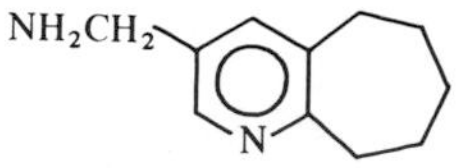		86%	m.p. 226–227°
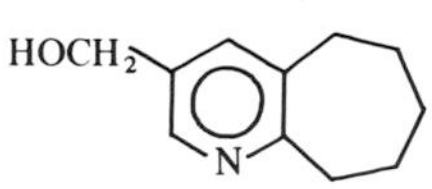		80%	m.p. 127–128°
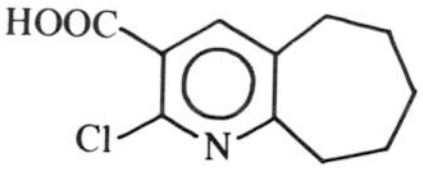		92%	m.p. 168–170°
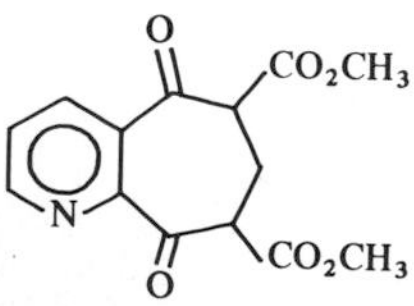	II.1.D	15%	m.p. 156–158°

NMR	IR	UV	Derivative (m.p.)	Reference
				190
	3270		Picrate (168–169°)	190
	1595, 1548, 1160 1036, 935, 699			280 281
	1602, 1570, 1195 1002, 930, 776			280 281
	1600, 1562, 1198 1074, 919, 708			280 281
	1624, 1560, 1159 1020, 915, 763			280 281
	3000, 1613, 1577 1472, 1202, 1032 932, 777			280 281
	3040, 1610, 1575 1473, 1192, 1072 910, 709			280 281
	1612, 1582, 1478 1227, 1025, 903 715			280 281
	1597, 1560, 1460 1170, 1150, 932 742			280 281
), 8.2(1)), 7.6(2) 6), 2.9(2)				73

TABLE II-14. *(CONTINUED)*

	Method of Preparation	Yield	b.p./m.p. (°C)
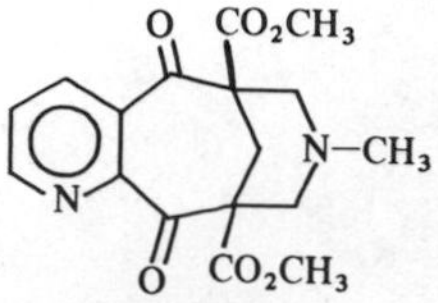		69%	m.p. 167–168°
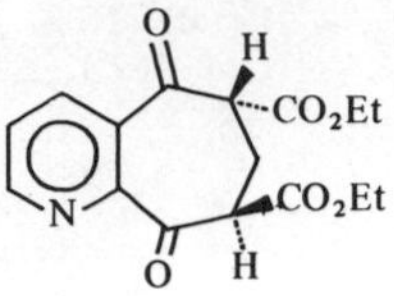	II.1.D	55%	m.p. 117–118°
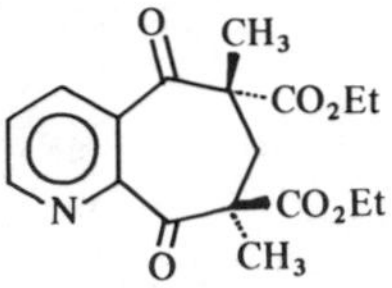	II.1.D	20%	m.p. 154–155°
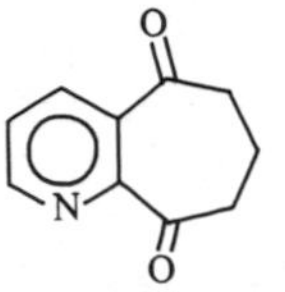		85%	m.p. 63–64°
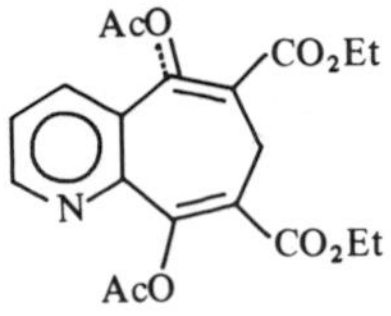		75%	m.p. 134–135°
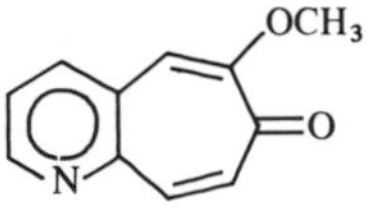	II.1.D	10%	m.p. 153°
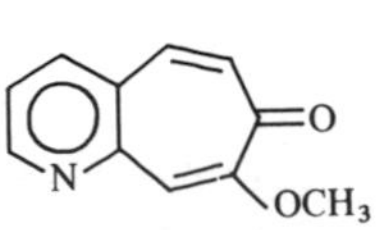	II.1.D	90%	
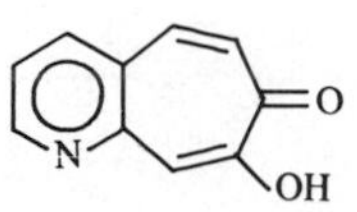	II.1.D		m.p. 167°
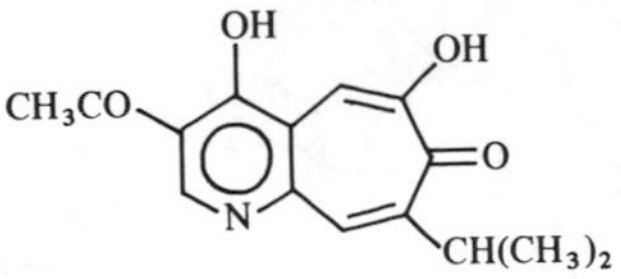			m.p. 290°

NMR	IR	UV	Derivative (m.p.)	Reference
				73
)(dd,1), 7.5(dd,1) 35(dd,1), 3.0(s,2) .6(OH), 4.4(q,4) 45(t,6)	1645 1615	204(4.11) 239(4.18) 281(4.35)		72
)(dd,1), 7.5(dd,1) 2(dd,1), 2.6(s,2) 3(m,4), 1.6(s,6) *(m,8)	1735 1710	213(4.33) 244(3.84) 275(sh)		72
35(dd,1), 7.5(dd,1) *5(dd,1), 2.9(m,2) 5(m,2), 2.4(m,2)	1690	212(4.08) 236(3.80) 270(sh)		72
3(dd,1), 7.45(dd,1) *(dd,1), 3.2(s,2) *(q,4), 2.3(s,6) 5(t,6)	1768 1710 1628	205(4.09) 236(4.44) 252(sh)		72
7(1), 7.97(1) 1(1), 7.15(1) 5(1), 6.87(1) 0(3)				75
9(1), 8.04(1) 6(1), 7.07(1) 7(1), 7.45(1) 3(3)				75
				75
				128

TABLE II-14.   *(CONTINUED)*

	Method of Preparation	Yield	b.p./m.p. (°C)
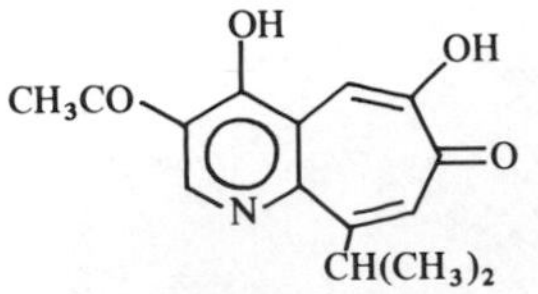			m.p. 169–169.7°
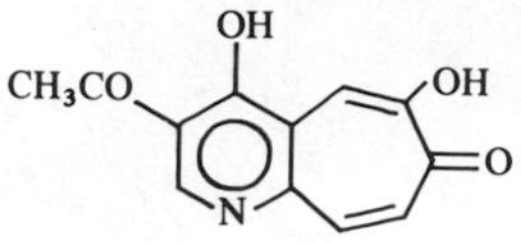			m.p. 280°
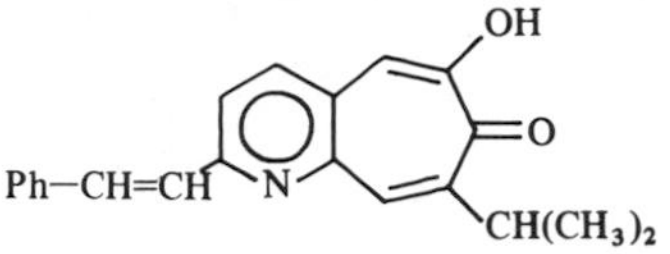			m.p. 142–142.5°
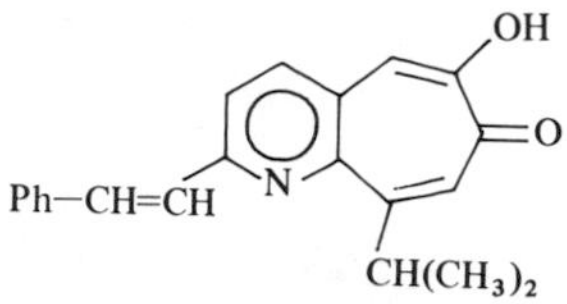			m.p. 182–183°
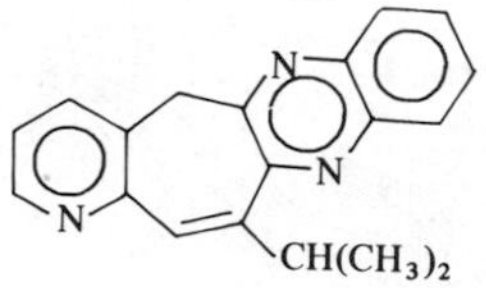		81%	m.p. 133–134°
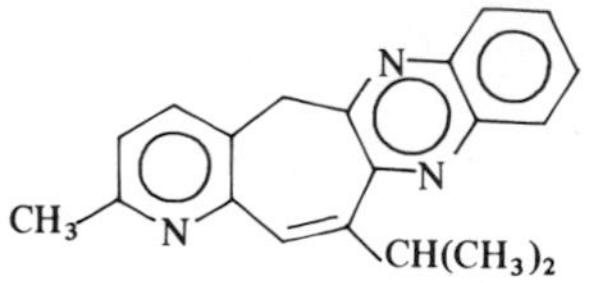		86%	m.p. 149.5–150.5°
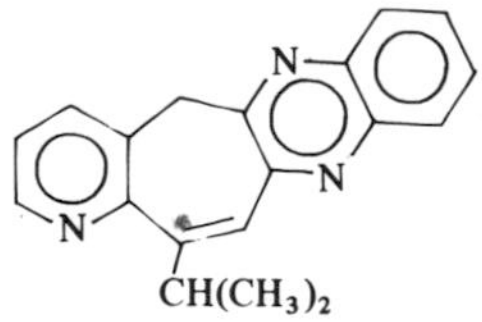		81%	m.p. 158.5–159°
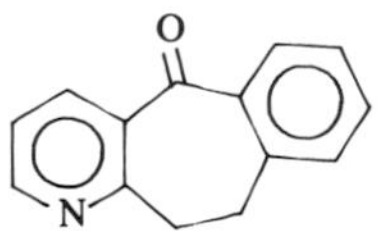	II.2.C	93%	m.p. 62–64°
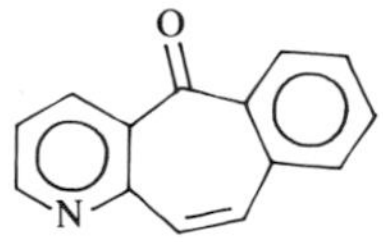		60%	m.p. 96–97°

NMR	IR	UV	Derivative (m.p.)	Reference
			Phenylhdrazone (249–254°)	128
				128
		355(4.5)		238
				238
		340(4.15)		238
		335(4.08)		238
		345(4.15)		238
(m,4), 7.6(m,4), (m,1), 8.30(dd,1), (m,1)	6.07 μ	276(10,150)	HCl (190–194°)	90 88
7.9(m,6), 8.5(m,1), (dd,1), 8.93(dd,1)	6.05 μ	226(21,600) 246(25,000) 305(11,850) 350(3,400)		90

TABLE II-14. (*CONTINUED*)

	Method of Preparation	Yield	b.p./m.p. (°C)
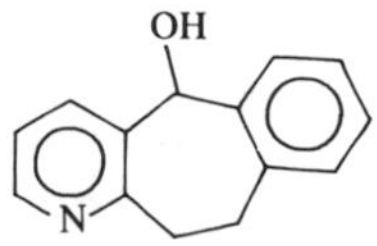		95%	m.p. 165–166°
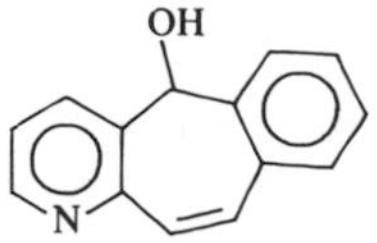		95%	m.p. 206–208°
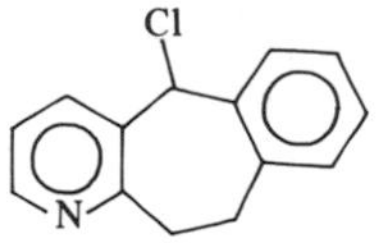		85%	m.p. 91–92°
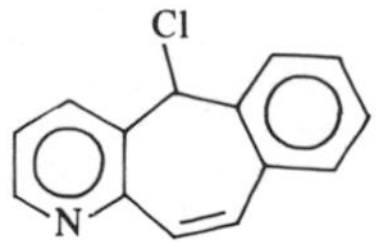		80%	m.p. 140–143°
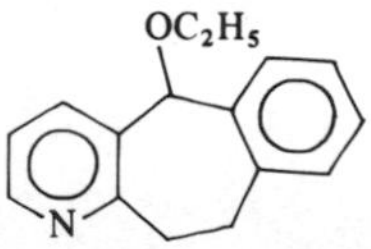		95%	b.p. 135° (0.33 mm) m.p. 56–57°
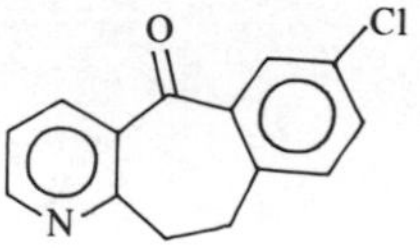	II.2.C	73%	m.p. 46–47°
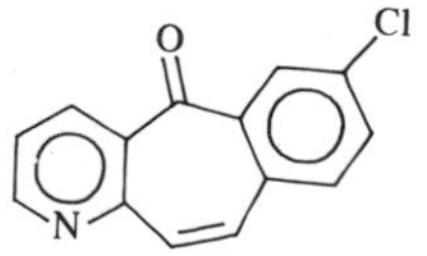		73%	m.p. 138–140°
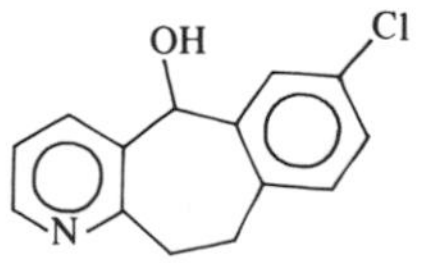		95%	m.p. 193–196°
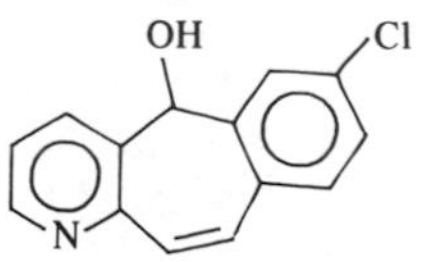		80%	m.p. 243–245°
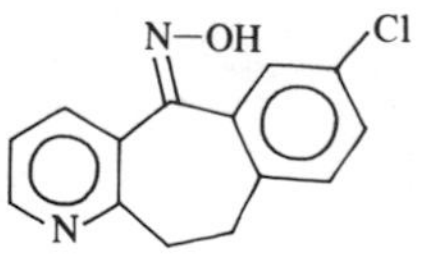		58%	m.p. 280–281°

NMR	IR	UV	Derivative (m.p.)	Reference
00(1), 3.35(*s*,4) 11(*s*,1), 6.9–7.7(*m*) 92(*dd*,1), 8.40(*m*,1)				90
14(*d*,1), 6.38(*d*,1) 0–8.0(*m*,5), 13(*d*,1), 8.52(*m*,1)				90
7–4.3(*m*,4), 03(*s*,1), 6.9–7.5(*m*,5) 68(*d*,1), 8.50(*m*,1)				90
08(*s*,1), 7.1–7.7(*m*,7) 84(*dd*,1), 8.60(*m*,1)				90
				90
				91
				91
				91
				91
				91

TABLE II-14. (CONTINUED)

	Method of Preparation	Yield	b.p./m.p. (°C)
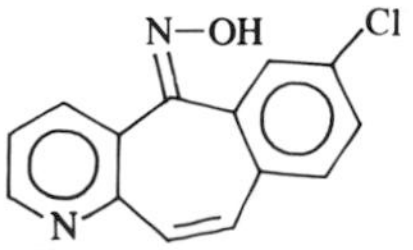		66%	m.p. 291–293°
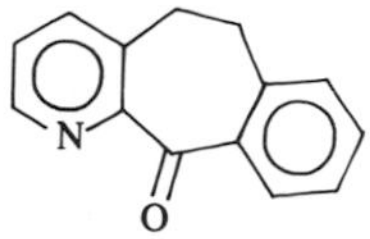	II.2.C	80%	m.p. 81–82°
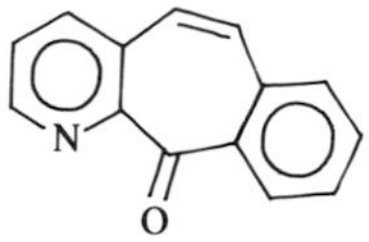		57%	m.p. 116–118°
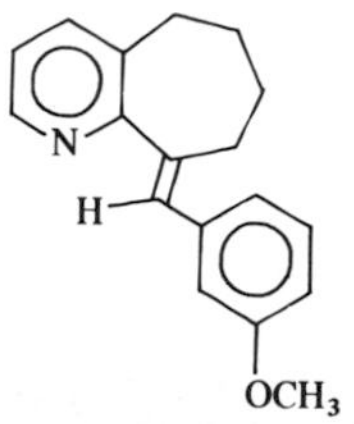	III.2.B	48%	m.p. 73.5–74°
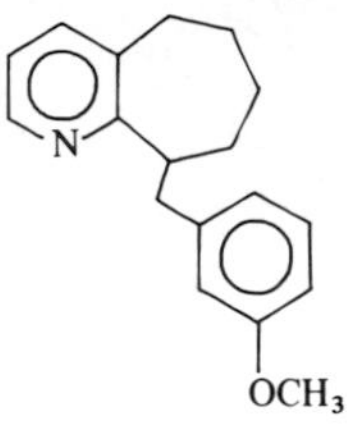		95%	m.p. 55.5–56.5°
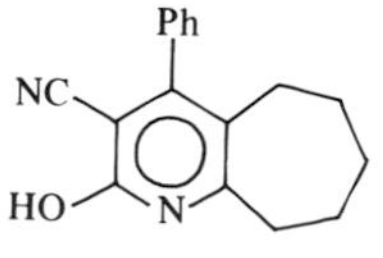	II.1.C	46%	m.p. > 300°
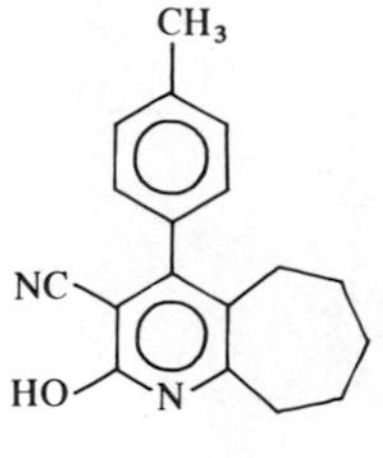	II.1.C	47%	m.p. > 260–262°

NMR	IR	UV	Derivative (m.p.)	Reference
				91
56(q,1), 8.60(q,1) 00(m,1), 3.08(s,4)	5.98 μ	280(10,200)		88
89(q,1), 8.83(q,1) 23(m,1), 6.91, 7.13	6.01 μ	226(21,500) 255(29,600) 306(10,300) 348(2,800)		88
54(dd,1), 61–7.6(m,7), 85(s,3) 57–3.04(m,4) 7–2.1(m,4)		270.5(4.17)		300
45(dd,1), 54–7.58(m,6), (s,3) 59–3.75(m,5) 2–2.15(m,6)			HCl (161°)	300
64(s,1), 7.50(s,5) 32(s,2), 2.14(s,2) 58(s,6)	2210, 1640 1590			318
49(s,1), 7.32(d,2) 1(d,2), 2.79(s,2) 86(s,3), 2.25(s,2) 7(s,6)	2210, 1641 1590			318

TABLE II-14.   (*CONTINUED*)

	Method of Preparation	Yield	b.p./m.p. (°C)
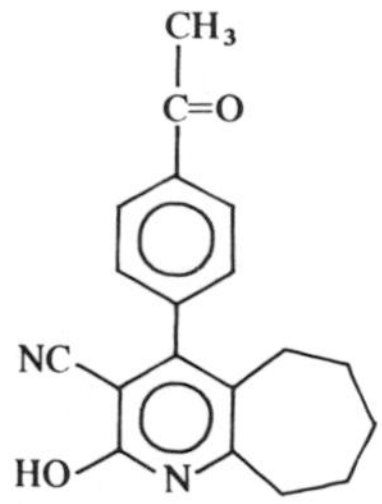	II.1.C	68%	m.p. > 272–273°
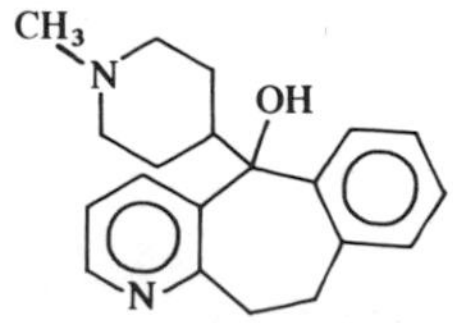		49%	m.p. 172–174°
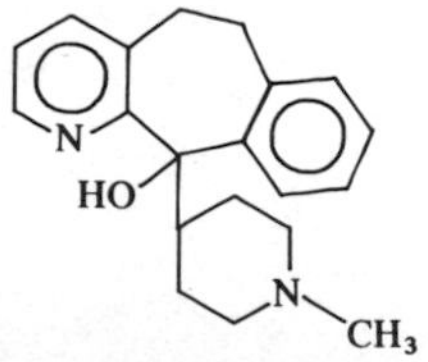		63%	m.p. 174–176°
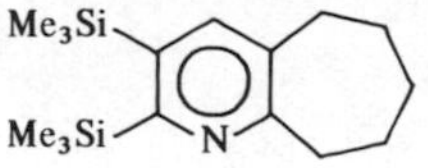	II.3.B	25%	Oil
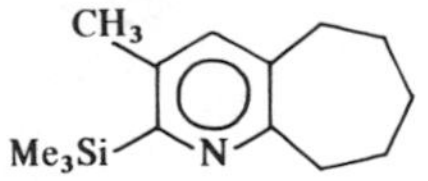	II.3.B.	66%	Oil
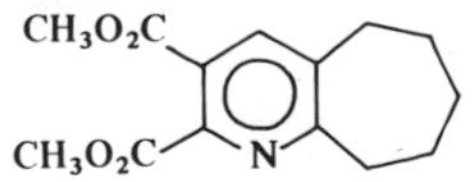	II.3.B	95%	m.p. 93–94°
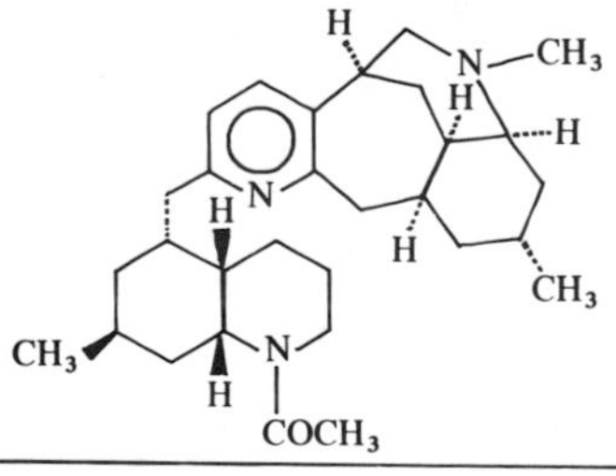			

NMR	IR	UV	Derivative (m.p.)	Reference
45(s,1), 7.24(d,2) 4(d,2), 3.82(s,3) 9(s,2), 2.27(s,2) 4(s,6)	2220, 1642 1594			318
				236
				236
				313
				313
				313
q) and q, J = 8 Hz)			276(8500) 279(13,400)	306

TABLE II-15.   3,4-CYCLOHEPTAPYRIDINES

	Method of Preparation	Yield	b.p./m.p. (°C)
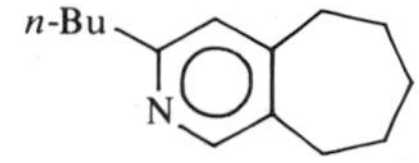 *n*-Bu structure	II.3.B	43%	
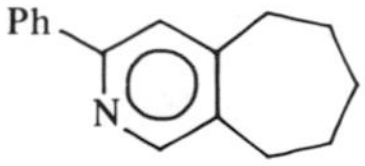 Ph structure	II.3.B	54%	
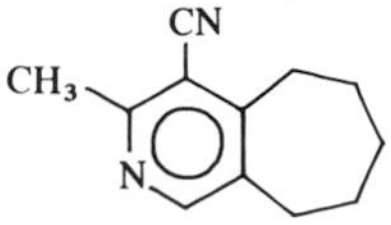 CN, CH₃ structure	II.2.E	72%	m.p. 112–116°
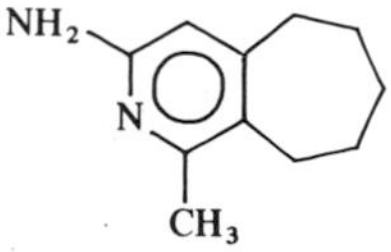 NH₂, CH₃ structure	II.7	77%	m.p. 86.5°
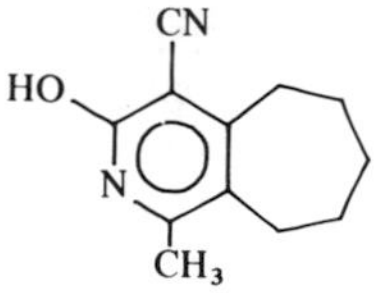 CN, HO, CH₃ structure	II.8	76%	m.p. 266.5–267°
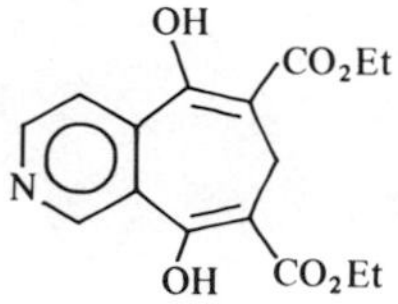 OH, CO₂Et, OH, CO₂Et structure	II.1.D	40%	m.p. 93–94°
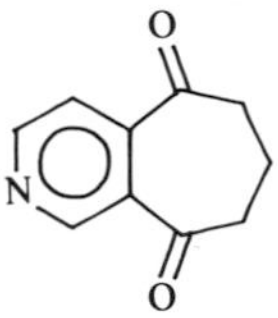 O, O structure		94%	m.p. 64–65°
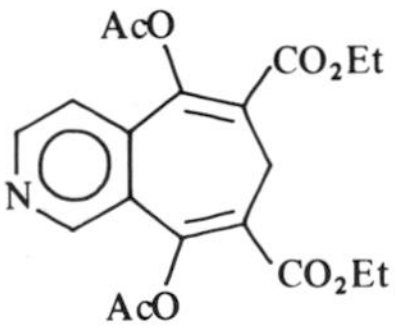 AcO, CO₂Et, AcO, CO₂Et structure		85%	
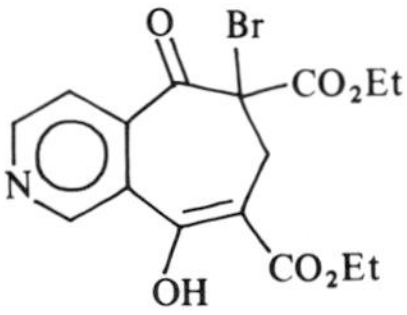 O, Br, CO₂Et, OH, CO₂Et structure			

NMR	IR	UV	Derivative (m.p.)	Reference
				146
				146
	2225		Picrate (191–195°)	107
56(2), 2.22(3) 5(2), 4.6(2) 83(1)	3300, 3460			170
	3322, 3145, 2252 1636, 1538, 833 780, 765	218(4.21) 244(3.85) 222(4.21) 252(3.73) 341(4.11)		180
(s,1), 8.75(d,1) (d,1), 2.95(s,2) 5(OH), 4.3(q,4) (t,6)	1645, 1610	206(4.08) 252(4.27) 272(4.18) 320(3.94)		72
(s,1), 9.0(d,1) 4(d,1), 2.85(m,4) 5(m,2)	1688	209(4.02) 228(3.69) 283(3.35)		72
2(s,1), 8.68(s,1) (d,1), 3.15(s,2) 4(q,4), 2.26(s,6) (t,6)	1775, 1710, 1620	212(sh) 241(4.45) 308(sh)		72
(s,1), 9.1(d,1) (d,1), 4.7(2), (1), 9.1(OH), (m,4), 1.1(m,6)	1740, 1720			72

TABLE II-15.   (*CONTINUED*)

	Method of Preparation	Yield	b.p./m.p. (°C)
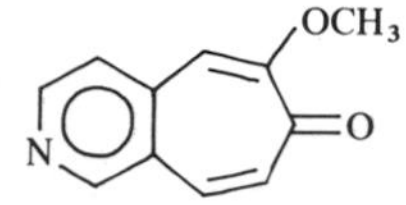	II.1.D		m.p. 180°
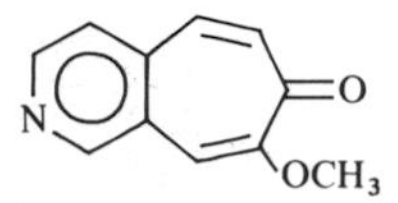	II.1.D		m.p. 145°
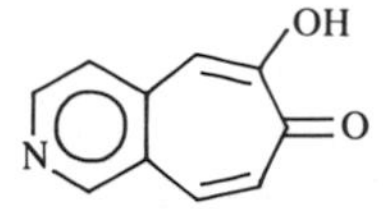	II.1.D		m.p. 230°
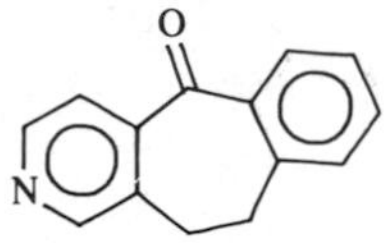	II.2.C	63%	m.p. 46–47°
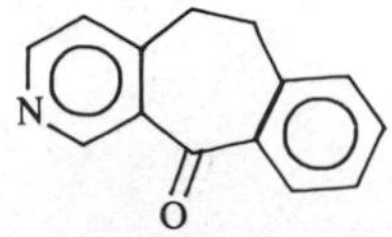	II.2.C	91%	m.p. 66–68°
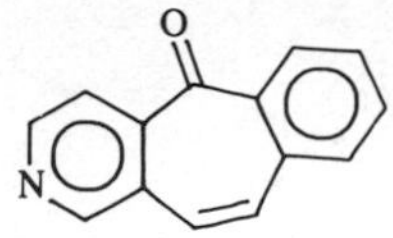		47%	m.p. 132–133°
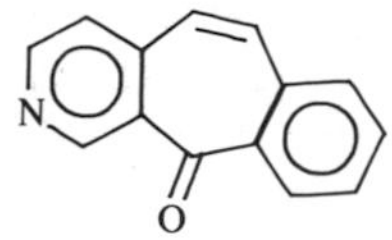		62%	m.p. 157–158°
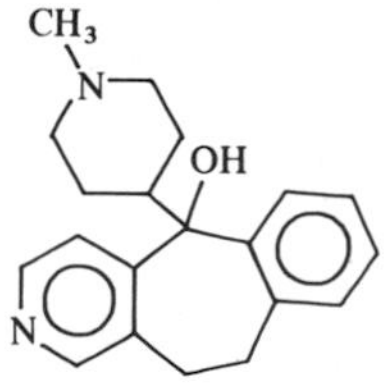		52%	m.p. 201–202°
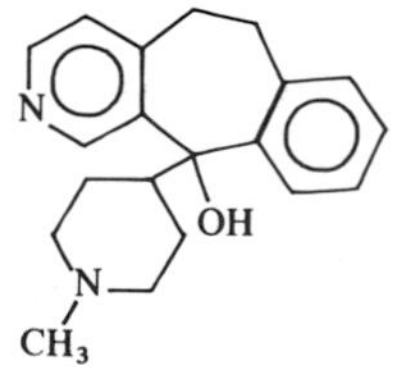		35%	m.p. 201–202°

NMR	IR	UV	Derivative (m.p.)	Reference
⁷(1), 8.72(1) ⁸(1), 7.68(1) ⁵(1), 6.90(1) ⁵(3)				76
(1), 8.71(1) ⁵(1), 7.58(1) (1), 7.05(1) ⁵(3)				76
⁵(1), 8.82(1) ⁵), 8.0(1) ⁵(1), 7.52(1)				76
(m,2), 7.65(m) (s,4)	6.09 μ	271(9950)		88
(d), 9.14(s) (m), 3.14(s,4)	6.09 μ	268(11,800)		88
s), 8.73(d) d), 8.25(m) 7.16	6.05 μ	235(25,600) 258(14,800) 310(12,400) 359(3,300)		88
d), 8.75(d) s), 8.24(m) 7.15	6.08 μ	233(27,000) 246(24,400) 310(13,700)		88
				236
				236

TABLE II-15.   (*CONTINUED*)

	Method of Preparation	Yield	b.p./m.p. (°C)
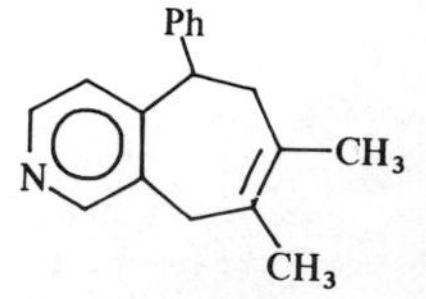	II.2.F	54%	m.p. 75°
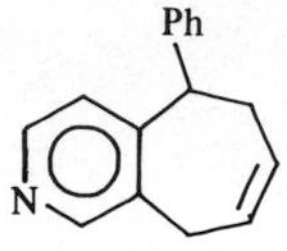	II.2.F	55%	m.p. 134–135°

NMR	IR	UV	Derivative (m.p.)	Reference
2(5,1), 8.24(*d*,1) 3–7.41(*m*,5), 4(*d*,1), 4.37(1*H*) 2–3.56(2*H*) 7–3.0(2*H*), 0(3*H*), 1.53(3*H*)	1595			302
2(*m*,2), 5–7.52(*m*,5) 4(*d*,1), 6–5.50(2*H*) 5(1*H*), 3.76(*d*,1) 2(*dd*,1), 5–2.87(2*H*)	1658 1591			302

TABLE II-16. CYCLOOCTAPYRIDINES

	Method of Preparation	Yield	b.p./m.p. (°C)
	II.1.B	50%	b.p. 59–60° (0.2 mm)
	II.1.B	55%	b.p. 73–75° (0.05 mm)
	II.1.A	62%	b.p. 150–151° (0.2 mm)
	II.1.B	58%	b.p. 95° (0.04 mm)
	II.1.B	53%	b.p. 100° (0.04 mm)
	II.3.C	50%	
	II.3.C	21%	
	II.1.C	22%	m.p. 295–296°
	II.1.C	24%	m.p. 258–260°

NMR	IR	UV	Derivative (m.p.)	Reference
			Picrate (150–151°)	62
			Picrate (160–161°)	62
				209
5(1), 2.59(3) 3(s)			Picrate (171–172°)	
8(1), 2.82(3) 3(2), 1.41(3)			Picrate (156–157°)	
5–7.4(m,10) 0(s,4) 3(s,12)		279(4.04) 312(3.31)		147
1(s,2), 5–7.2(m,8), 5(s,4), 2.64(s,6) 5(s,6)		276(4.11) 313(3.29)		147
50(s,1), 7.47(s,5) 4(s,2), 2.28(s,2) 3(s,8)	2210, 1640 1603			318
50(s,1), 7.25(d,2) 4(d,2), 2.80(s,2) 5(s,3), 2.26(s,2) 3(s,8)	2220, 1643 1610			318

TABLE II-16.　　(*CONTINUED*)

	Method of Preparation	Yield	b.p./m.p. (°C)
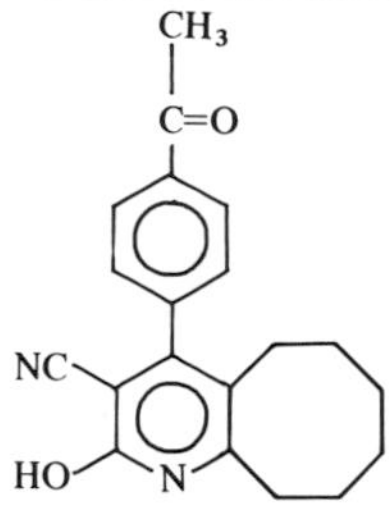	II.1.C	36%	m.p. 269–270°
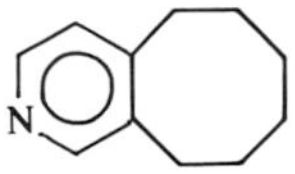	II.1.C	27%	b.p. 134–135° (11 mm)
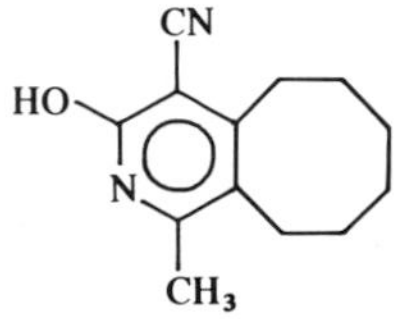	II.8	60%	m.p. 280–280.5°

NMR	IR	UV	Derivative (m.p.)	Reference
42(s,1), 7.25(d,2) (d,2), 3.81(s,3) (s,2), 2.30(s,2) (s,8)	2210, 1640 1606			318
				317
	3246, 3115, 2232 1629, 1529, 820 811, 790, 772	217(4.25) 242(3.82) 222(4.24) 249(3.67) 342(4.09)		180

TABLE II-17.   LARGER CYCLOALKAPYRIDINES

	Method of Preparation	Yield	b.p./m.p. ($°$C)
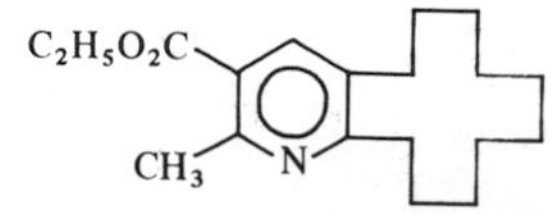 $C_2H_5O_2C$— [structure with CH_3, N]	II.1.B	66%	b.p. 142° (0.06 mm)
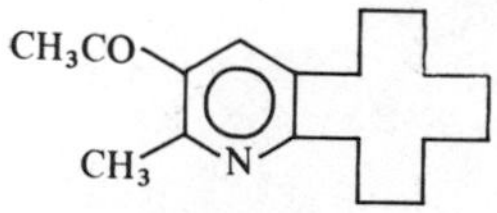 CH_3CO— [structure with CH_3, N]	II.1.B	63%	b.p. 135° (0.04 mm)
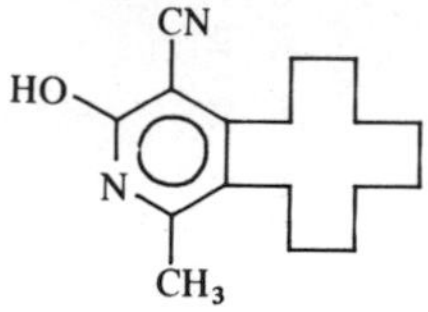 [structure with CN, HO, N, CH_3]	II.8	50%	m.p. 281.5–282°
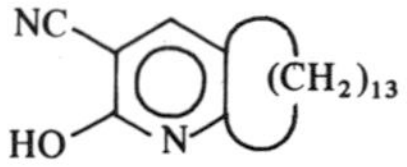 NC— [structure with HO, N, $(CH_2)_{13}$]	II.8		m.p. 247–248°
 HO— [structure with N, $(CH_2)_{13}$]		94%	m.p. 189–190°
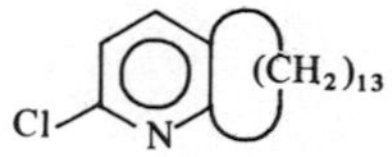 Cl— [structure with N, $(CH_2)_{13}$]		85%	
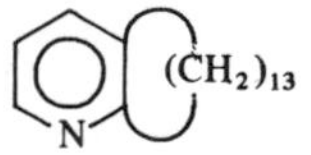 [structure with N, $(CH_2)_{13}$]			

NMR	IR	UV	Derivative (m.p.)	Reference
6(1), 2.80(3) 1(2), 1.40(3)			Picrate (137–138°)	65
4(1), 2.58(3) 2(3)			Picrate (190–191°)	65
	3311, 3156, 2237 1631, 1534, 829 778, 769	217(4.29) 239(3.83) 221(4.29) 249(3.59) 341(4.10)		180
				289
				289
			Picrate (130–131°)	289
			Picrate (194–195°)	289

TABLE II-18.   [2,3:4,5]DICYCLOPENTAPYRIDINE

	Method of Preparation	Yield	b.p./m.p. (°C)
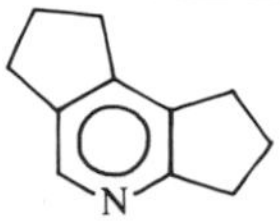	II.1.C		
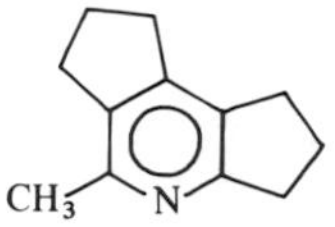	II.1.C		
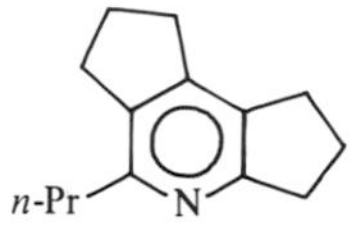	II.1.C		b.p. 104–106° (0.08 mm)
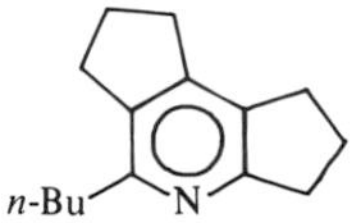	II.1.C		b.p. 94° (0.06 mm)
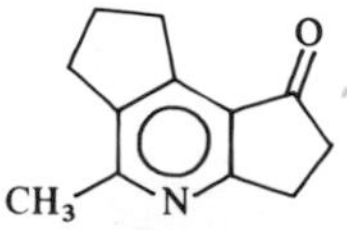	II.1.B	30%	m.p. 98–101°
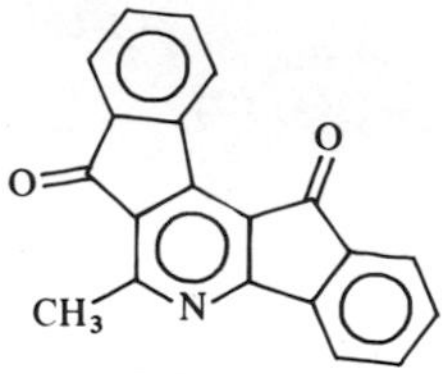			m.p. 263°
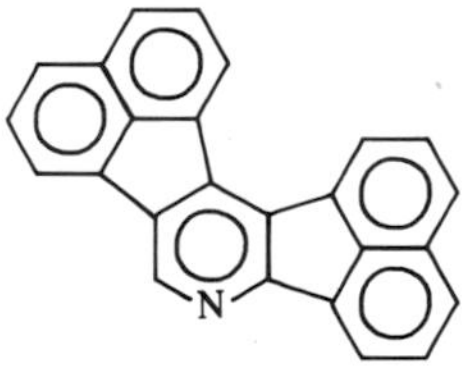	II.1.B	9%	m.p. > 300°

NMR	IR	UV	Derivative (m.p.)	Reference
(s,1)	2950, 2850, 1617	278(2804)		37
−2.70(8)	1580, 1470, 1440	274(2942)		
(quin, 2)	1395, 750	270(2927)		
(quin, 2)				
			Picrate (134°)	33
	1575	278	Chloroplatinate (187°)	290 291
	1605, 1585	275(3.81)	Chloroplatinate (184°)	290
	2959–2841	227(3.57)		35
	1460, 1379	284(4.03)		39
				64
				61
	1582, 1510, 1430 1358	332.5(4.63)		68

TABLE II-19.   [2,3:5,6]DICYCLOPENTAPYRIDINE

	Method of Preparation	Yield	b.p./m.p. (°C)
	II.1.A	66%	m.p. 87%
	III.2.C	18%	
	II.1.B	25%	m.p. 71–72°
	II.4		
	III.2.B		m.p. 215–217°
	III.2.B		m.p. 235–236°
	III.2.B		m.p. 185–186°
	II.1.C	80%	m.p. 203°

NMR	IR	UV	Derivative (m.p.)	Reference
30(s,1), 2.95(t,4)	2950, 2850, 1605	297(4875)		15
87(t,4)	1570, 1442, 1415	292(7200)		37
10($quin$, 4)	1305, 1230, 1215	287(7925)		
	730	282(6800)		
				194
			HCl (149–152°)	64
				153
				217
−7.1(m,14)				217
5(8), 2.35(9)				
−6.8(m,14)				217
(9), 2.92(8)				
				15

TABLE II-19. *(CONTINUED)*

	Method of Preparation	Yield	b.p./m.p. (°C)
	II.2.F	31%	m.p. 204–205°
	II.1.A		m.p. 332–334°
			m.p. 256°
		27%	m.p. 360°
			m.p. 395°
		56%	m.p. 370–372°
			m.p. 356–357°
	II.1.C		m.p. 292–293°

NMR	IR	UV	Derivative (m.p.)	Reference
9(6), 7.77(1) 7(2), 3.75(4)		237(35,800) 246(32,600) 270(6,400) 335(28,900) 343(32,500) 350(27,400)		113
				29
				29
				292
				293
				294
	1720			305
			Picrate 265–266°	305

TABLE II-20.   CYCLOPENTACYCLOHEXAPYRIDINES

	Method of Preparation	Yield	b.p./m.p. (°C)
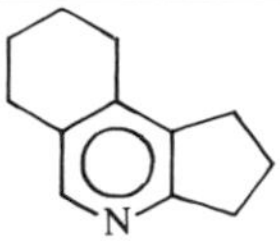	II.1.C		b.p. 108–110° (0.4 mm)
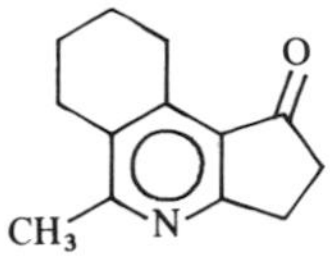	II.1.B	30%	m.p. 82.5–85°
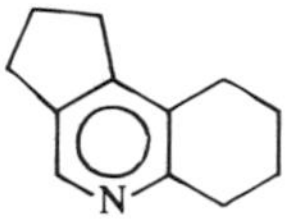	II.1.C		b.p. 108–110° (0.4 mm)
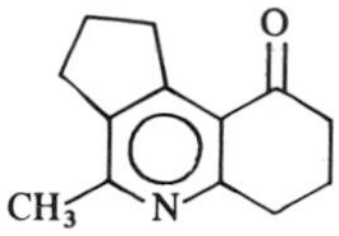	II.1.B	35%	b.p. 145–150° (0.1 mm) m.p. 73.5–75°
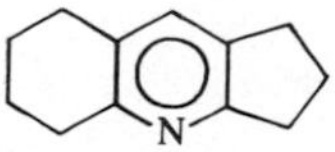	II.1.B	8%	b.p. 84° (0.5 mm)
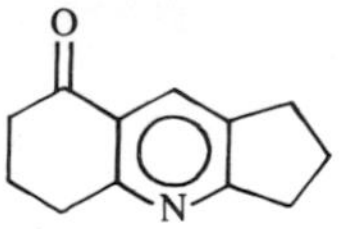	II.1.B	33%	m.p. 60–61°
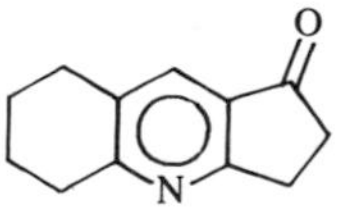	II.1.B	15–20%	m.p. 93.5–95.5°
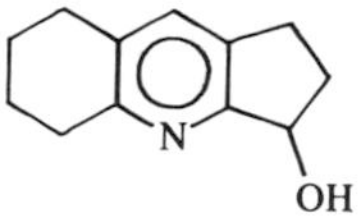	III.2.A		
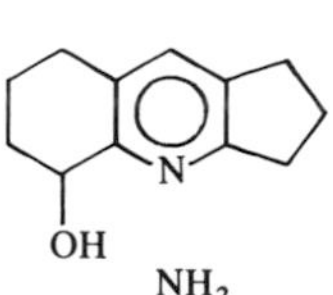	III.2.A		
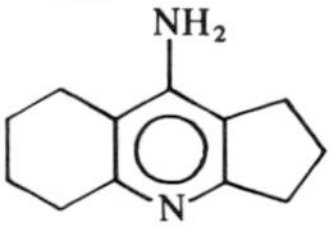	II.4		

NMR	IR	UV	Derivative (m.p.)	Reference
.06(s,1), 3.10–2.6(8) 09(quin,2) 80(m,4)	2870, 2830, 1600 1573, 1481, 1440 1402, 928, 820	276(4300) 272(4470)		37
			HCl (170–175°)	64
22(s,1) 3.10–2.5(8) 07(quin,2) 83(m,4)	2940, 2870, 1604 1581, 1470, 1412 1199, 925, 887 830	276(2840) 272(2985) 268(3205)		37
				64
15(s,1) 3.0–2.7(8) 06(quin,2) 84(m,4)	2935, 2860, 1610 1575, 1450, 1420 1227, 920, 750	295(4805) 289(6603) 285(7000) 281(6217)	Picrate (160–161°)	37 13, 11
				64
				64
				210
				210
				153

TABLE II-20. (*CONTINUED*)

	Method of Preparation	Yield	b.p./m.p. (°C)
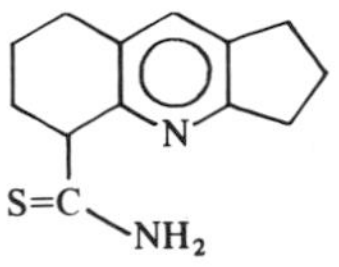	III.2.C		m.p. 118°
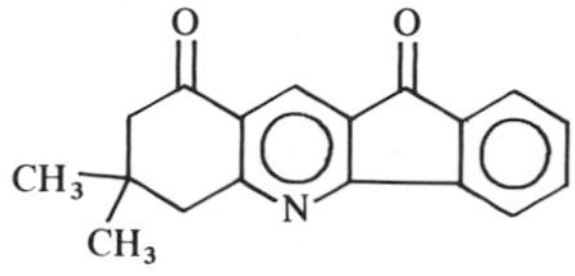			m.p. 154°
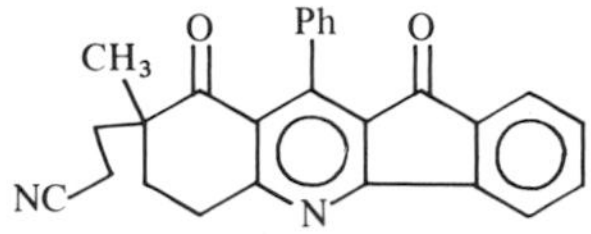	II.1.C		m.p. 169–170°
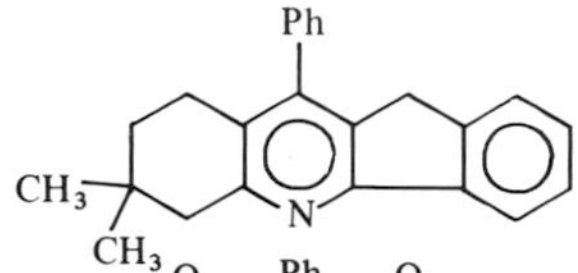			m.p. 142°
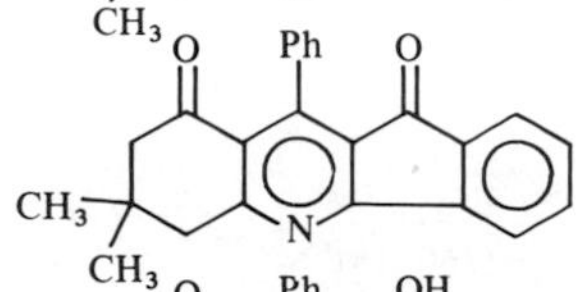	II.1.B		m.p. 193–194°
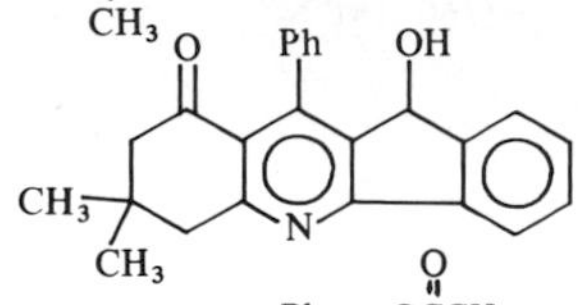			m.p. 189°
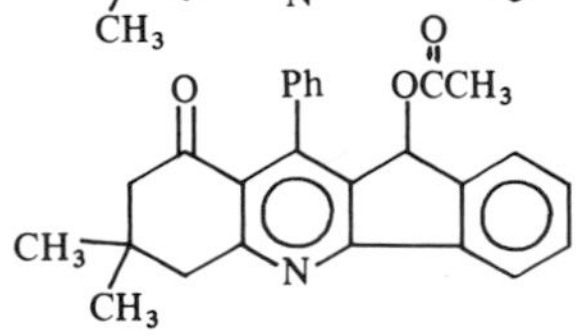			m.p. 171°
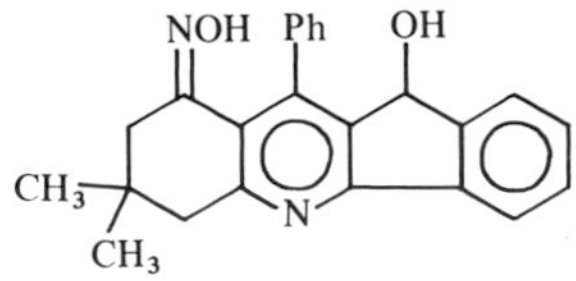			m.p. 261°
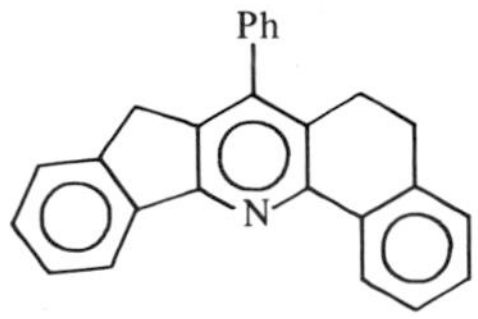	II.1.C	79%	m.p. 158°

NMR	IR	UV	Derivative (m.p.)	Reference
				224
				59
				295
	1548, 1494			232
	1603, 1689, 1558 1726			60
	1492, 1556, 1655 3384			232
				232
	1560, 1572			232
				312

TABLE II-21.   1,2,3,4,7,8,9,10-OCTAHYDROPHENANTHRIDINES

	Method of Preparation	Yield	b.p./m.p. (°C)
	II.1.C	54%	b.p. 110° (0.5 mm) m.p. 37–38°
	III.2.A		m.p. 140–141°
	III.2.A		m.p. 147.5–148.5°
	III.2.A	50%	m.p. 217.5–219.5°
	III.2.B	28%	m.p. 56–57°
	III.2.B	23%	m.p. 69–70.5°
	II.1.B		b.p. 172–176° (0.1 mm) m.p. 43–45°
	II.1.C	90%	m.p. 57°
	II.2.E	20%	b.p. 160–164° (3 mm)

NMR	IR	UV	Derivative (m.p.)	Reference
06(s,1), 85–2.50(8) 79(m,8)	2940, 2870, 1592 1470, 1415, 1325 1250, 932, 830 730	279(3064) 276(3321) 271(3423) 267(3013)	Picrate (181°)	37 34, 33
	3220			34
	3550			34
	3280, 3550(w)			34
				218
				218
			HCl (188–189°)	64
4–7.8(m,8) 8–7.6(s,3) 6–7.3(m,6) 3–6.9(m,2)	1590		Picrate 132°	64 33 38 41, 42
13(t,2) 7(m,8), 2.25(m,8) 6(q,2)	1580	218, 225, 230 275	Picrate (160.5–161.5°)	296 38 41, 42

TABLE II-21.　　(*CONTINUED*)

	Method of Preparation	Yield	b.p./m.p. (°C)
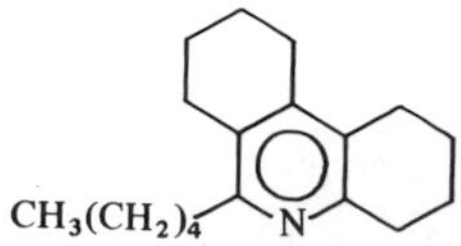 $CH_3(CH_2)_4$ structure	II.1.C		b.p. 174–175° (3 mm)
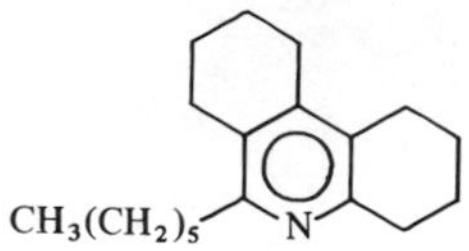 $CH_3(CH_2)_5$ structure	II.1.C		b.p. 190° (2.6 mm)
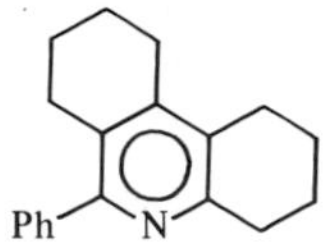 Ph structure	II.1.C	42%	m.p. 81–83%
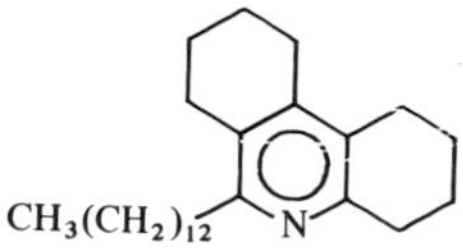 $CH_3(CH_2)_{12}$ structure	II.1.C		m.p. 43–44°
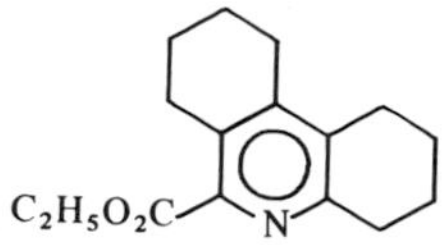 $C_2H_5O_2C$ structure	II.3.A.b	41%	

NMR	IR	UV	Derivative (m.p.)	Reference
9.3–8.9(*t*,3) 8.9–8.5(*m*,4) 8.5–7.8(*m*,10) 7.8–7.2(*m*,8) 7.2–6.9(*m*,2)	1575		Picrate (129–131°)	38 41, 42
9.4–8.9(*t*,3) 8.9–8.5(*m*,6) 8.5–7.8(*m*,10) 7.8–6.9(*m*,10)	1585		Picrate (134.5–136°)	38 41, 42
8.5–7.8(*m*,8) 7.8–7.2(*t*,6) 7.2–6.9(*m*,2) 2.8–2.0(*m*,5)	1570		Picrate (158–159°)	38 41, 42
	1578		Picrate (177–179°)	38 41, 42
2.5(*br*) 2.8(*br*) 4.7(*br*)				134

TABLE II-22.   1,2,3,4,5,6,7,8-OCTAHYDROACRIDINES

	Method of Preparation	Yield	b.p./m.p. (°C)
	II.1.A	95%	m.p. 69°
	II.1.B	26%	b.p. 175° (17 mm)
	II.7 II.1.B		b.p. 160–161° (4 mm)
	II.1.B	35%	m.p. 96–97°
	↲II.1.A	80%	m.p. 167°
	II.1.A	75%	m.p. 146°
		77%	
	III.2.A	91%	m.p. 127.5–128.5°
		90%	m.p. 156.5–158°
		55%	m.p. 147–149°
		58%	m.p. 143–144°
		68%	b.p. 86–87° (0.01 mm)

NMR	IR	UV	Derivative (m.p.)	Reference
02(s,1), 2.85(t,4)	2930, 2860, 1660	291(3980)	Picrate	37, 11, 19
70(t,4), 1.85(m,8)	1640, 1607, 1450,	286(5073)	(199–200°)	161, 13
	1250, 985, 936,	281(5520)		174, 15
	820, 710	277(4625)		18, 159
				297
			Picrate	174, 298
			(155°)	23, 175
				35
			HCl	64
			(184–185°)	
				15, 155
				15
				211
	1760		Picrate	211
	1260		(180–180.5°)	
			HCl	
			(179–179.5°)	
	3600–3500	276(4.33)	Picrate	211
			(135–136°)	212
			HCl	
			(202–203°)	
	700, 675			211
	1720, 3340			211
	3060, 3030		Picrate	211
			(m.p. 198°)	212
			HCl	
			(m.p. 202–203°)	

TABLE II-22. (*CONTINUED*)

	Method of Preparation	Yield	b.p./m.p. (°C)
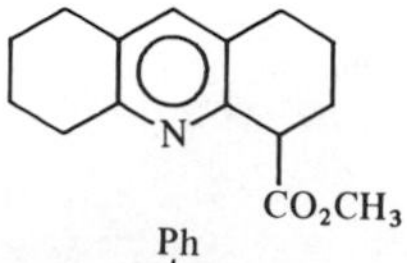	III.2.C	15%	
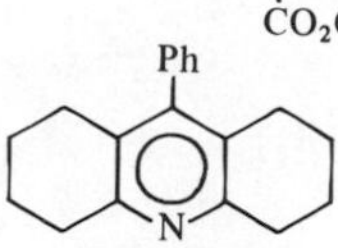	II.7		m.p. 176–177°
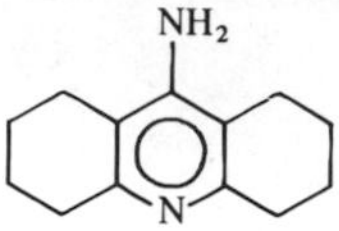		94%	m.p. 218–219° (dec)
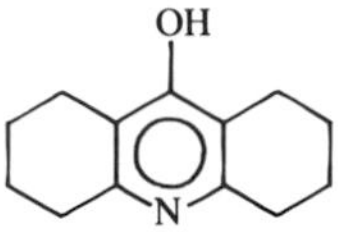		94%	m.p. 399–400° (dec)
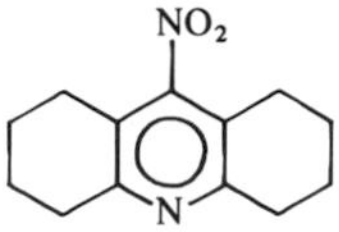		88%	m.p. 134–135°
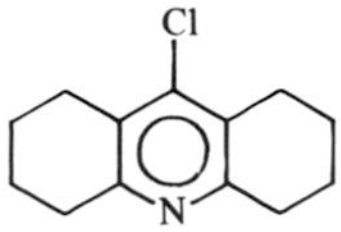		80%	m.p. 41–42°
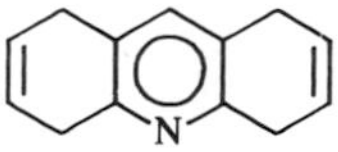	II.5.B	60%	m.p. 130.5°
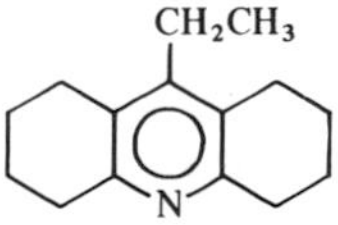	II.4	92%	b.p. 174–175° (4 mm)
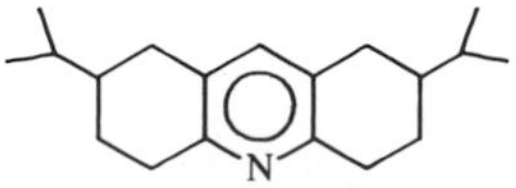	II.2.G	4.9%	m.p. 100–103°
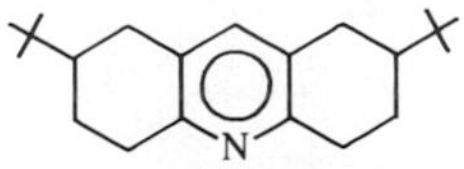	II.2.G	2.7%	m.p. 123–125°
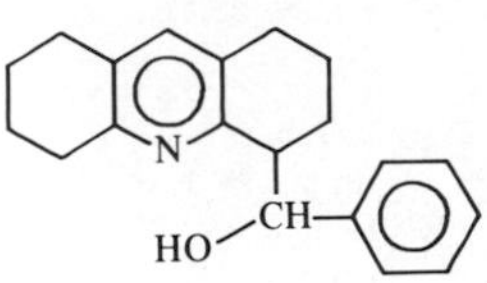	III.2.C	49%	m.p. 147–148°

NMR	IR	UV	Derivative (m.p.)	Reference
				194
			Picrate (175–176°)	174
	3480, 3300	249(2.92)	Picrate (189–190°) HCl (239.5–240.5°)	214
	1640, 3280, 3170	268(3.18)	Picrate (206–207°)	214
	1580, 1420			214
			Picrate (147–148.5°) HCl (230–231°)	214
.11(1), 5.9(q,4) .45 and 3.40	1662, 1710, 1601 1573, 888	283		164
			Picrate (192–193°)	148
				155
				155
	3340 3650 (dilute)		Picrate (161–162°)	219 220

TABLE II-22. (*CONTINUED*)

	Method of Preparation	Yield	b.p./m.p. (°C)
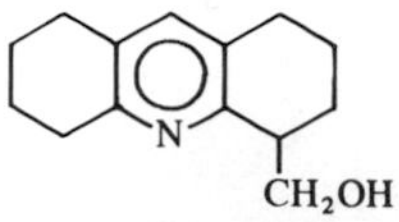 Structure	III.2.C	50%	m.p. 92–93°
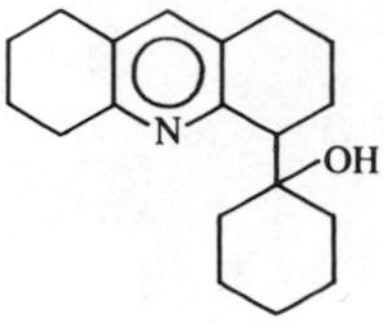 Structure	III.2.C	62%	m.p. 88–89°
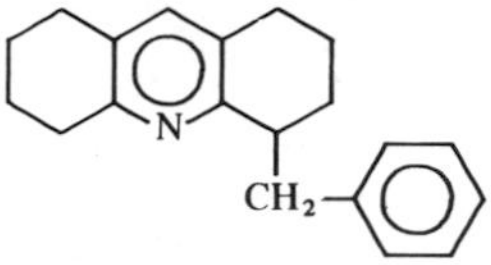 Structure	III.2.C		b.p. 142–146° (0.03 mm)
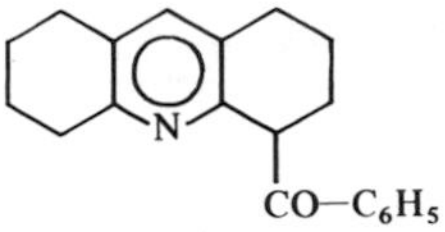 Structure	III.2.C	36%	m.p. 81.5–83°
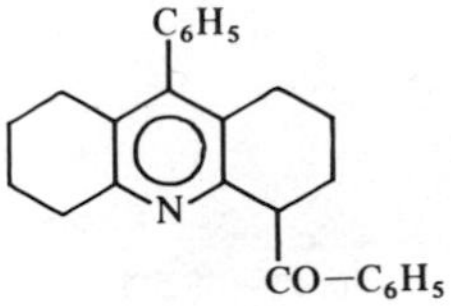 Structure	III.2.C	38%	m.p. 130–132°
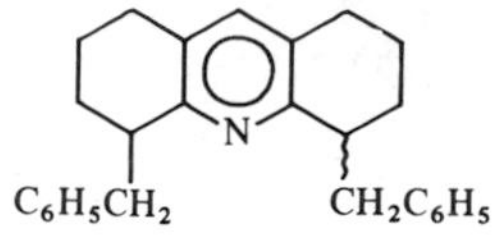 Structure	III.2.C	31% each isomer	m.p. 99–101° m.p. 79–81°
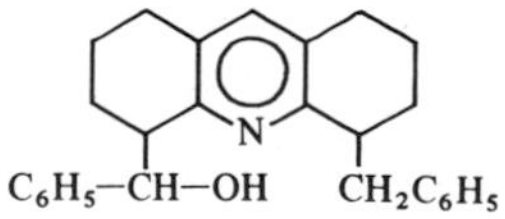 Structure	III.2.C	42%	
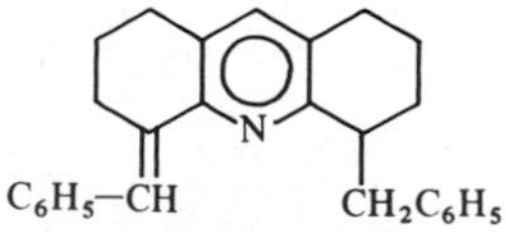 Structure		87%	m.p. 110–111°
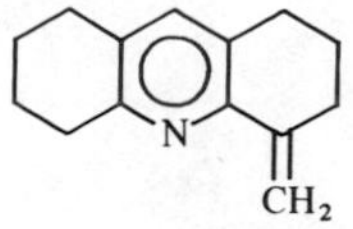 Structure		40%	b.p. 138–142° (1 mm)
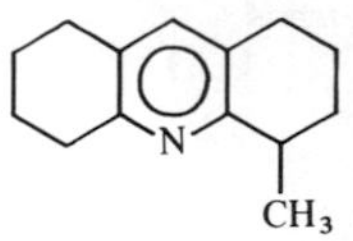 Structure		84%	m.p. 26–28°

NMR	IR	UV	Derivative (m.p.)	Reference
	3350		Picrate (132–133°)	219 220
	3350		Picrate (151–152°)	219 220
			Picrate (172–173°)	220
				222
				222
				221
				221
				221
	3080, 900, 1630		Picrate (157.5–159°)	299
8(d,3)			Picrate (144–145°)	299

TABLE II-22.   (*CONTINUED*)

	Method of Preparation	Yield	b.p./m.p. (°C)
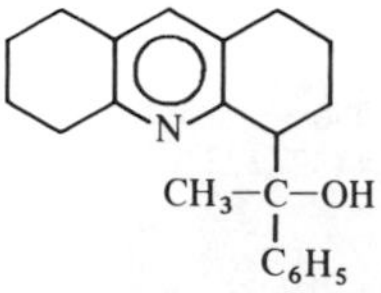	III.2.C	30%	m.p. 134–135°
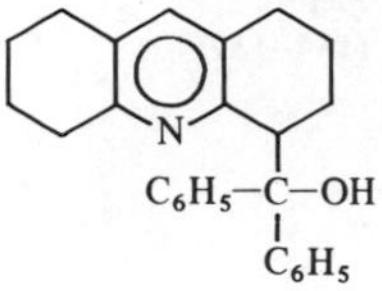	III.2.C	41%	m.p. 159–160°
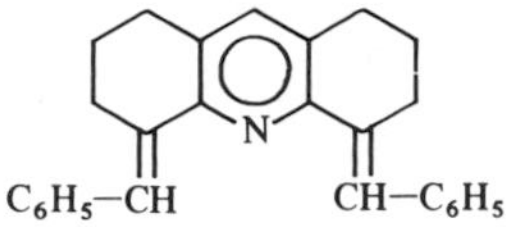	III.2.B	81.4%	m.p. 182–184°
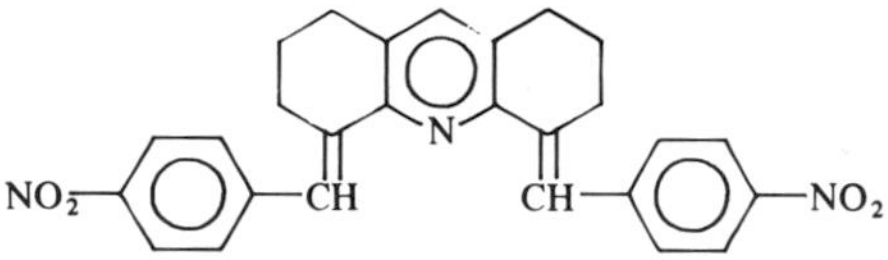	III.2.B	85%	m.p. 253°
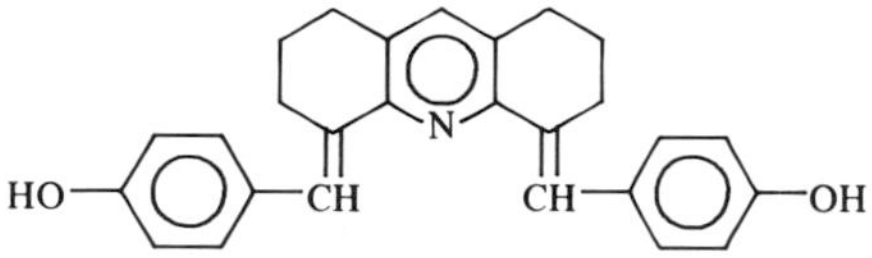	III.2.B	22%	m.p. 243°
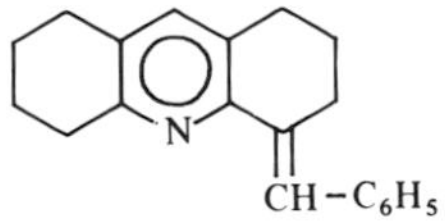	III.2.B	87%	m.p. 80–81°
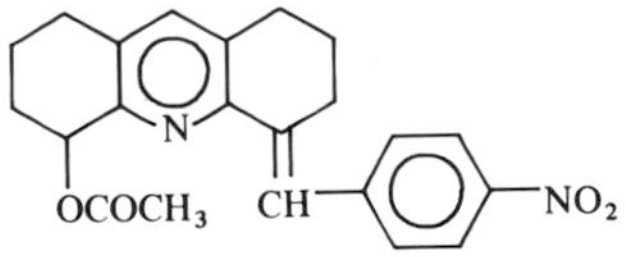	III.2.A	50%	m.p. 202–203°
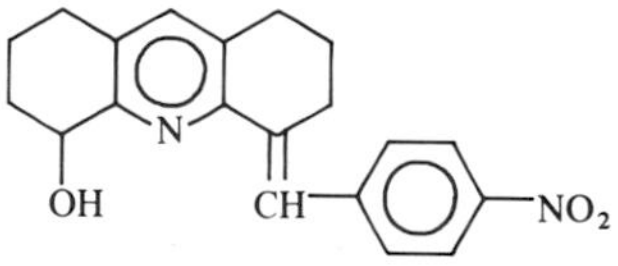		73%	m.p. 174–175°
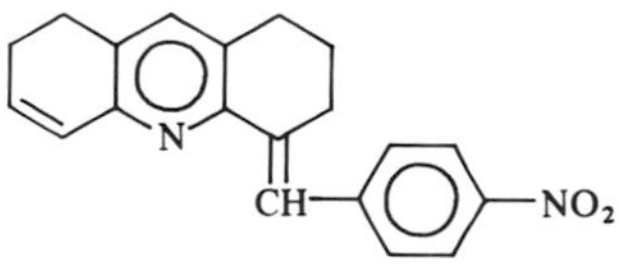		68%	

NMR	IR	UV	Derivative (m.p.)	Reference
	3250		Picrate (161°)	219
	3250			219
			Picrate (194–196°)	215 219
				216, 219
				216
			HCl (137°)	215
	1260, 1760, 1540 1360			213
	3530, 1360, 1540		Picrate (182.5–183°) HCl (189.5–190.5°)	213
			Picrate (176–177°) HCl (118–120°)	213

TABLE II-22.   (CONTINUED)

	Method of Preparation	Yield	b.p./m.p. (°C)
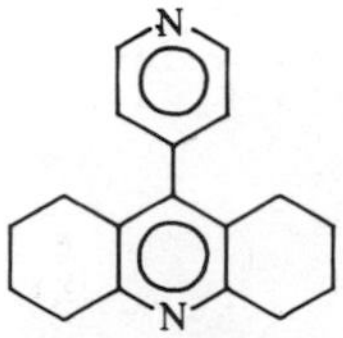	II.4	38%	m.p. 147–148°
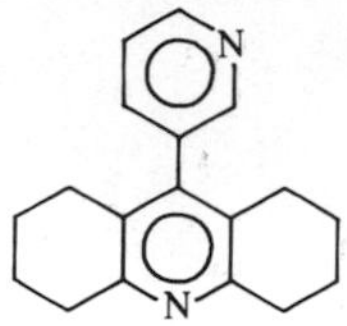	II.4	57%	m.p. 140–141°
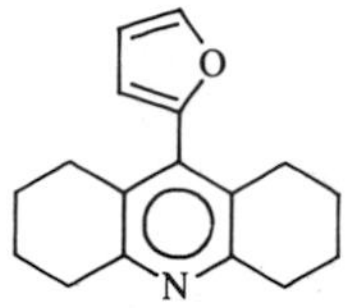	II.4		
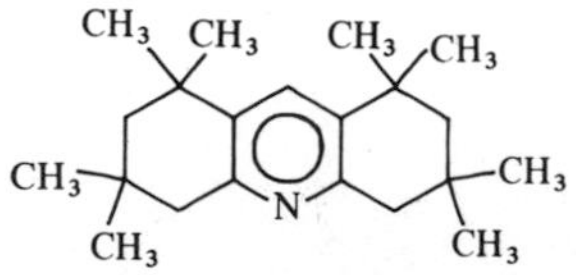	II.2.F	40%	m.p. 114–115°
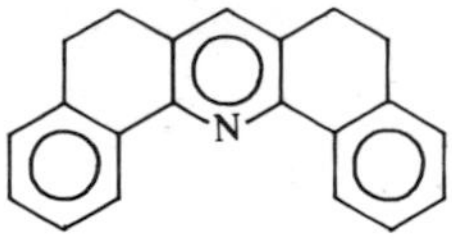	II.2.F	56%	m.p. 162–163°
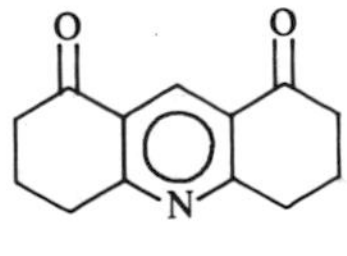	II.B	21%	m.p. 144°
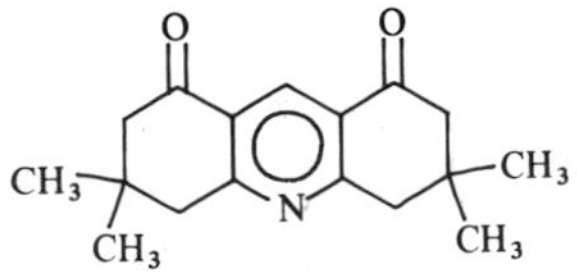	II.B	41%	m.p. 148°
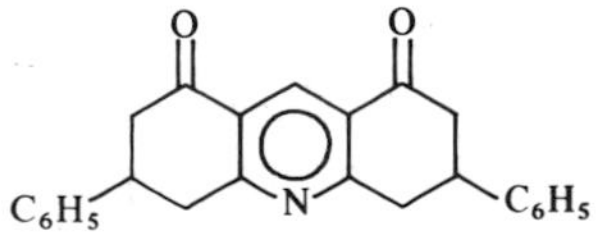	II.B	41%	m.p. 173°

NMR	IR	UV	Derivative (m.p.)	Reference
.62(2), 6.99(2) .75(8), 2.25(4) .83(4)			Picrate (253–254°)	151
.52(*m*,1) .31(*m*,1), 1.75(8) .25(4), 2.81(4)			Picrate (130°)	151
.53(1), 6.51(1) .33(1), 1.79(8) .57(4), 2.93(4)				151
.42(*s*,1), 2.58(*s*,4) .55(*s*,4), 1.28(*s*,12) .02(*s*,12)		215(6350) 271(6190) 276(7500) 280(7100) 288(6170)		113
				113
	2940, 1688 1583, 1418			43
.4(*s*,1) .5(*s*,4) .7(*s*,4) .1(*s*,12)	2940, 1685 1580, 1459 1417			43
.2(*s*,1) .6(*s*,10) .2(*m*,6) .6(*m*,4)	1705, 1590 1550, 1420 1226			43

TABLE II-22.   (*CONTINUED*)

	Method of Preparation	Yield	b.p./m.p. (°C)
	II.B	41%	m.p. 138°
	II.B	78%	
		61%	m.p. 174°
	II.1.C	95%	m.p. 151–152°C

NMR	IR	UV	Derivative (m.p.)	Reference
3.76(s,1) 3.15(m,4) 2.63(s,4) 1.50(m,20)	2920, 1685, 1590			43
	1605, 1580	220(3.83) 278(3.93)		43
3.38(s,1), 8.29(s,1) 3.09(s,2), 3.00(s,2) 2.65(s,2), 1.24(s,6) .06(s,6)		338(3.46) 246(sh,2.93)		43
				312

TABLE II-23.   ANNELATED CYCLOHEPTAPYRIDINES

	Method of Preparation	Yield	b.p./m.p. (°C)
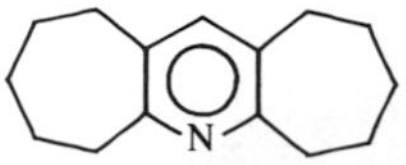	II.1.A	50%	m.p. 112°
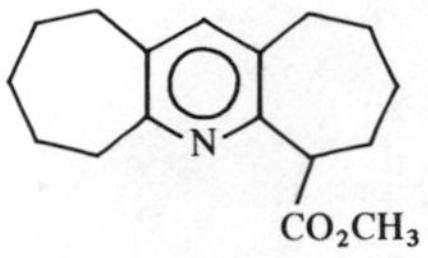	III.2.C	17%	
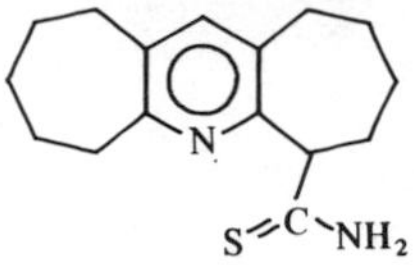		50%	
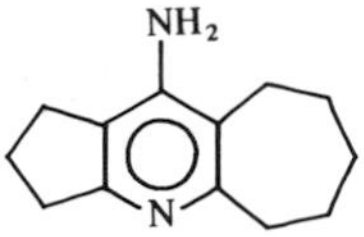	II.4		
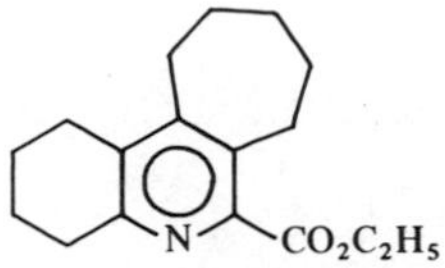	II.3.A.b	43%	
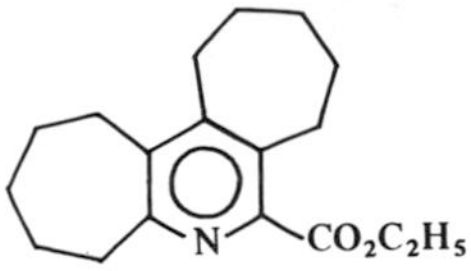	II.3.A.b	40%	
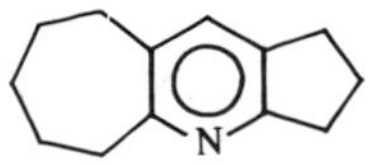	II.1.C	72.5%	b.p. 162–163° (17 mm)

NMR	IR	UV	Derivative (m.p.)	Reference
				15, 155
				194
			HCl (80–85°)	194
				153
nd 1.7				134
nd 1.8				134
				317

REFERENCES

1. H. Beschke, *Aldrichimica Acta,* **11,** 13 (1978).

2. R. W. Hoffman, "Dehydrobenzene and Cycloalkynes," Academic Press, New York, 1967, p. 275.

3. H. Pines and W. M. Stalick, "Base-Catalyzed Reactions of Hydrocarbons and Related Compounds," Academic Press, New York, 1977, p. 377.

4. R. G. Micetich, "Alkylpyridines and Arylpyridines," in R. A. Abramovitch, ed., *Pyridine and Its Derivatives, Supplement, Part Two,* Wiley, New York, 1974, p. 263.

5. See "Quinolines" and parts 1–4 of "Isoquinolines" in A. Weissberger and E. C. Taylor, eds., *The Chemistry of Heterocyclic Compounds,* Wiley, New York, 19XX.

6. F. Freeman, "The Chemistry of 1-Pyrindines," in A. R. Katritzky and A. J. Boulton, eds., *Advances in Heterocyclic Chemistry, Vol. 15,* Academic Press, 1973, p. 187.

7. (a) W. E. Billups, *Acc. Chem. Res.,* **11,** 245 (1978); (b) B. Halton, *Chem. Rev.,* **73,** 113 (1973).

8. C. Striegler, *J. Prakt. Chem.,* **86,** 241 (1912).

9. (a) E. Knoevenagel, *Ann. Chem.,* **281,** 33 (1894); (b) E. Knoevenagel, *Chem. Ber.,* **36,** 2180 (1903).

10. (a) H. Stobbe, *Chem. Ber.,* **35,** 3978 (1902); (b) H. Stobbe and H. Volland, ibid, **35,** 3973 (1902).

11. N. S. Gill, K. B. James, F. Lions, and K. T. Potts, *J. Am. Chem. Soc.,* **74,** 4923 (1952).

12. J. Epsztajn, W. E. Hahn, and B. K. Tosik, *Rocz. Chem.,* **44,** 431 (1970).

13. A. Risaliti and U. DeMartino, *Ann. Chim. (Rome),* **53,** 819 (1963).

14. J. Colonge, J. Dreux, and H. Delplace, *Bull. Soc. Chim. Fr.,* **1956,** 1635.

15. J. Colonge, J. Dreux, and H. Delplace, *Bull. Soc. Chim. Fr.,* **1957,** 447.

16. J. Colonge, J. Dreux, and M. Thiers, *Bull. Soc. Chim. Fr.,* **1959,** 1461.

17. G. Jones, R. K. Jones, and M. J. Robinson, *J. Chem. Soc., Perkin I,* **1973,** 968.

18. A. F. McKay, C. Podesva, E. J. Tarlton, and J. M. Billy, *Can. J. Chem.,* **42,** 10 (1964).

19. M. V. Moskovkina, V. A. Kaminskii, V. I. Vysotskii, and M. N. Tilichenko, *Khim. Geterotsik. Soedin.,* **1973,** 826; English translation: *Chemistry of Heterocyclic Compounds,* **1973,** 759.

20. V. A. Kaminskii, A. N. Saverchenko, and M. N. Tilichenko, *Zh. Org. Khim.,* **6,** 404 (1970); English translation: *J. Org. Chem. USSR,* **6,** 397 (1970).

21. T. I. Akimora, M. N. Tilichenko, V. V. Isakov, and T. A. Budina, *Zh. Org. Khim.,* **13,** 2095 (1977); English translation: *J. Org. Chem. USSR,* **13,** 1948 (1977).

22. G. E. Hall and J. Walker, *J. Chem. Soc.,* **1968,** 2237.

23. M. N. Tilichenko, M. A. Abramova, and M. E. Egorova, *Izvest. Vysshikh Ucheb. Zavedenii Khim. i Khim. Tekhnol.,* **3,** 130 (1960) [*Chem. Abstr.,* **54,** 17397 (1960)].

24. K. J. Clark, G. E. Fray, R. H. Jaeger, and R. Robinson, *Tetrahedron,* **6,** 217 (1959).

25. G. W. K. Cavill and D. L. Ford, *Aust. J. Chem.,* **13,** 296 (1960).

26. G. W. K. Cavill, D. L. Ford, and D. H. Solomon, *Aust. J. Chem.,* **13,** 469 (1960).

27. R. P. Thummel and D. K. Kohli, unpublished results.

28. E. Ya. Ozola and G. Ya. Vanag, *Khim. Geterotsik. Soedin.,* **1969,** 103; English translation: *Chemistry of Heterocyclic Compounds,* **1969,** 82.

29. G. Duburs and G. Vanag, *Lativijas PSR Zinatnu Acad. Vestis, Khim. Ser.,* **1962,** 119 [*Chem. Abstr.,* **59,** 5128 (1963)].

30. A. A. Usol'tsev, E. S. Karaulov, and M. N. Tilichenko, *Zh. Org. Khim.,* **13,** 84 (1977); English translation: *J. Org. Chem. USSR,* **13,** 77 (1977).

31. K. Hartke and R. Manusch, *Chem. Ber.,* **105,** 2584 (1972).

32. H. Junek and R. J. Schaur, *Monatsh. Chem.,* **99,** 89 (1968).

33. A. E. Chichibabin, *Bull. Soc. Chim. Fr.,* **1939,** 522.

34. W. E. Hahn and W. Koziolkiewicz, *Soc. Sci. Lodz., Acta Chim.,* **15,** 61 (1970).

35. A. Chichibabin and C. Barkovsky, *Comptes Rendus,* **212,** 914 (1941).

36. H. Chafetz and R. C. Anderson, *U.S. Pat.* 3,349,092 (1964) [*Chem. Abstr.,* **68,** 105026 (1968)].

37. R. P. Thummel and D. K. Kohli, *J. Org. Chem.,* **42,** 2742 (1977).

38. E. L. Patmore and H. Chafetz, *Am. Chem. Soc., Div. Petrol. Chem. Prepr.,* **17,** B27 (1972) [*Chem. Abstr.,* **80,** 3368p (1974)].

39. A. V. Upadysheva, E. P. Usova, I. A. Titova, and A. P. Zramenskaya, *Zh. Prikl. Khim.,* **44,** 1127 (1971); English translation: **44,** 1133 (1971).

40. N. Barbulescu, *Analele Univ., "C.I. Parhon" Bucaresti-Ser. Stiint nat.,* **13,** 101 (1956) [*Chem. Abstr.,* **53,** 1178 (1959)].

41. H. Chafetz and R. C. Anderson, *U.S. Pat.* 3,336,313 (1967) [*Chem. Abstr.,* **68,** 29621t (1968)].

42. H. Chafetz and E. L. Patmore, *U.S. Pat.* 3,408,351 (1968) [*Chem. Abstr.,* **70,** 19950b (1969)].

43. O. S. Wolfbeis and H. Junek, *Z. Naturforsch., B: Anorg. Chem. Org. Chem.,* **30B,** 249 (1975).

44. H. Antaki, *J. Chem. Soc.,* **1963,** 4877.

45. D. Gold and S.-S. V. Roessler, *Belg. Pat.* 858,391 (1978) [*Chem. Abstr.,* **89,** 163427w (1978)].

46. E. Breitmaier and E. Bayer, *Angew. Chem. Internat. Ed. Engl.,* **8,** 765 (1969).

47. J. H. Markgraf and W. L. Scott, *J. Chem. Soc., Chem. Commun.,* **1967,** 296.

48. J. H. Markgraf, R. J. Katt, W. L. Scott, and R. N. Shefrin, *J. Org. Chem.,* **34,** 4131 (1969).

49. R. P. Thummel and D. K. Kohli, *J. Heterocycl. Chem.,* **14,** 685 (1977).

50. P. Caluwe and T. G. Majewicz, *J. Org. Chem.,* **40,** 2566 (1975).

51. T. G. Majewicz and P. Caluwe, *J. Org. Chem.,* **41,** 1058 (1976).

52. T. Nozoe and K. Kikuchi, *Chem. Ind. (London),* **1962,** 358.

53. G. Stork, M. Ohaski, H. Kamachi, and H. Kakisawa, *J. Org. Chem.,* **36,** 2784 (1971).

54. W. Dammertz and E. Reimann, *Arch. Pharm.,* **310,** 172 (1977).

55. F. Bohlmann and R. Mayer-Mader, *Tetrahedron Lett.,* **1965,** 171.

56. F. Zymalkowski and M. Kothari, *Arch. Pharm.,* **303,** 667 (1970).

57. E. Reimann and R. Reitz, *Ann. Chem.,* **1976,** 610.

58. M. J. Vitola and V. E. Marquez, *J. Org. Chem.,* **42,** 2187 (1977).

59. E. I. Stankevich and G. Ya. Vanags, *Dokl. Akad. Nauk SSR,* **140,** 607 (1961); English translation: 934 (1961).

60. E. I. Stankevich and G. Ya. Yanag, *Zh. Obshch. Khim.,* **32,** 1146 (1962); English translation: *J. Gen. Chem. USSR,* **1962,** 1123.

61. V. Petrow, J. Saper, and B. Sturgeon, *J. Chem. Soc.,* **1949,** 2134.

62. E. Breitmaier and E. Bayer, *Tetrahedron Lett.,* **1970,** 3291.

63. A. Dornow and E. Neuse, *Arch. Pharm.,* **287,** 361 (1954).

64. C. Ruangsiyanand, H. J. Rimek, and F. Zymalkowski, *Chem. Ber.,* **103,** 2403 (1970).

65. G. Bouchon, K.-H. Spohn, and E. Breitmaier, *Chem. Ber.,* **106,** 1736 (1973).

66. D. K. Kohli, Ph.D. Dissertation, University of Houston, 1978, p. 49.

438 R. P. Thummel

67. A. C. W. Curran, *J. Chem. Soc., Perkin I*, **1976**, 975.
68. J. Tatsugi and M. Okumura, *Bull. Chem. Soc. Japan*, **51**, 1227 (1978).
69. W. A. Mosher, T. El-Zimaity, and D. W. Lipp, *J. Org. Chem.*, **36**, 3890 (1971).
70. W. A. Mosher and D. W. Lipp, *J. Org. Chem.*, **37**, 3190 (1972).
71. L. Neilands and G. Vanags, *Latvijas PSR Zinatnu Akad. Vestis, Kim. Ser.*, **1963**, 74 [*Chem. Abstr.*, **60**, 4102 (1964)].
72. G. Jones and R. K. Jones, *J. Chem. Soc., Perkin I*, **1973**, 26.
73. W. E. Hahn and C. Korzeniewski, *Soc. Sci. Lodz., Acta Chim.*, **17**, 193 (1972).
74. D. Binder, *Monatsh. Chem.*, **105**, 196 (1974).
75. C. Fugier, G. Quéguiner, and P. Pastour, *C. R. Acad. Sci. Paris*, **269**, 925 (1969).
76. G. Quéguiner, C. Fugier, and P. Pastour, *C. R. Acad. Sci. Paris*, **270**, 551 (1970).
77. J. J. Eisch and D. A. Russo, *J. Organometal. Chem.*, **14**, P13 (1968).
78. J. J. Eisch, H. Gopal, and D. A. Russo, *J. Org. Chem.*, **39**, 3110 (1974).
79. H. Pines and S. V. Kannan, *J. Chem. Soc., Chem. Commun.*, **1969**, 1360.
80. S. V. Kannan and H. Pines, *J. Org. Chem.*, **36**, 2304 (1971).
81. H. Pines, S. V. Kannan, and W. M. Stalick, *J. Org. Chem.*, **36**, 2308 (1971).
82. V. B. Reichert and A. Lechner, *Arzneim. Forsch.*, **15**, 36 (1965).
83. O. Périn-Roussel and P. Jacquignon, *C. R. Acad. Sci. Paris*, **278**, 279 (1974).
84. F. Ramirez and A. P. Paul, *J. Am. Chem. Soc.*, 77, 1035 (1955).
85. N. S. Prostakov, L. M. Kirillova, D. Phal'gumani, L. A. Shakhparonova, and V. P. Zvolinskii, *Khim. Geterotsik. Soedin.*, **1967**, 1068; English translation: *Chemistry of Heterocyclic Compounds*, **1967**, 832.
86. N. S. Prostakov, O. I. Sorokin, and A. Ya. Ismailov, *Khim. Geterotsik. Soedin.*, **1967**, 674; English translation: *Chemistry of Heterocyclic Compounds*, **1967**, 541.
87. A. C. Core and E. R. Kirch, *J. Pharm. Sci.*, **52**, 478 (1963).
88. F. J. Villani, P. J. L. Daniels, C. A. Ellis, T. A. Mann, and K. C. Wang, *J. Heterocycl. Chem.*, **8**, 73 (1971).
89. F. J. Villani, C. A. Ellis, M. D. Yudis, and J. B. Morton, *J. Org. Chem.*, **36**, 1709 (1971).
90. C. van der Stelt, P. S. Hofman, A. B. H. Funcke, and W. Th. Nauta, *Arzneim. Forsch.*, **18**, 756 (1968).
91. C. van der Stelt, P. S. Hofman, A. B. H. Funcke, and H. Timmerman, *Arzneim. Forsch.*, **22**, 133 (1972).
92. F. J. Villani, *U.S. Pat.* 3,357,986 (1967) [*Chem. Abstr.*, **69**, 27262y (1968)].
93. F. J. Villani, *U.S. Pat.* 3,366,635 (1968) [*Chem. Abstr.*, **69**, 10372m (1968)].
94. H. L. Lochte and A. G. Pitent, *J. Am. Chem. Soc.*, **82**, 469 (1960).
95. H. Suzumura, *Nippon Kagakukai, Bull.*, **34**, 1097 (1960).
96. H. L. Lochte and A. G. Pittman, *J. Org. Chem.*, **25**, 1462 (1960).
97. N. P. Shusherina, R. Ya. Levina, and K. Khua-min, *Zh. Obshch. Khim.*, **31**, 3477 (1961); English translation: *J. Gen. Chem. USSR*, **1961**, 3239.
98. N. P. Shusherina, K. Khua-min, and R. Ya. Levina, *Zh. Obshch. Khim.*, **33**, 3613 (1963); English translation: *J. Gen. Chem. USSR*, **1963**, 3545.
99. G. A. R. Kon and H. R. Nanji, *J. Chem. Soc.*, **1932**, 2426.
100. N. Nagano and M. Murakami, *Japan. Pat.*, 74 66,685 (1974) [*Chem. Abstr.*, **82**, 72800b (1975)].
101. T. Sakan, A. Fujino, F. Murai, A. Suzui, and Y. Butsugan, *Bull. Chem. Soc. Japan*, **33**, 712 (1960).
102. G. Simchen, *Chem. Ber.*, **103**, 389 (1970).

103. H. E. Schroeder and G. W. Rigby, *J. Am. Chem. Soc.*, **71**, 2205 (1949).

104. H. Junek and H. Aigner, *Z. Naturforsch.*, **25**, 1423 (1970).

105. G. Duaphin, B. Jamilloux, A. Kergomard, and D. Planat, *Tetrahedron*, **33**, 1129 (1977).

106. C. Jutz, H. G. Löbergin, and K. H. Trinkl, *Synthesis*, **1977**, 326.

107. C. Jutz, R. M. Wagner, A. Kraatz, and H. G. Löbering, *Ann. Chem.*, **1975**, 874.

108. L. A. Paquette and T. Kakihana, *J. Am. Chem. Soc.*, **93**, 174 (171).

109. A. G. Anastassiou, S. J. Girgenti, R. C. Griffith, and E. Reichmanis, *J. Org. Chem.*, **42**, 2651 (1977).

110. A. G. Anastassiou, E. Reichmanis, and S. J. Girgenti, *J. Am. Chem. Soc.*, **99**, 7392 (1977).

111. A. G. Anderson, Jr. and H. L. Ammon, *Tetrahedron Lett.*, **1966**, 2579.

112. D. L. Fishel and G. R. Newkome, *J. Am. Chem. Soc.*, **88**, 3654 (1966).

113. G. R. Newkome and D. L. Fishel, *J. Heterocycl. Chem.*, **4**, 427 (1967).

114. W. D. Crow, A. N. Khan, and M. N. Paddon-Row, *Aust. J. Chem.*, **28**, 1741 (1975).

115. W. D. Crow and M. N. Paddon-Row, *Aust. J. Chem.*, **28**, 1755 (1975).

116. W. D. Crow, A. N. Khan, M. N. Paddon-Row, and D. S. Sutherland, *Aust. J. Chem.*, **28**, 1763 (1975).

117. J. M. Riemann and W. S. Trahanovsky, *Tetrahedron Lett.*, **1977**, 1863.

118. J. M. Riemann and W. S. Trahanovsky, *Tetrahedron Lett.*, **1977**, 1867.

119. R. P. Thummel and D. K. Kohli, *J. Org. Chem.*, **43**, 4882 (1978).

120. R. P. Thummel and D. K. Kohli, *Tetrahedron Lett.*, **1979**, 143.

121. J. A. H. MacBride, *J. Chem. Soc., Chem. Commun.*, **1974**, 359.

122. J. W. Barton and R. B. Walker, *Tetrahedron Lett.*, **1975**, 569.

123. N. S. Prostakov, C. J. Matthew, and E. N. Sedykh, *Khim. Geterotsik. Soedin.*, **1967**, 1072; English translation: *Chemistry of Heterocyclic Compounds*, **1967**, 835.

124. N. S. Protakov, S. S. Moiz, A. T. Soldatenkov, V. P. Zvolinskii, and G. I. Cherenkova, *Khim. Geterotsik. Soedin.*, **1971**, 1398; English translation: *Chemistry of Heterocyclic Compounds*, **1971**, 1305.

125. P. Jacquignon and O. Perin-Roussel, *Collect. Czech. Chem. Commun.*, **41**, 1208 (1978).

126. S. H. Ruetman, *Synthesis*, **1975**, 382.

127. D. R. Carter and F. P. Boer, *Acta Cryst.*, **B30**, 2762 (1974).

128. K. Yamane, *Nippon Kagaku Zasshi*, **80**, 534 (1959) [*Chem. Abstr.*, **55**, 4500 (1961)].

129. W. Dittmar, J. Sauer, and A. Steigel, *Tetrahedron Lett.*, **1969**, 5171.

130. A. Demoulin, H. Gorissen, A.-M. Hesbain-Frisque, and L. Ghosez, *J. Am. Chem. Soc.*, **97**, 4409 (1975).

131. B. A. Kazanskii, G. Ya. Kondrat'eva and Yu. S. Dol'skaya, *Zh. Org. Khim.*, **6**, 2203 (1970); English translation: *J. Org. Chem. USSR*, **1970**, 2212.

132. L. B. Davies, P. G. Sammes, and R. A. Watt, *J. Chem. Soc., Chem. Commun.*, **1977**, 663.

133. Y. Butsugan, S. Yoshida, M. Muto, and T. Bito, *Tetrahedron Lett.*, **1971**, 1129.

134. P. B. J. Driessen, D. S. B. Grace, H. Hogeveen, and H. Jorritsma, *Tetrahedron Lett.*, **1976**, 2263.

135. R. J. Martens and H. J. den Hertog, *Tetrahedron Lett.*, **1962**, 643.

136. R. J. Martens and H. J. den Hertog, *Rec. Trav. Chim. Pays-Bas*, **83**, 621 (1964).

137. D. J. Berry, B. J. Wakefield, and J. D. Cook, *J. Chem. Soc.*, **1971**, 1227.

138. T. Kauffmann and F. -P. Boettcher, *Chem. Ber.*, **95**, 949 (1962).

139. T. Kauffmann, J. Hansen, K. Udluft, and R. Wirthwein, *Angew. Chem., Internat. Ed. Engl.*, **3**, 650 (1964).

440 R. P. Thummel

140. J. Kramer and R. S. Berry, *J. Am. Chem. Soc.*, **94**, 8336 (1972).
141. M. Mallet, F. Marsais, G. Queguiner, and P. Pastour, *C. R. Acad. Sci. Paris*, **275**, 1439 (1972).
142. J. D. Cook and B. J. Wakefield, *Tetrahedron Lett.*, **1967**, 2535.
143. K. Winterfeld and K. Nonn, *Chem. Ber.*, **100**, 2274 (1967).
144. Yu. I. Chumakov and N. B. Bulgakova, *Ukr. Khim. Zh.*, **36**, 514 (1970); English translation: *Ukranian Journal of Chemistry*, **1970**, 87.
145. T. Kametani, Y. Ichikawa, T. Suzuki, and K. Fukumoto, *Heterocycles*, **2** 171 (1974).
146. A. Naiman and K. P. C. Vollhardt, *Angew. Chem. Internat. Ed. Engl.*, **16**, 708 (1977).
147. C. Kaneko, S. Hayashi, and M. Ishikawa, *Rep. Inst. Med. Dent. Eng.*, **4**, 149 (1970) [*Chem. Abstr.*, **75**, 110169t (1971)].
148. N. S. Barbulescu, G. Badita, and M. N. Tilicenco, *Zh. Obshch. Khim.*, **33**, 4027 (1963); English translation: *J. Gen. Chem. USSR*, **1963**, 3968.
149. A. N. Saverchenko, V. A. Kaminskii, and M. N. Tilichenko, *Khim. Geterotsik. Soedin.*, **1973**, 384; English translation: *Chemistry of Heterocyclic Compounds*, **1973**, 355.
150. V. I. Vysotskii and M. N. Tilichenko, *Khim. Geterotsik. Soedin.*, **1969**, 751; English translation: *Chemistry of Heterocyclic Compounds*, **1969**, 560.
151. N. Barbulescu, F. Potmischil, and G. Badita, *Chem. Ber.*, **104**, 787 (1971).
152. N. Barbulescu and L. Ivan, *An. Univ. Bucuresti, Ser. Stiint. Natur. Chim.*, **15**, 47 (1966) [*Chem. Abstr.*, **70**, 87130j (1969)].
153. A. V. Upadysheva, N. D. Grigor'eva, and A. P. Znamenskaya, *Khim. Geterotsik. Soedin.*, **1977**, 1549; English translation: *Chemistry of Heterocyclic Compounds*, **1977**, 1240.
154. W. L. Bencze and M. J. Allen, *J. Am. Chem. Soc.*, **81**, 4015 (1959).
155. R. S. Monson, D. N. Priest, and J. C. Ullrey, *Tetrahedron Lett.*, **1972**, 929.
156. J. v. Braun, W. Gmelin, and A. Schultheiss, *Chem. Ber.*, **56B**, 1338 (1923).
157. J. v. Braun, A. Petzold, and J. Seemann, *Chem. Ber.*, **55**, 3779 (1922).
158. S. Shiotani, K. Sakai, and K. Mitsuhashi, *Yakagaku Zasshi*, **87**, 547 (1967) [*Chem. Abstr.*, **67**, 54022k (1967)].
159. J. Z. Ginos, *J. Org. Chem.*, **40**, 1191 (1975).
160. F. W. Vierhapper and E. L. Eliel, *J. Am. Chem. Soc.*, **96**, 2256 (1974).
161. F. W. Vierhapper and E. L. Eliel, *J. Org. Chem.*, **40**, 2729 (1975).
162. F. W. Vierhapper and E. L. Eliel, *J. Org. Chem.*, **40**, 2734 (1975).
163. H. Adkins and H. L. Coonradt, *J. Am. Chem. Soc.*, **63**, 1563 (1941).
164. A. J. Birch and H. H. Mantsch, *Aust. J. Chem.*, **22**, 1103 (1969).
165. T. Kato, H. Yamanaka, and T. Shimizu, *Yakugaku Zasshi*, **93**, 73 (1973) [*Chem. Abstr.*, **78**, 97453w (1973)].
166. G. N. Dorofeenko, V. I. Dulenko, and L. V. Dulenko, *Zh. Obshch. Khim.*, **34**, 3116 (1964); English translation: *J. Gen. Chem. USSR*, **1964**, 3155.
167. Yu. A. Zhdanov, G. N. Dorofeenko, V. A. Palchkov, and G. P. Safaryan, *Dokl. Akad. Nauk SSSR*, **155**, 1115 (1964); English translation: **1964**, 367.
168. G. N. Dorofeenko and V. I. Dulenko, *Dokl. Akad. Nauk SSSR*, **157**, 361 (1964); English translation: **1964**, 689.
169. G. I. Zhungietu and É. M. Perepelitsa, *Zh. Obshch. Khim.*, **36**, 1868 (1956); English translation: *J. Gen. Chem. USSR*, **1966**, 1850.
170. G. P. Safaryan and G. N. Dorofeenko, *Khim. Geterotsik. Soedin.*, **1976**, 1323; English translation: *Chemistry of Heterocyclic Compounds*, **1976**, 1097.
171. G. N. Dorofeenko, Yu. A. Zhdanov, and L. H. Étmetchenko, *Khim. Geterotsik. Soedin.*, **1969**, 781; English translation: *Chemistry of Heterocyclic Compounds*, **1969**, 576.

172. G. N. Dorofeenko, J. A. Shdanow, G. I. Shungijetu, and S. W. Kriwun, *Tetrahedron*, **22**, 1821 (1966).

173. G. N. Dorofeenko, G. V. Lazur'evskii, and G. I. Zhurngietu, *Dokl. Akad. Nauk SSSR*, **161**, 355 (1965); English translation: **1965**, 265.

174. S. S. Yang, W. Y. Haung, and S. M. Chen. *Chemistry (Taipei)*, **1962**, 65 [*Chem. Abstr.*, **59**, 3893 (1963)].

175. A. T. Balaban and N. S. Barbulescu, *Rev. Roum. Chim.*, **11**, 109 (1966).

176. N. Barbulescu, G. Nicolae, and V. Niculaita, *An. Univ. Bucuresti, Chim.*, **20**, 37 (1971) [*Chem. Abstr.*, **79**, 66146q (1973)].

177. N. Barbulescu and G. Nicolae, *Rev. Chim. (Bucharest)*, **22**, 368 (1971) [*Chem. Abstr.*, **75**, 129638s (1971)].

178. V. A. Kaminskii and M. N. Tilichenko, *Khim. Geterotsik. Soedin.*, **1974**, 1434; English translation: *Chemistry of Heterocyclic Compounds*, **1974**, 1263.

179. Yu. I. Chumakov and N. B. Bulgakova, *Khim. Geterotsik. Soedin.*, **1971**, 1533; English translation: *Chemistry of Heterocyclic Compounds*, **1971**, 1427.

180. F. Freeman and T. I. Ito, *J. Org. Chem.*, **34**, 3670 (1969).

181. F. Freeman, D. K. Farquhar, and R. L. Walker, *J. Org. Chem.*, **33**, 3648 (1968).

182. J. L. van der Baan and F. Bickelhaupt, *J. Chem. Soc., Chem. Commun.*, **1968**, 1661.

183. J. L. van der Baan and F. Bickelhaupt, *J. Chem. Soc., Chem. Commun.*, **1970**, 326.

184. V. Prelog and O. Metzler, *Helv. Chim. Acta*, **29**, 1170 (1946).

185. T. R. Kasturi, V. K. Sharma, and A. Srinivasan, *Tetrahedron*, **29**, 4103 (1973).

186. T. R. Kasturi and V. K. Sharma, *Tetrahedron*, **31**, 527 (1975).

187. J. W. Ducker and M. J. Gunter, *Aust. J. Chem.*, **28**, 581 (1975).

188. J. L. van der Baan and F. Bickelhaupt, *Tetrahedron*, **30**, 2447 (1974).

189. M. M. Robinson, *J. Am. Chem. Soc.*, **80**, 6254 (1958).

190. W. E. Hahn and J. Epsztajn, *Rocz. Chem.*, **37**, 403 (1963).

191. G. W. K. Cavill and A. Zeitlin, *Aust. J. Chem.*, **20**, 349 (1967).

192. John Wyeth and Brother, Ltd., *Brit. Pat.* 1,465,651 (1974).

193. D. E. Beattie, R. Crossley, A. C. W. Curran, G. T. Dixon, D. G. Hill, A. E. Lawrence, and R. G. Shepherd, *J. Med. Chem.*, **20**, 714 (1977).

194. R. Crossley, A. C. W. Curran, and D. G. Hill, *J. Chem. Soc., Perkin I*, **1976**, 977.

195. A. C. W. Curran and R. G. Shepherd, *J. Chem. Soc., Perkin I*, **1976**, 983.

196. John Wyeth and Brother, Ltd., *Brit. Pat.* 1,463,667 (1973).

197. John Wyeth and Brother, Ltd., *Brit. Pat.* 1,463,666 (1973).

198. John Wyeth and Brother, Ltd., *Brit. Pat.* 1,463,652 (1974).

199. John Wyeth and Brother, Ltd., *Brit. Pat.* 1,463,665 (1973).

200. John Wyeth and Brother, Ltd., *Brit. Pat.* 1,463,670 (1973).

201. John Wyeth and Brother, Ltd., *Brit. Pat.* 1,463,669 (1974) [*Chem. Abstr.*, **87**, 53108u (1977)].

202. John Wyeth and Brother, Ltd., *U.S. Pat.* 4,029,668 (1977).

203. John Wyeth and Brother, Ltd., *U.S. Pat.* 4,031,102 (1977).

204. John Wyeth and Brother, Ltd., *U.S. Pat.* 4,085,110 (1978).

205. W. E. Hahn and J. Epsztajn, *Rocz. Chem.*, **37**, 395 (1963).

206. W. E. Hahn and J. Epsztajn, *Rocz. Chem.*, **38**, 989 (1964).

207. J. Epsztajn, W. E. Hahn, and B. K. Tosik, *Rocz. Chem.*, **43**, 807 (1969).

208. E. Reimann and H. L. Ziegon, *Ann. Chem.*, **1976**, 1351.

209. J. Epsztajn, W. E. Hahn, and J. Z. Brezezinski, *Rocz. Chem.*, **49**, 123 (1975).

442 R. P. Thummel

210. W. E. Hahn, J. Epsztajn, B. Olejniczak, and S. Stasiak, *Rocz. Chem.*, **40**, 149 (1966) [*Chem. Abstr.*, **65**, 2213c (1966)].

211. G. A. Kimov, M. N. Tilichenko, and E. S. Karaulov, *Khim. Geterotsik. Soedin.*, **1969**, 297; English translation: *Chemistry of Heterocyclic Compounds*, **1969**, 226.

212. G. A. Klimov and M. N. Tilichenko, *Zh. Org. Khim.*, **2**, 1526 (1966); English translation: *J. Org. Chem. USSR*, **1966**, 1507.

213. I. V. Vigalok, I. E. Moisak, and N. V. Svetlakov, *Khim. Geterotsik. Soedin.*, **1969**, 175; English translation: *Chemistry of Heterocyclic Compounds*, **1969**, 133.

214. G. A. Klimov and M. N. Tilichenko, *Zh. Org. Khim.*, **36**, 1507 (1966); English translation: *J. Org. Chem. USSR*, **1966**, 1488.

215. M. N. Tilichenko and V. I. Vysotskii, *Zh. Obshch. Khim.*, **32**, 84 (1962); English translation: *J. Gen. Chem. USSR*, **1962**, 81.

216. V. I. Vysotskii and M. N. Tilichenko, *Khim. Geterotsik. Soedin.*, **1968**, 1080; English translation: *Chemistry of Heterocyclic Compounds*, **1968**, 785.

217. V. Baliah and R. Jeyaraman, *Indian J. Chem.*, **15B**, 798 (1977).

218. W. E. Hahn and W. Koziolkiewicz, *Lodz. Tow. Nauk. Wydz III, Acta Chim.*, **15**, 71 (1970) [*Chem. Abstr.*, **75**, 98419q (1971)].

219. V. A. Stonik, V. I. Vysotskii, and M. N. Tilichenko, *Khim. Geterotsik. Soedin.*, **1970**, 1542; English translation: *Chemistry of Heterocyclic Compounds*, **1970**, 1439.

220. M. N. Tilichenko, V. A. Stonik, and V. I. Vysotskii, *Khim. Geterotsik Soedin.*, **1968**, 570; English translation: *Chemistry of Heterocyclic Compounds*, **1968**, 422.

221. V. A. Stonik, V. I. Vysotskii, and M. N. Tilichenko, *Khim. Geterotsik. Soedin.*, **1972**, 673; English translation: *Chemistry of Heterocyclic Compounds*, **1972**, 611.

222. V. I. Vysotskii, V. A. Stonik, and M. N. Tilichenko, *Khim. Geterotsik. Soedin.*, **1972**, 984; English translation: *Chemistry of Heterocyclic Compounds*, **1972**, 895.

223. John Wyeth and Brother, Ltd., *Brit. Pat.* 1,463,668 (1977) [*Chem. Abstr.*, **87**, 68187t (1977)].

224. John Wyeth and Brother, Ltd., *U.S. Pat.* 4,000,142 (1976).

225. E. I. Stankevich and G. Ya. Vanag, *Khim. Geterotsik. Soedin.*, **1965**, 305; English translation: *Chemistry of Heterocyclic Compounds*, **1965**, 201.

226. T. Masamune, T. Saito, and G. Homma, *J. Fac. Sci., Hokkaido Univ., Ser. III*, **5**, 55 (1957) [*Chem. Abstr.*, **52**, 11062 (1958)].

227. T. Masamune and G. Homma, *J. Fac. Sci. Hokkaido Univ., Ser. III*, **5**, 64 (1957) [*Chem. Abstr.*, **52**, 14582 (1958)].

228. N. S. Protakov, K. M. S. Mokhomon, L. A. Gaivoronskaya, O. G. Kesarev, and A. A. Savina, *Khim. Geterotsik. Soedin.*, **1976**, 963; English translation: *Chemistry of Heterocyclic Compounds*, **1976**, 798.

229. N. S. Prostakov and A. Ya. Ismailov, *Dokl. Akad. Nauk Tadzh. SSR*, **15**, 29 (1972) [*Chem. Abstr.*, **77**, 164425h (1972)].

230. N. S. Prostakov, V. P. Shalimov, S. I. Manrikes, A. A. Savina, V. F. Zakharov, and V. P. Zvolinskii, *Khim. Geterotsik. Soedin.*, **1976**, 215; English translation: *Chemistry of Heterocyclic Compounds*, **1976**, 187.

231. N. S. Prostakov, M. E. Sintra, S. A. Soldatova, V. P. Shalimov, V. P. Zvolinskii, and A. A. Savina, *Khim. Geterotsik. Soedin.*, **1976**, 1231; English translation: *Chemistry of Heterocyclic Compounds*, **1976**, 1019.

232. E. I. Stankevich and G. Ya. Vanag, *Khim. Geterotsik. Soedin.*, **1965**, 750; English translation: *Chemistry of Heterocyclic Compounds*, **1965**, 508.

233. A. Schönberg and K. Junghans, *Chem. Ber.*, **95**, 2137 (1962).

234. A. Schönberg and K. Junghans, *Chem. Ber.*, **96**, 3328 (1963).

235. G. R. Newkome and J. M. Roper, *J. Org. Chem.,* **44**, 502 (1979).

236. F. J. Villani, P. J. L. Daniels, C. A. Ellis, T. A. Mann, K. -C. Wang, and E. A. Wefer, *J. Med. Chem.,* **15**, 750 (1972).

237. K. Yamane, *Nippon Kagaku Zasshi,* **80**, 1175 (1959).

238. K. Yamane, T. Matsumoto, and A. Yokoo, *Bull. Chem. Soc. Japan,* **36**, 1272 (1963).

239. J. F. Munshi and M. M. Joullié, *J. Heterocycl. Chem.,* **4**, 133 (1967).

240. T. Eguchi, *Bull. Chem. Soc. Japan,* **3**, 239 (1928) [*Chem. Abstr.,* **23**, 391 (1929)].

241. W. C. Thompson, *J. Am. Chem. Soc.,* **53**, 3160 (1931).

242. E. J. Poth, W. A. Schulze, W. A. King, W. C. Thompson, W. M. Slagle, W. W. Floyd, and J. R. Bailey, *J. Am. Chem. Soc.,* **52**, 1239 (1930).

243. W. C. Thompson and J. R. Bailey, *J. Am. Chem. Soc.,* **53**, 1002 (1931).

244. B. F. Armendt and J. R. Bailey, *J. Am. Chem. Soc.,* **55**, 4145 (1933).

245. R. W. Lackey and J. R. Bailey, *J. Am. Chem. Soc.,* **56**, 2741 (1934).

246. P. Arnall, *J. Chem. Soc.,* **1954**, 4040.

247. P. Arnall, *J. Chem. Soc.,* **1958**, 1702.

248. F. Runge, J. Freytag, and J. Kolbe, *Chem. Ber.,* **87**, 873 (1954).

249. Y. Hammouda and J. Le Men, *Bull. Soc. Chem. Fr.,* **1963**, 2901.

250. F. P. Guengerich, S. J. DiMari, and H. P. Broquist, *J. Am. Chem. Soc.,* **95**, 2055 (1973).

251. H. Auda, H. R. Juneja, E. J. Eisenbraun, G. R. Waller, W. R. Kayo, and H. H. Appel, *J. Am. Chem. Soc.,* **89**, 2476 (1967).

252. P. W. Jeffs, P. A. Luhan, A. T. McPhail, and N. H. Martin, *J. Chem. Soc., Chem. Commun.,* **1971**, 1466.

253. T. M. Capps, K. D. Hargrave, P. W. Jeffs, and A. T. McPhail, *J. Chem. Soc., Perkin II,* **1977**, 1098.

254. F. O. Snyckers, F. Strelow, and A. Wiechers, *J. Chem. Soc., Chem. Commun.,* **1971**, 1467.

255. P. A. Luhan and A. T. McPhail, *J. Chem. Soc., Perkin II,* **1972**, 2006.

256. K. V. Rao, K. Biemann, and R. B. Woodward, *J. Am. Chem. Soc.,* **85**, 2532 (1963).

257. A. J. Birch, D. N. Butler, and R. W. Rickards, *Tetrahedron Lett.,* **1964**, 1853.

258. G. P. Arsenault, *Tetrahedron Lett.,* **1965**, 4033.

259. D. E. Beattie, R. Crossley, A. C. W. Curran, D. G. Hill, and A. E. Lawrence, *J. Med. Chem.,* **20**, 718 (1977).

260. Tanbe Seiyaku Co., Ltd., *Japan. Pat.* 70 41,384 (1970) [*Chem. Abstr.,* **75**, 20215a (1971)].

261. Tanabe Seiyaku Co., Ltd., *Japan. Pat.* 72 00,811 (1972) [*Chem. Abstr.,* **76**, 140574j (1972)].

262. Scherico Ltd., *Belg. Pat.* 647,043 (1964) [*Chem. Abstr.,* **63**, 14829 (1965)].

263. L. E. Kholodov, N. M. Merzlyakova, E. A. Rudzit, and D. A. Kulikova, *Khim. Farm. Zh.,* **9**, 19 (1975) [*Chem. Abstr.,* **82**, 170630r (1975)].

264. D. Libermann, M. Moyeux, A. Rouaix, N. Rist, and F. Grumbach, *Compt. Rend.,* **244**, 402 (1957) [*Chem. Abstr.,* **51**, 6873 (1957)].

265. D. Libermann, N. Rist, F. Grumbach and S. Cals, *Bull. Soc. Chim. Biol.,* **39**, 1195 (1957) [*Chem. Abstr.,* **53**, 5496 (1959)].

266. Farbenfabriken Bayer A. G., *U.S. Pat.* 3,084,165 (1963) [*Chem. Abstr.,* **59**, 13956 (1963)].

267. R. J. Chorvat and R. Pappo, *Tetrahedron Lett.,* **1975**, 623.

268. Taisho Pharmaceutical Co., Ltd., *Japan. Pat.* 71 32,185 (1971) [*Chem. Abstr.,* **75**, 140701c (1971)].

444 R. P. Thummel

269. Taisho Pharmaceutical Co., Ltd., *Japan. Pat.* 71 31,861 (1971) [*Chem. Abstr.*, **75**, 140714j (1971)].

270. Pfizer Inc., *German Pat.* 2,746,244 (1978) [*Chem. Abstr.*, **89**, 24308v (1978)].

271. Shionogi and Co., Ltd., *Japan. Pat.* 75 111,076 (1975) [*Chem. Abstr.*, **84**, 135477r, (1976)].

272. Schering, A. G., *German Pat.* 2,449,030 (1976) [*Chem. Abstr.*, **85**, 32877n (1976)].

273. Merck and Co., Inc., *German Pat.* 2,020,762 (1970) [*Chem. Abstr.*, **74**, 99874a (1971)].

274. Osterreichische Stickstoffwerke A. G., *Austr. Pat.* 221,311 (1962) [*Chem. Abstr.*, **57**, 9824 (1962)].

275. Hoechst A. G., *German Pat.* 2,432,635 (1976) [*Chem. Abstr.*, **84**, 150525y (1976)].

276. Hoechst A. G., *German Pat.* 2,361,438 (1975) [*Chem. Abstr.*, **83**, 114227y (1975)].

277. American Cyanamid Co., *U.S. Pat.* 4,006,236 (1977) [*Chem. Abstr.*, **86**, 171283z (1977)].

278. R. P. Thummel and W. Nutakul, *J. Org. Chem.*, **43**, 3170 (1978).

279. (a) R. A. Finnegan, *J. Org. Chem.*, **30**, 1333 (1965). (b) A. Streitweiser, Jr., G. R. Zeigler, P. C. Mowry, A. Lewis, and R. G. Lawler, *J. Am. Chem. Soc.*, **90**, 1357 (1968).

280. E. M. Godar and R. P. Mariella, *J. Org. Chem.*, **25**, 557 (1960).

281. E. M. Godar and R. P. Mariella, *Appl. Spectrosc.*, **15**, 29 (1961).

282. C. B. Reese, *J. Am. Chem. Soc.*, **84**, 3979 (1962).

283. G. Bergson and A. M. Weidler, *Acta Chem. Scand.*, **16**, 2464 (1962).

284. N. S. Prostakov, A. V. Varlamov, G. A. Vasil'ev, O. G. Kesarev, and G. A. Urbina, *Khim. Geterotsik. Soedin.*, **1977**, 124; English translation: *Chemistry of Heterocyclic Compounds.*, **1977**, 105.

285. N. S. Prostakov, G. A. Urbina, L. A. Gaivoronskaya, V. P. Zvolinskii, and M. A. Galiullin, *Khim. Geterotsik. Soedin.*, **1977**, 1245; English translation: *Chemistry of Heterocyclic Compounds.*, **1977**, 1003.

286. M. M. Robison, *J. Am. Chem. Soc.*, **80**, 5481 (1958).

287. E. Godar and R. P. Mariella, *J. Am. Chem. Soc.*, **79**, 1402 (1957).

288. A. Ya. Ozola, E. I. Stankevich, S. V. Kalnin, and O. Ya. Neiland, *Khim. Geterotsik. Soedin.*, **1976**, 256; English translation: *Chemistry of Heterocyclic Compounds*, **1976**, 220.

289. V. Prelog and U. Geyer, *Helv. Chim. Acta*, **28**, 1677 (1945).

290. O. B. Edgar and D. H. Johnson, *J. Chem. Soc.*, **1958**, 3925.

291. I. Goodman, *J. Polymer Sci.*, **17**, 587 (1955).

292. G. Vanags, E. I. Stankevich, and E. Ya. Gren, *Zh. Obshch. Khim.*, **30**, 1620 (1960); English translation: *J. Gen. Chem. USSR*, **1960**, 1620.

293. G. Ya. Dubur and Ya. R. Uldrikis, *Khim. Geterotsik. Soedin.*, **1972**, 354; English translation: *Chemistry of Heterocyclic Compounds*, **1972**, 321.

294. G. Vanags and G. Ya. Dubur, *Zh. Obshch. Khim.*, **30**, 1898 (1960); English translation: *J. Gen. Chem. USSR*, **1960**, 1876.

295. D. Sveics, E. Stankevics, and O. Neilands, *Latv. PSR Zinat. Akad. Vestis, Kim. Ser.*, **1968**, 213 [*Chem. Abstr.*, **69**, 51964z (1968)].

296. Yu. S. Dol'skaya, G. Ya. Kondrat'eva, and B. Z. Bartkevich, *Izv. Akad. Nauk SSSR, Ser. Khim.*, **1978**, 1446; English translation: *Bulletin of Chemical Sciences*, **1978**, 1263.

297. V. I. Vysotskii and M. N. Tilichenko, *Khim. Geterotsik. Soedin.*, **1969**, 751; English translation: *Chemistry of Heterocyclic Compounds*, **1969**, 560.

298. M. N. Tilichenko and M. E. Egorova, *Uch. Zap. Saratovsk. Gos. Univ.*, **75**, 68 (1962) [*Chem. Abstr.*, **60**, 1697 (1964)].

299. V. A. Stonik, V. I. Vysotskii, and M. N. Tilichenko, *Khim. Geterotsik. Soedin.*, **1968**, 763; English translation: *Chemistry of Heterocyclic Compounds*, **1968**, 558.

300. W. Dammertz and E. Reimann, *Arch. Pharm. (Weinheim)*, **313**, 826 (1980).

301. W. Dammertz and E. Reimann, *Arch. Pharm. (Weinheim)*, **313**, 375 (1980).

302. G. Maas, *Chem. Ber.*, **112**, 3241 (1979).

303. T. Kusumi, K. Yoneda, and H. Kakisawa, *Synthesis*, **1979**, 221.

304. A. Naiman and K. P. C. Vollhardt, *Angew. Chem. Internat. Ed. Engl.*, **18**, 411 (1979).

305. J. N. Chattergea, S. C. Shaw, and S. N. Singh, *J. Indian Chem. Soc.*, **55**, 149 (1978).

306. W. A. Ayer, L. M. Browne, Y. Nakahara, M. Tori, and L. T. J. Delbaere, *Can. J. Chem.*, **57**, 1105 (1979).

307. M. Nitta, A. Sekiguchi, and H. Koba, *Chem. Lett.*, **1981**, 933.

308. J. A. H. MacBride and P. M. Wright, *Tetrahedron Lett.*, **22**, 4545 (1981).

309. R. Bernardi, T. Caronna, S. Morrocchi, and P. Traldi, *Tetrahedron Lett.*, **22**, 155 (1981).

310. I. Saito, K. Kanehira, K. Shimozono, and T. Matsuura, *Tetrahedron Lett.*, **21**, 2737 (1980).

311. J. I. Seeman, R. Galzerano, K. Curtis, J. C. Schug, and J. W. Viers, *J. Am. Chem. Soc.*, **103**, 5982 (1981).

312. A. R. Katritzky, A. M. El-Mowafy, G. Musumarra, K. Sakizadeh, C. Sana-Ullah, S. M. M. El-Shafie, and S. S. Thind, *J. Org. Chem.*, **46**, 3823 (1981).

313. D. J. Brien, A. Naiman, K. P. C. Vollhardt, *J. Chem. Soc., Chem. Commun.*, **1982**, 133.

314. L. A. Paquette, G. D. Ewing, S. V. Ley, H. C. Berk, and S. G. Traynor, *J. Org. Chem.*, **43**, 4712 (1978).

315. R. A. Abramovitch, W. D. Holcomb, and S. Wake, *J. Am. Chem. Soc.*, **103**, 1525 (1981).

316. K. Bogdanowicz-Wzwed, *Polish J. Chem.*, **52**, 295 (1978).

317. J. Epsztajn and R. M. K. Marcinkowski, *Polish J. Chem.*, **53**, 601 (1979).

318. K. Saito and S. Kambe, *Synthesis*, **1981**, 211.

319. S. Kanoktanaporn, J. A. H. MacBride, and T. J. King, *J. Chem. Research(s)*, **1980**, 204.

320. J. Koyama, T. Sugita, Y. Suzuta, and H. Irie, *Heterocycles*, **12**, 1017 (1979).

321. J. Epsztajn, A. Bieniek, and Z. Brzeziński, *Polish J. Chem.*, **54**, 341 (1980).

322. J. Bosch, J. Bonjoch, and I. Serret, *Heterocycles*, **14**, 1983 (1980).

323. D. L. Boger and J. S. Panek, *J. Org. Chem.*, **46**, 2179 (1981).

324. K. J. Gould, N. P. Hacker, J. F. W. McOmie, and D. H. Perry, *J. Chem. Soc., Perkin Trans. I*, **1980**, 1834.

325. F. A. L. Anet and C. R. Eves, *Can. J. Chem.*, **36**, 902 (1958).

326. E. Kleinman and C. H. Heathcock, *Tetrahedron Lett.*, **1979**, 4125.

CHAPTER III

Macrocyclic Pyridines

G. R. NEWKOME, V. K. GUPTA[‡]

Department of Chemistry, Louisiana State University, Baton Rouge, Louisiana

and

J. D. SAUER

Ethyl Corporation, Baton Rouge, Louisiana

I.	Nomenclature and Numbering.	448
II.	2,6-Pyridine Macrocycles	448
	1. Carbon Bridges	448
	2. Sulfur Bridges	458
	3. Nitrogen Bridges	458
	4. Carbon–Oxygen Bridges	459
	5. Carbon–Nitrogen Bridges	470
	6. Carbon–Sulfur Bridges	474
	7. Carbon–Nitrogen–Oxygen Bridges	476
	8. Carbon–Nitrogen–Sulfur Bridges	477
	9. Carbon–Nitrogen–Phosphorus Bridges	479
	10. Carbon–Oxygen–Phosphorus Bridges	480
	11. Carbon–Sulfur–Oxygen Bridges	480
III.	2,5-Pyridine Macrocycles	481
	1. Carbon and Carbon–Sulfur Bridges	481
	2. Carbon–Oxygen Bridges	484
IV.	2,4-Pyridine Macrocycles	485
	1. Carbon Bridges	485
	2. Carbon–Nitrogen Bridges	486
V.	2,3-Pyridine Macrocycles	487
	1. Carbon Bridges	487
	2. Carbon–Oxygen Bridges	488
	3. Carbon–Nitrogen Bridges	490
VI.	3,5-Pyridine Macrocycles	492
	1. Carbon Bridges	492
	2. Carbon–Oxygen Bridges	494
	3. Carbon–Sulfur Bridges	495
	4. Carbon–Nitrogen Bridges	496

‡ On leave from University of Delhi, Delhi, India, 1980–1983.

VII. 3,4-Pyridine Macrocycles   496
      1. Carbon Bridges   496
      2. Carbon–Nitrogen Bridges   496
VIII. 2,3; 5,6-Pyridine Macrocycles. Carbon Bridges   497
      Acknowledgment   498
 IX. Tables   499
      References   620

I. NOMENCLATURE AND NUMBERING

The conventional IUPAC nomenclature (1) has been applied to macrocyclic chemistry; however, as the number and type of incorporated subunit(s) and ring substituents increase, the practicality of the IUPAC system becomes more complicated and cumbersome, especially to the reader. Thus phane nomenclature (2–4) has been used to circumvent some of these problems. More recently Weber and Vögtle (5) have devised a Topology and Classification of Organic Neutral Ligands system; although this system needs to be expanded and modified, it should be consulted.

For the case in tabulation of ring substituents, as previously described (6), the atom adjacent to the pyridine (sub)ring will be designated as atom number one and all atoms in the largest continuous ring being numbered in succession with substituted positions taking preference when necessary.

II. 2,6-PYRIDINE MACROCYCLES

1. Carbon Bridges

[n]-(2,6)-Pyridinophanes (**III-1**) can be commonly constructed in at least three different ways: (a) cyclization of 2,6-disubstituted pyridines; (b) aromatization of 1,5-bifunctional cyclic compounds; and (c) pyridine ring synthesis accompanying simultaneous bridge formation. Kumada et al. (7, 8) accomplished a one-step synthesis of **III-1** by the cyclocoupling of di-Grignard reagents with 2,6-dichloropyridine in the presence of a nickel–phosphine catalyst. Further application of this procedure afforded **III-2** by use of the appropriate ethereal di-Grignard reagent.

$$(CH_2)_n \quad n = 6\text{–}10, 12$$
$$(10\text{–}33\%)$$

III-1   **III-2**

[7]-(2,6)-Pyridinophane (**III-3**) (9–11) has been synthesized by treatment of cyclododecane-1,5-dione (**III-4**) with hydroxylamine hydrochloride (9). The desired dione (**III-4**) was prepared (30%) from boraperhydrophenalene and one equivalent of acetic acid, followed by chromic acid oxidation.

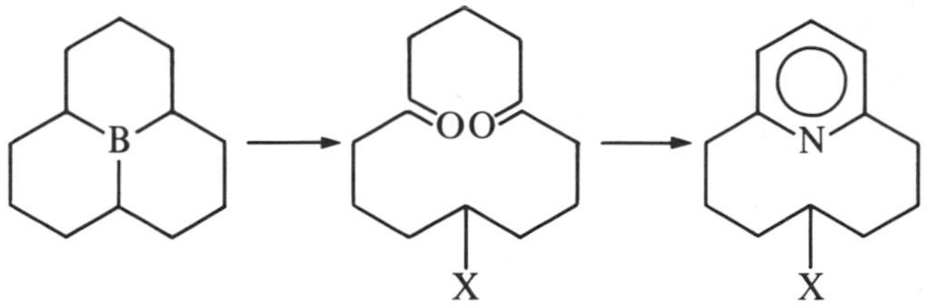

III-4   III-3

Büchi et al. (12) generated 2,6-pyridinophanes by a Stobbe condensation of cyclododecanone (**III-5**) with diethyl succinate to give initially an exocyclic carboxylic acid **III-6**, which was subsequently cyclized with either zinc chloride in acetic acid or preferably with polyphosphoric acid to the δ-keto-β,γ-unsaturated ester **III-7**. Upon acid hydrolysis, **III-7** underwent concomitant decarboxylation to give the α,β-unsaturated ketone **III-8**. Wolff–Kishner reduction to the bicyclo[10.3.0]pentadeca-1(12)-en-13-one (**III-8**) (13) gave two isomeric olefins, from which the trisubstituted olefin (**III-9**) was isolated as the major (70%) isomer. A subsequent Schmidt reaction on **III-9**, followed by dehydrogenation over 10% palladium on carbon at ca. 250°C afforded **III-10** and its 2,3-isomer, both in an overall 4% yield.

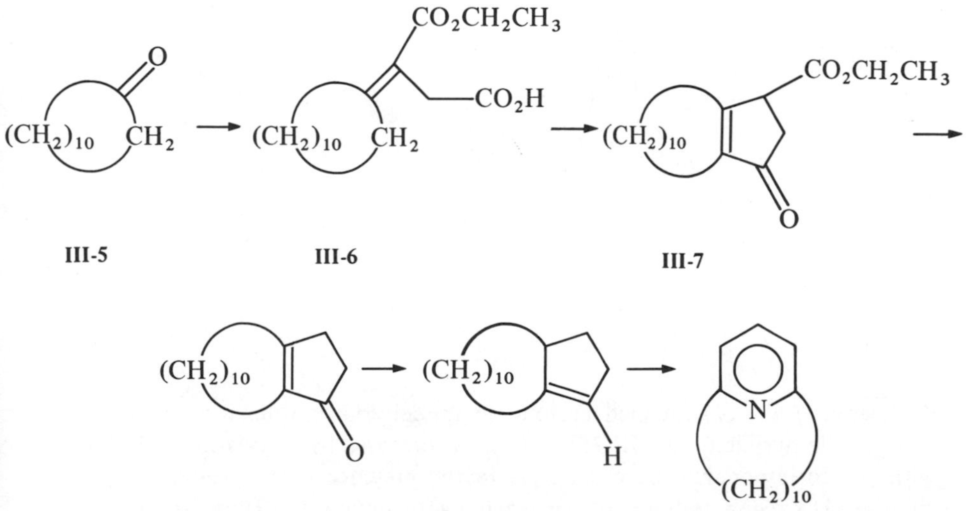

III-5   III-6   III-7

III-8   III-9   III-10

Conversion of **III-10** into (*dl*)-muscopyridine (**III-12**), was accomplished (12) by α-substitution *via* treatment of **III-10** *N*-oxide with acetic anhydride (14). Hydrolysis of the intermediate α-acetoxy derivative gave an alcohol **III-11a**, which was oxidized with chromium trioxide to afford **III-11b**. Alkylation of **III-11b**, followed by Wolff–Kishner reduction gave *dl*-**III-12**, which was resolved with di-*p*-toluoyl-L-tartaric acid (15).

III-11a

III-11b

III-12

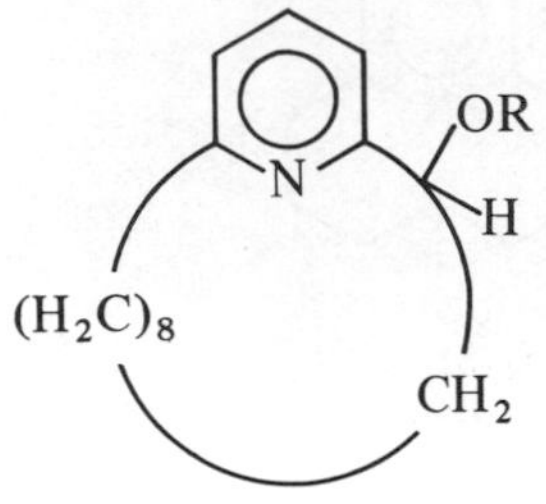

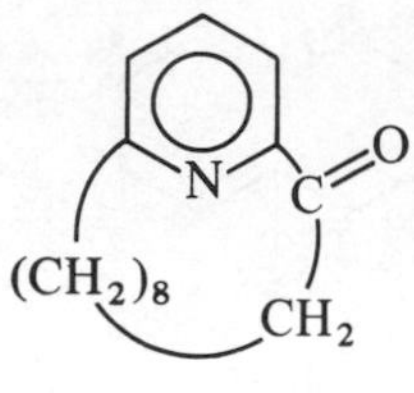

An alternative route to **III-12** involved the acid-catalyzed cyclization of **III-13** to give ketone **III-14**. A 1,6-conjugate addition of a methyl group followed by ring expansion, cyclization, and subsequent aromatization afforded *dl*-**III-12** (16).

III-13 **III-14**

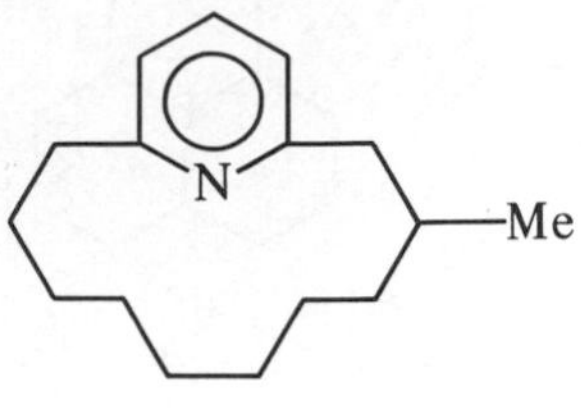

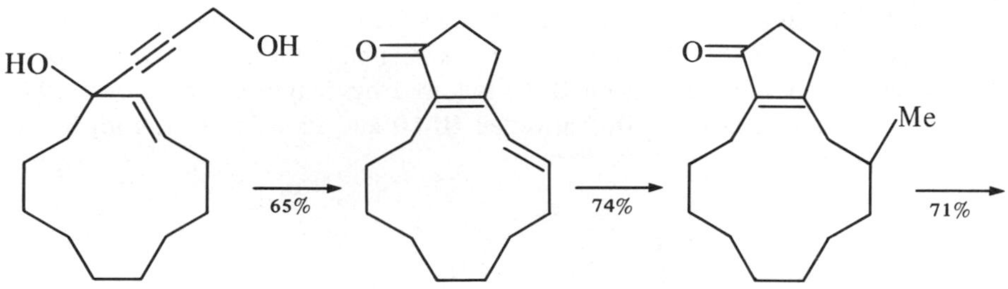

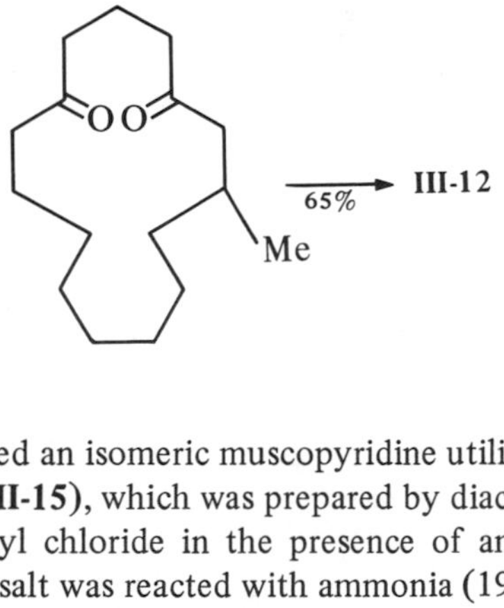

Balaban et al. (17) obtained an isomeric muscopyridine utilizing an intermediate – a bicyclic pyrylium salt (**III-15**), which was prepared by diacylation of isobutene with the corresponding diacyl chloride in the presence of anhydrous aluminium chloride (18). The pyrylium salt was reacted with ammonia (19) to give substituted pyridinophanes (**III-16**) in poor yield; however, the latter studies indicated the pyrylium salt to be dimeric (20). Thus, the initial macrocycle (17) was the dimer of **III-16**; traces (0.5%) of the desired monomer **III-16** were also isolated. [7]-(2,6)-Pyridinophane was prepared by a similar sequence (9, 21).

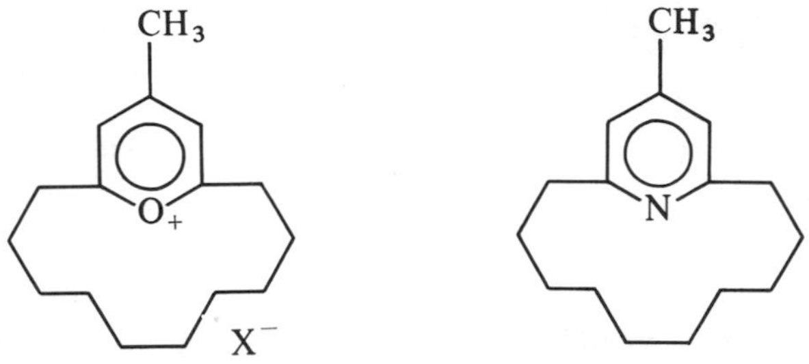

III-15   III-16

1-Methoxycyclododecene (**III-17**) was transformed to the activated intermediate **III-19**, *via* a Beckmann rearrangement of oxime **III-18**. Intramolecular trapping of the intermediate carbocation generated the dihydropyridine nucleus, which upon oxidation gave **III-16** (22).

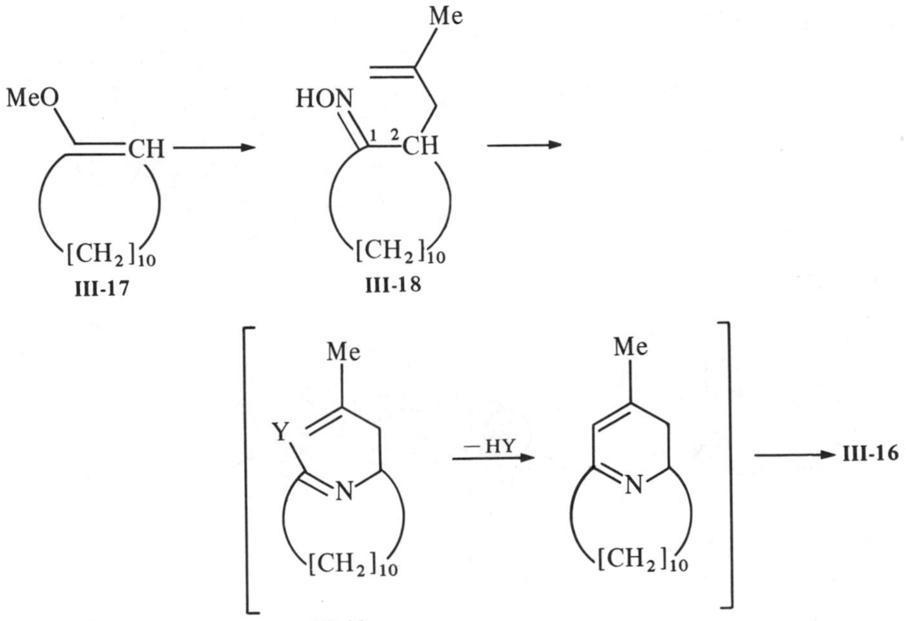

III-17   III-18

III-19

An alternate approach to these *C*-macrocycles was demonstrated by Isele and Schieb by the initial formation of a pyridinone nucleus (**III-21**) from an appropriately substituted diyne (**III-20**); then the terminal acetylenic groups were coupled by use of a copper catalyst (23). Reduction of the triple bonds and *O*-amination with *N*-chloroamine and sodium hydride gave **III-22**.

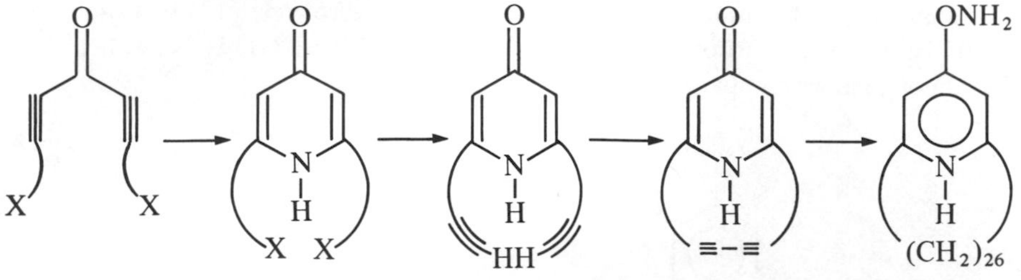

III-20   III-21   III-22

[2.2]-Cyclophanes (e.g., **III-23**) are generally prepared by carbon-carbon σ-bond formation using organometallic reagents. After the unsuccessful simple, one-step cyclocondensation of 2,6-pyridinecarboxyaldehyde with 2,6-lutidine (24, 25). Baker et al. in a now classic paper, described the synthesis of the first [2.2]-(2,6)-pyridinophane (**III-24**) by the cyclization of **III-25** through the action of either butyllithium in ether or phenyllithium in benzene–ether (24, 26). [2.2]-Metacyclo-2,6-pyridinophane was prepared in a similar manner (27). The treatment of 2,6-*bis*(bromomethyl)pyridine (**III-25**) with phenyllithium gave **III-24** in 25% yield (28). Cyclization of **III-26**, by means of sodium and tetraphenylethylene in THF gave an oligomeric mixture of 2,6-bridged pyridinophanes (29, 30). Kauffmann et al. modified this procedure by selective metalation, using the abstractability of the acidic α-methyl hydrogens, followed by copper transmetalation and oxidative coupling (31).

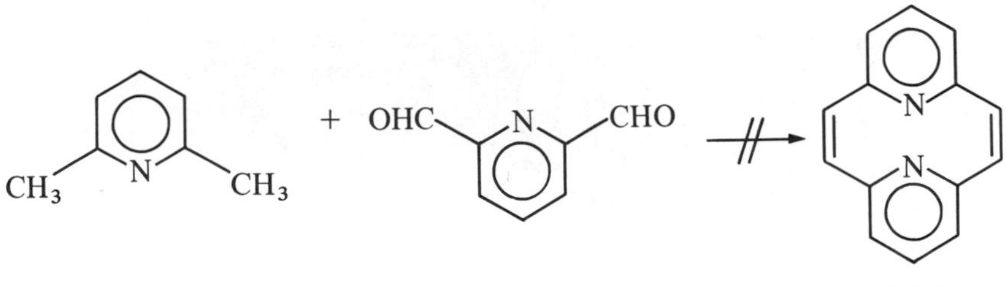

III-23

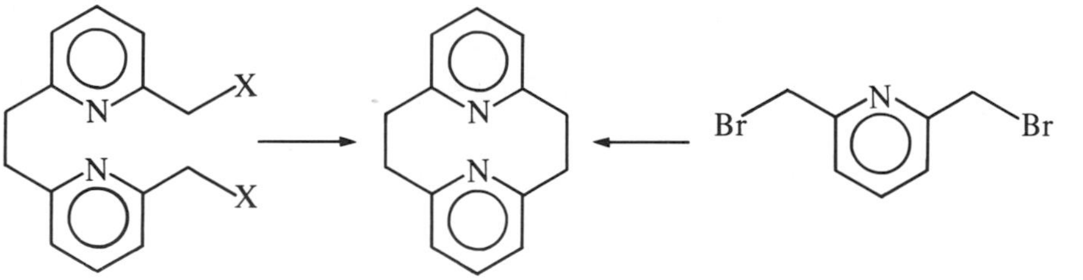

III-26   III-24   III-25

A ring-contractive, photochemical desulfurization of dithiaparacyclophanes (e.g., **III-27**) in the presence of trialkylphosphite (32) has provided an excellent route to the all carbon cyclophanes (**III-28**). Boekelheide et al. (33) converted **III-27** to the ring-contracted thioether **III-29** by a Stevens' rearrangement, subsequent conversion of **III-29** with Raney nickel gave **III-28**, whereas oxidation of

III-27 afforded the corresponding *bis*-sulfoxide **III-30**, which upon thermolysis afforded [2.2]-(2,6)-pyridinoparacyclophane-1,9-diene (**III-31**) (33, 34). Haenel (35) applied this later procedure to synthesize the related naphthalinopyridinophanes.

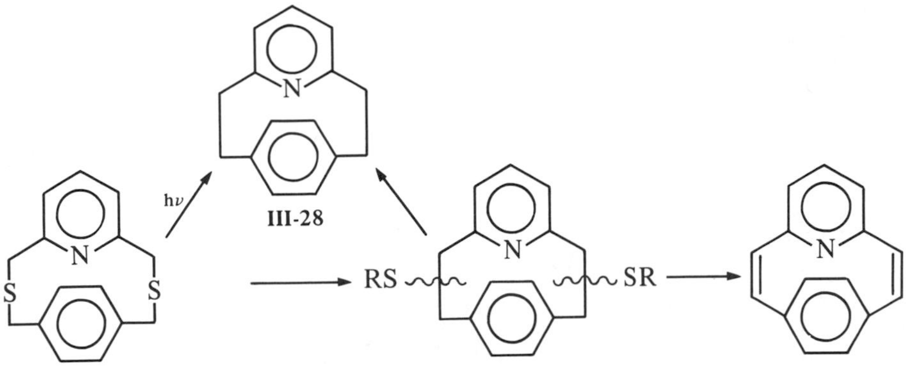

III-28

III-27

III-29
III-30 (S → O)

III-31

Fujita et al. (36, 37) constructed (86%) cyclophane **III-34** by the Wittig reaction of **III-33** and the *bis*-ylide **III-32** under high-dilution conditions, followed by catalytic hydrogenation of the unsaturated bridges with a Pd/C catalyst. Earlier a more exhaustive route to **III-34** was presented in order to establish the structure proof of *O*-methyllythranidine (**III-35**), an Lythraceous alkaloid (38).

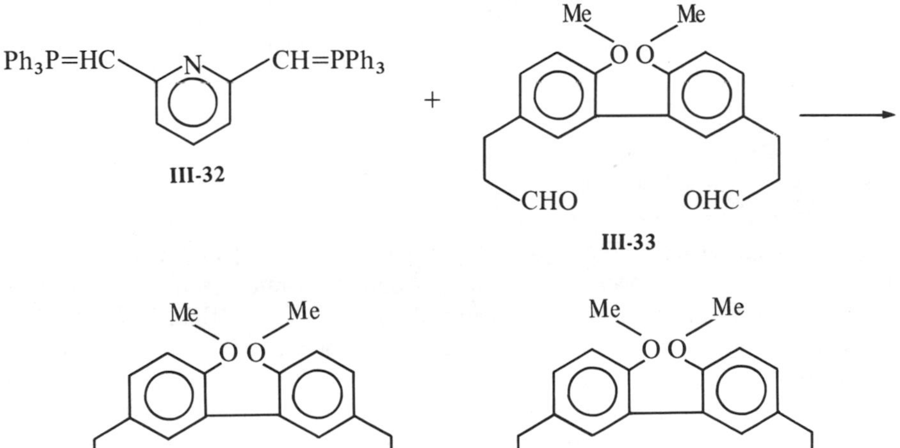

III-32

III-33

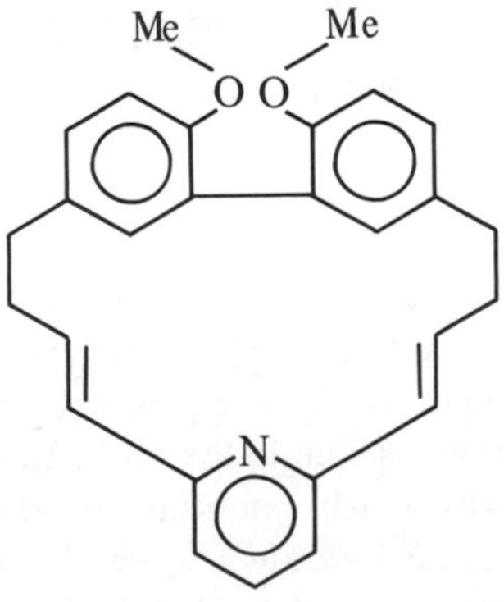

III-34

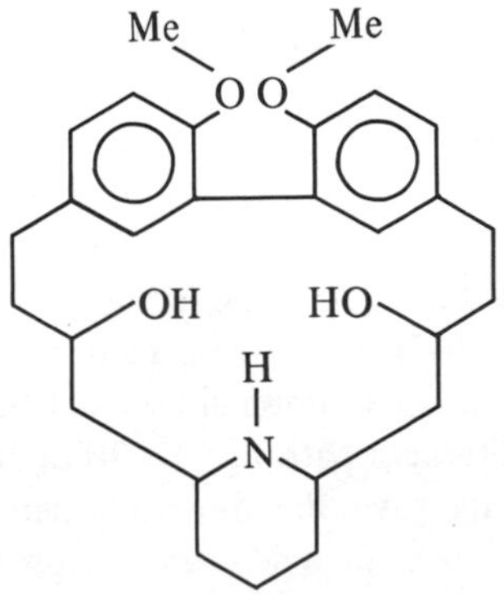

III-35

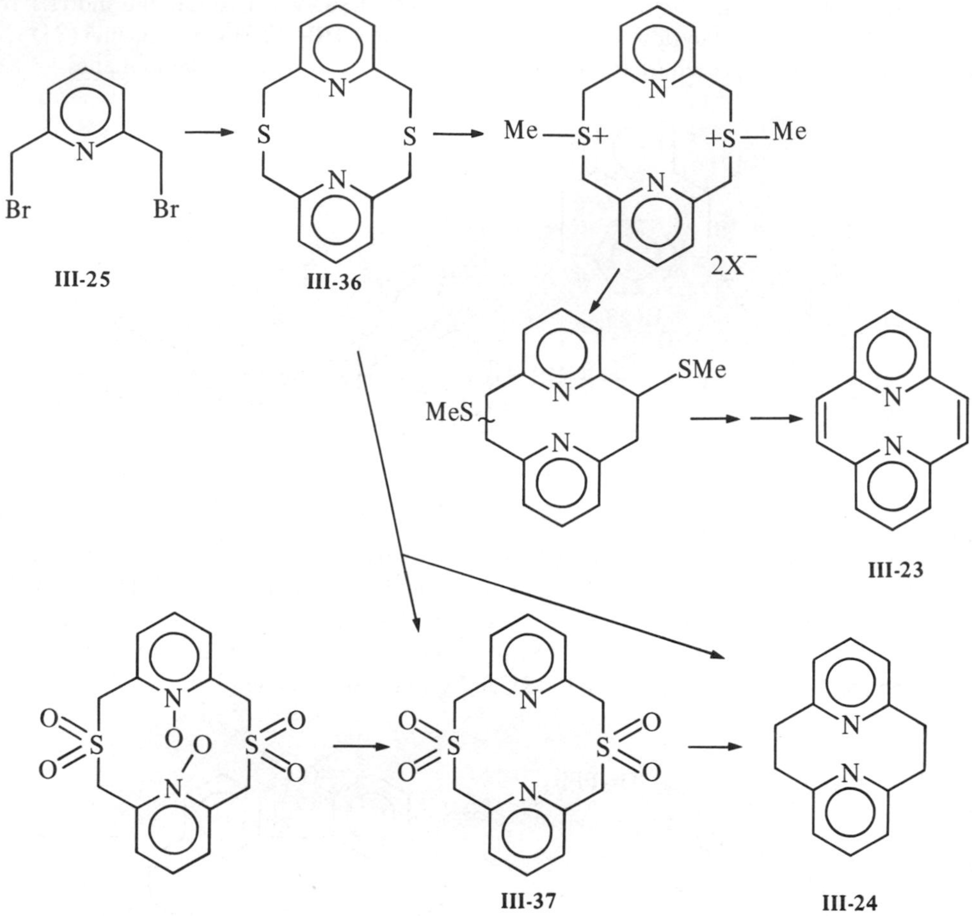

Various syntheses of these pyridinophanes from dithiacyclophanes by diverse ring-contractions have been devised. A two-step extrusion of sulfur by a Stevens' rearrangement, followed by a Hofmann elimination (e.g., **III-36** → **III-23**) (28), thermal extrusion of sulfur dioxide from the corresponding sulfone (**III-37** → **III-24**) (39), and photochemical desulfurization in the presence of trialkylphosphite (**III-36** → **III-24**) (40) are typical reaction routes to the carbon-bridged hetero-cyclophanes. For the production of polychlorinated derivatives of **III-24**, Martel et al. (41) utilized Boekelheide's work (28), but centered on the Ramberg–Bäcklund reaction (42) as the ring contractive procedure. Cooke (43) recently constructed [2.2]-(2,6)-pyridinophan-1-ene (**III-39**) *via* similar reaction sequence from **III-38**. In an alternate pathway, **III-38** upon treatment with trimethyloxonium hexafluoro-phosphate gave the *S*-methyl derivative (**III-40**) which smoothly underwent the Stevens' rearrangement in the presence of sodium hydride to give **III-41**. Facile *S*-oxidation of **III-41** and subsequent thermolysis of the *S*-oxide **III-42** gave **III-39** (43).

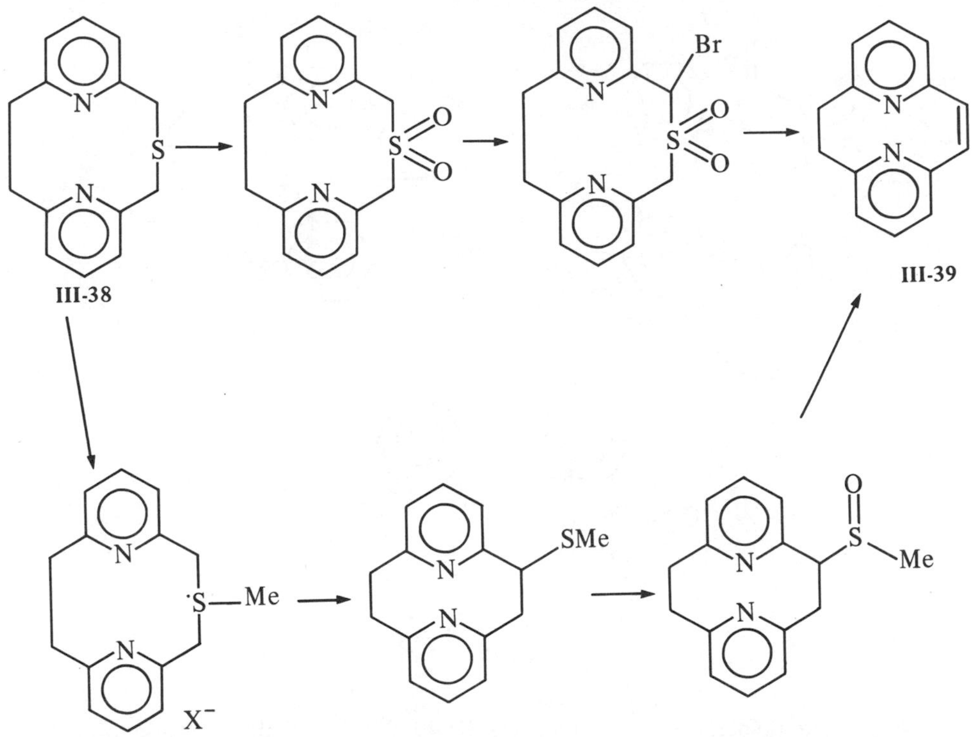

III-38

III-39

III-40   III-41   III-42

Carbonyl-bridged pyridine macrocycles have been reported (44). 2,6-Dibromo-pyridine was monolithiated with butyllithium in ether [2,6-dilithiopyridine is generated using THF solvent at $-100°C$ (45, 46)], then treated with 0.5 equiv. of methyl 2,6-pyridinedicarboxylate to give the diketone III-43. Base ketalization of III-43 with bromoethanol in the presence of excess lithium carbonate (47) afforded (80%) the diketal III-44, which after metal–halogen exchange, was treated with ethyl chloroformate or ethyl 2,6-pyridinedicarboxylate to give after hydrolysis the corresponding tri-(III-45) or tetraketone (III-46) (44). Modifications of this sequence gave III-47 and III-48.

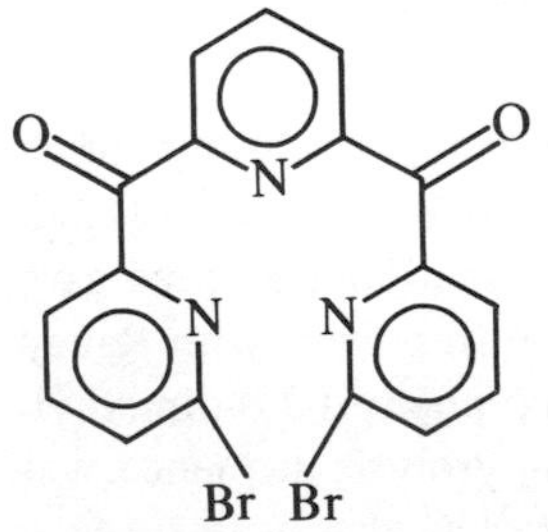

III-43 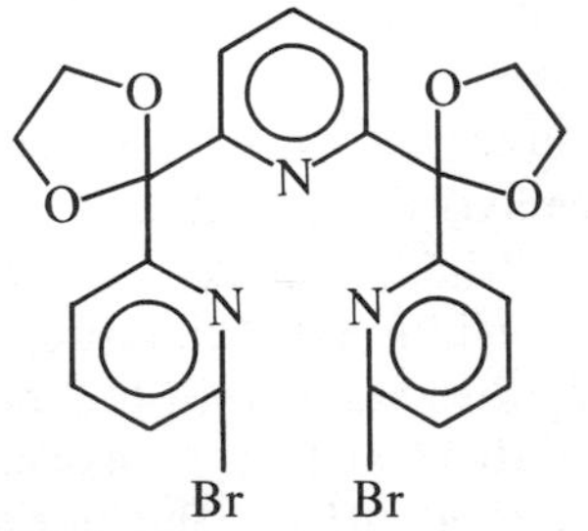   III-44 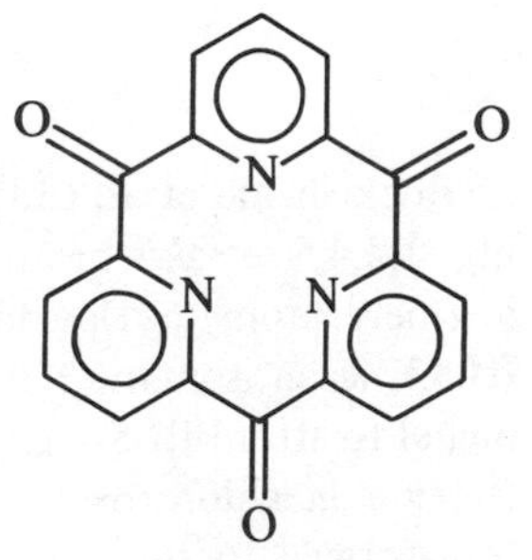   III-45

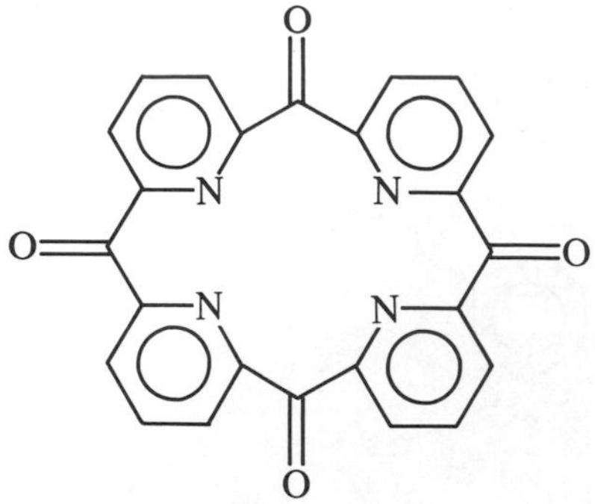

III-46

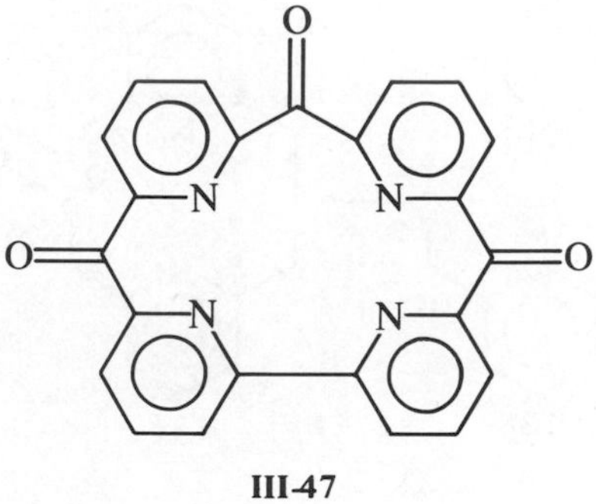

III-47

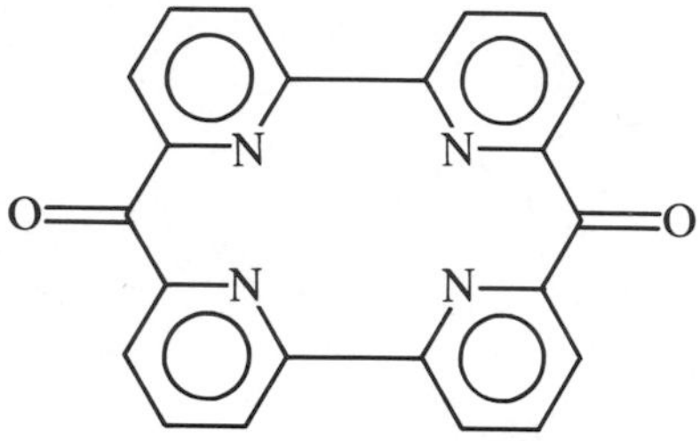

III-48

Attempted preparation of macrocycle **III-49** failed probably because of the enhanced acidity of the 2-methylenic hydrogens (48).

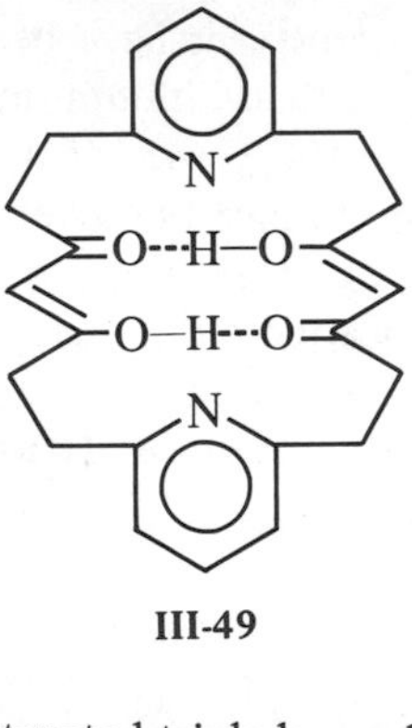

III-49

Boekelheide et al. (33) constructed triple-layered cyclophanes (**III-50**) by coupling 1,2,4,5-tetra*kis*(bromomethyl)benzene (**III-51**) with two equivalents of 2,6-*bis*(mercaptomethyl)pyridine (**III-52**). Standard sulfur contraction reactions gave **III-53**, as an isomeric mixture, which was desulfurized upon treatment with Raney nickel to afford **III-54**. **III-50**, when heated with 1-(2′-carboxyphenyl)-3,3-dimethyl-triazine in chloroform, gave **III-55** and by an oxidation-pyrolysis sequence was transformed to the desired tetraene **III-56**. This procedure has also been successfully applied to the construction of **III-57**.

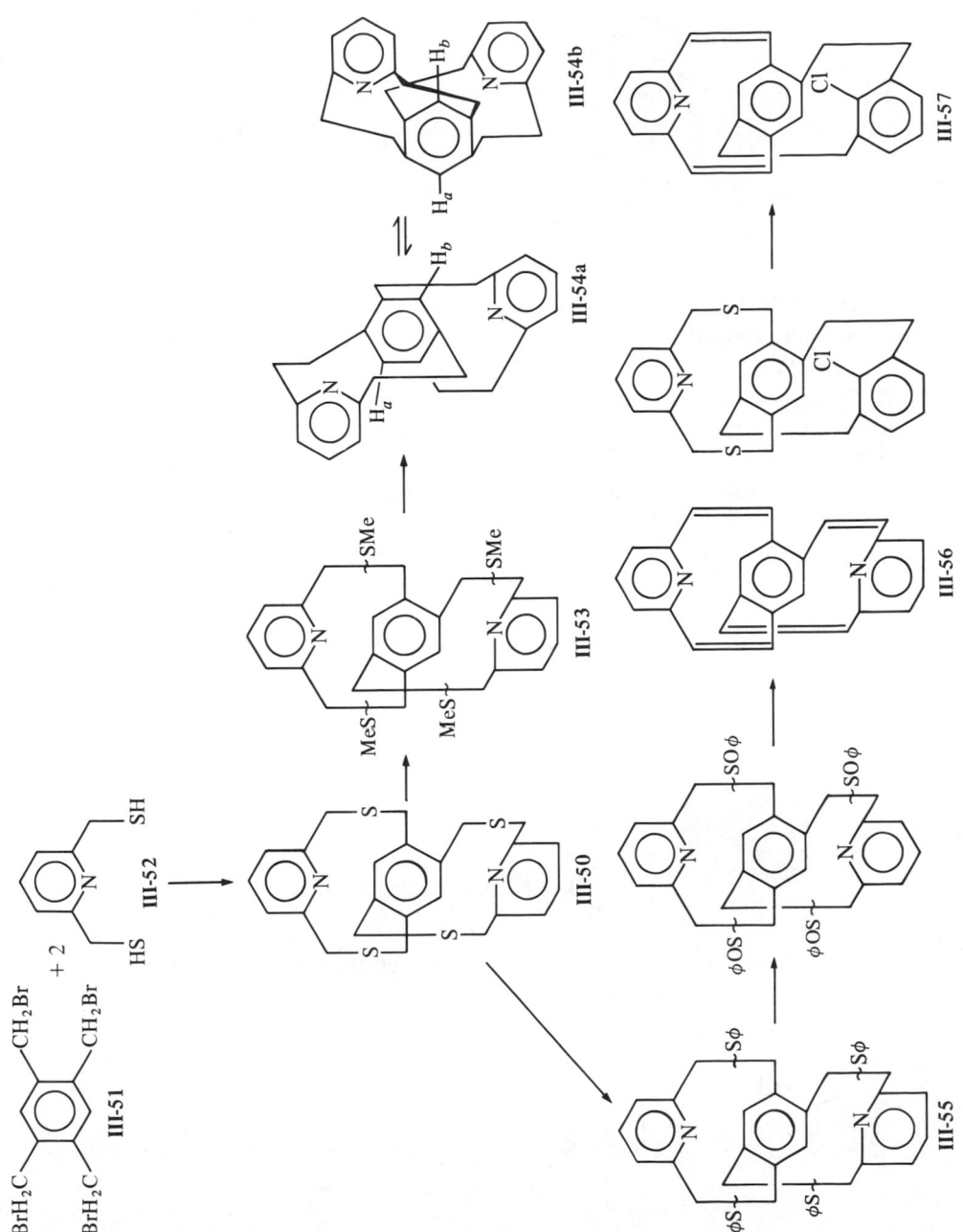

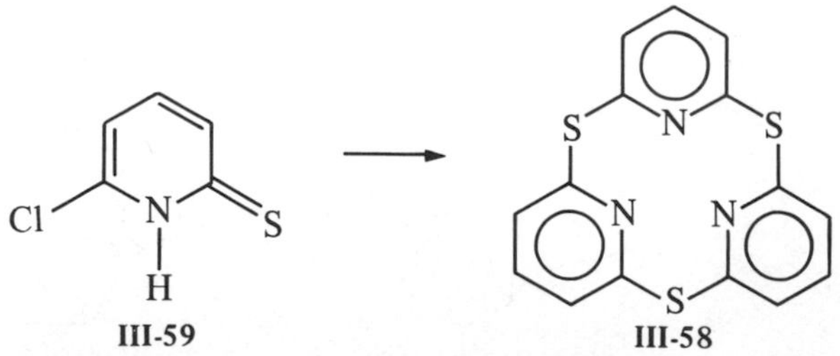

III-59   III-58

2. Sulfur-Bridges

The sulfur-bridged pyridine macrocycle **III-58** has been prepared by Undheim et al. (49) through the intramolecular cyclocondensation of 6-chloropyridin-2-thione (**III-59**) in the presence of P_2S_5 at 130°C. The x-ray crystal structure of **III-58** supports a nonplanar conformation (49, 50). Macrocycle **III-60** was obtained (45% overall) through a two-step sequence: 2,6-dichloropyridine was transformed into 2,6-dimercaptopyridine by treatment with sodium hydrogen sulfide in hot DMF, then oxidation of the dithiol with iodine in benzene and triethylamine under high dilution conditions (51).

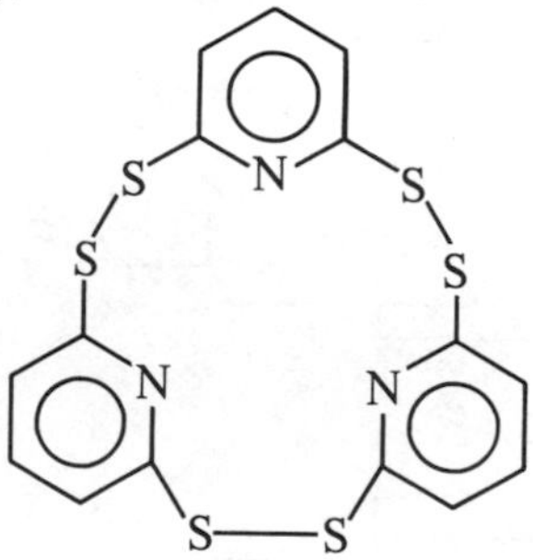

III-60

3. Nitrogen Bridges

A porphyrin-like, stable macrocycle **III-61**, possessing four pyridine rings was synthesized by a one-step template cyclization of 6,6'-dichloro-2,2'-dipyridine with ammonium tetrachlorozincate, followed by demetalation (52).

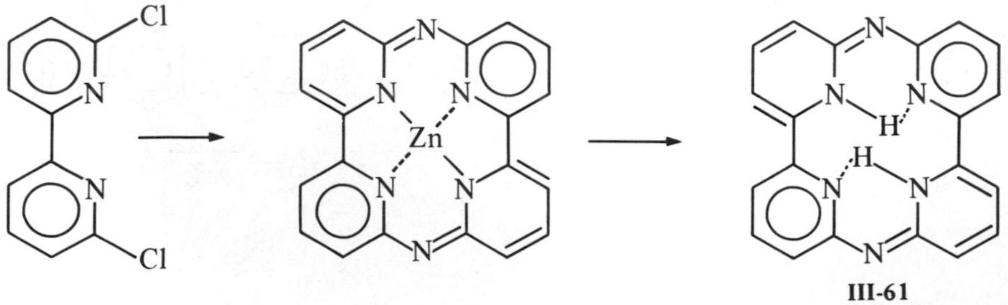

III-61

4.   Carbon–Oxygen Bridges

The carbon-heteroatom bridged heterocycles are divided into two major types: (1) those possessing a bridge-heteroatom isolated from the pyridine nucleus, and (2) those in which the bridge-heteroatom is directly adjacent to the pyridine nucleus.

Newkome and Robinson, in 1973, successfully synthesized **III-62** by the reaction of 2,6-*bis*(hydroxymethyl)pyridine with an activated *bis*-halide; yields of the smallest 22-member ring were high and steadily decreased to that of the 99-member ring (53). Cram et al. (54–56) similarly constructed macrocycles of type 1, for example, **III-63**, by the treatment of 2,6-*bis*(hydroxymethyl)pyridine with an appropriate polyethyleneglycol ditosylate. Pyridino-15-crown-5 (**III-63**; $n = 2$) was synthesized by others (57).

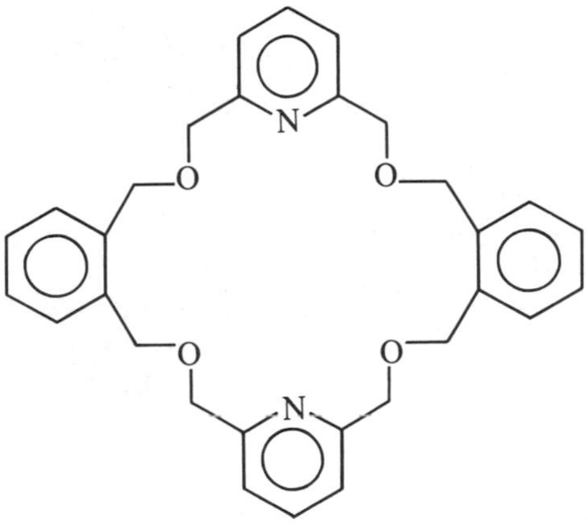

III-62

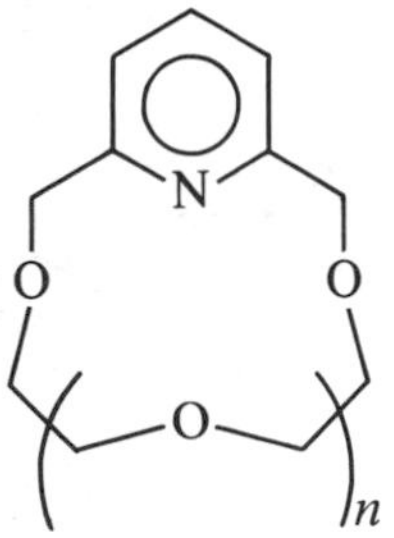

III-63

Utilizing an analogous procedure, a number of novel chiral pyridine macrocycles (**III-64–67**) were synthesized (55, 58) by the incorporation of the chiral binaphthyl moiety **III-68**, which has a variable dihedral angle ($60°$ to $120°$) as shown by Corey–Pauling–Koltun (CPK) models. This subunit (**III-68**) possesses a large barrier to racemization and thus imparts to the host the property of chiral recognition toward appropriate guest compounds. The absolute configuration of **III-68** has been established (59). Hosts **III-65b** and **III-66** exhibited chiral recognition toward phenylglycine and valine methyl ester hexafluorophosphate salts with higher recognition for the former; whereas, **III-64** and **III-65a** were less effective with the former guest. The major factors that influence the binding power and extent and direction of chiral recognition are probably the greater basicity and more rigid positioning of the heteroatom(s) of the pyridine versus that of the CH_2OCH_2 moieties, and the one-electron pair on nitrogen versus the two-electron pairs on oxygen (60, 61). These studies have been extended to other guest molecules (62).

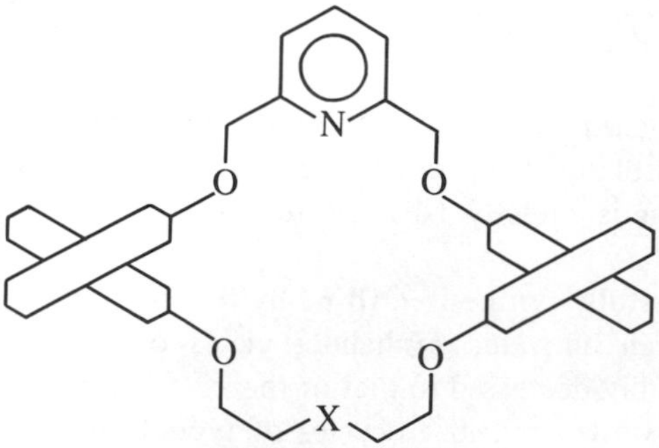

III-64 [X = CH$_2$]
III-66 [X = O]

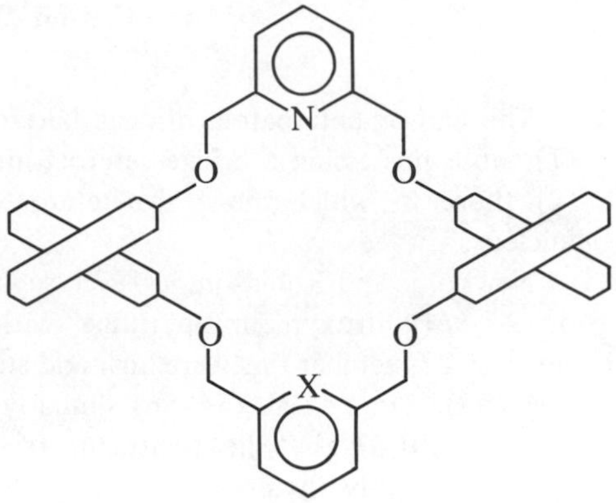

III-65a, X = C
    b, X = N

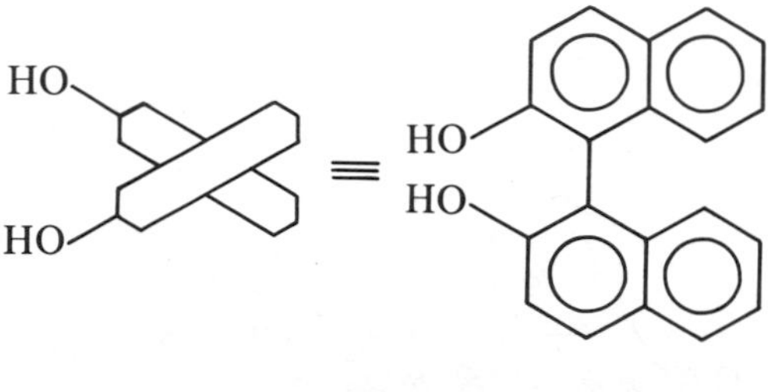

III-68

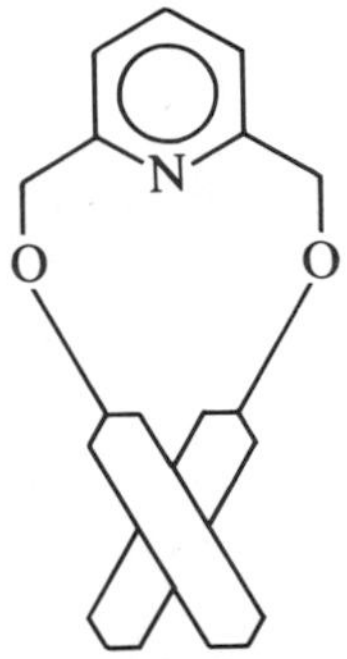

III-67

These chiral macrocycles serve as an optically active catalyst for asymmetric induction in the syntheses of new chiral centers, in the selective destruction of chiral centers, and in the interconversion of ligands attached at chiral centers.

Other related macrocycles, for example, **III-69**, were obtained during the synthesis of **III-70** (55). As a probe for binding ability in these *N*-macrocycles, the association constants, K_a, were conveniently determined using NMR spectroscopy with *tert*-butyl ammonium thiocyanate (55, 63). From association constants for **III-63** ($n = 3$) and **III-71** with alkali metal picrates, ΔG^0 values were ascertained (64). The pK_a values of the conjugate acids of the various macrocycles and the saturation constants (K$_S$) with H$_3$O$^+$, Na$^+$, K$^+$, NH$_4^+$, and Cs$^+$ were reported (55, 56).

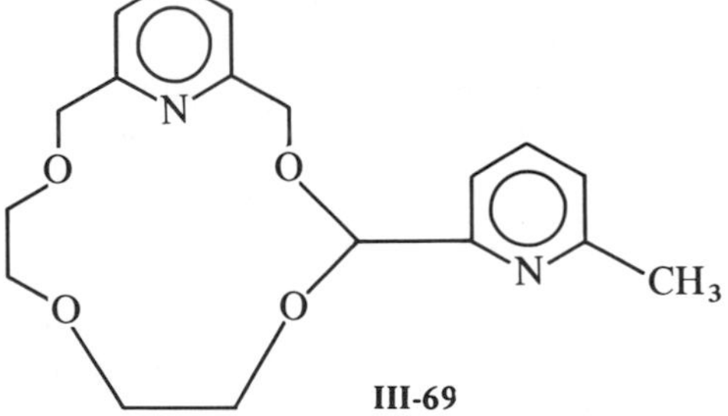

III-69

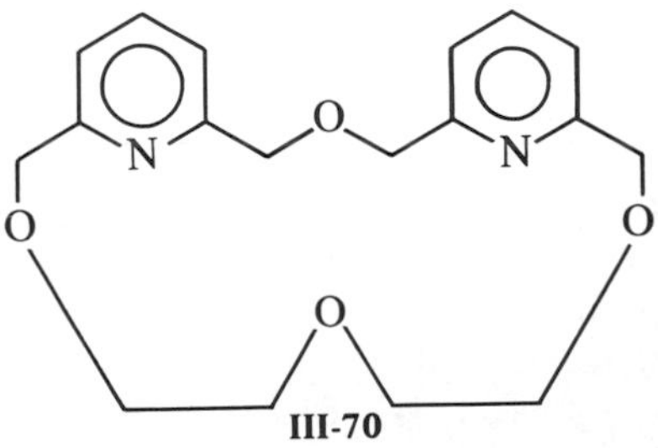

III-70

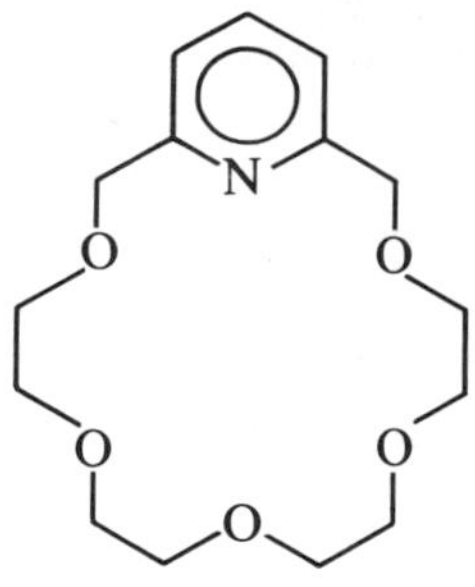

III-71

The x-ray single-crystal structure of the *tert*-butylammonium perchlorate complex of **III-63** ($n = 3$) showed that the RNH_3^+ forms hydrogen bonds with two oxygen and the basic pyridine *N*-atom; the complex has approximate *m* symmetry (65). Vögtle et al. (66–69) synthesized **III-72–III-76** by the standard procedure (55); these ligands differ in ring size and flexibility for subsequent evaluation of alkali metal ion complexation. Ligand **III-72** ($n = 1$) formed crystalline complexes (**III-77**) with salts or different alkali metal, ammonium, heavy metal, and rare earth ions, while **III-73** ($n = 1$) afforded a complex with only potassium thio-

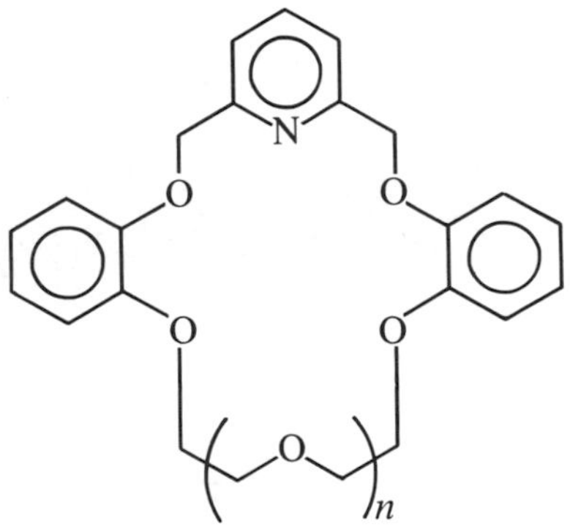

III-72 ($n = 0–2$)
III-73 [$N \rightarrow O$; ($n = 1, 2$)]

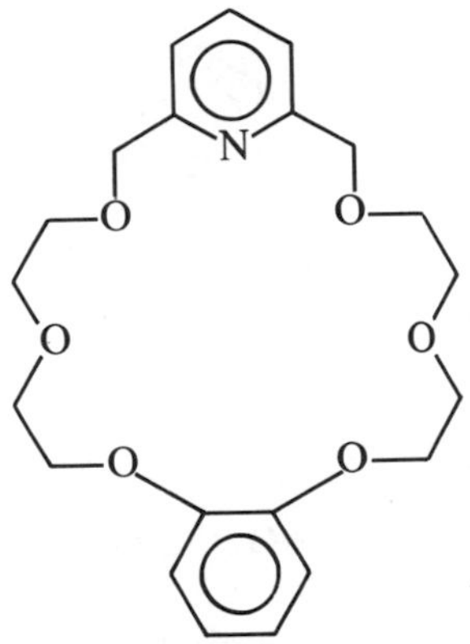

III-74

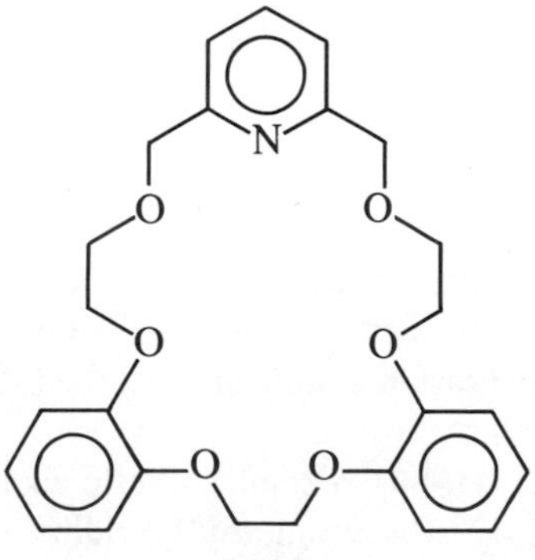

III-75

cyanate (66). These ligands also formed neutral stoichiometric crystalline complexes with CH acidic neutral molecules (67). The property of these pyridyl crowns as receptor cavities for acidic CH_3-, $-CH_2-$, and analogous heteroatom groups was demonstrated in **III-72** ($n = 1$) for the inclusion of volatile, highly toxic reagents, such as N,N-dimethylnitrosamine (68). A selective inclusion of aliphatic alcohols with **III-76** has been reported (69). A kinetic technique has been utilized to determine the relative complexing ability of **III-72** ($n = 1$) for p-$tert$-butylbenzenediazonium tetrafluoroborate (70). Whitney and Jaeger (71) prepared several deuterated analogs of **III-63** ($n = 3$), then evaluated the mass spectral fragmentation versus **III-63** ($n = 2, 3$) (57). In a study of natural-abundance ^{15}N NMR spectra of **III-72** ($n = 1$) and its KSCN complex, the trends indicated that the alkali metal ion complexation produces at least as large, and often larger, upfield shift of the pyridine nitrogen as does hydrogen bonding with methanol (72). Izatt et al. (73) evaluated the interaction of **III-63** ($n = 3$) and Ag^+, Cu^{2+}, and alkali and alkaline earth metal cations by titration calorimetry and various thermodynamic parameters (log K, ΔH and $T\Delta S$), observing that **III-63** ($n = 3$) complexes alkali and alkaline earth cations almost as strongly as 18-crown-6 and further formed complexes with Cu^{2+} as well (73).

III-77

III-76

The Hantzsch condensation has been used to prepare the 2,6-C,O-bridged pyridine macrocycles via the reaction of γ-bromoacetoacetate and the desired polyethylene glycol (74). After protecting the acidic β-dicarbonyl groups as the sodium chelate, nucleophilic substitution was carried out to afford **III-78**, which on subsequent Hantzsch condensation conditions afforded the 1,4-dihydro-2,6-pyridine crown ether **(III-79)**.

A pyridine–retronecate **(III-80)**, a semisynthetic alkaloid, has been constructed through the reaction of 2,6-bis(bromomethyl)pyridine with the dipotassium salt of retronecic acid (75).

III-78 ⟶

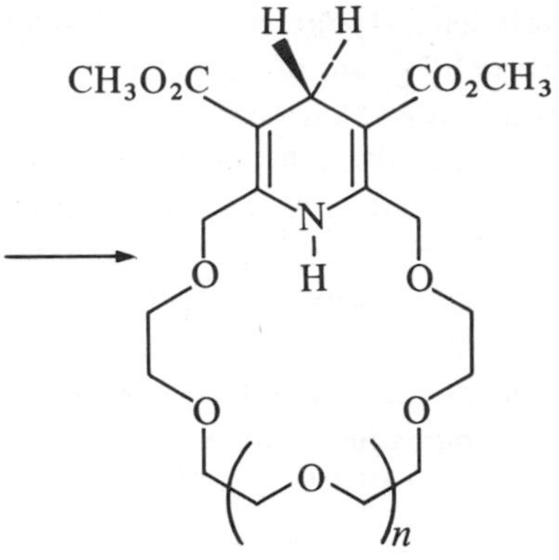

III-79 ($n = 1$–3)

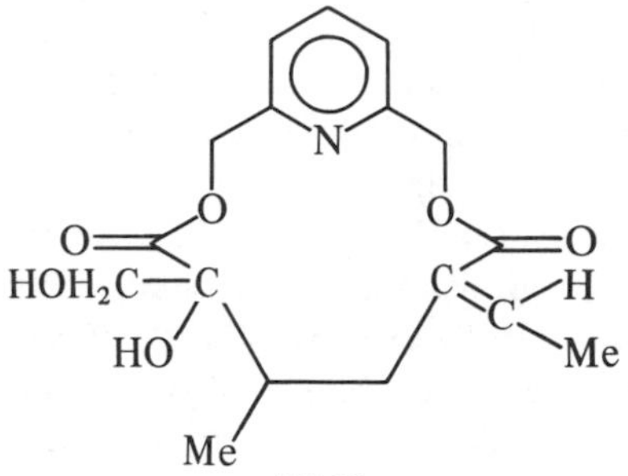

III-80

The most frequently used method for the synthesis of macrocyclic polyether diesters (e.g., **III-81**) has been the treatment of the very reactive *bis*-acid chloride and an appropriate glycol (76, 77) under high-dilution conditions. Treatment of the 2,6-pyridinedicarbonyl chloride with the appropriate oligoethylene glycols has generated the corresponding *bis*-lactone (77–81), whereas with α,ω-diols only the dimer **III-82b** was formed (77, 81). Host **III-81** gave strong complexes with Na$^+$, K$^+$, Ba^{2+}, and Ag$^+$ (log K = 4.3 to 4.9), while the analogous **III-63** ($n = 3$) formed stronger complexes with K$^+$, Ba^{2+}, and Ag$^+$ (82).

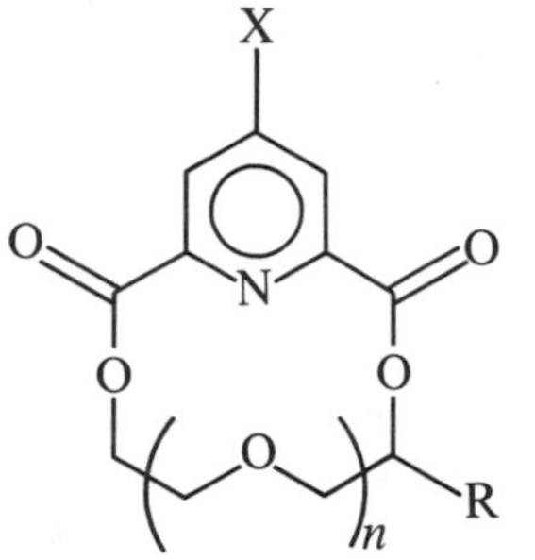

III-81 ($n = 2$–5)

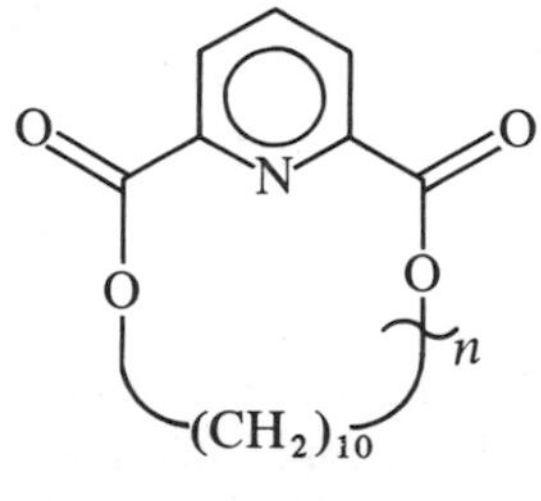

III-82a ($n = 1$)
III-82b ($n = 2$)

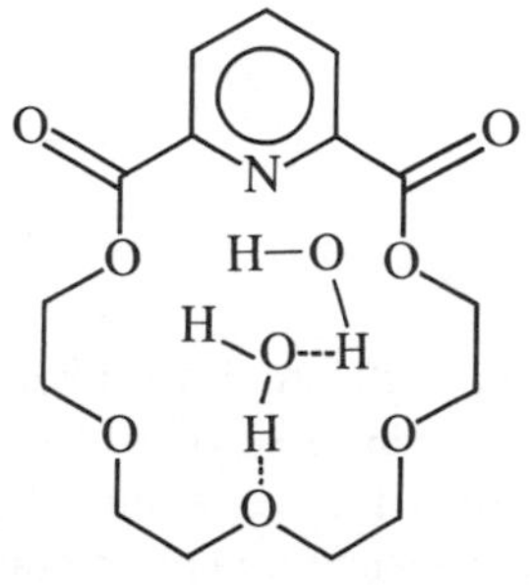

III-83

The x-ray analysis of **III-81** ($n = 3$) showed that all of the ring heteroatoms are about in the same plane and that the cavity is almost symmetrical (83, 84). Ligand **III-81** ($n = 3$) also formed a diaquo complex (**III-83**) in which the ring conformation is quite similar to the free ligand but the water hydrogen bonds to the ring oxygen remote to the pyridine nitrogen and *not* to the ring nitrogen atom (85). Due to the strong (2,6-) electron-withdrawing substituents on the pyridine subunit resulting in the diminished electron density of the pyridine *N*-electrons, the nitrogen atom is *not* the favored site of proton or guest attraction (86). Although the ammonium complex (**III-84**) is symmetrically hydrogen bonded to two ring oxygens and the nitrogen atom, the oxygen atoms are the preferred binding site and the nitrogen by default.

Magnetic nonequivalence for the protons labeled H_a and H_b in **III-84** was utilized for the computation of the energy of activation by means of temperature dependent ^{1}H NMR studies (82). These *bis*-lactones (e.g., **III-81**) have been subjected to complexation studies using ^{1}H NMR (82), calorimetric titrations, and membrane transport. From the free energy of activation (ΔG) data for the dissociation of the complex, the kinetic stabilities of the complexes with primary alkylammonium cations have been determined (82) and factors such as substituent effects, ring-size, and subring types have been studied. Rates of membrane transport of potassium nitrate with **III-81** ($n = 3$) were significant and decreased with order ($R = OCH_3 > H > Cl$), whereas rubidium nitrate favored ($n = 4$) and cesium nitrate ($n = 5$).

Results of calorimetric titrations suggested that **III-81** ($n = 3$) was an excellent host for Na^+, K^+, Rb^+ although the selectivity for K^+ over other ions was diminished (76, 87). Log K, ΔH, and $T\Delta S$ values for the complexation of **III-81** ($n = 3$) were also determined (88, 89). A review of these *bis*-lactones summarized the work thus far in the area (83).

III-84

III-85 ($n = 1, 2$)

Vögtle et al. (67) have isolated a crystalline neutral-molecule complex with 1:1 stoichiometry of **III-85** ($n = 1$) with nitromethane. A kinetic technique has been utilized to determine the relative complexing abilities of **III-81** for *p-tert*-butylbenzenediazonium tetrafluoroborate. Evidence was presented to support the involvement of the pyridine *N*-electrons in the complexation of this diazonium compound (67, 70).

Newkome et al. (90, 91) have constructed the second type of carbon–oxygen bridge *via* the direct nucleophilic displacement of the halo substituents of 2,6-dihalopyridines; 1:1 **(III-86)** and the dimeric 2:2 macrocycles have been isolated and characterized. Further application has been demonstrated in the preparation of tetraoxamuscopyridine **(III-86c)** (91). Similarly, nicotinic acid crown ethers **(III-87, III-88)** have been synthesized (92, 93). In these macrocycles (type 2), NMR Eu(fod) shift studies indicated that the predominant sites of metal coordination are the central bridging ethereal oxygens in **III-87** ($R = CN$; H), whereas when $R = CONMe_2$, the amide oxygen is the favored locus (92–94). Ligands **III-87** have been used as catalysts for aminolysis in an aprotic solvent and were found to be effective (95, 96).

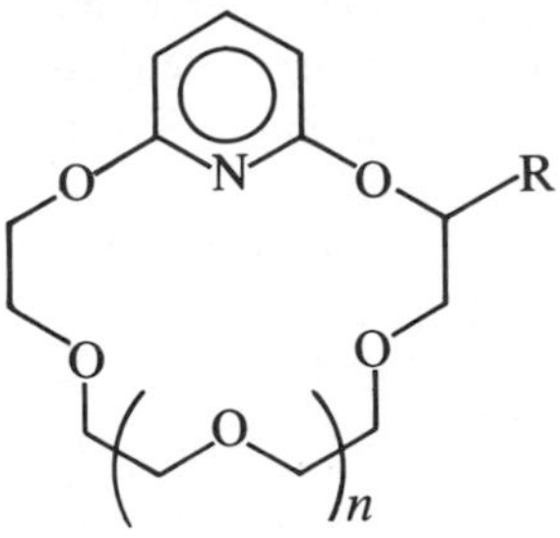

III-86a [$n = 1$; $R = H$]
III-86b [$n = 1$; $R = Me$]
III-86c [$n = 0$; $R = Me$]

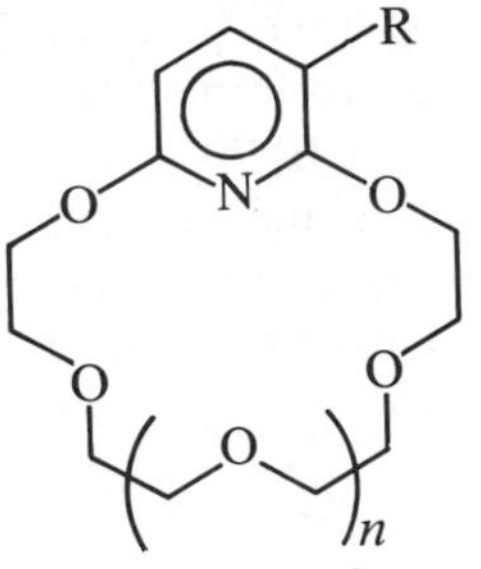

III-87 [$n = 1-3$]
  a, $R = CONMe_2$
  b, $R = CN$
  c, $R = H$

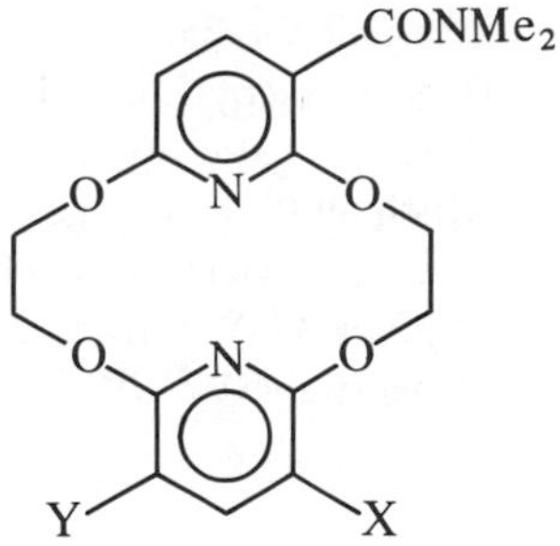

III-88 a, $X = H$; $Y = CONMe_2$
     b, $X = CONMe_2$; $Y = H$

Cram et al. (54, 55) synthesized numerous multiheteromacrocycles (**III-89–III-93**) which contained more than one pyridine subunit. Standard synthetic procedures from the appropriate diols were used. The complexing power, association constants [K_a] with *tert*-butylammonium thiocyanate, and pK_a values of these

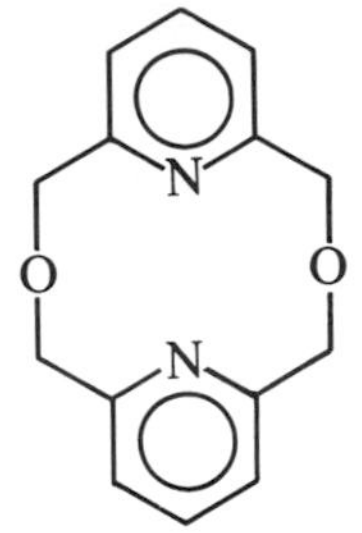

III-89

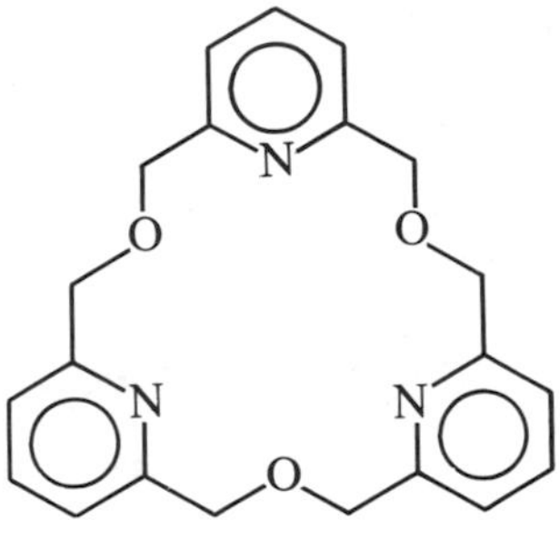

III-90

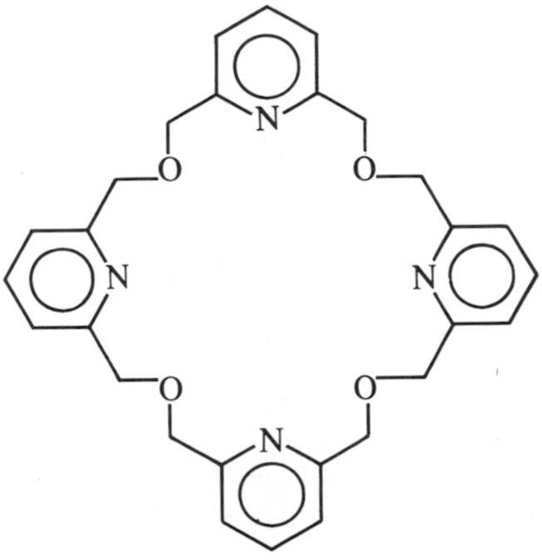

III-91

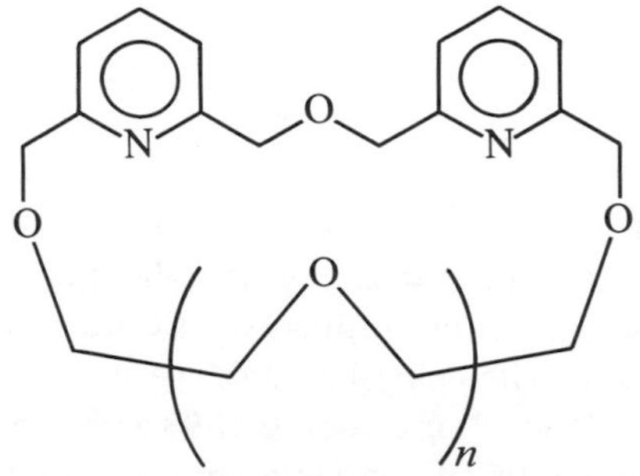

III-92 ($n = 0-5$)

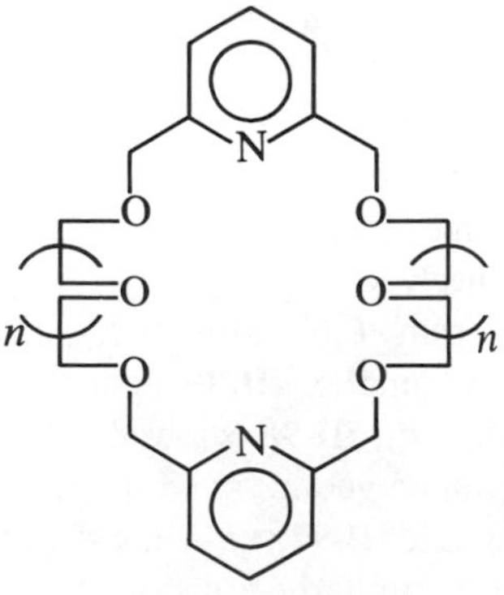

III-93 ($n = 0-3$)

macrocycles have been reported (55, 56, 63). Unlike in the *bis*-lactones (e.g., **III-81**), the pyridine nitrogens are the preferred site of hydrogen bonding as demonstrated by the complex derived from **III-90** and the *tert*-butylammonium ion in which all three nitrogens are hydrogen bonded to the ammonium ion (55).

Chiral macrocycles (**III-94, III-95**) were synthesized from chiral subunits (R,R)-$(+)$-N,N,N',N'-tetramethyltartramide and 1,2:5,6-di-*O*-isopropylidene-*D*-mannitol, respectively (97–99).

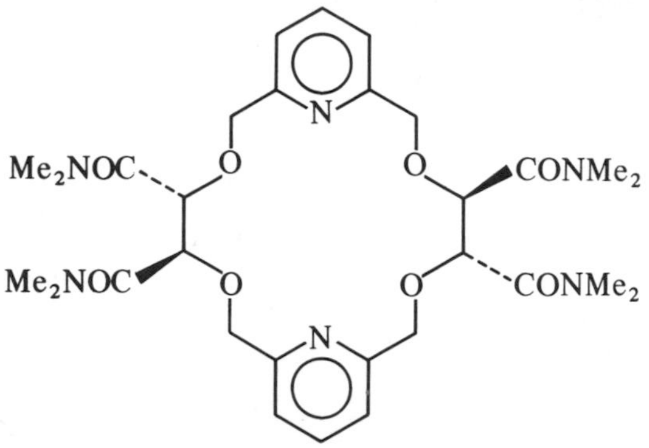

III-94

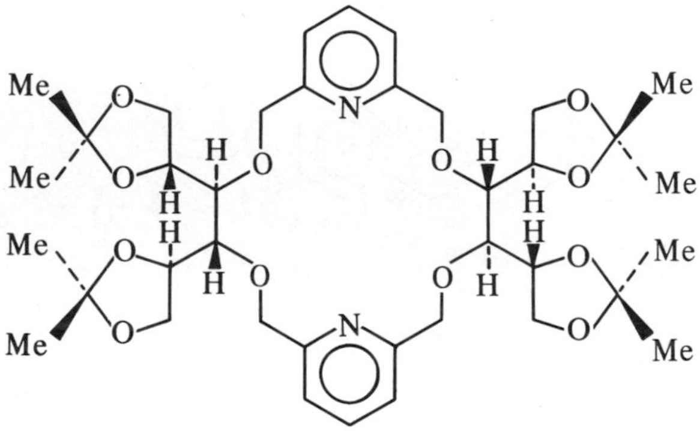

III-95

The synthesis of macrocyclic azaethers (**III-96**) has been achieved by cyclocondensation of 2,6-*bis*(hydroxymethyl)pyridine with α,α'-dibromo-*O*-xylene (53). Newkome et al. (100) prepared dipyridine crowns (**III-97**) by the reaction of ethyleneglycols with 6,6'-*bis*(chloromethyl)-2,2'-dipyridine, prepared in four steps from 6,6'-dibromo-2,2'-dipyridine (100a). Crown **III-97** ($n = 4$; $m = 1$) formed complex **III-98** upon treatment with cobalt(II) chloride; the x-ray crystal structure of **III-98** showed it to possess a pentacoordinated, distorted trigonal bipyramidal geometry with one coordination site being with an oxygen atom (100). Macrocycle **III-97** ($m = 2, n = 3$) formed a dinuclear complex **III-99** with copper(II) chloride; similarly a distorted trigonal bipyramidal configuration was found (101) at each bonding site.

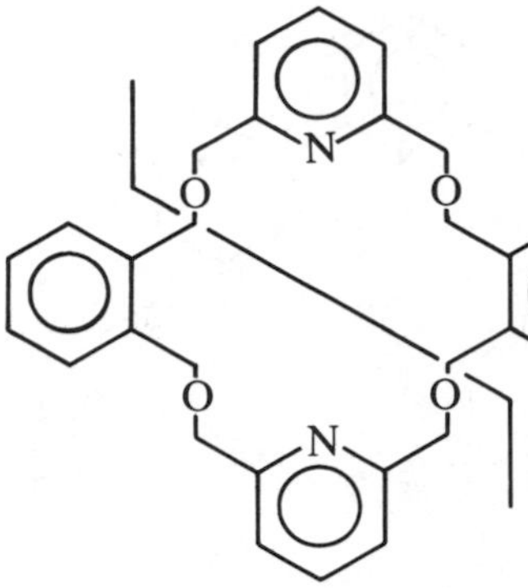

III-96 ($n = 1$–4)

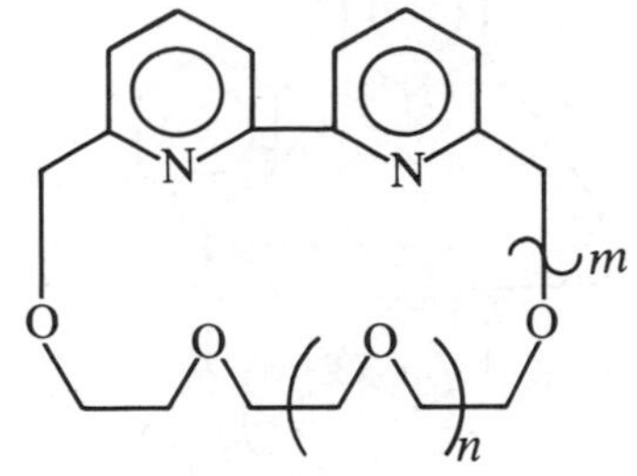

III-97 $m = 1, 3$
$n = 3, 4$

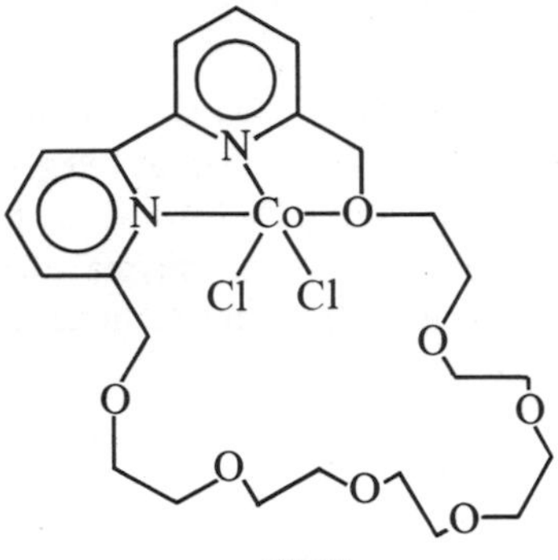

III-98

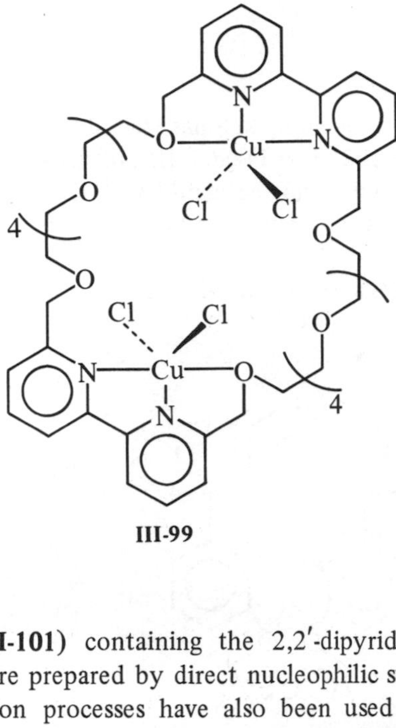

III-99

1:1- and 2:2-crown ethers (**III-100**–**III-101**) containing the 2,2′-dipyridine subunit having direct ring-oxygen bonds were prepared by direct nucleophilic substitution reactions (102). Novel contraction processes have also been used to prepare **III-100** ($n = 4$), in which either phosphorus (103) or carbon monoxide (104) is expelled. Variable temperature NMR studies of **III-100** ($n = 1$–4) support a conformational rotation of the larger rings from an *anti* to *syn* orientation at reduced temperatures ($< -70°C$). Only about 5% of **III-100** ($n = 3$) occupies the *anti* conformation at $-100°C$; an x-ray crystal structure established the *anti* conformation in the solid state (102).

Newkome et al. (105, 106) synthesized a series of spiro and ketonic macrocycles with specific cavities by application of the heteroatomic nucleophilic substitution procedure. The larger ketal macrocycles (e.g., **III-102**) were hydrolyzed to the

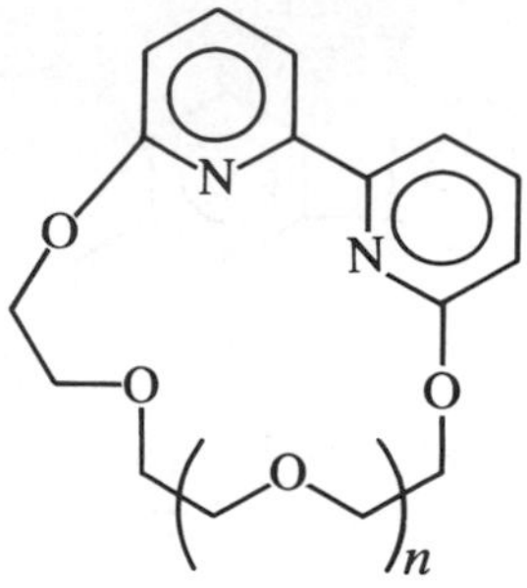

III-100 [n = 0–4]

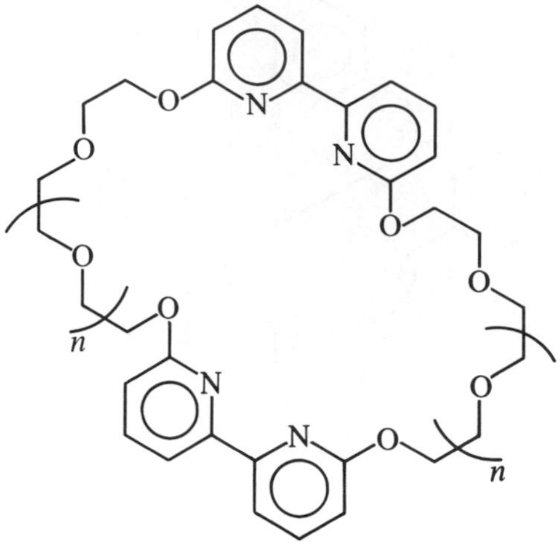

III-101 [n = 0–4]

corresponding ketones **(III-103–III-104)** in good yields. X-ray crystal structures of **III-102** (n = 1) and **III-103** (n = 1) indicated that the ketal macrocycles were globular and the main backbone of the molecule wraps around a central cavity in the shape somewhat like the seam of a tennis ball, whereas the ketonic macrocycle possessed a more open conformation. In both molecules, the imidate linkages all have a near 0° dihedral angle.

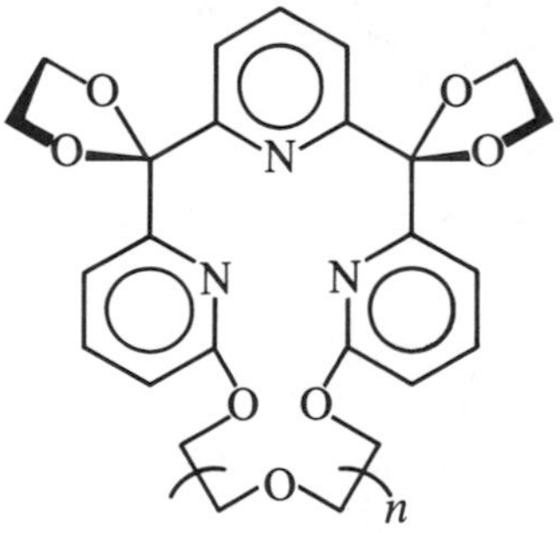

III-102 n = 0–5

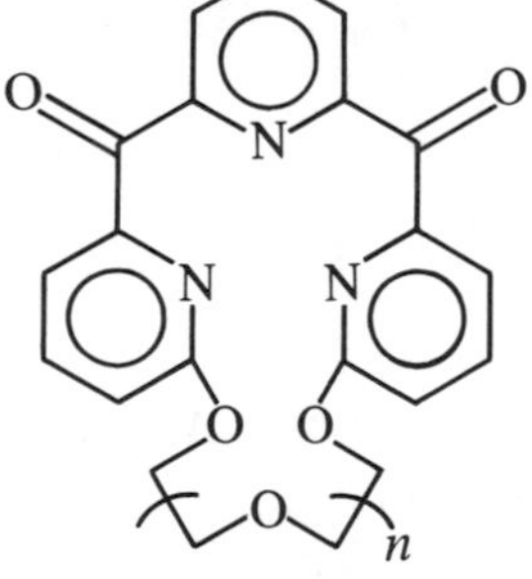

III-103 (n = 1–5)

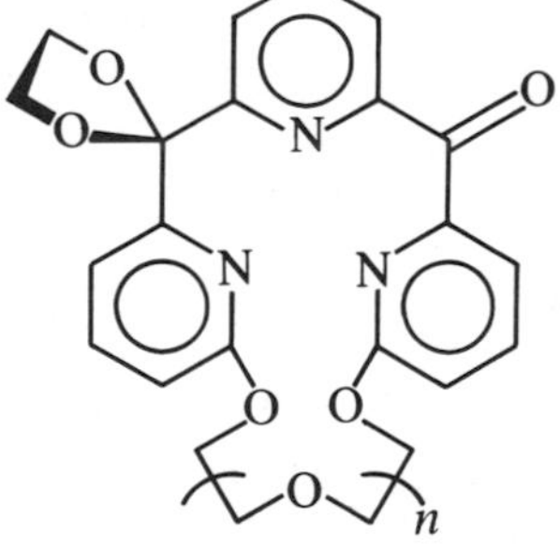

III-104 (n = 1, 4)

Other spiro and ketonic macrocycles **(III-105–III-108)** have been prepared (104, 107). The carbinol macrocycle **(III-109)** was obtained by reduction of the corresponding ketone with sodium borohydride. X-ray crystal studies of **III-107** (n = 5) revealed that it possesses a cavity containing a water molecule, while the small rings of **III-107** (n = 3) and **III-105** (n = 3) showed no evidence of included water (86, 104).

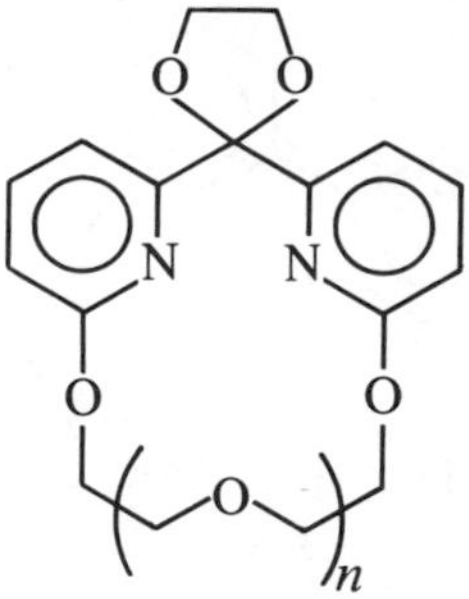

III-105 ($n = 1$–5)

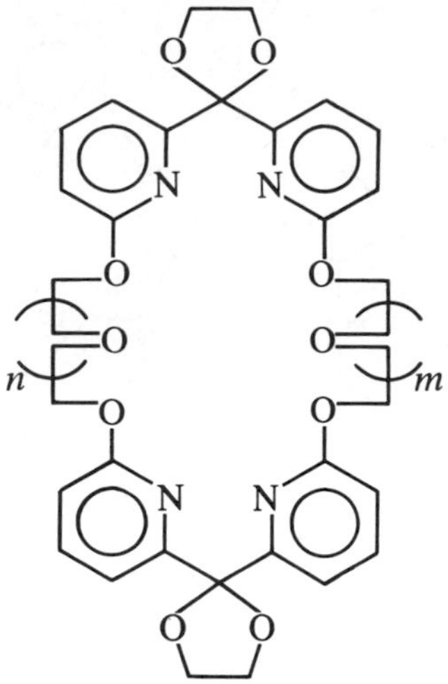

III-106 $n = 1; m = 2$
$n = m = 2$

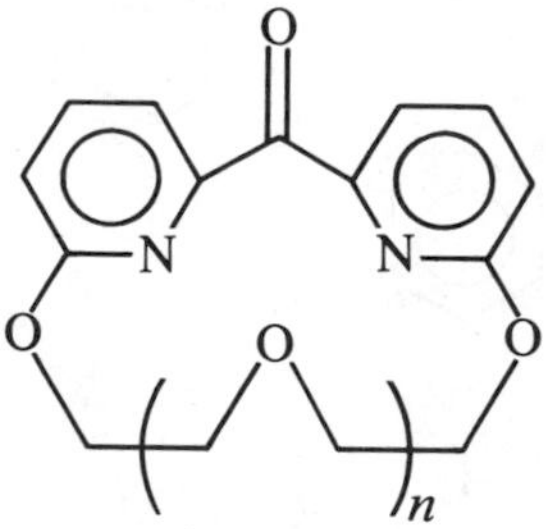

III-107 $n = 1$–5

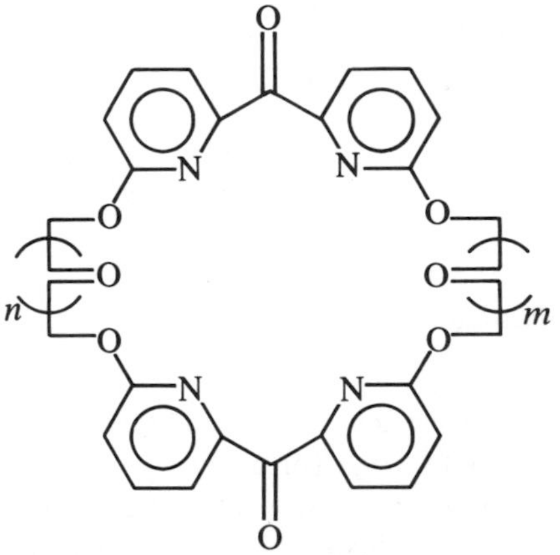

III-108 $n = 1; m = 2$
$n = m = 2$

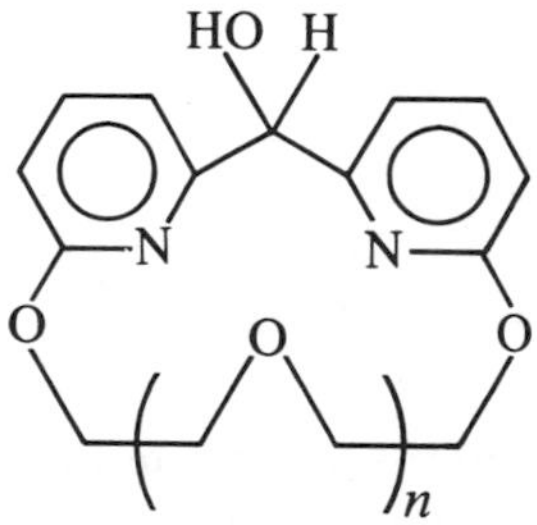

III-109 $n = 1$–5

The terpyridine crown ethers (**III-110**) were prepared in four steps from 2,6-dibromopyridine (108). Diketone **III-111** was easily transformed into **III-110** by treatment with methanolic ammonium acetate. X-ray analysis of **III-110** showed it to possess the *anti,anti* conformation **III-110a**, rather than the *syn,syn* conformation envisioned for complexation.

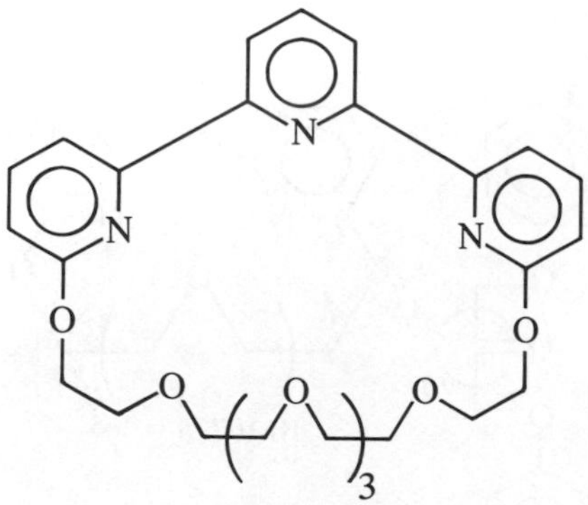

III-110

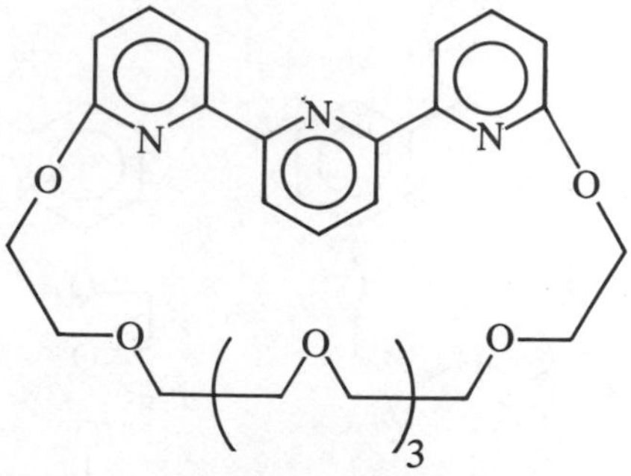

III-110a

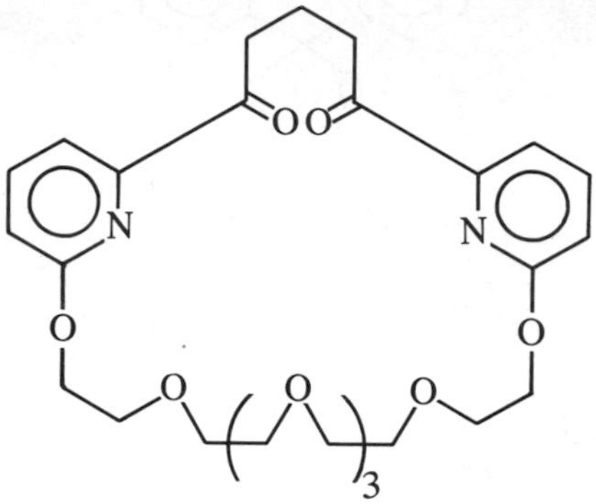

III-111

5.   Carbon–Nitrogen Bridges

The carbon–nitrogen bridged pyridine macrocycles have generally been produced by a Schiff-base condensation of either 2,6-diformyl (or acetyl)-pyridine with a *bis*-primary amine. Curry and Busch reported the first penta- and hexadentate macrocycles **(III-114** and **III-115**, respectively) in this series through the utilization of metal ion catalysis (109). The cyclocondensation process around a metal ion has become known as the "template effect" (110–114). Such metal ion intervention in a Schiff-base condensation has been utilized in the preparation of tetra-**(III-112)** (115–133), penta-**(III-114, III-116)** (109, 115, 130, 134–147), and hexadentate **(III-115)** (109, 148) pyridine macrocycles and their complexes. Catalytic reduction of the imine bonds in these unsaturated macrocycles gave the saturated tetra-**(III-113)** (118, 122, 125, 131, 149–153) and pentadentate (137) ligands.

Vögtle et al. (154–156) have synthesized a series of aza-bridged *bis*-lactams (e.g., **III-117)** *via* the reaction of 2,6-pyridinedicarbonyl chloride with diverse diamines under high-dilution conditions as described by Stetter and Marx (157). In a similar manner, aza-crowns with 2,2′-dipyridine were constructed (158). Condensation of

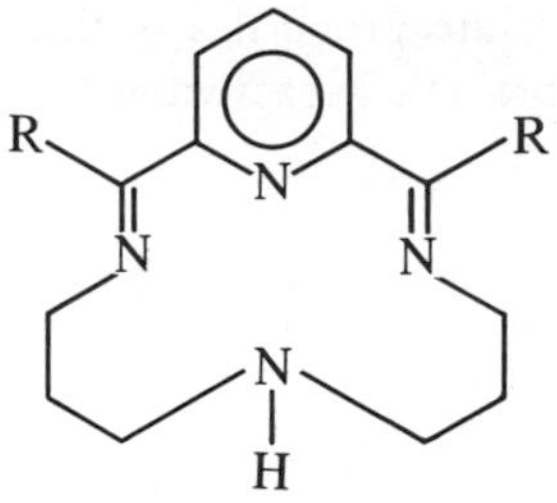

III-112

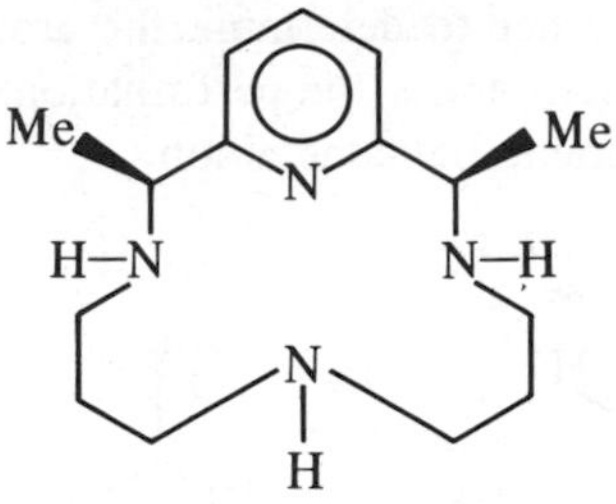

III-113 (meso)

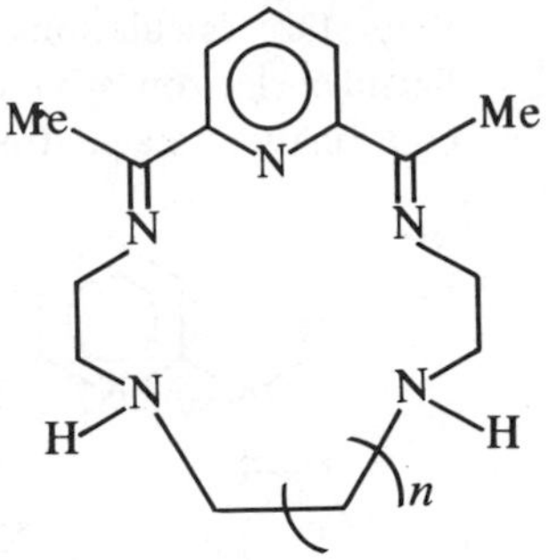

III-114 (n = 1-2)

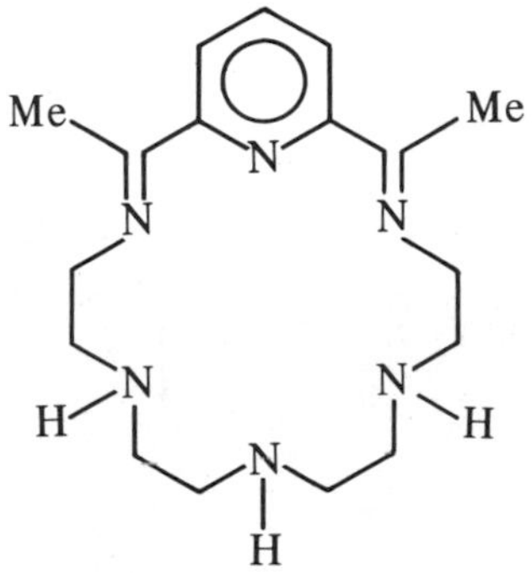

III-115

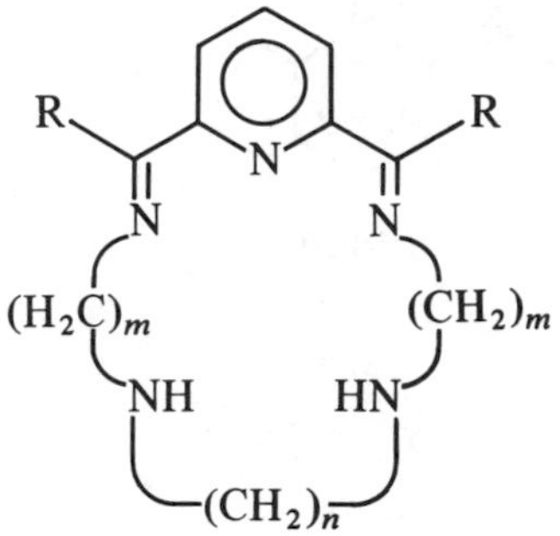

III-116   m = 3; n = 2
          n = 3; m = 2

2,6-dipicolinoyl dihydrazine with β-diketones has also been accomplished utilizing the template effect to give **III-118** (159).

Thermolysis of either a dicarbonyl compound (an imide) (160–166) or a dichloride (167) with 2,6-diaminopyridine gave **III-119** and **III-120**, respectively.

Bamfield and Mack (124, 168) directly constructed macrocycles as the complex, either by the interaction of 1,3,3-trichloroindolenine with pyridine in the presence of the metal salt or by heating the metal complex of 2,6-(1'-iminoisoindolin-3'-ylideneamino)pyridine with 1,3-diiminoisoindoline. Honeybourne (169) did the

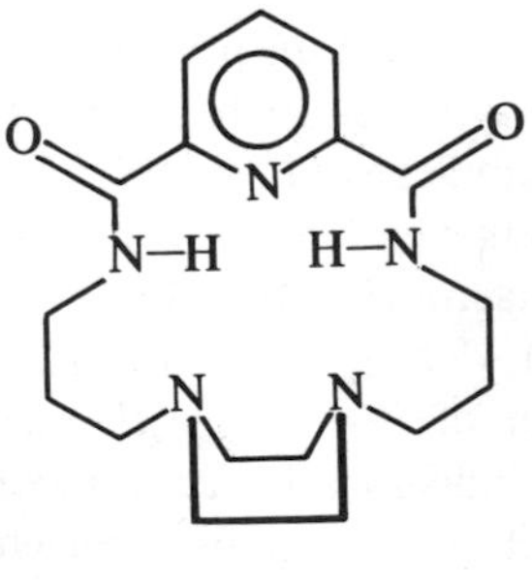

III-117

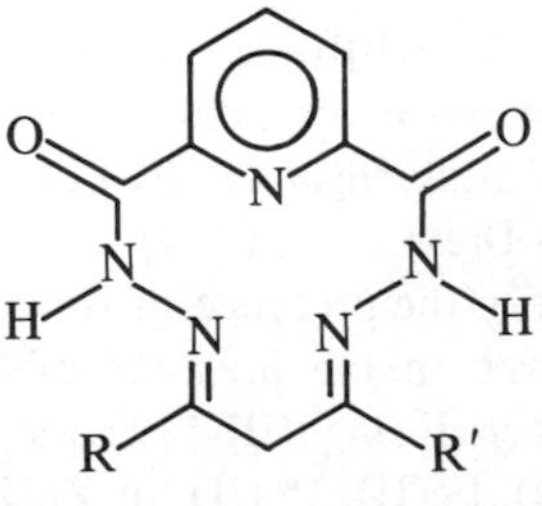

III-118

theoretical calculations in order to determine the ground state properties of the ligand π-electron system as to assess the perturbation of predicted reactivities by coordination to a central paramagnetic metal ion.

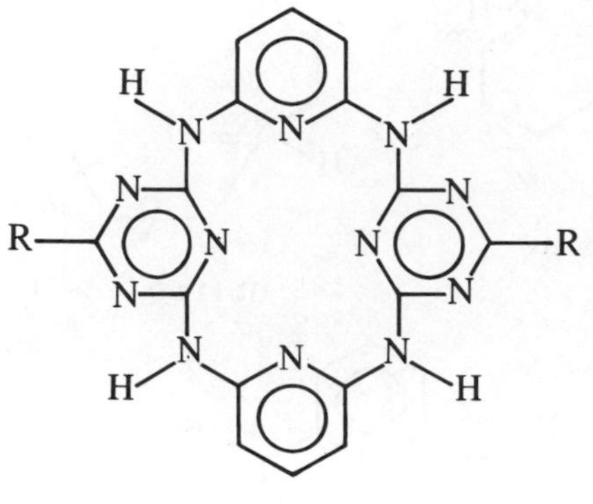

III-120

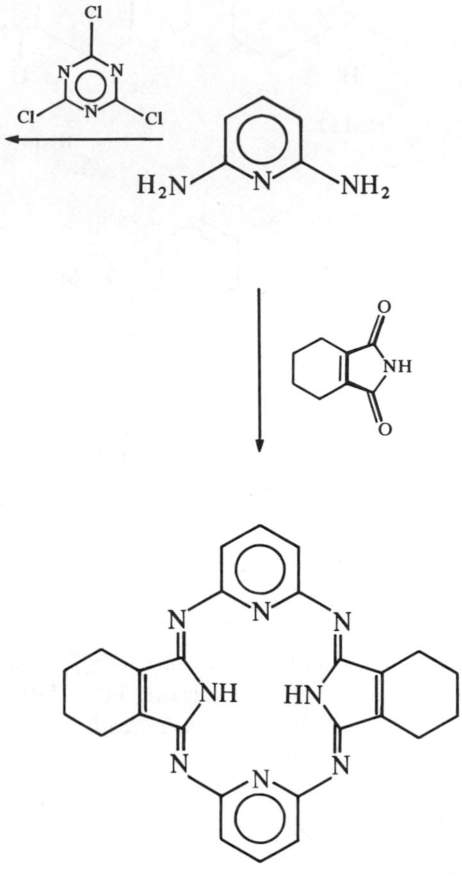

III-119

In the Schiff-base series, the use of linear polyfunctional *bis*-primary amines in the presence of suitable metal ions has lead to the synthesis of tetra-, penta-, and hexadentate ligands **(III-121, III-122)** and their complexes (123, 148, 170–173). When Drew et al. (174) condensed 2,6-diacetylpyridine with 3,3'-diaminodipropyl-amine in the presence of Ag$^+$ ions, a 28-membered macrocycle **III-123** was obtained; however, in the presence of Ni(II), Cu(II), Mn(II), Co(II), or Zn(II) salts the 14-membered ring **(III-112)** was generated. Treatment of **III-122** [Ba(II) salt] with Mn(II), Fe(II), Co(II), or Zn(II) caused **III-122** to undergo a ring-contraction from a 18- to 15-membered ring with the concomitant formation of a 5-membered imidazoline ring **(III-124)** (175).

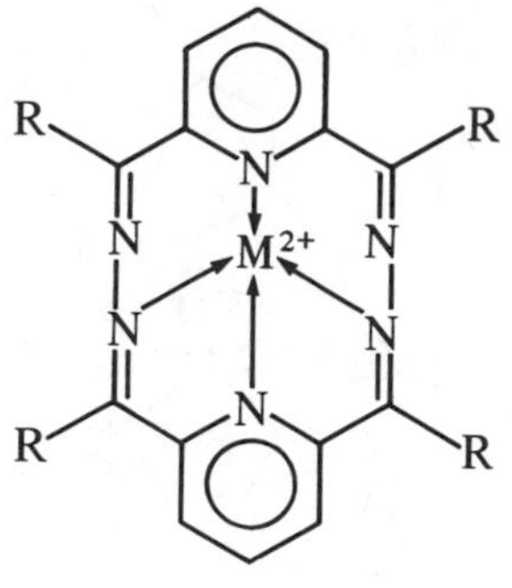

III-121

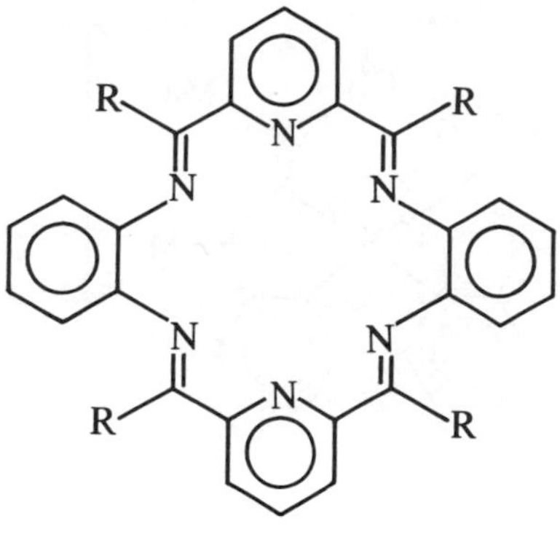

III-122

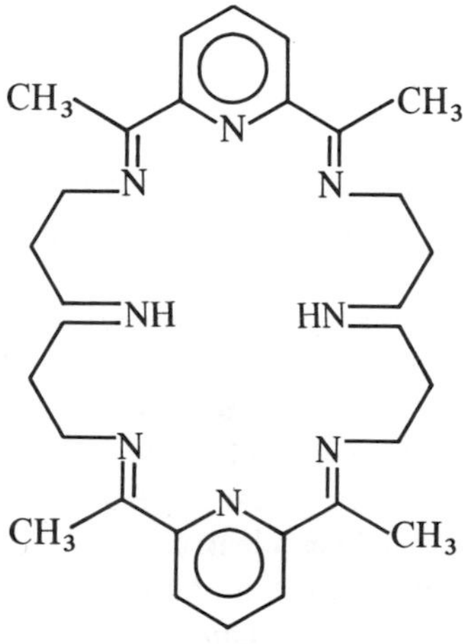

III-123

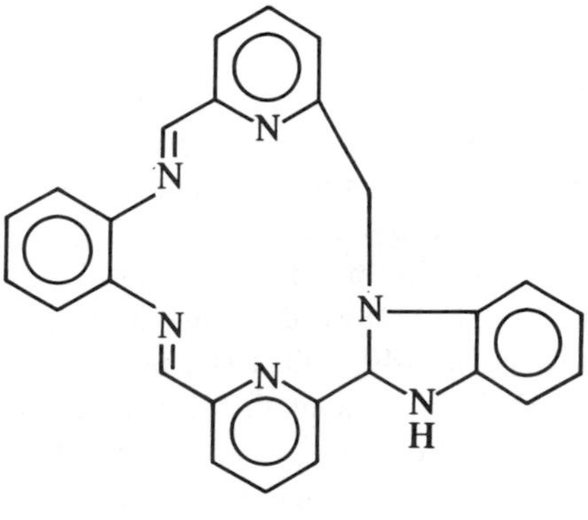

III-124

Incorporation of a 1,10-phenanthrolino (176) or 2,2'-dipyridino (177, 178) subunit has been accomplished by condensation of the dialdehyde (or ketone) with an appropriate diamine to give **III-125–III-127**. Complexes **III-128–III-129** of the dipyridine ligands have been realized (179–182).

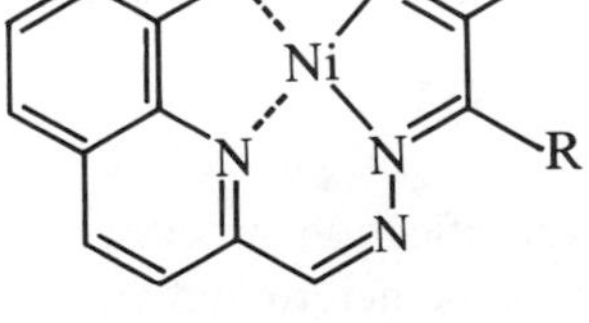

III-125

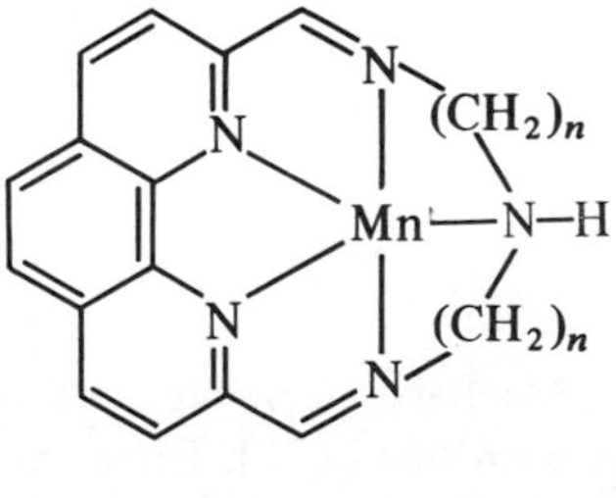

III-126 (n = 2, 3)

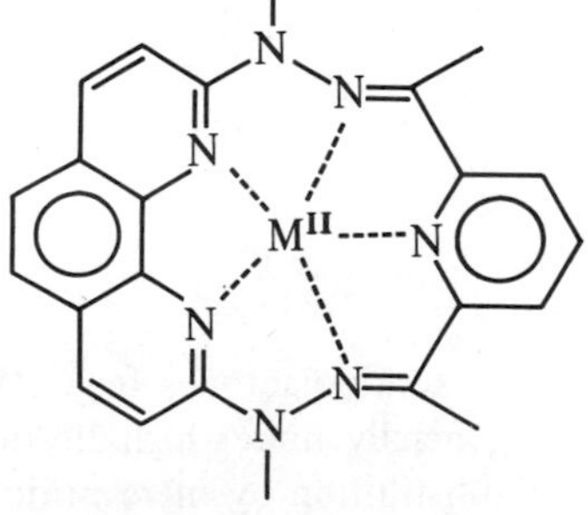

III-127 [M^{II} = Mn, Zn or Cd]

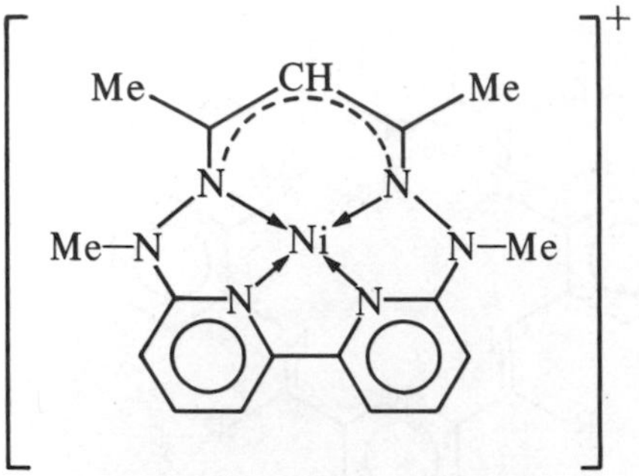

III-128 (X = IV, ClO$_4^-$, PF$_6^-$)

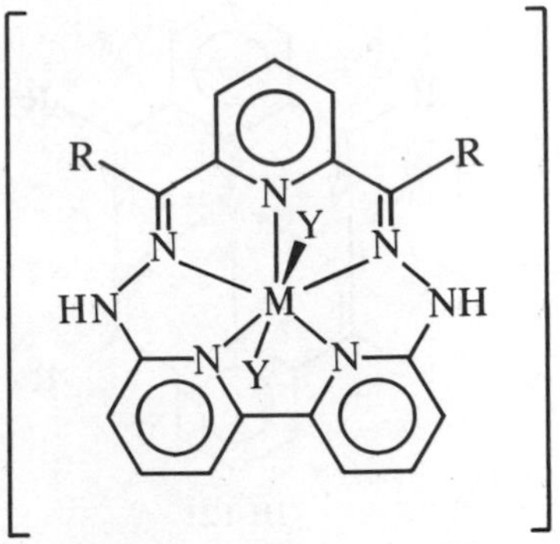

III-129 (M = Zn; Y = H$_2$O)

6. Carbon–Sulfur Bridges

Like the carbon–oxygen bridged macrocycles, two general types of carbon–sulfur bridges are possible: (1) isolated bridging sulfur atom(s), and/or (2) directly connected sulfur atoms to the subheterocyclic ring(s). Vögtle et al. (183) first demonstrated the construction of type 1 macrocycles by the reaction of 2,6-*bis*(bromomethyl)pyridine with dithioresorcinol to give **III-130.** Vögtle et al. (66, 154, 155, 184–189), Boekelheide et al. (28, 32–34), Martel and Rasmussen (39, 41), Galuszko (40, 190, 191), Bottino (192, 193), Bryan et al. (194), and Tsuge et al. (195) have all used this condensation procedure or some minor modifications thereof. Vögtle et al. (66, 154, 188, 196) have successfully condensed 2,6-pyridine-dithiol with an appropriate *bis*(halo)methylenic compound, thus demonstrating a route to the second type of sulfur-bridged macrocycle for example, **III-131** and **III-132.**

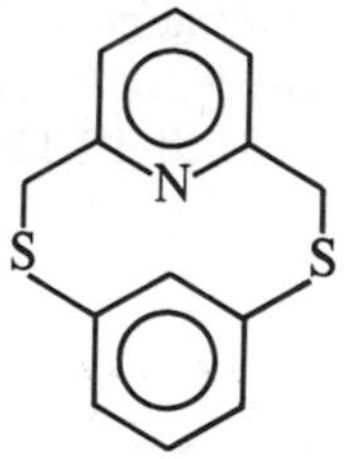

III-130 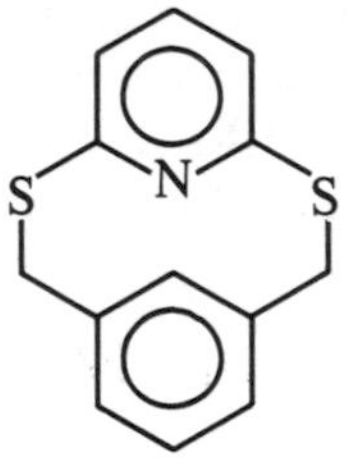 **III-131**

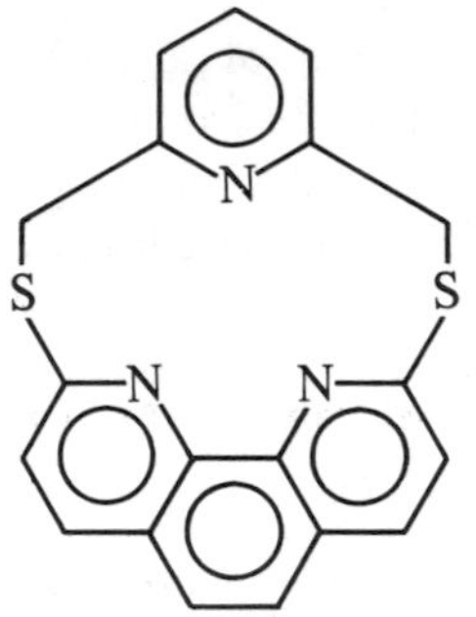

III-132

Bis-thiolactones (e.g., **III-133**) have been prepared by esterification processes generally under high-dilution conditions (77). Direct nucleophilic heteroaromatic substitution by mercaptide ion has been used to generate **III-134** and **III-135** (105) as well as the simpler sulfur-bridged systems **III-136** and **III-137** (197). Carbon-

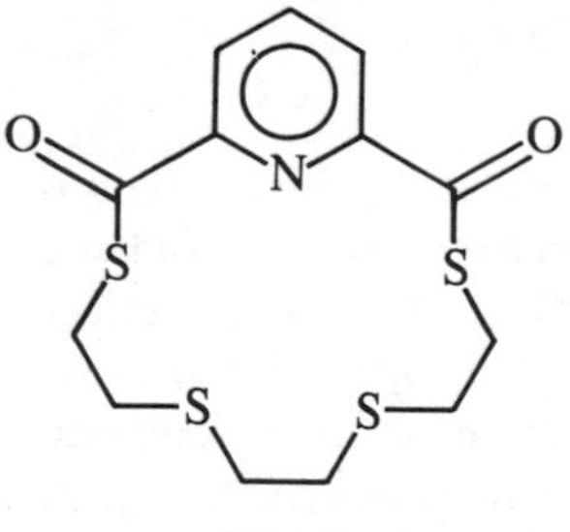

III-133

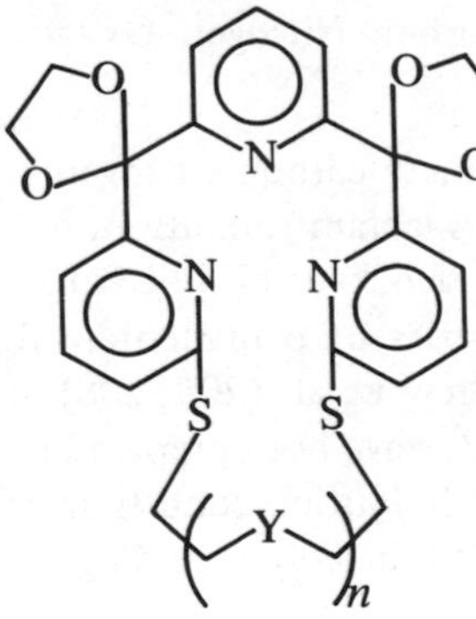

III-134 Y = S; n = 0–2
Y = O, n = 1

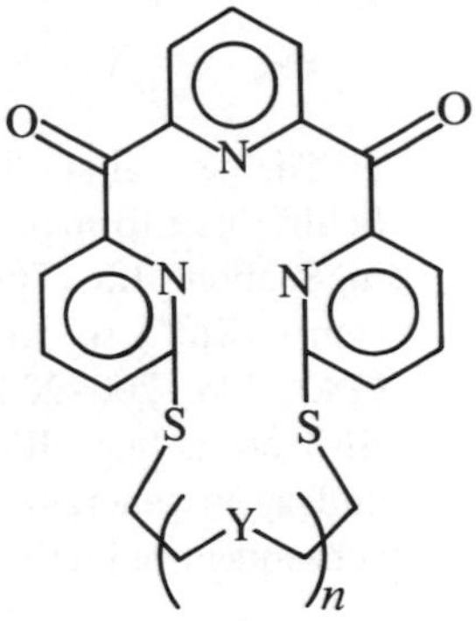

III-135 Y = S; n = 0–2
Y = O; n = 1

bridged cyclophanes have been derived from the intermediate carbon–sulfur bridged precursors, such as **III-138** (35, 43, 198).

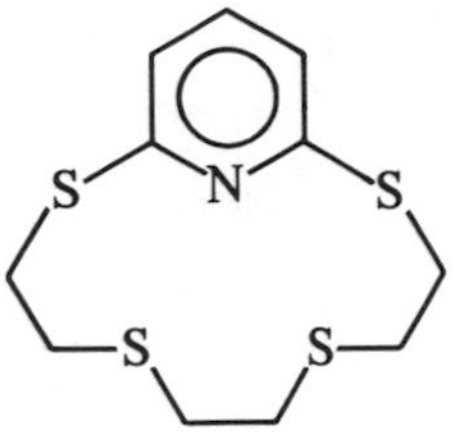

III-136

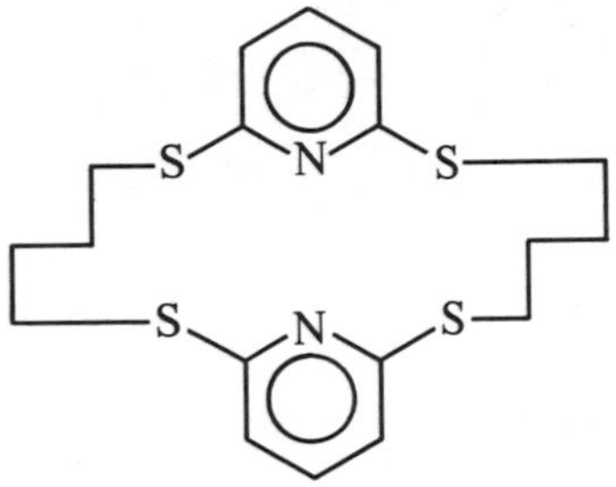

III-137

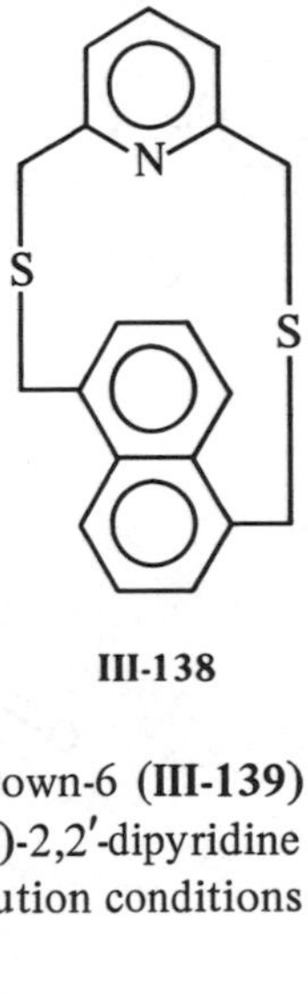

III-138

Recently, polypyridyl carbon–sulfur bridged analogs to 18-crown-6 (**III-139**) and larger oligomers have been prepared from 6,6′-*bis*(chloromethyl)-2,2′-dipyridine (100a) and 6,6′-*bis*(mercaptomethyl)-2,2′-dipyridine under high-dilution conditions (199).

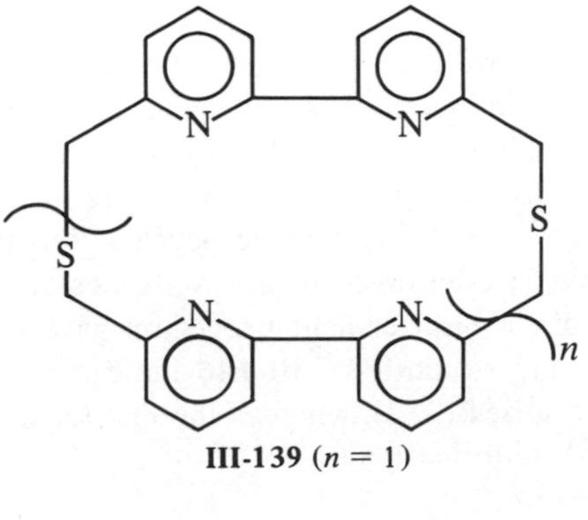

III-139 (n = 1)

7.   Carbon–Nitrogen–Oxygen Bridges

The general syntheses of these carbon–nitrogen bridged macrocycles are *via bis*-Schiff base formation and *bis*-lactam formation, both procedures were considered in Section II.5. Typically, Alcock et al. (115) have applied the template effect (Mn^{2+}, Zn^{2+}) to prepare a series of pentadentate ligands (**III-140**); Fenton et al. (143, 144, 200–204) and Drew et al. (205–208) have presented other examples. *Bis*-lactams (e.g., **III-141–143**) have been prepared by several obvious combinations and again generally under high-dilution conditions (66, 154, 156, 158, 209–210); subsequent reduction generated the corresponding *bis*-amines (e.g., **III-144**).

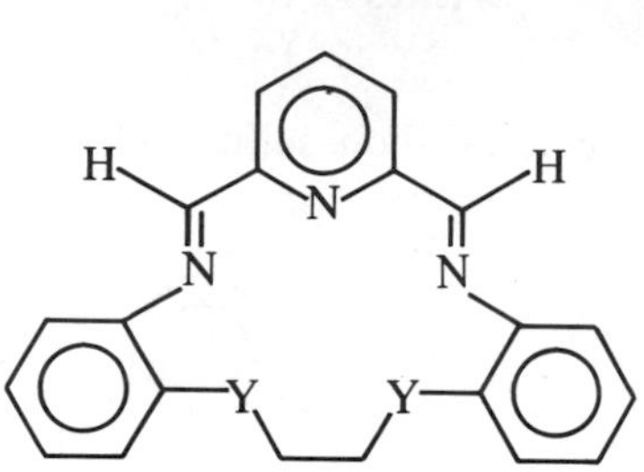

III-140 (Y = NH, O or S) 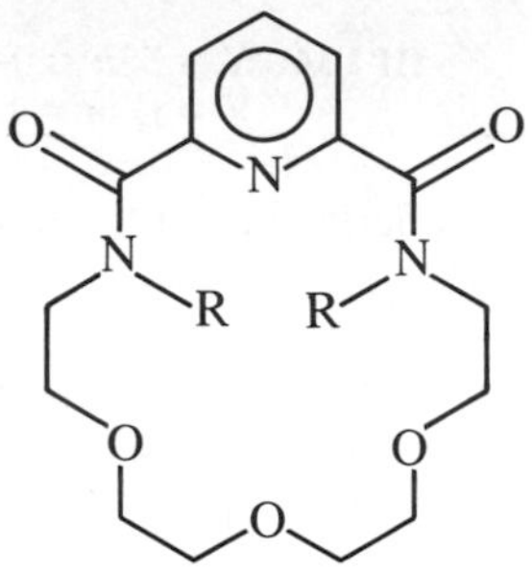   **III-141**

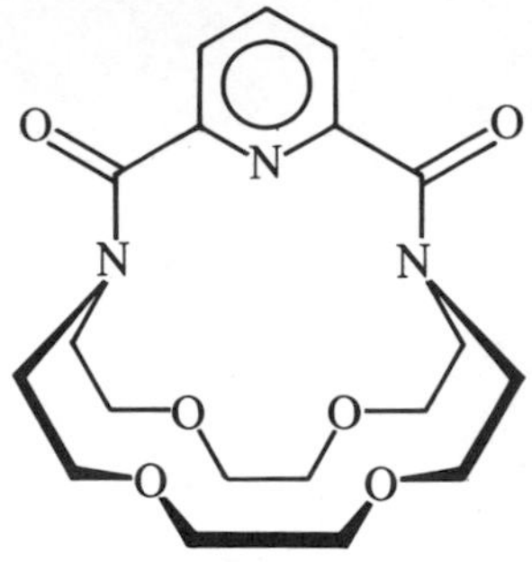

III-142

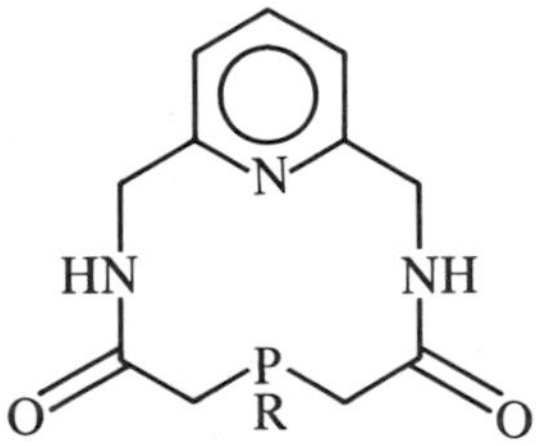

III-143
R = O; OCH$_2$CH$_2$O—,
—CH$_2$CH—,

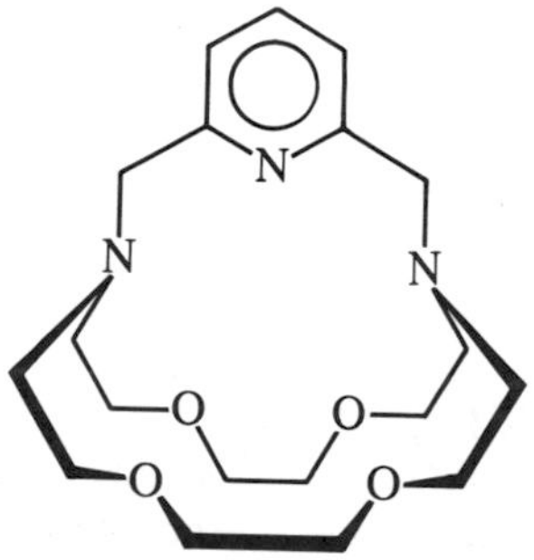

III-144

Bis-oximes (**III-145**) have been prepared by a simple nucleophilic substitution reaction with 2,6-*bis*(chloromethyl)pyridine and the appropriate dianion (211). Newkome et al. utilized direct nucleophilic heteroaromatic substitution to generate initially open-chained *bis*-tertiary amines, which were quaternized with *bis*-halides, then dealkylated to give the simple azomacrocycles. Repetition of this alkylation–dealkylation sequence gave entrance to a novel series of cryptands **III-147** (213). Application of this direct displacement procedure gave rise to **III-146** (212) and **III-148** (214). The x-ray structure of **III-146** indicated that the two bridge-head nitrogens are sp^2 hybridized (212), whereas the methylenic analog **III-149** has the anticipated *in-in* (sp^3) conformation (215).

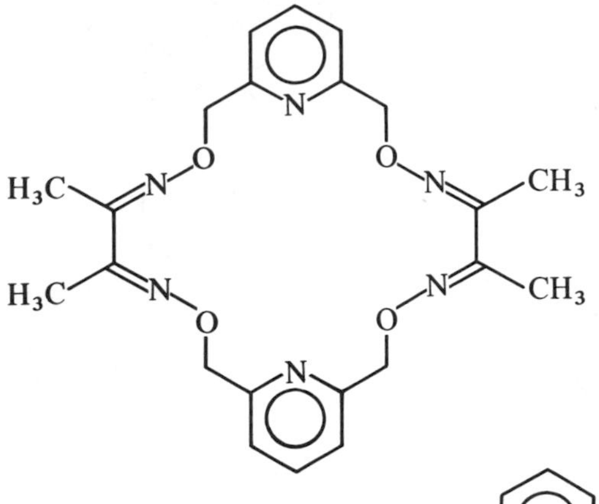

III-145

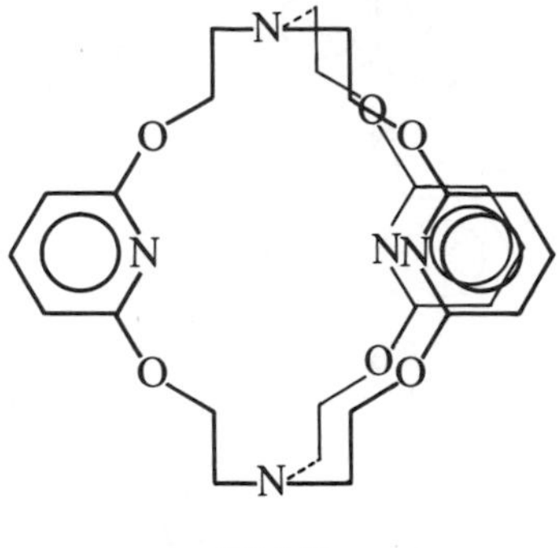

III-146

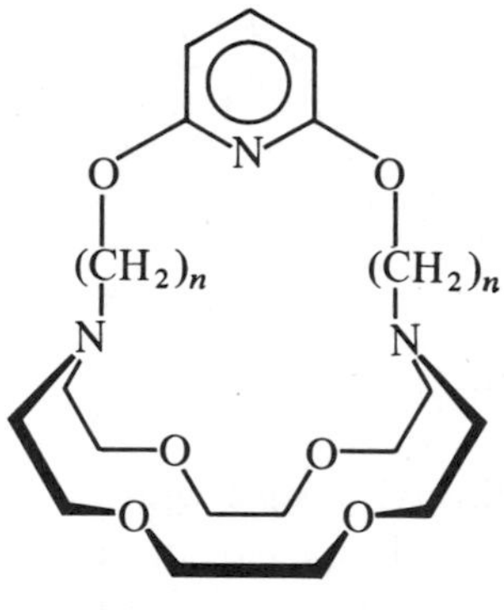

III-147 ($n = 2$–3)

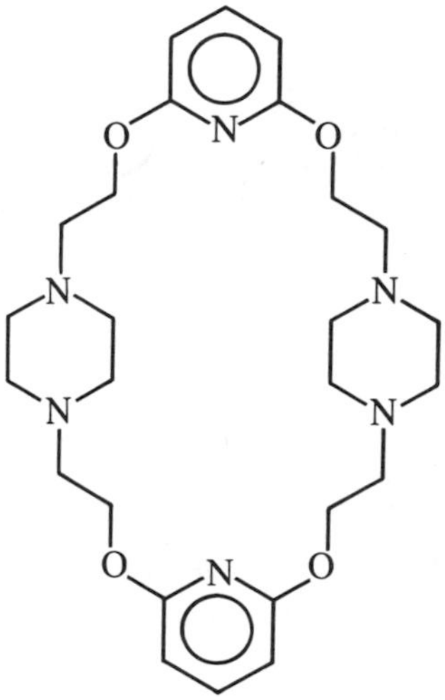

III-148

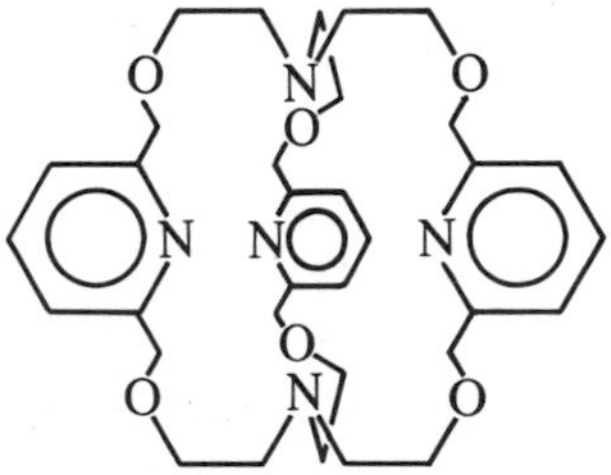

III-149

8. Carbon–Nitrogen–Sulfur Bridges

Vögtle (155) prepared **III-150** *via* standard nucleophilic substitution and *bis*-lactam (**III-151**) by an amidation procedure.

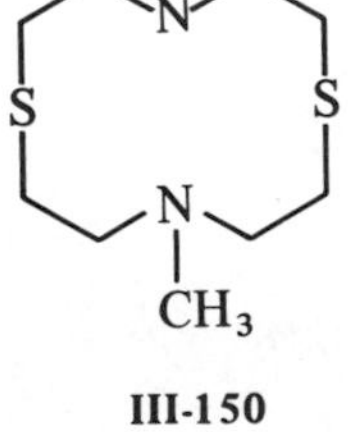

III-150

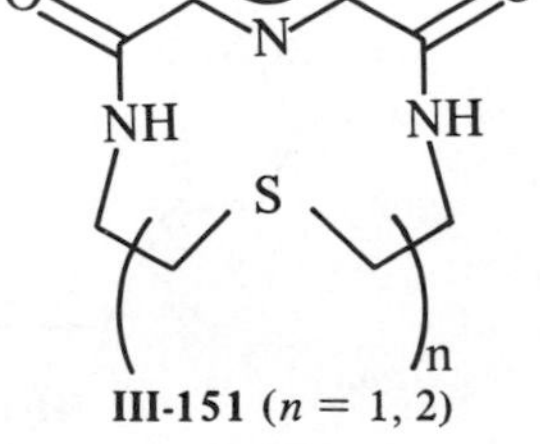

III-151 ($n = 1, 2$)

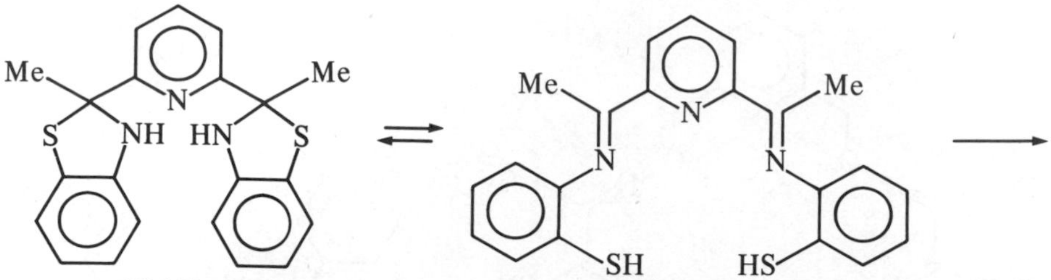

III-152a

III-152b

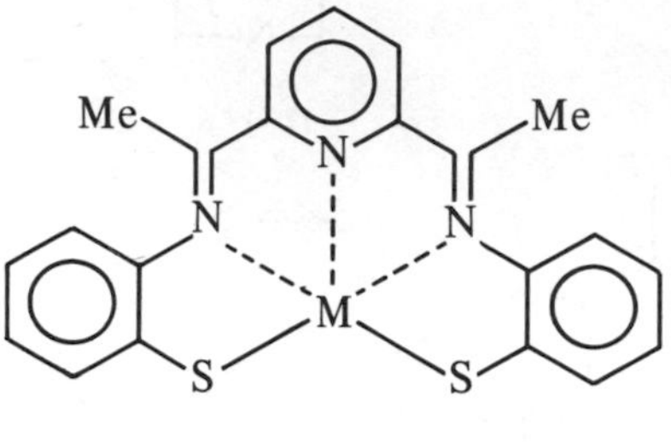

III-153

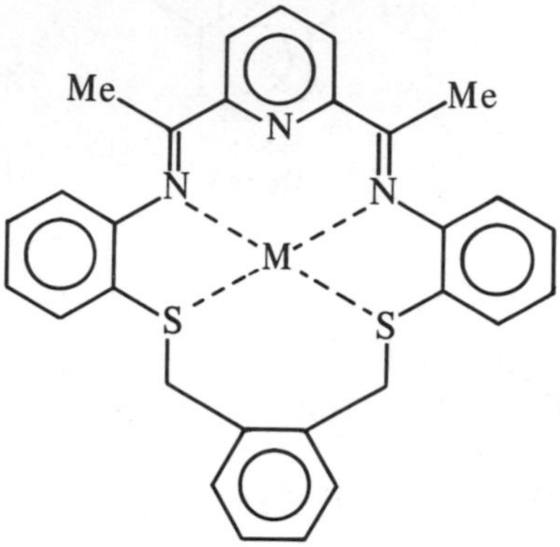

III-154

Lindoy (216) and Busch (110) have shown that 2,6-diacetylpyridine with 2-aminobenzenethiol gave the *bis*-benzothiazoline (**III-152**), which upon treatment with zinc or cadium acetate was transformed to the Schiff base complex **III-153** (217, 218). Macrocyclization was accomplished by subsequent treatment of this complex with α,α'-*bis*(bromomethyl)-*o*-xylene to give **III-154** (217). Drew et al. (143, 219) successfully applied Schiff-base chemistry to prepare **III-155** and **III-156**.

Macrocycles with three different subunits (e.g., **III-157**) can be constructed by heating 2,5-diamino-1,3,4-thiadiazole with an appropriate 1-iminoisoindolinylidene derivative (220).

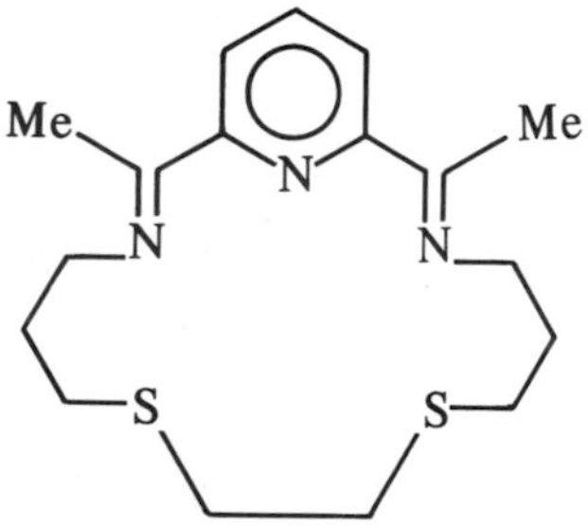

III-155

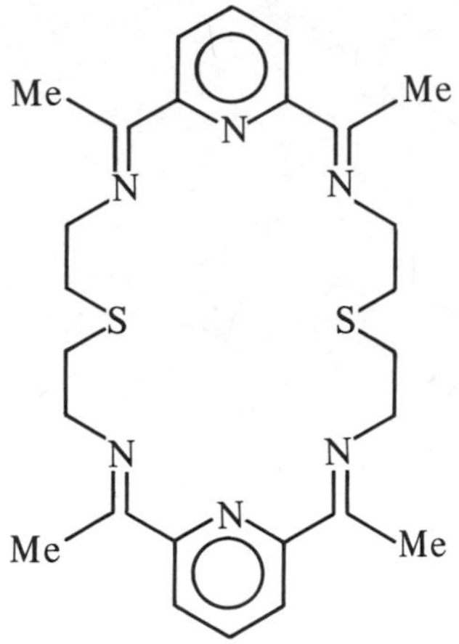

III-156

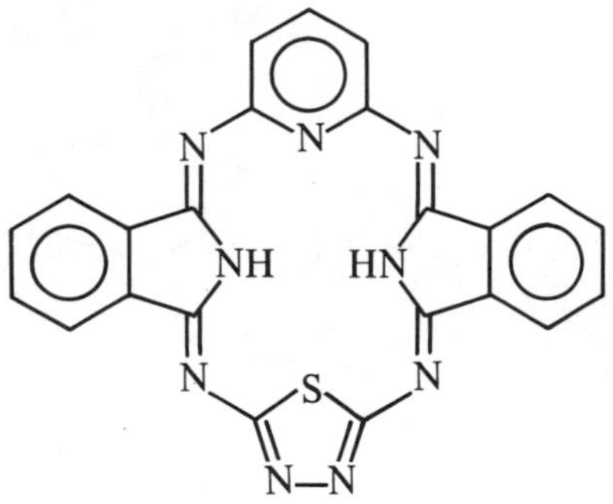

III-157

9. Carbon–Nitrogen–Phosphorus Bridges

Few phosphorus-bridged macrocycles containing a pyridine subring have been reported. Holm et al. (221) synthesized a most unusual six-coordinate complex **III-158** with nonoctahedral stereochemistry by a long but interesting sequence. The final encapsulation step was accomplished by the treatment of *tris*(2-aldoximo-6-pyridyl)phosphine **(III-159)** — metal ion (Fe^{2+}, Co^{2+}, Ni^{2+}, or Zn^{2+}) fluoro-borate complex with boron trifluoride etherate under homogeneous anaerobic reaction conditions to give **III-158**.

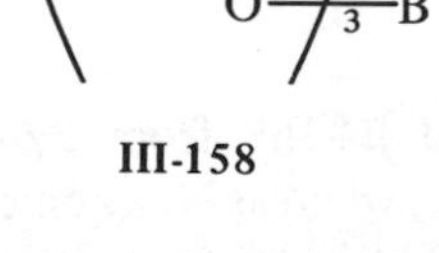

III-158

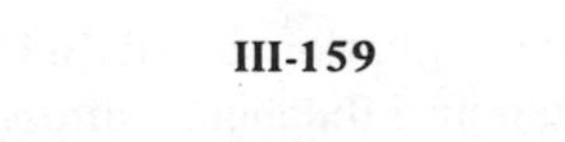

III-159

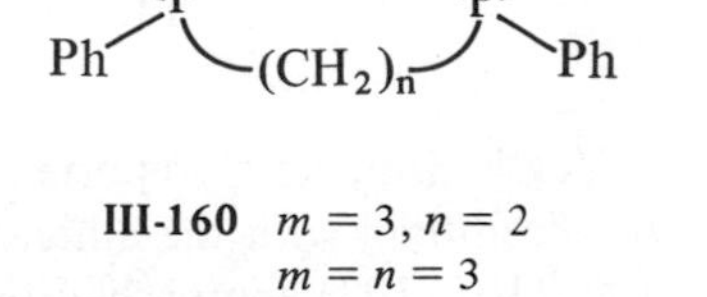

III-160 $m = 3, n = 2$
$m = n = 3$

Schiff-base chemistry has been used to synthesize the phosphorus-containing bridged macrocycles **III-160** (222) and **III-161** (223).

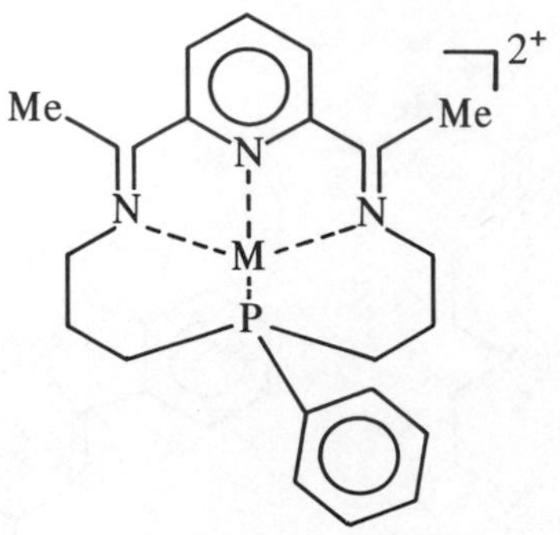

III-161

10. Carbon–Oxygen–Phosphorus Bridges

Newkome and Hager (103) reported the treatment of phenyl bis-(6-halo-2-pyridyl)phosphine with polyethylene glycolates to form the macrocyclic phosphines (**III-162**). Facile aerobic oxidation generated the corresponding P-oxides (**III-163**), which upon warming under basic conditions generated **III-100** ($n = 4$) by an unexpected contractive-phosphorus expulsion procedure (103).

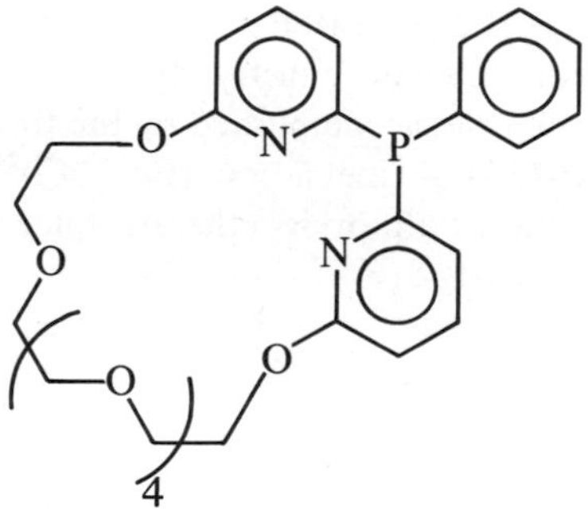

III-162
III-163 (P → O)

11. Carbon–Sulfur–Oxygen Bridges

Vögtle (66, 154) reported the formation of **III-164** and **III-165** from 2,6-pyridinedithiol with the ethereal terminal dihalide or ditosylate, whereas Newkome et al. (105, 197) approached the synthesis of direct ring substitution by the appropriate bis-mercaptide. An analogous route was utilized for the synthesis of the related dipyridine macrocycles (196). Disulfide products can also be generated under either reaction procedures. The α-methylenic-bridged analogs (**III-166**) have

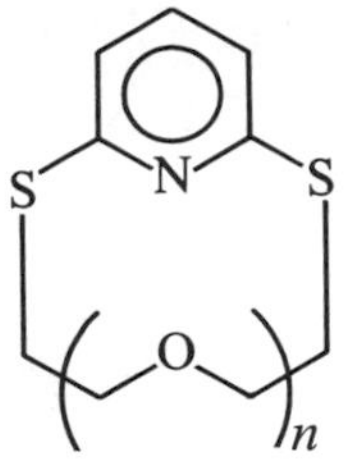

III-164 (*n* = 1–3)

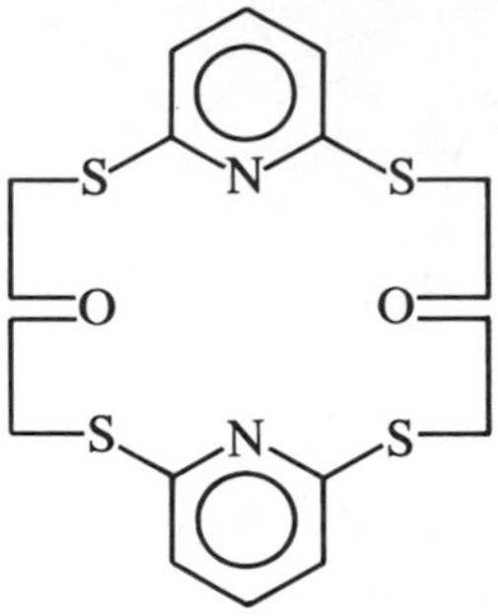

III-165

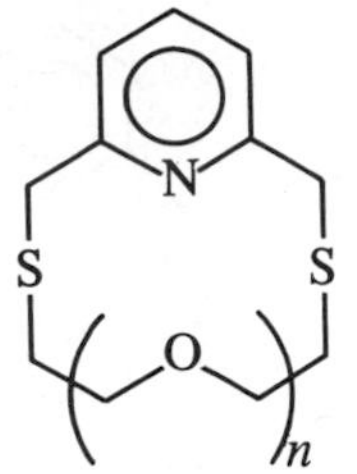

III-166 (*n* = 1–4)

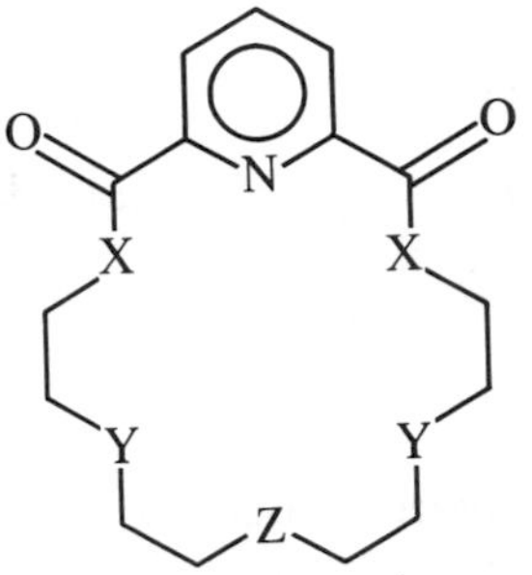

III-167 X = S; Y = Z = O
X = Y = O; Z = S
X = Z = O; Y = S

been prepared (66, 224) except under less rigorous reaction conditions. Synthesis of the *bis*-lactones (**III-167**) has been accomplished by a high-dilution reaction of the *bis*-acyl halide and diol or dithiol (70, 76, 81, 82).

III.   2,5-PYRIDINE MACROCYCLES

1.   Carbon and Carbon–Sulfur Bridges

Carbon-bridged [*n*](2,5)-pyridinophanes (**III-168**) were first constructed by Gerlach and Huber in 1968 (225) by the acid-catalyzed cyclization of *bis*(β-aminovinyl)diketones, to give ketones **III-169**, which were subsequently reduced under Wolff–Kishner conditions. Numerous reactions (226, 227) and conformational stability studies were carried out on the lower members of this series, especially [*n*] < 12 (225). The smallest bridged 2,5-pyridinophane yet reported possesses an 8-carbon atom bridge (225). (±)-[9](2,5)-Pyridinophane (**III-168**, *n* = 9) was resolved and shown to be thermally stable (225).

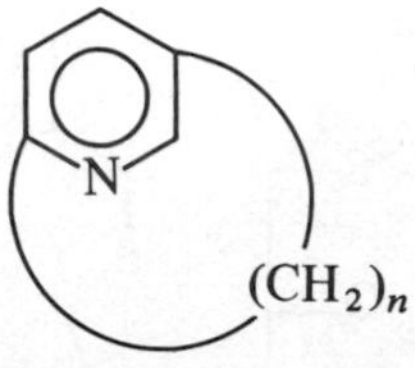

III-168 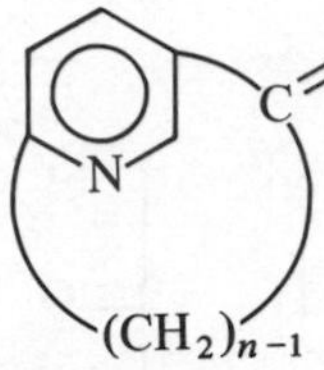   III-169

Stütz et al. (228) synthesized **III-170** by initial cyclization of hexadeca-1,14-diyne with copper acetate, then by heating the resultant cyclohexadeca-1,3-diyne with benzylamine at 170°C with a catalytic amount of Cu(I)Cl. Along with **III-170**, *N*-benzyl-[12](2,5)-pyrrolophane (**III-171**) was isolated.

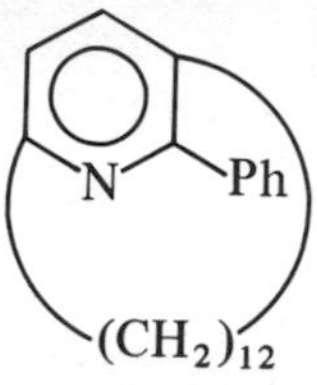

III-170

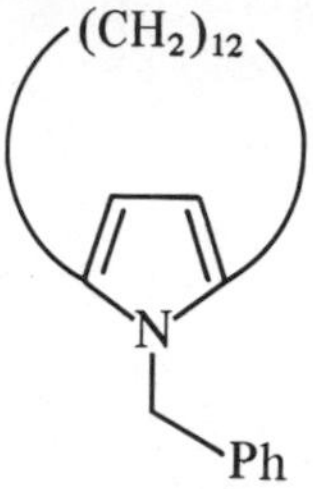

III-171

Bruhin and Jenny (229) prepared an isomeric mixture of **III-172** from the thermal 1,6-Hofmann elimination of (5-methyl-2-picolinyl)-trimethylammonium hydroxide. Crossed condensation reactions were also conducted from which **III-173** and **III-174** were isolated (229), when *p*-xylyltrimethylammonium hydroxide was used. Paudler et al. (230) reported the first mixed heterocyclophane **III-175** *via* a similar procedure; **III-172** and **III-176** were also isolated. The x-ray crystal structure of **III-175** showed that neither the pyridine nor furan ring was planar, and that the pyridine ring is bent in such a fashion that the unbridged carbon atoms of the pyridine and nitrogen atom are situated away from the center of the molecule (231).

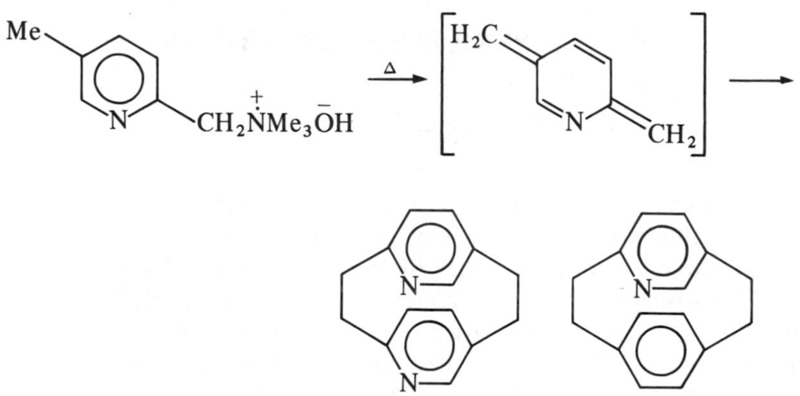

III-172

III-173

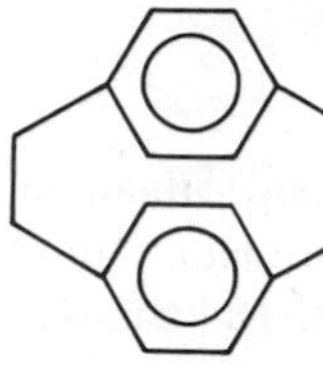

III-174

III-175

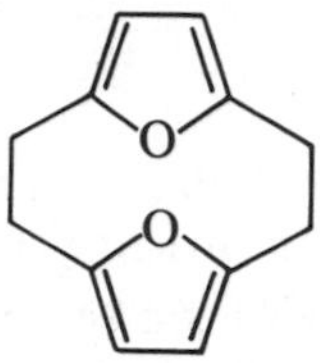

III-176

Application of the previously described S-ring contraction procedures of carbon–sulfur bridged macrocycles has been successfully utilized in the preparation of **III-173** from **III-177** (229, 232) and **III-178** (233). A new series of 2,5-pyridinophanes (**III-179**) was prepared by the coupling of 1,n-alkanedithiols ($n = 4, 6, 8$) with 5'-deoxy-2',5'-dichloro-3,4'-O-isopropylidenepyridoxine (234, 235). Iwata et al. (234, 234a) prepared **III-180** and **III-181**; no synthetic details were reported. Functionalized (2,5)-pyridinophanes (**III-181**) were resolved (234–237). Their catalytic activities for racemization of monosodium-L-glutamate (237, 238) and their nonenzymatic stereoselective transamination in the synthesis of chiral phenyl-alanine (237, 239, 240) were determined. The crystal structure and absolute configuration of a chiral vitamin B_6 analog (**III-181**; X = S, R = OH, R' = CH$_2$OH) have been reported (237, 241).

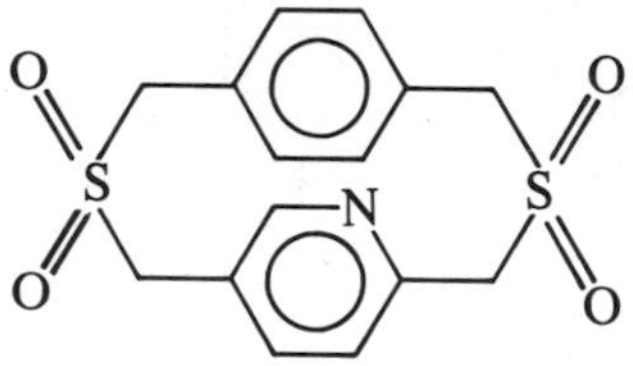

III-177

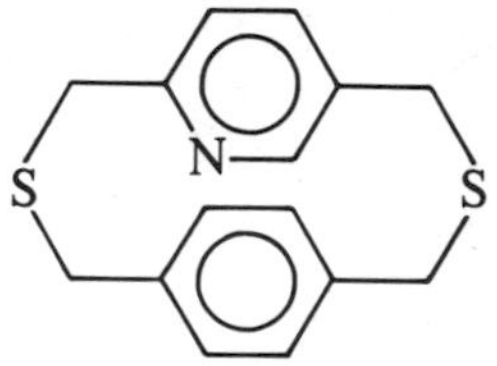

III-178

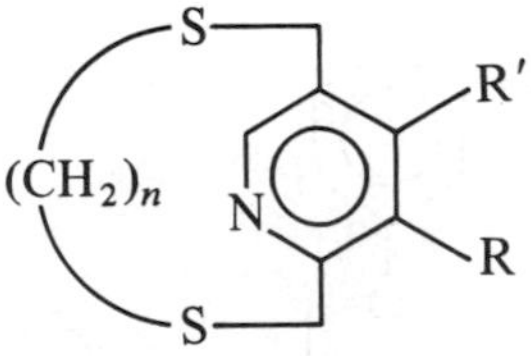

III-179

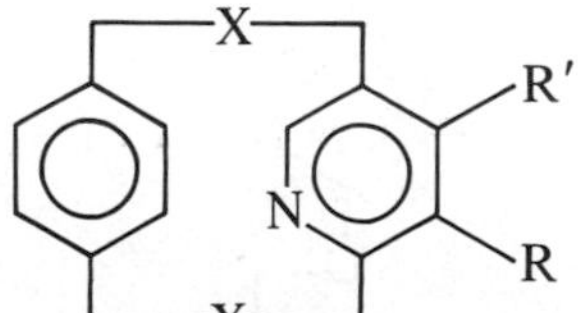

III-180 (X = S or None)

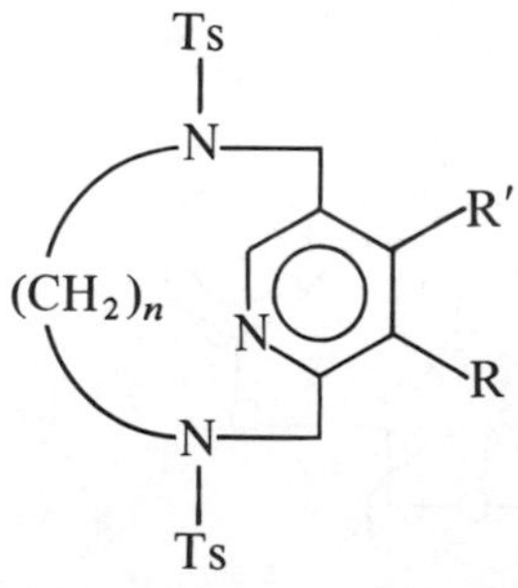

III-181

2.   Carbon–Oxygen Bridges

Newkome et al. (241a) reported the inclusion of a 6-oxanicotinate moiety into a series of lactone macrocycles. *Bis*-lactone **III-182** upon treatment with the disodium salt of diethylene glycol gave isomeric 2:2 macrocycles **III-184** and **III-185**, whereas, **III-182** with disodio pentaethylene glycolate, the major cyclic product was **III-183** ($n = 1$) along with traces of **III-183** ($n = 2$). An isomeric mixture of dimers (**III-186**) was also isolated.

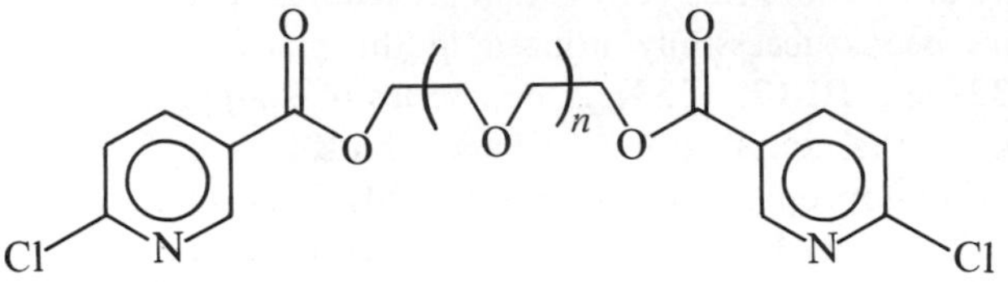

III-182 ($n = 1$–4)

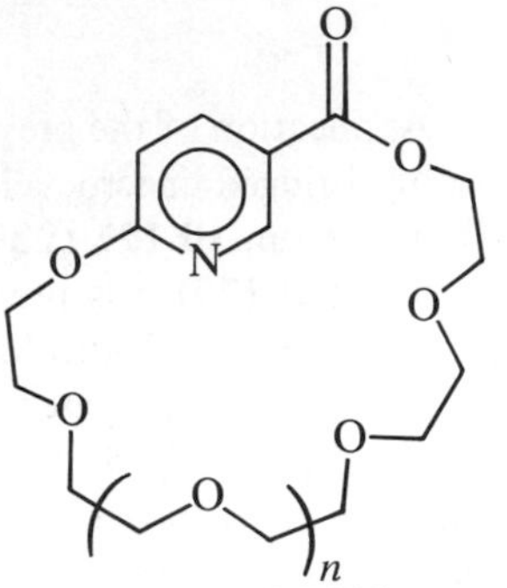

III-183
($n = 1, 2$)

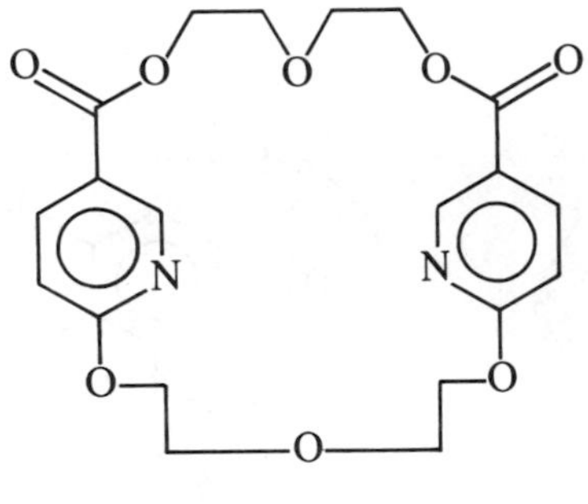

III-184

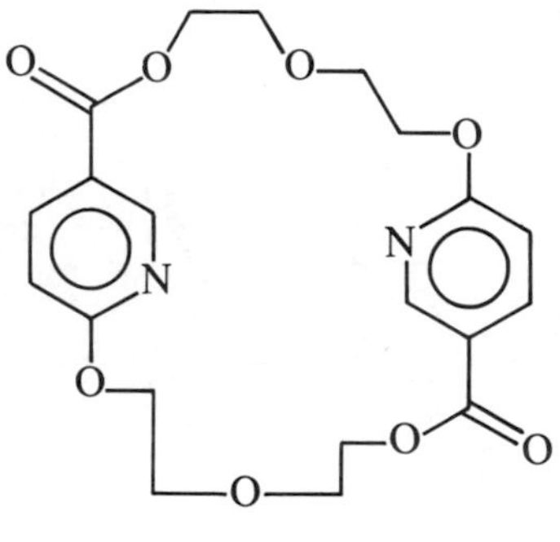

III-185

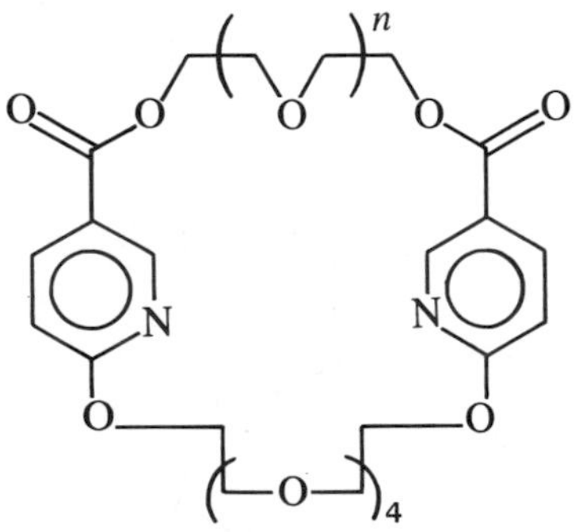

III-186a

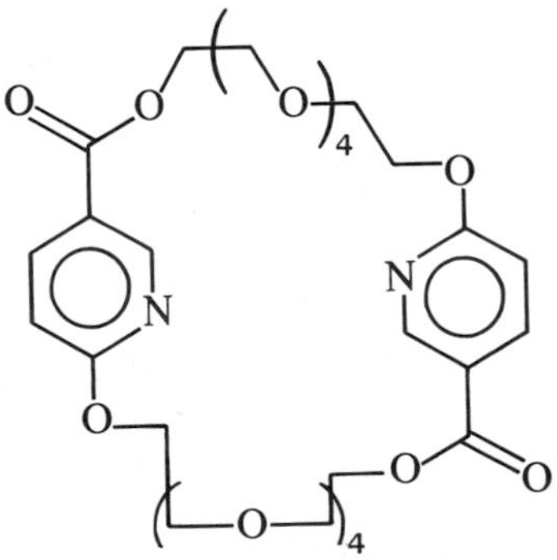

III-186b

IV. 2,4-PYRIDINE MACROCYCLES

1. Carbon Bridges

The synthesis of carbon-bridged [9](2,4)-pyridinophane **(III-188)** from 2-cyclododecenone **(III-187)** was first reported by Pagani et al. (242). The subheterocyclic ring was formed by initial treatment of **III-187** with ethyl cyanoacetate, then sequentially, hydrolytic-decarboxylation, reduction of the ketonitrile, and spontaneous cyclization of the resultant aminoalcohol to give the Δ^1-piperideine **(III-189)**, which catalytically dehydrogenated to give **III-188**. The ^{1}H NMR spectrum of **III-188** failed to exhibit the expected shielding effect of the π-electron cloud on the bridge methylene groups (242).

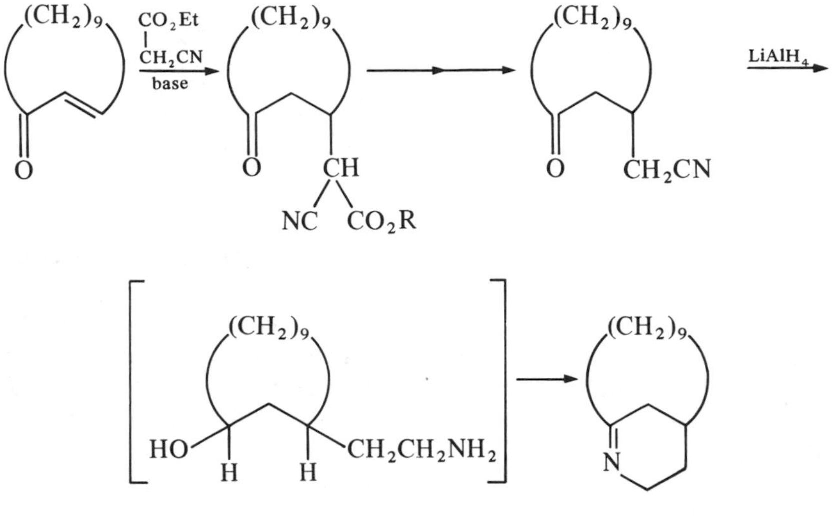

III-189

An alternate procedure to substituted [9](2,4)-pyridinophanes **(III-191)** was by the treatment of the corresponding pyrylium salt **(III-190)** with ammonium acetate (242, 243). The same pyrylium salt **(III-190)**, when reacted with hydrazine, generated the first [9](4,6)-pyridazinophane **(III-192)** (243).

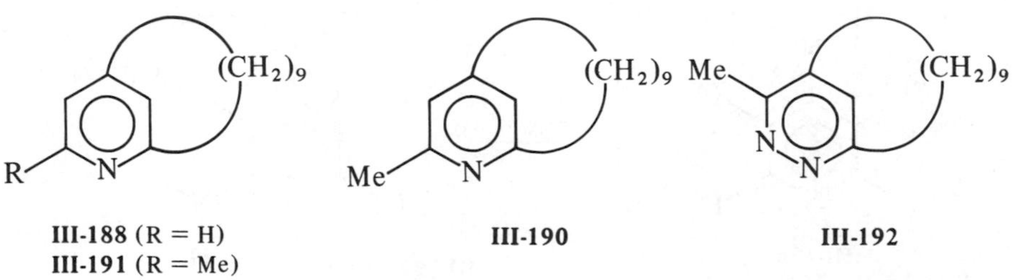

III-188 (R = H)
III-191 (R = Me)

III-190

III-192

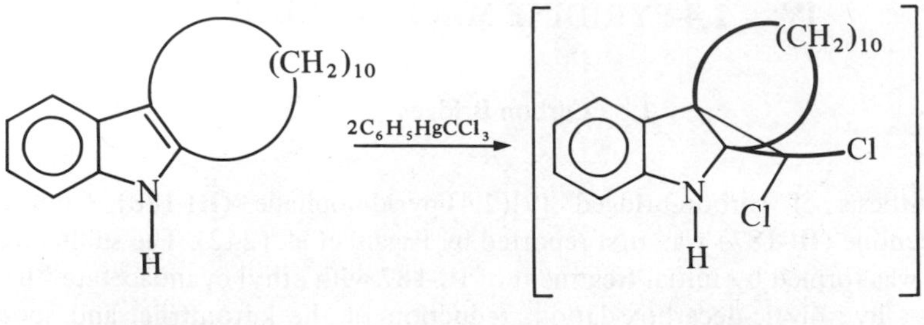

III-193　　　　　　　　**III-194**

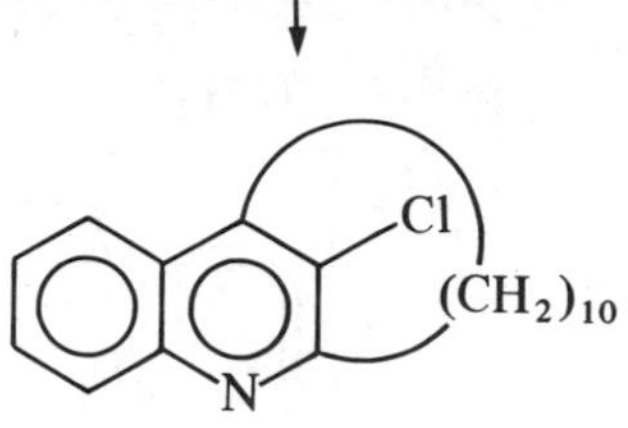

III-195

Parham et al. (244) prepared a large series of benzo[2,4]-pyridinophanes through a ring-expansion sequence. Treatment of indole **III-193** with phenyl(trichloromethyl)mercury afforded an intermediate **III-194**, which rearranged to the desired cyclophane **III-195**. This general procedure has been applied to the synthesis of numerous [n](2,4)pyridinophanes (244–247). Hydrodechlorination of **III-195** was easily accomplished by treatment with hydrazine and palladium on charcoal (248, 249).

2.　Carbon–Nitrogen Bridges

Treatment of 3,5-dichlorotrifluoropyridine with an appropriate long-chained α,ω-(primary) diamine generated initially **III-196**, which subsequently cyclized on heating to afford **III-197** (250).

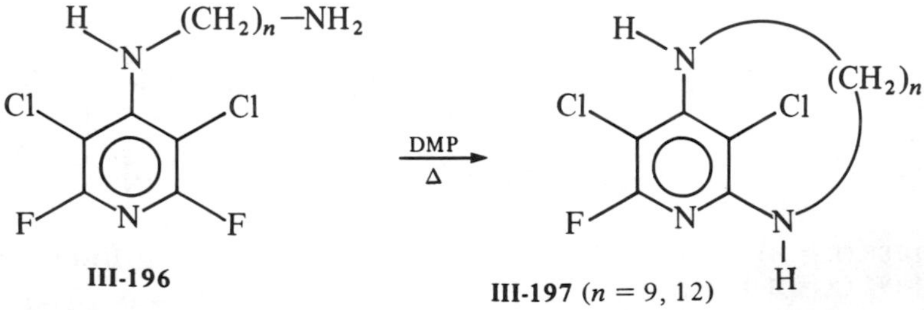

III-196　　　　　　**III-197** ($n = 9, 12$)

V.   2,3-PYRIDINE MACROCYCLES

1.   Carbon Bridges

The carbon-bridged macrocycles in this series were generally synthesized by the construction of the pyridine ring. When α-oxymethylene-cyclopentadecanone **(III-198)** was condensed with cyanoacetamide in the presence of base, the desired pyridine nucleus **(III-199)** was formed. The substituents on **III-199** were removed by standard methods to give **III-200** (251, 252). Analogous routes have been utilized to prepare **III-201**, **III-202** (253), and **III-203** (254). During the preparation of muscopyridine **(III-12)**, bicyclo[10.3.0]-pentadec-12-ene **(III-9)** when subjected to Schmidt reaction conditions, was transformed in part to the 2,3-isomeric macrocycle **(III-203)** (12). An explanation for the product distribution has been proposed (12).

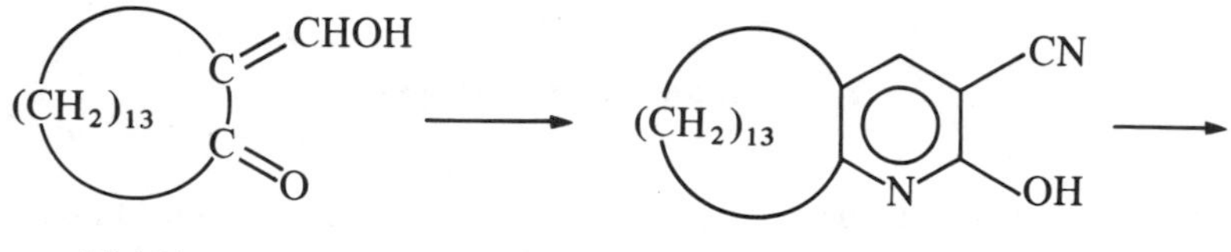

III-198   III-199

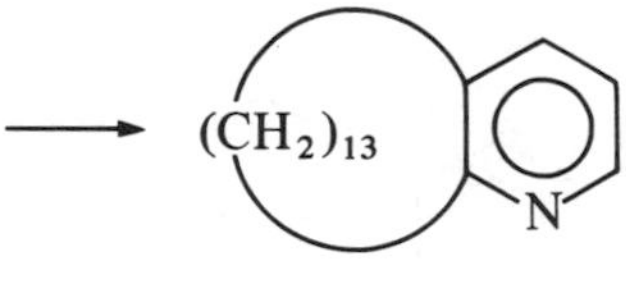

III-200

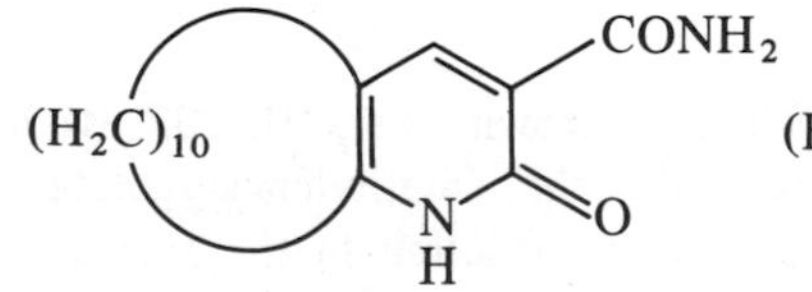

III-201 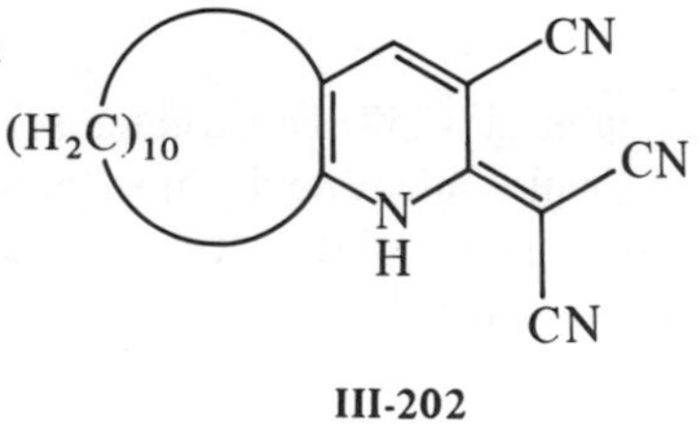   III-202

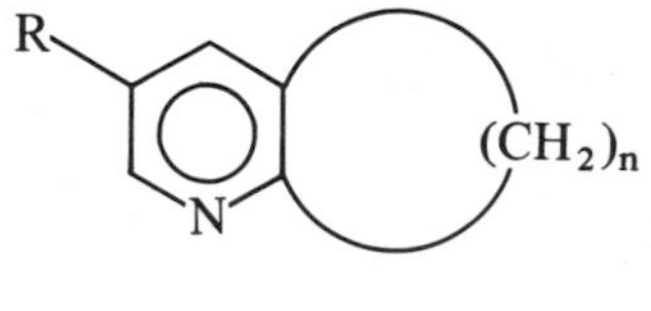

III-203

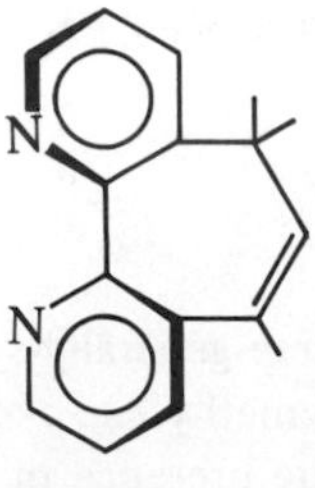

III-204 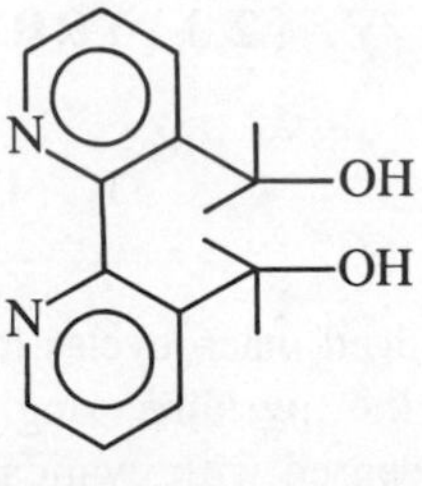   III-205

Inclusion of a dipyridine subunit in a mediocycle (III-204) has been demonstrated (255). Oxidation of o-phenanthroline and subsequent esterification of the resultant diacid gave 3,3'-di(carbomethoxy)-2,2'-dipyridine, which with excess methyllithium afforded a diol (III-205) that cyclized on the treatment with hot acid to give III-204.

2.   Carbon–Oxygen Bridges

Cesium carbonate has been utilized in the synthesis of 2,3-pyridine crown ethers (256). Thus when the dicesium salt of 2,3-dihydroxypyridine was treated with the appropriate dihalide, crown ethers (III-206) were obtained.

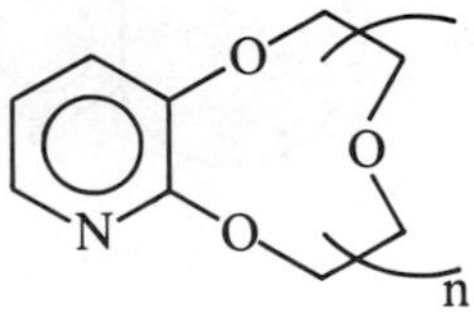

III-206 ($n = 3–5$)

Newkome et al. (257) constructed bis-lactone crown ethers (III-207, III-208) by subjecting 2-chloronicotinoyl chloride with disodium diethyleneglycolate. Transesterification and template cyclization procedures resulted in the formation of a 1:1-lactonic macrocycle (III-209).

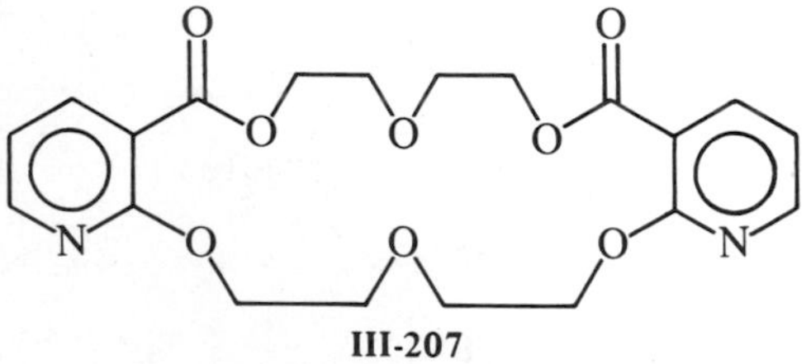

III-207

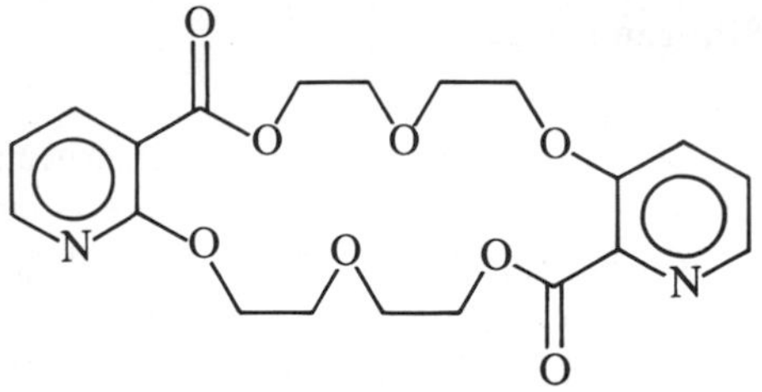

III-208 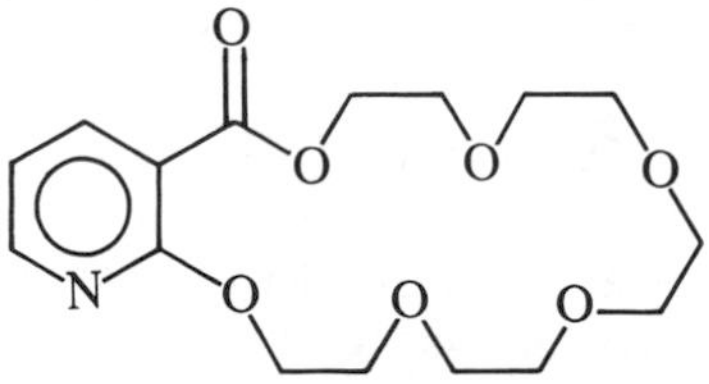   III-209

3,3'-Dicarbomethoxy-2,2'-dipyridine was reduced with sodium *bis*(2-methoxyethoxy)aluminium hydride to give **III-210**, which was condensed with an appropriate glycol ditosylate to afford **III-211** (258). A parallel sequence starting from **III-212** afforded good yields of **III-213** (259). Systems **III-204, III-211**, and **III-213** incorporate a bifunctional ligand, in that the molecule has more than one site for complexation and once the first metal ion is attached it will affect the coordination of a second metal ion. Such binding of two different metal ions was interpreted in terms of a simple model for allosteric effects (255, 286–260). Model **III-211** was used to evaluate the transport selectivity of Li^+, Na^+, and K^+; the ion transport rate followed the order $K^+ > Na^+ > Li^+$ across a $CHCl_3$ liquid membrane (261).

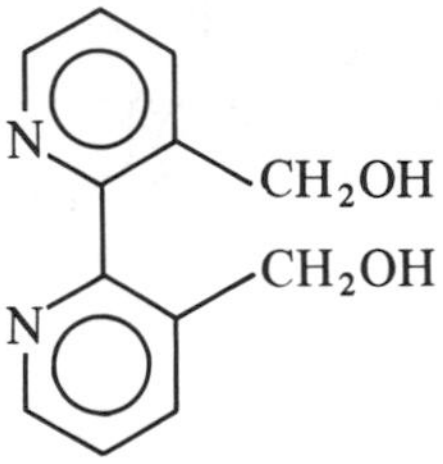

III-210 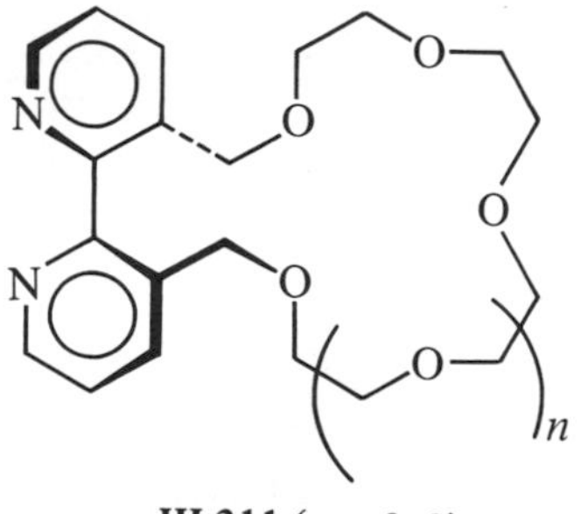   III-211 (n = 0–2) 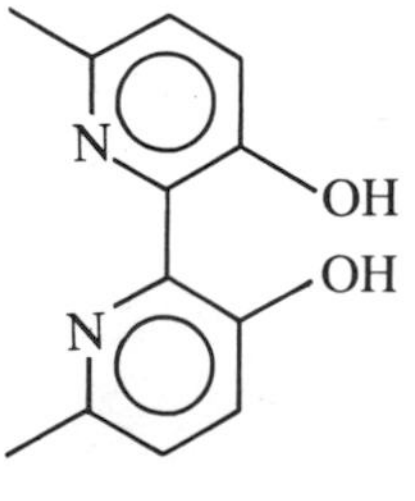   III-212

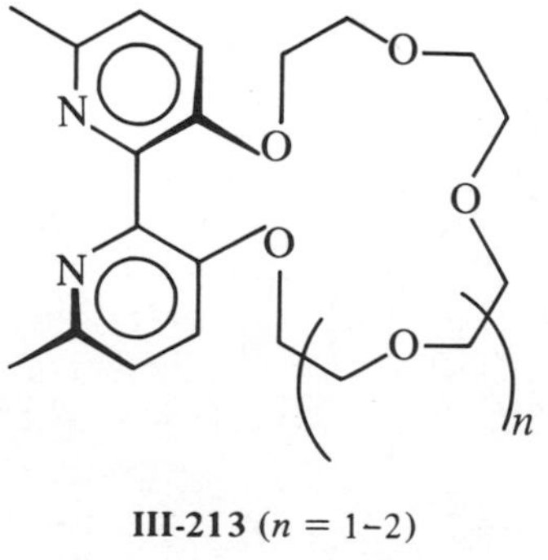

III-213 (n = 1–2)

3. Carbon–Nitrogen Bridges

Dziomko et al. (262) synthesized the metal complex of **III-214** by the transition metal ion coordination–templation of **III-215a**, followed by thermal intramolecular cyclization. Under similar conditions, 3-[(2′-amino-1′-naphthyl)-azo]-2-chloro-pyridine was cyclized to afford **III-216** (263). Condensation of **III-215b** with 2,2′-azodianiline in the presence of an equivalent of nickel acetate gave the Ni(II) complex **III-217** or **III-218** (264). Ligand **III-215** was easily generated by the diazotization of 2-chloro-3-aminopyridine with an appropriate pyrazole derivative (262).

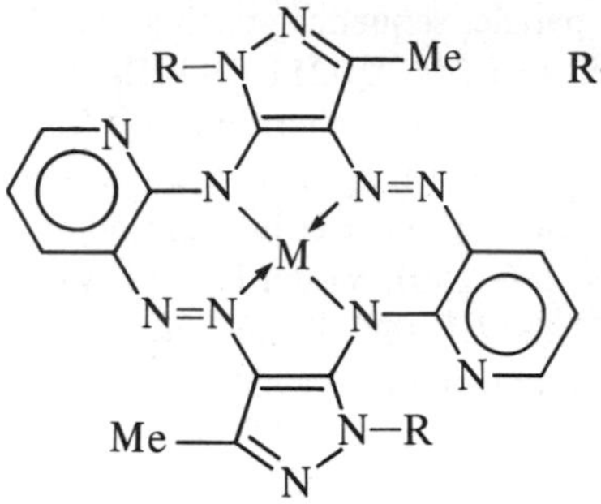

III-214

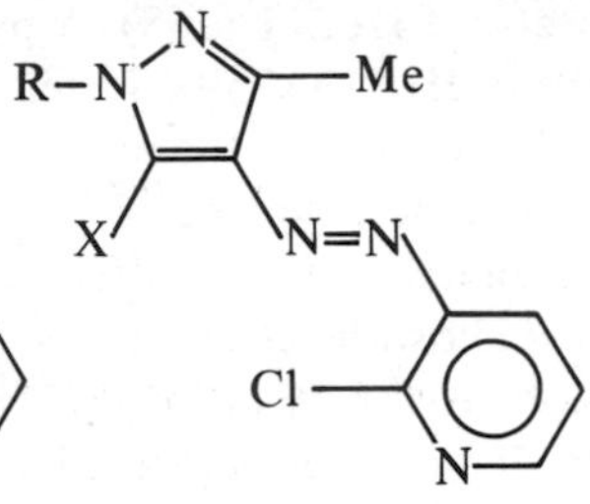

III-215 a; X = NH₂
        b; X = Cl

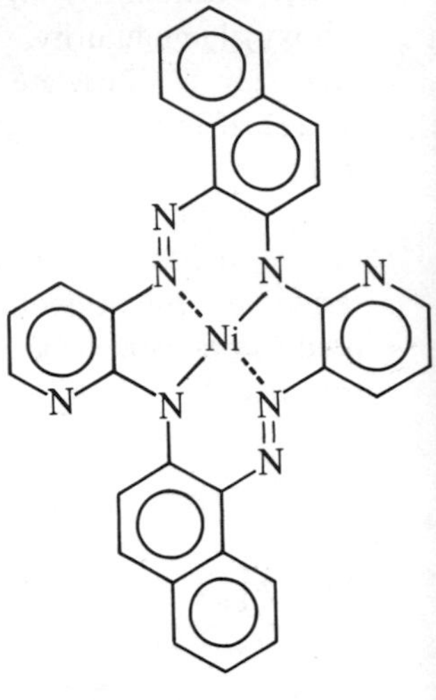

III-216

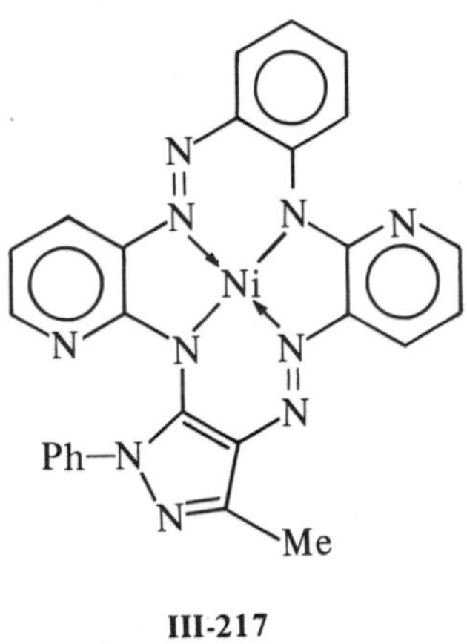

III-217

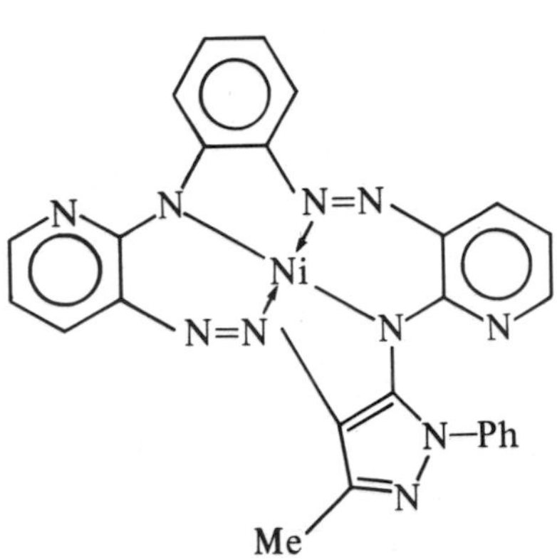

III-218

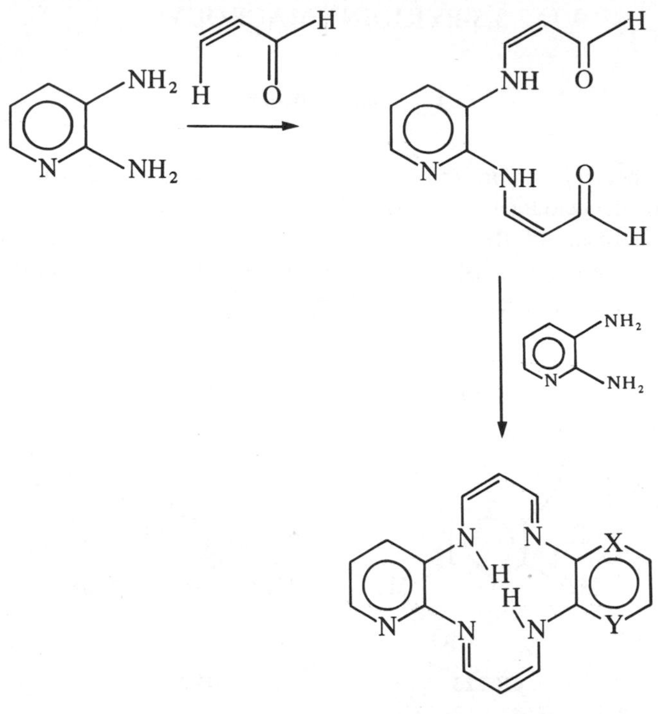

III-219 (X = N; Y = CH)
III-220 (X = CH; Y = N)

Reaction of 2,3-diaminopyridine with 2-substituted malonaldehyde or propynal afforded the symmetrical and unsymmetrical isomers **III-219** and **III-220**, respectively, even in the absence of a coordinating metal cation (265, 266). In the presence of a metal ion (e.g., Co²⁺ or Cu²⁺), the intermediate **III-221** was generated, then cyclized to the 14-membered macrocycle (**III-222**) (266). Several complexes of **III-219** and **III-220** have been reported (265, 266). The copper(II) and cobalt(II) complexes, **III-222**, have been utilized in the oxidation of ethylbenzene to acetophenone; semiconducting properties and thermal stabilities were reported (267). The EPR spectrum for the Co(II) complex **III-222** (X = N) has been reported (268).

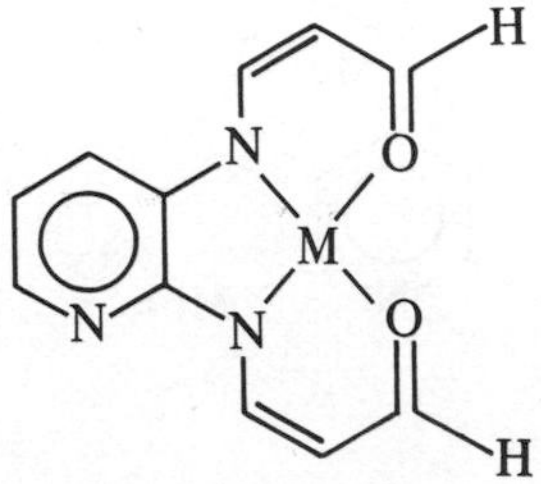

III-221 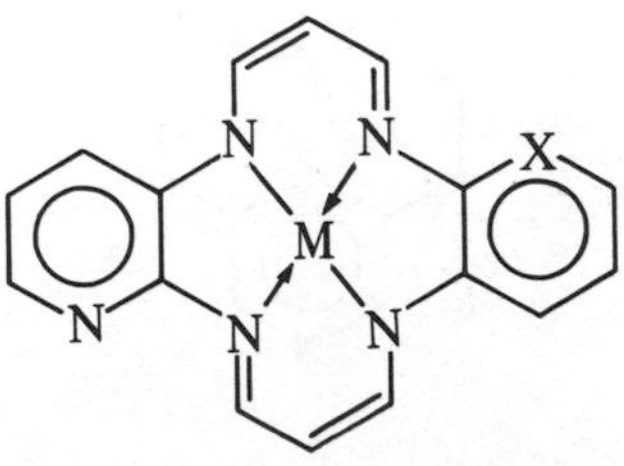   **III-222** (X = N or CH)

VI. 3,5-PYRIDINE MACROCYCLES

1. Carbon Bridges

The intermediary 3,5-bridged pyrylium salt **III-223** (269–271), prepared by diacetylation of cyclododecene, was transformed to **III-224** upon treatment with ammonia. Treatment of **III-223** with methylamine, aniline, or hydrazine gave the corresponding pyridinium salts (270). Europium-induced shift NMR studies of these macrocycles have been reported (272).

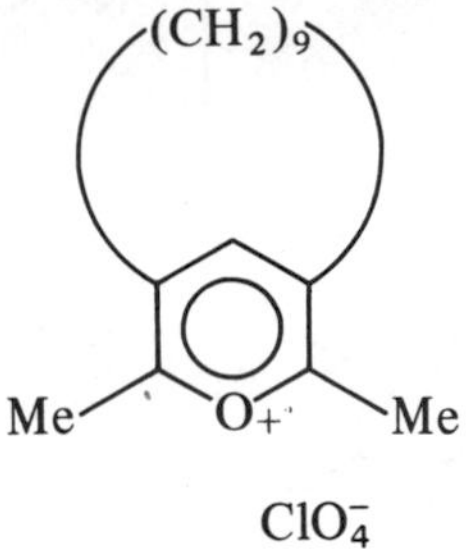

III-223 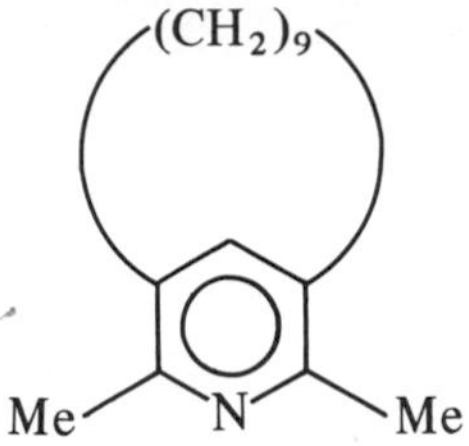   **III-224**

Metapyridinophanes (e.g., **III-225–226**) have been synthesized by a Wurtz coupling procedure of the substituted 3,5-*bis*-(halomethyl)pyridines (273, 274). A more tedious route to **III-229** from 5-carbethoxy-3-formyl-2,4,6-trimethylpyridine has been reported, in which a Wittig reaction was used in the initial stages and after routine functionalization, **III-227** was cyclized *via* a Wurtz cyclocoupling. Oxidation of **III-228** with ruthenium and molecular oxygen gave **III-229** [273], which photoisomerized to the substituted dehydropyrene **III-230**. A dark, thermal isomerization has been shown to be an equally facile process ($K_{1/2}$ MeOH = 8 sec at 17°C) (273).

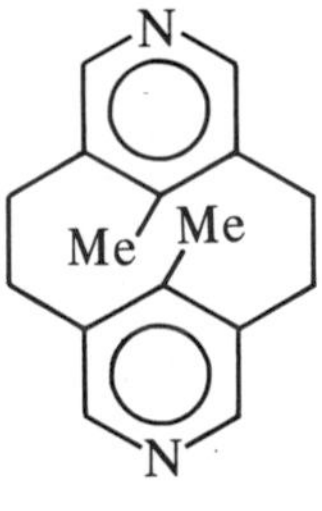

III-225 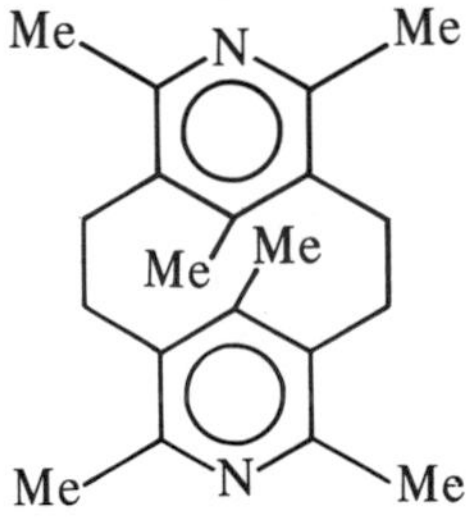   **III-226**

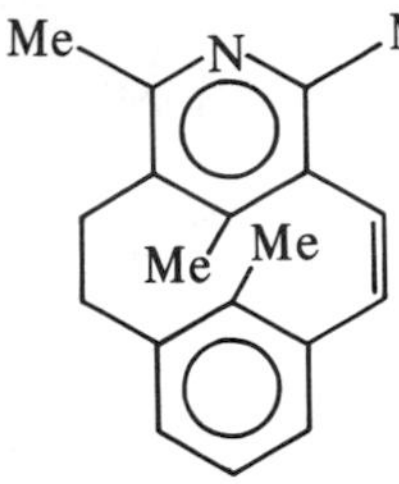

III-228

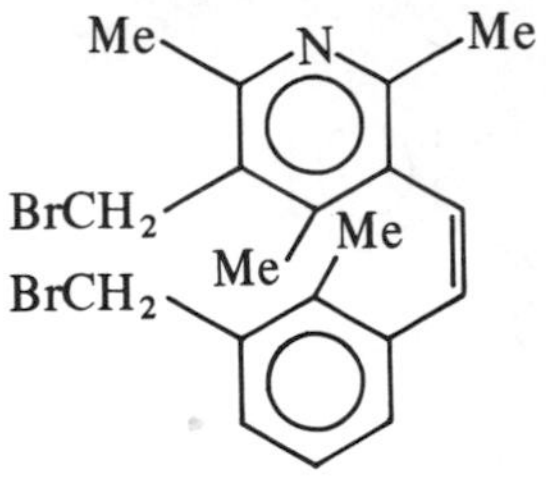

III-227

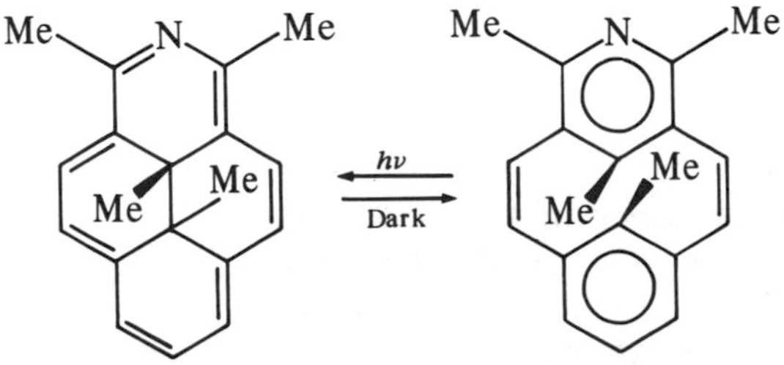

III-230

III-229

Jenny and Holzrichter (275, 276) constructed [2.2](3,5)-pyridinophane (**III-231**) by a double Wurtz coupling of 3,5-*bis*(chloromethyl)pyridine with sodium in the presence of tetraphenylethylene. Not only was the desired [2.2]-cyclophane isolated but also the homologous [2.2.2]- and [2.2.2.2]-macrocycles were obtained all in low yield.

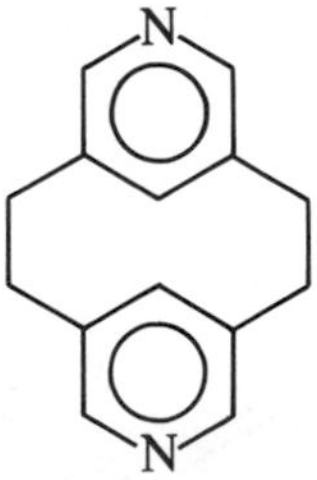

III-231

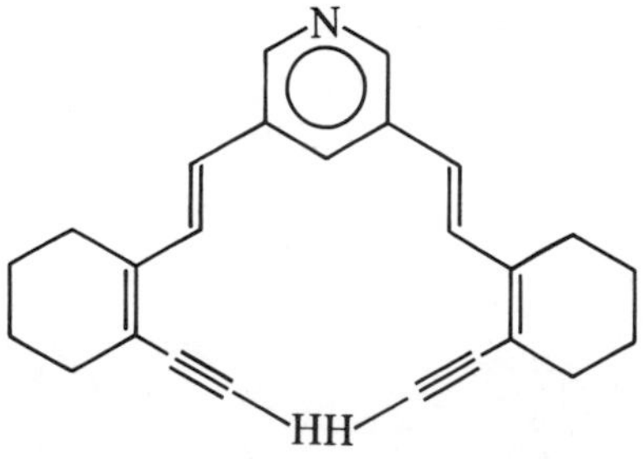

III-232

Sondheimer et al. in a series of papers (277–281) described the synthesis of heteroannulenes. The general mode of construction is described by the preparation of **III-233** (277). A di-Wittig reagent, prepared from 3,5-*bis*(bromomethyl)pyridine, was treated with 2 equiv. of an appropriate ynenealdehyde to afford the desired *trans*, *trans*-diene (e.g., **III-232**), as well as the other isomers. A copper-catalyzed cyclocoupling of **III-232** gave annulene **III-233**. 1,4-Reduction of **III-233** followed by *N*-trapping generated a novel series of aza[17]annulenes (**III-234**). A similar procedure has been applied to the synthesis of diatropic oxygen and sulfur analogs (280, 282).

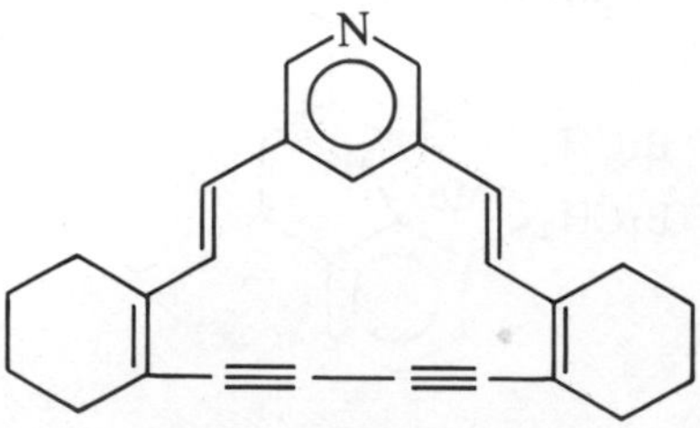

III-233

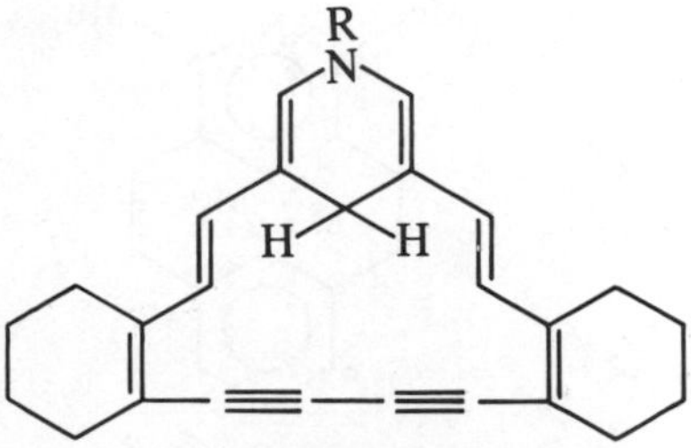

III-234

2. Carbon–Oxygen Bridges

The elegant application of the classical Hantzsch reaction to the synthesis of these carbon–oxygen *bis*-lactones has been reported by Kellogg et al. (74, 283). The transesterification of ethyl acetoacetate with various ethyleneglycols quantitatively afforded *bis*-ester **III-235**. Subsequent treatment of **III-235** with ammonium carbonate and formaldehyde then dehydrogenation of the dihydro intermediate gave monomer **III-236** and dimer **III-237** (74, 283). Substituents in the 4-position of the pyridine subunit were incorporated by use of aldehydes other than formaldehyde (74). Quaternization of **III-236** ($n = 3$) was accomplished by methylfluorosulfonic acid, then sodium perchlorate and reduction with sodium dithionite afforded the NADH model **III-238**, which undergoes facile isomerization to the 1,2-dihydro isomer **III-239** (74, 283). Kellogg et al. (284) devised an alternate route to these *bis*-lactones *via* the one-step condensation of the dicesium salt of pyridine-3,5-dicarboxylic acids with dihalides. Other alkali metal (Na^+, K^+, or Rb^+) carboxylates afforded low yields of **III-240** (284). Bradshaw et al. (76) also reported the

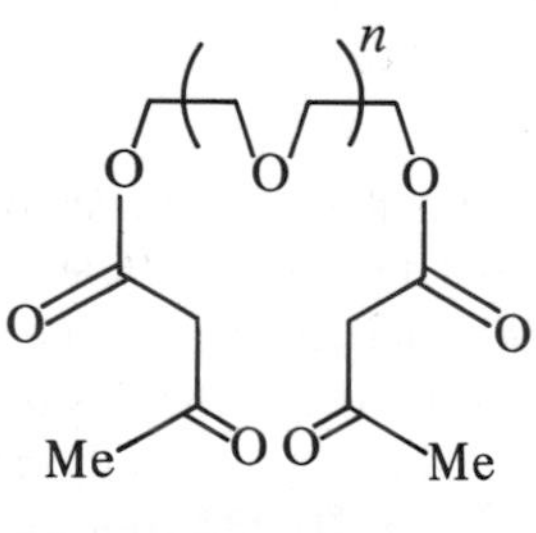

III-235 ($n = 0$–3)

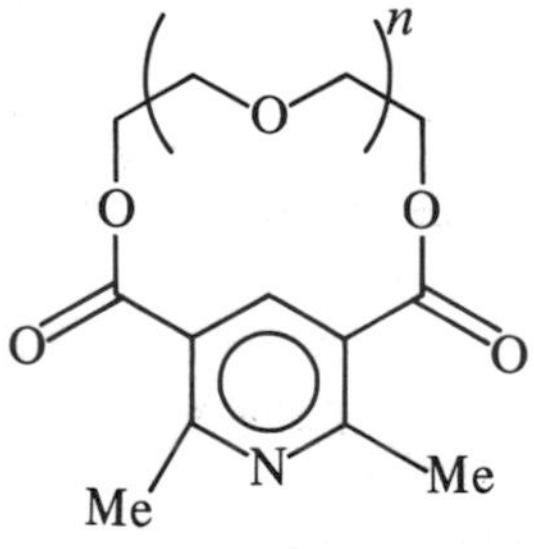

III-236 ($n = 2, 3$)

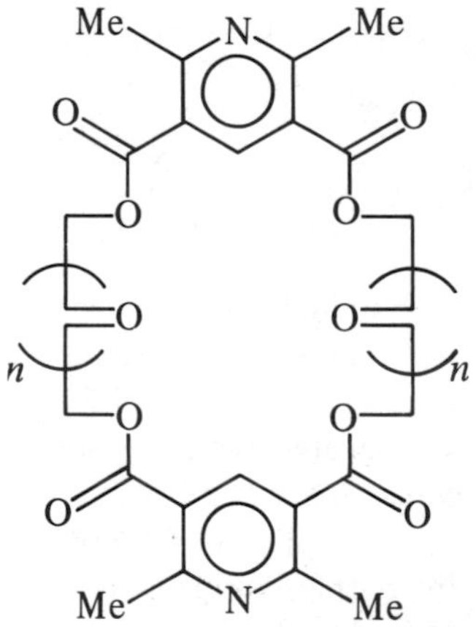

III-237 ($n = 0$–3)

preparation of **III-240**, which involved the condensation of the pyridinedicarbonyl chloride with an appropriate diol; yields by this procedure were generally low (< 10%) except for **III-240** ($n = 3$) (76).

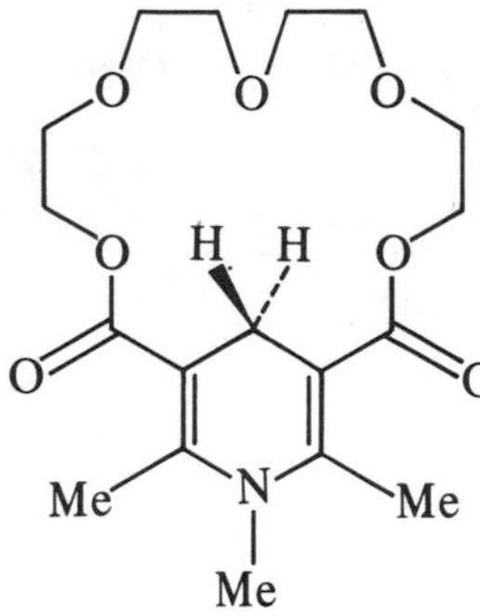

III-238

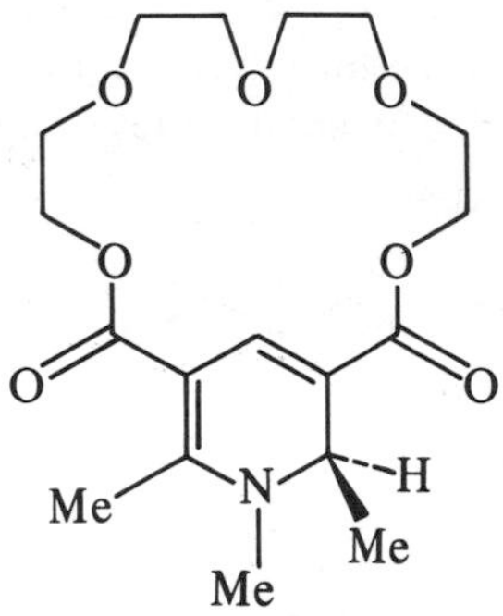

III-239

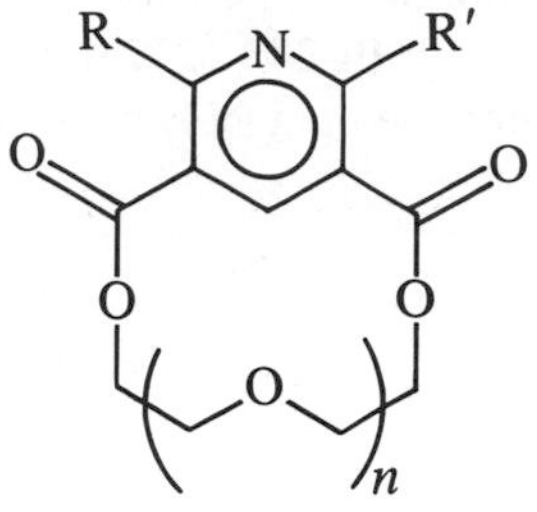

III-240 ($n = 2$–5)

Crown **III-238** formed a complex with sodium perchlorate; the x-ray crystal structure of this complex showed that the dihydropyridine ring possessed a boat conformation and tilted relative to the polyethylene glycol bridge (285). Kellogg et al. reported that **III-238** transferred hydride 2700 times faster than 1,2,6-trimethyl-3,5-*bis*-(carbethoxy)-1,4-dihydropyridine (286, 287). The asymmetric reductions with chiral 1,4-dihydropyridine crown ethers have also been reported (288, 289).

3. Carbon–Sulfur Bridges

2,11-Dithia[3]metacyclo[3](3,5)pyridinophane (**III-241**) has been synthesized by standard procedures and upon photolysis in the presence of triethylphosphite gave **III-242** and then **III-243** (290). Pyridinophanes **III-244** were converted to the corresponding **III-225** (274) by the standard *S*-oxidation-pyrolysis procedure considered earlier.

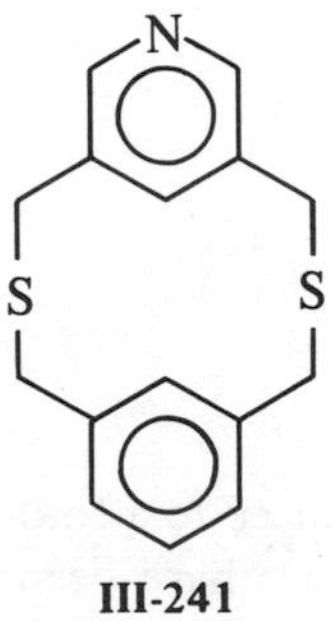

III-241

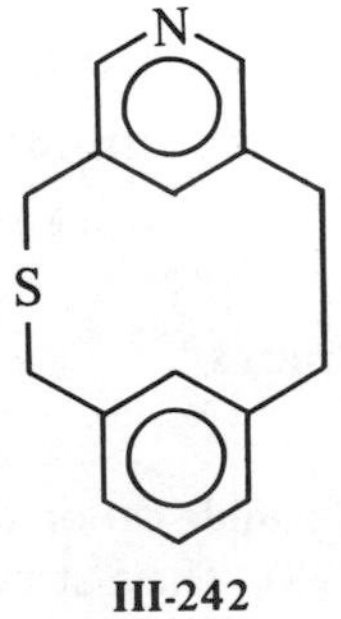

III-242

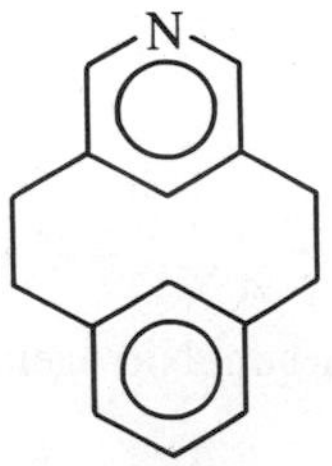

III-243

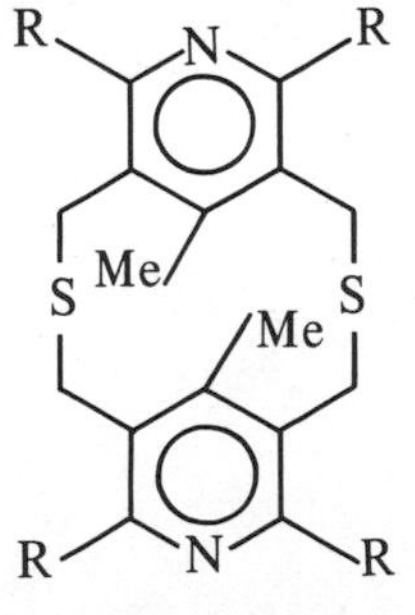

III-244 a R = H or Me
b S → O; N → O; R = H
c S → O; R = H

4.   Carbon–Nitrogen Bridges

The high-dilution cyclization of 3,5-pyridinedicarbonyl chloride with substituted diamines, following the procedure of Stetter (291), afforded entry into the *bis*-lactam **III-245** (292). An improved high-dilution procedure was recently devised to increase the yields of these cyclocondensation reactions (293).

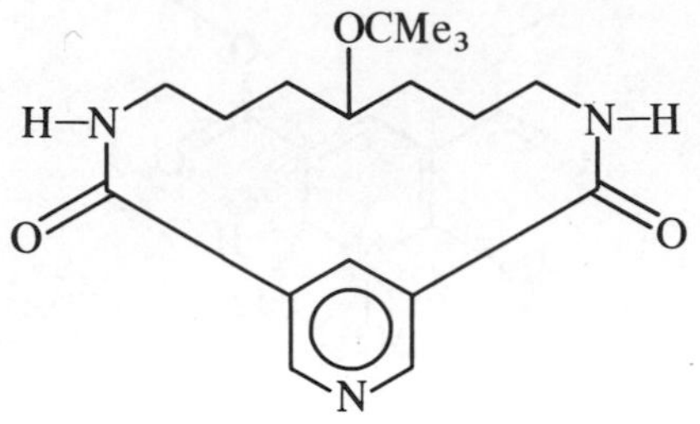

III-245

VII.   3,4-PYRIDINE MACROCYCLES

1.   Carbon Bridges

Freeman and Ito have reported the simple conversion of 2-acylcyclododecanones into substituted 5*H*-2-pyridinones, as well as 3,4-polymethylene pyridine (e.g., **III-247**) (294). The functionality can be removed by literature procedures (295).

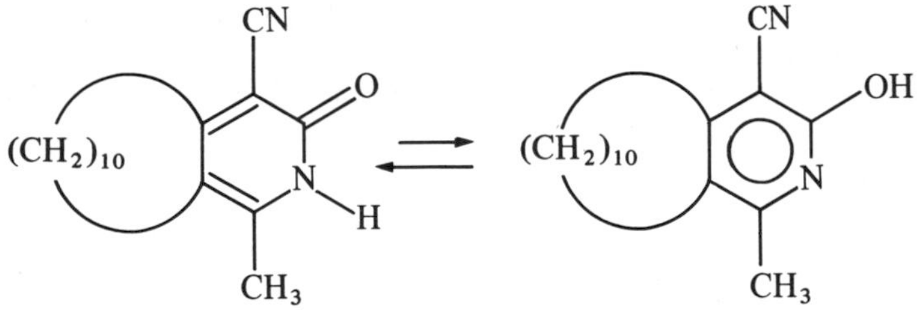

III-246   **III-247**

2.   Carbon–Nitrogen Bridges

Under basic conditions, the *N*-substituted nicotinamide cation cyclized to afford trimer **III-248** (296, 297) or tetramer **III-249** (298). Their structures have been established by unusual degradation techniques (296, 298).

VIII.   2,3;5,6-PYRIDINE MACROCYCLES. CARBON BRIDGES

Treatment of 3,5-*bis*(carbethoxy)-2,6-lutidine with 2 equiv. of LiAlH₄ gave 3-hydroxymethyl-5-carbethoxy-2,6-lutidine, which with thionyl chloride generated the corresponding chloromethyl derivative. Pyrolysis at 775°C of this halide formed **III-250**, which dimerized at 450°C under atmospheric pressures to give both possible dimers **III-251** and **III-252** (299). The above sequence was repeated and on gas-phase pyrolysis of the chloromethyl compounds at 750°C (10^{-3} mm) afforded cyclophanes **III-253** and **III-254** (299), respectively. The x-ray crystal structures of the polybridged cyclophanes have been determined (300, 301); the heteroatomic rings have a boat conformation in which the tips were inverted to the outside.

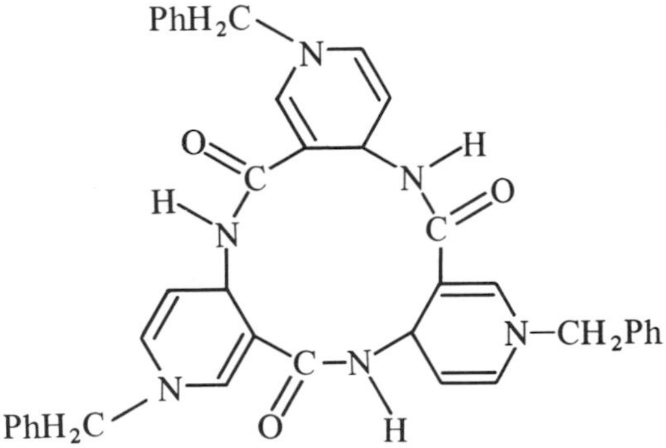

III-248

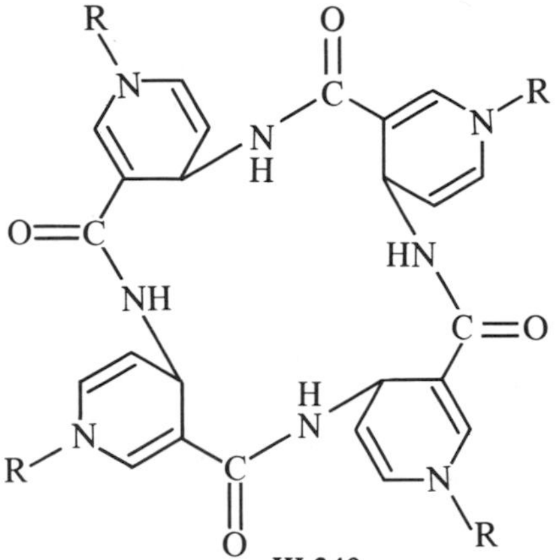

III-249

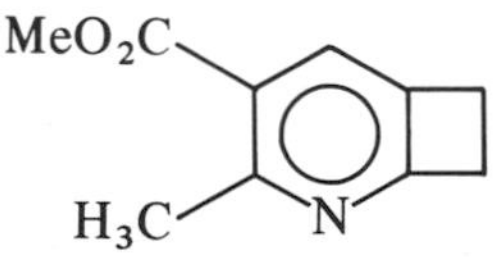

III-250

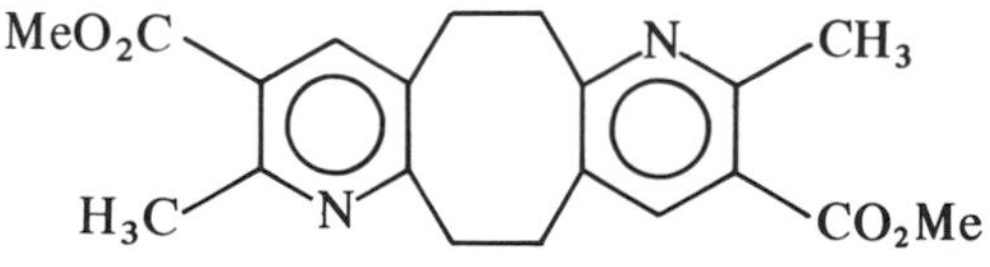

III-251

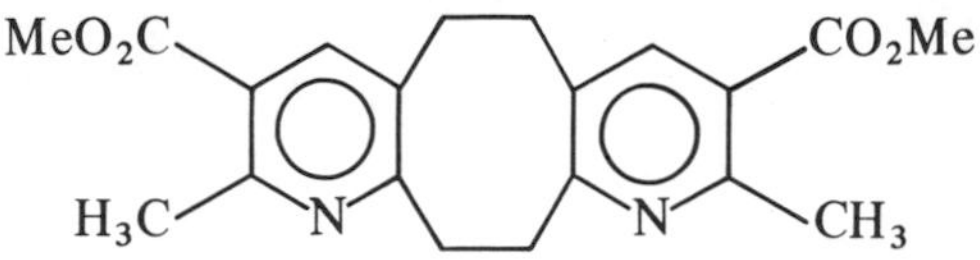

III-252

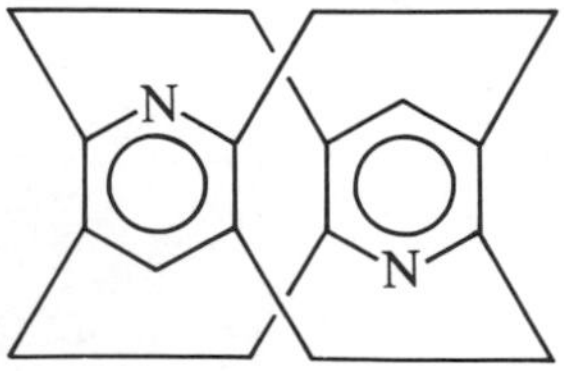

III-253

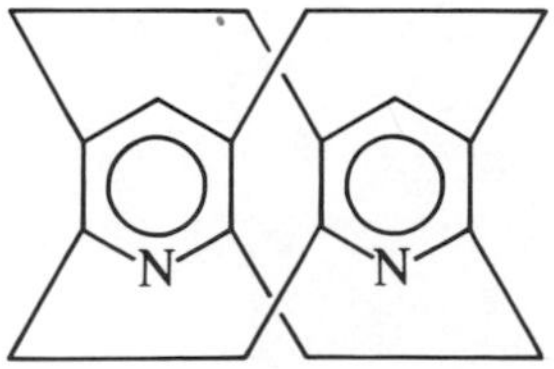

III-254

ACKNOWLEDGMENT

We wish to thank the National Science Foundation for partial support of this compendium.

IX.   TABLES

Notes for the following tables.

A = ^{1}H NMR
B = IR
C = UV
D = MS
E = ^{13}C NMR
F = PES
G = x ray
H = ^{15}N NMR
J = Mossbauer spectrum
K = ESR
( ) special letter = spectral data obtained on a complex.

Ring numbering: Bridged atom adjacent to the pyridine nucleus is denoted as position 1, followed by continuous numbering around the largest bridge or ring(s). Tabulated functionality can be located by this system.

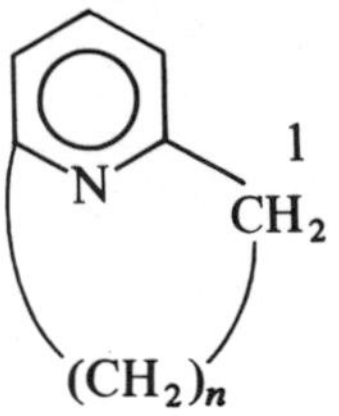

TABLE III-1.   2,6-PYRIDINE MACROCYCLES – CARBON BRIDGE

Compound	n	Substituents
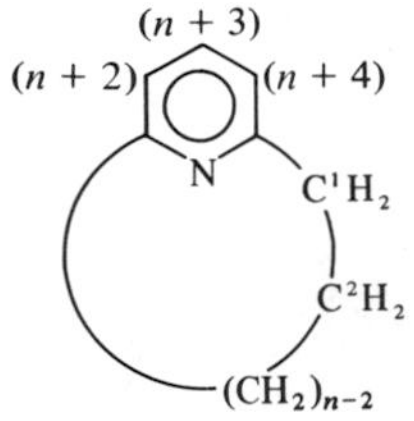	6	H
	7	H
		4-D
		1-CO_2Me
		1-OH
		1-(=O)
		1-(OMe)$_2$
		1-(=O); 2,2-(Me)$_2$
		2,2-(Me)$_2$
		1-(=CH$_2$)
		1-(=CMe$_2$)
		1-[=C(C$_6$H$_5$)$_2$]
		1-C(C$_6$H$_5$)$_2$OH
	8	H
	9	H
	10	H
		12-OH
		12-OAc
		1-OH
		1-OAc
		1-(=O)
		1-(=O); 2-Me
		1-(=O); 2,2-(Me)$_2$
		(±)-2-Me
		13-Me
		2,2-(Me)$_2$
		1,2,9,10-De(H)$_4$; N → O
	12	
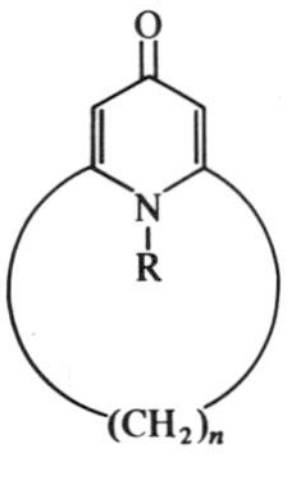	26	29-ONH$_2$, R = H
	26	R = H
	26	R = NH$_2$
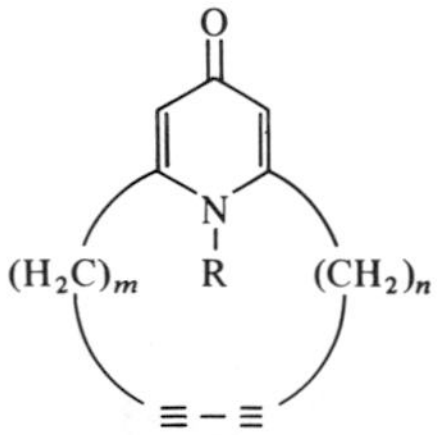	$m = n = 11$	R = H

Structure diagrams:

(n + 3), (n + 2), (n + 4) pyridine ring with N, C^{1}H$_2$, C^{2}H$_2$, (CH$_2$)$_{n-2}$

O=pyridone ring with N–R, (CH$_2$)$_n$

O=pyridone ring with N–R, (H$_2$C)$_m$ and (CH$_2$)$_n$, ≡ – ≡

Physical Data m.p. [b.p. (mm)], °C	Spectral Data Available	Metal Complex(es) General Comments	Reference
	A, C, D		7b
73 (3)]	A–C		7b, 9, 11, 21, 302
(7)]	A		9
0.03)]	A		9
-54.0 [95 (0.01)]	A		9
-34.5	A		9, 11
0.06)]	A		9
0.07)]	B–D		10
0.2)]	A–D		10
2.0)]	A–D		11
-118 (0.5)]	A–D		11
118	A–D		11
163	A, B		11
			7b
			7b
-16.6 [152–158 (3.7)]	B, C	N-Oxide (79–80.5°); π-crolonate (183–185°)	7b, 12, 396, 441
202	B, C	Subl: 125–130° (0.1)	12
			12
9	B, C		12
			12
8	B, C	DNP (191–192°)	12
	B	Picrolonate (113–115°)	12
-160 (0.36)]	B, C		12
-143 (2.2)]	B, C	Picrolonate (163–166°)	12, 16, 395, 407, 418
		$[\alpha]_D^{25}$ + 13.31°	8, 12, 15
105	B, C	Picrolonate [274° dec]	17, 22, 441
	A–D	Isolated < 1% yield	20
	B	Picrolonate [170–172° dec]	12
			303
			7b
			23
85	B		23
30	B		23
78	B		23

TABLE III-1.   *(CONTINUED)*

Compound	*n*	Substituents
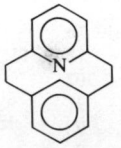		H
		H
		1(2), 7(8)-(SMe)$_2$
		1(2), 7(8)-[S$^+$(Me)$_2$]$_2$ 2BF$_4^-$
		16-H (BF$_4^-$)
		16N → O
		1,2,7,8-De(H)$_4$
		1,2,7,8-De(H)$_4$
		16-H (BF$_4^-$)
		1,2,7,8-De(H)$_4$; 16-BF$_3$
		4,15-(CH$_3$)$_2$
		4,15-(CH$_3$)$_2$; 1,2,7,8-De(H)$_4$
		4,5,14,15-(CH$_3$)$_4$
		4,5,14,15-(CH$_3$)$_4$; 1,2,7,8-De(H)$_4$
		11-(CH$_3$)
		11-(CH$_3$); 1,2,7,8-De(H)$_4$
		H
		1,9-(SMe)$_2$
		1,9-(SOMe)$_2$
		H
		R = OMe
		1,2,17,18-De(H)$_4$
		2,17-(OAc)$_2$

Physical Data m.p. [b.p. (mm)], °C	Spectral Data Available	Metal Complex(es) General Comments	Reference
	A	VTNMR study	27, 304, 438
.5–81.5	A, B, D, F^{29}	VTNMR study	33, 34, 302
–84			32
2–153	A, D		34
			34
9–183	A–C		34
5–167	A, B, D		34
7–158	A, C, D		33, 34
	G		305, 306
7–210	A, C		34
4–206	A, C		34
–66	A, C–F		33
3–144	A, C–F	SbF_5 complex	33
3–154	A, C, D, F		33
6–158	A, C, D, F		33
–80	A, D, F		33
0–121	A, D, F		33
0–112	A, C		35
			35
0–187			35
5	A, C		35
4.5–156.6	A, B	Lythraceous alkaloids	37, 38, 307
0–232	A, B		36
			36

TABLE III-1. *(CONTINUED)*

Compound	n	Substituents
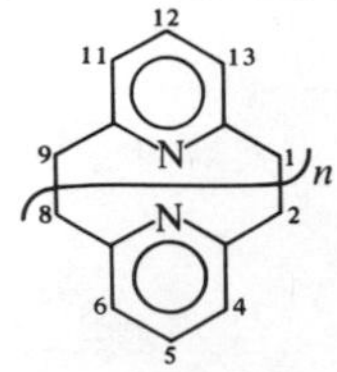	1	H
	1	*trans*-1,8-$(SMe)_2$
	1	1-(SMe)
	1	1,8(9)-$(SMe)_2$
	1	1-[(S(O)Me]
	1	1,8(9)-$(SMe_2)_2^+$
	1	1,2,8,9-De$(H)_4$
	1	1,2-De$(H)_2$
	1	1,1,2,2,8,8,9,9-$(Cl)_8$
	1	1,2,8,9-de$(H)_4$; 1,2,8,9-$(Cl)_4$
	1	1,2,8,9-de$(H)_4$; 1,2,8,9-$(-SCH_2\phi)_4$
	1	1,2-de$(H)_2$; 1,2,9-$(Cl)_3$; 8,8-$(OMe)_2$
	1	1,1,8,8-$(OMe)_4$; 2,9-$(Cl)_2$
	1	1,1,9,9-$(OMe)_4$; 2,8-$(Cl)_2$
	1	1,2-de$(H)_2$; 1,2,9-$(Cl)_3$; 8-$(X_S^S]$)
	1	1,2,8,9-$(Cl)_4$; 1,2,8,9-de$(H)_4$
	1	2,9-$(Cl)_2$; 1,1,8,8-$(OMe)_4$
	1	1,9-$(Cl)_2$; 2,2,8,8-$(OMe)_4$
	2	H
	3	H
	4	H
	5	H
	6	H
	7	H
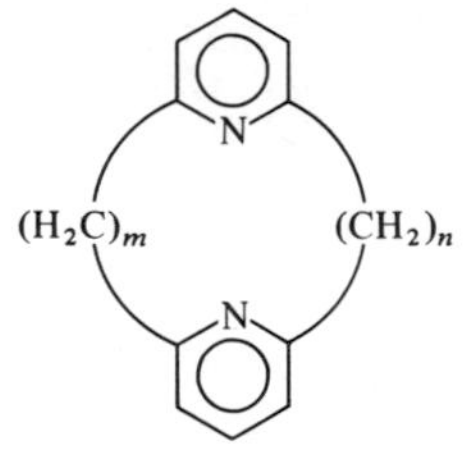	$n = 2; m = 3{-}6$	H
	$n = m = 10$	13,28-$(Me)_2$
	$n = m = 7$	3,5,15,17(=O)
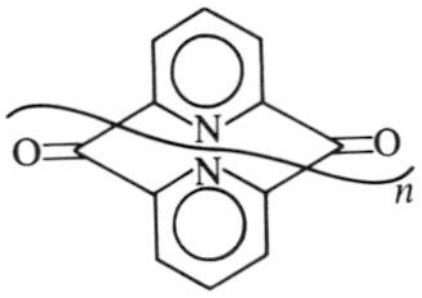	2	H
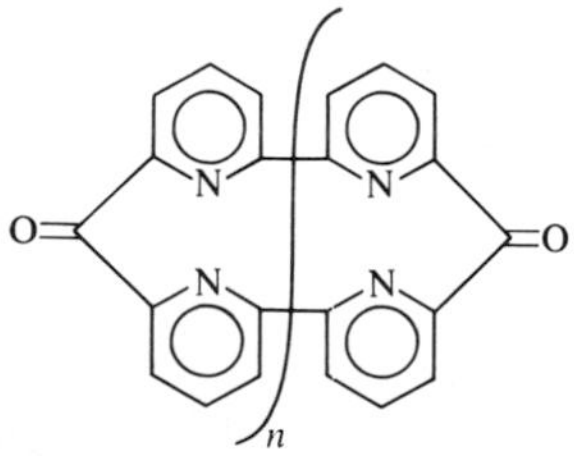	1	H
	2	H
	3	H
	4	H

Physical Data m.p. [b.p. (mm)], °C	Spectral Data Available	Metal Complex(es) General Comments	Reference
6–258	F^{308}, G^{309}		24, 28–31, 39, 40, 304, 308, 309
	A		27, 32, 310
4–235		Isomer A	28
4–126	A, B		43
7–168		Isomer B	35
4–164	A, B		43
			28
7.5–128	A, C		28, 41
2.5–182.5	A, B		43
9–362	A, D		41
4.5–256	A, C–E		41
	A, C, D		41
–178 (dec.)	A, C, D		41
–285 (dec.)			41
.5–275	A, D		41
	A, D		41
			311
	G		311
			311
–192	A		29
	A	Subl: 150–160° (0.01)	29, 31
–206	A		30
–159	A	Subl: 200–210° (0.01)	29
–161	A		29, 30
			29
			30
–105	B		20
		Attempted	48
–135	A, B		44
178	A, B		44
187	A, B		44
199 (dec.)	A		44
221	A		44

TABLE III-1.   *(CONTINUED)*

Compound	n	Substituents
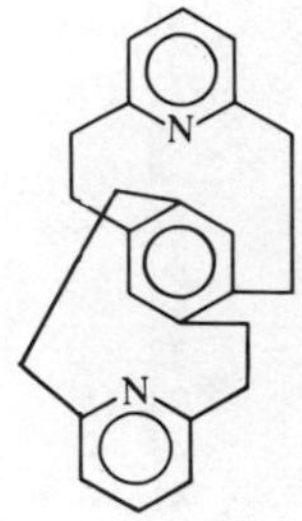		H de(H)$_8$
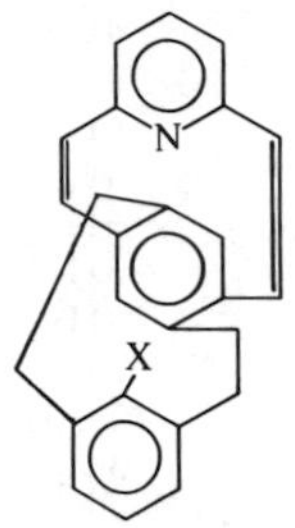		X = Cl
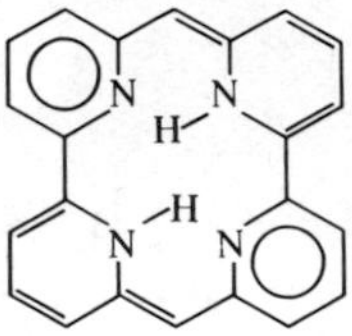		H
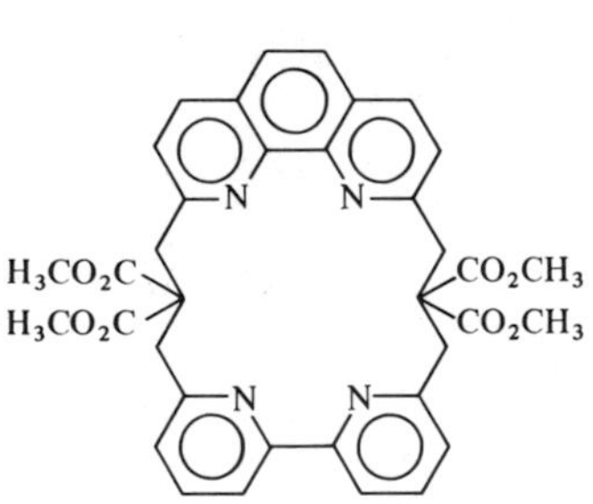		
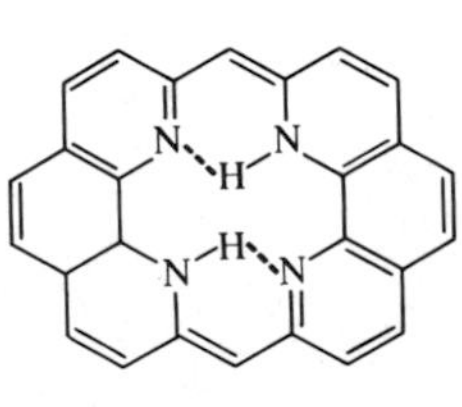		

Physical Data m.p. [b.p. (mm)], °C	Spectral Data Available	Metal Complex(es) General Comments	Reference
3–204	A, C, D, F, G^{312}		33 33, 312
–228	A, C, D		33
	B, C		313
	A, B		410
	$A–C$		439

TABLE III-1. *(CONTINUED)*

Compound 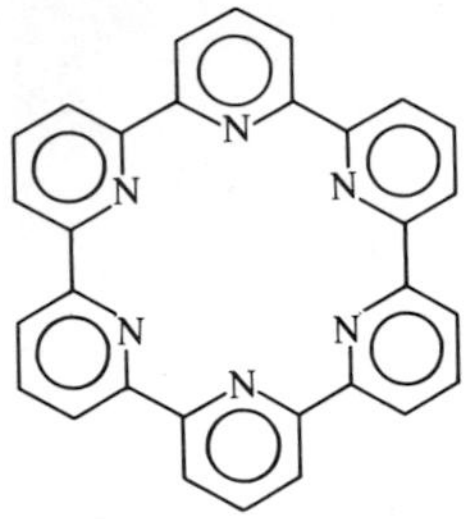	*n*	Substituents

Physical Data m.p. [b.p. (mm)], °C	Spectral Data Available	Metal Complex(es) General Comments	Reference
−295 (dec.)	*A, D, E*		460

TABLE III-2.   2,6-PYRIDINE MACROCYCLES – SULFUR BRIDGE

Compound	n	Substituents
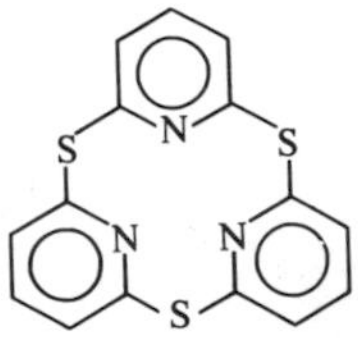		H
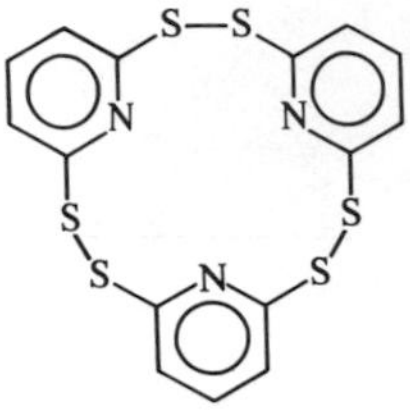		
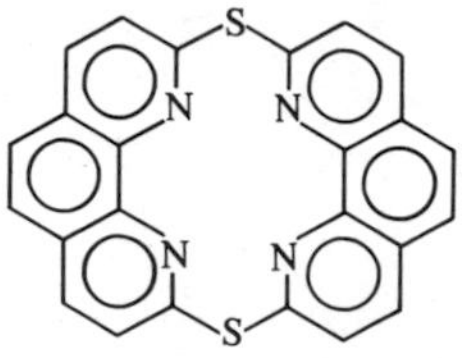		

TABLE III-3.   2,6-PYRIDINE MACROCYCLES – NITROGEN BRIDGE

Compound	n	Substituents
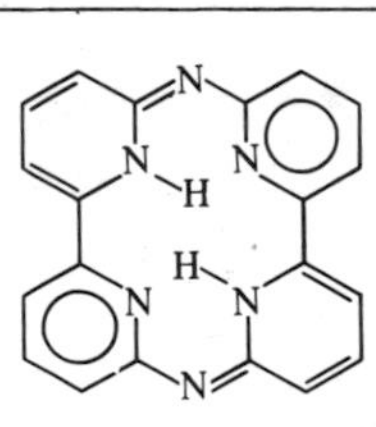		
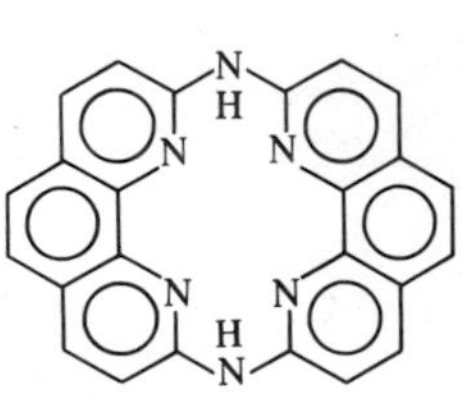		

Physical Data m.p. [b.p. (mm)], °C	Spectral Data Available	Metal Complex(es) General Comments	Reference
	G		49, 50
0–203 (dec)	A, D		51
	B, D		449

Physical Data m.p. [b.p. (mm)], °C	Spectral Data Available	Metal Complex(es) General Comments	Reference
452	A, B, D	Zn	52, 314, 440
450	A, B, D	Cu	432, 439

TABLE III-4.   2,6-PYRIDINE MACROCYCLES – CARBON, OXYGEN BRIDGE

Compound	n	Substituents
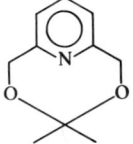	4	H
	1	R = H
	2	R = H
		R = H; 3,4;9,10-Dibenzo
	3	R = H
		R = H
		R = H
		R = H; 3,4;12,13-Dibenzo
		R = H; 3,4:12,13-Dibenzo; N → O
		R = CO_2Me; 1,4-diH_2
	4	R = H
		R = CO_2Me; 1,4-diH_2
		R = H; 3,4:15,16-dibenzo
		R = H; 3,4:15,16-dibenzo; N → O
		R = H; 3,4:9,10:15,16-tribenzo
		R = H; 9,10-benzo
		R = H; 6,7:12,13-dibenzo-
	5	R = CO_2Me; 1,4-diH_2
		R = H
	6	R = H
		H

Physical Data m.p. [b.p. (mm)], °C	Spectral Data Available	Metal Complex(es) General Comments	Reference
			7[b]
			55
(0.5 mm)]	A, D		55, 57
3–124	A, B, D		198
–41	A, D	pK$_a$4.8($\pm$ 0.2)Li, Na, K, Rb, Cs	54–56, 63, 64
	G	NH$_4^+$, [NH$_3$(t-Bu)]$^+$, [RNH$_3$]$^+$	65, 67, 71, 73
	D	Tetra-deuterated	57, 71
(dec)	A, H	K, Co, Na, NH$_4$, Ag, Pr, Rb, Ba, Hg ONNMe$_2$[m.p. 138–140°C], C$_6$H$_5$CH$_2$Cl [m.p. 145–155°C], epichlorohydrin [m.p. 125–135°C], (CH$_3$CO)$_2$O [m.p. 128–132, 138–143°C], 12[crown]4 [m.p. 170–174°]; MeCN [131°]; MeNO$_2$ [130°]; HCONH$_2$ [130°]; HCONMe$_2$ [117–125°]; MeCONMe$_2$ [65°]; DMSO [125°]; MeO$_2$CC≡CCO$_2$Me [85°]; (MeCO)$_2$ [125°]; (CH$_2$OH)$_2$ [125°]; Mesitylene [70°]	66–68, 70, 72
(dec)	A	K; MeCN [140°]; MeNO$_2$ [140–152°]; HCONH$_2$ [140°]; DMSO [174°]	66, 67
.5–116.5	A	Na$^+$	74, 442
			55
5–93.5	A	Na	74
100	A, B	MeCN [79–80°C][67]	67
–170	A, B		67
–104	A		69
67			69
100			69
100	A		74
			55
			55
	A		58

TABLE III-4. *(CONTINUED)*

Compound	n	Substituents
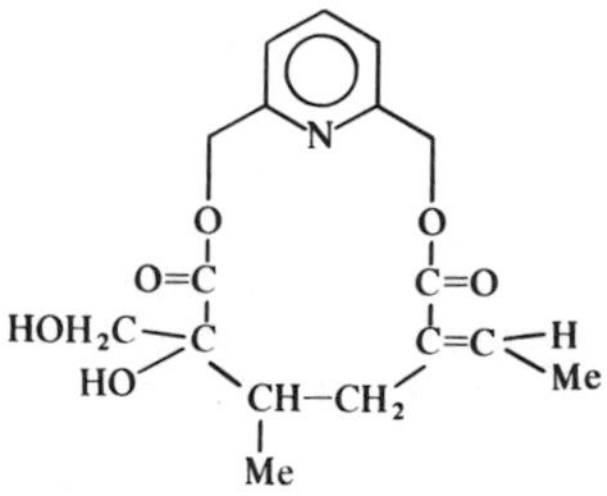		
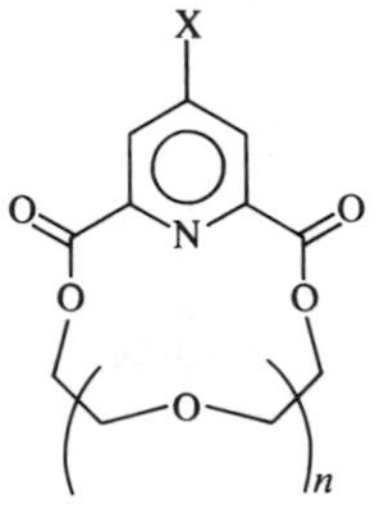	2	H
	2	3,10-(Me)$_2$
	2	3,10-(Et)$_2$
	3	H
	3	3,13-(Me)$_2$
	3	3,13-(Et)$_2$
	3	4,12-(Ph)$_2$
	3	X = (Cl)
	3	X = (Cl); 3,13-(Me)$_2$
	3	X = H; 3-(Et)
	3	X = (OMe)
	3	X = (OMe); 3,13-(Me)$_2$
	3	X = (OMe); 4,12-(Ph)$_2$
	3	X = H; 3,4; 12,13-dicyclohexano
	3	X = H; 3,4; 12,13-dibenzo
	3	X = H; 3,4-cyclohexano
	3	7,9-(Me)$_2$
	3	3,7,9,13-(Me)$_4$
	4	H
	4	3,16-(Me)$_2$
	4	3,16-(Et)$_2$
	4	X = (Cl)
	4	X = (OMe)
	4	X = H; 3,4; 15,16-dibenzo
	4	X = H; 3,4; 15,16-dibenzo; N → O

Physical Data m.p. [b.p. (mm)], °C	Spectral Data Available	Metal Complex(es) General Comments	Reference
	A, B, D	$[\alpha]_D + 51.9°$	75
9–140	A, B		76
55 (1 mm)]	A, B		78
70 (1.5 mm)]	A, B		79, 80, 87
5 (anh.)	A, B	Na, K, Ca, RNH_3^+	70, 77, 81, 85–89, 315
		Ba, K, Na, Ag, Pb, Rb, Cs, Ca, Sr	67, 316–318, 401
–66 (dihydrate)	G^{85}		85
	G^{319}	KSCN, RNH_3^+	76, 82, 319, 402
–83 (racemic)	A, B, E		78, 82, 320, 321, 402
(S, S)	A, B, E	$[\alpha]_D^{25}(S, S) = -13.7°$	320, 321, 401, 433
–92 (R, R)	A, B, E	$[\alpha]_D^{25}(R, R) = +13.2°$	320, 321, 401, 433
2–103 (R, S; meso)	A, B, E		320, 321
3.5–105.6	A, B		79, 80, 87
4.5–166 (racemic)	A, B		321, 402
–128 (S, S)	A, B	$[\alpha]_D^{25}(S, S) = +87°$	321, 433
7–128 (R, R)	A, B	$[\alpha]_D^{25}(R, R) = -91°$	321
9–102 (R, S; meso)	A, B		321
4–105	A	Li, Na, K, Rb, Cs, Ag, Mg, Ca, Sr, Ba, NH_4^+, RNH_3^+	82, 315
	A, B, E	$[\alpha]_D^{25} = -3.98°$	320, 401
88.5	A, B		80
–117	A	RNH_3^+	82, 315
99	A, B, E	$[\alpha]_D^{25} = -6.9°$	320, 321, 401
–142	A, B	$[\alpha]_D^{25} = +80.9°$	321
–138.5	(A, B)	K	80
		Na, K, CH_3NO_2 [85–90°]	67, 77
96.5	(A, B)	K, RNH_3^+	80
100	A, B		78, 82
5 (1 mm)]	A, B		78
–144.5	A, B	RNH_3^+	67, 70, 82, 87, 315
5 (1 mm)]	A, B		78
1 (1 mm)]	A, B		80
71	A		70, 76, 79, 82, 315
–123	A		70, 82, 315
		K	77
–170	A, B, D		67

TABLE III-4.   *(CONTINUED)*

Compound	n	Substituents
	5	H
	5	X = (Cl)
	5	X = (OMe)
	6	X = (Cl)
	6	X = (OMe)

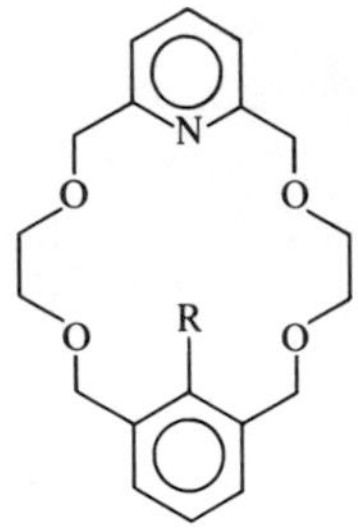

R = CO$_2$H
R = CO$_2$Me

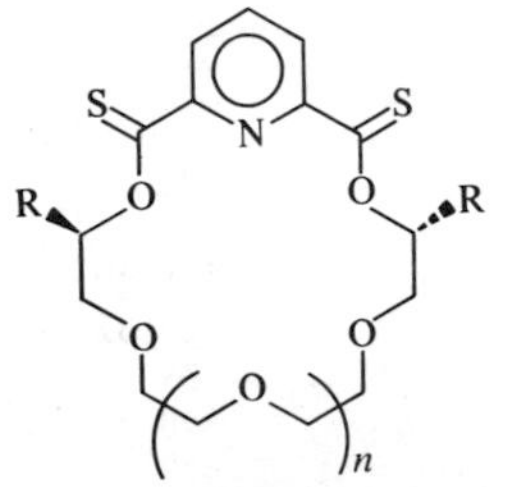

	n	
	0	R = H
	1	R = H
	2	R = H
	3	R = H
	1	R = CH$_3$

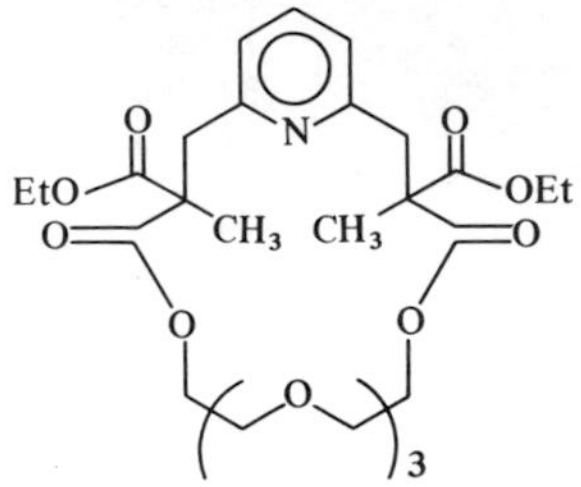

H

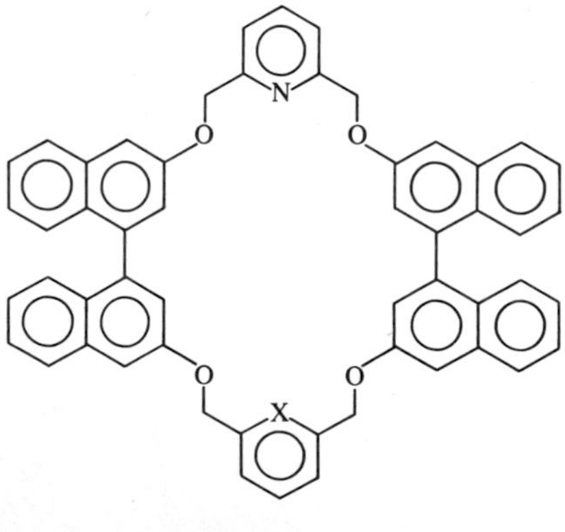

X = N

X = CH

Physical Data m.p. [b.p. (mm)], °C	Spectral Data Available	Metal Complex(es) General Comments	Reference
0–111	A, B		70, 76
–66			87
–73			87
–53			87
1			87
2–181	A, D		322
1	A, D		322
0–141	A, B, D, E	NaSCN, NaCLO$_4$	459
4–125	A, B, D, E	KSCN, KClO$_4$	459
4–125	A, B, D, E	CsSCN	459
	A, B, D, E		459
.5–64	A, B, D, E	KSCN	459
	A, B, D		323
5–298 (dec)	A, D	$[\alpha]^{25}_{589} - 250°$, $[\alpha]^{25}_{578} - 264°$, $[\alpha]^{25}_{546} - 319°$, $[\alpha]^{25}_{436} - 772°$	55, 58, 60–62, 419
	A, D	$[\alpha]^{25}_{589} - 269°$, $[\alpha]^{25}_{578} - 283°$, $[\alpha]^{25}_{546} - 339°$, $[\alpha]^{25}_{436} - 798°$	55, 58, 60–62, 419

TABLE III-4.   *(CONTINUED)*

Compound	*n*	Substituents

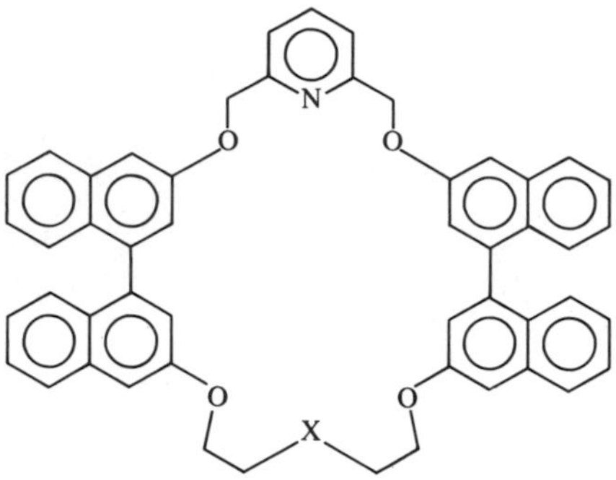

X = O

X = CH$_2$

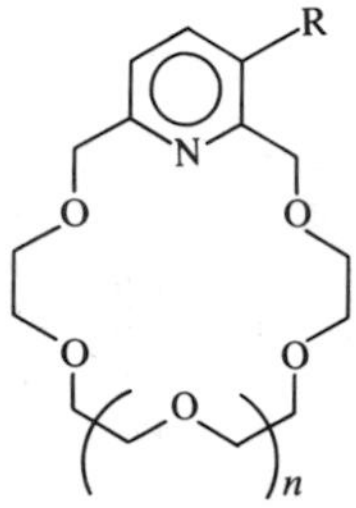

n = 1

R =   ;3,4; 12,13-dibenzo

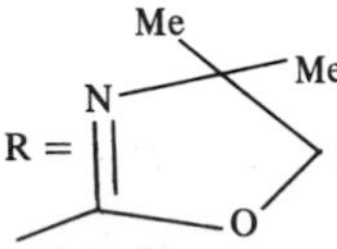

R = CO$_2$Et; 3,4; 12,13-dibenzo

n = 2, 3

R = N

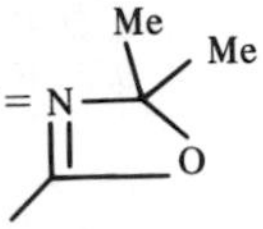

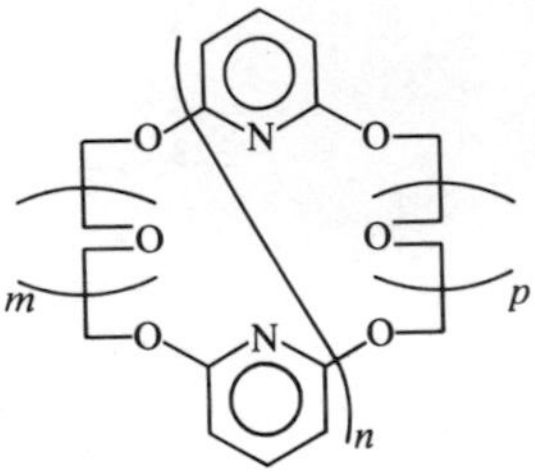

n = 0; *m* = 2		H
n = 0; *m* = 2		(±)-2-Me
n = 0; *m* = 3		H
n = 0; *m* = 3		17-CN
n = 0; *m* = 3		17-(C–NMe$_2$)
n = 0; *m* = 3		17-(CH$_2$NMe$_2$)
n = 0; *m* = 4		H
n = 0; *m* = 4		20-CN
n = 0; *m* = 4		20-(C–NMe$_2$)
n = 0; *m* = 4		20-(CH$_2$NMe$_2$)
n = 0; *m* = 4		20-CH$_2$OH
n = 0; *m* = 4		20-CH$_2$OMe
n = 0; *m* = 4		20-CH$_2$OEt
n = 0; *m* = 4		20-CH$_2$OPr
n = 0; *m* = 5		23-CN
n = 0; *m* = 5		23-(C–NMe$_2$)
n = 0; *m* = 5		23-(CH$_2$NMe$_2$)
n = 1; *m* = *p* = 0		H
n = 1; *m* = *p* = 0		6,15-(CONMe$_2$)$_2$
n = 1; *m* = 0; *p* = 1		H
n = 1; *m* = 1; *p* = 1		H
n = 1; *m* = *p* = 2		H
n = 1; *m* = *p* = 2		2,17(24)-(Me)$_2$

Physical Data m.p. [b.p. (mm)], °C	Spectral Data Available	Metal Complex(es) General Comments	Reference
	A	$[\alpha]_{578}^{25} - 242°, [\alpha]_{546}^{25} - 288°,$ $[\alpha]_{436}^{25} - 665°$	55, 58, 60, 61, 419
	A, D	$[\alpha]_{589}^{25} - 240°, [\alpha]_{578}^{25} - 250°,$ $[\alpha]_{546}^{25} - 301°, [\alpha]_{436}^{25} - 702°$	55, 58, 60–62, 419
–150	A, B, D		324
,5–125.5	A, B, D		324
	A, B, D		324
84	A, B		90, 91, 96
55	A, B, D, E		91
78	A, B, D	Na, K, Ba, Ag	90, 91, 95, 96
92	A, B, D		93
(0.3)]	A, B, D		92, 94
	A, B		92, 94
–160 (0.15)]	A, B, D		91, 94–96
(0.25)]	A, B, D		93
(0.4)]	A, B, D	Shift reagents	92–94
(0.15)]	A, B		92–94
	A, B, D		412
	A, B, D		412
	A, B, D		412
	A, B, D		412
(0.20)]	A, B, D		93
(0.3)]	A, B, D		92, 94
	A, B		92, 94
–216	A, B		90, 91, 325
(dec)	A		325
–95.5	$A–C$		90, 91
–112	$A–C$		90, 91, 96
–120	A		90, 91, 96
–110	A, B, D	Isomer A	91
	A, B, D	Isomer B	91
	A, B, D	Isomer C	91

TABLE III-4.   (*CONTINUED*)

Compound	n	Substituents
	$n = 1; m = 2; p = 3$	H
	$n = 1; m = p = 3$	H
	$n = 1; m = p = 4$	H
	$n = 1; m = p = 4$	$(CN)_2$
	$n = 2; m = p = 1$	H
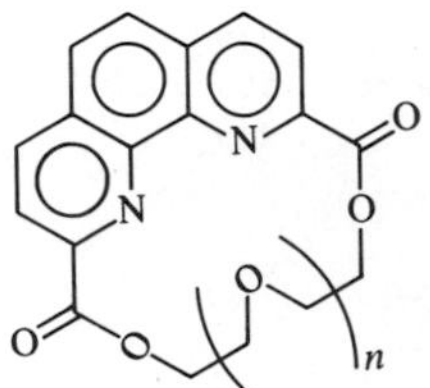	$n = 1$	H
	$n = 2$	H
	$n = 3$	H
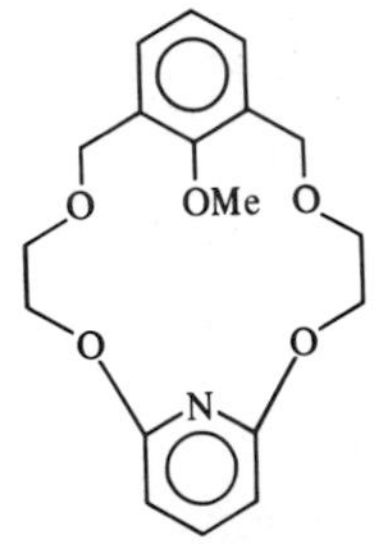		
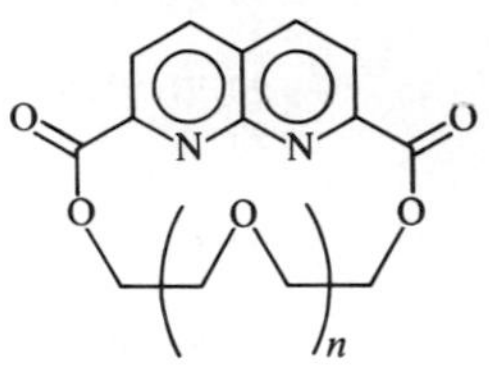	$n = 2$	H
	$n = 3$	H
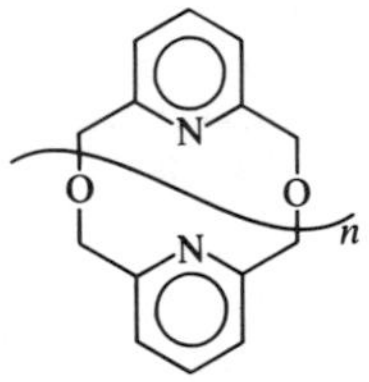	1	H
	2	H
	3	H
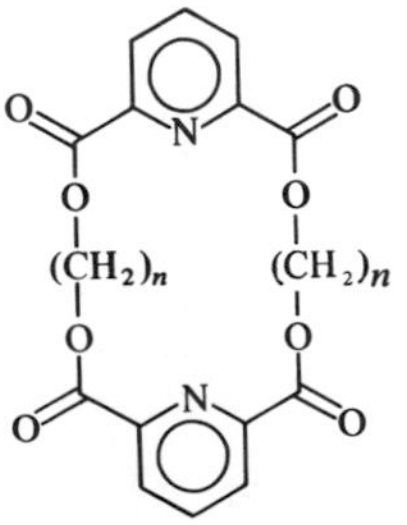	10	H

Physical Data m.p. [b.p. (mm)], °C	Spectral Data Available	Metal Complex(es) General Comments	Reference
72	A, B, D		91, 95
84	A, B, D		90, 91, 96
91	A, B, D		91, 95
−112°	A, B, D		93
.5−121.5	A, B		90, 91
−231	A, D, E		448
−201	A, D, E		448
−190	A, D, E		448
−71			435
−205	A, D, E		448
−177	A, D, E		448
−175 (dec)	A	pK_a 7.9(< 3), RNH_3^+	54−56, 63
−128	A	pK_a 5.3(3.7), RNH_3^+ tert-Butylammonium thiocyanate (1:1) (198−201°)	54−56, 63
−176 −173[55]	A	pK_a 4.8(> 3), RNH_3^+	54−56, 63
−158			77, 81

TABLE III-4.   *(CONTINUED)*

Compound	n	Substituents
		H
	$n = 0$	H 3,4:14,15-Dibenzo 3(R), 4(R), 14(R), 15(R)-(CONMe$_2$)$_4$ 1,6,12,17-(C=O)$_4$; 3,4; 14,15-dibenzo
	$n = 1$	3,14- or 4,15-
	$n = 1\text{–}3$	
	$n = 0$ $n = 1$ $n = 2$ $n = 3$ $n = 4$ $n = 5$	H H
	1 2 3 4	4,5:17,18-Dibenzo 4,5:17,18-Dibenzo 4,5:17,18-Dibenzo 4,5:17,18-Dibenzo
	1 2 3 4	H H H H

Physical Data m.p. [b.p. (mm)], °C	Spectral Data Available[a]	Metal Complex(es) General Comments[d]	Reference
~125		K, Co	326, 421
~148	A, D	pK_a 5.3(3.6), RNH_3^+	54–56, 63, 419
~186	A, D		
	A, D, E	$[\alpha]_D^{25} + 107°$	
~276			77
~149	A	$[\alpha]_D$ (CHCl$_3$) − 22°; K_a; $C_6H_5CH_2NH_3^+$ (SCN$^-$); t-BuNH$_3^+$; RNH$_3^+$	82, 99
			55
			55
	A	(Impure sample)	54–56
			55
			55
			55
			53
~143	A, C	NaSCN (195–196°)	53
130	C		53
109	C		53
105	C		53
	A, B, D	Co(SCN)$_2$	327
	A, B, D	CoCl$_2$, CuCl$_2$	327
	A, B, D	CuCl$_2$, PdCl$_2$	327
	A, B, D	CoCl$_2$	100

TABLE III-4.   (*CONTINUED*)

Compound	n	Substituents
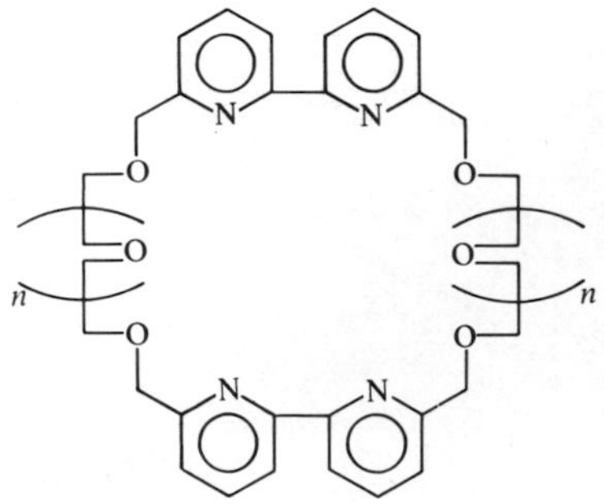	1 3 4 5	H H H H
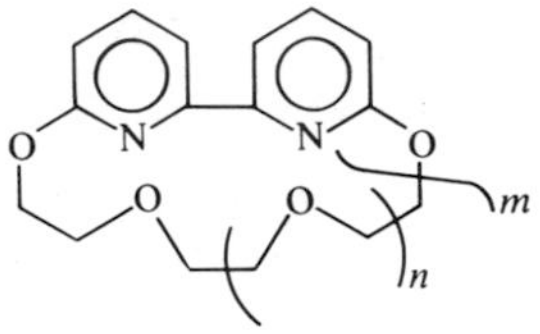	$n = 1; m = 1$ $n = 2; m = 1$ $n = 3; m = 1$ $n = 4; m = 1$ $n = 0; m = 2$ $n = 1; m = 2$ $n = 2; m = 2$ $n = 3; m = 2$ $n = 4; m = 2$	H H H H H H H H H
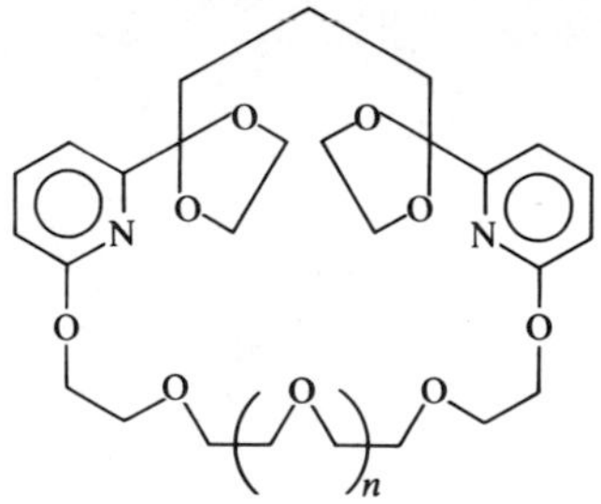	3	H
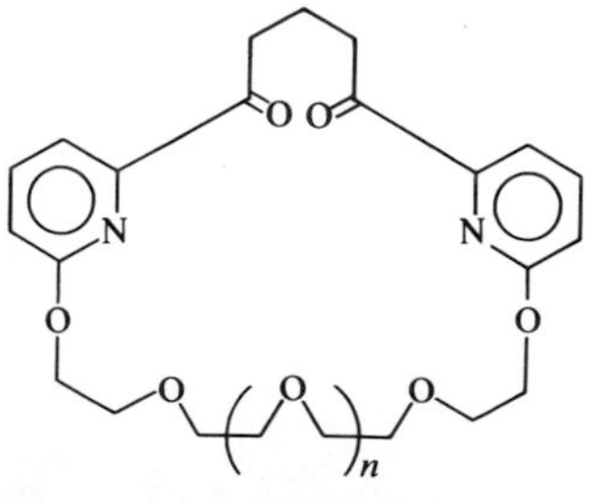	3	H

Physical Data m.p. [b.p. (mm)], °C	Spectral Data Available	Metal Complex(es) General Comments	Reference
5–147	A, B, D		
	A, B, D	$CuCl_2$, $CoCl_2$	327
	A, B, D	$CuCl_2$, $CoCl_2$, $PdCl_2$, $ZnCl_2$	101
	A, B, D	$CoCl_2$, $CuCl_2$	327
6–108	A–C		102
–101	A–C		102, 104
–92	A–C, G	Subl [80°C (1 mm)]	102
–43	A–C	Subl [80°C (1 mm)]	102, 103
3–105	A, B		102
6–148	A, B		102
1–122	A, B		102
–101	A, B		102
–72	A, B		102
	A, B, D		108
–100	A, B, D		108

TABLE III-4. (*CONTINUED*)

Compound	*n*	Substituents
	1 2 3 4 5	H H H H H
	1 2 3 4 5	H H H H H
	n = 1–4	H
	1 2 3 4 5 6	H H H H H H
	1 2 3 4 5 6	H H H H H H

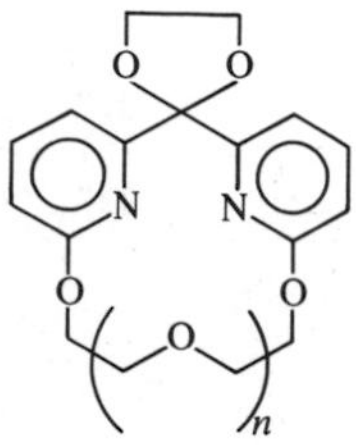

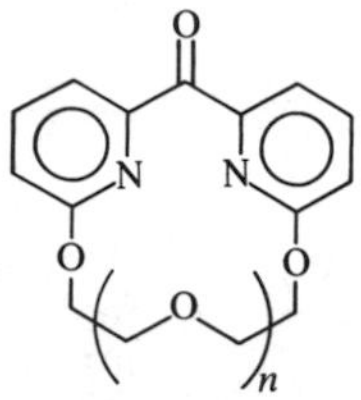

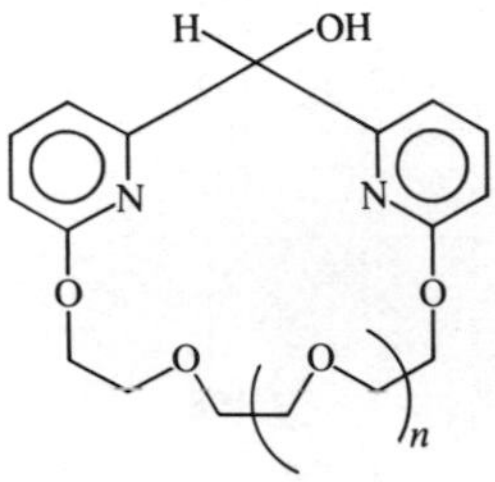

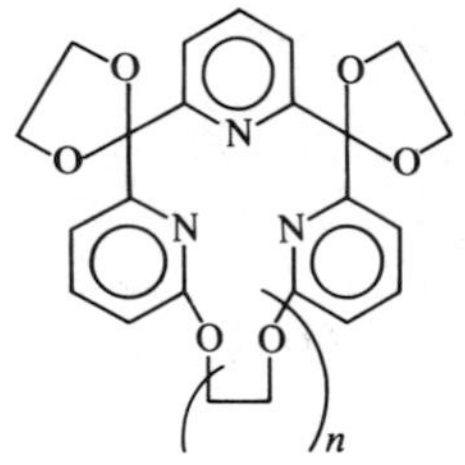

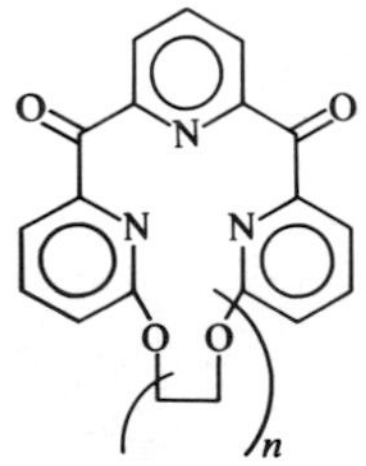

Physical Data m.p. [b.p. (mm)], °C	Spectral Data Available	Metal Complex(es) General Comments	Reference
–123.5	A, B, D		107
.5–120	A, B, D		107
72	A, B, D, [G]		104, 107
	A, B, D		104, 107
	A, B, D		104, 107
.5–122	A, B, D		107
5–96.5	A, B, D		107
–134	A, B, D, G		104, 107
70.5	A, B, D		104, 107
5–56	A, B, D, (G)	Neutral complex of H_2O	86, 104, 107
	A, B, D		107
9–211	A–E		44, 105
1–163	A–E,[106]		105, 106
2–135	A–E		105
8–131	A–D		105
–95	A–D		105
–87	A–D		105
		Not isolated	105
–121	A–E		105
–146	A–E, G[106]		105, 106
–120	A–D		105
77	A–D		105
63	A–D		105

TABLE III-4.   *(CONTINUED)*

Compound	*n*	Substituents
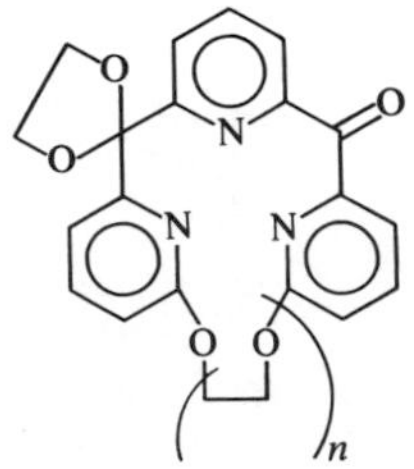	2 5	H H
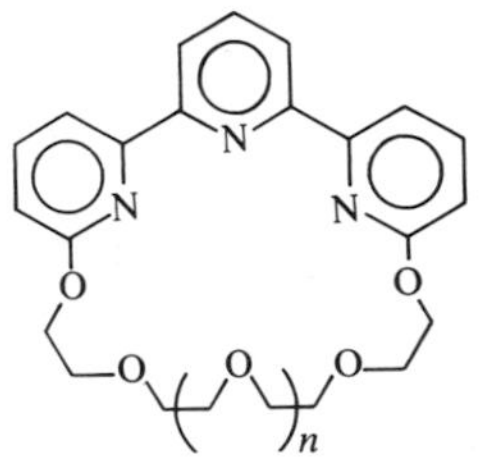	3	H
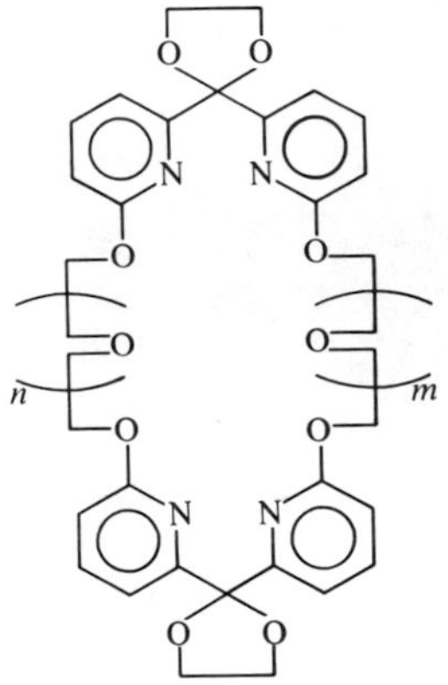	$n = 0, m = 0$ $n = 1, m = 1$ $n = 1, m = 2$ $n = 2, m = 2$ $n = 3, m = 3$ $n = 4, m = 4$ $n = 5, m = 5$	H H H H H H H
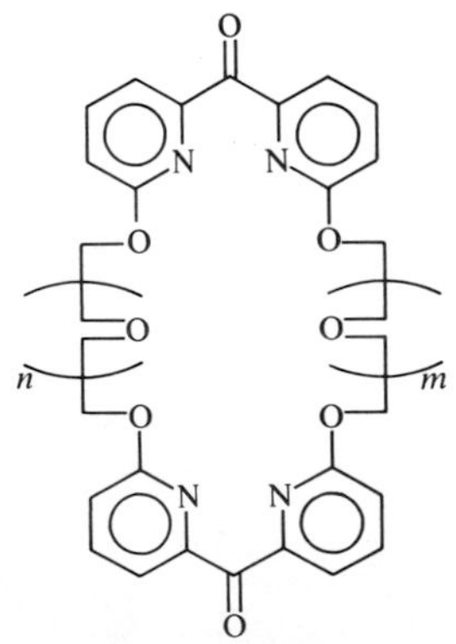	$n = 1, m = 2$ $n = 2, m = 2$	H H

Physical Data m.p. [b.p. (mm)], °C	Spectral Data Available	Metal Complex(es) General Comments	Reference
−144	A–D		105
−127	A–D		105
−128	A, B, D		108
	A, B, D	Complexes CHCl₃	107
	A, B, D		107
	A, B, D		107
	A, B, D		107
	A, B, D		107
	A, B, D		107
	A, B, D		107
−141	A, B, D		107
.5–170	A, B, D		107

TABLE III-4. *(CONTINUED)*

Compound 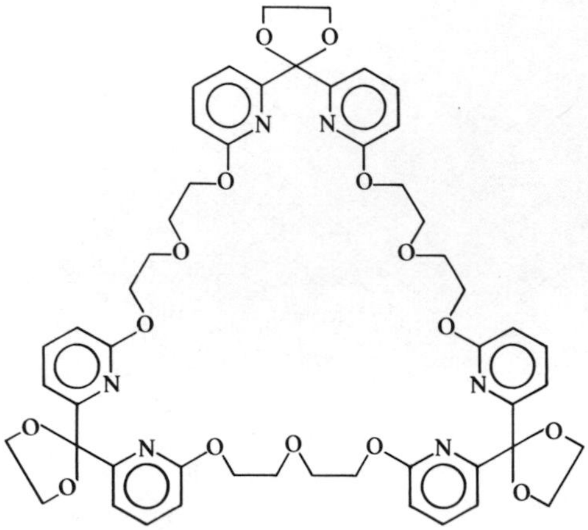	*n*	Substituents

Physical Data m.p. [b.p. (mm)], °C	Spectral Data Available	Metal Complex(es) General Comments	Reference
	A, B, D		107

TABLE III-5.   2,6-PYRIDINE MACROCYCLES – CARBON, NITROGEN BRIDGE

Compound	n	Substituents
	1 1	H 1,12-$(Me)_2$(abr: "B")
	2	1,12-$(Me)_2$; 3,4; 9,10-dibenzo 1,12-$(Me)_2$; 1,2,11,12-$(H)_4$ (abr:pyane N$_s$ 1,15-$(Me)_2$(abr: "A") 1,2,14,15-*tetra*(H)
	$m = n = 3$	H 1,11-$(Me)_2$(formerly cyp[149]; CR[118])

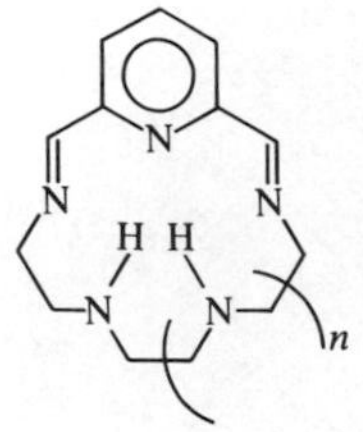

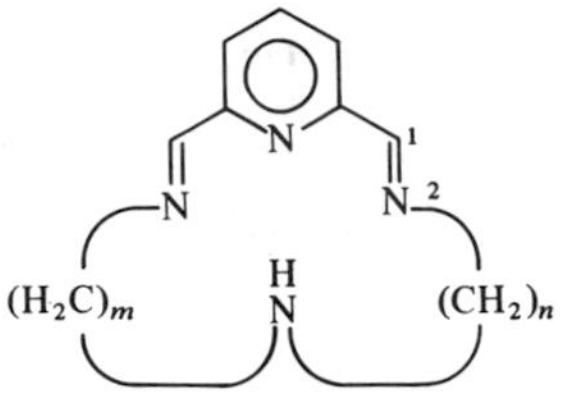

1,11-$(Me)_2$; 6-(CH_2CH_2OH)
1,11-$(Me)_2$; 6-$(CH_2)_3NH_2$
1,11-$(Me)_2$; 6-$CH_2CH_2N(Me))_2$

1,11-$(Me)_2$; 1,2,10,11-$(H)_4$
   (abr:CRH or CR + 4H)

1,11-$(Me)_2$; 1,2,10,11-
   $(H)_4$; 6-$CH_2CH_2N(Me)_2$
1,11-$(Me)_2$; 1,2,10,11-$(H)_4$; 6-$(CH_2CH_2O$
1,11-$(Me)_2$; 1,2-di(H)
1,6,11-$(Me)_3$(abr:N-MeCR)
1,11-$(Me)_2$; 5,6-de(H)

1,11-$(Me)_2$; 6-(H)
1,11-$(Me)_2$; 1,2-di(H); 6-(H)
1,11-$(Me)_2$; 1,2,10,11-*tetra*(H); 6-(H)

Physical Data m.p. [b.p. (mm)], °C	Spectral Data Available	Metal Complex(es) General Comments	Reference
		Fe, Mn, Zn	203, 328
	(B, C, J^{330})	Fe[(G)],[328] Mn, Zn	109, 134, 135, 138, 329, 330, 422, 430
	(B, C, G^{166})	Mn, Pb	130, 139, 143, 147, 166, 328, 331–333
	(B, C, G^{136})	Mg	136
		Mn, Zn[(G): Mn–(ClO$_4$)$_2$]	115, 331
	(B, C) (G^{334})	Fe, Co, Ni, Cu	137, 334, 422
		Fe	109, 453
		CA	425
	(B, C)	Zn	116
	$(A–C)$	Co	116–118, 126, 149, 335
		Mn, Cd	130, 132, 431
	$(A–C, K^{150},$		116, 119–121, 150, 336
	$G^{336, 337})$	Ni, electronic spectral studies	122, 337, 338, 427, 428
	(B, C, K)	Cu	119, 127, 339, 340, 430
	(B, C)	Zn	116, 118
	$(A–C)$	Ru	133
	$(A–C)$	Ni	131
		Ni, Cu, Co, Zn	341
	(B) , $(A–C)$	Ni	131, 342
	(B, C)	D isomer (131–134°)	122, 343, 430
		Ni, Electrocatalytic reduction of CO$_2$, pulse radiolysis	344, 345, 427, 428
	(B, C)	$Meso$ isomer (83–85°)	122
	$(A–C)$	From $meso$: Co, UO$_2$	118, 151, 346, 411
	$(A–C, K^{150}$ $G^{337})$	From $meso$: Ni	120, 121, 150, 152, 337, 338, 347
	(B, C, J)	Fe	125, 153, 348
	(B, C, K)	Cu	127
	$A–C$	Ni[ClO$_4$]$_3^-$ (diamagnetic); (ClO$_4$)$_2^-$ (paramagnetic)]; pK_a; Cu, Zn	128, 342 131, 349
25	$(A–C)$	Ni, Cu, Zn	131, 349
	$(A–C, K^{150})$	Ni	121, 150
	$(B–C)$ (G^{129})	Zn, Cu, [Co^{129}]	116, 129, 428
	$(A–C, K^{150})$	Ni	121, 150, 350, 393, 427
		Ni	350
		Ni	350
		Ni	350

TABLE III-5.   *(CONTINUED)*

Compound	n	Substituents
	$m = n = 2$	1,2,8,9-(H)$_4$; 2,5,8-(Ts)$_3$
	$m = 2; n = 0$	1,2,9,10-(H)$_4$
		1,2,9,10-(H)$_4$; 2,5,6,9-(CH$_2$COOH)$_4$
	$m = 2; n = 4$	1,11-(Me)$_2$(abr: 2,4-CR)
	$m = 3; n = 4$	1,12-(Me)$_2$(abr: 3,4-CR)
	$m = 3; n = 2$	1,10-(Me)$_2$(abr: 3,2-CR)
		1,14-(Me)$_2$
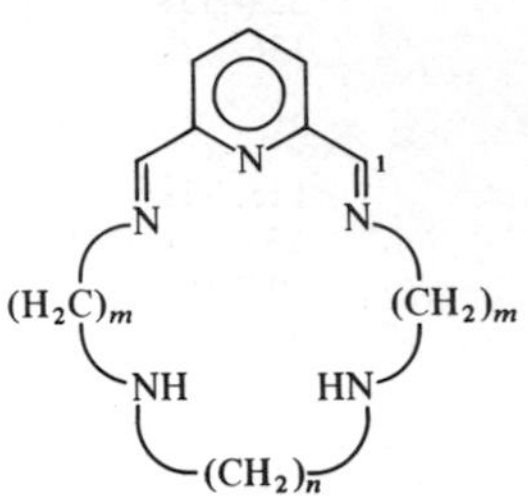	$m = 3; n = 2$	
	$m = 2; n = 3$	1,13-(Me)$_2$
		1,13-(Me)$_2$(abr: "C")
		1,13-(Me)$_2$; 1-(OMe); 2-(H)
		1,13-(Me)$_2$; 1-(OEt); 2-(H)
		1,2,12,13-*tetra*(H)
	$m = n = 2$	1,2,11,12-(H)$_4$
		1,2,11,12-(H)$_4$; 2,5,8,11-(CH$_2$COOH)$_4$
		1,2,11,12-(H)$_4$; 2,5,8,11-(Ts)$_4$

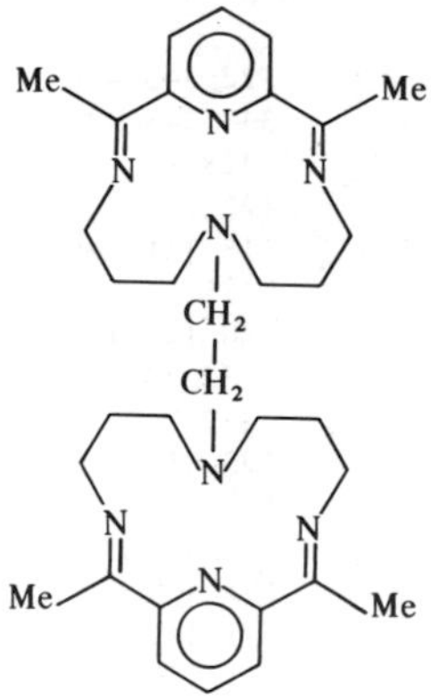

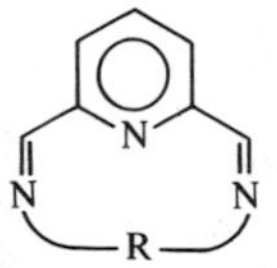

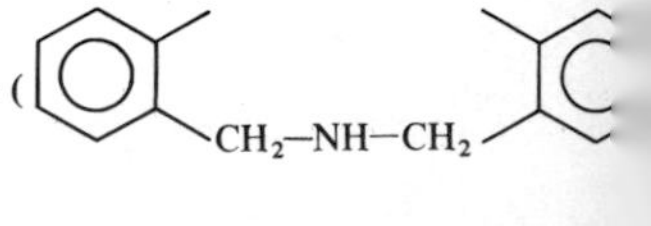

Physical Data m.p. [b.p. (mm)], °C	Spectral Data Available	Metal Complex(es) General Comments	Reference
-164	*A*		420
	A		420
	A, B	Cu, Pb, Cd, Co, Zn, Mg, Ca, Sr	420
	(*B, C*)	Ni, Cu	351, 428, 430
	(*B, C*)	Ni, Cu, Zn	116
	(*B, C*)	Attempted	116
	(*G*^{352})	Mn, Cd, Hg, Ag	141, 352
	(*G*^{166})	Mn	166
	(*A, B, C*)	Mn, Zn	130
	(*B, C*)	Pb	143
	(*A, B, D*)	Ca, Mg, Ba, Sr	144
	(*B, C*)	Ni	145
	(*B, C, G*)	Cd, Hg	146
	(*G*^{352})	Cd, Ag, Hg	352
		Fe, Mn, Zn	328
	(*B, C, G*^{328})	Mg, Fe, Mn	136, 138, 331
	(*G*^{353})	Ag	140–142, 328, 353
	(*G*^{352})	Cd	352
	(*A–C, G*^{166})	Mn	130, 166, 333
	(*B, C*)	Ni	145
	(*B, C*)	Ni	145
			434
	A		420
	A, B	Cu, Pb, Cd, Co, Zn, Mg, Ca, Sr	420
-161	*A*		420
	(*B, K*)	Cu	354
		Ni, Cu, Co	132

TABLE III-5.   (*CONTINUED*)

Compound	*n*	Substituents

$R = ($ phenyl $-N=N-$ phenyl $)$

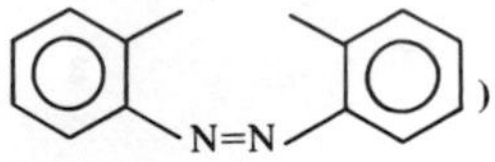

$R = ($ phenyl $-NH-$ phenyl $)$

$R = ($ phenyl $-CO-NH-$ phenyl $)$

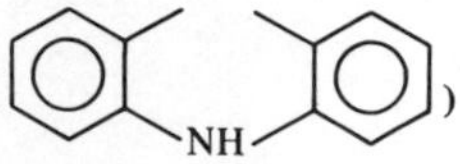

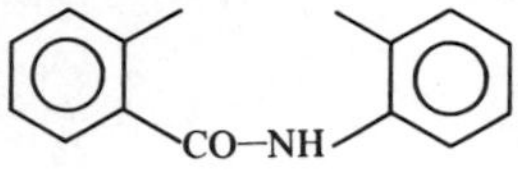

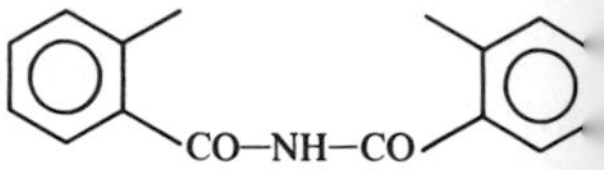

$R = ($ phenyl $-CO-NH-CO-$ phenyl $)$

$R = ($ phenyl $-NH-CH_2-CH_2-NH-)$

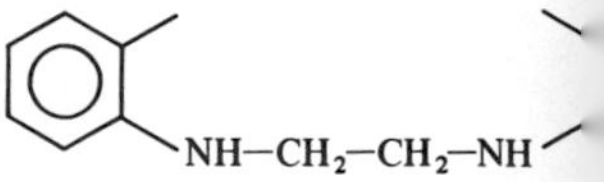

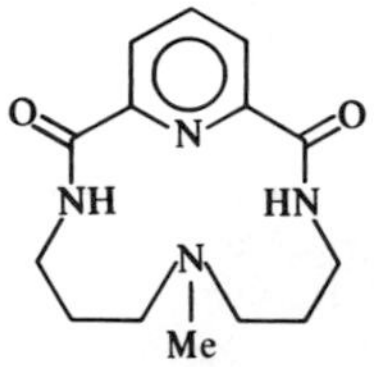

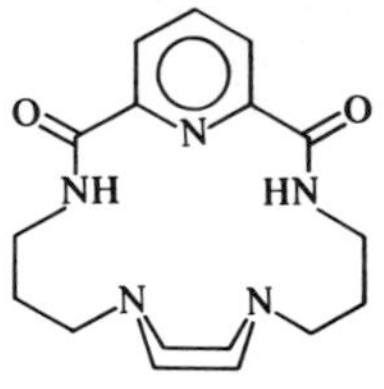

H

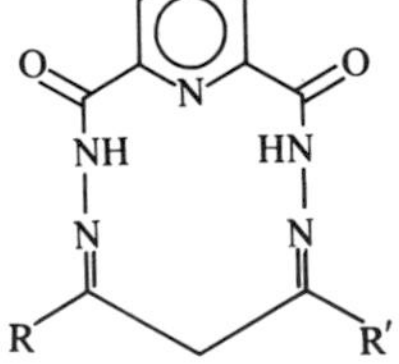

$R = R' = (CH_3)$
$R = (CH_2CH_3); R' = (CH_3)$
$R = (C_6H_5); R' = (CH_3)$
$R = R' = (CH_3); 5,6\text{-de}(H)_2; 7\text{-}(H)$

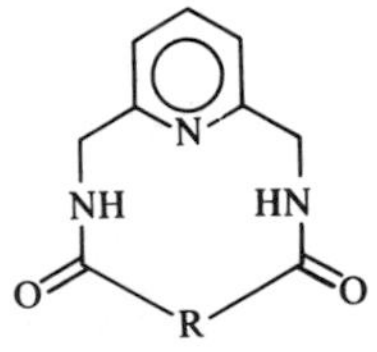

$R = (CH_2)$

Physical Data m.p. [b.p. (mm)], °C	Spectral Data Available	Metal Complex(es) General Comments	Reference
		Ni, Cu, Co	132
		Ni, Cu	132
		Ni, Cu, Co	132
		Ni, Cu, Co	132
		Mn, Zn	115, 331
166		Cu [m.p. 196–198° dec]	154, 443
228			154
	(B, C)	V	159
	(B, C)	V	159
	(B, C)	V	159
	(B)	Co, Ni, Cu	355
	D		156
	D		156

TABLE III-5.   *(CONTINUED)*

Compound	n	Substituents
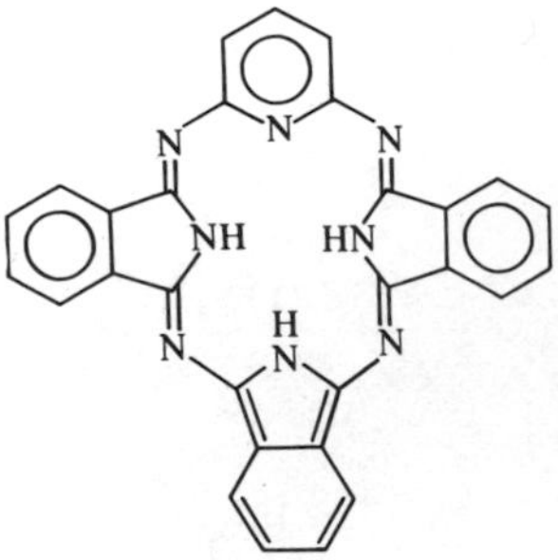		H
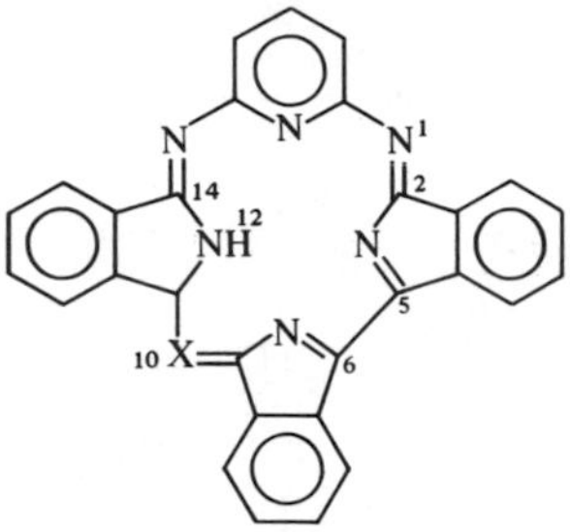		X = N X = C(cynophenyl); 11,12 -De(H)$_2$
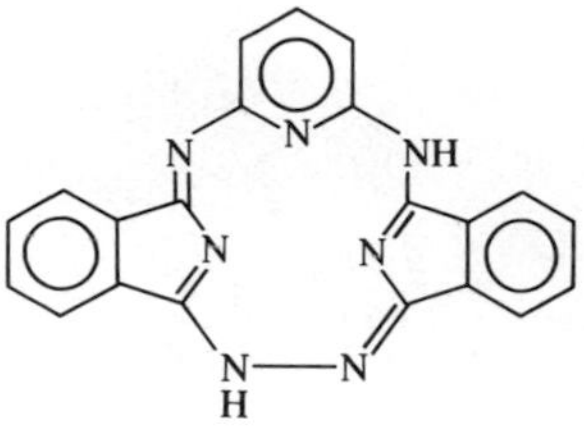		
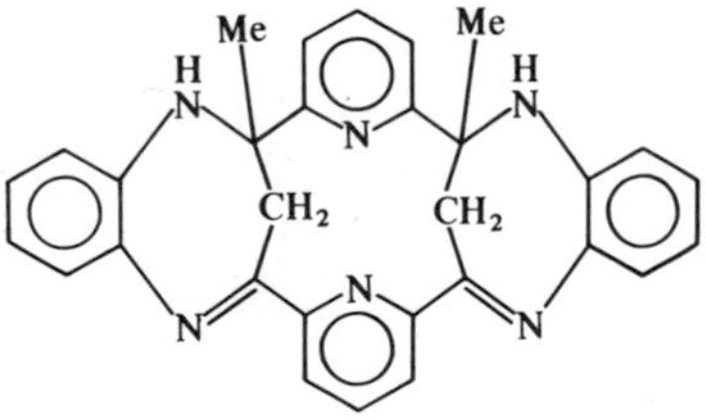		
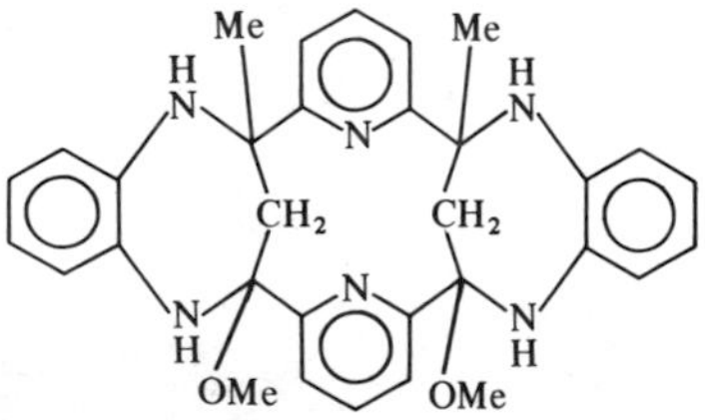		

Physical Data m.p. [b.p. (mm)], °C	Spectral Data Available	Metal Complex(es) General Comments	Reference
	(C, D)	Ni, Cu, Au	124
	C, D	Cu	168
	B, D	Cu	168
		Cu, Ni, Co	166
	A, B, D, G		148, 423, 424
	B, (G)	Cu	424

TABLE III-5.   (*CONTINUED*)

Compound	*n*	Substituents
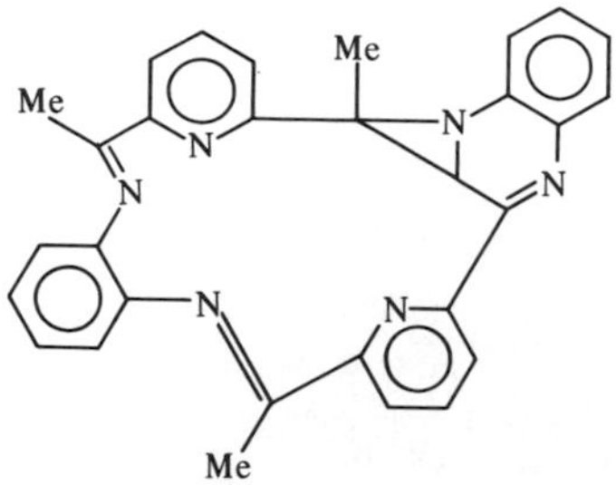		
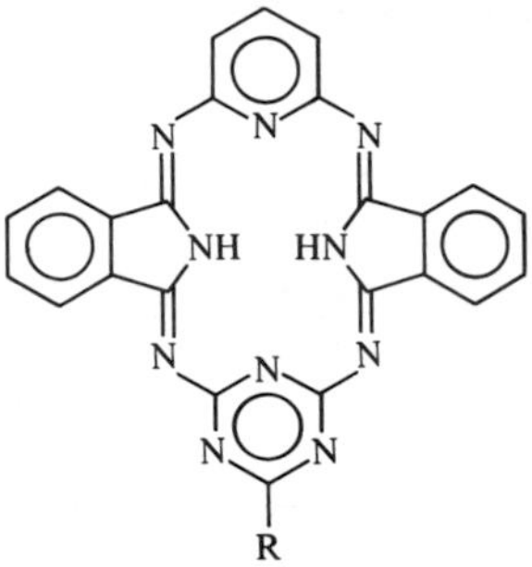		R = OH
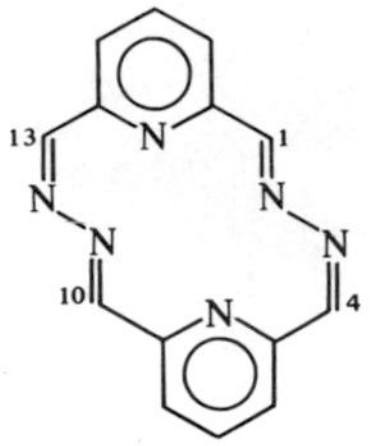		1,4,10,13-(Me)$_2$
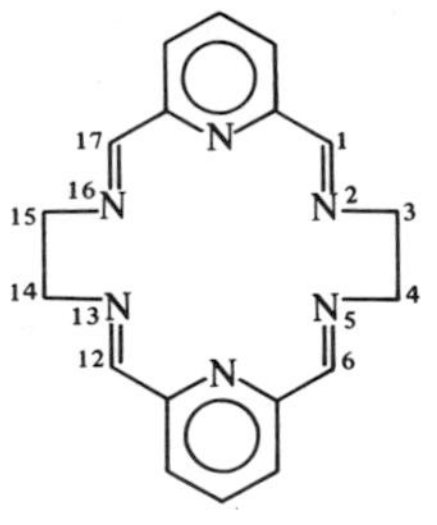		3,4:14,15-Dibenzo (abr: HADA) 1,6,12,17-(Me)$_4$; 3,4:14,15-dibenzo (abr:tmed) 1,6,12,17-(Me)$_4$
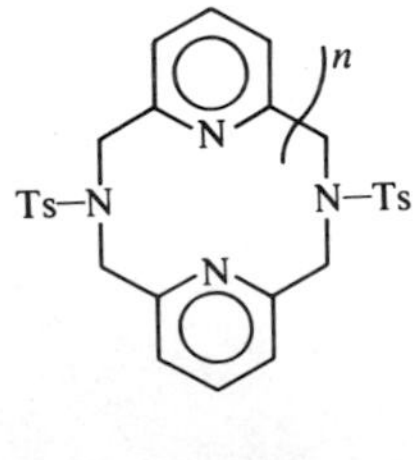	1 2	

Physical Data m.p. [b.p. (mm)], °C	Spectral Data Available	Metal Complex(es) General Comments	Reference
	B, (G)	Cu	424
		Cu, Co	165
	(B, C, G)	Fe	123, 445, 447
	(A–C)	Sc	171
		Zn	431
	(B, D, G)	Theoretical calculations	169
		Pb, Cd, Ca, Sr, Ba	172
/1 mm (subl)	*(B–D, K)*	Cu, La	148, 358a, 423
	(A, B, G)	Ca, Sr, Ba, Pb, Tl	170, 397, 426
	(A, B, G)	La	173, 356
20	*A, D*		409
20	*A, D*		409

TABLE III-5.   (*CONTINUED*)

Compound	*n*	Substituents

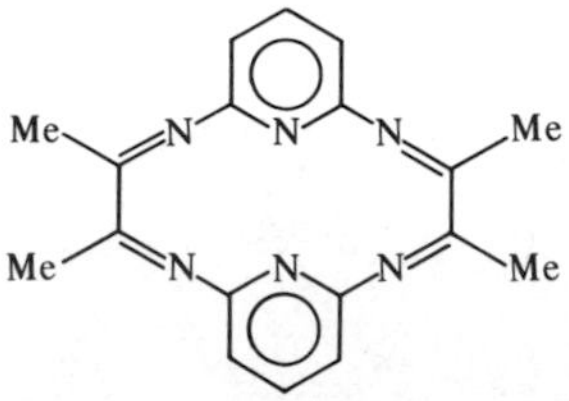

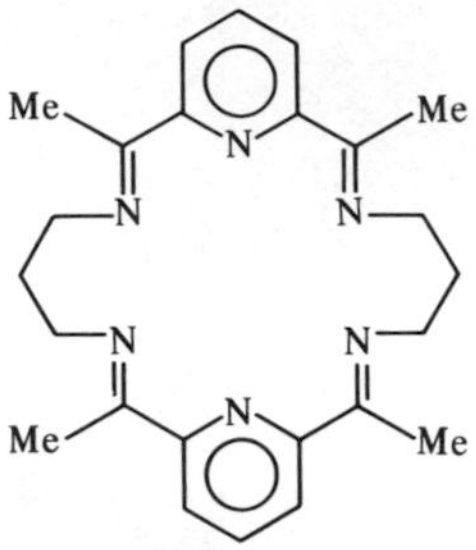

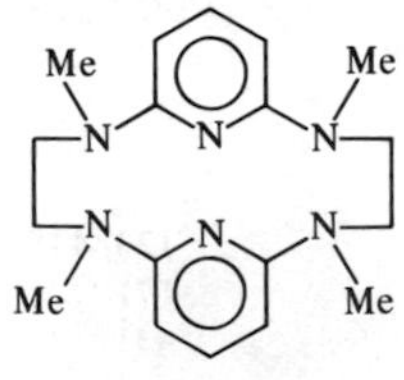

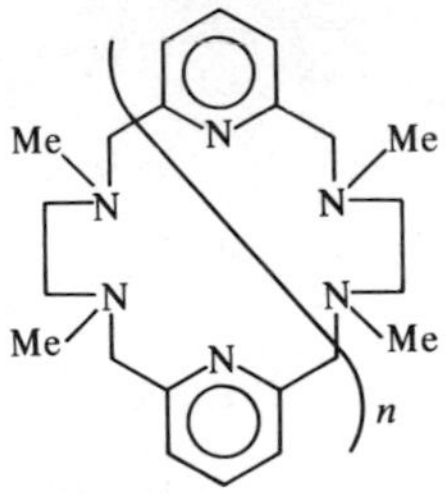	$n = 1$ $n = 2$	

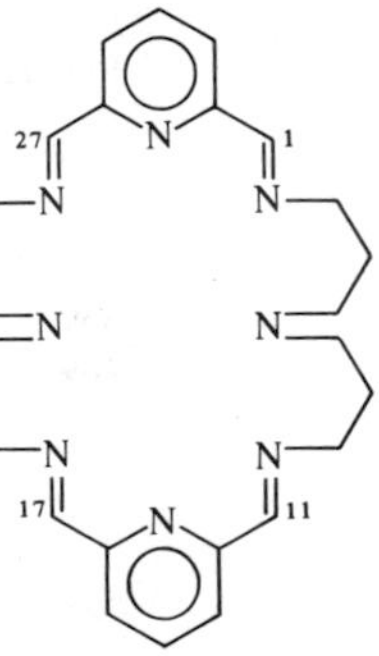

1,11,17,27-(Me)$_4$

Physical Data m.p. [b.p. (mm)], °C	Spectral Data Available	Metal Complex(es) General Comments	Reference
		Cu, Co, Ni	455
		Cu, Ba, Sr	452
	A, D		357
81	*A, D, [G]*	Co, Cu	357
	A, C		357
	(B, G)	Ag, Cu	174, 398

TABLE III-5.   (*CONTINUED*)

Compound	*n*	Substituents
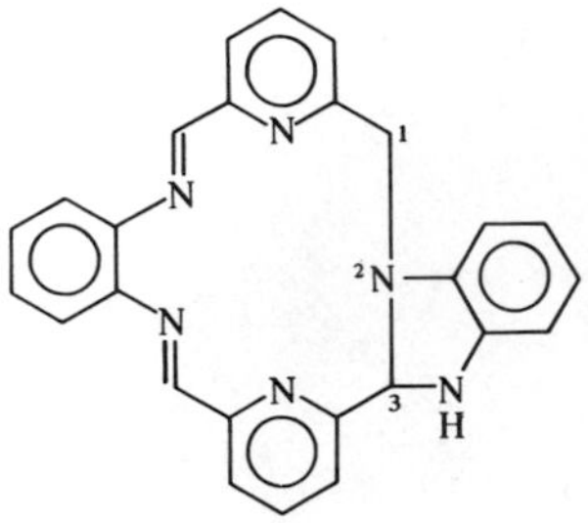		1-(OMe)
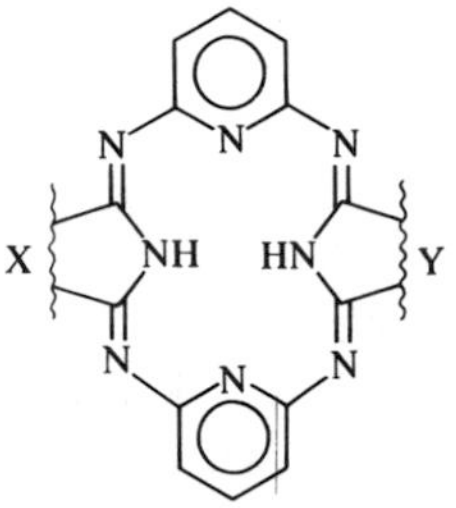		$X = Y =$ 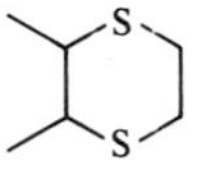
		$X = Y =$ 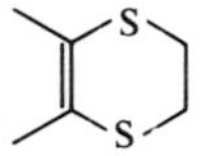
		$X = Y =$ 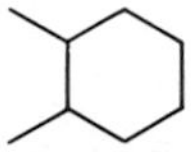
		$X = Y =$ 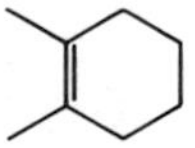
		$X = Y =$
		(abr: OAPI)
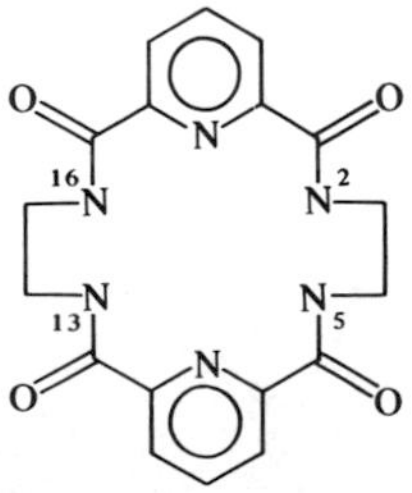		H 2,5,13,16-(Ts)$_4$

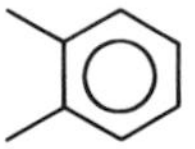

Physical Data m.p. [b.p. (mm)], °C	Spectral Data Available	Metal Complex(es) General Comments	Reference
	(B, D, G)	Mn, Fe, Co, Zn	175
			162
	C		160, 161
			162
	C		161, 163, 164
	A–D	Ge, theoretical calculations	169, 436
50			154, 155
			154

TABLE III-5. (*CONTINUED*)

Compound	n	Substituents

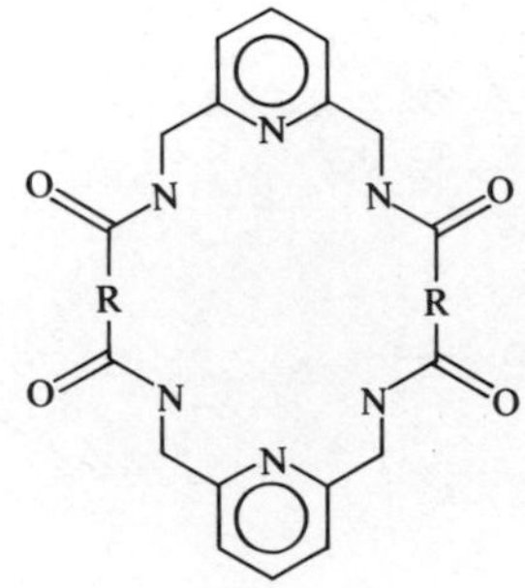

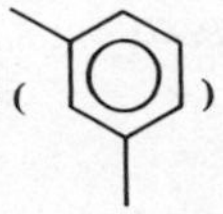

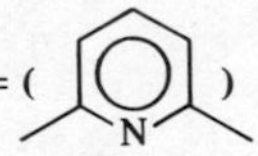

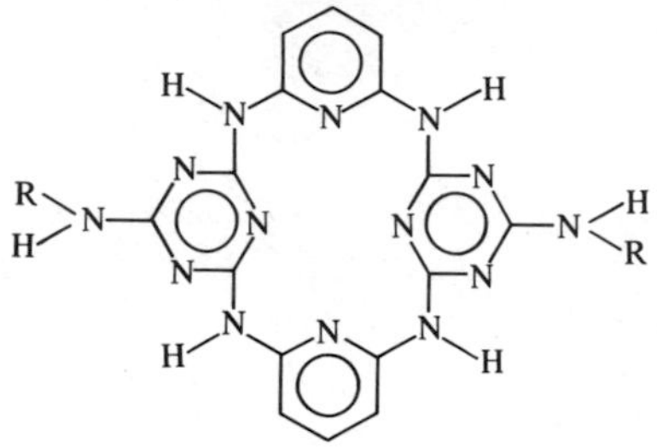

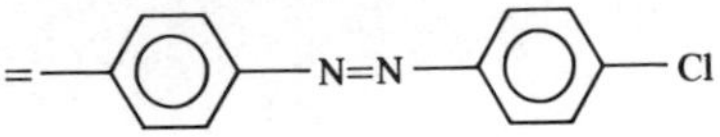

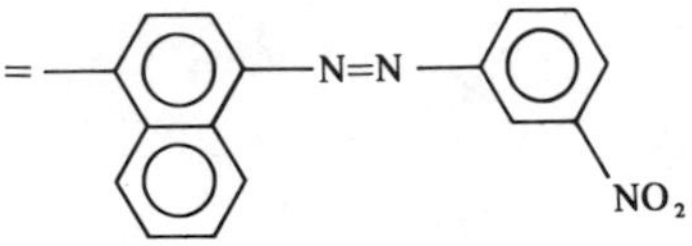

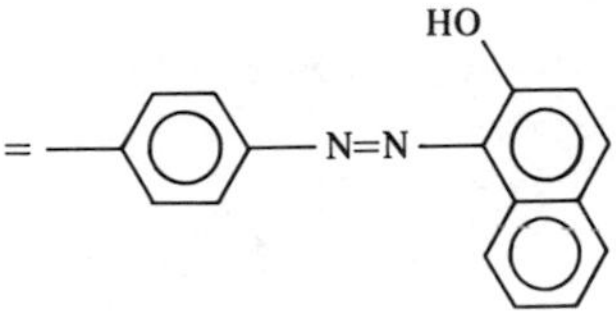

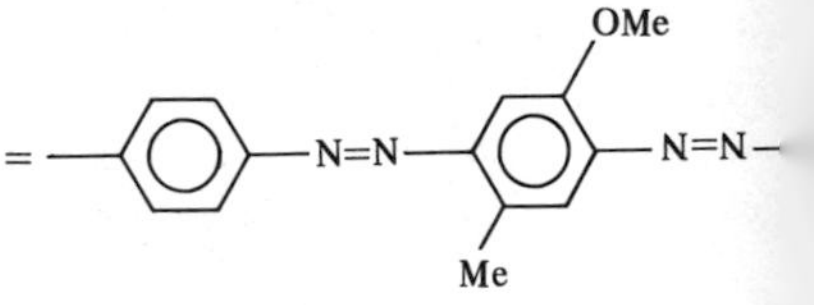

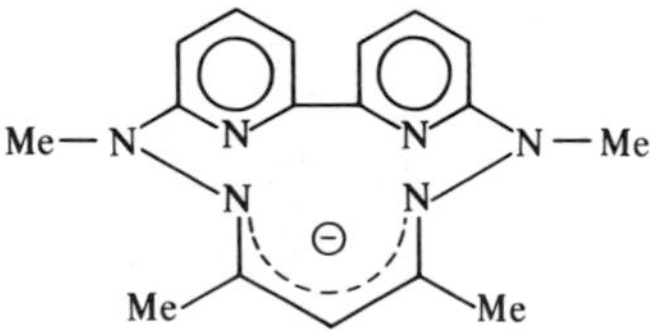

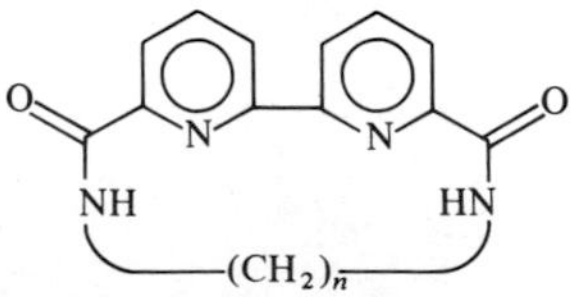

Physical Data m.p. [b.p. (mm)], °C	Spectral Data Available	Metal Complex(es) General Comments	Reference
300	*D*		156
20–322	*D*		156
	C	Cu	167
	C	Cu	167
	C	Cu	167
	C	Cu	167
	(A, B, C)	Ni	178
4–185	*A, D*		158

TABLE III-5. *(CONTINUED)*

Compound	n	Substituents

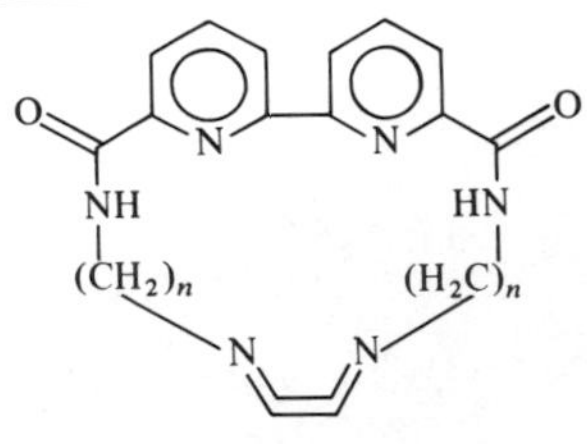

$n = 3$ H

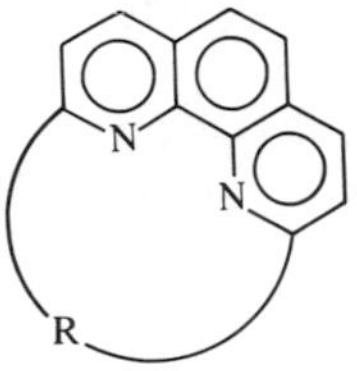

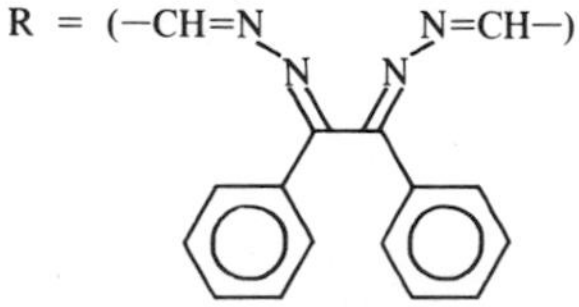

$R = (-CH=N \quad\quad N=CH-)$

$R = (-CH=N-(CH_2)_2-NH-(CH_2)_2-N=CH-)$
$R = (-CH=N-(CH_2)_3-NH-(CH_2)_3-N=CH-)$

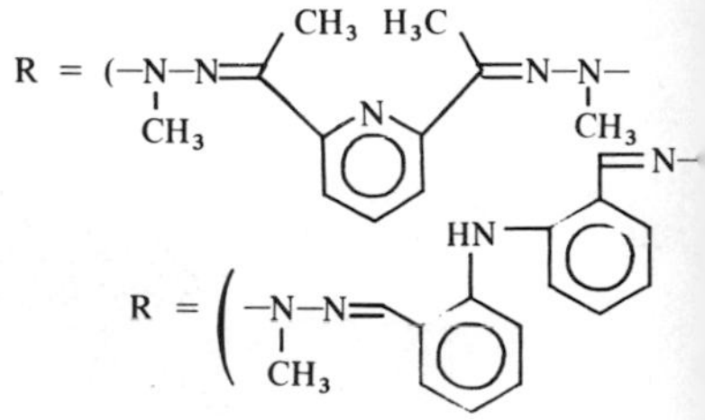

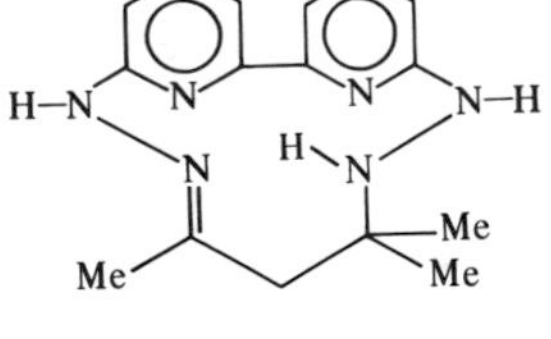

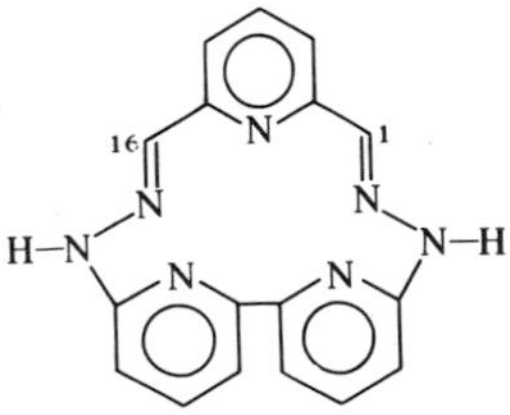

$R = (-N-N= \cdots =N-N-)$ (with CH₃ groups)

$R = \left(-N-N= \cdots \right)$

1,16-(Me)$_2$
1,3,14,16-(Me)$_4$
1,16-(OH)$_2$; 2,15-(H)$_2$
1,2,15,16-(H)$_4$; 1,3,14,16-(Me)$_4$

Physical Data m.p. [b.p. (mm)], °C	Spectral Data Available	Metal Complex(es) General Comments	Reference
7	A, D		158
	B	Ni, Co, Fe	176
	B	Mn	176
	B	Mn	176
	B, G	Mn, Zn, Cd	176
		Co	458
		Ni	390
	(G^{177})	Zn	177, 431
	(G^{180})	Fe	180, 431
	(G^{181})	Zn	181
	(A, B)	Zn, Cd	182

TABLE III-5. (*CONTINUED*)

Compound	n	Substituents
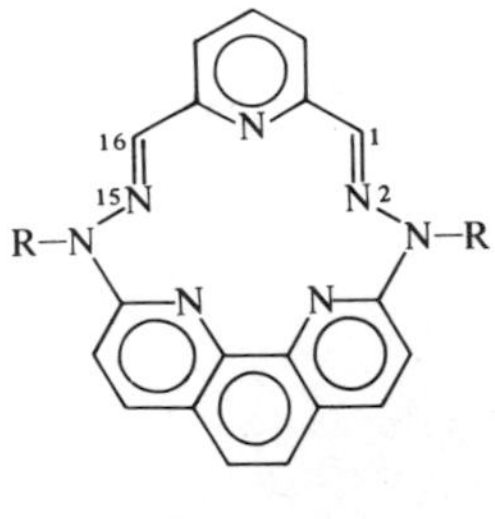		R = H
		1,16-(Me)$_2$; R $\neq$ H
		1,2,15,16-(H)$_4$; R = H
		1,16-(Me)$_2$; 1,2,15,16-(H)$_4$; R = H
		1,3,14,16-(Me)$_4$, R = H
		1,16-(Me)$_2$; R = CH$_2$CH$_2$OH
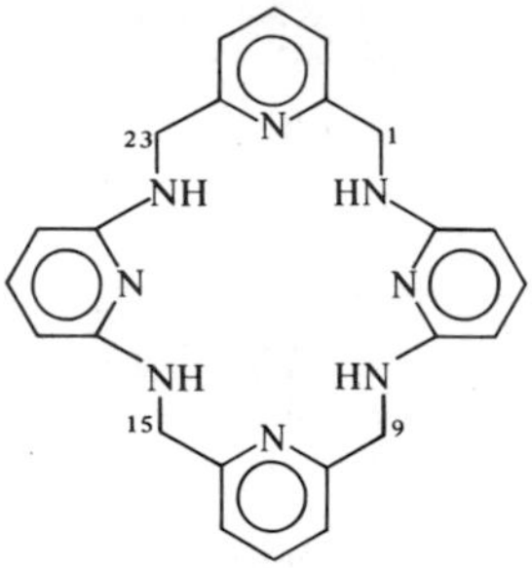		1,9,15,23-(C=O)$_4$
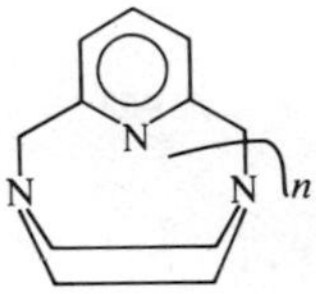	$n = 2$	
	$n = 3$	
	$n = 4$	
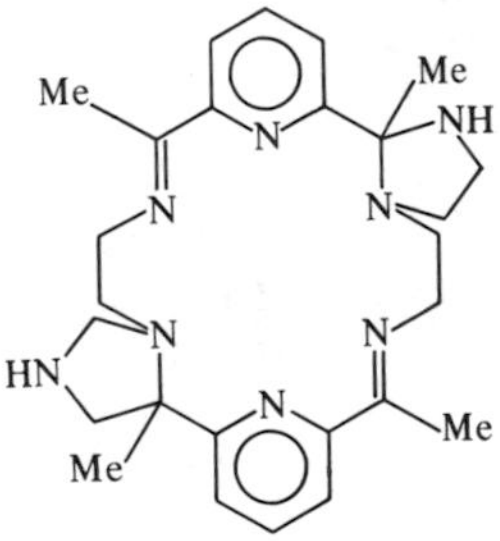		
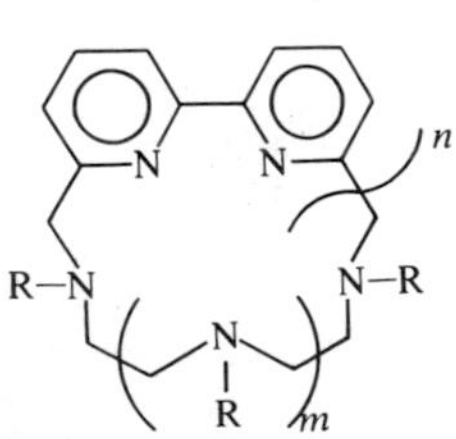	$m = 0; n = 2$	R = Tosyl
	$m = n = 1$	R = Tosyl
	$m = 2; n = 1$	R = Mesityl
	$m = n = 1$	R = H
	$m = 2; n = 1$	R = H

Physical Data m.p. [b.p. (mm)], °C	Spectral Data Available	Metal Complex(es) General Comments	Reference
	(A, G^{179})	Mn, Zn	179
	(A, G^{179})	Mn, Zn, Co	179, 456, 457
	(A, G^{179})	Mn, Zn	179
	(A, G^{179})	Mn, Zn	179
	(G^{180})	Fe, Ni	180, 437
		Mn	456
360			155
5 (dec)	A, B, D		357
	A, B		357
			357
	(B, G)	Ba, Sr, Ca	398
(dec)	A, D, G		409
20 (dec)	A, D		409
-177	A, D		409
	A, D		409
	A, D		409

TABLE III-5.   *(CONTINUED)*

Compound	n	Substituents
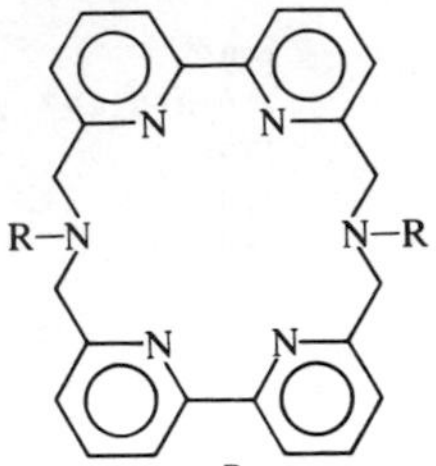		R = Tosyl R = H
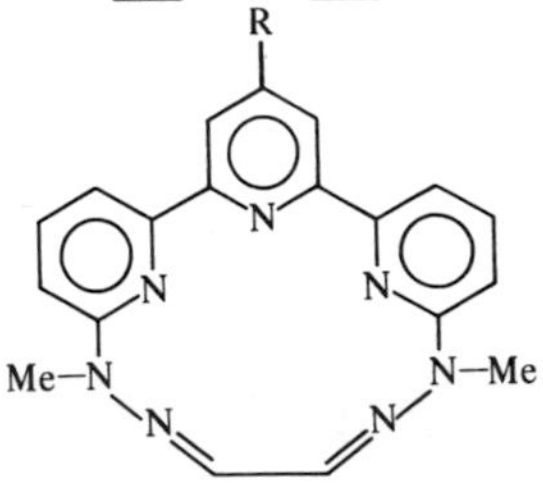		R = Ph R = H

Physical Data m.p. [b.p. (mm)], °C	Spectral Data Available	Metal Complex(es) General Comments	Reference
260	A, D		409
280	A, D		409
		Co	446
		Ni	454

TABLE III-6.   2,6-PYRIDINE MACROCYCLES – CARBON, SULFUR BRIDGE

Compound	n	Substituents
	2	H
	2	N → O
	3	H
	3	N → O
	4	N → O
	5	N → O
	6	N → O
	7	H
	7	N → O
	8	H
	8	N → O
	9	N → O
	10	N → O
	0	H
	1	H
	1	N → O
	1	5-sulfoxide
	2	H
	2	$1,12\text{-}(=O)_2$
		H
		H
		H
		N → O; 2, 10,18-*tris*(sulfone)

Physical Data m.p. [b.p. (mm)], °C	Spectral Data Available	Metal Complex(es) General Comments	Reference
	A–D	Cu	194
2–154	A	VTNMR	184, 302
–79	A		185, 302
7–109	A	VTNMR	184, 302
–99	A	VTNMR	184, 302
7–148	A	VTNMR	184, 189, 302
3–140	A	VTNMR	184, 302
	A–D	Cu	194
	A	VTNMR	184, 302
	A–D	Cu	194
–75	A	VTNMR	184, 302
7–120	A	VTNMR	184, 302
–55	A	VTNMR	184, 302
–77	A		155
2–163	A	Ag (m.p. 217–219°; A)	154, 155
		Hg (m.p. 198–200° dec; A)	154
		Ag, Hg, Au, Pd, Pt, Co	155
–152	A, D		189
–174	A		155
–133	A	Cd, Co, Ni	154, 155
5–138			77, 81
–153 (subl)	A	Zn	155
–173	A		183, 186, 302
–179	A, D		195
00	A, B		195

TABLE III-6.   (*CONTINUED*)

Compound	*n*	Substituents
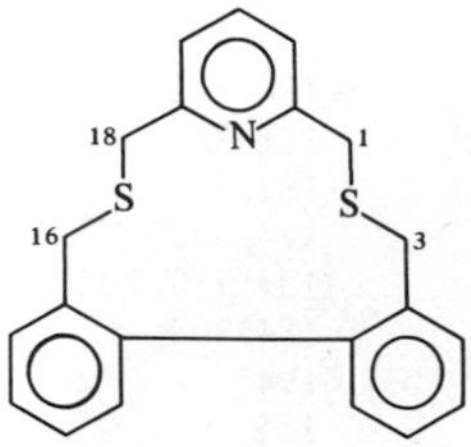		H 1-(*D*) 1-Me 1,18-(Me)$_2$ 2,17-*bis*(sulfone) 2,17-*bis*(sulfone); N → O
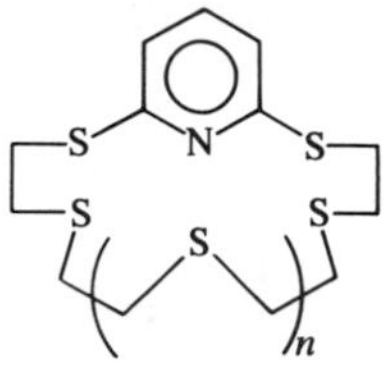	2	H
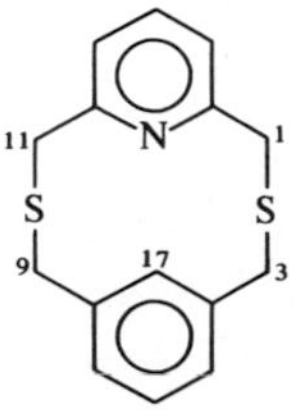		H 17-Me 17-F 17-NO$_2$
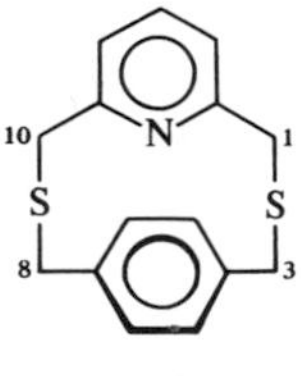		H 2,9-[SMe(BF$_4$)]$_2$ 2-Sulfone N → O; 2,9-*bis*(sulfone) N → O; 2-sulfoxide 2-sulfoxide; 9-sulfone N → O; 2,9-*bis*(sulfoxide) N → O; 2-sulfoxide; 9-sulfone 5,17-(CH$_3$)$_2$ 5,6,16,17-(CH$_3$)$_4$ 13-(CH$_3$)
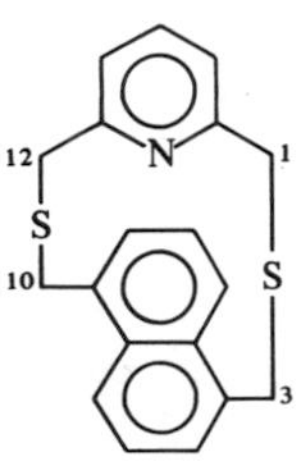		H 2,11-*bis*(sulfone)

Physical Data m.p. [b.p. (mm)], °C	Spectral Data Available	Metal Complex(es) General Comments	Reference
97–198	A, D		195
	A, D		195
28–129	A, D		195
2–104	A, D		195
300	A, B		195
			195
58.5–160.5	A, B, D		197
73–175	A		187
35–136	A		187
74–175	A		187
59–160			358
77–178	A, C, D		32–34
			34
28–230	A, B(D^{191})		190, 191
340	A, B, D(D^{191})		34, 190, 191
26–228 dec.	A, B(D^{191})		190, 191
250 dec.	A, B, (D^{191})		190, 191
20–250 (color change)	A(D^{191})	Subl: 220–245° (0.002)	190, 191
300	A, B(D^{191})		190, 191
0–161	A, D		33
6–138	A, C, D		33
73–175	A, D		33
5			35
			35

TABLE III-6.   *(CONTINUED)*

Compound	n	Substituents
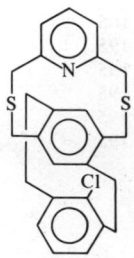		
		H
		15-Me
		15-OMe
		15-F
		15-Cl or Br
		H
		H
		H
		$(S\text{-}CH_3)_4$
		$(S\text{-}\phi)_4$
		$(S\text{-}\phi)_4$
		O

Physical Data m.p. [b.p. (mm)], °C	Spectral Data Available	Metal Complex(es) General Comments	Reference
9–220	*A, D, C*		33
–196	*A*		186, 304
9–131	*A*		188
5–208			188
2–144	*A*		186
		(attempted)	186
2–229	*A*		186
3–216	*A*	Fe, Co, Ni	155
2–253	*A, D*		33
	A, D		33
	A, D		33
	A, D		33

TABLE III-6.   *(CONTINUED)*

Compound	n	Substituents
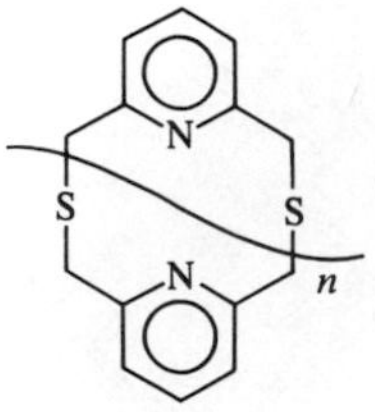	1	H
		$(N \rightarrow O)_2$
		$(N \rightarrow O)_2$; *bis*(sulfone)
		bis(sulfone)
		$[SMe(BF_4)]_2$
		$1,1,3,3,9,9,11,11$-$(Cl)_8$ *bis*-sulfone
	2	H
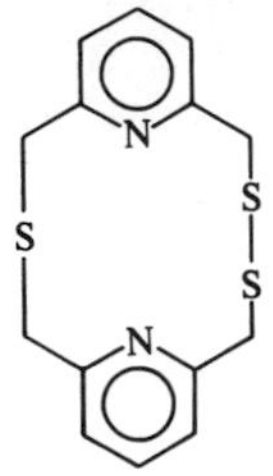		H
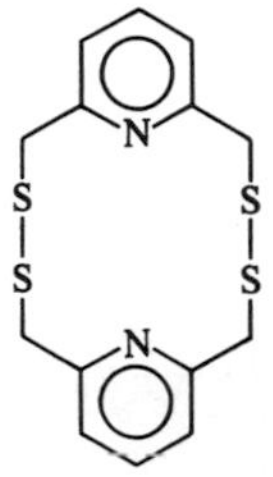		H
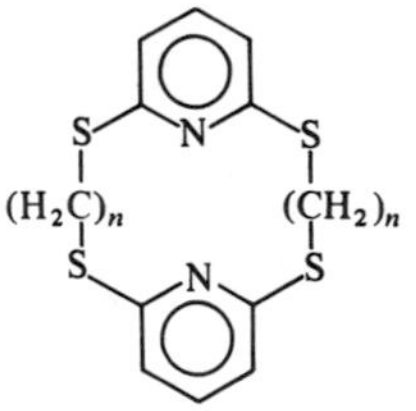	1	H
	4	H
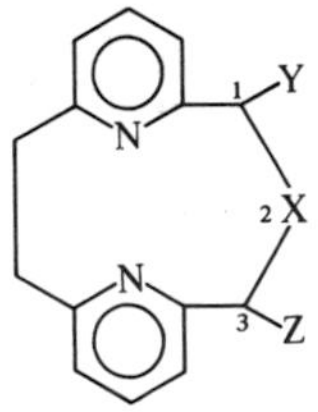		$Y = Z = H; X = S$
		$Y = Z = H; X = S; 6,7$-benzo
		$Y = Z = H; X = SO_2$
		$Y = Br; Z = H; X = SO_2$
		$Y = Z = Br; X = SO_2$
		$Y = Z = H; X = SO$
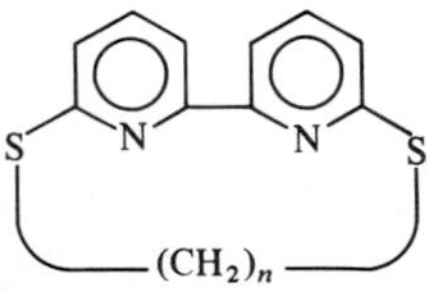		8

Physical Data m.p. [b.p. (mm)], °C	Spectral Data Available	Metal Complex(es) General Comments	Reference
20–222	A–D		39, 41, 187, 192, 302
30–230.5	[G]⁴⁰⁸		28, 40, 408
1 (dec.)	A		184
	A, D		39, 41
	A, C, D		39, 41
			35
2–356 (dec.)	C, D		41
5–188	A, D, C		40, 41
	A, D		40
0–152	A, D		40
5–197	A, [G]⁴⁰⁸	VTNMR	192, 408
	D		197
.5–141	A, B		43
2–166	A, D	VT(NMR)	198
3–233.5	A, B		43
5–215	A, B		43
4–205	A, B		43
4–135	A, B		43
			196

TABLE III-6.   (*CONTINUED*)

Compound	n	Substituents
	1 2	H H
	0 1 2	H H H
	0 1 2	H H H

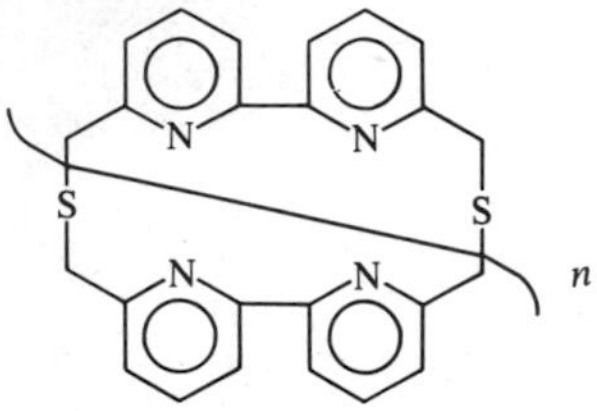

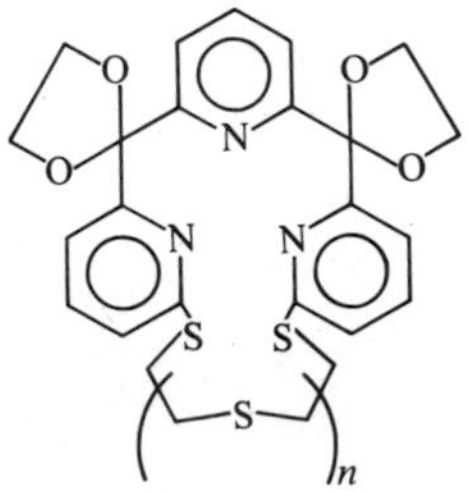

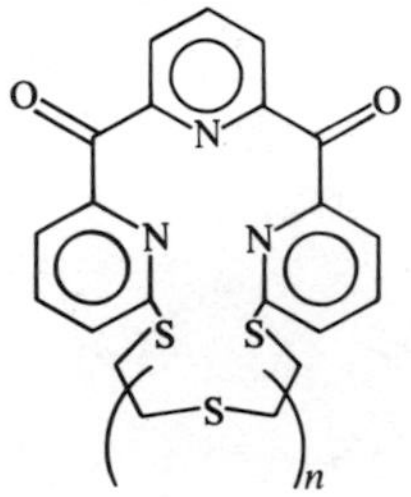

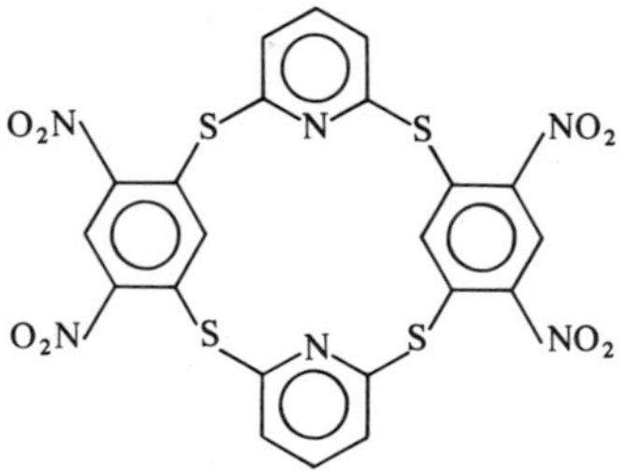

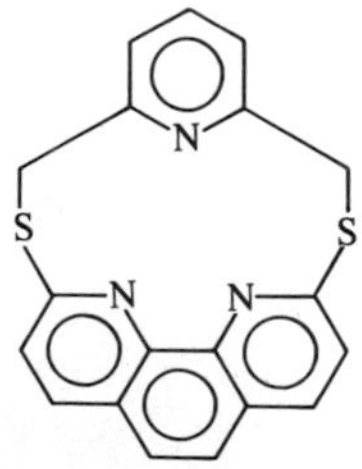

Physical Data m.p. [b.p. (mm)], °C	Spectral Data Available	Metal Complex(es) General Comments	Reference
235	A, B, D	VT(NMR)	199
173	A, B, D		
-237	A, B, C, D		188
-186	A, B, C, D		105
-147	A, B, C, D		105
201	A, B, C, D, G		105
160	A, B, C, D		105
130	A, B, C, D		105
	A, D		193, 359
	A		196

TABLE III-6. *(CONTINUED)*

Compound	n	Substituents

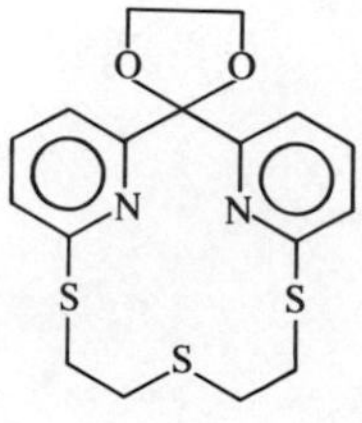

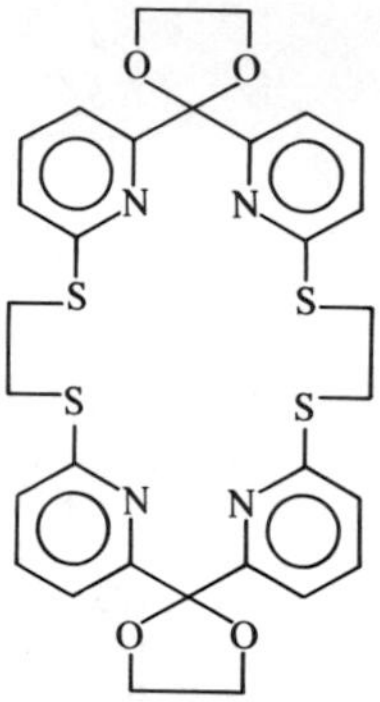

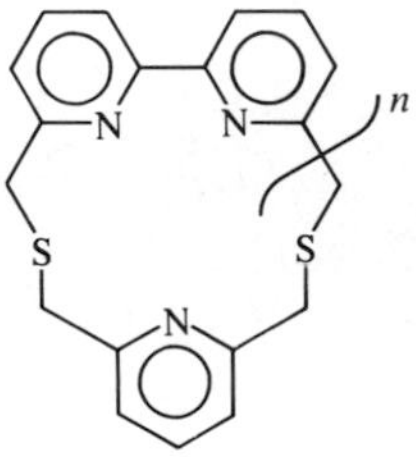

1
2

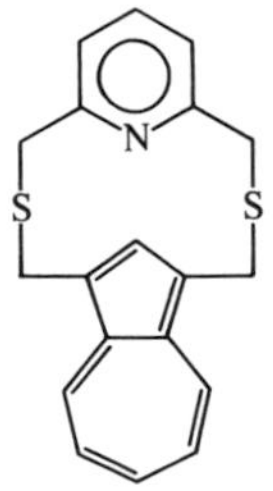

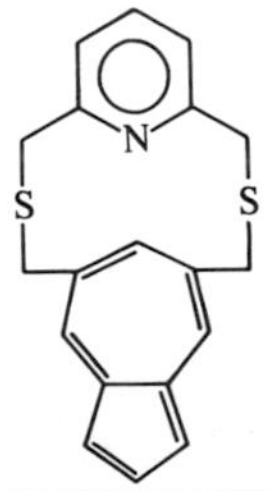

Physical Data m.p. [b.p. (mm)], °C	Spectral Data Available	Metal Complex(es) General Comments	Reference
	A, B, D		107
	A–B, D		107
(dec)	*A, D*		409
(dec)	*A, D*		409
			451
			451

TABLE III-7.　　2,6-PYRIDINE MACROCYCLES – CARBON, SELENIUM BRIDGE

Compound 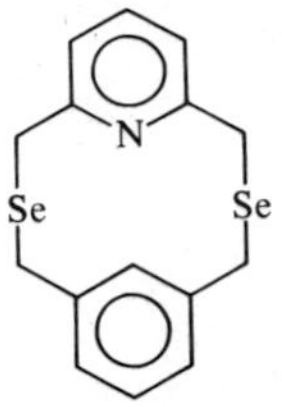	n	Substituents

Physical Data m.p. [b.p. (mm)], °C	Spectral Data Available	Metal Complex(es) General Comments	Reference
5–169.5	A		438

TABLE III-8. 2,6-PYRIDINE MACROCYCLES – CARBON, NITROGEN, OXYGEN BRIDGE

Compound	n	Substituents
	1	3,4:9,10-Dibenzo
	1	H
	1	1,12-(Me)$_2$
	1	1,12-(OCH$_2$CH$_3$)$_2$
	1	1,12-(OCH$_3$)$_2$
	1	1,12-(OCH$_2$CH$_2$CH$_2$CH$_3$)$_2$
	2	H
	2	H
	2	1,15-di(Me)$_2$
	2	1,15-di(Me)$_2$; 3,4,12,13-dibenzo
	0	X = O; Y = H
		X = H; Y = O
		2,11-(Tos)$_2$; X = Y = H
	1	X = O; Y = H
		X = H; Y = O
		2,11-(Tos)$_2$; X = Y = H
	2	X = O; Y = H
		X = H; Y = O
		2,11-(Tos)$_2$; X = Y = H
	2	H
	3	H
	2	H
	3	H
	2	H
	3	H

Physical Data m.p. [b.p. (mm)], °C	Spectral Data Available	Metal Complex(es) General Comments	Reference
		Mn, Zn	115, 331
	(A–D)	Mg, Mn, Cd, Hg	200, 202, 203
	(A–D)	Mg, Cd, Hg	200, 202, 205
	(A–D)	Mn(G^{207}), Fe, Zn, Cd, Co	205, 207
	(A, B, D)	Mn, Mg	203
			203
			203
	(B, G^{204})	Ba, Ca, Sr, Pb, Cd, Hg	143, 202, 204
		Pb, Ca, Ba, Sr	144, 201
	(B, C[G^{201}])	Pb, Ca, Ba, Sr, Cd, Hg	143–144, 201–202
	(B, C)	Pb	143–144
230	A, D		66, 154
212	D		156
177	A, D		66, 154
201	A, D		66, 154
251	D		156
185	A, D		66, 154
129	A, D		66, 154
	A, D, E		360
165	A, D		66, 154
	A, B, D		213
	A, B, D		213
209 (dec)	A, B, D		213
179 (dec)	A, B, D		213
233 (dec)	A, G		213
203 (dec)	A		213

TABLE III-8.   *(CONTINUED)*

Compound	n	Substituents
	2 3	H H
	$n = 0; m = 1$ $m = n = 1$ $m = n = 1$ $m = n = 1$ $m = 1; n = 2$ $m = n = 2$	$1,12\text{-}(C=O)_2$ $1,12\text{-}(C=O)_2$ H $1,12\text{-}(C=O)_2$; $6,7\text{-}(benzo)_2$ $1,12\text{-}(C=O)_2$ $1,13\text{-}(C=O)_2$ $1,13\text{-}(C=O)_2$
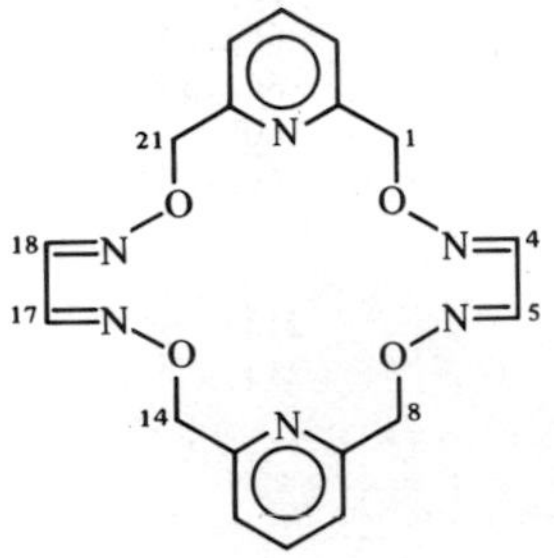		$4,5,17,18\text{-}(Me)_4$
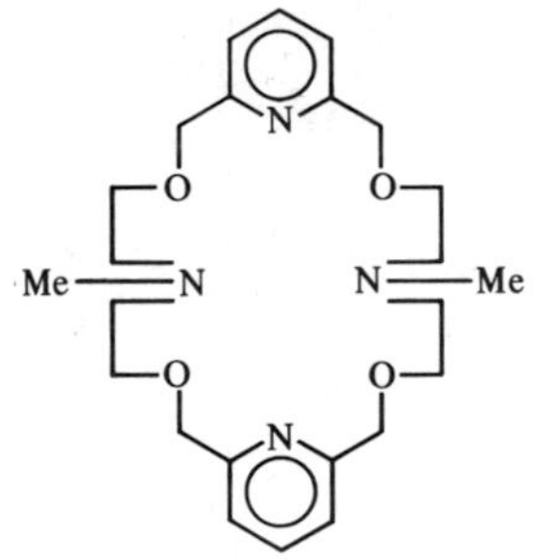		H
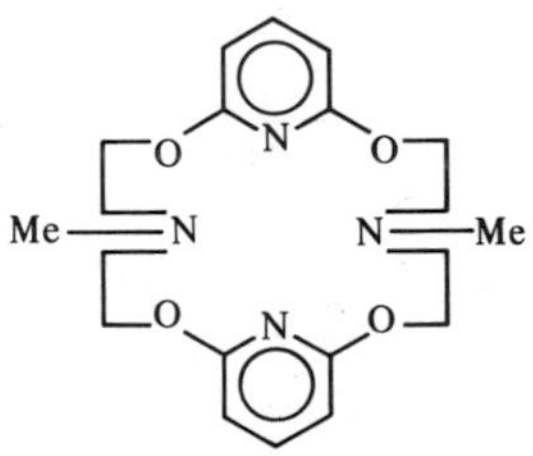		H

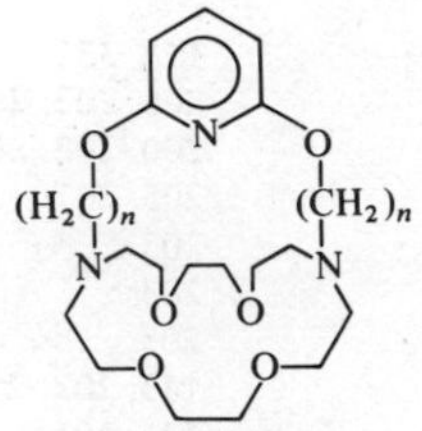

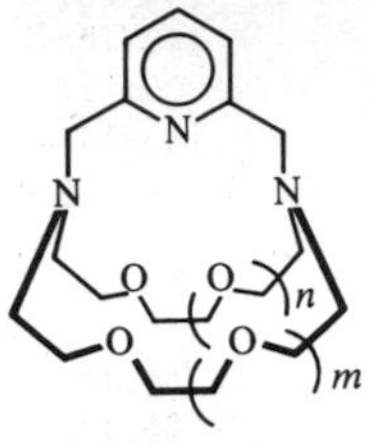

Physical Data m.p. [b.p. (mm)], °C	Spectral Data Available	Metal Complex(es) General Comments	Reference
	A		213
	A		213
-276			209
-186	*A, B*	Li, Na, K	209, 210
6	*A, D, E*	pK_a 8.31; Li, Na, K, Rb, Cs, Mg, Ca, Sr, Ba, [Na²³-NMR][361]	209, 210, 361
	A, D, E		362
	E		362, 363
	D		362
	D		362
	A, D		211
	A, B, D		357
	A, B, D		357

TABLE III-8. *(CONTINUED)*

Compound	n	Substituents
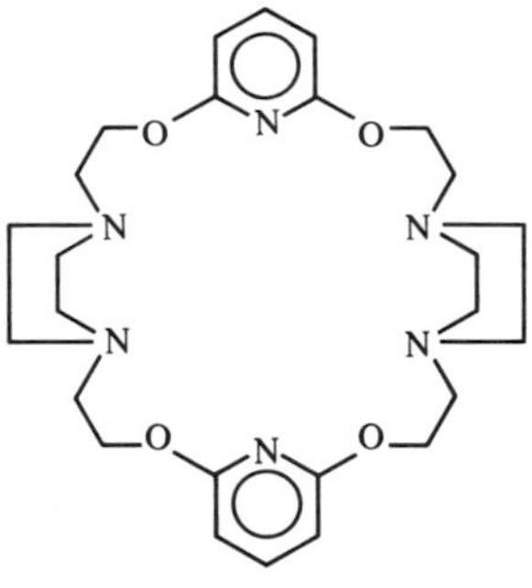		H
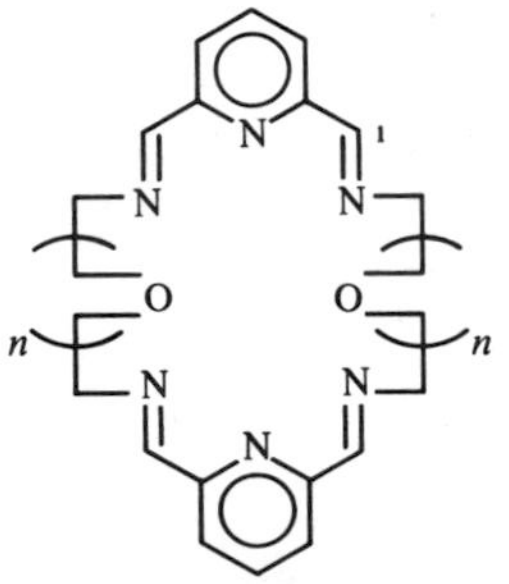	2	H 1,12,18-29-(Me)$_4$ 1,2,11,12,18,19,28,29-(*octa*-H) 1,2,11,12,18,19,28,29-(*octa*-H); 2,11,19,28-(Me)$_4$
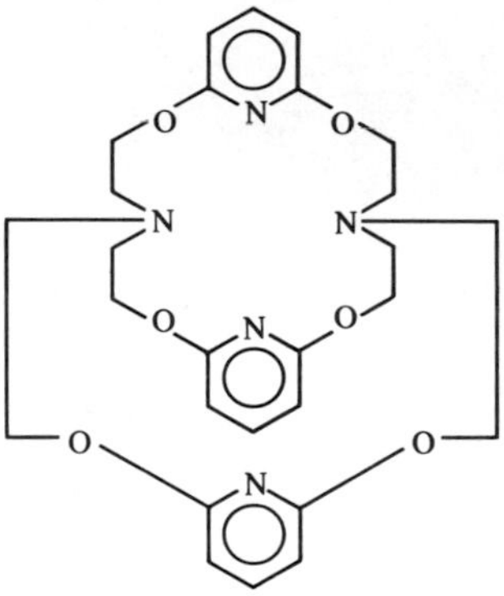		H
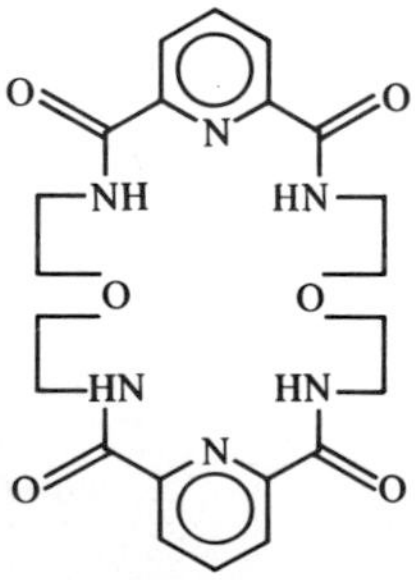		H

Physical Data m.p. [b.p. (mm)], °C	Spectral Data Available	Metal Complex(es) General Comments	Reference
(dec)	A, D, G	CoCl$_2$	214
	(A, B)	Cu, Zn, Ba	397
	(B, G)	Pb	143, 202, 208
	$[B, C, (G), K]$	Cu, Mg	364, 397, 399
	(B, C)	Pb, Ag, Co, Fe, Ni	205, 429
		Cu	206
		Cu	206
(dec)	$A–D, G, E$		212
–340 (subl)	A, D		66, 154

TABLE III-8.   (*CONTINUED*)

Compound	n	Substituents
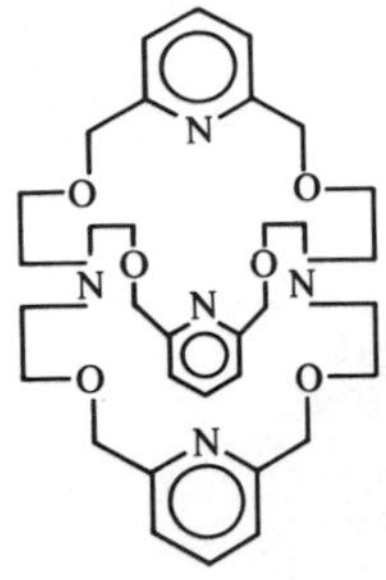		H
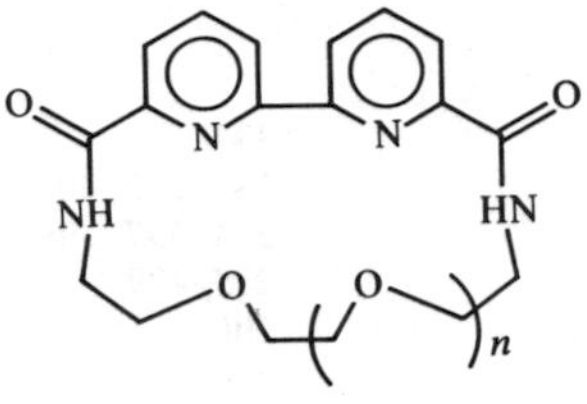	$n = 1$ $n = 2$	H H
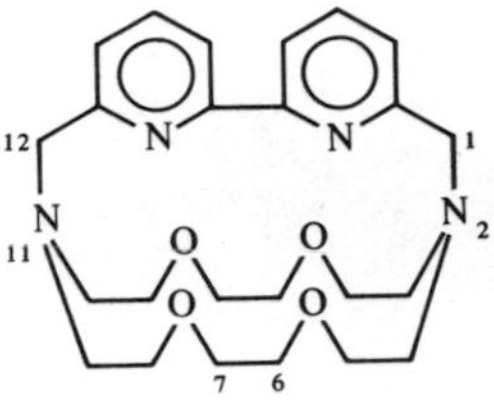		H 1,12-(=O)$_2$
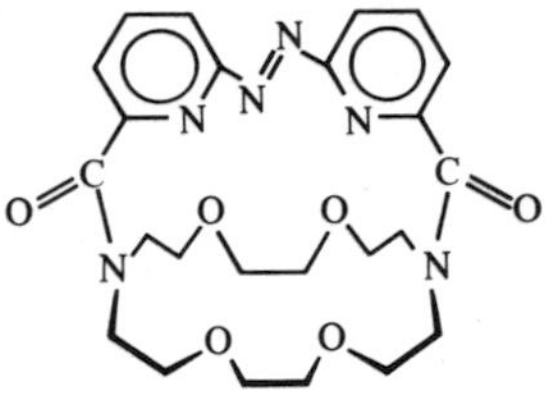		
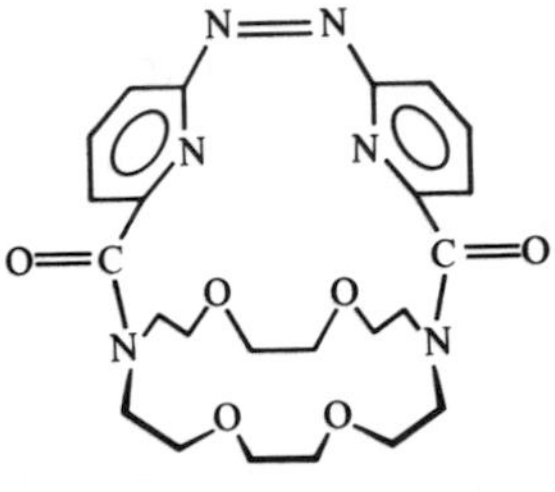		

Physical Data m.p. [b.p. (mm)], °C	Spectral Data Available	Metal Complex(es) General Comments	Reference
-119	A, B, D, E, G	Co, Cu	215
-228	A, D		158
-193	A, D		158
	A		158
266	A, D	Li, Na	158
49	B–D	Na, K, Cu, Ni, Co, Hg, Pb	405
			405

TABLE III-9. 2,6-PYRIDINE MACROCYCLES – CARBON, NITROGEN, SULFUR BRIDGE

Compound	n	Substituents
	1	5-Me
	1 2	1,9-(C=O)$_2$ 1,10-(C=O)$_2$
	1	3,4:9,10-Dibenzo
		1,11-di(Me)$_2$
		1,14-(Me)$_2$; 3,4:7,8:11,12-tribenzo

Physical Data m.p. [b.p. (mm)], °C	Spectral Data Available	Metal Complex(es) General Comments	Reference
69	*A*	Cu, Fe	155
-243	*A*		155
-236	*A*		155
		Mn, Zn	115, 331
	(B, C)	Pb	143
		Cu, Co, Ni, Zn	220
		Zn, Cd	217

TABLE III-9.   (*CONTINUED*)

Compound	*n*	Substituents

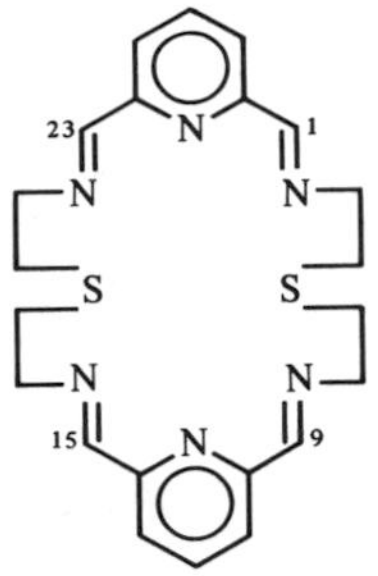

1,9,15,23-(Me)$_4$

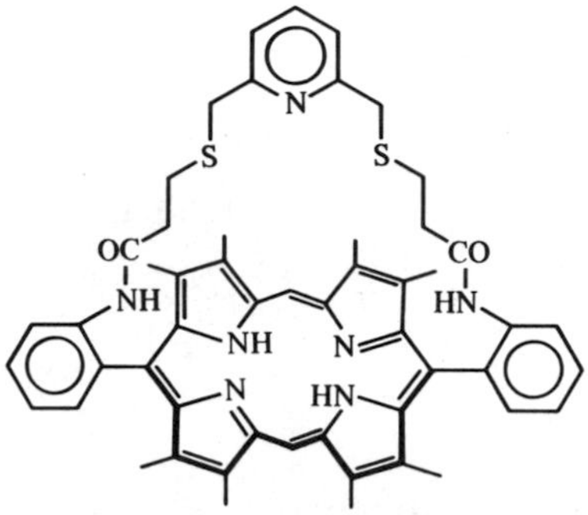

Physical Data m.p. [b.p. (mm)], °C	Spectral Data Available	Metal Complex(es) General Comments	Reference
	(*G, K*)	Cu	219, 399
	A, C, D, J, K	Cu–Fe	389

TABLE III-10.　2,6-PYRIDINE MACROCYCLES – CARBON, NITROGEN, PHOSPHORUS BRIDGE

Compound	n	Substituents
		1,11-(Me)$_2$
		1,11-(Me)$_2$; 1,2,10,11-(H)$_4$　(abr: pn_2-H$_4$) ["*meso*"]
		1,11-(Me)$_2$; 1,2,10,11-(H)$_4$; 6-S
	$m = 3; n = 2$	1,14-(Me)$_2$
	$m = n = 3$	1,15-(Me)$_2$
		(abr: PccBF)

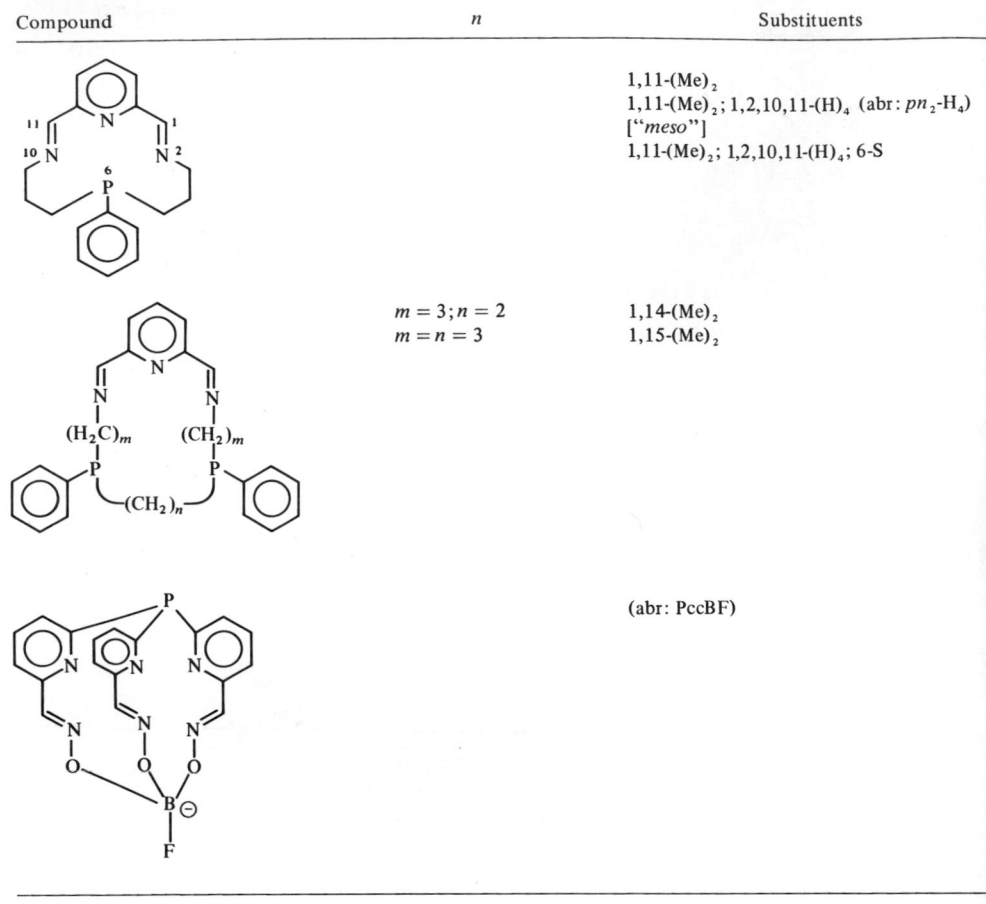

TABLE III-11.　2,6-PYRIDINE MACROCYCLES – CARBON–OXYGEN–PHOSPHORUS BRIDGE

Compound	n	Substituents
	$n = 3$	X = lone pair
	$n = 3$	X = O
	$n = 5$	X = lone pair
	$n = 5$	X = O

Physical Data m.p. [b.p. (mm)], °C	Spectral Data Available	Metal Complex(es) General Comments	Reference
		Ni	222
		Ni	222
	D		222
	(A–D)	Ag, Cd	223
	(A–D)	Ag, Cd	223
		Fe, Zn, Ni, Co	221
	(G)	Fe	365
	(G)	Ni	366

Physical Data m.p. [b.p. (mm)], °C	Spectral Data Available	Metal Complex(es) General Comments	Reference
7–108	A, B, D		108
7–138	A, B, D		108
1	A, B, D		103, 108
1	A, B, D		103, 108

TABLE III-12.   2,6-PYRIDINE MACROCYCLES – CARBON, SULFUR, OXYGEN BRIDGE

Compound	n	Substituents
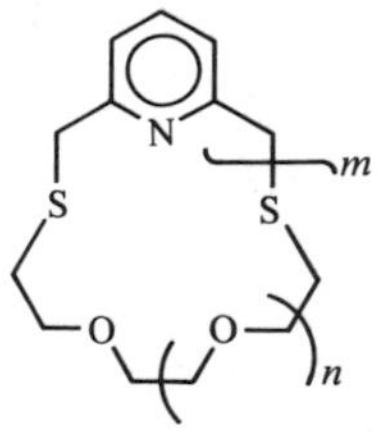	$m = 1; n = 0$	H
	$m = 1; n = 1$	H
	$m = 1; n = 2$	H
		$N \to O$
		$N \to O; bis$(sulfone)
	$m = 1; n = 2$	$1,15\text{-}(=O)_2$
	$m = 1; n = 3$	H
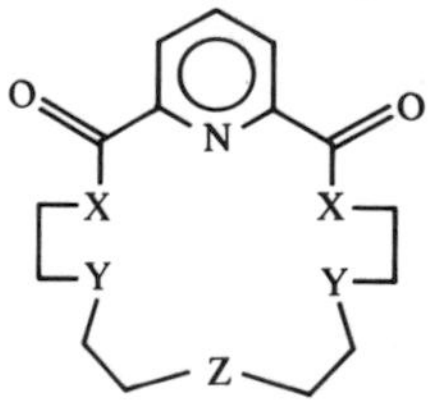		$X = S, Y = Z = O$
		$X = Y = O; Z = S$
		$X = Z = O; Y = S$
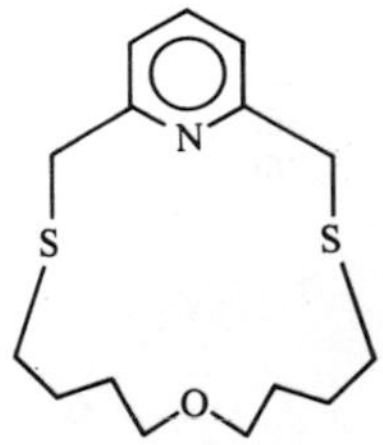		H
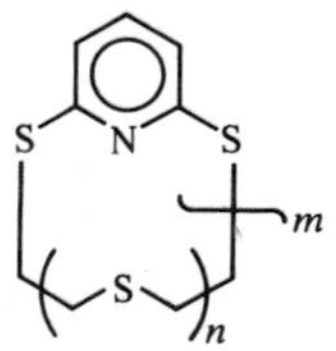	$n = 1; m = 1$	H
	$n = 2; m = 1$	H
	$n = 3; m = 1$	H
	$n = 1; m = 2$	H
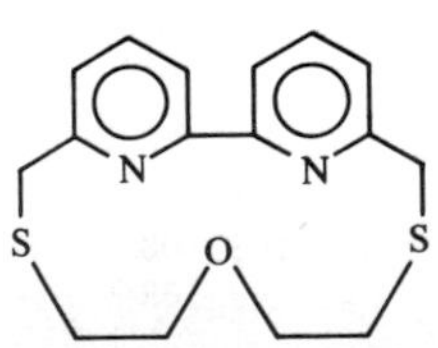		H

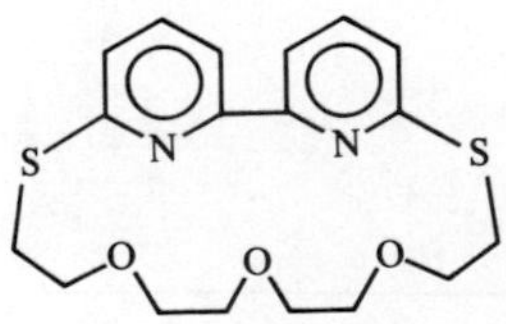

)

Physical Data m.p. [b.p. (mm)], °C	Spectral Data Available	Metal Complex(es) General Comments	Reference
3–135	A, D		66, 224
–91, 87–88[391]	A, B, D	Na, Co, Ni, Cu	66, 224, 391, 392
–59, 130–132[391]	A, (B), D	Na, Co, Cu, K, Ba, Ni	66, 224, 391, 392
1	A, (B), D	Na, K, NH$_4^+$, Ba	66, 224
8–201	A, D		66
6–148			77
			224
6–148			81
–97.5	A, B	RNH$_3^+$	70, 76, 82
7–129	A, B		76
.5–48.5	A, B, D	Co, Ni, Cu	491
–94; 110–112[66]; 9–121.5[197]	A, D		66, 154, 197
–77	A, D		66, 154
–76	A, B, D	Ag	66, 154
8–170	A, D		154
	A, B, D		327
5			196

TABLE III-12.   (*CONTINUED*)

Compound	*n*	Substituents
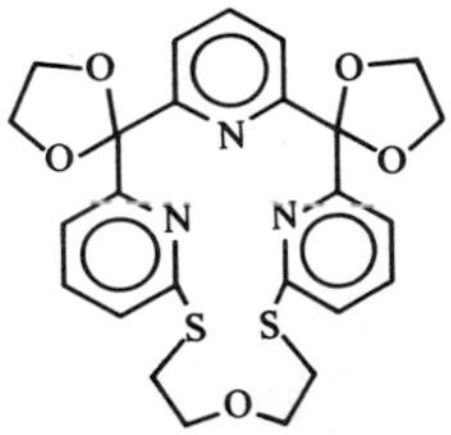		H
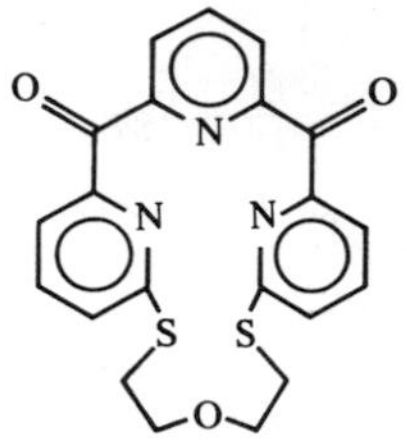		H
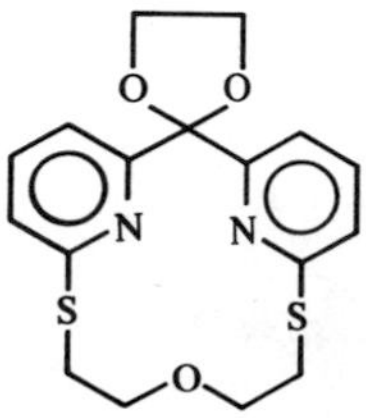		

Physical Data m.p. [b.p. (mm)], °C	Spectral Data Available	Metal Complex(es) General Comments	Reference
6–197	*A–D*		105
8–200	*A–D*		105
1–102	*A, B, D*		107

TABLE III-13.   2,5-PYRIDINE MACROCYCLES

Compound	n	Substituents
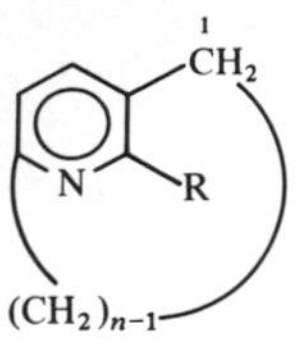	8	H
		1-(=O)
		1-OH
		1-OAc
	9	(±)–H
		(+)-H
		1-OH
		1-OAc
		1-(=O)
	10	H
		1-OH
		1-OAc
		1-(=O)
	11	H
		N → O
		1-OH
		1-OAc
		1-(=O)
		1-(=O); 2-(SeC$_6$H$_5$)
		1-(=O); 2,3-de(H)$_2$
		1-(=O); N → O
		1,11-(=O)$_2$
		11-(=O)
		11-(=NOH)
		11-OAc
		11-NH$_2$
		11-OAc; 1-(=O)
		11-OAc; 1-(=NOH)
		11-Cl; 1-(=O)
	12	H
		1-OH
		1-OAc
		1-(=O)
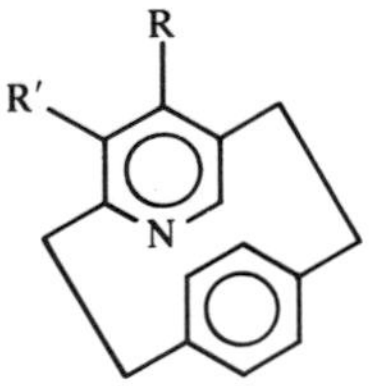		R = (Ph)
		R = R′ = H
		R = ⌢O Me
		R′ = O X Me
		R = CH$_2$OAc; R′ = OAc
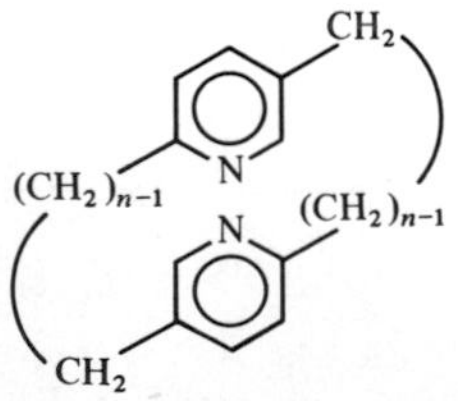	2	H
	8	1,13-(=O)$_2$
	9	1,14-(=O)$_2$
		H

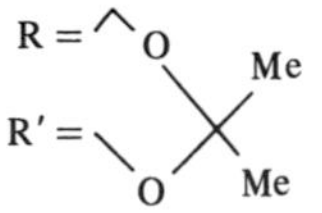

Physical Data m.p. [b.p. (mm)], °C	Spectral Data Available	Metal Complex(es) General Comments	Reference
-75 (0.01)]	A–C		225
-48 [105–110 (0.02)]	A–C		225
5–135 (0.02)]	A, B	Isomeric mixture	225
0–115 (0.01)]	A	Isomeric mixture	225
-81 (0.04)]	A–C		225
(0.01)]		$[\alpha]_D + 152°, [\alpha]_{365} + 1074°$	225
5–146 (0.03)]	A, B	Mixture	225
	A, B	Isomer B	225
-97	A, B	Isomer A	225
-59 [135–140 (0.01)]		Mixture	225
-72	A	Isomer A	225
-68	A	Isomer B	225
5–115 (0.03)]	A–C		225
-78 (0.01)]	A–C		225
5–160 (0.02)]	A, B	Mixture	225
5–130 (0.01)]	A	Mixture	225
-82 [140–150 (0.01)]	A–C		225
-95 (0.03)]	A–C		225
-56	A, C		227
0–145 (0.03)]	A, B	Mixture	225
0–115 (0.02)]	A	Mixture	225
37 [120–130 (0.03)]	A–C		225, 226
-76	A, B, D		226
	A, D		226
.5–131.5	A–D		227
64	A–D		227
64	A–D		227
-170			227
57	A, B, D		227
	A	[diHCl, m.p. 170–185°]	227
49	A, B, D		227
-138			227
	A, B, D		227
(0.01)]	A–C		225
)–150 (0.03)]	A, B		225
)–110 (0.02)]	A, B		225
48 [120–130 (0.02)]	A–C		225
	A		228
-258	A		229, 233
			234
			234
	A	4 isomers	229–230
-108	A–C		225
-148	A–C		225
-97.5	A–C		225

TABLE III-13. *(CONTINUED)*

Compound	n	Substituents
	12 12 12 12 12 12 12	1-(H) 1-(=O) 1-(NH$_2$) 1-(H); 2-(=O) 1-(H); 2-(=S) 12-(OAc); 2-(=O) 2,12-(=O)$_2$
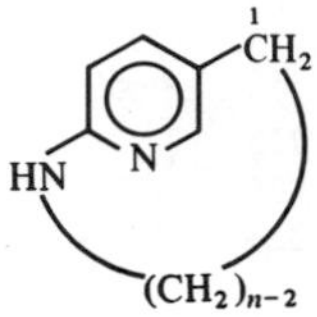	12	11-(=O)
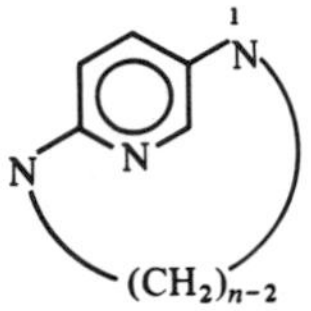	13	2,12-(=O)$_2$
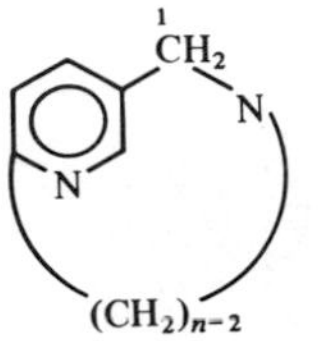	12	1-(=O)
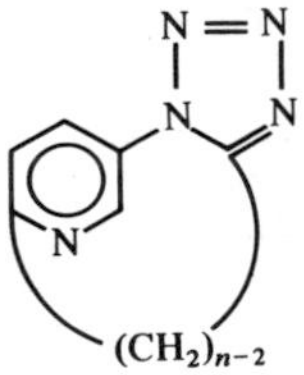	12	H
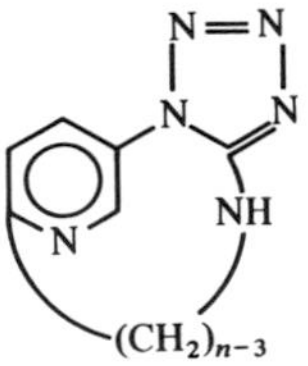	13	H
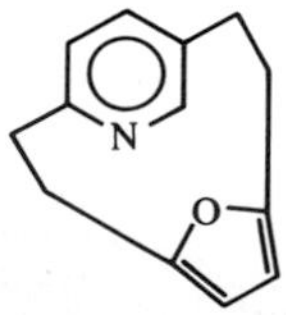		H

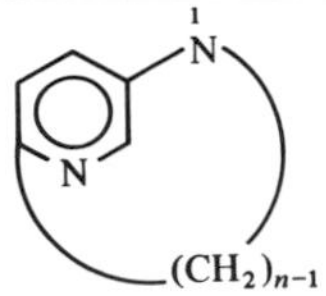

Physical Data m.p. [b.p. (mm)], °C	Spectral Data Available	Metal Complex(es) General Comments	Reference
-64	A, C, D		226
-54	A–D		226
0–160	A, D		226
1–164	A–D		226
3–164	A–D		226
8–141	A–D		227
9–165	A–D		227
5–148	A–D		227
5–220	A–D		227
3–146	A, B, D		226
3–121	A, C–E		226
–161	A–E		226
87	A, C, G[231]	No VTNMR changes	230, 231, 306

TABLE III-13.   *(CONTINUED)*

Compound	n	Substituents
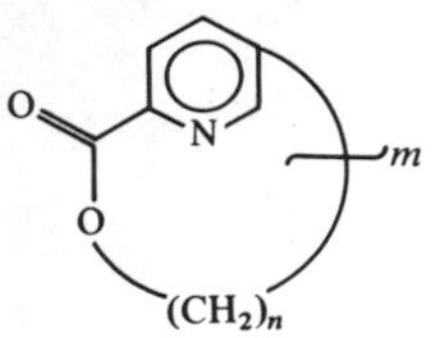	$9; m = 1$	H
	$m = 2$	H
	$12; m = 1$	H
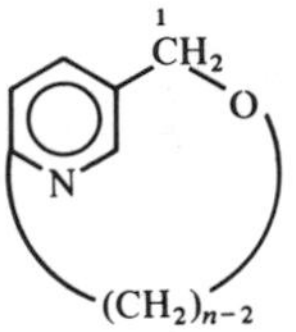	12	1-(=O)
	12	1-(=O); N → O
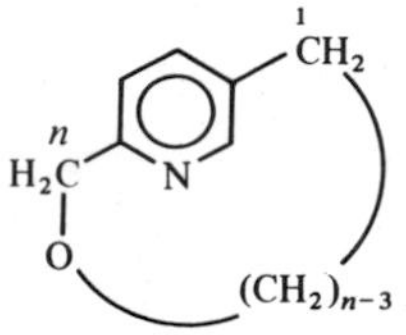	11	1-(=O)
	11	1-(=O); N → O
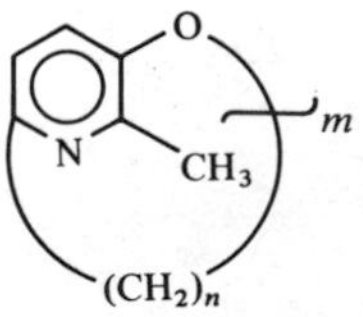	$7; m = 2$	2-(=O); 9,10,11,12,13,14-[carpaine] $(H)_6$
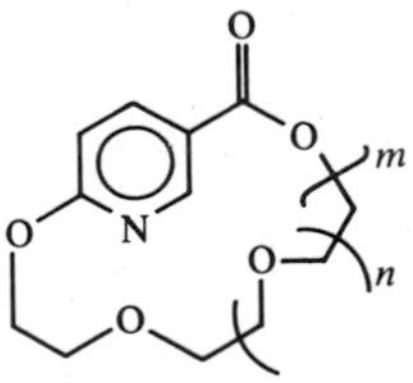	$0; m = 2$	H
	$m = 2$	H
	$3; m = 1$	H
	$m = 2$	H
	$4; m = 1$	H
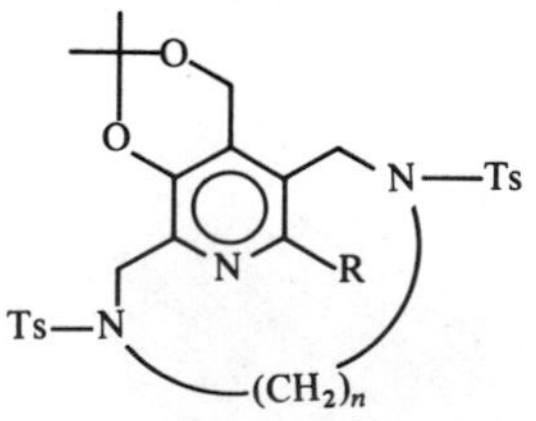	4	R = H
	5	R = H
	6	R = H
	7	R = H
	8	R = H
	9	R = H
	10	R = H
	5	R = Me
	6	R = Me
	7	R = Me
	8	R = Me
	9	R = Me
	10	R = Me
	12	R = Me

Physical Data m.p. [b.p. (mm)], °C	Spectral Data Available	Metal Complex(es) General Comments	Reference
)3–104	A, B		225
2–73	A, B		225
3–40	A, B		225
3–57	A–D		226
5–80	A–D		226
)–52	A–D		227
5–98	A–D		227
19–120 [subl: 120 (0.05)]	D	$[\alpha]_D^{12} + 24.7°; N,N'\text{-(Me)}_2$ [m.p. 79–81°]	367–369 370–372
1–142	A, B, D	syn isomer	241a
57–168	A, B, D	anti isomer	241a
il	A, B, D		241a
)–92	A, B, D		241a
il	A, B, D		241a
		Attempted	234a, 450
45–246 (dec)	A		234a, 450
4–225	A		234a, 450
4–226	A		234a, 450
8–209.5	A		234a, 450
9–160	A		234a, 450
1–143	A		234a, 450
c.	A		450
1–202	A		450
8–229	A		450
8–229	A		450
6–197	A		450
	A		450
	A		450

TABLE III-13. (*CONTINUED*)

Compound	n	Substituents
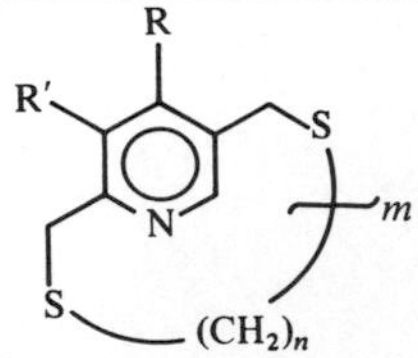 (structure)	4; $m = 1$	R = CHO; R' = OH
	$m = 1$	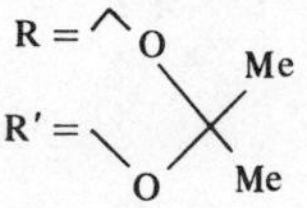 R = (acetonide group) Me R' = (acetonide group) Me
	5; $m = 1$	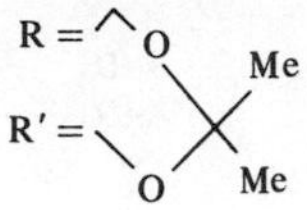 R = (acetonide group) Me R' = (acetonide group) Me
	$m = 1$	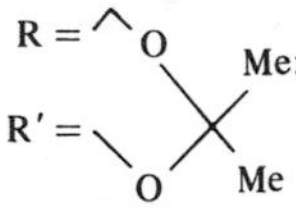 R = (acetonide group) Me; 5,5'-(Me)$_2$ R' = (acetonide group) Me
	$m = 1$	R = (CHO); 5,5'-(Me)$_2$ racemate R' = (OH)
	$m = 1$	R = (CHO); 5,5'-(Me)$_2$ (−) form R' = (OH)
	$m = 1$	R = (CHO) racemate R' = (OH)
	$m = 1$	R = (CH$_2$OH); R' = (OH)
	$m = 1$	R = (CH$_2$OH); 5,5'-(Me)$_2$ R' = (OH)
	$m = 1$	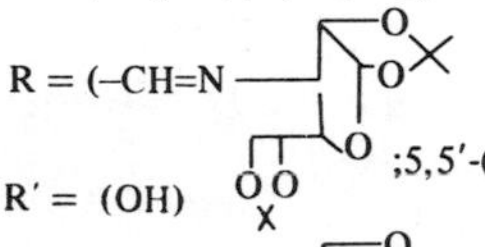 R = (−CH=N⎯ (sugar group)) ; 5,5'-(Me)$_2$ R' = (OH)
	$m = 1$	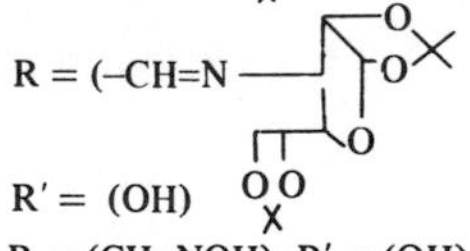 R = (−CH=N⎯ (sugar group)) R' = (OH)
	$m = 1$	R = (CH=NOH); R' = (OH)
	$m = 1$	R = (CH=NOH); 5,5'-(Me)$_2$ racemate R' = (OH)
	$m = 1$	R = (CH=NOH); 5,5'-(Me)$_2$ (−) form R' = (OH)
	$m = 1$	R = (CH$_2$NH$_2$); 5,5'-(Me)$_2$ racemate R' = (OH)
	$m = 1$	R = (CH$_2$NH$_2$); 5,5'-(Me)$_2$ (−) form R' = (OH)
	$m = 1$	R = (CH$_2$NH$_2$); 5,5'-(Me)$_2$ (+) form R' = (OH)
	$m = 1$	R = (CH$_2$NH$_2$); R' = (OH)
	6; $m = 1$	R = CHO; R' = OH
	$m = 1$	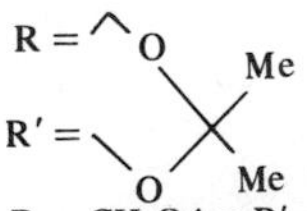 R = (acetonide group) Me R' = (acetonide group) Me
	$m = 1$	R = CH$_2$OAc; R' = OAc
	8; $m = 1$	R = CHO; R' = OH

Structure:

R' and R on pyridine ring with S-CH$_2$ linkages and (CH$_2$)$_n$ bridge, m repeat unit.

Physical Data m.p. [b.p. (mm)], °C	Spectral Data Available	Metal Complex(es) General Comments	Reference
1	A		234, 238, 403
5			234
9–111	A		236
9–140	A, C		404
7–148	B	Zn	404, 234b, 234c
	A–C	$[\alpha]_D^{24} - 422$, Zn	404, 234b, 234c
–92	A–C	(–) isomer$[\alpha]_D^{24} - 366°$	235–237, 239, 394, 403
2–172	A, B, G^{241}	$[\alpha]_D^{19} - 215°$	236, 237, 239, 241, 234b
280			404
280	B, C	$[\alpha]_D^{26} - 337°$	404
9–251^{236} 9–150^{237}	A–C	$[\alpha]_D^{26} + 236°$	236, 237
9–191	A, B	$[\alpha]_D^{23} - 291°$	235, 237, 239, 394
5–206			404
3–219	A–C	$[\alpha]_D^{23} - 363°$	404
4–155		Zn	404, 234b, 234c
4–175	A–C	$[\alpha]_D^{25} - 262°$, Zn	404, 234b, 234c
4–175		$[\alpha]_D^{21} + 262°$	404
3–144	A–C	$[\alpha]_D^{23} - 202°$	235, 237, 239, 240, 394
	A		234, 234b, 238, 403
			234
			234
	A		234, 238, 403

TABLE III-13.   *(CONTINUED)*

Compound	n	Substituents
	$m = 1$	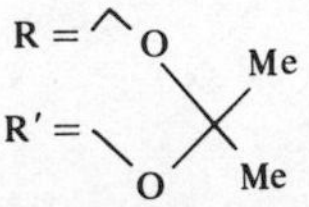 $R = \overset{\wedge}{}O$ Me $\quad$ $R' = \diagdown O$ Me
	$m = 1$	6-Me
	$m = 1$	$R = CH_2OAc; R' = OAc$
	$4; m = 2$	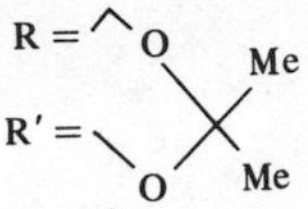 $R = \overset{\wedge}{}O$ Me $\quad$ $R' = \diagdown O$ Me
	$6; m = 2$	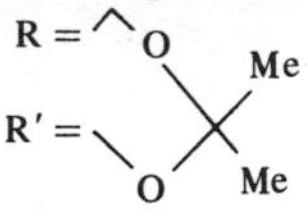 $R = \overset{\wedge}{}O$ Me $\quad$ $R' = \diagdown O$ Me

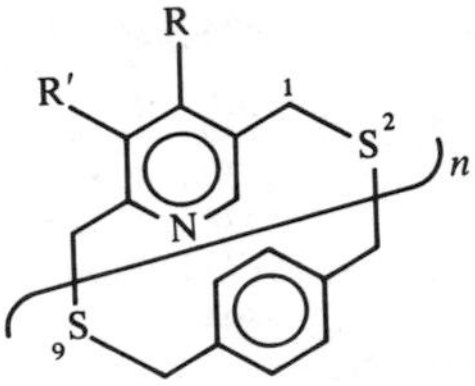	1	H
		2,9-*bis*(sulfone)
	1	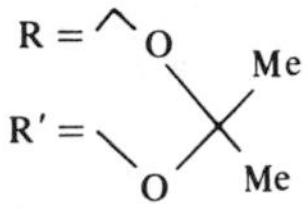 $R = \overset{\wedge}{}O$ Me $\quad$ $R' = \diagdown O$ Me
	1	$R = CH_2OAc; R' = OAc$
	1	$R = CHO; R' = OH$
	2	H
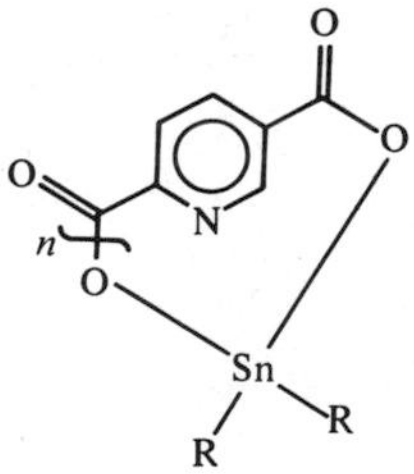	2	$R = Et$
		$R = i\text{-}Pr$

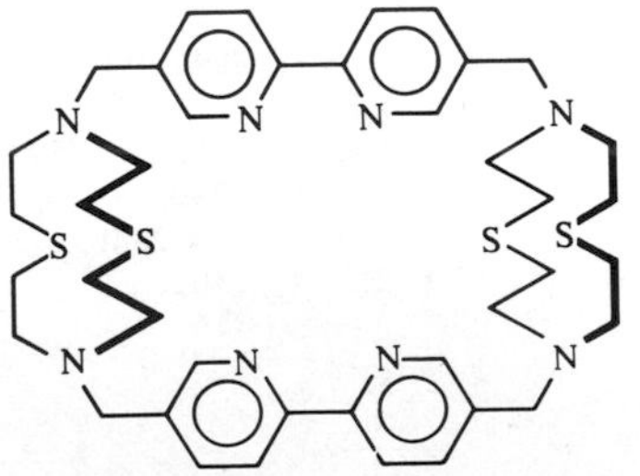

Physical Data m.p. [b.p. (mm)], °C	Spectral Data Available	Metal Complex(es) General Comments	Reference
			234
3	A		450
			234
			234
			234
205 dec)	A		233
			229, 233
			234
			234
219 (dec)	A		234, 238
218	A	Two isomers	233
) dec			373
ec			373
		Cu	414

TABLE III-13. (*CONTINUED*)

Compound	n	Substituents
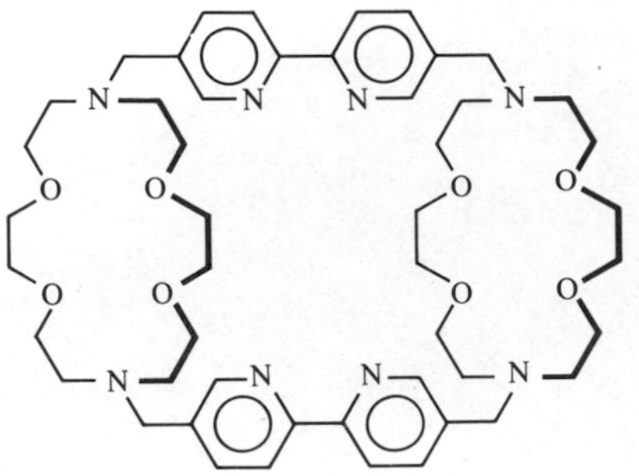		

TABLE III-14. 2,4-PYRIDINE MACROCYCLES

Compound	n	Substituents
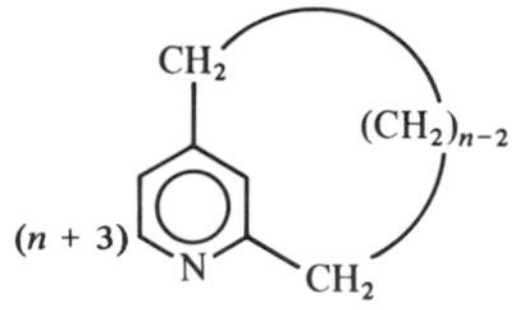	9	H 12-Me
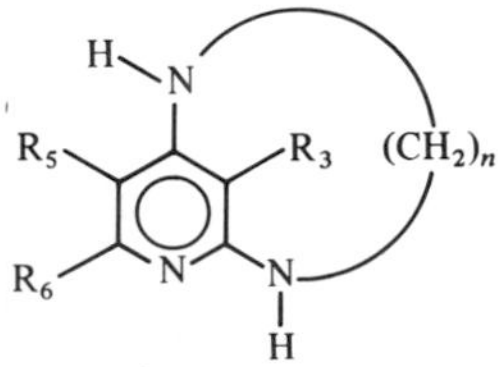	9 12	$R_3 = R_5 = (Cl)_2 ; R_6 = F$ $R_3 = R_5 = (Cl)_2 ; R_6 = F$

Physical Data m.p. [b.p. (mm)], °C	Spectral Data Available	Metal Complex(es) General Comments	Reference
		$^+H_3N-(CH_2)_7-NH_3^+$	414

Physical Data m.p. [b.p. (mm)], °C	Spectral Data Available	Metal Complex(es) General Comments	Reference
120 (0.3)]	A	MeI (127–128°)	242, 374
110 (0.2)]	A, B		243
	A, D		250
80 (3.5)]	A, D		250

TABLE III-15.   2,3-PYRIDINE MACROCYCLES

Compound	n	Substituents
	10	H 13-(CONH$_2$); 14(=O) 13-(CONH$_2$); 14(=C(CN)$_2$)
	13	H 17-Cl 17-OH 17-OH; 16-CN 15,18-(H)$_2$; 17,19-(OH)$_2$; 16-CN 15,17-(OH)$_2$; 16-CO$_2$CH$_2$CH$_3$ (H)$_6$ (cis) (H)$_6$ (trans)
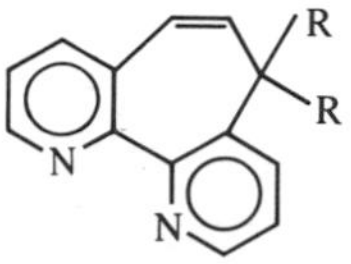		R = Me

		R = H
		R = CH(NH$_2$)-p (C$_6$H$_4$NO$_2$)
		R = CH-p(C$_6$H$_4$NO$_2$)
		R = CH(NH)-p-(C$_6$H$_4$NO$_2$)
		    Orn-Met-Ser-His-Trp-OH
		R = CH(NH)p-(C$_6$H$_4$NO$_2$)
		    Arg-Met-Ser-His-Trp-OH

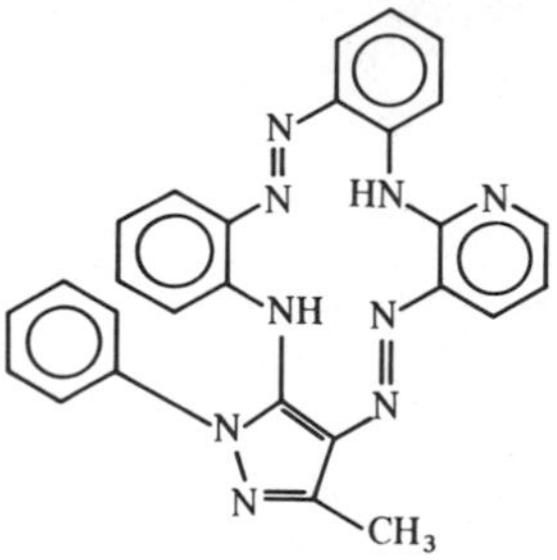		
		R = H; X = Y = CH R = H; X = N; Y = CH X = CH; Y = N R = (N=N-C$_6$H$_5$); X = N; Y = CH R = (NO$_2$); X = N; Y = CH R = (CN); X = N; Y = CH

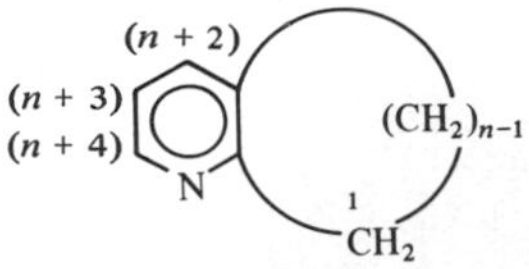

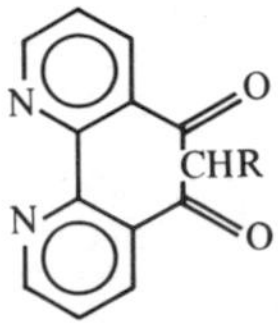

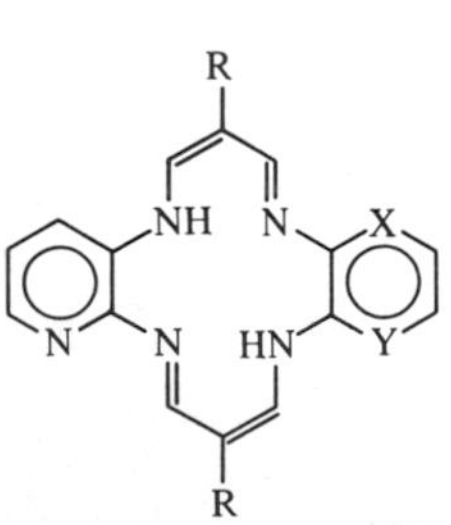

Physical Data m.p. [b.p. (mm)], °C	Spectral Data Available	Metal Complex(es) General Comments	Reference
5–175 (3.7)] 21.8–23.4	B, C	Picrate (154–155°)	12
	A, B		253
			253
5–127 (0.007)]		Picrate (137–138°)	251, 252
–131		Picrate (130–131°)	251
–190			251
–211			251
–248			251
–300			251
		Picrate (194–195°)	251
		Picrate (202–203°)	251
	A	Hg, Zn	255
			376
			376
			376
			376
			376
		Ni (two isomers)	264
–264	B–D	Co, Cu	266, 267
	B, D	Cu, Co	265–268
–328	B–D	Co, Ni, Cu	265
75	B–D	Ni	265
70	B–D	Ni	265

TABLE III-15.   (*CONTINUED*)

Compound	n	Substituents
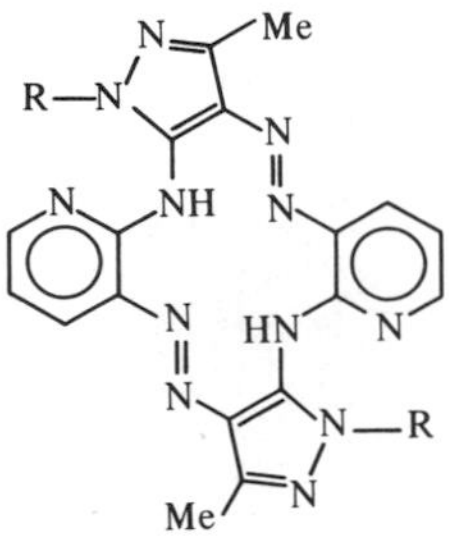		$R = Et$ $R = Ph$
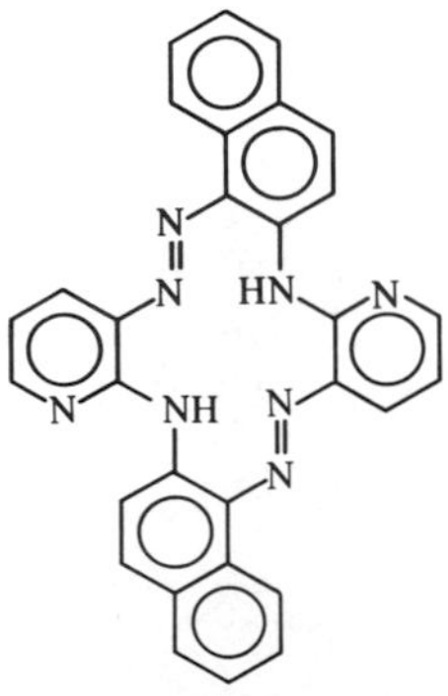		
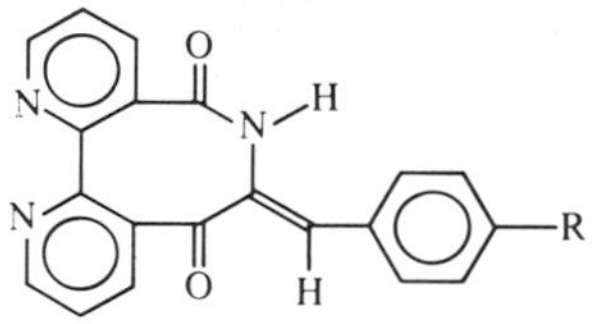		$R = NO_2$
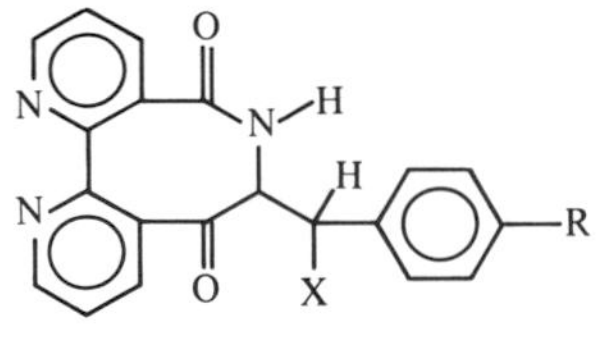		$R = NO_2$; $X = Gly\text{-}OEt$ $X = Tyr\text{-}OEt$ $X = His\text{-}OEt$ $X = Trp\text{-}OEt$ $X = Lys\text{-}CO_2CH_2C_6H_5$ $X = Orn\text{-}CO_2CH_2C_6H_5$ $X = Arg\text{-}CO_2CH_2C_6H_5$ $X = Orn\text{-}Trp\text{-}His\text{-}Ser\text{-}Met\text{-}OH$
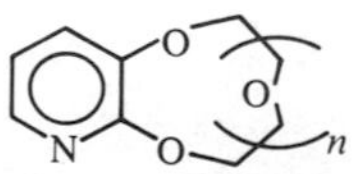	3 4 5	H H H

Physical Data m.p. [b.p. (mm)], °C	Spectral Data Available	Metal Complex(es) General Comments	Reference
	(A–D)	Ni	262
	(A–D)	Ni, Pd, Pt, Cu	262
		Ni, Cu	263
			377
			377
			377
			377
			377
			377
			377, 378
			377
			378
−65.5	A, C	Cs	256
8.5	A, C	Cs	256
	A, C	Cs	256

TABLE III-15.   (*CONTINUED*)

Compound	n	Substituents
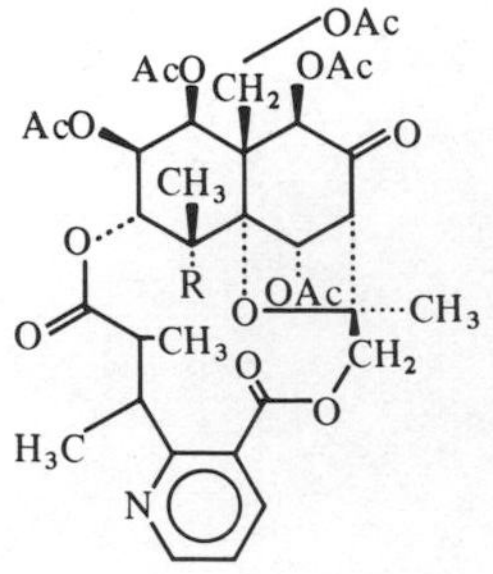		R = H R = OH
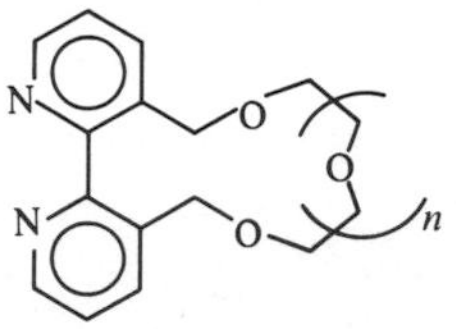	2 3 4	H H H
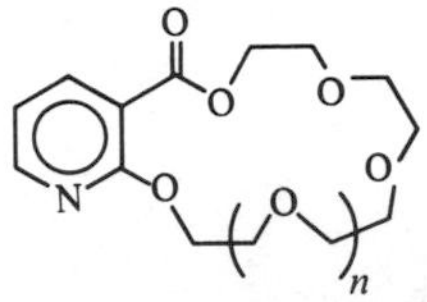	1	H
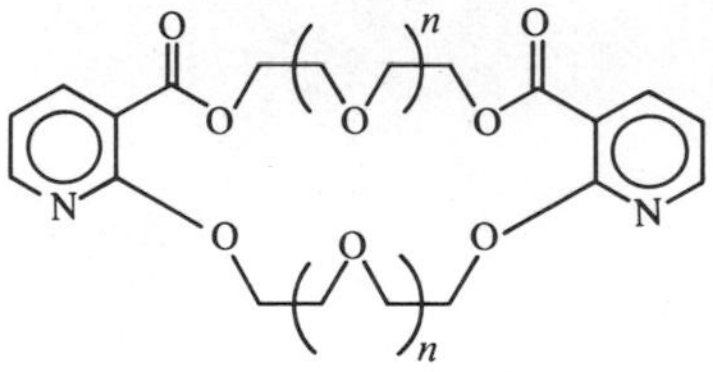	1	H
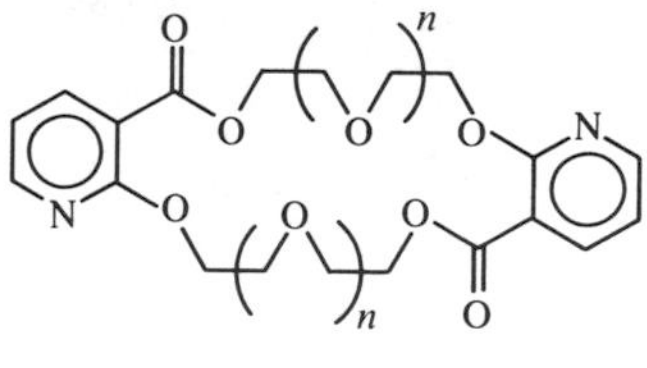	1	H
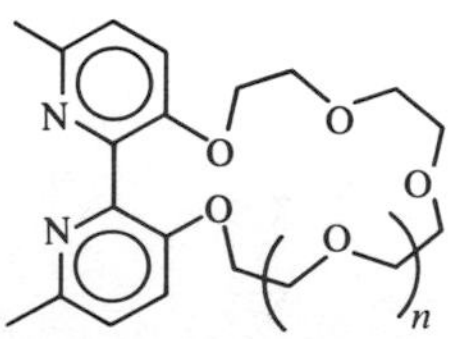	1 2	H H

Physical Data m.p. [b.p. (mm)], °C	Spectral Data Available	Metal Complex(es) General Comments	Reference
		Isolation and structure elucidation of alkaloids of Catha edulis, Euonymus europaeus, alatamine and wilfordine	379–382, 406, 415–417
3	A	Li, Na, K, NH₄⁺, Rb, Cs	258
	A, D	[W(CO)₄], Co, Zn, Hg, Ni, Li, Na, K, Rb, Cs, NH₄⁺	258, 260, 261
	A, D	[W(CO)₄], Co, Zn, Hg, Ni, Li, Na, K, Rb, Cs, NH₄⁺	258, 260, 261
	A, B, D		257
2.5–113	A, B, D		257
0.5–170	A, B, D		257
	A, D	[W(CO)₄], Li, Na, K, NH₄⁺, Rb, Cs	259
	A, D	[W(CO)₄], Li, Na, K, NH₄⁺, Rb, Cs	259

TABLE III-15.   *(CONTINUED)*

Compound 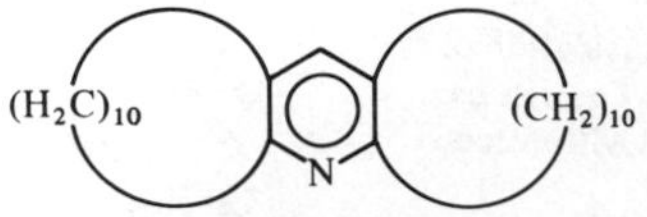	n	Substituents

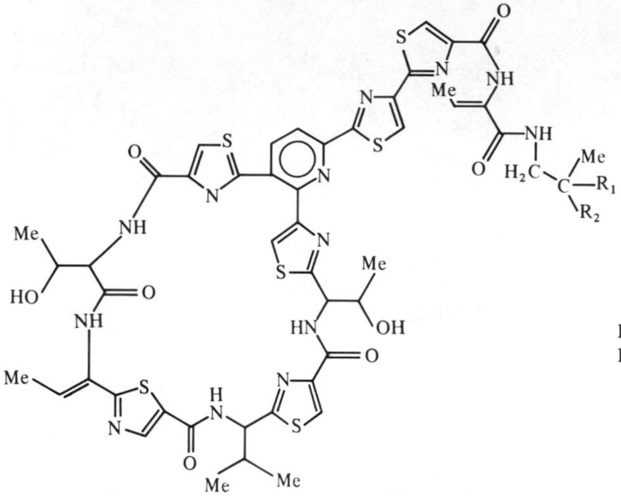

$R_1 = H; R_2 = OH$
$R_1, R_2 = O$

Physical Data m.p. [b.p. (mm)], °C	Spectral Data Available	Metal Complex(es) General Comments	Reference
			383
	A	Micrococcin P	375
	A		375
	A, E	Thiocillin I, II, III	413

TABLE III-16.   3,5-PYRIDINE MACROCYCLES

Compound	n	Substituents

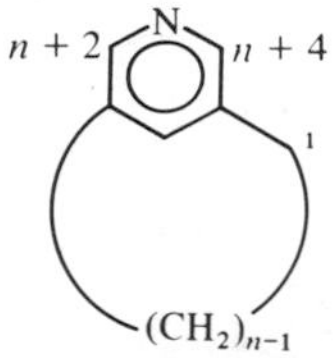	7	$9,11(Me)_2$
	9	$11,13\text{-}(Me)_2$
	9	$11,13\text{-}(Me)_2; 12\text{-}NH_2 \ (ClO_4^-)$
	9	$11,12,13\text{-}(Me)_3(ClO_4^-)$
	9	$11,13\text{-}(Me)_2; 12\text{-}C_6H_5(ClO_4^-)$
	9	$11,13\text{-}(Me)_2; 12\text{-}NHAc(ClO_4^-)$

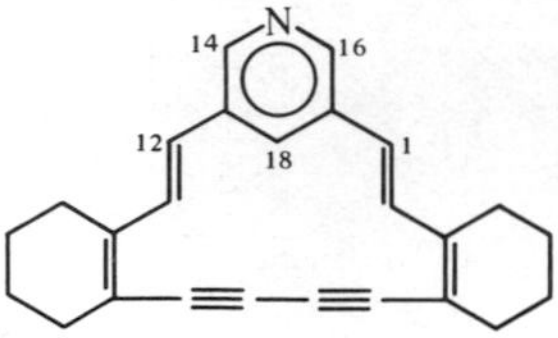

H
$15,18\text{-}(H)_2$
$15\text{-}(Me); 18\text{-}H$
$15\text{-}CO_2CH_2CH_3; 18\text{-}H$
$15\text{-}COCH_3; 18\text{-}CH_2CH_3$
$15\text{-}COCH_3; 18\text{-}H$
$15\text{-}CO_2CH_2CH_3; 18\text{-}CH_2CH_3$
$15,18\text{-}(Me)_2$
$15\text{-}Me; 18\text{-}CH_2CH_3$
$15\text{-}Me; 18\text{-}CH_2CH_2CH_3$
$15\text{-}Me; 18\text{-}n\text{-}Bu$
$15\text{-}H; 18\text{-}CH_3$
$15\text{-}H; 18\text{-}CH_2CH_3$

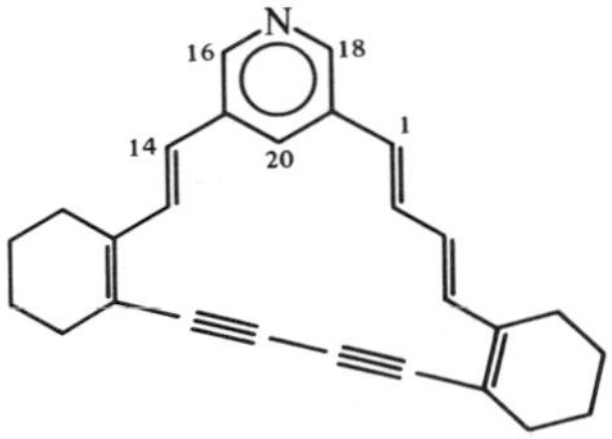

$20\text{-}H; 17\text{-}CO_2CH_2CH_3$

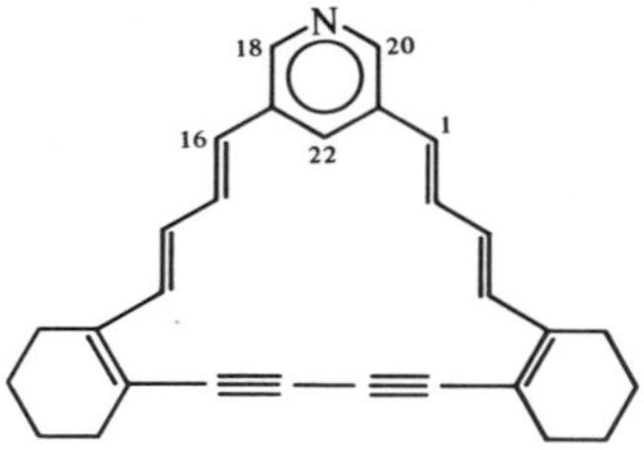

H
$22\text{-}H; 19\text{-}CO_2CH_2CH_3$

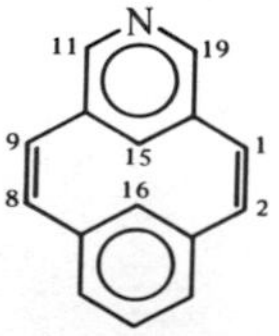

$11,13,15,16\text{-}(Me)_4$
$8,9\text{-}(H)_2; 11,13,15,16\text{-}(Me)_4$
$1,2,8,9\text{-}(H)_4$

Physical Data m.p. [b.p. (mm)], °C	Spectral Data Available	Metal Complex(es) General Comments	Reference
	A, E	Picrate (171–172°), Picrolonate (259°)	271, 272
	A		269
9	*A*		270, 384
6	*A, B*		270
4	*A, B*		270
8–199	*A, B, D*		384
9–250	*A, D*		277
p. (dec)	*A, D*	K (anion formation)	277–279
p. (dec)	*A, C, D*		277
p. (dec)	*A, D*		277
	A		280
o. (dec)	*A, D*		277
o. (dec)	*A*		280
(dec)	*B*		278
(dec)	*A*		278, 280
(dec)	*A*		278
(dec)	*A*		278
o. (dec)		K (anion formation)	279
o. (dec)		K (anion formation)	279, 280
. (dec)	*A, C*		385
. (dec)	*A*		281
o. (dec)	*A, C*		281
		HCl; $K_{1/2} \sim 8$ (MeOH)	273
1–89.6	*A–C*		273
–171 [subl: 60–65(0.03)]	*A*		290

TABLE III-16.   *(CONTINUED)*

Compound	n	Substituents
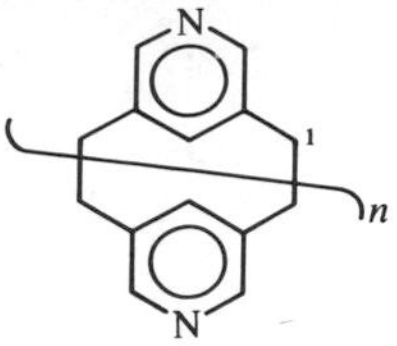	1	H
	2	H
		15,16-(Me)$_2$
		4,6,11,13,15,16-(Me)$_6$
	3	H
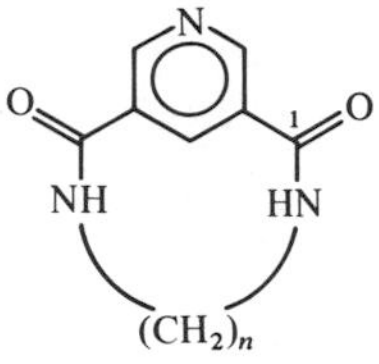	7	H
		6-OH
		6-O-*t*-Bu
		14-CH$_2$C$_6$H$_5$(Br$^-$)
		6-OH; 14-CH$_2$C$_6$H$_5$(Br$^-$)
		6,13-(OH)$_2$; 14-CH$_2$C$_6$H$_5$
		14-(2,6-Cl$_2$C$_6$H$_3$CH$_2$)(Br$^-$)
		6-OH; 14-(2,6-Cl$_2$C$_6$H$_3$CH$_2$)(Br$^-$)
		6-(=O)
		6-(=O); 14-(2,6-Cl$_2$C$_6$H$_3$CH$_2$)(Br$^-$)
		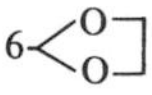 6$<$O$/$O
	5	H
	6	H
	8	H
	9	7-OH
		7-OH; 16-(2,6-Cl$_2$C$_6$H$_3$CH$_2$)(Br$^-$)
		7-(=O)
		7-(=O); 16-(2,6-Cl$_2$C$_6$H$_3$CH$_2$)(Br$^-$)
		7-(=O); 16-CH$_3$(I$^-$)
		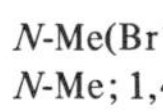 7$<$O$/$O
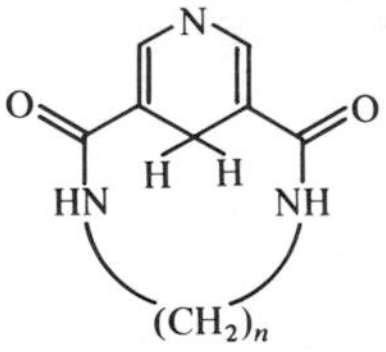	7	6-(=O); 14-(2,6-Cl$_2$C$_6$H$_3$CH$_2$)(Br$^-$)
		14-(2,6-Cl$_2$C$_6$H$_3$CH$_2$)
	9	7-(=O); 16-(2,6-Cl$_2$C$_6$H$_3$CH$_2$)
		7-(=O); 16-CH$_3$
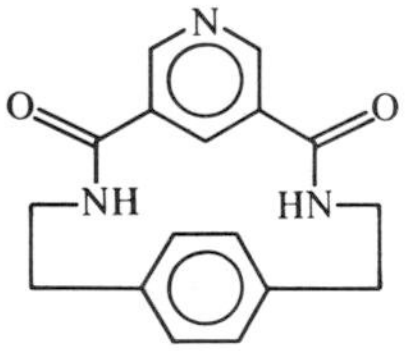		*N*-Me(Br$^-$)
		N-Me; 1,4-di(H)
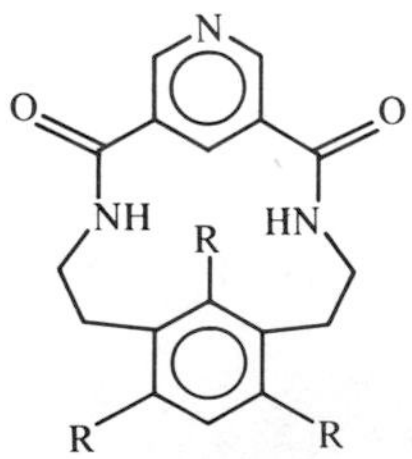		*N*-CH$_2$C$_6$H$_5$(Br$^-$); R = Me
		N-CH$_2$C$_6$H$_5$; 1,4-di(H); R = Me

Physical Data m.p. [b.p. (mm)], °C	Spectral Data Available	Metal Complex(es) General Comments	Reference
6–237	A, D		275, 276, 386
9–260	A, D		276
0–265 (impure)	A, D		274
4–246	A		273, 274
7–278	A, D		276
4–337 (dec)	A, B, D		292, 293, 442
6–318 (dec)	A, B, D		292, 293
0–312 (dec)	A, B, D		292
1–262 (dec)	A, B		292
9–241 (dec)	A–C		292
	A		292
8–290	A, B		293
4–255	A, B		293
3–316	A, B		293
7–259	A, B		293
6–267	A, B		293
6–238	A, B		293, 442
8–300	A, B		293, 442
1–343	A, B		293, 442
2–354	A, B		293
4–226	A, B		293
3–325	A, B		293
1–223	A, B		293
2–266	A, B		293
7–351 (dec)	A, B		293
5–268	A–C		293, 442
0–231 (dec)	A–C		293
5–217	A–C		293, 442
1–234	A–C		293
			387, 442
			387, 442
			387, 442
			387, 442

TABLE III-16.   *(CONTINUED)*

Compound	n	Substituents
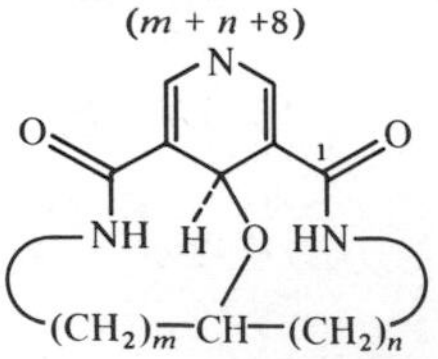 $(m + n + 8)$	$3; m = 3$	14-H
		15-OH; 16-H
	$4; m = 4$	16-H

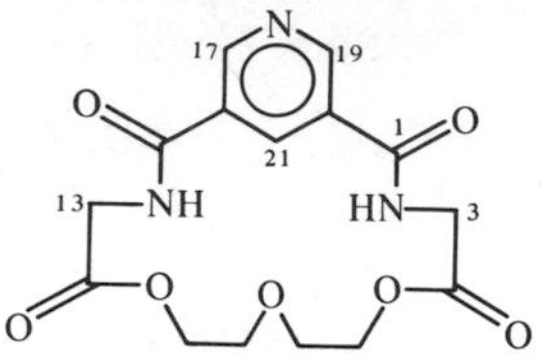	3,13-(CH$_3$)$_2$
	3,13-[(CH$_3$)$_2$CH-]
	3,13-[(CH$_3$)$_2$CH-]$_2$; 18-(CH$_3$)
	3,13-[(CH$_3$)$_2$CH-]$_2$; 18-(CH$_3$); 21-(H)$_2$
	3,13-(Bz)$_2$; 21-(H)$_2$; 18-(CH$_3$)
	2,3; 13,14-(◯)$_2$; 21-(H)$_2$; 18-(CH$_3$) 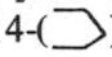

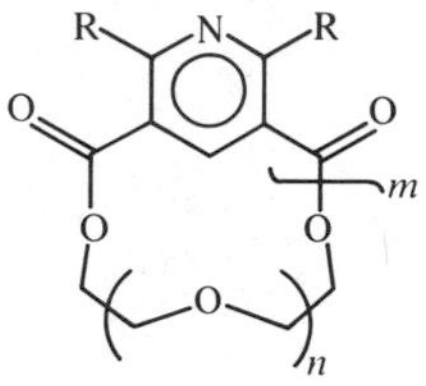	$0; m = 2$	R = Me
	$1; m = 2$	R = Me
	$2; m = 1$	R = H
		R = Me
	$m = 2$	R = H
	$3; m = 1$	R = H
		N-Me; dihydro
		R = Me
	$m = 2$	R = Me
	$4; m = 1$	H
	$5; m = 1$	H

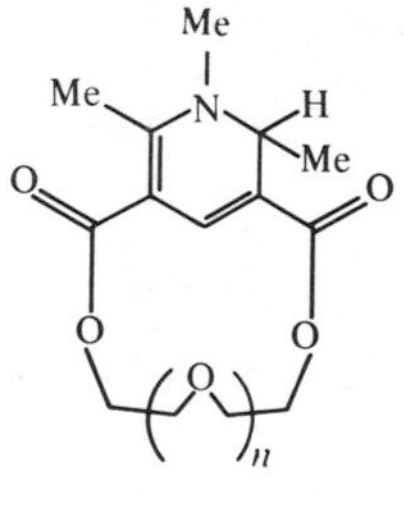	3	H

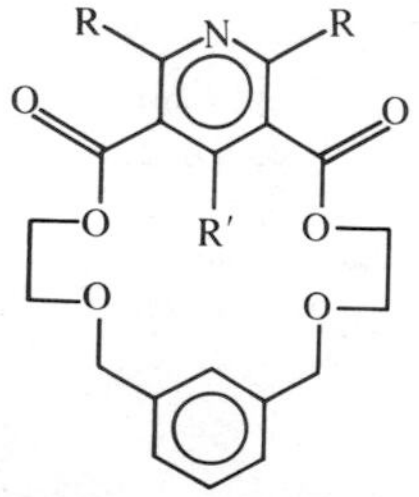	R = R′ = H
	R′ = H; R = Me
	R′ = H; R = H; N-Me(ClO$_4^-$)
	R = H; 1-Me; 1,4-diH$_2$
	R = R′ = Me
	R = Me; R′ = C$_6$H$_5$
	R = Me; R′ 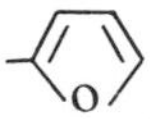 = ⟨◯$_O$⟩

Physical Data m.p. [b.p. (mm)], °C	Spectral Data Available	Metal Complex(es) General Comments	Reference
	C		293
	C		293
	C		293
.1–262.3 (dec)		$[\alpha]_D^{23} = -94.0°$	288, 442
.4–254.3 (dec)		$[\alpha]_D^{20} = -126.8°$	288, 442
		$[\alpha]_D^{22} = -159.6°$	288, 400, 442
.5–142.4	E	$[\alpha]_D^{25} = -133.9°$; Mg	288, 289, 442
	E	Mg	289, 442
	E	Mg	289, 442
–198	A–D		74, 87, 283
–126	A–D		74, 87, 283
–130 (137–137.5)[87]	A, B, D	Cs	74, 87, 283
–169	A–C		74, 283
5–96.5 (112–113)[87]	A, B	Cs, K, Rb, Na	76, 87, 284
	G	Na	285, 400
92	A–D	Me(ClO₄⁻) salt (m.p. 190–193°), Cs	74, 87, 283, 284, 286, 442
			87, 283
57 (54.5–55.5)[87]	A, B	Cs	76, 87, 284, 400, 442
51	A, B	Cs	76, 87, 284, 400
.5–133	A, C, D		74, 283
–128	A		74
.5–160	A, B		74
–214	A		74
–136	A		74
–181	A		74
.5–187.5	A		74
–180	A		74

TABLE III-16.   (*CONTINUED*)

Compound	n	Substituents
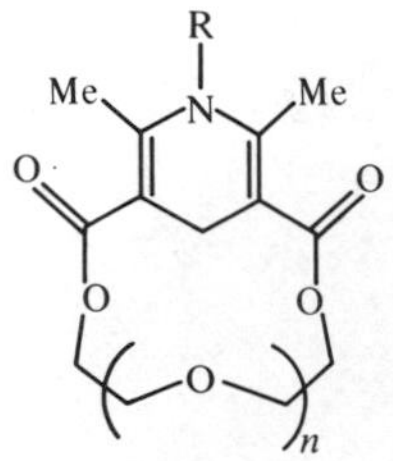	3	R = H R = Me
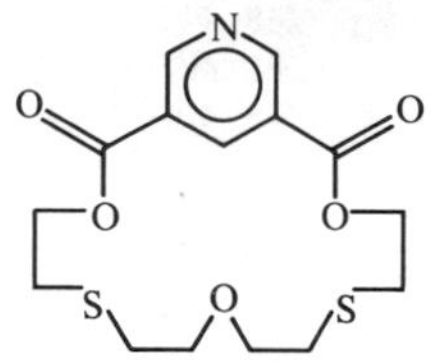		H
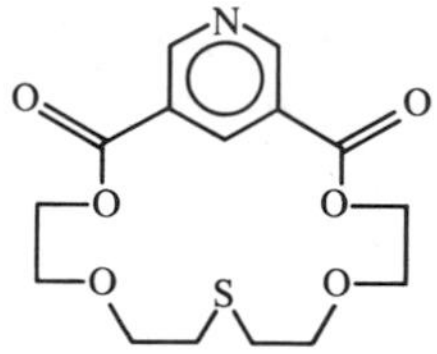		H
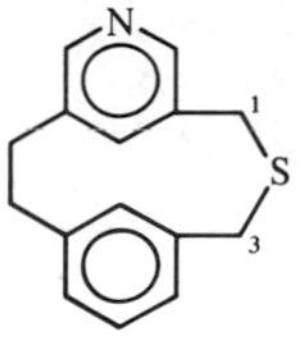	1	H
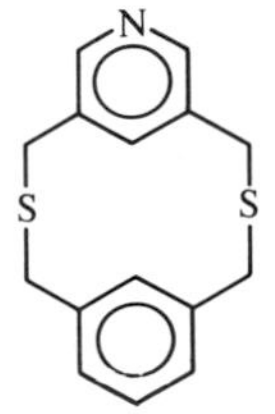		H
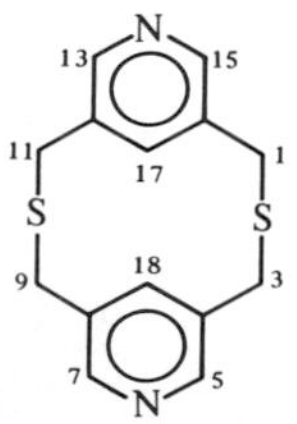		$17,18\text{-}(CH_3)_2$ $17,18\text{-}(CH_3)_2; 2,2,10,10\text{-}(O)_4$ $17,18\text{-}(CH_3)_2; 2,2,6,10,10,14\text{-}(O)_6$ $5,7,13,15,17,18\text{-}(CH_3)_6$

Physical Data m.p. [b.p. (mm)], °C	Spectral Data Available	Metal Complex(es) General Comments	Reference
–117	A, B		74, 442
–113	A–C	(NaClO$_4$ salts)	74, 283, 287, 400, 442
–104	A, B		76
–109	A, B		76
92 [subl: 55–60 (0.01)]	A		290
–178	A		290
–316	A, D		274
	A		274
	D		274
–299	A		274

TABLE III-16.　　(*CONTINUED*)

Compound	*n*	Substituents

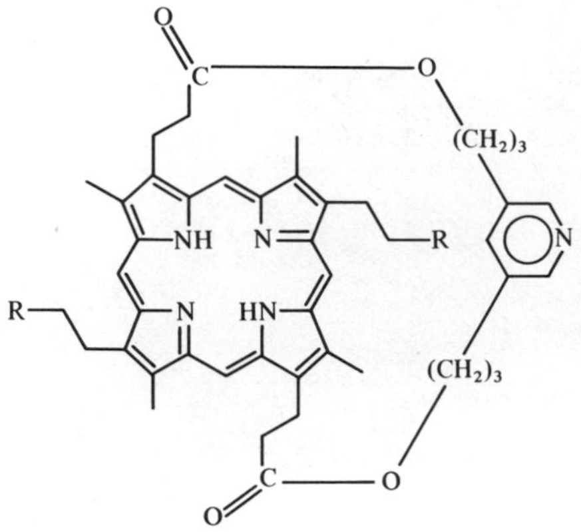

R = CO$_2$CH$_2$Ph
R = CO$_2$H
R = H

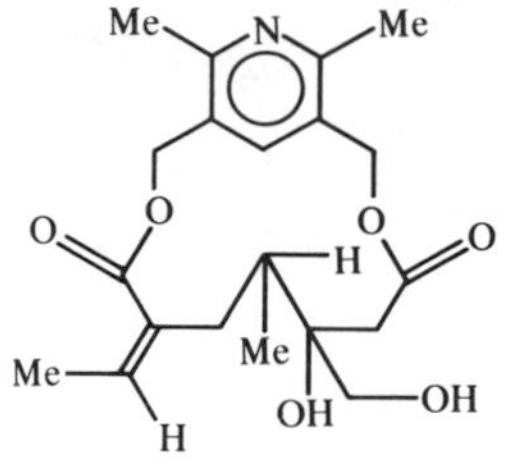

Physical Data m.p. [b.p. (mm)], °C	Spectral Data Available	Metal Complex(es) General Comments	Reference
			388
		Fe	388
			388
			444

TABLE III-17. 3,4-PYRIDINE MACROCYCLES

Compound	n	Substituents
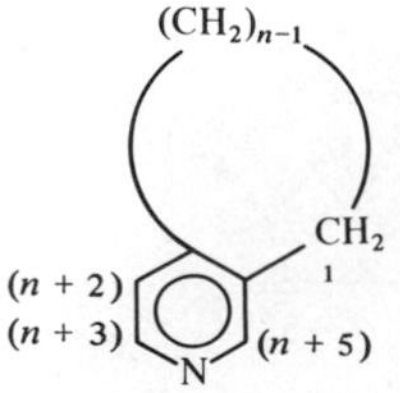 $(CH_2)_{n-1}$ $(n+2)$ $(n+3)$ CH_2 $(n+5)$ N	10	15-Me; 12-CN; 13-OH
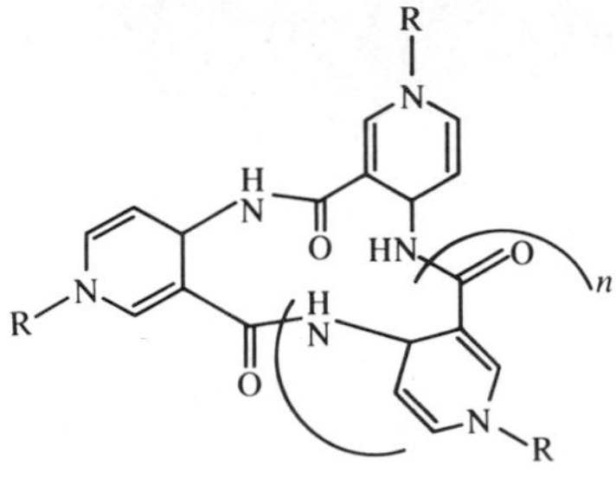	$n=1$	$R = CH_2\phi$
	$n=2$	$R = (CH_2)_2NH_2$
		$R = (CH_2)_3NH_2$
		$R = (CH_2)_3\phi$
		$R = (CH_2)_3C_6H_4OCH_3$
		$R = (CH_2)_3C_6H_3(OCH_3)_2$
		$R = CH_2OCH_2C_6H_3(OCH_3)_2$
		$R = (CH_2)_2NHCO_2C_2H_5$
		$R = (CH_2)_3NHCO_2C_2H_5$
		$R = (CH_2)_2NHSO_2C_6H_4CH_3$
		$R = (CH_2)_3NHSO_2C_6H_4CH_3$
		$R = (CH_2)_3N(CH_3)SO_2C_6H_4CH_3$
		$R = (CH_2)_2NHCOCF_3$
		$R = (CH_2)_3NHCOCF_3$
		$R = (CH_2)_2N(CO)_2C_6H_4$
		$R = (CH_2)_3N(CO)_2C_6H_4$

Physical Data m.p. [b.p. (mm)], °C	Spectral Data Available	Metal Complex(es) General Comments	Reference
5–282	B, C		294

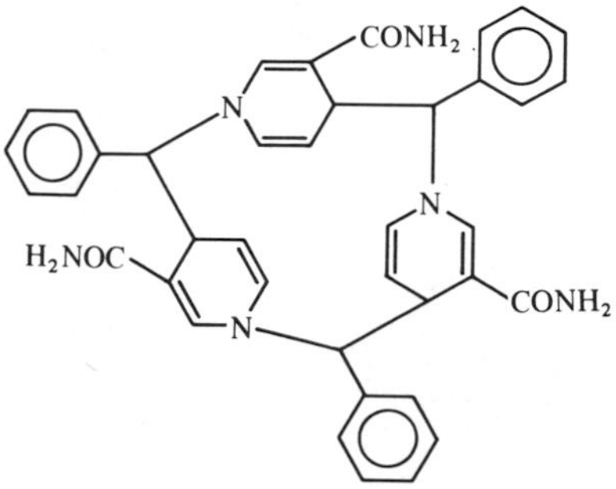		Original structural assignment[297]	296, 297
-231	A, C		298
	C		298
	A, C		298
	A, C		298
	C		298
	C		298
	A, C		298
	A, C		298
00	A, C		298
			298
			298
			298
-153	A, C		298
			298
-125	A, C		298

TABLE III-18.   2,3,5,6-PYRIDINE MACROCYCLES

Compound	n	Substituents

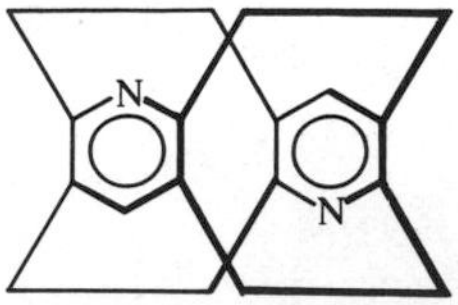

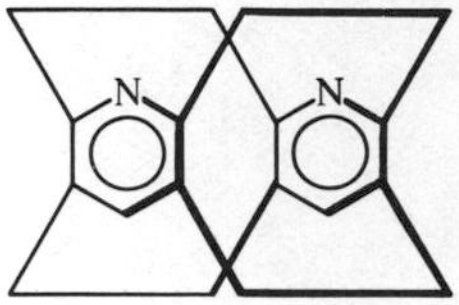

Physical Data m.p. [b.p. (mm)], °C	Spectral Data Available	Metal Complex(es) General Comments	Reference
50 (dec)	*A, C, G*	Photoelectron spectra	299, 299a, 300
10 (dec)	*A, C, G*	Photoelectron spectra	299, 299a, 301

REFERENCES

1. International Union of Pure and Applied Chemistry, *Nomenclature of Organic Chemistry*, Sections A and B, 2nd ed., London 1966; (b) "The Naming and Indexing of Chemical Compounds from Chemical Abstracts," Introduction to the Subject Index Vol. 54, Section 137, American Chemical Society, Washington, D.C., 1960; (c) A. M. Patterson, L. T. Capell, and D. F. Walker, "The Ring Index," 2nd ed., American Chemical Society, Washington, D.C., 1960.

2. (a) F. Vögtle and P. Neumann, *Tetrahedron Lett.,* **1969,** 5329; (b) Ibid., *Tetrahedron,* **26,** 5847 (1970).

3. Th. Kauffmann, *Tetrahedron,* **28,** 5183 (1972).

4. K. Hirayama, *Tetrahedron Lett.,* **1972,** 2109.

5. E. Weber and F. Vögtle, *Inorg. Chim. Acta,* **45,** L65 (1980).

6. G. R. Newkome, J. D. Sauer, J. M. Roper, and D. C. Hager, *Chem. Revs.,* **77,** 513 (1977).

7. (a) K. Tamao, S. -i. Kodama, T. Nakatsuka, Y. Kiso, and M. Kumada, *J. Am. Chem. Soc.,* **97,** 4405 (1975); (b) M. Kumada, unpublished data; (c) K. Tamao, S. Kodama, I. Nakajima, T. Nokatsuka, A. Minato, and M. Kumada, *Hukusokan Kagaku Toronkai Koen Yoshishu, 8th.,* **1975,** 174 [*Chem. Abstr.,* **85,** 5468y (1976)]; (d) Also see M. Kumada, in "Organotransition-Metal Chemistry," Y. Ishii and M. Tsutsui, eds., Plenum, New York, 1975, p. 211.

8. M. Kumada and K. Tamao, *Jap. Kokai,* 77 51,787 [*Chem. Abstr.,* **87,** 68170g (1977)].

9. S. Fujita and H. Nozaki, *Bull. Chem. Soc. Japan,* **44,** 2827 (1971).

10. S. Fujita, K. Imamura, and H. Nozaki, *Bull. Chem. Soc. Japan,* **46,** 1579 (1973).

11. S. Fujita, K. Imamura, and H. Nozaki, *Bull. Chem. Soc. Japan,* **45,** 1881 (1972).

12. K. Biemann, G. Buchi, and B. H. Walker, *J. Am. Chem. Soc.,* **79,** 5558 (1957).

13. M. Karpf and A. S. Dreiding, *Helv. Chim. Acta,* **59,** 1226 (1976).

14. V. Boekelheide and W. J. Linn, *J. Am. Chem. Soc.,* **76,** 1286 (1954).

15. H. Schinz, L. Ruzicka, U. Geyer, and V. Prelog, *Helv. Chim. Acta,* **29,** 1524 (1946).

16. H. Saimoto, T. Hiyamo, and H. Nozaki, *Tetrahedron Lett.,* **1980,** 3897.

17. A. T. Balaban, M. Gavat, and C. D. Nenitzescu, *Tetrahedron,* **18,** 1079 (1962).

18. A. T. Balaban and C. D. Nenitzescu, *Rev. Chemie. Acad. R. P. R.,* **6,** 269 (1961).

19. K. Dimroth, *Angew. Chem.,* **72,** 331 (1960).

20. U. K. Georgi and J. Retey, *Chem. Commun.,* **1971,** 32.

21. H. Nozaki, S. Fujita, and T. Mori, *Bull. Chem. Soc. Japan,* **42,** 1163 (1969).

22. P. Dubs and R. Stussi, *J. Chem. Soc., Chem. Commun.,* **1976,** 1021.

23. G. L. Isele and K. Scheib, *Chem. Ber.,* **108,** 2312 (1975).

24. W. Baker, K. M. Buggle, J. F. W. McOmie, and D. A. M. Watkins, *J. Chem. Soc.,* **1958,** 3594.

25. G. R. Newkome and D. L. Koppersmith, *J. Org. Chem.,* **38,** 4461 (1973).

26. W. Baker, *Chem. Brit.,* **1,** 250 (1965).

27. J. R. Fletcher and I. O. Sutherland, *Chem. Commun.,* **1969,** 1504.

28. V. Boekelheide and J. A. Lawson, *Chem. Commun.,* **1970,** 1558.

29. W. Jenny and H. Holzrichter, *Chimia,* **23,** 158 (1969).

30. W. Jenny and H. Holzrichter, *Chimia,* **22,** 306 (1968).

31. Th. Kauffmann, G. Beissner, W. Sahm, and A. Woltermann, *Angew. Chem. Internat. Ed.,* **9,** 808 (1970).

32. V. Boekelheide, I. D. Reingold, and M. Tuttle, *J. Chem. Soc., Chem. Commun.,* **1973,** 406.

33. I. D. Reingold, W. Schmidt, and V. Boekelheide, *J. Am. Chem. Soc.*, **101**, 2121 (1979).

34. V. Boekelheide, K. Galuszko, and K. S. Szeto, *J. Am. Chem. Soc.*, **96**, 1578 (1974).

35. M. W. Haenel, *Tetrahedron Lett.*, **1978**, 4007.

36. K. Fuji, K. Ichikawa, and E. Fujita, *Tetrahedron Lett.*, **1979**, 361.

37. K. Fuji, K. Ichikawa, and E. Fujita, *J. Chem. Soc., Perkins I*, **1980**, 1066.

38. E. Fujita, K. Fuji, and K. Tanaka, *J. Chem. Soc. (C)*, **1971**, 205.

39. H. J. J. -B. Martel and M. Rasmussen, *Tetrahedron Lett.*, **1971**, 3843.

40. K. Galuszko, *Rocz. Chem.*, **49**, 1597 (1975).

41. H. J. J. -B. Martel, S. McMahon, and M. Rasmussen, *Aust. J. Chem.*, **32**, 1241 (1979).

42. C. Y. Meyers, A. M. Malte, and W. S. Mathews, *J. Am. Chem. Soc.*, **91**, 7510 (1969).

43. M. P. Cooke, Jr., *J. Org. Chem.*, **46**, 1747 (1981).

44. G. R. Newkome, J. D. Sauer, P. K. Mattschei, and A. Nayak, *Heterocycles*, **9**, 1555 (1978).

45. G. R. Newkome and J. M. Roper, *J. Organometal. Chem.*, **186**, 147 (1980).

46. J. E. Parks, B. E. Wagner, and R. H. Holm, *J. Organometal. Chem.*, **56**, 53 (1973).

47. G. R. Newkome, J. D. Sauer, and G. L. McClure, *Tetrahedron Lett.*, **1973**, 1599.

48. A. H. Alberts and D. J. Cram, *J. Am. Chem. Soc.*, **101**, 3545 (1979).

49. K. R. Reistad, P. Groth, R. Lie, and K. Undheim, *J. Chem. Soc., Chem. Commun.*, **1972**, 1059.

50. P. Groth, *Acta Chem. Scand.*, **27**, 5 (1973).

51. F. Bottino, S. Cosentino, S. Cunsolo, and S. Pappalardo, *J. Heterocycl. Chem.*, **18**, 199 (1981).

52. S. Ogawa, T. Takeuchi, and S. Shiraishi, to be published.

53. G. R. Newkome and J. M. Robinson, *J. Chem. Soc., Chem. Commun.*, **1973**, 831.

54. M. Newcomb, G. W. Gokel, and D. J. Cram, *J. Am. Chem. Soc.*, **96**, 6810 (1974).

55. D. J. Cram, *U.S. Pat.* 4,080,337 (1978).

56. M. Newomb, J. M. Timko, D. M. Walba, and D. J. Cram, *J. Am. Chem. Soc.*, **99**, 6392 (1977).

57. R. R. Whitney and D. A. Jaeger, *Org. Mass Spectrometry*, **15**, 343 (1980).

58. E. P. Kyba, G. W. Gokel, F. de Jong, K. Koga, L. R. Sousa, M. G. Siegel, L. Kaplan, G. D. Y. Sogah, and D. J. Cram, *J. Org. Chem.*, **42**, 4173 (1977).

59. J. Jacques and C. Forquey, *Tetrahedron Lett.*, **1971**, 4617; E. P. Kyba, K. Koga, L. R. Sousa, M. G. Siegel, and D. J. Cram, *J. Am. Chem. Soc.*, **95**, 2692 (1973).

60. G. W. Gokel, J. M. Timko, and D. J. Cram, *J. Chem. Soc., Chem. Commun.*, **1975**, 444.

61. E. P. Kyba, J. M. Timko, L. J. Kaplan, F. de Jong, G. W. Gokel, and D. J. Cram, *J. Am. Chem. Soc.*, **100**, 4555 (1978).

62. S. C. Peacock, L. A. Domeier, F. C. A. Gaeta, R. C. Helgeson, J. M. Timko, and D. J. Cram, *J. Am. Chem. Soc.*, **100**, 8190 (1978).

63. J. M. Timko, R. C. Helgeson, M. Newomb, G. W. Gokel, and D. J. Cram, *J. Am. Chem. Soc.*, **96**, 7097 (1974).

64. L. J. Kaplan, G. R. Weisman, and D. J. Cram, *J. Org. Chem.*, **44**, 2226 (1979).

65. E. Maverick, L. Grossenbacher, and K. N. Trueblood, *Acta Cryst.*, **B35**, 2233 (1979).

66. E. Weber and F. Vögtle, *Chem. Ber.*, **109**, 1803 (1976).

67. F. Vögtle, W. M. Müller, and E. Weber, *Chem. Ber.*, **113**, 1130 (1980).

68. F. Vögtle and W. M. Müller, *Naturwissenschaften*, **67**, S.255 (1980).

69. E. Weber and F. Vögtle, *Angew. Chem. Internat. Ed.*, **19**, 1030 (1980).

70. R. A. Bartsch and P. N. Juri, *J. Org. Chem.*, **45**, 1011 (1980).

622 G. R. Newkome, V. K. Gupta, and J. D. Sauer

71. R. R. Whitney and D. A. Jaeger, *J. Heterocycl. Chem.*, **17**, 1093 (1980).

72. H. G. Förster and J. D. Roberts, *J. Am. Chem. Soc.*, **102**, 6984 (1980).

73. J. D. Lamb, R. M. Izatt, C. S. Swain, and J. J. Christensen, *J. Am. Chem. Soc.*, **102**, 475 (1980).

74. R. M. Kellogg, T. J. van Bergen, H. v. Doren, D. Hedstrand, J. Kooi, W. H. Kruizinga, and C. B. Troostwijk, *J. Org. Chem.*, **45**, 2854 (1980).

75. S. E. Drewes and A. T. Pitchford, *J. Chem. Soc., Perkin I*, **1981**, 408.

76. J. S. Bradshaw, R. E. Asay, G. E. Maas, R. M. Izatt, and J. J. Christensen, *J. Heterocycl. Chem.*, **15**, 825 (1978).

77. K. Frensch, G. Oepen, and F. Vögtle, *Ann. Chem.*, **1979**, 858.

78. J. S. Bradshaw and S. T. Jolley, *J. Heterocycl. Chem.*, **16**, 1157 (1979).

79. S. T. Jolley and J. S. Bradshaw, "Second Chemical Congress of the North American Continent," San Francisco, California, Aug. 25–29, 1980, No. 8.

80. S. T. Jolley and J. S. Bradshaw, *J. Org. Chem.*, **45**, 3554 (1980).

81. K. Frensch and F. Vögtle, *Tetrahedron Lett.*, **1977**, 2573.

82. S. L. Baxter and J. S. Bradshaw, *J. Heterocycl. Chem.*, **18**, 233 (1981).

83. J. S. Bradshaw, G. E. Maas, R. M. Izatt, and J. J. Christensen, *Chem. Rev.*, **79**, 37 (1979).

84. F. R. Fronczek, unpublished results, 1981.

85. G. R. Newkome, F. R. Fronczek, and D. K. Kohli, *Acta Cryst.*, **B37**, 2114 (1981).

86. G. R. Newkome, H. C. R. Taylor, F. R. Fronczek, T. J. Delord, D. K. Kohli, and F. Vögtle, *J. Am. Chem. Soc.*, **103**, 7376 (1981).

87. J. S. Bradshaw, R. E. Asay, S. L. Baxter, P. E. Fore, S. T. Jolley, J. D. Lamb, G. E. Maas, M. D. Thompson, R. M. Izatt, and J. J. Christensen, *Ind. Eng. Chem. Prod. Res. Dev.*, **19**, 86 (1980).

87a. J. S. Bradshaw, S. L. Baxter, J. D. Lamb, R. M. Izatt, and J. J. Christensen, *J. Am. Chem. Soc.*, **103**, 1821 (1981).

88. J. D. Lamb, R. M. Izatt, C. S. Swain, J. S. Bradshaw, and J. J. Christensen, *J. Am. Chem. Soc.*, **102**, 479 (1980).

89. R. M. Izatt, J. D. Lamb, R. E. Asay, G. E. Maas, J. S. Bradshaw, J. J. Christensen, and S. S. Moore, *J. Am. Chem. Soc.*, **99**, 6134 (1977).

90. G. R. Newkome, G. L. McClure, J. Broussard-Simpson, and F. Danesh-Khoshboo, *J. Am. Chem. Soc.*, **97**, 3232 (1975).

91. G. R. Newkome, A. Nayak, G. L. McClure, F. Danesh-Khoshboo, and J. Broussard-Simpson, *J. Org. Chem.*, **42**, 1500 (1977).

92. G. R. Newkome, T. Kawato, and A. Nayak, *J. Org. Chem.*, **44**, 2697 (1979).

93. G. R. Newkome and T. Kawato, *J. Org. Chem.*, **44**, 2693 (1979).

94. G. R. Newkome and T. Kawato, *Tetrahedron Lett.*, **1978**, 4643.

95. R. D. Gandour, D. A. Walker, A. Nayak, and G. R. Newkome, *J. Am. Chem. Soc.*, **100**, 3608 (1978).

96. J. C. Hogan and R. D. Gandour, *J. Am. Chem. Soc.*, **102**, 2865 (1980).

97. J. -M. Girodeau, J. -M. Lehn, and J. -P. Sauvage, *Angew. Chem. Internat. Ed.*, **14**, 764 (1975); *Angew. Chem.*, **87**, 813 (1975).

98. J. P. Behr, J. M. Girodeau, R. C. Hayward, J. -M. Lehn, and J. P. Sauvage, *Helv. Chim. Acta*, **63**, 2096 (1980).

99. D. A. Laidler and J. F. Stoddart, *J. Chem. Soc., Chem. Commun.*, **1976**, 979.

100. G. R. Newkome, D. K. Kohli, and F. Fronczek, *J. Chem. Soc., Chem. Commun.*, **1980**, 9.

100a. G. R. Newkome, W. E. Puckett, G. E. Kiefer, V. K. Gupta, Y. J. Xia, M. Coreil, and M. A. Hackney, *J. Org. Chem.*, **47**, 4116 (1982).

101. G. R. Newkome, D. K. Kohli, F. Fronczek, G. Chiari, B. Hales, and E. E. Case, *J. Am.*

Chem. Soc., **102,** 7608 (1980).

102. G. R. Newkome, A. Nayak, F. Fronczek, T. Kawato, H. C. R. Taylor, L. Meade, and W. Mattice, *J. Am. Chem. Soc.,* **101,** 4472 (1979).

103. G. R. Newkome and D. C. Hager, *J. Am. Chem. Soc.,* **100,** 5567 (1978).

104. G. R. Newkome and H. C. R. Taylor, *J. Org. Chem.,* **44,** 1362 (1979).

105. G. R. Newkome, A. Nayak, J. D. Sauer, P. K. Mattschei, S. F. Watkins, F. Fronczek, and W. H. Benton, *J. Org. Chem.,* **44,** 3816 (1979).

106. F. R. Fronczek, S. F. Watkins, and G. R. Newkome, *J. Chem. Soc., Perkin II,* **1981,** 877.

107. G. R. Newkome, H. C. R. Taylor, F. R. Fronczek, and T. J. Delord, *J. Org. Chem.,* **49,** in press (1984).

108. G. R. Newkome and D. C. Hager, unpublished results.

109. J. D. Curry and D. H. Busch, *J. Am. Chem. Soc.,* **86,** 592 (1964).

110. D. H. Busch, *Rec. Chem. Prog.,* **25,** 107 (1964).

111. D. St. C. Black and E. Markham, *Rev. Pure Appl. Chem.,* **15,** 109 (1965).

112. D. H. Busch, *Helv. Chim. Acta, Fasciculus extraordinarius Alfred Werner,* **1967,** 174.

113. L. F. Lindoy and D. H. Busch, *Prep. Inorg. React.,* **6,** 1 (1971).

114. D. St. C. Black and A. J. Hartshorn, *Coord. Chem. Rev.,* **9,** 219 (1972–1973).

115. N. W. Alcock, D. C. Liles, M. McPartlin, and P. A. Tasker, *J. Chem. Soc., Chem. Commun.,* **1974,** 727.

116. R. H. Prince, D. A. Stotter, and P. R. Woolley, *Inorg. Chim. Acta,* **9,** 51 (1974).

117. E. -i. Ochiai, K. M. Long, C. R. Sperati, and D. H. Busch, *J. Am. Chem. Soc.,* **91,** 3201 (1969).

118. K. M. Long and D. H. Busch, *Inorg. Chem.,* **9,** 505 (1970).

119. R. L. Rich and G. L. Stucky, *Inorg. Nucl. Chem. Lett.,* **1,** 61 (1965).

120. J. L. Karn and D. H. Busch, *Nature,* **211,** 160 (1966).

121. E. K. Barefield, F. V. Lovecchio, N. E. Tokel, E. Ochiai, and D. H. Busch, *Inorg. Chem.,* **11,** 283 (1972).

122. J. L. Karn and D. H. Busch, *Inorg. Chem.,* **8,** 1149 (1969).

123. V. L. Goedken, Y-ae. Park, S. -M. Peng, and J. M. Norris, *J. Am. Chem. Soc.,* **96,** 7693 (1974).

124. P. Bamfield and P. A. Mack, *J. Chem. Soc. (C),* **1968,** 1961.

125. D. P. Riley, P. H. Merrell, J. A. Stone, and D. H. Busch, *Inorg. Chem.,* **14,** 490 (1975).

126. K. M. Long and D. H. Busch, *J. Coord. Chem.,* **4,** 113 (1974).

127. L. F. Lindoy, N. E. Tokel, L. B. Anderson, and D. H. Busch, *J. Coord. Chem.,* **1,** 7 (1971).

128. T. J. Lotz and T. A. Kaden, *J. Chem. Soc., Chem. Commun.,* **1977,** 15.

129. C. -K. Poon, W. -K. Wan, and S. S. T. Liao, *J. Chem. Soc., Dalton,* **1977,** 1247.

130. J. C. Dabrowiak, L. A. Nafie, P. S. Bryan, and A. T. Torkelson, *Inorg. Chem.,* **16,** 540 (1977).

131. T. J. Lotz and T. A. Kaden, *Helv. Chim. Acta,* **61,** 1376 (1978).

132. D. St. C. Black and N. E. Rothnie, *Tetrahedron Lett.,* **1978,** 2835.

133. C. -K. Poon and C. -M. Che, *J. Chem. Soc., Chem. Commun.,* **1979,** 861.

134. S. M. Nelson and D. H. Busch, *Inorg. Chem.,* **8,** 1859 (1969).

135. S. M. Nelson, P. Bryan, and D. H. Busch, *Chem. Commun.,* **1966,** 641.

136. M. G. B. Drew, A. H. b. Othman, S. G. McFall, and S. M. Nelson, *J. Chem. Soc., Chem. Commun.,* **1975,** 818.

137. M. C. Rakowski, M. Rycheck, and D. H. Busch, *Inorg. Chem.,* **14,** 1194 (1975).

138. M. G. B. Drew, A. H. b. Othman, P. D. A. McIlroy, and S. M. Nelson, *J. Chem. Soc., Dalton,* **1975,** 2507.

139. F. A. Deeney and S. M. Nelson, *J. Phys. Chem. Solids,* **34,** 277 (1973).

140. M. G. B. Drew, J. Grimshaw, P. D. A. McIlroy, and S. M. Nelson, *J. Chem. Soc., Dalton,* **1976,** 1388.

141. M. G. B. Drew, A. H. b. Othman, S. G. McFall, P. D. A. McIlroy, and S. M. Nelson, *J. Chem. Soc., Dalton,* **1977,** 438.

142. M. G. B. Drew, A. H. b. Othman, and S. M. Nelson, *J. Chem. Soc., Dalton,* **1976,** 1394.

143. D. H. Cook, D. E. Fenton, M. G. B. Drew, A. Rodgers, M. McCann, and S. M. Nelson, *J. Chem. Soc., Dalton,* **1979,** 414.

144. D. H. Cook and D. E. Fenton, *J. Chem. Soc., Dalton,* **1979,** 266.

145. C. Cairns, S. G. McFall, S. M. Nelson, and M. G. B. Drew, *J. Chem. Soc., Dalton,* **1979,** 446.

146. M. G. B. Drew, S. G. McFall, and S. M. Nelson, *J. Chem. Soc., Dalton,* **1979,** 575.

147. M. D. Alexander, A. V. Heuvelen, and H. G. Hamilton, Jr., *Inorg. Nucl. Chem. Lett.,* **6,** 445 (1970).

148. R. W. Stotz and R. C. Stoufer, *Chem. Commun.,* **1970,** 1682.

149. E. -i. Ochiai and D. H. Busch, *Chem. Commun.,* **1968,** 905.

150. F. V. Lovecchio, E. S. Gore, and D. H. Busch, *J. Am. Chem. Soc.,* **96,** 3109 (1974).

151. E. -i. Ochiai and D. H. Busch, *Inorg. Chem.,* **8,** 1474 (1969).

152. E. -i. Ochiai and D. H. Busch, *Inorg. Chem.,* **8,** 1798 (1969).

153. A. C. Melnyk, N. K. Kildahl, A. R. Rendina, and D. H. Busch, *J. Am. Chem. Soc.,* **101,** 3232 (1979).

154. F. Vögtle, E. Weber, W. Wehner, R. Natscher, and J. Grutze, *Chem. Ztg.,* **98,** 562 (1974) [*Chem. Abstr.,* **82,** 72964h (1975)].

155. E. Weber and F. Vögtle, *Ann. Chem.,* **1976,** 891.

156. E. Buhleier, W. Wehner, and F. Vögtle, *Ann. Chem.,* **1978,** 537.

157. B. Stetter and J. Marx, *Ann. Chem.,* **607,** 59 (1957).

158. E. Buhleier, W. Wehner, and F. Vögtle, *Chem. Ber.,* **111,** 200 (1978).

159. S. Kher, S. K. Sahni, V. Kumari, and R. N. Kapoor, *Inorg. Chim. Acta,* **37,** 121 (1979).

160. V. F. Borodkin and R. D. Komarov, *Izv. Vyssh. Ucheb. Zaved., Khim. Khim. Tekhnol.,* **16,** 1304 (1973) [*Chem. Abstr.,* **79,** 137113q (1973)].

161. V. F. Borodkin and R. D. Komarov, *Izv. Vyssh. Ucheb. Zaved., Khim. Khim. Tekhnol.,* **16,** 1764 (1973) [*Chem. Abstr.,* **80,** 59924j (1974)].

162. V. F. Borodkin and R. D. Komarov, *USSR Pat.* 411,087; *Otkrytiya, Izobret., Prom. Obraztsy, Tovarnye Znaki.,* **51,** 77 (1974) [*Chem. Abstr.,* **80,** 108593m (1974)].

163. A. P. Snegireva and V. F. Borodkin, *Izv. Vyssh. Ucheb. Zaved., Khim. Khim. Tekhnol.,* **17,** 1364 (1974) [*Chem. Abstr.,* **82,** 97999m (1975)].

164. V. F. Borodkin, R. D. Komarov, and O. A. Aleksandrova, *Tr. Ivanov. Khim.-Teknol. Inst.,* **1972,** 141 [*Chem. Abstr.,* **79,** 115546f (1973)].

165. R. P. Smirnov, V. A. Gnedina, and V. F. Borodkin, *Tr. Vses. Mezhvuz. Nauch.-Tekh. Konf. Vop. Sin. Primen. Krasitelei,* **1970,** 17 [*Chem. Abstr.,* **76,** 14518f (1972)].

166. A. P. Snegireva and V. F. Borodkin, *Tr. Ivanov. Khim.-Teknol. Inst.,* **11,** 134 (1969) [*Chem. Abstr.,* **76,** 30246w (1972)].

167. V. F. Borodkin, V. A. Gnedina, and I. A. Grukova, *Izv. Vyssh. Ucheb. Zaved., Khim. Khim. Tekhnol.,* **16,** 1722 (1973) [*Chem. Abstr.,* **80,** 70791j (1974)].

168. P. Bamfield and P. A. Mack, unpublished data.

169. C. L. Honeybourne, *Tetrahedron,* **29,** 1549 (1973).

170. J. de O. Cabral, M. F. Cabral, W. J. Cummins, M. G. B. Drew, A. Rodgers, and S. M. Nelson, *Inorg. Chim. Acta,* **30,** L313 (1978).

171. W. Radecka-Paryzek, *Inorg. Chim. Acta,* **35,** L349 (1979).

172. M. G. B. Drew, J. de O. Cabral, M. F. Cabral, F. S. Esho, and S. M. Nelson, *J. Chem. Soc., Chem. Commun.,* **1979,** 1033.

173. W. Radecka-Paryzek, *Inorg. Chim. Acta,* **45,** L147 (1980).

174. M. G. B. Drew, S. G. McFall, S. M. Nelson, and C. P. Waters, *J. Chem. Research (S),* **1979,** 16; (*M*), **1979,** 360.

175. S. M. Nelson, F. S. Esho, M. G. B. Drew, and P. Bird, *J. Chem. Soc., Chem. Commun.,* **1979,** 1035.

176. M. M. Bishop, J. Lewis, T. D. O'Donoghue, and P. R. Raithby, *J. Chem. Soc., Chem. Commun.,* **1978,** 476.

177. Z. P. Hague, D. C. Liles, M. McPartlin, and P. A. Tasker, *Inorg. Chim. Acta,* **23,** L21 (1977).

178. J. Lewis and K. P. Wainwright, *J. Chem. Soc., Dalton,* **1978,** 440.

179. J. Lewis and T. O'Donoghue, *Inorg. Chim. Acta,* **30,** L339 (1978).

180. M. M. Bishop, J. Lewis, T. D. O'Donoghue, P. R. Raithby, and J. N. Ramsden, *J. Chem. Soc., Chem. Commun.,* **1978,** 828.

181. Z. P. Hague, M. McPartlin, and P. A. Tasker, *American Chemical Society Meeting, Hawaii,* **1979,** INOR 297.

182. J. Lewis, T. D. O'Donoghue, Z. P. Hague, and P. A. Tasker, *J. Chem. Soc., Dalton,* **1980,** 1664.

183. F. Vögtle, *Tetrehedron Lett.,* **1968,** 3623.

184. F. Vögtle and H. Risler, *Angew. Chem. Internat. Ed.,* **11,** 727 (1972).

185. F. Vögtle. *Tetrahedron,* **25,** 3231 (1969).

186. F. Vögtle and A. H. Effler, *Chem. Ber.,* **102,** 3071 (1969).

187. F. Vögtle and L. Schunder, *Chem. Ber.,* **102,** 2677 (1969).

188. F. Vögtle and P. Neumann, *Tetrahedron,* **26,** 5299 (1970).

189. E. Weber, W. Wieder, and F. Vögtle. *Chem. Ber.,* **109,** 1002 (1976).

190. K. Galuszko, *Rocz. Chem.,* **50,** 699 (1976).

191. K. Galuszko, *Rocz. Chem.,* **50,** 711 (1976).

192. F. Bottino and S. Pappalardo, *Chem. Lett.,* **12,** 1781 (1981).

193. F. Bottino and S. Pappalardo, *Org. Magn. Reson.,* **16,** 1 (1981).

194. P. S. Bryan and E. Doomes, *J. Coord. Chem.,* **6,** 97 (1976).

195. O. Tsuge and M. Okumura, *Heterocycles,* **6,** 5 (1977).

196. E. Buhleier and F. Vögtle, *Ann. Chem.,* **1977,** 1080.

197. G. R. Newkome, F. Danesh-Khoshboo, A. Nayak, and W. H. Benton, *J. Org. Chem.,* **43,** 2685 (1978).

198. G. R. Newkome, J. M. Roper, and J. M. Robinson, *J. Org. Chem.,* **45,** 4380 (1980).

199. G. R. Newkome and D. K. Kohli, *Heterocycles,* **15,** 739 (1981).

200. D. H. Cook, D. E. Fenton, M. G. B. Drew, S. G. McFall, and S. M. Nelson, *J. Chem. Soc., Dalton,* **1977,** 446.

201. D. E. Fenton, D. H. Cook, and I. W. Nowell, *J. Chem. Soc., Chem. Commun.,* **1977,** 274.

202. D. E. Fenton and R. Leonaldi, *Inorg. Chim. Acta,* **55,** L51 (1981).

203. D. H. Cook and D. E. Fenton, *Inorg. Chim. Acta,* **25,** L95 (1977).

204. D. E. Fenton, D. H. Cook, I. W. Nowell, and P. E. Walker, *J. Chem. Soc., Chem. Commun.,* **1978,** 279.

205. S. M. Nelson, M. McCann, C. Stevenson, and M. G. B. Drew, *J. Chem. Soc., Dalton,* **1979,** 1477.

206. M. G. Burnett, V. McKee, S. M. Nelson, and M. G. B. Drew, *J. Chem. Soc., Chem. Commun.*, **1980**, 829.

207. M. G. B. Drew, A. H. b. Othman, S. G. McFall, P. D. A. McIlroy, and S. M. Nelson, *J. Chem. Soc., Dalton*, **1977**, 1173.

208. M. G. B. Drew, A. Rodgers, M. McCann, and S. M. Nelson, *J. Chem. Soc., Chem. Commun.*, **1978**, 415.

209. W. Wehner and F. Vögtle, *Tetrahedron Lett.*, **1976**, 2603.

210. B. Tümmler, G. Maass, E. Weber, W. Wehner, and F. Vögtle, *J. Am. Chem. Soc.*, **99**, 4683 (1977).

211. W. Rasshofer, W. M. Muller, G. Oepen, and F. Vögtle, *J. Chem. Research (S)*, **1978**, 72; (M) **1978**, 1001.

212. G. R. Newkome, V. K. Majestic, F. R. Fronczek, and J. L. Atwood, *J. Am. Chem. Soc.*, **101**, 1047 (1979).

213. G. R. Newkome, V. K. Majestic, and F. R. Fronczek, *Tetrahedron Lett.*, **1981**, 3039.

214. F. R. Fronczek, V. K. Majestic, G. R. Newkome, W. H. Hunter, and J. L. Atwood, *J. Chem. Soc., Perkin II*, **1981**, 331.

215. G. R. Newkome, V. K. Majestic, and F. R. Fronczek, *Tetrahedron Lett.*, **1981**, 3055.

216. L. F. Lindoy, *Coord. Chem. Rev.*, **4**, 41 (1969).

217. L. F. Lindoy and D. H. Busch, *Inorg. Chem.*, **13**, 2494 (1974).

218. L. F. Lindoy, D. H. Busch, and V. Goedken, *J. Chem. Soc., Chem. Commun.*, **1972**, 683.

219. M. G. B. Drew, C. Cairns, A. Lavery, and S. M. Nelson, *J. Chem. Soc., Chem. Commun.*, **1980**, 1122.

220. N. A. Kolesnikov, V. F. Borodkin, and L. M. Fedorov, *Izv. Vyssh. Ucheb. Zaved., Khim. Khim. Tekhnol.*, **16**, 1084 (1973) [*Chem. Abstr.*, **79**, 105222h (1973)].

221. J. E. Parks, B. E. Wagner, and R. H. Holm, *J. Am. Chem. Soc.*, **92**, 3500 (1970); *Inorg. Chem.*, **10**, 2472 (1971).

222. J. Riker-Nappier and D. W. Meek, *J. Chem. Soc., Chem. Commun.*, **1974**, 442.

223. J. de O. Cabral, M. F. Cabral, M. G. B. Drew, S. M. Nelson, and A. Rodgers, *Inorg. Chim. Acta*, **25**, L77 (1977).

224. F. Vögtle and E. Weber, *Angew. Chem. Internat. Ed.*, **13**, 149 (1974).

225. H. Gerlach and E. Huber, *Helv. Chim. Acta*, **51**, 2027 (1968).

226. H. Reinshagen and A. Stutz, *Monatsh. Chem.*, **110**, 567 (1979).

227. H. Reinshagen, G. Schulz, and A. Stutz, *Monatsh. Chem.*, **110**, 577 (1979).

228. A. Stutz and H. Reinshagen, *Tetrahedron Lett.*, **1978**, 2821.

229. J. Brushin and W. Jenny, *Chimia*, **26**, 420 (1972); **25**, 238, 308 (1971).

230. C. Wong and W. W. Paudler, *J. Org. Chem.*, **39**, 2570 (1974).

231. J. L. Atwood, W. E. Hunter, C. Wong, and W. W. Paudler, *J. Heterocycl. Chem.*, **12**, 433 (1975).

232. F. Vögtle, *Angew. Chem. Internat. Ed.*, **8**, 274 (1969).

233. J. Brushin and W. Jenny, *Tetrahedron Lett.*, **1973**, 1215.

234. M. Iwata, H. Kuzuhara, and S. Emoto, *Chem. Lett.*, **1976**, 983.

234a. M. Iwata and H. Kuzuhara, *Chem. Lett.*, **1981**, 1749.

234b. Y. Tachibana, M. Ando, and H. Kuzuhara, *Chem. Lett.*, **1982**, 1765.

234c. Y. Tachibana, M. Ando, and H. Kuzuhara, *Chem. Lett.*, **1982**, 1769.

235. H. Kuzuhara, T. Komatsu, and S. Emoto, *Jpn. Kokai Tokkyo Koho*, **79**, 119, 485 [*Chem. Abstr.*, **92**, 128979t (1980)].

236. H. Kuzuhara, M. Iwata, and S. Emoto, *J. Am. Chem. Soc.*, **99**, 4173 (1977).

237. T. Komatsu, M. Ando. F. Sugawara, and H. Kuzuhara, *Koen Yoshishu-Tennen Yuki Kagokutsu Toronkai*, **22**, 627 (1979).

238. M. Iwata and H. Kuzuhara, *Chem. Lett.*, **1981**, 5.

239. H. Kuzuhara, T. Komatsu, and S. Emoto, *Tetrahedron Lett.*, **1978**, 3563.

240. H. Kuzuhara, T. Komatsu, and S. Emoto, *Jpn. Kokai Tokkyo Koko.*, 79 119,429 (1979) [*Chem. Abstr.*, **92**, 76571g (1980)].

241. T. Sakurai, H. Kuzuhara, and S. Emoto, *Acta Cryst.*, **B35**, 2984 (1979).

241a. G. R. Newkome, T. Kawato, F. R. Fronczek, and W. H. Benton, *J. Org. Chem.*, **45**, 5423 (1980).

242. A. Marchesini, S. Bradamante, R. Fusco, and G. Pagani, *Tetrahedron Lett.*, **1971**, 671.

243. S. Bradamante, G. Pagani, A. Marchesini, and U. M. Pagnoni, *Chim. Ind. (Milan)*, **55**, 962 (1973) [*Chem. Abstr;.* **80**, 95861v (1974)].

244. W. E. Parham. R. W. Davenport, and J. B. Biasotti, *Tetrahedron Lett.*, **1969**, 557.

245. W. E. Parham, R. W. Davenport, and J. B. Biasotti, *J. Org. Chem.*, **35**, 3775 (1970).

246. W. E. Parham, D. C. Egberg, and S. S. Salgar, *J. Org. Chem.*, **37**, 3248 (1972).

247. W. E. Parham, K. B. Sloan, and J. B. Biasotti, *Tetrahedron*, **27**, 5767 (1971).

248. W. L. Mosby, *Chem. Ind.*, **1959**, 1348.

249. P. N. Rylander, "Catalytic Hydrogenation over Platinum Metals," Academic Press, New York, 1967, Chapter 24.

250. D. Moran, M. N. Patel, N. A. Tahir, and B. J. Wakefield, *J. Chem. Soc., Perkin I*, **1974**, 2310.

251. V. Prelog and U. Geyer, *Helv. Chim. Acta*, **28**, 1677 (1945).

252. V. Prelog and S. Szpilfogel, *Helv. Chim. Acta*, **28**, 1684 (1945).

253. H. Junek, O. S. Wolfbeis, H. Sprintschnik, and H. Wolny, *Monatsh. Chem.*, **108**, 689 (1977).

254. E. Breitmaier and E. Bayer. *Tetrahedron Lett.*, **1970**, 3291.

255. J. Rebek, Jr. and J. E. Trend, *J. Am. Chem. Soc.*, **100**, 4315 (1978).

256. B. J. van Keulen, R. M. Kellogg, and O. Piepers, *J. Chem. Soc., Chem. Commun.*, **1979**, 285.

257. G. R. Newkome, T. Kawato, and W. H. Benton, *J. Org. Chem.*, **45**, 626 (1980).

258. J. Rebek, Jr., J. E. Trend, R. V. Wattley, and S. Chakravorti, *J. Am. Chem. Soc.*, **101**, 4333 (1979).

259. J. Rebek, Jr. and R. V. Wattley, *J. Heterocycl. Chem.*, **17**, 749 (1980).

260. J. Rebek, Jr., T. Costello, and R. V. Wattley, *Tetrahedron Lett.*, **1980**, 2379.

261. J. Rebek, Jr. and R. V. Wattley, *J. Am. Chem. Soc.*, **102**, 4853 (1980).

262. V. M. Dziomko, B. K. Berestevich, A. V. Kessenikh, Y. S. Ryabokobylko, and R. S. Kuzanyan, *Khim. Geterotsik. Soed.*, **1979**, 701.

263. V. M. Dziomko and U. Tomsons, *Tr. Vses. Nauchno Issled. Inst. Khim. Reakt. Osobo Chist. Khim. Veshchestv.*, **38**, 203 (1976).

264. V. M. Dziomko, U. A. Tomsons, R. S. Kuzanyan, and Yu. S. Ryabokobylko, *Zh. Vses. Khim. O-va.*, **22**, 115 (1977) [*Chem. Abstr.*, **86**, 189898p (1977)].

265. C. Reichhardt and W. Scheibelein, *Z. Natürforsch., Sect. B.*, **33**, 1012 (1978).

266. R. Müller and D. Wöhrle, *Makromol. Chem.*, **176**, 2775 (1975).

267. R. Müller and D. Wöhrle, *Makromol. Chem.*, **179**, 2162 (1978).

268. A. Pezeshk, *182nd ACS National Metting* (Inorg. Chem.), New York, New York Aug. 23–28, (1981), Nos. 153 and 236.

269. A. T. Balaban, *Tetrahedron Lett.*, **1968**, 4643.

628   G. R. Newkome, V. K. Gupta, and J. D. Sauer

270. A. T. Balaban, *Rev. Roumaine, Chim.*, **18**, 1609 (1973).

271. A. T. Balaban and I. I. Badilescu, *Rev. Roumaine, Chim.*, **21**, 1339 (1976).

272. A. T. Balaban, *Tetrahedron Lett.*, **1978**, 5055.

273. V. Boekelheide and W. Pepperdine, *J. Am. Chem. Soc.*, **92**, 3684 (1970).

274. K. Deuchert and S. Hünig, *Stud. Org. Chem. (Amsterdam)*, **3**, 202 (1979).

275. W. Jenny and H. Holzrichter, *Chimia*, **21**, 509 (1967).

276. W. Jenny and H. Holzrichter, *Chimia*, **22**, 139 (1968).

277. P. J. Beeby and F. Sondheimer, *J. Am. Chem. Soc.*, **94**, 2128 (1972).

278. P. J. Beeby and F. Sondheimer, *Angew. Chem. Internat. Ed.*, **11**, 833 (1972).

279. P. J. Beeby, J. M. Brown, P. J. Garratt, and F. Sondheimer, *Tetrahedron Lett.*, **1974**, 599.

280. J. M. Brown and F. Sondheimer, *Angew. Chem. Internat. Ed.*, **13**, 337 (1974).

281. P. J. Beeby and F. Sondheimer, *Angew. Chem. Internat. Ed.*, **12**, 410, 411 (1974).

282. J. M. Brown and F. Sondheimer, *Angew. Chem. Internat. Ed.*, **13**, 339 (1974).

283. T. J. van Bergen and R. M. Kellogg, *J. Chem. Soc., Chem. Commun.*, **1976**, 964.

284. O. Piepers and R. M. Kellogg, *J. Chem. Soc., Chem. Commun.*, **1978**, 383.

285. R. H. v.d. Veen, R. M. Kellogg, A. Vos, and T. J. v. Bergen, *J. Chem. Soc., Chem. Commun.*, **1978**, 923.

286. T. J. v. Bergen and R. M. Kellogg, *J. Am. Chem. Soc.*, **99**, 3882 (1977).

287. D. M. Hedstrand, W. H. Kruizinga, and R. M. Kellog, *Tetrahedron Lett.*, **1978**, 1255.

288. J. G. deVries and R. M. Kellogg, *J. Am. Chem. Soc.*, **101**, 2759 (1979).

289. P. Jouin, C. B. Troostwijk, and R. M. Kellogg, *J. Am. Chem. Soc.*, **103**, 2091 (1981).

290. J. Bruhin, W. Kneubuhler, and W. Jenny, *Chimia*, **27**, 277 (1973).

291. H. Stetter, L. Marx-Moll, and H. Rutzen, *Chem. Ber.*, **91**, 1775 (1958).

292. L. E. Overman, *J. Org. Chem.*, **37**, 4214 (1972).

293. D. C. Dittmer and B. B. Blidner, *J. Org. Chem.*, **38**, 2873 (1973).

294. F. Freeman and T. I. Ito, *J. Org. Chem.*, **34**, 3670 (1969).

295. V. Prelog and O. Metzler, *Helv. Chim. Acta*, **29**, 1170 (1946).

296. Y. Ohnishi, H. Minato, K. Okuma, and M. Kobayashi, *Chem. Lett.*, **1977**, 525.

297. D. C. Dittmer and J. M. Kolyer, *J. Org. Chem.*, **28**, 2288 (1963).

298. W. H. Guendel and W. Kramer, *Chem. Ber.*, **111**, 2594 (1978).

299. H. C. Kang and V. Boekelheide, *Angew. Chem. Internat. Ed.*, **20**, 571 (1981).

299a. Y. Zhong-zhi, E. Heilbronner, H. C. Kang, and V. Boekelheide, *Helv. Chim. Acta*, **64**, 2029 (1981).

300. A. H. Hanson, *Cryst. Struct. Comm.*, **10**, 313 (1981).

301. A. H. Hanson, *Cryst. Struct. Comm.*, **10**, 751 (1981).

302. H. Forister and F. Vögtle, *Angew. Chem. Internat. Ed.*, **16**, 429 (1977).

303. E. Doomes, 167th ACS Meeting, Los Angeles, April 1974, ORGN 124.

304. D. T. Hefelfinger and D. J. Cram, *J. Am. Chem. Soc.*, **93**, 4767 (1971).

305. L. H. Weaver and B. W. Matthews, *J. Am. Chem. Soc.*, **96**, 1581 (1974).

306. E. Fujita, K. Bessho, K. Fuji, and A. Sumi, *Chem. Pharm. Bull. Japan*, **18**, 2216 (1970).

307. E. Fujita and K. Fuji, *J. Chem. Soc.* **(C), 1971**, 1651.

308. F. Bernardi, F. P. Colonna, P. Dembech, G. Distefano, and P. Vivarelli, *Chem. Phys. Lett.*, **36**, 539 (1975).

309. N. B. Pahor, M. Calligaris, and L. Randaccio, *J. Chem. Soc., Perkin II*, **1978**, 38.

310. I. Gault, B. J. Price, and I. O. Sutherland, *Chem. Commun.*, **1967**, 540.

311. D. Taylor, *Aust, J. Chem.*, **31**, 1953 (1978).

312. A. W. Hanson, *Acta Cryst.*, **B33**, 2657 (1977).

313. S. Ogawa and S. Shiraishi, *American Chemical Society Meeting, Hawaii*, 1979, ORGN 108.

314. S. Ogawa and S. Shiraishi, *2nd IUPAC Symposium on Organic Synthesis, Jerusalem-Haifa, Israel*, Abstr. 79 (1978).

315. J. S. Bradshaw, G. E. Maas, R. M. Izatt, J. D. Lamb, and J. J. Christensen, *Tetrahedron Lett.*, **1979**, 635.

316. J. S. Bradshaw, G. E. Maas, M. D. Thompson, R. E. Asay, P. E. Fore, and J. D. Lamb, *First Symposium on Macrocyclic Compounds, Provo, Utah*, Aug. 1977, Abstr. 13.

317. J. D. Lamb, R. M. Izatt, P. A. Robertson, and J. J. Christensen, *J. Am. Chem. Soc.*, **102**, 2452 (1980).

318. J. D. Lamb, J. J. Christensen, J. L. Oscarson, B. L. Nielsen, B. W. Asay, and R. M. Izatt, *J. Am. Chem. Soc.*, **102**, 6820 (1980).

319. N. K. Dalley and S. B. Larson, *First Symposium on Macrocyclic Compounds, Provo, Utah*, Aug. 1977, Abstr. 25.

320. B. A. Jones, J. S. Bradshaw, and R. M. Izatt, *J. Heterocycl. Chem.*, **19**, 551 (1982).

321. J. S. Bradshaw, S. T. Jolley, and R. M. Izatt, *J. Org. Chem.*, **47**, 1229 (1982).

322. T. W. Bell, P. G. Cheng, M. Newcomb, and D. J. Cram, *J. Am. Chem. Soc.*, **104**, 5185 (1982).

323. G. R. Newkome and D. K. Kohli, unpublished results.

324. G. R. Newkome and C. R. Marston, *Tetrahedron* **39**, 2001 (1983).

325. G. R. Newkome and T. Kawato, *J. Am. Chem. Soc.*, **101**, 7088 (1979).

326. E. Weber, *Angew. Chem. Internat. Ed.*, **18**, 219 (1979).

327. G. R. Newkome and D. K. Kohli, *J. Am. Chem. Soc.*, **1983**, submitted.

328. M. G. B. Drew, A. H. b. Othman, W. E. Hill, P. McIlroy, and S. M. Nelson, *Inorg. Chim. Acta*, **12**, L25 (1975).

329. E. Fleischer and S. Hawkinson *J. Am. Chem. Soc.*, **89**, 720 (1967).

330. W. M. Reiff, G. J. Long, and W. A. Baker, Jr., *J. Am. Chem. Soc.*, **90**, 6347 (1968).

331. R. K. Boggess and W. D. Wiegele, *J. Chem. Ed.*, **55**, 156 (1978).

332. M. G. B. Drew, A. H. b. Othman, P. McIlroy, and S. M. Nelson, *Acta Cryst.*, **B32**, 1029 (1976).

333. M. G. B. Drew, A. H. b. Othman, S. G. McFall, and S. M. Nelson, *J. Chem. Soc., Chem. Commun.*, **1977**, 558.

334. M. G. B. Drew and S. Hollis, *Inorg. Chim. Acta*, **29**, L231 (1978).

335. K. Farmery and D. H. Busch, *Chem. Commun.*, **1970**, 1091.

336. E. B. Fleischer and S. W. Hawkinson, *Inorg. Chem.*, **7**, 2312 (1968).

337. R. Dewar and E. Fleischer, *Nature*, **222**, 372 (1969).

338. L. Y. Martin, C. R. Sperati, and D. H. Busch, *J. Am. Chem. Soc.*, **99**, 2968 (1977).

339. L. Fabbrizzi, M. Micheloni, and P. Pauletti, *J. Chem. Soc., Chem. Commun.*, **1978**, 833.

340. L. Fabbrizzi, M. Micheloni, and P. Paoletti, *J. Chem. Soc., Dalton*, **1979**, 1581.

341. H. Keypour and D. A. Stotter, *Inorg. Chim. Acta*, **33**, L149 (1979).

342. (a) Th. A. Kaden, *Chimia*, **30**, 207 (1976); (b) T. Lotz Dissertation, Basel, **1976**.

343. L. Sabatini and L. Fabbrizzi, *Inorg. Chem.*, **18**, 438 (1979).

344. B. Fisher and R. Eisenberg, *J. Am. Chem. Soc.*, **102**, 7361 (1980).

345. P. Morliere and L. K. Patterson, *Inorg. Chem.*, **20**, 1458 (1981).

346. R. H. Prince and D. A. Stotter, *Nature*, **249**, 286 (1974).

630　　　　　G. R. Newkome, V. K. Gupta, and J. D. Sauer

347. L. Fabbrizzi, M. Micheloni, and P. Paoletti, *J. Chem. Soc., Dalton,* **1980**, 134.

348. P. H. Merrell, V. L. Goedken, D. H. Busch, and J. A. Stone, *J. Am. Chem. Soc.,* **92**, 7590 (1970).

349. T. J. Lotz and Th. A. Kaden, *First Symposium on Macrocyclic Compounds, Provo, Utah,* Aug. 1977, Abstr. 14.

350. K. Mochizuki, T. Iko, and M. Fujimoto, *Bull. Chem. Soc. Japan,* **53**, 543 (1980).

351. H. Keypour and D. A. Stotter, *Inorg. Chim. Acta,* **19**, L48 (1976).

352. S. M. Nelson, S. G. McFall, M. G. B. Drew, A. H. b. Othman, and N. B. Mason, *J. Chem. Soc., Chem. Commun.,* **1977**, 167.

353. S. M. Nelson, S. G. McFall, M. G. B. Drew, and A. H. b. Othman, *J. Chem. Soc., Chem. Commun.,* **1977**, 370.

354. I. Murase, K. Hamada, and S. Kida, *Inorg. Chim. Acta,* **54**, L171 (1981).

355. V. B. Rana and M. P. Teotia, *Ind. J. Chem.,* **19A**, 267 (1980).

356. J. D. J. Backer-Dirks, C. J. Gray, F. A. Hart, M. B. Hursthouse, and B. C. Schoop, *J. Chem. Soc., Chem. Commun.,* **1979**, 744.

357. G. R. Newkome, V. K. Majestic, and F. R. Fronczek, *Inorg. Chim. Acta,* **77**, L47 (1983).

358. F. Vögtle, J. Grutze, R. Natscher, W. Wieder, E. Weber, and R. Grün, *Chem. Ber.,* **108**, 1694 (1975).

358a. W. Radecka-Paryzek, *Inorg. Chim. Acta,* **54**, L251 (1981).

359. F. Bottino, R. Fradullo, and S. Pappalardo, *Org. Mass Spectrom.,* **16**, 190 (1981).

360. E. Buhleier, K. Frensch, F. Luppertz, and F. Vögtle, *Ann. Chem.,* **1978**, 1586.

361. J. Grandjean, P. Laszlo, W. Offermann, and P. L. Rinaldi, *J. Am. Chem. Soc.,* **103**, 1380 (1981).

362. E. Buhleier, W. Wehner, and F. Vögtle, *Chem. Ber.,* **112**, 559 (1979).

363. E. Buhleier, W. Wehner, and F. Vögtle, *Chem. Ber.,* **112**, 546 (1979).

364. M. G. B. Drew, M. McCann, and S. M. Nelson, *J. Chem. Soc., Chem. Commun.,* **1979**, 481.

365. M. R. Churchill and A. H. Reis, Jr., *Chem. Commun.,* **1971**, 1307.

366. M. R. Churchill and A. H. Reis, Jr., *Chem. Commun.,* **1970**, 879.

367. H. Rapoport and H. D. Baldridge, Jr., *J. Am. Chem. Soc.,* **73**, 343 (1951).

368. H. Rapoport and H. D. Baldridge, Jr., *J. Am. Chem. Soc.,* **74**, 5365 (1952).

369. T. M. Smalberger, G. J. H. Rall, H. L. DeWaal, and R. R. Arndt, *Tetrahedron,* **24**, 6417 (1968).

370. M. Spiteller-Friedmann and G. Spiteller, *Monatsh. Chem.,* **95**, 1234 (1964).

371. J. L. Coke and W. Y. Rice, Jr., *J. Org. Chem.,* **30**, 3420 (1965).

372. G. J. H. Rall, T. M. Smalberger, and H. L. de Waal, *Tetrahedron Lett.,* **1967**, 3465.

373. D. A. Kochkin and I. B. Chekmareva, *Zhur. Obshechei Khim.,* **31**, 3010 (1961).

374. G. A. Pagani, *J. Chem. Soc., Perkin I,* **1974**, 2050.

375. B. W. Bycroft and M. S. Gowland, *J. Chem. Soc., Chem. Commun.,* **1978**, 256.

376. I. Nachev, *Symp. Pap. – IUPAC Int. Symp. Chem. Nat. Prod., 11th,* **1**, 111 (1978) [*Chem. Abstr.,* **92**, 76910s (1980)].

377. I. Nachev, *Symp. Pap. – IUPAC Int. Symp. Chem. Nat. Prod., 11th,* **1**, 65 (1978) [*Chem. Abstr.,* **92**, 42359d (1980)].

378. I. Nachev, *Symp. Pap. – IUPAC Int. Symp. Chem. Nat. Prod., 11th,* **1**, 117 (1978) [*Chem. Abstr.,* **92**, 76911t (1980)].

379. M. Pailer and K. Pfleger, *Monatsh. Chem.,* **107**, 965 (1976).

380. K. Yamada, Y. Shizuri, and Y. Hirata, *Tetrahedron,* **34**, 1915 (1978).

381. L. Crombie, W. M. L. Crombie, D. A. Whiting, and K. Szendrei, *J. Chem. Soc., Perkin I*, **1979**, 2976.

382. R. L. Baxter, W. M. L. Crombie, L. Crombie, D. J. Simmonds, D. A. Whiting, and K. Szendrie, *J. Chem. Soc., Perkin I*, **1979**, 2982.

383. Deutsche Goldund Silber-Scheideanstalt vorm. Roessler, *Belg. Pat.* 858,391 (1978) [*Chem. Abstr.*, **89**, 163427 (1978)].

384. D. J. Harris, G. Y.-P. Kan, T. Tschamber, and V. Snieckus, *Can. J. Chem.*, **58**, 494 (1980).

385. P. J. Beeby and F. Sondheimer, *Angew. Chem. Internat. Ed.*, **12**, 411 (1973).

386. W. Jenny and H. Holzrichter, *Chimia*, **22**, 247 (1968).

387. F. Rob, H. J. van Ramesdonk, J. W. Verhoeven, U. K. Pandit, and Th. J. de Boer, *Tetrahedron Lett.*, **21**, 1549 (1980).

388. A. R. Battersby and H. D. Hamilton, *J. Chem. Soc., Chem. Commun.*, **1980**, 117.

389. M. J. Gunter and L. M. Mander, *J. Org. Chem.*, **46**, 4792 (1981); *J. Am. Chem. Soc.*, **103**, 6784 (1981).

390. J. Lewis and K. P. Wainwright, *J. Chem. Soc., Chem. Commun.*, **1974**, 169.

391. S. Lu, M.S. Thesis, Southern University, Baton Rouge, Louisiana, May **1981**.

392. F. Vögtle, *Chimia*, **33**, 239 (1979).

393. P. Morliere and L. K. Patterson, *Inorg. Chim. Acta*, **64**, L183 (1982).

394. Insitute of Physical and Chemical Research, *Jpn. Kokai Tokkyo Koho JP 81,122385* and *81,122386*, 25th Sep. (1981) [*Chem. Abstr.*, **96**, 122831z, 122832a (1982)].

395. T. Hiyama, M. Shinoda, H. Saimoto, and H. Nozaki, *Bull. Chem. Soc. Jpn.*, **54**, 2747 (1981).

396. H. Nozaki, Y. Yamamoto, K. Oshima, K. Utimoto, and T. Hiyama, *3rd Symposium on Organic Synthesis, Madison, WI, June 15–20*, 1980, p. 241.

397. S. M. Nelson, C. V. Knox, M. McCann, and M. G. B. Drew, *J. Chem. Soc., Dalton*, **1981**, 1669.

398. M. G. B. Drew, J. Nelson, and S. M. Nelson, *J. Chem. Soc., Dalton*, **1981**, 1678.

399. M. G. B. Drew, S. M. Nelson, and J. Reedjk, *Inorg. Chim. Acta*, **64**, L189 (1982).

400. R. M. Kellogg and O. Piepers, *J. Chem. Soc., Chem. Commun.*, **1982**, 402.

401. B. A. Jones, J. S. Bradshaw, F. G. Morin, and D. M. Grant, *183rd ACS Meeting, Las Vegas, Nevada, March 28–April 2*, **1982**, ORGN. 25.

402. R. B. Davidson, R. M. Izatt, J. D. Lamb, J. J. Christenson, *183rd ACS Meeting, Las Vegas. Nevada, March 28–April 2*, **1982**, ORGN. 26.

403. Institute of Physical and Chemical Research, *Jpn. Tokkyo Koho, JP 81, 48, 207*, 14th Nov. (1981) [*Chem. Abstr.*, **96**, 200156z (1982)].

404. M. Ando, Y. Tachibana, and H. Kuzuhara, *Bull. Chem. Soc. Jpn.*, **55**, 829 (1982).

405. S. Shinkai, T. Minami, T. Kouno. Y. Kusano, and O. Manabe, *Chem. Lett.*, **1982**, 499.

406. D. Wu, L. Liu, and K. Chen, *Chem. Abstr.*, **97**, 6595x (1982).

407. K. Utimoto, S. Kato, M. Tanaka, Y. Hoshino, S. Fujikura, and H. Nozaki, *Heterocycles*, **18(Spec. Issue)**, 149 (1982).

408. G. R. Newkome, S. Pappalardo, and F. R. Fronczek, *J. Am. Chem. Soc.*, **105**, 5152 (1983).

409. G. R. Newkome, S. Pappalardo, V. K. Gupta, and F. R. Fronczek, *J. Org. Chem.*, **48**, 4848 (1983).

410. G. R. Newkome, G. E. Kiefer, and W. E. Puckett, *J. Org. Chem.*, **48**, 5112 (1983).

411. P. Zanello, R. Seeber, and A. Cinquantini, *Inorg. Chim. Acta*, **65**, L43 (1982).

412. G. R. Newkome and H.-W. Lee, *J. Org. Chem.*, **47**, 2800 (1982).

632   G. R. Newkome, V. K. Gupta, and J. D. Sauer

413. J. Shoji, T. Kato, Y. Yoshimura, and K. Tori, *J. Antibiotics,* **34**, 1126 (1981).

414. J.-M. Lehn, "Frontiers of Chemistry, Plenary and Keynote Lectures Presented at the 28th IUPAC Congress, Vancouver, British Columbia, 16–22 August 1981," K. J. Laidler, ed., Pergamon Press, Oxford, U.K., 1982, p. 265.

415. W. Da-gang, L. Ling, and C. Kung-chang, *Acta Botanica Yunnanica,* **3**, 471 (1981).

416. W. Da-gang, S. Xi-chang, and L. Feng, *Acta Botanica Yunnanica,* **1**, 29 (1979).

417. Institute of Botany; Institute of Medical Sciences, *K'o Hsueh T'ung Pao,* **22(10),** 458, 436 (1977) [*Chem. Abstr.,* **88**, 17707y (1978)].

418. S. Sakane, Y. Matsumura, Y. Yamamura, Y. Ishida, K. Maruoka, and H. Yamamoto, *J. Am. Chem. Soc.,* **105**, 672 (1983).

419. S. T. Jolley, J. S. Bradshaw, and R. M. Izatt, *J. Heterocycl. Chem.,* **19**, 3 (1982).

420. H. Stetter, W. Frank, and R. Mertens, *Tetrahedron,* **37**, 767 (1981).

421. E. Weber, *J. Org. Chem.,* **47**, 3478 (1982).

422. G. Ferraudi, *Inorg. Chem.,* **19**, 438 (1980).

423. J. de O. Cabral, M. F. Cabral, M. G. B. Drew, F. S. Esho, O. Haas, and S. M. Nelson, *J. Chem. Soc., Chem. Commun.,* **1982**, 1066.

424. J. de O. Cabral, M. F. Cabral, M. G. B. Drew, F. S. Esho, and S. M. Nelson, *J. Chem. Soc., Chem. Commun.,* **1982**, 1068.

425. H. Fujioka, E. Kimura, and M. Kodama, *Chem. Lett.,* **1982**, 737.

426. Y. Kawasaki and N. Okuda, *Chem. Lett.,* **1982**, 1161.

427. P. Morliere and L. K. Patterson, *Inorg. Chem.,* **21**, 1833 (1982).

428. L. Fabbrizzi, A. Lari, A. Poggi, and B. Seghi, *Inorg. Chem.,* **21**, 2083 (1982).

429. M. G. B. Drew, M. McCann, and S. M. Nelson, *Inorg. Chim. Acta,* **41**, 213 (1980).

430. A. G. Tomlinson, *Coord. Chem. Rev.,* **37**, 221 (1981).

431. M. N. Hughes, *Coord. Chem. Rev.,* **37**, 297 (1981).

432. S. Ogawa, T. Yamaguchi, and N. Gotoh, *J. Chem. Soc., Chem. Commun.,* **1972**, 577.

433. J. S. Bradshaw, B. A. Jones, R. B. Davidson, J. J. Christensen, J. D. Lamb, R. M. Izatt, F. G. Morin, and D. M. Grant, *J. Org. Chem.,* **47**, 3362 (1982).

434. E. Kimura, M. Kodama, and T. Yatsunami, *J. Am. Chem. Soc.,* **104**, 3182 (1982).

435. M. A. McKervey and T. O'Connor, *J. Chem. Soc., Chem. Commun.,* **1982**, 655.

436. M. Hanack, K. Mitulla, G. Pawlowski, and L. R. Subramanian, *J. Organometl. Chem.,* **204**, 315 (1981).

437. C. W. G. Ansell, J. Lewis, P. R. Raithby, J. N. Ramsden, and M. Schröder, *J. Chem. Soc., Chem. Commun.,* **1982**, 546.

438. H. Higuchi and S. Misumi, *Tetrahedron Lett.,* **23**, 5571 (1982).

439. S. Ogawa, *J. Chem. Soc., Perkin I,* **1977**, 214.

440. S. Ogawa and S. Shiraishi, *J. Chem. Soc., Perkin I,* **1980**, 2527.

441. S. Sakane, K. Maruoka, and H. Yamamoto, *Tetrahedron Lett.,* **24**, 943 (1983).

442. R. M. Kellogg. *Proceedings of the 26th OHOLO Conference, Zichron Yaacov, Israel, 22–25 March 1981,* p. 22.

443. H. Sigel and R. B. Martin, *Chem. Rev.,* **82**, 385 (1982).

444. S. E. Drewes, N. D. Emslie, A. T. Pitchford, and B. W. Wallace, *S. Afr. J. Chem.,* **35**, 115 (1982) [*Chem. Abstr.,* **98**, 89709g (1983)].

445. R. H. Petty and L. J. Wilson, *J. Chem. Soc., Chem. Commun.,* **1978**, 483.

446. E. C. Constable and J. Lewis, *Polyhedron,* **1**, 303 (1982).

447. R. H. Petty, B. R. Welch, L. J. Wilson, L. A. Bottomley, and K. M. Kadish, *J. Am. Chem. Soc.,* **102**, 611 (1980).

448. C. J. Chandler, L. W. Deady, J. A. Reiss, and V. Tzimos, *J. Heterocycl. Chem.*, **19**, 1017 (1982).

449. M. Hunziker and U. Hauser, *Heterocycles,* **19**, 2131 (1982).

450. M. Iwata and H. Kuzuhara, *J. Org. Chem.*, **48**, 1282 (1983).

451. Y. Fukazawa, J. Tsuchiya, M. Sobukawa, and S. Ito, *Heterocycles,* **20**, 147 (1983).

452. M. G. B. Drew, J. Nelson, F. Esho, V. McKee, and S. M. Nelson, *J. Chem. Soc., Dalton Trans.,* **1982**, 1837.

453. A. N. Scoville and W. M. Reiff, *Inorg. Chim. Acta,* **70**, 127 (1983).

454. E. C. Constable, J. Lewis, M. C. Liptrot, P. R. Raithby, and M. Schröder, *Polyhedron,* **2**, 301 (1983).

455. W. U. Malik, R. Bembi, and R. Singh, *Polyhedron,* **2**, 369 (1983).

456. C. W. G. Ansell, J. Lewis, P. R. Raithby, and T. D. O'Donoghue, *J. Chem. Soc., Dalton Trans.,* **1983**, 177.

457. C. W. G. Ansell, J. Lewis, M. C. Liptrot, and P. R. Raithby, and M. Schröder, *J. Chem. Soc., Dalton Trans.,* **1982**, 1593.

458. C. W. G. Ansell, J. Lewis, and P. R. Raithby, *J. Chem. Soc., Dalton Trans.,* **1982**, 2557.

459. B. A. Jones, J. S. Bradshaw, P. R. Brown, J. J. Christensen, and R. M. Izatt, *J. Org. Chem.,* **48**, 2635 (1983).

460. G. R. Newkome and H.-W. Lee, *J. Am. Chem. Soc.,* **105**, 5956 (1983).

CHAPTER IV

The Reviews of
Pyridine Chemistry – 1968-1982

G. R. NEWKOME

Department of Chemistry, Louisiana State University,
Baton Rouge, Louisiana

I.	Introduction and General Discussion	635
II.	General.	636
III.	Reactions of Pyridines	637
IV.	Synthesis of Pyridines	640
V.	Natural Products and Biochemical Aspects	642
VI.	Physical Aspects	648
VII.	Spectroscopy	649
VIII.	Dihydropyridines	650
IX.	Tetrahydropyridines	651
X.	Piperidines.	651
XI.	Didehydropyridines (Pyridynes)	652
XII.	Fused-Ring Pyridines.	652
XIII.	Pyridinium Compounds	652
XIV.	Pyridine N-Oxides	654
XV.	Pyridine Complexes	654
XVI.	Pyridinyl Radicals.	655
XVII.	Commercial Aspects	656

I.   INTRODUCTION AND GENERAL DISCUSSION

This review compiles most of the reviews of pyridine and its derivatives over the period 1968 through 1982. The survey is not restricted to English but rather whether the pertinent information is available from *Chemical Abstracts*. Under the broad topical subject area, the entries are arranged in chronological order, oldest first to the most recent. Most entries are listed in the order: title, authors, journal, language, number of references, and, in selected instances, chemical abstract reference for the more obscure listings, and a brief overview.

II. GENERAL

"Heteroaromatic Compounds – Pyrroles and Pyridines." K. Schofield, Butterworth, London, 1976, [Eng] 758 refs.

"Compounds Containing Six-Membered Rings with One Nitrogen Atom; Pyridine and Its Derivatives," D. M. Smith, in *Rodd's Chemistry of Carbon Compounds*, S. Coffey and M. F. Ansell, eds., Elsevier Scientific Publishing, New York, 1976, pp. 27–226, [Eng] 21 refs. Chapter 24 deals with the syntheses, reactions, limited physical data of individual compounds, polypyridyls. dehydropyridines, halogenated pyridines, and bipiperidyls.

"The Specific Synthesis of Pyridines and Oligopyridines," F. Krohnke, *Synthesis*, **1976**, 1 [Eng] 93 refs. This review describes the general synthetic routes to bi-, ter-, and oligopyridines; numerous specific procedures are included. A brief consideration of their complexation properties is included.

"Bicyclic Compounds Containing a Pyridine Ring: Cyclopolymethylenepyridines, Cycloalkenopyridines," N. Campbell, in *Rodd's Chemistry of Carbon Compounds*, S. Coffey and M. F. Ansell, eds., Elsevier Scientific Publishing, New York, **1976**, pp. 231–356 [Eng] 71 refs. Chapter 25 deals with only a few examples of 2,3- and 3,4-cyclopolymethylene pyridines.

"Construction of Synthetic Macrocyclic Compounds Possessing Subheterocyclic Rings, Specifically Pyridine, Furan and Thiophene," G. R. Newkome, J. D. Sauer, J. M. Roper, and D. C. Hager, *Chem. Revs.*, **77**, 513 (1977) [Eng] 500 refs.

"Pyridine as a Nonaqueous Solvent," J. M. Nigretto and M. Jozefowicz, *Chem. Non-Aqueous Solvents*, **5A**, 179 (1978) [Eng] 234 refs. [*Chem. Abstr.*, **91**, 79520e (1979).]

"Pyridines," D. M. Smith, in *Comprehensive Organic Chemistry*, P. G. Sammes, ed., Pergamon, Oxford, England, Vol. 4, 3–84 (1979), [Eng] 469 refs.

"Some 3,5-Bifunctional Pyridines and Related Pyridinophanes," K. Deuchert and S. Huenig, *Stud. Org. Chem.* (Amsterdam), **3**, 202 (1979) [Eng] 38 refs.

"Six-Membered Heterocyclic Systems," R. K. Smalley, *Aromat. Heteroaromat. Chem.*, **7**, 145 (1979) [Eng] 296 refs.

"Pyridine and Phenylpyridines," A. P. Kudchadker and S. A. Kudchadker, American Petroleum Institute Publ. (No. 710), **1979**, 52 pp. [Eng] 190 refs. An excellent general review of pyridine and phenylpyridine physical properties.

"Structure of Matter. Aromatic Heterocycles," M. Bartucz and M. Kajtar, *Termeszet. Vilaga*, **110**, 514 (1979) [Hung] 0 refs. [*Chem. Abstr.*, **92**, 215297 (1980).] Pyridine, in part, is reviewed.

"Chemical Features of *Alpha, Alpha*-Substituted Azines (as Exemplified by 2,6-Disubstituted Pyridines)," L. M. Yakhontov and D. M. Krasnokutskaya, *Izv. Sib. Otd. Akad., Nauk SSSR, Ser. Khim. Nauk*, **1980**, 56 [Russ] 97 refs. [*Chem. Abstr.*, **93**, 046254 (1980).] Pyridine azines are reviewed.

"Ion-exchange Resins of the Vinylpyridine Series," N. B. Galitskaya, V. B. Kargman, N. E. Kozhevnikova, and I. M. Todres, *Khim. Prom-st., Ser.: Okhr. Okruzhayushchei Sredy Ratsion Ispol'z. Prir. Resur.*, **1980**, 43 pp. [Russ] 232 refs. [*Chem. Abstr.*, **94**, 192973e (1981).]

"Advances in the Chemistry of α,α'-Disubstituted Pyridines," L. N. Yakhontov and D. M. Krasnokutskaya, *Usp. Khim.*, **50**, 1072 (1981) [Russ] 254 refs. [Eng. Trans.] *Russ. Chem. Rev.*, **50**, 565 (1981).

"Anhydrobases of the Pyridine Series (Review)", T. V. Stupnikova, B. P. Zemski, R. S. Sagitullin, and A. N. Kost, *Khim. Geterotsikl. Soedin.*, **1982**, 291 [Russ] 143 refs. [*Chem. Abstr.*, **97**, 23541u (1982).]

"Pyridine", H. Fujihara, *Yuki Gosei Kagaku Kyokaishi*, **40**, 672 (1982) [Japan] no refs. [*Chem. Abstr.*, **98**, 71832g (1983).]

"Pyridinecarboxylic Acids," V. V. Antonova, A. M. Bespalova, G. V. Krylova, and V. K. Promonenkov, Deposited Dic. **1981**, SPSTL 369 Khp–D81, 51 pp. [Russ] 177 refs. [*Chem. Abstr.*, **98**, 71834j (1982).]

"Pyridinophanes, Pyridinocrowns, and Pyridinocryptands," V. K. Majestic and G. R. Newkome, *Topics in Current Chemistry*, **106**, 79 (1982) [Eng] 152 refs.

III. REACTIONS OF PYRIDINES

"Alkenylation of 2-Hydroxy- and 2-Aminopyridines and Their Halo Derivatives," E. I. Fedorov and B. I. Mikhant'ev, *Tr. Voronezh. Gos. Univ.*, **73**, 60 (1969) [Russ] 6 refs. [*Chem. Abstr.*, **75**, 110109y (1971).]

"4-(2-Pyridylazo)resorcinol (PAR) as an Analytical Reagent: A Review," M. N. Desai and M. H. Gandhi, *Rec. Chem. Prog.*, **30**, 223 (1969) [Eng] 60 refs.

"Electrophilic Substitution in the Series of Six-Membered Nitrogen-Containing Heteroaromatic Compounds and Their *N*-Oxides," Zh. I. Aksel'rod and V. M. Berezovskii, *Usp. Khim.*, **39**, 1337 (1970) [Russ] 199 refs. [Eng. Trans.] *Russ. Chem. Rev.*, **39**, 627 (1970). Considerable attention is given to problems associated with the comparison of the reactivities of heteroaromatic compounds with the results of theoretical calculations through 1968.

"Radical Substitution of Aromatics and Heteroaromatics. II. Alkylation Reactions," H. J. M. Dou, G. Vernin, and J. Metzger, *Bull. Soc. Chim. Fr.*, **1971**, 4593 [Fr] 164 refs.

"Nucleophilic Reactions of 2-Aminothiazole and Related Heterocyclic Amines." C. D. Hurd, *Int. J. Sulfur Chem., Part A.* **1**, 97 (1971) [Eng] 30 refs. The structures of di- and tripyridylamines are reviewed and corrected. Nucleophilic reactions and basicities of *N*-heterocyclic amines, for example, 2-aminopyridines, are considered.

"Heteroaromatic Reactivity," J. H. Ridd, in *Physical Methods in Heterocyclic Chemistry*, A. R. Katritzky, ed., Academic Press, New York, Vol. 4, 1971, pp. 55–120 [Eng] 227 refs. Part II-A briefly considers the various modes of pyridyl substitution.

"Vinyl and Diene Monomers," E. C. Leonard, ed., Pt. 3, p. 1376, Wiley-Interscience, New York, 1971 [Eng] 2030 refs. Vinylpyridines are discussed.

"Hydrolysis of Some Pyridinium Aldoximes," I. Christenson, *Acta Pharm. Suec.*, **9**, 351 (1972) [Eng] 59 refs. [*Chem. Abstr.*, **78**, 28651x (1973).] A review of the author's papers in this area.

"Homolytic Arylation of Heterocycles," G. Vernin, H. J. M. Dou, and J. Matzger,

Bull. Soc. Chim. Fr., **1972,** 1173 [Fr] 213 refs. Homolytic arylation of pyridines, as well as other heterocycles, including the source of aryl radicals, substituents, solvents, and photolytic effects are reviewed.

"Electrochemical Properties of Nicotinamide and its Derivatives in Aqueous Solution. I. Critical Analysis of Available Data Relative to the Electrochemical Properties of Pyridine Derivatives in Aqueous Solution," D. Thevenot and R. Buvet, *J. Electroanal. Chem. Interfacial Electrochem.,* **39,** 429 (1972) [Fr] [*Chem. Abstr.,* **77,** 146840k (1972).] Electrochemical redox reactions and the potentiometric behavior of pyridine mixtures are reviewed. The conflicting theoretical explanations are evaluated.

"Isotopic Hydrogen Labeling of Heterocyclic Compounds by One-Step Methods," G. E. Calf and J. L. Garnett, *Adv. Heterocycl. Chem.,* **15,** 137 (1973) [Eng] 151 refs. Isotopic hydrogen labeling of pyridine.

"Polarography of Heterocyclic Compounds. I. Miscellaneous Data of the Polarographic Behavior of Heterocycles and Electrochemical Reduction of Heteroaromatic Compounds," Ya. P. Stradyn', V. P. Kadysh, and S. A. Giller, *Khim. Geterotsikl. Soedin.,* **1973,** 1587 [Russ] 150 refs; [Eng. trans.] *Chem. Heterocycl. Compd.,* **9,** 1435 (1975).

"Vapor-phase Oxidative Ammonolysis of Alkylpyridines and Quinoline," V. I. Trubnikov and E. S. Zhdanovich, *Vitam. Vitam. Prep.,* **1973,** 172 [Russ] 18 refs. [*Chem. Abstr.,* **80,** 36923u (1974).]

"Some Aspects of Azido-tetrazolo Isomerization," M. Tisler, *Synthesis,* **1973,** 123 [Eng] 107 refs. Pyridine azides.

"Synthesis and Chemical Transformation of Some Arylaromatic Azoacetylene Alcohols of Cyclic and Heterocyclic Series," T. A. Yagudeev, R. K. Karazhigitova, U. I. Urazaliev, and A. N. Nurgalieva, *Tr. Inst. Khim. Nefti Prir. Solei, Akad. Nauk Kaz. SSR,* **6,** 81 (1973) [Russ] 24 refs. [*Chem. Abstr.,* **80,** 145915p (1974).] 4-Piperidinol derivatives are considered.

"Catalytic Ammoxidation of 3-Picoline to 3-Cyanopyridine in Investigations on Nicotinic Acid and Nicotinamide Technology," E. Treszczanowicz, et al., *Przem. Chem.,* **52,** 167 (1973) [Pol] 7 refs. [*Chem. Abstr.,* **79,** 78470d (1973).]

"Mechanism of Contact Oxidation Reactions of Furan and Pyridine Compounds," S. Hillers and M. V. Shimanskaya, *Probl. Kinet. Katal.,* **15,** 47 (1973) [Russ] 30 refs. [*Chem. Abstr.,* **79,** 145472w (1973).]

"Oxidative Coupling *via* Organocopper Compounds," Th. Kauffmann, *Angew. Chem. Int. Ed. Engl.,* **13,** 291 (1974) [Eng] 94 refs. Oxidative coupling of pyridines.

"Dealkylation as a Means of Production of Light Pyridine Bases," A. P. Ivanovskii, A. M. Kut'in, and V. A. Shikhanov, *Vopr. Khim. Khim. Tekhnol.,* **34,** 140 (1974) [Russ] 31 refs. [*Chem. Abstr.,* **83,** 58550j (1975).] Vapor-phase, hydro- and oxidative dealkylations are reviewed.

"Base-Catalyzed Hydrogen Exchange," J. A. Elvidge, J. R. Jones, C. O'Brien, E. A. Evans, and H. C. Sheppard, *Adv. Heterocycl. Chem.,* **16,** 1 (1974) [Eng] 86 refs. For recent hydrogen–deuterium exchange studies of pyridine, see Part IV A.

"Advances in Homolytic Substitution of Heteroaromatic Compounds,"

F. Minisci and O. Porta, *Adv. Heterocycl. Chem.*, **16**, 123 (1974) [Eng] 142 refs. Various radical sources, homolytic alkylation, and arylation reactions of the pyridine nucleus are considered.

"Basic Isotopic Hydrogen Exchange of Heteroaromatic Compounds," N. N. Zatsepina and I. F. Tupitsyn, *Khim. Geterotsikl. Soedin.*, **1974**, 1587 [Russ] 148 refs; [Eng. trans.] *Chem. Heterocycl. Compd.*, **10**, 1397 (1976).

"Diels–Alder Condensation with 2-Pyrones and 2-Pyridones," N. P. Shusherina, *Usp. Khim.*, **43**, 1771 (1974) [Russ] 121 refs; [Eng. trans.] *Russ. Chem. Rev.*, **43**, 851 (1974). Endo-exo selectivity of this reaction is considered.

"The Reactions of Certain Aromatic Cations," S. V. Kriuan, O. F. Alferova, and S. Sayapina, *Russ. Chem. Rev.*, **43**, 835 (1974) [Eng] 220 refs.

"Diazotisation of Heterocyclic Primary Amines," R. N. Butler, *Chem. Revs.*, **75**, 241 (1975) [Eng] 202 refs. Diazotisation reaction of aminopyridines is described; Part IV A. 1–3.

"Structure and Reactivity of 3-Hydroxypyridine Derivatives During Electrophilic Substitution," K. M. Dyumaev and L. D. Smirnov, *Usp. Khim.*, **44**, 1788 (1975) [Russ] 117 refs; [Eng. trans.] *Russ. Chem. Rev.*, **44**, 823 (1975).

"Deuterium-Labeled Compounds," Y. Kawazoe and M. Maeda, *Kagaku No Ryoiki, Zokan*, **107**, 78 (1975) [Japan] 36 refs. [*Chem. Abstr.*, **84**, 134695v (1976).] Proton exchange of benzo derivatives is reviewed.

"The Willgerodt Reaction," E. V. Brown, *Synthesis*, **1975**, 358 [Eng] 172 refs. Numerous pyridyl ketones have been converted into amides *via* this reaction, as well as the Kinder modification. This conversion of α- or γ-alkylpyridines into thioamides and 2-(pyridyl)-benzothiazoles is discussed and examples tabulated.

"Ring Transformations in Nucleophilic Substitution Reactions of Halogen Derivatives of Azines," H. Poradowska and W. Czuba, *Wiad. Chem.*, **29**, 173 (1975) [Pol] 71 refs. [*Chem. Abstr.*, **84**, 3871e (1976).] Heteroaromatic ring transformations and reactions of halopyridines with nucleophiles are reviewed. The ANRORC mechanism is also discussed.

"A Special Case of Graebe–Ullmann Reaction in Polyphosphoric Acid: Decomposition of 1-(2'-Pyridyl)-1,2,3-benzotriazole Derivatives with Nitro Groups in the Pyridine Ring," P. Nantka-Namirski and J. Kalinowski, *Uniw. Adama Mickiewicza Poznaniu, Wydz. Mat., Fiz. Chem.*, [Pr.], *Ser. Chem.*, **18**, 259 (1975) [Eng] 21 refs. [*Chem. Abstr.*, **85**, 46447j (1976).]

"Certain Reactions of Pyridines and Related Heterocycles with Acetylenecarboxylic Acid Esters," R. M. Acheson, *Khim. Geterotsikl. Soedin.*, **1976**, 1011 [Russ] 45 refs; [Eng. trans.] *Chem. Heterocycl. Compd.*, **13**, 837 (1977).

"Electrophilic Substitution in a Series of Isomeric α-, β-, and γ-Hydroxypyridines," L. D. Smirnov and K. M. Dyumaev, *Khim. Geterotsikl. Soedin.*, **1976**, 1155 [Russ] 165 refs; [Eng. trans.] *Chem. Heterocycl. Compd.*, **12**, 955 (1976).

"Homolytic Substitution in the Aromatic and Heteroaromatic Series. Recent Progress in Heteroarylation," G. Vernin, *Bull. Soc. Chim. Fr.*, **1976**, 1257 [Fr] 164 refs. Comparison of the advantages and drawbacks of different methods used to produce heteroaromatic radicals is presented.

"Studies in the Heterocyclic Series. IX. Synthesis of Heterocyclic Thiols,"

C. O. Okafor, *Phosphorus and Sulfur,* **1**, 323 (1976) [Eng] 114 refs. Electrophilic and nucleophilic substitutions on a pyridine nucleus are reviewed.

"Energies and Alkylations of Tautomeric Heterocyclic Compounds Old Problems — New Answers," P. Beak, *Acct. Chem. Res.,* **10**, 186 (1977) [Eng] 54 refs. Pyridine–Pyridone tautomerization.

"Chemical Reactions of 2,3-Cycloalkenopyridines," H. Beschke, *Aldrichimica Acta,* **11**, 13 (1978) [Eng] 26 refs.

"New Synthetic Methods. 25. 4-Dialkylaminopyridines as Acylation Catalysts. 4. Pyridine Syntheses, 1. 4-Dialkylaminopyridines as Highly Active Acylation Catalysts," G. Hoefle, W. Steglich, and H. Vorbrueggen, *Angew. Chem.,* **90**, 602 (1978) [Ger] 125 refs. [Eng. trans.] *Angew. Chem. Internat. Edit.,* **17**, 569 (1978).

"Dealkylation of Pyridine Bases," P. Tomasik and K. Wojactynski, *Przem. Chem.,* **59**, 320 (1980) [Pol] 68 refs. [*Chem. Abstr.,* **93** 168025f (1980).]

"Cycloaddition Reactions of Pyridines," W. Sliwa, *Heterocycles,* **14**, 1793 (1980) [Eng] 115 refs. Includes: pyridinium *N*-methylides and *N*-iminoylides.

"Chemical Reactions of Newly Available Pyridines," H. Beschke, *Aldrichimica Acta,* **14**, 13 (1981) [Eng] 57 refs.

"Catalytic Hydrogenation of Pyridine and its Derivatives," D. Kreile, V. A. Slavinskaya, L. Krumina, A. E. Gutmanis, K. Ziemelis, and U. Tomsons, *Latv. P. S. R. Zinat. Akad. Vestis,* **1980**, 93 [Russ] 81 refs. [*Chem. Abstr.,* **94**, 47026 (1981).]

"Pyridine Ring Nucleophilic Recyclizations," A. N. Kost, S. P. Gromov, and R. S. Sagitullin, *Tetrahedron,* **37**, 3423 (1981) [Eng] 279 refs. The mechanism of ring cleavage and recyclization to hetero- and carbocycles is included.

"2,6-Di-*tert*-butylpyridine — An Unusual Base," B. Kanner, *Heterocycles,* **18**, 411 (1982) [Eng] 37 refs.

"Cycloaddition Reactions of Pyridine," W. Sliwa, *Wiad. Chem.,* **35**, 833 (1982) [Pol] 106 refs. [*Chem. Abstr.,* **97**, 55598j (1982).]

IV. SYNTHESIS OF PYRIDINES

"Syntheses de bases pyridiques à partir d'aldèhydes et de cétones: Mecanisme," J. Gelas, *Bull. Soc. Chim. Fr.,* **1967**, 3093 [Fr] 146 refs. Pyridines have been synthesized from aldehyde and ketone units; mechanistic aspects.

"Condensation of Oxazoles with Dienophiles — A New Method for the Synthesis of Pyridine Bases," M. Y. Karpeiskii and V. A. Florent'ev, *Russ. Chem. Rev.,* **38**, 540 (1969) [Eng] 39 refs.

"Synthesis of Heterocycles Using Acetylenic Compounds," K. Iwai, *Kagaku No Ryoiki, Zokan.,* **92**, 267 (1970) [Japan] 191 refs. [*Chem. Abstr.,* **73**, 10971w (1970).] Construction of substituted pyridines was considered.

"2-Aminopyridines," P. Arnall and G. R. Dace, *Mfg. Chem. Aerosol News,* **1970**, 21, 23–24, 26 [Eng] 123 refs. [*Chem. Abstr.,* **72**, 121247f (1970).] Production, physical data, chemistry, and practical uses of 2-aminopyridine derivatives are reviewed.

"Preparation of 3-Methylpyridine," H. J. Uebel and K. K. Moll, *Chem. Tech.* (Leipzig), **22**, 745 (1970) [Ger] 14 refs. [*Chem. Abstr.,* **74**, 141407c (1971).]

"Liquid-Crystalline Heterocycles," H. Schubert, *Wiss. Z. Martin-Luther Univ., Halle-Wittenberg, Math. Naturwiss. Reihe,* **19**, 1 (1970) [Ger]. [*Chem. Abstr.,* **75**, 98467d (1971).] Liquid–crystalline pyridinone derivatives are described.

"Preparation of Nitrogen Heterocycles with Fixed Labeling of Tritium (deuterium) by Isotope Exchange," Yu. M. Kapustin, *Nov. Metody Poluch. Radioaktiv. Prep., Sb Dokl. Simp.,* **1969** (published 1970), 363 [Russ] no refs. [*Chem. Abstr.,* **74**, 125276d (1971).] Preparation of labeled pyridines and their *N*-oxides by isotope exhange is reviewed.

"Pyridine. Purification and Tests for Purity," R. Lindauer and L. M. Mukherjee, *Pure Appl. Chem.,* **27**, 265 (1971) [Eng] 41 refs.

"Novel Pyridine Derivatives. Manufacture and Usage," P. Arnall and N. R. Clark, *Chem. Process* (London), **1971**, 9, 11–13, 15 [Eng] 35 refs. The industrial syntheses of pyridine and its derivatives are reviewed.

"New Method of Synthesizing 3-Alkylpyridines," T. Yoneyama, *Kagaku No Ryoiki,* **25**, 1050 (1971) [Japan] 3 refs. [*Chem. Abstr.,* **76**, 34024v (1972).] Synthesis of 3-alkylpyridines by the reaction of lithiotetrakis(*N*-dihydropyridyl)-aluminate with alkyl halides.

"Synthesis of Nitrogen Heterocycles by [4+2]-Cycloaddition from Nitrogen-Containing Heterodienes (Azadiene Synthesis)," K. N. Zellnin, I. P. Bezhan, and Z. M. Matveeva, *Khim. Geterotsikl. Soedin.,* **1972**, 579 [Russ] 110 refs; [Eng. trans.] *Chem. Heterocycl. Compd.,* **8**, 525 (1974).

"Make Pyridines Directly," Y. Kusunoki and H. Okazaki, *Hydrocarbon Process.,* **53**, 129 (1974) [Eng] 4 refs. Direct production of 2-picoline and 2-methyl-5-ethylpyridine from ethylene and ammonia in the presence of a palladium complex is reviewed.

"Recent Trends in Heteroaromatics," T. Sasaki, *Kagaku No Ryoiki,* **29**, 784 (1975) [Japan] 202 refs. [*Chem. Abstr.,* **85**, 21159d (1976).] "Recent" is only the years 1972–1974.

"Alkyl Pyridines from Ethylene," Y. Kusunoki and H. Okazaki, *Nikkakyo Geppo,* **28**, 603 (1975) [Japan] no refs. [*Chem. Abstr.,* **84**, 179932g (1976).]

"Synthesis of Pyridine Bases from Aldehydes and Ammonia," M. I. Farberov, V. V. Antonova, B. F. Ustavshchikov, and N. A. Titova, *Khim. Geterotsikl. Soedin.,* **1975**, 1587 [Russ] 21 refs; [Eng. trans.] *Chem. Heterocycl. Compd.,* **11**, 1349 (1976). Research accomplished at the Yaroslav Polytechnic Institute on the synthesis of pyridines is surveyed.

"Synthesis of Heterocycles by Means of the Ring Transformation Reactions of Monocyclic and Condensed 1,2-Oxazoles (Isoxazoles)," T. Nishiwaki, *Synthesis,* **1975**, 20 [Eng] 115 refs. Three procedures for the ring transformation of an isoxazole into a pyridine nucleus are reviewed.

"Phosphonic Acids of the Pyridine Series," K. Lewicka, *Wiad. Chem.,* **30**, 31 (1976) [Pol] 36 refs. [*Chem. Abstr.,* **84**, 164884b (1976).] Syntheses and chemistry of pyridinephosphonic and pyridylalkanephosphonic acids and their derivatives are reviewed.

"Direct Syntheses of Alkylpyridines," Y. Kusunoki and H. Okazaki, *Shokubai,* **18**, 26 (1976) [Japan] 19 refs. [*Chem. Abstr.,* **85**, 108460u (1976).]

"Synthesis of Heterocycles Using Pyridine Hydrohalides," R. Royer, *Khim. Geterotsikl. Soedin.*, **1977**, 579 [Russ] 63 refs; [Eng. trans.] *Chem. Heterocycl. Compd.*, **13**, 463 (1977). Reviews the formation of oxygen- and sulfur-containing heterocycles by a cyclization process in the presence of pyridinium halides.

"New Synthetic Reactions Based on the Onium Salts of Azaaromatic Compounds," T. Mukaiyama, *Heterocycles*, **6**, 1509 (1977) [Eng] no refs.

"Some Aspects of Direct Alkylation of Pyridine and Methyl Pyridines," C. V. Digiovanna, P. J. Cislak, and G. N. Cislak, *ACS Symp. Ser.*, **55**, 397 (1977) [Eng] 75 refs.

"Cobalt-Catalyzed Pyridine Syntheses from Alkynes and Nitriles," H. Bonnemann, *Angew. Chem. Int. Ed.*, **17**, 505 (1978) [Eng] 135 refs. Soluble organocobalt catalysts for the conversion of nitriles and alkynes to substituted pyridines are reviewed.

"γ-Piperidinones in Organic Synthesis," N. S. Prostakov and L. A. Gaivoronskaya, *Vsp. Khim.*, **47**, 859 (1978) [Russ] 415 refs; [Eng. trans.] *Russ. Chem. Rev.*, **47**, 447 (1978). Syntheses of γ-piperidinones and their application to the construction of other heterocycles are considered. Physiological activity is also surveyed.

"Synthesis of Nitrogen-Containing Heterocycles by the Imino Diels–Alder Reaction," J. I. Levin, *Heterocycles*, **12**, 949 (1979) [Eng] 65 refs.

"Catalytic Synthesis of Pyridine Bases," K. M. Akhmerov, D. Yusupov, and A. B. Kuchkarov, *Uzb. Khim. Zhur.*, **1978**, 62 [Russ] 21 refs. [*Chem. Abstr.*, **90**, 22691 (1979).]

"Catalytic Methods for the Production of Pyridine Bases," I. Lazdins and A. Avots, *Khim. Geterotsikl. Soedin.*, **1979**, 1011 [Russ] 234 refs. [*Chem. Abstr.*, **91**, 175087 (1979).]

"Synthesis of Pyridinecarboxylic Acids as Initial Products for Polymers," A. D. Kagarlitskii and B. V. Suvorov, *Tr. Inst. Khim. Nauk. Akad. Nauk. Kaz. SSR.*, **49**, 3 (1979) [Russ] 181 refs. [*Chem. Abstr.*, **92**, 147230 (1980).]

"Conversions of Primary Amino Groups into Other Functionality Mediated by Pyrylium Cations," A. R. Katritzky, *Tetrahedron*, **36**, 679 (1980) [Eng] 68 refs.

"Use of Transition Organometallic Compounds in Heterocyclic Synthesis," J. L. Davidson and P. N. Preston, *Adv. Heterocycl. Chem.*, **30**, 319 (1982) [Eng]. As a part of this review pyridine and derivatives are considered (pp. 376–381; 20 refs.).

V. NATURAL PRODUCTS AND BIOCHEMICAL ASPECTS

"Pyridine Alkaloids," W. A. Ayer and T. E. Habgood, *Alkaloids* (London), **11**, 459 (1968) [Eng] 227 refs. Review covers isolation of pyridine alkaloids and the biogenesis of nicotine, anabasine, and ricinine.

"Biosynthesis of the Piperidine Alkaloids," R. N. Gupta, *Lloydia*, **31**, 318 (1968) [Eng] 37 refs.

"Piperidine Alkaloids," H. W. Liebisch, *Biosyn. Alkaloids*, K. Mothes, Ed., **1969**, 275 [Ger] 242 refs.

"Total Synthesis of Natural Products Containing Heterocyclic Systems," W. Nagata, *Yuki Gosei Kagaku Kyokai Shi,* **26**, 732 (1968) [Jap] 53 refs. [*Chem. Abstr.,* **70**, 11851z (1969).]

"Biosynthesis of Some Alkaloids. Biosynthesis of Alkaloids with Pyridine, Piperidine, Tropan, Imidazole, and Purine Structures," D. Tarle and J. Petricic, *Farm. Glas.,* **24**, 341 (1968) [Croat] 17 refs. [*Chem. Abstr.,* **70**, 35017h (1969).]

"Natural Products with a Pyridine Structure and Their Biosynthesis," D. Gross, *Prog. Chem. Org. Nat. Prod.,* **28**, 109 (1970) [Ger] 28 refs.

"Pyrrolidine, Piperidine, Pyridine, and Imidazole Alkaloids," R. K. Hill, *Alkaloids,* S. W. Pelletier, ed., Van Nostrand Reinhold, New York, 1970, pp. 385–429 [Eng] 82 refs.

"Biochemical Application of UV/Visible Spectrophotometry: Determination of Enzyme Activity," R. Murton and F. E. Dunstan, *Instrum. News,* **20**, 14 (1970) [Eng]. [*Chem. Abstr.,* **73**, 105643x (1970).] A discussion of enzyme spectroscopy including pyridine nucleotide spectroscopy.

"Biochemistry of the Excess Synthesis of Pyridine Adenine Dinucleotides," R. V. Chagovets, A. G. Khalmuradov, and S. I. Shushevich, *Ukr. Biokhim. Zh.,* **42**, 191 (1970) [Ukrain] 74 refs. [*Chem. Abstr.,* **73**, 74542g (1970).] A review of pyridine adenine dinucleotides dynamics in liver after administration of substituted pyridines.

"Fusaric Acid [5-Butylpicolinic acid] and its Analogs," A. N. Kost, P. B. Terent'ev, and L. V. Modyanova, *Vestn. Mosk. Univ., Khim.,* **11**, 273 (1970) [Russ]. [*Chem. Abstr.,* **73**, 76950n (1970).] Biogenesis, physiological effects, methods of synthesis, and properties of fusaric acid are reviewed.

"Biosynthesis of the Alkaloidal Piperidine Ring," N. Ya. Lovkova, *Izv. Akad. Nauk SSSR, Ber. Biol.* **1970**, 451 [Russ] 35 refs. [*Chem. Abstr.,* **73**, 56275t (1970).]

"New Biological Reactions of NAD: Problems in the Synthesis and Utilization of Pyridine–Ribosyl Bond Energy in Mammalian Systems," Y. Nishizuka and O. Hayaishi, *J. Sci. Ind. Res.,* **29**, S15 (1970) [Eng]. [*Chem. Abstr.,* **74**, 73478z (1971).]

"Biochemical Evolution of Pyridine Coenzymes," A. G. Khalmuradov and S. I. Shusevich, *Vitaminy,* **5**, 29 (1970) [Russ] 50 refs. The evolutionary viewpoint, factual data on the chemistry and biochemistry of nicotinic acid and its derivatives are reviewed.

"Contributions of NMR Spectroscopy Towards the Elucidation of the Structure and Conformation of Pyridine, Piperidine, and Monoterpene Alkaloids," M. Vlassa, *Stud. Cercet. Chim.,* **18**, 1109 (1970) [Rom] 47 refs. [*Chem. Abstr.,* **74**, 125875y (1971).]

"Isolation, Structure, and Biosynthesis of Naturally Occurring Piperidine Compounds," D. Gross, *Prog. Chem. Org. Nat. Prod.,* **29**, 109 (1971) [Ger] 353 refs. Simple naturally occurring piperidines, alkyl (and dialkyl-), and heterocyclic substituted piperidines are reviewed.

"Pyrrolidine, Piperidine, and Pyridine Alkaloids," V. A. Snieckus, *Alkaloids* (London), **1**, 48 (1971) [Eng] 62 refs. Nicotine, conhydrine, skytanthine, and nupharolidine alkaloids are reviewed.

"Biosynthesis of Nicotinic Acid in Plants and Microbes," R. F. Dawson, D. R. Christman, and R. U. Byerrum, *Methods Enzymol.,* **18**, 90 (1971) [Eng] 44 refs.

"Regulation of Nicotinamide Metabolism," L. S. Dietrich, *Amer. J. Clin. Nutr.,* **24**, 800 (1971) [Eng] 50 refs. Pathways for NAD metabolism and effects of pyridine nucleotides on the activity of NMN phosphoribosyl transferase are reviewed.

"Pathways of Pyridine Coenzyme Biosynthesis and Mechanisms of their Intracellular Regulation," A. G. Khalmuradov and S. I. Shushevich, *Ukr. Biokhim. Zh.,* **43**, 774 (1971) [Russ] 174 refs. [*Chem. Abstr., 76*, 82583y (1972).]

"Biosynthesis (of Alkaloids). 1. General," R. B. Herbert, *Alkaloids* (London), **1**, 1 (1971) [Eng] 90 refs.

"Occurrence, Structure, and Biosynthesis of Natural Piperidine Compounds," D. Gross, *Fortschr. Chem. Org. Naturst.,* **29**, 1 (1971) [Ger] 353 refs.

"Syntheses of Carpyrinic Acid and Related Pyridines with Long Aliphatic Chains," G. Fodor, J.-P. Fumeaux, and V. Sankaran, *Synthesis,* **1972**, 464 [Eng] 34 refs. Syntheses of carpyrinic acid, dihydroprosopine [5-hydroxy-2-(11-hydroxydodecyl)-6-hydroxymethylpyridine], and several ant venoms [2-methyl-6-alkyl- and 6-alkenyl-pyridines] are described.

"Pyrrolidine, Piperidine, and Pyridine Alkaloids," V. A. Snieckus, *Alkaloid* (London), **2**, 33 (1972) [Eng] 72 refs. Review of *Lythrum, Nuphar,* and mono- and sesquiterpenoid alkaloids.

"Biosynthesis," J. Staunton, *Alkaloids* (London), **2**, 1 (1972) [Eng] 80 refs.

"Piperidine. Its Activity and Physiological Role in the Central Nervous System," Y. Kase and K. Miyata, *Farumishia,* **8**, 356 (1972) [Jap] 0 refs. [*Chem. Abstr., 78*, 154635h (1973).]

"Alkaloids," H. F. Hodson, *Annu. Rep. Progr. Chem. Sect. B.,* **68**, 493 (1972) [Eng] 80 refs. Pyridine alkaloids are considered.

"Nicotinic Acid," G.-S. Huang, *Petrochem.,* **12**, 47 (1973) [Eng] 7 refs.

"Alkaloids of Lythrum Plants," E. Fujita, *Farumashia,* **9**, 599 (1973) [Jap] 0 refs. [*Chem. Abstr., 80*, 80040c (1974).]

"Cinchona Alkaloids," M. R. Uskokovic and G. Grethe, *Alkaloids* (N.Y.), **14**, 181 (1973) [Eng] 54 refs.

"Alkaloids Derived from Pyridine, Piperidine, and Pyrrolidine," R. Danielak, *Chromatogr. Cienkowarstwowa Anal. Farm.,* **1973**, 152 [Pol] 24 refs. [*Chem. Abstr., 82*, 129302x (1975).]

"Pyrrolidine, Piperidine, and Pyridine Alkaloids," V. A. Snieckus, *Alkaloids* (London), **3**, 43 (1973) [Eng] 108 refs. Tobacco and mono- and sesquiterpenoid alkaloids are reviewed.

"Pyrrolidine, Piperidine, and Pyridine Alkaloids," V. A. Snieckus, *Alkaloids* (London), **4**, 50 (1974) [Eng] 113 refs.

"Synthesis and Antimicrobial Activity of Several Groups of Organic Compounds," M. N. Rottistrov, G. V. Kulik, E. M. Skrynik, T. V. Gorbonos, A. N. Bredikhina, and L. A. Taranova, *Fiziol, Aktiv. Veshchestva,* **5**, 123 (1973) [Russ] 17 refs. [*Chem. Abstr., 81*, 72907w (1974).] Antimicrobial activity of pyridine, in part, is reviewed.

"Reaction Mechanisms. III. Enzyme Mechanisms," M. Akhtar and D. C. Wilton, *Annu. Rep. Prog. Chem., Sect B,* **70**, 98 (1974) [Eng] 51 refs. [*Chem. Abstr.,* **82**, 120534s (1975).] Mechanisms of reactions of pyridine nucleotide-linked and flavine-linked enzymes are reviewed.

"Biologically Active Pyridine–Sulfur Compounds," W. Goebel, *Pharmazie,* **29**, 744 (1974) [Ger] 110 refs.

"Pyrrolidine, Piperidine, and Pyridine Alkaloids," V. A. Snieckus, *Alkaloids* (London), **5**, 56 (1975) [Eng] 66 refs.

"Pyridines and Reduced Pyridines of Pharmaceutical Interest," R. T. Coutts and A. F. Casy, *Chem. Heterocycl. Compt.,* **14**, 445 (1975) [Eng] 450 refs.

"Piperidine, A New Neuromodulator?," E. Giacobini, *Fed. Eur. Biochem. Soc. Meet.* (Proc)., **41**, 117 (1975) [Eng] 45 refs. [*Chem. Abstr.,* **84**, 177177y (1976).]

"Antibodies: Analytical Tools to Study Pharmacologically Active Compounds," J. J. Langone, H. von Vunakis, and L. Levine, *Acc. Chem. Res.,* **8**, 335 (1975) [Eng] 33 refs. Pyridines with biological activity are considered.

"Nuclear Magnetic Resonance of Alkaloids," T. A. Crabb, *Annu. Rep. NMR Spectrosc.,* **6A**, 249 (1975) [Eng] 338 refs.

"Pyrrolidine, Piperidine, and Pyridine Alkaloids," A. R. Pinder, *Alkaloids* (London), **6**, 54 (1976) [Eng] 28 refs.

"Biosynthesis of Ergot Alkaloids and Related Compounds," H. G. Floss, *Tetrahedron,* **32**, 873 (1976) [Eng] 334 refs.

"Neurobiology of Piperidine: Its Relevance to CNS Function," Y. Kase and T. Miyata, *Adv. Biochem. Psychopharmacol.,* **15**, 5 (1976) [Eng] 37 refs. [*Chem. Abstr.,* **86**, 101181r (1977).]

"Piperidine: A New Neuromodulator or a Hypogenic Substance?," E. Giacobini, *Adv. Biochem. Psychopharmacol.,* **15**, 17 (1976) [Eng] 111 refs. [*Chem. Abstr.,* **86**, 101182s (1977).]

"Isoprenoids and Alkaloids of Tobacco," C. R. Enzell, I. Wahlberg, and A. J. Aasen, *Progress in the Chemistry of Organic Products,* **34**, 1 (1977) [Eng] 307 refs. The structure, synthesis, biosynthesis, metabolism (in growing plants), postharvest reactions, and bacterial degradation of nicotine are considered; other pyridine derivatives from tobacco have also been reviewed.

"General Methods of Alkaloid Synthesis," R. V. Stevens, *Acc. Chem. Res.,* **10**, 193 (1977) [Eng] 35 refs.

"Pyrrolidine, Piperidine, and Pyridine Alkaloids," A. R. Pinder, *Alkaloids* (London), **7**, 35 (1977) [Eng] 33 refs.

"Phenothiazines with Piperidine Side Chains," C. D. Burrell, *Psychopharmacology* (N.Y.), **2**, 795 (1977) [Eng] 177 refs. [*Chem. Abstr.,* **87**, 78089b (1977).]

"Pyridine and Piperidine Alkaloids," J. D. Hunt and A. McKillop, "Rodd's Chemistry of Carbon Compounds," 2nd ed., S. Coffey and M. F. Ansell, eds., Elsevier, Amsterdam, **4G**, 115 (1978) [Eng] 202 refs.

"Microbiological Transformation of Pyridine Derivatives," A. N. Kost and L. V. Modyanova, *Khim. Geterotsikl. Soedin.,* **1978**, 1299 [Russ] 140 refs. [*Chem. Abstr.,* **90**, 2849r (1979).]

"Visualization of NAD(P)-Dependent Reactions," H. Moellering, A. W.

646 G. R. Newkome

Wahlefeld, and G. Michal, *Grundlagen Enzym. Anal.*, **1977**, 90 [Russ] 39 refs. [*Chem. Abstr.*, **90**, 18065 (1979).]

"(Labeled) Pyridine Nucleotides," K. Ueda, *Seikagaku Jikken Koza*, **6**, 413 (1977) [Japan] "many" refs. [*Chem. Abstr.*, **91**, 188748 (1979).] Tritium and ^{14}C pyridine nucleotide labeling is reviewed.

"Pyrrolidine, Piperidine, and Pyridine Alkaloids," A. R. Pinder, *Alkaloids* (London), **8**, 37 (1978) [Eng] 35 refs.

"Synthesis and Metabolism of Pyridine Nucleotides," E. R. Jaffe, *Hematology*, 2nd ed., McGraw-Hill, New York, 1977, pp. 146–148 [Eng] 14 refs. Erythrocyte pyridine nucleotide metabolism is reviewed.

"Enzyme Stereospecificities for Nicotinamides Nucleotides," K.-S. You, L. J. Arnold, Jr., W. S. Allison, and N. O. Kaplan, *Trends Biochem. Sci.*, **3**, 265 (1978) [Eng] 49 refs.

"Application and Potential of Coenzyme Labeling," T. Cronholm, T. Curstedt, and J. Sjovall, *Stable Isot., Proc. Int. Symp.*, **1978**, 169 [Eng] 28 refs. Pyridine nucleotide labeling is reviewed.

"Some Biological Transformations Involving Unsaturated Linkages: The Importance of Charge Separation and Charge Neutralization in Enzyme Catalysis," M. Akhta and C. Jones, *Tetrahedron*, **34**, 813 (1978) [Eng] 39 refs. Pyridine nucleotides charge separation in enzyme catalysis.

"Pyridine Nucleotide Independent Alcohol Dehydrogenase in Alkane Catabolizing Bacteria," H. Tauchert, W. Schoepp and H. Aurich, *Wiss. Z.*, **27**, 25 (1978) [Ger] 32 refs. [*Chem. Abstr.*, **90**, 116797 (1979).]

"Pyridine Coenzymes – Their Properties and Interaction with Apoenzymes," V. Velebuy and J. Kovar, *Chem. Listy*, **73**, 389 (1979) [Czech] 125 refs. [*Chem. Abstr.*, **90**, 199287 (1979).]

"Pyrrolidine, Piperidine, and Pyridine Alkaloids," A. R. Pinder, *Alkaloids* (London), **9**, 35 (1979) [Eng] 43 refs.

"Coenzymes," H. C. S. Wood, in *Comprehensive Organic Chemistry*, P. G. Sammes, ed., Pergamon, Oxford, Vol. 5, 489 pp. (1979) [Eng] 131 refs. NAD(P) is, in part, reviewed.

"Reconstruction of an Energy-Linked Reaction (Reduced Pyridine Nucleotide Transhydrogenation) in Fractionated *Escharichia Coli* Membranes with Purified ATPase," P. D. Bragg, *Methods Enzymol.*, **55**, 787 (1979) [Eng] 30 refs.

"Metabolic Pathway of Pyridine Nucleotides in Microorganisms," M. Kuwabara, *Hakko to Kogyo*, **37**, 955 (1979) [Japan] 69 refs. [*Chem. Abstr.*, **92**, 02882 (1980).]

"Studies of Purine Nucleoside Analogs," M. N. Preobrazhenskaya, I. A. Korbukh, V. N. Tolkachev, Ya. V. Dobrynin, and G. I. Vornovitskaya, *Nucleosides, Nucleotides Appl. Biol.*, **81**, 85 (1979) [Eng]. Pyrrolopyridine nucleosides are reviewed.

"Macrocyclic Piperidine and Piperidine Alkaloids in Carica Papaya," C.-S. Tang, *Trop. Foods: Chem. Nutr.* **1**, 55 (1979) [Eng] 40 refs. [*Chem. Abstr.*, **19**, 71656h (1979).]

"Physiological Importance of Pyridine Nucleotides with Special Consideration of the Energy Metabolism of Plants," G. Schilling, *Biol. Rundsch.*, **17**, 281 (1979) [Ger] 90 refs. [*Chem. Abstr.*, **92**, 90819 (1980).]

"Pyridine Nucleotide — Disulfide Oxidoreductases," A. Holmgren, *Experientia, Suppl. Dehydrogenases Requiring Nicotinamide Coenzymes,* **36**, 149 (1980) [Eng] 116 refs. [*Chem. Abstr.,* **92**, 193213 (1980).]

"Nicotinamide Adenine Dinucleotide Biosynthesis and Pyridine Nucleotide Cycle Metabolism in Microbial Systems," J. W. Foster and A. G. Moat, *Microbiol. Rev.,* **44**, 83 (1980) [Eng] 121 refs.

"Aminopyridines and Synaptic Transmission," S. Thesleff, *Neuroscience,* **5**, 1413 (1980) [Eng] 50 refs. [*Chem. Abstr.,* **93**, 179149a (1981).]

"4-Aminopyridine and Neuromuscular Transmission," L. H, D. J. Booij, *TGO, Tijdschr. Geneesmiddelenonderz.,* **5**, 697 (1980) [Eng] 45 refs. [*Chem. Abstr.,* **93**, 160744u (1980).]

"A Descriptive Evaluation of Quantitative Histochemical Methods Based on Pyridine Nucleotides," W. H. Outlaw, Jr., *Annu. Rev. Plant Physiol.,* **31**, 299 (1980) [Eng] 36 refs. [*Chem. Abstr.,* **93**, 64929n (1980).]

"Preparation of 3-Aminopyridine Adenine Dinucleotide and 3-Aminopyridine Adenine Dinucleotide Phosphate," B. M. Anderson and T. L. Fisher, *Methods Enzymol.,* **66**, 81 (1980) [Eng] 9 refs.

"Isolation and Analysis of Pyridine Nucleotides and Related Compounds by Liquid Chromatography," C. Bernofsky, *Methods Enzymol.,* **66**, 23 (1980) [Eng] 8 refs.

"Effect of Aminopyridines on Neuromuscular Transmission," P. Lechat, *Bull. Acad. Natl. Med.* (Paris), **164**, 139 (1980) [Fr] 15 refs. [*Chem. Abstr.,* **93**, 60749f (1980).]

"Regulation of Renal Ammoniagenesis During Acidosis: The Pyridine Nucleotide Hypothesis Revisited," H. G. Preuss, *Life Sci.,* **24**, 2293 (1980) [Eng] 62 refs. [*Chem. Abstr.,* **94**, 28312t (1980).]

"Physiologic Aspects of Pyridine Nucleotide Regulation in Mammals," C. Bernofsky, *Mol. Cell. Biochem.,* **33**, 135 (1980) [Eng] 116 refs. [*Chem. Abstr.,* **94**, 62317h (1980).]

"Pyridine Nucleotide-Linked Four-Electron Transfer Dehydrogenases," D. S. Feingold and J. S. Franzen, *Trends Biochem. Sci.,* **6**, 103 (1981) [Eng] 28 refs. [*Chem. Abstr.,* **95**, 37742 (1981).]

"Pyrrolidine, Piperidine, and Pyridine Alkaloids," A. R. Pinder, *Alkaloids* (London), **10**, 30 (1981) [Eng] 45 refs.

"Pyrrolidine, Piperidine, and Pyridine Alkaloids," A. R. Pinder, *Alkaloids* (London), **11**, 29 (1981) [Eng] 46 refs.

"Nicotinamines and Analogous Amino Acids, Enologenous Iron Carriers in Higher Plants," H. Ripperger and K. Schreiber, *Heterocycles,* **17**, 447 (1982) [Eng].

"Quality Evaluation of Spices, III. Analytical Techniques for the Determination of Piperine Pungent Principles in Black and White Pepper. Critical Appraisal," J. S. Pruthi, *Indian Oil Soap J.,* **36**, 167 (1970) [Eng] 55 refs. [*Chem. Abstr.,* **75**, 117168a (1971).]

VI. PHYSICAL ASPECTS

"Viscoelasticity of Polymers and Physical Chemistry of Their Solutions," C. Garbuglio, G. Ajroldi, and A. Mula, *Corsi Semin. Chim.*, **1968**, 241 [Ital] 8 refs. [*Chem. Abstr.*, **72**, 3779h (1970).] Review includes the properties of polymethylvinylpyridines.

"Prototropic Tautomerism of Heteroaromatic Compounds," A. R. Katritzky, *Chimia*, **24**, 134 (1970) [Eng] 33 refs. A general survey of heteroaromatic tautomerism (up to 1970) is compiled, including rules governing relative stabilities of tautomers.

"Liquid–Liquid Biphase Behavior in Aqueous Pyridine Bases," R. N. Paul, *Chem. Age India*, **21**, 716 (1970) [Eng] 24 refs. [*Chem. Abstr.*, **73**, 124022k (1970).] Liquid–liquid extraction procedures of pyridine bases is reviewed.

"Ion Exchangers from Pyridine and Quinoline Bases," V. M. Balakin and Z. Yu. Kokoshko, *Tr. Ural. Politekh. Inst.*, **184**, 11 (1970) [Russ] 104 refs. [*Chem. Abstr.*, **75**, 110746d (1971).]

"Use of the Hammett Equation for 2,5-Disubstituted Pyridine Systems with the Reaction Site at the 5-Position," P. Tomasik, *Pr. Nauk. Inst. Chem. Technol. Nafty Wegla Politech. Wroclaw.*, **1971**, 149 [Pol] 26 refs. [*Chem. Abstr.*, **78**, 42289s (1973).]

"Separation and Purification of Alkylpyridines. The Methods of Separation and Purification of Pyridine Homologs are Reviewed," L. Achremowicz, *Wiad. Chem.*, **25**, 491 (1971) [Pol] 182 refs. [*Chem. Abstr.*, **75**, 142590q (1971).] Pyridines from the pyrolysis of coal are described.

"Handbook of Molecular Dimensions. X-ray Bond Angles and Lengths," P. J. Wheatley, in *Physical Methods in Heterocyclic Chemistry*, A. R. Katritzky, ed., Academic Press, New York, Vol. 5, 1972, 598 pp. [Eng] 1367 refs. An indispensible reference guide to pyridine containing molecules.

"Ring Inversion in Some Six-Membered Heterocycles," V. M. Gittins, E. Wyn-Jones, and R. F. M. White, *Intern. Rotation Mol.*, **1974**, 425 [Eng] 123 refs.

"Gas–Liquid Chromatography of Pyridine and Quinoline Bases," V. M. Nabivach and L. A. Venger, *Zh. Anal. Khim.*, **30**, 604 (1975) [Russ] 75 refs. [*Chem. Abstr.*, **83**, 71148d (1975).]

"Conformational Analysis of Pentamethylene Heterocycles," J. B. Lambert and S. I. Featherman, *Chem. Rev.*, **75**, 611 (1975) [Eng] 117 refs. A survey of piperidine and its derivatives is well presented, specifically Part III.C.1.

"Tautomerism Problem and Spectroscopic Properties of Substituted *N*-Heterobenzenes," J. S. Kwiatkowski and M. Berndt, *Uniw. Adama Mickiewicza Poznaniu, Wydz. Mat., Fiz. Chem.*, [Pr], *Ser. Chem.*, **18**, 173 (1975) [Eng] 10 refs. [*Chem. Abstr.*, **84**, 120636t (1976).] The tautomerization of pyridines with emphasis on MO calculations and UV spectra is reviewed.

"Conformation of Piperidine and of Derivatives with Additional Ring Hetero Atoms," I. D. Blackburne, A. R. Katritzky, and Y. Takeuchi, *Acct. Chem. Res.*, **8**, 300 (1975) [Eng] 71 refs.

"Role of Sorbents in the Gas Chromatographic Analysis of Heterocyclic

Nitrogen-Containing Pharmaceuticals," L. P. Chekova, V. I. Trubnikov, and F. M. Shemyakin, *Farmatsiya* (Moscow), **24**, 77 (1975) [Russ] 83 refs. [*Chem. Abstr.*, **83**, 152408r (1975).] The role of stationary liquid phase and solid carriers in gas chromatography of pyridine, pyridinecarboxylic acids, and alkaloids is reviewed.

"A Critique of the Resonance Energy Concept with Particular Reference to Nitrogen Heterocycles Especially Porphyins," P. George, *Chem. Rev.*, **75**, 85 (1975) [Eng] 83 refs. Resonance energies of pyridine are included.

"The Tautomerism of Heterocycles," J. Elguero, C. Marzin, A. R. Katritzky, and P. Linda, "Advances in Heterocyclic Chemistry, Supplement 1," Academic Press, New York, 1976, 656 pp. [Eng]. Chapter 2 deals specifically with the tautomerism of substituted pyridines, as well as other six-membered ring heterocycles.

"Valence-Bond Isomers of Fluorine-Containing Aromatic Compounds. II," Y. Kobayashi, O. Yoshiro, and Y. Hanzawa, *Kagaku No Ryoiki*, **30**, 1112 (1976) [Japan] 87 refs. [*Chem. Abstr.*, **87**, 134843g (1977).]

"Compounds Containing a Six-Membered Ring with One Hetero Atom, Nitrogen; Introduction, Theoretical Description, and Spectroscopic Properties," M. H. Palmer, in *Rodd's Chemistry of Carbon Compounds*, S. Coffey and M. F. Ansell, eds., Elsevier Scientific Publishing, New York, 1976, pp. 1–26 [Eng]. Chapter 23 deals with the theoretical description and spectroscopic properties, for examples, NMR, chemical shifts and electron densities, ionization processes, electron impact phenomena, radical anions, and excited electronic states, of pyridines and its fused benzo relatives.

"High-Precision Coulometry and the Value of the Faraday," H. Diehl, *Anal. Chem.*, **51**, 318A–320A, 322A, 325A–326A, 328A–330A (1979) [Eng] 19 refs. Faraday constants of aminopyridine coulometry are reviewed.

"Phase Transition Temperatures of Some Mesogenic Organic Compounds," B. M. Volotin and R. S. Shishova, *Zhidk. Krist.*, **1979**, 285 [Russ] 36 refs. [*Chem. Abstr.*, **92**, 155931c (1980).] Organic liquid crystal transition is reviewed.

"Calculation of Diamagnetic Susceptibility of Organic Compounds. Comparison of Aromaticities of 2-Methoxypyridine and N-Methyl-2-pyridone," J. K. Lee, *Hwahak Kwa Kongop Ui Chinbo*, **20**, 5 (1980) [Korean] 29 refs. [*Chem. Abstr.*, **94**, 29603a (1981).]

VII. SPECTROSCOPY

"Contribution of NMR Spectroscopy Towards the Elucidation of the Structure and Conformation of Pyridine, Piperidine, and Monoterpene Alkaloids," M. Vlassa, *Stud. Cercet. Chim.*, **18**, 1109 (1970) [Rom] 47 refs. [*Chem. Abstr.*, **74**, 125875y (1971).]

"Ultraviolet Spectra of Heterocycles," W. L. F. Armarego, in *Physical Methods in Heterocyclic Chemistry*, A. R. Katritzky, ed., Academic Press, New York, Vol. 3, 1971, pp. 67–122 [Eng] 935 refs. Table IV in this review contains over 170 entries that deal with numerous substituted pyridines.

"Mass Spectrometry of Heterocyclic Compounds," G. Spiteller, in *Physical*

Methods in Heterocyclic Chemistry, A. R. Katritzky, ed., Academic Press, New York, Vol. 3, pp. 223–296 [Eng] 1971, 88 refs. Part V-C gives a brief survey of degradations in substituted pyridines.

"Nuclear Magnetic Resonance Spectra," R. F. M. White and H. Williams, in *Physical Methods in Heterocyclic Chemistry,* A. R. Katritzky, ed., Academic Press, New York, Vol. 4, 1971, pp. 121–235 [Eng] 608 refs. Part IV-A surveys the proton spectral data for substituted pyridine; the tables are especially useful. ^{13}C, ^{14}N, and ^{19}F spectra for several pyridines are also included.

"Infrared Spectroscopy in Heterocycles," A. R. Katritzky and P. J. Taylor, in *Physical Methods in Heterocyclic Chemistry,* A. R. Katritzky, ed., Academic Press, New York, Vol. 4, 1971, pp. 265–434 [Eng] 718 refs. Parts VII-A and V-B deal with infrared spectral data of substituted pyridines and pyridones, respectively.

"ESR Spectroscopy of Heterocyclic Radicals," B. C. Gilbert and M. Trenwith, in *Physical Methods in Heterocyclic Chemistry,* A. R. Katritzky, ed., Academic Press, New York, Vol. 4, 1974, pp. 95–145 [Eng] 268 refs.

"UV Photoelectron Spectra of Heterocyclic Compounds," E. Heilbronner, J. P. Maier, and E. Haselbach, in *Physical Methods in Heterocyclic Chemistry,* A. R. Katritzky, ed., Academic Press, New York, Vol. 4, 1974, pp. 1–52 [Eng] 165 refs.

"Nuclear Magnetic Resonance of Alkaloids," T. A. Crabb, *Annu. Rep. NMR Spectrosc.,* **6A**, 249 (1975) [Eng] 338 refs.

"Carbon-13 NMR Results in the Field of Nitrogen Heterocycles the Detection of Protonation Sites," H. Guenther, A. Gronenborn, U. Ewers, and H. Seel, *Nucl. Magn. Reson. Spectrosc. Mol. Biol.,* **11**, 193 (1978) [Eng] 10 refs. Pyridine protonation NMR is reviewed.

VIII.   DIHYDROPYRIDINES

"The Chemistry of Dihydropyridines," U. Eisner and J. Kuthan, *Chem. Rev.,* **72**, 1 (1972) [Eng] 611 refs. This excellent review covers the synthesis, structure, and physical and chemical properties of the five isomeric dihydropyridines.

"Use of Deuterium-Labeled Substances. 1,4-Dihydropyridine Derivative YC-93 as an Example," S. Higuchi and Y. Shiobara, *Shitsuryo Bunseki,* **27**, 193R (1979) [Japan] 10 refs. [*Chem. Abstr., 93,* 125226e (1980).]

"Syntheses with Dihydropyridines," S. Blechert, *Nachr. Chem., Tech. Lab.,* **28**, 651 (1980) [Ger] 15 refs. [*Chem. Abstr., 94,* 3880 (1980).]

"Development in Dihydropyridine Chemistry," J. Kuthan and A. Kurfurst, *Ind. Eng. Chem. Prod. Res. Dev.,* **21**, 191 (1982) [Eng] 1197 refs. Synthesis, structure, reactivity, properties, and practical applications of dihydropyridine derivatives.

"Recent Advances in the Chemistry of Dihydropyridines," D. M. Stout and A. I. Meyers, *Chem. Revs.,* **82**, 223 (1982) [Eng] 194 refs.

"Synthesis and Biological Activity of Substituted 3,5-Dicyanopyridines and Dihydropyridines," A. S. Petrovskii, V. K. Promonenkov, I. I. Naumova, and

Yu. A. Strepikheev, Deposited Dec., **1980**, SPSTL 257 khp-D80, [Russ] 27 pp., 134 refs. [*Chem. Abstr.*, **97**, 109785p (1982).]

"Asymmetric Reactions with Chiral Bridged 1,4-Dihydropyridine," R. M. Kellogg, *Stud. Org. Chem.* (Amsterdam), **10**, 22 (1982) [Eng] 19 refs.

IX. TETRAHYDROPYRIDINES

"3-Piperideines (1,2,3,6-Tetrahydropyridines)," M. Ferles and J. Pliml, *Adv. Heterocycl. Chem.*, **12**, 43 (1970) [Eng] 193 refs. This review considers the syntheses, properties, and reactions of 3-piperideines, as well as affords limited information on the naturally occurring and pharmaceutically important 3-piperideines.

"Chemistry and Pharmacology of Some Tetrahydropyridines Including Arecoline and Its Derivatives," R. T. Coutts and J. R. Scott, *Can. J. Pharm. Sci.*, **6**, 78 (1971) [Eng] 95 refs. A review on efficient means of preparing tetrahydropyridine derivatives.

X. PIPERIDINES

"Configuration of Piperidine Derivatives," A. Silhankova and M. Ferles, *Chem. Listy*, **61**, 1248 (1967) [Czech] 152 refs.

"Stereochemistry of the Quaternization of Piperidines," J. McKenna, *Top. Stereochem.*, **5**, 275 (1978) [Eng] 45 refs.

"Application of the Mannich Reaction for the Synthesis of Heterocyclic Systems. VI. Derivatives and Analogs of Bicyclo[4·3·1]decane. VI. Synthesis of 3,5-Piperidinedicarboxylic Acid Derivatives," W. L. Hahn and C. Korzeniewski, *Soc. Sci. Lvolz., Acta Chim.*, **16**, 113 (1971) [Eng] 28 refs. [*Chem. Abstr.*, **76**, 126724b (1972).]

"Synthesis and Study of Halogenated and Dienediyne Heterocycles and Enyne and Polyyne Cyclic Alcohols of Pyran and Piperidine Series," I. M. Azerbaev, T. G. Sarbaev, T. A. Yagudeev, A. N. Nurgalieva, Kh. Z. Zainullin, U. I. Urazaliev, N. N. Nurekeshova, Zh. Kh. Khimedenov, and U. Sh. Shunkarov, *Ts. Iust. Khim. Neftr, Priv. Solei Akad. Nauk Kaz. SSR.*, **6**, 58 (1973) [Russ] 62 refs. [*Chem. Abstr.*, **80**, 133133u (1974).]

"Conformational Analysis by Kinetic Methods. A. Critique. Theory and Experimental Development of Procedures Based on Very Fast Chemical Reactions," J. McKenna, *Tetrahedron*, **30**, 1555 (1974) [Eng] 29 refs.

"Butyrophenones and Diphenylbutylpiperidines," P. A. J. Janssen, *Med. Chem. Ser. Monogr.*, **4**, 129 (1974) [Eng], many refs. [*Chem. Abstr.*, **82**, 106074k (1975).]

"Conformation of Piperidine and of Derivatives with Additional Ring Hetero Atoms," I. D. Blackburne, A. R. Katritzky, and Y. Takeuchi, *Acc. Chem. Res.*, **8**, 300 (1975) [Eng] 71 refs.

"Butyrophenones and Diphenylbutylpiperidines," P. A. J. Janssen and W. F. M. van Bever, *Haudb. Exp. Pharmocol.*, **55**, 27 (1980) [Eng] many refs. [*Chem. Abstr.*, **94**, 57670 (1981).]

"A Review on Hindered Amines," R. Oda, *Kagaku* (Japan), **36**, 772 (1981) [Jpn] 7 refs. [*Chem. Abstr., 95*, 219432n (1981).]

"Photostabilizing Mechanisms of Hals: A Critical Review," J. Sedlar, J. Marchal, and J. Petriy, *Polym. Photochem.*, 2, 175 (1982) [Eng].

XI.   DIDEHYDROPYRIDINE (PYRIDYNES)

"The Hetarynes," Th. Kauffmann, *Angew. Chem. Int. Ed. Engl.*, **4**, 543 (1965) [Eng] 95 refs.

"Recent Aspects of the Chemistry of Arynes," S. Krisnappa, *J. Sci. Ind. Res.*, **29**, 538 (1970) [Eng] 59 refs. [*Chem. Abstr., 74*, 124992x (1971).] Proton acidity of heteroarynes and the relative stability of dehydropyridines are reviewed.

"Hetarynes. 15. Progress in the Hetaryne Field," Th. Kauffmann and R. Wirthwein, *Angew. Chem. Int. Ed. Engl.*, **10**, 20 (1971) [Eng] 72 refs. Generation of didehydropyridines is a part of this extensive review.

"Hetarynes," M. G. Reinecke, *Tetrahedron, 38*, 427 (1982) [Eng] 438 refs. Didehydropyridines are a portion (16 pp.) of this comprehensive review.

XII.   FUSED-RING PYRIDINES

"Thienopyridines," S. W. Schneller, *Internat. J. Sulfur Chem., Part B*, **7**, 309 (1972) [Eng] 40 refs.

"Studies in the Heterocyclic Series. VI, Chemistry and Applications of Isomeric Thiazolopyridines and Related Systems," C. O. Okafor, *Internat. J. Sulfur Chem., Part B*, 7, 121 (1972) [Eng] 165 refs.

"The Chemistry of 1-Pyridines," F. Freeman, in *Advances in Heterocyclic Chemistry*, A. R. Katritzky and A. J. Boulton, eds., Academic Press, New York, Vol. 15, 1973, pp. 187–231 [Eng] 163 refs. A review of a pyridine ring with a fused five-membered unsaturated ring.

"The 4-Oxopyranoazoles and 4-Oxopyranoazines," M. A. Khan, *Progr. Med. Chem., 9*, 117 (1973) [Eng] 33 refs.

"The Thienopyridines," J. M. Barker, in *Advances in Heterocyclic Compounds*, A. R. Katritzky and A. J. Boulton, eds., Academic Press, New York, Vol. 21, 65 (1977) [Eng] 150 refs. This review covers the literature through June 1975, and surveys in detail the syntheses reactions, and physical properties.

XIII.   PYRIDINIUM COMPOUNDS

"Novel Synthetic Utilization of Pyridinium *N*-Ylides," T. Sasaki, *Kagaku Kogyo*, **23**, 504 (1972) [Japan] 62 refs. [*Chem. Abstr., 76*, 140346m (1972).]

"Azomethine Ylides, Azomethine Imines, and Iminophosphoranes in Organic Synthesis," C. G. Struckwisch, *Synthesis*, **1973**, 469 [Eng] 63 refs. Pyridinum *N*-ylides.

"*N*-Acyl Pyridinium Salts and Their Corresponding Fused Derivatives," A. K. Sheinkman, S. I. Suminov, and A. N. Kost, *Usp. Khim.*, **42**, 1415 (1973) [Russ] 419 refs; [Eng. trans] *Russ. Chem. Rev.*, **1973**, 642.

"Heteroaromatic *N*-Imines," H. J. Timpe, *Adv. Heterocycl. Chem.*, **17**, 213 (1974) [Eng] 161 refs. Syntheses of pyridine *N*-imines (Part II.A) from either quaternary pyridinum salts, azides with pyridines, *via*, rearrangement reactions, or *N*-imines are considered. Physico-chemical and chemical properties are discussed.

"Hetarylation of Organic Compounds," A. K. Sheinkman, *Khim. Geterolsikl. Soedin.*, **1974**, 3 [Russ] 124 refs; [Eng. trans.] *Chem. Heterocycl. Compd.*, **10**, 1 (1975). Hetarylation by *N*-acyl salts of aromatic heterocycles is reviewed.

"Quaternization of Pyridine Bases with Alkyl Halides," L. Petrova, *Khim. Ind. (Sofia)*. **48**, 303 (1976) [Bulg] 42 refs. [*Chem. Anstr.*, **86**, 106242w (1977).] Effects of (a) substitutions in the pyridine ring, (b) the long chain alkyl halides, and (c) the solvent media on the quaternization process are reviewed.

"New Synthetic Methods. 14. Uses of Six-Membered Heteroaromatic Betaines in Synthesis," N. Dennis, A. R. Katritzky, and Y. Takeuchi, *Angew. Chem. Int. Ed.*, **15**, 1 (1976) [Eng] 41 refs. Cycloadditions of 3-hydroxy-1-alkylpyridinium betaines with electron-deficient olefins are reviewed.

"Structure and Reactivity of Cycloimmonium," G. Surpateanu, J. P. Catteau, P. Karafiloglou, and A. Lablache-Combier, *Tetrahedron,* **32**, 2647 (1976) [Eng] 178 refs. Pyridinium *N*-ylides and considered.

"Studies on Some Functional Polymers with a Poly(vinylpyridine) Matrix," N. Ise and T. Ohkubo, *Kyoto Daigaku Nippon Kagakuseni Kenkyusho Koenshu,* **33**, 49 (1976) [Japan] 18 refs. [*Chem. Abstr.*, **86**, 121797z (1977).]

"*o*-Mesitylenesulfonylhydroxylamine and Related Compounds — Powerful Aminating Reagents," Y. Tamura, J. Minamikawa, and M. Ikeda, *Synthesis*, **1977**, 1 [Eng] 107 refs. Table 2 in this review surveys the synthesis of *N*-amine salts of heteroaromatic amines, especially pyridine. Typical synthetic procedures are cited.

"Pseudobase Formation from Quaternary Pyridinium, Quinolinium, and Isoquinolium Cations," V. Simanek and V. Preininger, *Heterocycles,* **6**, 475 (1977) [Eng] 81 refs. Recent studies of addition reactions of the hydroxide ion with quaternary aromatic nitrogen compounds are presented.

"Photochemistry as a Tool in Heterocyclic Synthesis: From Pyridinium *N*-Ylides to Diazepines and Beyond," J. Streith, *Heterocycles,* **6**, 2021 (1977) [Eng] 42 refs. This review considers: (1) industrial applications of photoinduced heterocyclic synthesis *via* singlet oxygen and/or free radicals and (2) large bench-scale synthesis of 1,2-diazepines from pyridinium *N*-ylides.

'The Photochemistry of Aromatic *N*-Ylides Rearrangement and Fragmentation Patterns," J. Streith, *Pure Appld. Chem.*, **49**, 305 (1977) [Eng] 34 refs. Pyridinium *N*-ylides are considered.

"Quaternization of Heteroaromatic Compounds: Quantitative Aspects," J. A. Zoltewicz and L. W. Deady, *Adv. Heterocycl. Chem.*, **22**, 72 (1978) [Eng] 190 refs. Pyridine quarternization.

"Synthesis and Properties of *N*-Arylpyridinium Salts and Their Benzo Analogs," G. E. Tukhun, Yu. P. Andreichikov, and G. N. Dorofeenko, *Izv. Sev. -Kauk.*

Nauchn. Tsentra Vyssh, Shi., Estestv. Nauki, **1981,** 43 [Russ] 106 refs. [*Chem. Abstr.,* **96,** 103990j (1981).]

"Pyridinium Chlorochromate: A Versatile Oxidant in Organic Synthesis," G. Piancatelli, A. Scettri, and M. D'Auria, *Synthesis,* **1982,** 245 [Eng] 63 refs.

XIV.   PYRIDINE *N*-OXIDES

"Aromatic Amine Oxides," E. Ochiai, Elsevier, Amsterdam, 1967, 90 refs.

"Chemistry of the Heterocyclic *N*-Oxides," A. R. Katritzky and J. Lagowski, Academic Press, New York, 1971.

"Some Advances in the Reactions of Aromatic *N*-Oxides," H. Masatono, *Lect. Heterocycl. Chem.,* **1,** 51 (1972) [Eng] 58 refs.

"Aromatic Substitution via New Rearrangements of Heteroaromatic *N*-Oxides," R. Abramovitch and I. Shinkai, *Acct. Chem. Res.,* **9,** 192 (1976) [Eng] 56 refs.

"Pyridine *N*-Oxides," R. A. Abramovitch, *Lect. Heterocycl. Chem.,* **5,** S15 (1980) [Eng] 37 refs.

"Pyridine *N*-Oxides and their Derivatives," T. Talik, Z. Talik, H. Ban-Oganowska, and A. Puszko, *Pr. Nauk. Akad. Ekon. im. Oskara Langego Wroclaivin,* **191,** 97 (1982) [Pol] 279 refs. [*Chem. Abstr.,* **98,** 107082d (1983).]

XV.   PYRIDINE COMPLEXES

"The Coordination Chemistry of Aromatic Amine *N*-Oxide," R. G. Garvey, J. N. Nelson, and R. D. Ragsdale, *Coord. Chem. Rev.* **3,** 375 (1968) [Eng] 136 refs. Pyridine *N*-oxides are considered.

"Pyridine *N*-Oxide Complexes of Platinum(II)," M. Orchin and P. J. Schmidt, *Coord. Chem. Rev.,* **3,** 345 (1968) [Eng] 43 refs.

"The Electronic Spectra and Optical Activity of Phenanthroline and Dipyridyl Metal Complexes," S. F. Mason, *Inorg. Chim. Acta Rev.,* **2,** 89 (1968) [Eng] 84 refs.

"Cobalt and Nickel Complexes of Pyridine and Pyridine Derivatives," C. Guran, *Rev. Fiz. Chim., Ser. A.,* **6,** 264 (1969) [Rom] no refs. [*Chem. Abstr.,* **72,** 8747g (1970).] Complexation of pyridine and its derivatives with cobalt and nickel with special emphasis of structural and stereochemical problems is reviewed.

"The Chemistry of Complexes Containing 2,2'-Bipyridyl, 1,10-Phenanthroline, or 2,2',6',2''-Terpyridyl as Ligands," W. R. McWhinnie and J. D. Miller, *Adv. Inorg. Chem. Radiochem.,* **12,** 135 (1969) [Eng] 747 refs.

"Complexes of Pyridine Bases with Rhodium. Redox Behavior and Antibacterial Activity," R. D. Gillard, *Platinum Metals Rev.,* **14,** 50 (1970) [Eng] 6 refs. Acid adducts and stereochemical aspects of rhodium complexes are reviewed.

"Coordination Behavior of Some Chelating Ligands Containing Non- or Weakly Conjugated 2-Pyridyl Groups," W. R. McWhinnie, *Coord. Chem. Rev.,* **5,** 293 (1970) [Eng] 91 refs. Pyridine organic ligands, which are either bi- or ter-dentate, are reviewed.

"Organometallic Compounds," S. Tanuma, *Kotai Butsuri*, **5**, 404 (1970) [Japan] 4 refs. [*Chem. Abstr.*, **74**, 22616k (1971).] The structure and superconductivity of pyridine tantalum sulfide complexes are reviewed.

"Analytical Chemistry of Metal Complexes with Nitrogen-Containing Ligands of the 2,2'-Bipyridyl Type," A. T. Pilipenko and E. R. Falendysh, *Usp. Khim.*, **41**, 2094 (1972) [Russ] 249 refs; [Eng. trans.] *Russ. Chem. Rev.*, **41**, 991 (1972).

"Ternary Cu^{+2} Complexes: Stability, Structure, and Reactivity," H. Sigel, *Angew. Chem. Int. Ed. Engl.*, **14**, 394 (1975) [Eng] 86 refs.

"Applications of Heteroligand Complexes in the Analytical Chemistry of Neobium and Tantalum," S. V. Elinson, *Russ. Chem. Rev.*, **44**, 707 (1975) [Eng] 131 refs. Pyridine metal complexes.

"Equilibriums in Conplexes of *N*-Heterocyclic Molecules. III. Explanation for Classical Anomalies Among Complexes of 1,10-Phenanthrolines and 2,2'-Bipyridyls," R. D. Gillard, *Coord. Chem. Rev.*, **16**, 67 (1975) [Eng] 123 refs.

"Thermal Chemistry of Complexes of Halide Compounds of Elements with Pyridine," V. G. Tsvetkov, *Termodinam. Organ. Soedin.*, **1977**, 58 [Russ]. [*Chem. Abstr.*, **89**, 128603d (1978).]

"Coordination Compounds of the 3*d* Elements with Pyridine Derivatives as Chelate Ligands; The Effect of Steric and Electronic Factors on Their Structure," E. Uhlig, *Z. Chem.*, **18**, 440 (1978) [Ger] 93 refs. [*Chem. Abstr.*, **90**, 102848h (1979).]

"Electronic and Structural Properties of Some Bipyridylium + TCNQ and Related Complexes," D. D. Eley, G. J. Ansell, S. C. Wallwork, M. R. Willis, and J. Woodward, *Ann. N. Y. Acad. Sci.*, **313**, 417 (1978) [Eng] 53 refs.

"Metal Chelates of Pyridine 2-Aldehyde 2'-Pyridylhydrazone (paphyl) and Related Ligands," C. F. Bell, *Rev. Inorg. Chem.*, **1**, 133 (1979) [Eng] 46 refs.

"Absolute Configuration of (−)-*D-tris*(bipyridine)cobalt(III)," H. Kobayashi, *Kagaku* (Kyoto), **34**, 312 (1979) [Japan] 20 refs. [*Chem. Abstr.*, **93**, 36053d (1980).]

"Coordination by Halo/Nitro Substitutents Attached to Phenyl, Pyridyl, or Alkyl Groups," R. L. Dutta, S. K. Satapathi, and R. Sharma, *Indian J. Chem. Educ.*, **7**, 24 (1980) [Eng] 9 refs. [*Chem. Abstr.*, **95**, 79317b (1981).]

"The Studies on the New Reactions of Pyridine and Copper Chloride Complex and Lateral Metallation and Alkylation of Alkylpyridines," X. Zeng, X. Chen, D. Zou, H. Xu, Z. Ma, C. Zheng, Z. Yu, and Y. Gu, *Fundam. Res. Organomet. Chem., Proc. China-Jpn.-U.S. Trilateral Semin. Organomet. Chem. 1st. 1980*, **1982**, 29 [Eng] 17 refs. [*Chem. Abstr.*, **97**, 162710m (1982).]

XVI. PYRIDINYL RADICALS

"Organic Chemistry of Stable Free Radicals," A. R. Forrester, J. M. Hay, and R. H. Thomson, Academic Press, New York, 1968, p. 261 [Eng] 76 refs. Pyridinyl and viologen radicals are considered.

"Interaction in Stable Free Radicals. Molecular Complexes in Pyridinyl

Radicals," M. Itoh, *Yuki Gosei Kagaku Kyokai Shi*, **30**, 771 (1972) [Japan] 44 refs. [*Chem. Abstr.*, **78**, 3352n (1973).] Free radical interaction and related theory of pyridinyl radical dimers, pyridinyl diradicals, and pyridinyl cationic radicals are reviewed.

"Pyridyl Radical. *Ortho* Effect with Alkylbenzenes," G. Filippi, H. Jm. Dou, G. Vernin, and J. Metzger, *J. Heterocycl. Chem.*, **10**, 259 (1973) [Fr] 13 refs. The *ortho* effect on the reaction of pyridyl radicals with alkylbenzenes is reviewed.

"Pyridinyl Radicals in Biology," E. M. Kosower, in *Free Radicals in Biology, Volume II*, W. A. Pryor, ed., Academic Press, New York, 1976 [Eng] 107 refs. An excellent review of the chemical and physical properties of pyridinyl radicals, especially the major enzymes cofactors, the nicotinamide adenine nucleotides.

"Pyridinyl Paradigm Proves Powerful," E. M. Kosower, *ACS Symp. Ser.*, **69**, 447 (1978) [Eng] 18 refs.

XVII. COMMERCIAL ASPECTS

"Commercial Synthetic Pyridine Bases. 1. MEP [Methylethylpyridine] Manufacture, Chemistry and Uses," A. Nenz and M. Pieroni, *Hydrocarbon Process.*, **47**, 139 (1968) [Eng] 135 refs. 2-Methyl-5-ethylpyridine and 2-methyl-5-vinylpyridine are reviewed.

"Paraquat's Role in No-tillage Crop Production in the Southeastern United States," B. H. Tyler, *Proc. S. Weed Sci. Soc.*, **22**, 177 (1969) [Eng] 22 refs. [*Chem. Abstr.*, **72**, 65582x (1970).] Review of herbicides containing the bipyridinium group.

"Higher-Boiling Pyridines of Coal Tar," J. Vymetal, *Chem. Listy*, **64**, 65 (1970) [Czech] 54 refs. Review considers the various pyridines contained in coal tar and the separation of substituted anilines.

"Synthesis and Utilization of Pyridine Compounds," A. Paris, *Ind. Chim.* (Paris), **56**, (618) 9 (1969); **56** (619) 43 (1969) [Fr] 18 refs. Properties, industrial preparation, alkylation and arylation, oxidation and industrial uses of pyridines, as well as bipyridyl formation and preparation of piperidine and vinylpyridines are reviewed.

"Pyridine and its Homologs in the Atmosphere, I. Sampling and Analysis Methods," J. Neiser, V. Masek, and J. Pospisil, *Ropa Uhlie.*, **14**, 449 (1972) [Czech] 38 refs. [*Chem. Abstr.*, **77**, 143453a (1972).] Review of procedures to determine pyridine and derivatives in air.

"Pollution Analysis. Determination of Pyridine in the Atmosphere," O. Tada, *Kogai Bunseki Shishin.*, **3**, 49 (1972) [Japan] 5 refs. [*Chem. Abstr.*, **82**, 7264r (1975).]

"Coal-Tar Bases for Pesticide Productions," K. Wiszniowski, *Chemik*, **26**, 140 (1973) [Pol] 4 refs. [*Chem. Abstr.*, **79**, 49687z (1973).]

"Determination of Pyridine and its Homologs in the Atmosphere of Work Stations," J. Neiser, J. Branzovsky, J. Janik, J. Zvolsky, and V. Masek, *Koks, Smola, Gaz.*, **18**, 20 (1973) [Pol] 26 refs. [*Chem. Abstr.*, **79**, 9262a (1973).] Review on known titrimetric, spectrophotometric, gas-chromatographic, and polarographic methods to determine pyridines.

"Production of Fine Chemicals, Pharmaceuticals, and Insecticides from Light Pyridine Bases," S. K. Ray, H. S. Rao, and A. Lahiri, *J. Inst. Eng.* (India), **54**, 3 (1973) [Eng] 8 refs. [*Chem. Abstr.,* **81**, 93661b (1974).]

"Production and Use of Pyridine Bases in Poland," K. Wiszniowski, *Chemik,* **27**, 204 (1974) [Pol] 16 refs. [*Chem. Abstr.,* **81**, 135828u (1974).] Review of the recovery of pyridine from coke gas, tar, and petroleum.

"Pyridine and its Derivatives in Industry," A. Aydin, *Kim. Sanayi,* **22**, 131 (1974) [Turkish] 20 refs. [*Chem. Abstr.,* **83**, 147350m (1975).]

"Basic Monomers. Vinyl Pyridines and Aminoalkyl Acrylates and Methacrylates," L. S. Luskin, *Funct. Monomers: Their Prep., Polym., Appl.,* **2**, 555 (1974) [Eng] 672 refs. [*Chem. Abstr.,* **81**, 105993b (1974).]

"Extraction of Pyridine Bases from Coking By-products," Z. Kisza and A. Pilch-Kowalczyk, *Koks, Smola, Gaz,* **19**, 323 (1974) [Pol]. [*Chem. Abstr.,* **83**, 13007w (1975).]

"Pyridinecarboxylic Acid Thioamides, Thiocarbanilides, and Thiosemicarbazones (as Antituberculosis Drugs)," E. Krueger-Thiemer, H. Kroeger, H. J. Nestler, and J. K. Seydel, *Infektionskr. Ihre Erreger,* **4**, Pt. 3, 289 (1975) [Ger] 170 refs. [*Chem. Abstr.,* **86**, 50330d (1977).]

"Acrolein in the Gas Phase Synthesis of Pyridine Derivatives," H. Beschke and H. Friedrich, *Chem. -Ztg.,* **101**, 377 (1977) [Ger] 61 refs.

"Nicotinamide from β-Picoline," H. Beschke, H. Friedrich, H. Schaefer, and G. Schreyer, *Chem. -Ztg.,* **101**, 384 (1977) [Ger] 51 refs.

"Pyridine Derivatives as Drugs," A. Kleemann, *Chem. -Ztg.,* **101**, 389 (1977) [Ger] 62 refs. Synthetic procedures of pyridine derivatives with pharmaceutical activity.

"Catalytic Synthesis of Pyridine Bases," K. Akhmerov, D. Yusupov, and A. B. Kuchkarov, *Uzb. Khim. Zh.,* **1978**, 62 [Russ] 21 refs. [*Chem. Abstr.,* **90**, 22691 (1979).]

"Catalytic Methods of Obtaining Pyridine Bases," I. Lazdins and A. Avots, *Khim. Geterotsikl. Soedin.,* **1979**, 1011 [Russ], 234 refs.; [Eng. trans.] *Chem. Heterocycl. Compd.,* **15**, 823 (1980).

"4-Aminopyridine for Protecting Crops from Birds — A Current Review," J. Besser, *Proc. Verteb. Pest. Conf.,* **7**, 11 (1976) [Eng], 25 refs. [*Chem. Abstr.,* **90**, 017455 (1979).] A review of aminopyridine bird frightening agents.

"Toxicity of Pyrithiones," J. G. Black and D. Howes, *Toxical, Annu.,* **3**, 1 (1979) [Eng] 45 refs. The toxicity of hydroxypyridine-thione is reviewed.

"Solid-State Reactions in Organic Conductors and Their Technological Applications," S. Yoshimura and M. Murakami, *Ann. N.Y. Acad. Sci.,* **313**, 269 (1978) [Eng] 106 refs.

"Toxicological Cards. 152. 2,6-Dimethylpyridine," C. Morel, M. Gendre, and A. Cavigneaux, *Cah. Notes Doc.,* **99**, 319 (1980) [Fr] 17 refs. [*Chem. Abstr.,* **93**, 100592h (1980).]

"Pyridine and Pyridine Derivatives," G. L. Goe, in "Kirk-Othmer Encycl. Chem. Technol., 3rd Edit.," **19**, 454 (1982) [Eng] 127 refs. A general overview of the topic including manufacturing, substitution, oxidation, and chemistry.

Author Index

Numbers in parenthesis are reference numbers and indicate that the author's work is referred to although his name is not mentioned in the text. Numbers in *italics* show the pages on which the complete references are listed.

Aasen, A. J., 9(109), *198*
Abaev, G. N., 179(1633), *247*
Abdalla, M., 167(1400), *239*
Abd ElHalim, M. S., 167(1403), *239*
Abd El-Rahman, S., 167(1399), *239*
Abdesaken, F., 38(446), *209*
Abdulla, F., 124(1037), *228*
Abdulla, R. F., 123(1044a-b), *228*
Abdullina, R. M., 35(428, 429), *208*
Abdurakhmanov, A., 177(1581–1584), *246*
Abdusamatov, A., 14(183, 187, 193, 200–204), *200, 201*
Abe, Y., 95(734), *218*
Abel, D., 175(1523), *244*
Abjean, F., 70(600), *214*
Abramenko, P. I., 164(1384), *239*
Abramova, M. A., 259(23), 423(23), *436*
Abramovitch, R. A., 40(457), 42(469), 90(702), *209, 217*, 339(315), 345(315), *445*
Abu El-Haj, M. J., 126(1065), *229*
Achremowicz, L., 27(302), 204
Adachi, I., 71(616), 72(616), *214*
Adachi, J., 169(1430), *240*
Adachi, S., 181(1698), *250*
Adams, V. D., 159(1354), *238*
Adamus, M., 65(586), *213*
Adda, J., 35(426, 430), *208*
Adkins, H., 306, 306(163), *440*
Adler, B., 191(1773), *252*
Advani, B. G., 146(1191), *233*
Afifi, M. S., 14(191), *200*
Afonina, I. I., 137(1130, 1131), *231*
Afridi, A. S., 76(627), 79(667), 81(683), 82(683, 689), 91(707), *214, 216, 217*
Agui, H., 191(1775), *252*
Ah-Kow, G., 165(1391), *239*
Ahlers, K., 115(976), *226*
Ahmad, I., 10(119), *198*
Aigner, H., 169(1429), *240*, 282(104), *439*
Ajello, E., 71(613), *214*
Akano, M., 182(1703), *250*
Akhinyan, R. M., 6(60), *196*
Akhmerov, K. M., 177(1581–1584, 1586, 1587), 181(1692), *246, 249*

Akhnookh, Y., 167(1398, 1399), *239*
Akhtar, I. A., 80(679), *216*
Akhtar, M., 80(679), *216*
Akhtar, M. A., 170(1432), *240*
Akimora, T. I., 257(21), *436*
Akimova, T. L., 121(1024), *228*
Akopyan, A. N., 76(630, 631), *214*
Akramov, S. T., 15(206), *201*
Akulov, P. V., 24(284), *203*
Alajarin, M., 93(717), *217*
Alberola, A., 185(1722), *250*
Albert, A., 94(729), *218*
Albert, D. K., 29(360), *206*
Alberts, A. H., 456(48), *621*
Alcock, N. W., 476, 476(115), 533(115), 537(115), 569(115), 577(115), *623*
Aleksandrova, E. A., 190(1758), *251*
Aleksandrova, O. A., 545(164), *624*
Alekseeva, N. A., 28(366), *206*
Alessi, J. T., 74(624), *214*
Alexander, M. D., 533(147), *624*
Ali, E., 23(266), *203*
Aliev, A. G., 150(1292), *236*
Alimbaeva, P. K., 15(207), *201*
Ali-Zade, N. L., 113(935), *224*
Allen, M. J., 301(154), 345(154), *440*
Alles, H. U., 183(1711, 1712), *250*
Allison, W. S., 3(10), *195*
Alsuf'ev, V. A., 33(398), *207*
Alverez-Insua, A. S., 185(1723), *250*
Amakasu, T., 150(1291), *236*
Ammon, H. L., 285(111), 334, 334(111), 339(111), *439*
Anasstassiou, A. G., 285(109, 110), 335, 335(110), *439*
Andabu, M. B., 111(834), *221*
Andaburskaya, M. B., 167(1417), *240*
Anders, D. E., 31(380), *207*
Anderson, A. G., Jr., 285(111), 334, 334(111), 339(111), *439*
Anderson, L. B., 533(127), *623*
Anderson, R. C., 146(1195), 159(1354), 180(1674), *233, 238, 249*, 262(36), 263(41), 419(41), 421(41), *437*

Ando, K., 20(243), *202*
Ando, M., 483(237), 593(234b-c, 237, 404), *626, 627, 631*
Andreichikov, Y. P., 89(915), *224*
Andreyuk, E. I., 6(48), *196*
Anet, F. A. L., 325, 325(325), *445*
Angadiyavar, C. S., 168(1428), *240*
Angenot, L., 23(267), *203*
Anghelide, N., 126(1057), *229*
Anisimova, O. S., 162(1373), *239*
Ansell, C. W. G., 549(458), 551(437, 456, 457), *632, 633*
Antaki, H., 263(44), *437*
Antonova, V. V., 173(1472), 179(1628, 1635, 1640, 1641, 1645), 180(1673), 181(1687–1689), *242, 245, 247, 248, 249*
Anwar, M. M., 26(329), 27(331), *205*
Anwar, R. A., 7(63), 181(1697), *196, 250*
Aoi, I., 155(1311), *237*
Aoshima, S., 180(1678), *249*
Appel, H. H., 324(251), *443*
Arab, Al-, M. M., 76(636), 77(636), *215*
Arackal, T. J., 57(524, 525), *211*
Araki, E., 64(546), *212*
Areshidze, Kh. I., 181(1691), *249*
Ariesan, V., 148(1214), *234*
Armendt, B. F., 324, 324(244), *443*
Arnall, P., 324, 324(246, 247), *443*
Arndt, R. R., 591(369), *630*
Arnold, L. J., Jr., 3(10), *195*
Aronson, A., 12(161), *199*
Arques, A., 93(717), *217*
Arsenault, G. P., 325(258), *443*
Arthur, H. R., 14(196), *201*
Aryuzina, V. M., 112(849, 850), *221*
Asai, S., 66(595), *213*
Asakawa, M., 20(251), *202*
Asano, H., 13(176), *200*
Asao, T., 95(734), *218*
Asao, Y., 35(423), *208*
Asaoka, H., 55(504), *210*
Asay, B. W., 515(318), *629*
Asay, R. E., 463(76), 464(76, 89), 481(76), 495(76), 515(76, 87, 89, 316), 517(76, 87), 583(76), 611(76, 87), 613(76), *622, 629*
Asegawa, N., 180(1672), *249*
Ash, A. B., 171(1435a-b, 1436, 1437, 1438), *241*
Aslanov, K. A., 108(798), *220*
Astakhova, A. S., 108(803), *220*
Astakhova, L. G., 9(94), *197*
Atavin, A. S., 182(1708, 1709, 1710), *250*
Atmeh, R. F., 79(671), 124(1038), *216, 228*
Atwood, J. L., 476(212, 214), 482(231), 573(212, 214), 589(231), *626*

Aubouet, J., 138(1140), *231*
Auda, H., 324(251), *443*
Aue, D. H., 46(480), 184(1716), *210, 250*
Austin, D. J., 9(97), *198*
Austin, P. W., 148(1212), *234*
Avots, A., 172(1463), 177(1545, 1546), 179(1545, 1664, 1665), 179(1666), 181(1682–1684, 1686), *242, 244, 248, 249*
Awasthi, A. K., 170(1447), *241*
Ayer, W. A., 22(263), *203*, 395(306), *445*
Ayres, D. C., 115(970), *226*
Azawi, Al, F., 160(1359), *238*
Azimov, V. A., 112(849, 850), *221*
Azuma, K., 65(575), *213*

Babayan, G. S., 6(49), *196*
Backer-Dirks, J. D. J., 541(356), *630*
Baddar, F. G., 77(643), *215*
Bader, H. J., 87(884), 170(1446), 171(1445), *222, 241*
Badilescu, I. I., 492(271), 607(271), *628*
Badita, G., 42(465), *200*, 300(148, 151), 425(148), 431(150), *440*
Baetz, F., 38(447–449), 173(1516), *209, 243*
Bagal, L. I., 183(1713), *250*
Baidova, A. Y., 28(361), *206*
Baig, K. M., 7(63), *196*
Baikova, A. Y., 28(366, 368), *206*
Bailey, J. R., 323, 323(242, 243), 324, 324(244, 245), 339(243), *443*
Baker, F. S., 56(516), *211*
Baker, W., 452, 452(24, 26), 505(24), *620*
Baker, W. A., Jr., 533(330), *629*
Balaban, A. T., 86(699), 87(888, 889), 88(901, 903), 89(914, 917, 919), *217, 223, 224*, 309(175), 423(175), *441*, 450(17, 18), 492(269, 270), 501(17), 607(269–272), *620, 627, 628*
Balabanova, L. N., 182(1710), *250*
Balbaa, S. I., 14(191), *200*
Baldridge, H. D., Jr., 591(367, 368), *630*
Baldwin, J. E., 41(460), *209*
Baldwin, J. J., 130(1085), *230*
Baliah, V., 317(217), 411(217), *442*
Balicki, R., 107(777), 160(1358), 185(1725, 1730), *219, 238, 251*
Balitskaya, A. K., 6(55), *196*
Ball, G. J. H., 591(369, 372), *630*
Baltz, H., 177(1551), 179(1646), *244, 248*
Balyakina, M. V., 64(551), 148(1226), *212, 234*
Bamfield, P., 539(124, 168), *623, 624*
Banerjee, A. K., 6(30), *195*
Banetjee, K. S., 78(650), *215*
Banks, R. E., 40(456), 113(959), *209, 225*

Bapat, J. B., 88(903), *223*

Baranov, S. N., 87(895), *223*

Baraze, A., 188(1746–1748), *251*

Barbier, M., 13(173), *200*

Barbulescu, N., 42(465), 89(922, 924, 925), 121(1012, 1019, 1023), *209, 224, 227, 228,* 263(40), 300(151, 152), 309(176, 177), 431(150), *437, 440, 441*

Barbulescu, N. S., 300(148), 309(175), 423(175), 425(148), *440, 441*

Barczynski, P., 94(727), *217*

Barefield, E. K., 533(121), *623*

Barkovsky, C., 261(35), 409(35), 423(35), *437*

Barlow, M. G., 40(456), *209*

Barluenga, J., 174(1517), *243*

Baron, J., 26(314), *204*

Barrio, G., 168(1427), 176(1427), *240*

Barrio, J., 168(1427), 176(1427), *240*

Barsch, H., 177(1551), *244*

Barson, J. L., 11(129), *198*

Bartholomew, D. H., 84(855), 85(871), *222*

Bartkevich, B. Z., 190(1762), *252*, 419(296), *444*

Barton, J. W., 288(122), 337(122), *439*

Bartsch, R. A., 462(70), 464(70), 481(70), 513(70), 515(70), 517(70), 583(70), *621*

Bartulin, F. J., 140(1152), *232*

Barza, P., 194(1786), *252*

Basha, F. Z., 23(279), *203*

Battersby, A. R., 615(388), *631*

Battino, F., 511(51), *621*

Battiste, M. A., 44(476), *210*

Bauchon, G., 407(65), *437*

Baum, J., 37(443), 69(601), 70(607), *209, 214*

Baumann, H., 180(1671), *249*

Bautista, R. E., 34(401), *207*

Baxter, R. L., 17(222, 224), 19(222, 224), *201,* 603(382), *631*

Baxter, S. L., 463(82), 464(82,87), 481(82), 515(82, 87), 517(87), 523(82), 583(82), 611(87), *622, 623*

Bayer, E., 150(1273), 157(1335–1337), *236, 237, 238,* 264, 264(46), 267, 267(62), 339(62), 381(62), 403(62), *437, 487(254), 627*

Bayer, H. O., 126(1052), 145(1187), 148(1052), 150(1273), 157(1335–1337), *229, 236, 237, 238*

Baylis, A. B., 177(1557), *245*

Beattie, D. E., 313(193), 325(193, 259), *441, 443*

Becher, J., 144(1181–1184), *233*

Becker, K., 179(1643), *248*

Becker, S. E., 12(143), *199*

Beeby, P. J., 493(277–279, 281), 607(277–279,

281, 385), *628, 631*

Beekhuis, L. G. E., 131(1095), *230*

Begishuili, T. K., 6(42), *195*

Beilis, Y. I., 110(818), *220*

Beissner, G., 452(31), 505(31), *620*

Bekker, Z. E., 12(145), 13(145), *199*

Bekkerova, Z. R., 121(1022), *228*

Belegratis, K., 101(747), 156(1319–1321), *218, 237*

Bell, T. W., 517(322), *629*

Bell, W. H., 57(521), 139(1144), 190(1769), *211, 231, 252*

Bellamy, F. D., 44(473, 474), *209*

Bellingham, P., 114(947), *225*

Bellus, D., 51(491), *210*

Belova, T. P., 64(557), *212*

Bel'skii, F. I., 52(492), *210*

Bel'skii, I. F., 52(492), *210*

Belsky, I., 83(694), 86(694), *216*

Belyaeva, O. Ya., 192(1779), *252*

Bembi, R., 543(455), *633*

Bencze, W. L., 301(154), 345(154), *440*

Benes, J., 3(7), *194*

Ben'kovskii, V. G., 28(361, 366, 368), *206*

Bennett, G. B., 158(1347), *238*

Benson, R. H., 173(1481), *242*

Benton, W. E., 591(234a), *627*

Benton, W. H., 480(105, 197), 484(241a), 488(257), 527(105), 529(105), 557(197), 561(197), 563(105), 583(197), 585(105), 603(257), *623, 625, 627*

Benzie, R. J., 114(937–939), *224*

Beranger, S., 150(1282, 1284), *236*

Berestevich, B. K., 490(262), 601(262), 627

Berg, C. P., 3(2), *194*

Berg, W., 14(181), *200*

Berge, H., 110(816), *220*

Bergen, T. J. V., 489(286), 495(285, 286), 611(285, 286), *628*

Bergmann, F., 110(837), 111(837), *221*

Berg-Nielsen, K., 191(1774), *252*

Bergson, G., 334(283), *444*

Berk, H. C., 285(314), 345(314), *445*

Bernardi, F., 505(308), *628*

Bernardi, R., 347(309), *445*

Bernatek, E., 156(1323), *237*

Bernhard, R. A., 54(510), *211*

Bernofsky, C., 3(13), *195*

Berrie, A. H., 147(1208a), 148(1208a, 1209), 150(1294), *233, 236*

Berrouschot, H. D., 179(1643), *248*

Berry, D. J., 295(137), 297(137), *439*

Berry, R. S., 296, 296(140), 337(140), *440*

Berthet, D., 9(113), *198*

Bertini, V., 133(1112), *231*

Beruto, D., 177(1577), *245*

Beschke, H., 171(1454, 1455), 172(1458–1462, 1465, 1470, 1471, 1473, 1474), *241, 242, 254*(1), *436*

Bespalov, K. P., 179(1641, 1645), *248*

Besselievre, R., 23(272), *203*

Bessho, K., 503(306), 589(306), *628*

Bezborodov, A. M., 12(147, 148), *199*

Bezinger, N. N., 31(379), *207*

Bhaskaran, R., 12(142), *199*

Bialek, J., 115(974), *226*

Biasotti, J. B., 486(244, 245, 247), *627*

Bick, I. R. C., 8(74), *197*

Bickelhaupt, F., 132(1104), 148(1211), *230, 234,* 310, 310(182), 311, 311(188), *441*

Bicker, U., 40(452), *209*

Biehler, J. M., 111(841), *221*

Biellmann, J. F., 57(523), 108(812–814), *211, 220*

Biemann, K., 325(256), 443, 449(12), 450(12), 487(12), 501(12), 599(12), *620*

Bigalke, R. C., 13(177), *200*

Bilai, V. I., 6(54), 12(146–148), *196, 199*

Billups, W. E., 255(7), *436*

Billy, J. M., 257(18), 423(18), *436*

Bimecki, S., 78(660), *215*

Binder, D., 273, 273(74), 369(76), *438*

Biniecki, S., 83(688), *216*

Birch, A. J., 116(987), *226,* 306(164), 325(257), 425(164), *440, 443*

Bird, P., 545(175), *625*

Birkofer, L., 160(1357), *238*

Birnbaum, D., 6(25–28), *195*

Bischoff, C., 146(1197), 183(1714), *233, 250*

Bisenbaev, O., 6(50), *196*

Bishop, M. M., 549(176, 180), 551(180), *625*

Bito, T., 293(133), 341(133), *439*

Black, D. St. C., 535(132), 537(132), 623(132), *623*

Blidner, B. B., 496(293), 609(293), 611(293), *628*

Blount, J. F., 23(276), *203*

Blumbergs, P., 171(1435a, 1436), *241*

Boatman, S., 124(1043), *228*

Bobbitt, J. M., 133(1107), *230*

Boccato, G., 177(1553, 1573–1575), *245*

Bodem, G. B., 8(83), 9(83), *197*

Boehm, S., 73(622), *214*

Boekelheide, V., 450(14), 452, 452(28, 32, 33), 453(33, 34), 454, 454(28), 456, 456(33), 492(273), 497(299), 503(32–34), 505(28, 32), 507(33), 557(32–34), 559(33), 579(299), 607(273), 609(273), 619(299, 299a), *620, 621, 628*

Boell, W., 63(543–545), *211, 212*

Boennemann, H., 176(1524, 1528, 1529, 1532, 1533), *244*

Boer, F. P., 289(127), *439*

Boer, Th. J. de, 609(387), *631*

Boettcher, F., 6(25–28), *195*

Boettcher, F. P., 295, 295(138), *439*

Bogdanov, V. S., 190(1761), *252*

Bogdanowicz-Szwed, K., 133(1106), *230*

Bogdanowicz-Wzwed, K., 347(316), *445*

Boger, D. L., 291(323), *445*

Boggess, R. K., 533(331), 535(331), 537(331), 569(331), 577(331), *629*

Bogomolova, L. A., 12(146, 148), *199*

Bohleier, E., 480(196), *625*

Bohlmann, F., 266(55), *437*

Bollet, C., 28(365), *206*

Bol'shakov, G. F., 28(355), *206*

Bolt, A. J. N., 8(72), 10(72), *197*

Bomika, Z., 111(834), 167(1417), *221, 240*

Bonjoch, J., 264(322), *445*

Bonollo, L., 9(102), *198*

Borch, R. F., 126(1063), *229*

Borivoje, B., 183(1715), *250*

Borkhi, L. D., 115(971, 972), *226*

Borlai, O., 26(326), *205*

Borodkin, V. F., 478(220), 533(166), 539(166), 541(165), 545(160–164), 547(167), 577(220), *624, 626*

Bosch, J., 264(322), *445*

Bossert, F., 185(1721), *250*

Bota, A., 87(889), 89(919), *223, 224*

Botteghi, C., 84(863), 85(878), 118(999), *222, 227*

Bottino, F., 458(51), 561(192), 563(193, 359), *621, 623, 625, 630*

Bottomley, L. A., 541(447), *632*

Boucherle, A., 185(1729), *251*

Bouchon, G., 150(1272), *236,* 267(65), 341(65), 383(65), *437*

Boulton, A. J., 88(902), *223*

Bouquet, A., 23(269), *203*

Bowden, R. D., 40(455), 84(867), 85(867, 872–875), 114(948), 122(1028), 135(1118), 139(1146), 149(1241), *209, 222, 225, 228, 231, 232, 234*

Boyarintseva, O. N., 162(1373), *239*

Boyko, A. L., 34(413), *208*

Bozhkov, V. M., 3(9), *194*

Bozhkova, I. Z., 8(75), *197*

Bradamante, S., 485(243), 597(242, 243), *627*

Braden, R., 173(1506–1508), *243*

Bradshaw, J. S., 463(76, 78–80, 82, 83), 464(76, 82, 83, 87–89), 481(76, 82),

495(76), 515(76, 78–80, 82, 87–89, 315, 316, 320, 321, 401, 433), 517(76), *622, 623, 629, 632, 633*

Brandange, S., 13(175), *200*

Brandenburg, C. F., 26(298), 28(363), *204, 206*

Braun, J. v., 303, 303(156, 157), *440*

Braun, R. D., 110(815), *220*

Braz, R., 23(277), *203*

Bredereck, H., 186(1734), *251*

Breitmaier, E., 150(1272, 1273), 157(1335–1337, 1342), *236, 237, 238*, 264, 264(46), 267, 267(62, 65), 339(62), 341(65), 381(61), 383(65), 407(65), *437*, 487(254), *627*

Breitmaier, V., 403(62), *437*

Bremner, J. B., 8(74), *197*

Bren, V. A., 88(906), *223*

Brenner, M., 104(757), *218*

Brezezinski, J. Z., 315(209), 339(209), 403(209), *441*

Brien, D. J., 357(313), 395(313), *445*

Brinkmann, R., 176(1528, 1532), *244*

Brinkmeyer, R. S., 123(1044b), *228*

Brody, F., 118(998), 147(998), 166(998), *227*

Bronshtein, A. P., 24(284, 288), *203*

Broquist, H. P., 324(250), 443

Broussard-Simpson, J., 464(90), 519(90, 91), 521(90, 91), *622*

Brouwer, W. G., 170(1432), *240*

Brown, D., 29(370), *206*

Brown, D. J., 94(726), 159(1353), *217, 238*

Brown, E. V., 10(119), *198*

Brown, J. M., 493(279, 280, 282), 607(279, 280), *628*

Brown, M. R. W., 12(162), *199*

Brown, P. R., 517(459), *633*

Brown, R. T., 23(268), *203*

Browne, L. M., 22(263), *203*, 395(306), *445*

Brownlee, R. T. C., 86(700), *217*

Bruhin, J., 495(290), 613(290), *628*

Bruice, T. C., 108(806), *220*

Bruna, L., 78(649), *215*

Brunskill, J. S. A., 185(1726), *251*

Brus, G., 101(747), *218*

Brushin, J., 482, 482(229), 483(229, 233), 587(229, 233), 595(229, 233), 607(290), *626, 628*

Bryan, P. S., 533(130, 135), 535(130), 555(194), *623, 625*

Bryan, R. F., 19(236), *202*

Bryson, T. A., 125(1066), 127(1067), *229*

Buchardt, O., 92(710, 711), *217*

Buchi, G., 449, 449(12), 450(12), 501(12), 599(12), *620*

Buchi, W., 33(432), *208*

Buchmann, G., 178(1590), *246*

Budina, T. A., 121(1024), *228*, 257(21), *436*

Budzikiewicz, H., 12(168), 15(209), 18(228), *200, 201, 202*

Buerstell, H., 3(3), *194*

Buggle, K. M., 452(24), 505(24), *620*

Buhleier, E., 476(156, 158), 537(156), 547(156, 158), 549(158), 561(196), 563(196), 569(156, 360), 571(362–363), 575(158), 583(196), *624, 625, 630*

Bulatova, B. T., 28(361, 366), *206*

Bulgakova, N. B., 85(877, 879), *222*, 298(144), 310(179), *440, 441*

Bullock, E., 105(759), *218*

Bunkina, N. A., 24(288), *203*

Buratti, L., 99(100), *198*

Burdick, D., 9(110), *198*

Burg, B., 102(751), 103(751), *218*

Burger, B. V., 13(177), *200*

Burke, B. A., 16(221), *201*

Burlingame, A., 31(381), *207*

Burnett, M. G., 476(206), 573(206), *626*

Bursey, M. M., 186(1732), *251*

Burtzlaff, C., 191(1773), *252*

Buryak, A. L., 87(895), *223*

Buryan, P., 26(299), *204*

Busby, R. E., 56(516), *211*

Busch, D. H., 478, 478(110, 217, 218), 533(109, 117, 118, 120–122, 125–127, 335, 338, 348), 577(217), *623, 624, 626, 629, 630*

Buschi, G., 487(12), *620*

Buschmann, E., 99(744, 745), 100(744, 745), *218*

Bush, L. P., 9(84, 89, 96), 11(130), *197, 199*

Butler, D. N., 325(257), *443*

Butler, J. D., 48(483, 484), 52(483), *210*

Butsugan, Y., 14(178, 180), *200*, 281(101), 293(133), 341(133), 369(101), *438, 439*

Butt, A., 80(680), *216*

Butt, M. A., 80(679), *216*

Buttery, R. G., 33(395), 35(424), *207, 208*

Buyske, D. A., 9(114), *198*

Bycroft, B. W., 605(375), *630*

Bye, G., 53(500), *210*

Byerrum, R. U., 6(34), 11(137), *195, 199*

Bylicki, A., 28(350), *206*

Cabral, J. de O., 479(223), 539(423, 424), 541(170, 172, 423, 424), 581(223), *624, 625, 626, 632*

Cabral, M. F., 479(223), 539(423, 424), 541(170, 172, 423, 424), 581(223), *624, 625, 626, 632*

Cabrerizo, M. A., 167(1411), 185(1724), *240, 250*

Caccia, G., 85(878), 118(999), *222, 227*

Cairns, C., 478(219), 535(145), 579(219), *624, 626*

Callot, H. J., 108(84), *220*

Cals, S., 326(265), *443*

Caluwe, P., 117(995), 165(1394), *226, 239, 264(50, 51), 365(50), *437*

Camparini, A., 165(1389), *239*

Campazzi, G., 177(1578), *245*

Campbell, A., 84(855), 85(871), *222*

Campos, B., 34(402), *207*

Cane, S., 179(1656, 1657), *248*

Cantello, B. C. C., 149(1241), *234*

Capell, L. T., 459(1), *620*

Capps, T. M., 324(253), *443*

Carabateas, P. M., 107(774), *219*

Caramella, P., 71(611, 614), *214*

Carbateas, P. M., 167(1414), *240*

Carey, J. G., 84(866), 107(784), *219, 222*

Carlsen, P. H. J., 43(470, 471), *209*

Carlson, G. R., 79(675), *216*

Caronna, T., 347(309), *445*

Carpenter, A. P., Jr., 31(384), *207*

Carr, R. M., 168(1426), *240*

Carter, D. R., 289(127), *439*

Carter, G. B., 139(1144), *231*

Case, E. E., 525(101), *623*

Casey, W. J., 9(98), *198*

Castagnoli, N., 11(105), *198*

Catalano, A. W., 130(1081), *229*

Catalucci, E., 173(1503), *243*

Caude, M., 28(365), *206*

Caullet, C., 58(527–529), *211*

Cave, A., 23(269), *203*

Cavill, G. W. K., 14(186), *200*, 259(25, 26), 313, 313(191), 369(25, 191), *436, 441*

Celon, E., 64(554), *212*

Cerrini, S., 56(520), *211*

Chadwick, C., 122(1027), *228*

Chafetz, H., 146(1194–1196), 180(1674), *233, 249, 262*, 262(36, 38), 263(38, 41, 42), 419(38, 41, 42), 421(38, 41, 42), *437*

Chakravorti, S., 489(258), 603(258), *627*

Chalenko, V. G., 12(153), *199*

Chalk, A. J., 146(1200, 1201), *233*

Chamberlan, W. J., 9(110), *198*

Chamberlin, E. M., 65(587), *213*

Chambers, R. D., 115(979), *226*

Chandler, C. J., 521(448), *633*

Chandler, J. L. R., 3(14), *195*

Chandler, J. R., 3(17), *195*

Chandra, R., 180(1668), *249*

Chang, C. D., 177(1565), *245*

Chang, S. C., 74(624), 86(881), *214, 222*

Chaparro, B. F., 34(401), *207*

Chapple, C. L., 23(268), *203*

Charman, H. B., 178(1600, 1601), *246*

Chasin, L. A., 12(154), *199*

Chattergea, J. N., 359(305), 363(305), 413(305), *445*

Chatterjee, A., 11(139), 14(194, 195), *199, 201*

Chauvelier, J., 138(1139, 1140), *231*

Chauvin, M., 138(1140), *231*

Che, C. M., 533(133), *623*

Cheban, I. A., 6(48), *196*

Chebotareva, E. G., 182(1708), *250*

Chekhun, V. P., 64(557), *212*

Chekmareva, I. B., 115(965), *225*, 595(373), *630*

Chen, C. T., 171(1441), *241*

Chen, D. C., 171(1441), *241*

Chen, K., 603(406), *631*

Chen, S. M., 309(174), 423(174), 425(174), *441*

Chen, T. S., 105(759), *218*

Chen, Y. I., 171(1441), *241*

Cheng, C. C., 149(1263), 165(1392), *235, 239*

Cheng, P. G., 517(322), *629*

Chennat, T., 107(775), 219

Cherenkova, G. I., 289(124), 373(124), 375(124), *439*

Cherkosova, I. S., 65(574), *213*

Chermenskii, D. N., 12(148), *199*

Chernyavskaya, M. N., 25(311), *204*

Cherry, P. C., 33(393), *207*

Chiancone, F. M., 3(6), 9(102), *194, 198*

Chiari, G., 525(101), *623*

Chichibabin, A., 261(35), 409(35), 423(35), *437*

Chichibabin, A. E., 261(33), 409(33), 419(33), *437*

Chickenkova, L. G., 122(1031), *228*

Chindris, E., 148(1214), *234*

Chiraleu, F., 87(889), 88(903), 89(919), *223, 224*

Chirazi, M. A. A., 150(1286), *236*

Chistyakov, A. N., 115(973), *226*

Chiu, Y. Y. H., 23(278), *203*

Chivadze, G. O., 181(1691), *249*

Chorvat, R. J., 145(1188, 1189), *233, 327(267), *443*

Christ, B. G., 20(256), *202*

Christ, V., 13(177), *200*

Christensen, J. J., 462(73), 436(76, 83), 464(76, 83, 87–89), 481(76), 495(76),

513(73), 515(76, 87–89, 315, 317, 402), 517(76, 87, 459), 583(76), 611(76, 87), 613(76), *622, 629, 631*

Christman, D. R., 6(34), *195*

Chukhrii, F. N., 87(896), *223*

Chumak, A. D., 120(1001, 1007), 147(1007), 227

Chumakov, Y. I., 84(859, 861), 85(877, 879), 222

Chumakov, Yu., I., 298(144), 310(179), *440, 441*

Churchill, M. R., 581(365, 366), *630*

Cislak, F. E., 24(282), 179(1636), *203, 247*

Claramunt-Elguero, R. M., 88(903), *223*

Clardy, J., 22(259), *203*

Clark, B. A. J., 189(1754), *251*

Clark, K. J., 259(24), *436*

Clauson-Kaas, N., 48(489), 52(494), 53(499), *210*

Clement, R. A., 176(1525), *244*

Clemo, G. R., 27(301), *204*

Cliffton, M. D., 53(502), *210*

Cobb, R. L., 131(1090), *230*

Cohausz, G., 140(1151), *232*

Cojocaru, Z., 148(1214), *234*

Coke, J. L., 591(371), *630*

Colchester, J. E., 84(852, 856, 866), 107(784), 112(851), *219, 221, 222*

Collin, G., 147(1205, 1206), 179(1661), *233, 248*

Collins, E., 35(419), *208*

Cologne, J., 256(14, 15), 257(16), 339(16), 411(15), 423(15), 435(15), *436*

Colombini, C., 64(554), *212*

Colonna, F. P., 505(308), *628*

Coman, F., 194(1786), *252*

Constable, E. C., 553(446, 454), *632, 633*

Cook, D. H., 476(143, 144, 200, 203, 204), 478(143), 533(143, 203), 535(143, 144), 569(143, 144, 200, 203, 204), 573(143), 577(143), *624, 625*

Cook, G. L., 29(371–374), *206*

Cook, J. D., 295(137), 296(142), 297(137), *439, 440*

Cook, J. M., 77(642), *215*

Cook, P. D., 126(1059), *229*

Cook, S. T. M., 26(329), *205*

Cooke, M. P., Jr., 454, 454(43), 475(43), 561(43), *621*

Coonradt, H. L., 306, 306(162), *440*

Cooper, L. E., 179(1656, 1657), *248*

Coppock, B. M., 33(397), *207*

Corbier, B., 22(264), 120(264), *203*

Cordier, P., 129(1078), *229*

Core, A. C., 278(87), *438*

Coreil, M., 475(100a), *622*

Corran, J. A., 112(844, 851), 173(1480, 1484), *221, 242*

Correa, D. de B., 13(171), *200*

Corson, F. P., 41(461, 462), *209*

Cosentino, S., 458(51), 511(51), *621*

Cossey, A. L., 153(1326–1328, 1332), *237*

Costello, T., 489(260), 603(260), *627*

Crabtree, A., 148(1212), *234*

Craig, D., 178(1625), *247*

Cram, D. J., 459, 459(54–56, 58–62), 460(55, 56, 63, 64), 461(55), 465(54, 55), 466(55, 56, 63), 503(304), 505(43, 304), 513(54–56, 58, 63, 64), 517(55, 58, 60–62, 322), 519(55, 58, 60–62), 521(54–56, 63), 523(54–56, 63), 559(304), *621, 628, 629*

Crawford, L., 33(394), *207*

Creegan, F. J., 73(620), *214*

Crist, J. G., 177(1585), *246*

Crombie, L., 17(222–224), 19(222–224), *201, 603(381, 382), 631*

Crombie, W. M. L., 17(223, 224), 19(223, 224), *201, 603(381, 382), 631*

Crossley, R., 120(1004), *227, 313(193, 194), 325(193, 259), 341(194), 411(194), 425(194), 435(194), 441, 443*

Crow, W. D., 15(219), 16(219), 31(385), 42(467), *201, 207, 209, 286(114–116), 439*

Cummins, W. J., 541(170), *624*

Cunsolo, S., 458(51), 511(51), *621*

Cuong, L. D., 185(1728, 1729), *251*

Curran, A. C. W., 116(986), 120(1004, 1005), 157(1338–1340), 167(1418), *226, 227, 238, 240, 270, 270(67), 313, 313(193, 194), 314(195), 325(193, 259), 341(194), 411(194), 425(194), 435(194), 438, 441, 443*

Currie, R. B., 64(562), *212*

Curry, J. D., 533(109), *623*

Curtze, J., 122(1033), *228*

Czarnecki, J., 26(325), *205*

Czerwinska, E., 152(1298), *236*

Dabrowiak, J. C., 533(130), 535(130), 623

Da-gang, W., 603(415, 416), *632*

Dagne, E., 11(105), *198*

Dahm, F. L., 171(1454), *241*

Dalley, N. K., 515(319), *629*

Dammertz, W., 266(54), 349(301), 383(54), 393(300), *437, 445*

Daniels, P. J. L., 279(88), 322(236), 326(236), 389(88), 393(88), 395(236), 399(88, 236), *438, 443*

Danishefsky, S., 173(1514a, b), *243*

Danish-Khoshboo, F., 464(90, 91), 480(197), 519(90, 91), 521(90, 91), 557(197),

561(197), 583(197), *622, 625*
Darby, W. J., 6(36), *195*
Darlington, W. A., 12(164), *200*
Darragh, J. I., 141(1165, 1166), 179(1644), *232, 248*
Das, B. C., 12(198), 14(198), 15(198), *201*
Daugherty, H. W., 6(35), *195*
Daum, G., 114(945), *225*
Dauphin, G., 191(1771, 1772), *252*, 282(105), 283(105), 373(105), *439*
Dave, K. G., 133(1108), *230*
Davenport, R. W., 486(244, 245), *627*
David, L., 191(1772), *252*
David, S., 158(1350), *238*
Davidson, R. B., 515(402, 433), *631, 632*
Davies, L. B., 97(739), *218*, 292(132), 317(132), *439*
Davis, D. L., 11(130), *199*
Davis, N. R., 181(1697), *250*
Dawson, R. F., 6(34), *195*
Day, R. T., 126(1059), *229*
Deady, L. W., 521(448), *633*
De Almeida, M. E. L., 23(277), *203*
De Angelis, F., 55(514), *211*
Decora, A. W., 29(371–373), *206*
Deeb, A., 167(1400), *239*
Deeney, F. A., 533(139), *624*
Deguchi, T., 3(8), *194*
Dehnert, J., 153(1329), *237*
De Jong, F., 459(58, 61), 513(58), 517(58, 61), 519(58, 61), *621*
Delaveau, P., 23(272), *203*
Delboeve, L. T. J., 395(306), *445*
Del Carmen, M., 168(1427), 176(1427), *240*
Delord, T. J., 463(86), 515(86), 527(86), *622*
Delplace, H., 256(14, 15), 411(15), 423(15), 435(15), *436*
Del Rio-Estrada, C., 6(35), *195*
DeMartino, U., 256(13), 415(13), 423(13), *436*
Dembech, P., 505(308), *628*
Demidenko, E. I., 87(882), *222*
Demole, C., 8(69), 10(69), *197*
Demole, E., 8(69), 10(69), 9(113), *197, 198*
Demole, E. P., 8(71), *197*
Demoulin, A., 45(477), *210*, 291(130), *439*
Dempsey, W. B., 6(21, 23), *195*
DeMunno, A., 133(1112), *231*
Den Hertog, H. J., 294(135, 136), *439*
De Nie-Sarink, M. J., 108(804), *220*
Denisova, G. K., 179(1633, 1640), *247, 248*
Denzel, T., 150(1283), 164(1387), *236, 239*
Derbenev, V. V., 90(704, 705), *217*
Derijckere, A. M., 188(1749–1751), *251*
De Salles de Hys, L., 10(127), *198*

Desbois, M., 111(841, 842), 142(1168), *221, 232*
Deshapande, S. S., 78(650), *215*
Deuchert, K., 492(274), 495(274), 609(274), 613(274), *628*
Deumens, J. J. M., 131(1086, 1088, 1092, 1093, 1097, 1099, 1100), 139(1145), *232, 239*
De Valois, P. J., 35(421), *208*
De Voghel, G. J., 153(1325), *237*
DeVries, J. G., 495(288), 611(288), *628*
DeWall, H. L., 591(369, 372), *630*
Dewar, M. J. S., 60(532), *211*
Dewar, R., 533(337), *629*
Dewing, J., 139(1144), *231*
Dich, T. C., 84(863), *222*
Dickinson, E. M., 14(182), *200*
Dierkes, U., 140(1150), *232*
Dieterich, D., 173(1506–1509, 1511–1513), *243*
Dietz, K. P., 40(452), *209*
Diller, D., 110(837), 111(837), *221*
DiMari, S. J., 324(250), *443*
Dimroth, K., 86(698), *217*, 450(19), *620*
Dinculescu, A., 88(903), *223*
Dinkel, R., 178(1599), *246*
Dirinck, P., 34(411), *208*
Disler, E. N., 12(147), *199*
Distefano, G., 505(388), *628*
Dittmar, W., 77(644), 102(751, 754), 103(751, 754), 126(1054, 1055), *215, 218, 229*, 290(129), 373(129), *439*
Dittmer, D. C., 496(293, 297), 609(293), 611(293), 617(297), *628*
Divekar, P. V., 13(174), *200*
Dixon, G. T., 313(193), 325(193), *441*
Djerassi, C., 31(388), *207*
Dmitrieva, N. F., 108(803), *220*
Dodiuk, H., 83(694), 86(694), *216*
Dokshina, N. D., 31(377), *207*
Doktorova, L. I., 177(1548), 178(1589), *244, 246*
Doktorova, N. D., 60(535), *211*
Dolejs, L., 18(229), *202*
Dol'skaya, Yu. S., 181(1695), 190(1758–1768), *250, 251, 252*, 292(131), 339(131), 419(296), *439, 444*
Domeier, L. A., 459(62), 517(62), 519(62), *621*
Dominy, B. W., 126(1065), *229*
Domsa, K., 26(327), *205*
Dondoni, A., 93(713, 714), *217*
Donelson, D. M., 127(1067), *229*
Donetti, F., 177(1576), *245*
Doolittle, F. G., 31(380), *207*

Doomes, E., 501(303), 555(194), 569(303), 625, 628

Doren, H. v., 462(74), 494(74), 513(74), 611(74), 613(74), 622

Dorfman, L., 31(386), 207

Dorman, L. C., 78(651), 215

Dorn, H., 164(1386), 239

Dornow, A., 267(63), 437

Dorofeenko, G. N., 87(882, 885, 887, 892–894, 898, 899), 88(906), 89(915, 916, 918, 921), 90(703–705), 92(709), 93(718, 821, 822), 217, 222, 223, 224, 307(166–168), 309(172, 173), 371(171), 397(170), 440, 441

Dorrow, A., 343(63), 437

Dorsky, A. M., 111(840), 221

Doub, L., 148(1222), 149(1264, 1266, 1267), 234, 235

Douglas, A. W., 78(659), 215

Doya, M., 147(1202), 179(1649, 1650, 1660), 233, 248

Doyle, P., 149(1251), 235

Draghici, C., 87(889), 126(1057), 223, 229

Dreiding, A. S., 449(13), 620

Dreier, C., 144(1184), 233

Dresse, A., 23(267), 203

Dreux, J., 256(14, 15), 257(16), 339(16), 411(15), 423(15), 435(15), 436

Drew, M. G. B., 476(143, 200, 205–208), 478, 478(143, 219), 479(223), 533(136, 138, 143, 328, 332, 334), 535(140–143, 145, 146, 333, 352), 539(423, 424), 541(170, 172, 397, 423, 424), 543(174, 398), 545(175), 551(398), 569(143, 200, 205, 207), 573(143, 205, 206, 208, 364, 397, 399, 429), 577(143), 581(223), 623, 624, 625, 626, 629, 630, 631, 632

Drewes, S. E., 462(75), 615(444), 622, 632

Dreyer, D. L., 16(220), 201

Driessen, P. B. J., 39(450), 209, 293(134), 421(134), 435(134), 439

Drobinskaya, N. A., 64(555), 212

Dryanska, M., 177(1560), 245

Dubas-Sluyter, M. A. T., 161(1360), 238

Dubois, M., 23(267), 203

Dubravkova, L., 18(229), 202

Dubs, P., 136(1126), 231, 451(22), 501(22), 620

Dubur, G. Ya., 413(293, 294), 444

Duburs, G., 107(783), 108(809), 110(818), 111(834), 150(1271b), 167(1417), 219, 220, 221, 236, 240, 260(29), 413(29), 436

Duc, T. Q., 185(1728), 251

Ducker, J. W., 182(1704), 250, 311, 311(187), 441

Duffield, A. M., 18(230), 202

Duhamalieva, B. D., 6(55), 196

Dulenko, L. V., 307(166), 440

Dulenko, V. L., 87(899), 223, 307(166, 168), 440

Dumont, J. P., 35(426, 430), 208

Dunlap, R. B., 125(1066), 127(1067), 229

Dupre, M., 78(665), 215

Duran, F., 174(1518), 243

Durbin, D. E., 27(339), 205

Duschek, C., 191(1773), 252

Dutton, B. G., 84(855), 85(871), 222

Dvorko, G. F., 108(799–801), 220

Dyumaev, K. M., 50(928, 929), 224

Dzbanovskii, N. A., 115(972), 226

Dzhemilev, U. M., 107(786), 219

Dziembala, F., 25(310), 204

Dziomko, V. M., 490, 490(262–264), 599(264), 601(262, 263), 627

Easterly, J. P., 81(685), 216

Eberlein, W., 157(1334), 237

Ebisch, R., 188(1744), 251

Eckroth, D. R., 179(1627), 247

Edgar, O. B., 409(290), 444

Effenberger, F., 186(1734), 251

Effler, A. H., 555(186), 559(186), 625

Egberg, D. C., 486(246), 627

Egger, K. W., 190(1770), 252

Eggler, J., 173(1514b), 243

Egorova, M. E., 259(23), 423(23, 298), 436, 444

Eguchi, T., 323, 323(240), 443

Ehler, K. W., 129(1077), 134(1077), 229

Ehm, W., 153(1300), 236

Eicher, T., 38(445, 446), 69(603), 73(622), 76(633), 80(677), 209, 214, 216

Eichler, E., 148(1219, 1224), 234

Eigen, I., 25(293), 204

Eilhauer, H. D., 107(780), 219

Eilmes, J., 156(1324), 237

Eisch, J. J., 274, 274(77, 78), 339(77, 78), 369(77), 438

Eisenberg, R., 533(344), 629

Eisenbraun, E. J., 324(251), 443

Eisner, U., 107(775), 185(1720), 219, 250

Eizen, O. G., 31(378), 206

El-Agamey, A. G., 167(1407), 240

El Ashry, E. S. H., 79(669), 216

El-Bakoush, M. M. S., 189(1754), 251

Elderfield, R. C., 186(1735), 251

Eliel, E. L., 116(984), *226,* 305, 305(160–162),
    306, 306(161, 162), 423(161), *440*
Eliseeva, G. S., 6(46, 58, 62), *196*
Elkasaby, M., 120(1006), 167(1419, 1420),
    227, 240
El-Kerdaway, M., 167(1407), *240*
El-Kholy, I. E., 76(628), 77(640, 645),
    79(671), 83(691, 693), 102(1367), 124(1038),
    214, 215, 216, 228, 238
Ellanskaya, I. A., 6(54), *196*
Ellis, C. A., 279(88, 89), 322(236), 326(236),
    389(88), 393(88), 399(88), *438, 443, 448*
Ellis, P. D., 125(1066), 127(1067), *229*
El-Mowafy, A. M., 307(312), 349(312),
    417(312), 433(312), *445*
Eloy, F. G. F., 188(1749–1751), *251*
El-Rayyes, N. R., 77(641, 643), *215*
El-Shafie, S. M. M., 349(312), 417(312),
    433(312), *445*
Elvidge, J. A., 156(1322), *237*
Elwood, T. A., 186(1732), *251*
El'yanov, B. S., 65(572), *213*
El-Zimaity, T., 271(69), 351(69), 353(69),
    355(69), *438*
Emato, S., 483(234–236, 239–241), 587(234),
    593(234–236, 239), 595(234), *626, 627*
Emele, J. F., (104), *198*
Emmick, T. L., 124(1037), *228*
Emslie, N. D., 615(444), *632*
Endo, H., 20(243, 244), *202*
England, B. T., 94(726), *217*
Enikeeva, N. G., 34(417), *208*
Enomoto, S., 114(940), *225*
Enzell, C. R., 9(97, 109), *198*
Epsztajn, J., 88(902), 120(1003), *223, 227,*
    256(12), 312, 312(190), 314, 314(205, 206),
    315(207, 209), 316(210), 339(12, 205, 209),
    343(190, 210), 345(206), 383(12, 190, 207),
    385(190), 403(209), 405(317), 415(210), *436,*
    441, 442, 445
Erles, M., 116(993), *226*
Ernshaw, D. G., 29(370), *206*
Erofeev, N. S., 12(148), *199*
Escalier, J. C., 28(365), *206*
Esho, F. S., 539(423, 424), 541(172, 423,
    424), 545(175), *625, 632*
Essawy, A., 167(1400), *239*
Etheredge, S. J., 173(1514a), *243*
Etmetchenko, L. H., 371(171), *440*
Etmetchenko, L. N., 87(898), *223*
Eugster, E. H., 11(135), *198*
Evans, G., 149(1255), *235*
Evdokimova, G. A., 27(345), *205*
Eves, C. R., 325(325), *445*

Evstigneeva, R. P., 64(557), *212*
Eweiss, N. F., 88(904), *223*
Ewing, G. D., 285(314), 345(314), *445*

Fabbri, G., 177(1578), *245*
Fabbrizzi, L., 533(339, 343, 428), *629, 632*
Fahmy, A. F., 167(1399), *239*
Fales, H. M., 31(392), *207*
Fanghoenel, E., 188(1744), *251*
Faragher, R., 41(459), *209*
Farkas, R. L., 7(68), *197*
Farkh, Al-, Y. A., 77(641), *215*
Farmery, D. L., 149(1255), *235*
Farmery, K., 533(335), *629*
Farquhar, D. K., 149(1240), *234,* 310(181), *441*
Fattore, V., 177(1553), *245*
Fedeko, W., 56(520), *211*
Fedorov, L. M., 478(220), 577(220), *626*
Fedorov, V. O., 179(1667), 180(1668), *249*
Fedotova, O. V., 121(1008a), 122(1029–1031),
    227, 228
Fedyaeva-Basova, L. P., 122(1025), *228*
Fehlhaber, H. W., 22(265), *203*
Feiring, A. E., 116(983), *226*
Feix, G., 118(997), *227*
Felmeri, J., 149(1237), *234*
Feng, L., 603(416), *632*
Fenical, W., 22(257), *202*
Fenton, D. E., 476, 476(143, 144, 200–204),
    478(143), 533(143, 203), 535(143, 144),
    569(143, 144, 200–204), 573(143, 202),
    577(143), *624, 625*
Fenton, D. M., 178(1620, 1621), *247*
Ferenc, M., 81(686, 687), *216*
Ferraiolo, G., 177(1576, 1577), *245*
Ferraudi, G., 533(422), *632*
Ferrey, M., 72(617), *214*
Ferris, J. P., 130(1081), *229*
Ficini, J., 78(662, 663), *215*
Fields, E. K., 42(464), *209*
Fillion, H., 185(1728, 1729), *251*
Findlay, J. A., 20(280), *203*
Finnegan, R. A., 333(279a), *444*
Fischel, D. L., 286(212, 213), 413(113),
    431(113), *439*
Fischer, B., 533(344), *629*
Fischer, G., 86(699), *217*
Fishel, D. L., 187(1739–1741), *251*
Fisher, M. H., 78(659), *215*
Fisher, R. R., 125(1066), 127(1067), *229*
Fishinger, J. J., 78(659), *215*
Fleischer, E. 533(329, 336, 337), *629*
Fleming, T. L., 41(460), *209*
Fletcher, J. R., 452(27), 503(27), 505(27), *620*

Fleury, J. P., 111(841, 842), 142(1168), *221, 232*

Florent'ev, V. L., 60(531, 535, 536), 64(555, 556), *211, 212*

Floss, H. G., 12(215), *201*

Flowers, J. M., 9(114), *198*

Floyd, W. W., 323(242), *443*

Folkes, D. J., 34(415), *208*

Forbes, C. P., 22(188), *200*

Forcellese, M. L., 56(517, 518), *211*

Ford, D. L., 259(25, 26), 369(25), *436*

Fore, P. E., 515(87, 316), 517(87), 611(87), *622, 629*

Forister, H., 501(302), 503(302), *628*

Forman, M., 12(161), *199*

Forni, L., 173(1495–1497), *243*

Forquey, C., 459(59), 621

Forrester, R. B., 12(165), *200*

Forster, H. G., 462(72), 513(72), 561(302), *622, 628*

Foucaud, A., 72(617), *214*

Fowler, F. W., 73(621), *214*

Fradullo, R., 563(359), *630*

Frandsen, E. G., 144(1181–1184), *233*

Frank, B., 14(184), *200*

Frank, W., 535(420), *632*

Franke, G., 38(446), *209*

Fray, G. E., 259(24), *436*

Frederick, J. L., 115(978), *226*

Frederiks, J. C., 12(151), *199*

Freedman, F., 149(1240), *234*, 254(6), 310(180, 181), 397(180), 405(180), 407(180), *436, 441*, 496, 496(294), 617(294), *628*

Freeman, P. F. H., 148(1227), *234*

Freimiller, L. R., 121(1009), *227*

Frensch, K., 463(77, 81), 481(81), 515(77, 81), 521(77, 81), 523(77), 555(77, 81), 569(360), 583(77), *622, 630*

Fresneda, P. M., 93(717), *217*

Freytag, J., 324(248), *443*

Frick, H., 64(553), *212*

Friedman, S., 27(300), *204*

Friedmann, N., 36(435), *208*

Friedrich, H., 171(1454, 1455), 172(1458–1462, 1465, 1474), *241, 242*

Fritz, H., 110(830), *221*

Frohn, H. J., 115(976), *226*

Frolova, L. F., 6(55), *196*

Fronczek, F., 480(105), 523(100), 525(101, 102), 527(105, 106), 529(105), 563(105), 585(105), *622, 623*

Fronczek, F. R., 463(84–86), 476(212–215), 484(241a), 515(85, 86), 527(86), 541(409), 543(357), 551(357, 409), 553(409), 561(408),

565(409), 569(213), 571(213, 357), 573(212, 214), 575(215), 591(234a), *622, 626, 627, 630, 631*

Frost, C. M., 29(369), *206*

Frost, D. C., 102(756), *218*

Fruehauf, H. W., 102(752), *218*

Fuchs, O., 187(1737), *251*

Fuentes, L., 106(768), 166(1423, 1424), 167(1395), *219, 239, 240*

Fugier, C., 274(75, 76), 387(75), 399(76), *438*

Fuhr, K. H., 123(1044a), *228*

Fuid-Alla, H. M., 77(640), *215*

Fuji, K., 453(36–38), 503(36–38, 306, 307), 589(306), *621, 628*

Fujikura, S., 501(407), *631*

Fujimaki, M., 35(420), 54(508, 511, 512), *208, 210, 211*

Fujimori, T., 9(115), *198*

Fujino, A., 14(178-180), *200*, 281(101), 369(101), *438*

Fujioka, H., 533(425), *632*

Fujita, E., 453, 453(36–38), 503(36–38, 306, 307), 589(306), *621, 628*

Fujita, S., 449(9–11), 451(9, 21), 501(9–11, 21), *620*

Fujiwara, H., 142(1169), *232*

Fujiwara, T., 64(559, 566), *212*

Fujn, K., 78(664), *215*

Fukada, N., 182(1703), *250*

Fukazawa, Y., 565(451), *633*

Fuks, R., 189(1755), *251*

Fukumato, K., 298(145), 299(145), *440*

Fukunaga, K., 20(251), *202*

Funcke, A. B. H., 279(90, 91), 389(90), 391(90, 91), 393(91), *438*

Fusco, R., 485(242), 597(242), *627*

Fustero, S., 173(1517), *243*

Gadzhily, R. A., 150(1292), *236*

Gaeta, F. C. A., 459(62), 517(62), 519(62), *621*

Gaivoronskaya, L. A., 113(958), *225*, 319(228), 375(228), 379(285), *442, 445*

Galen, S. K., 50(490), 51(490), *210*

Galiulin, M. A., 379(285), *444*

Galloway, W. D., 10(117), *198*

Gal'pern, G. D., 31(379), *207*

Galuszko, K., 453(34), 454(40), 503(34), 505(40), 557(34, 190, 191), 561(40), *621, 625*

Gambacorta, A., 55(514), 56(517, 518, 520), *211*

Gamble, J. C., 159(1353), *238*

Gandour, R. D., 464(95, 96), 519(95, 96), 521(95, 96), *622*

Ganguly, S. N., 11(138), *199*

Ganzinger, D., 16(218), *201*

Garanin, V. I., 190(1764), *252*

Garbers, C. F., 13(177), *200*

Garkusha, G. A., 78(653), *215*

Garming, A., 140(1150, 1151), *232*

Garratt, P. J., 493(279), 607(279), *628*

Gasanova, S. M., 6(44), *196*

Gassenman, S., 150(1273), 157(1337, 1342), *236, 238*

Gatica, B. T., 140(1152), *232*

Gatoh, N., 511(432), *632*

Gault, I., 505(310), *628*

Gavat, M., 450(17), 501(17), *620*

Gavrilenko, M. N., 6(46), *196*

Gavuzzo, E., 56(520), *211*

Gegele, V. G., 26(322), *205*

Gegner, E., 144(1185), *233*

Geigy, J. R., 48(486–488), 53(497), *210*

Gelbein, A. P., 114(943, 944), *225*

Gelbke, P., 140(1150, 1151), *232*

Gellert, E., 16(217), *201*

Gemenden, C. W., 170(1432), *240*

Genet, J. P., 78(662, 663), *215*

Georgi, U. K., 451(20), 501(20), 505(20), *620*

Gepshtein, E. M., 25(308), *204*

Gerber, G. E., 7(63), *196*

Gerlach, H , 481, 481(225), 587(225), 591(225), *626*

Germaine, G. R., 12(160), *199*

Gewald, K., 149(1244), 163(1378), *235, 239*

Geyer, U., 407(289), *444*, 450(15), 487(251), 501(15), 599(251), *620, 627*

Ghiran, D., 148(1214), *234*

Gholson, R. K., 3(14, 17), *195*

Ghosez, L., 45(477), 174(1518), *210, 243*, 291, 291(130), *439*

Ghyczy, S., 27(338), *205*

Gianetti, G., 173(1497), *243*

Gibs, G. J., 116(980, 981), 117(994), *226*

Gilbertson, T. J., 8(82), *197*

Gilchrist, T. L., 41(459), *209*

Gill, N. S., 256(11), 339(11), 415(11), 423(11), *436*

Gillard, J. W., 8(74), *197*

Gilvarg, C., 12(152), *199*

Ginion, C., 23(267), *203*

Ginos, J. Z., 116(982), *226*, 304(159), 423(159), *440*

Girgenti, S. J., 285(109, 110), 335(110), *439*

Girgor'ev, N. P., 24(284, 288), *203*

Giri, V. S., 23(266), *203*

Gizatova, B. I., 28(352), *206*

Gladiali, S., 85(878), 118(999), *222, 227*

Gless, R. D., 111(836), *221*

Glushchenkova, E. V., 31(377), *207*

Glushkov, R. G., 192(1778–1780), *252*

Gmelin, W., 303(156), *440*

Gnedina, V. A., 541(165), 547(167), *624*

Godar, E., 341(287), *444*

Godar, E. M., 333, 333(280, 281), 341(280, 281), 343(280, 281), 385(280, 281), *444*

Godzhaev, S. P., 150(1292), *236*

Goedken, V., 478(218), *626*

Goedken, V. L., 533(348), 541(123), *623, 630*

Goel, O. P., 149(1265), *235*

Goeldner, M. P., 57(523), 108(812, 813), *211, 220*

Goetze, S., 73(619), *214*

Goi, H., 55(506), *210*

Gokel, G. W., 459(54, 58, 60, 61), 460(63), 465(54), 466(63), 513(54, 58, 63), 517(58, 60, 61), 519(58, 60, 61), 521(54, 63), 523(54, 63), *621*

Golab, J., 27(337), *205*

Gold, D., 263(45), *437*

Gollakota, K. G., 12(167), *200*

Golodova, K. G., 126(1047, 1049), 137(1132), *228, 231*

Golovina, N. I., 190(1760), *252*

Golovnya, R. V., 33(399, 400), 34(400, 417), 35(400, 428, 429), *207, 208*

Gompper, R., 175(1522), *244*

Goncharova, L. F., 12(147), *199*

Goodman, I., 409(291), *444*

Gopal, H., 274(78), 339(78), *438*

Gore, E. S., 533(150), *624*

Gori, G. B., 9(99), *198*

Gorissen, H., 45(477), *210*, 291(130), *439*

Gorrod, J. W., 11(106), *198*

Goshoev, M. G., 12(145), 13(145), *199*

Goto, G., 80(678), *216*

Gotor, V., 174(1517), *243*

Gottlieb, O. R., 13(171), 23(277), *200, 203*

Gould, K. J., 299(324), *445*

Govindachari, T. R., 23(275), *203*

Govorova, L. M., 115(973), *226*

Gowans, C. S., 6(29), *195*

Gowland, M. S., 605(375), *630*

Goy, J., 10(127), *198*

Grace, D. S. B., 39(450), *209*, 293(134), 421(134), 435(134), *439*

Graf, U., 12(150), *199*

Gramshaw, J. W., 34(415), *208*

Grandjean, J., 571(361), *630*

Granik, V. G., 192(1778–1780), *252*

Grant, D. M., 515(401, 433), *631, 632*
Granzer, E., 126(1056), *229*
Gray, C. J., 541(356), *630*
Grebenkin, A. P., 9(95), *197*
Green, J. L., 148(1215), *234*
Gregory, B., 105(759), *218*
Greibrokk, T., 52(495), *210*
Gren, E. Ya., 413(292), *444*
Greuter, H., 51(491), *210*
Gribov, B. G., 108(807), *220*
Griffith, R. C., 285(109), *439*
Grigg, R., 71(615), *214*
Grigoleit, G., 147(1204, 1205), 179(1661), *233, 248*
Grigor'ev, S. M., 25(305), *204*
Grigor'eva, F. N., 107(787), *219*
Grigor'eva, N. D., 95(733), 146(1198), 157(1343, 1344), *218, 233, 238*, 301(153), 411(153), 415(153), 435(153), *440*
Grimshaw, J., 535(140), *624*
Grinev, A. N., 162(1373), *239*
Grinsteins, E., 107(783), *219*
Grishin, O. M., 107(785), *219*
Groen, S. H., 131(1086, 1092, 1098, 1099), 139(1145), *230, 232*
Grohe, K., 128(1075), *229*
Gross, D., 3(4), 11(133), 14(181), *194, 199, 200*
Grossenbacher, L., 461(65), 513(65), *621*
Grossman, J., 73(621), *214*
Groth, P., 458(49, 50), 511(49, 50), *621*
Grout, R. J., 163(1377), *238*
Grudzien, J., 31(383), *207*
Grudzinskas, C. V., 126(1063), *229*
Gruenanger, P., 71(614), *214*
Gruett, M. D., 149(1246–1248), *235*
Grukova, I. A., 547(167), *624*
Grumbach, F., 326(264, 265), *443*
Grün, R., 557(358), *630*
Gruntz, U., 89(927), *224*
Grunwald, C., 11(130), *199*
Grushina, T. A., 148(1226), *234*
Grutze, J., 476(154), 537(154), 545(156), 555(154), 557(358), 569(154), 573(154), 583(154), *624, 630*
Gschwend, H. W., 142(1173), *232*
Guadagni, D. G., 35(424), *208*
Gubitz, G., 146(1199), *233*
Gudriniece, E., 132(1102), 133(1102), 150(1296), *230, 236*
Guendel, W. H., 496(298), 617(298), *628*
Guengerich, F. P., 324(250), *443*
Guerrera, F., 148(1234), *234*
Guertsen, G., 94(724), *217*

Gugunava, M. A., 50(928), *224*
Gulubov, A. Z., 8(75), *197*
Gunar, V. I., 65(572–574), 148(1226), *213, 234*
Gund, P., 78(659), *215*
Gundermann, K. D., 183(1711, 1712), *250*
Gundriniece, E., 148(1220, 1221), *234*
Gunter, M. J., 182(1704), *250*, 311, 311(187), *441*, 579(389), *631*
Gunther, M. A., 185(1722), *250*
Gupta, K. C., 171(1444), *241*
Gupta, V. K., 475(100a), 541(409), 553(409), 551(409), 565(409), *622, 631*
Gurevich, A. I., 149(1236, 1268), *234, 235*
Gurskaya, L. T., 40(454), *209*
Gurtovnik, P. F., 24(289), *203*
Gutkowski, B., 78(660), 83(688), *215, 216*
Gutzwiller, J. A. W., 112(848), *221*
Gyorfi, B., 9(85–87), *197*
Gyorgy, P., 6(38), *195*

Haarmann, W., 149(1260), *235*
Haas, O., 539(423), 541(423), *632*
Haber, R. R. G., 79(674, 676), *216*
Habermehl, G., 20(256), *202*
Habu, T., 34(408), *208*
Hacker, N. P., 299(324), *445*
Hackett, H. M., 31(384), *207*
Hackney, M. A., 475(100a), *622*
Haenel, M. W., 475(35), 503(35), 505(35), 557(35), *621*
Hager, D. C., 448(6), 480, 480(103), 525(103, 108), 529(108), *620, 623*
Hagino, Y., 3(1), *194*
Haglid, F., 133(1108), *230*
Hague, Z. P., 549(177, 182), *625*
Hahn, W. E., 120(1003), 181(1693), *227, 250*, 256(12), 261(34), 272(73), 312(190), 314, 314(205, 206), 315(207, 209), 316(210), 318, 318(218), 339(12, 205, 209), 343(190, 210), 345(206), 383(12, 190, 207), 385(73, 80), 387(73), 403(209), 415(210), 419(34, 218), *436, 437, 438, 441, 442*
Hajjar, Al-, F. H., 77(641, 643), 162(1367), *215, 238*
Hakr, J., 107(776), *219*
Hales, B., 525(101), *623*
Hall, G. E., 259(22), *436*
Halmekoski, J., 12(163), *200*
Halton, B., 255(7), *436*
Hamada, K., 535(354), *630*
Hamano, H., 150(1285), *236*
Hamilton, H. D., 615(388), *631*
Hamilton, H. G., Jr., 533(147), *624*

Hamilton, R., 173(1505), *243*

Hammouda, Y., 14(185), *200,* 324, 324(249), 369(249), *443*

Hamoud, H. S., 77(641), *215*

Hams, D. J., 607(384), *631*

Hanack, M., 545(436), *632*

Hanai, M., 169(1430), *240*

Handa, S. S., 8(73), *197*

Hanes, A., 87(889), *223*

Hanotier, J. D. V., 115(967, 968), *226*

Hanotier-Bridoux, M., 115(967, 968), *226*

Hansen, A. M., 76(633), 80(677), *214, 216*

Hansen, J., 295(139), *439*

Hanson, A. H., 497(300, 301), 619(300, 301), *628*

Hanson, A. W., 507(312), *629*

Hanson, C., 26(329), 27(331), *205*

Hara, K., 182(1705), *250*

Hara, Y., 62(542), *211*

Hardegger, E., 12(150), *199*

Harding, R. J., 35(418, 422), *208*

Hardt, P., 176(1535, 1538, 1539, 1541), *244*

Hardtmann, G. E., 148(1216, 1230–1232), *234*

Hardy, P. M., 7(64), *196*

Hargis, C. W., 177(1559), *245*

Hargrave, K. D., 324(253), *443*

Hariri, M., 146(1193), *233*

Harmetz, R., 139(1148, 1149), *232*

Harmony, J. A. K., 108(797), *220*

Harris, D. J., 106(764), *218*

Harris, E. E., 64(560, 562), 65(587), *212, 213*

Harris, E. F., 78(659), *215*

Harris, R. L. N., 153(1326–1328, 1332), *237*

Harris, T. M., 124(1040), *228*

Harrit, N., 92(710, 711), *217*

Hart, F. A., 541(356), *630*

Hart, N. K., 14(197), 15(197, 212, 213), 16(212), *201*

Hartke, K., 261(31), 355(31), 359(31), *437*

Hartmann, H., 126(1053, 1060, 1061), *229*

Haruna, Y., 6(22), *195*

Hasan, N. M., 62(541), *211*

Hasang, M., 80(680), *216*

Hassan, M., 167(1419, 1420), *240*

Hassner, A., 44(475), *209*

Hatada, T., 126(1058), *229*

Haug, P., 31(381), *207*

Haung, W. Y., 309(174), 423(174), 425(174), *441*

Hauptmann, S., 148(1228), 150(1228), *234*

Hauser, C. R., 79(668), 124(1043), 159(1355), *216, 228, 238*

Hauser, U., 511(449), *633*

Hawkinson, S., 533(329, 336), *629*

Hawthorne, J. O., 177(1585), *246*

Hayashi, S., 403(147), *440*

Hayashi, T., 167(1413), 168(1413), *240*

Hayashi, Y., 12(189), 14(189), *200*

Hayatsu, R., 27(342), *205*

Hayes, R., 71(615), *214*

Heber, D., 123(1036), *228*

Hecht, S. S., 9(112), 10(120), *198*

Heckel, E., 19(237), *202*

Heckman, M., 73(618), *214*

Hedstrand, D., 462(74), 494(74), 513(74), 611(74), 613(74, 287), *622, 628*

Hedstrand, D. M., 108(811), *220,* 495(287), *628*

Hefelfinger, D. T., 503(304), 505(304), 559(304), *628*

Hegedus, L. S., 136(1122), *231*

Heider, J., 157(1334), *237*

Heilbronner, E., 619(299a), *628*

Heinrich, E., 148(1210), *234*

Helbling, A. M., 13(172), *200*

Helgeson, R. C., 459(62), 460(63), 466(63), 513(63), 517(62), 519(62), 521(63), 523(63), *621*

Henderson, D. E., 31(384), *207*

Henderson, L. M., 3(1), *194*

Hennenberger, P., 25(296), *204*

Henriksen, L., 144(1184), *233*

Henskens, H. J. G., 131(1088), *230*

Henze, H. R., 53(498), *210*

Herak, M. J., 78(666), 79(670), *216*

Herbst, I., 155(1313, 1315), *237*

Herma, H., 146(1197), 183(1714), *233, 250*

Herzenberg, J., 177(1553), *245*

Hesbain-Frisque, A. M., 45(477), *210,* 291(130), *439*

Heuvelen, A. V., 533(147), *624*

Hibi, T., 155(1311), *237*

Hibino, S., 23(279), *203*

Highet, R. J., 31(387), 108(813), *207, 220*

Higuchi, H., 503(438), 567(438), *632*

Hiiragi, M., 191(1776, 1777), *252*

Hilgenberg, W., 3(3), *194*

Hill, D. G., 120(1004), *227,* 313(194), 325(193, 259), 341(194), 411(194), 425(194), 435(194), *441, 443*

Hill, G. T., 313(193), *441*

Hill, R. E., 6(19, 20), *195*

Hill, R. K., 11(134), *199*

Hill, W. E., 533(328), *629*

Hillert, G., 140(1151), *232*

Hindley, K. B., 124(1040), *228*

Hiraga, K., 80(678), *216*

Hirai, K., 184(1717, 1718), *250*

Hiraoka, N., 9(93), *197*
Hirata, Y., 18(226, 227, 231, 232, 233, 235), 185(1727), *202, 251,* 603(380), *630*
Hirota, K., 95(734), *218*
Hirota, T., 186(1736), *251*
Hirota, Y., 155(1311), *237*
Hirshfeld, H., 158(1350), *238*
Hitzel, V., 126(1056), 148(1056), *229*
Hiyamo, T., 450(16), 501(16, 395, 396), *620, 631*
Hobbs, M. E., 9(114), *198*
Hodgkin, J. H., 15(219), 16(219), 31(385), *201, 207*
Hodgson, R. L. 139(1143), *231*
Hodson, P. H., 12(164), *200*
Hoehn, H., 150(1283), 164(1387), *236, 239*
Hoerlein, G., 156(1317), *237*
Hoesh, L., 40(458), *209*
Hoffman, D., 9(112), 10(120), 11(107), *198*
Hoffman, R. W., 254(2), *436*
Hofman, P. S., 279(90, 91), 389(90), 391(90, 91), 393(91), *438*
Hogan, J. C., 464(96), 519(96), 521(96), *622*
Hogeveen, H., 39(450), *209,* 293(136), 421(134), 435(1134), *439*
Hokama, K., 187(1742, 1743), *251*
Holcomb, W. D., 42(469), *209,* 339(315), 345(315), *445*
Holland, J. W., 12(166), *200*
Hollis, S., 533(334), *629*
Hollitzer, O., 99(744), 100(744), *218*
Holm, R. H., 455(46), 479, 479(221), 581(221), *621, 626*
Holzrichter, H., 452(29, 30), 493, 493(275, 276), 505(29, 30), 609(275, 276), *620, 628*
Homeier, E. H., 178(1622), *247*
Homma, G., 319(226, 227), *442*
Honda, Y., 12(189), 14(189), 173(1493), *200, 243*
Honeybourne, C. L., 541(169), 545(169), *624*
Hong, P., 141(1158), 177(1542, 1543), *232, 244*
Horiguchi, Y., 105(761), *218*
Horn, U., 110(825), *221*
Hornby, J. C., 40(456), *209*
Horng, J. M., 171(1439–1441), *241*
Horsewood, P., 6(20), *195*
Horstmann, C., 15(209), *201*
Hoshino, Y., 501(407), *631*
Hosoda, K., 20(244), *202*
Hosokawa, G., 84(857), *222*
Hosokawa, T., 136(1124, 1125), *231*
Hossain, A. M. M., 115(970), *226*

Hotellier, F., 23(272), *203*
Houghton, E., 69(601), 70(605–607), *214*
Hozumi, T., 150(1276), 152(1276, 1297), 156(1318), *236, 237*
Hradetzky, F., 156(1320), *237*
Hruby, M., 81(686, 687), *216*
Huber, E., 481, 481(225), 591(225), *626*
Hubert, P., 34(404), 35(427), *207, 208*
Hueper, F., 14(184), *200*
Huff, R., 110(829), *221*
Huffman, R. W., 108(806), *220*
Hughes, C. G., 105(760), *218*
Hughes, G. J., 7(64), *196*
Hughes, M. N., 541(431), 549(431), *632*
Hughes, N., 147(1208a, 1294), 148(1208a, 1209), *233, 236*
Huguet, J., 58(527), *211*
Huisman, H. O., 161(1360), *238*
Hunig, S., 492(274), 495(274), 609(274), 613(274), *628*
Hunter, W. E., 482(231), 589(231), *626*
Hunter, W. H., 476(214), 573(214), *626*
Hunziker, M., 511(449), *633*
Huppatz, J. L., 153(1326–1328, 1332), *237*
Hursthouse, M. B., 541(356), *630*
Husson, A., 14(199), *201*
Husson, H. P., 14(199), *201*
Huyser, E. S., 108(797), *220*
Huys-Francotte, M., 135(1116), *231*
Hwang, T. M., 171(1441), *241*
Hynam, B. M., 163(1377), *239*

Ibragimov, I. I., 150(1292), *236*
Ibrahim, M. A., 120(1006), *227*
Ichikawa, K., 453(36, 37), 503(36, 37), *621*
Ichikawa, Y., 173(1485, 1490–1494), *242, 243,* 298(145), 299(145), *440*
Ichimoto, I., 78(664), *215*
Ichiyama, A., 3(18), *195*
Iddon, B., 115(979), *226*
Igon'kina, G. S., 31(376), *206*
Iida, T., 31(382), *207*
Iijima, I., 125(1046), *228*
Ikigler, A. A., 89(927), *224*
Ila, H., 154(1304–1306), 188(1306), *237*
Ilavsky, D., 163(1375, 1376), *239*
Il'in, G. S., 8(77), 9(85–92), *197*
Illi, V., 158(1346), *238*
Imae, Y., 12(158), *199*
Imai, T., 178(1622), *247*
Imamaura, J., 114(940), *225*
Imamura, K., 177(1561), *245,* 449(10, 11), 501(10, 11), *620*

Inaba, A., 144(1178), 189(1756, 1757), *233, 251*
Inglis, H. S., 85(870), 141(1165), *222, 232*
Inoue, H., 113(936), *224*
Inoue, I., 182(1705), *250*
Inoue, M., 114(940), *225*
Inoue, S., 55(504–506), *210*
Inukai, N., 133(1109), *230*
Inukai, T., 133(1110), *230*
Ioffe, I. I., 181(1682–1684), *249*
Ionas, G. P., 33(398), *207*
Ionava, L. V., 60(535), 64(555), *211, 212*
Iqbal, M., 56(516), *211*
Irie, H., 361(320), 363(320), *445*
Irikura, T., 164(1385), *239*
Irwin, W. J., 117(996), *226*
Isakov, V. V., 121(1024), *228, 257(21), 436*
Isakova, D. M., 6(58, 62), *196*
Isele, G. L., 138(1138), *231, 451, 451(23), 501(23), 620*
Ishida, Y., 501(418), *632*
Ishiguro, T., 149(1256), *235*
Ishikawa, F., 61(539), 64(565, 568, 569), *211, 212*
Ishikawa, M., 403(147), *440*
Isin, Zh. I., 166(1422), *240*
Islam, M., 66(597, 598), 67(597), *213*
Ismail, E. S., 167(1415), *240*
Ismailov, A. Y., 112(845), *221*
Ismailov, A Ya., 278(85, 86), 319(229), 375(229), 377(86), 379(86), *438, 442*
Ismatullaeva, M. G., 177(1587), 181(1692), *246, 249*
Ito, H., 171(1450, 1452, 1453), 179(1652), 180(1679), *241, 248, 249*
Ito, K., 148(1233), *234*
Ito, S., 565(451), *633*
Ito, T. I., 310, 310(180), 397(180), 405(180), 407(180), *441, 496, 496(294), 617(294), 628*
Itov, Z. L., 65(573), *213*
Ivakhnyuk, M. S., 137(1129), *231*
Ivan, L., 300(152), *440*
Ivanov, Ch., 177(1560), *245*
Ivanova, L. V., 123(1045), *223*
Ivanova, R. B., 108(807), *220*
Ivanova, T. S., 28(354), 29(354), *206*
Ivanovskii, A. P., 138(1135), 177(1547–1550, 1562, 1586), 179(1629, 1630, 1639, 1642), 180(1680), *231, 244, 245, 246, 247, 248, 249*
Iwanami, S., 66(591, 594), 124(1041), *213, 228*
Iwasawa, H., 158(1349), *238*

Iwata, M., 483(234a, 236), 587(234), 591(234a, 450), 593(234, 236, 238), 595(234, 450), *626, 627, 633*
Izatt, R. M., 462, 462(73), 463(76, 83), 464(76, 83, 87–89), 481(76), 495(76), 513(73), 515(76, 87–89, 315, 317, 318, 320, 321, 402, 433), 517(76, 87, 419, 459), 519(419), 583(76), 611(76, 87), 613(76), *622, 623, 629, 631, 632, 633*
Izman, G. V., 9(94), *197*
Izumi, T., 136(1123), *231*
Izumi, U., 101(746), *218*

Jackson, J. L., 71(615), *214*
Jacques, J., 459(59), *621*
Jacquignon, P., 275(83), 289(125), 373(83), 377(83), 379(125), *438, 439*
Jaeger, D. A., 459(57), 462, 462(57, 71), 513(57, 71), *621, 622*
Jaeger, R. H., 259(24), *436*
Jaffe, E. R., 3(5), *194*
Jagodzinski, T., 164(1381), 169(1431), *239, 240*
Jagt, J. C., 141(1164), *232*
Jahine, H., 149(1238), 167(1398, 1402), *234, 239*
Jallo, Al, H. N., 160(1359), 162(1367, 1368), *238*
James, K. B., 256(11), 339(11), 415(11), 423(11), *436*
Jamilloux, B., 191(1771, 1772), *252, 282(105), 283(105), 373(105), 439*
Janata, V., 78(649), 81(684), 150(1277), 152(1277), *215, 216, 236*
Janousek, Z., 135(1116), *231*
Janssen, P., 114(943, 944, 946), *225*
Janus, J. M., 114(941), *225*
Janz, G. J., 141(1159), *232*
Jarvik, M. E., 9(101), *198*
Jaworski, T., 41(463), 81(682), *209, 216*
Jeffreys, J. A. D., 170(1432), *240*
Jeffs, P. W., 324(252, 253), *443*
Jenner, P., 11(106), *198*
Jenny, W., 452(29, 30), 482, 482(229), 483(229, 233), 493(275, 276), 495(290), 505(29, 30), 587(229, 233), 595(229, 233), 607(290), 609(275, 276), 613(290), *620, 626, 628*
Jensen, H. B., 29(369, 370, 374), *206*
Jeroschewski, P., 110(816), *220*
Jeyaraman, R., 317(217), 411(217), *442*
Jiresch, W., 29(375), *206*
Jobour, Al-, N. H., 162(1369, 1370), *238*

Johns, S. R., 14(196, 197), 15(197, 212, 213), 16(212), *201*
Johnsgaard, M., 156(1323), *237*
Johnson, D. H., 409(290), *444*
Johnson, L. F., 31(388), *207*
Johnson, R. D., 11(136), *199*
Johnson, W. H., 9(97), *198*
Johnson, W. O., 149(1243), *234*
Joines, R. C., 74(624), *214*
Jolley, S. T., 463(78–80), 515(78–80, 87, 321), 517(87, 419), 519(419), 523(419), 611(87), *622, 623, 629, 632*
Jones, B. A., 515(320, 401, 433), 517(459), *629, 631, 632, 633*
Jones, G., 14(182), 31(391), *200, 207,* 257(17), 271, 271(72), 381(17), 387(72), 397(12), *436, 438*
Jones, R. A. Y., 57(526),*211*
Jones, R. K., 257(17), 271(72), 381(17), 387(72), 397(72), *436, 438*
Jones, R. L., 56(515, 519), *211*
Jongejan, H., 94(724), 95(730), *217, 218*
Joos, A., 118(997), *227*
Jorritsma, H., 39(450), *209,* 293(134), 421(134), 435(134), *439*
Jouin, P., 495(289), 611(289), *628*
Joullié, M. M., 323(239), *443*
Juenemann, W., 153(1329), *237*
Juneja, H. R., 324(251), *443*
Junek, H., 134(1113), 144(1179), 153(1301), 163(1379, 1380), 164(1388), 169(1429), 187(1738), *231, 233, 236, 239, 240, 251,* 261(32), 263(43), 282(104), 361(32), 363(32), 365(32), 431(43), 433(43), *437, 439,* 487(253), 599(253), *627*
Junghans, K., 321, 321(233, 234), 367(233, 234), *442*
Junjappa, H., 154(1304–1306), 188(1306), *237*
Junker, N., 9(97), *198*
Juri, P. N., 462(70), 481(70), 513(70), 515(70), 517(70), 583(70), *621*
Jusiak, L., 7(66), *197*
Just, E., 177(1579), *245*
Jutz, C., 150(1290), *236,* 283(106, 107), 285(107), 341(107), 359(107), 371(107), 383(107), 397(107), *439*

Kaden, T. A., 533(131), *623*
Kaden, Th. A., 533(349), 553(342), *629, 630*
Kadish, K. M., 541(447), *632*
Kadono, T., 7(67), *197*
Kadyrov, C. S., 109(819, 823), *220, 221*
Kaegi, H. H., 111(840), *221*
Kaempfer, I., 107(780), *219*

Kagasov, V. M., 27(343), 28(351), *205, 206*
Kaikaris, P. A., 108(803), *220*
Kaiser, E. M., 79(668), *216*
Kakihana, T., 284, 284(108), 337(108), *439*
Kakisawa, H., 71(612), *214,* 265(53), 285(303), 341(53), *437, 445*
Kalabin, G. A., 182(1710), *250*
Kalakutskii, B. T., 115(961), *225*
Kalesnikov, N. A., 478(220), *626*
Kallianos, A. G., 10(117), *198*
Kalnin, S. V., 349(288), 351(288), *444*
Kalnins, S., 150(1271a), *235*
Kaltenbronn, J. S., 148(1222), 149(1264, 1266, 1267), *234, 235*
Kalyanasundaram, R., 12(141), 199
Kamachi, H., 71(612), *214,* 265(53), 341(53), *437*
Kamada, M., 501(7b, 8), 513(7b), *620*
Kambe, S., 167(1413), 168(1413), *240,* 355(318), 357(318), 393(318), 395(318), 403(318), 405(318), *445*
Kamela, Z., 162(1374), *239*
Kametami, T., 109(820), 149(1261), 191(1776), *220, 235, 252*
Kametani, T., 298, 298(145), 299(145), *440*
Kaminskii, V. A., 89(923), 106(767), 121(1010, 1014, 1015, 1017, 1018, 1021, 1022), 181(1694), *219, 224, 227, 228, 250,* 257(19, 20), 300(149), 309(178), 423(19), *409, 436, 441*
Kan, G., 90(706), 106(765), *217, 218*
Kan, G. Y.-P., 607(384), *631*
Kanehira, K., 347(310), 349(310), *445*
Kaneko, C., 403(147), *440*
Kaneko, H., 9(115), *198*
Kang, H. C., 497(299), 579(299), 619(299, 299a), *628*
Kannan, S. V., 275, 275(79–81), 345(79, 80), 359(81), 373(81), *438*
Kano, H., 104(785), *218*
Kanoktanaporn, S., 333(319), *445*
Kantlehner, W., 135(1119), *231*
Kaplan, L., 459(58), 513(58), 517(58), 519(58), *621*
Kaplan, L. J., 459(61), 460(64), 513(64), 517(61), 519(61), *621*
Kaplan, N. O., 3(10, 12), *195*
Kapoor, R. N., 537(159), *624*
Kappe, T., 145(1190), 146(1193), 150(1286), 155(1313, 1315), *233, 236, 237*
Kapustin, M. A., 190(1764), *252*
Karacharov, N. N., 25(311), *204*
Karaulov, E. S., 182(1699, 1700), *250,* 316(211), 423(211), *442*

Karawya, M. S., 14(191), *200*
Karicheva, V. N., 31(379), *207*
Karlsson, K., 9(97), *198*
Karn, J. L., 533(120, 122), *623*
Karpeiskii, M. Y., 60(531, 535), 64(555, 556), *211, 212*
Karpf, M., 449(13), *620*
Kartavtsev, O. I., 108(792), *219*
Kasahara, A., 136(1120, 1121, 1123), *231*
Kascheres, A., 44(475), *209*
Kashima, C., 142(1171, 1172, 1174), *221, 232*
Kasturi, T. R., 132(1103, 1105), *230*, 311, 311(185, 186), *441*
Kasuga, R. J., 9(115), *198*
Katagiri, N., 17(225), 101(746), 153(1307), 155(1307), *201, 218, 237*
Katagiri, T., 142(1169, 1170), *232*
Kataoka, M., 64(547–549), 65(548), *212*
Kato, H., 35(420), 54(508, 511, 512), 69(602, 604), *208, 210, 211, 214*
Kato, M., 12(189), 14(189), *200*
Kato, S., 501(407), *631*
Kato, T., 17(225), 20(249), 39(451), 59(530), 78(658), 79(672), 85(876), 101(746, 748), 109(822), 111(833), 150(1275, 1276), 152(1276, 1297, 1299), 153(1307), 155(1307–1310, 1314), 156(1318), 167(1404, 1405), *201, 202, 209, 211, 215, 216, 218, 221, 222, 236, 237, 239, 240*, 306(165), *440*, 605(413), *632*
Kato, Y., 36(431), *208*
Katritzky, A. R., 76(627), 79(667), 81(683), 82(683, 689, 690), 86(700, 701), 87(701), 88(902–905), 89(920, 927), 91(707), 92(712), 93(713–715), 719), *214, 216, 217, 223, 224*, 307(312), 349(312), 417(312), 433(312), *445*
Katruk, E. A., 6(47, 48), *196*
Katt, R. J., 264(48), *437*
Kauffmann, T., 295(138, 139), *439*
Kauffmann, Th., 452, 452(31), 505(31), *620*
Kaufman, A., 178(1603), *246*
Kauppinen, V., 12(163), *200*
Kavan, I., 26(297), *204*
Kawagishi, N., 13(176), *200*
Kawahito, T., 173(1475), *242*
Kawai, Y., 27(333), *205*
Kawamata, J., 85(876), 111(833), 152(1299b), 155(1308), 167(1405), *221, 222, 236, 237, 240*
Kawamura, K., 55(505), *210*
Kawamura, N., 20(249, 250), 53(501), *202, 210*
Kawanisi, M., 173(1489), 193(1782), *242, 252*
Kawasaki, K., 191(1776, 1777), *252*

Kawasaki, T., 109(821), 167(1421), *220, 240*
Kawasaki, Y., 541(426), *632*
Kawato, T., 464(92–94), 484(241a), 488(257), 519(92–94, 325), 521(93), 525(102), 591(234a), 603(257), *622, 623, 627, 629*
Kawazu, M., 64(550), 65(575), *212, 213*
Kayo, W. R., 324(251), *443*
Kazanskii, B. A., 292(131), 339(131), *439*
Kellogg, R. M., 108(811), *220*, 462(74), 488(256), 489(286), 494, 494(74, 283, 284), 495, 495(285–289), 513(74, 442), 601(256), 609(442), 611(74, 283–286, 288, 289, 400, 442), 613(74, 283, 400, 442), *622, 627, 628, 631, 632*
Kendurkar, P. S., 171(1442, 1443), *241*
Kenji, I., 55(505), *210*
Kenny, D. H., 89(927), *224*
Kenny, J., 50(490), 51(490), *210*
Kergomard, A., 191(1771), *252*, 282(105), 283(105), 373(105), *439*
Kern, D., 140(1150, 1151), *232*
Kesarev, O. G., 319(228), 375(228), 377(228), *442, 444*
Kessenikh, A. V., 490(262), 601(262), *627*
Keypour, H., 535(351), *630*
Keszther, F., 10(122), *198*
Khakimdzhanov, S., 12(187), 14(187, 200, 201), *200, 201*
Khallik, A., 31(378), *206*
Khan, A. N., 286(114–116), *439*
Khan, M. A., 149(1254), 165(1390), *235, 239*
Khanna, L. A., 107(787), *219*
Kharchenko, V. G., 122(1030), *228*
Kharlampovich, G. D., 27(334), 28(351), *205, 206*
Khashab, Al-, A. Y., 76(636), 77(636), *215*
Kher, S. 537(159), *624*
Khidekel, M. L., 108(803, 807), *220*
Kholodov, L. E., 326(263), *443*
Kholodova, N. V., 89(915), *224*
Khomyakov, V. G., 115(971, 972), *226*
Khua-min, K., 281(97, 98), *438*
Khutornenko, G. A., 78(653), *215*
Kiang, A. K., 76(626), *214*
Kida, S., 535(354), *630*
Kiefer, G. E., 475(100a), 507(410), *622, 631*
Kigasawa, K., 191(1776, 1777), *252*
Kikuchi, K., 265(52), 365(52), 367(52), *437*
Kikuchi, M., 171(1451), *241*
Kildahl, N. K., 533(153), *624*
Kim, D., 23(279), 141(1167), *203, 232*
Kim, N. Yu., 166(1422), *240*
Kimov, G. A., 316(211), *442*
Kimura, E., 533(425), 535(434), *632*

Kimura, K., 12(155, 156), 20(252, 253), *199, 202*

King, S. B., 26(298), *204*

King, T. J., 71(615), *214, 333(319), 445*

King, W. A., 323(242), *443*

Kinlin, T. E., 34(405, 407), *207*

Kirch, E. R., 278(87), *438*

Kireev, G. V., 108(790, 792), *219*

Kirillova, M. A., 174(1520), *243*

Kirsanova, V. S., 25(306), *204*

Kirvun, S. V., 87(893, 895), *223*

Kisaki, T., 10(118), *198*

Kishida, Y., 184(1717, 1718), *250*

Kiso, Y., 448(7), *620*

Kita, K., 109(821), *220*

Kita, Y., 167(1421), *240*

Kitai, A., 6(39), *195*

Kitatsuji, E., 31(382), *207*

Kiva, E. A., 27(340), *205*

Kivilaakso, S., 12(163), *200*

Kiyoura, T., 131(1101), *230*

Kizawa, K., 34(402), *207*

Klar, H., 161(1364), *238*

Klasek, A., 18(230, 234), *202*

Kleeman, A., 173(1473), *242*

Klein, J. F. M., 131(1095, 1096), *230*

Kleinman, E., 325, 325(326), *445*

Klesment, I. R., 31(378), *207*

Klevenskaya, I. L., 6(45), *196*

Kliment'eva, N. I., 8(79), *197*

Klimov, G. A., 120(1002), 121(1008b), *227,* 316(212), 317(214), 359(212), 423(211, 212), 425(214), *442*

Klindukhova, T. K., 45(478), 138(1136), *210, 231*

Klingsberg, E., 106(764), *218*

Kloeden, D., 22(265), *203*

Klus, H., 10(126), *198*

Klyushnikova, T. M., 6(51), *196*

Knaus, G. N., 40(457), *209*

Kneubuhler, W., 495(290), 607(290), 613(290), *628*

Knoevengel, E., 256, 256(9), *436*

Knox, C. V., 541(397), 573(397), *631*

Ko, K., 177(1544), *244*

Koba, H., 369(307), *445*

Kobayashi, G., 126(1058), 170(1433), 171(1433), *229, 240*

Kobayashi, K., 496(296), 617(296), *628*

Kobayashi, S., 142(1171, 1172, 1174), *232*

Koblik, A. V., 89(916, 918), *224*

Koch, H., 161(1363), *238*

Kochkin, D. A., 595(373), *630*

Kocian, O., 116(993), *226*

Kodama, M., 533(425), 535(434), *632*

Kodama, S., 448(7), *620*

Koeda, T., 55(504), *210*

Koening, H., 63(543, 545), *211, 212*

Kofron, W. G., 124(1043), *228*

Koga, K., 173(1475), *242, 459(58, 59),* 513(58), 517(58), 519(58), *621*

Koga, M., 187(1743), *251*

Kogan, B. E., 27(340), *205*

Kogan, S. S., 29(356), *206*

Kogure, Y., 131(1101), *230*

Kohl, D. K., 157(1341), 165(1393), 181(1341), *238, 239,* 260(27), 262(37), 264(27, 37, 49), 267(37), 269(66), 288(120), 327(37), 332(37, 119), 333(120), 337(119, 120), 339(37), 369(37), 409(37), 411(37), 415(37), 419(37), 423(37), *436, 437, 439,* 463(85, 86), 475(199), 515(85, 86), 517(323), 523(100, 327), 525(101, 327), 527(86), 563(199), 583(327), *622, 623, 625, 629*

Kojima, H., 25(291), *203*

Kolbe, J., 324(248), *443*

Kole, M. M., 6(30), *195*

Kolesnikov, N. A., 577(220), *626*

Kolor, M. G., 35(425), *208*

Koloskova, N. M., 76(632), *214*

Kolosov, M. N., 149(1236, 1268), *234, 235*

Kolyer, J. M., 496(297), 617(297), *628*

Komarov, R. D., 545(160, 161, 162, 164), *624*

Komatsu, T., 483(235, 237, 239, 240), 593(235, 237, 239), *626, 627*

Komkov, I. P., 76(635), *215*

Komoto, M., 54(508, 511, 512), *210, 211*

Kon, G. A. R., 281(99), 371(99), *438*

Kondakova, L. V., 33(398), *207*

Kondo, M., 39(451), 59(530), 101(748), *209, 211, 218*

Kondrat'eva, G. Y., 57(552), 64(552, 571), 181(1695), 190(1758–1768), *212, 213, 250, 251, 252*

Kondrat'eva, G. Ya., 292(131), 339(131), 419(296), *439, 444*

Kondratov, V. K., 114(962), *225*

Konno, S., 109(822), *221*

Kononova, V. V., 140(1154), *232*

Kooi, J., 462(74), 494(74), 513(74), 611(74), 613(74), *622*

Kopay, C. M., 106(763), *218*

Kopf, J., 38(445), *209*

Koppersmith, D. L., 452(25), *620*

Korostova, S. E., 182(1710), *250*

Korshunov, M. A., 177(1547, 1549, 1550, 1562), 179(1629, 1630, 1639, 1642), 180(1680), *244, 245, 247, 248, 249*

Korte, D. E., 136(1122), *231*
Korth, H., 12(168), *200*
Korybut-Daszkiewicz, B., 41(463), *209*
Korzeniewski, C., 272(73), 385(73), 387(73), *438*
Korzhan, L. A., 25(311), *204*
Kosareva, M. A., 115(961), *225*
Koshigoe, T., 150(1285), *236*
Koshmina, N. V., 126(1048), 128(1072, 1073), 137(1127–1129), *228, 229, 231*
Kost, A., 12(140), *199*
Kost, A. N., 66(597, 598), 67(597), 150(1292), 164(1381–1283), 169(1431), *213, 236, 239, 240*
Kostyuk, V. A., 25(305, 307), *204*
Kosuge, T., 36(437, 438), 143(1177), *208, 233*
Kosuka, A., 109(820), *220*
Kotelev, V. V., 6(48), *196*
Kothari, M., 161(1361), *238, 266(56), 437*
Kotova, A. V., 28(362), *206*
Koudijis, A., 95(730), *218*
Kouno, T., 575(405), *631*
Kovalevskii, A. S., 87(893), *223*
Kowitz, W., 78(655), *215*
Koyama, J., 361(320), 363(320), *445*
Koyama, T., 186(1736), *251*
Koyasu, K., 177(1555), *245*
Kozerski, L., 152(1298), *236*
Kozikowski, A. P., 62(541), *211*
Koziolkiewicz, W., 181(1693), *250*
Kozlov, N. S., 40(454), *209*
Kozub, G. I., 108(803), *220*
Kozuka, A., 149(1261), *235*
Kozyrev, V. G., 84(858), *222*
Kraatz, A., 283(107), 341(107), 359(107), 371(107), 383(107), 397(107), *439*
Kramer, J., 296, 296(140), 337(140), *440*
Kramer, W., 496(298), 617(298), *628*
Krasnaya, Zh. A., 130(1079, 1080), 137(1133), 149(1245), 150(1289), *229, 233, 235, 236*
Kravtsova, V. N., 122(1031), *228*
Krechl, J., 163(1375, 1376), *238*
Kreidl, J., 149(1237), *234*
Kret, F., 27(336), *205*
Kretsch, M. J., 33(394), *207*
Kreutzberger, A., 175(1523), *244*
Krimm, H., 131(1087), *230*
Kriven'ko, A. P., 121(1008a), 122(1029–1031), *227, 228*
Krivun, S. V., 87(882), *222*
Kriwun, S. W., 309(172), *441*
Kroehnke, F., 107(788), 122(1033), 170(1434), *219, 228, 240*
Krow, G., 178(1626), *247*

Kruglov, E. A., 27(340), 28(368), *205, 206*
Kruizinga, W. H., 108(811), *220*, 462(74), 494(74), 495(287), 513(74), 611(74), 613(74, 287), *622, 628*
Krutosikova, A., 88(905), *223*
Kubelka, V., 163(1376), *239*
Kubo, M., 173(1504), *243*
Kubota, Y., 78(658), 155(1309–1310), *215, 237*
Kubrak, V. P., 28(354), 29(354), *206*
Kucherov, V. F., 130(1079, 1080), 137(1133), 149(1245), *229, 231, 235*
Kuchkarov, A., 177(1581–1584, 1587), *246*
Kuder, J. E., 130(1081), *229*
Kuebel, B., 73(619), 156(1317), *214, 237*
Kuhn, H., 10(126), 11(128), *198*
Kukhar, V. P., 134(1115), *231*
Kukharenko, T. A., 25(290), *203*
Kulakov, A. A., 24(289), *203*
Kulhanek, L., 26(321), *204*
Kulhankova, A., 26(315, 317, 320, 323, 324), *204, 205*
Kulikova, D. A., 326(263), *443*
Kulikovskii, A. V., 12(159), *199*
Kumada, M., 448, 448(7, 8), *620*
Kumar, A., 154(1304, 1305), *237*
Kumari, V., 537(159), *624*
Kung-Chang, C., 603(415), *632*
Kuniyoshi, M., 173(1475), *242*
Kuo, C. H., 112(847), 114(952, 953), *221, 225*
Kupchan, S. M., 19(236), 202
Kurakina, N. A., 78(653), *215*
Kurata, T., 54(511), *211*
Kurbatov, Y. V., 31(389), 108(789–796, 808, 810), 113(930–934), *207, 219, 220, 224*
Kurbatova, A. S., 31(389), 108(789, 791, 793–796), 113(930, 932), *219, 220, 224*
Kurihara, H., 126(1050, 1051), 127(1068), 150(1288), *228, 229, 236*
Kurilenko, V. M., 122(1025), *228*
Kurilo, G. N., 162(1373), *239*
Kurtev, B., 84(860), *222*
Kurz, J., 47(481), 159(1352), *210, 238*
Kusama, O., 191(1776, 1777), *252*
Kusano, Y., 575(405), *631*
Kushner, V. P., 3(9), *194*
Kusumi, T., 285(303), *439*
Kusunoki, Y., 178(1591, 1592, 1595–1598, 1604–1613, 1615–1619), *246, 247*
Kuthan, J., 107(776, 778, 779, 781, 782), 110(817), 163(1375, 1376), 185(1720), *219, 220, 239, 250*
Kut'in, A. M., 138(1135), 177(1547, 1549, 1550, 1562, 1586), 179(1629, 1630, 1639,

1642), 180(1680), *231, 244, 245, 246, 247, 248, 249*

Kutney, J. P., 31(388), *207*

Kutuzov, V. A., 12(153), *199*

Kutzner, H. J., 20(256), *202*

Kuzanyan, R. S., 490(262, 264), 599(264), 601(262), *627*

Kuz'min, V. I., 50(929), *224*

Kuz'mina, Z. F., 28(354), 29(354), *206*

Kuznetsov, V. I., 113(955, 956), *225*

Kuznetsova, M. A., 93(723), *217*

Kuzuhara, H., 483(234, 234a–237, 239–241), 587(234), 591(234a, 450), 593(234–239, 404), 595(234, 450), *626, 627, 631, 633*

Kvasnikov, E. I., 6(46, 51, 62), *196*

Kvitko, S. M., 149(1259), *235*

Kwiatkowski, S., 81(682), *216*

Kyba, E. P., 90(702), *217*, 459(58, 59, 61), 513(58), 517(58, 61), 519(61), *621*

Kyoshida, R., 193(1783), *252*

Lackey, R. W., 324(245), *443*

Laidler, D. A., 523(99), *622*

Lakhan, R., 60(534), *211*

Lamb, J. D., 462(73), 464(87, 89), 513(73), 515(87–89, 315–318, 402, 433), 517(87), 611(87), *622, 623, 629, 631, 632*

Lambert, B. F., 31(386), *207*

Lamberton, J. A., 14(196, 197), 15(212, 213), 16(212), *201*

Lamm, G., 149(1239), *234*

La Montagne, M. P., 171(1435a-b, 1436, 1438), *241*

Lan, S. J., 3(1), *194*

Lanchuk, A. Y., 28(362), *206*

Lang, K. F., 25(293), *204*

Lang, W. H., 177(1565), *245*

Lanum, W. J., 26(298), *204*

Lapoline, R., 22(260), *203*

Lari, A., 533(428), *632*

Larson, S. B., 515(319), *629*

Laszio, P., 571(361), *630*

Latham, D. R., 28(363), 29(357–359), *206*

Lattau, H., 128(1074), *229*

Latypova, F. N., 28(368), *206*

Latysheva, S. G., 6(32), *195*

Laukhina, L. L., 87(894), *223*

Laundon, R. D., 48(483, 484), 52(483), *210*

Lavery, A., 478(219), 579(219), *626*

Lavrinovics, E., 172(1463), 179(1666), *242, 249*

Lavrushin, V. F., 107(787), *219*

Lavy, J., 116(993), *226*

Lawrence, A. E., 313(193), 325(193, 259), *443*

Lawson, J. A., 452(28), 454(28), 505(28), *620*

Lazar, A., 187(1737), *251*

Lazdins, I., 172(1463), 179(1664–1666), *242, 248, 249*

Lazur'evskii, G. V., 87(886), *223*, 234–239, 309(173), *441*

Leach, M., 20(246, 247), *202*

Leanz, G., 12(152), *199*

Ley, S. V., 285(314), 345(314), *445*

Lezina, V. P., 50(928, 929), *224*

Lechner, A., 275(82), 347(82), *438*

Leddy, B. P., 89(927), *224*

Ledl, F., 36(439), *209*

Lee, H. W., 509(460), 519(412), *631, 633*

Lee, J. I., 171(1441), *241*

Leete, E., 8(76, 80–83), 9(83), 121(1034), *197, 228*

Lehmann, B., 97(735), *218*

Lehn, J. M., 597(414), *632*

Lei, J., 48(489), 49(489), *210*

Lelek, J., 26(299), *204*

Le Men, J., 14(185), *200*, 324(249), 369(249), *443*

Le Nguyen, N., 76(632), *214*

Leniewski, A., 150(1278), *236*

Lenke, D., 63(544), *212*

Lenzer, G., 77(639), *215*

Leonaldi, R., 476(202), 569(202), 573(202), *625*

Leonard, N. J., 168(1427), 176(1427), *240*

Leonov, I. D., 28(362), *206*

Leont'ev, V. B., 108(790, 792, 798), *219, 220*

LeRoux, M., 13(177), *200*

Lesher, G. Y., 149(1246–1248, 1269, 1270), *235*

Lesko, P. M., 22(260), *203*

Letyushova, N. N., 121(1016), *227*

Levi, M., 177(1560), *245*

Levina, R. Y., 77(637), *215*

Levina, R. Ya., 281(97, 98), *438*

Lewars, E. G., 105(760), *218*

Lewis, J., 88(905), 93(715), *217, 223, 547(178), 549(176, 180, 182, 390, 458), 551(179, 180, 437, 451, 456), 553(446, 454), 625, 632, 633*

Li, B. J., 177(1571), *245*

Li, L. H., 177(1571), *245*

Li, R. T., 134(1114), *231*

Liao, T. K., 149(1263), *235*

Libbey, L. M., 34(413), *208*

Libermann, D., 326(264, 265), *443*

Libert, M., 58(527–529), *211*

Lie, R., 458(49), 511(49), *621*

Liebman, A. A., 111(840), *221*
Liebscher, J., 126(1053, 1060, 1061), *229*
Liles, D. C., 476(115), 533(115), 537(115), 549(177), 569(115), 577(115), *623, 625*
Lin, A. J., 78(654), *215*
Lindoy, L. F., 478(216–218), 533(127), 577(217), *623, 626*
Lindsay, R. C., 34(409), *208*
Ling, L., 603(415), *632*
Ling, L. C., 33(395), 35(424), *207, 208*
Linn, W. J., 450(14), *620*
Lions, F., 256(11), 339(11), *436*
Lipkin, D., 97(735), *218*
Lipp, D. W., 271(69, 70), 351(69), 353(69), 355(69), *438*
Lipsch, J. M. J. G., 139(1145), *232*
Lipscomb, W. N., 23(278), *203*
Liptrot, M. C., 551(457), 553(454), *633*
Litsinova, L. A., 26(322), *205*
Litvinov, B. D., 24(289), *203*
Litvinova, S. A., 24(289), *203*
Liu, L., 603(406), *631*
Löbering, H. G., 283(106, 107), 341(107), 359(107), 371(107), 383(107), 397(107), *439*
Lochte, H. L., 280(94, 96), 324(94), 339(94), *438*
Loebering, H. G., 150(1290), *236*
Loh, H. H., 3(2), *194*
Lohacs, C., 126(1055), *229*
Lohaus, G., 77(644), *215*
Loidl, A., 55(507), *210*
Lomako, G. L., 40(454), *209*
Long, G. J., 533(330), *629*
Long, K. M., 533(117, 118, 126), *623*
Looker, J. H., 53(502), *210*
Lopatine, N. D., 114(954), *225*
Lora-Tamayo, M., 185(1722, 1723), *250*
Lorente, A., 166(1423), 167(1395, 1397), *239, 240*
Lorenzo, A., 93(717), *217*
Los, S. I., 6(53), *196*
Lovecchio, F. V., 533(121, 150), *623, 624*
Lovett, E. G., 97(735), *218*
Lovkova, M. Y., 8(77, 79), 9(85–88, 90, 92), *196, 197*
Lu, S., 583(391), *631*
Lucas, R. A., 31(386), *207*
Luhan, D. A., 324(252, 255), *443*
Lukin, E. V., 177(1580), *245*
Luk'yanets, E. A., 60(537), 62(537), *211*
Luning, B., 13(175), *200*
Luo, J. M., 171(1441), *241*
Luppertz, F., 569(360), *630*
Lutfullin, K. L., 14(192), *200*

Luttringer, J. P., 106(762), *218*
Lutz, T. J., 533(131, 349), *623, 630*
L'vova, S. D., 65(572–574), *213*
Lwowski, W., 42(466), *209*
Lynch, C. J., 9(99), *198*
Lyubopytova, N. S., 28(361, 368), *206*

Maas, G., 401(302), *445*
Maas, G. E., 463(76, 83), 464(76, 83, 89), 481(76), 495(76), 515(76, 87, 89, 315, 316), 517(76, 87), 583(76), 611(76, 87), *622, 629*
Maass, G., 476(210), 571(210), *626*
Macak, J., 26(299), *204*
MacBride, J. A. H., 288(121), 296(121), 333(319), 337(121, 308), *439, 445*
McCann, M., 476(143, 205, 208), 478(143), 533(143), 535(143), 541(397), 569(143, 205), 573(143, 205, 208, 364, 397, 429), 577(143), *624, 625, 626, 630, 631, 632*
McClure, G. L., 455(47), 464(90, 91), 519(90, 91), 521(90, 91), *621, 622*
McColl, I. S., 173(1480), *242*
Mac Donald, B., 102(756), 103(756), *218*
McDonald, F. R., 29(370–373), *206*
McDowell, C. A., 102(756), 103(756), *218*
McFall, S. G., 476(200, 201, 207), 533(136), 535(141, 145, 146, 333, 352), 543(174), 569(200, 207), *623, 624, 625, 626, 629, 630*
MacGee, J., 12(166), *200*
Machavariana, M. Z., 6(42), *195*
Machleidt, H., 157(1334), *237*
McIlhenny, H. M., 31(391), *207*
McIlroy, P., 533(328, 332), *629*
McIlroy, P. D. A., 476(207), 533(138), 535(140, 141), *623, 624, 626*
McInnes, A. G., 20(238–240), *202*
Mack, P. A., 539(124, 168), *623, 624*
McKay, A. F., 257(18), 423(18), *436*
MacKay, D., 47(482), *210*
McKay, J. F., 29(357–359), *206*
McKee, R. L., 148(1229), *234*
McKee, V., 476(206), 573(206), *626*
McKervey, M. A., 173(1505), *243, 521(435), 632*
Maclean, D. B., 22(262), *203*
McLean, S., 23(270, 271), 24(271), *203*
MacLeod, G., 33(397), *207*
McMahon, S., 454(41), 505(41), 561(41), *621*
McMillian, F. L., 108(797), *220*
McNulty, J., 126(1052), 148(1052), *229*
McNulty, P. J., 149(1242), *234*
McNutt, K. W., 6(36), *195*
McOmie, J. F. W., 299(324), *445*, 452(24), 505(24), *620*

McPartlin, M., 476(115), 533(115), 537(115), 549(177), 569(115), 577(115), *623, 625*
McPhail, A. T., 324(252, 253, 255), *443*
Maczynski, A., 25(303), 26(325), 27(335), 28(350), *204, 205, 206*
Madakyan, V. N., 113(958), *225*
Maeda, A., 193(1783), *252*
Maeda, I., 66(595), *213*
Maeda, K., 20(255), 136(1124), *202, 231*
Maehr, H., 20(246, 247), *202*
Maekh, S. K., 31(390), *207*
Maerkl, G., 139(1141), *231*
Maevskii, Y. V., 76(634, 635), 130(1082), *214, 215, 230*
Magar, N. G., 34(412), *208*
Magdesieve, N. N., 76(632), *214*
Maguire, J. H., 148(1229), *234*
Maia, J. G. S., 23(277), *203*
Maier, D. P., 86(881), *222*
Maier, W., 8(78), *197*
Mailey, E. A., 181(1685), *249*
Maing, I. Y., 35(425), *208*
Majestic, V. K., 476(212–215), 543(357), 551(357), 569(213), 571(213, 357), 573(212, 214), 575(215), *626, 630*
Majewicz, T. G., 117(995), 165(1394), *226, 239*, 264(50, 51), 365(50), *437*
Makarov, G. N., 24(284, 288), *203*
Makino, M., 36(437), *208*
Maksimov, Yu., V., 149(1259), *235*
Malarek, D. H., 111(840), *221*
Malik, W. U., 543(455), *633*
Malinowski, St., 179(1634), *247*
Mallet, M., 296(141), *440*
Malte, A. M., 454(42), *621*
Malysheva, N. V., 115(961, 962, 964, 977), *225, 226*
Mamaev, V. P., 129(1076), *229*
Manabe, O., 475(405), *631*
Manawwar, Z., 80(680), *216*
Manchin, P. J., 96(738), *218*
Mander, L. M., 579(389), *631*
Maned'yarov, G. M., 113(935), *224*
Maninger, I., 145(1190), *233*
Mann, D. F., 11(137), *199*
Mann, T. A., 279(88), 322(236), 326(236), 389(88), 393(88), 395(236), 399(88, 236), *438, 443*
Manrikes, S. I., 319(230), 377(230), *442*
Mansukhani, R., 158(1347), *238*
Mansurov, N. I., 6(33), *195*
Mantsch, H. H., 116(987), *226*, 306(164), 425(164), *440*
Manuel, M. F., 8(83), 9(83), *197*

Manusch, R., 261(31), 355(31), 359(31), *437*
Manzyuk, S. G., 6(33), *195*
Marascia, F. J., 73(620), *214*
Marchenko, N. B., 192(1778, 1780), *252*
Marchesini, A., 485(242, 243), 597(242, 243), *627*
Marcinkowski, R. M. K., 405(317), 435(317), *445*
Mardiev, Y. A., 113(935), *224*
Maretina, I. A., 174(1519, 1520), *243*
Mariella, R. P., 333(280, 281), 341(280, 281, 287), 343(280, 281), 385(280, 281), *444*
Marketz, H., 85(869), *222*
Markgraf, J. H., 264(47, 48), *437*
Markov, K., 177(1560), *245*
Markova, L. I., 122(1031), *228*
Markovac, A., 170(1435, 1436–1438), *241*
Markow, V. V., 25(311), *204*
Markus, G. A., 24(285), 25(305, 306), *203, 204*
Maronpot, R. R., 9(112), *198*
Marquez, V. E., 266(58), *437*
Mars, P., 114(941), *225*
Marsais, F., 296(141), *440*
Marsden, J. H., 139(1146), *232*
Marshall, W. D., 22(262), *203*
Marston, C. R., 519(324), *629*
Martel, H. J. J. B., 454(39, 41), 505(39, 41), 561(39, 41), *621*
Martens, R. J., 294(135, 136), *439*
Martes, M. P., 78(654), *215*
Martin, H. D., 73(618), *214*
Martin, L. Y., 533(338), *629*
Martin, N. H., 324(252), *443*
Martin, R. B., 537(443), *632*
Maruoka, K., 501(418, 441), *632*
Maruyama, T., 64(546), *212*
Marx-Moll, L., 496(291), *628*
Masamune, T., 319(226, 227), *442*
Masek, V., 24(287), 27(344), *203, 205*
Maslennikova, N. P., 121(1016), *227*
Mason, N. B., 535(352), *630*
Mason, R. B., 158(1347), *238*
Massoue, J. P., 28(365), *206*
Masuda, T., 62(542), *211*
Matasunaga, K., 193(1782), *252*
Matern, K., 147(1205), 179(1661), *233, 248*
Matskovskaya, E. S., 90(704), *217*
Matsuda, H., 184(1717, 1718), *250*
Matsuda, Y., 170(1433), 171(1433), *240*
Matsude, Y., 126(1058), *229*
Matsukubo, H., 69(602, 604), *214*
Matsumoto, H., 104(758), 148(1208b), *218, 233*

Matsumoto, M., 20(241, 242), *202*
Matsumoto, T., 322(238), 389(238), *443*
Matsumura, S., 13(169), 84(862), 85(862), *200, 222*
Matsumura, Y., 501(418), *632*
Matsuo, Y., 65(576, 578–582, 588), *213*
Matsushita, H., 9(115), *198*
Matsuura, A., 182(1702), *250*
Matsuura, I., 20(243, 244), *202*
Matsuura, K., 182(1701), *250*
Matsuura, T., 347(310), 349(310), *445*
Matsuzawa, T., 171(1449, 1451), *241*
Matthew, C. J., 289(123), 377(123), *439*
Matthews, B. W., 503(305), *628*
Mattice, W., 525(102), *623*
Mattschei, P. K., 455(44), 480(105), 505(44), 527(44, 105), 529(105), 563(105), 585(105), *621, 623*
Mauldin, R. K., 11(132), *199*
Maurer, B., 8(70), *197*
Maurer, K. H., 13(177), *200*
Maverick, E., 461(65), 513(65), *621*
Mavougou-Gomes, L., 52(496), *210*
May, E. L., 113(936), *224*
Mayer, J., 143(1175), *232*
Mayer-Mader, R., 266(55), *437*
Meade, L., 525(102), *623*
Meckl, H., 150(1280), *236*
Medeiros, R. W., 73(620), *214*
Medina, E., 13(170), *200*
Medvedeva, T. N., 12(147, 148), *199*
Meek, D. W., 479(222), 581(222), *626*
Meister, M., 53(499), *210*
Melius, P., 148(1215), *234*
Melling, J., 12(162), *199*
Melnyk, A. C., 533(153), *624*
Men'shikov, V. V., 164(1381–1383), 169(1431), *239, 240*
Mensler, K., 130(1085), *230*
Menyhart, J., 26(326, 327), *205*
Merrell, P. H., 533(125, 348), *623, 630*
Mertens, R., 535(420), *632*
Merzlyakova, N. M., 326(263), *443*
Metallidis, A., 144(1179, 1180), *233*
Metelli, R., 71(614), *214*
Metler, T., 138(1137), *231*
Metri, J., 167(1416), *240*
Metzger, C., 47(481), *210*
Metzler, O., 371(184), *441*, 496(295), *628*
Meyer, H., 159(1352), 185(1721), *238, 250*
Meyer, R. B., 129(1077), 134(1077), *229*
Meyers, A. I., 71(610), 102(750), 111(843), *214, 218, 221*
Meyers, C. Y., 454(42), *621*

Mezheritskii, V. V., 87(887), 89(921), *223, 224*
Mezhov, B. V., 89(916), *224*
Micetich, R. G., 254(4), *436*
Michau, J. D., 22(188), *200*
Micheloni, M., 533(339), *629*
Michener, E., 178(1626), *247*
Midorikawa, H., 157(1345), 167(1401, 1406), 168(1425), 182(1706), 186(1731), *238, 239, 240, 250, 251*
Mikaya, A. I., 180(1668, 1669), *249*
Mikhailova, L. N., 50(929), *224*
Mikhaleva, A. I., 182(1708–1710), *250*
Miki, H., 64(564, 570), 65(576, 578–582), 66(588, 589), *212, 213*
Mileva, P., 177(1560), *245*
Miller, S. I., 138(1137), *231*
Miller, S. L., 36(435), *208*
Mina, G. A., 150(1293), *236*
Minachev, Kh. M., 190(1764), *252*
Minami, N., 153(1307), 155(1307), *237*
Minami, T., 575(405), *631*
Minato, A., 448(7), *620*
Minato, H., 20(241), *202*, 496(296), 617(296), *628*
Minato, Y., 177(1567, 1569, 1570), 179(1647, 1648, 1651, 1658, 1659), 180(1677), *245, 248, 249*
Minozhedinova, N. S., 9(85–88, 90), *197*
Mirdzhalilova, S. M., 6(43), *196*
Mirianashvili, E. V., 6(61), *196*
Mironov, G. A., 33(399), 35(428, 429), *207, 208*
Mirzorakhimov, A., 6(33), *195*
Misawa, H., 127(1068), *229*
Mishima, H., 126(1050, 1051), *228, 229*
Mishrikey, M. M., 76(628), 77(640, 645), 79(671), 83(691, 693), 124(1038), *214, 215, 216, 228*
Misumi, S., 503(438), 567(438), *632*
Mitscher, L. A., 14(194), *201*
Mitsuhashi, K., 169(1430), *240*, 304(158), *440*
Mitulla, K., 545(436), *632*
Mitzusaki, S., 10(118), *198*
Miura, I., 6(19), *195*
Miwa, K., 25(291), *203*
Miwa, Y., 185(1727), *251*
Miyagawa, S., 180(1678), *249*
Miyake, A., 148(1233), *234*
Miyake, T., 177(1561), *245*
Miyamoto, N., 193(1782), *252*
Miyazaki, M., 125(1046), *228*
Miyazaki, S., 64(566, 569), *212*
Mkrtchyan, D. S., 6(49), *196*
Mochalov, V. V., 26(318), *204*

Modyanova, L. V., 12(140), *199*
Moenninghoff, H., 86(880), *222*
Moerck, R. E., 44(476), *210*
Moga-Iuga, M., 148(1214), *234*
Mohan, A. G., 178(1623), *247*
Moisak, I. E., 316(213), 429(213), *442*
Moiz, S. S., 289(124), 373(124), 375(124), *439*
Mokhoman, K. M. S., 319(228), 375(228), *442*
Mokhtar, H. M., 150(1274), *236*
Molina, P., 93(717), *217*
Moll, K. K., 173(1501), 177(1551, 1579), 179(1632, 1643, 1646), *243, 244, 245, 247, 248*
Mon, T. R., 33(395), *207*
Monich, N. V., 123(1045), *228*
Monson, R. S., 188(1745–1748), *251, 303(155), 423(155), 425(155), 435(155), *440*
Moore, G. J., 85(868), *222*
Moore, J. A., 73(620), 106(763), *214, 218*
Moore, L. P., 27(342), *205*
Moore, S. S., 464(89), 515(89), *622*
Morgan, D., 486(250), 597(250), *627*
Moraru, D., 194(1786), *252*
Morel, J., 175(1521), *244*
Mogan, M. E., 34(413), *208*
Morgan, P. J., 34(414), *208*
Mori, T., 451(21), 501(21), *620*
Morimoto, K., 78(656, 657), *215*
Morin, I. G., 515(401, 433), *631, 632*
Morisawa, Y., 64(547–549), 65(548), *212*
Morita, K., 184(1719), *250*
Morita, Y., 64(558, 559, 566), 66(590), *212, 213*
Moritz, K. L., 149(1258), *235*
Morliere, P., 533(345, 427), *629, 633*
Morris, G. F., 124(1043), 159(1355), *228, 238*
Morrocchi, S., 347(309), *445*
Mortimer, P. I., 110(832), *221*
Morton, J. B., 279(89), *438*
Mosby, W. L., 486(248), *627*
Mosher, W. A., 271(69, 70), 351(69), 353(69), 355(69), *438*
Mosiashvili, G. I., 6(61), *196*
Moskin, A. F., 178(1589), *246*
Moskovkina, M. V., 257(19), 423(19), *436*
Moskovkina, T. V., 121(1010), 122(1025, 1026), *227, 228*
Mothes, U., 12(215), *201*
Motomura, Y., 168(1425), *240*
Moyeux, M., 326(264), *443*
Mozzhuikhin, D. D., 108(807), *220*
Muchlstaedt, M., 172(1466), *242*
Muehlstaedt, M., 173(150), 179(1632, 1646), *243, 247, 248*

Mueller, A. K., 76(629), *214*
Mueller, E., 149(1260), *235*
Mueller, G. W., 57(524, 525), *211*
Mueller, R., 89(926), *224*
Muenzner, D., 128(1074), *229*
Mukhamedziev, M. M., 15(207), *201*
Mukherjee, M., 11(139), *199*
Mulders, E. J., 34(416), 36(436), *208*
Mullen, K., 113(959), *225*
Muller, A., 10(125), *198*
Muller, M., 461(67), *621*
Muller, R., 491(266, 267), 599(266, 267), *627*
Muller, W. M., 461(68), 462(67, 68), 464(67), 476(211), 513(67, 69), 515(67), 571(211), *621, 626*
Munshi, J. F., 323(239), *443*
Murahashi, S., 136(1124, 1125), *231*
Murai, F., 12(189), 14(178–180, 189), *200, 281(101), 369(101), *438*
Murai, H., 84(862), 85(862), *222*
Muraimukhamedov, K., 12(144), *199*
Murakami, K., 173(1500), *243*
Murakami, M., 124(1041), 133(1110, 1111), *228, 230, 231, 281(100), *438*
Murakami, M. I., 133(1109), *230*
Murakamo, M., 66(591, 594), *213*
Muralidhara, R., 34(405, 407), *207*
Murase, I., 535(354), *630*
Murata, A., 40(453), 139(1147), *209, 232*
Murray, D. G., 23(270, 271), 24(271), *203*
Murrell, W. G., 12(160), *199*
Musakin, A. P., 177(1580), *245*
Musumarra, G., 307(312), 349(312), 417(312), 433(312), *445*
Musumarva, J. G., 86(700), *217*
Muthard, D. A., 22(259), *203*
Muto, M., 293(133), 341(133), *439*
Mutterer, F., 110(825, 828, 829), *221*

Nacco, R., 84(863), *222*
Nachev, I., 599(376), 601(377, 378), *630*
Nadelson, J., 127(1069, 1070), *229*
Nadler, K., 15(121), *198*
Nafie, L. A., 533(130), 535(130), *623*
Nagai, S., 155(1311), *237*
Nagai, W., 185(1727), *251*
Nagano, N., 116(985), 133(1109–1111), *226, 230, 231, 281(100), *438*
Nagao, S., 84(864, 865), *222*
Nagiev, T. M., 113(935), *224*
Naguchi, M., 9(115), *198*
Naiman, A., 176(1536), *244, 288(304), 300(146), 357(313), 369(146), 395(313), 397(146), *440, 445*

Naiman, A. U., 176(1737), *244*
Naito, T., 60(538), 61(539), 64(565, 566, 568, 570), 65(577), 66(589, 590, 593), 111(839), *211, 212, 213, 221*
Nakagome, T., 191(1775), *252*
Nakahara, Y., 395(306), *445*
Nakajima, I., 448(7), *620*
Nakajima, K., 177(1558, 1572), 179(1653, 1662), *245, 248*
Nakajima, M., 12(189), 14(189), *200*
Nakamachi, A.,171(1448), 173(1500), *241, 243*
Nakamura, M., 34(408), *208*
Nakamura, S., 3(18), 171(1450), 179(1652), 180(1679), *195, 241, 248, 249*
Nakamura, T., 13(169), 84(857), 101(749), *200, 218, 222*
Nakamura, Y., 171(1452, 1453), *241*
Nakanishi, T., 181(1696), *250*
Nakano, K., 181(1690), *249*
Nakashima, T. T., 22(263), *203*
Nakatsuka, T., 448(7), *620*
Nakayama, S., 169(1430), *240*
Nanba, T., 186(1736), *251*
Nanji, H. R., 281(99), 371(99), *438*
Nantka-Namirski, P., 107(777), 149(1249, 1250), 160(1358), 185(1725, 1730), *219, 235, 238, 251*
Naplekova, N. N., 6(56), *196*
Narkevich, A. N., 93(721, 722), *217*
Narr, B., 149(1260), *235*
Nastasi, M., 106(762), *218*
Natasuki, R., 170(1433), 171(1433), *240*
Natscher, R., 476(154), 537(154), 545(154), 555(154), 557(358), 569(154), 573(154), 583(154), *624, 630*
Natsuki, R., 126(1058), *229*
Nauta, W. Th., 279(90), 389(90), 391(90), *438*
Nawata, Y., 20(243), *202*
Nayak, A., 455(44), 464(91, 92, 95), 480(105, 197), 505(44), 519(91, 92), 521(91, 95), 525(102), 527(44, 105), 529(105), 557(197), 561(197), 563(105), 583(197), 585(105), *621, 622, 623, 625*
Negievich, L., 107(785), *219*
Neidlein, R., 93(716, 720), *217*
Neilands, L., 271(71), 349(288), 351(288), 417(295), *438, 444*
Neilands, O., 150(1271a), *235*
Neiser, J., 27(344), *205*
Nelson, J., 543(398), *631*
Nelson, S. M., 476(143, 200, 205–208), 478(143, 219), 479(223), 533(134–136, 138, 139, 143, 322, 328), 535(140–143, 145, 146, 333, 352), 539(423, 424), 541(170, 172, 397,

423, 424), 543(174, 398), 545(175), 569(143, 200, 205, 207), 573(143, 205, 206, 208, 364, 397, 399, 429), 577(143), 579(219), 581(223), *624, 625, 626, 629, 630, 631, 632*
Nemchinskaya, V. L., 3(9), *194*
Nemcova, D., 78(649), 81(684), 150(1277), 152(1277), *215, 216, 236*
Nemec, J. W., 121(1009), *227*
Nemes, A., 187(1737), *251*
Nenitzescu, C. D., 450(17, 18), 501(17), *620*
Nenz, A., 178(1588), *246*
Neuhaeuser, S., 25(309), *204*
Neumann, P., 459(2), 559(188), 563(188), *620, 625*
Neunhoeffer, H., 37(444), 96(736, 737), 97(740, 741), 98(742), *209, 218*
Neurath, G. B., 9(108, 111), 10(108), *198*
Neurhoeffer, H., 102(752), *218*
Neuse, E., 267(63), 343(63), *437*
Newcomb, M., 459(54), 465(54), 513(54), 521(54), 523(54), *621*
Newkome, G. R., 187(1739–1741), *251*, 286(112, 113), 321(235), 413(113), *439, 443*, 448(6), 452(25), 455(44, 45, 47), 459(53), 463(85, 86), 464(90–95), 466(53), 475(100a, 198, 199), 476(212–215), 480(103, 105, 197), 484(241a), 488(257), 505(44), 507(410), 509(460), 513(108), 515(85, 86), 517(322, 323), 519(90–95, 324, 325, 412), 521(90, 91, 93, 95), 523(53, 100, 327), 525(101–104, 108, 327), 527(44, 86, 104–106), 529(105, 108), 541(409), 543(357), 551(357, 409), 553(409), 557(197), 561(197, 198, 408), 563(105, 199), 565(409), 569(213), 571(213, 357), 573(212, 214), 575(215), 583(197, 327), 585(105), 591(234a), 603(257), *620, 621, 622, 623, 625, 627, 630, 631*
Newomb, M., 459(56), 460(56, 63), 466(56, 63), 513(56, 63), 521(56, 63), 523(56, 63), *621*
Ng, C. Y., 3(1), *194*
Nguyen, T. T., 22(262), *203*
Nicholson, W. J., 113(959), *225*
Nickl, J., 149(1260), *235*
Nicolae, G., 89(922, 924, 925), *224, 309(176, 177), 441*
Nicoletti, R., 55(514), 56(517, 518, 520), *211*
Nicolson, A., 173(1502), *243*
Niculaita, V., 89(922), *224, 309(176), 441*
Nie, P. L., 88(902, 904), 93(713–715), *217, 223*
Niedermeyer, P., 188(1744), *251*
Nielsen, B. L., 515(318), *629*

Nigam, I. C., 114(942), *225*
Niida, T., 55(504, 506), *210*
Nikandrov, G. N., 177(1580), *245*
Nikitenko, A. V., 148(1220, 1221), *234*
Nikogosyan, V. G., 6(49), *196*
Nishi, M., 179(1638), *247*
Nishi, S., 147(1207), *233*
Nishikawa, S., 179(1647, 1648), 180(1677), *248, 249*
Nishimura, T., 127(1068), *229*
Nishio, K., 149(1252), *235*
Nishiwaki, T., 71(608), *214*
Nishizuka, Y., 3(18), *195*
Nistor, C., 148(1214), *234*
Nitta, M., 369(307), *445*
Nivazova, D. A., 108(794), *219*
Niwa, T., 55(506), 177(1654), 179(1654, 1658, 1659), 180(1677), *210, 248, 249*
Noda, M., 150(1275), 167(1404), *236, 239*
Noguchi, K., 177(1561), *245*
Nohara, A., 149(1256), *235*
Nokatsuka, T., 448(7), *620*
Nomura, K., 169(1430), *240*
Noravyan, A. S., 78(652), *215*
Norman, R. O. C., 168(1426), *240*
Norris, J. M., 541(123), *623*
Norris, K., 48(489), 49(489), 52(494), *210*
Norton, S. J., 78(661), *215*
Novelli, A., 168(1427), 176(1427), *240*
Novogorodova, N. Y., 31(390), *207*
Novoselov, V. S., 27(334), *205*
Nowell, I. W., 476(201, 204), 569(201), *625*
Nozaki, H., 449(9–11), 450(16), 451(9, 21), 501(9–11, 16, 21, 395, 396, 407), *620, 631*
Nozoe, T., 265(52), 367(52), *437*
Nuber, E., 587(225), *626*
Nuda, T., 55(505), *210*
Nulty, P. J., 145(1187), *233*
Numa, S., 173(1500), *243*
Numanov, I. U., 28(352, 353), *206*
Nunomura, N., 35(423), *208*
Nuralieva, Z. S., 15(207), *201*
Nurimov, E., 7(65), *196*
Nursten, H. E., 35(418, 422), *208*
Nutakul, W., 330(278), *444*
Nyiondi-Bonguen, E., 127(1071), *229*
Nyquist, E. B., 41(461), *209*
Nyu, K., 149(1261), *235*

Obayashi, A., 13(169), *200*
Obayashi, M., 184(1719), *250*
Oberkobusch, R., 147(1204, 1205), 179(1661), *233, 248*
O'Callaghan, C. N., 167(1396), *239*

Ochiai, E., 116(988), *226*, 533(117, 121, 149, 151, 152), *623, 624*
Ochiai, M., 184(1719), *250*
Ockels, W., 12(168), *200*
O'Conner, T., 521(435), *632*
O'Donoghue, T., 549(176, 180, 182), 551(179, 180, 456), *625, 633*
Oehl, R., 77(639), *215*
Oehldrich, J., 77(642), *215*
Oehlschlaeger, H., 150(1280), *236*
Oepen, G., 463(77), 476(211), 515(77), 521(77), 523(77), 555(77), 571(211), 583(77), *622, 626*
Offermann, W., 571(361), *630*
Offermanns, H., 172(1461, 1462), *241, 242*
Ogasawara, K., 149(1261), *235*
Ogata, M., 104(758), *218*
Ogawa, M., 111(839), *221*
Ogawa, S., 458(52), 507(313, 439), 511(52, 314, 432, 439), *621, 629, 632*
Ogawa, Y., 55(506), *210*
Ogino, T., 9(93), *197*
Ohanka, V., 110(817), *220*
Ohashi, M., 71(612), 150(1291), *214, 236, 265(53), 341(53), 437*
Ohata, K., 13(169), *200*
Ohl, H., 37(444), *209*
Ohl, J., 191(1773), *252*
Ohloff, G., 8(70), *197*
Ohnishi, Y., 496(296), 617(296), *628*
Ohta, M., 158(1349), *238*
Ohta, T., 143(1177), *233*
Okada, J., 181(1690), *249*
Okami, Y., 20(255), *202*
Okamoto, T., 143(1177), *233*
Okazaki, H., 178(1592, 1593, 1595–1598, 1602, 1604–1613, 1615–1619), *246, 247*
Oki, T., 6(39), *195*
Okubo, T., 84(862), 85(862), *222*
Okuda, N., 541(426), *632*
Okuma, K., 496(296), 617(296), *628*
Okumura, K., 116(985), *226*
Okumura, M., 270(68), 409(68), *438, 555(195), 557(195), 625*
Okumura, T., 7(67), *197*
Okytomi, T., 20(243, 244), *202*
Olejniczak, B., 316(210), 343(210), 415(210), *442*
Olekhnovich, E. P., 87(894), *223*
Olekhnovich, L. B., 87(892), *223*
Olemskaya, E. V., 179(1628), *247*
Oliver, D. W. H., 84(855), 85(871), *222*
Olson, G., 78(659), *215*
Olsson, K., 54(509), *210*

Omel'chenko, V. N., 149(1236, 1268), *234, 235*

Omori, E., 84(864, 865), *222*

Omoto, S., 55(506), *210*

Omura, H., 64(570), 65(577), 66(590), *212, 213*

Omura, Y., 61(539), 64(568), *211, 212*

Onishi, S., 64(558, 559), *212*

Onoprienko, V. V., 149(1236, 1268), *234, 235*

Oostveen, E. A., 94(728), *217*

Opalka, C. J., 149(1269, 1270), *235*

Opgenorth, H. J., 162(1371, 1372), *239*

Oppenheim, C., 113(959), *225*

Orakhovats, A., 84(860), *222*

Ornaf, R. M., 10(120), *198*

Ortiz, P. A., 34(401), *207*

Osbond, J. M., 67(599), *213*

Oscarson, J. L., 515(318), *629*

Oshima, K., 501(396), *631*

Oshuevea, N. A., 57(552), 64(552), *212*

Osman, S. F., 11(129), *198*

Osugi, J., 182(1705), *250*

Oszczapowicz, J., 27(337), *205*

Othman, A. H. B., 476(207), 533(136, 138, 328, 332), 535(141, 142, 333, 352), 569(207), *623, 624, 626, 629, 630*

Otroshchenko, O. S., 31(389), 108(790–793, 795, 796, 808, 810), 113(932, 933), 179(1663), *207, 219, 220, 224, 248*

Otto, H. H., 167(1412), *240*

Oude-Alink, B. A., 95(731, 732), 178(1624), *218, 247*

Outurquin, F., 165(1391), 175(1521), *239, 244*

Ovchinnikova, T. I., 179(1635), *247*

Overman, L. E., 188(1752, 1753), *251, 496(292), 628*

Ovsyannikova, V. N., 12(1002), *227*

Ozawa, I., 66(594), *213*

Ozerov, I. M., 31(376), *206*

Ozola, A., 150(1271), *235, 236*

Ozola, E. Ya., 260(28), 349(288), 351(288), 361(28), *436, 444*

Pachler, K. G. R., 13(177), *200*

Paddow-Row, M. N., 42(467), *209, 286(114–116), 439*

Padikkala, J., 13(173), *200*

Padwa, A., 43(470–472), *209*

Padyukova, N. S., 60(535, 537), 64(555, 556), *211, 212*

Pagani, G., 485(242, 243), 597(242, 243, 374), *627, 630*

Page, D. F., 149(1270), *235*

Pagnoni, U. M., 485(243), 597(243), *627*

Pailer, M., 10(124), 16(218), 29(375), *198, 201, 206,* 603(379), *630*

Pakrashi, S. C., 23(266), *203*

Palamar, A., 113(930, 932), *224*

Palchkov, V. A., 93(723), *217,* 307(167), *440*

Palecek, J., 107(778, 779, 781, 782), *219*

Palion, W. J., 179(1634), *247*

Pan, Y., 142(1173), *232*

Pandit, U. K., 108(804), *220,* 609(387), *631*

Panek, J. S., 291(323), *445*

Pankov, A. G., 177(1548), 178(1589), *244, 246*

Panosyan, A. K., 6(49), *196*

Papathanasopoulos, N., 149(1257), *235*

Pappalardo, S., 458(51), 511(51), 541(409), 551(409), 553(409), 561(192, 408), 563(193, 359), 565(409), *621, 623, 625, 630, 631*

Pappo, R., 145(1188, 1189), *233,* 327(267), *443*

Paquette, L. A., 284(108), 285(314), 337(108), 345(314), *439, 445*

Parello, J., 12(198), 14(198), 15(198), *201*

Parham, W. E., 486, 486(244–247), *627*

Paris, R. R., 23(269), *203*

Park, Y-ae, 541(123), *623*

Parkins, H., 16(221), *201*

Parks, J. E., 455(46), 479(221), 581(221), *621, 626*

Parks-Smith, D. G., 173(1482), *242*

Parliment, T. H., 35(425), *208*

Parrick, J., 56(516), 189(1754), *211, 251*

Partridge, M. W., 163(1377), *238*

Pastors, P., 151(1295), *236*

Pastour, P., 115(960), 117(960), 175(1521), *225, 244,* 274(75, 76), 296(141), 387(75), 399(76), *438, 440*

Pataraya, M. S., 6(61), *196*

Patchett, A. A., 78(659), *215*

Patel, A. N., 27(331), *205*

Patel, A. R., 171(1438), *241*

Patel, M. N., 486(250), 597(250), *627*

Patil, D. L., 34(412), *208*

Patmore, E. L., 146(1194, 1196), *233,* 262(38), 263(38), 363(42), 419(38, 42), 421(38, 42), *437*

Patterson, A. M., 459(1), *620*

Patterson, L. K., 533(345, 427), *629, 632*

Patts, K. T., 256(11), *436*

Paudler, W. W., 482(230, 231), 587(230), 589(230, 231), *626*

Paul, A. D., 275(84), 347(84), 349(84), *438*

Pauletti, P., 533(339), *629*

Paulmier, C., 165(1391), 174(1521), *239, 244*

Pavel, G. B., 120(1007), 147(1007), *227*

Pavel, G. V., 120(1001), 122(1032), *227, 228*

Pavlenko, M. G., 134(1115), *231*

Pawlowski, G., 545(436), *632*

Peacock, S. C., 459(62), 517(62), 519(62), *621*

Peck, R. L., 11(129), *198*
Pedersen, C., 92(710, 711), *217*
Pelcers, J., 111(834), 167(1417), *221, 240*
Pelosi, P., 133(1112), *231*
Peloso, A., 177(1576), *245*
Peltzing, K., 177(1551), *244*
Pelzing, K., 177(1579), *245*
Pendergast, W., 94(729), *218*
Peng, S. M., 541(123), *623*
Pepperdine, W., 492(273), 607(273), 609(273), *628*
Pera, M. H., 185(1728, 1729), *251*
Perepelitsa, E. M., 307(169), *440*
Perettie, D. J., 141(1160, 1161), *232*
Perin-Roussel, O., 275(83), 289(125), 373(83), 377(83), 379(125), *438, 439*
Pernemalm, P. A., 54(509), *210*
Perry, D. H., 299(324), *445*
Perveev, F. Ya., 126(1048), 128(1072, 1073), 137(1127–1132), *228, 229, 231*
Peschke, G., 12(157), *199*
Peters, K., 46(479), *210*
Peterson, D. A., 126(1063), *229*
Peterson, J. B., 48(489), 49(489), *210*
Peterson, J. V. B., 52(493, 494), *210*
Peterson, U., 14(184), *200*
Petrov, V. V., 115(969), 174(1519, 1520), *226, 243*
Petrova, L. K., 77(637), *215*
Petrow, V., 267(61), 365(61), 409(61), *437*
Petrus, A. M. J., 131(1098), *230*
Petty, R. H., 541(445, 447), *632*
Petzold, A., 303, 303(157), *440*
Pews, R. G., 41(461, 462), *209*
Pezeshk, A., 491(268), 599(268), *627*
Pfleger, K., 603(379), *630*
Philion, R. E., 149(1270), *235*
Phillips, J. N., 153(1326–1328, 1330–1332), *237*
Phillipson, J. D., 8(73), *197*
Pichler, H., 25(296), *204*
Picton, C., 6(59), *196*
Pidacks, C., 116(980, 981), *226*
Piepers, O., 488(256), 494(284), 601(256), 611(284, 400), 613(400), *627, 631, 728*
Pieroni, M., 177(1553, 1556), 178(1588), *245, 246*
Pierson, W. G., 31(386), *207*
Pinder, A. R., 11(442), *209*
Pines, H., 27(337), *205,* 254(3), 275(79–81), 345(79, 80), 359(81), 373(81), *438*
Pinhas, H., 150(1282, 1284), *236*
Pisarchuk, E. N., 6(51), *195*
Pitchford, A. T., 462(75), 615(444), *622, 632*
Pitel, D. W., 12(149), *199*

Pitent, A. G., 280(94), 339(94), *438*
Pittet, A. O., 34(405, 407), *207*
Pittmann, A. G., 280(96), 324(94), *438*
Pivnenko, N. S., 107(787), *219*
Planat, D., 191(1771), *252,* 282(105), 283(105), 373(105), *439*
Plattier, M., 22(264), 120(264, 1000), *203, 227*
Plattner, J. J., 111(836), *221*
Podesva, C., 257(18), 423(18), *436*
Poetsch, E., 126(1062), 154(1302), *229, 237*
Poggi, A., 533(428), *632*
Poletaeva, V. F., 12(143, 144), *199*
Polosin, V. M., 180(1668), *249*
Polumbrik, O. M., 108(799–802), *220*
Polyachenko, V. M., 140(1154), *232*
Polyak, O., 149(1237), *234*
Polyakova, T. I., 89(916, 918), *224*
Ponnamperuma, C., 36(434), *208*
Ponticelli, F., 165(1389), *239*
Ponticello, G. S., 130(1085), *230*
Ponticello, I., 144(1186), 148(1186), *233*
Poon, C. K., 533(133), *623*
Popov, Y. N., 27(340), 28(361, 366, 368), *205, 206*
Popova, N. S., 114(963), *225*
Porter, A. E. A., 96(738), *218*
Portmann, R., 153(1333), *237*
Portnova, T. V., 27(340), *205*
Poth, E. J., 323(242), *443*
Potier, P., 12(198), 14(198), 15(198), *201*
Potmischil, F., 42(462), 121(1012, 1019, 1023), *209, 227, 228,* 431(150), *440*
Potts, K. T., 37(443), 69(601), 70(605–607), *209, 214,* 339(11), 415(11), 423(11), *436*
Poulson, R. E., 29(369, 374), *206*
Pousset, J. L., 23(269, 272), *203*
Pouteau-Thouvenot, M., 13(173), *200*
Powell, R. G., 22(259), *203*
Powers, J. C., 144(1186), 148(1186), *233*
Pozdnyakova, T. M., 90(703), *217*
Prakasan, K. M., 12(167), *200*
Prakash, A., 113(959), *225*
Prasad, N. N., 12(142), *199*
Pratt, M. W. T., 26(329), 27(331), *205*
Predescu, I., 194(1786), *252*
Prelog, V., 371(184), 407(289), *441, 444,* 450(15), 487(251, 252), 496(295), 501(15), 599(251, 252), *620, 627*
Preobrazhenskii, N. A., 115(969), *226*
Price, B. J., 505(310), *628*
Priest, D. N., 188(1745), *251,* 303(155), 423(155), 425(155), 435(155), *440*
Prigoda, S. I., 12(153), *199*
Prilepo, A. Z., 12(153), *199*

Prince, R. H., 533(116, 346), 535(116), *623, 629*

Proevska, L., 84(860), *222*

Prokof'ev, E. P., 130(1079, 1080), 137(1133), 149(1245), *229, 231, 235*

Prokop, R. L., 53(502), *210*

Promonenkov, V. K., 179(1635), 180(1673), *247, 249*

Proshchina, N. G., 57(552), 64(552), *212*

Proskuryakov, V. A., 115(973), *226*

Proskuryakova, N. S., 12(146), *199*

Prostakov, N. S., 112(845, 846), 114(954–958), 172(1464), 179(1667), 180(1668, 1669), *221, 225, 242, 249*, 278(85, 86), 289(123, 124), 319(228–231), 337(86, 123, 230, 284), 373(124), 375(124, 228, 229), 379(86, 231, 285), *438, 439, 442, 444*

Puchnova, V. A., 60(537), 62(537), *211*

Puckett, W. E., 475(100a), 507(410), *622, 631*

Pufahl, K., 15(209), *201*

Pulverer, G., 12(168), *200*

Purrello, G., 150(1281), *236*

Pushkareva, I. D., 12(143), *199*

Pye, W. E., 23(279), *203*

Quadbeck-Seeger, H. J., 148(1213), *234*

Quan, P. M., 57(521), *211*

Quast, H., 46(479), *210*

Queguiner, G., 115(960), 117(960), *225*, 274(75, 76), 296(141), 387(75), 399(76), *438, 440*

Querchi, A., 71(611), *214*

Quick, J., 173(1515), *243*

Quijano-Rico, M., 34(401), *207*

Qureshi, S. A., 80(679), *216*

Radecka-Paryzek, W., 541(171, 173, 358a), *625, 630*

Radgus, A., 478(143), *624*

Radics, L., 20(280), *203*

Radtke, V., 149(1235), 155(1316), *234, 237*

Radzhan, P. K., 172(1464), 180(1669), *242, 249*

Rafia, F. K., 76(625), 77(645), *214, 215*

Rafla, F. K., 76(268), 83(691–693), 149(1254), 165(1390), *214, 216, 235, 239*

Raileanu, D., 126(1057), *229*

Raithby, P. R., 549(176, 180, 458), 551(180, 437, 456, 457), 553(454), *625, 632, 633*

Rajappa, S., 146(1191, 1192), *233*

Rakhmatullaev, T. U., 15(205, 206, 210, 211), *201*

Rakitin, I. I., 28(353), *206*

Rakovskii, V. E., 24(283), *203*

Rakowski, M. C., 533(137), *623*

Ramey, K. C., 178(1626), *247*

Ramirez, F., 275(84), 347(84), 349(84), *438*

Ramsden, C. A., 76(627), 79(667), 81(683), 82(683, 689), 88(904, 905), 91(707), *214, 216, 217, 223*

Ramsden, J. N., 549(180), 551(180, 437), *625, 632*

Rana, V. B., 537(355), *630*

Raninger, F., 76(629), *214*

Rao, G. S., 31(391), *207*

Rao, H. S., 115(966), *226*

Rao, K. V., 325(256), *443*

Raouf, A., 120(1006), 167(1419, 1420), *227, 240*

Rapoport, H., 111(836), 193(1784), *221, 252*, 591(367, 368), *630*

Rashidov, M. U., 14(204), *201*

Rasmussen, M., 454(39, 41), 505(39, 41), 561(39, 41), *621*

Rasshofer, W., 476(211), 571(211), *626*

Rastogi, R. R., 154(1304–1306), 188(1306), *237*

Rateb, L., 150(1274, 1293), *236*

Ravindranath, K. R., 23(275), *203*

Ray, A. B., 14(194, 195), *201*

Read, G., 3(174), *200*

Rebek, J., Jr., 488(255), 489(255, 258–261), 599(255), 603(258–261), *627*

Reczynska-Dutka, M., 160(1356), *238*

Redwan, D., 140(1150), *232*

Reedjk, J., 573(399), *631*

Rees, A. H., 105(760), *218*

Rees, C. W., 56(515, 519), *211*

Reese, C. B., 334, 334(282), *444*

Reichardt, C., 89(926), *224*, 491(265), 599(265), *627*

Reichert, V. B., 275(82), 347(82), *438*

Reichmanis, E., 285(109, 110), 335(110), *439*

Reichstein, T., 18(230), *202*

Reiff, H., 173(1056, 1509), *245*

Reiff, W. M., 533(330, 453), *629, 633*

Reim, H., 102(751, 753), 103(751, 753), *218*

Reimann, E., 161(1362, 1365), *238*, 266(54, 57), 315(208), 349(301), 383(54), 393(300), *437, 441, 445*

Reingold, I. D., 452(32, 33), 453(33), 456(33), 503(32, 33), 505(32), 507(33), 557(32), 559(33), *620, 621*

Reinshagen, H., 481(226, 227), 482(228), 587(226–228), 589(226, 227), 591(226, 227), *626*

Reis, A. H., Jr., 581(365, 366), *630*

Reiss, J. A., 521(448), *633*

Reistad, K. R., 458(49), 511(49), *621*
Reitz, R., 161(1362, 1365), *238,* 266(57), *437*
Remers, W. A., 116(980, 981), 117(994), *226*
Renard, M. L., 10(127), *198*
Rendina, A. R., 533(153), *624*
Renk, H. A., 139(1142), *231*
Rennie, R. A. C., 190(1769), *252*
Retey, J., 451(20), 501(20), 505(20), *620*
Rettig, A., 12(215), *201*
Revenko, E. A., 87(896), *223*
Reverberi, A., 177(1577), *245*
Reynolds, G. A., 74(624), 83(695, 696), 86(695, 696, 881), 87(900), 214, 216, 222, 223
Ricci, C., 3(11), *195*
Rice, W. Y., Jr., 591(371), *630*
Rich, R. L., 533(119), *623*
Richardson, M. T., 85(868), *222*
Richtzenhain, H., 114(943–945), *225*
Rickards, R. W., 325(257), *443*
Ried, W., 38(447–449), 127(1071), 173(1516), *209, 229, 243*
Riemann, J. M., 337(118), *439*
Riester, O., 150(1280), *236*
Rigby, G. W., 282(103), 347(103), *439*
Rigerte, B., 132(1102), 133(1102), *230*
Rigo, A., 177(1573–1575), *245*
Rigterink, H., 124(1039), *228*
Riker-Nappier, J., 479(222), 581(222), *626*
Riley, D. P., 533(125), *623*
Rimek, H. J., 150(1279), *236, 267*(64), 339(64), 409(64), 411(64), 415(64), 419(64), 423(64), *437*
Rinaldi, P. L., 571(361), *630*
Rinus, O., 167(1412), *240*
Risaliti, A., 256(13), 415(13), 423(13), *436*
Risler, H., 555(184), 561(184), *625*
Rist, N., 326(264, 265), *443*
Rob, I., 609(387), *631*
Robert, A., 72(617), *214*
Roberts, J. D., 462(72), 513(72), *622*
Robertson, P. A., 515(317), *629*
Robins, R. K., 126(1059), 129(1077), 134(1077), *229*
Robinson, J. M., 459(53), 466(53), 475(198), 513(198), 523(53), 561(198), *621, 625*
Robinson, M. J., 257(17), 381(17), *436*
Robinson, M. M., 312(189), 339(189), *441*
Robinson, R., 259(24), *436*
Robinson, W. E., 31(380), *207*
Robison, M. M., 31(386), *207,* 341(286), *444*
Roch, J., 149(1260), *235*
Rode, V. V., 25(307), *204*
Rodgers, A., 476(143, 208), 479(223),

533(143), 535(143), 541(170), 569(143), 573(143, 208), 577(143), 581(223), *624, 626*
Rodynyuk, I. S., 6(40), *195*
Roedig, A., 128(1075), 139(1141, 1142), *229, 231*
Roemer, A., 12(168), 18(228), *200, 202*
Roeraade, J., 9(97), *198*
Roessler, S. S. V., 263(45), *437*
Roger, S., 35(426), *208*
Rokovskii, V. E., 27(345), *205*
Romer, D., 121(1023), *228*
Ronsisvalle, G., 148(1234), *234*
Rooney, C. S., 130(1083, 1084), 148(1217–1219, 1224), *230, 234*
Rooney, J. J., 173(1505), *243*
Roos, J. P., 188(1753), *251*
Roper, J. M., 321(235), 448(6), 455(45), 475(198), 513(198), 561(198), *620, 621, 625*
Rosenburg, D. W., 65(587), *213*
Rosenmund, P., 77(639), *215*
Rosowsky, A., 149(1257), *235*
Ross, C. B., 178(1603), *246*
Ross, J. M., 173(1487, 1488), *242*
Ross, K. H., 46(479), *210*
Rosset, R., 28(365), *206*
Roth, H. J., 167(1408–1410), *240*
Rothnie, N. E., 535(132), 537(132), 623(132), *623*
Rouaix, A., 326(264), *443*
Rouillier, P., 22(264), 120(264), *203*
Rowe, J. M., 178(1601), *246*
Roy, S. C., 115(966), *226*
Roy, S. K., 115(966), *226*
Rozsa, H., 183(1715), *250*
Ruangsiyanand, C., 150(1279), *236, 267*(64), 339(64), 409(64), 411(64), 415(64), 419(64), 423(64), *437*
Ruby, P. R., 118(998), 147(998), 166(998), *227*
Rudnicki, A., 65(585, 586), *213*
Rudzit, E. A., 326(263), *443*
Ruetman, S. H., 289(126), 367(126), *439*
Rulko, F., 15(121, 208, 214), *198, 201*
Runge, F., 324(248), *443*
Rusanov, G. G., 26(318), *204*
Russo, D. A., 274(77, 78), 339(77, 78), 369(77), *438*
Rus'yanova, N. D., 115(961, 963, 964, 977), *225, 226*
Rutzen, H., 496(291), *628*
Ruyle, W. V., 78(659), *215*
Ruzicka, L., 450(15), 501(15), *620*
Ryabokobylko, Y. S., 490(262, 264), 599(262, 264), *627*
Ryall, R. W., 9(103), *198*

Rybakova, Z. P., 6(57), *196*
Rybina, G. I., 122(1031), *228*
Rycheck, M., 533(137), *623*
Rydon, H. N., 7(64), *196*
Rylander, P. N., 486(249), *627*
Rzabekova, N. L., 113(935), *224*

Saakin, A. M., 76(630), *214*
Saakyan, A. M., 76(631), *214*
Sabatini, L., 533(343), *629*
Sadighi, N., 57(526), *211*
Sadyknov, A. S., 31(389), *207*
Sadykov, A. S., 108(790–793, 796, 798, 808, 810), 113(932, 933), 179(1663), *219, 220, 224, 249*
Saehi, T., 20(243), *202*
Safaryan, A. A., 76(631), *214*
Safaryan, G. P., 87(885), 93(709), *217, 223,* 307(167), 308(170, 171), 397(170), *440*
Safranova, L. G., 6(41), 9(41), *195*
Sagitullin, R. S., 164(1381–1383), 169(1431), *239, 240*
Sahm, W., 452(31), 505(31), *620*
Sahni, S. K., 537(159), *624*
Saijo, S., 149(1252), *235*
Saikawa, H., 64(564), *212*
Saimoto, H., 450(16), 501(16, 395), *620, 631*
Saito, I., 347(310), 349(310), *445*
Saito, K., 55(506), 167(1413), 168(1413), *210, 240,* 355(318), 357(318), 393(318), 395(318), 403(318), 405(318), *445*
Saito, O., 136(1123), *231*
Saito, T., 136(1120), *231,* 319(226), *442*
Sakaguchi, M., 54(513), *211*
Sakaguchi, T., 109(821), 167(1421), *220, 240*
Sakai, K., 304(158), *440*
Sakai, Y., 171(1448), *241*
Sakakibara, H., 34(408), *208*
Sakakibara, J., 53(501), 84(867), 85(867), *210, 222*
Sakakibara, K., 173(1489), *242*
Sakamoto, T., 155(1314), *237*
Sakan, T., 12(189), 14(178–180, 189), *200,* 281(101), 369(101), *438*
Sakane, S., 501(418, 441), *632*
Sakizadeh, K., 307(312), 349(312), 417(312), 433(312), *445*
Sakurai, A., 157(1345), 167(1401, 1406, 1413), 168(1413, 1425), 182(1706), 186(1731), *238, 239, 240, 250, 251*
Sakurai, T., 483(241), *627*
Salbeck, G., 116(989), 156(1317), *226, 237*
Saleh, Al-, B., 40(456), *209*
Salgar, S. S., 486(246), *627*

Samatov, A., 14(183, 187), *200*
Samek, Z., 18(234), *202*
Sammes, M. P., 82(690), 92(708), *216, 217*
Sammes, P. G., 96(738), 97(739), *218,* 292(132), 371(132), *439*
Sammour, A., 120(1006), 167(1398, 1399, 1403, 1419, 1420), *227, 239, 240*
Samour, A. M., 167(1407), *240*
Samson, M., 176(1529), *244*
Samsonova, I. A., 6(25–28), *195*
Samtsevich, S. A., 6(52), *196*
Sana-Ullah, C., 307(312), 349(312), 417(312), 433(312), *445*
Sanchez, R. A., 36(435), *208*
Sandberg, E., 140(1153), *232*
Sanders, E., 78(661), *215*
Sanderson, A., 34(405, 407), *207*
Sanno, Y., 149(1256), *235*
Sano, E., 178(1509, 1605–1608), *246*
Sano, M., 64(658), *212*
Sano, T., 105(761), *218*
Santavy, F., 18(230, 234), *202*
Santhanam, K. S. V., 110(815), *220*
Saper, J., 267(61), 365(61), 409(61), *437*
Sartbaeva, U. A., 6(55), *196*
Sartori, P., 115(976), *226*
Sarukhanyan, F. G., 6(60), *196*
Sarychev, Y. F., 9(95), *197*
Sasakawa, T., 12(156), *199*
Sasaki, H., 20(243), *202*
Sasaki, M., 35(423), *208*
Sasakura, K., 23(274), *203*
Saskura, K., 194(1785), *252*
Sasse, K., 182(1707), *250*
Sassiat, P., 28(365), *206*
Sato, K., 150(1291), 155(1311), *236, 237*
Sato, T., 177(1558, 1572), 179(1653, 1662), *245, 248*
Sato, Y., 33(396), 36(431), *207, 208*
Sauer, J., 102(751, 753, 754), 103(751, 753, 754), *218,* 290(129), 373(129), *439,* 448(6), 455(44, 47), 480(105), 505(44), 527(44, 105), 529(105), 563(105), 585(105), *620, 621, 623*
Sauina, A. A., 377(230), *442*
Saunders, J. L., 9(84), *197*
Sausins, A., 110(818), *220*
Sauter, R., 186(1734), *251*
Saverchenko, A. N., 106(767), 121(1014, 1015, 1017, 1021, 1022), *219, 227, 228,* 257(20), 300(149), *436, 440*
Savina, A. A., 114(954), *225,* 319(228, 230, 231), 375(228), 379(231), *442*
Sayadyan, N. M., 6(49), *196*
Sayed, A. A., 149(1238), 167(1402), *234, 239*

Schaefer, H., 149(1244), 163(1378), 173(1470, 1471), *235, 239, 242*
Schaefer, V., 69(603), *214*
Schaefgen, L., 178(1625), *245*
Schaeren, S. F., 64(553), 65(584), *212, 213*
Schall, V., 139(1142), *231*
Schaller, H., 139(1141), *231*
Schamp, N., 34(411), *208*
Schaur, R. J., 261(32), 363(32), 365(32), *437*
Schefczik, E., 149(1253), *235*
Scheib, K., 138(1138), *231*, 451(23), 501(23), *620*
Scheibelein, W., 491(265), 599(265), *627*
Schenkluhn, H., 176(1528, 1533), *244*
Scheuermann, H., 162(1371, 1372), *238, 239*
Scheutzow, D., 139(1142), *231*
Schinz, H., 450(15), 501(15), *620*
Schmeltz, I., 9(110), 11(107), *198*
Schmelz, H., 167(1412), *240*
Schmidt, A., 153(1301), 163(1379), *236, 239*
Schmidt, H., 191(1773), *252*
Schmidt, H. W., 134(1113), *231*
Schmidt, R. R., 106(766), *218*
Schmidt, W., 452(33), 453(33), 456(33), 503(33), 507(33), 557(33), 559(33), *621*
Schmidtchen, F. P., 193(1784), *252*
Schmits, E., 40(452), *209*
Schneider, J. P., 129(1078), *229*
Schnekenburger, J., 123(1036), *228*
Schnitzer, A. M., 131(1090), *230*
Schnoes, H. K., 31(381), *207*
Schönberg, A., 321(233, 234), 367(233, 234), *442*, 541(356), *630*
Schoop, B. C., 541(356), *630*
Schott, H. H., 3(16), *195*
Schreiber, K., 15(209), *201*
Schreyen, L., 34(411), *208*
Schreyer, G., 171(1454), 173(1470, 1471, 1473, 1474), *241, 242*
Schrider, M. S., 193(1781), *252*
Schroeder, E., 161(1363), *238*
Schroeder, H. E., 282(103), 347(103), *439*, 551(437, 457), *632, 633*
Schroeder, M., 553(454), *633*
Schroeder, P. H., 77(646), *215*
Schroth, W., 86(699), *217*
Schubert, H., 128(1074), *229*
Schuendehuette, K. H., 149(1258), *235*
Schuep, W., 23(276), *203*
Schuette, H. R., 8(78), 14(181), *197, 200*
Schuetz, G., 87(884), 171(1445), *222, 241*
Schuler, W., 173(1470, 1471), *242*
Schultheiss, A., 303(156), *440*
Schultz, A. G., 23(273), *203*

Schultz, J. L., 27(341), *205*
Schulz, G., 481(227), 587(227), 589(227), 591(227), *626*
Schulze, W. A., 323(242), *443*
Schunder, L., 557(187), 561(187), *625*
Schweighardt, F. K., 27(300, 341), *204, 205*
Schweiss, D., 148(1222), 149(1267), *234, 235*
Scola, D. A., 133(1107), *230*
Scot, R. G., 27(342), *205*
Scott, A. I., 33(393), *207*
Scott, T. A., 6(59), *196*
Scott, W. L., 264(47, 48), *437*
Scoville, A. N., 533(453), *633*
Seada, M., 149(1238), *234*
Seaton, T., 40(455), 122(1028), 135(1118), *209, 228, 231*
Sednevets, L. G., 177(1548), *244*
Sedova, V. F., 129(1076), *229*
Sedykh, E. N., 289(123), 377(123), *439*
Seelye, R. N., 170(1432), *240*
Seemann, J., 303(157), *440*
Seghi, B., 533(428), *632*
Seidel, M. C., 126(1052), 148(1052), 149(1243), *229, 234*
Seifert, R. M., 35(424), *208*
Seitz, G., 86(880), *222*
Sekiguchi, A., 369(307), *445*
Selim, M. I. B., 167(1403), *239*
Semanov, G. N., 177(1580), *245*
Semenov, A. A., 140(1154), *232*
Semenov, S. S., 31(377), *207*
Senda, S., 95(734), *218*
Senkariuk, V., 187(1737), *251*
Serbousek, W. E., 53(502), *210*
Sergeeva, G. S., 157(1343, 1344), *238*
Serret, I., 264(322), *445*
Serverin, T., 36(439), *209*
Seto, M., 64(550), *212*
Seto, S., 155(1312), *237*
Sevcik, B., 78(649), *215*
Sevenet, T., 12(198), 14(198, 199), 15(198), *201*
Severin, D., 140(1151), *232*
Severin, T., 55(507), *210*
Sevoyan, A. G., 6(60), *196*
Shabamoto, T., 54(510, 513), *211*
Shafiee, A., 123(1035), *228*
Shakhmuradyan, S. B., 6(49), *196*
Shalimov, V. P., 319(230, 231), 377(230), 379(231), *442*
Shamma, M., 31(388), *207*
Shandala, M. Y., 76(636), 77(636), 162(1367, 1369, 1370), *215, 238*
Shanmugasundaram, E. R. B., 6(24), *195*

Shanmugasundaram, K. R., 6(24), *195*
Shaposhnikov, Y. K., 33(398), *207*
Sharma, D., 12(167), *200*
Sharma, V. K., 132(1103, 1105), *230*, 311(185, 186), *441*
Shaw, C. J. G., 56(516), *211*
Shaw, S. C., 359(305), 363(305), 413(305), *445*
Shcherbina, S. M., 6(54), *196*
Shcherbina, S. N., 12(146), *199*
Shcherev, A., 177(1560), *245*
Shchukina, M. N., 112(849, 850), *221*
Shdanov, J. A., 309(172), *441*
Shebaldova, A. D., 122(1031), *228*
Shefrin, R. N., 264(48), *437*
Sheinker, Yu. N., 192(1778), *252*
Shelepin, O. E., 93(721), *217*
Shepherd, R. G., 116(986), 120(1005), *226, 227*, 313(193), 314(195), 325(193), *441*
Sherif, O., 167(1402), *239*
Sherlock, M. H., 143(1175), *232*
Sherstyannykh, N. A., 9(94, 95), *197*
Sherstyuk, V. P., 84(859, 861), *222*
Shibanov, G. N., 93(723), *217*
Shibata, T., 85(876), 152(1299), *222, 236*
Shibata, U., 55(504), *210*
Shikhanov, V. A., 138(1135), 177(1547–1550, 1562, 1586), 179(1629, 1630, 1639, 1642), 180(1680), *231, 244, 245, 246, 247, 248, 249*
Shilina, S. G., 12(143, 145), 13(145), *199*
Shimada, S., 64(566), *212*
Shimamoto, N., 62(542), *211*
Shimanskaya, M. V., 177(1546), 181(1682–1684, 1686), *244, 249*
Shimizu, B., 22(264), 120(269, 1000), *203, 227*
Shimizu, T., 306(165), *440*
Shimo, N., 136(1124, 1125), *231*
Shimomura, H., 109(822), *221*
Shimozono, K., 347(310), 349(310), *445*
Shinkai, S., 575(405), *631*
Shinoda, M., 501(395), *631*
Shiotani, S., 304(158), *440*
Shiraishi, S., 20(254), *202*, 458(52), 507(313), 511(52, 314, 440), *621, 629, 632*
Shizuri, Y., 18(226, 227, 231, 232), *202*, 603(380), *630*
Shoji, J., 605(413), *632*
Shokhor, I. N., 183(1713), *250*
Shono, T., 12(189), 14(189), *200*
Shooley, J. N., 31(388), *207*
Shostakovskii, V. M., 52(492), *210*
Shreekrishna, M. G., 113(978), *226*
Shu, C. K., 34(406), *207*

Shudo, K., 143(1177), *233*
Shultz, J. L., 27(300), *204*
Shungijeto, G. I., 309(172), *441*
Shusherina, N. P., 77(637), *215*, 281(97, 98), *438*
Shvetsov, Y. A., 108(803), *220*
Shvo, Y., 83(694), 86(694), *216*
Sidorov, O. F., 26(318), *204*
Sieno, A., 20(251), *202*
Siegel, M. G., 459(58, 59), 513(58), 517(58), 519(58), *621*
Sigel, H., 537(443), *632*
Siggia, S., 31(384), *207*
Siklos, P., 26(327), *205*
Sikorska, T., 65(585), *213*
Sile, M., 172(1463), 177(1546), 179(1666), 181(1682–1684, 1686), *242, 244, 249*
Simada, S., 66(590), *213*
Simalty, M., 78(665), 87(883), *215, 222*
Simchen, G., 135(1117), *231*, 281(102), 371(102), *438*
Simmonds, D. J., 17(222, 224), 19(222, 224), *201*, 603(382), *631*
Simon, M. N., 36(433), *208*
Simoneit, B. R., 31(381), *207*
Simonovich, C., 79(676), *216*
Simons, D. M., 41(460), *209*
Simpson, B. D., 131(1090), *230*
Simpson, W. R. J., 158(1347, 1348), *238*
Sims, J. L., 9(96), *197*
Singer, P. P., 22(263), *203*
Singh, B., 158(1351), *238*
Singh, R., 543(455), *633*
Singh, S., 111(843), *221*
Singh, S. N., 359(305), 363(305), 413(305), *445*
Singh, U. P., 70(605, 606), *214*
Sinsheimer, J. E., 31(391), *207*
Sintra, M. E., 319(231), 379(231), *442*
Skala, V., 110(817), *220*
Skatteboel, L., 191(1774), *252*
Skelly, N. E., 81(685), *216*
Skrzecz, A., 27(335), *205*
Skura, J., 26(325), *205*
Skvortsova, G. G., 84(858), *222*
Slagle, W. M., 323(242), *443*
Slavyanova, O. V., 77(637), *215*
Sleeper, H. L., 22(257), *202*
Sliam, E., 87(889), *223*
Sloan, K. B., 486(247), *627*
Smagina, M. V., 6(56), *196*
Smalberger, T. M., 591(369, 372), *630*
Smets, F., 153(1325), *237*

Smirnov, L. D., 50(928, 929), *224*
Smirnov, R. P., 541(165), *624*
Smirnova, N. S., 122(1031), *228*
Smirnova, T. I., 179(1641, 1645), *248*
Smith, C. R., 22(259), *203*
Smith, D. G., 20(238–240), *202*
Smith, H., 149(1241), *234*
Smith, K., 34(414), *208*
Smith, R. E., 124(1043), 159(1355), *228, 238*
Smith, R. M., 19(236), *202*
Smolanoff, J., 44(472), *209*
Snegireva, A. P., 533(166), 539(166), 545(163), *624*
Snieckus, V., 90(706), 106(764, 765), *217, 218,* 607(384), *631*
Snyckers, F. O., 22(261), *203,* 324(254), *443*
Snyder, L. R., 28(364), *206*
Sobotta, R., 175(1522), *244*
Sobukawa, M., 565(451), *633*
Sochneva, E. O., 192(1778), *252*
Sogah, G. D. Y., 459(58), 513(58), 517(58), 519(58), *621*
Sokolov, D. V., 166(1422), *240*
Sokolovskaya, S. V., 76(634, 635), 130(1082), *214, 215, 230*
Soldatenkov, A. T., 172(1464), 179(1667), 180(1668, 1669), *242, 249,* 289(124), 373(124), 375(124), *439*
Soldatova, S. A., 319(231), 379(231), *442*
Soler, A., 93(717), *217*
Soliman, F. S. G., 150(1287), *236*
Soliman, G., 150(1293), *236*
Soliman, S. G., 146(1193), *233*
Solomon, D. M., 173(1514b), *243,* 259(26), *436*
Soma, K., 173(1485, 1490–1494), *242, 243*
Sominovitch, C., 79(694), *216*
Sommer, P., 160(1357), *238*
Somogyi, T., 187(1737), *251*
Sondheimer, F., 493(277–282), 607(277–281, 385), *628, 631*
Sone, M., 126(1058), *229*
Sonnemans, J., 114(941), *225*
Sonnenbichler, J., 19(237), *202*
Sono, E., 178(1591), *246*
Sonoda, A., 136(1124, 1125), *231*
Soroka, T. G., 93(721), *217*
Sorokin, O. I., 114(954), *225,* 278(85, 86), 377(86), 379(86), *438*
Soto, J. L., 106(768), 166(1423, 1424), 167(1395, 1397, 1411), 185(1722–1724), *219, 239, 240, 250*
Sotokin, O. I., 112(845), *221*

Sousa, L. R., 459(58, 59), 513(58), 517(58), 519(58), *621*
Spath, E., 10(122), *198*
Speckamp, W. N., 161(1360), *238*
Speer, R. J., 53(498), *210*
Spenser, I. D., 6(19, 20), 22(262), *195, 203*
Sperati, C. R., 533(117, 338), *623, 629*
Sperber, N., 50(490), 51(490), *210*
Speshilova, T. V., 28(352, 353), *206*
Spiegel, E., 46(479), *210*
Spies, H. S. C., 13(177), *200*
Spiridonova, G. I., 29(356), *206*
Spiteller, G., 13(170), *200,* 591(370), *630*
Spiteller-Friedmann, M., 591(370), *630*
Spohn, K. H., 150(1272), *236,* 267(65), 341(65), 383(65), 407(65), *437*
Springer, C. K., 50(490), 51(490), *210*
Sprintschnik, H., 487(253), 599(253), *627*
Spychala, S., 65(586), *213*
Srednevets, L. G., 178(1589), *246*
Sreenivasan, R., 146(1191, 1192), *233*
Srinivasan, A., 132(1103), *230,* 311(185), *441*
Srinivasan, R., 168(1428), *240*
Stacey, G. J., 149(1251), *235*
Stakheev, I. V., 6(32), *195*
Stalick, W. M., 254(3), 275(81), 359(81), 373(81), *436, 438*
Stanciu, T., 194(1786), *252*
Stanga, M., 173(1495), *243*
Stankevich, E. I., 150(1271a), *235,* 267(59), 319(225), 320(232), 349(288), 351(288), 365(59), 413(292), 417(59, 60, 232), *437, 442, 444*
Stankevics, B., 107(783), *219*
Stankevics, E., 150(1271b), *236,* 417(295), *444*
Stanton, D. W., 170(1432), *240*
Stasiak, S., 316(210), 343(210), 415(210), *442*
Staudinger, H., 3(16), *195*
Stedman, R. L., 9(110, 116), 10(116), *198*
Stefanescu, D., 194(1786), *252*
Stefanovich, L. I., 6(52), *196*
Steglich, W., 73(619), 99(744, 745), 100(744, 745), *214, 218*
Steigel, A., 102(751, 753, 754), 103(751, 753, 754), *218,* 290(129), 373(129) *439*
Steiner, D., 12(150), *199*
Stekhun, A. I., 28(354, 367, 368), 29(354), *206*
Stelzel, H. P., 150(1286, 1287), *236*
Stempel, A., 20(246, 247), 23(276), *202, 203*
Stensrud, T., 156(1323), *237*
Stepanova, M. I., 33(398), *207*
Stepanova, S. V., 65(572, 574), *213*
Stepanyan, M. L., 6(60), *196*
Stepanyuk, V. V., 6(46), *196*

Stephan, U., 8(78), *197*
Stetter, H., 496(291), 535(420), *628, 632*
Stevens, C. L., 171(1437), *241*
Stevens, R. V., 22(260), *203*
Stevenson, C., 476(208), 569(205), 573(205), *625*
Stijfs, P. A. M., 131(1093, 1094, 1099), *230*
Stobbe, H., 256, 256(10), *436*
Stoddart, J. F., 523(99), *622*
Stognii, L. P., 6(46), *196*
Stolz, G., 163(1380), *239*
Stone, J. A., 533(125, 340), *623, 630*
Stonik, V. A., 318(219–221), 319(222), 425(219, 220), 427(219, 220, 222, 299), 429(219), *442*
Storey, R. A., 115(979), *226*
Stork, G., 71(612), *214*, 265(53), 341(53), *437*
Stotter, D. A., 533(116, 346), 535(116, 351), *623, 629, 630*
Stotz, R. W., 539(148), 541(148), *623, 624*
Stoufer, R. C., 539(148), 541(148), *623, 624*
Stout, D. M., 102(750), *218*
Strautins, G., 181(1686), *249*
Streith, J., 106(762), *218*
Strelow, F., 22(261), *203, 324(254), 443*
Striegler, C., 256(8), 343(8), *436*
Strizhkova, A. S., 108(807), *220*
Strohschein, R. J., 158(1347, 1348), *238*
Strominger, J. L., 12(158), *199*
Struchkova, M. I., 64(557), *212*
Strzelecka, H., 78(665), 87(883), *215, 222*
Stucky, G. L., 533(119), *623*
Studeneer, A., 156(1317), *237*
Studentsov, O. V., 34(410), *208*
Studier, M. H., 27(342), *205*
Stuessi, R., 136(1126), *231*
Stuiber, D. A., 34(409), *208*
Sturgeon, B., 267(61), 365(61), 409(61), *437*
Stussi, R., 451(22), 501(22), *620*
Stutz, A., 481(226, 227), 482(228), 589(226, 227), 591(226, 227), *626*
Stytsenko, T. S., 130(1079, 1080), 137(1133), 149(1245), 150(1289), *229, 231, 235, 236*
Suassuna de Oliveira, E. N., 34(402), *207*
Subrahmanyam, G., 132(1103), *230*
Subramanian, L. R., 545(436), *632*
Sucrow, W., 161(1366), *238*
Sudenko, V. I., 6(62), *196*
Sugasawa, T., 23(274), 194(1785), *203, 252*
Sugawara, F., 593(237), *627*
Sugawara, I., 483(237), *627*
Sugi, H., 191(1776, 1777), *252*
Sugita, T., 361(320), 363(320), *445*
Sugiura, K., 18(226, 227, 233, 235), *202*

Sugiyama, N., 111(835), 142(1171–1174), *221, 232*
Sukhanyuk, B. P., 87(898), *223*
Sulser, H., 33(432), *208*
Suma, S., 171(1448), *241*
Sumi, A., 503(306), 589(306), *628*
Summons, R. E., 16(217), *201*
Sumnevich, V. G., 6(46), *196*
Sumskaya, A. I., 28(367), *206*
Sundberg, R. J., 104(757), *218*
Sunthankar, S. V., 148(1225), 149(1262), *234, 235*
Surovtseva, V. M., 12(148), *199*
Susan, A. B., 88(901), *223*
Suter, S. R., 104(757), *218*
Sutherland, D. S., 286(116), *439*
Sutherland, I. O., 452(27), 503(27), 505(27, 310), *620, 628*
Suto, J., 26(327), 27(346), *205*
Suvorov, B. V., 28(351), 115(965), *206, 225*
Suwalska, K., 65(586), *213*
Suwinski, J. W., 89(920), 93(719), *217, 224*
Suyama, K., 181(1696, 1698), *250*
Suzue, R., 6(22), *195*
Suzui, A., 281(101), 369(101), *438*
Suzuii, S., 14(178, 180), *200*
Suzuki, H., 40(453), 139(1147), *209, 232*
Suzuki, N., 173(1485, 1490–1494), *242, 243*
Suzuki, T., 298(145), 299(145), *440*
Suzumura, H., 280(95), 339(95), *438*
Suzuta, Y., 361(300), 363(320), *445*
Svab, A., 78(648, 649), *215*
Svanholt, K., 52(494), *210*
Sveics, D., 417(295), *444*
Svetlakov, N. V., 316(213), 429(213), *442*
Sviridova, L. A., 150(1292), *236*
Swain, C. S., 462(73), 464(88), 513(73), 515(87), *622*
Swierezek, R., 25(310), *204*
Swift, G., 84(853, 854), 173(1478–1480, 1483), *221, 222, 242*
Sybilska, D., 26(325), *204*
Sycheva, T. P., 112(849, 850), *221*
Szabo, L., 149(1237), *234*
Szafranski, A., 27(330), *205*
Szakolcai, A., 23(270), *203*
Szendrei, K., 17(223, 224), 19(223, 224), *201, 603(381), 631*
Szeto, K. S., 453(34), 503(34), 557(34), *621*
Szewczyk, J., 26(314), *204*
Szlompek-Nesteruk, D., 65(585, 586), *213*
Szpilfogel, S., 487(252), 599(252), *627*
Szulmajster, J., 12(154), *199*
Szychowski, J., 150(1278), *236*

Tabata, M., 9(93), *197*
Tabor, M. W., 12(166), *200*
Tachibana, Y., 593(234b, 235c, 404), *626, 631*
Tada, S., 84(862), 85(862), *222*
Taga, N., 125(1046), *228*
Tagano, T., 177(1654), 179(1654), *248*
Taguchi, M., 182(1701, 1702), *250*
Tahir, N. A., 486(250), 597(250), *627*
Taira, N., 17(225), *201*
Takabe, K., 142(1169, 1170), *232*
Takeda, K., 143(1177), 150(1291), *233, 236*
Takeda, S., 79(672), *216*
Takagaki, H., 66(592), *213*
Takaya, T., 102(750), *218*
Takahashi, K., 124(1041), *228*
Takahashi, M., 116(988), *226*
Takahashi, N., 20(251–254), 78(656, 657), *202
  215*
Takehara, M., 66(595), *213*
Takenaka, S., 171(1451), 173(1468, 1469),
  177(1555, 1563), *241, 242, 245*
Takeuchi, T., 458(52), 511(52), *621*
Takeshima, T., 182(1703), *250*
Talor, F., 139(1146), *232*
Tamano, A., 27(332, 333), 84(857), 101(749),
  147(1202), 179(1660), *205, 218, 222, 233,
  248*
Tamaki, E., 10(118), *198*
Tamao, K., 448(7, 8), 501(8), *620*
Tamazawa, K., 124(1041), *228*
Tambina, B., 78(666), 79(670), *216*
Tamono, A., 179(1649, 1650), *248*
Tamura, G., 20(243), *202*
Tamura, S., 20(252, 253), *202*
Tamura, Y., 109(821), 167(1421), *220, 240*
Tan, S. F., 76(626), *214*
Tanabe, O., 13(169), 84(862), 85(862), *206,
  222*
Tanabe, R., 116(988), *226*
Tanaka, J., 142(1169, 1170), *232*
Tanaka, K., 453(38), *621*
Tanaka, M., 501(407), *631*
Tanaka, S., 109(820), *220*
Tanaka, T., 125(1046), *228*, 503(38), *621*
Tanba, Y., 180(1676, 1681), *249*
Tani, H., 116(991), *226*
Tani, Y., 6(20, 23), *195*
Tanny, S. R., 73(621), *214*
Tapilov, Z., 181(1692), *249*
Taraz, K., 18(228), *202*
Tarlton, E. J., 257(18), 423(18), *436*
Tashiro, M., 187(1742), *251*
Tasker, P. A., 476(115), 533(15), 537(115),
  549(177, 182), 569(115), 577(115), *623, 625*

Tassi, D., 93(713, 714), *217*
Tatone, D., 84(863), 85(878), *222*
Tatsugi, J., 270(68), 409(68), *438*
Tatsunn, C., 78(664), *215*
Taub, D., 112(847), 114(952, 953), *221, 225*
Taurins, A., 134(1114), *231*
Taylor, E. C., 254(5), *436*
Taylor, H. C. R., 463(86), 515(86), 525(102,
  104), 527(86, 104, 107), 529(107), 531(107),
  565(107), 585(107), *622, 623*
Taylor, H. M., 123(1044a), 124(1037, 1042),
  228
Tedeschi, P., 165(1389), *239*
Teisseire, P., 22(264), 120(264), *203*
Teisseire, P. J., 120(1000), *227*
Tellert, K., 110(816), *220*
Teotia, M. P., 537(355), *630*
Terada, H., 179(1638), *247*
Teranishi, R., 33(395), *207*
Terashima, Y., 14(180), *200*
Terekhin, A. A., 122(1031), *228*
Terent'ev, P. B., 12(140), 66(597, 598),
  67(597), *199, 213*
Terent'ev, V. K., 25(306), *204*
Ter Heide, R., 35(421), *208*
Ter-Manuel'yants, E. E., 34(410), *208*
Terna, B., 60(534), *211*
Tertov, B. A., 89(916, 918), *224*
Tesarik, K., 26(297), 27(338), *204, 205*
Tescari, M., 173(1496), *243*
Teshigawara, T., 116(991), *226*
Teuber, H. J., 87(884), 170(1446), 171(1445),
  222, 241
Tewari, R. S., 170(1447), 171(1442–1444), *241*
Theander, O., 54(509), *210*
Thesing, J., 10(125), *198*
Thiers, M., 257(16), 339(16), *436*
Thind, S. S., 349(312), 417(312), 433(312),
  445
Thoma, J. A., 131(1088, 1094, 1096, 1097,
  1100), *230*
Thomas, D., 46(480), 184(1716), *210, 250*
Thomas, M. T., 106(763, 765), *218*
Thompson, M. D., 515(87, 316), 517(87),
  611(87), *622, 623, 629*
Thompson, N. E. S., 178(1624), *247*
Thompson, W. C., 323, 323(241, 242, 243),
  339(243), *443*
Thorne, R. L., 9(112), *198*
Thummel, R. P., 157(1341), 165(1393),
  181(1341), *238, 239*, 260(27), 262(37),
  264(37, 49), 267(37), 288(120), 327(37),
  330(278), 332(37, 119), 333(120), 337(119,
  120), 339(37), 369(37), 409(37), 411(37),

Thummel, R. P. (*Continued*)
  415(37), 419(37), 423(37), *436, 437, 439,*
  444
Thyagarajan, G., 113(936), *224*
Tilichenko, M. N., 89(923), 106(767),
  120(1001, 1002, 1007), 121(1008b, 1010,
  1011, 1013–1015, 1017, 1018, 1020–1022,
  1024), 122(1025, 1026, 1030, 1032),
  147(1007), 181(1694, 1700), *219, 224, 227,*
  228, 250, 257(19–21), 259(23), 300(149,
  178), 316(211, 212), 317, 317(214–216),
  318, 318(219–221), 319(222), 359(212),
  423(19, 23, 211, 212, 297, 298), 425(214,
  219, 220), 427(219, 220, 222, 229), 429(215,
  216, 219), *436, 440, 441, 442, 444, 445*
Timko, J. M., 459(56, 60–62), 460(56, 63),
  466(56, 63), 513(56, 63), 517(60–62),
  519(60–62), 521(56, 63), 523(56, 63), *621*
Timmer, R., 35(421), *208*
Timmermann, H., 279(91), 391(91), 393(91),
  438
Titova, I. A., 262(39), 409(39), *437*
Titova, N. A., 173(1472), 179(1628, 1631,
  1633, 1640, 1641, 1645), 180(1673),
  181(1687–1689), *242, 247, 248, 249*
Tobiki, H., 149(1252), *235*
Toda, H., 155(1312), *237*
Todhunter, E. N., 6(36), *195*
Togo, K., 66(595), 193(1783), *213, 252*
Tokai, N., 64(546), *212*
Tokareva, L. N., 28(362), *206*
Tokel, N. E., 533(121, 127), *623*
Tolmacheva, G. L., 28(352, 353), *206*
Tolstikov, G. A., 107(786), *219*
Tolysbsaev, B., 6(50), *196*
Tomasik, P., 27(336), 28(34), *205*
Tominage, Y., 126(1058), *229*
Tomko, J., 18(229), *202*
Tomlinson, A. G., 533(430), *632*
Tomsons, U. A., 490(263, 264), 599(264),
  601(263), *627*
Tonkous, V. V., 24(289), *203*
Tori, K., 605(413), *632*
Tori, M., 395(306), *445*
Torkelson, A. T., 523(130), 535(130), *623*
Tornetta, B., 148(1234), *234*
Torres, M., 113(957), *225*
Torsell, K., 12(216), *201*
Torssell, K., 14(190), *200*
Tosik, B. K., 256(12), 315(207), 339(12),
  383(12, 207), *436, 441*
Tosik, K., 120(1003), *227*
Tosunyan, A. O., 113(958), *225*
Toyoda, T., 23(274), 194(1785), *203, 252*
Trahanovsky, W. S., 337(118), *439*

Traldi, P., 347(309), *445*
Traynor, S. G., 285(314), 345(314), *445*
Tremper, A., 44(472), *209*
Trend, J. E., 488(255), 489(255, 258),
  599(255), 603(258), *627*
Trinkl, K. H., 150(1290), *236,* 283(106), *439*
Tritz, G. J., 3(15), *195*
Trofimov, B. A., 182(1708, 1709, 1710), *250*
Troostowijk, C. B., 462(74), 494(74), 495(289),
  513(74), 611(74, 289), 613(74), *622,*
  628
Troschuetz, R., 167(1408–1410), *240*
Triska, P., 163(1375–1376), *238*
Trubnikov, V. I., 115(969), *226*
Trueblood, K. N., 461(65), 513(65), *621*
Tschamber, T., 607(384), *631*
Tschesche, R., 22(265), *203*
Tselinskii, I. V., 183(1713), *250*
Tso, T. C., 10(123), *198*
Tsuboi, S., 188(1752), *251*
Tsuchida, H., 54(508, 511, 512), *210, 211*
Tsuchiya, J., 565(541), *633*
Tsuda, Y., 105(761), *218*
Tsuge, O., 144(1178), 187(1742, 1743),
  189(1756, 1757), *233, 251,* 555(195),
  557(195), *625*
Tsuji, K., 36(437, 438, 440), 143(1177), *208,*
  209, 233
Tsuruoka, T., 55(504, 505, 506), *210*
Tsuruta, Y., 177(1654), 179(1654), *248*
Tubiki, H., 191(1775), *252*
Tull, R. J., 139(1148, 1149), *232*
Tummler, B., 476(210), 571(210), *626*
Tupik, N. D., 6(53), *196*
Turchetto, E., 34(403), *207*
Turchin, K. F., 60(535), 64(555), *211, 212*
Turchl, I. J., 60(532), *211*
Turitsyna, N. F., 123(1045), *228*
Tuttle, M., 452(32), 503(32), 505(32), 557(32),
  620
Tvaruzek, P., 26(319), *204*
Tychin, G. A., 108(807), *220*
Tyler, W. P., 178(1625), *245*
Tyshchenko, A. A., 108(798), *220*
Tzimos, V., 521(448), *633*

Ubaev, K., 14(202), *201*
Uchida, A., 138(1137), *231*
Uchida, H., 158(1349), *238*
Uden, P. C., 31(384), *207*
Udluft, K., 295(139), *439*
Uebel, H. J., 172(1466), 173(1467, 1501),
  177(1551), 179(1632, 1646), *242, 243, 244,*
  247, 248
Ueda, F., 13(169), *200*

Ueda, K., 180(1678), *249*
Ueno, K., 64(565, 566, 568, 569, 570), 66(589, 590), *212, 213*
Uhlik, D. J., 6(29), *195*
Ujhidy, A., 26(326), *205*
Ulaste, V., 181(1682), *249*
Ulatowski, R., 25(310), *204*
Uldrikis, J., 108(809), *220*
Uldrikis, Ya. R., 413(293), *444*
Uldrikjis, J., 110(818), *220*
Ullrey, J. C., 188(1745), *251*, 303(155), 423(155), 425(155), 435(155), *440*
Ullrich, V., 3(16), *195*
Umezawa, H., 20(255), *202*
Uncuta, C., 89(917), *224*
Undheim, K., 52(495), 53(500), *210*, 458(49), 511(49), *621*
Uno, T., 150(1288), *236*
Upadysheva, A., 409(39), *437*
Upadysheva, A. V., 95(733), 146(1198), 157(1343, 1344), *218, 233, 238*, 262(39), 301(153), 411(153), 415(153), 435(153), *437, 440*
Uragami, M., 25(291), *203*
Urbina, G. A., 377(284), 379(285), *444*
Ushizawa, I., 17(225), *201*
Uskokovic, M. R., 112(848), *221*
Usol'tsev, A. A., 182(1699, 1700), *250*, 261(30), *436*
Usova, E. P., 262(39), 409(39), *437*
Ustavshchikov, B. F., 173(1472), 179(1628, 1631, 1633, 1635, 1640, 1641, 1645), 181(1687–1689), *242, 245, 247, 248, 249*
Usui, Y., 62(542), *211*
Utimoto, K., 501(396, 407), *631*

Vahldieck, J., 178(1590), *246*
Vaidya, S. D., 148(1225), 149(1262), *234, 235*
Valter, H., 106(766), *218*
Valtere, S., 148(1220), *234*
Vanag, G. Va., 260(28, 29), 319(225), 320(232), 361(28), *436, 442*
Vanags, G., 267(59), 271(71), 365(59), 413(29, 292, 294), 417(59, 232), *437, 438, 442, 444*
Van Allan, J. A., 74(624), 83(695, 696), 86(695, 696), *214, 216*
Van Allen, J. A., 86(881), 87(900), *222, 223*
Van Bergen, T. J., 462(74), 494(74, 293), 513(74), 611(74, 283), 613(74, 283), *622, 628*
Van der Baan, J. L., 132(1104), 148(1211), *230, 234*, 310(182), 311(188), *441*
Van der Gen, A., 22(258), *202*
Van Der Linde, L. M., 22(258), *202*
Van der Plas, H. C., 60(533), 74(623), 86(697),

94(724, 725, 727, 728), 95(730), 116(992), *211, 214, 216, 217, 218, 226*
Van der Steit, C., 279(90, 91), 389(90), 391(90, 91), 393(91), *438*
Van der Zalm, H., 131(1088, 1091), *230*
Van Dorsser, W., 23(267), *203*
Van Keulen, B. J., 488(256), 601(256), *627*
Van Leusen, A. M., 141(1164), *232*
Van Poelvoorde, E., 131(1091), *230*
Van Ramesolonk, H. J., 609(387), *631*
Van Ree, T., 22(188), *200*
Van Veldhuizen, A., 116(992), *226*
Van Wassenhove, F., 34(411), *208*
Varlamov, A. V., 114(955, 957), *225, 377(284), 444*
Vartanyan, S. A., 78(652), *215*
Vasil'ev, A. N., 182(1710), *250*
Vasil'ev, G. A., 114(954, 957), *225, 377(284), 444*
Vasserman, A. L., 89(921), *224*
Vatsala, T. M., 6(24), *195*
Vazhinskaya, I. S., 6(32), *195*
Veen, R. H. v.d., 495(285), 611(285), *628*
Veitch, J., 147(1206), *233*
Veksler, V. J., 53(503), *210*
Velter, H. J., 161(1366), *238*
Vereshchagin, A. L., 140(1154), *232*
Verhoeven, J. W., 609(387), *631*
Vernon, J. M., 168(1426), *240*
Verwiel, P. E. J., 20(245, 248), *202*
Veschambre, H., 191(1772), *252*
Vetrova, V. A., 179(1631, 1633), *247*
Vickroy, D. J., 11(131), *199*
Viehe, H. G., 135(116), 153(1325), *231, 237*
Vierhapper, F. W., 116(984), *226*, 305(160–162), 306(161, 162), *440*
Vigalok, I. V., 316(213), 429(213), *442*
Vilaplana, M. J., 93(717), *217*
Villani, F. J., 279(88, 89, 92, 93), 322(236), 326(236), 389(88), 393(88), 395(236), 399(88, 236), *438, 443*
Vinick, F. J., 142(1173), *232*
Vining, C. C., 13(174), *200*
Vining, L. C., 12(149), 20(238–240), *199, 202*
Vipond, P. W., 12(165), *200*
Viscontini, M., 13(172), *200*
Vissov, V. M., 87(891), *223*
Viswananthan, N., 23(275), *203*
Vitins, P., 190(1770), *252*
Vitola, M. J., 266(58), *437*
Vitzthum, O. G., 34(404), 35(427), *207, 208*
Vivarelli, P., 505(308), *628*
Vlasov, V. I., 87(890), *223*
Vlasova, T. F., 192(1778–1780), *252*

Vögtle, F., 448(5), 459(2), 461(66, 67, 68), 462(66–69), 463(77, 81, 86), 464(67), 476(66, 154, 156, 158, 209, 210, 211), 477(155), 480(66, 196), 481(66, 81, 224), 483(232), 501(302), 503(302), 513(66–69), 515(67, 77, 81, 86), 521(77, 81), 523(77), 527(86), 537(154, 156), 545(154, 155), 547(156, 158), 549(158), 551(155), 555(77, 81, 154, 155, 183–186, 189), 557(187, 358), 559(155, 186, 188), 561(184, 187, 196, 302), 563(188, 196), 569(66, 154, 156, 360), 571(209–211, 362, 363), 573(66, 154), 575(158), 577(155), 583(66, 77, 154, 196, 224), *620, 621, 622, 624, 625, 626, 628*

Vojtovic, K., 24(287), *203*

Volgina, I. V., 25(305, 307), *204*

Volke, J., 110(817), *220*

Volker, E. J., 106(763), *218*

Volkmann, R., 173(1514), *243*

Volkova, D. A., 6(48), *196*

Volland, H., 256(10), *436*

Vollering, M. C., 94(724), *217*

Vollhardt, K. P. C., 176(1536, 1537), *244*, 288(304), 300(146), 357(313), 369(146), 395(313), 397(146), *440, 445*

Vompe, A. F., 123(1045), *228*

Von Angerer, E., 76(633), 80(677), *214, 216*

Von Buelow, M. V., 23(277), *203*

Vonderwahl, R., 186(1733), *251*

Vondra, K., 107(782), *219*

Von Helden, J., 34(401), *207*

Von Philipsborn, G., 63(544), *212*

Von Schnering, H. G., 46(279), *210*

Vos, A., 495(285), 611(285), *628*

Vos, C., 20(245, 248), *202*

Voznyakovskaya, Yu. M., 6(44), *196*

Vukhrer, E. G., 6(43), *196*

Vullo, A. L., 150(128), *236*

Vymetal, J., 25(294, 295, 311), 26(315–317, 319–321, 323, 324, 328), 27(347, 348), *204, 205*

Vysatskii, V. I., 121(1010, 1011, 1013, 1020), 257(19), 227, 300(150), 317(215, 216), 318(219–221), 319(222), 423(19, 297), 425(219), 427(222, 299), 429(215, 216, 219), *436, 440, 442, 444, 445*

Wada, H., 18(226, 227, 233), *202*

Wada, M., 65(575), *213*

Wada, T., 155(1311), *237*

Wada, Y., 180(1677), 177(1654), 179(1654), 180(1677), *248, 249*

Wagner, B. E., 455(46), 479(221), 581(221), *621, 626*

Wagner, H., 19(237), *202*

Wagner, R. M., 283(107), 341(107), 371(107), 383(107), 397(107), *439*

Wah, H. K., 82(690), *216*

Wahlberg, I., 9(97, 109), *198*

Wahlberg, K., 12(216), *201*

Wainwright, K. P., 547(178), 549(390), *625, 631*

Wakatsuki, Y., 141(1155–1157), 176(1526, 1530, 1531, 1534), *232, 244*

Wake, S., 339(315), 345(315), *445*

Wakefield, B. J., 71(609), *214*, 295(137), 296(142), 297(137), *439, 440*, 486(250), 597(250), *627*

Wakisaka, K., 191(1776, 1777), *252*

Walba, D. M., 459(56), 460(56), 466(56), 513(56), 521(56), 523(56), *621*

Walker, B. H., 449(12), 450(12), 487(12), 501(12), 599(12), *620*

Walker, D. A., 464(95), 519(95), 521(95), *622*

Walker, D. F., 459(1), *620*

Walker, G., 168(1427), 176(1427), *240*

Walker, J., 259(22), *436*

Walker, P. E., 476(204), 569(204), *625*

Walker, R. B., 288(122), 337(122), *439*

Walker, R. L., 149(1240), *234*, 310(181), *441*

Wallace, B. W., 615(444), *632*

Waller, G. R., 11(136), 34(406), *199, 207*, 324(251), *443*

Walradt, J. P., 34(405, 407), *207*

Walter, J. A., 20(239), *202*

Walter, L. A., 50(490), 51(490), *210*

Waltermann, A., 505(31), *620*

Wanecki, F., 25(310), *204*

Wang, C. H., 171(1439–1441), *241*

Wang, C. S., 81(681, 685), *216*

Wang, D. Q., 177(1571), *245*

Wang, K. C., 279(88), 322(236), 326(236), 389(88), 393(88), 395(236), 399(88, 236), *438, 443*

Wang, P., 35(420), *208*

Wantuck, J. A., 178(1603), *246*

Warfield, A. H., 10(117), *198*

Waring, P., 159(1353), *238*

Warner, H. L., 149(1242, 1243), *234*

Wasson, B. H., 130(1084), 148(1217, 1218), *230, 234*

Wasson, B. K., 130(1083), *230*

Wat, C. K., 20(238, 240), *202*

Watanabe, A., 20(254), *202*

Watanabe, K., 33(396), 36(431), *207, 208*

Watanabe, M., 64(550), *212*

Watanabe, T., 64(547–549), 65(548), *212*

Watanabe, Y., 171(1449, 1451), 173(1468, 1469), 177(1555, 1563), *241, 242, 245*

Waters, C. P., 543(174), *625*

Watkins, D. A. M., 452(24), 505(24), *620*

Watkins, S. F., 480(105), 527(105, 106), 529(105), 563(105), 585(105), *623*

Watt, R. A., 97(739), *218,* 292(132), 371(132), *439*

Wattley, R. V., 489(258–261), 603(258–261), *627*

Weaver, L. H., 503(305), *628*

Weber, E., 448(5), 461(66, 67, 69), 462(66, 67, 69), 464(67), 476(66, 154, 210), 477(155), 480(66), 481(66, 224), 513(66, 67), 515(67), 523(326, 421), 537(154), 545(154, 155), 551(155), 555(154, 155, 189), 557(358), 559(155), 569(66, 154), 571(210), 573(66, 154), 577(155), 583(66, 154, 224), *621, 624, 626, 629, 630, 632*

Weber, J. H., 29(357, 358), *206*

Weber, L. D., 126(1063), *229*

Weeks, W. W., 9(89), *197*

Wefer, E. A., 322(236), 326(236), 359(236), 395(236), *443*

Wehinger, E., 185(1721), *250*

Wehner, W., 476(154, 156, 158, 209, 210), 537(154, 156), 545(154), 547(156, 158), 549(158), 555(154), 569(154, 156), 571(209, 210, 362, 363), 573(154), 575(158), 583(154), *624, 626*

Weidler, A. M., 334(283), *444*

Weigert, W., 173(1470, 1471), *242*

Weiner, M., 6(37), *195*

Weinreb, S. M., 23(279), 141(1167), *203, 232*

Weis, C. D., 110(825–827, 830, 831), *221*

Weisflog, J., 148(1228), 151(1228), *234*

Weisleder, D., 22(259), *203*

Weisman, G. R., 460(64), 513(64), *621*

Weiss, M. J., 116(980, 981), 117(994), *226*

Weissberger, A., 254(5), *436*

Welch, B. R., 541(447), *632*

Wendler, N. L., 112(847), 114(950, 952, 953), *221, 225*

Weng, S. J., 177(1571), *245*

Wenkert, E., 133(1108), *230*

Wentrup, C., 42(468), *209*

Werkhoff, P., 34(404), 35(427), *207, 208*

Werner, G., 96(737), 97(740, 741), 98(742), *218*

Werner, W., 116(990), *226*

Wessels, P. L., 13(177), *200*

Westphal, O., 118(997), *227*

Westwood, N. P. C., 102(756), 103(756), *218*

Weyer, R., 126(1056), 148(1056), *229*

Wharmby, M., 186(1735), *251*

Wheeler, W. R., 179(1636), *247*

Whincup, P. A. E., 173(1481), *242*

White, C. M., 27(300, 341), *204, 205*

White, J. W., 68(540), *211*

Whiting, D. A., 17(222, 223, 224), 19(222, 223, 224), *201*

Whitney, R. R., 459(57), 462(57, 71), 513(57, 71), 603(381, 382), *621, 622, 631*

Wiechers, A., 22(188, 261), *200, 203, 324(254), 443*

Wieder, W., 555(189), 557(358), *625, 630*

Wiegele, W. D., 533(331), 535(331), 537(331), 569(331), 577(331), *629*

Wihervaara, K., 6(31), *195*

Wilbert, G., 173(1487, 1488, 1498), *242, 243*

Wild, P., 107(788), 122(1033), *219, 228*

Wilder, P., 9(114), *198*

Wildman, W. C., 31(392), *207*

William, K. R., 115(979), *226*

Williams, G. L., 107(774), 167(1414), *219, 240*

Williams, H. W. R., 79(673), 130(1083, 1084), 148(1217–1219, 1224), *216, 230, 234*

Williams, T. H., 23(276), *203*

Wilson, L. J., 541(445, 447), *632*

Winecholt, R. L., 73(620), *214*

Winkler, T., 110(830), *221*

Winterfeldt, E., 123(1035), *228*

Winzerberg, K. N., 8(74), *197*

Wirth, R. K., 136(1122), *231*

Wirthwein, R., 295(139), *439*

Wisowaty, J. C., 125(1066), *229*

Wiszniowski, K., 24(286), *203*

Witerzens, P., 93(716, 720), *217*

Withycombe, D. A., 34(409), *208*

Witkiewicz, K., 15(208, 214), *201*

Wittek, P. J., 124(1040), 149(1263), *228, 235*

Witteveen, J. G., 22(258), *202*

Wittmann, H., 158(1346), *238*

Wobben, H. J., 35(421), *208*

Wöhrle, D., 491(266, 267), 599(266, 267), *627*

Woisa, K., 65(586), *213*

Wolf, F., 25(309), *204*

Wolf, U., 161(1366), *238*

Wolfbeis, O. S., 148(1223), 187(1738), *234, 251, 263(43), 431(43), 433(43), 437, 487(253), 599(253), 627*

Wolfe, J. F., 79(668), *216*

Wolny, H., 487(253), 599(253), *627*

Woltermann, A., 452(31), *620*

Wong, C., 482(230, 231), 587(230), 589(230, 231), 603(406), *626, 631*

Wong, L. L., 47(482), *210*

Wong, W. S., 76(626), *214*

Woodward, R. B., 325(256), *443*

Woolley, P. R., 533(116), 535(116), *623*

Work, S. D., 79(668), *216*
Woszczyk, A., 27(336), *205*
Woudenberg, M., 22(188), *200*
Wozniak, M., 116(992), *226*
Wren, J. J., 35(418, 422), *208*
Wright, D. J., 71(609), *214*
Wright, J. L. C., 20(238, 239, 240), *202*
Wright, P. M., 337(308), *445*
Wrobal, J. T., 150(1278), *236*
Wrtilek, I., 164(1388), *239*
Wu, T. T., 23(279), *203*
Wursch, J., 64(553), *212*
Wyrzykowska-Stankiewicz, D., 27(330), *205*

Xia, Y. J., 475(100a), *622*
Xi-chang, S., 603(416), *632*

Yablonskii, O. P., 178(1589), *246*
Yagodzin'ski, T., 164(1383), *239*
Yajiuma, I., 34(408), *208*
Yakhontov, L. N., 112(849, 850), *221*
Yakimchuk, I. G., 115(971), *226*
Yakimovich, S. I., 126(1049), *228*, 231
Yamada, K., 18(226, 227, 231–233, 235),
    187(1742), *202, 251*, 603(380), *630*
Yamagami, T., 64(558), *212*
Yamagishi, K., 173(1489), *242*
Yamaguchi, H., 170(1433), 171(1433), *240*
Yamaguchi, T., 116(991), *226*, 511(432), *632*
Yamaji, T., 173(1493), *243*
Yamamoto, H., 501(418), *632*
Yamamoto, K., 501(441), *632*
Yamamoto, M., 111(835), 142(1171, 1172,
    1174), *221, 232*
Yamamoto, T., 36(437), *208*
Yamamoto, Y., 39(451), 59(530), 79(672),
    101(748), 155(1314), *209, 211, 216, 218,
    237*, 501(396), *631*
Yamamura, H., 155(1311), *237*
Yamamura, Y., 501(418), *632*
Yamanaka, H., 109(822), 111(833), 127(1068),
    150(1276), 152(1276, 1297, 1299), 155(1308,
    1314), 167(1405), *221, 229, 236, 237, 240,*
    306(165), *440*
Yamane, K., 289(128), 322(237, 238),
    387(128), 389(127, 238), *439, 443*
Yamato, M., 186(1736), *251*
Yamazaki, H., 141(1155–1158), 176(1526,
    1530, 1531, 1534), 177(1542–1544), *232,
    244*
Yan, S. J., 165(1392), *239*
Yanag, G. Ya., 417(60), *437*
Yanai, T., 34(408), *208*
Yandovskii, V. N., 45(478), 138(1136), *210,
    231*

Yang, S. S., 309(174), 423(174), 425(174), *441*
Yanotovskii, M. Ts., 148(1226), *234*
Yarmchuk, L., 20(246), *202*
Yasakova, E. I., 12(143), *199*
Yasnikov, A. A., 107(785), *219*
Yasuda, H., 13(169), *200*
Yasuda, K., 84(867), 85(867), 173(1504), *222,
    243*
Yasuda, N., 66(592), *213*
Yasuda, S., 147(1207), 177(1567, 1569, 1570,
    1654), 179(1651, 1654), 180(1672, 1677),
    233, 245, 248, 249
Yasue, M., 20(249, 250), 53(501), *202, 210*
Yasumoto, T., 13(176), *200*
Yates, F. S., 88(903), *223*
Yatsunami, T., 535(434), *632*
Yazaki, T., 36(437), *208*
Yip, K. L., 92(708), *217*
Yokoo, A., 322(238), 389(238), *443*
Yokotsuka, T., 35(423), *208*
Yokoyama, M., 154(1303), 182(1703), *236,
    250*
Yokoyama, T., 25(291), *203*
Yoneda, K., 285(303), *439*
Yonehara, H., 20(251), *202*
Yoneyama, K., 20(254), *202*
Yorie, T., 155(1311), *237*
Yoshida, R., 66(595), *213*
Yoshida, S., 20(254), 78(656, 657), *202, 215,*
    293(133), 341(133), *439*
Yoshida, T., 177(1569), *245*
Yoshii, E., 31(382), *207*
Yoshikawa, T., 60(538), 61(539), 65(577),
    66(593), *211, 213*
Yoshimura, Y., 605(413), *632*
Yoshioka, K., 89(678), *216*
You, K. S., 3(10), *195*
Yudis, M. D., 279(89), *438*
Yukhnevich, A. D., 151(1296), *236*
Yuldashev, P. K., 14(192), *200*
Yunusov, S. Y., 14(183, 192, 193, 200–204),
    15, (206, 210, 211), 31(390), *200, 201, 207*
Yur'ev, V. P., 107(786), *219*
Yurkevich, E. A., 27(345), *205*
Yurkina, L. P., 114(962–964), 115(977), *225,
    226*
Yurugi, S., 62(542), *211*
Yusupov, D., 177(1583, 1584, 1587),
    181(1692), *246, 249*
Yusupova, N., 28(352), *206*

Zabunova, O., 177(1560), *245*
Zaher, H. A., 149(1238), 167(1402), *234, 239*
Zaichencko, Yu. A., 174(1519, 1520), *243*
Zaidi, N. A., 156(1322), *237*

Zaidis, E. G., 26(318), *204*
Zaika, E. I., 108(799–801), *220*
Zaika, O. I., 108(802), *220*
Zairov, F., 108(793), *219*
Zaitsev, A. A., 66(597, 598), 67(597), *213*
Zakharov, V. F., 113(955, 956), *225*, 319(230), 377(230), *442*
Zalewski, R., 28(349), *205*
Zalyalieva, S. V., 108(808, 810), 113(933, 934), *220, 224*
Zambelli, P., 173(1496), *243*
Zamudio, G. V., 34(401), *207*
Zankowska-Jasinska, W., 156(1324), 160(1356), 162(1374), *237, 238, 239*
Zapunnaya, K. V., 84(858), *222*
Zavadskaya, V. P., 28(367), *206*
Zech, W., 191(1773), *252*
Zefirov, N. S., 90(703), *217*
Zeitlan, A., 14(186), *200,* 313(191), 369(191), *441*
Zelenin, S. N., 108(803), *220*
Zenda, H., 36(437, 438), *208*
Zhamagortsyan, V. N., 78(652), *215*
Zhang, B. A., 177(1571), *245*
Zhardetskaya, N. K., 6(55), *196*
Zharkova, N. A., 27(343), *205*
Zhdanov, Y. A., 307(167), 371(171), *440*
Zhdanova, M. P., 88(906), 90(704, 705), 93(718), *217, 223*
Zhdanovich, E. S., 64(551), 115(965, 969), *212, 226*
Zhong-zhi, Y., 619(299a), *628*
Zhukova, N., 64(551), *212*
Zhukova, Z. N., 148(1226), *234*
Zhungietu, G. R., 87(886, 896, 897), *223,* 307(169), 309(173), *440, 441*
Zhuravlev, V. S., 50(928), *224*
Zhuravleva, I. L., 33(399), 34(417), 35(428, 429), *207, 208*

Zicha, B., 3(7), *194*
Zieba, U., 162(1374), *239*
Ziegeuner, G., 146(1199), *233*
Ziegler, E., 76(629), 101(747), 144(1179), 145(1190), 148(1223), 150(1286), 155(1313, 1315), 156(1319–1321), 158(1346), *214, 218, 233, 234, 236, 237, 238*
Ziegon, H. L., 315(208), *441*
Zielinski, H., 31(383), *207*
Ziemann, H., 173(1506, 1509, 1511–1513), *243*
Zilliken, F. W., 6(38), *195*
Ziyaev, A. A., 179(1663), *248*
Ziyaev, R., 14(193), *200*
Zlatkis, A., 27(339), *205*
Znamenskaya, A. P., 95(733), 146(1198), 157(1343, 1344), *218, 233, 238*
Znamenskava, A. P., 301(153), 411(153), 415(153), 435(153), *440*
Zolotareva, E. A., 12(148), *199*
Zolyomi, G., 187(1737), *251*
Zoorob, H. H., 167(1415, 1416), *240*
Zorko, F., 78(666), 79(670), *216*
Zramenskaya, A. P., 262(39), 409(39), *437*
Zubek, A., 164(1386), *239*
Zvezdina, E. A., 88(906), 90(704, 705), 93(718), *217, 223*
Zvolinskii, V. P., 114(954–956), *225,* 289(124), 319(230, 231), 373(124), 375(124), 377(230), 379(231, 285), *439, 442, 444*
Zvonkova, E. N., 64(557), *212*
Zymalkowski, F., 150(1279), 161(1361, 1364), *236, 238,* 266(56), 267(64), 339(64), 409(64), 411(64), 415(64), 419(64), 423(64), *437*
Zyul'kova, L. B., 181(1668), *249*
Zyuz'ko, A. S., 34(417), *208*

Subject Index

Page numbers followed by (R) denote Review reference; page numbers followed by (T) denote Table.

Aberrant biosynthesis, 8
Acridine:
  hydrogenation, 306
  preparation, 121 (T)
Acrolein-ammonia, 172 (T), 173
Actinidine, 12, 14, 31, 159, 281
β-Acylethylation, 256
Aldol condensation, 300
Alkaloids:
  biosynthesis, 644 (R)
  ergot, 645 (R)
  NMR, 645 (R), 649 (R)
  pyridine, 11–13
  syntheses of, 645 (R)
  tobacco, 7, 11, 645 (R)
  in tobacco leaf and smoke, 10
Alkylation, of pyridines, 642 (R)
Alkylpyridines:
  from coal, 25
  from ethylene, 641 (R)
  Oxidative ammonolysis, 638 (R)
  from petroleum, 28
  separation of, 648 (R)
  from shale, 29
  synthesis of, 641 (R)
3-Alkylpyridines, synthesis of, 641 (R)
Aminoazirines, 45, 291
β-Aminoenones, 264
Aminopyridines:
  from acetylenes, 137
  from amidines, 158, 159
  coulometry, 649 (R)
  from diazepine, 106
  diazotisation, 639 (R)
  from enamines, 129
  from imino esters, 158
  from malononitrile, 182
  neuromuscular transmission, 647 (R)
  from pyridines, 129
  review of, 647 (R)
1-Aminopyridines, 81
2-Aminopyridines, 640 (R)
  alkenylation, 637 (R)

  from azetinones, 47
  nucleophilic substitution, 637 (R)
4-Aminopyrimidine-5-carboxaldehyde, 264, 265
Ammoxidation:
  of 1,3-pentadiene, 139
  thienopyridines, 140
Anabasin:
  biogenesis, 642 (R)
  in tobacco, 10
Anabilysine, 7
Annelated cycloheptapyridines:
  physical data, 434, 435 (T)
  preparation, 434 (T)
Annelated pyridines:
  basicity, 332–333
  biologically active, 325–327
  carbocyclic, 253–445
  IR data, 333–334
  naturally-occurring, 323–325
  NMR data, 327–331
  physical properties, 327–335
  reactions of, 312–323
  syntheses of, 255–312
  tautomerization, 334–335
  UV data, 331–332
ANRORC mechanism, 639 (R)
Ant venoms, 644 (R)
Apoenzymes, 646 (R)
Aza[17]annulenes, 493
Azadibenzocycloheptadienones, 325, 326
Azadibenzocycloheptanes, 322
Azadienes, 290–293
  syntheses of, 641 (R)
2-Azaestradiol-3-methyl ether, 327
Azafluorenes, 283, 289, 319, 320, 321
Azafluorenones, 260, 267, 277, 278, 284, 320, 321
Azaindole, from aminopyrrole, 164
Azepines:
  oxidation of, 104
  photolysis of, 104
  ring-contraction, 104, 105
  silver-catalyzed rearrangement, 105

Azides, pyrolysis of, 189
Azido-tetrazolo isomerization, 638 (R)
Aziridino adducts, 102
Azocinyl dianion, 284
Azo compounds, 101
Azulenopyridines, 265

Basicity, annelated pyridines, 332–333
Bechmann rearrangement, 136, 137, 451
Benzo[g]guinolines, 300
Bicyclo[10.3.0]pentadeca-1(12)-en-13-one, 449
Biochemical aspects, 642 (R)
Birch reduction, 117
Bis-annelation, 255, 316–319
Bis-[1,8-diazafluorenylidene], 321
Bis-pyridinones, 81
Boekelheide rearrangement, 312, 313, 316, 450
Boraperhydrophenalene, 449
Boschniakine, 12
5-Butylpicolinic acid, 643 (R)
B6 vitamins, 3
    by microorganisms, 6

Cadaverine, thermolysis of, 139
Camptothecin, 23, 173
Cantleyine, 12
Carbenes:
    from alkylhalides, 55
    from bromoallenes, 57
    from CHCl₃, 55, 56
    with pyrroles, 55
    from sodium trichloroacetate, 56
Carbocyclic annelated pyridines, 253–445
    biologically active, 325–327
    naturally-occurring, 323–325
    physical properties, 327–335
    reactions of, 312–323
    synthesis of, 255–312
β-Carboline, 23
N-Carbomolymethylpyridinium chloride, 170
Carpyrinic acid, 644 (R)
Cathedulins, 17
Celestraceous alkaloids, 17, 18
Chichibabin condensation, 177, 261–263, 317, 450
Chiral binaphthyl, 459, 460
Chloropyridines, from cyanoacetamides, 153
Chromatography of pyridines, 648 (R)
Cinchona alkaloids, 644 (R)
D-(+)-Citronellal, 259
Claisen condensation, 271–274
Claisen rearrangement, 191, 287
Clitidine, 17
Coal, components, 24

Coal tar pyridines, 656 (R), 657 (R)
Cobalt complexes with pyridines, 654 (R), 655 (R)
Cobalt metallocycles, 141
Coenzyme labeling, 646 (R)
Condensation reactions, 255–274
    Aldol, 263, 300
    Aldol-Claisen, 271–274
    of β-amino-α,β-unsaturated carbonyls, 264–270
    chichibabin, 261–263, 317
    α-condensation, 314, 315, 317, 318
    with 1,5-dicarboxyls, 255, 309
    Friedlander, 264–270
    Hantzsch, 462
    Stobbe, 449
Configuration of, piperidines, 651 (R)
Conformation of:
    piperidines, 648 (R), 651 (R)
    pyridine macrocycles, 459, 467, 469
Conhydrine, 643 (R)
Copper complexes with pyridine, 655 (R)
Crops, protection against birds, 657
Crown ethers:
    with bis-lactones, 463, 464
    with dihydropyridines, 462, 463, 495
    with nicotinic acid, 463, 464
    with pyridines, 459
    reactions of, 469
Curtius rearrangement, 188
Cyanoethylation:
    of acetone, 131
    of methyl ethyl ketone, 131
Cyanopyridines:
    from dienes, 141, 143
    from ethylene and ammonia, 36
Cyclization reactions:
    acid-catalysed, 450
    azatrienes, 282–286
    cooligomerization, 299
    of cyanoesters, 132 (T)
    Diels-Alder reaction, 290–299
    of 1,5-dioxo compounds, 120 (T)
    electrophilic, 245, 276–282
    of five-carbon chain, 119
    Friedel-Crafts acylations, 278–279
    of imines, 190, 191
    metal catalysis, 136
    nitriles, 280–282
    nucleophilic, 274–275, 462, 467, 474, 475, 477, 480, 481
    of oxocarboxylic acids, 126
    photocyclization, 288, 300
    pyrolysis, 497
    rearrangements, 300–303

Schiff-base condensations, 470, 473, 476, 478, 479, 480, 490–491
thermal, 286–288
Cycloadditions, 290–300
with Azides, 102
3-hydroxy-1-alkylpyridinium, 653 (R)
photocyclizations, 300
of pyridines, 640 (R)
to pyridines, 641 (R)
pyridinium N-iminoylides, 640 (R)
pyridinium N-methylides, 640 (R)
Cycloalkanopyridines, larger than octa-:
physical data, 406–407 (T)
preparation of, 406 (T)
Cycloalkenopyridines, reactions of, 640 (R)
2,3-Cycloalkenopyridines, 267
2,3-Cyclobuta-1,8-naphthyridine, 264
Cyclobutapyridines:
physical data, 336–337 (T)
preparation of, 281, 336 (T)
2,3-Cyclobutapyridines, 286, 287
2,3-Cyclobutaquinoline, 264
Cyclocondensations, 261, 271, 452, 464, 466, 470, 471
intramolecular, 458
Cyclododecane-1,5-dione, 449
2-Cyclododecenone, 485
Cycloheptapyridines, 271–274, 279, 283, 285, 312, 315
2,3-Cycloheptapyridines, 257, 266, 299, 308
physical data, 380–395 (T)
preparation of, 380–395 (T)
3,4-Cycloheptapyridines:
physical data, 396–401 (T)
preparation of, 396–401 (T)
Cycloimmonium, 653 (R)
Cyclooctapyridines, 285
physical data, 402–405 (T)
preparation of, 402–405 (T)
Cyclotrimerization, 299
Cyclopentacyclohexapyridines:
physical data, 414–417 (T)
preparation of, 414–417 (T)
Cyclopentadienyl cobalt dicarbonyl, 299
Cyclopentapyridines, 257, 259, 261, 264, 274–277, 280, 281, 285, 292, 308, 310, 312–315, 323, 324
tuberculostatic activity, 326
2,3-Cyclopentapyridines, 257, 261, 280
physical data, 338–367 (T)
preparation of, 338–367 (T)
3, 4 Cyclopentapyridines, 290
physical data, 368–379 (T)

[2.2]-Cyclophanes, 452, 457, 493
Cyclopolymethylene pyridines, 636 (R)

Dealkylations, 638 (R), 640 (R)
Decarboxylation, 128
Decomposition, of Mannich bases, 256
Dehydrofusaric acid, 12
Dehydrogenases, 647 (R)
2,3-Dehydropyridine, 294, 295
3,4-Dehydropyridine, 295, 296
Desmosine, 7, 181
Desulfurization, 452, 454–456, 483, 495
Dewar pyridines, 39, 293, 294
4-Dialkylaminopyridines, acylation catalysts, 640 (R)
Diaminoazanaphthoquinones, 326
2,7-Diazabiphenylene, 288, 296
Diazafluorene, 321
Diazafluorenones, 321
Diazepines:
acid-catalyzed ring-contraction, 106
to pyridines, 105
1,2-Diazepines, 653 (R)
Diaziridines, 45
1,5-Dicarboxylic acids, 132
Dicyclobutapyridines, 288
[2,3 : 4,5]-Dicyclopentapyridine:
physical data, 408–409 (T)
preparation of, 408 (T)
[2,3 : 5,6]-Dicyclopentapyridine,
physical data, 410–413 (T)
preparation of, 410–413 (T)
Didehydropyridine, 652 (R)
Dieckmann cyclization, 273
Diels-Alder reactions, 290–299, 323
dienes with nitriles, 141
of pyridazines, 97–98
of pyridines, 642 (R)
2-pyridinones, 639 (R)
of pyrimidines, 97
2-pyrones, 639 (R)
Dienamines, 263
Dihydroangustine, 123
Dihydroazaphenanthridines, 309
Dihydroazepines, 104
Dihydroprosopine, 644 (R)
5,6-Dihydro-2-pyridine-7-one, 273
Dihydropyridines, 102, 650 (R)
from N-carbomethoxypyrrole, 74
dehydrogenation, 106, 108, 109 (T)
from dimethyl acetylenedicarboxylate, 184
via 1,3-dipolar addition, 73
from Hantzsch synthesis, 106

Dihydropyridines, (*Continued*)
  isotopic labeling, 650 (R)
  from oxazolones, 73
  oxidation, 106, 107 (T)
  syntheses of, 650 (R)
  thermal aromatization, 108
1,4-Dihydropyridines, 650 (R)
2,6-Dihydroxypyridine, 132–134
Diketene, 155 (T)
2,6-Dilithiopyridine, 455
1,5-Dimethyl-3,4-cyclopentapyridine, 259
Dipicolinic acid, 12
Dipyridine:
  azacrowns, 470, 473
  complexes, 654 (R), 655 (R)
  crown ethers, 466–468, 489
  mediocycles, 488
  thiacrowns, 474, 475, 480
2,6-Di-tert-butylpyridine, 640 (R)
Dithiaparacyclophanes, 452, 454–456, 483

Electrochemical oxidation, 110
  of furfuralamine, 52
Electrophilic cyclizations, 275–277
Enamines, with isocyanates, 144
Enolethers, with hydrazine, 129
Enzyme mechanism, 645 (R)
Ergot alkaloids, 645 (R)
2-Ethylpyridine:
  from catharanthine, 33
  from pseudo-catharanthine, 33
  from tabersonine, 33
Euonyminol, 17
Evonimine, 18
Evonine, 18
Extraction of pyridines, 648 (R)

Fabianine, 22
5-Fluoronicotinic acid, 8
Fontaphilline, 16
Friedlander condensation, 165, 264–270
  mechanism of, 267, 268
Frightening agents, 657 (R)
Fungicides, azanaphthoquinones, 327
Furarinolic acid, 12
Furfuryl amine, 51
Furoic acid, 51
Fusaric acid, 12, 643 (R)
Fused-ring pyridines, 652 (R)

Gentianine, 12, 15, 16
Gentiapicrin, 12, 15
Glutaconic acid, 132
Glutalic acid, 4

Glutarimides, 110, 111
Gould-Jacobs reaction, 191
Graebe-Ullmann reactions, 639 (R)
Guaipyridine, 22
Guareschi reaction, 147–149, 148 (T)
Guest-host interactions, 459, 462, 463, 465

Halfordine, oxidation of, 31
Halfordinols, 16
Halopyridines, with nucleophiles, 639 (R)
Hammett equation, 2,5-disubstituted pyridines,
  648 (R)
Hantzsch synthesis, 185, 255, 462, 494
Hetarylation, 653 (R)
Hetarynes, 652 (R)
Heteroaromatic reactivity, 637 (R)
Heteroarylation, 639 (R)
Hetero-cope rearrangements, 73
Heterocyclic thiols:
  Hofmann elimination, 454, 482
  synthesis of, 639 (R)
Hydrazonium salts, pyrolysis of, 187
Hydrogenation:
  of isoxazoles, 265, 269
  of quinolines and isoquinolines, 303–306
Hydrogen-deuterium exchange, 638 (R), 639 (R)
Hydrolysis:
  of 6-hydroxyactinidines, 313
  of uracils, 93, 95
Hydroxycotinine, from urine of smokers, 11
Hydroxypyridines:
  electrophilic substitution, 639 (R)
  from furfurylamines, 51
  from oxazoles, 64
  by oxidative ring-opening, 51
2-Hydroxypyridines:
  alkenylation, 637 (R)
  2-aminopyridines, 637 (R)
3-Hydroxypyridines, electrophilic substitution,
  639 (R)

Imidazole alkaloid biosynthesis, 643 (R)
Indicaine, 12
Indicainine, 12
Industrial pyridines, 657 (R)
Insecticides, tetrahydroquinolines, 327
Intramolecular cyclization, 490
Intramolecular cycloaddition, 287, 292, 293
Intramolecular cyclocondensation, 458
Intramolecular Diels-Alder reaction, 97
Ion exchange of pyridines, 648 (R)
Iridodial, 259
IR spectra, annelated pyridines, 333–334
Isocarbolines, 140

Isocarbostyrils, 136
Isodesmosine, 7, 18, 181
Isomerization, azido-tetrazolo, 638 (R)
Isoprenoids in tobacco, 645 (R)
Isoquinolines:
  halogenation of, 306, 307
  hydrogenation of, 303–306
  reduction of, 117
Isotopic labeling, 638 (R), 641 (R)
Isoxazoles:
  hydrogenation of, 265, 269
  hydrogenolysis of, 127
  to pyridines, 71
Isoxazolopyridine, from isoxazole, 165
IUPAC nomenclature, 448

Ketomethylpyridinium salts, 170
  reactions of, 171 (T)
Ketonetrimethylhydrazonium fluoroborates, 286
Khat, 17
Knoevenagel condensation, 132, 133, 147, 168, 310
Krohnke reaction, 170

Liquid chromatography, 647 (R)
Liquid crystals, 641 (R), 649 (R)
Lithium tetrachloropalladate, 136
Lutidine, from coal, 24–28
2,6-Lutidone, 81
Lutidylene, 298
Lycodine, 324
Lycopoduim alkaloids, 22
Lythraceous alkaloids, 453

Macrocyclic piperidines, 646 (R)
Macrocyclic pyridines, 636 (R)
  2,3-fused bridges, 487–492, 598–605 (T)
  2,3;5,6-fused bridges, 497, 618–619 (T)
  2,4-fused bridges, 485-486, 596–597 (T)
  2,5-fused bridges, 481-484, 586–595 (T)
  2,6-fused bridges, 448-481, 500–585 (T)
  3,4-fused bridges, 496, 616-617 (T)
  3,5-fused bridges, 492-496, 606-615 (T)
Malonyl chloride, 156 (T)
Mannich bases, 256
Mannich reaction, piperidine, 651 (R)
Mesitylenesulfonylhydroxylamine, 653 (R)
Metacyclo-(2,6)pyridinophanes, 452
α-Metallation, 313, 314, 315, 318, 452
Metapyridinophanes, 492
N-Methylcodine, 22
2,2-Methylenedicyclohexanone, 257, 258
Methylethylpyridine, 656 (R)
O-Methyl lythranidine, 453

N-Methylorthithine, 8
Methylpyridines, alkylation of, 642 (R)
2-Methylpyridines, preparation of, 102, 641 (R)
3-Methylpyridines, preparation of, 640 (R)
Muscopyridine, 450, 487

NAD, 3
  biological reactions, 643 (R)
  metabolism, 644 (R)
NADH, oxidation, 110
Naphthalinopyridinophanes, 453
Natural products, 642 (R)
Naucledine, 23
Naufoline, 23
Navenone-A, 22
Neurobiology, piperidine, 645 (R)
Neurospora, 3, 4
Neutral-molecule complexes, 459, 462–464, 468
Nickel complexes with pyridines, 654 (R)
Nicotinamide, 3, 5, 16
  from β-picoline, 657 (R)
  catalytic ammoxidation, 638 (R)
  electrochemical properties, 638 (R)
  metabolism of, 644 (R)
  nucleotides, 646 (R), 647 (R)
Nicotinamines, 647 (R)
Nicotine, 643 (R)
  biogenesis, 642 (R)
  content in tobacco, 9, 10
  variants, 8
Nicotinic acid, 644 (R)
  by baker's yeast, 6
  biosynthesis of, 3, 644 (R)
  catalytic ammoxidation, 638 (R)
  crown ethers, 462, 464, 484
  by fungus penicillium digitatum, 6
  by Gossypium barbadense, 6
  by microorganisms, 6
  from quinolines, 117
  by Rhodotorula glutinis, 6
Nicotinonitrile:
  crown ethers, 464
  from enamines, 130
  from ethylene and ammonia, 36
Niobium complexes, 655 (R)
Nitrene insertion, 40–42
NMR spectra:
  annelated pyridines, 327–331
  of macrocycles, 464, 467
  of pyridines, 649–650 (R)
Nomenclature:
  IUPAC, 448
  phane, 448

Norpyridoxine, from furfuralamine, 52
Nuclear magnetic resonance (NMR), alkaloids,
   645 (R)
Nucleophilic cyclizations, 274–275, 462, 267,
   474, 475
Nudifluorine, 11

Octahydroacridines, 121, 257–259, 262,
   263, 268, 269, 286, 300, 302,
   316–319
  physical data, 422–433 (T)
  preparation of, 422–433 (T)
Octahydrophenanthridines, 262, 263, 268, 318,
   319, 327
  physical data, 418–421 (T)
  preparation of, 418–421 (T)
Oligopyridines, 636 (R)
Organocobalt catalysts, 642 (R)
Organocopper compounds, oxidative coupling,
   638 (R)
Ortoleva-King reaction, 170
Oxazines:
  with active methylenes, 101
  with dienophiles, 99–100
Oxazoles:
  with dienophiles, 60–64, 640 (R)
  from nicotinic acid, 16
  with olefins, 60, 61
  ring transformations, 641 (R)
Oxazolone, 73
Oxidation:
  agents, 108 (T)
  NADH, 110
  vapor-phase, 262
N-oxidation, 312, 313, 316, 450
S-oxidation, 454
Oxidative coupling, 452
Oxidative dealkylation, 638
N-oxide acylation, 312, 313, 316
N-oxides, 450 (R)
  diastereomeric, 11
  reactions of, 312, 313, 316, 317,
   450
S-Oxides, reactions of, 454
4-Oxopyranoazoles, 652 (R)

Paphyl complexes, 655 (R)
Paraquat, 656 (R)
Pentachloropyridine, 135
Perfluorophenylnitrene, 40
Pesticides, 656 (R):
  octahydrophenanthridines,
   327

Pharmaceutical aspects of pyridine, 645 (R),
   657 (R)
Phenanthroline complexes, 654 (R),
   655 (R)
Phenylpyridines, 172, 636 (R)
Phosphorus expulsion, 480
Photocyclization, 288, 300
Picoline:
  from coal, 25–28
  from ethylene and ammonia, 641 (R)
  from N-hexylamine, 139
  from ketonitriles, 131
3-Piperideines, 651 (R)
Piperidine alkaloids, 645–647 (R)
  biosynthesis, 642–644 (R)
  isolation, 643 (R)
  NMR aspects, 643 (R)
  structure elucidation, 643 (R)
Piperidinediones, 110
Piperidines, 651 (R)
  conformational analysis, 648 (R)
  dehydrogenation, 108, 113, 114
  neurobiology, 645 (R)
  physiological role, 644 (R)
4-Piperidinol, 638 (R)
Piperidinones, 642 (R)
Piperine, 647 (R)
Poland, use of pyridine in, 657 (R)
Polarography, 638 (R)
Polymethylvinylpyridine, 648 (R)
Poly(vinyl)pyridines, 653 (R)
Porphyrin-like macrocycles, 458
Prototropic tautomerism, 648–649 (R)
Purification of, pyridine bases, 648 (R)
Purine nucleosides, 646 (R)
Pyrano-[3,4-b]-pyran, 80
Pyranopyridine, 80
Pyranoxazines, 80
Pyrazines, 101
  cycloaddition, 98
Pyrazolopyridine, from aminopyrazole,
   164
Pyridazines, Diels-Alder reaction, 97–98
(4,6)-Pyridazinophanes, 485 (R)
Pyridine adenine dinucleotides, biochemistry,
   643 (R)
Pyridine alkaloids, 11–13, 642–647 (R)
  actinidine, 259
  biological activities, 645 (R)
  biosyntheses, 8, 643 (R)
  biosynthesis, labeling, 8
  in castor fiber, 8
  metabolism, 8

O-methyl lythranidine, 453
NMR aspects, 643 (R)
retronecate, 462
Pyridine azides, 638 (R)
Pyridinecarboxylic acids, 637 (R)
  of polymers, 642 (R)
Pyridine coenzymes:
  biochemical evolution, 643 (R)
  biosynthesis, 644 (R)
Pyridine complexes, 654 (R)
2,6-Pyridine-crown ethers, 459–470
Pyridine cryptands, 477
2,6-Pyridinediol, 132–134
Pyridine-linked nucleotides, 645 (R)
  complexes, 654 (R)
  electrophilic substitution, 637 (R)
  isotopic exchange, 641 (R)
Pyridine nucleotides, 646–647 (R)
  biosynthesis, 3, 5
  metabolism, 644 (R)
  physiological aspects, 647 (R)
  spectroscopy of, 643 (R)
  from yeast S. cereoisiae, 3
Pyridine N-oxides, 654 (R)
Pyridinephosphonic acids, 641 (R)
Pyridine-pyridinone tautomerization, 640 (R)
Pyridine radicals, 650 (R)
Pyridines:
  from acetone oxime, 182
  from acetylenes, 137, 138, 141, 640 (R)
  from acetylenes and nitriles, 176 (T)
  from acrolein and ammonia, 171, 172 (T)
  from acyclic compounds, 118–194
  acylation catalysts, 640 (R)
  from aldehydes and ammonia, 177, 178,
    640 (R), 641 (R)
  alkenylation of, 637 (R)
  from alkylamines, 139
  alkylation of, 640 (R), 642 (R)
  alkylation reactions, 637 (R), 638 (R)
  from alkynes and nitriles, 645 (R)
  from N-allylimines, 190
  from amidinium salts, 175
  in amino acids, 3
  from aminofurans, 52
  from aminouracils, 186
  annelated carbocyclic, 253
  antimicrobial activity, 644 (R)
  atmospheric determinations, 656 (R)
  from azabutadienes, 46
  from azepines, 104–105
  from azetines, 46
  from azetinones, 73

  from azirines, 43, 44
  biochemical aspects, 642 (R), 645 (R)
  biogenesis, 3
  from cadaverine, 139
  from carbocyclics, 37
  from carbohydrates, 54
  catalytic hydrogenation, 640 (R)
  catalytic synthesis, 642 (R)
  from chloramine, 40
  chlorination, 40
  from coal, 24, 648 (R), 656 (R)
  commercial aspects, 656 (R)
  from δ-complexes of cyclobutadienes, 39
  from condensed rings, 113–118
  from cyanoacetate esters, 185 (T)
  cycloadditions, 640–641 (R)
  from cyclopentadiene and chloramine, 40
  from cyclopentanone and NH3, 40
  from cyclopropenylazides, 37
  dealkylations of, 638 (R), 640 (R)
  degradation of alkaloids, 31
  from diacetylenes, 146
  from diazepines, 105–106
  from diaziridines, 45
  from diaziridine imines, 46
  diazotization, 639 (R)
  from β-dicarbonyls, 142, 143, 147, 150 (T)
  from dicarboxylic acids, 132–135
  from β-dicarboxyls, 147–157
  from Diels-Alder reaction, 639 (R), 642 (R)
  from dienamides, 193
  from dienamines, 174
  from dienes with nitriles, 141
  from dienones, 138
  from dihydrofurans, 48
  from 2,3-dihydro-4H-pyran, 48
  from dihydropyridines, 46, 57, 102, 106–111
  from 1,5-dioxo compounds, 118–124 (T),
    449, 450, 469
  from diphenylcyclopropene thione, 74
  2,6-Disubstituted, 636 (R), 637 (R)
  3,5-Disubstituted, 636 (R)
  via electric discharge, 194
  electrochemical reduction of, 638 (R)
  electrophilic substitution, 637 (R), 639 (R)
  from enamines, 127, 128, 142, 152, 177, 193
  from enaminonitriles, 157
  from enynamines, 174
  in enzymes, 3
  ESR, 650 (R)
  exchange (H-D) studies, 638 (R), 639 (R),
    641 (R)
  from five-membered rings, 47

Pyridines: (*Continued*)
  from foods and beverages, 33
  from furanones, 48
  from furans, 48
  from furoic acids, 51
  general routes to, 636 (R)
  heteroarylation, 639 (R)
  from HMPTA, 188
  homolytic arylation, 637–638 (R)
  homolytic substitutions, 638–639 (R)
  from hydrazonium salt pyrolysis, 187
  from imines, 191
  industrial syntheses of, 641 (R), 656 (R)
  IR, 650 (R)
  isocyanides, 193
  isotopic hydrogen labeling, 638 (R), 641 (R)
  from isoxazoles, 47, 641 (R)
  from isoxazolium salts, 72
  from ketene *S,S*-acetals, 153, 154 (T)
  from ketenimines, 158
  from ketones, 640 (R)
  Knoevenagel reaction, 256
  liquid-crystalline, 641 (R)
  macrocyles containing, 636 (R)
  from malononitrile, 140, 153, 166, 182, 185 (T)
  from maltol glucosides, 53
  from mesoionic heterocyles, 70
  from mesoionic oxazolones, 68, 69
  from molecules with terminal unsaturation, 136–139
  monoterpenoid alkaloids, 12
  MS, 649 (R)
  natural products, 642 (R)
  from nature, 3
  from *N. Diderechii,* 24
  via nitrenes, 42
  from nitriles and dienes, 141
  from nitrite ylides, 37
  from 1-nitrobutane, 178
  NMR, 649–650 (R)
  nonaqueous solvent, 636
  nucleophilic recyclization, 640 (R)
  nucleophilic substitution, 639 (R)
  nucleotides, 3
  organocoppers, 638 (R)
  from oxazines, 99–100, 101 (T)
  from 1,2-oxazines, 41
  from oxazoles, 47, 60–68, 640–641 (R)
  oxidation, 638 (R)
  oxidative ammonolysis, 638 (R)
  oxidative couplings, 638 (R)
  from oxocarboxylic acids, 125–132
  from 1,5-pentanediol, 139
  from perfluoropiperidine, 113
  PES, 650 (R)
  from petroleum, 28
  from phenylazide, 42
  physical aspects, 648 (R)
  from piperidines, 113, 114
  from piperidone, 111
  polarography of, 638 (R)
  purification of, 641 (R)
  from pyrazines, 98–99
  from pyridazines, 97–98
  pyridinophanes, 636 (R)
  from pyridinopyrimidinones, 117
  from pyrimidines, 93–97, 129
  from pyrones, 75–79, 81
  from pyrroles, 55–57
  from pyrylium salts, 87–92, 451, 485, 492
  from quinoline, 113, 116 (T)
  radical substitution, 637–638 (R)
  reactions of, 637 (R)
  from reduced pyridines, 106–114
  review of, 635
  from ring-opening reaction, 117
  sesquiterpenoid alkaloids, 12
  from seven-membered heterocyles, 104–106
  from shale, 29
  from simulated Jovean atomosphere, 36
  from six-membered heterocyles, 75
    with three heteroatoms, 102–104
    with two heteroatoms, 101–102
  spectroscopy, 649 (R)
  synthesis of, 640 (R)
    with acetylenes, 640 (R)
    from aldehydes and ammonia, 641 (R)
    from alkynes and nitrites, 642 (R)
    with amino Diels-Alder reaction, 642 (R)
    with cobalt-catalysis, 642 (R)
    from ethylene and ammonia, 641 (R)
    from isoxazoles, 641 (R)
    with onium salts, 642 (R)
    with oxazoles, 640 (R)
    with transition organometallics, 642 (R)
  synthetic methods, 36
  tautomerization, 334–335, 648–649 (R)
  test for purity, 641 (R)
  from tetrahydrofurans, 48
  from tetrahydrofurfurylalcohol, 48
  from tetrahydropyridines, 111, 112
  from thia-[3,2-*a*]-pyridinium salts, 118
  from thiazolones, 68, 69
  from three-membered rings, 43
  in tobacco leaf and smoke, 9
  from triazenes, 188
  from triazines, 102–104

from α,β-unsaturated carbonyls, 166, 167 (T), 172
UV, 643 (R), 648–650 (R)
Pyridine-sulfur compounds, 645 (R)
Pyridinethiones, from thiocyanates, 144
Pyridinones:
  from acetylenic carbonyls, 161, 162
  from acyl azides, 188
  via aldol condensation, 130
  alkaloids, 17, 20
  from allenes and imines, 173
  from aminocrotonic acids, 152
  from 5-arylthiazoles, 72–73
  from azides, 189
  from cyanoacetamides, 153
  from D-glucaro-lactams, 55
  from diazetidines, 47
  from diaziridenes, 45
  from δ-dicarbonyls, 142, 143, 153, 161
  from dichloroketene, 174
  from diethynyl ketones, 138
  from dihydropyrrole and derivatives, 59
  from diketene, 153, 155 (T), 156
  from diphenylcyclopropenone, 71
  from enamides, 193
  from enamine oxidation, 100
  from enamines, 126, 130, 191, 192
  from enaminonitriles, 152
  from furfural, 49
  from gem-dichlorides, 130
  from glutarimides, 143
  from imidates, 146
  from isocyanates, 144, 189
  from isoxazoles, 71, 142, 143
  from ketene-esters, 158
  from δ-ketoamides, 145
  from ketoanilides, 160
  liquid crystalline, 641 (R)
  from malonyl chloride, 153, 158 (T)
  from maltol derivatives, 53–54
  metal catalysis, 136
  from oxazines, 101
  by oxidative ring-expansion, 49
  from oxime ethers, 123
  from oxocarboxylic acids, 125, 126 (T)
  from phenylhydrazones, 139
  from pyranodioxins, 80
  from pyrano-[3,4-b]-pyran, 80
  from pyrans, 83–85
  from pyrazines, 99
  from pyrones, 75, 80
  from pyrrolinediones, 57, 58
  from pyrylium salts, 83
  from tetracyclone, 40

from thioamides, 145
from 1,3,5-trioxo compounds, 123, 124
from δ-unsaturated amides, 145
from uracils, 95
from ynamines, 189
Pyridinium aldoximes, hydrolysis, 637 (R)
Pyridinium chlorochromate, 654 (R)
Pyridinium compounds, 652 (R)
Pyridinium halides:
  catalyst, 642 (R)
  chlorides, 7
Pyridinium N-imines, 653 (R)
Pyridinium N-iminoylides, cycloadditions, 640 (R)
Pyridinium N-methylides, cycloadditions, 640 (R)
Pyridinium N-ylides, 652–653 (R)
  to 1,2-diazepines, 105
Pyridinium salts:
  from pyrylium salts, 88–91
  reactions with nucleophiles, 92
Pyridinocrowns, 637 (R)
Pyridinocryptands, 637 (R)
Pyridinols:
  from acylfurans, 50–52
  alkaloids, 17, 20
2,3-Pyridino macrocyles, 487–491
  bis-lactones, 488, 489
  crown ethers, 488, 489
  Schiff bases, 490, 491
  substitution, physical properties, 598–605 (T)
2,3,5,6-Pyridino macrocyles, 7, 497
  substitution, physical properties, 618–619
2,4-Pyridino macrocyles, 485–486
  aza bridges, 486
  substitution, physical properties, 596–597 (T)
2,5-Pyridino macrocyles, 481–484
  bis-lactones, 484
  crown ethers, 484
  substitution, physical properties, 596–597 (T)
2,6-Pyridino macrocyles, 448–481
  with bis-lactams, 470, 476, 477
  bis-lactones, 463–464, 481
  bis-oximes, 476
  bis-thiolactones, 474, 476
  chiral, 460, 466
  NMR studies, 464
  porphyrin-like, 458
  with Schiff bases, 470–474, 476, 478–480
  substitution, physical properties, 500–585 (T)
  thiocrown ethers, 474–475, 477–480
3,4-Pyridino macrocyles, 496
  bis-lactams, 496
  substitution, physical properties, 616–617 (T)

3,5-Pyridino macrocycles:
  bis-lactams, 496
  bis-lactones, 494–495
  crown ethers, 494–495
  dihydro, 494–495
  substitution, physical properties, 606–615 (T)
  thiocrown ethers, 495
Pyridinophanes, 636–637 (R)
(2,3)-Pyridinophanes, 487–488
(2,4)-Pyridinophanes, 485, 486
(2,5)-Pyridinophanes, 481–483
(2,6)-Pyridinophanes, 448–457
(3,5)-Pyridinophanes, 492, 493, 495
Pyridinyl radicals, 655–656 (R)
Pyridoimidazole, 137
Pyrido-[2,3,-f]-morphan, 264
Pyrido-[3,2-d]-tropolone, 322
Pyridotropolones, 273, 289
Pyridoxal, 3, 5
  biosynthesis by E. coli, 6
Pyridoxal phosphate:
  by aerobic cellulose bacteria, 6
  by nonleguminous plants, 6
  by Rhizobium leguminosarum, 6
Pyridoxamine, 3, 5
Pyridoxine, (Vitamin B6), 3, 5, 6
  production and price, 68
  synthesis from oxazoles, 68
Pyridyl amides, Willgerodt reaction, 639 (R)
4-(2-Pyridylazo)resorcinol, 637 (R)
1-(2-Pyridyl)-1,2,3-benzotriazoles, 639 (R)
Pyridyl ketones:
  ketalization, 454
  Willgerodt reaction, 639 (R)
Pyridylthiols:
  electrophilic substitution, 639 (R)
  nucleophilic substitution, 639 (R)
2-Pyridylthione, 39
Pyridyne, 652 (R)
2,3-Pyridyne, 295
Pyrimidines:
  with aqueous acid, 95
  Diels-Alder reaction of, 96–97
  with dienophiles, 93, 96, 97
  nucleophilic substitution, 93, 94
  thermolysis, 97
Pyrindanes, 274–277, 280, 281, 308, 312–315, 323, 324
1-Pyrindines, 334, 652 (R)
Pyrolysis:
  of N-allylimines, 190
  of dihalides, 497
  of hydrazonium salts, 187
  of ketazines, 187

Pyrones:
  ammonolysis, 75
  to pyridines, 75
Pyrroles:
  with carbenes, 55
  electrochemical oxidation, 58
Pyrrolinediones, with diazoalkanes, 57
(2,5)-Pyrrolophanes, 482
Pyrrolopyridine nucleosides, 646 (R)
Pyrylium cations, 307–310, 450, 451, 485, 492
  with amines, 86–89, 91–92
  with ammonia, 86–89
  from 1,5-diketones, 83
  to pyridinium salts, 88–93
  from pyrones, 83
  reactions of, 642 (R)

Quaternization of pyridines, 653 (R)
Quinolines, 295
  Birch reduction, 117
  from coal, 25
  halogenation, 306–307
  hydrogenation, 303–306
  oxidation of, 113, 117
  oxidative ammonolysis, 638 (R)
  quinolinic acid, 4, 5
  Ramberg-Backlund reaction, 454
  reduction of, 116, 117

Rearrangement reactions, 300–303
  of allyl-2H-azirines, 43
  Beckmann, 451
  Boekelheide, 450
  Stevens, 452, 454, 455
Reduced acridines, from 1,5-dioxo compounds, 121
Reductions:
  of ketone macrocycles, 468
  Wolff-Kishner, 271
  see also Hydrogenation
Retronecic acid, 462
Review of:
  acrolein with pyridine, 657
  alkaloids:
    NMR, 645
    synthesis, 645
    tobacco, 645
  aminopyridines, 647, 657
  N-arylpyridinium salts, 653
  betaines in synthesis, 653
  chromatography of pyridines, 648
  coal tar pyridines, 656, 657
  commercial aspects of pyridines, 656
  complexes of dipyridines, 654, 655

complexes of pyridines, 654, 655
complexes of terpyridines, 654
configuration of piperidines, 651
conformational analysis, 648
conformation of piperidines, 651
Cu complexes of pyridines, 655
diazepins from pyridinium N-ylides, 653
didehydropyridines, 652
dihydropyridines, 650, 651
2,6-dimethylpyridine, 657
2,5-disubstituted pyridines, 648
enzyme mechanisms, 645
ergot alkaloids, 645
ESR spectra of pyridines, 650
extraction of pyridines, 648
fused-ring pyridines, 652
hetarylation, 653
hetarynes, 652
hindered amines, 652
industrial pyridines, 656
ion exchange of pyridines, 648
IR spectra of pyridines, 650
labeled pyridine nucleotides, 646
mesitylenesulfonylhydroxylamine, 653
methylethylpyridine, 656
MS spectra of pyridines, 649
NAD(H) reactions, 645
neurobiology, piperidine, 645
Ni(II) complexes of pyridines, 654
nicotinoamide nucleotides, 646, 647
nicotinoamide from β-picolines, 657
niobium complexes, 655 (R)
NMR of pyridines alkaloids, 645, 649, 650
4-oxopyranoazines, 652
4-oxopyranoazoles, 652
paraquat, 656
PES spectra of pyridines, 650
phenothiazines with piperidine, 645
3-piperidenines, 651
piperidine, neuromodular, 645
piperidine alkaloids, 645, 646
piperidines, 651
polymethylvinylpyridine, 648
polyvinylpyridines, 653
prototropic tautomerism, 648
Pt(II) complexes of pyridine N-oxides, 654
purification of pyridine bases, 648
pyridine, biological aspects, 645
pyridine alkaloids, 645–647
pyridinecarboxylic acid thioamides, 657
pyridinecarboxylic acid thiocarbanilides, 657
pyridinecarboxylic acid thiosemicarbazones, 657
pyridine complexes, 654, 655

pyridine-linked nucleotides, 645, 647
pyridine natural products, 642
pyridine nucleotides, 646
pyridine nucleotides, metabolism, 646
pyridine N-oxides, 654
pyridine, pharmaceutical aspects, 645
pyridine quaternization, 653
pyridine radicals, 650
pyridine-sulfur compounds, 645
pyridinium chlorochromate, 654
pyridinium N-imines, 653
pyridinium N-ylides, 652, 653
pyridinyl radicals, 655
pyridynes, 652
1-pyrindines, 652
pyrithiones, 657
quaternary pyridinium cations, 653
quaternization of pyridines, 653
quaternized piperidines, 651
of reviews, 635
rhodium complexes of pyridines, 654
ring inversion, 648
tantalium sulfide complexes of pyridine, 655
tautomerization of pyridines, 648, 649
tetrahydropyridines, 651
thiazolopyridines, 652
thienopyridines, 652
tobacco alkaloids, 645
UV spectra of pyridines, 649
vinylpyridines, 648, 657
viologen radicals, 655
x-ray data for pyridines, 648
Rhodium complexes with pyridine, 654 (R)
Ricinidine, 11
Ricinine, 11
  biogenesis, 642 (R)
Ring-contractions, 452–456, 483, 495
  of diazepines, 106
  CO-expulsion, 467
  metal-mediated, 472
  RPO-expulsion, 467, 680
Ring-expansions, 486
  with carbenes, 56–57
  with diazoalkanes, 57
  of pyrroles, 55
Ring inversion, 648 (R)
Ring transformations:
  from isoxazoles, 641 (R)
  isoxazoles into pyridines, 71
Rostratin, 16

Sceletium alkaloids, 22
Schiff-base condensation, 470, 472, 473, 476, 478–480, 490–491

Schmidt reaction, 449, 487
Selenolopyridines:
  from aminoselenophenes, 165
  from selenoles, 175
Sesbanine, 22
Sesquiterpenoid alkaloids, 16–22
Shale oil, components, 29
Steptonigrin, 23
Stevens' rearrangement, 452, 454
Stobbe condensation, 256, 449
Streptonigrin, 23, 325
Swertiamarin, 12, 16
Synthesis of pyridines, 640 (R)
Syphilobines, 22

Tantalum complexes with pyridines, 655 (R)
Tautomerization, 640 (R), 648–649 (R)
  pyridine, 334–335
Tecomanine, 31
Tecostidine, 12
Template cyclization, 458, 470, 471, 478, 479,
    488, 490, 491
Terpyridyl complexes, 654 (R)
  crown ethers, 469
Tetrahydroisoquinolines, 303–306
Tetrahydropyridines, 651 (R)
  dehydrogenation, 111, 112 (T)
  oxidation, 111
Tetrahydropyrimidines, 95
Tetrahydroquinolines, 117, 119, 120 (T), 264,
    281, 282, 285, 286, 291, 303–306,
    310–315, 324, 325, 327
Tetraoxamuscopyridine, 463
Tetrazaphenanthrene, 288
Thermal cyclizations, 286–288
Thermal Hofmann elimination, 482
Thermal rearrangement of tricyclic decanone
    oxime, 42
Thermolysis:
  of acetaldehyde and ammonia, 177
  of acrolein and ammonia, 172
  of acyl azides, 188
  of allyl-2H-azirines, 43
  of azides, 42
  cyclopentadiene and ammonia, 42
  of Diels-Alder adducts, 93, 96, 97
  of dihydropyridines, 108, 109 (T)
  of pyridinium acylimides, 92, 93
  of pyrimidines, 93, 95
  of pyrrole with CHCL$_3$, 56
  of pyrroles with sodium trichloroacetate, 56
  of tetrahydrofurfurylalcohol and NH$_3$, 48
Thiazolopyridines, 652 (R)
Thienopyridines, 652 (R)
Tobacco alkaloids, 7, 644–645 (R)
Tortuosamine, 22, 324
Transmetallation, 452
1,2,4-Triazenes, 291
Triazines, with dienophiles, 102–104, 103 (T)
Triphenylcyclopropane, 73
Tryptophan pathway, 4
Tuberculostatic activity:
  cyclopentapyridines, 326
  diaminoazanaphthoquinones, 326

Uracils, hydrolysis, 93, 95
UV spectra, annelated pyridines, 331–332

Vilsmeier reaction, 186
Vincarpine, 23
Vinyl pyridines, 636 (R), 637 (R), 653 (R),
    657 (R)
Viologen radicals, 655 (R)

Willgerodt reaction, 639 (R)
Wittig reaction, 453, 492, 493
Wolff-Kishner reduction, 449, 450, 481
Wurtz-couplng, 492, 493

X-ray data of pyridines, 648 (R)